FOODBORNE INFECTIONS AND INTOXICATIONS

Food Science and Technology
International Series

A complete list of books in this series appears at the end of this volume.

Foodborne Infections and Intoxications

Third Edition

Edited by

Hans P. Riemann and Dean O. Cliver

School of Veterinary Medicine
University of California, Davis

AMSTERDAM • BOSTON • HEIDELBERG • LONDON • NEW YORK • OXFORD
PARIS • SAN DIEGO • SAN FRANCISCO • SINGAPORE • SYDNEY • TOKYO

Academic Press is an imprint of Elsevier

Academic Press is an imprint of Elsevier
Radarweg 29, Po Box 211, 1000 AE Amsterdam, The Netherlands
84 Theobald's Road, London WC1X 8RR, UK
30 Corporate Drive, Suite 400, Burlington, MA 01803, USA
525 B Street, Suite 1900, San Diego, California 92101-4495, USA

First edition 1969
Second edition 1979
Third edition 2006

Library of Congress Catalog Number: 2005932980

British Library Cataloguing in Publication Data
A catalogue record for this book is available from the British Library

ISBN-13: 978-0-12-588365-8
ISBN-10: 0-12-588365-X

For information on all Academic Press publications
visit our web site at http://books.elsevier.com

Typeset by SPI Publisher Services, Pondicherry, India
Printed and bound in Italy

06 07 08 09 10 11 10 9 8 7 6 5 4 3 2 1

Cover image: The green objects are cells of *Escherichia coli* 0157:H7. The red matrix are beef muscle fibrils. The image is credited to Dr Pina Fratamico of the USDA-ARS, Eastern Regional Research Center, 600 E. Mermaid Lane, Wyndmoor, PA, USA.

Contents

Contributors

Carlos Abeyta (Ch. 5), Supervisory Microbiologist, US Food and Drug Administration, Pacific Regional Laboratory Northwest, 22201 23rd Dr. S.E., Bothell, WA 98021-4421

Sean Altekruse (Ch. 7), Policy Analysis and Formulation (OPPD), Food Safety and Inspection Service, USDA, 300 12th Street, S.W., Room 402, Washington, DC 20250-3700

Merlin S. Bergdoll [deceased] (Ch. 14), Food Research Institute, University of Wisconsin – Madison, 1925 Willow Drive, Madison, WI 53706-1187

Roberto A. Buffo (Ch. 18), Pasaje Bertres 251, 4000 – San Miguel de Tucuman, Argentina

Michael Casteel (Ch. 11), San Francisco Public Utilities Commission, Water Quality Laboratory, 1000 El Camino Real Millbrae, CA 94030

Fun S. Chu (Ch. 16), 16458 Denhaven Ct., Chino Hills, CA 91709-6176

Dean O. Cliver (Ch. 11), Department of Population Health & Reproduction, School of Veterinary Medicine, University of California, Davis, CA 95616-8743

Nathalie Corneau (Ch. 9), Health Canada, Tunney's Pasture Banning Research Center, Postal Locator 2204A2, Ottawa, Ontario K1A0L2

J. Cross (Ch. 12), c/o Dr. J. P. Dubey, Parasite Biology and Epidemiology Laboratory, Livestock and Poultry Sciences Institute, Beltsville Agricultural Research Center, Building 1040, Beltsville, MD 20705

Ivone Delazari (Ch. 19), Sadia Concordia SA Industria Comercio, Rua Fortunato Ferraz 659, 05093-901 Sao Paulo, Brasil

J. P. Dubey (Ch. 12), USDA, ARS, ANRI, Parasite Biology, Epidemiology & Systematics Lab., Beltsville Agricultural Research Center-East, Bldg. 1001, 10300 Baltimore Avenue, Beltsville, MD 20705-2350

Jeffrey M. Farber (Ch. 9), Health Canada, Tunney's Pasture Banning Research Center, Postal Locator 2204A2, Ottawa, Ontario K1A0L2

Pina Fratamico (Ch. 6), USDA-ARS, Eastern Regional Research Center, 600 E. Mermaid Lane, Wyndmoor, PA 19038

Daniel Y. C. Fung (Ch. 10), Department of Animal Science and Industry, Kansas State University, 225 Call Hall, Manhattan KS 66506-1600

Mansel W. Griffiths (Ch. 15), Department of Food Science, Ontario Agricultural College, University of Guelph, Guelph, Ontario, Canada N1G2W1

Maha N. Hajmeer (Chs 10 and 19), Department of Population Health and Reproduction, University of California, Davis, CA 95616-8743

Richard A. Holley (Ch. 18), Department of Food Science, University of Manitoba, Winnipeg, MB, Canada R3T 2N2

Keith Ito (Ch. 13), National Food Processors Association, 6363 Clark Avenue, Dublin, CA 94568

Eric A. Johnson (Ch. 17), Food Research Institute, University of Wisconsin, Madison, 1925 Willo Drive, Madison, WI 53706-1187

Vijay Juneja (Ch. 4), USDA-ERRC, 600 E. Mermaid Lane, Wyndmoor, PA 19038-8598

Philip Kass (Ch. 1), Department of Population Health and Reproduction, University of California, Davis, CA 95616

Charles Kaysner [retired] (Ch. 5), Supervisory Microbiologist, US Food and Drug Administration, Pacific Regional Laboratory Northwest, 22201 23rd Dr. S.E., Bothell, WA 98021-4421

Ronald G. Labbe (Ch. 4), University of Massachusetts, Department of Food Science, Amherst, MA 01003

Anna Lammerding (Ch. 2), Public Health Agency of Canada, Laboratory for Foodborne Zoonoses, Guelph, Ontario, Canada

Suzanne M. Matsui (Ch. 11), Division of Gastroenterology (111-GI), V.A. Palo Alto Healthcare System, 3801 Miranda Avenue, Palo Alto, CA 94304

Kåre Mølbak (Ch. 3), Statens Serum Institut, Artillerivej 5, DK 2300, Copenhagen S, Denmark

K. D. Murrell (Ch. 12), Parasite Biology and Epidemiology Laboratory, Livestock and Poultry Sciences Institute, Beltsville Agricultural Research Center, Building 1040, Beltsville, MD 20705

Truls Nesbakken (Ch. 8), The Norwegian School of Veterinary Science, Department of Food Safety and Infection Biology, PO Box 8146 Dep., 0033 Oslo, Norway

John Elmerdahl Olsen (Ch. 3), Department of Veterinary Microbiology, The Royal Veterinary and Agricultural University, Frederiksberg, Denmark

Franco Pagotto (Ch. 9), Health Canada, Tunney's Pasture Banning Research Center, Postal Locator 2204A2, Ottawa, Ontario K1A0L2

Nina Gritzai Parkinson (Ch. 13), National Food Processors Association, 6363 Clark Avenue, Dublin, CA 94568

Guillermo I. Perez-Perez (Ch. 7), Department of Microbiology, New York University School of Medicine, 550 First Avenue, New York, NY 10016

Hans P. Riemann (Chs 1 and 19), Department of Population Health and Reproduction, University of California, Davis, CA 95616

Riichi Sakazaki [deceased] (Ch. 5), c/o Asuka Jun-Yaku, 2-4 Kanda-Sudacho, Chiyoda-ku, Tokyo 101, Japan

Edward J. Schantz [deceased] (Ch. 17), Food Research Institute, University of Wisconsin, Madison, 1925 Willo Drive, Madison, WI 53706-1187

Heidi Schraft (Ch. 15), Department of Biology, Lakehead University, 955 Oliver Road, Thunder Bay, Ontario, Canada, P7B5E1

James L. Smith (Ch. 6), USDA-ARS, Eastern Regional Research Center, 600 E. Mermaid Lane, Wyndmoor, PA 190838

Ewen C. D. Todd (Ch. 2), National Food Safety and Toxicology Center, 165 Food Safety and Toxicology Building, Michigan State University, East Lansing, MI 48824-1314

Henrik Wegener (Ch. 3), Danish Zoonosis Centre, Danish Veterinary Institute, Bulowsvej 27, DK 1790 Copenhagen V, Denmark

Amy C. L. Wong (Ch. 14), Food Research Institute, University of Wisconsin – Madison, 1925 Willow Drive, Madison, WI 53706-1187

Preface to the third edition

A quarter of a century has passed since the second edition of *Foodborne Infections and Intoxications* was published. Significant discoveries and developments have taken place during this time, and many journal articles and several books dealing with foodborne pathogens have been published. Some important foodborne pathogens (e.g. noroviruses, enterohemorrhagic *Escherichia coli)* were unknown at the time of the second edition, and organisms such as *Yersinia* and *Campylobacter* were not conclusively proven to be foodborne.

In this third edition of *Foodborne Infections and Intoxications*, experts present updated accounts of the known characteristics of the most important foodborne pathogens, including their host ranges and the characteristics of the diseases they cause. The present volume also has a completely revised chapter on the epidemiology of foodborne diseases, with emphasis on investigation procedures, and a new chapter on risk assessment has been added. The chapter on the effects of food-processing procedures has been expanded to include a number of newer techniques, and the chapter on food safety presents a detailed discussion of hazard analysis-critical control points (HACCP) as a tool to assure safety. Four new chapters have been added, on *E. coli*, *Campylobacter* and related organisms, *Yersinia*, and *Listeria*, in addition to a chapter on other natural toxins (not including mycotoxins).

Much new information about the detection and identification of foodborne pathogens has been presented in books and articles in recent years. Still, about half the reported foodborne disease outbreaks in countries like the US have no identified agent. Without doubt many of these outbreaks are caused by viruses, which suggests a need for virus-detection procedures that can be applied by laboratories routinely charged with testing of suspect food samples. Since sampling and testing *per se* do not prevent foodborne disease outbreaks, there is also a need for research to develop effective interventions against common foodborne diseases and methods to assure the implementation of such interventions; the last two chapters of the book address this need.

There is, furthermore, a need for better setting of research priorities on foodborne diseases; some diseases, like human prion diseases, are so rare that even a 90 per cent reduction in incidence would have negligible public health significance. The chapter on risk assessment describes an important tool for setting priorities.

The editors especially thank the authors for contributing their vast expertise to this book.

Hans Riemann
Dean Cliver

Preface to the second edition

The organization and content of this edition follow the pattern of the first edition, but some changes have been made. The addition of a chapter on epidemiology provides information on the principles of foodborne disease transmission, the magnitude of the foodborne disease problem, and investigation of foodborne disease outbreaks. The chapters on laboratory methods and poisonous plants and animals have been deleted because these topics are adequately covered in other recent texts. A more extensive and detailed discussion of the foodborne infections and intoxications and their control has been provided. We hope that the changes will make the volume more useful as a reference work and textbook.

We are indebted to our co-authors for their willingness to make contributions. Special thanks are due to Ms Mary Jeanne Fanelli for excellent editorial assistance and to the staff of Academic Press for their cooperation.

Hans Riemann
Frank L. Bryan

Preface to the first edition

The broadness of the subject of this volume has made selectivity a necessity. Several types of food poisoning have not been included, such as those caused by toxic chemicals intentionally or unintentionally added by man. Among the naturally occurring agents only those considered most important have been included.

The content of this work could have been organized in several ways. We have chosen to group together those agents which must be present in food in a viable state in order to cause disease, although the disease which is provoked is often termed food poisoning (e.g. Salmonella food poisoning, perfingens food poisoning). The subdivision of the chapter on botulism into one on type E and another on types A, B, and F was motivated by the higher incidence of type E botulism in recent years and the accumulating literature dealing specifically with type E.

The purpose of the chapters on laboratory methods and food processing and preservation is to make the treatise more useful to readers who have special interests in these aspects of foodborne infections and intoxications. Since these chapters deal with a variety of organisms which are discussed in other chapters some repetition and scatter of information have resulted.

I am indebted to my co-authors for their willingness to make contributions without which this work could not have materialized. I also gratefully acknowledge their advice, understanding, and patience. I am indebted to Dr G. M. Dack for his encouragement and advice in connection with the planning of this volume. Special thanks are due to Mrs Patricia Akrabawi for excellent editorial assistance and to the staff of Academic Press for their cooperation.

Hans Riemann
January 1969

Dedication

While doing the food safety work that has occupied most of our combined 155 years of life, we have had the good fortune to deal with many exceptional scientists, some of whom are no longer with us.

We dedicate this book to the departed who have contributed to it, or would have, including Merlin S. Bergdoll, David A. A. Mossel, Riichi Sakazaki and Edward J. Schantz. Doctors Bergdoll, Sakazaki and Schantz were co-authors of chapters 14, 5, and 17, respectively. Dr Mossel was to have written the foreword to this book as did the late Zdeňek Matyáš to our recent *Foodborne Diseases,* second edition. The contributions of these five scientists to food safety worldwide have been huge. We are grateful to them, and to our younger colleagues who have completed these and other chapters, and have accepted the responsibility to continue this work.

We thank and salute them, one and all.

The editors

Incidence of Foodborne Disease

PART I

Epidemiology of foodborne diseases

1

Philip H. Kass and Hans P. Riemann

1 Introduction

Chemical, physical and biological agents transmitted by foods cause more than 200 recognized diseases in people (Bryan, 1982). Of these, infectious biological agents are the most important, causing the majority of foodborne disease. To put this figure in perspective, there are presently 412 known human infectious diseases, 118 of which are primarily found in humans but may also be found in animals, and 62 of which are principally animal infections but are also present in people. Among those diseases shared by humans and animals, 35 are widespread among animals, of which 12 are shared with livestock; 7 with non-human primates; several others with birds, fish or insects; and 2 with plankton (Morse, 1995). Only some of these infections are known to be foodborne, and the most important among these are the subjects of chapters in this book. It would be a grievous error, however, to believe that we have seen the entirety of infectious agents that are transmissible to people through food. So-called 'new' or 'emerging' foodborne agents have been discovered continuously over the years, from variants of well-known infectious agents such as *E. coli* to unprecedented infectious and non-reproductive agents (e.g. prion diseases such as Kuru and variant Creutzfeldt-Jakob syndrome). Even today, diseases that have traditionally been held to exist only in the realm of animals are crossing the species boundary to humans through routes that remain to be completely elucidated, such as avian influenza. Remarkably, even today no infectious cause is detected in approximately half of all reported foodborne disease outbreaks, making the discovery of 'new' agents virtually inevitable in the future.

Foodborne Infections and Intoxications 3e
ISBN-10: 0-12-588365-X
ISBN-13: 978-012-588365-8

Many of the infectious agents capable of causing foodborne diseases can be transmitted in ways other than via food or water. Agents transmitted by the fecal–oral route can cause infection through direct contact among hosts. Others, like *Coxiella burnetii*, the infectious agent of Q-fever, can be transmitted by the respiratory route, and botulism can be caused by wound infection with *Clostridium botulinum*. Some agents, though capable of causing infection by oral transmission, are seldom foodborne. Those agents that are most frequently foodborne are readily capable of occurring in enormous numbers in feces and hence in foods or water contaminated with feces (or other contaminated organic material); examples of these are *Clostridium perfringens*, *Staphylococcus aureus*, and *Bacillus cereus*.

Foodborne disease agents can be classified in different ways. The most common scheme is taxonomic combined with a classification based on mode of action; we have adopted this convention in grouping in the chapters of this book. Classification according to the source of the agent is largely useful only in situations where there can be only one or at most a few possible specific sources, such as in cassava and fugu poisoning. Another classification is based on clinical signs and symptoms of disease; under this scheme agents that use common themes in pathogenicity are grouped together, such as *Shigella*, *Yersinia,* and enteropathogenic *E. coli*. Alternatively, Bishai and Sears (1993) have distinguished foodborne disease organisms by the predominant clinical syndromes that they cause in the following (non-exhaustive) way:

- Nausea and vomiting (*S. aureus*, *B. cereus*, noroviruses, heavy metals, parasites)
- Non-inflammatory diarrhea (*C. perfringens*, *E. coli*, *Vibrio cholerae*)
- Inflammatory diarrhea (non-typhoidal *Salmonella*, *Shigella*, enteroinvasive *E. coli*, enterohemorrhagic *E. coli*, *Campylobacter*, *V. parahaemolyticus*, *Yersinia*, and other enteroinvasive pathogens)
- Neurological signs and symptoms (*C. botulinum*, ciguatera toxin, scombroid toxin, neurotoxic shellfish poisons, mushroom toxins, monosodium glutamate)
- Systemic and miscellaneous symptoms (*Listeria monocytogenes*, *Trichinella spiralis*, group A streptococci, hepatitis A virus, *Brucella* spp.).

It is clear that quite diverse agents can cause similar clinical signs and symptoms of foodborne diseases. This underscores the importance of laboratory identification of the agent in understanding the etiology, treatment and prevention of disease in individuals and in populations. Unfortunately, even with modern, sophisticated laboratory techniques, identification of the agent has not been accomplished in approximately half of the investigated foodborne disease outbreaks in the US to date. This failure of identification may occur because the agent is truly unknown (Mead *et al.*, 1999), because an inaccurate laboratory procedure has been applied, or because of mishandling of samples.

The focus of this chapter is on the epidemiology of foodborne diseases. Epidemiology has as its objective the study of the distributions of disease and health in populations, and how changes to these distributions are affected by causal determinants. The use of knowledge gained from this scientific discipline to effect changes in these distributions is the ultimate validation of epidemiological findings, as exemplified by the now legendary story of John Snow's removal of the handle of the Broad Street pump during a London cholera epidemic in the nineteenth century – an action that ultimately led to a dramatic decline in the area's disease morbidity (Snow, 1855). Snow's action demonstrated that fecal contamination of drinking water was a 'cause' of cholera many years before the causative agent, *Vibrio cholerae*, was identified, and indeed before the germ theory that microbial organisms could act as pathogenic agents was universally accepted.

2 Historical aspects

A few foodborne diseases, such as botulism, have been recognized and described since early historical times (Dolman, 1964). There is no doubt that the existence of foodborne diseases was recognized much earlier than the actual identification of pathogenic organisms, when humans learned by observation and/or experimentation which food items to avoid. Such proscriptions against certain foods, such as a ban on the consumption of pork in some religions, may too have been originally founded upon perceptive observations combined with rudimentary testing. In the middle of the 1800s, advances in scientific methods led to the identification of certain foodborne parasites, which formed the genesis of modern meat inspection procedures.

Since the dawn of the microbiological era, a great many microbial foodborne disease agents have been identified. By 1960, *Salmonella*, *Shigella*, *C. botulinum* and *S. aureus* were all well-known causes of foodborne diseases. *C. perfringens* and *B. cereus* were added to the list in the 1960s, followed by Norwalk virus in the 1970s; *Campylobacter*, *Yersinia*, 'new' strains of *E. coli* such as O157:H7, and *Cryptosporidium* were added in the 1980s; and *Cyclospora* in the 1990s.

The incidence of foodborne diseases in earlier times is completely unknown. When the human population was largely rural, most food was produced when and where it could be grown, raised or found. The unreliability and frequent unavailability of adequate supplies of food throughout an entire year necessitated storage of certain food items. This essential need led to development of preservation methods such as drying, salting, smoking, fermentation and, where climatic conditions permitted, refrigeration and freezing. Despite being shown later to have a sound scientific basis, food handlers' theory was sometimes better than their practice, and preserved foods undoubtedly sometimes caused food poisoning – which in turn led to further and safer refinements. There is little doubt that the advent of the use of heat in food preparation, delivered in the form of fire, helped to reduce the presence of pathogens (particularly in foods of animal origin), and hence the occurrence of foodborne illnesses.

3 Contemporary problems

3.1 Causes of foodborne diseases

Food itself does not normally cause disease in the short-term, except when it contains intrinsic toxins or allergenic components, or is consumed in toxic or physically incapacitating quantities. Although foods can admittedly be nutritionally deficient or contain substances known to be predictive of adverse long-term health impacts (e.g. excessive consumption of saturated fat), these effects fall instead within the realm of dietary-induced disease, and are not the subject of this book.

The 'web of causation' (MacMahon and Pugh, 1970) and the 'sufficient-component model of causation' (Rothman, 1976) are useful and complementary paradigms in unraveling the constellation of causes of foodborne diseases. The 'web of causation' dispelled the naïve but long-held and pervasive belief that the predominant determinant of disease in an individual was the presence of a specific agent. Instead, it diagrammatically illustrated the complex interplay between organism, host and environmental factors that inevitably occur before an individual's transition from a state of health to a state of disease. The 'sufficient-component model' postulates that different factors intrinsic to the host, organism and/or environment interact to cause disease; such factors are called 'component causes.' A set of minimally acting component causes – those that are minimally sufficient to initiate the transition of an individual from health to disease – are called 'sufficient causes'. It is important to note that there may be more than one, and indeed many, minimally sufficient sets of component causes that can lead to disease occurrence in a population, though only one sufficient cause exists for a single diseased individual. If a single disease is of infectious origin, then by definition every unique sufficient cause must include as a component cause the presence or influence of the infectious organism; such component causes that are common to all sufficient causes for a disease are designated 'necessary causes'.

Although single agents are obviously necessary causes, the mere presence of an agent in food or water may not be a component cause because the number of organisms (the dose) may be too low to cause infection or disease. Thc presence of other component causes besides the organism is inevitable, as illustrated in a study of *Salmonella* infection in poultry by Kinde *et al.* (1996). In this study, *Salmonella* from insufficiently treated urban sewage contaminated a stream that served as the only source of water for local wildlife. Wild animals subsequently became infected and carried the infection into proximate poultry houses, where they went searching for food. Through fecal and/or mechanical contamination of the environment, the poultry became infected, in turn increasing the risk of subsequent infection of humans through meat or eggs. Each component cause stipulated in the web of causation is the result of several antecedents, and the risk of human salmonellosis in this example could have been mitigated by removal of any single component cause, such as proper treatment of sewage, eliminating wildlife, or preventing access of wildlife to poultry houses. A distinct advantage of envisioning causation in this way is that it is not necessary fully to understand the causal mechanisms in their entirety to take preventive measures; elimination of a single component cause renders the set of component causes no longer sufficient. This approach has analogies to hazard analysis and critical control point methods (HACCP), where prevention is achieved through intervention at the critical control points.

3.2 Emerging foodborne diseases

3.2.1 Background and definitions

Mankind occupies a uniquely high position in the pecking order of nature. The single notable exception to this dominance is that people ostensibly can become the very victims of microbes and parasites that presumably occupy some of the lowest rungs of the evolutionary ladder. All humans are colonized almost from birth by a host of microbes, some of which are potentially pathogenic yet under normal circumstances do not cause disease. Through evolution, our species has acquired the mechanisms necessary to resist many different agents (Burnett and White, 1972). However, present-day exposures of people to a variety of foodborne pathogens readily occur over what formerly were secure geographical barriers, and at a rate so high that human evolution cannot keep pace. Furthermore, the absence (or near absence) of some agents in certain geographical areas leads to immunologically-naïve populations: resistance that would have otherwise been acquired in childhood is absent, leaving a highly susceptible population. For example, in some geographic regions hepatitis A virus infection is widespread, and children become brief shedders followed by active, lifelong immunity at an early age. In contrast, in other geographic areas hepatitis A infections are almost unknown, leaving a population that is very susceptible to infection from virus-contaminated foods. Demographics of human populations in many developing countries have dramatically shifted in only a few generations; with the gradual onset of urbanization and modernization, a further increase in the number of people with heightened disease susceptibility is to be expected.

Emerging foodborne diseases have been defined (Levine *et al.*, 1994) as diseases having one or more of the following characteristics:

- Clinical signs and symptoms differ from those of any diseases that preceded it
- Previously tolerated and acceptable conditions become intolerable
- A previously marginal population (afflicted with a certain disease) gains public voice
- New infection pathways, intermediate hosts, or reservoirs of pathogens evolve because of environmental or social changes.

Clearly an agent never before identified as foodborne also represents an emerging disease even if it may not be truly new (on an evolutionary scale), if it was accidentally overlooked in the past because its identification was never sought or because of inadequate and insensitive laboratory identification techniques. An emerging foodborne disease can also be attributed to a previously recognized foodborne agent when it appears in a population never before affected.

In industrialized countries, different patterns of foodborne disease outbreaks also appear to be emerging. Outbreaks that at one time were more commonly reported from smaller gatherings or cohorts, such as family picnics and church suppers, are now changing in frequency towards a greater occurrence of more diffuse and widespread outbreaks in larger populations (Tauxe, 1997). The globalization of food trade, large-batch production units, and increased consumption of 'fast' (i.e. ready-to-eat) food means that food that does not receive a terminal heat treatment may contribute to changes in outbreak patterns.

3.2.2 Changes in host susceptibility

Susceptibility to foodborne diseases may be altered for a number of reasons. Susceptibility increases as a result of impairment of the immune system caused by infection (especially AIDS), neoplasia, immune-mediated disease, immunosuppressive therapy used for cancer treatment or to prevent post-transplant organ rejection, and other medications that can alter the ecology of the gastrointestinal tract (notably antibiotics). Children of few years' age are considered more susceptible, and old age is also associated with a decrease in immune response; in addition, the elderly may have decreased gastric-acid secretion (Morris and Potter, 1997).

A 35-fold increase in the incidence of *Campylobacter* infections and a 280-fold increase in *Listeria* infections have been seen in AIDS patients; 5–10% of non-pregnant AIDS patients have developed *Toxoplasma gondii* encephalitis, and 10–20% of AIDS-associated diarrhea is due to *Cryptosporidium* infection (Morris and Potter, 1997).

While the AIDS epidemic may eventually be brought under control, the proportion of the US population that develops cancer, receives organ transplants or reaches old age will almost certainly continue to increase. In the US, white-male cancer incidence increased by 27% between 1973 and 1994; in white females the increase was 18%. Organ transplantation increased by 54% between 1988 and 1996. The proportion of US population over 74 years of age increased by 115% between 1950 and 1995; and of the 29 000 people in the US who reportedly died from diarrhea between 1979 and 1987, 51% were over 74 years old (Morris and Potter, 1997).

3.2.3 Food handling

Other factors contributing to changes in outbreak patterns may include decreased experience in food handling and preparation, and an increase in the number of meals taken outside the home and sometimes prepared by persons with limited training in and understanding of food safety. Less than 50 % of consumers are concerned about food safety. There are also an increasing number of women in the workforce living away from home and a greater number of single heads of households; this tends to limit the commitment to food preparation, and consumers seem to be more interested in convenience and saving time than in proper food handling and preparation (Collins, 1997). Furthermore, many consumers are not familiar with the properties of many of the new convenience foods, and the errors they commit in food preparation may occur because they have not absorbed information about how to handle food and protect themselves; that is, messages to the public about the importance of food safety may not have been delivered effectively (Bruhn, 1997). The situation may be exacerbated by the increasing number of vulnerable people and a shrinking public health infrastructure (Altekruse *et al.*, 1997).

The primary production of food occurs increasingly in large batches; this in itself may not pose an increased risk, but can result in widespread distribution of pathogens if and when they occur, and under conditions conducive to pathogen survival. Increasing imports of foods such as fruits and vegetables, grown and processed under undocumented conditions, may also lead to increased exposure to a myriad of bacteria and parasites (Beuchat and Ryu, 1997).

Industrially processed foods such as canned foods have, since the introduction of safe and calculated heat processes, had a very good safety record. However, different types of processing have been and continue to be implemented to provide even fresher, ready-to-use food in innovative packaging (Zink, 1997). These developments are driven by consumer appeal and competition; they are based on technical inputs from private and public laboratories, and require that distributors and users follow instructions on labels.

3.3 Incidence of foodborne disease

Information about the incidence of foodborne diseases comes from surveillance data usually collected in outbreaks; very little information is available on the incidence of sporadic cases. Outbreak reporting is admittedly incomplete; there is a considerable but unknown amount of underreporting. Attempts have been made (Bennett *et al.*, 1987; Todd, 1989) to estimate the degree of underreporting of foodborne diseases; reported annual cases in relation to the estimated number of cases ranged from about 13% for botulism to 0.01% for infections with *Vibrio* spp. (not *V. cholerae*). The total number of reported cases with known etiology was close to 11 000, while the total number of estimated cases was about 5 million. With this kind of uncertainty, reported numbers of foodborne disease outbreaks and cases may be more misleading than enlightening. However, published surveillance data do permit some comparisons of the magnitudes of incidences related to agent, year, season and other variables. Data compiled between 1985 and 1989 from 21 countries and presented by Todd (1994) indicate that salmonellosis was the most common foodborne disease in these countries except for Cuba, Denmark, Finland and Japan. *Staphylococcus aureus* intoxications ranked high in Cuba, Israel, Japan, Portugal, and Yugoslavia, while *Clostridium perfringens* infections were common in Denmark, Finland, Israel, and Sweden. The differences among countries probably reflect differences in the types of foods consumed (which can vary over time due to immigration and emigration), and differences in laboratory methods and surveillance systems. Reported foodborne disease outbreaks in the US showed no time trends through the periods 1983–1987 and 1988–1992 (Bean *et al.*, 1997), but there may be some trend in the relative frequency of isolation of different agents; *Clostridium perfringens*, *E. coli*, *Salmonella*, and hepatitis A seemed to be on the increase. The most striking feature of the data is that they show that no agent was discovered in between 54% and 64% of the investigated and reported presumptive foodborne outbreaks.

There are apparent seasonal trends in reported foodborne outbreaks in the US (Bean *et al.*, 1997). Outbreaks caused by bacteria peaked in May to August, while outbreaks caused by chemicals had a broader peak, from April to November. The peak for outbreaks caused by bacteria can probably be explained by better growth conditions during the warmer months; the reason for the peak in chemical outbreaks seems less clear. No seasonal trends were observed for parasitic or viral infections.

The 1988–1992 data for the US (Bean *et al.*, 1997) suggest that restaurants dominated among the places where foods contaminated by bacteria were eaten, with homes in second place. However, foods contaminated with bacteria were, in many instances, consumed at what was reported as 'other places'. For chemical food poisoning and parasitic infections, homes and restaurants ranked even. The interpretation of this information is uncertain because the number of meals consumed at home and at restaurants is unknown, and restaurant outbreaks may be more likely to be reported because more people are exposed.

The vehicles of transmission of foodborne diseases showed no apparent trends during 1983–1987 and 1988–1992 (Bean *et al.*, 1997). Beef, chicken, fruits and vegetables, 'other' fish and 'other' salads were all high on the list of vehicles, but the list is difficult to interpret because Chinese food and Mexican food are compared to individual food items such as ham, eggs and cheese. Furthermore, in 32–36% of reported outbreaks multiple vehicles were involved, and in 53–63% of reported outbreaks no vehicle was identified. Finally, the vehicles may not even be the sources of infection, but little information is available.

Data have been collected (Bean *et al.*, 1997) on what is called 'contributing factors' to foodborne disease outbreaks, and there is no apparent trend in these in the US for the period 1988–1992. Contributing factors are the same as component causes; with respect to bacterial food poisoning, improper holding temperature, inadequate cooking and poor personal hygiene were the leading causes; for chemical food poisoning, unsafe sources and – surprisingly – improper holding temperature were listed as leading causes; for parasitic infections no single cause was predominant; while for viral diseases the dominating cause was poor personal hygiene.

4 Epidemiological investigations of foodborne diseases

4.1 Foodborne disease surveillance

4.1.1 Passive surveillance

Although extensive data in the US are collected through local, state and national agencies, precise information about the epidemiology of foodborne diseases is scarce because it is difficult and expensive to obtain representative (and retrospective) data. Many countries have surveillance systems where outbreaks are reported, with an outbreak generally interpreted as two or more persons who become ill from the same food. Sporadic (single) cases are not reported except where there are specific mandatory requirements. Whether reporting actually occurs depends on the likelihood that afflicted individuals seek medical help, on the probability that the physician submits a sample to a laboratory, on the ability of the laboratory to detect the agent in question, and on the probability that the needed documentation is completed and submitted to a central agency that analyzes and publishes foodborne disease data. These successive sources of potential error mean that not only is there underreporting of disease incidence, but also the reporting is biased because severe cases or a high number of cases from a common source are more likely to be reported. For these reasons, foodborne disease incidence has been likened to the small part of an iceberg observed above water, with the predominant part acknowledged but unobservable. Laboratory diagnostic procedures have been substantially improved in recent years, but there is room for additional improvement because, as noted earlier, no causative agent is found in approximately 50% of reported foodborne outbreaks.

The surveillance alluded to above is also known as 'passive surveillance'. A few countries have had passive foodborne disease surveillance systems for more than half a century, and an increasing number of countries are using such systems.

Todd (1994) has presented an extensive review of surveillance systems existing throughout the world.

The objectives of foodborne disease surveillance are the following (Todd, 1994):

- Early warning of an illness (real or potential) that could affect a large number of members of a community
- Notifications by physicians of enteric or other specific diseases, that often are foodborne, to a reference laboratory
- Investigations of reports of foodborne illness and reporting of results on a regular basis
- Use of sentinel and special epidemiological studies to determine a more realistic level of morbidity caused by foodborne diseases (this type of activity is generally considered active surveillance).

Guzewich *et al.* (1997), Bryan *et al.* (1997a, 1997b) and Todd *et al.* (1997) have published a four-part critical review of foodborne disease surveillance. Part I (Guzewich *et al.*, 1997) describes the purpose and types of surveillance systems and networks. The listed components of foodborne disease surveillance are:

- Receiving notification of illnesses
- Investigating incidents and reporting findings
- Collating and interpreting data
- Disseminating information to effect control of current problems and provide guidance for prevention of disease.

This represents a fairly intricate system that requires well-coordinated activities at many levels, and its success depends upon the voluntary efforts of many participants. Unfortunately, foodborne disease investigations are sometimes poorly carried out, if at all, and the findings of investigations may be of insufficient quality for submitting reports and therefore remain in the office where the investigation was initiated. Recent reports on the incidence of foodborne diseases point to the difficulties in assessing the current status of foodborne morbidity and providing early warnings. Sample testing in the laboratory will help to overcome some of the deficiencies, as testing at the molecular level becomes more widespread.

Part II (Bryan *et al.*, 1997a) focuses on definitions and methods of tabulation of surveillance data, which can have a major influence on the way the data are analyzed and interpreted. For each step covered in disease investigation (time, space, dietary history, etc.), evaluations are made as to the value and limitation of data. The attempt of this part is to provide some degree of standardization, which is greatly needed.

Part III (Bryan *et al.*, 1997b) focuses on the food components, with collation of data listing vehicles, significantly important ingredients, places where foods were mishandled, methods of processing and preparation, and operations that contributed to outbreaks.

Part IV (Todd *et al.*, 1997) deals with the use of surveillance data, including:

- Developing new policies
- Evaluating effectiveness of programs
- Justifying food safety program budgets

- Modifying regulations
- Conducting hazard analysis and risk assessment, designing HACCP programs
- Informing the public about food safety
- Training food industry personnel
- Training public health officials
- Identifying new problems and research needs.

The goals of foodborne disease surveillance can only be fulfilled when the surveillance data reflect reality; this is presently not the case. Attempts have been made to estimate the degree of underreporting of foodborne diseases in the US (Bennett *et al.*, 1987; Todd, 1989), and these estimates have been revised by Mead *et al.* (1999) using more inclusive data sources. Mead *et al.* (1999) estimated that foodborne agents annually cause 76 million illnesses, 325 000 hospitalizations and 5000 deaths. The total number of foodborne illnesses reported through passive surveillance for 1993–1997 (Olsen *et al.*, 2000) was 86 058 with 29 deaths; about half of the outbreaks had an undetermined etiology. Table 1.1 shows a comparison between the estimates by Mead *et al.* (1999) and what was actually reported through passive surveillance

Table 1.1 Annual numbers of estimated and reported cases of foodborne illnesses per 100 000 persons in the United States

Agent	Estimation	Passive reporting
Norwalk virus	3274	0.09
Campylobacter	699	0.04
Salmonella, non-typhoid	478	0.32
C. perfringens	88	0.2
Giardia lamblia	71	0.003
S. aureus	66	0.1
T. gondii	40	–
Shigella	32	0.1
Y. enterocolitica	31	0.002
E. coli O157:H7	22	–
E. coli, enterotoxigenic	20	–
Rotavirus	14	–
Astrovirus	14	–
Cryptosporidium parvum	11	–
E. coli, *non-O157 STEC*	11	–
B. cereus	9.7	0.05
E. coli, other diarrhegnic	8.5	–
E. coli	0.23	
Cyclospora	5.2	–
Vibrio, other	1.8	0.01
Listeria	0.9	0.04
Brucella	0.3	0.007
V. vulnificus	0.02	–
Botulism	0.02	0.02
V. cholerae	0.02	0.0007
T. spiralis	0.02	0.007

Based on Mead *et al.* (1999) and Olsen *et al.* (2000).

(Olsen *et al.*, 2000). It is obvious that the incompleteness of passive reporting makes it impossible to use the data to represent the impact of foodborne disease on society. However, this should not be construed to imply that passive reporting is of no value: there is a lesson to be learned in each outbreak investigation when correctly performed, and passive surveillance will hopefully improve along the lines suggested by Guzewich, Bryan and Todd. It seems clear that an international standard of foodborne disease surveillance would be immensely valuable and make it possible to draw comparisons that are presently impossible. This in turn would provide new knowledge about causes of foodborne diseases. Because many of the activities in surveillance, especially those at the local level, are based on voluntary participation, it is important to maintain enthusiasm for the system. It is also important to improve the skills of those involved in conducting investigations and reporting results. The investigators must not only be familiar with the methodology of epidemiological investigation; they must also have knowledge of foodborne diseases and food production, processing and preparation.

4.1.2 Active surveillance

The inadequacy of passive surveillance gave impetus to establish sentinel studies, where the investigation of foodborne diseases can be performed in a more active fashion in limited geographical locations. Todd (1994) summarized some of these studies, which have been based on enrollment of local practitioners or have been epidemiological cohort studies where groups of people were interviewed about gastrointestinal disease syndromes at regular time intervals. These studies have yielded some surprising results in that disease incidence was much higher than expected.

In 1994 the Centers for Disease Control and Prevention (CDC) began implementing the Emerging Infections Program (EIP), in cooperation with selected state health departments, with foodborne diseases as a major component (CDC, 1996). The Active Surveillance Network (FoodNet) was established as collaborative effort among the CDC, the US Department of Agriculture (USDA), the Food and Drug Administration (FDA), and the EIP sites (Centers for Disease Control and Prevention, 1997). The components of FoodNet are:

- Survey of clinical laboratories
- Survey of physicians
- Survey of populations by interviewing residents
- Case-control studies.

FoodNet began collecting population-based active surveillance data on culture-confirmed cases of seven foodborne infections (*E. coli* O157:H7, *Campylobacter*, *Listeria*, *Salmonella*, *Vibrio*, and *Yersinia*) in five EIP sites. The 2001 preliminary data include *Campylobacter*, *E. coli* O157:H7, *Shigella*, *Vibrio*, *Yersinia enterocolitica*, *Cryptosporidium parvum*, *Cyclospora cayetanensis*, and hemolytic uremic syndrome (HUS) in nine sites in the US, representing 37.8 million persons (CDC, 2002). During 2001, 13 705 laboratory-diagnosed cases were reported; the overall incidences of the different agents/syndromes are shown in Table 1.2. There were considerable differences in incidences among the different sites. California had more frequent isolations of *Campylobacter* and *Shigella* than any other site, while the Minnesota site had

Table 1.2 Overall incidence per 100 000 persons of infections/syndrome in nine sites of active surveillance – United States, 2001[a]

Pathogen/syndrome	Incidence/100 000
Salmonella	15.1
Campylobacter	13.8
Shigella	6.4
E. coli O157:H7	1.6
Cryptosporidium	1.5
HUS[b]	0.9
Yersinia	0.4
Listeria	0.3
Vibrio	0.2
Cyclospora	0.1

[a] CDC (2002)
[b] Hemolytic uremic syndrome, usually caused by *E. coli* O157:H7.

the highest frequency of *E. coli* O157:H7, HUS, and *Cryptosporidium*. Among *Salmonella* isolates the dominating serotype was Typhimurium (15%), followed by Enteritidis (12%) and Newport (7%).

Not every isolation represents a case of foodborne disease, but Mead *et al.* (1999) have made estimates of the proportion of total cases that are foodborne. The estimates are 95% for non-typhoid *Salmonella* and 20% for *Shigella*. At the California site there were 14.3 isolations of *Salmonella* and 13.2 of *Shigella* per 100 000 persons; the adjusted numbers for foodborne cases become 13.6 isolations per 100 000 persons for *Salmonella* and 2.6 isolations per 100 000 persons for *Shigella*.

During the period 1996–2001, the incidence of most pathogens declined at the FoodNet sites. *Salmonella* declined by 15%, but there were differences among serotypes: Enteritidis and Typhimurium decreased, while Newport, Heidelberg, and Javiana increased. It has been suggested (CDC, 2002) that the decreases may be due to the implementation of egg quality assurance programs and improvements in hygienic manufacturing practices after the implementation of HAACP and performance standards in slaughterhouses.

Other active surveillance programs have been implemented. The Food Safety and Inspection Service (FSIS) *Salmonella* performance standards for slaughterhouses have been enforced since 1996, and programs directed at the primary production level and combined with intervention programs against *Salmonella* exist in Denmark and Sweden. Such programs are in principle not substantially different from control programs for tuberculosis or brucellosis, except that these two diseases affect not only humans but livestock as well.

4.2 Outbreak investigation

Outbreak investigations are at the core of foodborne disease surveillance, and the quality of these investigations is of utmost importance. The International Association of Milk, Food and Environmental Sanitarians (now the International

Association for Food Protection) published, in 1987, an extremely useful booklet on procedures for investigation of foodborne disease outbreaks; this booklet has been regularly updated, with the fifth edition appearing in 1999. Morse *et al.* (1994) have also described procedures, forms and tabulations used in investigation of foodborne diseases in New York State.

Although different authors and agencies promulgate or adopt somewhat different approaches to outbreak investigation, most outbreak investigations share the following key features:

- Receipt of an initial report or data
- Verification of the diagnosis
- Determination of whether an outbreak has occurred
- Search for additional data and cases
- Description of cases in terms of time, space and persons
- Formulation of hypotheses
- Further analytical, epidemiological, environmental, and laboratory studies
- Synthesis of findings with conclusions and recommendations
- Control measures
- Written reports.

When the outbreak is relatively confined in terms of time, space, and the size of the population at risk (as in some common source outbreaks), the preferred epidemiologic investigation method is the retrospective cohort study, where the fates of people who ate the putative food are compared to those of the people who did not. A cohort is an assemblage of individuals who share one or more characteristics that in turn define its membership – for example, being present at an event where a temporally defined foodborne outbreak occurred. In the special case of an outbreak investigation, we further define the cohort as being *closed* in the sense that there can be no immigration or emigration of its membership once the temporal definition of the cohort is established. Members of the cohort may be *censored*, meaning that they are lost to follow-up or are precluded from developing the disease due to competing causes. The validity of a cohort study rests on the assumption that censoring is: (a) unrelated to disease incidence, and (b) unrelated to the exposure(s) under study. Because the reasons for censoring are often unknown, the former assumption (a) is difficult to verify, so it is critical that the number of censored individuals be kept as minimal as possible. Conscientious and assiduous efforts at tracing back and establishing contact with all the members of the cohort are instrumental in preventing censoring bias.

Another assumption underlying the success of undertaking a cohort study to investigate a disease outbreak is that there is a distribution of exposure to the causative agent/source among the people in the cohort, allowing the comparison of disease incidence conditional on exposure status. The absence of a distribution, which could occur if everyone in the cohort consumed the putative source of an infectious organism, precludes the determination of comparative measures of exposure-specific disease incidence. The mere finding that all diseased individuals consumed a particular food or drink is not necessarily sufficient to implicate that consumable. To illustrate this point, suppose at a dinner gathering of people a particular salad was contaminated.

If everyone ate the contaminated salad, and 75% of the individuals developed the disease, then the salad would seemingly be implicated. Suppose, however, that everyone also drank the same water and ate several of the other foods offered: 75% of these individuals would also develop disease. Thus, there would be no epidemiologic basis by which investigators could distinguish the potential responsible food or drink from incidental ones without ancillary information.

The aforementioned example illustrates the need for a coherent way to evaluate data collected in the course of doing cohort studies. Suppose that we undertake a cohort study of an outbreak of an intestinal disease at a single-day gathering and collect retrospective information about all foods and drinks consumed by the individuals present. For simplicity, we shall assume that there was no loss to follow-up of any of the members of the cohort, that any consumption of a food is considered to be a positive food intake exposure (obviating the need retrospectively to measure quantities of items consumed), and that there is no misclassification of an individual's disease status. Individuals can then be cross-classified by their binary exposure and by their binary disease status.

The assumptions above are sufficient to allow the calculation of the proportion of individuals in the cohort that develop disease (*incidence proportion*), both crudely and conditional on exposure status. Although a cohort study is by definition non-experimental, the unexposed group in a cohort study is analogous, though not identical in construct, to the control group in an experimental study. However, it is important to appreciate the distinction between the two study types: in an experimental study exposure is typically randomized to ensure comparability between the study groups, while in a non-experimental study subjects typically select their own exposure for reasons unknown but that may be causally important. To illustrate, an individual on antibiotics may be both more susceptible to developing enteric disease, and also specifically avoid certain foods, making such foods appear risk-protective. The absence of randomization, then, can be a serious impediment to cohort study validity. The reasons why individuals select their own exposures are examples of confounders – variables that lead to a biased (i.e. invalid) statistic. Although it is possible to obtain valid statistics derived from incidence proportions through analytic control of confounding, this subject is beyond the scope of this chapter; a more thorough discussion of confounding may be found in Rothman and Greenland (1998).

The incidence proportion among individuals exposed to a particular food can be designated by the following notation:

$$P(D \mid E)$$

where P = disease probability (i.e. the proportion of individuals that get the disease), D = development of disease, and E = the conditional status of being exposed. Similarly, we can write the incidence proportion among unexposed individuals as:

$$P(D \mid \overline{E})$$

where $\overline{\mathrm{E}}$ = the conditional status of being unexposed. A parameter that is causally interpretable (in the absence of any biases) as the proportionate change in the average risk of disease moving from non-exposed to exposed status is the *incidence proportion ratio (IPR)*, denoted by:

$$\frac{P(D \mid E)}{P(D \mid \overline{E})}$$

When based on observed data, the IPR statistic is extremely useful in distinguishing causal foods from non-causal foods in investigating disease incidence during an outbreak. If a particular food was not contaminated, and its consumption was independent of any food that was contaminated, then the incidence proportion of disease among those who ate the food would be expected to be approximately equal to the incidence proportion of disease among those who never ate the food, and the IPR would be approximately equal to 1.0. Formulae for variance estimators and confidence intervals of the crude IPR, and tests of the null hypothesis that the crude IPR = 1 are available (Rothman and Greenland, 1998); these formulae generally require the construction of contingency tables cross-classified by disease and exposure frequencies.

It is also possible to take the difference between the exposure-specific incidence proportions, rather than the ratio above, leading to the calculation of the *incidence proportion difference* (IPD):

$$IPD = P(D \mid E) - P(D \mid \overline{E})$$

Under the null hypothesis of no effect of a consumed food, the expected IPE = 0. Although the IPD can vary between –1 and 1, in practice it is unlikely that foods protect against foodborne disease, hence the IPD should fall between 0 and 1. Although the magnitude of the IPD is constrained by the incidence proportion among the unexposed (a characteristic from which the IPR does not suffer), virtually all foodborne diseases would have such a low background incidence during the finite period of an outbreak investigation that this should not be an impediment to causal analysis. As with the IPR, the IPD has its own formulae for variance estimation, a confidence interval, and a test of the null hypothesis that the IPD = 0 (Rothman and Greenland, 1998).

When it is not possible clearly to enumerate the constituents of a cohort during an outbreak, particularly those that did not develop disease and hence were not reported, or when the cohort is so large that gathering retrospective information on everyone in the cohort is impractical, the epidemiologic design of choice is generally the case-control study. While these can be used to study foodborne disease outbreaks when cohort studies are not feasible, they are better suited to studying sporadic disease, and will be discussed in more detail in the following section.

4.3 Investigation of sporadic cases of foodborne diseases

4.3.1 Measures of effect

From the standpoint of causal identification of substances contributing to foodborne disease, the ability to conduct a cohort study offers a distinct advantage over other study designs; notably, the ability to estimate exposure-specific risks and their conjunctive effect measures. However, the conditions under which a cohort study can be conducted are highly restrictive: there must be a way of inventorying the members of the cohort defined by time and space; the period of the outbreak must be relatively brief, with a rapid rise in the incidence of the disease, followed by a relatively quick return to an endemic level; censoring must be minimal and unrelated to exposure and disease status; and the morbidity must be high enough for the epidemic to be recognized.

The vast majority of diseases transmitted by food or water, however, fail to fulfill most, if not all, of these criteria. Most foodborne disease that comes to the attention of public health officials occurs in a sporadic and seemingly random temporal and spatial pattern. Such sporadic incidence is consonant with either the occurrence of etiologically unrelated and isolated cases or with a paucity of cases eventually diagnosed and reported from one or more unrecognized epidemics. The latter represents an extreme case of censoring in which almost all diseased individuals are unaccounted for and hence lost to follow-up. What led these individuals from an epidemic to become ill is immaterial, and no outbreak investigation of them is possible; they cannot be practically studied as part of a larger cohort because none could be enumerated. Furthermore, it often takes a considerable amount of detective work to identify the vehicle/source outbreaks that do not involve a common source in a restricted time and space, especially if the incubation period of the agent is long or the food is distributed through different channels at different periods of time.

The solution to how to study determinants of sporadic foodborne infection requires an appreciation that so long as exposure-specific risks are unnecessary to know, case-control studies offer efficient alternatives to conducting cohort studies while still yielding ratio measures of effect. Indeed, it would be a mistake to consider case-control studies as distinctly different from cohort studies; the designs are distinguished by whether sampling of the population at risk occurs, as it does in case-control studies, or whether the population at risk is inventoried in its entirety, as in cohort studies. Due to the necessity of sampling based on outcome status (conventionally diseased or non-diseased subgroups), case-control studies are unable to provide estimates of disease incidence without ancillary information not usually collected or readily available for foodborne illnesses. By conditioning on outcome status, investigators can only measure the probability distribution of exposure(s) in the respective study groups (e.g. the proportion of cases or controls that are exposed to a particular food):

$$P(E|D) \text{ and } P(E|\overline{D})$$

where D and $\overline{D}$ represent cases and controls, respectively.

It is important to recognize that these two distributions (and their ratio), where exposure is an 'outcome' of disease rather than a cause, are not quantities of intrinsic causal interest. It is illogical to consider how disease status can 'affect' the distribution of prior exposures because of the inverse temporal relationship. The distributions have intrinsic value, nevertheless, when transformed into exposure odds, defined as the probability of each level of exposure divided by its converse probability, such that the numerator and the denominator of the exposure odds sum to one:

$$P(E|D)/(1 - P(E|D)) = P(E|D)/P(\overline{E}|D), \text{ and}$$
$$P(E|\overline{D})/(1 - P(E|\overline{D})) = P(E|\overline{D})/P(\overline{E}|\overline{D})$$

The first statement is called the exposure odds among cases, and the second is called the exposure odds among controls. The ratio of these two odds is the well-known exposure (case-control) odds ratio:

$$\frac{P(E|D)/P(\overline{E}|D)}{P(E|\overline{D})/P(\overline{E}|\overline{D})}$$

Like its components, the exposure odds ratio has no natural causal interpretation. However, it is possible to show using Bayes theorem that the exposure odds ratio is algebraically and numerically equivalent to the incidence (disease) odds ratio:

$$\frac{P(D|E)/P(\overline{D}|E)}{P(D|\overline{E})/P(\overline{D}|\overline{E})}$$

Although the exposure odds ratio and the incidence odds ratio are interpretively different, the significance of their algebraic equivalence should not be minimized: it provides the fundamental basis for the legitimate use of case-control studies in causal inference.

Odds ratios derived from case-control studies are difficult to literally interpret because, unlike probabilities that have a domain of 0 to 1 inclusive, the domain of odds lies in the interval of 0 to infinity. However, as we will show, most case-control studies of foodborne disease can be designed so that the exposure odds ratios can be interpreted as other, more easily understood ratio measures of effect. The most important – and controversial – issue in designing a case-control study of sporadic foodborne disease is how the controls are selected.

4.3.2 Cumulative incidence case-control studies

For the reasons presented above, it is not always practical to conduct a cohort study in the investigation of a foodborne disease outbreak. Although a cohort may in fact exist and can be enumerated, censoring can prevent the calculation of exposure-specific incidence proportions, and hence any derivative effect measures. When a common source is suspected of transmitting the totality of short-term disease in the cohort, it is still possible to test this hypothesis using a *cumulative incidence case-control study*.

As in other case-control studies, cases with foodborne disease are compared to controls that did not develop foodborne disease. Unlike other types of case-control studies, however, controls are sampled from those members of the cohort who remained disease-free throughout the entire duration of the outbreak. By definition, then, controls in this type of case-control study are not eligible to become cases because they are only selected after the outbreak is over.

Cornfield (1951) demonstrated that the exposure odds ratio from a cumulative incidence case-control study has attractive statistical properties when the conditional incidence of disease is assumed to be rare (i.e. less than 5%). Under this restriction, the case-control odds ratio remains quantitatively equivalent to the incidence odds ratio, but now the incidence odds is approximately equal to the incidence proportion, i.e.

$$\frac{P(D)}{P(\bar{D})} \cong P(D)$$

This property, ubiquitously known in the epidemiologic lexicon as the *rare disease assumption*, leads to the powerful conclusion that the case-control odds ratio may be interpreted as the incidence proportion ratio. It is important to recount that it is not sufficient for this assumption to hold overall (unconditionally); it must also be met marginally and jointly conditional on all levels of exposures (foods) and strata of confounders (e.g. age groups, gender, HIV status etc.). Unfortunately, it would be surprising if this assumption was not violated in an outbreak investigation; depending on the dose of a foodborne pathogen, the morbidity associated with consumption of a contaminated food could be quite high. Therefore, caution must be exercised in the interpretation of the odds ratio, so as not to confuse odds with concepts such as probability, risk or likelihood.

Validity in a cumulative incidence case-control study of a foodborne outbreak depends on an assumption of random sampling – i.e. the enrolled cases being representative of all cases, and controls being representative of all non-diseased individuals. If the reasons why individuals cannot be located or refuse to participate are somehow related to the foods that they ate, then bias can result. Although a minimum of one control should be located for each enrolled case (for reasons of statistical efficiency), this may not always be possible when morbidity is high. Nevertheless, as long as resources exist for obtaining information in an outbreak investigation, there is no virtue in not making an effort to locate as many controls (and cases) as possible: approximately half of all outbreak investigations fail to implicate a cause, underscoring the difficulty in successfully undertaking retrospective studies. Investigators should also be cognizant that cases may be more motivated to recall a complete dietary history during an outbreak compared to controls. To the extent that investigators can obtain documentation of all consumables available during an outbreak, such a listing can be used to establish prior food intake during interviews to minimize lack of recall, rather than asking individuals to volunteer what they remember eating. Although some error in recall is to be expected, it would not be surprising to have some controls recall eating food found to be contaminated – not because of error in recall or high endemic level of disease, but because the quantity of a food eaten may serve as a proxy for the number of pathogenic organisms consumed. Similarly, cases may not recall eating a contaminated food, but could have still been exposed to an infectious dose through unrecognized cross-contamination. Though investigations such as this assume heterogeneity of the at-risk population, they generally assume homogeneity of distribution of the pathogen in the suspect food. Exceptions, in which the infectious agent or toxin is localized in part of a lot of food, are common and will surely complicate epidemiological analysis.

4.3.3 Incidence density case-control studies

Although the prototypical epidemiologic study of foodborne disease is the outbreak investigation, most cases that occur in large populations are seemingly unrelated and sporadic. The study of such cases is not performed for the purpose of determining a contaminated food that is in all likelihood long discarded. Instead, research into the determinants of disease in sporadic cases is performed for enhancing public health through disease prevention. Such work has as its underlying goal the development of a better understanding of what actions individuals took that placed themselves at heightened risk of foodborne disease, and what host characteristics they had that predisposed them to succumbing to infection or intoxication. To illustrate this concept, consider that much of the poultry sold to consumers in the US is contaminated with enteric pathogens including *Campylobacter* and *Salmonella*. Yet most people who consume poultry do not develop illness from it. The salient issue is not about whether poultry is contaminated, and if so with what pathogen, but rather is a contrast of what individuals who ate poultry did or did not do that influenced whether they developed disease or not. When endemic contamination of consumer foods is ubiquitous in large populations, the identification of such behaviors can lead to preventive health strategies that if adopted and implemented can profoundly affect disease morbidity. Such strategies can include education, identification of individuals at heightened risk, improved food hygiene at the post-harvest level, avoidance of certain drugs that affect host defenses, etc.

Public health officials typically collect information from reported cases about risk factors related to foodborne disease, though they generally lack the resources necessary to conduct the controlled studies that could identify the characteristics and behaviors that contributed as component causes to the development of disease. Such studies are difficult to perform because, unlike the situation of an outbreak investigation where a cohort can often be clearly defined, sporadic cases arise in large populations that are dynamic in membership and defy enumeration short of through a census. Clearly, cohort and cumulative incidence case-control studies are inappropriate for studying sporadic cases.

Instead, researchers can rely on an alternative study design: the *incidence density case-control study*. Over an extended but closed period of time, cases are recruited and studied, retrospectively, prospectively or both (ambispectively), from a large population. Case ascertainment may occur at local health agencies (where mandatory reporting occurs), at hospitals (where a confirmatory disease diagnosis is made) or at laboratories (where isolation of organisms or toxins occurs). To be sure, such cases can hardly be described as typical of all cases that occur but remain undiagnosed: study cases are generally more severely affected, more likely to utilize medical services, and preferentially patronize medical personnel who are willing to obtain a diagnosis instead of solely treating symptoms. This select group of patients may or may not be representative of the larger but unrecognized body of individuals with foodborne disease with respect to behaviors and host characteristics. In general, if disease severity is associated with the presence of risk factors, then the problem of lack of representativeness and generalizability is offset by the inclusion of individuals in a study that will make it easier to identify such risk factors. If severity and the other imponderable factors that lead an individual to seek diagnosis and treatment are unassociated with risk factors, the lack of representativeness is no longer an issue.

Just as cases are recruited throughout the study period, controls are selected from the population at risk during the identical period. This method of sampling controls distinguishes incidence density sampling from cumulative incidence sampling: in the former, controls are eligible to become cases after being sampled as controls, and would be included in the study separately as both a case and as a control. Although controls can be randomly selected from throughout the study period, it is preferable to select controls at approximately the same time that cases occur, creating matched sets of, typically, one case and one or more controls. When a matched analysis is performed using stratified analysis or conditional logistic regression, any confounding by time will be controlled for. This study design also yields case-control odds ratios but, unlike cumulative incidence case-control studies, these odds ratios are interpretable as *incidence rate ratios*: the proportionate change in the incidence rate of disease moving from unexposed to exposed status. The gravity of this interpretation should be appreciated – it allows the case-control odds ratio to estimate a readily understood measure of proportionate change in the disease rate even when the disease is not rare. In other words, even in the absence of the rare disease assumption, an incidence density case-control study need not be interpreted in terms of relative odds. Furthermore, when the disease is rare, the incidence density case-control odds ratio is a superior estimator of the incidence proportion ratio, compared to the cumulative incidence case-control odds ratio (Greenland and Thomas, 1982).

Control selection in incidence density case-control studies of foodborne disease is neither trivial nor uncontroversial. Controls are sampled to reflect the exposure distribution in the source population (at risk) of cases; *selection bias* arises when there is a disparity in this distribution between the control and source populations. When cases emanate from a large population and are obtained from a disease registry, it is often efficient to take a primary sample of the source population. The latter can be performed through standard survey protocols, such as random digit dialing, neighborhood interviews, etc. For example, one study of determinants of sporadic salmonellosis was performed by obtaining cases from county health departments, which maintain vital information on patients with reportable diseases (Kass *et al.*, 1992). Controls were selected through random digit dialing of the counties that reported to their respective health departments that provided cases. Although this method of sampling is relatively straightforward, caution must be exercised that the method of obtaining controls is unrelated to the distribution of risk factors in the source population. For example, if individuals in lower socioeconomic strata are less educated about proper food preparation techniques and are also less likely to own a telephone, then a random digit dialing survey would not capture such individuals and hence would underestimate the prevalence of improper food-handling skills.

An alternative control sampling procedure is often employed when cases are obtained from hospitals. Although it is uncontroversial to locate cases with a foodborne illness in this way, it is unclear what the source population of such cases is; indeed, such a population may be only a hypothetical construct. Although some hospitals have a monopoly on medical care for a defined region, it is also common for other hospitals to serve as referral centers for patients living both far and near from them. Patients at such hospitals may not necessarily be representative of all patients in the source population, even if they are a census of cases that are seen at those hospitals. In this scenario of a hospital-based study, a key requirement for control selection is that the control, had she or he developed the foodborne disease under study, would have entered the hospital and been diagnosed with the disease through exactly the same mechanism as the cases were. This underlying tenet of control selection is important to adhere to because the reasons why certain individuals patronize specific hospitals could in some unknown or unspecified way be related to the risk factors under study. For example, if hormone replacement therapy is via immune modulation, a possible risk factor for enteric infection, and one region hospital is recognized for its treatment of postmenopausal women, then patients under such therapy may preferentially attend that hospital over another. For this reason, controls are typically selected from the same hospitals as cases, and from patient diagnostic groups that would be more likely to also be seen at the same hospital for gastrointestinal foodborne illness. To illustrate this point, it is plausible that, if patients had an infectious respiratory disease at a hospital, if they later developed an infectious gastrointestinal foodborne disease they would present themselves to the same hospital. In contrast, a group of cancer patients would be ill-advised as a control group, because such patients often gravitate to certain tertiary care facilities known for prowess in treating such conditions, and such patients may be atypical of the source population with respect to their distribution of risk factors.

There are additional considerations when selecting a control group for a hospital-based case-control study. It is essential to recognize that, regardless of what disease or group of disease diagnoses is used to constitute such a control group, the risk factors under study must *not* be determinants of these control diagnoses. Were this to occur, the distribution of the exposure(s) in the controls would be unrepresentative of the source population, and in fact would be spuriously closer to the distribution of the exposure(s) in the cases, leading to case-control odds ratios that were biased towards the null value of 1.0. As an illustration of this, consider a case-control study to determine whether individuals taking oral antibiotics are at higher risk of developing *Campylobacter, Salmonella*, and *E. coli* infections. While the case definition is obvious with confirmatory stool cultures, the control definition is less transparent. Patients who had been treated for other infectious diseases would pointedly not be appropriate as controls because such individuals are likely to be have been exposed to oral antibiotics to an extent clearly higher than that of the source population of cases. Failure to recognize such associations would result in effect measures not only being biased towards the null, but could even lead to factors that are harmful appearing to be protective (or *vice versa*).

This admonition should never be construed to mean that controls should be selected because they are unexposed; any such study would be fatally flawed from its inception. The key central theme is that controls should not be *selected* on the basis of their exposure status, or of a proxy of exposure. In the above example, treatment for infectious disease was a proxy for antibiotic use. On the other hand, if individuals treated for closed fractures and ligament or tendon damage were selected as controls (under the assumption that these conditions are not treated with antibiotics), then even if these controls were taking antibiotics for other ancillary medical reasons they could remain in the study as controls because their control selection did not depend on the other reasons.

Findings from such case-control studies should be interpreted with respect not only to the odds ratios or incidence rate ratios, but also to baseline (unexposed) rates. The reason for this is that even factors that exert large *proportional* effects on disease rates may be of negligible public health importance if their baseline disease rates are low and the factors are rare in the population at risk. In other words, overall morbidity in a population is a function of not only the relative effect of a factor, but also its prevalence in the population. For this reason, weaker risk factors can exert a greater influence on disease incidence if they are relatively common than stronger risk factors that are relatively rare. It makes little sense to conduct studies of risk factors if their effect on morbidity is small or they are not subject to mitigation through education or public health interventions.

Bibliography

Altekruse, S. F., M. L. Cohen and D. L. Swerdlow (1997). Emerging foodborne diseases. *Emerg. Infect. Dis.* **3**, 285–293.

Bean, N. H., J. S. Goulding, M. T. Daniels and F. J. Angulo (1997). Surveillance for foodborne disease outbreaks – United States, 1988–1992. *J. Food Prot.* **60**, 1265–1286.

Bennett, J. V., S. D. Holmberg, M. F. Rogers and S. L. Solomon (1987). Infections and parasitic diseases. In: R. W. Amler and H. B. Dull (eds), *Closing the Gap: The Burden of Unnecessary Illness*, pp. 102–114. Oxford University Press, New York, NY.

Beuchat, L. R. and J.-H. Ryu (1997). Produce handling and processing practices. *Emerg. Infect. Dis.*, **3**, 459–465.

Bishai, W. R. and C. L. Sears (1993). Food poisoning syndromes. *Gastroenterol. Clin. N. Am.* **22**, 579–608.

Bruhn, C. M. (1997). Consumer concerns: motivating to action. *Emerg. Infect. Dis.* **3**, 511–515.

Bryan, F. L. (1982). *Diseases transmitted by foods.* Centers for Disease Control, Atlanta.

Bryan, F. L., J. J. Guzewich and E. C. D. Todd (1997a). Surveillance of foodborne disease. II. Summary and presentations of descriptive data and epidemiological patterns; their value and limitations. *J. Food Prot.* **60**, 567–578.

Bryan, F. L., J. J. Guzewich and E. C. D. Todd. (1997b). Surveillance of foodborne disease. III. Summary and presentation of data on vehicles and contributory factors; their value and limitations. *J. Food Prot.* **60**, 701–714.

Burnet, M. and D. O. White (1972). *Natural History of Infectious Disease*, 4th edn. Cambridge University Press, London.

CDC (Centers for Disease Control and Prevention) (1996). *CDC's Emerging Infections Program*. CDC, Atlanta.

CDC (Centers for Disease Control and Prevention) (1997). Foodborne diseases active surveillance network, 1996. *Morbid. Mortal. Weekly Rep.* **46**, 258–261.

CDC (Centers for Disease Control and Prevention) (2002). Preliminary FoodNet data on the incidence of foodborne illnesses – selected sites, United States. *Morbid. Mortal. Weekly Rep.* **51**, 325–329.

Collins, J. E. (1997). Impact of changing consumer lifestyles on the emergence/reemergence of foodborne pathogens. *Emerg. Infect. Dis.* **3**, 258–261.

Cornfield, J. (1951). A method of estimating comparative rates from clinical data. Applications to cancer of the lung, breast, and cervix. *J. Natl Cancer Inst.* **11**, 1269–1275.

Dolman, G. E. (1964). Botulism as a world health problem. In: K. H. Lewis and K. J. Cassell Jr (eds), *Botulism*, pp. 5–32. US Department of Health, Education, and Welfare, Public Health Service, Cincinnati, OH.

Food Safety and Inspection Service, US Department of Agriculture (2004). FSIS News Release, 5 August 2004.

Greenland, S. and D. C. Thomas (1982). On the need for the rare disease assumption in case-control studies. *Am. J. Epidemiol.* **116**, 547–553.

Guzewich, J. J., F. L. Bryan and E. C. D. Todd (1997). Surveillance of foodborne disease. I. Purposes and types of surveillance systems and networks. *J. Food Prot.* **60**, 555–566.

International Association of Milk, Food and Environmental Sanitarians (1999). *Procedures to Investigate Foodborne Illness*, 5th edn. IAMFES, Des Moines, IA.

Kass, P. H., T. B. Farver, J. J. Beaumont *et al.* (1992). Disease determinants of sporadic salmonellosis in four northern California counties. A case-control study of older children and adults. *Ann. Epidemiol.* **2**, 683–696.

Kinde, H. L., D. M. Read, R. P. Chin *et al.* (1996). Sewage effluent: likely source of *Salmonella enteritidis* phage type 4 infection in a commercial layer flock in Southern California. *Avian Dis.* **40**, 672–676.

Levine, R., T. Awerbuch, U. Brinkman *et al.* (1994). The emergence of new diseases. *Am. Scientist* **82**, 52–59.

MacMahon, B. and T. Pugh (1970). *Epidemiology, Principles and Methods*. Little, Brown and Company, Boston, MA.

Mead, P. S., L. Slutsker, V. Dietz *et al.* (1999). Food-related illness and death in the United States. *Emerg. Infect. Dis.* **5**, 607–625.

Morris, J. G. and M. Potter (1997). Emergence of new pathogens as a function of changes in host susceptibility. *Emerg. Infect. Dis.* **3**, 435–441.

Morse, D. L., G. S. Birkhead and J. J. Guzewich (1994). Investigating foodborne diseases. In: Y. H. Hui, J. R. Gorman, K. D. Murrell and D. O. Cliver (eds), *Foodborne Disease Handbook*, Vol. 1, pp. 547–603. Marcel Dekker, New York, NY.

Morse, S. S. (1995). Factors in the emergence of infectious diseases. *Emerg. Infect. Dis.* **1**, 7–15.

Olsen, S. J., L. C. Mackinnon, J. S. Goulding *et al.* (2000). Surveillance of foodborne disease outbreaks – United States, 1993–1997. *Morbid. Mortal. Weekly Rep.* **49**, 1–51.

Rothman, K. J. (1976). Causes. *Am. J. Epidemiol.* **104**, 587–592.

Rothman, K. J. and S. Greenland (1998). *Modern Epidemiology*, 2nd edn. Lippincott-Raven, Philadelphia, PA.

Snow, J. (1855). *On the Mode of Communication of Cholera*, 2nd edn. Churchill, London. Reproduced in *Snow on Cholera*, Commonwealth Fund, New York, 1936. Reprinted 1965 by Hafner, New York, NY.

Tauxe, R. V. (1997). Emerging foodborne diseases: an evolving public health challenge. *Emerg. Infect. Dis.* **3**, 425–434.

Todd, E. C. D. (1989). Preliminary estimates of cost of foodborne diseases in the United States. *J. Food Prot.* **52**, 595–601.

Todd, E. C. D. (1994). Surveillance of foodborne diseases. In: Y. H. Hui, J. R. Gorman, K. D. Murrell and D. O. Cliver (eds), *Foodborne Disease Handbook*, Vol. 1, pp. 461–536. Marcel Dekker, New York, NY.

Todd, E. C. D., J. J. Guzewich and F. L. Bryan (1997). Surveillance of foodborne disease. IV. Dissemination and uses of surveillance data. *J. Food Prot.* **60**, 715–723.

Zink, D. L. (1997). The impact of consumer demands and trends on food processing. *Emerg. Infect. Dis.* **3**, 467–469.

Microbial food safety risk assessment

2

Anna M. Lammerding and Ewen C. D. Todd

1 Introduction

Foodborne illness has major consequences worldwide, in terms of human health and financial burden to society and the food industry (Buzby *et al.*, 1996; Todd, 1997; Buzby and Frenzen, 1999; Frenzen *et al.*, 1999; Sockett and Todd, 2000). In the US, it has been estimated that about 76 million illnesses, 325 000 hospitalizations and 5020 deaths are caused by foodborne agents every year (Mead *et al.*, 1999). Unknown agents account for more than 62 million (81%) of these cases. Of the 14 million cases in which the agent is identified, Norwalk-like viruses are estimated to cause 9.3 million illnesses, and *Salmonella, Listeria* and *Toxoplasma* are responsible for most of the 1510 deaths attributable to an etiological agent.

In recent years, many new food safety pathogens have been recognized and new food safety issues have arisen, including, for example, hamburger meat, apple cider, alfalfa sprouts and well-water contaminated with enterohemorrhagic *Escherichia coli*; coleslaw, soft cheeses, pâtés and hot dogs contaminated with *Listeria monocytogenes*; tomatoes, sprouts and melons contaminated with *Salmonella enterica*; mesclun lettuce, basil and raspberries contaminated with *Cyclospora cayetanensis*; *Cryptosporidium parvum* in water and produce; and antimicrobial resistance in several genera of pathogens (Smith and Fratamico, 1995; Beuchat, 1996; Altekruse *et al.*, 1997; Como-Sabetti, 1997; Tauxe, 1997; Van Beneden *et al.*, 1999; Millar *et al.*, 2002; White *et al.*, 2002). Better laboratory detection methods, targeted surveillance and better outbreak recognition have improved our ability to identify and track foodborne disease. However, there is a general consensus

Foodborne Infections and Intoxications 3e
ISBN-10: 0-12-588365-X
ISBN-13: 978-012-588365-8

that over the last two decades there has been a real increase in foodborne illness, although more recently there may be evidence of a decline in some foodborne diseases, at least in the US (Vugia *et al.*, 2003). Despite exemplary improvements in the food industry at large, and unprecedented efforts of public and private organizations, disease-causing microorganisms are capable of adapting to new niches, new vehicles of transmission and new hosts, acquiring novel resistance and virulence mechanisms along the way. Microbiological hazards in food and water continue to impose a significant burden on public health and the economics of food production.

The recent statistics of foodborne disease indicate the extent of the problem and the need for safe preparation and handling of foods (Farber and Todd, 2000). However, although the risks are apparent, there is a need to better define the extent and severity of foodborne disease, especially for food products in commerce, together with a re-examination of current food safety priorities and practices. In 1994, the US Council for Agricultural Science and Technology published a report advocating the use of risk assessment techniques to establish priorities for managing and improving food safety (Foegeding and Roberts, 1994). The recommendations were revisited in 1998 with even stronger recommendations to base food safety policy on risk assessment and risk management practices (Foegeding and Roberts, 1998). It is recognized that there is no such thing as zero risk, and that it is not feasible to reduce all risks for all foods. However, food industry and regulatory risk managers should identify those risks that have the largest impact on public health, and implement programs to reduce the level of risk to the minimum that is practical, technologically feasible and socially acceptable. Recently, a risk-based approach to controlling foodborne disease has been proposed by the International Commission on Microbiological Specifications in foods (ICMSF, 2002). This includes setting specific public health goals relating to reduction of a disease by a certain time, and food safety objectives to meet these by industry.

What is 'risk'? Risks, and their assessment, are common in our everyday lives. Generally, we think about risk as the likelihood, or probability, of some adverse event or situation occurring. We may be able to quantify it in broad terms, such as one-in-a-million chance if we think it is a rare possibility. However, that thought process should also address not only how likely it is that something will go wrong, but also how bad it will be if it does go wrong (Kaplan and Garrick, 1981). For perspective, a common example might be that although the likelihood of falling in the home is greater than that of having an automobile accident, the latter gives us more concern, partly because we read or hear about the more serious accidents more frequently.

Evaluations of risks associated with foodborne hazards in the past have most often been general considerations of the hazard, routes of exposure, handling practices and/or consequences of exposure. Quantifying any of these elements is challenging, since many factors influence the risk of foodborne disease. These complicate interpretations of data about the prevalence, numbers and behavior of microorganisms, and confound the interpretation of human health statistics. Consequently, policies, regulations, and other types of decisions concerning food safety hazards have been largely based on subjective information and observations. However, advances in science combined with increased consumer awareness, global trade and recognition of the economic and social impacts of microbial foodborne illness have moved us toward the threshold of using quantitative risk assessment to help determine food safety priorities.

This chapter focuses on aspects relevant to microbial food safety risk assessment. However, the principles and practices discussed also apply to assessment of the risk from all types of foodborne hazards, including all biological, chemical, and physical hazards.

2 Movement away from traditional methods to evaluate the safety of food

2.1 The need for change

Traditional visual or organoleptic tests to detect visible contamination and deterioration of raw foods, and end-product laboratory testing for pathogens or indicator organisms in prepared foods, does not necessarily deliver reasonable assurances of safety. This is because of the practical limitations of sampling enough of the food product, the time needed to obtain test results, and the lack of sensitivity and/or specificity of current methods (ICMSF, 2002). Pathogens are infrequently present in foods, and if they are they are usually present only in very low numbers. However, most bacterial agents can multiply under conditions of temperature abuse and/or survive improper cooking, and some can cause illness even if only a few cells are ingested. Thus, a more practical risk-based approach is required to make management decisions relative to the safety of a product up to the moment of consumption.

2.2 Regulation

An important motivation for the adoption of formal quantitative risk assessment is the demand for rational, scientifically valid regulations to replace the traditional, subjective approaches that were developed many years ago. For example, to move away from prescriptive regulations about the numbers of sinks, or cleanliness of walls in food-processing operations, to an 'outcome-based' approach focused on those aspects of food processing, preparation and handling critical to reducing or eliminating pathogens. Thus, a system is needed that quantifies the relationship between regulatory requirements and public health outcomes and will help to establish priorities for risk management.

2.3 HACCP

Given current systems of primary food production, it is difficult to prevent raw materials contaminated with potentially pathogenic microorganisms from entering the food supply. Food safety risk management strategies must include steps in the processing of foods that reduce or destroy microbes, prevent recontamination of product, and improve retail, food service and consumer practices to prevent the mishandling of foods. The Hazard Analysis Critical Control Point (HACCP) approach is one such strategy that can be applied to the entire food chain, from production to consumption. Key steps in the processing, distribution, marketing and preparation of a food that are critical to the safety of the product (i.e. critical control points or 'CCPs') are monitored and controlled. However, HACCP is a risk-based management system that has been developed on the basis of qualitative assessments of hazards, their impacts and their control. The efficacy of HACCP is limited by its inability to quantify the potential combined result of multiple control-point deviations, and to relate the operation of a HACCP system to a measurable public health outcome (Buchanan and Whiting, 1998a).

2.4 Global food trade

Perhaps providing the greatest impetus for food safety risk assessment is international trade. The worldwide recognition of the importance of international commerce and the need to facilitate trade and ensure the quality and safety of food led to the establishment of the Joint FAO/WHO Food Standards Programme and the Codex Alimentarius Commission (CAC) in 1962 (Lupien and Kenny, 1998). Standards for foods in international trade are developed and agreed to by consensus among the Codex member countries. The Uruguay Round of the General Agreement on Tariffs and Trade (GATT), which was replaced by the World Trade Organization (WTO) in 1995, declared that formal risk assessment must be the basis for food standards and the resolution of non-tariff trade issues (Horton, 2001). Thus, it is important to develop uniform approaches to risk assessment.

3 Managing risk through risk assessment

Structured scientific risk assessment processes were introduced within US federal regulatory agencies during the late 1970s as a means of standardizing the basis for decision-making. These were driven by the need for regulatory action in situations where large numbers of people were, or could be, exposed to relatively low levels of chemical substances that had been identified as hazardous to health, but only under conditions of relatively intense exposures (Rodricks, 1994). The actual decision-making processes and the communication of information from the science-based risk assessment activities were defined as risk management and risk communication (NRC, 1983, 1994). The three elements, risk assessment, risk management and risk communication, are collectively referred to as risk analysis. This approach is widely used in fields such as environmental health, toxicology, cancer research, and chemical and nuclear industries. More recent is the use of risk analysis in microbial food safety.

It is noted that within each discipline, and even among workers within a field, some differences exist in terminology and approaches to categorizing activities and evaluating the risk parameters. However, the basic framework described by NRC underlies most developments in this field. At the international level, Codex is responsible for defining risk assessment principles and practices for all foodborne hazards, and for promoting consistency and clarity in the establishment of Codex standards through the use of risk analysis. Through a series of consultations, international discussion, debate and consensus, a framework, principles and guidelines have evolved for the application of risk analysis for food safety hazards (FAO/WHO, 1995, 1997, 1998). Table 2.1 lists the current definitions of terms for risk analysis, adopted in 1999 (CAC, 1999).

Table 2.1 Definitions of risk analysis terminology for foodborne hazards (CAC, 1999)

Term	Definition
Hazard	**A biological, chemical or physical agent in, or condition of, food with the potential to cause an adverse health effect**
Risk	A function of the probability of an adverse health effect and the severity of that effect, consequential to a hazard(s) in food
Risk analysis	A process consisting of three components: risk assessment, risk management and risk communication
Risk assessment	A scientifically based process consisting of the following steps: (i) hazard identification, (ii) hazard characterization, (iii) exposure assessment, and (iv) risk characterization
Quantitative risk assessment	A risk assessment that provides numerical expressions of risk and indication of the attendant uncertainties
Qualitative risk assessment	A risk assessment based on data, which, while forming an inadequate basis for numerical risk estimations, nonetheless when conditioned by prior expert knowledge and identification of attendant uncertainties permits risk ranking or separation into descriptive categories of risk
Hazard identification	The identification of biological, chemical and physical agents capable of causing adverse health effects and which may be present in a particular food, or group of foods
Hazard characterization	The qualitative and/or quantitative evaluation of the nature of the adverse health effects associated with the hazard; for the purpose of microbiological risk assessment, the concerns relate to microorganisms and/or their toxins
Dose–response assessment	The determination of the relationship between the magnitude of exposure (dose) to a chemical, biological or physical agent and the severity and/or frequency of associated adverse health effects (response)
Exposure assessment	The qualitative and/or quantitative evaluation of the likely intake of biological, chemical and physical agents via food as well as exposures from other sources if relevant
Risk characterization	The process of determining the qualitative and/or quantitative estimation, including attendant uncertainties, of the probability of occurrence and severity of known or potential adverse effects in a given population based on hazard identification, hazard characterization and exposure assessment
Risk estimate	Output of risk characterization
Transparent	Characteristics of a process where the rationale, the logic of development, constraints, assumptions, value judgments, decisions, limitations and uncertainties of the expressed determination are fully and systematically stated, documented and accessible
Uncertainty analysis	A method to estimate the uncertainty associated with model inputs, assumptions and structure/form
Risk management	The process of weighing policy alternatives in the light of results of risk assessment and, if required, selecting and implementing appropriate control options, including regulatory measures
Risk communication	The interactive exchange of information and opinions concerning risk and risk management among risk assessors, risk managers, consumers and other interested parties

Definitions and usage of the term 'risk' vary. Frequently, the word 'risk' is used to reflect the probability or likelihood of an event, i.e. the probability that an identified hazard will cause harm. However, more precisely, as indicated earlier, risk is a function of both the probability of an adverse outcome and the impact of that adverse event attributable to a hazard. 'Severity' is frequently used to describe the seriousness of the effect(s) of the hazard. For microbial hazards, estimates of severity include but are not limited to the number and impact of the symptoms, the proportion of cases hospitalized, the case/fatality ratio, the long-term effect of sequelae, and the duration of illness.

3.1 Risk assessment

Risk assessment is the scientific evaluation of the probability of occurrence and severity of known or potential adverse health effects resulting from exposure to a hazard. The result is an estimate of the likelihood and impact of an adverse effect on human health produced by a likely level of exposure to a hazardous agent.

Risk assessments are typically divided into four distinct steps: hazard identification, exposure assessment, dose–response assessment, and risk characterization (NRC, 1983). This basic framework has been modified by introducing the term 'hazard characterization' – a step that includes a dose–response assessment if the data are available, but allows more subjective assessments of the consequences of exposure if a dose–response assessment cannot be carried out (Figure 2.1).

3.2 Risk management

Risk management includes consideration of policy alternatives in light of the results of risk assessment and the selection and implementation of appropriate control options. Industry and regulators should strive for production and processing systems that ensure that all food is safe and wholesome. However, it is acknowledged that complete freedom from hazards is an unattainable goal.

The primary goal of the management of risks associated with food in international trade is to protect public health by controlling such risks as effectively as possible. The decision-making process takes into consideration the scientific evidence within the context of other factors and/or values. These may include cost–benefit assessment; social, cultural and ethical concerns; and political and legal issues. The extent to which decision-makers take into account additional factors is dependent on the risk issue and its context. For example, in the resolution of trade disputes between countries or in establishment of international food standards, decisions should be principally based upon the scientific assessment of the risk of human illness, and of the degree to which there are viable strategies to reduce the risk.

In the establishment of national priorities in relation to public health, managers will generally consider in their deliberations the values of regulation, surveillance or research; public perceptions of the risk and other social concerns; and the economics of different management options. The food industry will generally need to consider the business costs in making decisions. Risk management strategies for foods may range from promulgating regulatory standards and issuing microbiological guidelines, to

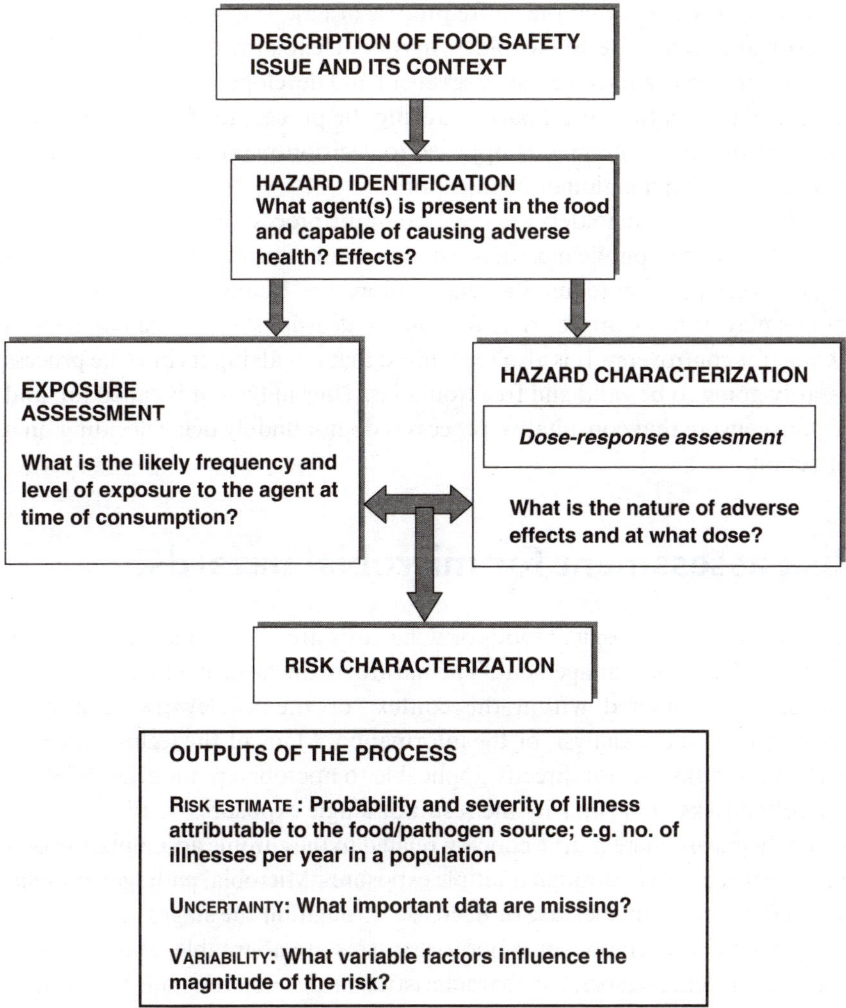

Figure 2.1 Steps of microbial food safety risk assessment (adapted from Lammerding and Fazil, 2000).

product labeling and consumer education. Options may include the implementation of interventions at any part of the food chain to prevent, reduce or eliminate contamination of food or limit opportunities for microbial growth. Finally, banning a food from the marketplace is also an action that can be taken if the risk from consuming a product is unacceptably high and no other feasible management options exists.

3.3 Risk communication

Risk communication emphasizes the need to exchange information and opinion interactively among risk assessors, risk managers and other interested parties. Interested parties, or stakeholders in the risk issue, may include, for example, growers, producers, processors, food workers, and consumer groups and associations.

Increasingly, it is realized that the entire process of assessing and managing risks should be open and interactive. Stakeholders may be able to provide specialized information and insights relevant to the risk assessment and development of management strategies. The participation of all parties during the process tends to increase the acceptability of the final outcome, as opposed to decision-making with no external input or no apparent explanation or justification.

The extent to which stakeholders are involved in the process, at what stage, and in what kind of forum (e.g. public meetings, written communications) will depend on the issue. Not every decision requires extensive processes. Limited deliberations and consultations may be more appropriate for routine decisions with little impact and little potential for controversy. It is also recognized that not all inputs into the process are necessarily going to be valid and free from bias. Thus judgment is called for, and managers must ensure that consultative processes do not unduly delay deciding on a course of action.

4 Risk assessment for microbial hazards

Chemical, physical and biological foodborne hazards are all considered within the realm of food safety risk management. The nature of the hazard will dictate what factors must be considered within the context of the framework, and what approaches are used in the analysis of the information. Many of the techniques used to assess chemical risks are not directly applicable to microbial pathogens (ICMSF, 1998). Microbial risks are primarily the result of single exposures. Unlike toxic or carcinogenic chemicals, there is little concern related to the chronic accumulation of a pathogen or microbial toxin through multiple exposures. Microbial pathogen populations are dynamic, and may increase or decrease throughout the stages of food processing, handling and preparation. Microorganisms are adaptable, and they may acquire or lose virulence-associated characteristics, or develop resistance to antimicrobial control measures. The risk assessment should consider microbiological growth, survival and death in foods, and the complexity of the interaction (including sequelae) between human and hazard following consumption. For many pathogens, the potential for further spread of the organisms from an infected host to others is also an important risk factor.

Biological agents of concern to public health include pathogenic strains of bacteria, molds, viruses, helminths, protozoa and algae, and certain toxic products they may produce. Bacteria that produce toxins, such as *Clostridium botulinum* and *Staphylococcus aureus*, require consideration of bacterial growth/inactivation characteristics as well as the resistance of the toxin to heat and other inactivation processes. Parasites and viruses are infective agents that do not grow in foods, and therefore reduction by decontamination or inactivation steps is an important consideration in an assessment. There are also two major groups of non-bacterial microbial toxins that can occur in foods: mycotoxins, produced by certain molds, and seafood toxins, synthesized by certain species of marine phytoplankton, mainly dinoflagellates. Since mycotoxins are normally ingested in small quantities over long periods of time, their effect is more like chemical agents producing chronic illness, and they are normally treated as chemicals in the risk assessment process. In contrast, most of the seafood toxins cause acute effects, some with long-term sequelae, and perhaps should be included with the microorganisms and their products in risk assessments. However, there are other toxins, e.g. those produced by cyanobacteria (blue-green algae) in fresh or brackish water, that may cause both acute and chronic effects. In addition, the emergence of new foodborne diseases, such as hemorrhagic colitis and hemolytic uremic syndrome caused by *Escherichia coli* O157:H7, may have resulted from the incorporation of new genetic material from other known pathogens. As the field of microbial food safety risk assessment advances, there will be a need to develop novel techniques to assess or quantify the exchange of genetic material among pathogens, and the development of resistance to antimicrobial agents (Salisbury *et al.*, 2002). However, for the most part, quantitative microbial risk assessments so far have assumed that all strains of the pathogen being assessed have similar characteristics, including their virulence.

5 Conducting a risk assessment

5.1 Purpose of the risk assessment and approaches to the process

Formal assessments require time, expertise and data. The decision to conduct a risk assessment should be made carefully. Not all food safety issues or management decisions require risk assessments, and in some cases not all elements of a risk assessment are needed. For example, there may be circumstances where only an exposure assessment is appropriate. Risk assessment is just one tool for decision-makers, and food safety managers should consider this within an overall sound decision-making framework.

Detailed, comprehensive assessments are generally recommended when significant decisions must be made about implementation of new regulations, or to resolve major safety or trade disputes. However, in addition to estimating risk, a systematic analysis of any issue can be useful for many different types of problems. An important benefit can be to identify significant data gaps about any one issue, and hence set priorities for research funding.

The work begins by identifying and understanding what information the risk manager needs in order to make a risk management decision and to help others understand that decision (ILSI, 1996; Lammerding and Fazil, 2000). Risk assessments can be developed to help rank hazards in relation to the impact they have on public health. This is most often used as a means of setting priorities when multiple health concerns are competing for limited resources. A second category is more detailed analyses of entire 'farm-to-fork' pathways, in order to identify likely sources of a pathogen and evaluate the impact that various activities associated with the production, transportation, marketing and consumption of a food product have on the final consumer risk. These types of assessments have been described as product/ pathogen pathway analyses, and Process Risk Models (Cassin *et al.*, 1998a, 1998b).

A detailed focus on only one phase of the food process is also useful to determine what interventions, if any, at that stage could help prevent foodborne illness. The food safety concern may be about a specific food produced by a specific process, or an entire commodity group – such as the risk of salmonellosis from all eggs produced in a country or region. The assessment may focus on a specific pathogen, or consider all microbial hazards associated with a foodstuff.

Stating the reason for doing the assessment at the beginning of the work will help to define the context and parameters of the assessment, and should reflect the importance of the activity. Setting guidelines for value judgment and policy choices, when assessors need to make decisions about selection or suitability of data, for example, is a risk manager's responsibility, and should be carried out together with the risk assessor(s). The output form and possible output alternatives of the risk assessment should also be defined at the onset. Finally, choosing the most suitable approach to a specific risk assessment problem will depend on the problem being investigated, the time needed for the work and the availability of data.

Within the general guidelines, different approaches can be taken to assemble and analyze relevant information. In very general terms, a risk assessment may be qualitative or quantitative (or degrees in between); quantitative assessments may be point-estimate analyses or probabilistic (stochastic) analyses. The focus of the assessment may be only to derive a measure of the risk (e.g. for risk ranking), or it may be to help understand what factors in the production, distribution and consumption of a food contribute most significantly to a large risk, what data gaps exist, and what management strategies may be most effective.

Descriptive *evaluations* of a pathogen/food concern generally describe the background of the issue by providing a review of existing literature, and possibly providing suggestions for risk management. This has been the traditional approach to health hazard evaluations, and can be of value if the management question is simple and a clear-cut decision recommended. However, such descriptive reports generally lack an estimation of the magnitude of the risk, which, by definition, is the purpose of formal risk assessment.

Qualitative risk assessments include descriptive components but evaluate exposure and dose–response information in a categorical manner so that some estimate of likelihood and impact can be made. A qualitative estimate of risk may be derived using a ranking system to describe the probabilities of exposure and of becoming ill, such as negligible, low, medium or high. Impact or severity of illness can be similarly ranked. Such a process has been used to rank risks associated with various seafood products (Huss *et al.*, 2000). However, qualitative statements and measurements must be precisely defined to avoid misinterpretations. Ross and Sumner (2002) provide an example of this within a spreadsheet software format. The user of the software selects from qualitative statements and/or provides quantitative data concerning factors that will affect the food safety risk to a specific population, arising from a specific food product and hazard. The spreadsheet converts the qualitative inputs into pre-defined numerical values and combines them with the quantitative inputs to generate indices of public health risk. Sumner and Ross (2002) applied the system to rank 10 seafood hazard/product combinations on a scale of 0 (no risk) to 100 (all meals are lethal).

When time, resources and/or data are limited, a simplified approach may be appropriate. Such an process may also provide a preliminary type of assessment to determine if a potential risk is important enough to warrant more detailed analysis (van Gerwen *et al.*, 2000). However, at this time, no specific guidelines exist for qualitative or simple quantitative types of assessments, other than adherence to the general framework and principles of microbial risk assessment.

If the risk manager seeks information on the expected number of acute salmonellosis cases per 100 000 population associated with eggs or poultry, or the risk of salmonellosis per egg or chicken meal, quantitative risk assessment is required. This is a mathematical analysis of information, and gives a numerical estimation of risk. Until recently, the most common method for quantitative assessments was to use single values or point-estimates as inputs – for example, combining means, the fiftieth percentiles (medians) or the ninety-fifth percentiles (as an example of a worst-case scenario). The result of a point estimate assessment is an average, or worst-case, estimate of the risk. This approach has limitations in producing realistic outputs, particularly for diverse, dynamic and variable biological systems. This is particularly true if many worst-case scenarios are combined for an overall estimate of risk, because it leads to a value for the risk that is highly unlikely to occur. An example is the early assessment of Peeler and Bunning (1994), which was criticized for its use of worst-case scenarios by Cassin *et al.* (1996). The alternative to point-estimate analyses is probabilistic risk assessment, which uses the entire range of possible values for each input variable, described by unique distribution curves. The combined output calculated from the individual distributions of input factors is a distribution of the risk in a population, or for any random meal.

The difference between a point estimate and a probability distribution to describe an input is illustrated in Figure 2.2. The example shows the concentration of a pathogen in a unit of food. It can readily be seen that there is a substantial loss of information when a single point is used to describe an entire set of reported data. The

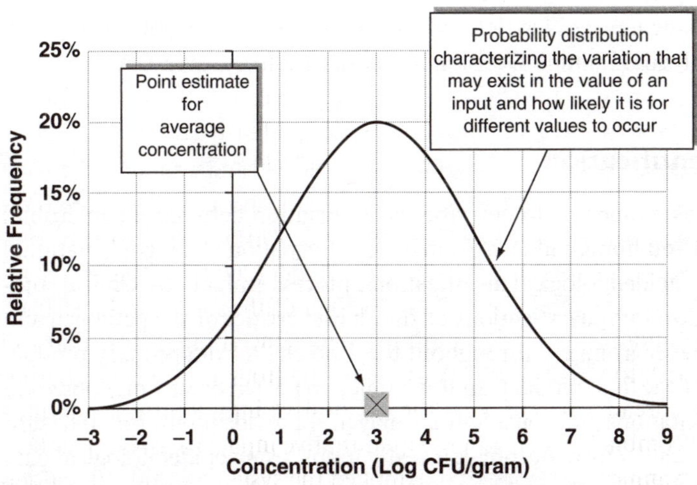

Figure 2.2 Comparison of point estimate and probability distribution for a data set.

point-estimate specifies the value that a parameter could take, while the probability distribution specifies the range of values that could occur, as well as how frequently different values occur. Probability distributions are assigned based on actual laboratory or observational data, on knowledge of the underlying biological phenomena, or on expert opinion if no other information is available (Vose, 2000).

The importance of acknowledging the range of possible values is underlined by recognizing that it is unlikely that microbial risks to human health are uniformly distributed, or that 'average' occurrences or events are likely to cause significant problems (Potter, 1994). Consideration of the extremes, or tails, of how likely or unlikely it is that such events will occur and who might be affected should be part of sound risk management deliberations (Thompson and Graham, 1996). This is especially true when trying to consider the possibility of an outbreak occurrence where often it is the case that a failure or loss of control happens at more than one stage of the food chain – for example, temperature abuse during retail storage of raw meat, allowing growth of a pathogen, combined with cross-contamination in the home, or undercooking.

Probabilistic risk assessments can be evaluated using analytical techniques, but only for very simple models. More typically, these types of assessments require the application of Monte Carlo analysis – a numerical technique that is especially suited to computer applications. Monte Carlo analysis is based on randomly selecting a single number from each of the probability distributions assigned to the input parameters. These values are used to calculate the mathematical solution defined by the risk assessment model, and the end result is stored. This sequence is repeated several thousand times (iterations), with a different set of numbers for the inputs selected at each iteration. Values that are more likely to occur, according to the defined probability distribution, are selected more frequently. Within a complex model there are several distributions that will be combined – for example, distributions of specific pathogens in cattle, in the slaughterhouse, and at various steps in processing, storage, growth, cooking and serving. The ranges and frequencies of all the individual input parameters are combined to generate the output of interest. A simplified illustration of a Monte Carlo model is shown in Figure 2.3. Commercial software for Monte Carlo simulations include @Risk (Palisade Corp., Newfield, NY), Crystal Ball (Decisioneering, Inc., Denver, CO) and Analytica (Lumina Decision Systems, Inc., Los Gatos, CA).

5.2 Hazard identification

The first step in risk assessment is to describe the association between the microbial pathogen(s) in a food and human illness. Sources of information can include national surveillance databases, epidemiological investigations, process evaluations, clinical studies, animal studies, laboratory investigations of the characteristics of the pathogen and its interaction with the environment throughout the food chain from primary production to consumption of the final product, and/or studies on related organisms, foods and conditions. Expert elicitations and consultations may also provide input. Initial hazard information typically results from outbreak investigations, using epidemiological data and laboratory identification of the pathogen in human specimens and in the samples of food consumed. Laboratory-based surveillance also provides valuable information for

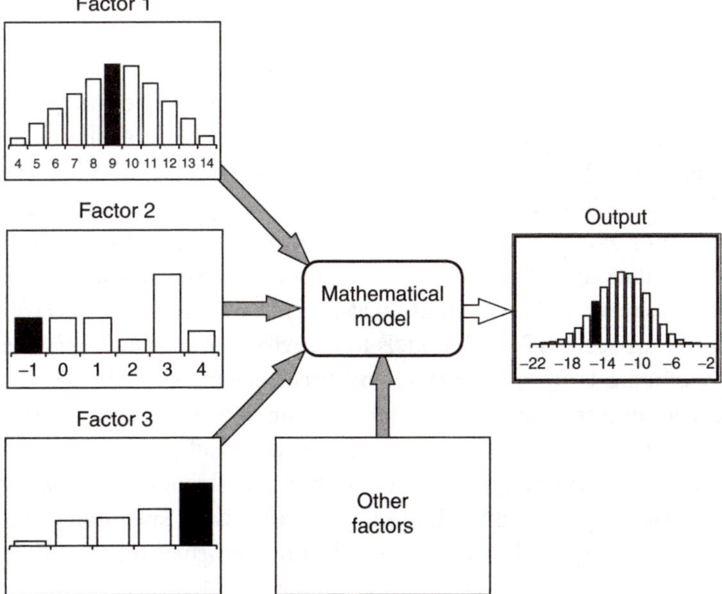

Figure 2.3 Illustration of a Monte Carlo simulation model for probabilistic quantitative risk assessment. In each iteration of the model, a single point value is randomly selected from the distributions defined for each input factor (bold columns in the graphs for each factor). For the exposure assessment, factors include prevalence and concentration of the pathogen in the food at different stages in the production, and consumption data such as frequency and amount eaten. The dose-response assessment would also be a major factor in a full assessment. These values are used in the calculation of the single output value in say, the probability of a contaminated serving or the number of people ill. The next iteration selects a new set of single values from each distribution, calculating a new output value. Repeated thousands of times, the distribution of the output is constructed from the individual input distributions, and the mathematical relationship between the parameters.

risk assessment, including characterizing the stages of food production, processing and handling, and also the populations that may be at greatest risk.

A hazard identification step is primarily a brief qualitative description of the risk issue, providing a background and rationale for the assessment. When the scope of the issue is broad – for example, all microbial hazards associated with the consumption of a meat product – the hazard identification stage can be used to identify the most significant organism(s), typically the most resistant or virulent pathogen that would be of most concern in the food. This allows narrowing the focus for more detailed analysis.

5.3 Exposure assessment

The exposure assessment is an estimate of the likelihood of ingesting a pathogen and the likely number (or dose) of the organism at the time of consumption. The nature of the pathogen and/or microbial toxin, the type of food, the scope of the assessment and the availability of relevant data will dictate the parameters that must be considered. The most important elements of an exposure assessment are data on the prevalence and concentration of a pathogen in the final product, and the relevant consumption data for that product (Jaykus, 1996; ICMSF, 1998; Lammerding and Fazil, 2000).

Factors affecting the presence and level of the agent, from the source of contamination (including interaction with the environment, soil and water) up to the point of consumption, should be considered. The exposure unit should be considered as the unit that could potentially result in illness, and for most biological agents in food this is normally considered to be a single-meal serving size. Rarely are data available at the exact point of consumption, and hence estimations must be derived from what is known about the contamination of the product at an earlier stage, and what factors will affect the level of contamination at the time of ingestion. This will require the construction of scenarios and models to describe and predict the range of possible exposures. In a quantitative assessment, this stage involves the derivation of mathematical equations to describe the relationship between model parameters. All potential sources of entry of the hazard into the food product should be evaluated, and consideration given to the verifiable effectiveness of existing control measures. In assessing individual data sets, the sensitivity, specificity and validity of sampling and testing methods used to collect the information should be taken into account.

Data on the prevalence and/or concentration of microorganisms in food are usually collected for raw materials before processing, during production, or in a finished product either before or after distribution to the retail level. Microbial populations may either increase through growth or decrease through inactivation or dilution with uncontaminated material during a formulation step. Most often, numbers of pathogens are too low to detect easily, or to detect without using non-quantitative enrichment. Microbial pathogen levels may be kept low, for example, by proper time/temperature controls during food processing, but they can substantially increase under abusive conditions such as improper food storage or inadequate cooking temperatures. Types of processing, the storage environment and its temperature, the relative humidity of the environment, and the gaseous composition of the atmosphere influence the survival and growth of microorganisms in foods. Other relevant factors include pH, moisture content or a_w, nutrient content, the presence of antimicrobial substances, and competing microflora. The time for transit from production to retail, and the length of storage by the retailer and consumer, are also critical factors influencing growth of the pathogens and need to be estimated. A final product inactivation step can be heat processing in a package, or cooking. Key elements that are typically unknown in a process, but which may greatly influence final concentrations of pathogens, are the consumer storage time and temperature for ready-to-eat foods.

Predictive food microbiology uses microbiological, mathematical and statistical information to develop models that describe the growth and decline of microbes under specified conditions, and can be used to predict population changes relative to changes in specific parameters (Ross and McMeekin, 1994; Whiting and Buchanan, 1994; van Gerwen and Zwietering, 1998). Such parameters include temperature, pH, a_w, and the growth medium (i.e. in a food or nutritive broth). Pioneering work in predictive microbiology in the 1920s resulted in thermal death time calculations using D and z values to predict the safe processing conditions for different sizes of various canned or pasteurized food products. These models have been extended to irradiated foods and are still in use today. Advances in this field provide valuable techniques for estimating exposure for microbial risk assessment. Whiting and Buchanan (1994) describe three levels of predictive models. The primary level models quantify colony-forming units/ml, production of toxin and other metabolic products, as well as absorbance and impedance, over time. The growth rate of a pathogen in broth can be calculated, for example, with the Gompertz function, one of the most widely used primary level models, to describe the lag, exponential growth and stationary phases. Secondary-level models show how other conditions affect the primary one – for example, how growth is increased or decreased by changes in pH or a_w. Buchanan and Whiting (1996) describe natural log-quadratic equations to measure growth in broth of different pathogens (*Aeromonas hydrophila, Bacillus cereus, Escherichia coli* O157:H7, *Listeria monocytogenes, Shigella flexneri, Salmonella enterica*, and *Staphylococcus aureus*) under different environmental conditions. Tertiary-level models depend on computer software to turn primary- or secondary-level models into packages for modelers. Using such software can test the potential effects of changing conditions on the growth of pathogens. One of the most widely used is the Pathogen Modeling Program of the USDA, originally developed by Buchanan and Whiting (1998b). This can be used for generating exposure assessment data to predict, for example, when a pathogen will grow to levels that could cause human infections.

Additional factors that are considered in exposure assessment are the distribution of the agent in the food, consumer handling practices, consumption patterns, and host demographics. Consumption patterns relate to socioeconomic and cultural backgrounds, ethnicity, seasonality, age differences, regional differences, and consumer preferences and behavior. A high-risk population is a segment of the population that has an increased likelihood of exposure to a hazard, an increased likelihood of illness due to exposure to a hazard, and/or an increased likelihood that the illness resulting from exposure to a hazard will be severe or life threatening (Gerba *et al.*, 1996).

Typical meals involving the food in question can be determined from national surveys or small population studies. Data for specific target groups, like infants or the aged, are useful. There may also be cultural, social, economic or demographic factors that might influence estimation of consumption patterns and practices. When risk assessments are conducted for international trade considerations, differences in exposure data between countries and regions must be considered. Between nations, there will be some differences in pathogen prevalence and concentration attributable to underreporting due to the existing national control programs, as well as a real variation due to geographic and ecological differences. Food distribution systems can vary from one country to the next, including, for example, greater or lesser temperature control during storage, or differences in transit times. More typically, for most pathogen–food combinations in exposure assessments, prevalence and concentration data will be so limited that worldwide data have to be used. Consumption data, however, are national or regional statistics.

5.4 Hazard characterization and dose–response assessment

Hazard characterization is a qualitative or quantitative description of the consequences upon exposure to a pathogen or its toxin in a food. The specific response or adverse effect that is being measured may be infection (intestinal colonization without symptoms of illness), acute mild to acute severe illness, chronic complications such as reactive arthritis, and/or death. The dose–response assessment specifically refers to a mathematical relationship that translates the number of organisms ingested (e.g. colony- or plaque-forming-units) into a probability of an adverse outcome.

The likelihood of a pathogen causing illness is dependent on (a) the characteristics of the organism itself – e.g. virulence factors, resistance to gastric acidity and the host's immune response; (b) the susceptibility of the host – e.g. immunocompetence or nutritional status; and (c) the characteristics of the food – e.g. a food with high fat content will protect the organism from gastric acidity.

Two distinct hypotheses have been proposed for the nature of the dose–response relationship for foodborne pathogens. The first is that there exists a threshold number of organisms, or minimum infectious dose, that must be ingested before any infection or illness occurs. The second hypothesis is that each pathogen cell has an equal capacity to cause an infection or illness. Thus there is no threshold number, and the probability of infection increases as the levels of the biological agent increase. For example, it has been estimated that a single cell of *Shigella* spp., a pathogen noted for its high infectivity, has a probability of 0.005 of causing an infection. Another way of expressing this concept is that if 1000 people each consumed one *Shigella* spp. cell, five individuals in the group would become infected (ICMSF, 1998).

The nature of dose–response relationships is currently a widely debated area. The few data that are available are based on feeding studies and information from outbreak investigations. The limited number of controlled human feeding studies that have been conducted have usually involved very small numbers of healthy adult males, which makes extrapolations to general or at-risk populations very limited (Teunis *et al.*, 1996, 1999; Buchanan *et al.*, 2000). Another issue is that most of these studies were conducted many decades ago. Ethical considerations make it unlikely that many such studies will be conducted in the future. Information from experimental feeding trials with animal models must be cautiously interpreted when extrapolating to the human population, although they have been used for years in chemical risk assessments. Data collected from foodborne outbreak investigations can be invaluable, but unfortunately information such as numbers of the pathogen in the food, estimates of amount of food consumed and the proportion of people infected or ill among all individuals exposed is often not obtained or reported. Another approach that has been taken is to use monitoring data about the prevalence and concentrations of a specific pathogen in a specific foodstuff, and compare those data with the incidence of related illness in the population exposed to that food (Buchanan *et al.*, 1997). However, for most pathogens, sufficient data about their levels in foods are currently not available, nor does the reporting of most foodborne illnesses provide accurate statistics or attribution to a food. Finally, experts' opinions may help to arrive at some consensus about dose–response relationships (Martin *et al.*, 1995).

This does not substitute for scientific data, but, given the difficulty of collecting experimental data on harmful pathogens, expert judgment may provide an option for estimating relative risk to the population.

5.5 Risk characterization

The final step in risk assessment combines the information generated in hazard identification, exposure assessment and hazard characterization to produce a complete picture of the assessed risk. Risk managers and risk assessors at the beginning of the work should have defined the form of the risk assessment outcome, and the types of questions to be answered in a risk characterization.

The *risk estimate* should reflect the range of contamination of a food product, factors that might affect growth or inactivation of the pathogen, and the variability of the human response to the microbial pathogen. Risk characterization should also provide insights about the nature of the risk which are not captured by a simple qualitative or quantitative statement of risk. Such insights include, for example, a description of the most important factors contributing to the average risk, and a discussion of gaps in data and knowledge. The risk assessor may also include a comparison of the effectiveness of alternative methods of risk reduction for consideration by the risk manager.

Of necessity, the modeling of complex systems requires some simplification and the use of assumptions. These assumptions should be clearly stated and should be understood by the risk manager. Every effort should be made to compare the results produced against independent observed data if such data are available. At the very least, the assessment should undergo rigorous peer review to ensure that the results are reasonable and plausible, and that the data, models and assumptions were appropriate. The conclusions from a risk assessment should be defensible and reproducible.

5.6 Uncertainty and variability

Risk assessments should be based on scientific evidence and knowledge. In many cases, data may be lacking or conflicting, or give a wide range of values for any one measurement. Two factors, then, have to be considered: the uncertainty and the variability of the data. These are distinct and should not be confused. The degree of confidence in the final estimation of risk depends on the uncertainty factors identified in the previous steps. High-quality quantitative data are preferable to qualitative information, but even these are rarely complete, especially in a farm-to-fork food process. Assumptions will be made during the course of the assessment, based on informed judgment. Recognition and acceptance of a degree of uncertainty are fundamental to an estimation of risk.

The basis of uncertainty is two-fold. First, there is uncertainty regarding the quantity and quality of the information used in the assessment. Secondly, there is uncertainty regarding the validity of the assumptions made during the process. Both aspects influence the degree of confidence in the risk estimation. Scientific and statistical uncertainties include those that might arise in the evaluation and extrapolation of information obtained from epidemiological, microbiological and laboratory animal studies. Uncertainties arise whenever attempts are made to use data concerning the occurrence of certain phenomena obtained under one set of conditions to make estimations or predictions about phenomena likely to occur under other sets of conditions for which data are not available. One example is to use dose–response data for one pathogen (which already include a certain amount of uncertainty) as surrogate information applied to another related pathogen. More information can help to reduce uncertainty – for example, use of human outbreak data to limit the upper and lower bounds of the dose–response equation, or simply more research.

Variability, on the other hand, represents diversity or heterogeneity in a well-characterized population or phenomena. Biological variation includes the differences in virulence that exist in microbiological populations, and in the degree of susceptibility within the human population. Unlike factors that we are uncertain about, more research or further measurements will not reduce the variability in risk assessment data.

It is important to recognize the respective influences of uncertainty and variability on the results of a risk assessment and on the risk management decisions pursued. If the output of interest, i.e. the risk estimate, is influenced by uncertainty in a parameter, the management decision may be to do more research or collect more data to better characterize and understand that factor, and to add the new information into the assessment. However, if a risk mitigation decision is required under circumstances where uncertainty is significant and additional data are not readily obtainable, then a conservative (or 'safe') strategy may be warranted, with the understanding that more information would allow a better decision. If the variability in one or more parameters results in a large risk estimate, then better control of these processes or factors may be needed to reduce the risk. In a descriptive, qualitative assessment, the assessor will use descriptive estimates or language to highlight those factors that are uncertain and/or variable and have an important influence on the estimation of risk. Point-estimate quantitative assessments are limited because the use of single representative numbers excludes the variances in the data. Probabilistic assessments incorporate variability and uncertainty, and techniques are available to calculate quantitatively their respective influences on the risk estimate (Vose, 2000).

5.7 Modeling mitigation strategies

In its application to microbial food safety, an important benefit of risk assessment is the understanding derived from a systematic analysis of factors that may affect the presence of a pathogen in the food, and those related to the pathogen/host interaction. The approach is valuable in focusing researchers as well as managers on aspects for which insufficient data currently exist and, potentially, on steps in the food chain where interventions may be most effective.

The effectiveness of plausible interventions can be tested with risk assessment models, and specifically those constructed quantitatively using Monte Carlo computer modeling techniques (Cassin *et al.*, 1998b). Incorporating new data or hypothetical targets for pathogen reduction can create different scenarios, and the risk estimate can be recalculated with the new information or assumptions. The outcome of the recalculation allows comparison of proposed interventions against the current situation, in terms of how much risk reduction might be achieved. This application is very advantageous if examining multi-step processes and models. As an example, a risk assessment for *E. coli* O157:H7 in ground-beef hamburgers, which modeled the pathway from farm to human illness, was used to identify major factors that were important to the risk estimate for consumption of this product in North America (Cassin *et al.*, 1998a). A number of these factors were proposed as potential points for intervention. Hypothetical values were incorporated into the model for different stages – for example, reduced levels of *E. coli* O157:H7 entering a slaughterhouse by theoretically excluding cattle shedding high numbers. Other mitigations tested were the effect of stricter temperature controls in retail store meat cabinets, and some degree of consumer education to encourage better cooking of ground-beef patties in the home. Re-simulation of the model using these alternative values allowed a comparison of the risk reduction for each strategy. Clearly, the more accurate the initial assessment data, and the more realistic the assumptions about implementation and outcomes of interventions, the more realistic the calculated risk reduction estimates will be. Nevertheless, as quantitative risk assessments improve, the ranking of interventions based on their impact on the number of predicted illnesses will become important in assigning resources to improve food safety. This type of 'what-if?' experimentation with risk assessment models can provide more information for making risk management decisions, and insights for future research or data collection. These can be done relatively quickly once the basic model has been set up.

6 Applications of microbial risk assessment techniques for food safety

The field of microbial food safety risk assessment is relatively new, although advancing rapidly. Hence, only a limited number of examples are available in the literature at this time. Substantially more work is available on quantification of waterborne microbial risks, and the reader is directed to Haas *et al.* (1999) for further information and examples.

Quantifying microbial risks, or simply quantifying potential consumer exposure to pathogens in foods, was really not considered feasible until the mid-1990s. An example of a simple semi-quantitative risk estimation process was published by Todd (1996), regarding the risk of salmonellosis from eggs. The first efforts fully to quantify potential consumer exposures to pathogens in a food used point-estimate values for bacterial concentrations and prevalences, together with survival kinetics and consumption information, for *L. monocytogenes* (Peeler and Bunning, 1994) and *B. cereus* (Notermans *et al.*, 1997) in pasteurized milk. Farber *et al.* (1996) combined quantitative values for a dose–response relationship with exposure estimates to provide a measure of the risk of listeriosis associated with the consumption of pâté and soft cheeses. The work by Cassin *et al.* (1998a) on *E. coli* O157:H7 in ground beef was one of the first probabilistic risk assessments that modeled all the stages of the food production chain. Increasingly, probabilistic methods have been adopted to estimate risks associated with a number of different pathogen/food combinations. Table 2.2 lists some of the

Table 2.2 Examples of published microbial food safety risk assessments[a]

Pathogen	Commodity	Scope/focus	Reference
Bacillus cereus	Cooked, chilled vegetable	Retail and consumer phase model to estimate exposure; identify knowledge gaps	Nauta *et al.* (2003)
	Cooked rice	Effects of storage and holding temperatures on risk of emetic disease	McElroy *et al.* (1999)
Escherichia coli O157:H7	Ground beef	Farm to consumer's risk of illness; evaluation of possible interventions in food chain	Cassin *et al.* (1998a)
	Ground beef	Estimation of public health impact in USA; input into regulatory decision-making	USDA-FSIS (2001)
Campylobacter jejuni	Ground and fresh beef	Retail to consumer; health impact due to antibiotic-sensitive vs antibiotic-resistant strains	Anderson *et al.* (2001)
	Poultry meat	Farm to consumer risk of illness; effect of food handling practices and pre-consumer interventions	Rosenquist *et al.* (2003)
Listeria monocytogenes	Milk	PE[b] values for raw milk contamination, survival of pasteurization, and consumer exposure	Peeler and Bunning (1994)
	Pâté, soft cheese	PE exposure and dose-response	Farber *et al.* (1996)
	Ready-to-eat meat, smoked fish	Dose-response model based on food survey and epidemiological data in a population	Buchanan *et al.* (1997)
	Raw milk soft cheese	Raw milk contamination through cheese processing to consumer risk	Bemrah *et al.* (1998); Sanaa *et al.* (2004)
	Smoked or gravad fish	Retail to consumer illness; factors influencing risk estimation	Lindqvist and Westöö (2000)
	Meat and poultry	Processing to consumer; effectiveness of sanitation of food contact surfaces and other mitigation strategies	Gallagher *et al.* (2003)
	Ready-to-eat foods	Retail to consumer risk	Chen *et al.* (2003)
	Ready-to-eat foods	Retail to consumer risk, relative risk ranking of products to prioritize focus of regulatory actions	FDA/FSIS/CDC (2003)
Mycobacterium paratuberculosis	Milk	Raw milk and pasteurization phases to estimate consumer exposure; comparison of probabilistic vs PE estimations.	Nauta and van der Giessen (1998)

Table 2.2 Examples of published microbial food safety risk assessments[a]—*cont'd*

Pathogen	Commodity	Scope/focus	Reference
Salmonella enterica	Shell eggs	Semi-quantitative	Todd (1996)
	Pasteurized liquid egg[c]	Shell eggs to human illness	Whiting and Buchanan (1997)
	Poultry meat, frozen	Retail to consumer illness	Brown *et al*. (1998)
	Poultry meat, fresh	Post-processing to consumer risk	Oscar (1998)
	Whole broiler chicken	Post-processing to consumer risk, effects of interventions	FAO/WHO (2002)
	Shell eggs and egg products[c]	Layer flocks to human illness via different pathways of egg preparation and consumption; impact of interventions	USDA-FSIS (1998; for overview see Hope *et al*., 2002); FAO/WHO (2002b)
	Turkey meat	Frozen to human illness; risk management options for catering establishments	Bemrah *et al*. (2003)

[a] Probabilistic risk assessments unless otherwise noted
[b] PE, point-estimates
[c] *Salmonella enterica* subsp. Enteritidis

examples published in the literature or available on websites. The same methods can also be applied when the focus of the investigation is on only one component of a risk assessment – either exposure or the dose–response assessments.

Already, a number of diverse food safety concerns are being studied using the structured approach defined by the risk assessment process. Additionally, the process and techniques are typically used to quantify likely risk reductions for any one intervention implemented somewhere in the food chain, and/or to identify areas where information is lacking and more research is needed.

7 Summary

Risk assessment is a systematic process for the collection, organization and analysis of data that can be brought to bear to assist in rational, objective decision-making processes. As a structured inquiry into the hazard, exposure and dose–response parameters, the risk assessment document itself serves as a database of relevant information and a record of assumptions and decisions. The assessment can be readily updated as better information is acquired or a food system is changed.

It is acknowledged that assessors are faced with many data gaps and knowledge constraints when constructing quantitative risk models. Currently, there is a concerted international effort under the auspices of the World Health Organization and the Food and Agriculture Organization to begin to resolve exposure and dose–response issues by examining all available data for *Salmonella* spp., *Listeria monocytogenes*, *Vibrio* spp., *Campylobacter* and *E. coli* O157:H7 collected from around the world. The ones published so far are for *Salmonella* in eggs and broiler chickens, *L. monocytogenes* in ready-to-eat foods and *Enterobacter sakazakii* in powdered infant formula with the full technical reports and interpretive summaries (FAO/WHO, 2002a, b; 2004a, b; 2005). Such efforts will help to provide much-needed insights and parameters for microbial risk assessments. Furthermore, it might be anticipated that with increasing awareness of the information needs for risk assessment combined with the capabilities of advanced laboratory techniques, better information will be made available from food research, food surveillance and outbreak investigations in the future.

Despite its limitations, the quantitative approach is useful to gain insights and to make some inferences with relevance to risk management (Linqvist *et al.*, 2002). Through rigorous problem definition, incorporation of all available data, and identification of variability and uncertainties, the risk assessment approach is a useful tool to improve the level of knowledge and the decision-making processes in food safety. The process can be used to examine risk factors from production through consumption, and to improve our understanding of key issues through model development and highlighting critical data gaps. There is a need for more information from human and animal surveillance systems; for improved outbreak investigation with quantification of pathogens found in suspected food vehicles; for greater knowledge regarding the prevalence and concentration of pathogens in foods and their ingredients throughout the food chain, the impact of competing flora and existing and proposed intervention procedures, the effect of consumer storage and preparation practices, and the consumption patterns of specific foods at the local and national level; and details of high-risk and normal populations relative to the food consumed and pathogens ingested.

It is also important to consider the value of different types of risk assessments. A complex, data- and resource-intensive assessment is not always a necessary or an appropriate process for a food safety risk management issue. A comprehensive risk assessment typically takes considerable time, and is thus not always an option for decision makers. Also, more important than the assessment itself is translating and communicating the components and results of a risk assessment adequately to non-scientific audiences, for purposes of discussion and decision-making. In particular, probabilistic analyses, incorporating uncertainty and variability in the risk estimation, and resulting in a broad range of values (e.g. numbers of illnesses per population) are generally difficult to understand by non-risk assessors. As a basis for policy decisions, managers typically prefer median or mean figures that are more readily understood.

For a number of reasons, then, there is also a need to consider and develop alternate risk assessment methods, simplified risk models and analytical techniques that are aligned with the established principles of food safety risk assessment. Although structured and systematic, risk assessment should be a flexible tool to use as is appropriate for the task at hand.

As the field advances, development of risk assessments will not only lead to better management of risk to reduce human foodborne illness; it will also improve our understanding of processes and interactions in the food chain, such as the ecology, physiology and biovariability of microbial pathogens, and the nature and variability of the pathogen–host relationships that lead to foodborne illness.

Bibliography

Altekruse, S. F., M. L. Cohen and D. L. Swerdlow (1997). Emerging foodborne diseases. *Emerg. Infect. Dis.* **3**, 285–293.

Anderson, S. A., R. W. Y. Woo and L. M. Crawford (2001). Risk assessment of the impact on human health of resistant *Campylobacter jejuni* from fluoroquinolone use in beef cattle. *Food Control* **12**, 13–25.

Bemrah, N., M. Sanaa, M. H. Cassin *et al.* (1998). Quantitative risk assessment of human listeriosis from consumption of soft cheese made from raw milk. *Prev. Vet. Med.*, **37**, 129–145.

Bemrah, N., H. Bergis, C. Colmin *et al.* (2003). Quantitative risk assessment of human salmonellosis from the consumption of a turkey product in collective catering establishments. *Intl J. Food Microbiol.* **80**, 17–30.

Beuchat, L. (1996). Pathogenic microorganisms associated with fresh produce. *J. Food Prot.* **59**, 204–216.

Brown, M. H., K. W. Davies, C. M. P. Billon *et al.* (1998). Quantitative microbiological risk assessment: principles applied to determining the comparative risk of salmonellosis from chicken products. *J. Food Prot.* **61**, 1446–1453.

Buchanan, R. L. and R. C. Whiting (1996). Risk assessment and predictive microbiology. *J. Food Prot.* **59** (Suppl.), 31–36.

Buchanan, R. L. and R. C. Whiting (1998a). Risk assessment: a means for linking HACCP and public health. *J. Food Prot.* **61**, 1531–1537.

Buchanan, R. L. and R. C. Whiting (1998b). USDA Pathogen Modeling Program, Release 5.1. USDA Agricultural Research Service, Wyndmoor, PA.

Buchanan, R. L., W. G. Damert, R. C. Whiting and M. van Schothorst (1997). Use of epidemiologic and food survey data to estimate a purposefully conservative dose–response relationship for *Listeria monocytogenes* and incidence of listeriosis. *J. Food Prot.* **60**, 918–922.

Buchanan, R. L., J. L. Smith and W. L. Long (2000). Microbial risk assessment: dose–response relations and risk characterization for food safety. *Intl J. Food Microbiol.* **58**, 159–172.

Buzby, J. C. and P. D. Frenzen (1999). Food safety and product liability. *Food Policy* **24**, 637–651.

Buzby, J. C., T. Roberts, J. C.-T. Lin and J. M. MacDonald (1996). *Bacterial Foodborne Disease, Medical Costs & Productivity Losses.* USDA Economic Research Service, Washington, DC. Agricultural Economic Report Number 741.

CAC (Codex Alimentarius Commission) (1999). *Principles and Guidelines for the Conduct of a Microbiological Risk Assessment.* Food and Agriculture Organization, Rome. CAC/GL-30.

Cassin, M. H., G. M. Paoli, R. S. McColl and A. M. Lammerding (1996). A comment on 'Hazard assessment of *Listeria monocytogenes* in the processing of bovine milk', letter to the Editor. *J. Food Prot.* **59**, 341–343.

Cassin, M. H., A. M. Lammerding, E. C. D. Todd *et al.* (1998a). Quantitative risk assessment for *Escherichia coli* O157:H7 in ground beef hamburgers. *Intl J. Food Microbiol.* **41**, 21–44.

Cassin, M. H., G. M. Paoli and A. M. Lammerding (1998b). Simulation modeling for microbial risk assessment. *J. Food Prot.* **61**, 1560–1566.

Chen, Y. H., E. H. Ross, V. N. Scott and D. E. Gombas (2003). *Listeria monocytogenes*: low levels equal low risk. *J. Food Prot.* **66**, 570–577.

Como-Sabetti, K. (1997). Outbreaks of *E. coli* O157:H7 associated with eating alfalfa sprouts – Michigan and Virginia, June–July 1997. *Morbid. Mortal. Weekly Rep.* **46**, 741–745.

FAO/WHO (1995). *Application of Risk Analysis to Food Standards Issues.* Report of the Joint FAO/WHO Expert Consultation. WHO, Geneva. WHO/FNU/FOS/95.3.

FAO/WHO (1997). *Risk Management and Food Safety.* Report of a Joint FAO/WHO Expert Consultation. FAO, Rome. FAO Food and Nutrition Paper No. 65.

FAO/WHO (1998). *The Application of Risk Communication to Food Standards and Safety Matters.* Report of a Joint FAO/WHO Expert Consultation. FAO, Rome. FAO Food and Nutrition Paper No. 70.

FAO/WHO (2002a). *Interpretive Summary. Risk Assessments of Salmonella in Eggs and Broiler Chickens. Microbiological Risk Assessment Series 1.* World Health Organization and the Food and Agriculture Organization of the United Nations, Geneva, Switzerland.

FAO/WHO (2002b). *Risk Assessments of Salmonella in Eggs and Broiler Chickens. Microbiological Risk Assessment Series 2.* World Health Organization and the Food and Agriculture Organization of the United Nations, Geneva, Switzerland.

FAO/WHO (2004a) *Interpretive Summary. Risk Assessments of* Listeria monocytogenes *in ready-to-eat foods. Microbial Risk Assessment Series 7.* World Health Organization and the Food and Agriculture Organizations of the United Nations, Geneva, Switzerland.

FAO/WHO (2004b) *Risk Assessments of* Listeria monocytogenes *in ready-to-eat foods. Microbial Risk Assessment Series 5.* World Health Organization and the Food and Agriculture Organizations of the United Nations, Geneva, Switzerland.

FAO/WHO (2005) Enterobacter sakazaki *and other microorganisms in powdered infant formula: meeting report. Microbial Risk Assessment Series 6.* World Health Organization and the Food and Agriculture Organizations of the United Nations, Geneva, Switzerland.

Farber, J. M., W. H. Ross and J. Harwig (1996). Health risk assessment of *Listeria monocytogenes* in Canada. *Intl J. Food Microbiol.* **30**, 145–156.

Farber, J. M. and E. C. D. Todd (2000). *Safe Handling of Foods.* Marcel Dekker, New York, NY.

FDA/FSIS/CDC (2003). FDA Center for Food Safety and Applied Nutrition, USDA Food Safety and Inspection Service, Centers for Disease Control and Prevention. *Quantitative assessment of the relative risk to public health from foodborne* Listeria monocytogenes *among selected categories of ready-to-eat foods* (available at http://www.foodsafety.gov/~dms/lmr2-toc.html).

Foegeding, P. M. and T. Roberts (eds). (1994). *Foodborne Pathogens: Risks and Consequences.* Council for Agricultural Science and Technology, No. 122. Ames, IA.

Foegeding, P. M. and T. Roberts (eds). (1998). *Foodborne Pathogens: Review of Recommendations*. Council for Agricultural Science and Technology, No. 22. Ames, IA.

Frenzen, P. D., T. L. Riggs, J. C. Buzby *et al.* and the FoodNet Working Group (1999). *Salmonella* cost estimate updated using FoodNet data. *Food Review* **22(2)**, 10–15.

Gallagher, D. L., E. D. Ebel and J. R. Kause (2003). FSIS/USDA. *Draft FSIS risk assessment for* Listeria *in ready-to-eat meat and poultry products*. (available at http://www.fsis.usda.gov/OPHS/lmrisk/DraftLm22603.pdf).

Gerba, C. P., J. B. Rose and C. N. Haas (1996). Sensitive populations: who is at the greatest risk? *Intl J. Food Microbiol.* **30**, 113–123.

Haas, C. N., J. B. Rose and C. P. Gerba (1999). *Quantitative Microbial Risk Assessment*. John Wiley & Sons, New York, NY.

Horton, L. R. (2001). Risk analysis and the law: international law, the World Trade Organization, Codex Alimentarius and national legislation. *Food Addit. Contamin.* **18**, 1057–1067.

Hope, B. K., R. Baker, E. D. Edel *et al.* (2002). An overview of the *Salmonella* Enteritidis risk assessment for shell eggs and egg products. *Risk Analysis* **22**, 203–218.

Huss, H. H., A. Reilly and P. K. Ben Embarek (2000). Prevention and control of hazards in seafood. *Food Control* **11**, 149–156.

ICMSF (International Commission on Microbiological Specifications for Foods) (1998). Potential application of risk assessment techniques to microbiological issues related to international trade in food and food products. *J. Food Prot.* **61**, 1075–1086.

ICMSF (International Commission for Microbiological Specifications for Foods) (2002). *Microorganisms in Foods 7: Microbiological Testing in Food Safety Management*. Kluwer Academic/Plenum Publishers, New York, NY.

ILSI Risk Science Institute Pathogen Risk Assessment Working Group (1996). A conceptual framework to assess the risks of human disease following exposure to pathogens. *Risk Analysis* **16**, 841–848 (rev. 2000; contact ILSI-RSI, 1126 Sixteenth St. NW. Washington, DC 20036-4810).

Jaykus, L. (1996). The application of quantitative risk assessment to microbial food safety risks. *Crit. Rev. Microbiol.* **22**, 279–293.

Kaplan, S. and B. J. Garrick (1981). On the quantitative definition of risk. *Risk Analysis* **1**, 11–27.

Lammerding, A. M. and A. Fazil (2000). Hazard identification and exposure assessment for microbial food safety risk assessment. *Intl J. Food Microbiol.* **58**, 147–157.

Lindqvist, R. and A. Westöö (2000). Quantitative risk assessment for *Listeria monocytogenes* in smoked or gravad salmon and rainbow trout in Sweden. *Intl J. Food Microbiol.* **58**, 181–196.

Lindqvist, R., S. Sylven and I. Vagsholm (2002). Quantitative microbial risk assessment exemplified by *Staphylococcus aureus* in unripened cheese made from raw milk. *Intl J. Food Microbiol.* **78**, 155–170.

Lupien, J. R. and M. F. Kenny (1998). Tolerance limits and methodology: effect on international trade. *J. Food Prot.* **61**, 1571–1578.

Martin, S. A., T. S. Wallsten and N. D. Beaulieu (1995). Assessing the risk of microbial pathogens: application of judgement-encoding methodology. *J. Food Prot.* **58**, 529–535.

McElroy, D. M., L. A. Jaykus and P. M. Foegeding (1999). A quantitative risk assessment for *Bacillus cereus* emetic disease associated with consumption of Chinese-style rice. *J. Food Safety* **19**, 209–229.

Mead, P. S., L. Slutsker, V. Dietz *et al.* (1999). Food-related illness and death in the United States. *Emerg. Infect. Dis.* **5**, 607–625.

Millar, B. C., M. Finn, L. Xiao *et al.* (2002). *Cryptosporidium* in foodstuffs – an emerging aetiological route of human foodborne illness. *Trends Food Sci. Technol.* **13**, 168–187.

Nauta, M. J. and J. W. B. van der Giessen (1998). Human exposure to *Mycobacterium paratuberculosis* via pasteurized milk: a modeling approach. *Vet. Record* **143**, 293–296.

Notermans, S., J. Dufrenne, P. Teunis *et al.* (1997). A risk assessment study of *Bacillus cereus* present in pasteurized milk. *Food Microbiol.* **14**, 143–151.

NRC (National Research Council) (1983). *Risk Assessment in the Federal Government: Managing the Process.* National Academy Press, Washington, DC.

NRC (National Research Council) (1994). *Science and Judgment in Risk Assessment.* National Academy Press, Washington, DC.

Oscar, T. P. (1998). The development of a risk assessment model for use in the poultry industry. *J. Food Safety* **18**, 371–381.

Peeler, J. T., and V. K. Bunning (1994). Hazard assessment of *Listeria monocytogenes* in the processing of bovine milk. *J. Food Prot.* **57**, 689–697.

Potter, M.E. (1994). The role of epidemiology in risk assessment: a CDC perspective. *Dairy Food Environ. Sanit.* **14**, 738–741.

Rodricks, J. V. (1994). Risk assessment, the environment, and public health. *Environ. Health Perspect.* **102**, 258–264.

Rosenquist, H., N. L. Nielsen, H. M. Sommer *et al.* (2003). Quantitative risk assessment of human campylobacteriosis associated with thermophilic *Campylobacter* species in chickens. *Intl J. Food Microbiol.* **83**, 87–103.

Ross, T. and T. A. McMeekin (1994). Predictive microbiology – a review. *Intl J. Food Microbiol.* **23**, 241–264.

Ross, T. and J. Sumner (2002). A simple, spreadsheet-based, food safety risk assessment tool. *Intl J. Food Microbiol.* **77**, 39–53.

Salisbury, J. G., T. J. Nicholls, A. M. Lammerding *et al.* (2002). A risk analysis framework for the long-term management of antibiotic resistance in food-producing animals. *Intl J. Antimicrob. Agents* **20**, 159–170.

Sanaa, M., L. Coroller and O. Cerf (2004). Risk assessment of listeriosis linked to the consumption of two soft cheeses made from raw milk: Camembert of Normandy and Brie of Meaux. *Risk Analysis* **24**, 389–399.

Smith, J. L. and P. M. Fratamico (1995). Factors involved in the emergence and persistence of foodborne diseases. *J. Food Prot.* **58**, 696–708.

Sockett, P. N. and E. C. D. Todd (2000). The economic cost of foodborne disease. In: *The Microbiological Safety and Quality of Food*, pp. 1563–1588. Aspen Publishers, Gaithersburg, MD.

Sumner, J. and T. Ross (2002). A semi-quantitative seafood safety risk assessment. *Intl J. Food Microbiol.* **77**, 55–59.

Tauxe, R. V. (1997). Emerging foodborne diseases: an evolving public health challenge. *Emerg. Infect. Dis.* **4**, 1–12.

Teunis, P. F. M., O. G. van der Heijden, J. W. B. van der Geissen and A. H. Havelaar (1996). *The Dose–Response Relation in Human Volunteers for Gastro-intestinal Pathogens*. National Institute of Public Health and Environment, report 2845500002. Bilthoven, The Netherlands.

Teunis, P. F. M., N. J. D. Nagelkerke and C. N. Haas (1999). Dose–response models for infectious gastroenteritis. *Risk Analysis* **19**, 1251–1260.

Thompson, K. M. and J. D. Graham (1996). Going beyond the single number: using probabilistic risk assessment to improve risk management. *Human Ecol. Risk. Assess.* **2**, 1008–1034.

Todd, E. C. D. (1996). Risk assessment of cracked eggs in Canada. *Intl J. Food Microbiol.* **30**, 125–143.

Todd, E. C. D. (1997). Epidemiology of foodborne diseases: a worldwide view. *World Health Stat. Q.* **50**, 30–50.

USDA-FSIS (1998). The *Salmonella* Enteritidis Risk Assessment Team. Salmonella *Enteritidis Risk Assessment: Shell Eggs and Egg Products. Final Report* (available at http://www.fsis.usda.gov/ophs/risk/index.htm).

USDA-FSIS (2001). *Draft Risk Assessment of the Public Health Impact of* Escherichia coli *O157:H7 in Ground Beef* (available at http://www.fsis.usda.gov/OPPDE/rdad/FRPubs/00-023NReport.pdf).

Van Beneden, C., W. Keene, R. Strang *et al.* (1999). Multinational outbreak of *Salmonella enterica* serotype Newport infections due to contaminated alfalfa sprouts. *J. Am. Med. Assoc.* **281**, 158–162.

van Gerwen, S. J. C. and M. H. Zwietering (1998). Growth and inactivation models to be used in quantitative risk assessments. *J. Food Prot.* **61**, 1541–1549.

van Gerwen, S. J. C., M. C. te Giffel, K. van't Riet *et al.* (2000). Stepwise quantitative risk assessment as a tool for characterization of microbiological food safety. *J. Appl. Bacteriol.* **88**, 938–951.

Vose, D. (2000). *Risk Analysis: A Quantitative Guide*, 2nd edn. John Wiley & Sons, New York, NY.

Vugia, D., J. Hadler, S. Chaves *et al.* (2003). Preliminary FoodNet data on the incidence of foodborne illnesses – selected sites, United States, 2002. *Morbid. Mortal. Weekly Rep.* **52**, 340–343.

White, D. G., S. Zhoa, S. Simjee *et al.* (2002). Antimicrobial resistance of foodborne pathogens. *Microbes Infect.* **4**, 405–412.

Whiting, R. C. and R. L. Buchanan (1994). IFT scientific status summary: microbial modeling. *Food Technol.* **48**, 113–120.

Whiting, R. C., and R. L. Buchanan (1997). Development of a quantitative risk assessment model for *Salmonella enteritidis* in pasteurized liquid eggs. *Intl J. Food Microbiol.* **36**, 111–125.

Foodborne Infections

Salmonella infections

3

Kåre Mølbak, John E. Olsen and Henrik C. Wegener

1 Historical aspects and contemporary problems

Bacteria of the genus *Salmonella* are widespread and important causes of foodborne infections in man, and are the most frequent etiologic bacterial agents of foodborne disease outbreaks. Serotypes Typhi (*S.* Typhi), *S.* Paratyphi and *S.* Sendai are highly adapted to man. *S.* Typhi and *S.* Paratyphi have humans as their main reservoir, and enteric fever (typhoid and paratyphoid fever) as their most important clinical manifestation. Enteric fever continues to be an important cause of morbidity and mortality in developing countries.

The non-typhoid *Salmonella* serotypes, which include some 2500 different serotypes, are widely distributed in nature, including in the gastrointestinal tracts of mammals, reptiles, birds, and insects. Most clinical infections of humans are transmitted from healthy carrier animals to humans through food. The main clinical manifestation of human infection with non-typhoid *Salmonella* is an acute gastrointestinal illness and, less frequently, septicemia. Non-typhoid *Salmonella* was among the earliest of the so-called emerging pathogens. In particular two *Salmonella* serotypes, *S.* Enteritidis and *S.* Typhimurium, became major causes of foodborne illness in the 1980s and 1990s, with an important impact on public health and the economy in industrialized countries.

Foodborne Infections and Intoxications 3e
ISBN-10: 0-12-588365-X
ISBN-13: 978-012-588365-8

1.2 Historical aspects

Water and milk were found to be vehicles of the etiologic agent of enteric fever by epidemiological evidence several years before the agent itself was identified (Budd, 1874). The organism, now named *S. enterica* serotype Typhi, was discovered in 1880 (Eberth, 1880) and isolated on culture media in 1884 (Gaffky, 1884). In 1885, the bacteriologist Theobald Smith, who worked in the US Department of Agriculture, isolated *S.* Choleraesuis from porcine intestine, and the genus *Salmonella* was named after D. E. Salmon, his laboratory chief.

The first report of a laboratory-confirmed outbreak of foodborne salmonellosis described an episode in which 58 persons in 25 different families who had eaten beef developed acute gastroenteritis; one died (Gärtner, 1888). Gärtner isolated the 'Gärtner-bacillus' from the infected cow from which the meat came, and from organs of the fatal case. (Kauffmann determined that the 'Gärtner-bacillus' from this outbreak was serotype Enteritidis, but other outbreaks of the 'Gärtner-bacillus' were of serotype Dublin and possibly other serotypes; Kauffmann, 1930.) Mice, rabbits, guinea pigs, and goats were affected when inoculated with the bacillus. In the following years, several outbreaks of salmonellosis affecting man or animals were reported, and the old concept of 'meat poisoning' was linked with the etiologic agent *Salmonella*. Human salmonellosis occurred primarily among individuals who ate meat from ill animals, mainly cattle, but also pigs or goats.

There are only limited historical data on the prevalence of *Salmonella* in healthy animals, carcasses or meat; but it is likely to have been much lower than found in industrialized countries in the 1980s and 1990s. For example, bacteriological testing in connection with meat inspection in Copenhagen central abattoir revealed *Salmonella* Typhimurium on only two occasions between 1922 and 1941 (Hansen, 1942). A total of 662 human cases of *Salmonella* infections were found in Denmark in a 6-year period from 1936 to 1942; 79 % were serotype Typhimurium and 6 % serotype Enteritidis (Harhoff, 1948). Harhoff concludes: 'Although hens' eggs are used so very often, practically speaking they are never seen as a cause of infection because hens are very resistant to *Salmonella* infections (leaving out *S. pullorum* and *S. gallinarum*, specific to hens and nonpathogenic to human beings).' At that time, contaminated ducks' eggs caused most egg-associated outbreaks; meat from ill animals (mainly cattle) was another common cause of foodborne outbreaks, and large milkborne epidemics were seen when the milk came from a cow with *Salmonella* mastitis. Cheese made from raw milk was also known to be a source of *Salmonella* infection, whereas broiler chickens were an uncommon source of foodborne outbreaks (Harhoff, 1948). A devastating event occurred in Alvesta, Sweden, in 1955, when meatborne outbreak of *S.* Typhimurium affected some 9000 individuals and caused 90 deaths (Bengtsson *et al.*, 1955). This outbreak prompted an early implementation of *Salmonella* control in Sweden.

In several European countries, rodenticides based on cultures of *S.* Enteritidis were used as rat baits. Human disease was associated with this use, for example among persons who handled the rodenticides or accidentally ingested it (Threlfall *et al.*, 1996; Painter *et al.*, 2004).

While *S.* Typhi became an enormous problem in the US in the early industrial era, the disease burden associated with non-typhoid *Salmonella* was low before World

War II (Tauxe, 1999). Improvements in sanitation nearly eliminated *S.* Typhi as a cause of indigenous infections in the US and other developed countries. Decades later, non-typhoid *Salmonella* infections began to increase in importance – a trend that may have peaked near 1990 (Figure 3.1).

The development of serotyping was fundamental for the understanding of the epidemiology of *Salmonella* infections. While *S.* Typhi was easy to recognize by biochemical tests, it was, for example, impossible to distinguish between *S.* Typhimurium and *S.* Paratyphi B on the basis of fermentation of sugars or other biochemical properties. Many bacteriologists considered *S.* Typhimurium and *S.* Paratyphi B to be identical. However, it was an enigma as to why infections with seemingly identical bacteria often did not share pathological, clinical and epidemiological features. In outbreaks where the source was identified to be a human reservoir, the clinical illness was often septicemia. However, when the agent came from an animal reservoir, the infection resulted in gastroenteritis, but the strain could infect laboratory mice that often rapidly died from septicemia (Gärtner, 1922). This and several other questions were resolved at the end of the 1920s when White and Kauffmann, through the use of improved serological methods, succeeded in designing a classification system for *Salmonella*. The foundation for this serotyping scheme was the discovery of the flagellar H antigen and the thermostable somatic O antigen by Weil and Felix (Weil and Felix, 1918) and the phase-shift in the H antigen (Andrewes, 1922).

Salmonella serotypes were regarded as different species, and were named according to the disease they caused (e.g. *S.* Enteritidis, *S.* Typhi, *S.* Paratyphi, *S.* Abortus equi, and *S.* Bovismorbificans), or the animals from which they were isolated (Kelterborn, 1967). For example, *S.* Gallinarum and *S.* Pullorum were important pathogens in poultry, *S.* Choleraesuis an important swine pathogen, and *S.* Typhimurium got its name because it was originally isolated from ill laboratory mice (Löffler, 1892). A limited number of serotypes were named after the person who isolated it (e.g. *S.* Virchow). Currently, each new antigenically distinguishable type is typically named after the geographic place at which it was first isolated. Most of the important

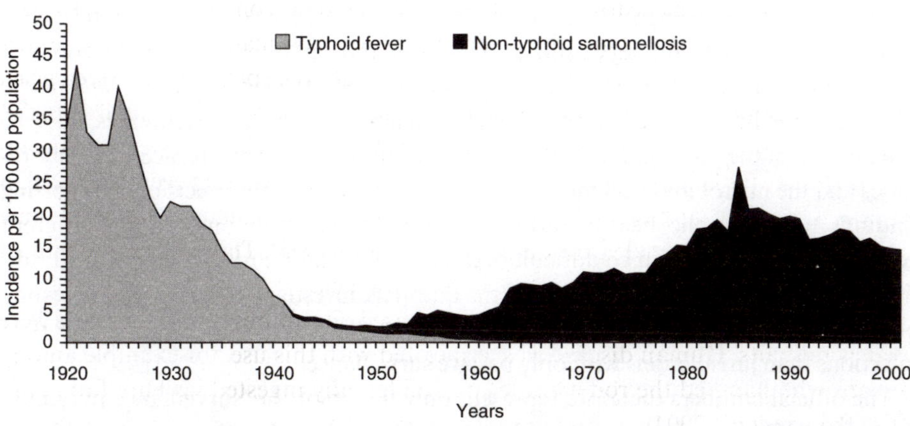

Figure 3.1 The fall and rise of reported *Salmonella* infections in the United States, 1920–2001. *Source:* Centers for Disease Control and Prevention (CDC), National Notifiable Diseases Surveillance Data.

Salmonella serotypes are now considered to belong to a single species, *Salmonella enterica* (Brenner *et al.*, 2000, Popoff, 2001). Kelterborn provides a historical account of the *Salmonella* serotypes up to 1965 (Kelterborn, 1967). The current classification of *Salmonella* is further described in section 2 of this chapter.

The Kauffmann–White scheme, which at present includes more than 2500 serotypes, is the most successful bacterial typing scheme in history. *Salmonella* serotypes are often correlated with clinical severity, reservoir, and occurrence of resistance. Information about the distribution of different serotypes, as well as subtypes, in different animal species, foods and man may be used to quantify the relative importance of different sources of *Salmonella*-infections (Altekruse *et al.*, 1993; Hald *et al.*, 2004). In the future, the Kauffmann–White scheme is likely to continue as the foundation for *Salmonella* surveillance and epidemiological studies.

1.3 Contemporary problems

In the past two to three decades, the incidence of non-typhoid *Salmonella* infections in humans has shown a marked increase. In this section we will review data on the incidence of *Salmonella* infections in humans, discuss the temporary problems due to the global emergence of foodborne non-typhoid *Salmonella* infections, mention specific issues due to antimicrobial drug-resistance in *Salmonella*, and summarize international surveillance.

1.3.1 Incidence of *Salmonella* infections in humans

Most developed countries have laboratory-based surveillance of *Salmonella* infections. In addition, many countries have systems for recording outbreaks and notification systems where clinicians submit data on patients with *Salmonella* infections to national public health institutions. Official numbers of *Salmonella* infections are commonly derived from the laboratory-based surveillance, where clinical microbiology laboratories report positive findings and, in some countries, submit *Salmonella* isolates to national reference laboratories for serotyping and other characterization. These data are pivotal for measuring trends over time and detecting outbreaks; however, the figures from the official reporting systems do not measure the burden of illness, and the degree of surveillance differs among countries. Moreover, the reported incidence is a composite measure of several factors. These factors include the true incidence of *Salmonella* infections, the health-care seeking behavior of patients with gastroenteritis, and the likelihood that the physician will request a stool culture. Furthermore, access to laboratories and the microbiological methods in place vary, as does the precision in reporting findings to the public health authorities. Finally, comparisons among different geographical locations can be difficult because public health jurisdictions with a tradition of active case-searching as part of the outbreak investigations or extensive testing of contacts of known patients or food-handlers are likely to report higher numbers of infections than jurisdictions with only passive surveillance.

The official numbers therefore represent only the tip of the 'surveillance pyramid'. Although this tip is likely to include the most severe cases, there is a significant amount of illness that is not recorded in the official systems. In population-based

Table 3.1 Reported and estimated numbers of human Salmonellosis cases per 100 000 population

Country	Reported	Estimated	Reference
England	70	220	Wheeler *et al.*, 1999
United States	14	520	Mead *et al.*, 1999
The Netherlands	24	340	Van Pelt *et al.*, 2003

studies in England and Wales (Wheeler *et al.*, 1999), the US (Mead *et al.*, 1999) and the Netherlands (de Wit *et al.*, 2001a, 2001b; van Pelt *et al.*, 2003) the true burden of disease was investigated. The studies suggest that for every reported *Salmonella* infection, between 3.8 and 38 persons actually fell ill (Table 3.1). The differences between the three studies may partly be due to different methodologies, but it is interesting that there is less variation between estimated incidences in the general population than the reported figures indicate. This suggests that the sensitivity of the surveillance is a critical factor when attempts are made to compare numbers between countries. It is currently estimated that there are 1.4 million infections of non-typhoidal *Salmonella* in the US each year (Mead *et al.*, 1999).

1.3.2 The global emergence of foodborne non-typhi *Salmonella* infections

There are marked differences regarding the sensitivity of the *Salmonella* surveillance among different countries; still, surveillance is crucial for examining trends and for the detection of outbreaks. This is in particular true if the surveillance includes serotyping. Available data suggest that the incidence of foodborne non-typhi *Salmonella* infections has increased during the last couple of decades. For several reasons, this increase is not likely to be an artifact. If the increase could be explained by increased awareness and surveillance, the relative distribution between the serotypes would, by and large, remain constant. This has not been the case; several countries have experienced decreases in the incidence of *S*. Typhi, while the non-typhoid serotypes have lately increased both proportionally and in absolute numbers (Tauxe, 1999; Figure 3.1). Moreover, important changes in the serotype distribution among the non-typhoid types, as well as distribution of subtypes (e.g. phage types), is incompatible with the hypothesis that increased diagnostic activity is the major cause of the increasing incidence of foodborne *Salmonella* (Olsen et al., 2001a; Herikstad *et al.*, 2002; van Pelt *et al.*, 2003).

In countries with data from food animals or foods, parallel trends can be observed between serotype distribution in human infections and the prevalence and distribution of *Salmonella* in animals and food (Wegener *et al.*, 2003). Thus, the emergence of *Salmonella* in poultry production and later in egg production has been related to the industrialization and the globalization of poultry and egg production (St. Louis *et al.*, 1988; Rodrigue *et al.*, 1990; Tauxe, 1999). It has been hypothesized (Baumler *et al.*, 2000) that the epidemic of salmonellosis in humans was triggered by *S*. Enteritidis filling the ecological niche vacated by the avian pathogens *S*. Gallinarum and *S*. Pullorum, although this hypothesis remains debated (Riemann *et al.*, 2000; Ward *et al.*, 2000). It is furthermore thought that this emergence was amplified by international trade of live animals – a trade that was part of the post-World War II efforts to ensure food security and protein security in the developed countries. In countries such as Sweden, where measures against *Salmonella* were taken many years ago, indigenous *Salmonella* infections are lower than in other countries of Europe (Wegener *et al.*, 2003).

With the exception of an outbreak in the 1950s, human *Salmonella* infections were infrequent in Denmark up to the late 1970s. In the period 1980–2000, Denmark was affected by three distinct epidemics (Figure 3.2). By comparing the trends in humans with the data from food animals and food, it can be shown that in the first epidemic broiler chickens were the main reservoir; in the second, pigs; and in the third, shell eggs (Wegener *et al.*, 2003).

Table 3.2 illustrates some of the factors that have contributed to the increasing incidence of non-typhoid *Salmonella* infections in the industrialized countries. The table shows that the current problems are complex, and suggests that factors in primary food production, in plants for food processing and in consumer behavior, and a changing demography, all contribute. However, the quantitatively most important factor in the spread of *Salmonella* in the modern food production system is international trade of live animals colonized with *Salmonella*. This international trade, which serves to supply the food animal production with breeder animals to improve the genetic stock, is an especially efficient way of disseminating *Salmonella* vertically throughout the mass production system. A particularly good example of this is the pandemic of *S*. Enteritidis. Several years ago, the pandemic spread of *S*. Enteritidis in the human populations of Europe and the Americas was associated with the current intensive egg production (Rodrigue *et al.*, 1990). In 1995, *S*. Enteritidis had become the most frequently isolated serotype in 35 countries, followed by *S*. Typhi (12 countries) and *S*. Typhimurium (8 countries). The mean national proportion of *Salmonella* isolates that were *S*. Enteritidis was 36.3 %, with the highest proportions in Europe (58.6 %) and the Americas (42.6 %) (Herikstad *et al.*, 2002). Figure 3.3 shows the relative increase in the incidence of human infections with *S*. Enteritidis in several countries. *S*. Enteritidis may colonize the ovaries of modern breeds of layer

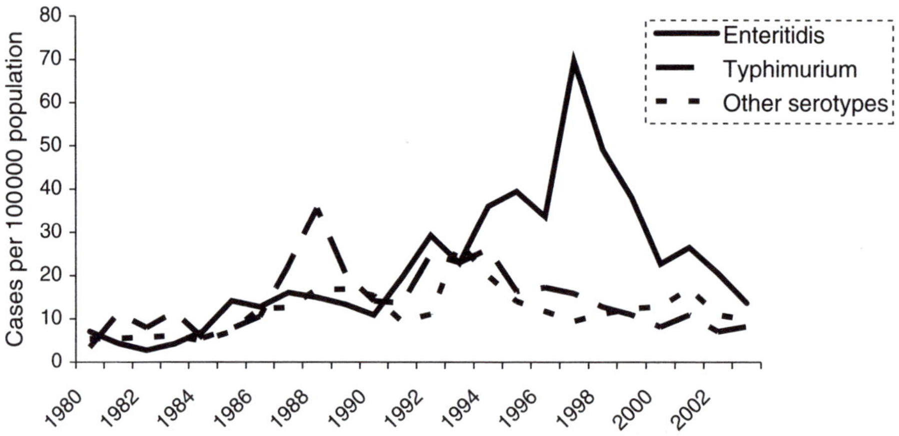

Figure 3.2 Incidence of human *Salmonella* infections in Denmark, 1980–2003, by serotype. *Source:* Statens Serum Institut.

Table 3.2 Some causes (contributing factors) of the emergence of non-typhoid Salmonellosis in industrialized countries

Part of food chain	Causes	Effects
Animal feeding	Large-scale production of compound feed with geographically diverse ingredients	Widespread contamination, changes in intestinal ecology
	Antimicrobial growth promoters	Change in intestinal ecology, selection of Gram negative bacteria
	Antimicrobial treatment of sick animals	Selection and spread of resistant *Salmonella*
Animal production systems	Trading animals for rearing or breeding	Introduction of *Salmonella* in breeding stocks, widespread distribution of infected animals
	Long-distance transport of animals	Transport-induced stress enhances shedding and spread of *Salmonella*
	Large-scale production systems	Amplification of infection and contamination
Food processing	Large-scale production of minimally processed and ready-to-eat products	Amplification of contamination, outbreaks affecting large populations
Retailers and restaurants	Faulty storage, less familiarity with food handling	Cross-contamination, increase in numbers of *Salmonella* during storage
Consumption	Less familiarity with food preparation among consumers, increasing numbers of immunosuppressed consumers	Increased incidence of salmonellosis especially in the vulnerable segments of the population

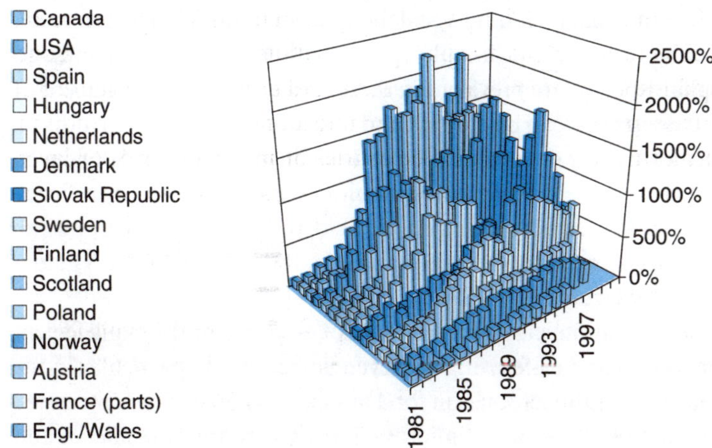

Figure 3.3 The global epidemic of *Salmonella* Enteritidis. The figure shows the relative change in the incidence of human infections with *S.* Enteritidis in 15 different countries. *Source:* World Health Organization.

hens, and thus has the potential for vertical transmission from breeders to layers and then to eggs sold for human consumption (St Louis, 1988; Humphrey, 1994, 1999; Saeed *et al.*, 1999). Outbreak investigations indicate that a principal cause of *S.* Enteritidis infection is the consumption of raw or undercooked table eggs, or dishes contaminated with raw eggs (Coyle *et al.*, 1988; Mishu *et al.*, 1994; Angulo and

Swerdlow, 1999). This hypothesis has furthermore been corroborated by the results of case-control studies of risk factors for sporadic *S.* Enteritidis infections (Mølbak and Neimann, 2002, and references therein). While it is encouraging that a recent decline in the incidence of *S.* Enteritidis has been reported, this serotype remains the most important one in terms of the public health burden, morbidity, and mortality in industrialized countries.

1.3.3 Emergence of antimicrobial drug resistance in *Salmonella*

The increasing prevalence of antimicrobial drug resistance in *S.* Typhi and in several non-typhoid *Salmonella* serotypes is a concern for public health, as is the increasing number of classes of antimicrobials involved.

In areas where typhoid fever is endemic, use and misuse of over-the-counter antimicrobial drugs in humans contributed to the 1972 emergence of resistance in *S.* Typhi (Anonymous, 1972; Paniker and Vimala, 1972; Anderson, 1975; Rowe *et al.*, 1997). In Tajikistan in 1998, contamination of a municipal water system with a multidrug resistant *S.* Typhi caused a massive outbreak, with some 10 000 cases and 100 deaths. The outbreak strain was resistant to ampicillin, chloramphenicol, nalidixic acid, streptomycin, sulfisoxazole, tetracycline, and trimethoprim-sulfamethoxazole (Mermin *et al.*, 1999). In the US, 25 % of 293 patients were infected with *S.* Typhi isolates that were resistant to one or more microbial agent; and 17 % were resistant to five or more agents, including ampicillin, chloramphenicol, and trimethoprim-sulfamethoxazole. Quinolone resistance (nalidixic acid resistance) was found in 7 %. Patients with multidrug resistant or nalidixic acid resistant *S.* Typhi were likely to report travel outside the US, in particular to the Indian subcontinent (Ackers *et al.*, 2000). Multidrug resistant *S.* Typhi is also common in South East Asia, including Thailand and Vietnam (Connerton *et al.*, 2000; Swaddiwudhipong and Kanlayanaphotporn, 2001). Strains resistant to quinolones and/or cephalosporins are prevalent in these areas, and the management of patients infected with these strains can be difficult and costly.

The appropriate breakpoint for determining ciprofloxacin-resistance in *Salmonella* is being debated, and it is of concern that fluoroquinolones may have reduced efficacy in the treatment of patients infected with nalidixic acid-resistant strains, although these strains are determined to be susceptible according to the current NCCLS guidelines (Aarestrup *et al.*, 2003; Crump *et al.*, 2003).

In non-typhoid *Salmonella*, antimicrobial agents are not essential for the management of most cases; however, they can be lifesaving for severe infections, in particular in the case of septicemia. The use of antimicrobials in food animals has no doubt contributed to the emergence and dissemination of antimicrobial drug resistant *Salmonella*, and once a drug-resistant *Salmonella* has been established in food production establishments it may be disseminated even in the absence of selective pressure from drug use (Glynn *et al.*, 1998; Angulo *et al.*, 2000; Fey *et al.*, 2000; Bager and Helmuth, 2001). Complex groups of genes may encode for resistance with insignificant cost to the fitness of the bacterium. Drug resistant *Salmonella* types have a selective advantage when transferred to the gut ecosystem in animals or humans treated with antimicrobials; therefore the emergence of resistance and increased transmission are often linked. Table 3.3 summarizes epidemiological data from the UK regarding 'successful' multidrug-resistant

Table 3.3 Emergence and disappearance of *Salmonella* Typhimurium in the United Kingdom

Peak year	Cases in peak year	Phage type	R-type[a]	Confirmed cases, 1967–2000
1967	511	29	ACKNSSuTFu	2364
1983	400	204	CSSuT	3649
1986	294	204c	ACGKSSuTTm	1614
1989	83	193	ACKSSuT	200
1996	4006	104	ACSSuT+	22 220

[a] A, ampicillin; C, chloramphenicol; Fu, furazolidone; G, gentamycin; K, kanamycin; N, neomycin; S, streptomycin; Su, sulfonamides; T, tetracycline; +, common acquisition of resistance to other antimicrobial drugs.

Salmonella serotypes that in the period since the mid-1960s have caused considerable morbidity and mortality in humans.

In non-typhoid *Salmonella*, resistance is most common in *S*. Typhimurium and several less frequently reported serotypes, including *S*. Hadar, *S*. Heidelberg, *S*. Newport, *S*. Saintpaul, *S*. Senftenberg and *S*. Virchow (Seyfarth *et al*., 1997; Threlfall *et al*., 1999a, 2003; Breuil *et al*., 2000; Swartz, 2002; Logue *et al*., 2003). The primary reservoir of these serotypes is food animals, and they often show multidrug resistance encoded by large transferable plasmids. In *S*. Typhimurium DT104, which is usually resistant to at least five drugs (ampicillin, chloramphenicol, streptomycin, sulfonamides, and tetracycline), the genes encoding for resistance are found on two large integrons on the bacterial chromosome. *S*. Typhimurium DT104 is a recent example of a multidrug-resistant *Salmonella* clone that has spread to several continents (Tauxe, 1999; Threlfall, 2000). *S*. Newport is an emerging multidrug resistant *Salmonella* in the US; it is often resistant to at least amoxicillin/clavulanate, ampicillin, cefoxitin, ceftiofur, cephalothin, chloramphenicol, streptomycin, sulfamethoxazole, and tetracycline. In addition, it may have decreased susceptibility to ceftriaxone, thereby complicating therapy for serious *Salmonella* infections. The primary reservoir of this type is cattle, and outbreaks have been caused by meat, raw milk, and cheese from raw milk, as well as from direct contact with animals (Anonymous, 2002a).

The prevalence of antimicrobial drug resistance is low in *S*. Enteritidis. This serotype does not readily acquire the mobile genetic elements that often encode for resistance in other *Salmonella* serotypes. However, an increase in quinolone (nalidixic acid) resistance in *S*. Enteritidis is a cause of concern, in particular because quinolone resistance may be associated with reduced efficacy of early empirical treatment with ciprofloxacin or other fluoroquinolones (Mølbak *et al*., 2002). Quinolone resistance is caused by a mutation on the bacterial chromosome and is not transferable. The prevalence of drug resistance is also low in *Salmonella* serotypes that traditionally are confined to feral reservoirs or reptiles, where the selective pressure from antimicrobial drugs is limited.

The human health consequences of resistance are reduced efficacy of treatment, delayed treatment, and limitation of therapeutic choice. Resistance also leads to increased transmission and risk of the horizontal spread of resistance genes. The overall impact may be an increased risk of outbreaks – in particular in hospital settings or among persons treated with antimicrobial drugs – and increased human morbidity and mortality (Helms *et al.*, 2002; Travers and Barza, 2002). Table 3.4 summarizes some of the clinical–biological mechanisms that may explain an increased morbidity and mortality associated with drug resistance in *Salmonella*.

Table 3.4 Consequences for human health of the emergence of antimicrobial resistant *Salmonellae*

Causes	Effects
Selective advantage in patients treated with antimicrobial drugs	Increased transmission, places like hospitals at high risk (Barza and Travers, 2002)
Risk of treatment failure in patients with extraintestinal infection	Reduced efficacy of early empirical treatment
Drug resistance genes located on mobile genetic elements	Horizontal transmission of resistance
Limited choices of effective antimicrobial drugs	Increased cost of treatment
Increase virulence due to co-selection of virulence factors or improved fitness of bacterial strains	Increased risk of admission to hospital or blood stream infection

1.3.4 National and international initiatives for surveillance and control of *Salmonella* infections

New types of *Salmonella* often emerge in several countries at more or less the same time, and multi-state or international outbreaks often call for a coordinated response. For this reason, several national and international networks are currently addressing the problem of emerging *Salmonella* infections. An important objective is to improve and enhance surveillance, including serotyping, of *Salmonella*.

The WHO Global Salm-Surv is a worldwide network for the surveillance of salmonellosis and other foodborne diseases. The mission is to reduce the global burden of foodborne disease by strengthening national and regional foodborne disease surveillance and response systems. The most important activities include international training courses, an electronic discussion group, external quality assurance programs, reference testing, and a country databank with summary information on the most common *Salmonella* serotypes (Petersen *et al.*, 2002).

Enter-net is an international surveillance network for human gastrointestinal infections, with a focus on *Salmonella* and verocytotoxin-producing *Escherichia coli*. This network conducts international surveillance, and involves the countries of the European Union in addition to several other countries. The database is built on a one-line-per-case-patient principle, and data are submitted from the member countries to the Enter-net hub on a regular basis. This allows the timely recognition of outbreaks and analysis of trends (Threlfall *et al.*, 1999b; Lindsay *et al.*, 2002).

Three networks based in the US contribute to improved *Salmonella* surveillance, outbreak response and research:

- PulseNet is a national network of public health laboratories that perform pulsed field gel electrophoresis (PFGE) on *Salmonella* and other bacteria that might be foodborne (Tauxe, 1998; Swaminathan *et al.*, 2001). This network permits the rapid comparison of PFGE patterns ('fingerprints') through an electronic database at the Centers for Disease Control and Prevention (CDC). Closely related or identical patterns may suggest a common source of infection. PulseNet is helpful in detecting and investigating outbreaks, particularly those that involve many states. Bacterial subtyping with the PulseNet protocol is now undertaken in several other countries, and extension of this collaboration will greatly improve the national and international public health infrastructure and food safety.

- The Foodborne Diseases Active Surveillance Network (FoodNet) is the principal foodborne disease component of CDC's Emerging Infections Program (EIP). FoodNet is a collaborative project of the CDC, the EIP sites, the US Department of Agriculture, and the Food and Drug Administration. The project consists of active surveillance for foodborne diseases, and related epidemiologic studies are designed to help public health officials better understand the epidemiology of foodborne diseases in the US (Angulo *et al.*, 1998; Mead *et al.*, 1999). Australia has established a similar surveillance network, OZFOODNET (Dalton and Unicomb, 2001).
- The US National Antimicrobial Resistance Monitoring System (NARMS) for Enteric Bacteria was established in 1996, within the framework of the CDC's Emerging Infections Program's Epidemiology and Laboratory Capacity Program and the Foodborne Diseases Active Surveillance Network. In 2002, 28 state and local public health laboratories participated in NARMS. The main activity of NARMS is to monitor antimicrobial resistance of enteric bacteria, including *Salmonella* isolated from humans. Currently, approximately 108 million persons (40 % of the US population) reside within the NARMS surveillance sites. NARMS is a collaboration between the CDC, the US Food and Drug Administration (Center for Veterinary Medicine), and the US Department of Agriculture (Food Safety and Inspection Service and Agricultural Research Services).

The Nordic countries have established programs for monitoring resistance in bacteria from food animals, food, and humans. Some of these integrated programs, such as DANMAP, also include surveillance for consumption of antimicrobials (Bager, 2000).

2 Characteristics of *Salmonella*

2.1 Classification

2.1.1 Characterization of *Salmonella* by biochemical and conventional phenotypic methods

The family *Enterobacteriaceae* consists of Gram-negative, facultative anaerobic, non-spore-forming rods. *Salmonella* Liegnières 1900 conforms to the general definition of the family. Members of the genus are (usually) motile by peritrichous flagella, reduce nitrates to nitrites, (usually) produce gas from glucose, (usually) produce hydrogen sulfide on triple-sugar iron agar, and (usually) grow on citrate as the sole carbon source. They are indole- and urease-negative and (usually) lysine- and ornithine-decarboxylase-positive and sucrose-, salicin-, inositol-, and amygdalin-negative.

Phenylalanine and tryptophan are not oxidatively deaminated, and lipase and deoxyribonuclease are not produced (Le Minor, 1984).

The genus consists of two species: *S. enterica* (Le Minor *et al.*, 1982, 1986) and *S. bongori* (Reeves *et al.*, 1989). The Judicial Commission of the International Committee of Systematic Bacteriology, however, has not accepted that *S.* Typhi, the cause of typhoid fever in humans, is reduced to a serotype (Wayne, 1994), and it has been suggested that the species name *S. typhi* be maintained in a genus consisting of *S. enterica*, *S. typhi* and *S. bongori* (Euzeby, 1999). The committee has now ruled that the type species of *Salmonella* is *S. enterica* and that the epithet *enterica* takes priority over all other epithets that may have been applied to this species (Anonymous, 2005). Recently it has been argued that the name *S. choleraesuis* has priority over *S. enterica* (Yabuuchi and Ezaki, 2000), but the species name *S. enterica* will be used throughout this chapter, in line with Brenner *et al.* (2000). *S. enterica* is diverse, and consists of six subspecies. The important zoonotic salmonellae almost exclusively belong to the first subspecies, ssp. *enterica*. The subspecies names and the biochemical criteria for identification of *Salmonella* species and subspecies are listed in Table 3.5.

Serotyping by slide agglutination based on O-, H- and Vi-antigens according to the Kauffmann–White scheme is an important tool for classification of *Salmonella*. More than 2500 serotypes are recognized by the WHO Collaborating Centre for *Salmonella* Serotyping, at the Institut Pasteur, Paris (Popoff, 2001). An internationally accepted nomenclature for *Salmonella* has emerged based on the recommendations of the WHO Collaborating Centre. According to this, genus, species and subspecies designations are italicized, whereas the serotype designations are not. The first letter of the serotype designation is capitalized. The use of only the genus and the serotype name, e.g. *S.* Typhimurium, is accepted as a convenient and short form (Brenner *et al.*, 2000). Tables 3.6 and 3.7 list the most frequent *Salmonella* serotypes isolated from humans in 17 European countries (Enter-net countries) and the US, 1998–2002. *S.* Enteritidis ranks as number one in Europe, whereas Typhimurium is the most common serotype in the US.

Common serotypes, such as *S.* Enteritidis and *S.* Typhimurium, may be further subdivided by phage typing. The phage typing systems of Anderson *et al.* (1977) and Ward *et al.* (1987) have become the internationally accepted, and the Health Protection Agency, UK, serves as the WHO reference center for *Salmonella* phage typing. Phage types are numbered consecutively and given a two-letter prefix. *S.* Typhimurium phage types are named DTs for definitive types, whereas *S.* Enteritidis phage types are PTs – examples are *Salmonella* Typhimurium DT104 and *Salmonella* Enteritidis PT4. The tracking of the recent global spread of *S.* Typhimurium DT104 has documented the continued value of phage typing and antimicrobial resistance testing as a tool to investigate trends and patterns of *Salmonella* epidemiology (Tauxe, 1999; Threlfall, 2000). Thus far, the advancement of molecular techniques has not replaced the traditional phenotypic methods for epidemiological characterization of *Salmonella*. The combined use of phenotypic and genotypic typing methods, such as serotyping, phage typing, antimicrobial resistance testing, pulsed field gel electrophoresis (PFGE) and plasmid profiling, allows very detailed comparison of strains of *Salmonella*. As will be evident from the sections that follow, this has greatly improved our understanding of the epidemiology of foodborne salmonellosis.

Table 3.5 Characteristics differentiating *Salmonella* species and subspecies

Species	Sub-species	Type strain	Dulcitol	ONPG	Malonate	Gelatinase	Sorbitol	Culture w/KCN	L(+)-tartrate[a]	Galacturo-nate	Gamma-glutamyl transferase	Beta-glucoro-nidase	Mucate	Salicine	Lactose	Lysed by phage O1	Usual reservoir
S. enterica	*enterica*	ATCC 13312	+	–	–	+	–	+	–	+*	d	+	–	–	–	+	Warm-blooded animals
	salamae	CIP 8229	+	–	+	+	+	–	–	+	+	d	+	–	–	+	
	arizonae	ATCC 13314	–	+	+	+	+	–	–	–	–	–	+	–	– (75 %)	–	Cold-blooded animals
	diari-zonae	CIP 8231	–	+	+	+	+	–	–	+	+	+	–	+	–	–	
	houte-nae	CIP 8232	–	–	–	+	+	+	–	+	+	–	–	+	–	–	
	indica	CIP 2501	d	d	–	+	–	–	–	+	+	d	+	–	d	+	
S. bongeri	–	CIP 8233	+	+	–	–	+	+	–	+	+	–	+	–	–	d	

+ = 90 % or more positive reactions; – = 90 % or more negative reactions; d = different reactions in different serovars.
[a] d-tartrate; *Typhimurium; d, Dublin.
Based on Ezaki *et al.* (2000) and Popoff (2001).

Table 3.6 The 20 serotypes of *Salmonella* isolated most frequently from humans in 17 European countries, 1998–2002. (*Source:* Enter-net *Salmonella* database, 1998–2002)

Rank	Serotype	Frequency	Percentage
1	*S.* Enteritidis	218 552	57.7
2	*S.* Typhimurium	68 615	18.1
3	*S.* Hadar	10 479	2.8
4	*S.* Virchow	7067	1.9
5	*S.* Infantis	4246	1.1
6	*S.* Brandenburg	3050	0.8
7	*S.* Newport	2931	0.8
8	*S.* Agona	2669	0.7
9	*S.* Heidelberg	2478	0.7
10	*S.* Derby	2382	0.6
11	*S.* Braenderup	2047	0.5
12	*S.* Blockley	2002	0.5
13	*S.* Typhi	1982	0.5
14	*S.* Bovismorbificans	1862	0.5
15	*S.* Stanley	1732	0.5
16	*S.* Panama	1546	0.4
17	*S.* Paratyphi B	1399	0.4
18	*S.* Montevideo	1381	0.4
19	*S.* Bredeney	1365	0.4
20	*S.* Oranienburg	1298	0.3
	All others	39 469	10.4
	Total	378 552	100.0

Table 3.7 The 20 serotypes of *Salmonella* most frequently isolated from humans, 1998–2002 (US Centers for Disease Control and Prevention)

Rank	Serotype	Frequency	Percentage
1	*S.* Typhimurium	38 297	23.3
2	*S.* Enteritidis	28 547	17.4
3	*S.* Newport	15 312	9.3
4	*S.* Heidelberg	9316	5.7
5	*S.* Javiana	5822	3.5
6	*S.* Montevideo	3857	2.4
7	S. Muenchen	3784	2.3
8	*S.* Oranienburg	3049	1.9
9	*S.* Thompson	2733	1.7
10	*S.* Infantis	2708	1.7
11	*S.* Agona	2629	1.6
12	*S.* Saintpaul	2500	1.5
13	*S.* Braenderup	2327	1.4
14	*S.* Hadar	2051	1.3
15	*S.* Paratyphi B var. Java	1915	1.2
16	*S.* Typhi	1763	1.1
17	*S.* Poona	1530	0.9
18	*S.* Mississippi	1497	0.9
19	*S.* Berta	1202	0.7
20	*S.* Stanley	946	0.6
	All Others	32 259	19.7
	Total	164 044	100.0

2.1.2 Characterization of *Salmonella* by molecular methods – molecular typing

Typing methods are probably applied more extensively to *Salmonella* than to any other zoonotic pathogen, and the majority of the molecular methods developed for characterization of bacteria has been applied to *Salmonella* both for epidemiological investigation and for research. The choice of method depends on the serotype and specific circumstances, and often more than one method is used to improve the quality of the typing.

Plasmid profiling was originally the most popular method for molecular typing of *Salmonella*. It offers excellent discriminatory power for many serotypes, where plasmids of varying sizes occur frequently (Olsen, 2000). It can be particularly useful in the investigation of foodborne disease outbreaks, but is less useful for large-scale, population-based studies covering extended periods of time because the plasmid genotype can be relatively labile due to loss or uptake of plasmids.

Chromosomally-based typing methods primarily consist of restriction endonuclease methods, where the chromosome is cut into fragments of various sizes. The results are subsequently analyzed by interpretation of the patterns produced by gel electrophoresis (DNA fingerprints), or in combination with hybridization to DNA-blots to highlight variation in selected loci by use of hybridization probes (RFLP-pattern). RFLP-based methods, such as ribotyping (Grimont and Grimont, 1986) and IS*200* typing (Stanley *et al.*, 1991) were previously used extensively with *Salmonella*, but are now uncommon.

Separation of large restriction fragments, generated by rare cutters (restriction enzymes with few recognition sites in the chromosome), by pulsed field gel electrophoresis (PFGE) is currently the molecular 'gold standard' for *Salmonella* typing, and good standardized protocols have been developed for use with *Salmonella* (Swaminathan *et al.*, 2001). PFGE profiling effectively subdivides most serotypes, where other methods fail to do so (Olsen, 2000). However, discrimination within clones, such as phage type DT104, is less efficient (Liebana *et al.*, 2002). Figure 3.4 shows an example of a typing result with PFGE. Lack of stability of DNA during the analysis is an obstacle with some serotypes (Stanley *et al.*, 1995; Baggesen *et al.*,

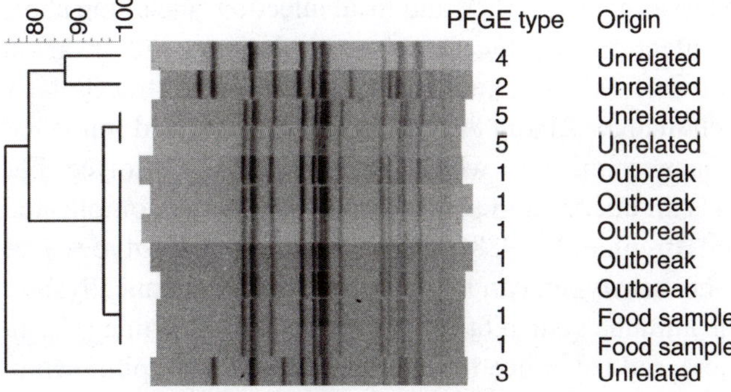

Figure 3.4 PFGE profiles of *Salmonella* Typhimurium isolates from a restaurant outbreak in Denmark (Ethelberg *et al.*, 2004). Isolates from patients visiting the restaurant and isolates obtained from food samples in the restaurant had indistinguishable PFGE profiles (PFGE type 1). The UPGMA tree shows the relationship between the outbreak strain and four other common PFGE types in the outbreak period.

1996), but substitution with HEPES buffer for the normal Tris-borate buffer during preparation may help overcome this problem (Koort *et al.*, 2002).

Typing methods based on PCR technology are increasingly used for characterization of bacteria, but not so commonly with *Salmonella*. This is possibly because the typing schemes are already well developed, and it has been difficult to obtain a higher discrimination than is possible by the traditional methods – for example, phage typing has been shown to be more discriminatory than RAPD (random amplification of polymorphic DNA) for typing of *S*. Typhimurium and *S*. Enteritidis (Hilton *et al.*, 1996). More promising is the use of PCR as an alternative to serotyping. This would eliminate the need for carefully standardized sera, and allow for 'serotyping' outside of reference laboratories. Amplification of phase 1 and phase 2 flagellin genes of 264 serotypes and digestion with two restriction enzymes yielded 116 patterns. Of these patterns, 80 % were specifically associated with one antigen. However, flagellin–antigen RFLP did not precisely match the diversity evidenced by flagellar–antigen agglutination (Dauga *et al.*, 1998), probably because some serotypes include bacteria that are not clonally related. Recently, a multiplex PCR method has been developed to identify the most common phase-2 antigens (Echeita *et al.*, 2002). Antibody-based typing may be preserved as the gold standard, but it is very likely that PCR-based methods will take over for everyday 'serotyping'.

PCR-based typing is also being developed as a substitute for DNA fingerprinting. In an attempt to automate typing analysis, a fluorescent-based AFLP (Amplified Fragment Length Polymorphism) typing system employing capillary electrophoresis was tested. This method performed equally well as PFGE, and distinguished 48 types among 97 strains of *Salmonella*. Strains from a large *S*. Typhimurium outbreak were included, and showed identical profiles apart from small diffezrences in intensity of signals (Lindstedt *et al.*, 2000).

2.2 Virulence factors and their genetic basis

Non-typhoid *Salmonella* cause infections in humans and a broad range of animals. They give rise to a variety of disease syndromes including enterocolitis, bacteremia/septicemia (some times referred to as enteric fever), and focal infections including abortion. During the course of infection, *Salmonella* colonizes the intestine. Invasion happens rapidly after ingestion of bacteria, and bacteria can be seen intracellularly less than 15 minutes post-challenge (Zhang *et al.*, 2003). The localized intestinal infection causes diarrhea, the mechanism of which is beginning to be resolved. For most serotypes, bacteremia is an uncommon and transient event in uncomplicated salmonellosis (Mandal and Brennand, 1988). The host-adapted serotypes (see below), and in rare cases other serotypes, can however propagate systemically; and eventually kill the host if the immune system fails to eliminate the bacterium (Uzzau *et al.*, 2000). Mice show increased tolerance to high numbers of *S*. Typhimurium when lipid A is detoxified by inactivation of the *waaN* gene (Khan *et al.*, 1998), indicating that death occurs due to an overload of endotoxin.

S. Typhimurium has been the model organism for studies of salmonellosis, and a great deal of our current understanding of virulence factors springs from studies of this serotype. It must, however, be remembered that serotypes differ in disease association. For example some serotypes have a narrow host range (Table 3.8), and the genetic background for this is still not understood in detail (Uzzau *et al.*, 2000). *Salmonella* normally infects animals via the fecal–oral route. Alternative routes, such as lung infections, occur in animals (Fedorka-Cray *et al.*, 1995), and presumably can happen in humans. It is not known whether different virulence factors are active during alternative routes of infection. Also, it is currently not known if the infection progresses through the same mechanism in all mammals including humans.

2.2.1 Overview of virulence factors and their genetic background

More than 4 % of the genetic information in *S.* Typhimurium has been associated with fatal disease in a mouse model, indicating that *Salmonella* are highly specialized pathogens (Bowe *et al.*, 1998). The virulence genes are scattered throughout the chromosome and in some serotypes on large virulence-associated plasmids. Large genetic elements, termed *Salmonella* pathogenicity islands (SPIs), are essential for virulence. A schematic presentation of virulence factors and their role in salmonellosis is shown in Figure 3.5.

2.2.2 Virulence factors involved in *Salmonella*-induced diarrhea

The precise mechanism by which *Salmonella* cause diarrhea is not fully understood. *Salmonella* harbors an enterotoxin with homology to the enterotoxins in *E. coli* and *Vibrio cholerae*, but this enterotoxin produced *in vitro* does not induce fluid secretion in a rabbit ileal loop model (Wallis *et al.*, 1986), and disruption of the gene (*stn*) does not influence occurrence of diarrhea in cattle (Wallis *et al.*, 1999).

Most of our understanding of the function of individual virulence factors and their role in development of diarrhea comes from studies of the effect on cell cultures and from investigations using cattle intestinal loop models. In cattle, *S.* Typhimurium causes fibrino-purulent necrotizing enteritis with heavy infiltration of neutrophils. The pathological lesions are mostly located in the terminal ileum and the cranial part of the colon (Zhang *et al.*, 2003). The symptoms are similar to those reported from non-human primates and infected humans (Rout *et al.*, 1974; McGovern and Slavutin, 1979).

Table 3.8 Host ranges of some *Salmonella* serotypes

Group	Serotype	Normal host	Zoonotic importance	Virulence plasmid (size kb)
Host restricted	Typhi	Humans	No	None
	Gallinarum	Birds	No	85
	Abortus ovis	Sheep/goat	No	50–59
	Typhisuis	Pigs	No	None
Host-adapted	Dublin	Cattle	Yes	83
	Choleraesuis	Pigs	Yes	50
Ubiquitous	Typhimurium	Many	Yes	90
	Enteritidis	Many	Yes	54

Adapted from Gulig *et al.* (1993) and Uzzau *et al.* (2000).

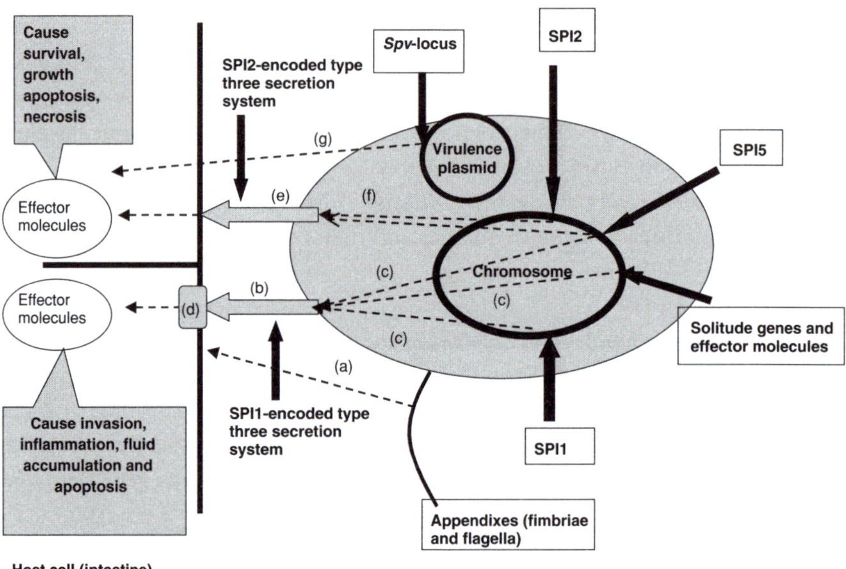

Figure 3.5 Schematic presentation of *Salmonella* virulence genes and their role in development of salmonellosis. The virulence genes involved in development of diarrhea and systemic disease are shown separately, although the invasion is the same in the two disease syndromes, and effector molecules in some cases are shared in the two infection types. The initial contact is believed to be mediated through fimbriae (a). Genes from the *Salmonella* pathogenicity island 1 (SPI1) encode a type-three secretion system (b). This needle complex injects effector molecules encoded from SPI1, SPI5 and from solitude genes into the intestinal cell (c), leading to uptake of the bacterium, fluid secretion, inflammation and delayed apoptosis. Some of the effector molecules are located in the cell membrane of the host cell (d) and are essential for translocation of the other effector molecules and for virulence. Genes from SPI2 are essential for survival and growth inside professional phagocytes during the systemic phase of infection. SPI2 encodes a type-three secretion system (e), which translocates effector molecules into the *Salmonella*-containing vacuole in the infected cell (f). Genes encoded from large virulence plasmids are essential for systemic disease in non-typhi serotypes that are normally associated with systemic disease (g). The plasmids encode spv-genes (salmonella plasmid virulence), which modify the cytoskeleton of phagocytic cells.

The approximately 40-kb *Salmonella* pathogenicity island 1 (SPI1) and the effector molecules secreted by this system are essential for fluid accumulation and influx of neutrophils (Wallis *et al.*, 1999), and mediate cytoskeleton rearrangements, which lead to the membrane ruffling and uptake of the bacterium (Zhou *et al.*, 2001). The SPI1-genes encode a type-three secretion system, which forms a needle complex. It secretes effector molecules into the membrane (SipB, SipC and SipD) and cytosol of the host cell (SopB, SopE2, SipA, SopD, SopA, AvrA, SptP, SlrP) (reviewed by Zhang *et al.*, 2003). Sip-molecules, AvrA and Sptp, are encoded from SPI1, while the remaining effector molecules are encoded outside this element. SopA, SopB (encoded from SPI5) and SopD are essential for *S*. Dublin-induced fluid accumulation in cattle (Jones *et al.*, 1998), and SopB, SopE and SipA are essential for the uptake of bacteria (Zhou *et al.*, 2001).

SopB and SopE have polyphosphatase activity and are believed to alter phospatidylinositol levels in the cell, which antagonize closure of chloride channels, thus suggesting a secretory mechanism behind the diarrhea. This assumption is supported by studies using intestinal loops in pigs, where *S*. Typhimurium-induced secretion is reduced by 40 % when 5-HT_3 (secretonin) receptors are blocked before the challenge (Grøndahl *et al.*, 1998). However, the evidence for the secretion as the important mechanism in *Salmonella* diarrhea is mostly circumstantial, and other models have been proposed. The fluid accumulation in intestinal loops is significantly reduced when influx of neutrophils is prevented (Wallis *et al.*, 1990), suggesting that neutrophil-induced damage or signaling is significant in development of diarrhea. The inflammation leading to necrosis of the intestinal mucosa has also been implicated as an important cause of fluid and protein loss during *Salmonella* infection (Santos *et al.*, 2002). It has been suggested that the inflammation is not just caused by the invasion of the bacterium in the intestine, but also by induction of signaling molecules (reviewed by Zhang *et al.*, 2003). Effector molecules secreted by the SPI1-encoded, type-three secretion system induce a pro-inflammatory signaling event, which may play an important role in the development of inflammation, the attraction of neutrophils, damage to the mucosa, and increased secretion. In accordance with this assumption, early events in diarrhea in cattle are characterized as an acute neutrophilic inflammatory response associated with increased expression of IL-8, GRO (growth-related oncogenes) a and g, GCP2 (granulocyte-chemotactic protein 2) IL-1β, IL-1Ra and IL-4 (Santos *et al.*, 2002).

2.2.3 Virulence genes involved in systemic infection

Salmonellae that cause systemic infection are taken up by phagocytic cells in the lamina propria. From there they are transported to systemic sites by exploiting the 'normal' cell trafficking system. It has been suggested that macrophages are mainly responsible for the transport of bacteria, but peripheral dendritic cells also contain salmonellae during the infection and may help *Salmonella* to reach the draining lymph node (Yrlid *et al.*, 2001). Many organs, including the spleen, liver, lungs, lymph nodes, and reproductive tissue, are culture-positive in animals suffering from systemic salmonellosis. Dissemination and multiplication of *Salmonella* seems to take place in CD18-containing phagocytes, at least for *S*. Typhimurium infection in mice (Vazquez-Torres *et al.*, 1999). Whether this is general to all *Salmonella* infections remains to be clarified.

Salmonellae survive inside macrophages *in vitro* (Fields *et al.*, 1986), in an acidified vacuole (Rathman *et al.*, 1996) termed the *Salmonella*-containing vacuole (*svc*), to indicate that it is not a normal phagosome (Brumell *et al.*, 2002). The survival was early suggested to be due to inhibition of phagosome–lysosome fusion (Buchmeier and Heffron, 1991), but conflicting evidence has been presented (Carrol *et al.*, 1979; Oh *et al.*, 1996). Genes encoded in *Salmonella*-pathogenicity island 2 (SPI2) are essential for systemic salmonellosis in animal models (Hensel *et al.*, 1998; Jones *et al.*, 2001). SPI2, which is absent in *S. bongori*, encodes a type-three secretion system that translocates effector molecules into the cytosol of the infected cell (Figure 3.5). The effector molecules are not well characterized (Hensel, 2000), but are known to affect phagosome–lysosome formation (Uchiya *et al.*, 1999) and cause exclusion of NADPH oxidase from the *Salmonella*-containing vacuole (Vazquez-Torres *et al.*, 2000). Gene(s) of SPI2 together with *spv* genes (see below) and *ompR* are responsible for late death of macrophages (Monack *et al.*, 2001). Interestingly, SPI2 is only induced *in vivo* when *Salmonella* is located intracellularly in the phagocyte, and the genes are preferentially expressed in acidic phagosome environments (Cirillo *et al.*, 1998). *In vitro*, expression can be stimulated by Mg^{2+} deprivation and phosphate starvation (Deiwick *et al.*, 1999).

In addition to the SPI2-induced macrophage killing mentioned above, *Salmonella* kills macrophages *in vitro* by a caspase-1, but not caspase-3, dependent mechanism, which results in activation of IL-1β (Brennan and Cookson, 2000). The mechanism has been described as apoptosis (Monack *et al.*, 1996; Hersh *et al.*, 1999), but it stimulates a pro-inflammatory response (Brennan and Cookson, 2000) and has features of necrosis (Watson *et al.*, 2000). The killing requires expression of *sipB* of SPI1 (Hersh *et al.*, 1999). The reaction seems essential for virulence following oral challenge, as LD_{50} is 3 log units higher in caspase-1 deficient mice than in wild-type mice (Monack *et al.*, 2000).

Salmonella plasmid-virulence (*spv*) genes are essential for systemic infections caused by certain non-typhoid serotypes of ssp. I (Table 3.8). The *spv* operon encodes five highly conserved genes (*spvR*, *A*, *B*, *C*, and *D*), irrespective of the serotype (Gulig *et al.*, 1993). *SpvB* and *spvC* are the virulence-enhancing genes (Matsui *et al.*, 2001). The operon is present on large, 54–100-kb serotype-associated plasmids (SAPs).

The *spv*-genes allow the bacterium to grow at systemic sites (Hackett *et al.*, 1986) while the entero-pathogenicity is not influenced (Wallis *et al.*, 1995). Recent studies have shown that the product of the major structural gene of the *spv* operon, *spvB*, is a mono(ADP-ribosyl)transferase and is hence related to the large group of toxins that inactivate host proteins by ADP-ribosylation (Otto *et al.*, 2000). The target molecule is actin, and ADP-ribosylation results in a block of conversion of G actin to F actin, leading to destabilization of the cytoskeleton (Lesnick *et al.*, 2001; Tezcan-Merdol *et al.*, 2001). The specific mode of action is essential for the pathogenesis in the mouse model (Lesnick *et al.*, 2001).

Other genes are essential during the systemic phase of infection. A two-component system, PhoP/PhoQ, controls expression of a number of genes entitled *pag* (*C–P*) for *phoP*-activated genes (Belden and Miller, 1994). Genes of this regulon encode a defensine-resistant phenotype (Miller *et al.*, 1990) and influence macrophage processing of *Salmonella* antigens (Wick *et al.*, 1995). Analysis of human gene expression by use of micro-arrays indicates a role of this system in macrophage cell deaths during *S.* Typhimurium infection (Detweiler *et al.*, 2001). Heat-shock proteins are induced during intracellular growth of *Salmonella* (Buchmeier and Heffron, 1990), and the heat-shock protein *htrA* (Johnson *et al.*, 1991) and the stress-associated chaperone and protease *clpXP* (Yamamoto *et al.*, 2001) have been identified as virulence genes.

The major host responses to systemic infection include INF-γ stimulation of phagosome–lysosome formation and oxygen-dependent killing mechanisms, such as NADPH coupled respiratory burst, nitric oxide production, and production of superoxide (Kagaya *et al.*, 1989; van Diepen *et al.*, 2002). Protection against oxidative killing is therefore essential for virulence in *Salmonella.* The majority of *S.* Typhimurium mutants with an increased ability to propagate in macrophages inhibit NO production (Eriksson *et al.*, 2000), and strains that are hyper-susceptible to superoxide are avirulent (van Diepen *et al.*, 2002).

2.2.4 The host-specific phenotypes of some serotypes of *Salmonella*

The mechanism behind the host-specific/host-adapted phenotype seen with some serotypes (Table 3.8) is not fully understood. The host-specific infection is characterized by a selective ability of the serotype to propagate at systemic sites in the preferred host (Barrow *et al.*, 1994). The host-adapted serotypes are capable of doing this in more than one host; but they too are almost exclusively isolated from only one host. Host adaptation and host specificity are not due to a superior ability to invade the intestine of the preferred host, nor to a higher ability to enter and grow inside phagocytic cells (*in vitro*) of this host (Uzzau *et al.*, 2001; Paulin *et al.*, 2002; Chadfield *et al.*, 2003), and the oxidative response induced in the macrophage of the preferred host is not different from the response induced by other serotypes (Chadfield and Olsen, 2001). The majority of the serotypes that show host adaptation or host specificity carries a serotype-associated plasmid (SAP), but this has not been shown to be important for adaptation or specificity (Uzzau *et al.*, 2000).

S. Gallinarum has been shown to be less cytotoxic to avian macrophages than other host-specific and host-adapted serotypes (Chadfield *et al.*, 2003). Compared to *S.* Typhimurium and *S.* Enteritidis, it further induces a cytokine response, with less stimulation of pro-inflammatory cytokine IL-6 (Kaiser and Lamont, 2001). This indicates that a stealth-like technique is implicated in the host-specific infection, but it does not explain the propagation to full-blown infection. It has never been possible to isolate mutants or recombinant strains that have changed phenotype from that of host adaptation/restriction to either become adapted/restricted to a different host or broaden their host range. A likely explanation is that host specificity is caused by a regulatory mechanism of many genes rather than the presence or absence of one or a few genes. It is likely that host adaptation mechanisms will only be unraveled when full genome-expression analyses from intracellular *Salmonella* are performed, for example by using micro-array analysis.

3 Nature of the infection in man and animals

3.1 Epidemiology of non-typhoid *Salmonella*

Man most commonly acquires non-typhoid *Salmonella* as a foodborne infection, with raw or undercooked eggs, poultry, meat and unpasteurized milk being common vehicles of infection; also, other foods cross-contaminated during preparation, storage or serving may be involved. Increasingly common are infections from contaminated uncooked vegetables, fruits, etc. Direct or indirect contact with animals colonized with *Salmonella* is another source of infection, including contact during visits to petting zoos and farms (Friedman *et al.*, 1998).

Reptiles kept as pets, such as turtles, iguanas, other lizards and snakes, are often identified as non-food sources of infection. *Salmonella* belonging to subspecies II, III (*S.* Arizonae), IV, V, and VI are often found in cold-blooded animals, and are considered to be reptile-associated. Examples of subgroup I reptile-associated serotypes are *S.* Paratyphi var Java (*S.* Java), *S.* Poona (D'Aoust *et al.*, 1990; Anonymous, 2000a), *S.* Marina (Mermin *et al.*, 1997), and several others (Woodward *et al.*, 1997). Table 3.9 summarizes reptile-associated serotypes (Ackman *et al.*, 1995; Woodward *et al.*, 1997). Reptile-associated serotypes or other environmental *Salmonella* may be the cause of foodborne outbreaks, in particular from fruits, vegetables and spices contaminated by *Salmonella* from feral reptiles or other animals. This may explain recent outbreaks of *S.* Poona in the US (Anonymous, 2002b; see also Table 3.12).

Table 3.9 Reptile-associated *Salmonella* subspecies and serotypes

Reptile association	Serotypes
Reptiles are the source of more than 50 % of different serotype isolates (CDC 1981–1990)	Amoutive, Anecho, Anjona, Aqua, Banana, Belem, Blukwa, Chicago, Chichiri, Claibornei, Eastbourne, Elisabetville, Gatuni, Gombe, Guarapiranga, Hull, Hvittingfoss, Ilala, Inglis, Kintambo, Kisarawe, Koketime, Kua, Lome, Maracaibo, Mikawasima, Nottingham, Oslo, Patience, Ramatgan, Redlands, Soahanina, Soerenga, Tamale
Reptiles are the source of more than 50 % of different subspecies isolates (CDC 1981–1990)	Subspecies IIIa (*S. enterica* subspecies *arizonae*) Subspecies IIIb (*S. enterica* subspecies *diarizonae*) Subspecies IV (*S. enterica* subspecies *houtenae*)
Serotypes associated with reptiles in other reports	Abaetetuba, Anatum, Cerro, Chester, Braenderup, Ealing, Florida, Infantis, Irumu, Jangwani, Java, Javiana, Jangwani, Litchfield, Manhattan, Matadi, Miami, Monschaui, Montevideo, Muenchen, Muenster, Newport, Oranienburg, Panama, Poano, Pomona, Poona, Rubislaw, Saintpaul, Stanley, Tilene, Urbana, Uganda, Wassenaar Subspecies VI (*S. enterica* subspecies *indica*)

Modified and extended from Ackman *et al.* (1995).

The distribution of a few serotypes shows geographical restrictions with respect to human infections. Examples of such serotypes include *S*. Javiana, *S*. Mississippi, *S*. Rubislaw and *S*. Bareilly, which are found in limited areas of the Eastern and South Eastern US (Helfrick *et al.*, 2001). *S*. Weltewreden is widespread throughout South and South East Asia, and also seems to be a serotype restricted to the Pacific regions of the US. Infections with such serotypes may be caused by direct or indirect contact with a feral reservoir – for example, reptiles or amphibians. It is also possible that infection with such restricted serotypes is associated with a local and limited production of food items. It is important to underscore that these serotypes cause only a fraction of the total number of human *Salmonella* infections.

Although waterborne transmission of non-typhoid *Salmonella* has been reported (Angulo *et al.*, 1997; Taylor *et al.*, 2000), waterborne outbreaks of *Salmonella* infections are uncommon in industrialized countries. *Salmonella* was not implicated as the cause of recent waterborne-disease outbreaks (including recreational water) in the US (Barwick *et al.*, 2000).

Person-to-person transmission is uncommon among healthy adults in developed countries but may be of significance in certain settings, such as institutions, hospitals, nursing homes etc., where it may be difficult to keep hygienic standards high and the population is particularly vulnerable to infection (Lyons *et al.*, 1980; Seals *et al.*, 1983; Stone *et al.*, 1993). In some of these outbreaks the involved strains have been multiresistant, making it a challenge to disentangle secondary person-to-person transmission from foodborne transmission or spread from contaminated surfaces (Olsen *et al.*, 2001b). Person-to-person and nosocomial transmissions remain problems in the less-developed countries (Riley et al., 1984; Kumar *et al.*, 1995; Nair *et al.*, 1999).

3.1.1 Age- and gender-specific incidence

The incidence of non-typhoid *Salmonella* is usually highest in infants and young children. The incidence rate of some serotypes increases slightly in young adults in their twenties. There is also a slight increase among elderly persons (Figure 3.6). In the US, the overall yearly reported incidence ranged from 19 to 13 per 100 000 population from 1987 to 1997, whereas the incidence in infants was as high as 122 per 100 000 (Olsen *et al.*, 2001a).

Among infants and young children, the incidence is usually higher in boys than in girls. In contrast, the isolation rate for women (10.2 per 100 000) exceeded that for men (8.8 per 100 000) in the period 1987–1997 (Olsen *et al.*, 2001a). This gender difference is particularly marked in fertile women (Figure 3.6). At the extremes of age, the incidence in males is again higher than in females. In the US, an increasing median age of infection is associated with a female dominance that, in particular, is found for serotypes that affect middle-aged adults. The biological or epidemiological basis for the age- and gender-specific incidence is not clear. Health-seeking behavior, physician's behavior, age- and gender-specific exposures and transmission patterns, as well as hormonal and immunological factors, may be important (Reller *et al.*, 2002).

The reptile-associated *Salmonella* types disproportionately affect infants and young children, in particular boys, and they often cause severe infections, including bacteremia and meningitis (Ackman *et al.*, 1995; Mermin *et al.*, 1997; Ward, 2000).

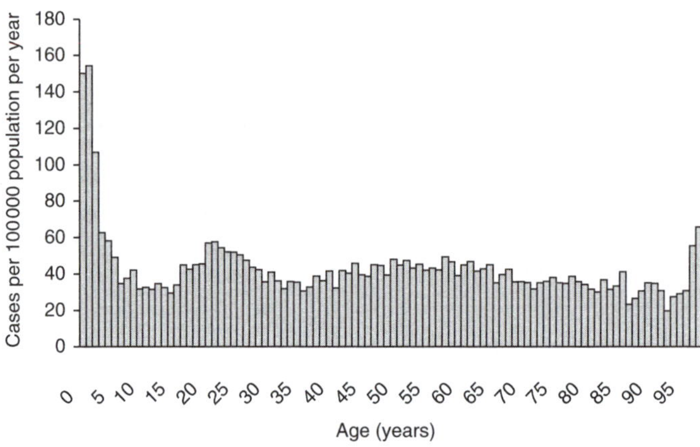

Figure 3.6 The age specific incidence of human *Salmonella* infections in Denmark, 1999–2003. In the first peak, among children below 2 years of age, the incidence was highest in boys (relative rate 1.15; 95 % CI 1.01–1.30). Around the second peak, among individuals between 20 and 25 years of age, the incidence was higher in females than in males (relative rate 1.17; 95 % CI 1.02–1.33) (n = 11 900 episodes).
Source: Statens Serum Institut.

3.1.2 Seasonality

Non-typhoid *Salmonella* infections have a marked seasonality. In the northern hemisphere most infections are reported around August (Figure 3.7), whereas *Salmonella* peaks in March in Australia (Hall *et al.*, 2002). Although this general pattern is consistent for most serotypes and countries, some serotypes demonstrate slight variations. For example, in the US *S.* Hadar peaks earlier, whereas *S.* Javiana has an unusually marked seasonal variation (Helfrick *et al.*, 2001; Olsen *et al.*, 2001a).

3.1.3 Infective dose and other factors influencing infection

Salmonella infections are acquired by the fecal–oral route, although the swallowing of contaminated aerosols may cause infections in rare situations (Rubenstein and

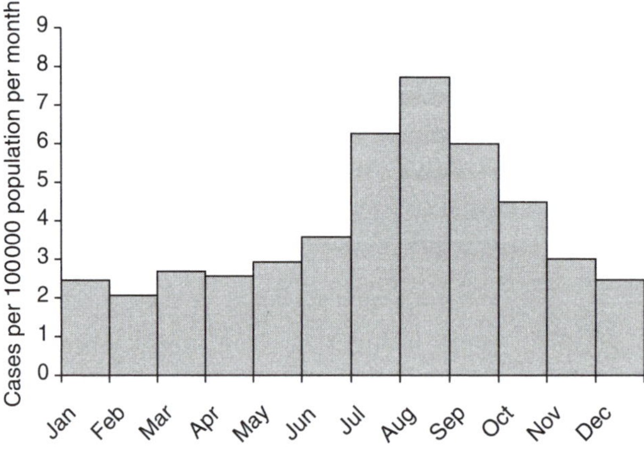

Figure 3.7 The monthly incidence of human infections with *Salmonella* in Denmark, 1999–2003.
Source: Statens Serum Institut.

Fowler, 1955; Grunnet and Hansen, 1978; Fannin *et al.*, 1985). Most persons are probably exposed to *Salmonella* from time to time, either from contaminated foods or from environmental sources. Under certain conditions, this exposure leads to clinical infection, subclinical infection or asymptomatic carriage. The number of organisms ingested, the vehicle of infection and several host factors are important in determining the outcome of exposure. In addition, some serotypes or strains of *Salmonella* may be more successful in infecting humans than others.

Data on the number of *Salmonella* organisms required to cause disease come from volunteer studies and investigation of outbreaks in which the numbers of bacteria in the food vehicles have been determined. Volunteer studies are not generalizable because of the limited number of subjects included; moreover, the physical and chemical properties of the vehicle may not represent relevant food vehicles, and laboratory-passaged strains may be less virulent than those found in food animals. Furthermore, higher doses of *Salmonella* may be required to cause disease in healthy volunteers than in populations at high risk.

Between 1936 and 1970, nine studies were performed on volunteers administered a variety of *Salmonella* serotypes. The attack rate increased with increasing inoculum size, and the number of organisms required varied with the host specificity of the serotype. For example, ingestion of less than 10^5 organisms of *S.* Typhi resulted in disease, whereas ingestion of 10^{10} *S.* Pullorum, an organism that is highly adapted to fowl, was required to produce gastroenteritis in man (Blaser and Newman, 1982). In a review of epidemiological data, Glynn and Bradley (1992) concluded that there was evidence of a correlation between dose and severity for several common serotypes, including Typhimurium, Enteritidis, Infantis, Newport and Thompson. In the investigation of a single-source outbreak, the effects of ingested *S.* Enteritidis dose on incubation period and on the severity and duration of illness were estimated in a cohort of 169 persons who developed gastroenteritis after eating hollandaise sauce made from shell eggs. The cohort was divided into three groups based on self-reported dose of sauce ingested. As dose increased, the median incubation period decreased and greater proportions reported body aches and vomiting. Increased dose was also associated with increases in median weight loss in kilograms, maximum daily number of stools, subjective rating of illness severity, and the number of days of confinement to bed. In this outbreak, ingested dose was an important determinant of the incubation period, symptoms and severity of acute salmonellosis (Mintz *et al.*, 1994).

Experiences from outbreaks have shown that under certain circumstances very small inocula have been sufficient to cause disease. In an outbreak of *S.* Typhimurium infection from contaminated chocolate, ≤ 10 *S.* Typhimurium per 100 g of chocolate was found. The authors concluded that the infectious dose was fewer than 10 organisms (Kapperud *et al.*, 1990). In an outbreak spread by paprika and paprika-powdered potato chips contaminated by a variety of serotypes (Saintpaul, Rubislaw and Javiana), the infective dose was estimated at 4–45 organisms with an attack rate of only 1 in 10 000 exposed persons (Lehmacher *et al.*, 1995). Finally, a massive outbreak of *S.* Enteritidis was linked to commercially distributed ice cream made from a liquid premix that had been transported in tanker trucks used previously to haul liquid raw eggs. The highest level of product contamination documented in this outbreak was only 6 organisms per half-cup (65 g) serving of ice cream (Hennessy *et al.*, 1996). In these three outbreaks, the estimated concentration of *Salmonella* is likely to have been similar to that of the product at the time of consumption. These studies confirm that low-level contamination of food by *Salmonella*, and thus extremely small infectious doses, can cause disease in humans. On this basis, single-hit models now play a more prominent role in dose–response assessment than models with a threshold (Teunis and Havelaar, 2000; Haas, 2002).

Ingested *Salmonella* must pass the acid barrier of the stomach, which is a first line of defense against enteric infections. Most strains of *Salmonella* survive poorly at normal gastric pH (< 1.5), but survive well at pH ≥ 4.0, and may have an adaptive acid-tolerance response that might promote survival at low pH. If the *Salmonella* are suspended in a lipophilic vehicle, the kill from the acid barrier is reduced, and consequently the infectious dose is reduced. Examples of such vehicles include ice cream, several other types of desserts, some sauces, and chocolate.

An important factor that increases the risk of infection is reduced gastric acidity due, for example, to achlorhydria, antacid use or gastric surgery. Neonates and infants may be at high risk of infection because of their relative achlorhydria and the buffering capacity of breast milk or formula. In addition, high-iron infant formula may increase the risk of *Salmonella* infection (Haddock *et al.*, 1991). This may contribute to transmission of *Salmonella* to neonates and infants from infected family members, or within neonatal departments or infant care settings.

Therapy with antimicrobial drugs has been shown to be a risk factor for *Salmonella* infections. Recent administration of antimicrobials may provide a relative advantage for the Gram-negative flora, a so-called competitive effect. In addition, if the *Salmonella* is resistant to the drug, it has a selective advantage compared with other bacteria in the gut (Barza and Travers, 2002). In the investigation of a large milkborne outbreak of multidrug resistant *S.* Typhimurium, it was found that smaller quantities of milk transmitted the disease to persons who took antibiotics during the incubation period (Ryan *et al.*, 1987).

Chronic illness, including immunosuppressive disease, malignant disease and diabetes, may also decrease the number of bacteria needed to cause infection.

In conclusion, in both non-typhoidal *Salmonella* gastroenteritis and *S.* Typhi infection, low bacterial inocula may cause disease, in particular in association with deficiencies in host defenses or a food vehicle that protects the bacteria from the gastric acid.

3.1.4 Incubation period

The classical symptoms of non-typhoid *Salmonella* infection as an acute gastrointestinal illness are indistinguishable from those due to many other gastrointestinal pathogens. Within 6–48 hours after ingestion of the contaminated food, diarrhea, often with fever and abdominal cramps, occurs. Most patients develop symptoms 24–48 hours after exposure, but with ingestion of a high dose the incubation period may be as short as a few hours. In other situations, patients may initially be subclinically colonized, and symptoms develop as late as 10 days after exposure.

3.1.5 Symptoms in the acute phase of illness

Most patients develop a gastrointestinal illness with acute diarrhea as the main symptom. Other common symptoms include abdominal pain or cramps, fever, chills, nausea, vomiting, pain in the joints, headache, myalgia, and general malaise. Diarrhea may vary in volume and frequency. In most cases the stools are loose and of moderate volume. Blood may occur in the stools, and this is probably an important criterion for physicians to request a stool culture. Weight loss is a common feature of the infection. In a case-control study of *S.* Enteritidis infections, self-reported symptoms included diarrhea (99.8 %), abdominal pains (89.7 %), fever (82.2 %), pains in joints (63.7 %), headache (59.6 %), nausea (59.5 %), vomiting (42.7 %) and blood in stools (22.7 %). Although diarrhea usually lasts for less than a week, the median duration of sickness in this study was 13 days (interquartile range 8–17 days) (Mølbak *et al.* 2002). In uncomplicated cases, fever usually resolves within 48–72 hours. It is generally impossible to differentiate clinically between infection with non-typhoid *Salmonella* and other foodborne agents, although viral gastroenteritis due to calicivirus have vomiting as a more important symptom (> 50 % patients) and is usually of shorter duration than *Salmonella*-infection and other bacterial foodborne infections.

Salmonella and other enteric bacterial pathogens can cause a syndrome of pseudoappendicitis, in which patients have severe abdominal pain (Saphra and Winter, 1957; Fisker *et al.*, 2003).

Between 3 % and 7 % of immunocompetent persons infected with *S.* Typhimurium or Enteritidis have positive blood cultures (Saphra and Winter, 1957; Fisker *et al.*, 2003). The risk of bloodstream infection is higher for some of the rare serotypes. *S.* Dublin is a highly invasive serotype; but *S.* Choleraesuis, *S.* Oranienburg, and in some studies *S.* Virchow, are also frequently isolated from blood. It has been suggested that there is an almost inverse relationship between invasiveness and the incidence of the zoonotic serotypes (Schønheyder and Ejlertsen, 1995). The proportion of blood isolates is higher in patients of extreme ages, in particular in the elderly.

Immunosuppression and other chronic underlying illness, including inflammatory bowel disease, organ transplantation (Mussche *et al.*, 1975), malignancy (Han *et al.*, 1967; Wolfe *et al.*, 1971), and malnutrition are risk factors for severe gastroenteritis as well as bloodstream infection. Patients with HIV/AIDS are at a particularly high risk for invasive *Salmonella* infection (Sperber and Schleupner, 1987; Levine *et al.*, 1991).

A subset of patients develops a primary septicemia without prominent gastrointestinal symptoms. The infectious dose, the *Salmonella* serotype, and host factors are important determinants of this syndrome, which is associated with a high risk of death. For example, in an outbreak caused by homemade ice cream, all eight affected persons became severely ill 8–18 hours after they had eaten the ice cream. A previously healthy 13-year-old boy died 37 hours after exposure, his mother and four younger siblings were transferred to intensive care units in hospitals in adjoining states, and the remaining two adult males were hospitalized locally. The ice cream contained 10^6 *Salmonella*/g and, according to food histories, the fatal illness occurred in the boy who had eaten the largest amount of ice cream (10^9 organisms) (Taylor *et al.*, 1984a). A similar outbreak has been reported from Denmark (Anonymous, 2000b).

3.1.6 Complications

Non-typhoidal *Salmonella* gastroenteritis is often described as a self-limiting disease without any complications. However, carefully conducted epidemiological studies suggest that complications are experienced by a significant proportion of patients. For example, in a study of 3328 patients, extraintestinal disease was present in 135 (4.1 %), and complications in 233 (7.0 %) patients included 27 unnecessary appendectomies (Fisker *et al.*, 2003).

Complications are often related to bloodstream infection, and include endocarditis or arterial infections. Localized infections include abdominal, soft-tissue and urogenital infections (Miller *et al.*, 1995). In addition, persistence of bowel symptoms commonly occurs after bacterial gastroenteritis and is responsible for considerable morbidity and health care costs (Neal *et al.*, 1997).

3.1.7 Sequelae

Sequelae or late-onset complications are characterized by reactive symptoms. Between 2 % and 15 % of episodes of *Salmonella* gastroenteritis are followed by symptoms of reactive arthritis. Multiple joints are usually involved, most commonly the knee, ankle, and wrist. Symptoms can range from mild arthralgias to severe arthopathy, and can last for variable periods. Most cases of reactive arthritis are self-limiting, but severe arthopathy may be incapacitating. As for some other bacterial gastrointestinal infections, reactive symptoms are correlated with HLA-B27 antigen. Nearly all patients who develop arthritis report a febrile diarrheal episode, and studies have indicated an association between the severity of the initial episode of enteritis and the risk of developing reactive arthritis. Arthritis occurs at an average of 10 days after the onset of diarrhea.

Reiter's syndrome is a triad of arthritis, conjunctivitis and urethritis. A subset of patients with reactive symptoms will develop the full triad, whereas others may experience conjunctivitis, iritis, or bursitis without arthritis. Erythema nodosum is another well-known reactive consequence.

3.1.8 Case fatality

The mortality after uncomplicated *Salmonella* gastroenteritis is low. Acute phase mortality was estimated to be 1.3 % in the US (Cohen and Tauxe, 1986) and 1.2 % in Denmark (Fisker *et al.*, 2003). The average case-fatality rate reported to FoodNet was lower, at 0.7 % (Mead *et al.*, 1999). This figure, after a correction of underreporting of deaths, was used for a recent assessment of mortality due to foodborne infections in the US. However, this approach assumes that deaths are limited to the acute phase of infection. Furthermore, the confounding effect of comorbidity was not taken into account. Recently, new estimates of the excess mortality associated with infections with *Salmonella* were obtained in a population-based registry study including 26 974 case-patients. The 1-year mortality risk was 3.1 %, 2.9 times higher than a matched sample of the Danish population, suggesting that the mortality after *Salmonella* infections may be underestimated (Helms *et al.*, 2003).

3.1.9 Carrier state

The mean duration of carriage of non-typhoid *Salmonella* is 4–5 weeks after resolution of the gastroenteritis (Buchwald and Blaser, 1984). The duration of culture positivity varies with the age of the patient and the serotype. For some less common serotypes, such as *S.* Panama, Muenchen and Newport, more than 20 % are still positive at 20 weeks, whereas 90 % of *S.* Typhimurium patients are negative at 9 weeks. A higher proportion of infants than adults has prolonged shedding, but the delayed clearance in neonates and infants does not result in permanent carriage. Treatment with antimicrobials may lead to longer excretion (Aserkoff and Bennett, 1969). A chronic carrier state is defined as the persistence of *Salmonella* in stool or urine for periods greater than a year. It is estimated that between 0.2 % and 0.6 % of patients with non-typhoid *Salmonella* infections develop chronic carriage, which is less than for *S.* Typhi.

3.2 Epidemiology of *S.* Typhi and *S.* Paratyphi

Enteric fever is a severe systemic illness characterized by fever and abdominal symptoms. Enteric fever caused by *S.* Typhi is referred to as typhoid fever, but a similar but less severe syndrome is caused by *S.* Paratyphi A, B or C, and is called paratyphoid fever. *S.* Paratyphi may also cause a gastrointestinal illness resembling non-typhoid *Salmonella* gastroenteritis. In the pre-antibiotic era the case fatality rate of typhoid fever was high (10 %–20 %), and mortality continues to be high in some areas of Africa and Asia. Inadequate access to health care, delayed treatment, and antimicrobial resistance contribute to high case fatality rate. However, the fatality may be reduced to less than 1 % with prompt and relevant antibiotic therapy.

S. Typhi and *S.* Paratyphi are highly adapted to man. Often the infection is acquired by food or water contaminated with human excreta; direct person-to-person transmission is rare, but anal–oral transmission of *S.* Typhi has been demonstrated (Dritz and Braff, 1977). The worldwide annual incidence of typhoid fever is estimated to be about 17 million cases, with approximately 600 000 deaths (Chin, 2000). Areas with high disease burdens include South and East Asia, Africa south of the Sahara, and Latin America. This situation is related to growth of the population, increased urbanization, inadequate sanitation and insufficient water supply, and crowding in homes and settlements. The situation is worsened by an overburdened health care system and an increasing number of people with HIV/AIDS. In Indonesia typhoid fever is among the five major causes of death (Edelman and Levine, 1986). By contrast, the incidence of typhoid fever is low in industrialized countries. *S.* Typhi infection became an enormous problem in the US in the industrial era, but the incidence of typhoid fever has since decreased; it was 1 case per 100 000 population in 1955 and 0.2 cases in 1966 (Figure 3.1), but has since remained fairly stable with fewer than 500 cases annually. This progress is likely to be related to improved food-handling practices and water treatment, although improved management of chronic carriers (for example by long-term fluoroquinolone treatment) may also have contributed to the decline. Contaminated food or water remains a source of outbreaks. Usually, water-borne transmission involves the ingestion of fewer bacteria and, consequently, has a longer incubation period and a lower attack rate than foodborne transmission.

In developed countries, enteric fever is often associated with foreign travel or immigration. The risk of acquiring *S.* Typhi and *S.* Paratyphi is particularly associated with travel to destinations in South-East Asia, the Indian subcontinent, South America and Africa.

3.2.1 Age- and gender-specific incidence

In endemic areas, the incidence of *S.* Typhi infection is often highest in preschool children or children 5–19 years old. Adults have acquired substantial immunity from previous exposures. In industrialized countries the median age of infection is in young adults, and the infection is usually found more frequently in males than in females. This pattern reflects that most *S.* Typhi infections in industrialized countries are imported. Immunosuppressed patients, including those with HIV/AIDS, transplant patients, and those with diseases that include biliary or urinary tract abnormalities, are at higher risk of typhoid fever.

3.2.2 Infective dose

The infective dose is believed to be smaller for *S.* Typhi than for *S.* Paratyphi. As for non-typhoid *Salmonella*, susceptibility to infection is increased in individuals with achlorhydria and underlying illness such as HIV infection. Low bacterial inocula may cause disease, in particular in association with deficiencies in host defenses or if a food vehicle protects the bacteria from the gastric acid. As for non-typhoid *Salmonella*, it is thought that clinical severity depends on the number of bacteria ingested. However, a re-analysis of the relationship between challenge dose and severity of disease suggests a need for caution in interpreting this relationship (Glynn *et al.*, 1995).

3.2.3 Incubation period

Typhoid fever is characterized by an insidious onset, and the clinical picture may be mild or atypical. Owing to these characteristics the incubation period may be difficult to determine, but it ranges from 3 days to 1 month, with a usual range of 8–14 days. *S.* Paratyphi may cause *Salmonella* gastroenteritis resembling non-typhoid *Salmonella*, and in this case the incubation period is 1–10 days.

3.2.4 Acute illness

Typhoid and paratyphoid fever are severe systemic illnesses characterized by fever and abdominal symptoms. The symptoms may be relatively non-specific, and many cases are atypical. In contrast to patients with acute abdominal illness caused by other Gram-negative bacteria, symptoms may be present for weeks. Some patients have a prodromal phase with mild diarrhea and abdominal pains.

The symptoms during the first week are headache, malaise and rising fever, and may include constipation or non-productive cough. In the second week of illness, fever is a classic sign but is not always present, and the pattern of the fever is of little practical use in the diagnosis. Some patients may look dazed and apathetic and have sustained fever. Between 20 % and 40 % patients have abdominal pains. Some patients have diarrhea at presentation, including AIDS patients and infants. Other common symptoms or signs include chills, anorexia, a slightly distended abdomen, and splenomegaly. Among Caucasian patients, 25 % to 50 % have rose spots of 2–4 mm in diameter on the trunk; these rose spots fade under pressure. The spots are caused by bacterial embolization, and may also be found in invasive non-typhoid *Salmonella* infections. Bradycardia is common in adults during the first 2 weeks of disease, but this sign is not a reliable diagnostic sign because it occurs in less than 50 % of patients.

In the third week, the patients become more dazed and ill. Without treatment, the high fever persists and a delirious confused state sets in (typhoid state). Abdominal distension is pronounced, and abdominal sounds are scanty. Diarrhea is common, with liquid, foul green-yellow stools (pea-soup diarrhea). Pulmonary symptoms are also often present. Death may occur from septicemia, myocarditis, intestinal hemorrhage or perforation. Patients who survive into the fourth week slowly improve, but intestinal complications may still occur.

Paratyphoid fever shares the features of typhoid fever, although the infection is milder and of shorter duration. *S.* Paratyphi may cause *Salmonella* gastroenteritis resembling non-typhoid *Salmonella* infection.

The definitive diagnosis of enteric fever requires the isolation of *S.* Typhi or *S.* Paratyphi from the patients. Cultures of blood (early in the disease), stool (after the first week), urine, rose spots, bone marrow, and gastric and intestinal secretions can all be useful. Bone marrow cultures provide the best bacteriological conformation, even in patients who have received antimicrobial drugs. Most serological tests, including the classical Widal test, have limited sensitivity and specificity, and are of little diagnostic value.

3.2.5 Complications

Between 10 % and 20 % of patients treated with antimicrobial drugs suffer a relapse after initial recovery. A relapse is usually milder and shorter than the initial illness.

Common complications of enteric fever are intestinal hemorrhage and perforation; the recognition of these complications may be difficult. The liver, gallbladder, and pancreas may also be affected. Other complications include focal infections, an often-fatal toxic myocarditis, and degeneration of skeletal muscle fibers. A dazed, confused state, characterized by disorientation, delirium, and restlessness, is characteristic of late-stage typhoid fever. These symptoms may occasionally be present in the early stage of the disease, and patients may be misdiagnosed as suffering from neurological or psychiatric illness. A number of other neurological or psychological symptoms may develop during convalescence. Subclinical, disseminated intravascular coagulation occurs commonly in typhoid fever.

Carriers of *S.* Typhi have been reported to have an increased risk of hepatobiliary cancer. Gallstones increase the risk of gallbladder cancer and are almost always present in typhoid carriers, so the causal route of this condition is debatable (Welton *et al.*, 1979; Mellemgaard and Gaarslev, 1988; Dutta *et al.*, 2000).

3.2.6 Carrier state

About 10 % of untreated typhoid patients discharge bacteria for 3 months after the onset of symptoms, and between 1 % and 4 % of patients with *S.* Typhi infection develop chronic carriage. Considerably fewer patients with *S.* Paratyphi become chronic carriers. The frequency of chronic carriage is higher in women and in persons with biliary abnormalities. Pathology studies suggest that most carriers of *S.* Typhi have gallstones, and this is thought to be the locus of the carrier state. Concurrent infection with *Schistosoma* spp. is another risk factor for *S.* Typhi carriage. In the absence of urinary tract pathology, urinary carriage is rare after the third month. Serology for the Vi antigen is useful for distinguishing chronic carriage from acute infection.

Fluoroquinolones are excreted in the bile, and are excellent for the management of chronic carriers, even in the presence of biliary pathology. A recent decline in carrier rates may be associated with the shift in therapeutic choice from chloramphenicol or co-trimoxazole to fluoroquinolones.

S. Typhi carriers working as food handlers are important sources of infection, as illustrated by the classical case of 'Typhoid Mary'. The legendary Typhoid Mary Mallon spent the last 15 years of her life (1915–1930) in quarantine in a New York City Hospital as a result of her chronic carriage of *S.* Typhi and her intractable affinity for working as a professional cook despite order from the health department. A total of 26 cases and 3 deaths were attributed to Mary Mallon (Burrows, 1954). Even today, in industrialized countries, many indigenous cases may be traced to chronic carriers who contaminate food.

4 Prevalence of *Salmonella* in foods and feeds – principles for control

Salmonella can occur in a wide range of different food products. It is reasonable to assume that all products contaminated with *Salmonella* at the point of consumption have the potential to cause human disease. Traditionally, meat and eggs have constituted the major vehicles of foodborne human salmonellosis in the industrialized world; however, in recent years a number of new and surprising vehicles of *Salmonella* infections have been recognized. Control of salmonellosis should involve all stages in the 'feed-to-food' chain, and ideally be conducted as an integrated effort, where control measures in each stage of production are coordinated with efforts in other stages to optimize the effect. Methods for detecting *Salmonella* in feedstuffs, food animals, environment and foods are pivotal to the development and implementation of successful control efforts.

4.1 Poultry

Poultry may acquire *Salmonella* from various sources, including parent birds, feedstuffs, rodents, wild birds, and other vehicles. Clinical disease is uncommon, with the exception of infection caused by the host-adapted serovar Pullorum/Gallinarum, but most infections are of importance as potential sources of food poisoning in man. A wide range of *Salmonella* serotypes have the ability to colonize poultry:

S. Typhimurium, *S.* Enteritidis, *S.* Hadar, *S.* Virchow, *S.* Infantis and, recently, *S.* Paratyphi B var. Java have all been frequently isolated from poultry in several countries world-wide.

4.1.1 Occurrence in poultry

Salmonella can frequently be isolated from most species of live poultry, such as broilers, turkeys, ducks, and geese. The levels of *Salmonella* in poultry can vary depending on the country, the nature of the production system, and the specific control measures in place. In Sweden, poultry and poultry products are almost free from *Salmonella* due to a decade-long effort to achieve this. In contrast, in most other countries with intensive poultry production a high proportion of poultry are infected. A survey from the US showed that up to 60 % of broiler flocks and 36 % of carcasses after slaughter harbored *Salmonella* (Bailey *et al.*, 2002). The European Union annual survey of the occurrence of zoonotic agents in feed, animals, and food from 2000 (Anonymous, 2002c) showed that *Salmonella* occurrence in broiler flocks ranged from 0.1 % in Sweden to more than 30 % of flocks in some of the other countries reporting surveillance data. Countries adhering to the EU-approved control programs (Table 3.10) generally had a lower occurrence of *Salmonella* compared to other countries not adhering to the program.

Salmonella can also frequently occur in turkey flocks, and with especial frequency in commercial flocks of geese and ducks, where levels exceeding 90 % have been reported. *Salmonella* can also be isolated from wild ducks and geese, but at lower levels (Fallacara *et al.*, 2001).

4.1.2 Occurrence in poultry products

Poultry products may be contaminated at several stages during the slaughter process, either from feces during evisceration, or by cross-contamination from contact with contaminated products or surfaces of the production line. Particular 'hot spots' for contamination in poultry slaughter include defeathering, evisceration and cutting, whereas chilling in a water bath reduces the load of *Salmonella* but may facilitate cross-contamination from contaminated to uncontaminated products

Table 3.10 *Salmonella* surveillance according to the EU Zoonosis Directive 92/117/EEC

Production unit	Sampling frequency	Sample material
Central rearing stations (broiler and table egg sector)	Chickens at 1 day of age	10 samples of crate material, 20 dead/destroyed chickens
	4 weeks	30 dead chickens and 60 fecal samples
	2 weeks before moving birds	12 × 5 fecal samples
Breeder (hatching egg production) broiler and table egg sector	Every 2 weeks	50 dead chickens or meconium from 250 chickens taken from the hatchery

(Corry *et al.*, 2002; Fluckey *et al.*, 2003; Northcutt *et al.*, 2003). A close correlation between levels of *Salmonella* in broiler flocks and in carcasses after slaughter has been documented in some studies (Figure 3.8) (Wegener *et al.*, 2003), whereas other studies have not found such an association (Heyndrickx *et al.*, 2002). The explanation of this difference may be variations in pre- and post-harvest monitoring programs.

4.1.3 Pre-harvest control

Preventing the introduction of *Salmonella* into poultry flocks is a prerequisite to keeping birds *Salmonella*-free. There are several well-documented sources of *Salmonella* infection in poultry production. The significance of different risk factors varies between different production systems, countries and regions.

Wild fauna provide a reservoir for *Salmonella*, and are consequently a potential source for transmission of infection to domestic animals. Birds of all species, rodents, foxes, badgers, and other animals have been shown to be sources of *Salmonella* (Edel *et al.*, 1976; de Boer *et al.*, 1983; Euden, 1990; Evans and Davies, 1996). An association between the level of occurrence of *Salmonella* in husbandry systems and the level of *Salmonella* in wild birds in the same region has sometimes been demonstrated, suggesting that transmission of *Salmonella* between domesticated and wild birds can go in both directions.

Insects, for example litter beetles and flies, have been shown to be risk factors for re-introduction of *Salmonella* into poultry houses after depopulating, cleaning and restocking the premises (Edel *et al.*, 1973; Davies and Wray, 1995). Humans (such as farm staff, veterinarians, and visitors) and domesticated animals (such as cats and dogs) can also serve as vectors of *Salmonella* introduction into food animal flocks. Finally, *Salmonella* may be transmitted by airborne spread of aerosols from, for

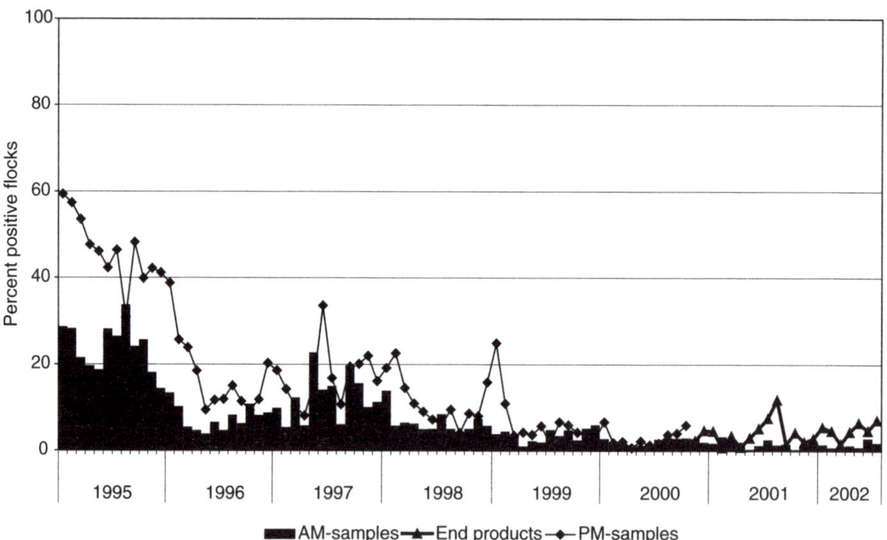

Figure 3.8 Prevalence of *Salmonella* in Danish broiler flocks and carcasses/end products after slaughter. AM, ante mortem; PM, post mortem. *Source:* Annual Report on Zoonoses in Denmark 2002, Ministry of Food, Agriculture and Fisheries

instance, manure, human waste dumps and contaminated water (Fannin *et al.*, 1985; Hardman *et al.*, 1991).

A major source of *Salmonella* infection in herds or flocks of food animals is the introduction of infected animals. This is very well documented for poultry, where vertical transmission is a main mode of introduction of *S.* Enteritidis (Humphrey, 1994, 1999; Saeed *et al.*, 1999). The highly integrated and multi-layered pyramids of poultry production, where a few breeding flocks are supplying very large numbers of producers with animals, provide an excellent opportunity for vertical transmission followed by propagation of *Salmonella* in food animals. Environmental contamination in hatcheries can be a key factor in the spread of *Salmonella* (Skov *et al.*, 1999).

Feedstuffs can harbor *Salmonella*, and contaminated feedstuffs are a potential source of *Salmonella* in food animals. In many places feedstuffs are heated (pelleted) to prevent *Salmonella* introduction. Contaminated water is a potent source of *Salmonella* in food animals.

Applying the all-in-all-out type production principles and ensuring that houses are totally emptied of animals, cleaned and disinfected before new animals are introduced is an important factor in the prevention of *Salmonella* transmission from one group of animals to another (Skov *et al.*, 1999). This can eliminate *Salmonella* in the environment and thus prevent infection of newly introduced animals. Apart from thoroughness of cleaning, the time for which the premises are kept vacant between depopulation and repopulation can be important – the longer the vacant period, the lower the risk that residual *Salmonella* will survive in the environment.

Feeding is an important tool in ensuring optimal gut health. Incorporation of organic acids, especially formic acid and propionic acids, in the feed has been shown to reduce the incidence of *Salmonella* in broilers, layers and swine (Wingstrand *et al.*, 1997).

Vaccination to boost immunity to *Salmonella* infection and colonization has been tried in most animal species. It is most widely used in poultry, where live attenuated vaccines have been applied on a very large scale in some countries in Europe. Hassan and Curtiss (1997) showed that chickens, vaccinated orally with an avirulent live *S.* Typhimurium strain and subsequently challenged with wild-type *S.* Typhimurium or *S.* Enteritidis, were partially protected from infection, compared to non-vaccinated chickens. Vaccination does not offer complete protection of the birds against infection; but it may, under some circumstances and for some serotypes, increase the animals' resistance to infection/colonization, and thus reduce the numbers of animals in a flock harboring *Salmonella* or the levels of *Salmonella* shed by individual animals.

Competitive exclusion (CE) consists of the early establishment of an adult intestinal microflora in young animals to prevent colonization by *Salmonella*. Undefined mixed bacterial cultures from the crop or intestinal content of adult chicks have been used to protect newly hatched chicks against colonization with *Salmonella*. Attempts to develop defined cultures have been less successful, and so far no commercial products of defined cultures are available. Studies have shown that CE cultures provide some level of protection against colonization with a wide range of serovars in broilers. However, CE treatment does not offer complete protection. It should be used in combination with other measures, such as vaccination and the use of acid-treated feed. The application of CE treatment for *Salmonella* control in other animal species is not well investigated.

Elimination of *Salmonella* from the animal gut with bacteriophages has been studied, and *Salmonella*-reducing effects have been observed when high doses of lytic phages have been applied (Berchieri *et al.*, 1991). So far, this concept has not led to development of commercial products.

Antibiotics should never be used to control latent infections with *Salmonella*, due to the risk of antimicrobial resistance development. Antibiotics may be used to control clinical salmonellosis, where animal welfare reasons dictate it.

Under some conditions, dietary carbohydrates may cause changes in the intestinal flora of chicks and selectively promote the propagation of the part of the gut flora that inhibits colonization with *Salmonella* (Fernandez *et al.*, 2002). The effects observed in experimental studies vary, and more research is needed.

Complete elimination of the infection from broiler flocks that are already infected is virtually impossible. Infected poultry tend to clear the infection with increased age; but in the current intensive production system the birds are slaughtered too early in life to take advantage of this natural clearance.

During catching and transportation to the slaughterhouse, birds may become infected from contaminated catching machines and/or transport crates if these have not been adequately cleaned and dried (Slader *et al.*, 2002).

The Danish *Salmonella* control program is an example of a successful pre-harvest control program, based on detection and slaughter of infected flocks and application of farm hygiene. This program is described in detail elsewhere (Wegener *et al.*, 2003). By implementation of intensive monitoring (Table 3.11) and a top-down eradication program, the flock prevalence of *Salmonella* has been reduced from nearly 80 % in the early 1990s to less than 2 % in 2002 (Figure 3.9).

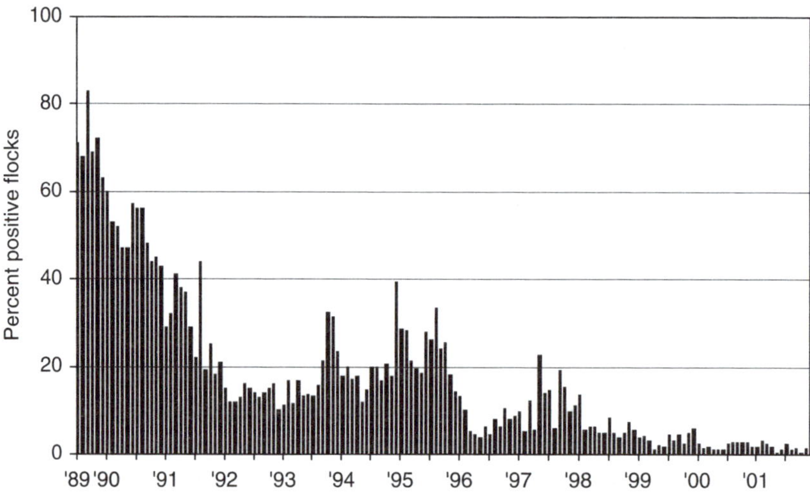

Figure 3.9 Prevalence of *Salmonella* in Danish broiler flocks following implementation of a top-down eradication strategy in 1989. The program has been regularly evaluated and revised. *Source:* Danish Veterinary and Food Administration.

Table 3.11 *Salmonella* surveillance in the Danish broiler and table egg production, 1999

Production unit	Sampling frequency	Sample material
Central rearing stations (broiler and table egg sector)	Chickens day 1	10 samples of crate material, 20 dead/destroyed chickens[a]
	1 week	40 dead chickens
	2 weeks	20 dead chickens
	4 weeks	30 dead chickens and 60 fecal samples[a]
	8 weeks	60 fecal samples
	2 weeks prior to moving birds	12×5 fecal samples and 60 blood samples[a, b]
Breeders (hatching egg production, broiler and table egg sector)	Every 2 weeks	50 dead chickens or meconium from 250 chickens taken from the hatchery[a, c]
	Every 4 weeks	60 fecal samples and 60 samples of blood or eggs[d]
Hatchery	Every week	Wet 'dust'
Rearing (table egg production)	Chickens day 1	10 samples of crate material and 20 dead/destroyed chickens
	Every 3 weeks	5 × 2 drag swab samples or 60 fecal samples
	Every 12 weeks	5 × 2 drag swab samples or 12 × 5 fecal samples, and 60 blood samples[b]
Table egg production	Every 9 weeks for eggs sold to packers	5 × 2 drag swab samples[e] or fecal samples, and 60 eggs
	Every 6 months for eggs sold at the barn door	Fecal and egg samples

Data: Danish Veterinary and Food Administration.
[a] Requirements of the EU Zoonosis Directive (92/117/EEC).
[b] Samples taken by the district veterinary officer
[c] Samples taken by the district veterinary officer every 8 weeks
[d] Samples taken by the district veterinary officer every 3 months
[e] The drag swab samples are collected by 15 cm pieces of tube gauze mounted on the foot wear during house inspection; the method has the same or better sensitivity as 60 fecal samples.

4.1.4 Post-harvest control

The poultry slaughter process involves several stages that are critical to *Salmonella* contamination:

- *Scalding* reduces the total microbial load of the carcass; but as contaminated and non-contaminated birds may share the same water, the risk of cross-contamination during scalding is high. The temperature of the scalding water is very important to reduce *Salmonella* contamination. Temperatures between 50°C and 60°C are usually applied. Frequent replacement of scalding water or use of countercurrent-scalding or spray-scalding systems may help to reduce microbial loads and the risk of cross-contamination.

- *Defeathering* squeezes out fecal material from the feathers as well as from the gut by the action of the rubber fingers that mechanically remove the feathers. The machines are difficult to clean and easily become contaminated with *Salmonella* during the slaughter of infected flocks early in the day, then pass on the contamination to subsequent non-infected flocks throughout the remainder of the day.
- *Evisceration* with modern automated evisceration machines frequently ruptures the intestines, causing fecal leakage to occur. Fecal contamination of the inner and outer surfaces of the carcass during evisceration is an important mode of *Salmonella* contamination if birds were colonized with *Salmonella* at the time of slaughter.
- *Rinse/chill* is used to remove visible fecal contamination and to chill the carcass by immersion in water or by spraying. The immersion process has been shown in several studies to be a risk factor for cross-contamination. Spray systems are less prone to causing cross-contamination, but may be less efficient in removing already established contamination. In some countries, decontaminants (e.g. organic acids, sodium triphosphate, sodium hypochlorite) may be added to the water to reduce microbial contamination. In a study of 250 chickens slaughtered in a commercial plant in Puerto Rico, James *et al.* (1992) found that the level of Enterobacteriaceae was reduced from $\log_{10}$ 2.70 before evisceration, to 2.50 before chilling, and 1.56 after chilling; however, *Salmonella* was found on 24 % of the carcasses before evisceration, 28 % after evisceration and 49 % after chilling. This shows that the microbial load was reduced but the proportion of *Salmonella*-contaminated carcasses increased.
- *Cutting and/or packaging* and any subsequent steps where the product comes into contact with hands, utensils or surfaces may result in cross-contamination.

Reduction of the bacterial load of the chicken carcass or cuts of chicken can be effectively achieved by irradiation; however, organoleptic properties may change and consumers, at least in some regions of the world, have a low degree of acceptance of irradiated foods (Lewis *et al.*, 2002).

Storing at refrigerated temperatures below 5°C throughout the chain of distribution, storage and retail sale is important, because *Salmonella* can multiply in chicken meat at temperatures exceeding 6°C (Oscar, 2002).

Handling during preparation and cooking in the home or in professional food establishments is of critical importance for the prevention of salmonellosis. Gorman *et al.* (2002) found that *Salmonella* and other pathogens could be spread from fresh chicken to hands and food-contact surfaces in the domestic kitchen, such as the dish cloth, refrigerator handle, oven handle, counter top or draining board, during preparation of a traditional Sunday roast-chicken lunch.

4.2 Shell eggs

Contaminated shell eggs are probably the most common cause of human salmonellosis worldwide (Herikstad *et al.*, 2002). *Salmonella* can contaminate eggs on the shell as well as internally. Shell eggs are found to be much more frequently contaminated on the shell compared to contamination of the white/yolk. Furthermore, surface contamination is associated with many different serotypes, while infection of the white/yolk primarily is associated with one serotype only, *S.* Enteritidis (Humphrey, 1994). This is because *S.* Enteritidis is capable of infecting the reproductive organs of the laying hen (the ovaries), and from this site can contaminate the interior of the egg before the shell is deposited (Humphrey *et al.*, 1991). Experimentally-infected laying hens can produce eggs with a relatively high level of contamination/infection. In a study by Bichler *et al.* (1996), 26.5 % of eggs from experimentally-infected hens were positive for *Salmonella* on the exterior, while only 2.9 % were positive for *S.* Enteritidis internally. *Salmonella* usually occurs in low numbers inside the egg (< 100), but the levels of *Salmonella* may become very high (> 10^8 CFU/ml) if the contaminated eggs have been stored at elevated temperatures (> 12°C) for extended periods of time (> 20 days) (Humphrey and Whitehead, 1993).

Salmonella can occur in all types of shell egg production. In a monitoring program in Denmark, *Salmonella* antibodies have been found in layers from organic, free-range, barn-yard, deep-litter and battery-type production (Figure 3.10). Surveys of naturally contaminated shell eggs for *Salmonella* usually detect a very low prevalence of contaminated/infected eggs. A survey of 5700 eggs from the UK detected 0.6 % positive in the contents, a survey of 13 000 eggs from Denmark found that 0.02 % of the eggs were infected internally, and two surveys from USA comprising 1400 and 1200 eggs respectively detected no positive samples (Baker *et al.*, 1980; Humphrey *et al.*, 1991; Bager, 1996; Schutze *et al.*, 1996). In the same period, *S.* Enteritidis from eggs was a major source of foodborne salmonellosis in these countries. This apparent

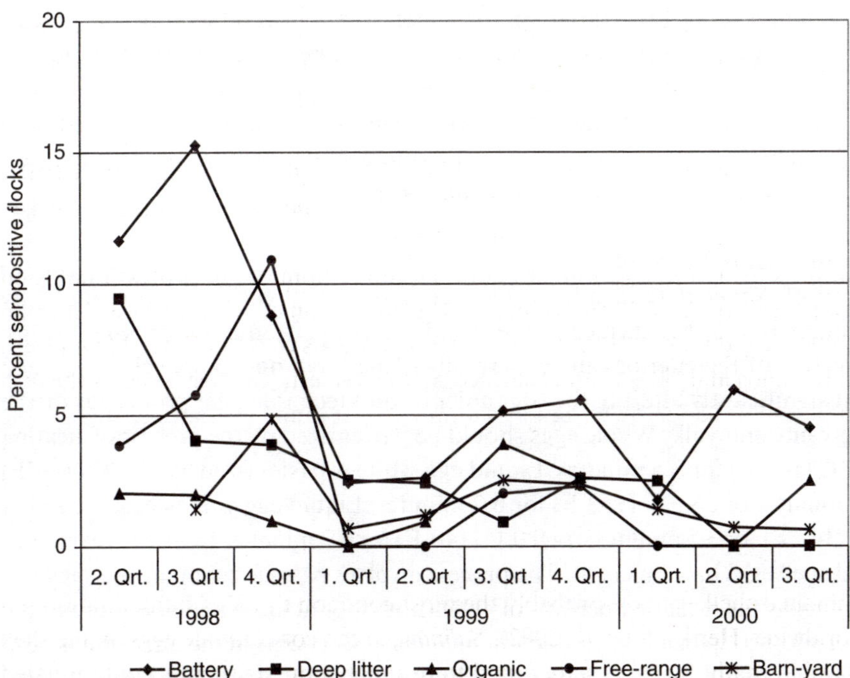

Figure 3.10 *Salmonella* in different shell egg production systems in Denmark 1998–2001. *Source:* Danish Zoonosis Centre and Danish Veterinary and Food Administration.

disagreement may be best explained by differences in the handling of dishes where raw eggs are used (e.g. cakes and desserts, mayonnaise, ice cream). Storage for longer periods of time at room temperatures may occur before the food is shared by many persons, for instance at a party (the term 'wedding cake outbreak' refers to large *S.* Enteritidis outbreaks associated with contaminated cream cakes).

Eggs of other species, such as ducks, may also harbor *Salmonella*. A survey from Thailand showed that 12.4 % and 11 % of duck eggs were infected on the surface and in the interior, respectively (Saitanu *et al.*, 1994).

4.2.1 Pre-harvest control

An effective way to control *Salmonella* in eggs is by preventing the vertical spread of *Salmonella* Enteritidis between different generations of birds. The principle is a top-down approach, eliminating *Salmonella* from the top of the production pyramid and downwards by test and slaughter. At present there are no effective tools to eliminate *Salmonella* from infected flocks of hens, so infected birds have to be eradicated and replaced with *Salmonella*-free birds. *Salmonella* cannot be eliminated from infected birds, but management, hygiene, and the use of vaccines or competitive exclusion cultures may reduce the level of infection in a flock of laying birds and thereby reduce the number of infected eggs produced. Vaccination with attenuated live *Salmonella* vaccines to enhance resistance to infection is used in many countries specifically to reduce *S.* Enteritidis in shell egg production (Nassar *et al.*, 1994; Parker *et al.*, 2001; Woodward *et al.*, 2002).

Molting is a process whereby a second laying cycle is induced in layers after completion of a laying cycle (after approximately 1 year of production). The process involves starvation (several weeks) and sometimes a period of complete light restriction. Apart from the animal welfare issues relating to this practice, it has been shown that *Salmonella* shedding and levels of *Salmonella* in shell-eggs produced from infected and molted hens is enhanced. Furthermore, birds are more sensitive to infection during the starvation period (Holt *et al.*, 1994).

4.2.2 Post-harvest control

The level of *Salmonella* in infected/contaminated shell eggs can be reduced by decontamination of the surface of the shell by heating, irradiation or other means. The infection of the interior can be thermally reduced but not completely eliminated without significantly altering the organoleptic and technological properties of the raw egg white and yolk. Whole eggs should be pasteurized by the process of heating to 60.0°C (140°F) for 3.5 minutes. Liquid egg white is pasteurized at 56.7°C (134°F) for 3.5 minutes or 55.6°C (132°F) for 6.2 minutes. Liquid egg yolk is pasteurized at 61.1°C (142°F) for 3.5 minutes or 60.0°C (140°F) for 6.2 minutes. In recent years, pasteurized shell eggs have emerged. The process involves pasteurization of the shell egg in water baths for extended periods of time (hours) to achieve a temperature in the center of ~55°C. Up to a 7-log reduction in *Salmonella* counts in the interior has been documented by this method. The major limitation to the widespread use of these post-harvest control methods is the relatively high cost involved (~2 cents per egg). Irradiation of eggs has been approved in the US (3 kGy), and studies indicate that this approach may achieve the same effect as thermal inactivation while retaining the desired organoleptic and technical qualities (e.g. 'whipability') of the egg (Wong *et al.*, 1996).

The pandemic of *S.* Enteritidis, notably phage type 4, is caused by the vertical spread of *Salmonella* from a few central breeding establishments, which supply most countries in the world with parent birds of layers. The inability to eliminate *S.* Enteritidis from these high-yielding lines of breeding birds and the industry's unwillingness to eradicate infected birds of valuable genetic lineages have significantly hampered the effective control of the global *S.* Enteritidis problem. This unfortunate situation has resulted in hundreds of millions of cases of human salmonellosis worldwide since the early 1970s, and the situation continues.

4.3 Pigs

Pork and pork products have in recent years increasingly become recognized as important sources of human salmonellosis (Nielsen and Wegener, 1997). Pigs are colonized by *Salmonella* on the farm, and pork is subsequently contaminated during slaughter or subsequent processing. Surveys from different countries have detected *S. enterica* in a high proportion of pig herds in some countries: 11.4 % in Denmark, 23 % in the Netherlands, and 46.7 % in the US (Harris *et al.*, 1997; van der Wolf *et al.*, 1999; Christensen *et al.*, 2002). The occurrence of *Salmonella* in pigs at slaughter has been found to vary greatly between studies, from 39.9 % in a survey from the USA to 0.1 % in a survey from Norway (Hurd *et al.*, 2002; Sandberg *et al.*, 2002). A high proportion of carcasses may be contaminated immediately after slaughter. Tamplin *et al.* (2001) reported 73 % in a study, but the level of contamination is reduced markedly after chilling of the carcass; only 0.7 % of the carcasses were positive at the chilled carcass stage. This is consistent with findings for *E. coli* and *Campylobacter*, that the chill step is a potentially important pathogen reduction step in meat processing. Prevalence of *Salmonella* in offal such as tongues and livers is significantly higher than on the surface of carcasses. A study by von Altrock *et al.* (2000) found that 1 % of carcasses, 2.7 % of tongues and 5.3 % of livers harbored *Salmonella* in a survey of several pig slaughterhouses in Germany. Surveys of pork at retail have found *Salmonella* prevalence in the range of 1–3 % (Wegener *et al.*, 2003), with some high-risk products (such as minced meat) reaching as high as 13 % (Oosterom *et al.*, 1985). A number of pork-associated outbreaks of human salmonellosis have confirmed the significance of pork as a source of human salmonellosis (Maguire *et al.*, 1993; Wegener and Baggesen, 1996; Mertens *et al.*, 1999; Mølbak *et al.*, 1999).

Control of *Salmonella* in pork can be implemented both on the farm and at slaughter and processing. Pre-harvest control consists of monitoring of *Salmonella* at the herd level, and implementation of *Salmonella* reduction measures in infected herds through hygiene, separation of animals, feeding strategy, and strict control over *Salmonella* in the supply chain of pigs for breeding and fattening. During transportation, holding and slaughter, pigs from infected herds should be separated from pigs from *Salmonella*-free herds to reduce the risk of cross-contamination. Post-harvest control requires the implementation of food safety assurance programs, such as

HACCP, and optimal hygiene and storage conditions throughout the stages of processing, handling and distribution. The pathogen level on carcasses and products can be reduced by various methods, such as hot water or steam treatment, irradiation, etc. Denmark is the only country with a nationwide 'feed-to-food' control program of *Salmonella* in pork (Mousing *et al.*, 1997; Wegener *et al.*, 2003). It is based on routine testing and classification of slaughter pig herds and subsequent slaughter of pigs according to the 'inherent risk' as measured by the test program, and will be described in some detail in the following sections.

4.3.1 Pre-harvest control

The Danish program for pre-harvest control includes serological testing of monthly blood samples of pigs from breeding and multiplying herds. If a specified cut-off level is reached, bacteriological confirmatory testing is carried out. Further, if the serological reactions exceed a specified high level, all movement of animals is restricted. Slaughter pig herds are monitored continuously by serological testing of 'meat juice' (meat juice is drip fluid released from meat samples after freezing and thawing) (Nielsen *et al.*, 1998). Meat samples for testing are collected at the slaughter line, and the number of samples and frequency of sampling are determined by the size of the herd. Approximately 700 000 slaughter pigs are currently tested each year out of a production of 22 million (Figures 3.11, 3.12). The herds are categorized into three levels based on the proportion of seropositive meat juice samples during the last 3 months. Owners of level-2 and level-3 herds are encouraged to seek advice on how to reduce the *Salmonella* problem in the herd (e.g. through changes in feeding, hygiene and other management practices). Furthermore, the payment from the slaughterhouse is reduced by 2 % and 4 % respectively for level-2 and level-3 animals (Figure 3.12).

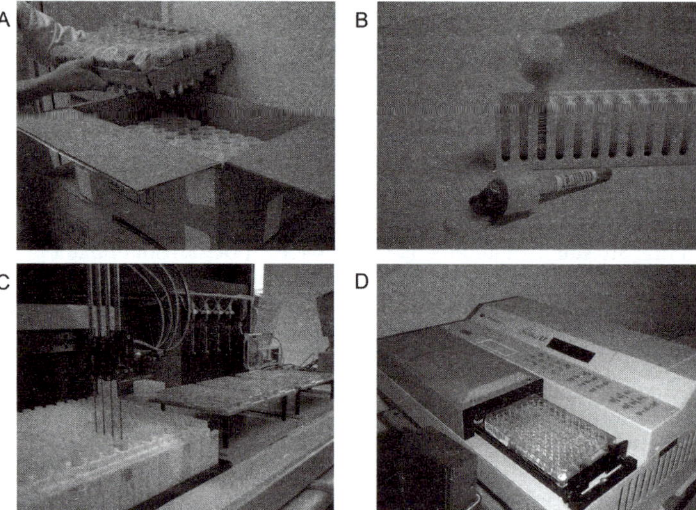

Figure 3.11 (A) Receipt of pork samples from the slaughterhouse. Each tube is labelled with a barcode indicating herd of origin. Samples are frozen overnight. (B) The tube is entered in a rack with the barcode facing outward. Meat juice sieves into the tube from the container during thawing. (C) Withdrawal of meat juice from tube and transfer to microtiter tray. (D) ELISA analysis, reading and transfer of results to central database.

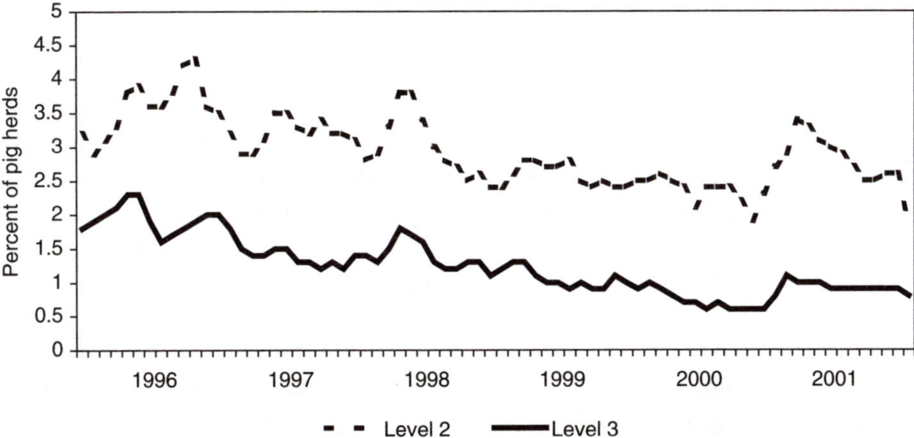

Figure 3.12 Prevalence of *Salmonella* in Danish pig herds as determined by continuous serological testing of all commercial pig herds (*n* > 700 000 samples tested per year). The herds are categorized in three levels based on the proportion of seropositive meat juice samples during the last three months. Owners in level 2 and 3 are encouraged to seek advice on how to reduce the *Salmonella* problem in the herd (e.g. feeding, hygiene, management). Furthermore, pigs from level-3 herds can only be slaughtered in special slaughterhouses under special hygienic precautions.

4.3.2 Post-harvest control

In Denmark, pigs from herds in levels 1 and 2 (see above) are slaughtered without any special precautions. Pigs from level-3 herds can only be slaughtered in special slaughterhouses under special hygienic precautions. Carcasses from level-3 herds are tested bacteriologically after slaughter, and if the level of contamination exceeds a certain limit then all carcasses from the particular herd undergo heat treatment or other risk-reducing processing. All slaughterhouses undertake routine bacteriological testing of carcasses according to a sampling plan, which ensures that testing is representative of the national swine production (> 30 000 samples per year). Slaughterhouses that exceed a certain predetermined level of *Salmonella* in the routine monitoring of carcasses are obliged to investigate and reduce the contamination problem to an acceptable level. This program has been further described by Feld *et al.* (2000), Hald and Andersen (2001), and Nielsen and Wegener (1997). The prevalence of swine herds in levels 2 and 3 has been steadily reduced since the program began. Bacteriological testing has indicated that herd infection levels were reduced by some 50% (from 14.7 % to 7.2 % in small herds and 22.2 % to 11.4 % in large herds) between 1993 and 1998 (Christensen *et al.*, 2002). During the same period, the level of *Salmonella* contamination in pork end products as determined by the routine monitoring program was reduced from approx. 3 % to less than 1 % (Wegener *et al.*, 2003).

4.4 Cattle, beef, and dairy products

Cattle may be asymptomatically infected with *Salmonella* and, via the gastrointestinal content, beef may be contaminated during slaughter and processing, and milk during milking. *Salmonella* Dublin, which is highly pathogenic to humans, is strongly associated with cattle (host-adapted, Table 3.8). This makes cattle an important target for *Salmonella* control efforts.

Salmonella surveillance and control programs, if any, are generally conducted at the slaughterhouse level. At the herd level, cattle are often examined only if there are clinical indications. However, more countries are considering the implementation of *Salmonella* surveillance and control programs at the herd level. Possibilities for control depend very much on farm facilities and production form.

In general, movement and purchase should be kept to a minimum. Most infection is introduced into *Salmonella*-free herds by purchasing infected cattle. A herd should be kept as a closed herd if possible. If stock is purchased or returned from the market, the animals should be kept in quarantine for 4 weeks. Testing animals for *Salmonella* before and after purchase can be used as a preventive measure. During quarantine in particular, good hygiene should be practiced (changing boots, clothes, and equipment) and disinfection should take place.

Surface water should be fenced off and pastures not occupied for 4–5 weeks after the application of manure, slurry or sludge. Effective rodent and bird control should be carried out, and contact with wildlife avoided. Farm personnel and visitors should wear clean, protective clothing and disinfected boots. Footbaths with active disinfectants should be available when entering the herd's area.

Wet areas should be avoided, and dry and clean areas provided for all animals – calves in particular. In calf-rearing units, an all-in/all-out system should be adopted, allowing for cleaning and disinfection between batches.

Wildlife and water are potential reservoirs of *Salmonella*. The *Salmonella* organism is a hardy bacterium, which can survive in the environment for extended periods of time, depending on climatic conditions, thereby posing a risk for exposure and subsequent infection.

The levels of *Salmonella* reported in beef vary, but are usually lower than in chicken. McEvoy *et al.* (2003) tested cattle and carcasses for the presence of *Salmonella* spp. at a commercial Irish abattoir over a 12-month period. *Salmonella* was isolated from 2 % of fecal, 2 % of rumen and 7.6 % of carcass samples. *Salmonella* was most frequently isolated from samples taken during the period August to October. *S.* Dublin was isolated from 72 % of positive samples. *S.* Agona and *S.* Typhimurium DT104 were each isolated from 14 % of positive samples. Ransom *et al.* (2002) found that 6.7 % of beef carcasses tested in the US were positive for *Salmonella*. In comparison, 63.3 % of the hides investigated in the same study were positive for *Salmonella*. In the US, infections with multidrug-resistant *S.* Newport have emerged in recent years; the primary reservoir is thought to be bovine (Anonymous, 2002a).

Pasteurization of milk effectively kills *Salmonella*, but consumption of unpasteurized milk and milk products is a well-documented risk factor for human salmonellosis. Furthermore, inadequately pasteurized milk as well as milk and milk products contaminated after pasteurization are recognized sources of human disease (Tacket *et al.*, 1985; Ryan *et al.*, 1987; Anonymous, 1997, 1998; Villar *et al.*, 1999; De Valk *et al.*, 2000).

4.5 New food vehicles of transmission

An array of new food vehicles in foodborne disease transmission has been identified in recent years. Traditionally, the food implicated in a foodborne outbreak was undercooked meat, poultry, eggs, or unpasteurized milk. Now, additional foods previously thought safe are considered to be hazardous.

The new food vehicles share several features. Contamination typically occurs early in the production process, rather than just before consumption. Because of consumer demand and the global food market, ingredients from many countries may be combined in a single dish, which makes the specific source of contamination difficult to trace. The foods have fewer barriers to microbial growth, such as added salt, sugar or preservatives. Because the food has a short shelf life, it may often be eaten or thrown away by the time the outbreak is recognized. Therefore, efforts to prevent contamination at the source are very important.

Increasing numbers of reported foodborne outbreaks are being traced back to fresh produce (Mahon *et al.*, 1997; Ooi *et al.*, 1997; Mohle-Boetani *et al.*, 1999). A series of outbreaks recently investigated by the CDC has linked *Salmonella* to fresh fruits and vegetables harvested in the US and elsewhere (Table 3.12). Various possible points of contamination have been identified, including contamination during production and harvest, initial processing and packing, distribution, and final processing (Table 3.13). Untreated or contaminated water seems to be a particularly likely source of contamination. Water used for spraying, washing, and maintaining the appearance of produce must be microbiologically safe. Chlorination of water used in the processing of fresh produce appears to be a particularly important control point (Anonymous, 1999; Weissinger *et al.*, 2000). As illustrated in Table 3.12, bean sprouts are a common vehicle, not only in the US but also elsewhere (Puohiniemi *et al.*, 1997; Anonymous, 1999; Van Beneden *et al.*, 1999). Several treatments can cause reductions in *Salmonella* populations on sprout seeds; however, the treatment may not eliminate the pathogen (Weissinger and Beuchat, 2000; Stan and Daeschel, 2003).

Table 3.12 Foodborne *Salmonella* outbreaks in the US traced to fresh produce, 1990–2003

Year	Serotype	Vehicle	Cases	States involved
1990	Chester	Cantaloupes	245	30
	Javiana	Tomatoes	174	4
1991	Poona	Cantaloupes	> 400	23
1993	Montevideo	Tomatoes	84	3
1995	Stanley	Alfalfa sprouts	242	17
	Hartford	Orange juice	63	21
	Newport	Alfalfa sprouts	133	8
1997	Saphra	Cantaloupes	24	1
1998	Havana	Alfalfa sprouts	18	2
1999	Typhimurium	Clover seed sprouts	112	1
	Muenchen	Alfalfa sprouts	157	7
2003	Kottbus	Alfalfa sprouts	31	3

Table 3.13 Contamination sources during production and processing of produce

Production/processing phase	Event	Contamination sources
Production and harvest	Growing, picking	Irrigation water, manure
	Bundling	Lack of field sanitation
Initial processing	Washing, waxing	Wash water, handling
	Sorting, boxing	Wash water, handling
Distribution	Trucking	Ice, dirty trucks
Final processing	Slicing, squeezing	Wash water, handling
	Shredding, peeling	Cross-contamination

Based on *in vitro* data, the US Food and Drug Administration recommends chemical disinfection of raw sprout seeds to reduce enteric pathogens contaminating the seed coats (Gill *et al.*, 2003; Thomas *et al.*, 2003).

4.6 Feed

Salmonella is a common contaminant of animal feeding stuffs. Both vegetable- and animal-derived feed constituents can harbor *Salmonella* and may lead to contamination of the finished feed. Pelleting of compound feed, which generates a sufficiently high temperature in the product (> 82°C for 15 seconds), is an effective method of reducing the level of contamination. In the feed mill, hygiene at all stages of processing is critical to avoid post-pelleting contamination. *Salmonella* can contaminate at several locations in the production environment, particularly in areas that become moist from condensation, with resulting growth and contamination of passing feedstuffs.

In a survey in the Netherlands, Veldman *et al.* (1995) investigated the rate of contamination with *Salmonella* in 360 samples of poultry feeds and feed components; 10 % were contaminated. Mash foods, mostly used for layer-breeders, were far more frequently (21 %) contaminated than pelleted feeds (1.4 %). The rate of contamination of fishmeal was 31 %, meat and bone meal 4 %, tapioca 2 %, and maize grits 27 %. Twenty-eight serotypes of *Salmonella* were isolated.

On 30 swine farms in the US, *Salmonella* was isolated from 36 of 1264 (2.8 %) feed and feed ingredient samples and from 14 of 30 (46.7 %) farms (Harris *et al.*, 1997). Thirteen *Salmonella* serotypes and two untypeable isolates were cultured. Six different herd characteristics (lack of bird-proofing; using farm-prepared feed for finishing-age pigs rather than purchased feed; and housing pigs in facilities other than total confinement in the growing, finishing, gestating, and breeding stages of production) were associated with an increased risk of recovery of *Salmonella* organisms from at least one feed or feed ingredient. Sanitation and pest-control measures continue to be important.

Dried pet foods, such as dried pig ears, may contain *Salmonella*, and this can result in human disease; recent outbreaks of human salmonellosis have been traced to contaminated dried pig ears used as pet treats in Canada (Clark *et al.*, 2001).

5 Foods most often associated with human infections

There is no direct association between the prevalence of *Salmonella* in a given animal reservoir or vehicle and its relative contribution to human illness. Several other factors influence whether exposure to *Salmonella* causes a clinical infection in man. Strain-to-strain variation, the numbers of organisms, the physical and chemical properties of the vehicle, and host factors are all significant. The prevalence of *Salmonella* in eggs is usually much lower than in broiler chickens and meat. Nonetheless, eggs are a commonly recognized source of *Salmonella* outbreaks. This predicament is related to the fact that eggs are often served raw or undercooked and may be pooled during the preparation of dishes, thereby exposing a large number of persons. This is particularly so if the handling and storage of the given dish allows for multiplication of the strain.

Salmonella is usually introduced into a kitchen environment from contaminated eggs, poultry or meat. Following this introduction, a number of other factors are important, including the hygiene behavior of the food handlers, temperature abuse, storage of food items, risk of cross-contamination, and any final heat treatments before serving. If hygiene barriers break down in a kitchen or during serving (e.g. at a buffet meal), parts of the environment or the utensils used in food preparation may eventually be contaminated with *Salmonella*. In this situation the original source may no longer be recognized, and cases of illness may be associated with the consumption of a number of different and apparently unrelated food items. *Salmonella* carriers among the staff may in this situation serve as a continuous source of outbreaks (Ethelberg *et al.*, 2004).

To identify and evaluate the sources of human *Salmonella* infections, investigation of foodborne outbreaks is very important. Additional information is available from case-control studies of sporadic infections with *Salmonella*. More recently, systematic typing of strains collected at various points in the 'farm-to-fork' chain and risk-assessment approaches have proven to be informative.

6 Epidemiology of Salmonellosis

6.1 Surveillance and outbreak investigation

Surveillance and outbreak investigation serve several purposes. Investigation of foodborne-disease outbreaks provides information that supports prevention and control measures in the food industry and helps to identify critical control points to reduce contamination.

Outbreak investigations have proven to be a critical means of identifying new subtypes of *Salmonella* and new vehicles, as well as maintaining awareness of contemporary problems such as *S.* Enteritidis. The pathogen is not identified in many outbreaks because of incomplete or delayed laboratory examination. Prompt and thorough investigation of foodborne outbreaks is needed for a timely identification of etiologic agents, sources and vehicles. By analyzing data on foodborne disease outbreaks, epidemiologists can evaluate trends and determine the most common food vehicles and common errors in food handling. This information provides the basis for regulatory and other changes to improve food safety.

In the US, summaries of the data reported to the Foodborne Disease Outbreak Surveillance system have been published for 1983–1987 (Bean *et al.*, 1990), 1988–1992 (Bean *et al.*, 1996), and 1993–1997 (Olsen *et al.*, 2000). In 1993–1997, a total of 878 (32 %) of 2751 outbreaks reported to the CDC had a known etiology. *Salmonella* was the most commonly identified agent, responsible for 357 (13%) outbreaks, 32 610 (38 %) outbreak-associated cases, and 13 (45 %) outbreak-associated deaths. Although *S.* Typhimurium is the most common *Salmonella* serotype in the US (Table 3.7), *S.* Enteritidis was the most frequently reported cause of foodborne disease outbreaks, accounting for 196 of them. *S.* Enteritidis also resulted in more deaths than any other pathogen. Of *S.* Enteritidis outbreaks with a known vehicle, about 80 % were associated with eggs or dishes containing eggs. Table 3.14 summarizes the most common vehicles of transmission in foodborne *Salmonella* outbreaks in the US in the period 1993–1997 (Olsen *et al.*, 2000). Food vehicles were classified as individual food items (e.g. milk or eggs) or as food categories (e.g. ice cream or Mexican food). Homemade ice cream containing milk and eggs is listed under 'ice cream' rather than 'milk' or 'eggs'. Taken together, eggs, ice cream, baked foods and other dishes or salads that might have been prepared with or contaminated from raw eggs were the most common vehicles. Beef and other meats were also commonly incriminated. However, compared with previous years, vegetables and fruits were increasingly associated with foodborne transmission of *Salmonella*. Examples included alfalfa sprouts

Table 3.14 Reported outbreaks of foodborne salmonellosis by vehicles of transmission, US, 1993–1997

Vehicle	Number of outbreaks	% outbreaks with single vehicle
Beef	14	11.4
Pork/ham	5	4.1
Chicken	6	4.9
Turkey	6	4.9
Other or unknown meat	6	4.9
Shellfish	1	0.8
Milk	3	2.4
Cheese	1	0.8
Eggs	17	13.8
Ice cream	11	8.9
Other dairy	1	0.8
Baked foods	12	9.8
Potato salad	1	0.8
Poultry, fish, and egg salad	5	4.1
Other salad	13	10.6
Chinese food	1	0.8
Mexican food	5	4.1
Non-dairy beverage	2	1.6
Multiple vehicles	61	
Unknown vehicle	184	

(Mahon *et al.* 1997), orange juice (Cook *et al.*, 1998), tomatoes (Cummings *et al.*, 2001) and melons (Anonymous, 2002b); see also Table 3.12.

England and Wales also have outbreak-reporting surveillance, which has been described in a series of publications (Djuretic *et al.*, 1996; Ryan *et al.*, 1997; Evans *et al.*, 1998; Gillespie *et al.*, 2001, 2003; Kessel *et al.*, 2001; Smerdon *et al.*, 2001; Long *et al.*, 2002; Meakins *et al.*, 2003).

The contamination of ready-to-eat products constitutes a particular problem. For the consumer, thorough cooking and appropriate hygienic measures to avoid cross-contamination during transport, storage, preparation and serving may reduce the risk of infection due to low concentrations of *Salmonella* in raw meat or raw shell eggs. However, there is no terminal treatment that precludes infection from a ready-to-eat product. This is exemplified by outbreaks due to contaminated commercial ice cream (Hennessy *et al.*, 1996), pork fillet (Mølbak *et al.*, 1998), potato chips (Lehmacher *et al.*, 1995), chocolate (Kapperud *et al.*, 1990) and salad vegetables (Long *et al.*, 2002; Ward *et al.*, 2002). Such products may cause severe morbidity and mortality, in particular when they are introduced into vulnerable populations.

Foodborne *S.* Typhi and *S.* Paratyphi are usually transmitted by food items contaminated with the feces or urine of patients and carriers. Contamination may occur when night soil is used as fertilizer, or by the hands of carriers. Important vehicles include shellfish taken from sewage-contaminated beds (Earampamoorthy and Koff, 1975; Stroffolini *et al.*, 1992), as well as raw fruits and vegetables and contaminated milk or milk products. In an outbreak in England, dairy cows excreted *S.* Paratyphi B in milk and feces, and it has been suggested that livestock may serve as a reservoir for paratyphoid (Thomas and Harbourne, 1972).

In industrialized countries, typhoid and paratyphoid fevers have now been controlled by modern water and sewerage systems, pasteurization of milk, and shellfish sanitation programs. However, even in the US and Europe, *S.* Typhi still has the potential to cause large outbreaks. Some examples of food vehicles include Mexican food (Taylor *et al.*, 1984b), shrimp salad (Lin *et al.*, 1988), orange juice (Birkhead *et al.*, 1993), pork meat served at a village festival (Pradier *et al.*, 2000), and imported tropical fruit (Katz *et al.*, 2002). Typhoid carriers among food handlers were found in several of these investigations, and food contamination by a carrier was the most plausible source. Hamburgers, döner kebabs and food from street vendors were likely vehicles in a large outbreak of *S.* Paratyphi B among European tourists returning from a holiday resort in Turkey (Fisher *et al.*, 1999). Again, a carrier is the most likely explanation of such an outbreak.

6.2 Case-control studies of sporadic infections

Data from outbreaks are helpful in determining food vehicles associated with human *Salmonella* infections. However, information from outbreaks may not represent sporadic cases of illness. Case-control studies are studies where data on relevant exposures are obtained from case-patients as well as asymptomatic (uninfected) controls. The odds of exposure for a given food item are calculated for both case-patients and controls, and the ratio between these odds is a measure of the relative risk of infection associated with a given food item. Well-conducted case-control studies are important sources of information because it is possible to estimate the relative role of several different food exposures over a long period of time in a representative sample of culture-confirmed cases of *Salmonella* infections. However, case-control studies also have their limitations.

Salmonella infections may be associated with the consumption of many different food items. Even in a large study, the statistical power to determine the role of each item may be small. Some food exposures (meat, poultry, and eggs) may be a very common part of the usual diet of both cases and controls. It may be hard to get exact information on the various events that led from exposure to this food item to illness. The size of the exposure window (i.e. the most likely incubation period), the brand of the product, the time of purchase and the handling after purchase are some of the parameters that are of importance; but it may be difficult to obtain the necessary information from a questionnaire.

Case-control studies are subject to a number of biases, including recall bias due to memory lapses, and possibly selection bias. The methodological problems are relevant for the use of case-control data collected as a part of outbreak investigations, but this issue becomes graver when the aim is to quantify the role of many different foods as risk factors. In an outbreak investigation, the aim is to identify a single or possibly a few important vehicles, and the 'signal' of these vehicles will usually be strong even in a relatively small study. This is particularly the case when the investigators have a strong hypothesis to test.

At least 20 case-control studies were published between 1985 and 2004. Most of these studies examined risk factors for *S*. Enteritidis, and associated illness with exposure to eggs and dishes prepared with raw eggs. Other studies have examined a broader range of serotypes, or *S*. Typhimurium. Table 3.15 summarizes the main findings in these studies.

Table 3.15 Food items and other risk factors identified in case-control studies of foodborne Salmonellosis

Serotype	No. of cases / controls	Risk factors	Location	Reference
Reptile-associated	42/28	Reptile in household	New York State, US	Ackman *et al.*, 1995
All serotypes	209/209	Undercooked eggs (*S*. Ent.); foreign travel; gastro-duodenal disease; antacids	North Thames, UK	Banatvala *et al.*, 1999
All serotypes	85/164, only children	Lack of breastfeeding; antibiotics; low social class; chronic non-infectious intestinal disease	Italy	Borgnolo *et al.*, 1996
S. Enteritidis phage type 4	157/196	Raw egg, runny eggs, sandwiches with mayonnaise; precooked chicken	England	Cowden *et al.*, 1989
S. Enteritidis phage type 4	19/19	Eggs	Wales	Coyle *et al.*, 1988

Table 3.15 Food items and other risk factors identified in case-control studies of foodborne Salmonellosis—*cont'd*

Serotype	No. of cases / controls	Risk factors	Location	Reference
S. Enteritidis	105/105	Undercooked eggs; storage of eggs more than 2 weeks after purchase; diarrhea in household, especially infants	France	Delarocque-Astagneau *et al.*, 1998
S. Typhimurium	101/101, only children	Undercooked ground beef; antibiotics; diarrhea in household	France	Delarocque-Astagneau *et al.*, 2000
S. Enteritidis & *S.* Typhimurium	106/212	Undercooked eggs; chicken (*S.* Tm.); restaurant food (*S.* Tm.); travel (*S.* Ent.); hamburger (*S.* Ent.)	Minnesota, US	Hedberg *et al.*, 1993
All serotypes	94/226	Imported poultry purchased abroad; catered food (*S.*Tm.); foreign travel	Norway	Kapperud *et al.*, 1998
All serotypes	120/265	Undercooked chicken; foreign travel; diabetes; hormonal therapy; antibiotics	Northern California, USA	Kass *et al.*, 1992
S. Enteritidis	182/345	Chicken outside the home	FoodNet sites, US	Kimura *et al.*, 2004
All serotypes	965/256	Undercooked eggs; puppies, kittens or turtles in household	Southwest Germany	Kist and Freitag, 2000
All serotypes	115/115	Chronic medical conditions; inconsistent handwashing	Louisiana, US	Kohl *et al.*, 2002
S. Enteritidis	455/507	Foreign travel; undercooked eggs; eggs from battery layers	Denmark	Mølbak and Neimann, 2002
S. Enteritidis	45/316	Eggs	New York State, US	Morse *et al.*, 1994
All serotypes	99/99	Raw eggs; handling free range eggs	Wales	Parry *et al.*, 2002
S. Enteritidis	58/98	Undercooked eggs; restaurant food	Southern California, US	Passaro *et al.*, 1996
All serotypes	223/223	Foreign travel; undercooked eggs (*S.* Ent); medications other than antacids	Switzerland	Schmid *et al.*, 1996
All serotypes	60/60	Turtle in household (study focused on pet ownership)	Puerto Rico	Tauxe *et al.*, 1985
S. Enteritidis	35/59	Undercooked eggs; restaurant food	Wisconsin, US	Trepka *et al.*, 1999

Limited numbers of case-control studies have attempted to identify risk factors for typhoid fever. In a study in Indonesia, consumption of food from street vendors was associated with an increased risk, whereas hand washing with soap after using the toilet was protective (Velema *et al.*, 1997). Another study found that living in a house without a water supply from the public network and with open sewers was associated with increased risk. Never or rarely washing hands before eating, and being unemployed or in a part-time job, were also risk factors (Gasem *et al.*, 2001). In Pakistan, ice cream, eating food from a street vendor in the summer months, taking antimicrobials before the onset of disease, and drinking water at the worksite were independently associated with an increased risk (Luby *et al.*, 1998).

Case-control studies and their advantages and limitations are also discussed in Chapter 1.

6.3 Risk assessment and other quantitative approaches to evaluate the risks of *Salmonella* infections from foods

The principles of risk assessments are discussed in Chapter 2. Most risk assessments focus on a specific combination of a pathogen and a source/commodity. Data needed for risk assessments are obtained from studies of the prevalence of *Salmonella* in food animals and foods, outbreak investigations, epidemiological studies, and laboratory work.

In Denmark, the relative contribution of the most important food sources has been determined by systematic bacterial typing of *Salmonella* collected from different parts of the food chain, including food animals, carcasses, and food (Wegener *et al.*, 2003; Hald *et al.*, 2004). The findings have been compared to the type distribution of human *Salmonella* isolates. Some subtypes of *Salmonella* are limited to a certain animal reservoir or food item, and the proportion of this subtype in humans mirrors the relative importance of this reservoir or food item. This subtype serves, in other words, as a marker of how 'successful' the reservoir is in causing human infections, and it is then possible to estimate the contribution of this reservoir as a source of infection with other subtypes. This approach is subject to some limitations. Only the role of animals or foods under surveillance can be determined, and the data will therefore only be valid for these sources. This approach is better suited to countries with a systematic, 'farm-to-table' surveillance than to countries where *Salmonella* isolates from animals are limited to clinical isolates or collected in small-scale surveys. It is also important to have data on imported foods to quantify their role as a source of infection, and it is essential to be able to exclude human cases that became infected with *Salmonella* during foreign travel.

This approach has proven to be an important means of evaluating trends and determining priorities. This is appreciated by scientists and decision-makers, as well as the food industry and the consumers (Wegener *et al.*, 2003). Figure 3.13 shows Danish *Salmonella* statistics with a quantification of the three major sources: poultry, pigs, and eggs. Human *Salmonella* infections in Denmark showed three epidemic

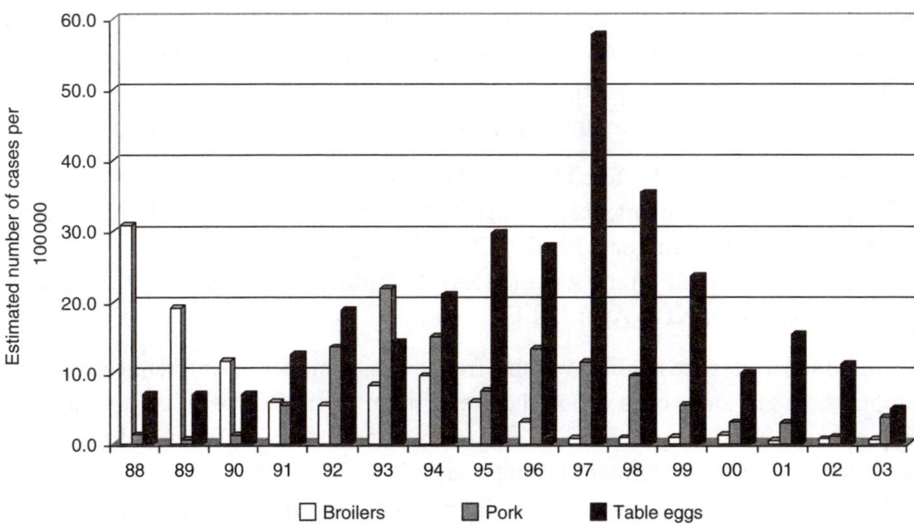

Figure 3.13 Estimated number of *Salmonella* infections attributed to the three major sources, Denmark, 1988–2003. For explanation of the method, see Wegener *et al.* (2003) and Hald *et al.* (2004). *Source:* Annual Report of Zoonoses in Denmark, 2003.

peaks (Figure 3.2) corresponding to these three sources. Successful interventions, mainly at the pre-harvest levels, led to a substantial decline in the overall incidence as well as the contribution from each of these sources.

7 Principles of detection of Salmonella

7.1 Conventional detection methods

Detection of *Salmonella* in samples where there are low initial cell numbers, or where the cells are stressed due to physical or chemical injury, requires a three-stage procedure involving pre-enrichment in non-selective broth, enrichment in selective broth, and subsequent detection on selective and indicative (differential) agar media. This procedure is applicable to most environmental and food samples, and also to fecal and organ samples from animals without symptoms of disease. Detection of *Salmonella* in clinical samples from animal or human patients with diarrhea, where the number of viable *Salmonella* cells in the sample is high and where time is a critical factor, can with reasonable confidence be carried out by direct culture on selective indicative (differential) agar media.

A large variety of different media for isolation and detection of *Salmonella* are available, and many different combinations of media and culture conditions have been developed to isolate *Salmonella* from different types of products and samples. It is, however, important to note that there is no single method that is optimal to all types of samples. Finding the best method for a given type of samples may require extensive literature studies, and often comparative testing of different methods.

Procedures are available from a number of standard-setting bodies such as the ISO, AOAC, IDF, NMKL, etc. The choice of method may be dictated by regulation or requirements from trading partners.

The culture media most commonly used for conventional detection are:

- For pre-enrichment – Buffered Peptone Water (BPW) or Lactose Broth (LB)
- For selective enrichment, Rappaport–Vassiliadis Broth (RV), Selenite Cystine Broth (SC), or Tetrathionate Broth (TB)
- For plating, Brilliant Green Agar (BGA), Bismuth Sulfite Agar (BSA), Hektoen Agar (HA) and/or XLD Agar (XLD).

All these substrates are readily available from a large number of different commercial suppliers. At any stage of isolation the use of more than one type of culture medium may enhance sensitivity. Table 3.16 shows combinations of different culture media prescribed by different standard setting bodies.

Table 3.16 Combinations of pre-enrichment, enrichment and plating media prescribed by different standard-setting agencies for isolation of *Salmonella* from food

Agency	Pre-enrichment	Enrichment	Plating
ISO (International Organization for Standardization)	BPW	RV, TB	XLD or other media
APHA (American Public Health Association)	LB	SC, TB	SSagar, BSA, HA
AOAC/FDA (Association of Analytical Communities/ US Food and Drug Administration)	LB, TSB, nutrient broth	SC, TB	BGA, HA, XLD, BSA
IDF (International Dairy Federation)	BPW, distilled water + brilliant green	TB, SC	BGA, BSA
BSI (British Standards)	BPW	RV, SC	BGA or othe media
AFNOR (The French Standards Association)	BPW	RV, SC	BGA, XLD, HA, deoxycholate-citrate-lactose agar
NMKL (Nordic Committee on Food Analysis)	BPW	RV	BGA, XLD

Key: BGA, brilliant green agar; BPW, buffered peptone water; BSA, bismuth sulfite agar; HA, Hektoen enteric agar; LB, lactose broth; SC, selenite cystein agar; SS, salmonella shigella agar; TB, tetrathionate broth; XLD, xylose lysine deoxycholate agar.

7.2 Rapid detection methods

Extensive research and development of rapid *Salmonella* detection methods have been carried out, and many different methods are commercially available today. However, while detection times have been reduced to 18–24 hours, none of the so-called rapid methods can detect *Salmonella* in a food product instantly, or even within a normal 8-hour working day. This is still a strongly desired goal by developers. The critical obstacle to achieving faster detection times is the need to culture *Salmonella* until cell numbers reach a range of 10^4–10^5 cells per ml in order to achieve sufficient numbers of cells in the small volume of sample (10–20 μl) commonly employed in modern detection methods such as PCR or ELISA.

The principles for rapid detection of foodborne pathogens will not be described in detail in this chapter. The techniques have been reviewed by Ibrahim *et al.* (1985), Blackburn (1993), and Olsen *et al.* (1995). However, some of the methods available for rapid detection of *Salmonella* will be summarized. Rapid detection methods offer detection times as low as 18 hours, and may be more convenient and less labor-intensive than conventional methods. The materials needed for the tests are usually more expensive than those used in conventional tests.

7.2.1 Detection of antibodies to *Salmonella* by enzyme immunoassay (EIA)

The detection of antibodies to *Salmonella* by EIA offers a sensitive and cost-effective method for mass screening of animal flocks/herds for indications of a past/present *Salmonella* infection. The limitation of the method is that the immune response of the individual animal is not elicited before 1–2 weeks after infection takes place, so the infection may have been cleared from the animal while the antibodies are still present. Thus the method offers a convenient tool for flock or herd screening, but is less useful for testing of individual animals. The method has been applied in *Salmonella* pre-harvest control programs for layers, broilers, and pigs in a number of countries, including Denmark, the Netherlands and Germany (Nielsen *et al.*, 1998; Wiuff *et al.*, 2000; van Winsen *et al.*, 2001; Zamora and Hartung, 2002). A number of commercial kits are available for testing poultry, cattle, and pigs. An obvious advantage of this method is that it can be automated and no incubation is required to increase the numbers of bacterial cells (Figure 3.11).

7.2.2 Immunological detection of *Salmonella* antigens by EIA

The EIA is a well-established technique for assaying antigens. Antibodies labeled with an enzyme are bound to *Salmonella* antigens, and the level of antigen present is determined by enzymatic conversion of a substrate, usually resulting in a color change which can be read visually or by a spectrophotometer. One of the reagents is usually bound to a solid matrix, such as the surface of a microtiter plate well. Swaminathan and Feng (1994) have published a good review of this technique. Some of the commercially available EIA kits for *Salmonella* antigen detection are *TECRA*®, from TECRA® Diagnostics Australia (*TECRA*® International Pty Ltd) and the Salmonella-Tek™ ELISA test system (from BioMérieux S.A.). Furthermore, BioMérieux S.A. has developed a fully automated *Salmonella* EIA to be used in the VIDAS® system. The BIOLINE® *Salmonella* ELISA Test (Bioline ApS) for detection of *Salmonella* in food and feeds has recently been granted AOAC approval.

The EIAs rely on the standard cultural procedures for pre-enrichment and selective enrichment to provide enough *Salmonella* cells for detection. EIA technology that enables detection at an earlier stage of resuscitation and/or culture can provide even more rapid results. A number of such assays have been commercialized. The Foss Electric EiaFoss® *Salmonella* method (Foss Electric A/S) uses a combination of immunocapture to concentrate cells and automated EIA testing. The assay is completed within 18 hours. A dipstick-based assay has been developed by TECRA® to detect *Salmonella* in foods, which utilizes an antibody-coated dipstick to capture *Salmonella*. The dipstick is transferred to EIA reagents to detect *Salmonella*. This assay is complete within 22 hours.

7.2.3 DNA-based detection of *Salmonella*

Despite the many expectations, DNA methods have not replaced standard culture methods for *Salmonella* detection. On the contrary, when used, DNA methods are often applied for species identification in enrichment broths or even colonies on agar plates. The essential principle of nucleic acid-based detection methods is the specific formation of double-stranded nucleic acid molecules from two complementary, single-stranded molecules under defined physical and chemical conditions. When performed *in vitro*, this is termed hybridization. In diagnostic assays, one of the strands is produced in the laboratory in the form of probe molecules, DNA fragments or oligonucleotides, or primers for polymerase chain reaction (PCR), while nucleic acids from the target organism provide the other strand. (See Olsen *et al.*, 2000, for a review of probes and PCR detection of foodborne pathogens.)

As with other pathogenic bacteria, DNA probes preceded PCR methods for detection of *Salmonella*. Currently, the use of DNA probes is limited to colony confirmation tests and *in situ* hybridization analysis, while all the new developments apply PCR. It is, however, quite possible that DNA arrays will revive the use of traditional DNA probes for detection of bacteria. The DNA-based procedures have specific advantages compared to the culture method in cases of atypical phenotypic characteristics of *Salmonella*.

In-situ hybridization has been used in studies of *Salmonella* invasion and pathogenicity (the FISH technique) (Licht *et al.*, 1996; Nordentoft *et al.*, 1997). In this system, fluorescent-labeled probes are hybridized to fixed tissue sections, whereby the *in situ* location of *Salmonella* can be studied. Similar techniques have been used to study stress-induced changes in *Salmonella* ribosomes (Tolker-Nielsen and Molin, 1996) and in studies of other enteric bacteria, such as *E. coli* (Poulsen *et al.*, 1994).

Polymerase chain reaction (PCR)-based detection of Salmonella was developed by Widjojoatmodjo *et al.* (1991), using primers isolated from replication genes. Rahn *et al.* (1992) used primers selected from the sequence of the *invA* gene of *Salmonella*. In a testing of 630 strains, 575 of which belonged to 102 specified serotypes, and a collection of non-*Salmonella*, four strains of two serotypes of subspecies I gave false negative results. False replicons were observed from non-*Salmonella* strains, but not of the expected size. Members of subspecies III are deficient in the *invABC* operon (Galan and Curtiss, 1991); it is not known whether this deficiency in the genes will influence the PCR detection of this subspecies. The detection limit of the assay was shown to be 300 organisms, or 27 pg of chromosomal DNA.

The *invA*-directed PCR method has been combined with other PCR methods to obtain either further specification or multiplex detection. In a safety control for chicken, a PCR method with *invA*-directed primers was used successfully as a general *Salmonella* detection method (Mahon *et al.*, 1994). *InvA*-PCR has also been included in multiplex PCR methods with simultaneously detection of *E. coli*, fecal coliforms, *Salmonella* and pathogenic *Vibrio* using colorimetric DNA–DNA hybridization for detection of amplicons (Brasher *et al.*, 1998), and in a multiplex method where a common enrichment broth and a common PCR-reaction mix were used to detect 13 different foodborne pathogenic bacteria in the same reaction (Wang *et al.*, 1997). Such methods show the potential for practical every day use of PCR methods in food microbiology. In another testing where enrichment in tetrathionate broth was used both as a source of DNA for *invA* based capillary PCR and as a part of a traditional culture method, all 53 samples from commercial chicken flocks were also subjected to delayed (5 days) secondary enrichment procedure (DES). The PCR method detected 26 out of 30 culture-positive samples, and was positive in another 9, 3 of which were positive by DES. Both methods showed suboptimal sensitivity compared to DES (Carli *et al.*, 2001), which illustrates that for both PCR and culture methods short detection time and improved sensitivity are difficult to achieve at the same time.

New primers continue to be tested, and PCR methods have now become commercially available. The choice of primers will probably be determined by commercial considerations rather than performance of the different primers. The Dupont Company has developed a homogeneous fluorescence assay that can be combined with different PCR methods, and which can quantify PCR products. From 100 to 1 000 000 CFU could be detected quantitatively in a trial with a commercial *Salmonella* PCR assay (BAX-system). When used for colony confirmation, 83 of 84 strains gave positive signals. The reason for the negative reaction of a strain of *S.* Havana must be related to the primer sequence used (Tseng *et al.*, 1997).

7.2.4 DNA arrays as a diagnostic tool for *Salmonella*

DNA arrays are biochips, on which oligonucleotides or PCR products are bound to a solid support in a manner that enables hybridization results to be scored automatically. Usually glass or plastic forms the solid support, and probes (sample DNA or RNA) are fluorescence-labeled. A charge-coupled device (CCD) scanner or a scanning confocal microscope is used to capture the image. A description of the technological developments behind the DNA chips has been reviewed elsewhere (Southern, 1996; Marshall and Hodgson, 1998; Ramsay, 1998), and will not be covered here.

DNA chips can be used to detect organisms by hybridization in a way that is technically similar to colony-hybridization, but considerably faster and with many different probes applied simultaneously.

As the number of fully sequenced organisms keeps growing, sequences from different genes can be incorporated into ready-made DNA discs. While traditional DNA hybridization works with one probe at a time, thousands of probes can be placed on a DNA chip (Braxton and Bedilion, 1998). This leaves room for probes for all the relevant pathogenic bacteria in veterinary medicine and food hygiene, for sub-typing of selected bacteria by oligonucleotides that span selected mutations, and for determination of the presence of the common antibiotic resistance genes.

The first DNA chips for use in routine food microbiology and clinical microbiology are commercially available, for example from GeneScan. The food microbiology array consists of probes for *Campylobacter*, *Salmonella* and *Listeria monocytogenes*, and is used in combination with a multiplex PCR method for amplification of the relevant DNA. DNA chips, however, face some of the same problems of other DNA methods. A major problem is still to capture the relevant bacteria or nucleic acid from the samples to allow sensitive detection by direct hybridization.

DNA methods have so far mainly been applied in research. Strains are often isolated and purified by traditional culture methods. This use is referred to as a 'colony or culture confirmation' test. In food microbiology, PCR methods in particular must be expected to be used increasingly for rapid and reliable screening for *Salmonella* in samples of mixed cultures. Detection of the pathogens directly in clinical and food samples is possible, but this has only been reported in very few investigations. Major problems concerning the viability of detected cells (Bej *et al.*, 1991), the presence of enzyme inhibitors in the samples (Rossen *et al.*, 1992) and the accessibility of the target organisms remain to be solved before this method can be routinely applied. Once these problems have been solved, the DNA methods can be turned into truly rapid methods that provide the microbiologist with an answer within 1–2 hours.

DNA discs will continue to be developed on a commercial basis, enabling automated identification of bacteria. With improved methods for accessing nucleic acid in complex samples, this will change our routine diagnostic methods. The capacity of the discs will allow typing to be performed simultaneously with studies of, for example, single nucleotide polymorphism (SNP), speeding up typing analysis. However, like other DNA methods, a master plate must be established and the isolate obtained for further characterization; the value is mostly in the ability rapidly to screen out negative samples.

7.2.5 Sampling

Apart from the actual disease status of animal populations in a country or region (e.g. prevalence or incidence), the sampling protocol plays a large part in what can eventually be concluded from our efforts (Lo Fo Wong and Dahl, 2001). These factors include sample type, sample size, sample volume, sample frequency, and sample location. The sample size depends on the expected prevalence of the pathogen or disease under investigation, the purpose of sampling, and the sensitivity and specificity of the applied test. The sample type depends to a great extent on the diagnostic methods available for the analysis of the samples, and their sensitivity and specificity. Generally, there is an increasing sensitivity (i.e. the probability of culturing at least one viable organism) with increasing sample volume (Davies *et al.*, 1999; Hurd *et al.*, 1999a). The sample frequency is of importance when the occurrence of the pathogen under investigation is expected to change over time, due to dynamics in the population, management factors, seasonality or, as with *Salmonella*, intermittent shedding (Hurd *et al.*, 1999b). The choice of sample location, whether certain organs of the animal or specific places in the environment, will inherently affect the success of finding the organism.

Sampling in general, as a means to assure the safety of a given batch of food, is discussed in Chapter 19.

Bibliography

Aarestrup, F. M., C. Wiuff, K. Mølbak and E. J. Threlfall (2003). Is it time to change fluoroquinolone breakpoints for *Salmonella* spp.? *Antimicrob. Agents Chemother*. **47**, 827–829.

Ackers, M. L., N. D. Puhr, R. V. Tauxe and E. D. Minz (2000). Laboratory-based surveillance of *Salmonella* serotype Typhi infections in the United States: antimicrobial resistance on the rise. *J. Am. Med. Assoc*. **283**, 2668–2673.

Ackman, D. M., P. Drabkin, G. Birkhead and P. Cleslak (1995). Reptile-associated salmonellosis in New York State. *Pediatr. Infect. Dis. J.* **14**, 955–959.

Altekruse, S., J. Koehler, F. Hickman-Brenner *et al.* (1993). A comparison of *Salmonella enteritidis* phage types from egg-associated outbreaks and implicated laying flocks. *Epidemiol. Infect*. **110**, 17–22.

Anderson, E. S. (1975). The problem and implications of chloramphenicol resistance in the typhoid bacillus. *J. Hyg. (Lond)*. **74**, 289–299.

Anderson, E. S., L. R. Ward, M. J. Saxe and J. D. de Sa (1977). Bacteriophage-typing designations of *Salmonella typhimurium. J. Hyg. (Lond)*. **78**, 297–300.

Andrewes, F. W. (1922). Studies in group agglutination. I. The *Salmonella* group and its antigenic structure. *J. Path. Bact*. **25**, 505–525.

Angulo, F. J. and D. L. Swerdlow (1999). Epidemiology of human *Salmonella enterica* serovar Enteritidis infections in the United States. In: A. M. Saeed, R. K. Gast and M. E. Potter (eds), Salmonella enterica *serovar Enteritidis in Humans and Animals. Epidemiology, Pathogenesis, and Control*, pp. 33–41. Iowa State University Press, Ames, IA.

Angulo, F. J., S. Tippen, D. J. Sharp *et al.* (1997). A community waterborne outbreak of salmonellosis and the effectiveness of a boil water order. *Am. J. Public Health* **87**, 580–584.

Angulo, F. J., A. C. Voetsch, D. Vugia *et al.* (1998). Determining the burden of human illness from food borne diseases. CDC's emerging infectious disease program Food Borne Diseases Active Surveillance Network (FoodNet). *Vet. Clin. North Am. Food Anim. Pract*. **14**, 165–172.

Angulo, F. J., K. R. Johnson, R. V. Tauxe and M. L. Cohen (2000). Origins and consequences of antimicrobial-resistant nontyphoidal *Salmonella*, implications for the use of fluoroquinolones in food animals. *Microb. Drug. Resist*. **6**, 77–83.

Anonymous. (1972). Typhoid fever. *Mexico. Morbid. Mortal. Weekly Rep*. **21**, 177–178.

Anonymous. (1997). Preliminary report of an international outbreak of *Salmonella anatum* infection linked to an infant formula milk. *Euro Surveill*. **2**, 22–24.

Anonymous. (1998). Defective pasteurisation linked to outbreak of *Salmonella typhimurium* definitive phage type 104 infection in Lancashire. *Commun. Dis. Rep. Weekly* **8**, 335–338.

Anonymous (1999). Microbiological safety evaluations and recommendations on sprouted seeds. National Advisory Committee on Microbiological Criteria for Foods. *Intl J. Food Microbiol*. **52**, 123–153.

Anonymous (2000a). Baby dies of *Salmonella poona* infection linked to pet reptile. *Commun. Dis. Rep. Weekly* **10**, 161.

Anonymous (2000b). *Salmonella enteritidis*, Denmark. *Weekly Epidemiol. Rec.* **75**, 54–55.

Anonymous (2002a). Outbreak of multidrug-resistant *Salmonella newport*–United States, January–April 2002. *Morbid. Mortal. Weekly Rep.* **51**, 545–548.

Anonymous (2002b). Multistate outbreaks of *Salmonella* serotype Poona infections associated with eating cantaloupe from Mexico – United States and Canada, 2000–2002. *Morbid. Mortal. Weekly Rep.* **51**, 1044–1047.

Anonymous (2002c). *EU Trends and Sources of Zoonotic Infections in the European Union 2000*. CRL-E, BGVV, Berlin.

Anonymous (2005). Opinion 80. Judicial Commission of the International Committee on systematics of prokaryotes. *Int. J. Systematic and Evol. Microbiol.* **55,** 519-520.

Aserkoff, B. and J. V. Bennett (1969). Effect of antibiotic therapy in acute salmonellosis on the fecal excretion of salmonellae. *N. Engl. J. Med.* **281**, 636–640.

Bager, F. (1996). Investigation of Salmonella prevalence in shell-eggs [in Danish]. *Zoonose-Nyt* **3**, 7–8.

Bager, F. (2000). DANMAP, monitoring antimicrobial resistance in Denmark. *Intl J. Antimicrob. Agents* **14**, 271–274.

Bager, F. and R. Helmuth (2001). Epidemiology of resistance to quinolones in *Salmonella. Vet. Res.* **32**, 285–290.

Baggesen, D. L., H. C. Wegener and J. P. Christensen (1996). Typing of *Salmonella enterica* serovar Saintpaul, an outbreak investigation. *APMIS* **104**, 411–418.

Bailey, J. S., N. A. Cox, S. E. Craven and D. E. Cosby (2002). Serotype tracking of *Salmonella* through integrated broiler chicken operations. *J. Food. Prot.* **65**, 742–745.

Baker, R. C., J. P. Goff and J. E. Timoney. (1980). Prevalence of salmonellae on eggs from poultry farms in New York State. *Poultry Sci.* **59**, 289–292.

Banatvala, N., A. Cramp, I. R. Jones and R. A. Feldman (1999). Salmonellosis in North Thames (East), UK, associated risk factors. *Epidemiol. Infect.* **122**, 201–207.

Barrow, P. A., M. B. Huggins and M. A. Lovell (1994). Host specificity of *Salmonella* infection in chickens and mice is expressed in vivo primarily at the level of the reticuloendothelial system. *Infect. Immun.* **62**, 4602–4610.

Barwick, R. S., D. A. Levy, G. F. Graun *et al.* (2000). Surveillance for waterborne-disease outbreaks – United States, 1997–1998. *Morbid. Mortal. Weekly Rep. CDC Surveill. Summ.* **49**, 1–21.

Barza, M. and K. Travers (2002). Excess infections due to antimicrobial resistance, the 'Attributable Fraction'. *Clin. Infect. Dis.* **34** (Suppl. 3), S126–130.

Baumler, A. J., B. M. Hargis and R. M. Tsolis (2000). Tracing the origins of *Salmonella* outbreaks. *Science* **287**, 50–52.

Bean, N. H., P. M. Griffin, J. S. Goulding and C. B. Ivey (1990). Foodborne disease outbreaks, 5-year summary, 1983–1987. *Morbid. Mortal. Weekly Rep. CDC Surveill. Summ.* **39**, 15–57.

Bean, N. H., J. S. Goulding, C. Lao and F. J. Angulo. (1996). Surveillance for foodborne-disease outbreaks – United States, 1988–1992. *Morbid. Mortal. Weekly Rep. CDC Surveill. Summ.* **45**, 1–66.

Bej, A. K., M. H. Mahbubani and R. M. Atlas (1991). Amplification of nucleic acids by polymerase chain reaction (PCR) and other methods and their applications. *Crit. Rev. Biochem. Mol. Biol.* **26**, 301–334.

Belden, W. J. and S. I. Miller. (1994). Further characterization of the PhoP regulon, identification of new PhoP-activated virulence loci. *Infect. Immun.* **62**, 5095–5101.

Bengtsson, E., P. Hedlund, A. Nisell and H. Nordenstam (1955). An epidemic due to *Salmonella typhimurium* (Breslau) occurring in Sweden in 1953. *Acta Med. Scand.* **153**, 1–20.

Berchieri, A. Jr, M. A. Lovell and P.A. Barrow (1991). The activity in the chicken alimentary tract of bacteriophages lytic for *Salmonella typhimurium. Res. Microbiol.* **142**, 541–549.

Bichler, L. A., K. V. Nagaraja and D. A. Halvorson (1996). *Salmonella enteritidis* in eggs, cloacal swab specimens, and internal organs of experimentally infected White Leghorn chickens. *Am. J. Vet. Res.* **57**, 489–495.

Birkhead, G. S., D. L. Morse, W. C. Levine *et al.* (1993). Typhoid fever at a resort hotel in New York, a large outbreak with an unusual vehicle. *J. Infect. Dis.* **167**, 1228–1232.

Blackburn, C. W. (1993). Rapid and alternative methods for the detection of salmonellas in foods. *J. Appl. Bacteriol.* **75**, 199–214.

Blaser, M. J. and L. S. Newman (1982). A review of human salmonellosis, I. Infective dose. *Rev. Infect. Dis.* **4**, 1096–1106.

Borgnolo, G., F. Barbone, G. Scornavacca *et al.* (1996). A case-control study of *Salmonella* gastrointestinal infection in Italian children. *Acta Paediatr.* **85**, 804–808.

Bowe, F., C. J. Lipps, R. M. Tsolis *et al.* (1998). At least four percent of the *Salmonella typhimurium* genome is required for fatal infection of mice. *Infect. Immun.* **66**, 3372–3377.

Brasher, C. W., A. DePaola, D. D. Jones and A. K. Bej (1998). Detection of microbial pathogens in shellfish with multiplex PCR. *Curr. Microbiol.* **37**, 101–107.

Braxton, S. and T. Bedilion (1998). The integration of microarray information in the drug development process. *Curr. Opin. Biotechnol.* **9**, 643–649.

Brennan, M. A. and B. T. Cookson (2000). *Salmonella* induces macrophage death by caspase-1-dependent necrosis. *Mol. Microbiol.* **38**, 31–40.

Brenner, F. W., R. G. Villar, F. J. Angulo *et al.* (2000). *Salmonella* nomenclature. *J. Clin. Microbiol.* **38**, 2465–2467.

Breuil, J., A. Brisabois, I. Casin *et al.* (2000). Antibiotic resistance in salmonellae isolated from humans and animals in France, comparative data from 1994 and 1997. *J. Antimicrob. Chemother.* **46**, 965–971.

Brumell, J. H., P. Tang, M. L. Zaharik and B. B. Finlay (2002). Disruption of the *Salmonella*-containing vacuole leads to increased replication of *Salmonella enterica* serovar typhimurium in the cytosol of epithelial cells. *Infect. Immun.* **70**, 3264–3270.

Buchmeier, N. A. and F. Heffron (1990). Induction of *Salmonella* stress proteins upon infection of macrophages. *Science* **248**, 730–732.

Buchmeier, N. A. and F. Heffron (1991). Inhibition of macrophage phagosome-lysosome fusion by *Salmonella typhimurium. Infect. Immun.* **9**, 2232–2238.

Buchwald, D. S. and M. J. Blaser (1984). A review of human salmonellosis, II. Duration of excretion following infection with nontyphi *Salmonella. Rev. Infect. Dis.* **6**, 345–356.

Budd, W. (1874). *Typhoid Fever. Its Nature, Modes of Spreading, and Prevention.* Arno Press, New York, NY.

Burrows, W. (1954). *Textbook of Microbiology.* Saunders, Philadelphia, PA.

Carli, K. T., C. B. Unal, V. Caner and A. Eyigor (2001). Detection of salmonellae in chicken feces by a combination of tetrathionate broth enrichment, capillary PCR, and capillary gel electrophoresis. *J. Clin. Microbiol.* **39**, 1871–1876.

Carrol, M. E., P. S. Jackett, V. R. Aber and D. B. Lowrie (1979). Phagolysosome formation, cyclic adenosine 3′, 5′-monophosphate and the fate of *Salmonella typhimurium* within mouse peritoneal macrophages. *J. Gen. Microbiol.* **110**, 421–429.

Chadfield, M. and J. Olsen (2001). Determination of the oxidative burst chemiluminescent response of avian and murine-derived macrophages versus corresponding cell lines in relation to stimulation with *Salmonella* serotypes. *Vet. Immunol. Immunopathol.* **80**, 289–308.

Chadfield, M. S., D. J. Brown, S. Aabo *et al.* (2003). Comparison of intestinal invasion and macrophage response of *Salmonella* Gallinarum and other host-adapted *Salmonella enterica* serovars in the avian host. *Vet. Microbiol.* **92**, 49–64.

Chin, J. (2000). *Control of Communicable Diseases Manual.* American Public Health Association, Washington, DC.

Christensen, J., D. L. Baggesen, B Nielsen and H. Stryhn (2002). Herd prevalence of *Salmonella* spp. in Danish pig herds after implementation of the Danish *Salmonella* Control Program with reference to a pre-implementation study. *Vet. Microbiol.* **88**, 175–188.

Cirillo, D. M., R. H. Valdivia, D. M. Monack and S. Falkow (1998). Macrophage-dependent induction of the *Salmonella* pathogenicity island 2 type III secretion system and its role in intracellular survival. *Mol. Microbiol.* **30**, 175–188.

Clark, C., J. Cunningham, R. Ahmed *et al.* (2001). Characterization of *Salmonella* associated with pig ear dog treats in Canada. *J. Clin. Microbiol.* **39**, 3962–3968.

Cohen, M. L. and R. V. Tauxe (1986). Drug-resistant *Salmonella* in the United States, an epidemiologic perspective. *Science* **234**, 964–969.

Connerton, P., J. Wain, T. T. Hien *et al.* (2000). Epidemic typhoid in Vietnam, molecular typing of multiple-antibiotic-resistant *Salmonella enterica* serotype typhi from four outbreaks. *J. Clin. Microbiol.* **38**, 895–897.

Cook, K. A., T. E. Dobbs, W. G. Hlady *et al.* (1998). Outbreak of *Salmonella* serotype Hartford infections associated with unpasteurized orange juice. *J. Am. Med. Assoc.* **280**, 1504–1509.

Corry, J. E., V. M. Allen, W. R. Hudson *et al.* (2002). Sources of *Salmonella* on broiler carcasses during transportation and processing, modes of contamination and methods of control. *J. Appl. Microbiol.* **92**, 424–432.

Cowden, J. M., D. Lynch, C. A. Joseph *et al.* (1989) Case-control study of infections with Salmonella enteritidis phage type 4 in England. *Br. Med.* J. **299**, 771–773.

Coyle, E. F., S. R. Palmer, C. D. Ribero *et al.* (1988). *Salmonella enteritidis* phage type 4 infection, association with hen's eggs. *Lancet* **2**, 1295–1297.

Crump, J. A., T. J. Barrett, J. I. Nelson and F. J. Angulo (2003). Reevaluating fluoroquinolone breakpoints for *Salmonella enterica* serotype Typhi and for non-Typhi salmonellae. *Clin. Infect. Dis.* **37**, 75–81.

Cummings, K., E. Barrett, J. C. Mohle-Boetani *et al.* (2001). A multistate outbreak of *Salmonella enterica* serotype Baildon associated with domestic raw tomatoes. *Emerg. Infect. Dis.* **7**, 1046–1048.

Dalton, C. B. and L. E. Unicomb (2001). Sentinel foodborne disease surveillance in Australia, priorities and possibilities. *Environ. Health* **1**, 10.

D'Aoust, J. Y., E. Daley, M. Crozier and A.M. Sewell (1990). Pet turtles, a continuing international threat to public health. *Am. J. Epidemiol.* **132**, 233–238.

Dauga, C., A. Zabrovskaia and P. A. Grimont (1998). Restriction fragment length polymorphism analysis of some flagellin genes of *Salmonella enterica. J. Clin. Microbiol.* **36**, 2835–2843.

Davies, P. R., J. A. Funk, M. G. Nichols *et al.* (1999). Effects of some methodologic factors on detection of *Salmonella* in swine feces. *Proceedings of the 3rd International Symposium on Epidemiology and Control of Salmonella in Pork, 5–7 August, 1999, Washington, DC, USA.*

Davies, R. H. and C. Wray (1995). Contribution of the lesser mealworm beetle (*Alphitobius diaperinus*) to carriage of *Salmonella enteritidis* in poultry. *Vet. Record* **137**, 407–408.

de Boer, E., W. M. Seldam and H. H. Stigter (1983). *Campylobacter jejuni*, *Yersinia enterocolitica* and *Salmonella* in game and poultry [in Dutch]. *Tijdschr. Diergeneeskd* **108**, 831–836.

Deiwick, J., T. Nikolaus, S. Erdogan and M. Hensel (1999). Environmental regulation of *Salmonella* pathogenicity island 2 gene expression. *Mol. Microbiol.* **31**, 1759–1773.

Delarocque-Astagneau, E., J. C. Desenclos, P. Bouvet and P. A. Grimont (1998). Risk factors for the occurrence of sporadic *Salmonella enterica* serotype Enteritidis infections in children in France, a national case-control study. *Epidemiol. Infect.* **121**, 561–567.

Delarocque-Astagneau, E., C. Bouillant, V. Vaillant *et al.* (2000). Risk factors for the occurrence of sporadic *Salmonella enterica* serotype Typhimurium infections in children in France, a national case-control study. *Clin. Infect. Dis.* **31**, 488–492.

Detweiler, C. S., D. B. Cunanan and S. Falkow (2001). Host microarray analysis reveals a role for the *Salmonella* response regulator phoP in human macrophage cell death. *Proc. Natl. Acad. Sci. USA* **98**, 5850–5855.

de Valk, H., E. Delarocque-Astagneau, G. Colomb *et al.* (2000). A community-wide outbreak of *Salmonella enterica* serotype Typhimurium infection associated with eating a raw milk soft cheese in France. *Epidemiol. Infect.* **124**, 1–7.

de Wit, M. A., M. P. Koopmans, L. M. Kortbeek *et al.* (2001a). Gastroenteritis in sentinel general practices, The Netherlands. *Emerg. Infect. Dis.* **7**, 82–91.

de Wit, M. A., M. P. Koopmans, L. M. Kortbeek *et al.* (2001b). Sensor, a population-based cohort study on gastroenteritis in the Netherlands, incidence and etiology. *Am. J. Epidemiol.* **154**, 666–674.

Djuretic, T., P. G. Wall, M. J. Ryan *et al.* (1996). General outbreaks of infectious intestinal disease in England and Wales 1992 to 1994. *Commun. Dis. Rep. CDR Rev.* **6**, 57–63.

Dritz, S. K. and E. H. Braff (1977). Sexually transmitted typhoid fever. *N. Engl. J. Med.* **296**, 1359–1360.

Dutta, U., P. K. Garg, R. Kumar and R. K. Tandon (2000). Typhoid carriers among patients with gallstones are at increased risk for carcinoma of the gallbladder. *Am. J. Gastroenterol.* **95**, 784–787.

Earampamoorthy, S. and R. S. Koff (1975). Health hazards of bivalve-mollusk ingestion. *Ann. Intern. Med.* **83**, 107–110.

Eberth, C. J. (1880). Die Organismen in den Organen bei Typhus abdominalis. *Virchows Arch. f. Path. Anat.* **81**, 58–73.

Echeita, M. A., S. Herrera, J. Garaizar and M. A. Usera (2002). Multiplex PCR-based detection and identification of the most common *Salmonella* second-phase flagellar antigens. *Res. Microbiol.* **153**, 107–113.

Edel, W., M. van Schothorst, P. A. M. Guinée and E. R. Kampelmacher (1973). Mechanisms and prevention of *Salmonella* infections in animals. In: B. C. Hobbs and J. H. B. Christian (eds), *The Microbiological Safety of Food*, pp. 247–256. Academic Press, London.

Edel, W., M. van Schothorst and E. H. Kampelmacher (1976). Epidemiological studies on *Salmonella* in a certain area ('Walcheren project'). I. The presence of Salmonella in man, pigs, insects, seagulls and in foods and effluents. *Zentralbl. Bakteriol. [Orig A]* **235**, 475–484.

Edelman, R. and M. M. Levine (1986). Summary of an international workshop on typhoid fever. *Rev. Infect. Dis.* **8**, 329–349.

Eriksson, S., J. Bjorkman, S. Boerg *et al.* (2000). *Salmonella typhimurium* mutants that downregulate phagocyte nitric oxide production. *Cell. Microbiol.* **2**, 239–250.

Ethelberg, S., M. Lisby, M. Torpdahl *et al.* (2004). Prolonged restaurant-associated outbreak of multidrug-resistant *Salmonella* Typhimurium among patients from several European countries. *Clin. Microbiol. Infect.* **10**, 904–910.

Euden, P. R. (1990). *Salmonella* isolates from wild animals in Cornwall. *Br. Vet. J.* **146**, 228–232.

Euzeby, J. P. (1999). Revised *Salmonella* nomenclature, designation of *Salmonella enterica* (*ex* Kauffmann and Edwards 1952) Le Minor and Popoff 1987 sp. nov., nom. rev. as the neotype species of the genus *Salmonella* Lignieres 1900 (approved lists 1980), rejection of the name *Salmonella choleraesuis* (Smith 1894) Weldin 1927 (approved lists 1980), and conservation of the name *Salmonella typhi* (Schroeter 1886) Warren and Scott 1930 (approved lists 1980). Request for an opinion. *Intl J. Syst. Bacteriol.* **49 Pt 2**, 927–930.

Evans, H. S., P. Madden, C. Douglas *et al.* (1998). General outbreaks of infectious intestinal disease in England and Wales, 1995 and 1996. *Commun. Dis. Public Health* **1**, 165–171.

Evans, S. and R. Davies (1996). Case control study of multiple-resistant *Salmonella typhimurium* DT104 infection of cattle in Great Britain. *Vet. Record* **139**, 557–558.

Ezaki, T., Y. Kawamura and E. Yabuuchi (2000). Recognition of nomenclatural standing of *Salmonella typhi* (Approved Lists 1980), *Salmonella enteritidis* (Approved Lists 1980) and *Salmonella typhimurium* (Approved Lists 1980), and conservation of the specific epithets *enteritidis* and *typhimurium*. Request for an opinion. *Intl J. Syst. Evol. Microbiol.* **50**, 945–947.

Fallacara, D. M., C. M. Monahan, T. Y. Morishita and R. F. Wack (2001). Fecal shedding and antimicrobial susceptibility of selected bacterial pathogens and a survey of intestinal parasites in free-living waterfowl. *Avian Dis.* **45**, 128–135.

Fannin, K. F., S. C. Vana and W. Jakubowski (1985). Effect of an activated sludge wastewater treatment plant on ambient air densities of aerosols containing bacteria and viruses. *Appl. Environ. Microbiol.* **49**, 1191–1196.

Fedorka-Cray, P. J., L. C. Kelley, T. J. Stabel *et al.* (1995). Alternate routes of invasion may affect pathogenesis of *Salmonella typhimurium* in swine. *Infect. Immun.* **63**, 2658–2664.

Feld, N. C., L. Ekeroth, K. O. Gradel *et al.* (2000). Evaluation of a serological *Salmonella* mix-ELISA for poultry used in a national surveillance programme. *Epidemiol. Infect.* **125**, 263–268.

Fernandez, F., M. Hinton and B. Van Gils (2002). Dietary mannan-oligosaccharides and their effect on chicken caecal microflora in relation to *Salmonella* Enteritidis colonization. *Avian Pathol.* **31**, 49–58.

Fey, P. D., T. J. Safranek, M. E. Rupp *et al.* (2000). Ceftriaxone-resistant salmonella infection acquired by a child from cattle. *N. Engl. J. Med.* **342**, 1242–1249.

Fields, P. I., R. V. Swanson, C. D. Haidaris and F. Heffron (1986). Mutants of *Salmonella typhimurium* that cannot survive within the macrophage are avirulent. *Proc. Natl. Acad. Sci. USA* **83**, 5189–5193.

Fisher, I., A. Lieftucht and K. Mølbak (1999). Paratyphoid fever after travel to Turkey – update 16 September 1999. *Euro Surveill. Weekly* **3**, 38.

Fisker, N., K. Vinding, K. Mølbak and M. K. Hornstrup (2003). Clinical review of nontyphoid *Salmonella* infections from 1991 to 1999 in a Danish county. *Clin. Infect. Dis.* **37**, e47–52.

Fluckey, W. M., M. X. Sanchez, S. R. McKee *et al.* (2003). Establishment of a microbiological profile for an air-chilling poultry operation in the United States. *J. Food Prot.* **66**, 272–279.

Friedman, C. R., C. Torigian, P. J. Shillam *et al.* (1998). An outbreak of salmonellosis among children attending a reptile exhibit at a zoo. *J. Pediatr.* **132**, 802–807.

Gaffky, G. T. A. (1884). Zur Aetiologie des Abdominaltyphus. *Mitt. a. d. Kaiserl. Ges. Amte* **2**, 372–420.

Galan, J. E., and R. Curtiss III (1991). Distribution of the invA, -B, -C, and -D genes of *Salmonella typhimurium* among other *Salmonella* serovars, invA mutants of *Salmonella typhi* are deficient for entry into mammalian cells. *Infect. Immun.* **59**, 2901–2908.

Gärtner, A. (1888). Über die Fleischvergiftung in Frankenhausen a. Kyffhäuser und den Erreger derselben. Correspondenz-Blätter des allgem. ärztl. *Vereins von Thüringen* **17**, 573–600.

Gärtner, W. (1922). Kann der Paratyphus B abdominalis in klinischer, patologisk-anatomischer, epidemiologischer und bakteriologischer Hinsicht der sog. Gastroenteritis paratyphosa B abgetrennt werden? *Zbl. Bakt. Orig.* **87**, 486–525.

Gasem, M. H., W. M. Dolmans, M. M. Keuter and R. R. Diohomoeljanto (2001). Poor food hygiene and housing as risk factors for typhoid fever in Semarang, Indonesia. *Trop. Med. Intl. Health* **6**, 484–490.

Gill, C. J., W. E. Keene, J. C. Mohle-Boetani *et al.* (2003). Alfalfa seed decontamination in a *Salmonella* outbreak. *Emerg. Infect. Dis.* **9**, 474–479.

Gillespie, I. A., G. K. Adak, S. J. O'Brien *et al.* (2001). General outbreaks of infectious intestinal disease associated with fish and shellfish, England and Wales, 1992-1999. *Commun. Dis. Public Health* **4**, 117–123.

Gillespie, I. A., G. K. Adak, S. J. O'Brien and F. J. Bolton (2003). Milkborne general outbreaks of infectious intestinal disease, England and Wales, 1992–2000. *Epidemiol. Infect.* **130**, 461–468.

Glynn, J. R. and D. J. Bradley (1992). The relationship between infecting dose and severity of disease in reported outbreaks of *Salmonella* infections. *Epidemiol. Infect.* **109**, 371–388.

Glynn, J. R., R. B. Hornick, M. M. Levine and D. J. Bradley (1995). Infecting dose and severity of typhoid, analysis of volunteer data and examination of the influence of the definition of illness used. *Epidemiol Infect.* **115**, 23–30.

Glynn, M. K., C. Bopp, W. Dewitt *et al.* (1998). Emergence of multidrug-resistant *Salmonella enterica* serotype typhimurium DT104 infections in the United States. *N. Engl. J. Med.* **338**, 1333–1338.

Gorman, R., S. Bloomfield and C. C. Adley (2002). A study of cross-contamination of food-borne pathogens in the domestic kitchen in the Republic of Ireland. *Intl J. Food Microbiol.* **76**, 143–150.

Grimont, F. and P. A. Grimont (1986). Ribosomal ribonucleic acid gene restriction patterns as potential taxonomic tools. *Ann. Microbiol.* **137B**, 165–175.

Grøndahl, M. L., G. M. Jensen, C. G. Nielsen *et al.* (1998). Secretory pathways in *Salmonella* Typhimurium-induced fluid accumulation in the porcine small intestine. *J. Med. Microbiol.* **47**, 151–157.

Grunnet, K. and J. C. Hansen (1978). Risk of infection from heavily contaminated air. *Scand. J. Work. Environ. Health* **4**, 336–338.

Gulig, P. A., H. Danbara, D. G. Guiney *et al.* (1993). Molecular analysis of spv virulence genes of the *Salmonella* virulence plasmids. *Mol. Microbiol.* **7**, 825–830.

Haas, C. N. (2002). Conditional dose-response relationships for microorganisms, development and application. *Risk Analysis* **22**, 455–463.

Hackett, J., I. Kotlarski, V. Mathan *et al.* (1986). The colonization of Peyer's patches by a strain of *Salmonella typhimurium* cured of the cryptic plasmid. *J. Infect. Dis.* **153**, 1119–1125.

Haddock, R. L., S. N. Cousens and C. C. Guzman (1991). Infant diet and salmonellosis. *Am. J. Public Health* **81**, 997–1000.

Hald, T. and J. S. Andersen (2001). Trends and seasonal variations in the occurrence of *Salmonella* in pigs, pork and humans in Denmark, 1995–2000. *Berl. Munch. Tierarztl. Wochenschr.* **114**, 346–349.

Hald, T., D. Vose, H. C. Wegener and T. Koupeev (2004). A Bayesian approach to quantify the contribution of animal-food sources to human salmonellosis. *Risk Analysis* **24**, 255–269.

Hall, G. V., R. M. D'Souza and M. D. Kirk (2002). Foodborne disease in the new millennium, out of the frying pan and into the fire? *Med J. Aust.* **177**, 614–618.

Han, T., J. E. Sokal and E. Neter (1967). Salmonellosis in disseminated malignant diseases. A seven-year review (1959–1965). N. *Engl. J. Med.* **276**, 1045–1052.

Hansen, A. C. (1942). Die beim Hausgeflügel in Dänemark festgestellten *Salmonella*-typen. *Zbl. Bakt. Orig.* **149**, 222–235.

Hardman, P. M., C. M. Wathes and C. Wray (1991). Transmission of salmonellae among calves penned individually. *Vet. Record* **129**, 327–329.

Harhoff, N. (1948). *Gastroenteritisbaciller af Salmonellagruppen i Danmark*. Nyt Nordisk Forlag – Arnold Busck, Copenhagen.

Harris, I. T., P. J. Fedorka-Cray, J. T. Gray *et al.* (1997). Prevalence of *Salmonella* organisms in swine feed. *J. Am. Vet. Med. Assoc.* **210**, 382–385.

Hassan, J. O. and R. Curtiss III (1997). Efficacy of a live avirulent *Salmonella typhimurium* vaccine in preventing colonization and invasion of laying hens by *Salmonella typhimurium* and *Salmonella enteritidis. Avian Dis.* **41**, 783–791.

Hedberg, C.W., M. J. David, K. E. White *et al.* (1993). Role of egg consumption in sporadic *Salmonella enteritidis* and *Salmonella typhimurium* infections in Minnesota. *J. Infect. Dis.* **167**, 107–111.

Helfrick, D. L., S. J. Olsen, R. D. Bishop *et al.* (2001). *An Atlas of Salmonella in the United States* US Department of Health and Human Services, Centers for Disease Control and Prevention, Atlanta, GA.

Helms, M., P. Vastrup, P. Gerner-Smidt and K. Mølbak (2002). Excess mortality associated with antimicrobial drug-resistant *Salmonella typhimurium. Emerg. Infect. Dis.* **8**, 490–495.

Helms, M., P. Vastrup, P. Gerner-Smidt and K. Mølbak (2003). Short and long term mortality associated with foodborne bacterial gastrointestinal infections, registry based study. *Br. Med. J.* **326**, 357.

Hennessy, T. W., C. W. Hedberg, L. Slutsker *et al.* (1996). A national outbreak of *Salmonella enteritidis* infections from ice cream. The Investigation Team. *N. Engl. J. Med.* **334**, 1281–1286.

Hensel, M. (2000). *Salmonella* pathogenicity island 2. *Mol. Microbiol.* **36**, 1015–1023.

Hensel, M., J. E. Shea, S. R. Waterman *et al.* (1998). Genes encoding putative effector proteins of the type III secretion system of *Salmonella* pathogenicity island 2 are required for bacterial virulence and proliferation in macrophages. *Mol. Microbiol.* **30**, 163–174.

Herikstad, H., Y. Motarjemi and R. V. Tauxe (2002). *Salmonella* surveillance, a global survey of public health serotyping. *Epidemiol. Infect.* **129**, 1–8.

Hersh, D., D. M. Monack, M. R. Smith *et al.* (1999). The *Salmonella* invasin SipB induces macrophage apoptosis by binding to caspase-1. *Proc. Natl. Acad. Sci. USA.* **96**, 2396–2401.

Heyndrickx, M., D. Vandekerchove, L. Herman *et al.* (2002). Routes for salmonella contamination of poultry meat, epidemiological study from hatchery to slaughterhouse. *Epidemiol. Infect.* **129**, 253–265.

Hilton, A. C., J. G. Banks and C. W. Penn (1996). Random amplification of polymorphic DNA (RAPD) of *Salmonella*, strain differentiation and characterization of amplified sequences. *J. Appl. Bacteriol.* **81**, 575–584.

Holt, P. S., R. J. Buhr, D. L. Cunningham and R. E. Porter Jr (1994). Effect of two different molting procedures on a *Salmonella enteritidis* infection. *Poultry Sci.* **73**, 1267–1275.

Humphrey, T. J. (1994). Contamination of egg shell and contents with *Salmonella enteritidis*, a review. *Intl J. Food Microbiol.* **21**, 31–40.

Humphrey, T. J. (1999). Contamination of eggs and poultry meat with *Salmonella enterica* serovar Enteritidis. In: Salmonella enterica *serovar Enteritidis in humans and animals. Epidemiology, Pathogenesis, and Control*, pp. 183–192. Iowa State University Press, Ames, IA.

Humphrey, T. J. and A. Whitehead (1993). Egg age and the growth of *Salmonella enteritidis* PT4 in egg contents. *Epidemiol. Infect.* **111**, 209–219.

Humphrey, T. J., A. Whitehead, A. H. Gawler *et al.* 1991. Numbers of *Salmonella enteritidis* in the contents of naturally contaminated hens' eggs. *Epidemiol Infect.* **106**, 489–496.

Hurd, H. S., T. J. Stabel and S. A. Carlson (1999a). Sensitivity of various fecal sample collections techniques for detection of *Salmonella* Typhimurium in finish hogs. *Proceedings of the 3rd International Symposium on Epidemiology and Control of* Salmonella *in Pork, 5–7 August, 1999, Washington, DC, USA.***11

Hurd, H. S., W. D. Schlosser and E. D. Ebel (1999b). The effect of intermittent shedding on prevalence estimation in populations. *Proceedings of the 3rd International Symposium on Epidemiology and Control of* Salmonella *in Pork, 5–7 August, 1999, Washington, DC, USA.*

Hurd, H. S., J. D. McKean, R. W. Griffith *et al.* (2002). *Salmonella enterica* infections in market swine with and without transport and holding. *Appl. Environ. Microbiol.* **68**, 2376–2381.

Ibrahim, G. F., M. J. Lyons, R. A. Walker and G. H. Fleet (1985). Rapid detection of salmonellae by immunoassays with titanous hydroxide as the solid phase. *Appl. Environ. Microbiol.* **50**, 670–675.

James, W. O., J. C. Prucha, R. L. Brewer *et al.* (1992). Effects of countercurrent scalding and postscald spray on the bacteriologic profile of raw chicken carcasses. *J. Am. Vet. Med. Assoc.* **201**, 705–708.

Johnson, K., I. Charles, G. Dougan *et al.* (1991). The role of a stress–response protein in *Salmonella typhimurium* virulence. *Mol. Microbiol.* **5**, 401–407.

Jones, M. A., M. W. Wood, P. B. Mullan *et al.* (1998). Secreted effector proteins of *Salmonella dublin* act in concert to induce enteritis. *Infect. Immun.* **66**, 5799–5804.

Jones, M. A., P. Wigley, K. L. Page *et al.* (2001). *Salmonella enterica* serovar Gallinarum requires the *Salmonella* pathogenicity island 2 type III secretion system but not the Salmonella pathogenicity island 1 type III secretion system for virulence in chickens. *Infect. Immun.* **69**, 5471–5476.

Kagaya, K., K. Watanabe and Y. Fukasawa (1989). Capacity of recombinant gamma interferon to activate macrophages for *Salmonella*-killing activity. *Infect. Immun.* **57**, 609–615.

Kaiser, M. G. and S. J. Lamont (2001). Genetic line differences in survival and pathogen load in young layer chicks after *Salmonella enterica* serovar enteritidis exposure. *Poultry Sci.* **80**, 1105–1108.

Kapperud, G., S. Gustavsen, I. Hellesnes *et al.* (1990). Outbreak of *Salmonella typhimurium* infection traced to contaminated chocolate and caused by a strain lacking the 60-megadalton virulence plasmid. *J. Clin. Microbiol.* **28**, 2597–2601.

Kapperud, G., J. Lassen and V. Hasseltvedt (1998). *Salmonella* infections in Norway, descriptive epidemiology and a case–control study. *Epidemiol. Infect.* **121**, 569–577.

Kass, P. H., T. B. Farver, J. J. Beaumont *et al.* (1992). Disease determinants of sporadic salmonellosis in four northern California counties. A case–control study of older children and adults. *Ann. Epidemiol.* **2**, 683–696.

Katz, D. J., M. A. Cruz, M. J. Trepka *et al.* (2002). An outbreak of typhoid fever in Florida associated with an imported frozen fruit. *J. Infect. Dis.* **186**, 234–239.

Kelterborn, E. (1967). Salmonella *Species*. Dr. W. Junk N. V., Den Haag.

Kessel, A. S., I. A. Gillespie, S. J. O'Brien *et al.* (2001). General outbreaks of infectious intestinal disease linked with poultry, England and Wales, 1992–1999. *Commun. Dis. Public Health* **4**, 171–177.

Khan, S. A., P. Everest, S. Servos *et al.* (1998). A lethal role for lipid A in *Salmonella* infections. *Mol. Microbiol.* **29**, 571–579.

Kimura, A. C., V. Reddy, R. Marcus *et al.* and Emerging Infections Program FoodNet Working Group (2004). Chicken consumption is a newly identified risk factor for sporadic Salmonella enterica serotype Enteritidis infections in the United States, a case–control study in FoodNet sites. *Clin. Infect. Dis.* **38** (Suppl. 3), S244–S252.

Kist, M. J., and S. Freitag (2000) Serovar specific risk factors and clinical features of *Salmonella enterica* ssp. *enterica* serovar Enteritidis, a study in South–West Germany. *Epidemiol. Infect.* **124**, 383–392.

Kohl, K. S., K. Rietberg, S. Wilson and T. A. Farley (2002). Relationship between home food–handling practices and sporadic salmonellosis in adults in Louisiana, United States. *Epidemiol. Infect.* **129**, 267–276.

Koort, J. M., S. Lukinmaa, M. Rantala *et al.* (2002). Technical improvement to prevent DNA degradation of enteric pathogens in pulsed field gel electrophoresis. *J. Clin. Microbiol.* **40**, 3497–3498.

Kumar, A., G. Nath, B. D. Bhatia *et al.* (1995). An outbreak of multidrug resistant. *Salmonella typhimurium* in a nursery. *Indian Pediatr.* **32**, 881–885.

Lehmacher, A., J. Bockemuhl and S. Aleksic (1995). Nationwide outbreak of human salmonellosis in Germany due to contaminated paprika and paprika-powdered potato chips. *Epidemiol. Infect.* **115**, 501–511.

Le Minor, L. (1984). The genus Salmonella Lignières 1900. In: N. R. Krieg and J. G. Holt (eds), *Bergey's Manual of Systematic Bacteriology*, Vol. 1, pp. 427–458. Williams and Williams, Baltimore, MD.

Le Minor, L., M. Veron and M. Popoff (1982). A proposal for *Salmonella* nomenclature. *Ann. Microbiol.* (*Paris*) **133**, 245–254.

Le Minor, L., M. Y. Popoff, B. Laurent and D. Hermant (1986). Characterization of a 7th subspecies of *Salmonella, S. choleraesuis* subsp. *indica* subsp. nov. *Ann. Inst. Pasteur Microbiol.* **137B**, 211–217.

Lesnick, M. L., N. E. Reiner, J. Fierer and D. G. Guiney (2001). The *Salmonella* spvB virulence gene encodes an enzyme that ADP-ribosylates actin and destabilizes the cytoskeleton of eukaryotic cells. *Mol. Microbiol.* **39**, 1464–1470.

Levine, W. C., J. W. Buehler, N. H. Bean and R. V. Tauxe (1991). Epidemiology of nontyphoidal *Salmonella* bacteremia during the human immunodeficiency virus epidemic. *J. Infect. Dis.* **164**, 81–87.

Lewis, S. J., A. Velasquez, S. L. Cuppett and S. R. McKee (2002). Effect of electron beam irradiation on poultry meat safety and quality. *Poultry Sci.* **81**, 896–903.

Licht, T. R., K. A. Krogfelt, P. S. Cohen *et al.* (1996). Role of lipopolysaccharide in colonization of the mouse intestine by *Salmonella typhimurium* studied by *in situ* hybridization. *Infect. Immun.* **64**, 3811–3817.

Liebana, E., L. Garcia-Migura, C. Clouting *et al.* (2002). Multiple genetic typing of *Salmonella enterica* serotype typhimurium isolates of different phage types (DT104, U302, DT204b, and DT49) from animals and humans in England, Wales, and Northern Ireland. *J. Clin. Microbiol.* **40**, 4450–4456.

Lin, F. Y., J. M. Becke, C. Groves *et al.* (1988). Restaurant-associated outbreak of typhoid fever in Maryland, identification of carrier facilitated by measurement of serum Vi antibodies. *J. Clin. Microbiol.* **26**, 1194–1197.

Lindsay, E. A., A. J. Lawson, R. A. Walker *et al.* (2002). Role of electronic data exchange in an international outbreak caused by *Salmonella enterica* serotype Typhimurium DT204b. *Emerg. Infect. Dis.* **8**, 732–734.

Lindstedt, B. A., E. Heir, T. Verdunt and G. Kapperud (2000). Fluorescent amplified fragment length polymorphism genotyping of *Salmonella enterica* subsp. enterica serovars and comparison with pulsed field gel electrophoresis typing. *J. Clin. Microbiol.* **38**, 1623–1627.

Löffler, F. (1892). Über Epidemien unter den im hygienischen Institute zu Greifswald gehaltenen Mäusen und über die Bekämpfung der Feldmausplage. *Zbl. Bakt.* **11**, 129–141.

Lo Fo Wong, D. M. A. and J. Dahl (2001) The influence of method of analysis and test characteristics on measures of association in epidemiological studies. PhD thesis. Royal Veterinary and Agricultural University, Copenhagen, Denmark.

Logue, C. M., J. S. Sherwood, P. A. Olah *et al.* (2003). The incidence of antimicrobial-resistant *Salmonella* spp on freshly processed poultry from US Midwestern processing plants. *J. Appl. Microbiol.* **94**, 16–24.

Long, S. M., G. K. Adak, S. J. O'Brien and I. A. Gillespie (2002). General outbreaks of infectious intestinal disease linked with salad vegetables and fruit, England and Wales, 1992–2000. *Commun. Dis. Public Health* **5**, 101–105.

Luby, S. P., M. K. Faizan, S. P. Fisher-Hoch *et al.* (1998). Risk factors for typhoid fever in an endemic setting, Karachi, Pakistan. *Epidemiol. Infect.* **120**, 129–138.

Lyons, R. W., C. L. Samples, H. N. DeSilva *et al.* (1980). An epidemic of resistant *Salmonella* in a nursery. Animal-to-human spread. *J. Am. Med. Assoc.* **243**, 546–547.

Maguire, H. C., A. A. Codd, V. E. Mackay *et al.* (1993). A large outbreak of human salmonellosis traced to a local pig farm. *Epidemiol. Infect.* **110**, 239–246.

Mahon, B. E., A. Ponka, W. N. Hall *et al.* (1997). An international outbreak of *Salmonella* infections caused by alfalfa sprouts grown from contaminated seeds. *J. Infect. Dis.* **175**, 876–882.

Mahon, J., C. K. Murphy, P. W. Jones and P. A. Barrow (1994). Comparison of multiplex PCR and standard bacteriological methods of detecting *Salmonella* on chicken skin. *Lett. Appl. Microbiol.* **19**, 169–172.

Mandal, B. K. and J. Brennand. (1988). Bacteraemia in salmonellosis, a 15 year retrospective study from a regional infectious diseases unit. *Br. Med. J.* **297**, 1242–1243.

Marshall, A. and J. Hodgson (1998). DNA chips, an array of possibilities. *Natl Biotechnol.* **16**, 27–31.

Matsui, H., C. M. Bacot, W. A. Garlington *et al.* (2001). Virulence plasmid-borne spvB and spvC genes can replace the 90-kilobase plasmid in conferring virulence to *Salmonella enterica* serovar Typhimurium in subcutaneously inoculated mice. *J. Bacteriol.* **183**, 4652–4658.

McEvoy, J. M., A. M. Doherty, J. J. Sheridan *et al.* (2003). The prevalence of *Salmonella* spp. in bovine faecal, rumen and carcass samples at a commercial abattoir. *J. Appl. Microbiol.* **94**, 693–700.

McGovern, V. J. and L. J. Slavutin (1979). Pathology of salmonella colitis. *Am. J. Surg. Pathol.* **3**, 483–490.

Mead, P. S., L. Slutsker, V. Dietz *et al.* (1999). Food-related illness and death in the United States. *Emerg. Infect. Dis.* **5**, 607–625.

Meakins, S. M., G. K. Adak, B. A. Lopman and S. J. O'Brien (2003). General outbreaks of infectious intestinal disease (IID) in hospitals, England and Wales, 1992–2000. *J. Hosp. Infect.* **53**, 1–5.

Mellemgaard, A. and K. Gaarslev (1988). Risk of hepatobiliary cancer in carriers of *Salmonella typhi. J. Natl Cancer Inst.* **80**, 288.

Mermin, J., B. Hoar and F. J. Angulo (1997). Iguanas and *Salmonella marina* infection in children, a reflection of the increasing incidence of reptile-associated salmonellosis in the United States. *Pediatrics* **99**, 399–402.

Mermin, J. H., R. Villar, J. Carpenter *et al.* (1999). A massive epidemic of multidrug-resistant typhoid fever in Tajikistan associated with consumption of municipal water. *J. Infect. Dis.* **179**, 1416–1422.

Mertens, P. L., J. F. Thissen, A. W. Houben and F. Sturmans (1999). An epidemic of *Salmonella typhimurium* associated with traditional salted, smoked, and dried ham [in Dutch]. *Ned. Tijdschr. Geneeskd.* **143**, 1046–1049.

Miller, S. I., W. S. Pulkkinen, M. E. Selsted and J. J. Mekalanos (1990). Characterization of defensin resistance phenotypes associated with mutations in the phoP virulence regulon of *Salmonella typhimurium. Infect. Immun.* **58**, 3706–3710.

Miller, S. I., E. L. Hohmann and D. A. Pegues (1995). *Salmonella* including *Salmonella* Typhi. In: G. L. Mandell, J. E. Bennett and R. Dolin (eds), *Mandell, Douglas and Bennett's Principles and Practice of Infectious Diseases*, pp. 2013–2033. Churchill Livingstone, New York, NY.

Mintz, E. D., M. L. Cartter, J. L. Hadler *et al.* (1994). Dose–response effects in an outbreak of *Salmonella enteritidis. Epidemiol. Infect.* **112**, 13–23.

Mishu, B., J. Koehler, L. A. Lee *et al.* (1994). Outbreaks of *Salmonella enteritidis* infections in the United States, 1985–1991. *J. Infect. Dis.* **169**, 547–552.

Mohle-Boetani, J. C., R. Reporter, S. B. Werner *et al.* (1999). An outbreak of *Salmonella* serogroup Saphra due to cantaloupes from Mexico. *J. Infect. Dis.* **180**, 1361–1364.

Mølbak, K. and J. Neimann (2002). Risk factors for sporadic infection with *Salmonella enteritidis*, Denmark, 1997–1999. *Am. J. Epidemiol.* **156**, 654–561.

Mølbak, K., N. H. Eriksen, T. Hald and D. L. Baggesen (1998). Outbreak of *Salmonella* Manhattan infection in Denmark with international implications. *Euro Surveill. Weekly* **2**, 9.

Mølbak, K., D. L. Baggesen, E. M. Aarestrup *et al.* (1999). An outbreak of multidrug-resistant, quinolone-resistant *Salmonella enterica* serotype typhimurium DT104. *N. Engl. J. Med.* **341**, 1420–1425.

Mølbak, K., P. Gerner-Smidt and H. C. Wegener (2002). Increasing quinolone resistance in *Salmonella enterica* serotype Enteritidis. *Emerg. Infect. Dis.* **8**, 514–515.

Monack, D. M., B. Raupach, A. E. Hromockyj and S. Falkow (1996). *Salmonella typhimurium* invasion induces apoptosis in infected macrophages. *Proc. Natl. Acad. Sci. USA* **93**, 9833–9838.

Monack, D. M., D. Hersh, N. Ghori *et al.* (2000). *Salmonella* exploits caspase-1 to colonize Peyer's patches in a murine typhoid model. *J. Exp. Med.* **192**, 249–258.

Monack, D. M., W. W. Navarre and S. Falkow (2001). *Salmonella*-induced macrophage death, the role of caspase-1 in death and inflammation. *Microbes Infect.* **3**, 1201–1212.

Morse, D. L., G. S. Birkhead, J. Guardino *et al.* (1994). Outbreak and sporadic egg-associated cases of *Salmonella enteritidis*, New York's experience. *Am. J. Public Health* **84**, 859–860.

Mousing, J., P. T. Jensen, C. Halgaard *et al.* (1997). Nation-wide *Salmonella enterica* surveillance and control in Danish slaughter swine herds. *Prev. Vet. Med.* **29**, 247–261.

Mussche, M. M., N. H. Lameire and S. M. Ringoir (1975). *Salmonella typhimurium* infections in renal transplant patients. Report of five cases. *Nephron* **15**, 143–150.

Nair, D., N. Gupta, S. Kabra *et al.* (1999). *Salmonella senftenberg*, a new pathogen in the burns ward. *Burns* **25**, 723–727.

Nassar, T. J., H. M. al-Nakhli and Z. H. Ogaily (1994). Use of live and inactivated *Salmonella enteritidis* phage type 4 vaccines to immunise laying hens against experimental infection. *Rev. Sci. Tech.* **13**, 855–867.

Neal, K. R., J. Hebden and R. Spiller (1997). Prevalence of gastrointestinal symptoms six months after bacterial gastroenteritis and risk factors for development of the irritable bowel syndrome, postal survey of patients. *Br. Med. J.* **314**, 779–782.

Nielsen, B. and H. C. Wegener (1997). Public health and pork and pork products, regional perspectives of Denmark. *Rev. Sci. Tech.* **16**, 513–524.

Nielsen, B., L. Ekeroth, F. Bager and P. Lind (1998). Use of muscle fluid as a source of antibodies for serologic detection of *Salmonella* infection in slaughter pig herds. *J. Vet. Diagn. Invest.* **10**, 158–163.

Nordentoft, S., H. Christensen and H. C. Wegener (1997). Evaluation of a fluorescence-labelled oligonucleotide probe targeting 23S rRNA for *in situ* detection of *Salmonella* serovars in paraffin-embedded tissue sections and their rapid identification in bacterial smears. *J. Clin. Microbiol.* **35**, 2642–2648.

Northcutt, J. K., M. E. Berrang, J. A. Dickens *et al.* (2003). Effect of broiler age, feed withdrawal, and transportation on levels of coliforms, *Campylobacter, Escherichia coli* and *Salmonella* on carcasses before and after immersion chilling. *Poultry Sci.* **82**, 169–173.

Oh, Y. K., C. Alpuche-Aranda, E. Berthiaume *et al.* (1996). Rapid and complete fusion of macrophage lysosomes with phagosomes containing *Salmonella typhimurium. Infect. Immun.* **64**, 3877–3883.

Olsen, J. E. (2000). Molecular typing of *Salmonella.* In: C. Wray and A. Wray (eds), Salmonella *in Domestic Animals*, pp. 429–466. CABI Publishing, Wallingford.

Olsen, J. E., S. Aabo, W. Hill *et al.* (1995). Probes and polymerase chain reaction for detection of food–borne bacterial pathogens. *Intl J. Food Microbiol.* **28**, 1–78.

Olsen, S. J., L. C. MacKinnon, J. S. Goulding *et al.* (2000). Surveillance for food-borne–disease outbreaks–United States, 1993–1997. *Morbid Mortal Weekly Rep. CDC Surveill. Summ.* **49**, 1–62.

Olsen, S. J., R. Bishop, F. W. Brenner *et al.* (2001a). The changing epidemiology of *Salmonella*, trends in serotypes isolated from humans in the United States, 1987–1997. *J. Infect. Dis.* **183**, 753–761.

Olsen, S. J., E. E. DeBess, T. E. McGivern *et al.* (2001b). A nosocomial outbreak of fluoroquinolone-resistant *Salmonella* infection. *N. Engl. J. Med.* **344**, 1572–1579.

Ooi, P. L., K. T. Goh, K. S. Neo and C. C. Ngan (1997). A shipyard outbreak of salmonellosis traced to contaminated fruits and vegetables. *Ann. Acad. Med. Singapore* **26**, 539–543.

Oosterom, J., R. Dekker, G. J. de Wilde *et al.* (1985). Prevalence of *Campylobacter jejuni* and *Salmonella* during pig slaughtering. *Vet. Q.* **7**, 31–34.

Oscar, T. P. (2002). Development and validation of a tertiary simulation model for predicting the potential growth of *Salmonella typhimurium* on cooked chicken. *Intl J. Food Microbiol.* **76**, 177–190.

Otto, H., D. Tezcan-Merdol, R. Girisch *et al.* (2000). The spvB gene–product of the *Salmonella enterica* virulence plasmid is a mono(ADP-ribosyl)transferase. *Mol. Microbiol.* **37**, 1106–1115.

Painter, J. A., K. Mølbak, J. Sonne-Hansen *et al.* (2004). *Salmonella*-based rodenticides and public health. *Emerg. Infect. Dis.* **10**, 985–987.

Paniker, C. K. and K. N. Vimala (1972). Transferable chloramphenicol resistance in *Salmonella typhi. Nature* **239**, 109–110.

Parker, C., K. Asokan and J. Guard-Petter. (2001). Egg contamination by *Salmonella* serovar enteritidis following vaccination with Delta-aroA *Salmonella* serovar typhimurium. *FEMS Microbiol. Lett.* **195**, 73–78.

Parry, S. M., S. R. Palmer, J. Slader *et al.* and South East Wales Infectious Disease Liaison Group (2002). Risk factors for salmonella food poisoning in the domestic kitchen – case-control study. *Epidemiol. Infect.* **129**, 277–285.

Passaro, D. J., R. Reporter, L. Mascola *et al.* (1996). Epidemic *Salmonella enteritidis* infection in Los Angeles County, California. The predominance of phage type 4. *West. J. Med.* **165**, 126–130.

Paulin, S. M., P. R. Watson, A. R. Benmore *et al.* (2002). Analysis of *Salmonella enterica* serotype-host specificity in calves, avirulence of *S. enterica* serotype gallinarum correlates with bacterial dissemination from mesenteric lymph nodes and persistence *in vivo*. *Infect. Immun.* **70**, 6788–6797.

Petersen, A., F. M. Aarestrup, F. J. Angulo *et al.* (2002). WHO Global Salm-Surv external quality assurance system (EQAS), an important step toward improving the quality of *Salmonella* serotyping and antimicrobial susceptibility testing worldwide. *Microb. Drug Resist.* **8**, 345–353.

Popoff, M. Y. (2001). *Antigenic Formulas of the* Salmonella *Serovars*. WHO Collaborating Center for Reference and Research on Salmonella, Institut Pasteur, Paris.

Poulsen, L. K., F. Lan, C. S. Kristensen *et al.* (1994). Spatial distribution of *Escherichia coli* in the mouse large intestine inferred from rRNA *in situ* hybridization. *Infect. Immun.* **62**, 5191–5194.

Pradier, C., O. Keita-Perse, E. Bernard *et al.* (2000). Outbreak of typhoid fever on the French Riviera. *Eur. J. Clin. Microbiol. Infect. Dis.* **19**, 464–467.

Puohiniemi, R., T. Heiskanen and A. Siitonen (1997). Molecular epidemiology of two international sprout-borne *Salmonella* outbreaks. *J. Clin. Microbiol.* **35**, 2487–2491.

Rahn, K., S. A. De Grandis, R. C. Clarke *et al.* (1992). Amplification of an invA gene sequence of *Salmonella typhimurium* by polymerase chain reaction as a specific method of detection of *Salmonella*. *Mol. Cell Probes* **6**, 271–279.

Ramsay, G. (1998). DNA chips, state-of-the art. *Natl Biotechnol.* **16**, 40–44.

Ransom, J. R., K. E. Belk, R. T. Bacon *et al.* (2002). Comparison of sampling methods for microbiological testing of beef animal rectal/colonal feces, hides, and carcasses. *J. Food Prot.* **65**, 621–626.

Rathman, M., M. D. Sjaastad and S. Falkow (1996). Acidification of phagosomes containing *Salmonella typhimurium* in murine macrophages. *Infect. Immun.* **64**, 2765–2773.

Reeves, M. W., G. M. Evins, A. A. Heiba *et al.* (1989). Clonal nature of *Salmonella typhi* and its genetic relatedness to other salmonellae as shown by multilocus enzyme electrophoresis, and proposal of *Salmonella bongori* comb. nov. *J. Clin. Microbiol.* **27**, 313–320.

Reller, M., R. Tauxe and K. Mølbak (2002). Excessive *Salmonella*-related morbidity in women; US surveillance data, 1968–2000. *40th Annual Meeting of the Infectious Society of America, Chicago, October 24–27, 2002.*

Riemann, H., P. Kass and D. Cliver (2000). *Salmonella enteritidis* epidemic. *Science* **287**, 1754–1755; author reply 1755–1756.

Riley, L. W., B. S. Ceballos, L. R. Trabulsi *et al.* (1984). The significance of hospitals as reservoirs for endemic multiresistant *Salmonella typhimurium* causing infection in urban Brazilian children. *J. Infect. Dis.* **150**, 236–241.

Rodrigue, D. C., R. V. Tauxe and B. Rowe (1990). International increase in *Salmonella enteritidis*, a new pandemic? *Epidemiol. Infect.* **105**, 21–27.

Rossen, L., P. Norskov, K. Holmstrom and O. F. Rasmussen (1992). Inhibition of PCR by components of food samples, microbial diagnostic assays and DNA-extraction solutions. *Intl J. Food Microbiol.* **17**, 37–45.

Rout, W. R., S. B. Formal, G.J. Dammin and R. A. Gianella (1974). Pathophysiology of *Salmonella* diarrhea in the Rhesus monkey, intestinal transport, morphological and bacteriological studies. *Gastroenterology* **67**, 59–70.

Rowe, B., L. R. Ward and E. J. Threlfall (1997). Multidrug-resistant *Salmonella typhi*, a worldwide epidemic. *Clin. Infect. Dis.* **24** (Suppl. 1), S106–109.

Rubenstein, A. D. and R. N. Fowler (1955). Salmonellosis of the newborn with transmission by delivery room resuscitators. *Am. J. Public Health* **45**, 1109–1114.

Ryan, C. A., M. K. Nickels, N. T. Hargrett-Bean *et al.* (1987). Massive outbreak of antimicrobial-resistant salmonellosis traced to pasteurized milk. *J. Am. Med. Assoc.* **258**, 3269–3274.

Ryan, M. J., P. G. Wall, G. K. Adak *et al.* (1997). Outbreaks of infectious intestinal disease in residential institutions in England and Wales 1992–1994. *J. Infect.* **34**, 49–54.

Saeed, A. M., D. Thiagarajan and E. K. Asem (1999). Mechanism of transovarian transmission of *Salmonella enterica* serovar Enteritidis in laying hens. In: A. M. Saeed (ed.). Salmonella Enterica *serovar Enteritidis in Humans and Animals*, pp. 193–211. Iowa State University Press, Ames, IA.

Saitanu, K., J. Jerngklinchan and C. Koowatananukul (1994). Incidence of salmonellae in duck eggs in Thailand. *Southeast Asian J. Trop. Med. Public Health* **25**, 328–331.

Sandberg, M., P. Hopp, J. Jarp and E. Skjerve (2002). An evaluation of the Norwegian *Salmonella* surveillance and control program in live pig and pork. *Intl J. Food Microbiol.* **72**, 1–11.

Santos, R. L., S. Zhang, R. M. Tsolis *et al.* (2002). Morphologic and molecular characterization of *Salmonella typhimurium* infection in neonatal calves. *Vet. Pathol.* **39**, 200–215.

Saphra, I. and J. W. Winter (1957). Clinical manifestations of salmonellosis in man; an evaluation of 7779 human infections identified at the New York *Salmonella* Center. *N. Engl. J. Med.* **256**, 1128–1134.

Schmid, H., A. P. Burnens, A. Baumgartner and J. Oberreich (1996). Risk factors for sporadic salmonellosis in Switzerland. *Eur. J. Clin. Microbiol. Infect. Dis.* **15**, 725–732.

Schønheyder, H. C. and T. Ejlertsen (1995). Survey of extraintestinal nontyphoid *Salmonella* infections in a Danish region. Inverse relation of invasiveness to frequency of isolation. *Apmis* **103**, 686–688.

Schutze, G. E., H. A. Fawcett, M. J. Lewno *et al.* (1996). Prevalence of *Salmonella enteritidis* in poultry shell eggs in Arkansas. *South Med. J.* **89**, 889–891.

Seals, J. E., P. L. Parrott, J. E. McGowan Jr and R. A. Feldman (1983). Nursery salmonellosis, delayed recognition due to unusually long incubation period. *Infect. Control* **4**, 205–208.

Seyfarth, A. M., H. C. Wegener and N. Frimodt-Moller (1997). Antimicrobial resistance in *Salmonella enterica* subsp. *enterica* serovar typhimurium from humans and production animals. *J. Antimicrob. Chemother.* **40**, 67–75.

Skov, M. N., O. Angen, M. Chriel *et al.* (1999). Risk factors associated with *Salmonella enterica* serovar typhimurium infection in Danish broiler flocks. *Poultry Sci.* **78**, 848–854.

Slader, J., G. Domingue, G. Jorgensen *et al.* (2002). Impact of transport crate reuse and of catching and processing on *Campylobacter* and *Salmonella* contamination of broiler chickens. *Appl. Environ. Microbiol.* **68**, 713–719.

Smerdon, W. J., G. K. Adak, S. J. O'Brien *et al.* (2001). General outbreaks of infectious intestinal disease linked with red meat, England and Wales, 1992–1999. *Commun. Dis. Public Health* **4**, 259–267.

Southern, E. M. (1996). DNA chips, analysing sequence by hybridization to oligonucleotides on a large scale. *Trends Genet.* **12**, 110–115.

Sperber, S. J. and C. J. Schleupner (1987). Salmonellosis during infection with human immunodeficiency virus. *Rev. Infect. Dis.* **9**, 925–934.

Stan, S. D. and M. A. Daeschel (2003). Reduction of *Salmonella enterica* on alfalfa seeds with acidic electrolyzed oxidizing water and enhanced uptake of acidic electrolyzed oxidizing water into seeds by gas exchange. *J. Food Prot.* **66**, 2017–2022.

Stanley, J., C. S. Jones and E. J. Threlfall (1991). Evolutionary lines among *Salmonella enteritidis* phage types are identified by insertion sequence IS200 distribution. *FEMS Microbiol. Lett.* **66**, 83–89.

Stanley, J., N. Baquar and A. Burnens (1995). Molecular subtyping scheme for *Salmonella panama. J. Clin. Microbiol.* **33**, 1206–1211.

St Louis, M. E., D. L. Morse, M. E. Potter *et al.* (1988). The emergence of grade A eggs as a major source of *Salmonella enteritidis* infections. New implications for the control of salmonellosis. *J. Am. Med. Assoc.* **259**, 2103–2107.

Stone, A., M. Shaffer and R. L. Sautter (1993). *Salmonella poona* infection and surveillance in a neonatal nursery. *Am. J. Infect. Control* **21**, 270–273.

Stroffolini, T., G. Manzillo, R. DeSana *et al.* (1992). Typhoid fever in the Neapolitan area, a case-control study. *Eur. J. Epidemiol.* **8**, 539–542.

Swaddiwudhipong, W. and J. Kanlayanaphotporn (2001). A common-source waterborne outbreak of multidrug-resistant typhoid fever in a rural Thai community. *J. Med. Assoc. Thai.* **84**, 1513–1517.

Swaminathan, B. and P. Feng (1994). Rapid detection of food-borne pathogenic bacteria. *Ann. Rev. Microbiol.* **48**, 401–426.

Swaminathan, B., T. J. Barrett, S. B. Hunter and R. V. Tauxe (2001). PulseNet, the molecular subtyping network for foodborne bacterial disease surveillance, United States. *Emerg. Infect. Dis.* **7**, 382–389.

Swartz, M. N. (2002). Human diseases caused by foodborne pathogens of animal origin. *Clin. Infect. Dis.* **34** (Suppl. 3), S111–122.

Tacket, C. O., L. B. Dominguez, H. J. Fisher and M. L. Cohen (1985). An outbreak of multiple-drug-resistant *Salmonella enteritis* from raw milk. *J. Am. Med. Assoc.* **253**, 2058–2060.

Tamplin, M. L., I. Feder, S. A. Palumbo *et al.* (2001). *Salmonella* spp. and *Escherichia coli* biotype I on swine carcasses processed under the hazard analysis and critical control point-based inspection models project. *J. Food Prot.* **64**, 1305–1308.

Tauxe, R. V. (1998). New approaches to surveillance and control of emerging foodborne infectious diseases. *Emerg. Infect. Dis.* **4**, 455–456.

Tauxe, R. V. (1999). *Salmonella* Enteritidis and *Salmonella* Typhimurium DT104, successful subtypes in the modern world. In: W. M. Scheid, W. A. Craig and J. M. Hughes (eds), *Emerging Infections 3*, pp. 37–52. ASM Press, Washington, DC.

Tauxe, R.V., J. G. Rigau-Perez, J. G. Wells and P. A. Blake (1985). Turtle-associated salmonellosis in Puerto Rico. Hazards of the global turtle trade. *J. Am. Med. Assoc.* **12**, 237–239.

Taylor, D. N., C. Bopp, K. Birkness and M. L. Cohen (1984a). An outbreak of salmonellosis associated with a fatality in a healthy child, a large dose and severe illness. *Am. J. Epidemiol.* **119**, 907–912.

Taylor, J. P., W. X. Shandera, T. G. Betz *et al.* (1984b). Typhoid fever in San Antonio, Texas, an outbreak traced to a continuing source. *J. Infect. Dis.* **149**, 553–557.

Taylor, R., D. Sloan, T. Cooper *et al.* (2000). A waterborne outbreak of *Salmonella* Saintpaul. *Commun. Dis. Intell.* **24**, 336–340.

Teunis, P. F. and A. H. Havelaar (2000). The Beta Poisson dose–response model is not a single-hit model. *Risk Analysis* **20**, 513–520.

Tezcan-Merdol, D., T. Nyman, U. Lindberg *et al.* (2001). Actin is ADP-ribosylated by the *Salmonella enterica* virulence-associated protein SpvB. *Mol. Microbiol.* **39**, 606–619.

Thomas, G. W. and J. F. Harbourne (1972). *Salmonella paratyphi* B infection in dairy cows. II. Investigation of an active carrier. *Vet. Record* **91**, 148–150.

Thomas, J. L., M. S. Palumbo, J. A. Farrar *et al.* (2003). Industry practices and compliance with US Food and Drug Administration guidelines among California sprout firms. *J. Food Prot.* **66**, 1253–1259.

Threlfall, E. J. (2000). Epidemic *Salmonella typhimurium* DT 104 – a truly international multiresistant clone. *J. Antimicrob. Chemother.* **46**, 7–10.

Threlfall, E. J., A. M. Ridley, L. R. Ward and B. Rowe (1996). Assessment of health risk from *Salmonella*-based rodenticides. *Lancet* **348**, 616–617.

Threlfall, E. J., I. S. Fisher, L. R. Ward *et al.* (1999a). Harmonization of antibiotic susceptibility testing for *Salmonella*, results of a study by 18 national reference laboratories within the European Union-funded Enter-net group. *Microb. Drug. Resist.* **5**, 195–200.

Threlfall, E. J., L. R. Ward and B. Rowe (1999b). Resistance to ciprofloxacin in non-typhoidal salmonellas from humans in England and Wales – the current situation. *Clin. Microbiol. Infect.* **5**, 130–134.

Threlfall, E. J., I. S. Fisher, L. Berghold *et al.* (2003). Antimicrobial drug resistance in isolates of *Salmonella* enterica from cases of salmonellosis in humans in Europe in 2000, results of international multi-centre surveillance. *Euro. Surveill.* **8**, 41–45.

Tolker-Nielsen, T. and S. Molin (1996). Role of ribosome degradation in the death of heat-stressed *Salmonella typhimurium. FEMS Microbiol. Lett.* **142**, 155–160.

Travers, K. and M. Barza (2002). Morbidity of infections caused by antimicrobial-resistant bacteria. *Clin. Infect. Dis.* **34** (Suppl. 3), S131–134.

Trepka, M. J., J. R. Archer, S. F. Altekruse *et al.* (1999). An increase in sporadic and outbreak-associated *Salmonella enteritidis* infections in Wisconsin, the role of eggs. *J. Infect. Dis.* **180**, 1214–1219.

Tseng, S. Y., D. Macool, V. Elliott *et al.* (1997). An homogeneous fluorescence polymerase chain reaction assay to identify *Salmonella. Anal. Biochem.* **245**, 207–212.

Uchiya, K., M. A. Barbieri, K. Funato *et al.* (1999). A *Salmonella* virulence protein that inhibits cellular trafficking. *Embo J.* **18**, 3924–3933.

Uzzau, S., D. J. Brown, T. Wallis *et al.* (2000). Host adapted serotypes of *Salmonella enterica. Epidemiol. Infect.* **125**, 229–255.

Uzzau, S., G. S. Leori, V. Petruzzi *et al.* (2001). *Salmonella enterica* serovar-host specificity does not correlate with the magnitude of intestinal invasion in sheep. *Infect. Immun.* **69**, 3092–3099.

Van Beneden, C. A., W. E. Keene, R. A. Strong *et al.* (1999). Multinational outbreak of *Salmonella enterica* serotype Newport infections due to contaminated alfalfa sprouts. *J. Am. Med. Assoc.* **281**, 158–162.

van der Wolf, P. J., J. H. Bongers, A. R. Elbers *et al.* (1999). *Salmonella* infections in finishing pigs in The Netherlands, bacteriological herd prevalence, serogroup and antibiotic resistance of isolates and risk factors for infection. *Vet. Microbiol.* **67**, 263–275.

van Diepen, A., T. van der Straaten, S. M. Holland *et al.* (2002). A superoxide-hypersusceptible *Salmonella enterica* serovar Typhimurium mutant is attenuated but regains virulence in p47(phox–/–) mice. *Infect. Immun.* **70**, 2614–2621.

van Pelt, W., M. A. de Wit, W. J. Wannet *et al.* (2003). Laboratory surveillance of bacterial gastroenteric pathogens in The Netherlands, 1991–2001. *Epidemiol. Infect.* **130**, 431–441.

van Winsen, R. L., A. van Nes, D. Keuzenkamp, H. A. Urlings *et al.* (2001). Monitoring of transmission of *Salmonella enterica* serovars in pigs using bacteriological and serological detection methods. *Vet. Microbiol.* **80**, 267–274.

Vazquez-Torres, A., J. Jones-Carson, A. J. Baumler *et al.* (1999). Extraintestinal dissemination of *Salmonella* by CD18-expressing phagocytes. *Nature* **401**, 804–808.

Vazquez-Torres, A., J. Jones-Carson, P. Mastroni *et al.* (2000). Antimicrobial actions of the NADPH phagocyte oxidase and inducible nitric oxide synthase in experimental salmonellosis. I. Effects on microbial killing by activated peritoneal macrophages *in vitro*. *J. Exp. Med.* **192**, 227–236.

Veldman, A., H. A. Vahl, G. J. Bonggreve and D. C. Fuller (1995). A survey of the incidence of *Salmonella* species and *Enterobacteriaceae* in poultry feeds and feed components. *Vet. Record* **136**, 169–172.

Velema, J. P., G. van Wijnen, P. Bult *et al.* (1997). Typhoid fever in Ujung Pandang, Indonesia – high-risk groups and high-risk behaviours. *Trop. Med. Intl Health* **2**, 1088–1094.

Villar, R. G., M. D. Macek, S. Simons *et al.* (1999). Investigation of multidrug-resistant *Salmonella* serotype typhimurium DT104 infections linked to raw-milk cheese in Washington State. *J. Am. Med. Assoc.* **281**, 1811–1816.

von Altrock, A., A. Schutte and G. Hildebrandt (2000). Results of the German investigation in the EU Project '*Salmonella* in Pork (Salinpork) – 2. Investigations in a slaughterhouse. *Berl. Munch. Tierarztl. Wochenschr.* **113**, 225–233.

Wallis, T. S., W. G. Starkey, J. Stephen *et al.* (1986). Enterotoxin production by *Salmonella typhimurium* strains of different virulence. *J. Med. Microbiol.* **21**, 19–23.

Wallis, T. S., A. T. Vaughan, G. I. Clarke *et al.* (1990). The role of leucocytes in the induction of fluid secretion by *Salmonella typhimurium*. *J. Med. Microbiol.* **31**, 27–35.

Wallis, T. S., S. M. Paulin, J. S. Plested *et al.* (1995). The *Salmonella dublin* virulence plasmid mediates systemic but not enteric phases of salmonellosis in cattle. *Infect. Immun.* **63**, 2755–2761.

Wallis, T. S., M. Wood, P. Watson *et al.* (1999). Sips, Sops, and SPIs but not stn influence *Salmonella* enteropathogenesis. *Adv. Exp. Med. Biol.* **473**, 275–280.

Wang, R. F., W. W. Cao and C. E. Cerniglia (1997). A universal protocol for PCR detection of 13 species of foodborne pathogens in foods. *J. Appl. Microbiol.* **83**, 727–736.

Ward, L. (2000). *Salmonella* perils of pet reptiles. *Commun. Dis. Public Health* **3**, 2–3.

Ward, L. R., J. D. de Sa and B. Rowe (1987). A phage-typing scheme for *Salmonella enteritidis. Epidemiol. Infect.* **99**, 291–294.

Ward, L. R., J. Threlfall, H. R. Smith and S. J. O'Brien (2000). *Salmonella enteritidis* epidemic. *Science* **287**, 1753–1754; author reply 1755–1756.

Ward, L. R., C. Maguire, M. D. Hampton *et al.* (2002). Collaborative investigation of an outbreak of *Salmonella enterica* serotype Newport in England and Wales in 2001 associated with ready-to-eat salad vegetables. *Commun. Dis. Public Health* **5**, 301–304.

Watson, P. R., S. M. Paulin, P. W. Jones and T. S. Wallis (2000). Interaction of *Salmonella* serotypes with porcine macrophages *in vitro* does not correlate with virulence. *Microbiology* **146**, 1639–1649.

Wayne, L. G. (1994). Actions of the judicial commission of the international committee on systematic bacteriology on request for opinions published between January 1985 and July 1993. *Intl J. Syst. Bacteriol.* **44**, 177–178.

Wegener, H. C. and D. L. Baggesen (1996). Investigation of an outbreak of human salmonellosis caused by *Salmonella enterica* ssp. *enterica* serovar Infantis by use of pulsed field gel electrophoresis. *Intl J. Food Microbiol.* **32**, 125–131.

Wegener, H. C., T. Hald, D. L. Wong *et al.* (2003). *Salmonella* control programs in Denmark. *Emerg. Infect. Dis.* **9**, 774–780.

Weil, E. and A. Felix. 1918. Über die Doppelnatur der Rezeptoren beim Paratyphus. *Berl. Klin. Wochenschr.* **31**, 986–988.

Weissinger, W. R. and L. R. Beuchat (2000). Comparison of aqueous chemical treatments to eliminate *Salmonella* on alfalfa seeds. *J. Food Prot.* **63**, 1475–1482.

Weissinger, W. R., W. Chantarapanont and L.R. Beuchat (2000). Survival and growth of *Salmonella baildon* in shredded lettuce and diced tomatoes, and effectiveness of chlorinated water as a sanitizer. *Intl J. Food Microbiol.* **62**, 123–131.

Welton, J. C., J. S. Marr and S. M. Friedman (1979). Association between hepatobiliary cancer and typhoid carrier state. *Lancet* **1**, 791–794.

Wheeler, J. G., D. Sethi, J. M. Cowden *et al.* (1999). Study of infectious intestinal disease in England, rates in the community, presenting to general practice, and reported to national surveillance. The Infectious Intestinal Disease Study Executive. *Br. Med. J.* **318**, 1046–1050.

Wick, M. J., C. V. Harding, N. J. Twesten *et al.* (1995). The phoP locus influences processing and presentation of *Salmonella typhimurium* antigens by activated macrophages. *Mol. Microbiol.* **16**, 465–476.

Widjojoatmodjo, M. N., A. C. Fluit, R. Torensma *et al.* (1991). Evaluation of the magnetic immuno PCR assay for rapid detection of *Salmonella. Eur. J. Clin. Microbiol. Infect. Dis.* **10**, 935–938.

Wingstrand, A., J. Dahl, L. K. Thomsen *et al.* (1997). Influence of dietary administration of organic acids and increased feed structure on *Salmonella* Typhimurium infection in pigs. *Proceedings of the IIth International Symposium on Epidemiology and Control of* Salmonella *in Pork., Copenhagen, Denmark, 20–22 August 1997.*

Wiuff, C., E. S. Jauho, H. Stryhn *et al.* (2000). Evaluation of a novel enzyme–linked immunosorbent assay for detection of antibodies against *Salmonella*, employing a stable coating of lipopolysaccharide-derived antigens covalently attached to polystyrene microwells. *J. Vet. Diagn. Invest.* **12**, 130–135.

Wolfe, M. S., D. B. Louria, D. Armstrong and A. Blevins (1971). Salmonellosis in patients with neoplastic disease. A review of 100 episodes at Memorial Cancer Center over a 13-year period. *Arch. Intern. Med.* **128**, 546–554.

Wong, Y. C., T. J. Herald and K. A. Hochmeister (1996). Comparison between irradiated and thermally pasteurized liquid egg white on functional, physical, and microbiological properties. *Poultry Sci.* **75**, 803–808.

Woodward, D. L., R. Khakhria and W. M. Johnson (1997). Human salmonellosis associated with exotic pets. *J. Clin. Microbiol.* **35**, 2786–2790.

Woodward, M. J., G. Gettinby, M. F. Breslin *et al.* (2002). The efficacy of Salenvac, a *Salmonella enterica* subsp. *Enterica* serotype Enteritidis iron-restricted bacterin vaccine, in laying chickens. *Avian Pathol.* **31**, 383–392.

Yabuuchi, E. and T. Ezaki. (2000). Arguments against the replacement of type species of the genus *Salmonella* from *Salmonella choleraesuis* to '*Salmonella enterica*' and the creation of the term 'neotype species', and for conservation of *Salmonella choleraesuis. Intl J. Syst. Evol. Microbiol.* **50**, 1693–1694.

Yamamoto, T., H. Sashinami, A. Takaya *et al.* (2001). Disruption of the genes for ClpXP protease in *Salmonella enterica* serovar Typhimurium results in persistent infection in mice, and development of persistence requires endogenous gamma interferon and tumor necrosis factor alpha. *Infect. Immun.* **69**, 3164–3174.

Yrlid, U., M. Svensson, A. Hakansson *et al.* (2001). *In vivo* activation of dendritic cells and T cells during *Salmonella enterica* serovar Typhimurium infection. *Infect. Immun.* **69**, 5726–5735.

Zamora, B. M. and M. Hartung (2002). Chemiluminescent immunoassay as a microtiter system for the detection of *Salmonella* antibodies in the meat juice of slaughter pigs. *J. Vet. Med. B. Infect. Dis. Vet. Public Health* **49**, 338–345.

Zhang, S., R. A. Kingsley, R. L. Santos *et al.* (2003). Molecular pathogenesis of *Salmonella enterica* serotype typhimurium-induced diarrhea. *Infect. Immun.* **71**, 1–12.

Zhou, D., L. M. Chen, L. Hernandez *et al.* (2001). A *Salmonella* inositol polyphosphatase acts in conjunction with other bacterial effectors to promote host cell actin cytoskeleton rearrangements and bacterial internalization. *Mol. Microbiol.* **39**, 248–259.

Clostridium *perfringens* gastroenteritis

4

R. G. Labbe and V. K. Juneja

1 Historical aspects and contemporary problems

The relationship between *C. perfringens* (earlier called *Clostridium welchii*) and diarrhea was recognized at the end of the nineteenth century (Andrews, 1899), after which the association fell into obscurity until the mid-1940s. By contrast, the association of *C. perfringens* with gas gangrene became generally recognized during World War I, and the agent remains today the most common cause of such infections.

Contemporary descriptions of outbreaks of food poisoning due to *C. perfringens* date from 1943, when children became ill after eating school meals containing gravy heavily contaminated with anaerobic, spore-forming bacilli, including *C. perfringens* (Knox and MacDonald, 1943). Shortly thereafter, McClung (1945) in the US described four outbreaks of food poisoning resulting from consumption of chicken that had been steamed the previous day. Symptoms were typical of those associated with *C. perfringens* foodborne illness, and the organism itself was isolated from the cooked chickens. The work of Hobbs and her colleagues (1953) further established the role of *C. perfringens* in foodborne illness. They described variations in outbreak strains, which were only weakly β-hemolytic but produced so-called 'heat-resistant' (HR) spores. These were defined as surviving 100°C for 1 hour. However, the notion that only strains with these characteristics were associated with foodborne illness was dispelled in the 1960s, when British (Sutton and Hobbs, 1968), American (Hall *et al.*,

Foodborne Infections and Intoxications 3e
ISBN-10: 0-12-588365-X
ISBN-13: 978-012-588365-8

1963) and Canadian (Hauschild and Thatcher, 1967) investigators described the isolation of outbreak strains that were β-hemolytic and produced 'heat-sensitive' (HS) spores. Thus, the phenotypic characteristics, levels of β-hemolysis and heat resistance, are currently of little diagnostic value.

In the 1970s, great strides were made in determining the presumptive mode of action of the enterotoxin, following its purification by Hauschild and Hilsheimer (1971) and Stark and Duncan (1972). Today, *C. perfringens* remains a principal agent of foodborne illness. In countries where its role in foodborne illness is sought, it consistently ranks number two or three as the etiological agent of foodborne illness, this in the face of more trendy 'emerging' foodborne illnesses that periodically capture popular attention. Because of its spore-forming ability, widespread occurrence and rapid growth rate, it is regularly implicated in illnesses associated with food prepared well in advance of consumption, such as in food-service settings.

2 Characteristics of the organism

2.1 Taxonomy

C. perfringens is a Gram-positive, spore-forming, rod-shaped bacterium that is encapsulated and non-motile. The G-C content is from 24–27 moles %. Based on analyses of 16S rRNA, its closest relative is *Clostridium pasteurianum* (Canard *et al.*, 1992a). Spores, usually located subterminally, are formed only reluctantly and in specially formulated media, although apparently more readily in the intestine (see below). Important cultural characteristics (used in confirmation of isolates) include the reduction of nitrate, liquefaction of gelatin, and fermentation of lactose. The production of lecithinase (α-toxin) has also been used for classification purposes, and is demonstrated by a pearly opalescence surrounding colonies grown on egg yolk agar which can be inhibited by type A antitoxin (Nagler reaction). A detailed description of the cultural characteristics of *C. perfringens* was provided by Willis (1969).

The ability to sporulate is an important property of *C. perfringens* because: (a) it is a factor in classification; (b) spores can survive cooking procedures and resume vegetative cell growth if proper conditions are present; and (c) high levels of enterotoxin formation are associated with sporulation. These aspects are discussed below in more detail.

2.2 Sporulation of *C. perfringens*

The presence of sporulating cells in laboratory media can be readily seen by phase contrast microscopy (Figure 4.1A). The refractile appearance is due to the spores' low water content. The mature spore (Figure 4.1B) shows characteristic structures, including the core cortex, spore coat layers and subcoat region. The spore cortex is composed of peptidoglycan, a carbohydrate–peptide polymer. Its hydrolysis occurs during germination by spore-lytic enzymes (Gombas and Labbe, 1985) located in the coat/subcoat region (Labbe *et al.*, 1978).

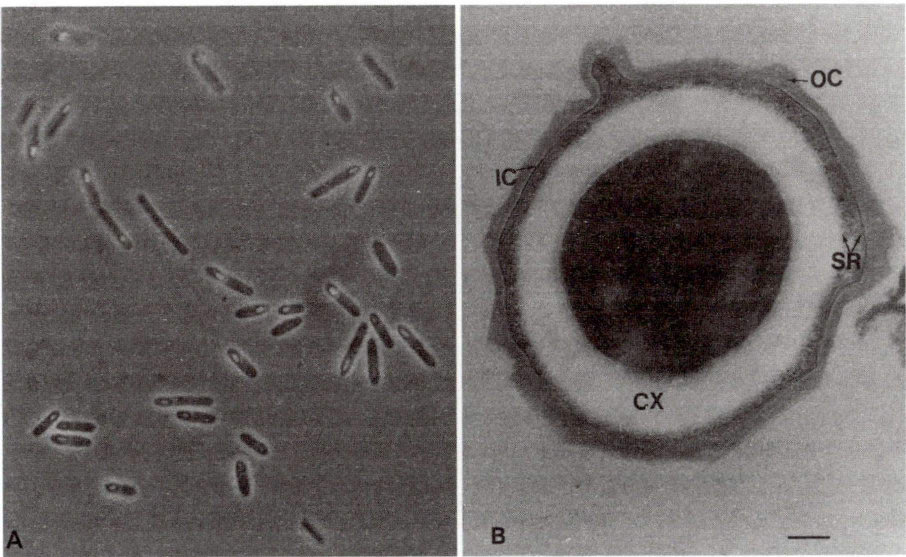

Figure 4.1 Spores of *C. perfringens*: (A) phase contrast micrograph of sporulating cells, showing mature, refractile spores; (B) ultrastructure of spore, showing outer coat (OC), inner coat (IC), subcoat region (SR), cortex (CX), and core (C). Reproduced from Labbe *et al.* (1978) with permission of the National Research Council of Canada.

The bacterial sporulation process is divided into stages, I–VII. Figure 4.2 shows various stages of the process in *C. perfringens*. Stage II is the earliest microscopically visible stage, i.e. the formation of an asymmetric forespore membrane. Engulfment of the forespore is associated with Stage III, and spore maturation with Stage VI. Stage VII corresponds to the release of the mature spore.

2.2.1 Factors affecting sporulation

Because of its association with enterotoxin formation, a good deal of work has centered on the factors that affect sporulation, which is highly strain dependent. Unfortunately, as stated above, most strains sporulate only reluctantly in laboratory media, and even then only to moderate levels. Low levels of spores, sufficient for culture carriage, are usually produced in commercially available cooked-meat medium (CMM). However the levels of sporulation are insufficient for high *C. perfringens* enterotoxin (CPE) yields, or for obtaining adequate spore crops for additional studies.

Over a period of many years, several sporulation media have been devised. The medium of Duncan and Strong (1968), or minor modifications of it (Labbe and Rey, 1979; Harmon and Kautter, 1986; Craven, 1988; Juneja *et al.*, 1993), has been the most widely used. Adjuncts, such as certain methylxanthines – caffeine, theobromine, and theophylline (Sacks and Thompson, 1977; Labbe and Nolan, 1981; Juneja *et al.*, 1993) – and bile salts (Heredia *et al.*, 1991) can also enhance the level of sporulation. Optimal conditions are often determined empirically, and a specific protocol and medium may not apply for all strains. The recommended temperature for routine use is 37°C since certain strains possess little amylase activity above 40°C (Garcia-Alvarado *et al.*, 1992a), and starch is the principal carbohydrate in the

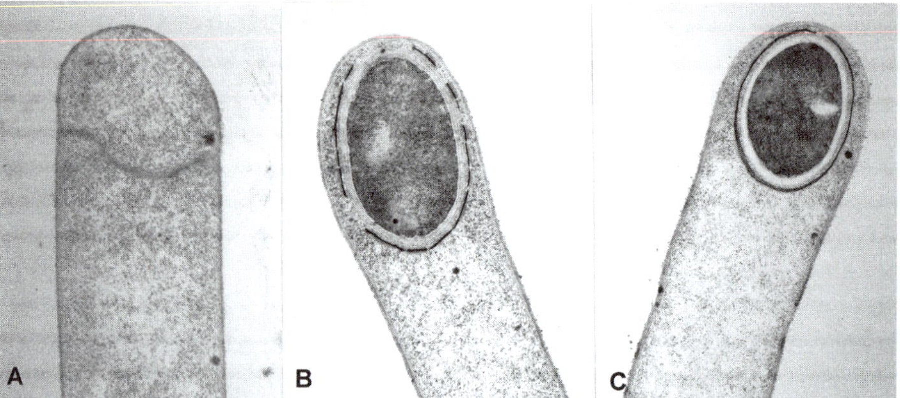

Figure 4.2 *C. perfringens* at intermediate stages of sporulation: (A) stage II, forespore septum formation; (B) stage V, deposition of spore coats; (C) stage VI, maturation of spore. A and B, from Labbe, unpublished; C from Duncan *et al.* (1973); with permission of the American Society for Microbiology.

Duncan–Strong medium. Solid sporulation media have also been described (Uemura, 1978a; Stelma *et al.*, 1985;) but not widely adopted. Since sporulation and enterotoxin formation in the small intestine occur during pathogenesis in foodborne outbreaks (see below), it is perhaps not surprising that human intestinal contents (Skjelkvale and Uemura, 1977a) and bovine feces (Stelma *et al.*, 1985) enhance the sporulation process.

Sacks and Thompson (1978; Sacks, 1983) developed a defined sporulation medium which has proven useful for metabolic studies of the sporulation process (Dillon and Labbe, 1989; Decaudin and Tholozan, 1996), for purification of extracellular sporulation products (Park and Labbe, 1991a), or for radioactive labeling of CPE (Granum and Skjelkvale, 1981; Loffler and Labbe, 1986). As with complex sporulation media, not all strains respond with a high percentage of sporulating cells. This medium and other defined media (Riha and Solberg, 1971; Ting and Fung, 1972; Sebald and Costilow, 1975; Goldner *et al.*, 1985) are based on an earlier formulation by Boyd *et al.* (1948) for vegetative cell growth.

2.2.2 Enterotoxin formation during sporulation

The clinical symptoms associated with foodborne illness due to *C. perfringens* are due to an intracellular enterotoxin produced in the small intestine during sporulation of ingested vegetative cells. CPE and the mature spore are released from the mother cell together. These events can be reproduced in the laboratory, and the pattern of CPE formation in laboratory media is now well established (Labbe, 1980; Stringer, 1985). The sequence of events is presumably similar during sporulation in the intestine. *In vitro*, heat-resistant spores can be detected 3–4 hours after inoculation of a sporulation medium with vegetative cells. Maximum spore levels are reached within 7–8 hours, with levels as high as $3–4 \times 10^7$/ml possible. After approximately 10 hours the toxin and mature spore are released from the sporangium, at which point the CPE is detectable in the culture fluid. The kinetics of these events generally corresponds to the time required for onset of symptoms in clinical cases. CPE is not secreted outside the cell, and is therefore not an exotoxin in the classical sense. In fact the CPE gene (*cpe*) does not encode a 5′ signal peptide usually associated with toxin secretion (Czeczulin *et al.*, 1993).

During exponential cell growth, no CPE can be detected. However about 1 hour after the end of exponential growth (T_0), the induction of CPE mRNA and CPE protein occurs (Melville *et al.*, 1994), and production of these rises sharply early in the stationary phase. The early induction of CPE mRNA supports the earlier conclusions of Duncan *et al.* (1972a), who isolated spore mutants and showed that mutants blocked at Stage 0 failed to make CPE while those blocked at later stages were able to do so. It appears that CPE gene expression is induced after Stage 0 and before Stage III (forespore formation); Stage II mutants have not been isolated.

Some strains of *C. perfringens* that produce high concentrations of enterotoxin form an inclusion body during sporulation (Duncan *et al.*, 1973; Takumi *et al.*, 1991; Figure 4.3). It is composed of enterotoxin, and possesses biological activity (Loffler and Labbe, 1983, 1985). This aggregated CPE is in addition to the soluble toxin present in the sporangium. Since harsh conditions are necessary for its dissolution, it probably plays a minor role in human illness.

Although significant accumulation of CPE in *C. perfringens* cells occurs only during the sporulation process, very low levels of CPE have been detected in vegetative cells of this organism (Granum *et al.*, 1984). While a 48-kDa CPE-related protein has been shown to exist in vegetative cells (Ryu and Labbe, 1993), Czeczulin *et al.* (1993) did in fact detect trace levels of a 35-kDa CPE-immunoreactive species in lysates of vegetative cells, though no biological activity was detected. Further, cell lysates from recombinant *cpe*-positive *Escherichia coli* were shown by quantitative Western blotting analysis to contain moderate amounts of CPE whose synthesis was driven by a clostridial promoter (Czeczulin *et al.*, 1993). Taken together, the results suggest that, while sporulating cells may not be required for *cpe* expression, sporulation does result in high level *cpe* expression.

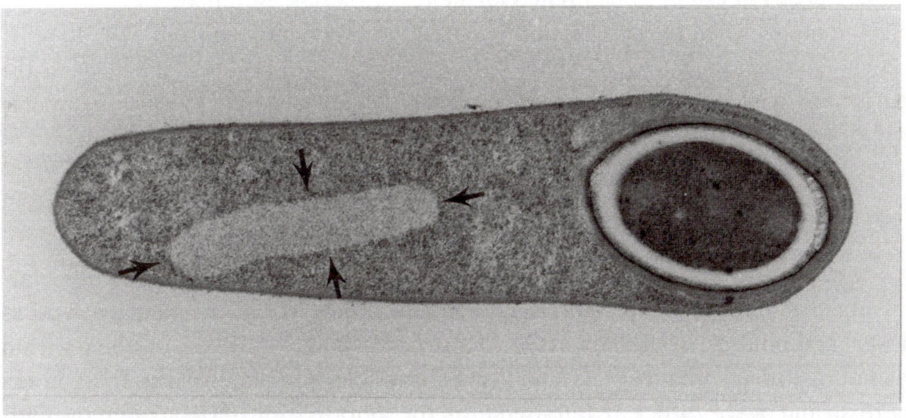

Figure 4.3 Electron micrograph of sporulating cells of ent$^+$ *C. perfringens*. Arrows indicate inclusion body; from Labbe, unpublished.

CPE can represent a significant portion (20% or more) of total cell lysate protein (Labbe, 1981; Garcia-Alvarado *et al.*, 1992b; Czeczulin *et al.*, 1993). The levels are distinctly strain- and nutrient-dependent (Labbe and Rey, 1979). Nevertheless, the function of CPE remains unknown. Earlier proposals that CPE may serve as a structural spore coat protein (Frieben and Duncan, 1973) now appear unlikely since relatively small amounts of CPE are associated with the intact spore, perhaps as a result of having been 'trapped' during spore maturation (Ryu and Labbe, 1993). In addition, many enterotoxin-negative (ent^-) strains are able to sporulate. At this point it seems that the *cpe* gene is activated by transcriptual factors which also control sporulation genes, and that the *cpe* gene itself plays no role in sporulation. Thus the functions of, and an explanation for, the high levels of CPE formed during sporulation remain elusive.

2.3 Spore heat resistance

The ability of spores of *C. perfringens* to survive heat-processing procedures contributes to its role in virtually all outbreaks of foodborne illness caused by this organism. Spores vary widely in their heat resistance (Roberts, 1968). As mentioned above, an original proposition that only the HR spore strains are involved in foodborne outbreaks has been disproven. Although spore heat-resistance determinations of outbreak strains are not routinely performed, it is reasonable to assume that the HR biotype is more frequently involved. For example, Ando *et al.* (1985) demonstrated that the D_{95} (D value is the time required to inactivate 90% of the population) values for the HR group were between 17.6 and 63 minutes, compared with between 1.3 and 2.8 minutes for the HS spore strains. Using serological tests, the authors also demonstrated that few HS spore strains (1 of 17) were enterotoxin-positive (ent^+) compared to HR spore strains (17 of 20). Similar results were obtained using biological assays (ileal loop tests; Sunagawa *et al.*, 1987). D-values for various strains of *C. perfringens* spores have been summarized elsewhere (Labbe, 1989). However it should be pointed out that one important extrinsic factor in spore heat resistance is the temperature at which sporulation occurs; increased heat resistance is associated with higher sporulation temperatures. For example, D_{95} values for spores of one ent^+ strain increased from 116 to 200 minutes when the sporulation temperature was increased from 37°C to 43°C (Garcia-Alvarado *et al.*, 1992b). Other factors affecting spore heat resistance include the menstruum in which inactivation was performed, and the medium in which the spores were grown (Weiss and Strong, 1967; Roberts, 1968; Adams, 1973; Craven, 1990).

2.4 Spore germination

In contrast to the reluctance of *C. perfringens* to undergo sporulation, its heat-activated spores germinate readily in common, complex laboratory media, as well as in meat and poultry. In aqueous systems, the process can readily be monitored on the basis of loss of absorbance over time. As in the case with heat activation, HR and HS spore strains have different requirements for germination. HR strains respond to potassium ion alone as the germinant, whereas HS strains respond to a mixture of L-alanine, inosine, and $CaCl_2$ in the presence of CO_2 (Oka *et al.*, 1983).

Another difference between HR and HS strains is their requirement for heat activation of spores (Roberts, 1968; Ando *et al.*, 1985). For example, only 0.13–3.5% of HR spore strains grew without heat activation, compared to 30–50% of HS strains. A temperature/time of 75°C/10 minutes effectively activates spores without injury. Heat activation not only promotes germination of spores, but also increases sporulation and enterotoxin formation of certain strains when applied following successive passages through a sporulation medium, presumably by selecting for a spore-forming population (Uemura *et al.*, 1973; Tsai and Riemann, 1974; Skjelkvale and Duncan, 1975; Uemura, 1978b).

2.5 Recovery of spores from thermal injury

Cassier and Sebald (1969) first reported that the addition of lysozyme to plating media could increase the recovery of heat-injured *C. perfringens* spores. This observation was confirmed and extended by others (Duncan *et al.*, 1972b; Adams, 1973, 1974). Lysozyme enhanced recovery of both injured HR and HS spore types, with recovery rates of between 1% and 8% in the presence of 1 μg of lysozyme/ml of plating medium, compared with between 0.0001% and 0.0007% in its absence. A similar though less dramatic phenomenon occurred when concentrated *C. perfringens* culture supernatant fluids (CSF) were used in place of lysozyme (Duncan *et al.*, 1972b). The CSF contains a protein (Tang and Labbe, 1987) that, like lysozyme, acts at the level of germination and replaces the native spore-lytic enzyme (Gombas and Labbe, 1985) that is inactivated by heat. A similar effect of lysozyme on heat-injured spores has since been noted for certain strains of *Clostridium botulinum* (Peck *et al.*, 1993). In addition to lysozyme being a recognized food additive, certain vegetables contain lysozyme-like activity, which can similarly promote the recovery of heated non-proteolytic *C. botulinum* spores (Stringer and Peck, 1996).

In the case of gamma irradiation, there is again a difference between HR and HS spore strains, the latter being less radiation resistant (Roberts, 1968; Gombas and Gomez, 1978). Prior irradiation sensitized the spores to subsequent heating, although the reverse was not true.

2.6 Effect of temperature on vegetative cell growth

Under proper conditions, *C. perfringens* displays the shortest generation time of any bacterial pathogen. Meat and poultry items subjected to improper holding times and temperatures easily fulfill these conditions. Spores that occur naturally in such products can survive heat treatments that provide the heat activation for germination. Generation times of 7–8 minutes in ground beef have been reported (Willardsen *et al.*, 1978, 1979; Labbe and Huang, 1995).

The adverse implications for food safety of an organism with a generation time of less than 8 minutes are noteworthy when one considers that a >100 000-fold increase in population could occur within 2 hours under optimal conditions. Not surprisingly, the maximum and minimum growth temperatures for *C. perfringens* vary depending on the strain, pH, growth medium, and presence of other microorganisms. Autoclaved ground beef is a better growth medium than the commonly-used laboratory medium, thioglycollate broth (Willardsen *et al.*, 1978). The generally accepted temperature range for growth of *C. perfringens* is 15–50°C. Growth at 15°C in vacuum-packaged, cooked ground beef occurred only after several days under anaerobic conditions, and not at all in cooked ground beef held aerobically (Juneja *et al.*, 1994a). Similar results were obtained by others using other types of meat and poultry (Craven, 1980). Unlike other foodborne pathogens, *C. perfringens* grows best at relatively high temperatures, typically 43–46°C. As mentioned above, generation times of less than 10 minutes are possible. The upper temperature limit is near 50°C. In one study none of three retail isolates grew above 49°C, yet Willardsen *et al.* (1978) reported a generation time of 30.8 minutes for an eight-strain composite of ent^+ and ent^- strains grown at 51°C, while in another study only a 1 log increase occurred within 12 hours at 51.1°C in roast beef (Brown and Twedt, 1972).

Thermal adaptation by vegetative cells of *C. perfringens* has been reported in which more than a two-fold increase in D_{57} value occurred when the growth temperature was increased from 37°C to 45°C (Roy et al., 1981). Likewise, a sublethal heat shock at 55°C for 30 minutes increased the heat tolerance of vegetative cells of an ent^+ strain two- to three-fold, and was maintained for 2 hours (Heredia *et al.*, 1997). This suggests that mild reheating of previously cooked foods (perhaps containing low numbers of surviving cells) could result in growth by this organism. This 'acquired thermotolerance' may also explain the 'Phoenix Phenomenon' reported by Shoemaker and Pierson (1976). In that case, the number of vegetative cells, when exposed to a temperature of 52°C, fell during the first few hours but subsequently returned to initial levels. Craven (1980) has reviewed the earlier body of literature devoted to the effect of temperature on the growth of *C. perfringens* in meat and poultry.

2.7 Growth during cooling

Juneja and co-workers developed models to predict the germination, outgrowth and lag (GOL), and exponential growth rates (EGR) of *C. perfringens* from spores at temperatures applicable to the cooling of cooked meat (Juneja *et al.*, 1999) and chicken (Juneja and Marks, 2002). *C. perfringens* growth from spores was not observed at a temperature of < 15°C or > 51°C, for up to 3 weeks. It was found that the use of the logistic function provided a better prediction of relative growth than the use of the Gompertz function. From the parameters of the Gompertz or logistic function, the growth characteristics, germination, out growth, and lag times and exponential growth rate were calculated. These growth characteristics were subsequently described by Ratkowsky functions, using temperature as the independent variable. The standard errors and confidence intervals were computed on the predictions of relative growth for a given temperature by applying multivariate statistical procedures. Closed-form equations developed that allow prediction of growth for a general cooling scenario (Juneja *et al.*, 2001). The predictive models should aid in evaluating the safety of cooked product after cooling, and thus with the disposition of products subject to cooling deviations.

Multiplication of germinated spores or vegetative cells may occur during refrigeration of cooked foods. During refrigeration of large cuts of cooked meat, the internal temperatures can pass through the growth range of *C. perfringens* for sufficiently long to permit significant growth. For example during refrigeration of cooked, deboned turkey meat or 3.8–9.1 liters (1.0–2.4 gallons) of cooked turkey stock, temperatures were within the growth range of *C. perfringens* for 7–9 hours and 10 hours respectively (Bryan *et al.*, 1971; Bryan and McKinley, 1974). This highlights the need to reduce the size of meat portions for cooking, or to reduce large volumes of meat and gravies after cooking.

The ability of *C. perfringens* to grow over a wide, relatively high temperature range has prompted additional recent studies on the kinetics of growth during cooling. In the case of *C. perfringens*, negligible growth of a three-strain composite of spores occurred in cooked beef cooled from 54.4°C to 7.2°C in 15 hours or less, although a several-log increase did occur if the cooling period was extended to 18 hours (Juneja *et al.*, 1994b). Similarly, no growth from spores occurred within 150 minutes in cooked, ground beef inoculated at 60°C and cooled to 15°C at a falling rate of 15°C/h (Shigehisa *et al.* 1985). Also, refrigerated, aerobically- or anaerobically-packaged cooked, ground beef containing 10^3 *C. perfringens*/g and temperature-abused at 28°C for 5 hours did not support the growth of *C. perfringens* (Juneja *et al.*, 1994a). It appears that short periods of temperature abuse of small portions of refrigerated meat containing *C. perfringens* do not necessarily result in a hazardous level of cells.

2.8 Effect of low temperature on vegetative cells

A number of reports highlight the sensitivity of *C. perfringens* vegetative cells to low temperature, and current laboratory protocols specify holding conditions for suspect foods in outbreaks if such foods are to be held before analysis (Labbe and Harmon, 1992). For example, cell numbers fell by 80–90% within 50 minutes when held in 0.1% peptone at 4°C (Traci and Duncan, 1974). Yet Juneja *et al.* (1994c) detected no loss in viability within a 6-day storage period for two ent^+ strains held at 4°C in cooked ground turkey. Further work is needed in this area to determine whether food provides a protective effect as compared to laboratory media.

In the case of frozen storage, more than 90% of cells were inactivated following storage at –17.7°C for 30 days in a starch paste, and even more in buffer (Strong *et al.*, 1966). Other studies of *C. perfringens* in food products held at –17.7°C for 48 hours, –23°C for 14 days, or –20°C for 17 days resulted in survival rates of 0.1–6% (Canada *et al.*, 1964; Traci and Duncan, 1974; Fruin and Babel, 1977). Not surprisingly, spores were significantly more resistant in these challenge studies.

The phenomenon of cold shock lethality also highlights the inadvisability of using cold buffer in washing vegetative cell suspensions in experimental work involving *C. perfringens* if viability is to be maintained.

2.9 Other factors affecting growth and survival

Although an anaerobe, *C. perfringens* readily grows in tubes of complex media exposed to air. Once initiated, growth proceeds rapidly. The organism will grow well at an oxidation–reduction potential (E_h) of +200 mV with an upper limit of near +300 mV (Pearson and Walker, 1976). The larger the inoculum and the more active the organism's metabolic state, the higher the limiting E_h will be. Intense reducing conditions are established by the organism during logarithmic growth, reaching as low as –400 mV (Tabatabi and Walker, 1970).

C. perfringens is not especially tolerant of low water activity. Limiting levels of 0.97 with sucrose as the humectant, 0.93 with glycerol (Kang *et al.*, 1969), and 0.96 with glucose (Strong *et al.*, 1970) have been reported.

Like most microorganisms, *C. perfringens* initiate growth most readily under neutral pH conditions, although excellent growth occurs at values between 6.0 and 7.0, a pH range similar to that of most meat and poultry products. Growth is severely limited at pH 5.0 and 8.3 (Smith, 1972). A pH of 5.0 is easily reached in carbohydrate-rich media. Cells will slowly die off at such low pH levels. Likewise, in cases of foodborne illness some inactivation of cells occurs during stomach passage due to the low pH levels encountered. As with other environmental insults, the resistance depends in part on the metabolic state of the cells – log phase cells being more susceptible than stationary phase cells (Hauschild *et al.*, 1967; Fisher *et al.*, 1970; Sutton and Hobbs, 1971; Wrigley *et al.*, 1995). In foodborne outbreaks, cells are ingested along with large volumes of food, which provides some protective, buffering effect against stomach acidity. It has been reported that acid exposure (pH 2.0, 30 minutes) enhances enterotoxin formation (Wrigley *et al.*, 1995), although others have been unable to produce this effect (De Jong *et al.*, 2002).

The association of *C. perfringens* with meat as a vehicle prompted much work on the effects of curing salts and other food additives. A level of 7.8% NaCl was required to prevent the growth of most strains, although some inhibition occurred at levels of 5–6% NaCl (Gough and Alford, 1965; Mead, 1969a; Roberts and Derrick, 1978; Craven, 1980). No growth of a mixture of vegetative cell strains occurred in the presence of 400 μg/ml of sodium nitrite at 20 or 30°C (Gibson and Roberts, 1986). As the concentration of sodium chloride was increased from 3% to 6%, the level of nitrite required for inhibition of growth decreased from 300 to 25 μg/ml (Roberts and Derrick, 1978). Gibson and Roberts (1986) determined the conditions required to inhibit growth of mixed strains of *C. perfringens* in laboratory media. These were (at pH 6.2) storage at 15°C, 1% NaCl, and 50 μg/ml of sodium nitrite. At 20°C, higher concentrations of at least one curing agent were required for inhibition. They concluded that growth of this organism was prevented by levels of curing salts used commercially, providing that the pH was 6.2 or below. This dependence on pH suggests that the nitrous acid was the effective agent, an observation reported by others (Labbe and Duncan, 1970; Riha and Solberg, 1975a). Levels that are inhibitory to vegetative cells will not necessarily prevent spore germination, but act at the level of outgrowth of spores (Labbe and Duncan, 1970).

Heating in laboratory media enhances the inhibitory effect of sodium nitrite, a phenomenon termed the 'Perigo effect' (Perigo and Roberts, 1968; Riha and Solberg, 1975b). For example, only one-twentieth to one-fortieth of the amount of nitrite required to inhibit seven strains of *C. perfringens* was needed if the nitrite was autoclaved with the test medium. The effect was also noted in a heated pork–nitrite system (Ashworth *et al.*, 1973).

Klindworth *et al.* (1979) reported that butylated hydroxyanisole (BHA) (between 50 and 150 ppm) inhibited growth of *C. perfringens* in thioglycollate broth. Both BHA and sorbic acid had a synergistic inhibitory effect when used in conjunction with nitrite or esters of para-hydroxybenzoic acid (Robach *et al.*, 1978; Klindworth *et al.*, 1979).

Finally, it should be noted that cured meat products are rarely vehicles for outbreaks of *C. perfringens* food poisoning. In the period 1978–1997, only three such outbreaks (each with a ham vehicle) were reported to the US Centers for Disease Control and Prevention (CDC, 1990, 1996 and 2000). The heat treatment applied to such products, the presence of curing agents, the low initial spore load, and the relative sensitivity of the surviving (and perhaps injured) spore population to the curing agents contribute to the safety of these products.

2.10 Molecular genetics of *C. perfringens*

The development of genetic methods for the analysis of *C. perfringens* has accelerated dramatically in recent years due to the advent of techniques such as pulsed field gel electrophoresis (PFGE) and the availability of restriction enzymes with rare cutter sites or G+C-rich recognition sequences. The physical map of the 3.6-Mb chromosome of the type A strain CPN50 now comprises about 100 markers (Katayama *et al.*, 1995; Cole and Canard, 1997). Several toxins produced by other type strains have also been cloned and sequenced, including α, ε, ι and κ, as well as sialidase and hyaluronidase (Rood and Cole, 1991; Rood and Lyristis, 1995). The complete *cpe* gene from two ent$^+$ strains has also been cloned (Czeczulin *et al.*, 1993; Cornillot *et al.*, 1995). The availability of *E. coli–C. perfringens* shuttle plasmids and vectors (Roberts *et al.*, 1988; Sloan *et al.*, 1992; Bannum and Rood, 1993; Rood, 1997), reporter systems (Phillips-Jones, 1993; Matsushita *et al.*, 1994; Bullifent *et al.*, 1995), methods for electroporation-induced transformation (Scott and Rood, 1989; Allen and Blaschek, 1990; Chen *et al.*, 1996) and other genetic methods facilitated work in this area that has shown that CPE is regulated at the transcriptional level. Although only a minority of *C. perfringens* isolates possess the *cpe* gene, it appears that most if not all naturally *cpe*-negative isolates produce factors involved in transcriptional regulation of CPE expression (Czeczulin *et al.*, 1996).

The association of elevated CPE levels with sporulation suggests that growth phase-specific *cpe* regulatory factors may be involved in CPE expression. The *cpe* gene was found at the same chromosomal locus in three strains associated with food poisoning, whereas it was plasmid-bound in 20 strains from domestic livestock (Cornillot *et al.*, 1995). Similarly, Collie and McClane (1998) reported that the *cpe* gene was located on an extrachromosomal element in the case of strains isolated from non-foodborne gastrointestinal diseases, and confirmed its chromosomal location in all food poisoning isolates examined. Why food poisoning would be caused only by those isolates carrying chromosomal *cpe* genes is unknown, but Sarker *et al.* (2000) have shown that the spores and vegetative cells of chromosomal *cpe* isolates have a higher degree of heat resistance than isolates having *cpe* on a plasmid. The presence of the *cpe* gene on a 6.3-kb transposon has been demonstrated (Brynestad *et al.*, 1997). This could explain the low prevalence of the *cpe* gene in natural isolates. The presence of *cpe*-positive and -negative isolates with identical or nearly identical PFGE patterns in a single outbreak also suggests the involvement of a moveable genetic element for the *cpe* gene (Ridell *et al.*, 1998). The existence of two types of promoter regions was reported by Melville *et al.* (1994). The cloning and sequencing of the entire CPE structural gene was reported by Katayama *et al.* (1995), and the complete genome sequence of *C. perfringens* was published in 2002 (Shimizu *et al.*, 2002). The molecular biology of the *C. perfringens* genome in general, and the regulation of the CPE production in particular, has been reviewed elsewhere (Cole and Canard, 1997; Melville *et al.*, 1997; McClane, 2001).

3 Nature of the disease

3.1 Disease states associated with *C. perfringens*

Emerging epidemiological evidence incriminates CPE in human and veterinary diseases apart from foodborne illness. Table 4.1 summarizes these and provides related references. The focus here is on food poisoning due to CPE produced by type A strains, although CPE has also been detected in types C and D (Skjelkvale and Duncan, 1975; Uemura and Skjelkvale, 1976; Markovic *et al.*, 1993). The essential role of CPE in foodborne illness was shown unequivocally by elegant experiments reported by Sarker *et al.* (1999). Through the construction of *cpe*-knockout mutants, they demonstrated inability of such mutants to induce either intestinal fluid accumulation or histopathological damage while full virulence was restored when the mutant was complemented with a shuttle plasmid carrying the wild type *cpe* gene.

3.2 Symptoms associated with foodborne illness

The classic symptoms of *C. perfringens* type A food poisoning are diarrhea with lower abdominal cramps. Vomiting is not common, and fever is rare. Symptoms typically occur 8–24 hours after ingestion of temperature-abused foods containing large numbers of vegetative cells of the organism. Mortality is low, and such cases have been associated with elderly patients. Symptoms subside within 1–2 days, although cramps can continue a little longer. The brevity of the events is probably due to the normal turnover of intestinal cells to which the toxin is bound, as well as to removal of unbound CPE by the diarrheal process which it itself induced. Antibiotic therapy

Table 4.1 Veterinary and human diseases associated with *C. perfringens* enterotoxin, apart from foodborne illness[a]

Illness	References
Antibiotic-associated diarrhea of humans	Borriello *et al.*, 1984; Borriello, 1985; Hatheway, 1990
Chronic non-foodborne diarrhea	McClane *et al.*, 1988a
Acute sporadic (non-foodborne) human diarrhea	Larson and Borriello, 1988; Brett, 1994; Mpamugo *et al.*, 1995
Sudden infant death syndrome	Lindsay *et al.*, 1993; Mach and Lindsay, 1994; Lindsay, 1996
Animal diarrhea	Collins *et al.*, 1989; Citino, 1995

[a] Modified from McClane, 1994.

is not recommended because the symptoms are self-limited. Dehydration, especially in the elderly, is the most important concern.

There have been isolated reports of foodborne illness caused by *C. perfringens* with short incubation periods (< 8 hours; Robinson and Messer, 1969; Sanders and Hutchenson, 1974; Roach and Sienko, 1992). This suggests a possible role for preformed toxin, or perhaps ingestion of sporulating cells. The ability of this organism to produce detectable CPE in foods has been demonstrated in the laboratory (Mead, 1969b; Dework, 1972; Naik and Duncan, 1977a; Craven *et al.*, 1981). However, the presumed unpalatable nature of such foods argues against outbreaks with such short incubation periods.

Surprisingly, most serum samples taken from Brazilians, Americans and Koreans showed antibody against CPE (Uemura *et al.*, 1974; S. Ryu, personal communication), and 5 of 10 volunteers given purified CPE orally developed anti-CPE titers (Skjelkvale and Uemura, 1977a). However, circulating antibody does not neutralize CPE in the intestinal lumen (Niilo, 1971). More recently it has been shown that CPE, like staphylococcal enterotoxin A, is a superantigen, able to stimulate large numbers of T cells (Bowness *et al.*, 1992).

A more severe type of foodborne illness, caused by *C. perfringens* type C, occurred in northern Germany after World War II. Described as enteritis necroticans, it is an often fatal illness associated with hemorrhagic necrosis of the jejunum caused by β toxin. The same syndrome was later associated with natives of the highlands of Papua New Guinea (Walker, 1985), where it is termed 'pig-bel' due to its association with pig feasting. About 70% of the natives were found to be carriers of the type C strain (Lawrence *et al.*, 1979). Malnutrition is an important risk factor, leading to levels of proteases too low to destroy the toxin. In addition, food containing protease inhibitors, such as trypsin inhibitors, may prevent the breakdown of the β toxin in the gut (Lawrence, 1988). More recently, enteritis necroticans due to *C. perfringens* type C was described among refugee children at an evacuation site in Thailand, although no association with any food was established (Johnson *et al.*, 1987).

3.3 Volunteer and animal studies

Early human feeding studies involved the administration of whole cultures in food vehicles. In one study, 21 of 24 volunteers became ill when fed cultures in soup (Dische and Elek, 1957). In later studies, cells were grown for 3 hours at 46°C in beef stew before administration. Of volunteers given strains previously shown to produce enterotoxin in the rabbit ileum, 61% had diarrhea, whereas none given rabbit-negative strains became ill (Strong *et al.*, 1971). When purified CPE was administered, one of two volunteers given 8 mg and all of five volunteers given 10 mg developed classical symptoms of food poisoning (Skjelkvale and Uemura, 1977b).

A number of animal species have also been used to demonstrate that CPE is the factor responsible for illness or in mode-of-action studies. These include pigs (Collins *et al.*, 1989), lambs (Hauschild *et al.*, 1970; Niilo, 1971), sheep (Niilo, 1971, 1972; Niilo *et al.*, 1971), calves (Niilo, 1973a, 1973b), chickens (Niilo, 1974, 1976), dogs (Bartlett *et al.*, 1972), rats (McDonel and Asano, 1975; McDonel *et al.*, 1978), rabbits

(Duncan and Strong, 1969, 1971; Hauschild *et al.*, 1971a; Niilo 1971; McDonel and Duncan, 1975, 1977; McDonel *et al.*, 1978; Stringer *et al.*, 1982; Sherman *et al.*, 1994), guinea pigs (Niilo, 1971), mice (Yamamoto *et al.*, 1979; Miwantani *et al.*, 1978), and monkeys (Duncan and Strong, 1971; Hauschild *et al.*, 1971b; Uemura *et al.*, 1975).

3.4 Mode of action of *C. perfringens* enterotoxin

3.4.1 Histopathological effects

An enterotoxin responsible for food poisoning was demonstrated by Duncan and Strong (1969) by its ability to induce fluid accumulation in ligated ileal loops of rabbits. Exposure of the rat ileum to CPE demonstrated a reversal in transport from absorption in controls to secretion of fluid, sodium and chloride. Glucose absorption was inhibited, while transport of potassium and bicarbonate was unaffected (McDonel, 1974).

A further understanding of effects at the cellular and molecular level came about with the availability of purified CPE and the use of cultured cell lines. Histological studies of CPE-treated ileal loops of rabbits revealed desquamation of epithelial cells at villous tips (McDonel *et al.*, 1978), indicating that the brush border microvillous membrane is the initial site of action of CPE. Intestinal damage can occur rapidly, within 15–30 minutes of toxin treatment (Sherman *et al.*, 1994). The normal, folded configuration of the brush border is lost, and large quantities of membrane and cytoplasm are lost to the lumen. The toxin also causes 'bleb' formation in rabbit small intestinal cells (Figure 4.4) and in cultured Vero (African green monkey kidney) cells (McClane and McDonel, 1979). The rabbit terminal ileum was most notably affected

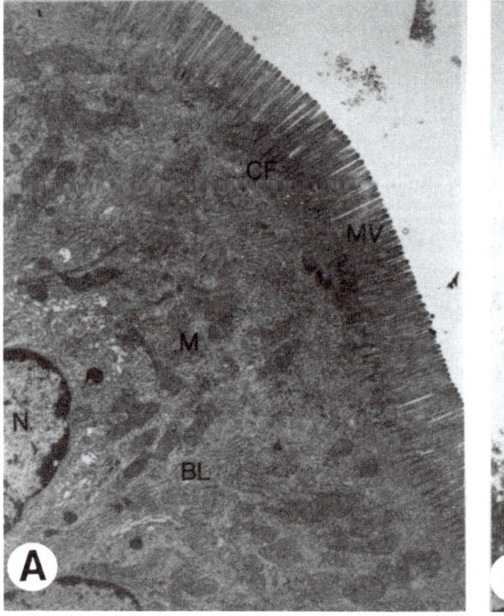

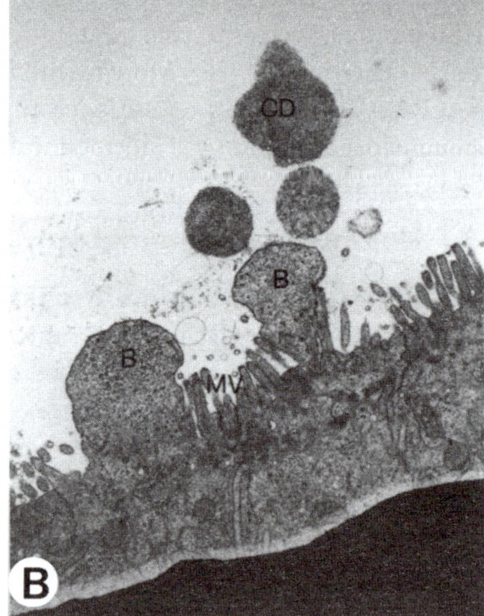

Figure 4.4 Effect of *C. perfringens* enterotoxin (CPE) on rat ileum: (A) control showing normal villus cell morphology – MV, microvilli; (B) treated with CPE for 90 min. showing bleb formation – B, bleb; CD, cytoplasmic bleb. Reprinted from McDonel (1979) with permission of the American Society for Clinical Nutrition.

by CPE; the colon was surprisingly unaffected (McDonel and Duncan, 1977), even though CPE binds well to rabbit colonic cells (McDonel, 1980).

3.4.2 Molecular and cellular effects

Work describing the tissue damage induced by CPE was followed by a description of its action at the cellular level. The assumption is that CPE's cytotoxic effects on mammalian cells cause the tissue damage associated with CPE treatment in rabbit small intestine. While intestinal, HeLa and liver cell lines have been used in such studies (McDonel, 1980; Skjelkvale *et al.*, 1980; Jarmund and Telle, 1982; Matsuda *et al.*, 1986), Vero cells in particular have provided a detailed understanding of CPE action. By contrast, Chinese hamster ovary (CHO) cells are insensitive to CPE (Wieckowski *et al.*, 1994). Structure and function mapping studies of CPE have identified the N-terminal region as involved with insertion and cytotoxicity and the C-terminal end with binding (Figure 4.5; McClane, 1997).

McClane and co-workers established the sequence of at least four events in CPE action (Figure 4.6). An initial calcium-independent step involves the binding of CPE to a 50-kDa membrane protein (Wnek and McClane, 1983; Sugii and Horiguchi, 1988). More recently, the members of the claudin family have been identified as functional CPE receptors (Katahira *et al.*, 1997; Furuse *et al.*, 1998). The claudin family includes several transmembrane proteins, which play important structural roles in epithelial tight junctions. Unlike *E. coli* labile toxin and *Vibrio cholera* enterotoxin, no role has been identified for ganglioside GM_1 from rabbit brush border membranes in the binding of CPE (Wnek and McClane, 1986). Binding is followed by a small (90-kDa) then large (160-kDa) complex formation. The latter causes the membrane to lose its normal permeability properties, initially to molecules of molecular weight < 200, such as ions and amino acids (Matsuda *et al.*, 1986; McClane, 1994) – perhaps by the formation of pore structures, although the precise mechanism remains unclear. The inhibition of DNA, RNA and protein synthesis appears to be a secondary consequence. Subsequent calcium-dependent events include permeability changes for larger molecules of up to 3–5 kDa, and morphological damage such as bleb formation. The change in permeability leads to the loss of macromolecular precursors and of viability, and presumably the

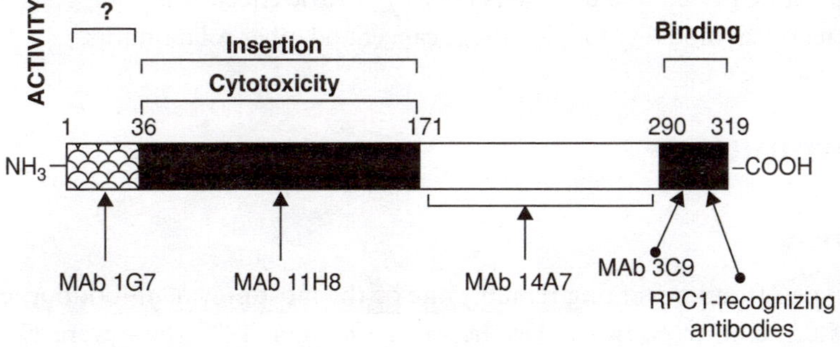

Figure 4.5 Map of CPE functional regions involved in insertion, cytotoxicity and binding. Also shown (arrows) are sequences required for presentation of epitopes recognized by various monoclonal antibodies (MAb). Reprinted from McClane (1997) with permission of the American Society for Microbiology Press.

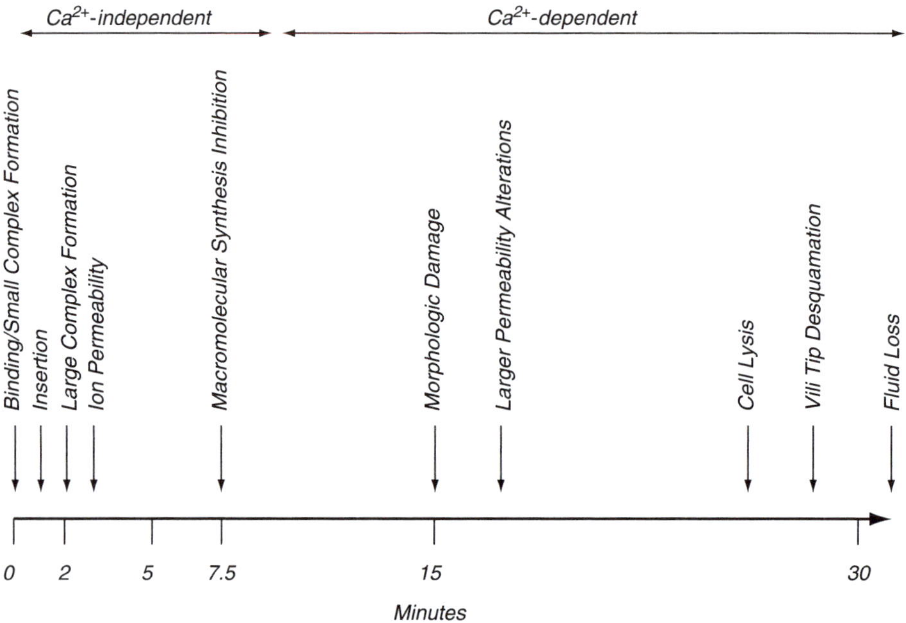

Figure 4.6 Time line for known events in CPE action, showing Ca^{2+}-dependent and -independent actions. Reprinted from McClane (1997) with permission of the American Society for Microbiology Press.

morphological damage seen in CPE-treated intestinal epithelial cells including, as noted above, intestinal cell lysis and desquamation of intestinal cells from villous tips. The net secretion of fluids and electrolytes would correspond to the clinical symptoms associated with *C. perfringens* foodborne illness. The events associated with CPE action on mammalian cells have been described in detail by McClane (2001).

A potential medical application of CPE for treating pancreatic cancer cells was recently suggested (Michel *et al.*, 2001). Such cells over-express claudin-4, a class of transmembrane proteins located within the tight junctions of cells. As mentioned above, a CPE receptor on intestinal cells has been identified as a member of the claudin family (Katahira *et al.*, 1997). *In vitro* and *in vivo* treatment of pancreatic cancer cell lines and tissues with CPE led to a dose-dependent cytotoxic effect, which suggests a promising new treatment modality for pancreatic cancer and other solid tumors.

4 Epidemiology

4.1 Incidence

C. perfringens type A food poisoning remains one of the most prevalent foodborne diseases in Western countries. In the US, between 1983 and 1997 there were 121 confirmed outbreaks, with 9316 cases and a total of 13 deaths (Bean and Griffin, 1990; Bean *et al.*, 1997; CDC, 1990, 2000), placing *C. perfringens* third behind *Salmonella* and *Staphylococcus aureus* in number of confirmed outbreaks and second in terms of number of cases (of bacterial etiology). However, the great majority of cases go unreported. Todd has estimated that there are actually 250 000–652 000 cases of *C. perfringens* type A food poisoning each year in the US alone, with an annual cost of $123 million (Todd, 1989a, 1989b; Mead *et al.*, 1999). The number of cases per outbreak is typically larger than that for *S. aureus*. In the US, for the period 1973–1992 the average number of cases/outbreak for *C. perfringens* was about 75, compared to 45 for *S. aureus* (CDC, 1990, 1996, 2000).

As mentioned above, the incidence is widely under-reported due to the perceived mild nature of the illness. In addition, there are differences among countries in their surveillance arrangements. For example, in England and Wales, where it is a notifiable disease, there were an average of 53 outbreaks (1156 cases) per year for the period 1983–1993 (R. Gilbert, personal communication) – almost five times more than in the US where, as with most foodborne diseases, *C. perfringens* type A food poisoning is not a notifiable illness. A clearer picture of the incidence of *C. perfringens* foodborne illness, at least in the case of Europe, should emerge with the undertaking of the WHO surveillance program (World Health Organization, 1995).

Food service establishments such as restaurants, hospitals, factories, prisons, schools, and caterers are the most likely sources of the illness, without any seasonal prevalence. In such institutions food is often prepared well in advance and held for later serving, allowing the possible growth in temperature-abused foods. This is one reason for the large case-to-outbreak rate noted above. Another reason for this is that public health officials are unaware of (or not motivated to investigate) outbreaks, due to the relatively mild symptoms in patients, who may not seek medical attention.

4.2 Food vehicles

Meat and poultry have long been recognized as primary vehicles for *C. perfringens* type A food poisoning. The most recent US data also include several outbreaks in which the vehicles were apparently Mexican foods (CDC, 1990, 1996, 2000). These products are known to contain cells or spores of *C. perfringens* (see below). The latter can survive a cooking procedure and then germinate and grow during slow cooling, with rapid growth possible between 37 and 50°C. Although commercially prepared foods are rarely involved in *C. perfringens* foodborne illness, there was a recent outbreak in a British hospital in which vacuum-packaged pork was implicated. The item had been packaged at a commercial, meat-packaging facility and improperly cooled. The increasing popularity of vacuum-packaged meat products suggests this will not be the last incident in which this type of product is the vehicle.

4.3 Reservoirs

4.3.1 Feces

C. perfringens has been found in the intestinal contents of virtually every animal species examined, although the level in normal human feces, usually 10^3–10^4/g, is relatively small compared to other strict anaerobes, such as *Bacteriodes*. Up to 39 different *Clostridium* species have been isolated from human feces, with

C. perfringens, usually enterotoxin-negative, being the most common isolate (George and Finegold, 1985). Normal stools may contain more than one strain (Klotz, 1965; Hall and Hauser, 1966). In foodborne illness, the outbreak strains are ingested and present in feces at higher levels than are resident strains (Sutton and Hobbs, 1968), an observation adopted as a criterion for a foodborne outbreak (see below). However certain population groups have since been found to carry relatively high fecal spore levels of *C. perfringens*, of the order of 10^7–10^9/g. These were healthy adults in Japan, and the elderly in Japan and in the UK (Nakagawa and Nishida, 1969; Akama and Otani, 1970; Stringer, 1985). Such high counts have also been found in patients with infectious or chronic diarrhea not associated with food poisoning (Jackson *et al.*, 1986; Larson and Borriello, 1988). George and Feingold (1985) have presented a detailed review of the clostridial flora in the human gastrointestinal tract.

4.3.2 Environment

C. perfringens type A is part of the microflora of the soil. It has been detected at levels of several thousand per gram in virtually all soil samples examined (Taylor and Gordeon, 1940; Smith and Gardner, 1949). Types B, C, D, and E are obligate parasites, mostly of domestic animals, and do not persist in soils, perhaps because these types are unable to compete with other microflora (Smith and Williams, 1984). The organism can also be found in low numbers in marine sediments, probably representing recent pollution (Matches *et al.*, 1974; Saito, 1990). Of samples of hospital kitchen dust, 90% also contained *C. perfringens* (McKillop, 1959). The potential of flies as a vector was noted by Hobbs *et al.* (1953), who isolated *C. perfringens* from blowflies at several locations associated with food preparation.

The association of *C. perfringens* with the intestinal tract has led to its use as an indicator of fecal pollution of water (Olivieri, 1982; Sorensen *et al.*, 1989). It has similarly been used as an indicator of movement of sewage sludge on the ocean floor, where extreme conditions preclude the use of traditional indicators of water pollution such as coliforms and fecal streptococci (Hill *et al.*, 1993). Membrane filtration protocols have been developed for analysis of water for this organism (Armon and Payment, 1988; Bezirtzoglou *et al.*, 1996).

4.3.3 Food

The older literature contains many surveys of the incidence of *C. perfringens* in raw and processed foods. In the case of meat and poultry, the median number of samples containing *C. perfringens* is near 50%, with levels < 10^2/g (Hall and Angelotti, 1965; Roberts, 1972; Foster *et al.*, 1977; Smart *et al.*, 1979; Bauer *et al.*, 1981; Lin and Labbe, 2003). Variations in reported levels are due in part to the fact that early surveys focused on HR spore strains (Hall and Angelotti, 1965), or did not distinguish between the spore and vegetative-cell states (Ladiges *et al.*, 1974; Foster *et al.*, 1977; Ali and Fung, 1991). Not surprisingly, the incidence of *C. perfringens* is less in processed foods including meats (Strong *et al.*, 1963; Hall and Angelotti, 1965; Nakamura and Kelley, 1968). Although not commonly reported as vehicles of foodborne illness due to *C. perfringens*, the body surface and alimentary canal of most fish of several species harbor *C. perfringens* (Hobbs *et al.*, 1965; Taniguti and Zenitani, 1969). Of particular concern are herbs and spices. The latter are well known for harboring bacterial spores, including *C. perfringens*, which have been detected at levels of 10^1–10^4/g (Powers *et al.*, 1975; Erickson and Deibel, 1978; DeBoer *et al.*, 1985; Neut *et al.*, 1985; Smith, 1985). In food service settings, such products can be added to large amounts of cooked foods.

Reviews on the incidence of *C. perfringens* in foods were largely conducted in the 1960s and 1970s, and have been summarized elsewhere (Bryan, 1969; Walker, 1975; Stringer, 1985). Unfortunately, such early surveys did not identify the enterotoxin-producing potential of isolates. More recent surveys indicate that *C. perfringens* enterotoxin-positive (CPE-positive) strains are uncommon in retail foods. Use of genetic probes for the enterotoxin gene and immunoassays for the enterotoxin (CPE) reveal an incidence ranging from 0 to 17% from a variety of products (Saito, 1990; Daube *et al.*, 1996; Miwa *et al.*, 1998; Lin and Labbe, 2003; Wen and McCabe, 2004). In the case of healthy animals, the incidence of cpe^+ strains is somewhat higher, ranging from 6% to 22% (Van Damme-Jongsten *et al.*, 1989; Tschirdewahn *et al.*, 1991).

4.4 Control and prevention

It is not possible to prevent carriers from handling food, because most people harbor *C. perfringens* in their intestinal tracts. Since the organism is present in animals, it is not surprising that *C. perfringens* is found in raw meat and poultry. The spores will also survive indefinitely in dust and in environmental nooks. Cooking at temperatures not exceeding 100°C will allow the survival of spores of *C. perfringens*. The cooking process drives off oxygen, creating near anaerobic conditions in foods such as rolls of cooked meat, pies, stews and gravies, and in poultry carcasses. Therefore, prevention must be concerned not only with destruction but also with the multiplication of vegetative cells in cooked foods, and this is the most practical way of preventing *C. perfringens* foodborne illness. Improper cooling after cooking has been identified as a principal cause of outbreaks in the US and UK (Bryan, 1969; Roberts, 1982). Government agencies and public health officials have issued guidelines specifying the cooling parameters for food. For example the US Food and Drug Administration recommends that all foods be cooled from 60°C to 5°C in 6 hours or less (US FDA, 1997). Another contributing factor is inadequate reheating of cooked, chilled foods. Proper reheating to an internal temperature of 75°C before serving is important if the organism has had an opportunity to multiply because of previous temperature abuse. Public health agencies recommend holding cooked foods at or below 4°C (40°F) or above 60°C (140°F) (US FDA, 1997). Such temperatures will prevent the growth of *C. perfringens* in hazardous foods. Other recommended procedures include reducing the size of large portions of meat to hasten cooling, and ensuring proper cooling capacity of refrigeration facilities. In the case of the food service industry, certain trade organizations offer short courses that describe recommended food-handling procedures in detail.

5 *C. perfringens* enterotoxin

5.1 Biochemical characteristics

CPE was initially purified independently by Hauschild and Hilsheimer (1971) and Stark and Duncan (1972). Its amino acid sequence was subsequently determined (Van Damme-Jongsten *et al.*, 1989; Czeczulin *et al.*, 1993; Cornillot *et al.*, 1995); it is composed of a single polypeptide with a highly conserved sequence of 319 amino acids, an M_r of 35 317, and a pI of pH 4.3 (McClane, 2001). CPE has a unique amino acid sequence except for some limited homology with a non-neurotoxic protein made by *C. botulinum* (McClane, 1996). Analysis of deletion mutants indicates that the minimum size of CPE fragment that retains cytotoxicity comprises amino acids 45–319 (Kokai-kun and McClane, 1997a, 1997b). The toxin aggregates from solution at concentrations above 0.5 mg/ml at 4°C, perhaps due to its high hydrophobic amino acid content (43%) (Granum and Stewart, 1993).

Activation of biological activity occurs following exposure to trypsin and chymotrypsin, which results in loss of 25 and 36 amino acids from the N-terminus respectively, thus suggesting the activation of CPE in the small intestine during foodborne illness (Granum *et al.*, 1981; Granum and Richardson, 1991). CPE is easily inactivated by heat, with a 90% loss of biological activity within 1 minute at 60°C (Granum and Skjelkvale, 1977). However, serological activity was lost more slowly, 20% remaining after heating at 60°C for 20 minutes. About 12% of this loss could be restored by treatment with urea for 1 hour (Naik and Duncan, 1978). A restoration of biological activity, during holding at 20°C, occurred following heating at 55°C for 1 minute (which caused a 70% loss of activity). Surprisingly, CPE was less heat-stable in chicken gravy than in phosphate buffer (both pH 6.1) (Bradshaw *et al.*, 1982). CPE stored at 37°C for 7 days or at –20°C for 28 days lost biological but not serological activity (McDonel and McClane, 1981). However, neither biological nor serological activity was affected by storage at 4°C for several weeks in 0.02 M phosphate buffer, pH 6.8 (Granum *et al.*, 1981). Of the original activity of CPE, 5–15% was lost each time it was thawed, due to protein precipitation (denaturation). Lyophilization is the preferred method of preparation for long-term storage.

Maximum UV absorption by CPE occurs at 276 nm, at which the extinction coefficient $E_{1\,cm}^{1\%}$ is 1.33 mg^{-1} cm^2 (Granum and Whitaker, 1980). In 0.02-M phosphate buffer, pH 6–8, its maximum solubility at 25°C is 3.94 mg/ml.

Demonstration of a single band following sodium dodecyl sulfate-polyacrylamide gel electrophoresis (SDS-PAGE) is routinely used to verify that a protein has been purified to homogeneity. In the case of CPE, it has been repeatedly shown that SDS causes aggregation of the toxin, resulting in multiple, higher molecular-weight bands (Enders and Duncan, 1977; Granum, 1982; Salinovich *et al.*, 1982; Reynolds *et al.*, 1986; Horiguchi *et al.*, 1987; McClane *et al.*, 1988b; Czeczulin *et al*, 1993; Ryu and Labbe, 1993). This anomalous aggregation does not occur at CPE concentrations below 15 μg/ml or under non-denaturing conditions, i.e. in the absence of SDS (Ryu and Labbe, 1989).

5.2 Purification

A number of purification protocols for CPE have been described. That using ammonium sulfate precipitation of sporulating cell extracts followed by gel filtration is relatively simple and has been widely adopted (Granum and Whitaker, 1980; McDonel and McClane, 1988). The soluble, intracellular protein concentration of certain strains can comprise 20% or more of CPE (Labbe and Rey, 1979). The use of high CPE-producing strains, together with growth at 43°C, permits CPE purification to homogeneity within 48 hours (Heredia *et al.*, 1994).

Several workers have identified a minor band in following PAGE (non-denaturing) of purified CPE (Sakaguchi *et al.*, 1973; Enders and Duncan, 1976; Bartholomew and Stringer, 1983). This band was shown to possess biological and serological activity similar to but weaker than CPE. The MW but not the charge was the same as that of CPE. The presence of this band may have been due to endogenous protease activity (Loffler and Labbe, 1983; Park and Labbe, 1991b). The use of the protease inhibitors EDTA and phenylmethylsulfonylfluoride (PMSF) and low temperatures effectively prevented such artifacts (Park and Labbe, 1990; Heredia *et al.*, 1994).

6 Detection of the organism and enterotoxin

6.1 Criteria for outbreaks

A number of criteria have been proposed for establishing an outbreak of *C. perfringens* type A food poisoning. These include: (a) > 10^6 spores/g feces from ill individuals; (b) > 10^5 cells/g in incriminated food ; (c) the presence of the same serotype of *C. perfringens* in both contaminated food and feces; (d) the presence of the same serotype in all ill individuals in an outbreak; or (e) detection of enterotoxin in feces of ill individuals. The usefulness of each is described below.

6.2 Toxin typing

In addition to CPE, *C. perfringens* produces at least 11 extracellular toxins (McDonel, 1979; Smith and Williams, 1984), although each isolate only produces a specific subset of these. Recently a novel toxin, termed beta-2 (β-2) toxin, has been identified (Gibert *et al.*, 1997). It has no significant amino acid sequence homology with β-toxin but does have similar biological activity. β-2 producing strains have been associated with intestinal disorders in animals (Gibert *et al.*, 1997; Herholz *et al.*, 1999).

Strains are divided into five types based on the production of four extracellular toxins: α, β, ε, and ι (Table 4.2). In principle all types produce α-toxin (lecithinase, phospholipase C), although occasionally a lecithinase-negative strain is isolated from outbreaks (Skjelkvale *et al.*, 1979; Brett, 1994). Neutralization of biological activity in the skin of mice or guinea pigs by specific antisera is the traditional method of toxin typing (Oakley and Warrack, 1953; Sterne and Batty, 1975; Smith and Williams, 1984). More recently, the use of the PCR test for this purpose has been

Table 4.2 Distribution of the major lethal toxins among types of *C. perfringens*

Type	Lethal toxin			
	Alpha	Beta	Epsilon	Iota
A	+	–	–	–
B	+	+	+	–
C	+	+	–	–
D	+	–	+	–
E	+	–	–	+

described (Daube *et al.*, 1994; Moller and Ahrens, 1996; Songer and Meer, 1996; Uzal *et al.*, 1997, Yamagishi *et al.*, 1997; Yoo *et al.*, 1997). The chromosomal organization of these toxin types is relatively constant, which suggests that phenotypic differences such as production of extracellular toxins reflect minor changes at the chromosomal level or are due to the acquisition of extrachromosomal genes (Canard *et al.*, 1992a, 1992b). Since all foodborne isolates are type A, it is not necessary to determine the toxin type of isolates from outbreaks.

6.3 Serological typing

It is unusual to find the same serotype of *C. perfringens* in a population selected at random. However, after a common-source outbreak the same serotype can be recovered from the stools of most patients. The capsular polysaccharides provide the basis for the serological typing scheme. Although well over 100 serological types exist, the capsular polysaccharides from only a few serotypes have been isolated and purified, and their sugar compositions, and more recently their complete primary structures, determined (Cherniak *et al.*, 1983; Kalelkar *et al.*, 1997). The procedure for serotyping has been described by Stringer *et al.* (1982)

The first 13 serotypes used were prepared in the 1950s by Hobbs and co-workers (Hobbs *et al.*, 1953) at the Food Hygiene Laboratory (London), and reactive isolates were termed 'Hobbs types', a term still used today in the literature – e.g. Hobbs serotype 9 (H9). Eventually a number of antisera were prepared from outbreak strains. Stringer *et al.* (1990) were able to identify a specific serotype in 77% of 646 isolates from foodborne outbreaks in the UK using 75 antisera. The addition of 34 American and 34 Japanese antisera raised the percentage of isolates typeable by only 2%, presumably because the additional antisera were raised against strains from a variety of sources – e.g. environmental or fecal from healthy individuals. Thus only strains from confirmed outbreaks should be used for antiserum production. Nevertheless, the development of an effective serological typing scheme has been hindered by the antigenic heterogeneity of strains of *C. perfringens*, so that isolated outbreak strains are often non-typeable. In addition, the necessary extensive collection of antisera is not generally available outside the UK. This has led to the use of other criteria for confirmation of outbreaks.

6.4 Other typing and subtyping methods

Other schemes for typing *C. perfringens* isolates include the use of bacteriocins. Mahony (1974) described a bacteriocin-typing protocol using 10 different bacteriocins. Watson and co-workers (Watson *et al.*, 1982; Watson, 1985) described the selection of 20 bacteriocins in an evaluation of bacteriocin typing in the laboratory investigation of *C. perfringens* food-poisoning outbreaks. They found that a much greater proportion of strains from food-poisoning outbreaks were bacteriocinogenic compared with isolates from human and animal infections, various foods, and the environment. However, as a practical matter, this technique requires a laboratory supply of bacteriocins.

A proposal to classify *C. perfringens* on the basis of chemotypes based on the composition of the capsular polysaccharide was made by Paine and Cherniak (1975). Strain differentiation based on plasmid analysis has also been proposed (Mahony *et al.*, 1987; Eisgruber, 1997; Eisgruber *et al.*, 1995): 70–80% of outbreak strains have been found to possess one or more plasmids (Mahony *et al.*, 1987; Phillips-Jones *et al.*, 1989) while, by comparison, 67% of freshly isolated, meat-associated strains have been found to carry one or more plasmids. As in the case of serotyping, these results limit the usefulness of plasmid profiling as the sole technique for distinguishing fecal or food-poisoning strains, since the predominant strains may not possess a plasmid (Mahony *et al.*, 1992).

Pons *et al.* (1993, 1994) proposed enzyme electrophoretic typing (zymotyping) as a marker for epidemiological analysis. They found that all strains from human or animal isolates could be typed by multilocus enzyme or esterase electrophoretic typing with reagents that are not specific for *C. perfringens*. However the same group of workers (Leflon-Guibot *et al.*, 1997) later concluded that random amplified polymorphic DNA (RAPD) PCR analysis is more rapid, less fastidious and less time-consuming than zymotying. The specific usefulness of RAPD in investigations of foodborne outbreaks remains to be confirmed. Other molecular methods of detecting *C. perfringens* include a 16S rDNA-based PCR method (Wang *et al.*, 1994); a PCR procedure (Baez and Juneja, 1995a; Fach and Popoff, 1997); and nested (Miwa *et al.*, 1996, 1998), duplex (Kanakaraj *et al.*, 1998; Augustynowicz *et al.*, 2002), and multiplex PCR (Daube *et al.*, 1996; Songer and Meer, 1996; Meer and Songer, 1997). The last is a useful alterative to the standard *in vivo* typing methods since it detects the genes for the major toxins.

Pulsed field gel electrophoresis (PFGE) has shown substantial discriminatory power for subtyping *C. perfringens* associated with outbreak and retail foods. Epidemiologically unrelated isolates from outbreaks (Ridell *et al.*, 1998; Maslanka *et al.*, 1999; Schalch *et al.*, 1999), non-outbreak retail isolates (Lin and Labbe, 2003) or non-foodborne human gastrointestinal disease isolates (Collie *et al.*, 1998) have been shown to have unique banding patterns. It is expected that the ability of PFGE to identify clonal relationships of isolates will assist epidemiological investigations of foodborne disease outbreaks caused by *C. perfringens*.

Ribotyping has been shown to be useful for classification of strains below the species level, and more easily interpreted than plasmid profiling (Schalch *et al.*, 1997; Kilic *et al.*, 2002) or PFGE (Schalch *et al.*, 2003). It should be noted that in at least one report endogenous nucleases of isolates limited the usefulness of PFGE (Schalch *et al.*, 2003). As with many molecular methods, and unlike serological typing, the procedure requires specialized techniques and equipment and expensive commercial reagents. Clearly, a comparative evaluation of the several proposed procedures for characterization of outbreak strains of *C. perfringens* will be required.

6.5 Isolation and identification of the organism

6.5.1 Enumeration

The procedure selected for enumeration of *C. perfringens* depends upon whether the sample is from routine sampling or from a suspected outbreak. In the former case, low numbers would be expected and a most probable number (MPN) procedure is the most appropriate (see below).

The media developed early for enumerating viable cells of *C. perfringens* were iron-sulfite agar together with one or more selective antibiotics (reviewed by Walker, 1975). The plating medium currently recommended in North America and elsewhere (International Organization for Standardization [ISO 7937:1997]) is tryptose-sulfite-cycloserine (TSC) agar, in which *C. perfringens* appears as black colonies. Its usefulness and superiority over other media for enumeration of *C. perfringens* in food and feces was demonstrated by comparative studies conducted by Hauschild *et al.* (1977, 1979). The egg-yolk-free variation is recommended, because it is just as effective and simpler. Incubation is at 37°C in an anaerobic chamber. Detailed procedures for its use are available elsewhere (British Standards Institute, 1996, BS 5763; ISO 7937:1997; *US FDA Bacteriological Analytical Manual*, 1995). In the past, neomycin blood agar has been widely used as a plating medium in the UK, since hemolysis is readily visible. As mentioned above, it was earlier proposed that food outbreak strains were weakly hemolytic. However, neomycin can inhibit food poisoning strains of this organism when used at a concentration of 100 μg/ml or greater (Spencer, 1969).

An elevated fecal spore count (after heating at 75°C for 20 minutes) is also a diagnostic criterion for implicating *C. perfringens* in foodborne outbreaks. As mentioned in section IV above, *C. perfringens* is a normal inhabitant of the human intestine. Normal fecal specimens contain between 10^3 and 10^4 spores/g (Hauschild *et al.*, 1974; Harmon *et al.*, 1986). Outbreak stools contain > 10^6 spores/g (Hauschild *et al.*, 1979; Harmon *et al.*, 1986). Only a minimal reduction in spore count occurred following frozen storage, which offers a convenient alternative if analysis is delayed.

More recently, two new plating media have been proposed. One, BCP (*Bacillus cereus*/*C. perfringens*), a blood-free egg-yolk medium, was shown to be equal or superior to TSC for unstressed or cold-shocked vegetative cells (Hood *et al.*, 1990). Significantly, pyruvate was found to promote full growth of several injured strains, presumably by degrading hydrogen peroxide, an effect noted earlier by Harmon and Kautter (1977). Unfortunately, a commercial egg yolk emulsion was not a satisfactory substitute for yolks from fresh eggs.

Most recently, Gubash and Ingham (1997) developed an egg-yolk-free, surface plating medium called bismuth-iron-sulfite-cycloserine (BISC) agar for selective isolation of *C. perfringens* from feces and foods. The advantages over TSC agar were an elimination of the need to prepare pour or overlay agar plates, and the consistent black or dark grey colonies. Without an overlay, pure cultures of *C. perfringens* on TSC agar can appear as white colonies. Significantly, neither TSC nor BISC was as effective as blood agar in recovering cold- or heat-stressed vegetative cells. TSC, BCP, and BISC all required confirmation of isolated colonies (see below).

An MPN procedure is called for in the case of routine sampling where low numbers are expected. Iron milk medium (IMM), consisting of pasteurized whole milk with 2% iron powder, is simple to prepare, inexpensive, and relatively sensitive (Abeyta and Wetherington, 1994). Selection is based on the rapid growth of *C. perfringens* at 45°C, and the typical 'stormy fermentation' reaction, during which acid from lactose fermentation rapidly coagulates the milk and is followed by fractionation of the curd into a spongy mass. If results are read after 18 hours of incubation, confirmatory tests (see below) are not necessary (Abeyta *et al.*, 1985a). Similar counts of *C. perfringens* were obtained from the environment and food samples using TSC medium or the IMM method (St John *et al.*, 1982; Abeyta *et al.*, 1985b).

Neut *et al.* (1985) demonstrated the advantages of a lactose sulfite (LS) broth compared to TSC agar as part of a MPN procedure for the enumeration of low numbers of *C. perfringens* without the need of confirmatory tests. The superior LS medium was itself apparently surpassed in sensitivity by a thioglycollate-based medium supplemented with selective antibiotics and other adjuncts (Smith, 1985). This liquid medium, termed Rapid Perfringens Medium, recovered a greater number of organisms from inoculated and naturally-contaminated samples than did LS medium.

The succession of proposed media for MPN procedures is reminiscent of the situation in the case of selective plating media in the 1970s, by which time a number had evolved. The situation was resolved by an international comparative study, a measure for which MPN methods for enumeration of *C. perfringens* would seem to be a timely candidate.

6.5.2 Confirmatory tests

None of the plating methods mentioned above is completely selective for *C. perfringens*. The possible (though unlikely) presence of other sulfite-reducing clostridia requires the confirmation of a representative number (about 10) of isolates. The recommended media for this purpose in North America and Australia are tubes of motility-nitrate (MN) and lactose-gelatin (LG) (Labbe and Harmon, 1992; Roberts *et al.*, 1995). Reported counts are obtained by multiplying the total count by the fraction of isolates that are confirmed as Gram-positive, non-motile rods, and are positive for nitrate reduction, lactose fermentation, and gelatin liquefaction. The use of TSC agar for enumeration together with MN and LG for confirmation has been adopted as official first action by AOAC International. However, the ISO and British Standard methods recommend the use of lactose–sulfite medium alone as a confirmatory test (British Standards Institute, 1996; International Organization for Standardization, 1997). More recently, lactose–sulfite has been shown to be inferior to MN and LG media for confirmatory purposes (Eisgruber *et al.*, 2000).

Traditionally in the UK, confirmation procedures have included streaking on lactose–egg-yolk medium to check for lecithinase activity (Hobbs *et al.*, 1971). This is seen as a zone of opalescence surrounding the colonies. Type A antitoxin is applied to half the plate to neutralize lecithinase produced by *C. perfringens* (Nagler reaction); occasional lecithinase-negative isolates would not be detected by this test.

6.6 Detection of the enterotoxin

6.6.1 Serological assays

The direct detection of CPE in outbreak stools is a useful technique for confirmation of a foodborne outbreak due to *C. perfringens*. As mentioned in section 4 above, elderly and institutionalized patients may carry elevated spore levels. In such cases, detection of CPE may be the only definitive evidence of *C. perfringens* etiology in gastroenteritis cases. Over the years a number of serological methods have been proposed, including electroimmunodiffusion, reversed passive hemagglutination, fluorescent antibody, and single- and double-gel diffusion (Labbe, 1989). However, currently the most widely used are an ELISA and reverse passive latex agglutination (RPLA), which have sensitivities of 2–4 ng/g feces (Berry *et al.*, 1988; Labbe, 1989). Stools from healthy individuals contain undetectable levels of CPE (Berry *et al.*, 1988) compared with outbreak stool samples which contain > 1 μg/g (Bartholomew *et al.*, 1985; Berry *et al.*, 1988; Birkhead *et al.*, 1988). A more recent ELISA method was described with a detection limit of 1 pg/ml of purified CPE (Uemura *et al.*, 1992). However, its sensitivity in testing fecal samples, which typically contain less, was not determined.

Two kits for serological detection of CPE are available commercially; an ELISA (Tech Lab, Inc.), and RPLA (Oxoid). In cases where the enterotoxigenicity of isolates is to be determined, the instructions included with commercial CPE detection kits may be misleading by implying that a single sporulation medium will induce sporulation in all isolates. As mentioned above, *in vitro* sporulation is highly strain- and medium-dependent, and diarrheal strains often fail to sporulate *in vitro* (Kokai-kun *et al.*, 1994).

If costs of commercial test kits are an issue, R. Labbe has found an older, serologi cal assay method, counterimmunoelectrophoresis (Naik and Duncan, 1977b), to be simple and useful in cases where high sensitivity is not critical (Garcia-Alvarado *et al.*, 1992c).

6.6.2 Western blotting

Western immunoblots can specifically identify 35-kDa CPE using specific antiserum. The method avoids the presumptive identification of a 48-kDa CPE-related protein as CPE (Ryu and Labbe, 1993). In addition, although ELISA and RPLA have proven very reliable, an occasional false positive cannot be excluded (Notermans *et al.*, 1984; Bartholomew *et al.*, 1985). Thus Western blotting is the method of choice for verifying the enterotoxin-producing potential of clinical, food and environmental isolates. Detection of 10 ng of CPE is possible (Kokai-kun *et al.*, 1994). The drawbacks of the method are the additional effort involved and the need to obtain a moderate level of sporulation in cultures in order to detect CPE in cell extracts.

6.6.3 Gene-probe methods

The difficulty in obtaining consistent sporulation of isolates of *C. perfringens* in order to detect CPE by serological methods led to the development of methods for the direct detection of the *cpe* gene in vegetative cells. As indicated in section 4 above, these assays include the use of PCR as well as radioactive or digoxigenin (dig)-labelled gene probes (Van Damme-Jongsten *et al.*, 1990; Tschirdewahn *et al.*, 1991; Kokai-kun *et al.*, 1994; Baez and Juneja, 1995b). However, isolates identified as *cpe*-positive by such assays can only be regarded as potentially enterotoxigenic, since it is not known whether they also possess the regulatory factors required for CPE expression.

Van Damme-Jongsten *et al.* (1989, 1990) found that in the feces of non-symptomatic animals, only 6% of strains possessed the *cpe* gene. In the case of food and fecal isolates from 186 outbreaks, they reported that only 60% hybridized with synthetic DNA probes, which suggested that strains that cannot produce the enterotoxin had been isolated. More recently Ridell *et al.* (1998) found a somewhat higher percentage (86–88%) of food poisoning isolates were *cpe*-positive when analyzed by the polymerase chain reaction.

6.6.4 Other methods

As mentioned above, Vero cells have been widely used in CPE mode-of-action studies. Biological assays for the enterotoxin have also been developed using this cell line. Activity is measured as inhibition of plating efficiency (McDonel and McClane, 1981, 1988) or killing of Vero cells (Mahony *et al.*, 1989). However, in the case of stool samples, the cells are at least 10-fold less sensitive than ELISA or RPLA methods. Additionally, problems with reproducibility and non-specific cytotoxin reactions were noted (Berry *et al.*, 1988).

Other methods (yet to be widely adopted) for implicating *C. perfringens* or its enterotoxin in foodborne outbreaks include pyrolysis mass spectrometry (Sisson *et al.*, 1992), immunomagnetic separation-ELISA (Cudjoe *et al.*, 1991), and DOT-ELISA (Mehta *et al.*, 1989).

The sensitivity of *C. perfringens* to cold shock led Harmon and Kautter (1976) to develop a method for estimating previously viable cells by measuring lecithinase (alpha toxin) activity in implicated food. A hemolysis indicator plate assay was found to be dose-dependent on the number of cells above a threshold of about 10^4 CFU/ml (Foegeding and Busta, 1980). However, others have found lecithinase activity to be time-, temperature-, and strain-dependent (Nakamura *et al.*, 1969), and repressed in carbohydrate-supplemented media (Wrigley, 1990). This technique would be of little value in those rare outbreaks involving lecithinase-negative strains (Pinegar and Stringer, 1977; Brett, 1994). Thus the procedure may be of value only when patients' stools are unavailable and when suspect food has undergone prolonged cold storage. Accordingly, the alpha toxin method has been adopted as official first action by AOAC for the examination of outbreak foods in which the presence of large numbers of vegetative cells, which may no longer be viable, is suspected.

Bibliography

Abeyta, C. Jr and J. Wetherington (1994). Iron milk medium for recovering *Clostridium perfringens* from shellfish: collaborative study. *AOAC Intl.* **77**, 351–356.

Abeyta, C. Jr, A. Michalovski and M. Wekell (1985a). Differentiation of *Clostridium perfringens* from related clostridia in iron milk medium. *J. Food Prot.* **48**, 130–134.

Abeyta, C. Jr, M. Wekell and J. Peeler (1985b). Comparison of media for enumeration of *Clostridium perfringens* in foods. *J. Food Sci.* **50**, 1732–1735.

Adams, D. (1973). Inactivation of *Clostridium perfringens* at ultrahigh temperatures. *Appl. Microbiol.* **26**, 282–287.

Adams, D. (1974). Requirement for and sensitivity to lysozyme by *Clostridium perfringens* spores heated at ultra high temperatures. *Appl. Microbiol.* **27**, 797–801.

Akama, K. and S. Otani (1970). *Clostridium perfringens* as the flora in the intestine of healthy persons. *Jpn J. Med. Sci. Biol.* **23**, 161–175.

Ali, M. and D. Fung (1991) Occurrence of *Clostridium perfringens* in ground beef and ground turkey evaluated by three methods. *J. Food Safety* **11**, 197–203.

Allen, S. and H. Blaschek (1990). Factors involved in the electroporation-induced transformation of *Clostridium perfringens. FEMS Microbiol. Lett.* **70**, 217–220.

Ando, X., T. Tsuzuki, H. Sunagawa and S. Oka (1985). Heat resistance, spore germination and enterotoxigenicity of *Clostridium perfringens. Microbiol. Immunol.* **29**, 317–326.

Andrews, F. (1899). On an attack of diarrhea in the wards of St Bartholomew's Hospital, probably caused by infections of rice pudding with *Bacillus enteritidis sporogenes. Lancet* **i**, 8–12.

Armon, R. and P. Payment (1988). A modified m-CP medium for the enumeration of *Clostridium perfringens* from water samples. *Can. J. Microbiol.* **34**, 78–79.

Ashworth, J., L. Hargreaves and B. Jarvis (1973). The production of an antimicrobial effect in pork heated with sodium nitrite under simulated commercial pasteurization conditions. *J. Food Technol.* **8**, 177–181.

Augustynowicz, E., A. Gzyl and J. Slusarczyk (2002). Detection of enterotoxigenic *Clostridium perfringens* with a duplex PCR. *J. Med. Microbiol.* **51**, 169–172.

Baez, L. and V. Juneja (1995a). Detection of enterotoxigenic *Clostridium perfringens* in raw beef polymerase chain reaction. *J. Food Prot.* **58**, 154–159.

Baez, L. and V. Juneja (1995b). Nonradioactive colony hybridization assay for detection and enumeration of enterotoxigenic *Clostridium perfringens* in raw beef. *Appl. Environ. Microbiol.* **61**, 807–810.

Bannum, T. and J. Rood (1993). *Clostridium perfringens – Escherichia coli* shuttle vectors that carry single antibiotic resistance determinants. *Plasmid* **229**, 233–235.

Bartholomew, B. and M. Stringer (1983). Observation on the purification of *Clostridium perfringens* type A enterotoxin and the production of a specific antiserum. *FEMS Microbiol. Lett.* **18**, 43–48.

Bartholomew, B., M. Stringer, G. Watson and R. Gilbert (1985). Development and application of an enzyme-linked immunosorbent assay for *Clostridium perfringens* type A enterotoxin. *J. Clin. Pathol.* **38**, 222–228.

Bartlett, M., H. Walker and R. Ziprin (1972). Use of dogs as an assay for *Clostridium perfringens* enterotoxin. *Appl. Microbiol.* **23**, 196–197.

Bauer, F., J. Carpenter and J. Reagan (1981). Prevalence of *Clostridium perfringens* in pork during processing. *J. Food Prot.* **44**, 279–283.

Bean, N. and P. Griffin (1990). Foodborne disease outbreaks in the United States, 1973–1987, pathogens, vehicles and trends. *J. Food Prot.* **53**, 804–817.

Bean, N., J. Goulding, M. Daniels and F. Angulo (1997). Surveillance for foodborne disease outbreaks – US 1988–1992. *J. Food Prot.* **60**, 1265–1286.

Berry, P., J. Rodhouse, S. Hughes *et al.* (1988). Evaluation of ELISA, RPLA, and Vero cell assays for detecting *Clostridium perfringens* enterotoxin in fecal specimens. *J. Clin. Pathol.* **41**, 458–461.

Bezirtzoglou, E., D. Dimitriou and A. Panagiou (1996). Occurrence of *Clostridium perfringens* in river water by using a new procedure. *Anaerobe* **2**, 169–173.

Birkhead, G., R. Vogt, E. Heun *et al.* (1988). Characterization of an outbreak of *Clostridium perfringens* food poisoning by quantitative fecal culture and fecal enterotoxin measurement. *J.. Clin. Microbiol.* **26**, 471–474.

Borriello, S. (1985). Newly discovered clostridial disease of the gastrointestinal tract, *Clostridium perfringens* enterotoxin-associated diarrhea and neutropenic enterocolitis. In: S. Borriello (ed.), *Clostridia in Gastrointestinal Disease*, pp. 223–230. CRC Press, Boca Raton, FL.

Borriello, S., H. Larson, A. Welch *et al.* (1984). Enterotoxingenic *Clostridium perfringens*, a possible cause of antibiotic-associated diarrhea. *Lancet* **i**, 305–307.

Bowness, P., P. Moss, H. Tranter *et al.* (1992). *Clostridium perfringens* enterotoxin is a superantigen reactive with human T cell receptors Vb6.9 and Vb22. *J. Exp. Med.* **176**, 893–896.

Boyd, M., M. Logan and A. Tytell (1948). The growth requirement of *Clostridium perfringens* BP6K. *J. Biol. Chem.* **174**, 1013–1025.

Bradshaw, J., G. Stelma, V. Jones *et al.* (1982). Thermal inactivation of *Clostridium perfringens* enterotoxin in buffer and in chicken gravy. *J. Food Sci.* **47**, 914–916.

Brett, M. (1994). Outbreaks of food poisoning associated with lecithinase-negative *Clostridium perfringens. J. Med. Microbiol.* **41**, 405–407.

British Standards Institute (1996). BS 5763. Methods for microbiological examination of food and animal feeding stuffs. Part. 9. Examination of *Clostridium perfringens*.

Brown, D. and R. Twedt (1972). Assessment of the sanitary effectiveness of holding temperatures on beef cooked at low temperatures. *Appl. Microbiol.* **24**, 599–603.

Bryan, F. (1969). What the sanitarian should know about *Clostridium perfringens* foodborne illness. *J. Milk Food Technol.* **32**, 381–389.

Bryan, F. and T. McKinley (1974). Prevention of foodborne illness by time–temperature control of thawing, cooking, chilling, and reheating turkeys in school lunch kitchens. *J. Milk Food Technol.* **37**, 420–425.

Bryan, F., T. McKinley and B. Mixon (1971). Use of time–temperature evaluations in detecting the responsible vehicle and contributing factors of foodborne disease outbreaks. *J. Milk Food Technol.* **34**, 576–580.

Brynestad, S., B. Synstad and P. E. Granum (1997). The *Clostridium perfringens* enterotoxin gene is on a transposable element in type A human food poisoning strains. *Microbiology* **143**, 2109–2115.

Bullifent, H., A. Moir and R. Titball (1995). The construction of a reporter system and use for the investigation of *Clostridium perfringens* gene expression. *FEMS Microbiology Lett.* **131**, 99–105.

Canada, J., D. Strong and L. Scott (1964). Response of *Clostridium perfrigens* spores and vegetative cells to temperature variations. *Appl. Microbiol.* **12**, 273–276.

Canard, B., T. Garnier, B. LaFay, R. Christen, and S. T. Cole (1992a). Phylogenetic analysis of the pathogenic anaerobe *Clostridium perfringens* using the 16S rRNA nucleotide sequence. *Intl J. System Bacteriol.* **42**, 312–314.

Canard, B., B. Saint-Joanis and S. Cole (1992b). Genomic diversity and organization of virulence genes in the pathogenic anerobe *Clostridium perfringens. Mol. Microbiol.* **6**, 1421–1429.

Cassier, M. and M. Sebald (1969). Germination lysozyme-dépendent des spores de *Clostridium perfringens* ATCC 3624 après treatment thermique. *Ann. Inst. Pasteur* **117**, 312–324.

CDC (Centers for Disease Control) (1990). Foodborne disease outbreaks, 5-year summary, United States, 1983–1987. *Morbid. Mortal. Weekly Rep.* **39**(SS01), 15–23.

CDC (Centers for Disease Control and Prevention) (1996). Surveillance for foodborne disease outbreaks – United States, 1988–1992. *Morbid. Mortal. Weekly Rep.* **45**(SS05), 1–55.

CDC (Centers for Disease Control and Prevention) (2000). Surveillance for foodborne disease outbreaks – United States, 1993–1997. *Morbid. Mortal. Weekly Rep.* **49**(SS01), 1–51.

Chen, C.-K., C. Boucle and H. Blaschek (1996). Factors involved in the transformation of previously non-transformable *Clostridium perfringens* type B. *FEMS Microbiol. Lett.* **140**, 185–191.

Cherniak, R., K. Dayalu and R. Jones (1983). Analysis of the common polysaccharide antigens from the cell envelopes of *Clostridium perfringens* type A. *Carbohydr. Res.* **119**, 171–190.

Citino, S. (1995). Chronic, intermittent *Clostridium perfringens* enterotoxicosis in a group of cheetahs (*Acinonyx jubatus jubatus*). *J. Zoo Wild. Med.* **26**, 279–285.

Cole, S. and B. Canard (1997). Structure, organization and evolution of the genome of *Clostridium perfringens*. In: J. Rood, B. McClane, G. Songer and R. Titball (eds), *Molecular Biology and Pathogenesis of the Clostridia*, pp. 49–63. Academic Press, London.

Collie, R. and B. McClane (1998). Evidence that the enterotoxin gene can be episomal in *Clostridium perfringens* isolates associated with non-foodborne human gastrointestinal diseases. *J. Clin. Microbiol.* **36**, 30–36.

Collie, R., J. Kokai-kun and B. McClane (1998). Phenotypic characterization of enterotoxigenic *Clostridium perfringens* isolates from non-foodborne human gastrointestinal diseases. *Anaerobe* **4**, 69–79.

Collins, J., M. Bergeland, D. Bouley *et al.* (1989). Diarrhea associated with *Clostridium perfringens* type A enterotoxin in neonatal pigs. *J. Vet. Diagn. Invest.* **1**, 351–353.

Cornillot, E., B. Saint-Joanis, G. Daube *et al.* (1995). The enterotoxin gene (*cpe*) of *Clostridium perfringens* can be chromosomal or plasmid-borne. *Mol. Microbiol.* **15**, 639–647.

Craven, S. (1980). Growth and sporulation of *Clostridium perfringens* in foods. *Food Technol.* **34(4)**, 80–87.

Craven, S. (1988). Increased sporulation of *Clostridium perfringens* in a medium prepared with the prereduced anaerobically sterilized techniques or with carbon dioxide or carbonate. *J. Food Protect.* **51**, 700–706.

Craven, S. (1990). The effect of the pH of the sporulation environment on the heat resistance of *Clostridium perfringens* spores. *Curr. Microbiol.* **22**, 233–237.

Craven, S., L. Blankenship and J. McDonel (1981). Relationship of sporulation, enterotoxin formation and spoilage during growth of *Clostridium perfringens* type A. *Appl. Environ. Microbiol.* **41**, 1184–1191.

Cudjoe, K., L. Thorsen, T. Sorensen *et al.* (1991). Detection of *Clostridium perfringens* type A enterotoxin in fecal and food samples using immunomagnetic separation (IMS)-ELISA. *Intl J. Food Microbiol.* **12**, 313–322.

Czeczulin, J., P. Hanna and B. McClane (1993). Cloning, nucleotide sequencing, and expression of the *Clostridium perfringens* enterotoxin gene in *Escherichia coli. Infect. Immun.* **61**, 3429–3439.

Czeczulin, J., R. Collie and B. McClane (1996). Regulated expression of *Clostridium perfringens* enterotoxin in naturally cpe-negative type A, B, and C isolates of *C. perfringens. Infect. Immun.* **64**, 3301–3309.

Daube, G., B. China, P. Simon *et al.* (1994). Typing of *Clostridium perfringens* by *in vitro* amplification of toxin genes. *J. Appl. Bacteriol.* **77**, 650–655.

Daube, G., B. Simon, C. Limbourg *et al.* (1996). Hybridization of 2659 *Clostridium perfringens* isolates with gene probes for seven toxins (α, β, ε, ι, θ, μ and enterotoxin) and sialidase. *Am. J. Vet. Res.* **57**, 496–501.

DeBoer, E., W. Spiegelenberg and F. Janssen (1985). Microbiology of spices and herbs. *Ant. van Leeuw.* **51**, 435–438.

Decaudin, M. and J.-L. Tholozan (1996). A comparative study of the conditions of growth and sporulation of three strains of *Clostridium perfringens* type A. *Can. J. Microbiol.* **42**, 298–304.

DeJong, A. R. Beumer and F. Rombouts (2002). Optimizing sporulationn of *Clostridium perfringens. J. Food Prot.* **65**, 1457–1462.

Dework, F. Jr (1972). Sporulation of *Clostridium perfringens* type A in vacuum sealed meats. *Appl. Microbiol.* **24**, 834–836.

Dillon, M. and R. Labbe (1989). Stimulation of growth and sporulation of *Clostridium perfringens* by its homologous enterotoxin. *FEMS Microbiol. Lett.* **59**, 221–224.

Dische, F. and S. Elek (1957). Experimental food poisoning by *Clostridium welchii. Lancet* **i**, 71–74.

Duncan, C. and D. Strong (1968). Improved medium for sporulation of *Clostridium perfringens. Appl. Microbiol.* **16**, 82–89.

Duncan, C. and D. Strong (1969). Ileal loop fluid accumulation and production of diarrhea in rabbits by cell-free products of *Clostridium perfringens. J.Bacteriol.* **100**, 86–94.

Duncan, C. and D. Strong (1971). *Clostridium perfringens* type A food poisoning. I. Response of the rabbit ileum as an indication of enteropathogenicity of strains of *Clostridium perfringens* in monkeys. *Infect. Immun.* **3**, 167–170.

Duncan, C., D. Strong and M. Sebald (1972a). Sporulation and enterotoxin production by mutants of *Clostridium perfringens. J. Bacteriol.* **110**, 378–391.

Duncan, C., R. Labbe and R. Reich (1972b). Germination of heat and alkali-altered spores of *Clostridium perfringens* type A by lysozyme and an initiation protein. *J. Bacteriol.* **109**, 550–559.

Duncan, C., G. King and W. Frieben (1973). A paracrystalline inclusion body formed during sporulation of enterotoxin producing strains of *Clostridium perfringens. J. Bacteriol.* **114**, 845–859.

Eisgruber, H (1997). Plasmid profiling for strain differentiation of *Clostridium perfringens* from food poisoning cases and outbreaks. *Arch. Lebensmittel* **48**, 10–14.

Eisgruber, H., M. Wiedmann and A. Stolle (1995). Use of plasmid profiling as a typing method for epidemiologically related *Clostridium perfringens* isolates from food poisoning cases and outbreaks. *Lett. Appl. Microbiol.* **20**, 290–294.

Eisgruber, H., B. Schalch, B. Sperner and A. Stolle (2000). Comparison of four routine methods for the confirmation of *Clostridium perfringens* in food. *Intl J. Food Microbiol.* **57**, 135–140.

Enders, G. and C. Duncan (1976). Anomalous aggregation of *Clostridium perfringens* entertoxin under dissociating conditions. *Can. J. Microbiol.* **22**, 1410–1414.

Enders, G. and C. Duncan (1977). Preparative polyacrylamide gel electrophoresis purification of *Clostridium perfringens* enterotoxin. *Infect. Immun.* **17**, 425–429.

Erickson, J. and R. Deibel (1978). New medium for rapid screening and enumeration of *Clostridium perfringens* in foods. *Appl. Environ. Microbiol.* **36**, 567–571.

Fach, P. and M. Popoff (1997). Detection of enterotoxigenic *Clostridium perfringens* in food and fecal samples with a duplex PCR and the slide latex agglutination test. *Appl. Environ. Microbiol.* **63**, 4232–4236.

Fisher, L., D. Strong and C. Duncan (1970). Resistance of *Clostridium perfringens* to varying degrees of acidity during growth and sporulation. *J. Food Sci.* **35**, 91–95.

Foegeding, P. and F. Busta (1980). *Clostridium perfringens* cells and phospholipase C activity at constant and linearly rising temperatures. *J. Food Sci.* **45**, 918–924.

Foster, J., J. Fowler and W. Ladiges (1977). A bacteriological study of raw ground beef. *J. Food Prot.* **40**, 790–795.

Frieben, W. and C. Duncan. (1973). Homology between enterotoxin protein and spore structural protein in *Clostridium perfringens* type A. *Eur. J. Biochem.* **39**, 393–401.

Fruin, J. and F. Babel (1977). Changes in the population of *Clostridium perfringens* type A frozen in a meat medium. *J. Food Prot.* **40**, 622–625.

Furuse, M., K. Fujita, T. Hiiragi *et al.* (1998). Claudin-1 and -2: novel integral membrane proteins localizing at tight junctions with no sequence similarity to occludin. *J. Cell Biol.* **141**, 1539–1550.

Garcia-Alvarado, J., M. Rodriguez and R. Labbe (1992a). Influence of elevated temperature on starch hydrolysis by enterotoxin-positive and enterotoxin-negative strains of *Clostridium perfringens* type A. *Appl. Environ. Microbiol.* **58**, 326–330.

Garcia-Alvarado, J., R. Labbe and M. Rodriguez (1992b). Sporulation and enterotoxin production by *Clostridium perfringens* type A at 37 and 43°C. *Appl. Environ. Microbiol.* **58**, 1411–1414.

Garcia-Alvarado, J., R. Labbe and M. Rodriguez (1992c). Isolation of inclusion bodies from vegetative *Clostridium perfringens*, partial purification of a 47 kDa inclusion protein. *J. Appl. Bacteriol.* **73**, 157–162.

George, W. and S. Finegold. (1985). Clostridia in the human gastrointestinal flora. In: S. P. Borriello (ed.), *Clostridia in Gastrointestinal Disease*, pp. 1–38. CRC Press, Boca Raton, FL.

Gibert, M., C. Jolivet-Renaud and M. Popoff (1997). Beta2 toxin, a novel toxin produced by *Clostridium perfringens. Gene* **203**, 65–73.

Gibson, A. and T. Roberts (1986). The effect of temperature, sodium chloride, sodium nitrite and storage temperature on the growth of *Clostridium perfringens* and fecal streptococci in laboratory media. *Intl J. Food Microbiol.* **3**, 195–210.

Goldner, S., M. Solberg and L. Post (1985). Development of a minimal medium for *Clostridium perfringens* by using an anaerobic chemostat. *Appl. Environ. Microbiol.* **50**, 202–206.

Gombas, D. and R. Gomez (1978). Sensitization of *Clostridium perfringens* spores to heat by gamma irradiation. *Appl. Environ. Microbiol.* **36**, 403–407.

Gombas, D. and R. Labbe (1985). Purification and properties of spore lytic enzymes from *Clostridium perfringens* type A spores. *J. Gen. Microbiol.* **131**, 1487–1496.

Gough, B. and J. Alford (1965). Effect of curing agents on the growth and survival of food poisoning strains of *Clostridium perfringens. J. Food Sci.* **30**, 1025–1029.

Granum, P. (1982). Inhibition of protein synthesis by a tryptic polypeptide of *Clostridium perfringens* type A enterotoxin. *Biochim. Biophys. Acta* **708**, 6–11.

Granum, P. and M. Richardson (1991). Chymotrypsin treatment increases the activity of *Clostridium perfringens* enterotoxin. *Toxicon* **29**, 898–900.

Granum, P. and R. Skjelkvale (1977). Chemical modification and characterization of enterotoxin from *Clostridium perfringens* type A. *Acta Pathol. Microbiol. Immunol. Scand. Sect. B.* **85**, 89–94.

Granum, P. and R. Skjelkvale (1981). Endogenous radiolabeling of enterotoxin from *Clostridium perfringens* type A in a defined medium. *Appl. Environ. Microbiol.* **42**, 596–598.

Granum, P. and G. Stewart (1993). Molecular biology of *Clostridium perfringens* enterotoxin. In: M. Sebald (ed.), *Genetics and Molecular Biology of Anaerobic Bacteria*, pp. 235–247. Springer Verlag, New York, NY.

Granum, P. and J. Whitaker (1980). Improved method for purification of enterotoxin from *Clostridium perfringens* type A. *Appl. Environ. Microbiol.* **39**, 1120–1122.

Granum, P., J. Whitaker and R. Skjelkvale (1981). Trypsin activation of enterotoxin from *Clostridium perfringens* type A. Fragmentation and some physicochemical properties. *Biochim. Biophys. Acta* **668**, 325–332.

Granum, P., W. Telle, O. Olsvik and A. Stavn (1984). Enterotoxin formation by *Clostridium perfringens* during sporulation and vegetative growth. *Intl J. Food Microbiol.* **1**, 43–49.

Gubash, S. and L. Ingham (1997). Comparison of a new bismuth-iron-sulfite-cycloserine agar for isolation of *Clostridium perfringens* with trypose-sulfite-cycloserine and blood agars. *Zbl. Bakt.* **285**, 397–402.

Hall, H. and R. Angelotti (1965). *Clostridium perfringens* in meat and meat products. *Appl. Microbiol.* **13**, 352–356.

Hall, H. and G. Hauser (1966). Examination of feces from food handlers for salmonellae, shigellae, enteropathogenic *Escherichia coli*, and *Clostridium perfringens*. *Appl. Microbiol.* **14**, 928–932.

Hall, H., R. Angelotti, K. Lewis, and M. Foter (1963). Characteristics of *Clostridium perfringens* strains associated with food and food-borne disease. *J. Bacteriol.* **85**, 1094–1103.

Harmon, S. and D. Kautter (1976). Estimating population levels of *Clostridium perfringens* in foods, based on alpha toxin. *J. Milk Food Technol.* **39**, 107–113.

Harmon, S. and D. Kautter (1977). Recovery of clostidia on catalase-treated plating media. *Appl. Environ. Microbiol.* **33**, 762–770.

Harmon, S. and D. Kautter (1986). Improved medium for sporulation and enterotoxin production by *Clostridium perfringens. J. Food Prot.* **49**, 706–711.

Harmon, S., D. Kautter and C. Hatheway (1986). Enumeration and characterization of *Clostridium perfringens* spores in the feces of food poisoning patients and normal controls. *J. Food Prot.* **49**, 23–28.

Hatheway, C. (1990). Toxigenic clostridia. *Clin. Microbiol. Rev.* **3**, 66–98.

Hauschild, A. and R. Hilsheimer (1971). Purification and characteristics of the enterotoxin of *Clostridium perfringens* type A. *Can. J. Microbiol.* **17**, 1425–1433.

Hauschild, A. and F. Thatcher (1967). Experimental food poisoning with heat susceptible *Clostridium perfringens* type A. *J. Food Sci.* **32**, 467–471.

Hauschild, A., R. Hilsheimer and F. Thatcher (1967). Acid resistance and infectivity of food poisoning *Clostridium perfringens. Can J. Microbiol.* **13**, 1401–1406.

Hauschild, A., L. Nilo and W. Dorwood (1970). Response of the ligated intestinal loops in lambs to an enteropathogenic factor of *Clostridium perfringens* type A. *Can. J. Microbiol.* **16**, 339–343.

Hauschild, A., R. Hilsheimer and C. Rogers (1971a). Rapid detection of *Clostridium perfringens* enterotoxin in a modified ligated intestinal loop technique in rabbits. *Can. J. Microbiol.* **17**, 1475–1476.

Hauschild, A., M. Walcroft and W. Campbell (1971b). Emesis and diarrhea induced by enterotoxin of *Clostridium perfringens* type A in monkeys. *Can. J. Microbiol.* **17**, 1141–1143.

Hauschild, A., R. Hilsheimer and D. Griffith (1974). Enumeration of fecal *Clostridium perfringens* spores in egg yolk free tryptose-sulfite-cycloserine agar. *Appl. Microbiol.* **27**, 527–530.

Hauschild, A., R. Gilbert, S. Harmon *et al.* (1977). ICMSF methods studies. VIII. Comparative study for the enumeration of *Clostridium perfringens* in foods. *Can. J. Microbiol.* **23**, 884–892.

Hauschild, A., P. Desmarchelier, R. Gilbert *et al.* (1979). ICMSF methods studies. XII. Comparative study for the enumeration of *Clostridium perfringens* in feces. *Can. J. Microbiol.* **25**, 953–963.

Heredia, N., R. Labbe, M. Rodriguez and J. Garcia-Alvarado (1991). Growth, sporulation, and enterotoxin production by *Clostridium perfringens* type A in the presence of human bile salts. *FEMS Microbiol. Lett.* **84**, 15–22.

Heredia, N., J. Garcia-Alvarado and R. Labbe (1994). Improved rapid method for production and purification of *Clostridium perfringens* type A enterotoxin. *J. Microbiol. Methods* **20**, 87–91.

Heredia, N., G. Garcia, R. Luevanos *et al.* (1997). Elevation of the heat resistance of vegetative cells and spores of *Clostridium perfringens* type A by sublethal heat shock. *J. Food. Prot.* **60**, 998–1000.

Herholz, C., R. Miserez, J. Nicolet *et al.* (1999). Prvelence of β2-toxigenic *Clostridium perfringens* in horses with intestinal disorders. *J. Clin. Microbiol.* **37**, 358–361.

Hill, R., I. Knight, M. Anikis and R. Colwell (1993). Benthic distribution of sewage sludge indicated by *Clostridium perfringens* at a deep-ocean dump site. *Appl. Environ. Microbiol.* **59**, 47–51.

Hobbs, B., M. Smith, C. Oakley *et al.* (1953). *Clostridium welchii* food poisoning. *J. Hygiene* **51**, 75–101.

Hobbs, G., D. Cann, B. Wilson and J. Shewan (1965). The incidence of organisms of the genus *Clostridium* in vacuum-packed fish in the United Kingdom. *J. Appl. Bacteriol.* **28**, 265–270.

Hobbs, G., K. Williams and A. Willis (1971). Basic methods for the isolation of clostridia. In: D. Shapton and R. Board (eds), *Isolation of Anaerobes*, pp. 1–23. Academic Press, London.

Hood, A., A. Tuck and C. Dane (1990). A medium for the isolation, enumeration and rapid presumptive identification of injured *Clostridium perfringens* and *Bacillus cereus. J. Appl. Bacteriol.* **69**, 359–372.

Horiguchi, Y., T. Akai and G. Sakaguchi (1987). Isolation and function of a *Clostridium perfringens* enterotoxin fragment. *Infect. Immun.* **55**, 2912–2915.

International Organization for Standardization (1997). Microbiology of food and animal feeding stuffs – horizontal method for enumeration of *Clostridium perfringens* – colony count. ISO 7937.

Jackson, S., D. Yip-Chuck, J. Clark and M. Brodsky (1986). Diagnostic importance of *Clostridium perfringens* enterotoxin analysis in recurring enteritis among the elderly, chronic care psychiatric patients. *J. Clin. Microbiol.* **23**, 748–751.

Jarmund, T. and W. Telle (1982). Binding of *Clostridium perfringens* enterotoxin to hepatocytes, small intestinal epithelial cells and Vero cells. *Acta Pathol. Microbiol. Immunol. Scand. Sect. B* **90**, 377–381.

Johnson, S., P. Echeverria, D. Taylor *et al.* (1987). Enteritis necroticans among Khmer children at an evacuation site in Thailand. *Lancet* **ii**, 496–500.

Juneja, V. and H. Marks (2002). Predictive model for growth of *Clostridium perfringens* during cooling of cooked cured chicken. *Food Microbiol.* **19**, 313–327

Juneja, V., J. Call and A. Miller (1993). Evaluation of methylxanthines and related compounds to enhance *Clostridium perfringens* sporulation using a modified Duncan and Strong medium. *J. Rapid Methods Autom. Microbiol.* **2**, 203–218.

Juneja, V., B. Marmer and A. Miller. (1994a). Growth and sporulation potential of *Clostridium perfringens* in aerobic and vacuum-packaged cooked beef. *J. Food Prot.* **57**, 393–398.

Juneja, V., O. Snyder and M. Cygnarowicz-Provost (1994b). Influence of cooling rate on outgrowth of *Clostridium perfringens* spores in cooked ground beef. *J. Food Prot.* **57**, 1063–1067.

Juneja, V., R. Whiting, H. Marks and O. Snyder (1999). Predictive model for growth of *Clostridium perfringens* at temperatures applicable to cooling of cooked meat. *Food Microbiol.* **16**, 335–349.

Juneja, V., J. Novak, H. Marks and D. Gombas (2001). Growth of *Clostridium perfringens* from spore inocula in cooked cured beef, development of a predictive model. *Innovat. Food Sci. Emerg. Technol.* **2**, 289–301.

Kalelkar, S., J. Glushka, H. van Halbeek *et al.* (1997). Structure of the capsular polysaccharide of *Clostridium perfringens* Hobbs 5 as determined by NMR spectroscopy. *Carbohydr. Res.* **299**, 119–128.

Kanakaraj, R., D. Harris, J. G. Songer and B. Bosworth (1998). Multiplex PCR assay for the detection of *Clostridium perfringens* in feces and intestinal contents of swine feed. *Vet. Microbiol.* **63**, 39–28.

Kang, C., M. Woodburn, A. Pagenkopf and R. Cheney (1969). Growth, sporulation, and germination of *Clostridium perfringens* in media of controlled water activity. *Appl. Microbiol.* **18**, 798–805.

Katahira, J., N. Inoue, Y. Horiguchi *et al.* (1997). Molecular cloning and functional characterization of the receptor of *Clostridium perfringens* enterotoxin. *J. Cell Biol.* **136**, 1239–1247.

Katayama, S.-I., B. Dupay, T. Garnier and S. Cole (1995). Rapid expansion of the physical and genetic map of the chromosome of *Clostridium perfringens* CPN50. *J. Bacteriol.* **177**, 5680–5685.

Kilic, U., B. Schalch and A. Stolle (2002). Ribotyping of *Clostridium perfringens* from industrially produced ground meat. *Appl. Microbiol.* **34**, 238–243.

Klindworth, K., P. Davidson, C. Brekke and A. Branen (1979). Inhibition of *Clostrdium perfringens* by butylated hydroxyanisole. *J. Food Sci.* **44**, 564–567.

Klotz, A. (1965). Application of fluorescent antibody techniques to detection of *Clostridium perfringens. Public Health Rep.* **80**, 305–311.

Knox, R. and E. MacDonald (1943). Outbreaks of food poisoning in certain Leicester institutions. *Med. Offr* **69**, 21–22.

Kokai-kun, J. and B. McClane (1997a). The *Clostridium perfringens* enterotoxin. In: J. Rood, G. Songer, B. McClane and R. Titball (eds), *Molecular Biology and Pathogenesis of the Clostridia*, pp. 325–358. Academic Press, London.

Kokai-kun, J. and B. McClane (1997b). Deletion analysis of the *Clostridium perfringens* enterotoxin. *Infect. Immun.* **65**, 1014–1022.

Kokai-kun, J. J. G.Songer, J. Czeczulin *et al.* (1994). Comparison of Western immunoblots and gene detection assays for identification of potentially enterotoxigenic isolates of *Clostridium perfringens. J. Clin. Microbiol.* **32**, 2533–2539.

Labbe, R. (1980). Relationship between sporulation and enterotoxin production in *Clostridium perfringens. Food Technol.* **34(4)**, 88–90.

Labbe, R. (1981). Enterotoxin formation by *Clostridium perfringens* type A in a defined medium. *Appl. Environ. Microbiol.* **41**, 315–317.

Labbe, R. (1989). *Clostridium perfringens*. In: M. Doyle (ed.), *Foodborne Bacterial Pathogens*, pp. 191–234. Marcel Dekker, New York, NY.

Labbe, R. and C. Duncan (1970). Growth from spores of *Clostridium perfringens* in the presence of sodium nitrite. *Appl. Microbiol.* **19**, 353–359.

Labbe, R. and S. Harmon (1992). *Clostridium perfringens*. In: C. Vanderzant and D. Splittstoesser (eds), *Compendium of Methods for the Microbiological Examination of Foods*, 3rd edn, pp. 623–635. American Public Health Association, Washington, DC.

Labbe, R. and T. Huang (1995). Generation times and modeling of enterotoxin-positive and enterotoxin-negative strains of *Clostridium perfringens* in laboratory media and ground beef. *J. Food Prot.* **58**, 1303–1306.

Labbe, R., and J. Nolan (1981). Stimulation of *Clostridium perfringens* enterotoxin formation by caffeine and theobromine. *Infect. Immun.* **34**, 50–54.

Labbe, R. and D. Rey (1979). Raffinose increases sporulation and enterotoxin production by *Clostridium perfringens* type A. *Appl. Environ. Microbiol.* **37**, 1196–1200.

Labbe, R., R. Reich and C. Duncan (1978). Alteration in ultrastructure and germination of *Clostridium perfringens* type A following extraction of spore coats. *Can. J. Microbiol.* **24**, 1526–1536.

Ladiges, W., J. Foster and W. Ganz (1974). Incidence and viability of *Clostridium perfringens* in ground beef. *J. Milk Food Technol.* **37**, 622–626.

Larson, H. and S. Borriello (1988). Infectious diarrhea due to *Clostridium perfringens. J. Infect. Dis.* **157**, 390–391.

Lawrence, G. (1988). Necrotizing enteritis and *Clostridium perfringens. J. Infect. Dis.* **151**, 803–804.

Lawrence, G., P. Walker, J. Garap and M. Avusi (1979). The occurrence of *Clostridium welchii* type C in Papua New Guinea. *Papua New Guinea Med. J.*, **22**, 69–73.

Leflon-Guibout, V., J.-L. Pons, B. Heym and M.-H. Nicolas-Chanoine (1997). Typing of *Clostridium perfringens* strains by use of random amplified polymorphic DNA (RAPD) system in comparison with zymotyping. *Anaerobe* **3**, 245–250.

Lin, Y. T. and R. Labbe (2003). Enterotoxigenicity and genetic relatedness of *Clostridium perfringens* isolates from retail food. *Appl. Environ. Microbiol.* **69**, 1642–1646

Lindsay, J. (1996). *Clostridium perfringens* type A enterotoxin, more than just explosive diarrhea. *Crit. Rev. Microbiol.* **22**, 257–277.

Lindsay, J., A. Mach, M. Wilkinson *et al.* (1993). *Clostridium perfringens* type A cytotoxic enterotoxin as trigger for death in the sudden infant death syndrome. *Curr. Microbiol.* **27**, 51–59.

Loffler, A. and R. Labbe (1983). Intracellular proteases during sporulation and enterotoxin formation by *Clostridium perfringens* type A. *Curr. Microbiol.* **8**, 187–190.

Loffler, A. and R. Labbe (1985). Isolation of an inclusion body from sporulating, enterotoxin-positive *Clostridium perfringens. FEMS Microbiol. Lett.* **27**, 143–147.

Loffler, A. and R. Labbe (1986). Characterization of a parasporal inclusion body from sporulating, enterotoxin-positive *Clostridium perfringens* type A. *J. Bacteriol.* **165**, 542–548.

Mach, A. and J. Lindsay (1994). Activation of *Clostridium perfringens* cytotoxic enterotoxin in vivo and *in vitro*, role in triggers for sudden infant death. *Curr. Microbiol.* **28**, 261–267.

Mahony, D. (1974). Bacteriocin susceptibility of *Clostridium perfringens*, a provisional typing scheme. *Appl. Microbiol.* **28**, 172–176.

Mahony, D., M. Stringer, S. Borriello and J. Mader (1987). Plasmid analysis as a means of strain differentiation in *Clostridium perfringens. J. Clin. Microbiol.* **25**, 1333–1335.

Mahony, D., E. Gilliatt, S. Dawson *et al.* (1989). Vero cell assay for rapid detection of *Clostridium perfringens* enterotoxin. *Appl. Environ. Microbiol.* **55**, 2141–2143.

Mahony, D., R. Ahmed and S. Jackson (1992). Multiple typing techniques applied to a *Clostridium perfringens* food poisoning outbreak. *J. Appl. Bacteriol.* **72**, 309–314.

Markovic, L (1993). In vitro effect of *Clostridium perfringens* enterotoxin on vero cells and in vivo effect in animals. *Acta Veterinaria (Beograd)* **43**, 191–198.

Maslanka, S., J. Kerr, G. Williams *et al.* (1999). Molecular subtyping of *Clostridium perfringens* by pulsed-field gel electrophoresis to facilitate foodborne disease outbreak investigations. *J. Clin. Microbiol.* **37**, 2209–2214.

Matches, J., J. Liston and D. Curran (1974). *Clostridium perfringens* in the environment. *Appl. Microbiol.* **28**, 655–660.

Matsuda, M., K. Ozutsumi, H. Iwashi and N. Sugimoto (1986). Primary action of *Clostridium perfringens* type A enterotoxin on Hela and Vero cells in the absence of extracellular calcium, rapid and characteristic changes in membrane permeability. *Biochem. Biophys. Res. Commun.* **141**, 704–710.

Matsushita, C., O. Matsushita, M. Koyama and A. Okabe (1994). A *Clostridium perfringens* vector for the selection of promoters. *Plasmid* **31**, 317–319.

McClane, B (1994). *Clostridium perfringens* enterotoxin acts by producing small molecule permeability alterations in plasma membranes. *Toxicology* **87**, 43–67.

McClane, B (1996). An overview of *Clostridium perfringens* enterotoxin. *Toxicon* **34**, 1335–1343.

McClane, B. (1997). *Clostridium perfringens*. In: M. Doyle, L. Beuchat and T. Montville (eds), *Food Microboiology:Fundamentals and Frontiers*, pp. 305–326. American Society for Microbiology Press, Washington, DC.

McClane, B (2001). *Clostridium perfringens*. In: M. Doyle, L. Beuchat and T. Montville (eds), *Food Microbiology, Fundamentals and Frontiers*, 2nd edn, pp. 351–372. American Society for Microbiology Press, Washington, DC.

McClane, B. and J. McDonel (1979). The effect of *Clostridium perfringens* enterotoxin on morphology, viability and macromolecular synthesis in Vero cells. *J. Cell Physiol.* **99**, 191–200.

McClane, B., P. Hanna and A. Wnek (1988a). *Clostridium perfringens* type A enterotoxin. *Microbiol. Pathog*. **4**, 317–323.

McClane, B., A. Wnek, K. Hulkower and P. Hanna (1988b). Divalent cation involvement in the action of *Clostridium perfringens* type A enterotoxin, early events in enterotoxin action are divalent cation dependent. *J. Biol. Chem*. **263**, 2423–2435.

McClung, L. (1945). Human food poisoning due to growth of *Clostridium perfringens* (*welchii*) in freshly cooked chicken, preliminary note. *J. Bacteriol*. **50**, 229–231.

McDonel, J. (1974). *In vivo* effects of *Clostridium perfringens* enteropathogenic factors on the rat ileum. *Infect. Immun*. **10**, 1156–1162.

McDonel, J. (1979). The molecular mode of action of *Clostridium perfringens* enterotoxin. *Am. J. Clin. Nutr*. **32**, 210–218.

McDonel, J. (1980). Binding of *Clostridium perfringens* ^{125}I-enterotoxin to rabbit intestinal cells. *Biochemistry* **21**, 4801–4807.

McDonel, J. and T. Asano (1975). Analysis of unidirectional sodium fluxes during diarrhea induced by *Clostridium perfringens* enterotoxin in the rat terminal ileum. *Infect. Immun*. **2**, 274–280.

McDonel, J. and C. Duncan (1975). Histopathological effects of *Clostridium perfringens* enterotoxin in the rabbit ileum. *Infect. Immun*. **12**, 1214–1218.

McDonel, J. and C. Duncan (1977). Regional localization of activity of *Clostridium perfringens* typer A enterotoxin in the rabbit ileum, jejunum, and duodenum. *J. Infect. Dis*. **136**, 661–666.

McDonel, J. and B. McClane (1981). Highly sensitive assay for *Clostridium perfringens* enterotoxin that uses inhibition of plating efficiency of Vero cells grown in culture. *J. Clin. Microbiol*. **13**, 940–946.

McDonel, J. and B. McClane (1988). Production, purification and assay of *Clostridium perfringens* enterotoxin. *Methods Enzymol*. **165**, 94–103.

McDonel, J., L. Chang, J. Pounds and C. Duncan (1978). The effects of *Clostridium perfringens* enterotoxin on rat and rabbit ileum. *Lab. Investig*. **39**, 210–218.

McKillop, E. (1959). Bacterial contamination of hospital food with special reference to *Clostridium welchii* food poisoning. *J. Hygiene* **57**, 31–35.

Mead, G. (1969a). Combined effect of salt concentration and redox potential of the medium on the initiation of vegetative growth by *Clostridium welchii. J. Appl. Bacteriol*. **32**, 468–470.

Mead, G. (1969b). Growth and sporulation of *Clostridium welchii* in breast and leg muscle of poultry. *J. Appl. Bacteriol*. **32**, 86–91.

Mead, P., L. Slutsker, V. Dietz *et al.* (1999). Food-related illness and death in the United States. *Emerg. Infect. Dis*. **5**, 607–625.

Meer, R. and J. G. Songer (1977). Multiplex polymerase chain reaction assay for genotyping *Clostridium perfringens. Am. J. Vet. Res*. **58**, 702–706.

Mehta, R., K. Narayan and S. Notermans (1989). DOT-enzyme linked immunosorbent assay for detection of *Clostridium perfringens* type A enterotoxin. *Intl J. Food Microbiol*. **9**, 45–50.

Melville, S., R. Labbe and A. Sonenshein (1994). Expression from the *Clostridium perfringens cpe* promotor in *C. perfringens* and *Bacillus subtilis. Infect. Immun*. **62**, 5550–5558.

Melville, S., R. Collie and B. McClane (1997). Regulation of enterotoxin production in *C. perfringens*. In: J. Rood, B. McClane, G. Songer and R. Titball (eds), *Molecular Biology and Pathogenesis of the Clostridia*, pp. 471–487. Academic Press, London.

Michel, P., M. Buchholz, M. Rolke *et al* (2001). Claudin-4, a new target for pancreatic cancer treatment using *Clostridium perfringens* enterotoxin. *Gastroenterology* **121**, 678–684.

Miwa, N., T. Nishina, S. Kubo and K. Fujikura (1996). Nested PCR for the detection of low levels enterotoxigenic *Clostridium perfringens* in animal feces and meat. *J. Vet. Med. Sci.* **58**, 197–203.

Miwa, N., T. Nishina, S. Kubo *et al.* (1998). Amount of enterotoxigenic *Clostridium perfringens* in meat detected by nested PCR. *Intl J. Food Microbiol.* **42**, 195–200.

Miwantani, T., T. Honda, S. Arai and A. Nakayama (1978). Mode of lethal activity of *Clostridium perfringens* type A enterotoxin. *Jpn J. Med. Sci. Biol.* **31**, 221–223.

Moller, K. and P. Ahrens (1996). Comparison of toxicity neutralization, ELISA and PCR tests for typing of *Clostridium perfringens* and detection of the enterotoxin gene by PCR. *Anaerobe* **2**, 103–110.

Mpamugo, O., T. Donovan and M. Brett (1995). Enterotoxigenic *Clostridium perfringens* as a cause of sporadic diarrhea. *J. Med. Microbiol.* **43**, 442–445.

Naik, H. and C. Duncan (1977a). Enterotoxin formation in foods by *Clostridium perfringens* type *A. J. Food Safety* **1**, 7–18.

Naik, H. and C. Duncan (1977b). Rapid detection and quantitation of *Clostridium perfringens* enterotoxin by counterimmunoelectrophoresis. *Appl. Environ. Microbiol.* **34**, 125–128.

Naik, H. and C. Duncan (1978). Thermal inactivation of *Clostridium perfringens* enterotoxin. *J. Food Prot.* **41**, 100–103.

Nakagawa, M. and S. Nishida (1969). Heat resistance and toxigenicity of *Clostridium perfringens* strains in normal intestines of Japanese. *Jpn J. Microbiol.* **13**, 133–138.

Nakamura, M., and K. Kelly. 1968. *Clostridium perfringens* in dehydrated soups and sauces. *J. Food Sci.* **33**, 424–425.

Nakamura, M., J. Schulze and W. Cross (1969). Factors affecting lecithinase activity and production in *Clostridium welchii. J. Hygiene* **67**, 153–162.

Neut, C., J. Pathak, C. Romond and H. Beerens (1985). Rapid detection of *Clostridium perfringens*, comparison of lactose sulfite broth with tryptose-sulfite-cycloserine agar. *J. Assoc. Off. Anal. Chem.* **68**, 881–883.

Niilo, L. (1971). Mechanism of action of the enteropathogenic factor of *Clostridium perfringens* type A. *Infect. Immun.* **3**, 100–106.

Niilo, L. (1972). The effect of enterotoxin of *Clostridium welchii* (perfringens) on the systemic blood pressure of sheep. *Res. Vet. Sci.* **13**, 503–505.

Niilo, L. (1973a). Fluid secretory response of bovine Thiry jejunal fistula to enterotoxin of *Clostridium perfringens. Infect. Immun.* **1**, 1–4.

Niilo, L. (1973b). Effect on calves of the intravenous injection of the enterotoxin of *Clostridium perfringens* type A. *J. Comp. Pathol.* **83**, 265–269.

Niilo, L. (1974). Response of the ligated intestinal loops in chicken to the enterotoxin of *Clostridium perfringens. Appl. Microbiol.* **28**, 889–891.

Niilo, L. (1976). The effect of enterotoxin of *Clostridium welchii* (*perfringens*) type A on fowls. *Res. Vet. Sci.* **20**, 225–226.

Niilo, L., A. Hauschild and W. Dorward (1971). Immunization of sheep against experimental *Clostridium perfringens* enteritis. *Can. J. Microbiol.* **17**, 391–395.

Notermans, S., C. Heuvelman, H. Beckers and T. Uemura (1984). Evaluation of the ELISA as a tool in diagnosing *Clostridium perfringens* enterotoxin. *Zentralbl. Bakteriol. Hyg. Abt. 1 Orig. B.* **179**, 225–234.

Oakley, C. and G. Warrack (1953). Routine typing of *Clostridium welchii. J. Hyg. Camb.* **51**, 102–107.

Oka, S., T. Tsuzuki and Y. Ando (1983). Relationship between heat resistance and germination of spores of *Clostridium perfringens. J. Food Hyg. Soc. Jpn* **24**, 390–395.

Olivieri, V. (1982). Bacterial indicators of pollution. In: W. Pipes (ed.), *Bacterial Indicators of Pollution*, pp. 21–41. CRC Press, Boca Raton, Fl.

Paine, G. and R. Cherniak (1975). Composition of the capsular polysaccharides of *Clostridium perfringens* as a basis for their classification by chemotypes. *Can. J. Microbiol.* **21**, 181–185.

Park, K. B. and R. Labbe (1990). Proteolysis of *Clostridium perfringens* type A enterotoxin during purification. *Infect. Immun.* **58**, 1999–2001.

Park, K. B. and R. Labbe (1991a). Isolation and characterization of extracellular proteases of *Clostridium perfringens* type A. *Curr. Microbiol.* **23**, 215–219.

Park, K. B. and R. Labbe (1991b). Purification and characterization of intracellular proteases of *Clostridium perfringens* type A. *Can. J. Microbiol.* **37**, 19–27.

Pearson, M. and H. Walker (1976). Effect of oxidation reduction potential upon growth and sporulation of *Clostridium perfringens. J. Milk Food Technol.* **39**, 421–425.

Peck, M. W., D. A. Fairbarn and B. M. Lund (1993). Heat resistance of spores of non-proteolytic *Clostridium botulinum* estimated on medium containing lysozyme. *Lett. Appl. Microbiol.* **16**, 126–131.

Perigo, J. and T. A. Roberts (1968). Inhibition of clostridia by nitrite. *J. Food Technol.* **3**, 91–95.

Phillips-Jones, M. (1993). Bioluminescence (*lux*) expression in the anaerobe *Clostridium perfringens. FEMS Microbiol. Lett.* **106**, 265–270.

Phillips-Jones, M., L. Iwanejko and M. Longden (1989). Analysis of plasmid profiling as a method for rapid differentiation of food-associated *Clostridium perfringens* strains. *J. Appl. Bacteriol.* **67**, 243–254.

Pinegar, J. and M. Stringer (1977). Outbreaks of food poisoning attributed to lecithinase-negative *Clostridium welchii. J. Clin. Pathol.* **30**, 491–492.

Pons, J.-L., B. Picard, P. Niel *et al.* (1993). Esterase electrophoretic polymorphism of human and animal strains of *Clostridium perfringens. Appl. Environ. Microbiol.* **59**, 496–501.

Pons, J.-L., M.-L. Combe and G. Leluan (1994). Multilocus enzyme typing of human and animal strains of *Clostridium perfringens. FEMS Microbiol. Lett.* **121**, 25–30.

Powers, E., R. Lawyer and Y. Masuoka (1975). Microbiology of processed spices. *J. Milk Food Technol.* **38**, 683–687.

Reynolds, D., H. Tranter and P. Hambleton (1986). Scaled-up production and purification of *Clostridium perfringens* type A enterotoxin. *J. Appl. Bacteriol.* **60**, 517–525.

Ridell, J., J. Bjorkroth, H. Eisgruber *et al.* (1998). Prevalence of the enterotoxin gene and clonality of *Clostridium perfringens* strains associated with food poisoning outbreaks. *J. Food Prot.* **61**, 240–243.

Riha, W. and M. Solberg (1971). Chemically defined medium for the growth of *Clostridium perfringens. Appl. Microbiol.* **22**, 738–739.

Riha, W. and M. Solberg (1975a). Chemically defined medium for growth of *Clostridium perfringens. Appl. Microbiol.* **22**, 738–739.

Riha, W. and M. Solberg (1975b). *Clostridium perfringens* inhibition by sodium nitrite as a function of pH, inoculum size, and heat. *J. Food Sci.* **40**, 439–444.

Roach, R. and D. Sienko (1992). *Clostridium perfringens* outbreak associated with minestrone soup. *Am. J. Epidemiol.* **136**, 1288–1291.

Robach, M., F. Ivey and C. Hickey (1978). System for evaluating clostridial inhibition in cured meat products. *Appl. Environ. Microbiol.* **36**, 210–212.

Roberts, D. (1972). Observations on procedures for thawing and spit-roasting frozen dressed chickens, and post-cooking care and storage, with particular reference to food poisoning bacteria. *J. Hygiene* **70**, 565–588.

Roberts, D. (1982). Factors contributing to outbreaks of food poisoning in England and Wales 1970–1979. *J. Hygiene* **89**, 491–498.

Roberts, D., W. Hoover and M. Greenwood (1995). *Practical Food Microbiology*, 2nd edn. Public Health Service, London.

Roberts, I., W. M. Holmes and P. Hylemon (1988). Development of a new shuttle plasmid system for *Escherichia coli* and *Clostridium perfringens. Appl. Environ. Microbiol.* **54**, 268–270.

Roberts, T. A. (1968). Heat and radiation resistance and activation of spores of *Clostridium welchii. J. Appl. Bacteriol.* **31**, 133–144.

Roberts T. A. and C. Derrick (1978). The effect of curing salts on the growth of *Clostridium perfringens* (*welchii*) in laboratory medium. *J. Food Technol.* **13**, 349–353.

Robinson, J. and M. Messer (1969). *Clostridium perfringens* food poisoning –Texas. *Morbid. Mortal. Weekly Rep.* **18**, 20–21.

Rood, J. (1997). Genetic analysis in *Clostridium perfringens*. In: L.J. Rood, B. McClane, G. Songer and R. Titball (eds), *Molecular Biology and Pathogenesis of the Clostridia*, pp. 65–72. Academic Press, London.

Rood, J. and S. Cole (1991). Molecular genetics and pathogenesis of *Clostridium perfringens. Microbiol. Rev.* **55**, 621–648.

Rood, J. and M. Lyristis (1995). Regulation of extracellular toxin production in *Clostridium perfringens. Trends Microbiol.* **3**, 192–196.

Roy, R., F. Busta and D. Thompson (1981). Thermal inactivation of *Clostridium perfringens* after growth at several constant and linearly rising temperature. *J. Food Sci.* **46**, 1586–1590.

Ryu, S. and R. Labbe (1989). Coat and enterotoxin-related proteins in *Clostridium perfringens* spores. *J. Gen. Microbiol.* **135**, 3109–3118.

Ryu, S. and R. Labbe (1993). A 48 kilodalton enterotoxin-related protein from *Clostridium perfringens* vegetative and sporulating cells. *Intl J. Food Microbiol.* **17**, 343–348.

Sacks, L. (1983). Influence of carbohydrates on growth and sporulation of *Clostridium perfringens* in a defined medium with or without guanosine. *Appl. Environ. Microbiol.* **46**, 1169–1175.

Sacks, L. and P. Thompson (1977). Increased spore yields of *Clostridium perfringens* in the presence of methylxanthines. *Appl. Environ. Microbiol.* **34**, 189–193.

Sacks, L. and P. Thompson (1978). Clear defined medium for the sporulation of *Clostridium perfringens. Appl. Environ. Microbiol.* **35**, 405–410.

Saito, M. (1990). Production of enterotoxin by *Clostridium perfringens* derived from humans, animals, foods, and the natural environment in Japan. *J. Food Prot.* **53**, 115–118.

Saito, M., M. Matsumoto and M. Funabashi (1992). Detection of *Clostridium perfringens* enterotoxin gene by the polymerase chain reaction amplification procedure. *Intl J. Food Microbiol.* **17**, 47–55.

Sakaguchi, G., T. Uemura and H. Riemann (1973). Simplified method for purification of *Clostridium perfringens* enterotoxin. *Appl. Microbiol.* **26**, 762–767.

Salinovich, O., W. Mattice and E. Blakeney Jr (1982). Effects of temperature, pH and detergents on the molecular conformation of the enterotoxin of *Clostridium perfringens. Biochim. Biophys. Acta* **707**, 147–153.

Sanders, S. and R. Hutchenson Jr (1974). *Clostridium perfringens* – Tennessee. *Morbid. Mortal. Weekly Rep.* **23**, 19–20.

Sarker, M., R. Carman and B. McClane (1999). Inactivation of the gene (*cpe*) encoding *Clostridium perfringens* enterotoxin eliminates the ability of two cpe-positive *C. perfringens* type A human gastrointestinal disease isolates to affect rabbit ileal loops. *Mol. Microbiol.* **33**, 946–958.

Sarker, M., R. Shivers, S. Sparks *et al.* (2000). Comparative experiments to examine the effects of heating on vegetative cells and spores of *Clostridium perfringens* isolates carrying plasmid enterotoxin genes versus chromosomal enterotoxin genes. *Appl. Environ. Microbiol.* **60**, 3234–3240.

Schalch, B., J. Bjorkroth, H. Eisgruber *et al.* (1997). Ribotyping for strain characterization of *Clostridium perfringens* isolates from food poisoning cases and outbreaks. *Appl. Environ. Microbiol.* **63**, 3992–3994.

Schalch, B., B. Sperner, H. Eisgruber and A. Stolle (1999). Molecular methods for the analysis of *Clostridium perfringens* relevant to foods. *FEMS Immunol. Microbiol.* **24**, 281–286.

Schalch, B., H. Bader, R. Schau *et al.* (2003). Molecular typing of *Clostridium perfringens* from a foodborne disease outbreak in a nursing home, rybotyping versus pulsed field gel electrophoresis. *J. Clin. Microbiol.* **41**, 892–899.

Scott, P. and J. Rood (1989). Electroporation mediated transformation of lysostaphin-treated *Clostridium perfringens. Gene* **82**, 327–333.

Sebald, M. and R. Costilow (1975). Minimal growth requirements for *Clostridium perfringens* and isolation of auxotrophic mutants. *Appl. Microbiol.* **29**, 1–6.

Sherman, S., E. Klein and B. McClane (1994). *Clostridium perfringens* type A enterotoxin induces tissue damage and fluid accumulation in rabbit ileum. *J. Diarrheal Dis. Res.* **12**, 200–207.

Shigehisa, T., T. Nakagami and S. Taji (1985). Influence of heating and cooling rates on spore germination and growth of *Clostridium perfringens* in media and in roast beef. *Jpn J. Vet. Sci.* **47**, 259–267.

Shimizu, T., K. Ohtani, H. Hirakawa *et al.* (2002). Complete genome sequence of *Clostridium perfringens*, an anaerobic flesh-eater. *Proc. Natl. Acad. Sci. USA* **99**, 996–1001.

Shoemaker, S. and M. Pierson (1976). 'Phoenix phenomenon' in the growth of *Clostridium perfringens. Appl. Environ. Microbiol.* **32**, 803–807.

Sisson, P., J. Kramer, M. Brett and N. Lightfoot (1992). Application of pyrolysis mass spectrometry to the investigation of outbreaks of food poisoning and non-gastrointestinal infection associated with *Bacillus* species and *Clostridium perfringens. Intl J. Food Microbiol.* **17**, 57–66.

Skjelkvale, R. and C. Duncan (1975). Enterotoxin formation by different toxigenic types of *Clostridium perfringens. Infect. Immun.* **11**, 563–575.

Skjelkvale, R. and T. Uemura (1977a). Detection of enterotoxin in feces and anti-enterotoxin in serum after *Clostridium perfringens* food poisoning. *J. Appl. Bacteriol.* **42**, 355–363.

Skjelkvale, R. and T. Uemura (1977b). Experimental diarrhea in human volunteers following oral adminstration of *Clostridium perfringens* enterotoxin. *J. Appl. Bacteriol.* **43**, 281–286.

Skjelkvale, R., M. Stringer and J. Smart (1979). Enterotoxin production by lecithinase-positive and lecithinase-negative *Clostridium perfringens* isolated from food poisoning outbreaks and other sources. *J. Appl. Bacteriol.* **47**, 329–339.

Skjelkvale, R., H. Tolleshaug and T. Jarmud (1980). Binding of enterotoxin from *Clostridium perfringens* type A to liver cells in vivo and in vitro. The enterotoxin causes membrane leakage. *Acta Pathol. Microbiol. Immunol. Scand. Sect. B* **88**, 95–102.

Sloan, J., T. Warner, P. Scott *et al.* (1992). Construction of a sequenced *Clostridium perfringens – Escherichia coli* shuttle plasmid. *Plasmid* **27**, 207–219.

Smart, J., T. A. Roberts, F. Stringer and N. Shan (1979). The incidence and serotypes of *Clostridium perfringens* on beef, pork, and lamb carcasses. *J. Appl. Bacteriol.* **46**, 377–383.

Smith, L. (1972). Factors involved in the isolation of *Clostridium perfringens. J. Milk Food Technol.* **35**, 71–76.

Smith, L. and M. Gardner (1949). Vegetative cells of *Clostridium perfringens* in soil. *J. Bacteriol.* **58**, 407–411.

Smith, L. and B. Williams (1984). *The Pathogenic Anaerobic Bacteria*, pp. 1–34. Charles Thomas Publishing Co., Springfield, IL.

Smith, M. (1985). Companion of rapid perfringens medium and lactose sulfite medium for detection of *Clostridium perfringens. J. Assoc. Off. Anal. Chem.* **68**, 807–808.

Songer, J. G. and R. Meer (1996). Genotyping of *Clostridium perfringens* by polymerase chain reaction is a useful adjunct to diagnosis of clostridial enteric disease in animals. *Anaerobe* **2**, 197–203.

Sorensen, D., S. Eberl and R. Dicksa (1989). *Clostridium perfringens* as a point source indicator in non-point polluted streams. *Water Res.* **23**, 191–197.

Spencer, R. (1969). Neomycin-containing media in the isolation of *Clostridium botulinum* and food poisoning strains of *Clostridium welchii. J. Appl. Bacteriol.* **32**, 170–174.

Stark, R. and C. Duncan (1972). Purification and biochemical properties of *Clostritium perfringens* type A enterotoxin. *Infect. Immun.* **6**, 662–673.

Stelma, G., C. Johnson and D. Shah (1985). Detection of enterotoxin in colonies of *Clostridium perfringens* by a solid phase enzyme-linked immunosorbent assay. *J. Food Prot.* **48**, 227–231.

Sterne, M. and I. Batty (1975). Criteria for diagnosing clostridial infection. *In* M. Sterne (ed.), *Pathogenic Clostridia*, pp. 79–122. Butterworths, London.

St John, W., J. Matches and M. Wekell (1982). Use of iron milk medium for enumeration of *Clostridium perfringens. J. Assoc. Off. Anal. Chem.* **65**, 1129–1133.

Stringer, M. (1985). *Clostridium perfringens* type A food poisoning. In: S. Borrellio (ed.), *Clostridia in Gastrointestinal Disease*, pp. 117–143. CRC Press, Boca Raton, FL.

Stringer, M., G. Watson and R. Gilbert (1982). *Clostridium perfringens* type A, serological typing and methods for the detection of enterotoxin. In: J. Corry, D. Robert and F. Skinner (eds), *Isolation and Identification Methods for Food Poisoning Organisms*, pp. 111–135. Academic Press, London.

Stringer, M., P. Turnbull and R. Gilbert (1990). Application of serological typing to the investigation of outbreaks of *Clostridium perfringens* food poisoning, 1970–1978. *J. Hygiene* **84**, 443–456.

Stringer, S. and M. Peck (1996). Vegetable juice aids the recovery of heated spores of non-proteolytic *Clostridium botulinum. Lett. Appl. Microbiol.* **23**, 407–411.

Strong, D., J. Canada and B. Griffiths (1963). Incidence of *Clostridium perfringens* in frozen chicken gravy. *J. Food Sci.* **29**, 479–482.

Strong, D., K. Weiss and L. Higgins (1966). Survival of *Clostridium perfringens* in starch pastes. *J. Am. Diet. Assoc.* **49**, 191–194.

Strong, D., E. Foster and C. Duncan (1970). Influence of water activity on the growth of *Clostridium perfringens. Appl. Microbiol.* **19**, 980–987.

Strong, D., C. Duncan and G. Perna (1971). *Clostridium perfringens* type A food poisoning. II. Response of the rabbit ileum as an indication of enteropathogenicity of stains of *Clostridium perfringens* in human beings. *Infect. Immun.* **3**, 171–178.

Sugii, S. and Y. Horiguchi (1988). Identification and isolation of the binding substance for *Clostidium perfringens* enterotoxin on Vero cells. *FEMS Microbiol. Lett.* **52**, 85–90.

Sunagawa, H., T. Tsuzuki, K. Takeshi *et al.* (1987). Enterotoxigenicity of heat-resistant and heat-sensitive strains of *Clostridium perfringens. Jpn J. Vet. Sci.* **49**, 853–858.

Sutton, R. and B. Hobbs (1968). Food poisoning caused by heat-sensitive *Clostridium welchii.* A report of five recent outbreaks. *J. Hygiene* **66**, 135–142.

Sutton, R. and B. Hobbs (1971). Resistance of vegetative cells of *Clostridium welchii* to low pH. *J. Med. Microbiol.* **4**, 539.

Tabatabi, L. and H. Walker (1970). Oxidation–reduction potential and growth of *Clostriduim perfringens* and *Pseudomonas fluorescens. Appl. Microbiol.* **20**, 441–446.

Takumi, K., A. Takeoka, T. Koga and Y. Endo (1991). Relationship between sporulation, enterotoxin production, and inclusion body formation in *Clostridium perfringens* NCTC 8798. *J. Gen. Appl. Microbiol.* **37**, 341–353.

Tang, S. and R. Labbe (1987). Mode of action of *Clostridium perfringens* initiation protein (spore-lytic enzyme). *Ann. Inst. Pasteur* **138**, 597–608.

Taniguti, T. and B. Zenitani (1969). Incidence of *Clostridium perfringens* in fishes. I. On the application of the LAS medium to detection of *Clostridium perfringens. J. Food Hyg. Soc. Jpn* **10**, 199–203.

Taylor, A. and W. Gordon (1940). A survey of the types of *Clostridium, welchii* present in soil and in the intestinal contents of animals and man. *J. Pathol. Bacteriol.* **50**, 271–276.

Ting, M. and D. Fung (1972). Chemically defined medium for growth and sporulation of *Clostridium perfringens. Appl. Microbiol.* **24**, 755–759.

Todd, E. (1989a). Preliminary estimates of costs of foodborne disease in the United States. *J. Food Prot.* **52**, 595–601.

Todd, E. (1989b). Cost of acute bacterial foodborne disease in Canada and the United States. *Intl J. Food Microbiol.* **9**, 313–326.

Traci, P. and C. Duncan (1974). Cold shock lethality and injury in *Clostridium perfringens. Appl. Microbiol.* **28**, 815–821.

Tsai, C. and H. Riemann (1974). Relation of enterotoxigenic *Clostridium perfringens* type A to food poisoning. 1. Effect of heat activation on the germination, sporulation and enterotoxigenesis of *Clostridium perfringens. J. Formosan Med. Assoc.* **73**, 653–659.

Tschirdewahn, B., S. Notermans, K. Wernars and F. Untermann (1991). The presence of enterotoxigenic *Clostridium perfringens* strains in feces of various animals. *Intl J. Food Microbiol.* **14**, 175–178.

Uemura, T. (1978a). Sporulation and enterotoxin production by *Clostridium perfringens* on solidified sporulation media. *J. Food Hyg. Soc. Jpn* **19**, 462–467.

Uemura, T. (1978b). Incidence of enterotoxigenic *Clostridium perfringens* in healthy humans in relation to the enhancement of enterotoxin productjion by heat treatment. *J. Appl. Bacteriol.* **44**, 411–419.

Uemura, T. and R. Skjelkvale (1976). An enterotoxin produced by *Clostridium perfringens* type D. Purification by affinity chromatography. *Acta Pathol. Microbiol. Immunol. Scand. Sect. B* **84**, 414–420.

Uemura, T., G. Sakaguchi and H. Riemann (1973). In vitro production of *Clostridium perfringens* enterotoxin and its detection by reversed passive hemagglutination. *Appl. Microbiol.* **26**, 381–385.

Uemura, T., C. Genigeorgis, H. Riemann and C. Franti (1974). Antibody against *Clostridium perfringens* type A enterotoxin in human sera. *Infect. Immun.* **9**, 470–471.

Uemura, T., G. Sakaguchi, T. Itoh *et al.* (1975). Experimental diarrhea in cynomogus monkeys by oral administration with *Clostridium perfringens* type A viable cells or enterotoxin. *Jpn J. Med. Sci. Biol.* **28**, 165–177.

Uemura, T., S. Yoshitake, D. Hu and T. Kajikawa (1992). A highly sensitive ELISA for *Clostridium perfringens. Lett. Appl. Microbiol.* **15**, 23–25.

US Food and Drug Administration (1995). *Bacteriological Analytical Manual*, pp. 16.01–16.06. AOAC International, Gaithersburg, MD.

US Food and Drug Admin., Div. Retail Food Prot (1997). *Food Code*, p. 55. US Dept Health Services, Public Health Service, Washington, DC.

Uzal, F., J. Plumb, L. Blackall and W. Kelly (1997). PCR detection of *Clostridium perfringens* producing different toxins in feces of goats. *J. Appl. Microbiol.* **25**, 339–344.

Van Damme-Jongsten, M., K. Werners and S. Notermans (1989). Cloning and sequencing of the *Clostridium perfringens* enterotoxin gene. *Antonie van Leeuw* **56**, 181–190.

Van Damme-Jongsten, M., J. Rodhouse, R. Gilbert and S. Notermans (1990). Synthetic DNA probes for detection of enterotoxigenic *Clostridium perfringens* strains isolated from outbreaks of food poisoning. *J. Clin. Microbiol.* **28**, 131–133.

Walker, H. (1975). *Clostridium perfringens. Crit. Rev. Food Sci. Nutr*. **7**, 71–104.

Walker, P. D. (1985). Pig-bel. In: S. Borriello (ed.), *Clostridia in Gastrointestinal Disease*, pp. 93–116. CRC Press, Boca Raton, FL.

Wang, R. E., W.-W. Cao, W. Franklin *et al* (1994). A 16S rDNA-based PCR method for rapid and specific detection of *Clostridium perfringens* in food. *Mol. Cellul. Probes* **8**, 131–138.

Watson, G. (1985). The assesment and application of a bacteriocin typing scheme for *Clostridium perfringens. J. Hygiene* **94**, 69–79.

Watson, G., M. Stringer, R. Gilbert and D. Mahony (1982). The potential of bacteriocin typing in the study of *Clostridium perfringens* food poisoning. *J. Clin. Pathol.* **35**, 1361–1365.

Weiss, K. and D. Strong (1967). Some properties of heat resistant and heat sensitive strains of *Clostridium perfringens. J. Bacteriol.* **93**, 21–26.

Wen, Q. and B. McClane (2004). Detection of enterotoxigenic *Clostridium perfringens* type A isolates in American retail foods. *Appl. Environ. Microbiol.* **70,** 2685–2691.

Wieckowski, E., A. Wnek and B. McClane (1994). Evidence that an ~50 kDa mammalian plasma membrane protein with receptor-like properties mediates the amphiphilicity of specifically bound *Clostridium perfringens* enterotoxin. *J. Biol. Chem.* **269**, 10838–10848.

Willardsen, R., F. Busta, C. Allen and L. Smith (1978). Growth and survival of *Clostridium perfringens* during constantly rising temperatures. *J. Food Sci.* **43**, 470–475.

Willardsen, R., F. Busta and C. Allen (1979). Growth of *C. perfringens* in three different beef media and fluid thioglycollate medium at static and constantly rising temperature. *J. Food Prot*. **42**, 144–148.

Willis, A. T. (1969). *Clostridium welchii*. In: *Clostridia of Wound Infections*, pp. 41–156. Butterworths, London.

Wnek, A. and B. McClane (1983). Identification of a 50,000 molecular weight protein from rabbit brush border membranes that binds *Clostridium perfringens* enterotoxin. *Biochem. Biophys. Res. Commun.* **112**, 1099–1105.

Wnek, A. and B. McClane (1986). Comparison of receptors for *Clostridium perfringens* type A and cholera enterotoxins in isolated rabbit intestinal brush border membranes. *Microbiol. Path.* **1**, 89–100.

World Health Organization (1995). *WHO Surveillance Program for Control of Foodborne Infections and Intoxications in Europe* (ed. K. Schmidt). Federal Institute for Health Protection of Consumers and Veterinary Medicine, Berlin.

Wrigley, D. (1990). Phospholipase C production and *Clostridium perfringens* growth in soy meal/ground beef mixtures. *J. Food Safety* **10**, 295–306.

Wrigley, D., H. Hanwella and B. Thon (1995). Acid exposure enhances sporulation of certain strains of *Clostridium perfringens. Anaerobe* **1**, 263–269.

Yamagishi, T., K. Sugitani, K. Tanishima and S. Nakamura (1997). Polymerase chain reaction test for differentiation of five toxin types of *Clostridium perfringens. Microbiol. Immunol.* **41**, 295–299.

Yamamoto, K., I. Ohishi and G. Sakaguchi (1979). Fluid accumulation in mouse ligated intestine inoculated with *Clostridium perfringens* type A enterotoxin. *Appl. Environ. Microbiol.* **37**, 181–186.

Yoo, H. S., S. U. Lee, K. Y. Park and Y. H. Park (1997). Molecular typing and epidemiological survey of prevalence of *Clostridium perfringens* types by multiplex PCR. *J. Clin. Microbiol.* **35**, 228–232.

Vibrio infections

5

Riichi Sakazaki[1], *Charles Kaysner, and Carlos Abeyta, Jr*

1 *Vibrio parahaemolyticus*

1.1 Historical aspects

Fujino (1951) and Fujino *et al.* (1953) found a vibrio associated with a large outbreak of food poisoning occurring near Osaka, Japan. Later, Takikawa (1958) described another outbreak of gastroenteritis caused by an organism similar to the one isolated by Fujino *et al.* (1953). In 1960 an explosive epidemic of gastroenteritis occurred along the Pacific coast of Japan, and the Ministry of Health and Welfare established a committee to study this epidemic. Subsequently, a vibrio similar to the organism described by Fujino *et al.* and Takikawa was identified as the etiologic agent.

1.2 Characteristics

1.2.1 Classification and phenotypic characteristics

The organism was given different names, but the most widely accepted name, *Vibrio parahaemolyticus*, was suggested by Sakazaki *et al.* (1963). *V. parahaemolyticus* has a single polar, sheathed flagellum in broth culture, but a young culture on the surface of solid medium may have unsheathed, peritrichous flagella. The vibrio is halophilic, it grows in media containing 1–8 % sodium chloride, but not in media without salt. Many strains swarm on the surface of agar media. *V. parahaemolyticus*

[1] deceased

Foodborne Infections and Intoxications 3e
ISBN-10: 0-12-588365-X
ISBN-13: 978-012-588365-8

Table 5.1 Phenotypic characteristics of *Vibrio* species associated with human disease

Species	Oxidase	Nitrate reduction	Growth in peptone water with (%) NaCl: 0	6	8	10	Swarming	Susceptibility to O/129 (μg/ml): 10	150	Flagellation on solid medium	Indole	Voges-Proskauer	Urease	Lysine decar-boxy-lase	Ornithine decar-boxylase	Arginine dehy-drolase	ONPG	Gas from glucose	Acid from: L arabi-nose	Cellobiose	Lactose	Sucrose	Salicin	Growth on TCBS agar
V. cholerae	+	+	+	d	–	–	–	S	S	M	+	+	–	+	+	–	+	–	–	–	–	+	–	Y
V. mimicus	+	+	+	d	–	–	–	S	S	M	+	–	–	+	+	–	+	–	–	–	d	–	–	G
V. s para-haemolyticu	+	+	–	+	+	–	d	R	S	P	+	–	d	+	+	–	–	–	d	–	–	–	–	G
V. vulnificus	+	+	–	+	–	–	–	S	S	M	+	–	–	+	D	–	+	–	–	+	+	D	+	G,Y
V. fluvialis	+	+	–	+	d	–	–	R	S	P	d	–	–	–	–	+	d	–	+	d	–	+	–	G
V. firnissii	+	+	–	+	d	–	–	R	S	P	d	–	–	–	–	+	d	+	+	d	–	+	–	G
V. alginolyticus	+	+	–	+	+	+	+	R	S	P	+	+	–	+	d	–	–	–	–	–	–	+	–	Y
V. metch-nikovii	–	–	–	d	–	–	–	S	S	M	d	+	–	d	–	d	d	–	–	–	D	+	–	Y
V. hollisae	+	+	–	d	–	–	–	R	S	M	d	–	–	–	–	–	–	–	+	–	–	–	–	–
V. cincinna-tiensis	+	+	–	+	–	–	–	R	S	M	–	+	–	d	–	–	+	–	+	+	–	+	+	Y
V. carchaniae	+	+	–	+	–	–	–	R	S	M	+	d	–	+	–	–	–	–	–	d	–	D	–	G,Y
V. damsela[a]	+	+	–	+	–	–	–	S	S	M	–	+	–	d	–	+	–	–	–	–	–	–	–	G

+, 90–100 % strains positive; –, 0–10 % strains positive; d, 11–89 % strains positive; S, susceptible; R, resistant; M, monotrichous; P, peritrichous; Y, yellow; G, green.

[a] *V. damsela* is presently classified in the genus *Photobacterium*.

is susceptible to the vibrio-static agent O/129 (2,4-diamino-6,7-diisopropyl pteridine) at a concentration of 150 μg/ml but is resistant to 10 μg/ml. The phenotypic characteristics of *V. parahaemolyticus* are summarized in Table 5.1, together with those of other human pathogenic vibrios.

Although the flagellins of the polar and peritrichous flagella of the same strain differ in their immunological properties (Shinoda *et al.*, 1974), antigens made from intact flagellae of different strains of *V. parahaemolyticus* are serologically identical (Sakazaki *et al.*, 1968a; Terada, 1968). An antigenic scheme for *V. parahaemolyticus*, in which 11 O groups and 41 K antigens were recognized, was established by Sakazaki *et al.* (1968a); later, the number of K antigens was expanded to 70.

1.2.2 Ability to survive and grow in the environment

V. parahaemolyticus grows at temperatures between 10° and 44°C, but fails to grow at 4°C. Low temperatures not only arrest multiplication but also cause a rapid initial decrease in numbers of viable cells, although survival for several weeks occurs in refrigerated seafood. Under optimal conditions the generation time of *V. parahaemolyticus* in the exponential phase is 9–13 minutes.

V. parahaemolyticus is found in estuarine environments throughout the world, and can be isolated from estuarine waters during the summer but not in the winter. However, the vibrio can be isolated from the sediment when the temperature is less than 10°C, and it probably survives in the sediment during the winter. It is unable to survive in ocean water. *V. parahaemolyticus* adheres to chitin through production of chitinase; in this way it can colonize zoo-plankton and the surface of shellfish. It also colonizes the digestive tract of shellfish. *V. parahaemolyticus* is often found during the summer in fresh water or streams, but especially in brackish water. It is unknown whether *V. parahaemolyticus* enter into a viable but non-culturable state in cold environments; such a state has been reported for *V. cholerae* and *V. vulnificus*.

1.3 Nature of infection

1.3.1 Clinical manifestations

V. parahaemolyticus produces gastroenteritis in humans, and signs and symptoms usually occur approximately 12 hours after ingestion of contaminated food. The outstanding features are severe abdominal pain and diarrhea. The spectrum of manifestations may be very mild with only a few loose stools, or may be severe with stools containing blood and mucus. Recovery and disappearance of vibrio from stools are usually complete within a few days. Person-to-person transmission has not been observed. The mortality rate of *V. parahaemolyticus* infection is very low. *V. parahaemolyticus* is occasionally isolated from infected skin or soft tissue lesions of fish handlers, but its etiologic role in these infections is unknown.

The minimal infectious dose of *V. parahaemolyticus* in man varies with strain and individual, but was 10^5–10^8 CFU in human volunteer experiments (Takikawa, 1958; Sanyal and Sen, 1974).

1.3.2 Virulence factors

Kato *et al.* (1965) found that strains of *V. parahaemolyticus* isolated from patients with gastroenteritis were hemolytic whereas those from seafish and sea environment were predominantly non-hemolytic on a modified brain heart infusion agar. A thermostable extracellular substance is responsible for the hemolytic reaction. This hemolysis is called the Kanagawa reaction. Sakazaki *et al.* (1986b) found that 96 % of 2720 strains from patients with gastroenteritis were Kanagawa-positive, while only 1 % of 650 environmental strains were Kanagawa-positive. Feeding Kanagawa-negative strains to 15 adult human volunteers failed to induce clinical signs (Sakazaki *et al.*, 1968b). The hemolysin responsible for the Kanagawa reaction is a thermostable direct hemolysin (TDH). It is probable that TDH plays a role in the pathogenesis of the vibrio gastroenteritis.

Some outbreaks of gastroenteritis may be associated with Kanagawa-negative *V. parahaemolyticus*; Honda *et al.* (1988) demonstrated that those Kanagawa-negative strains produced a TDH-related hemolysin (TRH) but not TDH. Shirai *et al.* (1990) and Kishishita *et al.* (1992) also found that not only TDH-positive but also TRH-positive vibrios were associated with gastroenteritis. Further studies have shown that many clinical strains produce both hemolysins, or at least possess the genes that encode them.

A variety of pili and other potential colonization factors have been described for *V. parahaemolyticus*, but substantial studies have indicated that candidate adhesins are lacking.

1.3.3 Toxins

The TDH is a 21-kDa protein that is not affected by heating at 100°C at pH 6.0. It is a pore-forming toxin (Honda *et al.*, 1992) and expresses hemolytic activity, cytotoxicity, enterotoxicity, and cardiotoxicity. In the hemolytic reaction the TDH is strongly active against erythrocytes of dogs, mice, rats, and humans, weakly active against those of rabbits and sheep, and inactive against horse erythrocytes. Nishibuchi *et al.* (1992) reported that a TDH-positive strain caused fluid accumulation in ligated ileal loops in rabbits whereas its TDH-negative mutant failed to do so. It was indicated that G_{T1b} is the intestinal receptor for TDH, but work by Yoh *et al.* (1995) suggested the presence of other unknown receptors.

Although TRH is similar to TDH in biological, immunological, and physicochemical properties, it is thermolabile and differs in activity on erythrocytes (Honda *et al.*, 1988). The TRH is linked epidemiologically to gastroenteritis (Nishibuchi, 1990), but the activity of this toxin is still uncertain.

1.4 Prevalence in food

Coastal fish and shellfish are usually contaminated with *V. parahaemolyticus* during the summer, and may contain high levels of the organism. The vibrios on the surface of fish or shellfish do not proliferate when the temperature is kept below 10°C, but at 20°C the number of vibrios increases very rapidly. Asakawa *et al.* (1974) reported that about 10–100 cells of *V. parahaemolyticus* were found on the surface of coastal fish just after landing, but if they were kept at atmospheric temperatures in the summer the numbers became more than 10^6 within a few hours. Vibrios are not found on market fish in the winter.

Oysters are often contaminated with *V. parahaemolyticus*. In Japan, they are seldom implicated in vibrio gastroenteritis because they are usually eaten during the cold season. However, in the US this species is a leading cause of gastroenteritis associated with raw shellfish consumption, particularly during the summer months (Hlady, 1997). Boiled or roast fish may sometimes be incriminated when the food has been contaminated from raw fish.

1.5 Foods most often associated with human infection

Gastroenteritis due to *V. parahaemolyticus* is almost always associated with seafood. Raw fish meat and shellfish are the most important sources of the disease in Japan, where the high incidence is without doubt due to the national custom of eating raw fish and fish products. However, the lack of correlation of the Kanagawa reaction between isolates from patients and from implicated seafoods is one of the puzzling *V. parahaemolyticus* problems.

Seafoods responsible for illness vary with local eating habits in different countries. Any kind of seafish served as 'sushi' and 'sashimi' can possibly cause gastroenteritis. The vibrio may be killed in 0.5 % acetic acid within several minutes, but raw fish or shellfish treated with vinegar, which is commonly done in Japan, often transmit vibrio gastroenteritis. Vibrio gastroenteritis has sometimes been associated with consumption of raw vegetables which have been contaminated with the vibrio through kitchen utensils.

V. parahaemolyticus infection may be less important in European countries. Infection is unlikely in those countries where people are not in the habit of eating raw fish meat. Nevertheless, cases of gastroenteritis due to this vibrio have been reported in East European countries, the UK and Africa, in addition to the US. In these countries seafoods are usually cooked shortly before consumption, but crab and shrimp – the seafoods most often associated with vibrio infection – are usually handled after cooking, which may result in cross-contamination from other sources. It is probable that *V. parahaemolyticus* is a frequent cause of diarrheal disease in developing countries where people have a poor water supply and poor sanitary conditions. In such countries, waterborne infection with the vibrio could be considered.

1.6 Principles of detection

Although several selective agar media have been devised, TCBS (thiosulfate citrate bile salts sucrose) agar is the recommended medium for isolation of *V. parahaemolyticus*. This species forms green colonies on TCBS agar, since it does not ferment sucrose. However, routine use of TCBS agar for plating of stool specimens may not be cost-effective. MacConkey agar containing additional 0.5 % NaCl is convenient for routine culturing of diarrheal stools.

Enrichment culture is used for detection of the vibrio in foods and from marine sources. Polymyxin salt broth with a pH of 7.6 and containing 2 % NaCl and 50 µg/ml of polymyxin B may yield selective growth of *V. parahaemolyticus*. Alkaline peptone water has also been used in many laboratories for enrichment. It should be noted that some factor(s) in shellfish may inhibit growth of vibrios. It is recommended that shellfish be cut into small pieces but not homogenized. After the pieces have been put into the enrichment broth it should be shaken vigorously and the pieces removed from the broth with forceps. However, enrichment culturing of seafood that is suspected of causing an outbreak may not be helpful, because most of the isolates from the food will be Kanagawa-negative – in contrast to isolates from patients.

The colonial appearance of *V. parahaemolyticus* on TCBS agar is so typical that provisional identification of isolates from stool specimens may be made directly on the plate. However, isolates from marine sources must be further examined in order to differentiate them from related organisms. The addition of 1% NaCl to media for biochemical tests is essential for positive reactions. Miyamoto *et al.* (1989) developed a rapid and sensitive assay for *V. parahaemolyticus* in seafoods using a fluorogenic method.

Wagatsuma agar has been used to test for the Kanagawa reaction. However, the test with Wagatsuma agar is difficult to perform due to problems with the preparation of this specialized medium. Honda *et al.* (1989) reported an enzyme-linked immunosorbent assay for detection of THD-producing vibrios, and Bej *et al.* (1999) described a multiplex PCR procedure to detect both hemolysin genes.

Several molecular approaches to detect Kanagawa-positive vibrios have been developed. DNA probes and oligonucleotide probes specific for the genes encoding TDH (*tdh*) or TRH (*trh*) were described by Nishibuchi *et al.* (1985, 1986) and Yamamoto *et al.* (1992). In their methods, however, the probes also hybridize with *tdh* genes in some strains of *V. cholerae* non-O1, *V. hollisae* and *V. mimicus*. Lee and Pan (1993) reported a polymerase chain reaction technique using oligonucleotide primers derived from the nucleotide sequence of the *tdh* gene. Also, Bej *et al.* (1999) described a multiplex PCR procedure to detect both hemolytic genes.

Strains capable of producing urease have been found to be associated with many clinical cases. On the west coast of the US, the majority of Kanagawa-positive strains were urease positive (Kaysner *et al.*, 1994). Upon further investigation these common serogroups were also found to contain both the *tdh* and *trh* genes. Subsequently, clinical isolates from other areas of the US, and also Asia, were found to be urease positive. *V. parahaemolyticus* possessing the *trh* gene nearly always produces urease (Okuda *et al.*, 1997).

Serotyping of *V. parahaemolyticus* isolates can be performed by slide agglutination tests using O and K antisera. In outbreaks of *V. parahaemolyticus* infections it is, however, seldom that the serovar identified in the patients is also detected in the incriminated seafood. Therefore, serotyping of isolates from seafoods and marine sources is generally not helpful. Several serotypes have been found in a majority of patients, i.e. serogroups O4:K12 and O1:K56 on the west coast of the US. It has recently been found that clinical strains in Asian countries are predominantly serogroup O3:K6 (DePaola *et al.*, 2003). This serogroup also caused the largest reported oyster-associated outbreak in the US, in 1998 (Daniels *et al.*, 2000).

2 *Vibrio vulnificus*

2.1 Historical aspects

Hollis *et al.* (1976) studied a biogroup of marine vibrio referred to as the 'lactose-fermenting' or 'Lac$^+$' vibrio. This vibrio was subsequently given the scientific name *Vibrio vulnificus* by Farmer (1979).

2.2 Characteristics

2.2.1 Classification and phenotypic characteristics

This vibrio can grow on/in ordinary media containing 1–6 % NaCl, but not at 0 % or 8 %. Most strains are encapsulated. This species is divided into two biogroups; biogroup 1 involves both clinical and environmental strains and is indole-positive, while biogroup 2 includes only strains that are pathogenic to eels and is indole-negative. Most strains do not ferment sucrose, but an occasional strain may be fermentative. Cellobiose is fermented by all strains, differentiating this species from *V. parahaemolyticus*. For phenotypic characteristics of *V. vulnificus* see Table 5.1.

Serologically, 18 O groups were defined in this species by Shimada and Sakazaki (1984). Recent studies found that multiple strains may be present in one sample of shellfish, based on pulsed field gel electrophoresis and ribotype patterns (Tamplin *et al.*, 1996).

2.2.2 Virulence factors

Although a variety of potential virulence factors, such as capsule (Kreger *et al.*, 1981), cytotoxin (Gray and Kreger, 1987), collagenase (Smith and Merkel, 1982), siderophore (Simpson and Oliver, 1983) and protease (Kothary and Kreger, 1985) have been reported in clinical and environmental strains of *V. vulnificus*, the role of the capsule in the pathogenesis of human infection appears to be clearly established. Encapsulated cells are virulent to mice, are resistant to the bacteriocidal activity of human serum and to phagocytosis by macrophages, and are able to grow in iron-deficient media; these characteristics distinguish them from unencapsulated cells (Tamplin *et al.*, 1985). In individuals, especially those with hepatic cirrhosis, hepatoma or hemochromatosis, the vibrio induces septicemia. It has been suggested that lack of complement, functional defects of the reticuloendothelial system or the presence of free Fe^{2+} in serum, or all three, are important factors in development of septicemia. However, Biosca *et al.* (1993) suggested that the production of capsule by strains of biotype 2 is not associated with illness in eels.

The cytolysin (hemolysis), proteases, and collagenase are considered tissue-damaging factors in wound infections with *V. vulnificus*, but the production of these factors has been found in strains that are either virulent or avirulent in mice (Morris *et al.*, 1987).

2.2.3 Ability to survive and grow in the environment

V. vulnificus is widely distributed in the estuarine environment, but it can be isolated only during the warmer months. Although unable to be cultured in the cold season, the vibrios are likely to be present but are in an apparently viable but non-culturable state (Oliver and Wanucha, 1989).

2.3 Nature of infection

2.3.1 Human infections

In humans this species is associated with two disease syndromes, primary septicemia and wound infection. Septicemia caused by *V. vulnificus* is very serious, with a case/fatality rate of about 50 %. Progression of illness can be very rapid, from asymptomatic to death within 24 hours. The important risk factor is pre-existing hepatic disease, especially cirrhosis, and about 75 % of patients with primary septicemia caused by *V. vulnificus* have hepatic disease leading to increased levels of iron in serum (Blake *et al.*, 1979).

Wound infection with *V. vulnificus* usually develops after trauma and exposure to marine environment, and progresses rapidly. The infection is most commonly present as a cellulitis with a case/fatality rate of 7 %. Approximately one-third of the patients with wound infection may have underlying disease.

Other infectious diseases from which *V. vulnificus* have been isolated include pneumonia and endometritis (Tison and Kelly, 1984). *V. vulnificus* has also been recovered from the stools of patients with diarrhea (Johnston *et al.*, 1986), but its etiological role in diarrhea has not been proven. Observations of large series of patients with *V. vulnificus* infections in Taiwan, Korea and the US have been published (Park *et al.*, 1991; Chuang *et al.*, 1992; Hlady, 1997).

It has been suggested that biogroup 2 is not associated with human infection. However, Veenstra *et al.* (1992) reported the implication of a strain of biogroup 2 in a patient with septicemia.

2.3.2 Reservoirs and transmission

Principal reservoirs for *V. vulnificus* are coastal seawater and brackish water. During summer months, filterfeeding shellfish, such as oysters and clams, can contain high levels of this species. *V. vulnificus* is not sewage-associated.

2.4 Prevalence in foods

Oliver (1981) described that oysters contaminated with *V. vulnificus* and held for 24 hours on ice showed a 3-log decrease in cells. At the most common storage temperature of 8°C there can be up to a 2-log increase, and at 20°C even higher increases. Oliver (1981) also found that homogenization of oyster meat may lead to release of lethal factors that are very detrimental to the vibrios when combined with cold storage.

2.5 Foods most often associated with human infections

Foodborne *V. vulnificus* infections have been associated almost exclusively with raw seafoods, especially oysters.

2.6 Principles of detection

V. vulnificus grows well on TCBS agar. Blood agar is most commonly used in the clinical laboratory, since blood and wounds are the usual sources of this organism. Two agar media, cellobiose-polymyxin B-colistin agar (CPC; Massad and Oliver, 1987) and sodium dodecyl sulfate-polymyxin B-sucrose agar (Bryant *et al.*, 1987) have been developed for the isolation and enumeration of *V. vulnificus* in shellfish and environmental samples. A recent modification of CPC leaving out the polymyxin B, called CC agar (Hoy and Dalsgaard, 1998), has also been used effectively for isolation. The samples require enrichment using alkaline peptone water before plating. *V. vulnificus* has been detected by PCR (Brauns *et al.*, 1991; Hill *et al.*, 1991). A labeled DNA probe to detect *V. vulnificus* and a sandwich enzyme-linked immunosorbent assay for cytotoxin detection in environmental samples were reported by Wright *et al.* (1993), Parker and Lewis (1995) and DePaola *et al.* (1997). Additionally, DNA probes have been used to differentiate *V. vulnificus* from the phenotypically similar *V. parahaemolyticus* (Wright *et al.*, 1993; Kaysner *et al.*, 1994; Gooch *et al.*, 2001).

3 *Vibrio cholerae*

3.1 Historical aspects

Cholera was originally endemic in the delta of the River Ganges in eastern India, but before 1960 it had expanded into six pandemics that affected most of the world. There was another focus of a cholera-like disease, which was called El Tor cholera, in Cerebes, Indonesia; this disease has been spreading to large parts of the world since 1961 and the expansion is continuing.

The causative agent of the seventh cholera pandemic is a different serovar of the same serogroup (O1) as the original cholera vibrio, but in late 1992 the Indian subcontinent experienced an epidemic of cholera caused by a new type of O group designated O139. This is the first serogroup other than O1 to have caused a large-scale epidemic. There have been no reports of spread of this serogroup to other continents (Albert, 1994).

3.2 Characteristics

3.2.1 Classification and phenotypic characterization

So-called NAG (non-agglutinable) vibrios or NCV (non-cholera vibrios) are now classified into *V. cholerae*, in which the cholera vibrio is assigned to O group 1 (Sakazaki *et al.*, 1967).

V. cholerae has a single polar, sheathed flagellum. Many strains of *V. cholerae* non-O1 are encapsulated. There are reports, although rare, of cholera-like diarrheal illnesses and septicemic infections caused by non-O1/O139 serogroups.

Growth occurs between pH 6.0 and 9.6, and between 15°C and 42°C. Growth is supported by peptone water without additional salt. Phenotypic characteristics of

V. cholerae are shown in Table 5.1. *V. mimicus* is a closely related species (Table 5.1) reportedly causing human gastroenteritis and sharing many characteristics with *V. cholerae*, including the ability to grow in media without added NaCl (Davis *et al.*, 1981); it can be differentiated from *V. cholerae* on TCBS agar by its lack of sucrose fermentation.

V. cholerae O1 is divided into two biovars, the classical and El Tor (eltor), but the classical biovar has become exceedingly rare. The differential characteristics of the two biovars are shown in Table 5.2. The differentiation was originally based on hemolytic activity on sheep red blood cells, but the hemolysis test does not always give reproducible results. A modified CAMP test in which the synergistic hemolysis is associated with eltor biovar and non-O1 strains was described by Lesmanna *et al.* (1994). Non-O1 serogroups are predominantly hemolytic. *V. cholerae* is inhibited by the vibriostatic compound O/129, but O1 strains resistant to this compound are increasingly found; O139 strains are resistant to O/129.

V. cholerae is divided into a number of O groups. So far, 140 O groups have been established (Sakazaki *et al.*, 1970; Shimada *et al.*, 1994). There are three O antigenic variants of *V. cholerae* O1 named Ogawa, Inaba and Hikojima, based on their three O antigen factors. The Ogawa strains are the original form, and the Inaba strains are mutants that have lost an Ogawa-specific factor. The rare Hikojima variants are regarded as a stable intermediate form between the Ogawa and Inaba variants (Sakazaki and Tamura, 1971). The conversion from Ogawa to Inaba is irreversible.

3.2.2 Virulence factors

3.2.2.1 Colonization factor

The involvement of a variety of pili has been suggested for intestinal colonization by *V. cholerae* O1 and O139,but a long filamentous pilus is essential for colonization (Taylor *et al.*, 1987). This pilus is named toxin-coregulated pilus (TCP). Outer membrane proteins (OMP) also appear important for colonization. One possible colonization factor of OMP is the accessory colonization factor (acf) (Peterson and Mekalanos, 1988). An adhesin called OmpU was shown to mediate adherence to human epithelial cells (Sperandio *et al.*, 1995). Franzon *et al.* (1993) reported that hemagglutinin is a possible colonization factor. The capsular polysaccharide mediates adherence to epithelial cells in *V. cholerae* non-O1 (Johnson *et al.*, 1991). Such a capsule

Table 5.2 Biovars of *Vibrio cholerae* O1

Test	Biovar	
	Classical	El tor
Hemolysis of sheep erythrocytes	–	+[a]
Voges-Proskauer	–	+
Chicken erythrocyte agglutination	–	+
Polymyxin B, 50 IU	–	+
Mukerjee's phage IV	Susceptible	Resistant

[a] Recent isolates are sometimes negative.

would also promote septicemia caused by non-O vibrios in susceptible hosts. Motility may help *V. cholerae* reach the intestinal mucosa.

3.2.2.2 Detachment

The ability to detach from host cells could be important for allowing bacteria to leave mucosal cells and reattach to newly formed mucosal cells. *V. cholerae* produces mucinase that degrades different types of protein. It allows *V. cholerae* to detach from cultured human intestinal epithelial cells.

3.2.2.3 Cholera toxin (CT)

Cholera toxin, a heat-labile protein, is responsible for the massive diarrhea characteristics of cholera. CT consists of one subunit A and five identical subunits B. Excreted CT attaches through subunit B to the surface of mucosal cells by binding G_{MI} gangliosides. Once CT is bound, subunit A is released and translocated into the host cell. The intracellular target of CT is adenyl cyclase, which mediates the transformation of ATP to cyclic AMP (c-AMP) with subsequent increase in intracellular levels of c-AMP. The increased intracellular c-AMP concentration leads to increased chloride secretion by intestinal crypt cells and to decreased NaCl-coupled absorption by villus cells, resulting in water flow into the lumen and diarrhea (Finkelstein *et al.*, 1992). CT or imunologically related enterotoxins are also demonstrated in some non-O1 strains. The presence of the CT gene or production of CT in vitro is a trait used by health officials to assess the possible virulence of strains isolated from environmental samples or food products.

Two new enterotoxins of *V. cholerae* were described by Fasano *et al.* (1991) and Trucksess *et al.* (1993). One of them, zona occludens toxin (Zot), increases the permeability of the small intestine mucosa by disrupting the intracellular tight junctions that bind mucosal cells together. The other is an accessory cholera toxin (Ace), which causes significant fluid accumulation in rabbit ileal loops. Genes encoding CT (*ctx*) are located on a transposon, and only strains that aquire this transposon produce toxin. The *zot* and *ace* genes are close to the *ctx* on the chromosome and may be on the same transposon, thus suggesting that *ctx*, *zot* and *ace* genes constitute a virulence cassette. On the other hand, Sanyal *et al.* (1983) reported a toxin in CT-negative O1 strains. This toxin was named 'new cholera toxin', and was shown to cause diarrhea in human volunteers (Saha and Sanyal, 1990).

The hemolysin of *V. cholerae* is a cytotoxin that causes fluid accumulation in rabbit ileal loops. In contrast to the fluid produced with CT; however, the accumulated fluid produced in response to hemolysin is invariably bloody and contains mucus (Ichinose *et al.*, 1987). Occasional strains of *V. cholerae* also produce other toxins, including a Shiga toxin in O1 strains (O'Brien *et al.*, 1984) and the thermostable enterotoxin (ST) of *E. coli* in some non-O1 strains (Morris, 1990). Genes encoding the THD-related hemolysin (TRH) produced by some Kanagawa-negative strains of *V. parahaemolyticus* were found on a plasmid in some non-O1 strains (Honda *et al.*, 1986).

Environmental O1 isolates outside of epidemic areas are almost always CT-negative. However, those CT-negative strains, as well as non-O1 strains, may cause not only diarrhea but also extraintestinal infections.

3.2.3 Ability to survive and grow in the environment

V. cholerae O1 does not survive in fresh water for more than 7–10 days, particularly during the warm season; its survival in seawater is longer. *V. cholerae* produces chitinase and it preferably colonizes the surface of shellfish and zooplanktons, therefore the estuarine environment represents an ideal setting for survival and persistence of *V. cholerae*. Spira *et al.* (1981) reported that water hyacinths from Bangladesh waters were shown to be colonized by *V. cholerae* and to promote its growth. Colwell and Huq (1994) noticed a survival strategy of *V. cholerae* in the environment in which it remained viable but was not culturable. The Rugose variant of *V. cholerae* is also assumed to have a resistant form (Morris *et al.*, 1993).

3.3 Nature of the infection

3.3.1 Clinical manifestations

In typical cholera there is profuse diarrhea, with large volumes of so-called rice-water stool passed painlessly. This can amount to twice the body weight within 4–6 days. Gastric disturbances, particularly subacidity and gastrectomy, are risk factors in severe cholera. If untreated there will be prostration with symptoms of severe dehydration, and death can occur very quickly after onset of symptoms. The disease is easily treated by monitoring body fluid loss combined with rehydration.

Cholera is usually a disease of the lower socioeconomic groups because of their poor hygienic standards. Most patients have either a mild diarrhea or no symptoms at all. The ratio of severe to mild asymptomatic cases is between 1:5 and 1:10 for classical cholera, but only 1:25 to 1:100 for eltor cholera. However, during the recent epidemic in Peru there were over 70 % severe cases and a 60 % case/fatality rate (Pan American Health Organization, 1993). In addition, it was reported that the vibrios isolated in South and Central America can be clearly distinguished genetically from strains causing the seventh pandemic. The World Health Organization estimates that there are greater than 150 000 cases of cholera per year on multiple continents.

3.3.2 Reservoirs and transmission

Excretion of the vibrios by infected persons usually lasts for only a few days, and carriers are rarely found to harbor the vibrios for a long period. Cholera is more likely to occur in families with asymptomatically infected breast-fed infants, who themselves are protected against the illness by maternal antibodies. Domestic animals in epidemic areas often carry O1 vibrios (Sanyal *et al.*, 1974).

Cholera is not spread by direct contact. The most important mode of spread is through the environment, particularly contaminated water (Hughes *et al.*, 1982). Raw shellfish are also important sources of infection. Not only the surface but also the digestive tracts of shellfish are colonized by *V. cholerae*; the digestive tract is infected through ingestion of zooplankton in which the vibrios were absorbed. *V. cholerae* persists for many weeks in shellfish, and can thus maintain its lifecycle without continuous contamination with human feces containing the vibrio.

3.4 Prevalence in foods and water

The presence of CT-positive strains of *V. cholerae* is not always associated with fecal contamination from cholera patients. Water may also become contaminated with *V. cholerae* during household storage. In developing countries, beverages are potential vehicles of cholera transmission. Beverages containing ice may be specifically incriminated because the ice is made from contaminated municipal water. Chlorination of public water systems is an effective means of controlling epidemics.

V. cholerae survives for between 2 and 14 days on most foods, and survival is increased when foods are cooked before contamination. A short survival period is common in acidic foods such as fruits, whereas survival may be several weeks in cooked and raw vegetables. A pH of 5.0 or lower has been found to be very detrimental to *V. cholerae* survival.

3.5 Foods and water most often associated with infection

A significantly increased risk of infection is associated with use of contaminated water for food preparation, bathing or washing. Bathing in contaminated surface water may be particularly risky in Moslem communities, since it is common to rinse the mouth with water. Seafoods, particularly shellfish, may acquire the vibrio from environmental sources, and may serve as a vehicle in epidemic cholera.

3.6 Principles of detection

3.6.1 Isolation

Numerous agar media have been devised for the isolation of *V. cholerae*, but TCBS agar is probably the most widely used; on this medium *V. cholerae* appears as yellow colonies due to sucrose fermentation. For enrichment culture alkaline peptone water (pH 8.6–9.0) is used; it supports good growth of *V. cholerae*, but the incubation period is best limited to 8 hours to prevent overgrowth with other organisms. Stool specimens should be collected as soon as possible, and inoculated onto isolation agar plates. For formed feces, enrichment culture may be necessary. Water samples are passed through a membrane filter and the filter disk is then placed in alkaline peptone water. Moore's swabs have also been used successfully to isolate *V. cholerae*. Food samples are blended in alkaline peptone water and then treated the same way as water samples. Shellfish should be cut into small pieces but not homogenized. After putting the pieces into enrichment broth and shaking vigorously, the pieces are removed from the broth because factors in those seafoods may inhibit the vibrios. DePaola *et al.* (1987) recommended incubation of a separate portion of the sample in alkaline peptone water at 42°C. For the detection of vibrios that are in a dormant state, Huq *et al.* (1990) reported a method using fluorescence microscopy.

The rapid diagnosis of cholera is of great importance; therefore suspect colonies from foods related to an outbreak should be tested first for agglutination directly with *V. cholerae* O1 and O139 antisera.

3.6.2 Direct antigen detection methods

There are enough O antigens in rice-water stools of cholera patients to be agglutinated with O antibodies. A coagglutination test using monoclonal antibodies to an O1-specific epitope has been developed recently. An antigen capture test using a colloidal gold-based colorimetric immunoassay has been reported by Hasan *et al.* (1994).

3.6.3 Toxin assays

Various modifications of an ELISA using purified G_{MI} ganglioside receptor as the capture molecule are now commonly used to assay CT. Almeida *et al.* (1990) reported, however, that the latex agglutination assay to detect CT was less complicated and less time-consuming than the ELISA. In addition, a variety of molecular approaches to toxin detection have been developed. PCR has been used for the detection of the *ctx* gene (Fields *et al.*, 1992), and was shown to be more sensitive than a bead ELISA for detecting CT in stool specimens (Ramamurthy *et al.*, 1993).

Acknowledgment

The passing of Riichi Sakazaki, an internationally known microbiologist and a genuinely gracious scientist, saddens us. I, Charles Kayser, first met Dr Sakazaki in 1973, when he toured the US to provide information to many public health laboratories on the identification of *V. parahaemolyticus*. This was after the first major outbreak in the US. Over the years and having had the opportunity to meet him at various scientific meetings, he was always the gracious individual and always interested to hear of work in other laboratories. This excellent scientist will be missed.

Bibliography

Albert, J. M. (1994). *Vibrio cholerae* O139 Bengal. *J. Clin. Microbiol.* **32**, 2345–2349.

Almeida, R. J., F. W. Hickman-Brenner, E. G. Sowers *et al.* (1990). Comparison of a latex agglutination assay and an enzyme-linked immunosorbent assay for detecting cholera toxin. *J. Clin. Microbiol.* **28**, 28 130.

Asakawa, Y., S. Akahane and M. Noguchi (1974). Quantitative studies on pollution with *Vibrio parahaemolyticus* during distribution of fish. In: T. Fujino, G. Sakaguchi, R. Sakazaki and Y. Takeda (eds), *International Symposium on* Vibrio parahaemolyticus, pp. 97–103. Saikon Publishing Co., Tokyo.

Bej, A. K., D. P. Patterson, C. W. Brasher *et al.* (1999). Detection of the total and hemolysin-producing *Vibrio parahaemolyticus* in shellfish using multiplex PCR ampliications of *tl, tdh*, and *trh*. *J. Microbiol. Methods* **36**, 215–225.

Biosca, E. G., H. Llorens, E. Garay and C. Amaro (1993). Presence of a capsule in *Vibrio vulnificus* biotype 2 and its relationship to virulence for eels. *Infect. Immun.* **61**, 1611–1618.

Blake, P. A., M. H. Merson, R. E. Weaver *et al.* (1979). Disease caused by a marine vibrio. Clinical characteristics and epidemiology. *N. Engl. J. Med.* **300**, 1–5.

Brauns, L. A., M. C. Hudson and J. D. Oliver (1991). Use of the polymerase chain reaction in detection of culturable and nonculturable *Vibrio vulnificus* cells. *Appl. Envir. Microbiol.* **57**, 2651–2655.

Bryant, R. G., J. Jarvis and J. M. Janda (1987). Use of sodium dodecyl sulfate-polymyxin B-sucrose medium for isolation of *Vibrio vulnificus* from shellfish. *Appl. Environ. Microbiol.* **53**, 1556–1559.

Chuang, Y.-C., C.-Y. Yuan, C.-Y. Liu *et al.* (1992). *Vibrio vulnificus* infection in Taiwan: report of 28 cases and review of clinical manifestations and treatment. *Clin. Infect. Dis.* **15**, 271–276.

Colwell, R. R. and A. Huq (1994). Vibrios in the environment: viable but not culturable *Vibrio cholerae*. In: I. K. Wachsmuth, P A. Blake and O. Olsvik (eds), Vibrio cholerae *and Cholera: Molecular to Global Perspectives*, pp. 117–133. ASM Press, Washington, DC.

Daniels, N., L. Mackinnon, R. Bishop *et al.* (2000). *Vibrio parahaemolyticus* infections in the United States, 1973–1998. *J. Infect. Dis.* **181**, 1661–1666.

Davis, B. R., G. R. Fanning, J. M. Madden *et al.* (1981). Characterization of biochemically atypical *Vibrio cholerae* strains and designation of a new pathogenic species, *Vibrio mimicus. J. Clin. Microbiol.* **14**, 631–639.

DePaola, A., C. A. Kaysner and R. M. McPherson (1987). Elevated temperature method for recovery of *Vibrio cholerae* from oysters (*Crassostrea gigas*). *Appl. Envir. Microbiol.* **53**, 1181–1182.

DePaola, A., M. L. Motes, D. W. Cook *et al.* (1997). Evaluation of alkaline phosphatase-labeled DNA probe for enumeration of *Vibrio vulnificus* in Gulf Coast oysters. *J. Microbiol. Methods* **29**, 115–120.

DePaola, A. J., C. A. Ulaszek, B. J. Kaysner *et al.* (2003). Molecular, serologic, and virulence characteristics of *Vibrio parahaemolyticus* isolated from environmental, food and clinical sources in North America and Asia. *Appl. Environ. Microbiol.* **69**, 33–357.

Farmer, J. J. III (1979). *Vibrio ('Beneckea') vulnificus*, the bacterium associated with sepsis, septicemia and the sea. *Lancet* **ii**, 903.

Fasano, A., B. Bandry, D. W. Pumplin *et al.* (1991). *Vibrio cholerae* produces a second enterotoxin, which affects intestinal tight junctions. *Proc. Natl Acad. Sci. USA* **88**, 5242–5246.

Fields, P. I., T. Popovic, I. K. Wachsmuth and O. Olsvik (1992). Use of polymerase chain reaction for detection of toxigenic *Vibrio cholerae* O1 strains in Latin American cholera epidemic. *J. Clin. Microbiol.* **30**, 2118–2121.

Finkelstein, R. A., M. Boesman-Finkelstein, Y. Chang and C. C. Hase (1992). *Vibrio cholerae* hemagglutinin/protease, colonial variation, virulence, and detachment. *Infect. Immun.* **60**, 472–478.

Franzon, V. L., A. Barker and P. Manning (1993). Nucleotide sequence encoding the mannose-fucose-resistant hemagglutinin of *Vibrio cholerae* O1 and construction of a mutant. *Infect. Immun.* **61**, 3032–3037.

Fujino, T. (1951). Bacterial food poisoning. *Saisin Igaku* **6**, 263–271.

Fujino, T., Y. Okuno, D. Nakada *et al.* (1953). On the bacteriological examination of shirasu food poisoning. *Med. J. Osaka Univ.* **4**, 299–304.

Gooch, J. A., A. DaPaola, C. A. Kaysner and D. L. Marshall (2001). Evaluation of two direct plating methods using non-radioactive probes for enumeration of *Vibrio parahaemolyticus* in oysters. *Appl. Environ. Microbiol.* **67**, 721–724.

Gray, L. D. and A. S. Kreger (1987). Mouse skin damage caused by cytolysin from *Vibrio vulnificus* and *Vibrio vulnificus* infection. *J. Infect. Dis.* **158**, 236–241.

Hasan, J. A. K., A. Huq, M. L. Tamplin *et al.* (1994). A novel kit for rapid detection of *Vibrio cholerae* O1. *J. Clin. Microbiol.* **32**, 249–252.

Hill, W. E., S. P. Keasler, M. W. Trucksess *et al.* (1991). Polymerase chain reaction identification of *Vibrio vulnificus* in artificially contaminated oysters. *Appl. Environ. Microbiol.* **56**, 707–711.

Hlady, W. G. (1997). Vibrio infections associatd with raw oyster consumption in Florida, 1981–1994. *J. Food Prot.* **60**, 353–357.

Hoi, L. and I. Dalsgaard (1998). Improved isolation of *Vibrio vulnificus* from seawater and sediment with cellobiose-colistin agar. *Appl. Environ. Microbiol.* **64**, 1721–1724.

Hollis, D. G., R. E. Weaver, C. N. Baker and C. Thornsberry (1976). Halophilic vibrio species isolated from blood cultures. *J. Clin. Microbiol.* **3**, 425–431.

Honda, T., M. Nishibuchi, T. Miwatani and J. B. Kaper (1986). Demonstration of a plasmid-borne gene encoding a thermostable direct hemolysin in *Vibrio cholerae* non-O1 strains. *Appl. Environ. Microbiol.* **52**, 1218–1220.

Honda, T., Y. Ni, M. Yoh and T. Miwatani (1988). Production of monoclonal antibodies against thermostable direct hemolysin of *Vibrio parahaemolyticus* and the application of the antibodies for enzyme-linked immunosorbent assay. *Med. Microbiol. Immunol.* **178**, 245–253.

Honda, T., Y. Ni, M. Yoh *et al.* (1989). Production of monoclonal antibodies against thermostable direct hemolysin of *Vibrio parahaemolyticus* and application of the antibodies for enzyme-linked immunosorbent assay. *Med. Microbiol. Immunol.* **178**, 245–253.

Honda, T., Y. Ni, T. Miwatani *et al.* (1992). The thermostable direct hemolysin of *Vibrio parahaemolyticus* is a pore-forming toxin. *Can. J. Microbiol.* **38**, 1175–1180.

Hughes, J. M., J. M. Boyce, R. J. Levine *et al.* (1982). Epidemiology of eltor cholera in rural Bangladesh, importance of surface water in transmission. *Bull. WHO* **60**, 395–401.

Huq, A., R. R. Colwell, R. Rahman *et al.* (1990). Detection of *Vibrio cholerae* O1 in the aquatic environment by fluorescent-monoclonal antibody and culture methods. *Appl. Environ. Microbiol.* **56**, 2370–2373.

Ichinose, Y., K. Yamamoto, N. Nakasone *et al.* (1987). Enterotoxicity of El Tor-like hemolysin of non-O1 *Vibrio cholerae. Infect. Immun.* **55**, 1090–1093.

Johnson, G., J. Holmgren and A. M. Svennerholm (1991). Identification of a mannose-binding pilus on *Vibrio cholerae* El Tor. *Microb. Pathog.* **11**, 433–441.

Johnston, J. M., S. F. Becker and L. M. McFarland (1986). Gastroenteritis in patients with stool isolates of *Vibrio vulnificus. Am. J. Med.* **80**, 336–338.

Kato, T., Y. Obara, H. Ichinoe *et al.* (1965). Grouping of *Vibrio parahaemolyticus* (biotype 1) by hemolytic reaction [in Japanese]. *Shokuhin Eisei Kenkyyu* **15**, 83–86.

Kaysner, C. A., C. Abeyta Jr, P. A. Trost *et al.* (1994). Urea hydrolysis can predict potential pathogenicity of *Vibrio parahaemolyticus* strains in the Pacific Northwest. *Appl. Environ. Microbiol.* **60**, 3020–3022.

Kishishita M., N. Matsuoka, K. Kumagai *et al.* (1992). Sequence variation in the thermostable direct hemolysin (*trh*) gene of *Vibrio parahaemolyticus. Appl. Environ. Microbiol.* **58**, 2449–2453.

Kothary, M. H. and A. S. Kreger (1985). Production and partial characterization of an elastokytic protease of *Vibrio vulnificus. Infect. Immun.* **50**, 534–540.

Kreger, A. S., L. DeChatelet and P. Shirley (1981). Interaction of *Vibrio vulnificus* with human polymorphonuclear leukocytes, association of virulence with resistance to phagocytosis. *J. Infect. Dis.* **144**, 244–248.

Lee, C. and S.-F. Pan (1993). Rapid and specific detection of the thermostable direct hemolysin gene in *Vibrio parahaemolyticus* by the polymerase chain reaction. *J. Gen. Microbiol.* **139**, 3225–3131.

Lesmana, M., D. Subekti, P. Tjaniad and G. Pazzaglia (1994). Modified CAMP test for biogrouping *Vibrio cholerae* O1 strains and distinguishing them from strains of *Vibrio cholerae* non-O1. *J. Clin. Microbiol.* **32**, 235–237.

Massad, G. and J. D. Oliver (1987). New selective and differential medium for *Vibrio cholerae* and *Vibrio vulnificus. Appl. Environ. Microbiol.* **53**, 2262–2264.

Miyamoto, T., Y.-I. Seu, H. Miwa and S. Hatano (1989). A fluorogenic assay for the rapid detection of some vibrio species including *Vibrio parahaemolyticus* in foods. *Jpn J. Food Hygiene* **30**, 534–541.

Morris, G. J., A. C. Wright, D. M. Roberts *et al.* (1987). Identification of environmental *Vibrio vulnificus* isolates with a DNA probe for the cytotoxin-hemolysin gene. *Appl. Environ. Microbiol.* **53**, 193–195.

Morris J. G. Jr (1990). Non-O1 *Vibrio cholerae*, a look at the epidemiology of an occasional pathogen. *Epidemiol. Rev.* **12**, 179–191.

Morris, J. G. Jr, J. A. Johnson, E. W. Rice *et al.* (1993). *Vibrio cholerae* O1 can assume a rugose survival form which resists chlorination, but retains virulence for humans. In: *Abstracts of the Twenty-ninth Joint Conference on Cholera and Related Diseases, National Institutes of Health, Bethesda*, pp. 177–182. NIH.

Nath, G. and S. C. Sanyal (1992). Emergence of *Vibrio cholerae* O1 resistant to vibriostatic agent O/129. *Lancet* **340**, 366–367.

Nishibuchi, M. (1990). Molecular epidemiological evidence for association of thermostable direct hemolysin (TDH) and TDH-related hemolysin of *Vibrio parahaemolyticus* with gastroenteritis. *Infect. Immun.* **58**, 3568–3573.

Nishibuchi, M., M. Ishibashi, Y. Takeda and J. B. Kaper (1985). Detection of the thermostable direct hemolysin gene and related DNA sequences in DNA colony hybridization test. *Infect. Immun.* **49**, 481–486.

Nishibuchi, M., W. E. Hill, G. Zon *et al.* (1986). Synthetic oligodeoxyribonucleotide probes to detect Kanagawa phenomenon-positive *Vibrio parahaemolyticus. J. Clin. Microbiol.* **23**, 1091–1093.

Nishibuchi, M., A. Fasano, R. G. Russell and J. B. Kaper (1992). Enterotoxigenicity of *Vibrio parahaemolyticus* with and without genes encoding thermostable direct hemolysin. *Infect. Immun.* **60**, 3539–3545.

O'Brien, A. D., M. E. Chen, R. K. Holmes *et al.* (1984). Environmental and human isolates of *Vibrio cholerae* and *Vibrio parahaemolyticus* produce a *Shigella dysenteriae* 1 (Shiga)-like cytotoxin. *Lancet* **i**, 77–78.

Okuda, J., M. Ishibashi, S. L. Abbott *et al.* (1997). Analysis of thermostable direct hemolysin (*tdh*) gene and the tdh-related hemolysin (*trh*) genes in urease-positive strains of *Vibrio parahemolyticus* isolated on the west coast of the United States. *J. Clin. Microbiol.* **35**, 1965–1971.

Oliver, J. D. (1981). Lethal cold stress of *Vibrio vulnificus* in oysters. *Appl. Environ. Microbiol.* **41**, 710–717.

Oliver, J. D. and D. Wanucha (1989). Survival of *Vibrio vulnificus* at reduced temperatures and elevated nutrients. *J. Food. Safety* **10**, 79–86.

Pan American Health Organization (1993). *Cholera in the Americas*. Pan American Health Organization, Washington, DC.

Park, S. D., H. S. Shon and N. J. Joh (1991). *Vibrio vulnificus* septicemia in Korea, clinical and epidemiological findings in seventy patients. *J. Am. Acad. Dermatol.* **24**, 397–403.

Parker, J. W. and D. H. Lewis (1995). Sandwich enzyme-linked immunosorbent assay for *Vibrio vulnificus* hemolysin to detect *V. vulnificus* in environmental specimens. *Appl. Environ. Microbiol.* **61**, 476–486.

Peterson, K. M. and J. J. Mekalanos (1988). Characterization of the *Vibrio cholerae* ToxR regulon, identification of novel genes involved in intestinal colonization. *Infect. Immun.* **56**, 2822–2829.

Ramamurthy, T., A. Pal, P. K. Bag *et al.* (1993). Detection of cholera toxin gene in stool specimens by polymerase chain reaction, comparison with bead enzyme-linked immunosorbent assay and culture method for laboratory diagnosis of cholera. *J. Clin. Microbiol.* **31**, 3068–3070.

Saha, S. and S. C. Sanyal (1990). Immunobiological relationships of the enterotoxins produced by cholera toxin gene-positive (CT^+) and -negative (CT^-) strains of *Vibrio cholerae* O1. *J. Med. Microbiol.* **32**, 33–37.

Sakazaki, R. and K. Tamura (1971). Somatic antigen variation in *Vibrio cholerae. Jpn J. Med. Sci. Biol.* **24**, 93–100.

Sakazaki, R., S. Iwanami and H. Fukumi (1963). Studies on the enteropathogenic, facultatively halophilic bacteria, *Vibrio parahaemolyticus*. I. Morphological, cultural and biochemical properties and its taxonomical position. *Jpn J. Med. Sci. Biol.* **16**, 161–188.

Sakazaki, R., C. Z. Gomez and M. Sebald (1967). Taxonomical studies of the so-called NAG vibrios. *Jpn J. Med. Sci. Biol.* **20**, 265–280.

Sakazaki, R., S. Iwanami and K. Tamura (1968a). Studies on the enteropathogenic, facultatively halophilic bacteria. *Vibrio parahaemolyticus*. II. Serological characteristics. *Jpn J. Med. Sci. Biol.* **21**, 313–324.

Sakazaki, R., K. Tamura, T. Kato *et al.* (1968b). Studies on the enteropathogenic, facultatively halophilic bacteria, *Vibrio parahaemolyticus*. III. Enteropathogenicity. *Jpn J. Med. Sci. Biol.* **21**, 325–331.

Sakazaki, R., K. Tamura, C. Z. Gomez and R. Sen (1970). Serological studies on the cholera group of vibrios. *Jpn J. Med. Sci. Biol.* **23**, 13–20.

Sanyal, S. C. and P. C. Sen (1974). Human volunteer study on the pathogenicity of *Vibrio parahaemolyticus*. In: T. Fujino, G. Sakaguchi, R. Sakazaki and Y. Takeda (eds), *International Symposium on* Vibrio parahaemolyticus, pp. 227–235. Saikon Publishing Co., Tokyo.

Sanyal, S. C., S. J. Singh, I. C. Tiwari *et al.* (1974). Role of household animals in maintenance of cholera infection in a community. *J. Infect. Dis.* **130**, 575–579.

Sanyal, S. C., K. Alam, P. K. B. Neogi *et al.* (1983). A new cholera toxin. *Lancet* **i**, 1337.

Shimada, T. and R. Sakazaki (1984). On the serology of *Vibrio vulnificus. Jpn J. Med. Sci. Biol.* **37**, 241–246.

Shimada, T., E. Arakawa, K. Itoh *et al.* (1994). Extended serotyping scheme for *Vibrio cholerae. Curr. Microbiol.* **28**, 175–178.

Shinoda, S., T. Miwatani, T. Honda and Y. Takeda (1974). Antigenicity of flagella of *Vibrio parahaemolyticus*. In: F. Fujino, G. Sakaguchi, R. Sakazaki and Y. Takeda (eds), *International Symposium on* Vibrio parahaemolyticus, pp. 193–197. Saikon Publishing Co., Tokyo.

Shirai, H., H. Ito, T. Hirayama *et al.* (1990). Molecular epidemiologic evidence for association of thermostable direct hemolysin (TDH) and TDH-related hemolysin of *Vibrio parahaemolyticus* with gastroenteritis. *Infect. Immun.* **58**, 3568–3573.

Simpson, L. M. and J. D. Oliver (1983). Siderophore production by *Vibrio vulnificus. Infect. Immun.* **41**, 644–649.

Smith, G. C. and J. R. Merkel (1982). Collagenolytic activity of *Vibrio vulnificus*, potential contribution to its invasiveness. *Infect. Immun.* **35**, 1155–1156.

Sperandio, V., A. Giron, W. D. Silveira and J. B. Kaper (1995). The OmpU outer membrane protein, a potential adherence factor of *Vibrio cholerae. Infect. Immun.* **63**, 4433–4438.

Spira, W. M., A. Huq, Q. S. Ahmed and Y. A. Saeed (1981). Uptake of *Vibrio cholerae* biotype eltor from contaminated water by water hyacinth (*Eichornia crassipes*). *Appl. Environ. Microbiol.* **42**, 550–553.

Takikawa, I. (1958). Studies on pathogenic halophilic bacteria. *Yokahama Med. Bull.* **2**, 313–322.

Tamplin, M. L., S. Specter, G. E. Rodrick and H. Friedman (1985). *Vibrio vulnificus* resists phagocytosis in the absence of serum opsonins. *Infect. Immun.* **49**, 715–718.

Tamplin, M. L., J. K. Jackson, C. Buchreiser *et al.* (1996). Pulsed-field gel electrophoresis and ribotype profiles of clinical and environmental *Vibrio vulnificus* isolates. *Appl. Environ. Microbiol.* **62**, 3572–3580.

Taylor, R. K., V. L. Miller, D. B. Furlong and J. J. Mekalanos (1987). Use of *phoA* gene fusions to identify a pilus colonization factor coordinately regulated with cholera toxin. *Proc. Natl Acad. Sci. USA* **84**, 2833–2837.

Terada, Y. (1968). Serological studies on *Vibrio parahaemolyticus*. II. Flagellar antigens [in Japanese]. *Jpn J. Bacteriol.* **23**, 767–771.

Tison, D. L. and M. T. Kelly (1984). Vibrio species of medical importance. *Diagn. Microbiol. Infect. Dis.* **2**, 263–276.

Trucksess, M., J. E. Galen, J. Michalski *et al.* (1993). Accessory Cholera Enterotoxin (Ace). *Proc. Natl Acad. Sci. USA* **90**, 5267–5271.

Veenstra, J., P. J. Rietra, C. P. Stoutenbeek *et al.* (1992). Infection by an indole-negative variant of *Vibrio vulnificus* transmitted by eels. *J. Infect. Dis.* **166**, 209–210.

Wright, A. C., G. A. Miceli, W. L. Landry *et al.* (1993). Rapid identification of *Vibrio vulnificus* on nonselective media with an alkaline phosphate-labeled oligonucleotide probe. *Appl. Environ. Microbiol.* **59**, 541–546.

Yamamoto, K., T. Honda and T. Miwatani (1992). Enzyme-labeled oligonucleotide probes for detection of the genes for thermostable direct hemolysin (TDH) and TDH-related hemolysin (TRH) of *Vibrio parahaemolyticus. Can. J. Microbiol.* **38**, 410–416.

Yoh, M., N. Morinaga, M. Noda and T. Honda (1995). The binding of *Vibrio parahaemolyticus* ^{125}I-labeled thermostable direct hemolysin to erythrocytes. *Toxicon* **33**, 651–657.

Escherichia coli infections

6

Pina M. Fratamico and James L. Smith

1 Introduction

In the second edition of *Foodborne Infections and Intoxications* (Riemann and Bryan, 1979), Bryan (1979) devoted about four pages to a discussion of enterotoxigenic and enteroinvasive strains of *Escherichia coli* as foodborne enteric pathogens. The knowledge of *E. coli* strains that can cause gastrointestinal illness has changed drastically since 1979, as shown by the fact that an entire chapter of this edition (the third) is devoted to foodborne pathogenic *E. coli*.

In 1895 Escherich first described *E. coli*, and subsequently, the organism was recognized as a normal inhabitant of human and animal intestinal tracts (Doyle and Padhye, 1989; Neill *et al.*, 1994). The organism is not a major bowel bacterium and, other than its role in providing a source of vitamins in some animals, the function of *E. coli* in the ecology and physiology of the bowel is unclear (Sussman, 1985). The pathogenic role of *E. coli* was obscured by its commensal status, and the ability of the organism to cause disease was not suspected for a number of years (Doyle and Padhye, 1989; Neill *et al.*, 1994). *Escherichia coli* was recognized as a cause of urinary tract infections in the early 1920s (Bettelheim, 1992), and the cause of infantile gastroenteritis was attributed to strains described as enteropathogenic *E. coli* in the late 1940s (Levine, 1987).

Foodborne Infections and Intoxications 3e
ISBN-10: 0-12-588365-X
ISBN-13: 978-012-588365-8

Human *E. coli* strains can be broadly grouped as commensal strains, extraintestinal pathogenic strains, and intestinal pathogenic strains (Russo and Johnson, 2003). Extraintestinal pathogenic *E. coli* acquired genes conferring the ability to cause disease outside of the gastrointestinal tract, and include strains that cause urinary tract, abdominal, and pelvic infections, in addition to septicemia, meningitis, and endocarditis. *E. coli*, however, can cause infection of surgical wounds, in addition to infections in nearly every organ and anatomical site. Intestinal pathogenic strains have acquired virulence factors conferring the ability to cause gastrointestinal diseases, including enteritis and colitis.

Currently, many *E. coli* strains are known to act as intestinal pathogens, varying in the mechanisms by which they produce diarrhea. The mechanisms include the production of toxins and/or of host cell attachment factors, and the invasion of colonic mucosal cells (Table 6.1; Guerrant and Thielman, 1995). Usually, a given infection involves more than one virulence factor. At the present time, at least six different categories of diarrheic *E. coli* are known (Table 6.1). A new category called cell-detaching *E. coli* (CDEC), defined by the ability of the bacteria to cause detachment of tissue culture cells from solid supports and by the production of α-hemolysin, pyelonephritis-associated pili and cytotoxic necrotizing factor 1 (CNF-1), has been identified (Fábrega *et al.*, 2002; Okeke *et al.*, 2002). The CDEC may also possess virulence factors found in other categories of *E. coli* (Clarke, 2001; Fábrega *et al.*, 2002; Okeke *et al.*, 2002). *E. coli* strains that induce the formation of actin stress fibers and activation of DNA synthesis in cell cultures, leading to the formation of multi-nucleated giant cells, have been referred to as necrotoxigenic *E. coli* (NTEC). There are two toxins involved, CNF-1 in NTEC1 strains, and cytotoxic necrotizing factor-2, CNF-2, in NTEC2 strains. Many NTEC also produce cytolethal distending toxins (De Rycke *et al.*, 1999; Mainil *et al.*, 2003). Interestingly, the gene that encodes CNF-1 in NTEC1 strains is found on a pathogenicity island that also has genes that

Table 6.1 *Escherichia coli* strains that cause gastrointestinal illness

Strain	Type of diarrhea	Virulence factors	Genetic coding
Enteroinvasive (EIEC)	Acute dysenteric	Cell invasion and intracellular multiplication	Plasmid and chromosomal
Diffusely adherent (DAEC)	Watery, in children	Both fimbrial and non-fimbrial adhesins	Plasmid and chromosomal
Enteroaggregative (EAEC)	Persistent	Aggregative adherence and heat-stable enterotoxin	Plasmid
Enterotoxigenic (ETEC)	Acute watery	Adherence and heat-labile toxins	Plasmid and chromosomal
Enteropathogenic (EPEC)	Acute and/or persistent	Localized and attaching and effacing adherence	Plasmid and chromosomal
Enterohemorrhagic (EHEC)	Bloody, with or without sequelae including hemolytic uremic syndrome	Attaching and effacing adherence, enterohemolysin, and Shiga-like toxins	Phage, plasmid, and chromosomal

Modified from Guerrant and Thielman (1995).

encode for α-hemolysin and P-fimbriae (De Rycke *et al.*, 1999); therefore, NTEC1 may be related to CDEC. Due to the promiscuity of genetic transfer in Gram-negative organisms (Clarke, 2001), it is probable that additional diarrheic *E. coli* strains will emerge. The characteristics of the six major categories of diarrheagenic *E. coli* are discussed below.

2 Characteristics of *Escherichia coli*

2.1 Classification and biochemical characteristics

Escherichia coli is in the family Enterobacteriaceae. The organism is a Gram-negative, non-spore-forming, straight rod (1.1–1.5 μm × 2.0–6.0 μm) arranged in pairs or singly; is motile by means of peritrichous flagella or may be non-motile; and may have capsules or microcapsules. *Escherichia coli* is a facultatively anaerobic, chemo-organotrophic microorganism with an optimum growth temperature of 37 °C (Brenner, 1984; Ørskov, 1984). It is oxidase negative, catalase positive, fermentative (glucose, lactose, D-mannitol, D-sorbitol, arabinose, maltose), reduces nitrate, and is β-galactosidase positive. Approximately 95 % of *E. coli* strains are indole and methyl red positive, but are Voges-Proskauer and citrate negative (Doyle and Padhye, 1989). Interestingly, DNA relatedness indicates that *E. coli* and *Shigella* form a single species, and it is difficult to separate the two biochemically (Brenner, 1984). *Shigella* strains were placed in a different genus in the 1940s to distinguish them from non-pathogenic *E. coli*; however, a number of studies based on genetic typing and DNA sequencing have shown that *Shigella* strains fall within *E. coli* (Lan and Reeves, 2002). The separate nomenclature is maintained for medical and epidemiological purposes.

Serotyping and serogrouping of *E. coli* is useful for subdividing the species into serovars. Serological typing in *E. coli* involves serological identification of three surface antigens: O (somatic lipopolysaccharide), K (capsular) and H (flagellar). Although the numbers of the different *E. coli* O, K, and H antigens reported in the literature vary, Ørskov and Ørskov (1992) state that there are 173 O antigens, 80 K antigens and 56 H antigens. Mol and Oudega (1996) have suggested that the fimbrial (F) surface antigens should be a fourth component of serological testing. Determining the serogroup (O antigen) and serotype (O and H, and often K antigens) is an important means of defining the various pathogenic strains of *E. coli*, since certain serotypes are associated with the various categories of diarrheagenic *E. coli*. For example, *E. coli* serotype O157:H7 is an enterohemorrhagic *E. coli*, and *E. coli* serogroup O124 is an enteroinvasive strain. *E. coli* serotyping is important for making the proper diagnosis and for performing foodborne outbreak and epidemiological investigations. However, novel *E. coli* serotypes frequently emerge as intestinal pathogens and, when recognized, are included in a specific category of diarrheagenic *E. coli*. Thus serotyping alone cannot be relied on for categorizing a strain of *E. coli*, and the identification of specific virulence characteristics/genes must also be performed (Barlow *et al.*, 1999).

2.2 *Escherichia coli* in foods: growth and survival

Since *E. coli* is an inhabitant of the gastrointestinal tract of humans and animals, it is expected to be present in the environment, water and food. The organism can be present in the environment due to animal defecation or contamination with untreated human sewage, in foods of animal origin such as meat or milk, in produce from manured land or from plots irrigated with fecally-contaminated water, in cooked or uncooked foods prepared by infected food handlers, and in water contaminated with human sewage or animal waste.

E. coli strains do not grow under refrigeration conditions; however, the organism can survive for weeks at 4°C or –20°C. The limits of temperature for growth of *E. coli* are 7–46°C, and the optimum growth temperature is approximately 37°C (Bell and Kyriakides, 1998). The heat resistance of *E. coli* is similar to that of other enteric bacteria; however, heat sensitivity is affected by the food environment and exposure of the organism to prior stress or growth conditions. *E. coli* generally grows within the pH range of 4.4–9.0, at an a_w of at least 0.95, and at NaCl levels of less than 8.5 % (Bell and Kyriakides, 1998). Many studies related to the growth and survival of *E. coli* have been performed with *E. coli* O157:H7. These are discussed in Section 8.

2.3 *Escherichia coli* as an indicator organism

As a normal member of the gastrointestinal tract of humans and animals, the presence of *E. coli* in the environment, water or food suggests fecal contamination. The presence of *E. coli* in a food implies that enteric pathogens may also be present. However, studies have shown that the presence or absence of fecal pathogens cannot be correlated with detection of *E. coli* (Pierson and Smoot, 2001). The presence of the organism in heat-processed foods may represent process failure, post-processing contamination from equipment or personnel, or contact with raw product (Pierson and Smoot, 2001). At best, the presence of *E. coli* in food or water is an indication of uncleanliness and careless handling.

3 Enteroinvasive *Escherichia coli* (EIEC)

The enteroinvasive *E. coli* strains are a cause of bacillary dysentery. These strains are biochemically and genetically related to *Shigella*, cause disease symptoms similar to those of *Shigella*, and have *Shigella*-like characteristics such as the lack of motility, the inability to ferment lactose and failure to decarboxylate lysine. The EIEC are able to invade HeLa cells and to induce keratoconjunctivitis in the guinea pig eye (Sérény test) (Hale *et al*., 1997; Nataro and Kaper, 1998). In most patients an EIEC infection results in watery diarrhea; however, in a few cases stools may contain blood and mucus. The infective dose of EIEC is several logs higher than that of *Shigella* (Hale *et al*., 1997), indicating that person-to-person spread is uncommon, although it has been reported (Harris *et al.*, 1985).

The site of EIEC infection is the colonic mucosa. The bacterial cells attach to the epithelial cells of the colon with subsequent penetration of the enterocytes via endocytosis. The endocytic vacuole is lysed followed by intracellular multiplication of the bacterial cells. There is directional movement of EIEC through the cytoplasm mediated by the attachment of cellular actin to one pole of the bacterial cell; actin aids in propelling the bacteria into adjacent epithelial cells (Nataro and Kaper, 1998). Thus there is cell-to-cell spread of EIEC without entrance into the extracellular milieu.

A 140-MDa plasmid (pINV) encodes the genes necessary for EIEC to invade, multiply and survive within the colonic enterocytes. The *ipa* (invasion plasmid antigen) genes present on the plasmid encode the Ipa proteins, IpaA–IpaD, which are necessary for the invasive phenotype (Hale *et al.*, 1997; Nataro and Kaper, 1998). The EIEC plasmid has virulence genes identical to those present on the *Shigella* 120–140-MDa invasion plasmid. The sequences of three genes, *ipgD, mxiC* and *mxiA*, in the invasion region of the virulence plasmid of *Shigella* and EIEC were analyzed to determine the evolutionary relationships of the pINV plasmids (Lan *et al.*, 2001). Two distinct forms of the plasmid were identified in *Shigella* species, pINV A and pINV B, and the EIEC strains had plasmids identical to those found in *Shigella* strains. Furthermore, Wang *et al.* (2001a) found that 12 of the 33 O-antigen forms in *Shigella* were identical to those of *E. coli* strains.

Food- and waterborne outbreaks of EIEC have occurred; however, they are not common in industrialized nations. Implicated foods include French soft cheeses, potato salad, and guacamole (Gordillo *et al.*, 1992; Hale *et al.*, 1997; Willshaw et al., 2000). Suspect foods are generally cooked foods that are not reheated after being handled by infected food workers, but raw foods may also be involved. There has been one report of an outbreak in an institution due to person-to-person transfer of EIEC (Harris *et al.*, 1985). Traveler's diarrhea has also been associated with EIEC infection (Hale *et al.*, 1997). Infections by EIEC do not appear to be important contributors to morbidity in developed countries; such infections are probably more important in developing countries, particularly among young children. Similar to *Shigella*, the reservoir for EIEC is the human intestinal tract.

A virulence antigen-specific, monoclonal antibody-based, enzyme-linked immunoassay has been used to detect *Shigella* and EIEC (Pal *et al.*, 1997). In addition, DNA probes and primers have been developed and used to detect *Shigella* and EIEC by hybridization or by the polymerase chain reaction (PCR) respectively (Houng *et al.*, 1997). López-Saucedo *et al.* (2003) described a multiplex PCR that differentiated enterotoxigenic *E. coli*, enteropathogenic *E. coli*, Shiga toxin-producing *E. coli* and EIEC in stool samples. A PCR assay using enriched stool samples from children with acute diarrhea was more sensitive than stool culture or colony hybridization for detection of *Shigella* and EIEC (Dutta *et al.*, 2001). While it is possible to differentiate between the various diarrheagenic *E. coli* strains via multiplex PCR, it is difficult to differentiate between *Shigella* and EIEC.

4 Diffusely adherent *Escherichia coli* (DAEC)

The category known as diffusely adherent *E. coli* (DAEC) is poorly characterized, and the involvement of DAEC in diarrhea remains controversial. When bacteria attach uniformly to the surface of HeLa or HEp-2 cells the adherence is termed diffuse; whereas in localized adherence the bacteria adhere in groups at one or a few sites on the cell surface (Scaletsky *et al.*, 1984). Bilge *et al.* (1989) designated a fimbrial adhesin, F1845, as responsible for diffuse HEp-2 cell adhesion by diarrheic *E. coli* isolates. A DNA probe targeting the *daaC* gene, associated with expression of the F1845 fimbriae, has been developed to detect DAEC (Bilge *et al.*, 1989). The *daaC* gene can be found either on the bacterial chromosome or on a plasmid. The probe is specific for this gene; however, it is rather insensitive, which suggests that other adhesins are also responsible for the diffuse-adherent pattern (Willshaw *et al.*, 2000). A second putative adhesin that mediates the diffuse adherence phenotype, designated AIDA-I (adhesin involved in diffuse adherence), is a 100-kDa cell surface protein (Benz and Schmidt, 1992). Some strains of DAEC induce finger-like projections, which jut from the surface of epithelial cells (HEp-2 and Caco-2 cells). These projections enclose the bacterial cells, embedding and protecting them from gentamicin; however, the bacteria are not intracellular (Cookson and Nataro, 1996). The role of these finger-like projections in pathogenesis is unknown.

Some strains of DAEC are not diarrheic; however, if infected with a diarrheic strain the patient has fever and vomiting, and stools are watery and mucoid. Diarrhea caused by DAEC occurs in developing countries, mainly in children between 48 and 60 months of age; infants are rarely affected (Cookson and Nataro, 1996; Nataro and Kaper, 1998). Nursing infants may be protected against DAEC because human milk proteins have been shown to inhibit the adherence of DAEC (Nascimento de Araújo and Giugliano, 2000). Jallat *et al.* (1993) demonstrated that most of the *E. coli* strains isolated from stools of diarrheic infants, children and adults in a French hospital were DAEC (100/262 were diffusely adherent on HEp-2 cells). Only one-third of the DAEC strains hybridized with the F1845 *daaC* probe, however, indicating that DAEC strains are quite heterogenous. In a study of 24 diarrheic children, Poitrineau *et al.* (1995) found that vomiting but not diarrhea was significantly associated with the presence of DAEC in their stools. Those children carrying F1845 DNA probe-positive DAEC had approximately three times longer hospital stays than children harboring other DAEC types. Strains in the DAEC category apparently vary in the level of pathogenicity.

5 Enteroaggregarive *Escherichia coli* (EAEC)

Escherichia coli strains that do not secrete labile toxin, stable toxin or Shiga toxin and adhere in an aggregative or 'stacked brick' (AA phenotype) adhesion pattern to HEp-2 cells are known as enteroaggregative *E. coli* (EAEC) (Nataro *et al.*, 1995; Law and Chart, 1998). This definition may include both pathogenic and non-pathogenic strains. A number of diarrheic outbreaks have been caused by EAEC and associated with food or drinking water; however, the organism was seldom isolated from the suspect vehicle (Cobeljic *et al.*, 1996; Itoh *et al.*, 1997; Smith *et al.*, 1997; Nataro *et al.*, 1998; Okeke and Nataro, 2001). Strains of EAEC have been isolated from foods such as formula from baby feeding bottles (Morais *et al.*, 1997) and tabletop sauces such as guacamole from Mexican-style restaurants (Adachi *et al.*, 2002a).

EAEC strains are a common cause of persistent diarrhea in children in developing countries; however, disease caused by EAEC is probably underreported and underdiagnosed as a cause of childhood diarrhea in industrialized countries (Okeke and Nataro, 2001). Protein components of human milk inhibit the adhesion of EAEC to HeLa cells (Nascimento de Araújo and Giugliano, 2000). It is likely that infants are protected against EAEC diarrhea while they are nursing. In adults, EAEC has been reported as a causative agent of diarrhea in individuals who travel to developing countries (Adachi *et al.*, 2001, 2002b; Okeke and Nataro, 2001) and in HIV-infected individuals (Okeke and Nataro, 2001). Other immunocompromised populations are also probably susceptible to EAEC-induced diarrhea.

The diarrhea induced by EAEC is watery and often protracted, and is associated with abdominal pain. Borborygmus (rumbling due to gas), low-grade fever, vomiting and dehydration may occur. Gross mucus and blood may be present in the stools, and up to one-third of patients may have grossly bloody stools (Nataro *et al.*, 1998; Okeke and Nataro, 2001). Histologically, a thick mucous gel is present on the intestinal mucosa, and there are necrotic lesions in the ileal mucosa (Eslava *et al*, 1998). The inflammatory cytokines IL-8 and IL-1, produced during EAEC infection, induce mucosal inflammation (Okeke and Nataro, 2001; Greenberg *et al.*, 2002). Infection with EAEC may also be asymptomatic. EAEC infections may lead to malnutrition and growth retardation in infants and children (Nataro *et al.*, 1998; Steiner *et al.*, 1998). Oral hydration is an effective therapy (Law and Chart, 1998; Nataro *et al.*, 1998; Smith and Cheasty, 1998).

A three-stage model has been proposed for EAEC pathogenesis: Stage I involves initial adherence to the intestinal mucosa and mucous layer; stage II involves enhanced production of mucus, leading to a thick EAEC-encrusted biofilm on the mucosal surface; and stage III involves elaboration of cytotoxin(s), which result in intestinal secretion and damage to the intestinal mucosa (Nataro *et al.*, 1998; Okeke and Nataro, 2001). The thick cover on the mucosal surface may promote tenacious colonization and lead to malnutrition. The persistent diarrhea seen in EAEC-infected patients may be due to the inability of individuals with pathogen-induced malnutrition to repair the damage done to the intestinal mucosa. Pathogenicity of EAEC has been modeled using tissue culture (Nataro *et al.*, 1996) and gnotobiotic piglets (Tzipori *et al.*, 1992).

5.1 Virulence factors

Information concerning virulence factors in EAEC is limited and confusing. There appear to be several types of fimbriae involved with aggregative attachment, and while the pathology suggests that a toxin is involved in EAEC diarrhea, it is not clear that the toxin(s) responsible have been identified. A study of iron utilization has shown that EAEC are able to utilize heme or hemoglobin as the sole iron source and produce siderophores at a level similar to that of *Shigella* and enterohemorrhagic *E. coli* (Okeke *et al.*, 2004). Most strains possessed genes associated with multiple iron utilization systems, which may provide EAEC with a competitive advantage over other bacteria that are negative for these systems. Analysis of EAEC strains isolated from Mongolian children with diarrhea has shown that AggR (transcriptional activator)-positive strains that caused diarrhea were more likely to possess several other EAEC virulence genes than AggR-negative strains. Furthermore, the isolation of AggR-positive EAEC was significantly higher in the diarrheal group than in controls; thus, AggR may serve as a marker for virulent EAEC strains (Sarantuya *et al.*, 2004).

5.1.1 Attachment

Using biopsies from normal patients, Knutton *et al.* (1992) demonstrated attachment of EAEC to the colonic (44/44 biopsy samples) and the ileal mucosa (36/44 of biopsy samples). None of the EAEC strains attached to the duodenal mucosa. Knutton *et al.* (1992) suggested that EAEC is a large-bowel pathogen that colonizes the colon by adhesion mediated by fimbriae. A number of EAEC adherence factors have been demonstrated. Nataro *et al.* (1992) described aggregative adherence fimbriae I (AAF/I), and Czeczulin *et al.* (1997) characterized AAF/II. However, by using DNA probes for AAF/I and AAF/II, Czeczulin *et al.* (1997) found that only a minority of EAEC strains possessed these fimbriae. Other aggregative fimbriae have been described by Knutton *et al.* (1992) and Collinson *et al.* (1992). In addition, afimbrial adhesins are expressed by some strains of EAEC (Okeke and Nataro, 2001).

5.1.2 Putative toxins

The EAST1 toxin, a plasmid-mediated, low molecular weight, heat-stable toxin with an *in vitro* mode of action similar to that of the ETEC heat-stable toxin, was first demonstrated in EAEC (Savarino *et al.*, 1991). The *astA* gene encoding EAST1 was found in approximately 41 % of EAEC strains (Okeke and Nataro, 2001). It is not clear, however, that EAST1 has a role in EAEC-induced diarrhea *in vivo* (Navarro-Garcia *et al.*, 1998). A more detailed discussion of EAST1 can be found in Section 6.

A high molecular weight (108-kDa), heat-labile protein toxin was found in EAEC (Narvarro-García *et al.*, 1998). A partially purified preparation induced tissue damage, inflammation and secretion of mucus in isolated rat jejunum. Eslava *et al.* (1998) reported on the genetic cloning, sequencing and characterization of the 108-kDa toxin. The toxin gene (*pet*) is located on the 65-MDa EAEC virulence plasmid, which also contains the genes for the aggregative phenotype, AA. The plasmid-encoded toxin (Pet) appears to belong to the autotransporter class of secreted proteins, and is highly homologous to other autotransporter proteins such as the EspP protease of EHEC and the cryptic protein EspC of enteropathogenic *E. coli* (Eslava *et al.*, 1998). Okeke and Nataro (2001) reported that 18–44 % of EAEC isolates possess Pet. Morabito *et al.* (1998) isolated EAEC strains that were involved in an outbreak of HUS. These strains were unusual in that they produced Stx2, had the AA phenotype, and possessed the *astA* gene for EAST1 but lacked the EHEC genes, *eaeA, hly* and *katP*.

5.2 Heterogeneity of EAEC strains

EAEC form a very heterogeneous group comprising more than fifty O serogroups (Chart *et al.*, 1997). Serotyping is not useful for identifying EAEC strains. Types of fimbriae vary from bundles of fine filaments (Knutton *et al.*, 1992) and thin fimbriae (Collinson *et al.*, 1992) to bundle forming fimbriae (Nataro *et al.*, 1992; Czeczulin *et al.*, 1997); however, not all EAEC express fimbriae (Chart *et al.*, 1997). In addition, not all strains of EAEC produce the EAST1 or Pet toxins (Savarino *et al.*, 1996; Eslava *et al.*, 1998). Nataro *et al.* (1995) also demonstrated that EAEC strains are heterogeneous in their ability to induce diarrhea in adult volunteers. The heterogeneity of EAEC renders identification of strains and diagnosis of EAEC-induced illnesses difficult.

5.3 Diagnosis of EAEC infections

The most definitive identification of EAEC is the demonstration of adhesion to HEp-2 cells (Law and Chart, 1998; Miqdady *et al.*, 2002). However, the technique is only suitable for use in research laboratories. It is time-consuming and cumbersome, and the type of adherence can be easily misinterpreted.

Utilizing the adhesion-associated region of the adherence plasmid of EAEC, Baudry *et al.* (1990) and DebRoy *et al.* (1994) developed DNA probes for the identification of EAEC strains. However, neither probe identified all EAEC having the AA phenotype. The probe developed by Baudry's group detected 70.5 % (43/61) of isolates, whereas the probe developed by DebRoy and coworkers detected 93.4 % (57/61).

Schmidt *et al.* (1995) developed a PCR assay based on the probe of Baudry *et al.* (1990). Of 50 EAEC strains (positive in the HEp-2 adherence assay), 88 % (44/50) were positive with the Baudry probe, and 86 % (43/50) were positive using the PCR. Both probes reacted with less than 1 % (4/418) of other *E. coli* strains tested (Schmidt *et al.*, 1995). Thus the probes and PCR tests were quite specific but did not detect all EAEC that showed the typical AA phenotype. Therefore, the HEp-2 adherence assay is the only reliable method for identification of EAEC strains; however, the test does not distinguish pathogenic from non-pathogenic strains of EAEC.

A multiplex PCR using three plasmid-borne genes (aggregative adherence (AA) probe, *aap* and *aggR*) was tested on 28 AA-positive *E. coli* isolated from diarrheic patients and detected 23/28 (82 %) of the strains (Cerna *et al.*, 2003). It appears that the only reliable technique for detection of EAEC is the determination of the AA phenotype using a tissue culture assay.

Clearly, additional studies on EAEC and the determination of the type of adherence factors involved in the aggregative type of adherence are needed. It is probable that the aggregative pattern is mediated by different types of adhesion molecules. The literature indicates that not all EAEC strains cause diarrhea. Is the aggregative pattern important for virulence *in vivo*, or is it merely a diagnostic tool to detect a certain type of *E. coli*? Thus, tests to identify pathogenic EAEC and differentiate diarrheic from non-diarrheic strains of EAEC are needed. The role of the EAST1 and Pet toxins as *in vivo* agents of pathogenesis in EAEC strains is uncertain. Generation of mutants lacking the expression of these toxins or cloning of EAEC virulence factors into laboratory strains of *E. coli* and use in cell culture assays, animal models (Law and Chart, 1998) and human volunteers may be useful in understanding the virulence of EAEC. Growth retardation and other growth deficits appear to be related to EAEC infections (Nataro *et al.*, 1998; Steiner *et al.*, 1998). Does malnutrition predispose to EAEC infection? Does EAEC-induced mucosal damage lead to malabsorption of nutrients? Does EAEC-induced stimulation of mucus formation impose a barrier to intestinal absorption of nutrients? Animal studies should clarify the putative role of EAEC in malnutrition.

6 Enterotoxigenic *Escherichia coli* (ETEC)

6.1 The disease

In developed countries ETEC strains are an uncommon cause of diarrhea; however, in developing countries they are a major cause of diarrhea, with high morbidity and mortality in infants, young children and the elderly. In addition, the organism is the primary cause of traveler's diarrhea in visitors to developing countries (Cohen and Giannella, 1995; O'Brien and Holmes, 1996).

Newborn and young domestic animals (calves, lambs, and pigs) are susceptible to ETEC-induced diarrhea; however, ETEC do not cause disease in adult animals (Gyles, 1992). The ETEC colonize the small intestine of both human adults and children by attaching to the enterocytic brush border with the aid of bacterial adherence factors; however, the organisms do not invade or damage intestinal cells (Cohen and Gianella, 1995). The ETEC secrete toxin(s), which lead to the production of a non-inflammatory watery diarrhea. Blood, mucus, and leukocytes are absent in stools. The infected individuals may show nausea and mild to moderate abdominal cramping, but without fever (Neill *et al.*, 1994; Cohen and Gianella, 1995). Diarrhea may be prolonged in children, leading to severe dehydration, and mortality can be high. Serious malnutrition may result in children infected with ETEC. Traveler's diarrhea in adults is usually a mild self-limited illness lasting 1–5 days (Neill *et al.*, 1994; Cohen and Gianella, 1995).

ETEC infection is transmitted through the ingestion of contaminated food or water. An ETEC-infected food handler with poor personal hygiene can contaminate food and water. Also, ETEC-containing sewage can contaminate potable water (Black *et al.*, 1981). Studies with human volunteers indicate that the infective dose is approximately 10^8 organisms (Levine *et al.*, 1977), therefore person-to-person transmission of ETEC infection is not likely under most circumstances. Humans appear to be the major reservoir for ETEC, and there are no animal reservoirs, although young animals are susceptible to ETEC infection (Doyle and Padhye, 1989).

Antibiotic treatment is generally not recommended for most cases of traveler's diarrhea, since antibiotic use may lead to antibiotic resistance in ETEC and may also change the intestinal flora. If the diarrhea is severe or prolonged, the infection can be treated with trimethoprim/sulfmethoxazole, and rehydration therapy may be required. Mild cases of ETEC-induced traveler's diarrhea can be treated with antidiarrheal drugs to limit fluid accumulation and intestinal mobility (Berkow, 1992; Cohen and Giannella, 1995).

The best prophylaxis against ETEC infection and other gastrointestinal diseases in infants is breastfeeding (Black and Lanata, 1995; Pickering *et al.*, 1995). In a study involving Bangladeshi children, Clemens *et al.* (1997) found that exclusive breastfeeding of infants (children less than 1 year of age) was protective against severe ETEC-induced illness. This protective effect was not observed with breastfed children during the second and third years of life. Breastfeeding provided significantly greater protection against diarrhea induced by *Vibrio cholerae* as compared to ETEC-induced diarrhea (Clemens *et al.*, 1997). In ETEC-infected infants and children, oral rehydration is an effective therapy, and rehydration prior to the occurrence of severe diarrheic dehydration can be lifesaving. Antimotility agents and antibiotic therapy are not recommended for use in ETEC-infected children. In developing countries, antibiotic resistance is common, and the use of antibiotics often does not remove the offending organisms from the gut (Pickering *et al.*, 1995). Similarly, antimotility agents interfere with peristaltic removal of the pathogen. Vaccines for use in humans to control ETEC infections are not available (O'Brien and Holmes, 1996).

6.2 Foodborne outbreaks

Water and food have been implicated in outbreaks of ETEC (CDC, 1994; Table 6.2). Foods implicated in outbreaks include salads, dipping sauces and ready-to-eat items, including hot dogs, cold roast beef, cold turkey, and Brie cheese (foods that are served raw or foods that are cooked but served cold). Although ETEC does not appear to be a major cause of diarrheic foodborne outbreaks in the United States, many disease cases may go unrecognized due to the non-availability of laboratory tests for identification of ETEC strains. Beatty *et al.* (2004) reported results of an 8-year study on the incidence of outbreaks due to ETEC in the US and on cruise ships. Sixteen outbreaks due to ETEC occurred from 1996 to 2003, and *E. coli* O169:H41 was the serotype identified in 10 of the outbreaks. The vehicle of infection was identified in 11 of the outbreaks, and included drinking water, ice, various vegetables and salads, enchiladas, tacos, tortilla chips, quesadillas, fajitas, chicken, lasagna, and catfish.

6.3 Basis of pathogenicity

The ETEC cause diarrheal illness by adherence and colonization of the intestinal mucosa, and the synthesis and release of at least one member of two groups of enterotoxins – heat-labile toxins (LT) and heat-stable toxins (ST) (Cohen and Giannella, 1995). ETEC strains may produce an LT only or an ST only, or both an LT and an ST.

6.3.1 The LT enterotoxins

Some features of the LT enterotoxins of ETEC are listed in Table 6.3. Polyclonal antibodies raised against a particular LT-I toxin neutralize other LT-Is and cholera toxin (CT) produced by *V. cholerae*, but do not neutralize the activity of LT-II toxins.

Table 6.2 Foodborne outbreaks in which ETEC strains were implicated

Reference	Toxin	Number ill	Comments
Ryder *et al.*, 1976	ST	55	Outbreak occurred in a children's hospital in the United States; baby formula was implicated
Merson *et al.*, 1976	ST/LT LT ST	8 12 5	Outbreak occurred in travelers attending a meeting in Mexico City; salads containing raw vegetables were implicated
Hobbs *et al.*, 1976	ST	67	Outbreak occurred on a cruise ship; food suspected
Kudoh *et al.*, 1977	ST	129	Two separate outbreaks in two different locations in Japan (food eaten not stated); outbreaks associated with eating lunch
Danielsson *et al.*, 1979	LT	60	Outbreak occurred in Swedish restaurant; shrimp and mushroom salads were implicated
	LT	2	Incident occurred in a Swedish home; cold, shop-sliced roast beef implicated
Lumish *et al.*, 1980	LT	349	Outbreak occurred on two separate trips of same cruise ship; drinking water and crabmeat cocktail were implicated
Taylor *et al.*, 1982	ST/LT	415	Outbreak occurred in a restaurant in Wisconsin; Mexican food was implicated (food items included sauces, garnish, flour tortillas, and guacamole)
Wood *et al.*, 1983	LT	282	Outbreak occurred in a hospital in Texas; associated with eating in hospital cafeteria; no specific food was implicated
Riordan *et al.*, 1985	ST/LT ST	27	Outbreak occurred at a cold buffet at a school in England; curried turkey mayonnaise was implicated
MacDonald *et al.*, 1985	ST	45	Clusters of outbreaks occurred at office parties in Washington, DC; French Brie cheese was implicated; cheese from the same plant (same brand and lot) was implicated in outbreaks in Illinois (75 cases), Wisconsin (35 cases), Georgia (10 cases) and Colorado (4 cases). The same brand of cheese caused outbreaks in Denmark, the Netherlands and Sweden
CDC, 1994	LT, ST	47	Outbreak occurred on a plane from North Carolina to Rhode Island; garden salad was implicated
	LT, ST	97	Outbreak occurred at a buffet served in a mountain lodge in New Hampshire; tabouleh salad was implicated
Mitsuda *et al.*, 1998	ST1b	>600	Outbreak occurred with school lunches at four elementary schools in Japan; tuna paste implicated – raw carrots, onions and cucumbers that were part of the tuna paste were probably contaminated with ETEC

Table 6.2 Foodborne outbreaks in which ETEC strains were implicated—*cont'd*

Reference	Toxin	Number ill	Comments
Roels *et al.*,1998	ST	372–645	Outbreak occurred at a banquet in Milwaukee, Wisconsin; pan-fried spiced potatoes were implicated
	ST and LT/ST	97	Outbreak occurred on a cruise ship; drinking ship's tap water and/or beverages with ice were implicated
Daniels *et al.*, 2000	ST, LT and LT/ST	19	Outbreak occurred on a cruise ship; beverages with ice and ice water were implicated
	ST, LT and LT/ST	197	Outbreak occurred on a cruise ship; unbottled water and beverages with ice were implicated
Huerta *et al.*, 2000	ST, LT and LT/ST	229	Outbreak occurred at military posts and civilian communities in the Golan Heights, Israel; drinking water was implicated
Naimi *et al.*, 2003	ST, LT and LT/ST	77	Outbreak occurred in a restaurant in Minnesota; parsley from Mexico was implicated

Modified from Fratamico *et al.* (2002).

Table 6.3 Toxins of enterotoxigenic *E. coli*

Toxin vs heat[a]	Variants	Preferred receptor	Structure	Mode of action
Heat-labile (LT)	LTh-I LTp-I	GM1	One A unit (~240 amino acids) combined with five B units (100 amino acids each)	Activation of adenylate cyclase with stimulation of intestinal Cl– secretion
	LT-IIa	GD1b		
	LT-IIb	GD1b		
Heat-stable (ST)	STIa (STp)	Guanylate cyclase STaR (on human enzyme)	18-amino acid peptide	Activation of guanylate cyclase with stimulation of Cl^- secretion
	STIb (STh)	Guanylate cyclase STaR (on human enzyme)	19-amino acid peptide	
	STII	Unknown	48-amino acid peptide	Mode of action unknown; stimulation of intestinal HCO_3^- secretion

Modified from O'Brien and Holmes (1996)

[a] After 30 minutes at 100 °C, LT toxin activity is lost; ST toxin activity is retained.

Thus, there is a close relationship between CT and LT-I. Antibodies to a particular LT-II neutralize LT-IIs but not LT-I or CT (O'Brien and Holmes, 1996). The LTh-I variant is of human origin, whereas, the LTp-I is of porcine origin. The CT-encoding genes are carried on the genome of a filamentous bacteriophage maintained as an integrated prophage (Davis and Waldor, 2003); however, the genes for LT-I are plasmid-mediated. The genes for LT-II are chromosomally encoded (O'Brien and Holmes, 1996). LT-I causes a milder disease in humans than CT; however, LT-II does not appear to be involved in human disease and is found primarily in animal isolates (O'Brien and Holmes, 1996).

Similar to CT, the LTs are oligomeric peptides consisting of one A polypeptide subunit non-covalently bound to five B polypeptides. The A subunits of LT-I and LT-II have approximately 50 % amino acid identity; whereas, the B subunits of LT-I and LT-II are approximately 10 % identical (O'Brien and Holmes, 1996). The B subunits of LT-I are ring structures that bind to the ganglioside GM1 on the host cell. The A subunit undergoes proteolytic nicking to produce A1 and A2 fragments. The A1 fragment is linked to A2 by a disulfide bond, and the A2 fragment is bound to the B subunits. After binding to the host cell, the toxin is endocytosed and translocated through the cell (Butterton and Calderwood, 1995; Nataro and Kaper, 1998). The cellular target of LT is adenylate cyclase located on the membrane of intestinal epithelial cells. The A1 fragment exhibits ADP-ribosyl transferase activity. ADP ribosylation of a GTP-binding protein mediates activation of adenylate cyclase, with a resultant increase in cyclic AMP within the intestinal mucosa. The net result is the stimulation of chloride secretion and a decrease in sodium absorption. The increased luminal ion content leads to a loss of fluid and electrolytes with production of a watery diarrhea (Butterton and Calderwood, 1995; Nataro and Kaper, 1998).

For CT activity proteolytic nicking of the A subunit is necessary (Butterton and Calderwood, 1995), whereas nicking is not necessary for enzymatic activity of the LT-I A subunit (Grant *et al.*, 1994). Nicking does, however, enhance the biological and enzymatic activity of LT. Tsuji *et al.* (1997) constructed an ETEC mutant in which the nicking region of the A-subunit of LT was deleted. The mutant had less diarrheal activity with decreased induction of cyclic AMP; thus, nicking of the A-subunit of LT-I appears to be necessary for optimum activity of the toxin. Other bacteria that produce LT-like toxins include *Klebsiella, Enterobacter, Aeromonas, Plesiomonas, Campylobacter* and *Salmonella* (Cohen and Gianella, 1995).

6.3.2 The ST enterotoxins

Certain characteristics of the ST enterotoxins are shown in Table 6.3. These monomeric toxins are subdivided into STI (STa) and STII (STb) families. Two toxins are found in the STI family (Table 6.3), including one of porcine origin – STp, an 18-amino-acid peptide. The other, STh, is of human origin and consists of 19 amino acids. Three intramolecular disulfide bonds are present in the STI peptides. STI molecules are heat and acid stable, are not denatured by detergents, are water and methanol soluble, and are resistant to proteases. With disruption of the disulfide bonds, the toxins become inactive (Cohen and Gianella, 1995; O'Brien and Holmes,

1996; Nataro and Kaper,1998). The STI peptides are poorly antigenic, and must be conjugated to a carrier protein to prepare antisera for diagnostic purposes. Genes for both STI and STII are located predominately on plasmids, but some are on transposable elements. Genes encoding LT, ST, colonization factors, colicin and antibiotic resistance may be present on the same plasmid (Cohen and Gianella, 1995; O'Brien and Holmes, 1996; Nataro and Kaper, 1998).

STI is synthesized in the bacterial cell cytoplasm as a precursor protein consisting of a PRE-region (amino acid residues 1–19), a PRO-region (amino acid residues 20–54) and a MATURE-region (amino acid residues 55–72). The precursor protein is translocated across the inner membrane into the periplasmic space via Sec proteins of the type II secretion pathway. The PRE-region acts as a signal protein in the translocation process and is cleaved during or after translocation. In the periplasmic space, the protein consists of the PRO- and MATURE-regions (Okamoto and Takahara, 1990). The PRO-region is then cleaved, disulfide bonds are formed in the mature 18-amino acid toxin, and the toxin is translocated across the outer membrane in an unknown manner (Yamanaka *et al.*, 1994, 1997). Yamanaka *et al.* (1998) have shown that an outer membrane protein, TolC, is involved in some manner in the translocation of periplasmic STI across the outer membrane into the external environment.

The major receptor for STI is guanylate cyclase C in the membrane of enterocytes in the small intestine. Binding of the enzyme by STI leads to an accumulation of cyclic GMP, and secretion of chloride and water into the intestinal lumen (Cohen and Gianella, 1995; O'Brien and Holmes, 1996; Nataro and Kaper, 1998). Guanylin is a mammalian hormone which aids in the regulation of fluid and electrolyte absorption in the gut. STs and guanylin are homologous and bind to the same receptor on intestinal epithelial cells (Rabinowitz and Donnenberg, 1996).

The STI toxins produce a reversible short-term effect that is quick acting (within 5 minutes) and is mediated by guanylate cyclase. However, the biological effect of CT and LT is prolonged with a lag phase of about 1 hour, is reversible, and is mediated by activation of adenylate cyclase (Cohen and Gianella, 1995). LT-I and CT bind to adenylate cyclase from various tissues, whereas STI binds only to intestinal guanylate cyclase (Gyles, 1992).

STII appears to be primarily found in ETEC strains isolated from pigs. While STI is methanol soluble, STII is methanol insoluble. STII is a larger peptide than STI (5.1 kDa and 2 kDa, respectively), and does not cross-react immunologically with STI (O'Brien and Holmes, 1996). STII induces secretion of bicarbonate ions and water into the intestinal lumen, and increases the intracellular Ca^{2+} in intestinal cells (O'Brien and Holmes, 1996). STII does not appear to contribute to human disease (Salyers and Whitt, 1994); however, a few cases of STII-induced human diarrhea have been reported (Lortie *et al.*, 1991; Okamoto *et al.*, 1993). Reviews by Dubreuil (1997) and Nair and Takeda (1998) discuss various aspects of the heat-stable enterotoxins.

Other bacteria, including *Citrobacter freundii, Yersinia enterocolitica* and non-O1 *Vibrio cholerae*, produce toxins similar to STI (Smith, 1988; Chaudhuri *et al.*, 1998). An STIa-containing plasmid from ETEC could be transferred to species of *Shigella, Salmonella, Klebsiella, Enterobacter, Edwardsiella, Serratia* and *Proteus* with stable maintenance of the plasmid and expression of toxin (Smith, 1988).

6.3.3 EAST1 (EAEC heat-stable toxin 1)

EAST1 is a heat-stable enterotoxin present in enteroaggregative *E. coli* (EAEC) strains (Savarino *et al.*, 1993). The gene for EAST1, *astA*, was found to be present in strains of ETEC isolated from both humans and animals (Yamamoto and Echeverria, 1996; Yamamoto and Nakazawa, 1997). The gene is present in piglet strains of ETEC producing STI, LT, or STI and LT, and was associated with adhesin factor K88. The K88 and EAST1 genes are on separate plasmids (Yamamoto and Nakazawa, 1997). In LT-, STI-, or LT- and STI-producing ETEC strains from humans, however, colonization factor antigens (CFA/I, CFA/II or CFA/IV) were associated with EAST1, and all genes were located on the same plasmid (Yamamoto and Echeverria, 1996).

Savarino *et al.* (1996) detected the EAST1 gene in 100 % of 75 *E. coli* O157:H7 strains, in 47 % of 227 EAEC, in 41 % of 149 ETEC, in 22 % of 65 EPEC strains, and in 13 % of 70 DAEC strains, utilizing an *astA* DNA probe. In addition, *astA* was present in non-diarrhea-producing *E. coli* strains. Rich *et al.* (1999) found no correlation between the severity of diarrheic symptoms due to EAEC infection and the presence of EAST1. Thus, the *astA* gene appears to be common in *E. coli*; however, the significance of the EAST1 toxin in the pathogenesis of ETEC is unclear at the present time.

EAST1 is a low molecular weight, cysteine-rich, 38-amino-acid polypeptide enterotoxin that is plasmid encoded, partially heat stable (63 % of activity remained after 65 °C for 15 minutes) and protease sensitive (Savarino *et al.*, 1991; O'Brien and Holmes,1996; Ménard and Dubreuil, 2002). The EAST1 toxin does not cross-react serologically with STI; however, EAST1 shows homology with the receptor-binding domains of STI. Presumably the enterotoxin associates with the same receptor binding site of guanylate cyclase as STI, leading to cyclic GMP secretion (Savarino *et al.*, 1991; O'Brien and Holmes, 1996; Ménard and Dubreuil, 2002). Thus, EAST1 appears to be a member of the STI family of heat-stable enterotoxins and produces diarrhea by secreting cyclic GMP (Savarino *et al.*, 1993). A review by Ménard and Dubreuil (2002) summarizes various aspects of EAST1.

6.3.4 Colonization factors

The attachment of the ETEC strains to host cells is an important initial step in pathogenesis. The colonization factor antigens (CFAs), including CFAs I, II, and IV, are major adherence factors in human strains and are found only in diarrhea-causing ETEC (Salyers and Whitt, 1994). The CFA genes are plasmid encoded, and CFAs, ST and LT may be encoded by the same plasmid (Cohen and Giannella, 1995; Mol and Oudega, 1996). CFAs CFAII and CFAIV are further divided into CS (coli surface) antigens (Nirdnoy *et al.*, 1997). In addition to CFAI, CFAII, and CFAIV, a number of other CFAs have been described (Cassels and Wolf, 1995; Gaastra and Svennerholm, 1996; Grewal *et al.*, 1997; Ricci *et al.*, 1997). The CFA structures may be fimbrial rods, flexible fibrils, helical fibrils or curly fibrils (Cassels and Wolf, 1995). In 241 ETEC strains isolated from Mexican children, 46 % of the strains possessed a CFA. Of LT/ST strains, 65 % expressed a CFA, whereas 50 % of ST and 25 % of LT strains expressed a CFA (López-Vidal *et al.*, 1990). López-Vidal *et al.* (1990) found that children infected with ETEC lacking CFAs (I, II or IV) had diarrhea similar to those infected with CFA-containing ETEC. The fact that ETEC-lacking CFAs can cause diarrhea suggests that unidentified colonization factors were responsible for diarrhea in those ETEC strains.

Lyte *et al.* (1997) demonstrated that norepinephrine stimulated growth and K99 pilus-mediated adhesion of ETEC. ETEC strains that express K99 pili are pathogenic for lambs, calves and pigs. The distal two-thirds region of the small intestine is preferentially colonized by ETEC. This area is highly innervated with adrenergic nerves, which produce norepinephrine at terminals present in the mucosal lining (Lyte *et al.*, 1997). While the K99 adhesion pili are virulence factors necessary for colonization of ETEC in animals (Parry and Rooke, 1985), it is likely that intestinal norepinephrine also stimulates induction of CFAs in human strains of ETEC.

6.4 ETEC and the immune response

Both purified CT and LT (or the B subunit of the toxins) have been used as oral adjuvants, since they are potent mucosal immunogens. LT is less toxic than CT, and can be used at levels that do not induce diarrhea (Baqar *et al.*, 1995). CT induces a TH_2 (T-helper cells involved in the humoral immune response) response with production of IL-4 (interleukin 4) and IL-5 (interleukin 5) cytokines; in addition, the immunoglobulins IgA, IgG_1 and IgE are produced. LT, however, induces a mixed TH_1 and TH_2 response with production of IFN-γ (interferon-gamma), IL-4 and IL-5 cytokines. LT induces an IgA, IgG_1, IgG_2 and IgG_{2b} antibody response profile (Takahashi *et al.*, 1996). The IgE response induced by CT indicates that its use as a mucosal adjuvant can lead to immediate-type allergic hypersensitivity, and the use of LT would be more desirable for this purpose. Oral administration of LT and heat-killed *Campylobacter jejuni* stimulated both local and systemic *Campylobacter*-specific IgA and IgG in non-human primates (Baqar *et al.*, 1995). Co-administration of LT with oral inactivated influenza vaccine to mice showed increased antiviral serum IgG and mucosal IgA as compared to use of the vaccine alone (Katz *et al.*, 1997). The use of LT as an oral adjuvant to increase the secretion of secretory IgA on mucosal surfaces appears to be a viable option in the control of human gastrointestinal and pulmonary diseases.

In developing countries, ETEC infections decrease as individuals become older – which suggests the development of protective immunity against ETEC infections. Therefore, it should be possible to develop vaccines against ETEC. Recent studies have indicated that the oral administration of killed ETEC cells combined with recombinant cholera-toxin B subunit to children or adults provided significant protection against ETEC infection (Savarino *et al.*, 1999; Cohen *et al.*, 2000; Qadri *et al.*, 2000). However, the development of an effective ETEC vaccine with broad protective powers is difficult due to the large number of different intestinal adherence factors expressed by ETEC strains (Nataro and Kaper, 1998).

Mason *et al.* (1998) constructed a synthetic gene coding for *E. coli* LT-B subunit for use in transgenic potatoes. Feeding mice with raw tubers that expressed the LT-B subunit protein resulted in high levels of serum and mucosal anti-LT-B immunoglobulins. Plant-derived vaccine antigens, particularly produced in raw edible fruits, should prove useful in protecting children and adults against diarrheic diseases (Walmsley and Arntzen, 2000).

6.5 Detection of ETEC

There are no serological or biochemical markers to differentiate toxin-producing strains from non-toxigenic ETEC strains. Therefore, it is necessary to detect the toxins produced by ETEC strains. Nataro and Kaper (1998) described a number of molecular diagnostic techniques that can be used to detect LT and ST. A multiplex PCR allowing simultaneous detection of the ETEC LTI and STII genes was used to detect the pathogen in skim milk and porcine stool (Tsen *et al.*, 1998). Monoclonal antibodies were produced against ETEC colonization factors and used to determine the prevalence of ETEC possessing the different colonization factors in children with diarrhea in Argentina (Viboud et al., 1993). López-Saucedo *et al.* (2003) described a single multiplex polymerase chain reaction that could be used to detect diarrheic *E. coli*, including enterotoxigenic *E. coli*. A DNA colony hybridization assay, including a pooled-toxin (STp, STh, and LT) probe assay and individual probe assays to detect toxins and a number of different colonization factors, was developed to detect and characterize ETEC (Steinsland *et al.*, 2003).

7 Enteropathogenic *Escherichia coli* (EPEC)

7.1 Disease and epidemiology

Strains in the EPEC category are an important cause of infantile diarrhea in developing countries where water quality and hygiene are poor. The EPEC strains cause infections with high morbidity and mortality and are a threat to infants and young children worldwide. It is estimated that EPEC cause at least 117 million diarrheal episodes per year in developing countries (not including China) (Clarke *et al.*, 2002). Outbreaks of diarrhea due to EPEC are rare in developed countries; however, outbreaks have occurred in day-care centers and pediatric wards (Vallance and Finlay, 2000). Transmission occurs primarily by the fecal–oral route, and contaminated hands, food and fomites serve as sources of infection. EPEC generally affects children less than 2 years of age, and especially infants less than 6 months of age. The diarrhea is self-limiting in most cases; however, in severe cases it can be prolonged, with wasting and failure to thrive (Fagundes-Neto and Scaletsky, 2000). Acute EPEC infections are manifested by profuse watery, mucoid (but non-bloody) diarrhea, often accompanied by vomiting and fever, and in severe cases death may result (Vallence and Finlay, 2000; Willshaw *et al.*, 2000; Clarke *et al.*, 2002). In adult volunteers, 10^8–10^{10} CFU are necessary to induce diarrhea; however, it is probable that the infectious dose in children is lower (Clarke *et al.*, 2002). The incubation period for EPEC infection in children is unknown. Breastfeeding is protective, and infants generally become infected following weaning due to preparation of weaning foods with contaminated water. Oral hydration is the treatment of choice in mild cases, and parenteral rehydration is needed in severe cases. Children may suffer several diarrheal episodes each year due to EPEC, and no vaccines are currently available (Willshaw *et al.*, 2000; Clarke *et al.*, 2002). The reservoir for EPEC strains is the human gastrointestinal tract, and there is no evidence of zoonotic infections with human EPEC serotypes.

7.2 Basis of pathogenicity

The small-bowel epithelium is the site of EPEC infection (Vallance *et al.*, 2002). EPEC strains bind loosely to the surface of small-bowel epithelial cells in a localized adherence pattern, and inject virulence factors into the cells. Disease is the result of the translocated bacterial virulence factors interacting with components of the host cells and altering the host-cell signaling pathways (Vallance and Finlay, 2000).

The EPEC adherence factor (EAF) plasmid is necessary for localized adherence. Densely packed three-dimensional clusters of bacteria adhering to the surface of tissue-culture cells is characteristic of localized adherence. The bundle-forming pilus (BFP), encoded by the EAF plasmid, is responsible for localized adherence, and is required for full virulence. BFP mutants are impaired in their ability to induce diarrhea (Frankel *et al.*, 1998; Vallance and Finlay, 2000; Donnenberg and Whittam, 2001).

After initial attachment (localized adherence) of EPEC to the intestinal-cell membrane, proteins are secreted; this results in intimate bacterial attachment and the formation of cuplike pedestals on the microvilli on which the bacteria rest with the accumulation of polymerized filamentous actin, α-actinin, talin, ezrin and myosin light chain. These lesions are referred to as 'attaching and effacing' (A/E), and have been observed *in vitro* and *in vivo* (Donnenberg and Whittam, 2001). The A/E pathology is mediated by genes located on a 35-kb pathogenicity island, the locus of enterocyte effacement (LEE), which comprises 41 open reading frames. The G+C content of the LEE region is 38.4 %, in contrast to the *E. coli* chromosome, which has a G+C content of 50.8 % (Frankel *et al.*, 1998). The LEE genes are separated into three domains: Tir (translocated intimin receptor) and the intimin outer membrane protein; EspA-D, encoding secreted proteins and their chaperones; and a region encoding a type-III secretion system, which translocates bacterial proteins directly into the host cell (Frankel *et al.*, 1998; Donnenberg and Whittam, 2001). The Tir protein, encoded by LEE, is translocated via the type-III secretion system and is inserted into the host-cell plasma membrane. The Esps are involved in the translocation process. The inserted Tir is phosphorylated and then acts as the receptor for the intimin outer membrane protein. Intimin is the product of the *eae* gene located downstream of the *tir* gene in the LEE locus. Intimin is essential for intimate adherence and A/E formation (DeVinney *et al.*, 1999a, 1999b; Donnenberg and Whittam, 2001). Thus, EPEC strains insert their own receptor (Tir) for the intimin adhesin protein, with resultant A/E lesion formation (DeVinney *et al.*, 1999a).

A number of mechanisms have been proposed to explain how EPEC cause diarrhea; however, none of the proposed mechanisms have been studied in enough detail to elucidate the diarrheic mechanism (Nataro and Kaper, 1998; Vallance and Finlay, 2000). Thus, it is not clear how an EPEC infection actually triggers diarrhea.

Savarino *et al.* (1996) detected the EAST1 gene, *astA*, in 22 % of 65 EPEC strains; however, the significance of EAST1 toxin in these strains is unknown. Some EPEC strains that produce the A/E lesion have a gene (*lifA*) that encodes the toxin lymphostatin. Lymphostatin inhibits lymphocyte activation and selectively inhibits the production of IL-2, IL-4, IL-5 and IFN-γ. In addition, the toxin inhibits proliferation of lymphocytes (Klapproth *et al.*, 2000). The expression of lymphostatin may suppress the immune response against the bacteria and thereby prolong the infection, which would enhance the spread of the organism to other individuals.

7.3 Animal models

EPEC is a human pathogen, and does not infect most laboratory animals (Vallance and Finlay, 2000). However, A/E lesion-inducing *E. coli* strains have been isolated from rabbits (REPEC). The REPEC strains infect the small bowel of weanling rabbits and produce a disease similar to that caused by human EPEC. The rabbits suffer diarrhea and weight loss (DeVinney *et al.*, 1999a; Milon *et al.*, 1999). The pattern of adherence of REPEC is diffuse rather than localized as with human EPEC; however, the REPEC LEE-encoded secreted proteins are similar to those of human EPEC (Tauschek *et al.*, 2002). *Citrobacter rodentium* produces A/E lesions in mice but, unlike human EPEC and REPEC, *C. rodentium* colonizes the large bowel rather than the small bowel (Higgins *et al.*, 1999). The organism induces a TH1 response with production of interleukins, tumor necrosis factor alpha and gamma interferon. In addition, *C. rodentium* induces intestinal epithelial cell hyperplasia rather than diarrhea (Higgins *et al.*, 1999).

7.4 Detection of EPEC

Both A/E and the localized adherence phenotype of EPEC can be determined by the use of HEp-2 or HeLa cells (Nataro and Kaper, 1998). A fluorescence actin-staining assay has been used to detect A/E. This assay involves staining of actin that accumulates under the attached EPEC, using fluorescein-labeled phalloidin. Genotypic assays based on the use of DNA probes and the PCR have been described to evaluate the three major characteristics of EPEC: A/E (detection of the *eae* gene), presence of the EAF plasmid (EAF or *bfpA* gene probes), and the lack of Shiga toxin genes (use of gene probes or PCR primers targeting *stx*, discussed in Section 8 (Nataro and Kaper, 1998). López-Saucedo *et al.* (2003) described a multiplex PCR method for the detection of diarrheagenic strains of *E. coli*, including EPEC (gene targets for EPEC were *bfpA* and *eaeA*). A single multiplex PCR reaction could distinguish between EPEC, ETEC, EIEC and Shiga toxin-producing *E. coli* based on amplification of specific virulence genes.

8 Enterohemorrhagic *Escherichia coli* (EHEC)

Enterohemorrhagic *E. coli* (EHEC) were first identified as human pathogens in 1982, after the occurrence of outbreaks of hemorrhagic colitis due to consumption of undercooked hamburgers contaminated with *E. coli* O157:H7. The term 'EHEC' refers to *E. coli* serogroups including O26, O111, O103, O104, O118, O145 (with various H antigen types) and others that share the same clinical, pathogenic and epidemiologic features with *E. coli* O157:H7, the EHEC serotype that is responsible for the greatest proportion of disease cases. Since 1982 numerous outbreaks have been documented, and it is estimated that *E. coli* O157:H7 is responsible for greater than 73 000 cases of illness and 61 deaths each year in the United States (Mead *et al.*, 1999).

8.1 Disease characteristics

E. coli O157:H7 can cause an asymptomatic infection; however, it usually leads to a mild non-bloody diarrhea or an acute grossly bloody diarrhea termed hemorrhagic colitis (HC) (Ryan *et al.*, 1986; Griffin, 1995; Su and Brandt, 1995; Mead and Griffin, 1998; Stephan *et al.*, 2000). The incubation period usually ranges from 3 to 8 days, but can be as short as 1–2 days. Illness in patients with non-bloody diarrhea is less severe, and these individuals are less likely to develop systemic sequelae or to die. HC is marked by an acute onset of severe abdominal cramps, followed by a progression of watery to bloody diarrhea that lasts for 4–10 days. The cecum and the ascending colon are the predominantly affected areas. Fever is usually absent or low-grade, stools are usually free of white blood cells, and about half of the patients have vomiting. With hemorrhagic colitis there may be elevation of blood leukocytes, edema with 'thumb printing,' hemorrhage of the lamina propria, superficial ulceration, pseudomembrane formation, and necrosis of the superficial colonic mucosa.

In 2–7 % of patients, predominantly infants and children, *E. coli* O157:H7 infection can lead to hemolytic uremic syndrome (HUS) – a severe post-diarrheal systemic complication and the leading cause of acute renal failure in children in the United States (Mead and Griffin, 1998). The disorder is characterized by microangiopathic hemolytic anemia, thrombocytopenia and renal insufficiency (Su and Brandt, 1995). Central nervous system complications may occur in 30–50 % of patients. The production of Shiga toxins (Stx) by EHEC strains plays a large role in the pathogenesis of hemorrhagic colitis and HUS. Damage and death of endothelial cells through the action of Shiga toxins, which bind to specific receptors on endothelial cells, result in the deposition of platelets and fibrin, leading to abnormal white blood cell adhesion, reduced blood flow in small vessels of the affected organs, increased coagulation, and thrombus formation (O'Loughlin and Robins-Browne, 2001). As red blood cells and platelets pass through the narrowed blood vessels they are mechanically damaged, with resultant hemolytic anemia and thrombocytopenia. Histological changes in the kidney include capillary wall thickening, endothelial cell swelling, and thrombosis of capillaries in the glomeruli, resulting in necrosis of kidney tissue with complete occlusion of renal microvessels. Approximately 50 % of patients with HUS require dialysis; about 3–5 % die; and about 5 % develop chronic renal failure, stroke and other major sequelae (Mead and Griffin, 1998).

8.2 Basis of pathogenicity

Virulence factors of *E. coli* O157:H7 and other typical EHEC strains include the production of one or more types of Shiga toxins, intestinal colonization and the production of A/E lesions as occurs with EPEC and mediated by genes located on the LEE locus, and the presence of a plasmid of approximately 60 MDa (pO157) (LeBlanc,

2003). In the US, EHEC serotype O157:H7 is the most common cause of HC and HUS; however, numerous other serotypes produce Shiga toxins and have also caused HC and HUS. Thus, *E. coli* strains that produce Shiga toxins are referred to as Shiga toxin-producing *E. coli* (STEC). The most important virulence factors in the pathogenesis of EHEC infection are the Shiga toxins (Stx), which have also been referred to as verotoxins or verocytotoxins because of their cytopathogenic effect on Vero cells (African green monkey kidney cells). Cytotoxicity assays consisting of addition of serial dilutions of the samples to Vero cell monolayers, followed by examination of the cells by microscopy for cytotoxic effects, have been used for detection of the Shiga toxins. Shiga toxin is produced by *Shigella dysenteriae* type 1, and the gene is found on the chromosome. Shiga toxin 1 differs by only one amino acid from Shiga toxin, thus Stx1 was previously called Shiga-like toxin I. In *E. coli*, Shiga toxin 1 (Stx1) and Shiga toxin 2 (Stx2) are encoded on the genomes of temperate bacteriophages. Variants of Shiga toxin 1 and Shiga toxin 2 have been identified (Table 6.4; Schmidt *et al.*, 2000; Schmidt, 2001Zhang *et al.*, 2002; ; Bürk *et al.*, 2003; Leung *et al.*, 2003). An STEC strain may produce Stx1, Stx2, or a combination of one or both toxins and one of the variants. Stx1c is associated with *E. coli* found in sheep, and the Stx2d variant was idenfied in an *E. coli* strain of bovine origin (Brett *et al.*, 2003; Bürk *et al.*, 2003). *E. coli* strains that produce Stx2e are responsible for edema disease in swine (Cornick *et al.*, 1999). The gene, stx_{2e}, was believed to be chromosomally encoded; however, it was later found to be encoded on the genome of a Shiga toxin 2e-converting bacteriophage in an ONT(non-typable):H$^-$ *E. coli* strain isolated from a patient with diarrhea (Muniesa *et al.*, 2000).

Shiga toxins are composed of a single A polypeptide and a B-pentamer that binds to the eukaryotic cell receptor, globotriaosylceramide (Gb_3), expressed on epithelial and endothelial cells; the receptor for Stx2e, on the other hand, is globotetraosylceramide (Gb_4). After binding of the B-pentamer to the glycolipid receptors,

Table 6.4 *E. coli* Shiga toxins

Toxin type	Percent identity with relevant Stx	
	A subunit	B subunit
Stx1[a]	–	–
Stx1c	97	97
Stx1d	93	92
Stx2	–	–
Stx2c	100	97
Stx2d	99	97
Stx2e	93	84
Stx2f	63	57
Stx2g (Vt2g)		[b]

[a] The A and B subunits of Stx1 are 55 % and 57 % identical to those of Stx2, respectively; Stx1 is 98 % identical to Stx of *Shigella dysenteriae*, differing by only one amino acid in the A subunit

[b] The nucleic acid sequence of the *vt2g* A subunit showed 63 to 95 % similarity to other *vt* gene A subunit sequences; the nucleic acid sequence of *vt2g* B subunit showed 77 % to 91 % similarity to other *vt* gene B subunit sequences (Leung *et al.*, 2003).

the holotoxin is internalized by endocytosis via clathrin-coated pits. It is transported through the golgi network to the endoplasmic reticulum and the nuclear membrane. In the golgi, the 32-kDa A subunit is cleaved by a calcium-sensitive serine protease to an active 28-kDa peptide (A_1) and a 4-kDa peptide (A_2), but the fragments remain associated by a disulfide bond. In the endoplasmic reticulum, the disulfide bond linking the two peptides is reduced and the A1 fragment is translocated into the cytoplasm. In the cytoplasm the A1 subunit, an N-glycosidase, acts on the 60S ribosomal subunit, removing a single adenine residue from the 28S rRNA of eukaryotic ribosomes. Aminoacyl-tRNA then no longer binds to ribosomes, resulting in an irreversible inhibition of protein synthesis in eukaryotic cells. Recent studies have shown that Stxs can modulate the expression of chemokines and cytokines by human epithelial and endothelial cells, contributing to the inflammatory responses and the pathology of EHEC diseases (Cherla *et al.*, 2003; Matussek *et al.*, 2003) Furthermore, Stxs induce apoptosis (programmed cell death) in some cell types, which may contribute to the development of bloody diarrhea and HUS (Cherla *et al.*, 2003).

Similar to the EPEC, *E. coli* O157:H7 and other EHEC possess the LEE locus and produce A/E lesions in the intestinal mucosa (LeBlanc, 2003). The *E. coli* O157:H7 LEE locus of *E. coli* O157:H7 strain EDL933 consists of 41 genes that are found in the same order and orientation as those in the EPEC O127:H6 strain. The average nucleotide identity between the LEE locus in the two strains is approximately 94 %. The *esc* genes that encode the type-III secretion system are highly conserved; however, there is less similarity between the other genes including *eae*. The *eae* genes of EPEC and *E. coli* O157:H7 (strain 933) share 87 % identity, with high conservation in the N-terminal region but variability in the C-terminal 280 amino acid region of intimin, which is involved in binding to enterocytes and Tir (Frankel *et al.*, 1998). Both the EHEC and EPEC Tir bind intimin; however, the EHEC Tir is not tyrosine-phosphorylated, indicating that tyrosine phosphorylation is not required for A/E lesion formation (DeVinney *et al.*, 1999c). At least 14 variants of intimin, including alpha 1, alpha 2, beta 1, beta 2, gamma 1, *eae*-xi and others have been identified using intimin type-specific PCR assays, and it has been suggested that different intimins in EPEC and EHEC may explain the different host tissue cell tropism (small bowel vs large bowel, respectively) (Frankel *et al.*, 1998; Blanco *et al.*, 2004).

Additional factors contributing to the virulence of EHEC are encoded on a *ca.* 60-MDa virulence plasmid. This plasmid is heterogenous and varies in size among EHEC strains, even within *E. coli* O157:H7 strains (LeBlanc, 2003). The complete DNA sequences of two pO157 virulence plasmids from strains EDL933 and RIMD 0509952 have been published (Burland *et al.*, 1998; Makino *et al.*, 1998). The plasmid from strain EDL 933 was a 92-kb F-like plasmid composed of 100 open reading frames, including the plasmid-encoded genes for the EHEC hemolysin (operon *ehxCABD*), which belong to the RTX family of exoproteins; KatP, a periplasmic catalase-peroxidase that functions to protect the bacterium against oxidative stress; a serine protease (EspP) that may contribute to the cytotoxic activity and tissue destruction by EHEC; and a gene cluster related to the type-II secretion pathway of Gram-negative bacteria (Etp system), composed of a cluster of 13 genes, *etpC* through *etpO*. Another unusually large ORF of 3169 amino acids showed strong sequence similarity within the first 700 amino acids to the N-terminal activity-containing domain of the large clostridial toxin (LCT) gene family in *Clostridium difficile* that includes ToxA and ToxB. Tatsuno *et al.* (2001) reported that *toxB* was required for adherence of *E. coli* O157:H7 to epithelial cells. One study showed that the EHEC enterohemolysin may contribute to the development of HUS through the production of IL-1β from human monocytes, and may mediate translocation of Stxs that stimulate the production of IL-1β (Taneike *et al.*, 2002).

The low infectious dose of *E. coli* O157:H7 corresponds to the organism's ability to tolerate acid environments, and the organism possesses at least three acid resistance systems (Lin *et al.*, 1996; see 'Growth and survival', below). The TAT (twin arginine translocation) system may be another virulence factor of EHEC (Pradel *et al.*, 2003). Deletion of the *tatABC* genes encoding the TAT system of *E. coli* O157:H7, involved in export of proteins across the cytoplasmic membrane, resulted in a decrease in secretion of Stx1 and abolished the synthesis of flagella. A cytolethal-distending toxin (*cdt*) gene cluster was identified in *E. coli* O157:H7 (6 % of isolates examined) and in sorbitol-fermenting *E. coli* O157:H- (87 % of isolates) isolated from patients with diarrhea and HUS (Janka *et al.*, 2003). Their studies suggested that *cdt* may have been acquired by phage transduction.

E. coli O157:H7 releases membrane vesicles into the culture medium, which were shown to contain DNA encoding the *eae*, stx_1, stx_2 *and uidA* genes (Kolling and Matthews, 1999). Furthermore, *E. coli* O157:H7 vesicles facilitated the transfer of virulence and antibiotic resistance genes to other enteric bacteria, and the genes were expressed in the recipient bacteria (Yaron *et al.*, 2000). Thus vesicle formation may be a mechanism for transport and transfer of genetic material and Shiga toxins. There is no evidence that *E. coli* O157:H7 is invasive *in vivo*; however, Matthews *et al.* (1997) showed that the pathogen invaded certain cell lines, including RPMI-4788 (human), MAC-T (bovine mammary secretory) and MDBK (bovine kidney, but not HeLa cells *in vitro*. The organism demonstrated both localized and diffuse adherence to the cells, and microtubules were required for invasion. The authors suggested that the ability to invade bovine mammary cells might be important in asymptomatic carriage of *E. coli* O157:H7 in cattle. Oelschlaeger *et al.* (1994) showed that *E. coli* O157:H7 invaded human ileocaecal (HCT-8) and bladder (T24) cell lines but not INT 407 intestinal cells. These investigators also reported that microfilaments were involved in internalization.

Quorum sensing is a phenomenon through which small signaling molecules, termed autoinducers, provide a means for cell–cell communication in response to cell population density. Quorum sensing may control virulence in *E. coli* O157:H7 by influencing transcription of genes in the LEE operon (Sperandio *et al.*, 1999; Anand and Griffiths, 2003). Sperandio *et al.* (1999) suggested that intestinal colonization of EHEC might be regulated by quorum-sensing signals produced by non-pathogenic *E. coli* present in the intestinal flora. Using *E. coli* K12 DNA arrays, hybridization patterns of cDNA from RNA extracted from *E. coli* O157:H7 and from its isogenic *luxS* (gene involved in synthesis of autoinducer 2, AI-2) mutant showed up-regulation of 235 genes and down-regulation of 169 genes in the wild type strain compared to the mutant (Sperandio *et al.*, 2001). Up-regulated genes included those involved in chemotaxis, the SOS response and the synthesis of flagella and Stx. Thus, quorum sensing in *E. coli* O157:H7 is a global regulatory system. In a subsequent publication, Sperandio *et al.* (2003) reported that a molecule termed AI-3, and not the AI-2 molecule, is the signal involved in quorum sensing in EHEC. It was demonstrataed that AI-3, whose synthesis also depends on LuxS, cross-talks with the mammalian hormone epinephrine, indicating that quorum sensing in *E. coli* O157:H7 may also involve bacterium–host communication.

8.3 Treatment and vaccines

Although there is no established therapy for *E. coli* O157:H7 infection, several vaccines are being developed and other promising regimens evaluated. The use of antibiotics in the treatment of infection is controversial, since antimicrobial therapy may increase the risk of development of HUS (Mølbak *et al.*, 2002). Antibiotics may induce the expression of the Shiga toxins, and/or bacterial injury caused by the antibiotic may result in increased release of preformed toxins. Mulvey *et al.* (2002) suggested that administration of an Stx-binding agent, Synsorb-Pk, given in combination with antibiotics may potentially absorb sufficient amounts of toxin to prevent uptake into the circulatory system. This assumption requires testing in humans or in an appropriate animal model.

A number of vaccine protocols for use in cattle and humans are being investigated (Horne *et al.*, 2002). A plant cell-based intimin vaccine tested in mice showed the development of an intimin-specific mucosal immune response and a reduced duration of shedding of *E. coli* O157:H7 (Judge *et al.*, 2004). This plant-based vaccine system is being explored for oral administration to cattle to decrease shedding of the pathogen. Vaccination of cattle with type-III secreted proteins reduced the duration of shedding and numbers of *E. coli* O157:H7 in feces (Potter *et al.*, 2004). Additionally, the prevalence of the organism in cattle was reduced in a clinical trial conducted under conditions of natural exposure in a feedlot setting. A vaccine consisting of liposomes incorporating monophosphoryl lipid A and antigens from an *E. coli* O157:H7 lysate induced IgG and IgA serum-antibody and mucosal-antibody responses in immunized mice (Tana *et al.*, 2003). A number of other vaccine strategies, including toxoid and O-specific polysaccharide-protein conjugate vaccines, for prevention of EHEC disease are under investigation (Keusch *et al.*, 1998; Konadu *et al.*, 1998). Treatments based on use of Synsorb Pk, a synthetic analog of Shiga toxin receptor Gb_3, bound to a calcinated diatomaceous material called Chromosorb are undergoing clinical trials (Takeda *et al.*, 1999; Trachtman and Christen, 1999). Other proposed treatments include administration of recombinant bacteria expressing a Shiga toxin receptor mimic; humanized Shiga toxin-neutralizing monoclonal antibodies; pooled bovine colostrum containing antibodies to Shiga toxins, intimin, and the EHEC hemolysin; and bovine lactoferrin and its peptides; however, further studies are needed to elucidate the effects of these therapies *in vivo* (Shin *et al.*, 1998; Huppertz *et al.*, 1999; Paton *et al.*, 2001; Yamagami *et al.*, 2001).

8.4 Infectious dose

Analyses of foods implicated in disease outbreaks have revealed that the infectious dose for EHEC is less than 50 organisms (Tilden *et al.*, 1996; Tuttle *et al.*, 1999). The calculated number of *E. coli* O157:H7 found in raw ground beef patties implicated in an outbreak that occurred in the western US in November 1992 to February 1993 was 1.5 organisms per gram, or 67.5 per patty (Tuttle *et al.*, 1999). In an outbreak that occurred in Australia associated with EHEC O111:H^-, the contaminated, fermented sausages contained fewer than one *E. coli* O111:H^- per 10 g (Paton *et al.*, 1996). The occurrence of waterborne outbreaks, and outbreaks associated with visiting farms and petting zoos, in addition to person-to-person transmission of EHEC infection, provides further evidence of a low infective dose (Crump *et al.*, 2002; O'Donnell *et al.*, 2002; Olsen *et al.*, 2002).

8.5 Antibiotic resistance

Over the past 20 years there has been an increase in antibiotic resistance observed in *E. coli* O157:H7 isolates (Schroeder *et al.*, 2002a; Wilkerson and van Kirk, 2004). Resistance to tetracycline was the most common resistance found in bovine and human *E. coli* O157:H7 isolates, followed by resistance to streptomycin and ampicillin (Wilkerson and van Kirk, 2004). Mizan *et al.* (2002) found that antibiotic resistance plasmids could readily be transferred from a commensal *E. coli* strain to *E. coli* O157:H7 in bovine rumen fluid, at a frequency exceeding that observed for mating in LB broth. Antimicrobial resistance also appears to be widespread in non-O157 STEC (DeCastro *et al.*, 2003; Schroeder *et al.*, 2002b). Schroeder *et al.* (2002a) suggested that selection pressure imposed by the use of antimicrobials, including tetracycline derivatives, sulfa drugs, and penicillins in human and veterinary medicine, is resulting in the selection of antimicrobial resistant strains of STEC.

8.6 Animal models

A number of animal species have been evaluated as models of EHEC infection; however, in no animal system can the entire spectrum of the disease processes observed in humans be replicated. EHEC do not normally cause disease in cattle, but colostrum-deprived neonatal calves develop diarrhea 18 hours following inoculation with 10^{10} CFU of bacteria. The calves can become colonized with EHEC at levels greater than or equal to 10^6 CFU/g of intestinal tissue or feces, and may develop A/E lesions in the small or large intestine (Dean-Nystrom, 2003). Intimin was required for colonization of EHEC O157:H7 in newborn calves (Dean-Nystrom *et al.*, 1998). Ritchie *et al.* (2003) used an infant rabbit model to study the role of stx_2, *eae* and *tir* in EHEC pathogenesis. EHEC derivatives with deletions in the *tir* and *eae* genes did not colonize or form A/E lesions, or cause inflammation and diarrhea. The stx_2 gene increased the severity and duration of diarrhea, but was not involved in attachment. Intragastric inoculation of Stx2 induced diarrhea and inflammation. Gnotobiotic piglets infected orally with Stx2-producing *E. coli* O157:H7 and O26:H11 developed gastrointestinal illness and thrombotic microangiopathy in the kidneys, which is typically seen in humans with HUS (Gunzer *et al.*, 2002). Comparing an *E. coli* O157:H7 strain with a mutation in the *eaeA* gene with the wild-type strain, Tzipori *et al.* (1995) showed that intimin (the product of *eaeA*) facilitated attachment to cells and affected the site of intestinal colonization. Pigs injected intramuscularly with Stx1 developed vascular damage and necrosis in the intestines and brain, similar to that which develops in humans with EHEC disease (Dykstra *et al.*, 1993). The greyhound dog model is being investigated, due to similarities noted between EHEC disease and a condition in greyhounds called idiopathic cutaneous and renal glomerular vasculopathy (Fenwick and Cowan, 1998). Dogs with this illness exhibit renal changes similar to those seen in humans with HUS, and it is suspected that the disease in dogs is caused by Shiga toxin-producing *E. coli*. The disease was observed in dogs from which *E. coli* O157:H7 was isolated and in dogs administered Stx1 or Stx2 by intravascular inoculation. Baboons administered Stx1 by intravenous infusion develop renal failure and damage to the gastrointestinal mucosa (Melton-Celsa and O'Brien, 2003). The kidney lesions that develop are similar to those seen in kidneys of patients with HUS. A macaque monkey model is being developed in which the animals infected with *E. coli* O157:H7 acquire diarrhea and A/E lesions. Several mouse models have been developed to study EHEC pathogenesis (Melton-Celsa and O'Brien, 2003), and ferrets are being investigated as a model system to study EHEC-mediated HUS (Woods *et al.*, 2002). Oral inoculation of chicks with *E. coli* O157:H7 showed that colonization occurred in the cecum and colon, and A/E lesions were detected in two out of seven chicks (Beery *et al.*, 1985; Sueyoshi and Nakazawa, 1994). Use of appropriate animal models will advance our understanding of the pathophysiology of EHEC-induced HC and HUS, and provide information that will help to prevent, control and treat EHEC infection.

8.7 Growth and survival

Although there was some variation among strains, the minimum and maximum growth temperatures for *E. coli* O157:H7 studied in brain–heart infusion broth were 10° and 45 °C, respectively, and several strains grew slowly at 8 °C (Palumbo *et al.*, 1995). Numerous studies have been conducted addressing the effects of environmental stresses and food production processes on the growth, survival and inactivation of *E. coli* O157:H7. *E. coli* O157:H7 has no unusual heat resistance; however, the thermal resistance can be influenced by a number of factors, including pH, growth conditions and growth phase of the cells, and the method of heating. Juneja *et al.* (1999) found that increasing the concentration of sodium pyrophosphate decreased the heat resistance; whereas increasing the concentration of NaCl increased resistance. The amount of fat in ground beef influences thermal tolerance of *E. coli* O157:H7. The *D* values for beef containing 2 % fat and 3 % fat were 4.1 and 5.3 minutes respectively at 57.2°C, and 0.3 and 0.5 minutes respectively at 62.8°C (Line *et al.*, 1991).

López-González *et al.* (1999) showed that irradiation doses approved for red meats were sufficient to reduce the level of *E. coli* by several log values; however, the radiation temperature, oxygen permeability of the packaging material, and the medium in which the organism was irradiated influenced the D_{10} values. Treatment consisting of dry heat in combination with an irradiation dose of 2.0 kGy was effective in eliminating *E. coli* O157:H7 from inoculated alfalfa and mung bean seeds, whereas a dose of 2.5 kGy was needed to eliminate the pathogen from radish seeds (Bari *et al.*, 2003). Irradiation did not affect the germination percentage for alfalfa seeds or the length of the sprouts, but it did decrease the lengths of the radish and mung bean sprouts.

Although there is some variability among strains, *E. coli*, O157:H7 is relatively acid-tolerant compared to other foodborne pathogens. The pathogen can grow at pH levels ranging from 4.4 to 9.0, and can survive for extended periods in foods at pH levels of 3.5–5.5. *E. coli* O157:H7 survived for up to 2 months, with only a 100-fold reduction in cell numbers during fermentation, drying and storage of fermented sausage; for 5–7 weeks in mayonnaise at 5°C; and for 10–31 days in apple cider at 8°C (Glass *et al.*, 1992; Zhao *et al.*, 1993, 1994). Application of warm (20°C) and hot (55°C) acetic, citric, and lactic acid sprays did not appreciably reduce the levels of *E. coli* O157:H7 on raw beef (Brackett *et al.*, 1994). Three systems of acid tolerance have been characterized in *E. coli* O157:H7: an acid-induced, oxidative system requiring the Rpos alternate-sigma factor; an acid-induced, arginine-dependent system; and a glutamate-dependent system (Lin *et al.*, 1996). Thus several acid resistance systems function in *E. coli* O157:H7, which permit survival under different acid-stress conditions, and once induced remain active during prolonged periods of cold storage.

E. coli O157:H7 survived in inoculated tap and bottled spring and mineral water for up to 300 days or more (Warburton *et al.*, 1998), and for 14 days at <15°C in farm water stored outdoors, demonstrating the potential that farm water might serve as a vehicle for transfer of the organism in a herd (McGee *et al.*, 2002). The organism survived for 77, >226 and 231 days in manure-amended autoclaved soil stored at 5°C, 15°C and 21°C, respectively (Jiang *et al.*, 2002). It persisted for 25–41 days in fallow soils, 47–96 days on rye roots, and 92 days on alfalfa roots (Gagliardi and Karns, 2002). Persistence of the organism was not affected by the presence of manure, whereas the presence of clay increased persistence.

8.8 Sources of foodborne cases and outbreaks

Cattle are the major reservoir for *E. coli* O157:H7, and transmission of the pathogen from cattle to humans occurs via contaminated food or water. A survey conducted to determine the distribution and prevalence of *E. coli* O157:H7 in cattle in four major feeder-cattle states in the US showed that 10.2 % of fecal samples (out of 10 662 samples tested) and 13.1 % of water or water-tank sediment samples were positive, with over 60 % of feedlots having at least one positive water or water sediment sample (Sargeant *et al.*, 2003). Elder *et al.* (2000) found an overall prevalence of *E. coli* O157:H7 or O157:NM, in the feces and hides of fed cattle presented for slaughter at meat processing plants in the midwestern US, of 28 % (91/327) and 11 % (38/355) in feces and on hides, respectively – a prevalence much higher than previous studies had reported. The authors suggested that the isolation methods employed and time of year that the samples were collected were the likely reasons for the differences in results. Cattle harboring *E. coli* O157:H7 are generally disease-free; however, the organism causes fatal ileocolitis in newborn calves. Tolerance to infection by *E. coli* O157:H7 is likely due to lack of Gb_3, the Shiga toxin receptor, in the bovine gastrointestinal tract (Pruimboom-Brees *et al.*, 2000). The duration of shedding of *E. coli* O157:H7 in cattle is about 30 days (Sanderson *et al.*, 1999), and the principal site of colonization is the terminal rectum (Grauke *et al.*, 2002; Naylor *et al.*, 2003). *E. coli* O157:H7 has also been isolated from deer, pigs, horses, goats, sheep, cats, dogs, rabbits, poultry, and rats, and from birds such as ravens, doves, and seagulls (Meng *et al.*, 2001; Feder *et al.*, 2003). *E. coli* O157:H7 strains administered to young adult sheep persisted longer in the gastrointestinal tract of the animals than did EPEC or ETEC strains that were included in the same inoculum (Cornick *et al.*, 2000). The ability of EHEC to persist in ruminants may explain how this reservoir is maintained. The pathogen was isolated from houseflies collected from a school in Japan at which a disease outbreak occurred (Kobayashi *et al.*, 1999). Feeding experiments showed that the bacteria were harbored in the fly's intestine and were shed for at least 3 days after feeding, indicating proliferation of the organism in the fly.

Foods of bovine origin, including ground beef, raw milk and roast beef, have been associated with *E. coli* O157:H7 infection; however, goat cheese, venison jerky, and environmental contamination with sheep feces have also been linked with outbreaks (Meng *et al.*, 2001; Ogden *et al.*, 2002). An outbreak affecting 732 individuals and with 55 cases of HUS and 4 deaths occurred in late 1992 and early 1993 in the western US and Canada (Bell *et al.*, 1994). Inadequately cooked ground beef served at multiple outlets of the same fast-food restaurant chain was implicated as the cause of infection. Numerous other food vehicles, such as apple cider, mayonnaise, pea salad, cantaloupe, lettuce, hard salami, and alfalfa and radish sprouts, have also been linked to outbreaks (Meng *et al.*, 2001). Fruits and vegetables are likely contaminated with cattle manure during harvesting and processing. Due to the ability of *E. coli* O157:H7 to tolerate acidic environments, the organism can survive in foods of low pH such as apple cider or fermented products. White radish sprouts were implicated as the vehicle of infection in a large outbreak that occurred in Japan in 1996 that involved 9578 individuals, many of whom were schoolchildren, with 90 cases of HUS and 11 deaths (Bettelheim, 1997; Michino *et al.*, 1999). Contaminated recreational water, well water, groundwater and municipal water systems have also been linked to outbreaks. In Walkerton, Ontario, Canada, in May 2000, an estimated 2300 individuals became seriously ill and 7 died due to exposure to drinking water contaminated with *E. coli* O157:H7 from cattle excrement from a nearby farm that had washed into the town's wells during a flood weeks earlier. Outbreaks resulting from visits to agricultural fairs and petting zoos have also occurred, likely due to exposure to animals – in particular cattle – and the farm environment (CDC, 2001; Crump *et al.*, 2003). Additonally, person-to-person transmission of HC or HUS due to *E. coli* O157:H7 infection occurs in nursing homes, day-care centers, and between family members (Al-Jader *et al.*, 1999; Carter *et al.*, 1987).

8.9 Diagnosis and methods for detection, isolation and identification of EHEC

Routine diagnosis of *E. coli* O157:H7 infection involves isolation of the pathogen, from stools of patients presenting with bloody diarrhea or HUS, on sorbitol MacConkey agar (SMAC). However, since *E. coli* O157:H7 also causes non-bloody diarrhea, it has been recommended that non-bloody stools from patients with diarrhea should also be cultured. Demonstration of the Shiga toxins in fecal filtrates is also useful for diagnosis; however, non-O157 STEC may also cause HC or HUS. Therefore, isolation and identification of the causative organism is necessary for epidemiological purposes. Different strategies for detection and isolation of *E. coli* O157:H7 in foods have been described (Fratamico *et al.*, 2002; Deisingh and Thompson, 2004). *E. coli* O157:H7 does not ferment sorbitol within 24 hours, thus colonies are colorless on SMAC. Sorbitol-negative colonies can be picked and characterized for responses to other biochemical parameters, for the presence of the O157 and H7 antigens, and for the presence of Shiga toxin genes or for the production of the Shiga toxins (Su and Brandt, 1995). Commercially available selective and differential agar media for isolation of *E. coli* O157:H7 include Rainbow® agar O157 (Biolog), CHROMagar® O157 (Hardy Diagnostics), BCM® O157:H7 Agar (Biosynth) and Fluorocult® *E. coli* O157:H7 Agar (Merck). Modifications of SMAC medium have resulted in agars with increased selectivity for *E. coli* O157:H7 or the ability to differentiate it from colonies of other organisms. These include CT-SMAC (which contains potassium tellurite and cefixime) or CR-SMAC (in which cefixime and rhamnose are added to SMAC), and SMA-BCIG containing the substrate for β-glucuronidase, 5-bromo-4-chloro-3-indoxyl-β-D-glucuronic acid cyclohexylammonium salt (BCIG). Presumptive positive colonies can be tested for the presence of the O157 and H7 antigens using commercially available latex agglutination kits or antisera, including the RIM® *E. coli* O157:H7 Latex Test (Remel) and the ImmunoCard STAT! *E. coli* O157 (Meridian Diagnostics) test.

Except for the production of Shiga toxins, there are generally no phenotypic markers (such as the inability to ferment sorbitol) that are shared by all non-O157 EHEC and that can be utilized as a strategy in the development of differential media to distinguish them from non-pathogenic *E. coli*. Thus, the incidence of disease caused by non-O157 EHEC may be underestimated due to the lack of adequate methods for detection of these pathogens. Methods for detection of non-O157 EHEC are described below.

Conventional methods rely on culture and agar media to grow, isolate and enumerate viable *E. coli* O157:H7. In recent years, however, numerous companies have developed methods for detection of *E. coli* O157:H7 and other pathogens that are specific, faster and often more sensitive than conventional methods. Immunoassays and genetic-based assays such as the polymerase chain reaction (PCR) are examples of rapid methods. Immunoassays rely on binding of an antibody to a specific bacterial antigen. A number of immunoassay formats, including enzyme-linked immunosorbent or fluorescent assays, lateral flow immunoassays and latex agglutination assays, for detection of *E. coli* O157:H7 are commercially available (Feng, 2001; Fratamico *et al.*, 2002).

Immunomagnetic separation (IMS) with magnetic particles coated with antibodies specific for *E. coli* O157 is available from Dynal, Inc., or the particles can be purchased from a number of vendors and the beads then coated with the desired antibody. Use of IMS on food enrichments and other complex matrices results in concentration of the target bacteria and sequestering from non-target organisms and other matrix components that interfere with subsequent detection systems (Fratamico and Crawford, 1999). IMS has been used in conjunction with plating onto selective agars and the PCR, and has been incorporated in commercially available kits including the PATH *IGEN™ E. coli* O157 test (Igen International) and the EHEC-Tek™ for *E. coli* O157:H7 (Organon Teknika) for detection of the pathogen in foods (Fratamico and Crawford, 1999). Shiga toxins can be detected in bacterial culture supernatants, in food enrichments and in stool samples, using immunoassays – including colony immunoblot assays, latex agglutination, and antibody capture or toxin receptor-mediated enzyme linked immunosorbent assays. Immunoassay-based kits for the detection of Stx1 and Stx2 include the VTEC RPLA toxin detection kit (Oxoid), the Premier EHEC® (Meridian Diagnostics), and the RIDASCREEN® Verotoxin kit (r-biopharm). Nucleic acid-based detection systems, including PCR and DNA hybridization, rely on discrimination of *E. coli* O157:H7 from closely related organisms based on unique DNA or RNA sequences. Numerous PCR-based methods targeting genes including stx_1, stx_2, *eae*, *uidA*, *hly*, *fliC*, *rfbE* and others have been described (Deisingh and Thompson, 2004). PCR assays have also been performed in a multiplex format in which more than one sequence is amplified simultaneously in a single reaction (Fratamico *et al.*, 2000; Campbell *et al.*, 2001). Real-time PCR assays employing fluorogenic probes to visualize amplification of target sequences during the reaction have been applied for detection of *E. coli* O157:H7 (Ibekwe and Grieve, 2003; Sharma and Dean-Nystrom, 2003). Commercially available PCR-based assays include the BAX® *E. coli* O157:H7 (Dupont Qualicon), the Probelia PCR for *E. coli* O157:H7 (BioControl Systems), and the TaqMan® *E. coli* O157:H7 detection kit (Applied Biosystems). More recently, seven specific genes of *E. coli* O157:H7 were detected using oligonucleotide arrays (Liu *et al.*, 2003). Biotin-labeled target DNA was obtained by incorporation of biotin-16-dUTP during multiplex PCR, and this was followed by hybridization to probes that were spotted onto glass slides and staining with streptavidin-Cy3. Call *et al.* (2001) used a combination of immunomagnetic capture and multiplex PCR, followed by detection of the PCR products using a microarray, to detect *E. coli* O157:H7 in rinse fluid from chickens at a level of 55 CFU/ml. Several other types of methods have also been described (Deisingh and Thompson, 2004).

8.10 Evolution of *Escherichia coli* O157:H7

The H7 flagellar gene (*fliC*) and *eae* gene sequences of *E. coli* O157:H7 and O55:H7 are nearly identical (Reid *et al.*, 1999). Additionally, multilocus enzyme electrophoresis analyses have indicated that *E. coli* O157:H7 evolved from a progenitor strain with serotype O55:H7 (Feng *et al.*, 1998). Through phylogenetic analyses based on enzyme allele profiles, a model for the stepwise emergence of *E. coli* O157:H7 was formulated. The immediate *E. coli* O55:H7 ancestor that ferments sorbitol and expresses β-glucuronidase, evolved from an EPEC-like ancestor. The O55:H7 'clone' acquired the LEE pathogenicity island and a mutation at −10 in the β-glucuronidase gene *uidA*. Acquisition of the stx_2 gene by transduction by a toxin-converting bacteriophage resulted in the Stx2-producing *E. coli* O55:H7 strain. A second mutation then occurred in the *uidA* gene at +92, and the O antigen changed from O55 to O157, possibly by lateral transfer and recombination of a region of the *rfb* locus containing the *rfbE* gene that was homologous to the perosamine-synthetase gene of *Vibrio cholerae*. The EHEC virulence plasmid was acquired at this stage. Two distinct lines then evolved from this progenitor: a non-motile sorbitol$^+$ and β-glucuronidase$^+$ Stx2-producing strain (O157:H$^-$), and a sorbitol-negative β-glucuronidase$^+$ lineage that acquired the stx_1 gene (O157:H7, stx_1^+, stx_2^+). The latter lineage lost β-glucuronidase activity, producing the immediate ancestor of the O157:H7 'clone' that has spread worldwide. Comparison of *gnd* gene sequences (located adjacent to the *E. coli* O antigen gene cluster, also known as the *rfb* locus) showed that *gnd* co-transferred with the adjacent *rfb* locus into *E. coli* O157 and O55 in distantly separated lineages, and also that intragenic recombination may have contributed to allelic variation in this region of the O157 chromosome (Tarr *et al.*, 2000). L. Wang *et al.* (2002) sequenced the *E. coli* O55 O-antigen genes and flanking sequences to understand how the shift from O55 to O157 occurred. They identified two recombination sites, one within the *galF* gene and the other between the *hisG* and *amn* genes, providing evidence for the recombination event proposed for the evolution of the *E. coli* O157:H7 clone.

8.11 Genomic analysis of *Escherichia coli* O157:H7

The genome of *E. coli* O157:H7 EDL933 has been sequenced, providing information on the evolution of this organism (Perna *et al.*, 2001). In addition, the sequence data help in the identification of genes associated with virulence and in the development of methods for detection of the organism. Comparison of the O157:H7 genome to that of *E. coli* K12 revealed that both share a similar backbone sequence of *ca.* 4.1 Mb; however, 1.34 Mb of DNA in *E. coli* O157:H7 is missing in K12, and 0.53 Mb of DNA in K12 is missing in O157:H7 (Perna *et al.*, 2001; Sperandio, 2001). The *E. coli* O157:H7 and K12 genomes consist of 5416 and 4405 genes respectively. Both O157:H7 and K12 are punctuated by hundreds of islands or DNA segments of up to 88 kb in length, designated K-islands in K12 and O-islands in O157:H7. Approximately 26 % of the *E. coli* O157:H7 genes are encoded within O-islands of diverse sizes, which are not found in K12. Only 40 % of the O-island genes can be assigned a function. Putative virulence genes are encoded on nine large islands, while smaller islands encode genes involved in synthesis of fimbriae, in iron uptake and utilization, and in survival in different environments. Sequences related to known bacteriophages were identified in 18 multigenic regions. Together the sequence data indicate that there is a high level of diversity between the O157:H7 and K12 genomes, and that the islands were probably acquired through horizontal gene transfer from other organisms. Taylor *et al.* (2002) showed that there was considerable variability in the presence, number and location of O-islands encoding tellurite resistance within *E. coli* O157:H7 strains. Furthermore,

Shaikh and Tarr (2003) demonstrated that *E. coli* O157:H7 genomes possessed novel truncated bacteriophages and multiple stx_2 bacteriophage-insertion sites. Several antibiotics promoted excision of bacteriophages, and bile salts attenuated excision.

8.12 Genetic fingerprinting and outbreak investigation

Molecular typing methods are used to determine the genetic relatedness of bacterial isolates, to aid in epidemiologic investigations of foodborne disease outbreaks. Phenotypic methods such as serotyping, phage typing or multilocus enzyme electrophoresis are gradually being replaced by genetic fingerprinting techniques such as ribotyping, random amplified polymorphic DNA, and pulsed field gel electrophoresis (PFGE). The PFGE technique involves isolation of intact DNA, followed by digestion with restriction enzymes and analysis of the digestion products (typically 10–20 products ranging in size from 10 to 800 kb) that are separated by agarose gel electrophoresis with programmed variations in the direction and duration of the electric field. During an outbreak in 1993 caused by hamburgers contaminated with *E. coli* O157:H7 served at a fast-food restaurant chain, PFGE was used to determine the genetic relatedness of clinical and food isolates. All of the isolates associated with the multistate outbreak had identical phage type and PFGE patterns (Barrett *et al.*, 1994). It was determined that use of a standardized subtyping method would allow rapid comparison of isolates from different parts of the country and determination of a common source of infection, to prevent further spread of infection. Thus, a national molecular subtyping network for foodborne disease surveillance, known as PulseNet, was established by the Centers for Disease Control and Prevention in collaboration with the Association of Public Health Laboratories in 1996 (Swaminathan *et al.*, 2001). Laboratories participating in PulseNet perform PFGE on outbreak isolates from humans and/or the suspected food and enter the PFGE patterns into an electronic database to allow rapid comparison of the fingerprint patterns by the CDC. All 50 state public health laboratories, local public health laboratories, Food and Drug Administration and USDA Food Safety and Inspection Service laboratories participate in PulseNet, which is playing an integral role in the surveillance and investigation of outbreaks of foodborne illness caused by *E. coli* O157:H7 (CDC, 2002). PulseNet has been expanded and currently tracks non-typhoidal *Salmonella*, *Shigella*, *Listeria monocytogenes* and *Campylobacter*. Noller *et al.* (2003) reported that a subtyping technique known as multilocus variable-number tandem repeat analysis (MLVA), which targets short tandem repeats in the DNA at multiple loci, had a sensitivity equal to that of PFGE; however, specificity was superior to that of PFGE, since MLVA differentiated strains with identical PFGE types.

8.13 Importance of non-O157 STEC/EHEC

Over 200 STEC (also referred to as verocytotoxin-producing *E. coli* or VTEC) serotypes have been identified; 100 or more non-O157 O:H serotypes of STEC have been responsible for cases and outbreaks of HC and HUS (Nataro and Kaper, 1998; World Health Organization, 1998; Karmali, 2003; Table 6.5), and strains in this

Table 6.5 Outbreaks caused by non-O157 EHEC

Country	Serotype	Number affected	Vehicle of transmission	Reference
France	O103:H2	6	Not known	Mariani-Kurkdjian *et al.*, 1993
Italy	O111:NM	9	Not known	Caprioli *et al.*, 1994 7 (suspected)
United States	O104:H21	11 (confirmed)	Milk	CDC, 1995
Australia	O111:H$^-$	22	Uncooked, semi-dry fermented sausage	Paton *et al.*, 1996
Japan	O118:H2	126	Salad	Hashimoto *et al.*, 1999
Australia	O113:H21	3	Not known	Paton *et al.*, 1999
United States	O111:H8	58	Ice, salad bar	CDC, 2000
United States	O121:H19	11	Swimming water	McCarthy *et al.*, 2001
Ireland	O26:H11	4	Not known	McMaster *et al.*, 2001
Germany	O26:H11	6 (non-bloody diarrhea)	'Seemerrolle' beef	Werber *et al.*, 2002 5 (asymptomatic)
Germany	O26:H11	3	Not known	Misselwitz *et al.*, 2003

subset of STEC are referred to as EHEC. The term 'EHEC' refers to STEC serotypes that share the same clinical, pathogenic and epidemiologic features with *E. coli* O157:H7. Several important non-O157:H7 EHEC belong to *E. coli* serogroups O26, O103, O111, O113, O121 and O145 (Nataro and Kaper, 1998).

In Australia, Latin America and many European countries, non-O157 STEC serotypes are more prevalent than *E. coli* O157:H7 (Nataro and Kaper, 1998; Elliott *et al.*, 2001; Werber *et al.*, 2002; Karmali, 2003). Non-O157 STEC strains may account for 20–70 % of STEC infections throughout the world (World Health Organization, 1998). For example, *E. coli* O121:H19 was linked to an outbreak of HUS at a lake in Connecticut (McCarthy *et al.*, 2001); *E. coli* O111:H$^-$ was responsible for an outbreak in Australia involving 21 cases with 1 fatality, linked to consumption of a locally-produced semi-dry fermented sausage (Paton *et al.*, 1996); and *E. coli* O111:H8 caused a disease outbreak indistinguishable from that caused *E. coli* O157:H7 in attendees at a youth camp (Brooks *et al.*, 2004). *E. coli* O26:H11 caused a multistate outbreak in Germany involving 11 case subjects and was linked to a certain type of beef referred to as *Seemerrolle* (Werber *et al.*, 2002).

Although non-O157 STEC have caused HC and HUS, infections with some STEC strains may result in asymptomatic or mild diarrhea. It is likely that these strains do not possess all of the virulence factors of *E. coli* O157:H7, and further studies are needed to assess this possibility. In an outbreak linked to STEC O26:H11 in Germany, diarrhea was non-bloody and five persons remained asymptomatic (Werber *et al.*, 2002). Friedrich *et al.* (2002) found that STEC possessing different stx_2 variants differed in their capacity to produce HUS. Strains harboring stx_{2c} caused HUS, whereas strains harboring stx_{2d} or stx_{2e} did not, although all strains caused diarrhea. Pradel *et al.* (2002) used a subtractive genomic hybridization technique to identify virulence genes specific to an *E. coli* O91:H21 strain that caused HUS. A comparison to the DNA of strains that were not associated with human illness showed that the O91:H21 strain possessed fragments corresponding to previously identified unique sequences in *E. coli* O157:H7, and sequences from *Shigella flexneri* and enteropathogenic and STEC plasmids; in addition, the strain possessed three copies of the stx_2 gene. The authors suggested that highly pathogenic STEC strains acquire virulence genes by lateral gene transfer to a larger degree than do strains with lesser virulence.

An estimation of the true incidence of disease caused by the non-O157 STEC is complicated by the need to detect the presence of the Shiga toxins or of the *stx* genes. Unlike *E. coli* O157:H7, most non-O157 STEC cannot be detected based on selective and differential media that will detect the lack of the ability to ferment sorbitol or lack of β-glucuronidase activity. In addition to plating stool specimens from patients seen at the Inova Fairfax Hospital in Fairfax, Virginia, onto SMAC, Park and coworkers (2002) used a commercially available assay to detect the presence of Shiga toxins in the stool. *E. coli* O157:H7 was found in 45 out of 65 patient stool specimens, and non-O157 STEC strains were found in 20 specimens. The serotypes of the strains were O26:H12, O45:H2, O103:H2, O111:NM, O153:H2, O88:H25, O145:NM and O96:H9. Thus, they recommended that assays for the Shiga toxins be incorporated into clinical protocols for testing bloody stool specimens. This recommendation is also underscored by The Centers for Disease Control and Prevention.

A number of the O antigen gene clusters that contain genes involved in the synthesis of the O antigens of the different *E. coli* serogroups have recently been sequenced, and PCR assays targeting unique sequences within these regions have been used to detect specific *E. coli* serogroups (Fratamico *et al.*, 2003; Wang *et al.*, 2001b). For example, the *E. coli wzx* (O antigen flippase) and *wzy* (O antigen polymerase) genes are suitable targets for serogroup-specific PCR assays. Multiplex PCR assays targeting a serogroup specific region within the O antigen gene cluster, in addition to the *stx* or other virulence gene have also been reported (Wang, G. *et al.*, 2002). A plating medium consisting of washed sheep's blood agar and containing mitomycin C enhanced the ability to detect enterohemolysin-producing O157:H7 and non-O157 STEC strains (Sugiyama *et al.*, 2001). Magnetic particles linked with antibodies specific for *E. coli* O26, O103, O111 and O145 are commercially available (Dynal). Other methods for detection of STEC, including immunological and DNA-based methods, have been described (Karch *et al.*, 1999; Bettelheim, 2003).

Bibliography

Adachi, J. A., Z.-D. Jiang, J. J. Mathewson *et al.* (2001). Enteroaggregative *Escherichia coli* as a major etiologic agent in Traveler's diarrhea in 3 regions of the world. *Clin. Infect. Dis.* **32**, 1706–1709.

Adachi, J. A., C. D. Ericsson, Z.-D. Kiang *et al.* (2002a). Natural history of enteroaggregative and enterotoxigenic *Escherichia coli* infection among US travelers to Gluadalajara, Mexico. *J. Infect. Dis.* **185**, 1681–1683.

Adachi, J. A., J. J. Mathewson, Z.-D. Jiang *et al.* (2002b). Enteric pathogens in Mexican sauces of popular restaurants in Guadalajara, Mexico and Houston, Texas. *Ann. Intern. Med.* **136**, 884–887.

Al-Jader, L., R. L. Salmon, A. M. Walker *et al.* (1999). Outbreak of *Escherichia coli* O157 in a nursery, lessons for prevention. *Arch. Dis. Child.* **81**, 60–63.

Anand, S. K. and M. W. Griffiths (2003). Quorum sensing and expression of virulence in *Escherichia coli* O157, H7. *Intl J. Food Microbiol.* **85**, 1–9.

Baqar, S., A. L. Bourgeois, P. J. Schultheiss *et al.* (1995). Safety and immunogenicity of a prototype oral whole-cell killed *Campylobacter* vaccine administered with a mucosal adjuvant in non-human primates. *Vaccine* **13**, 22–28.

Bari, M. L., E. Nazuka, Y. Sabina *et al.* (2003). Chemical and irradiation treatments for killing *Escherichia coli* O157:H7 on alfalfa, radish, and mung bean seeds. *J. Food Prot.* **66**, 767–774.

Barlow, R.S., R. G. Hirst, R. E. Norton *et al.* (1999a). Novel serotype of enteropathogenic *Escherichia coli* (EPEC) as a major pathogen in an outbreak of infantile diarrhoea. *J. Med. Microbiol.* **48**, 1123–1125.

Barrett, T. J., H. Lior, J. H. Green *et al.* (1994). Laboratory investigation of a multistate food-borne outbreak of *Escherichia coli* O157:H7 by using pulsed-field gel electrophoresis and phage typing. *J. Clin. Microbiol.* **32**, 3013–3017.

Baudry, B., S. J. Savarino, P. Vial *et al.* (1990). A sensitive and specific DNA probe to identify enteroaggregative *Escherichia coli*, a recently discovered diarrheal pathogen. *J. Infect. Dis.* **161**, 1249–1251.

Beatty, M. E., C. A. Bopp, J. G. Wells *et al.* (2004). Enterotoxin-producing *Escherichia coli* O169:H41, United States. *Emerg. Infect. Dis.* **10**, 518–521.

Beery, J. T., M. P. Doyle and J. L. Schoeni (1985). Colonization of chicken cecae by *Escherichia coli* associated with hemorrhagic colitis. *Appl. Environ. Microbiol.* **49**, 310–315.

Bell, B. P., M. Goldoft, P. M. Griffin *et al.* (1994). A multistate outbreak of *Escherichia coli* O157:H7-associated bloody diarrhea and hemolytic uremic syndrome from hamburgers – the Washington experience. *J. Am. Med. Assoc.* **272**, 1349–1353.

Bell, C. and A. Kyriakides (1998). *E. coli – Practical Approach to the Organism and its Control in Foods*. Blackie Academic and Professional, New York, NY.

Benz, I. and M. A. Schmidt (1992). Isolation and serologic characterization of AIDA-I, the adhesin mediating the diffuse adherence phenotype of the diarrhea-associated *Escherichia coli* strain 2787 (O126:H27). *Infect. Immun.* **60**, 13–18.

Berkow, R. (ed.) (1992). *The Merck Manual of Diagnosis and Therapy*, 16th edn. Merck Research Laboratories, Rahway, NJ.

Bettelheim, K. A. (1992). The genus *Escherichia*. In: A. Balows, H. G. Trüper, M. Dworkin *et al.* (es), *The Prokaryotes*, Vol. III, 2nd edn, pp. 2696–2736. Springer-Verlag, New York, NY.

Bettelheim, K. A. (1997). *Escherichia coli* O157 outbreak in Japan, lessons for Australia. *Aust. Vet. J.* **75**, 108.

Bettelheim, K. A. (2003). Non-O157 verotoxin-producing *Escherichia coli*, a problem, paradox, and paradigm. *Exp. Biol. Med.* **228**, 333–344.

Bilge, S. S., C. R. Clausen, W. Lau and S. L. Moseley (1989). Molecular characterization of a fimbrial adhesin, F1845, mediating diffuse adherence of diarrhea-associated *Escherichia coli* to HEp-2 cells. *J. Bacteriol.* **171**, 4281–4289.

Black, R. E. and C. F. Lanata (1995). Epidemiology of diarrheal diseases in developing countries. InL M. J. Blaser, P. D. Smith, J. I. Ravdin *et al.* (eds), *Infections of the Gastrointestinal Tract*, pp. 13–36. Raven Press, New York, NY.

Black, R. E., M. M. Merson, B. Rowe *et al.* (1981). Enterotoxigenic *Escherichia coli* diarrhoea, acquired immunity and transmission in an endemic area. *Bull. WHO* **59**, 263–268.

Blanco, M., J. E. Blanco, A. Mora *et al.* (2004). Serotypes, virulence genes, and intimin types of Shiga toxin (Verotoxin)-producing *Escherichia coli* isolates from cattle in Spain and identification of a new intimin variant gene (*eae*-ξ). *J. Clin. Microbiol.* **42**, 645–651.

Brackett, R. E., Y.-Y. Hao and M. P. Doyle (1994). Ineffectiveness of hot acid sprays to decontaminate *Escherichia coli* O157:H7 on beef. *J. Food Prot.* **57**, 198–203.

Brenner, D. J. (1984). Family 1. Enterobacteriaceae. In: N. R. Krieg (ed.), *Bergey's Manual of Systematic Bacteriology*, Vol. 1, pp. 408–420. Williams & Wilkins, Baltimore, MD.

Brett, K. N., V. Ramachandran, M. A. Hornitzky *et al.* (2003). stx_{1c} is the most common Shiga toxin 1 subtype among Shiga toxin-producing *Escherichia coli* isolates from sheep but not among isolates from cattle. *J. Clin. Microbiol.* **41**, 926–936.

Brooks, J. T., D. Bergmire-Sweat, M. Kennedy *et al.* (2004). Outbreak of Shiga toxin-producing *Escherichia coli* O111:H8 infections among attendees of a high school cheerleading camp. *Clin. Infect. Dis.* **38**, 190–198.

Bryan, F. L. (1979). Infections and intoxications caused by other bacteria. In: H. Riemann and F. L. Bryan (eds), *Food-borne Infections and Intoxications*, 2nd edn, pp. 211–297. Academic Press, New York, NY.

Bürk, C., R. Dietrich, G. Acar *et al.* (2003). Identification and characterization of a new variant of Shiga toxin 1 in *Escherichia coli* ONT, H19 of bovine origin. *J. Clin. Microbiol.* **41**, 2106–2112.

Burland, V., Y. Shao, N. T. Perna *et al.* (1998). The complete DNA sequence and analysis of the large virulence plasmid of *Escherichia coli* O157:H7. *Nucl. Acids Res.* **26**, 4196–4204.

Butterton, J. R. and S. B. Calderwood (1995). *Vibrio cholerae* O1.*In* M. J. Blaser, P. D. Smith, J. I. Ravdin *et al.* (eds), *Infections of the Gastrointestinal Tract*, pp. 649–670. Raven Press, New York, NY.

Call, D. R., F. J. Brockman and D. P. Chandler (2001). Detecting and genotyping *Escherichia coli* O157:H7 using multiplexed PCR and nucleic acid microarrays. *Intl J. Food Microbiol.* **67**, 71–80.

Campbell, G. R., J. Prosser, A. Glover and K. Killham (2001). Detection of *Escherichia coli* O157:H7 in soil and water using multiplex PCR. *J. Appl. Microbiol.* **91**, 1004–1010.

Caprioli, A., I. Luzzi, F. Rosmini *et al.* (1994). Community-wide outbreak of hemolytic-uremic syndrome associated with non-O157 verocytotoxin-producing *Escherichia coli*. *J. Infect. Dis.* **169**, 208–211.

Carter, A. O., A. A. Borczyk, J. A. Carlson *et al.* (1987). A severe outbreak of *Escherichia coli* O157:H7-associated hemorrhagic colitis in a nursing home. *N. Engl. J. Med.* **317**, 1496–1500.

Cassels, F. J. and M. K. Wolf (1995). Colonization factors of diarrheagenic *E. coli* and their intestinal receptors. *J. Indust. Microbiol.* **15**, 214–226.

CDC (Centers for Disease Control and Prevention) (1994). Foodborne outbreaks of enterotoxigenic *Escherichia coli* – Rhode Island and New Hampshire, 1993. *Morbid. Mortal. Weekly Rep.* **43**, 87–89.

CDC (Centers for Disease Control and Prevention) (1995). Outbreak of acute gastroenteritis attributable to *Escherichia coli* serotype O104:H21 – Helena, Montana, 1994. *Morbid. Mortal. Weekly Rep.* **44**, 501–503.

CDC (Centers for Disease Control and Prevention) (2000). *Escherichia coli* O111:H8 outbreak among teenage campers – Texas, 1999. *Morbid. Mortal. Weekly Rep.* **49**, 321–324.

CDC (Centers for Disease Control and Prevention) (2001). Outbreaks of *Escherichia coli* O157:H7 infections among children associated with farm visits – Pennsylvania and Washington, 2000. *Morbid. Mortal. Weekly Rep.* **50**, 293–297.

CDC (Centers for Disease Control and Prevention) (2002). Multistate outbreak of *Escherichia coli* O157:H7 infections associated with eating ground beef – United States, June–July 2002. *Morbid. Mortal. Weekly Rep.* **51**, 637–639.

Cerna, J. F., J. P. Nataro and T. Estrada-Garcia (2003). Multiplex PCR for detection of three plasmid-borne genes of enteroaggregative *Escherichia coli* strains. *J. Clin. Microbiol.* **41**, 2138–2140.

Chart, H., J. Spenser, H. R. Smith and B. Rowe (1997). Identification of enteroaggregative *Escherichia coli* based on surface properties. *J. Appl. Microbiol.* **83**, 712–717.

Chaudhuri, A. G., J. Bhattacharya, G. G. Nair *et al.* (1998). Rise of cytosolic Ca^{2+} and activation of membrane-bound guanylyl cyclase activity in rat enterocytes by heat-stable enterotoxin of *Vibrio cholerae* non-O1. *FEMS Microbiol. Lett.* **160**, 125–129.

Cherla, R. P., S.-Y. Lee and V. L. Tesh (2003). Shiga toxins and apoptosis. *FEMS Microbiol. Lett.* **228**, 159–166.

Clarke, S.C. (2001). Diarrhoeagenic *Escherichia coli* – an emerging problem? *Diag. Microbiol. Infect. Dis.* **41**, 93–98.

Clarke, S. C., R. D. Haigh, P. P. E. Freestone and P. H. Williams (2002). Enteropathogenic *Escherichia coli* infection, history and clinical aspects. *Br. J. Biomed. Sci.* **59**, 123–127.

Clemens, J. D., M. R. Rao, J. Chakraborty *et al.* (1997). Breastfeeding and risk of life-threatening enterotoxigenic *Escherichia coli* diarrhea in Bangladeshi infants and children. *Pediatrics* **100(6)**, E2.

Cobeljic, M., B. Miljkovic-Selimovic, D. Paunovic-Todosijevic *et al.* (1996). Enteroaggregative *Escherichia coli* associated with an outbreak of diarrhoea in a neonatal ward. *Epidemiol. Infect.* **117**, 11–16.

Cohen, D., N. Orr, M. Haim *et al.* (2000). Safety and immunogenicity of two different lots of the oral, killed enterotoxigenic *Escherichia coli*-cholera toxin B subunit vaccine in Israeli young adults. *Infect. Immun.* **68**, 4492–4497.

Cohen, M. B. and R. A. Gianella (1995). Enterotoxigenic *Escherichia coli*. In: M. J. Blaser, P. D. Smith, J. I. Ravdin *et al.* (eds), *Infections of the Gastrointestinal Tract*, pp. 691–707. Raven Press, New York, NY.

Collinson, S. K., L. Emödy, T. J. Trust and W. W. Kay (1992). Thin aggregative fimbriae from diarrheagenic *Escherichia coli*. *J. Bacteriol.* **174**, 4490–4495.

Cookson, S. T. and J. P. Nataro (1996). Characterization of HEp-2 cell projection formation induced by diffusely adherent *Escherichia coli. Microb. Pathog.* **21**, 421–434.

Cornick, N. A., I. Matisse, E. Samuel *et al.* (1999). Edema disease as a model for systemic disease induced by Shiga toxin-producing *E. coli. Adv. Exp. Med. Biol.* **473**, 155–161.

Cornick, N. A., S. L. Booher, T. A. Casey and H. W. Moon (2000). Persistent colonization of sheep by *Escherichia coli* O157:H7 and other *E. coli* pathotypes. *Appl. Envrion. Microbiol.* **66**, 4926–4934.

Crump, J. A., A. C. Sulka, A. J. Langer *et al.* (2002). An outbreak of *Escherichia coli* O157:H7 infections among visitors to a dairy farm. *N. Engl. J. Med.* **347**, 555–560.

Crump, J. A., C. R. Braden, M. E. Dey *et al.* (2003). Outbreaks of *Escherichia coli* O157 infections at multiple county agricultural fairs, a hazard of mixing cattle, concession stands and children. *Epidemiol. Infect.* **131**, 1055–1062.

Czeczulin, J. R., S. Balepur, S. Hicks *et al.* (1997). Aggregative adherence fimbriae II, a second fimbrial antigen mediating aggregative adherence in enteraggregative *Escherichia coli. Infect. Immun.* **65**, 4135–4145.

Daniels, N. A., J. Naimann, A. Karpati *et al.* (2000). Traveler's diarrhea at sea, three outbreaks of waterborne enterotoxigenic *Escherichia coli* on cruise ships. *J. Infect. Dis.* **181**, 1491–1495.

Danielsson, M.-L., R. Möllby, H. Brag *et al.* (1979). Enterotoxigenic-enteric bacteria in foods and outbreaks of food-borne diseases in Sweden. *J. Hygiene Camb.* **83**, 33–40.

Davis, B. M. and M. K. Waldor (2003). Filamentous phages linked to virulence of *Vibrio cholerae. Curr. Opin. Microbiol.* **6**, 35–42.

Dean-Nystrom, E. A. (2003). Bovine *Escherichia coli* O157:H7 infection model. *Methods Mol. Med.* **73**, 329–338.

Dean-Nystrom, E. A., B. T. Bosworth, H. W. Moon and A. D. O'Brien (1998). *Escherichia coli* O157:H7 requires intimin for enteropathogenicity in calves. *Infect. Immun.* **66**, 4560–4563.

DebRoy, C., B. D. Bright, R. A. Wilson *et al.* (1994). Plasmid-coded DNA fragment developed as a specific gene probe for the identification of enteroaggregative *Escherichia coli. J. Med. Microbiol.* **41**, 393–398.

DeCastro, A. F. P., B. Guerra, L. Leomil *et al.* (2003). Multidrug-resistant Shiga toxin-producing *Escherichia coli* O118:H16 in Latin America. *Emerg. Infect. Dis.* **9**, 1027–1028.

Deisingh, A. K. and M. Thompson (2004). Strategies for the detection of *Escherichia coli* O157:H7 in foods. *J. Appl. Microbiol.* **96**, 419–429.

De Rycke, J., A. Milon and E. Oswald (1999). Necrotoxic *Escherichia coli* (NTEC), two emerging categories of human and animal pathogens. *Vet. Res.* **30**, 221–233.

DeVinney, R. A. Gauthier, A. Abe and B. B. Finlay (1999a). Enteropathogenic *Escherichia coli*, A pathogen that inserts its own receptor into host cells. *Cell. Mol. Life Sci.* **55**, 961–976.

DeVinney, R., D. G. Knoechel and B. B. Finlay (1999b). Enteropathogenic *Escherichia coli*, cellular harassment. *Curr. Opin. Microbiol.* **2,** 83–88.

DeVinney, R., M. Stein, D. Reinscheid *et al.* (1999c). Enterohemorrhagic *Escherichia coli* O157:H7 produces Tir, which is translocated to the host cell membrane but is not tyrosine phosphorylated. *Infect. Immun.* **67**, 2389–2398.

Donnenberg, M. S. and T. S. Whittam (2001). Pathogenesis and evolution of virulence in enteropathogenic and enterohemorrhagic *Escherichia coli. J. Clin. Invest.* **107**, 539–548.

Doyle, M. P. and V. V. Padhye (1989). *Escherichia coli*. In: M. P. Doyle (ed.), *Foodborne Bacterial Pathogens*, pp. 235–281. Marcel Dekker, New York, NY.

Dubreuil, J. D (1997). *Escherichia coli* STb enterotoxin. *Microbiology* **143**, 1783–1785.

Dutta, S., A. Chatterjee, P. Dutta *et al.* (2001). Sensitivity and performance characteristics of a direct PCR with stool samples in comparison to conventional techniques for diagnosis of *Shigella* and enteroinvasive *Escherichia coli* infection in children with acute diarrhoea in Calcutta, India. *J. Med. Microbiol.* **50**, 667–674.

Dykstra, S. A., R. A. Moxley, B. H. Janke *et al.* (1993). Clinical signs and lesions in gnotobiotic pigs inoculated with Shiga-like toxin I from *Escherichia coli. Vet. Pathol.* **30**, 410–417.

Elder, R. O., J. E. Keen, G. R. Siragusa *et al.* (2000). Correlation of enterohemorrhagic *Escherichia coli* O157 prevalence in feces, hides, and carcasses of beef cattle during processing. *Proc. Natl. Acad. Sci. USA* **97**, 2999–3003.

Elliott, E. J., R. M. Robins-Browne, E. V. O'Loughlin *et al.* and Contributors to the Australian Paedriatic Surveillance Unit (2001). Nationwide study of haemolytic uraemic syndrome, clinical, microbiological, and epidemiological features. *Arch. Dis. Child.* **85**, 125–131.

Eslava, C., F. Navarro-García, J. R. Czeczulin *et al.* (1998). Pet, an autotransporter enterotoxin from enteraggregative *Escherichia coli. Infect. Immun.* **66**, 3155–3163.

Fábrega, V. L. A., A. J. P. Ferreira, F. R. da Silva Patrício *et al.* (2002). Cell-detaching *Escherichia coli* (CDEC) strains from children with diarrhea, identification of a protein with toxigenic activity. *FEMS Microbiol. Lett.* **217**, 191–197.

Fagundes-Neto, U. and Scaletsky, I. C. A. (2000). The gut at war, the consequences of enteropathogenic *Escherichia coli* infection as a factor of diarrhea and malnutrition. *São Paulo Med. J.* **118**, 21–29.

Feder, I., F. M. Wallace, J. T. Gray *et al.* (2003). Isolation of *Escherichia coli* O157:H7 from intact colon fecal samples of swine. *Emerg. Infect. Dis.* **9**, 380–383.

Feng, P. (2001). Development and impact of rapid methods for detection of foodborne pathogens. In: M. P. Doyle, L. R. Beuchat and T. J. Montville (eds), *Food Microbiology, Fundamentals and Frontiers*, pp. 775–796. ASM Press, Washington, DC.

Feng, P., K. A. Lampel, H. Karch and T. S. Whittam (1998). Genotypic and phenotypic changes in the emergence of *Escherichia coli* O157:H7. *J. Infect. Dis.* **177**, 1750–1753.

Fenwick, B. W. and Cowan, L. A. (1998). Canine model of the hemolytic uremic syndrome. In: J. B. Kaper and A. D. O'Brien (eds), Escherichia coli *O157:H7 and other Shiga Toxin-producing* E. coli *Strains*, pp. 268–277. American Society for Microbiology, Washington, DC.

Frankel, G., A. D. Phillips, I. Rosenshine *et al.* (1998). Enteropathogenic and enterohemorrhagic *Escherichia coli*, More subversive elements. *Mol. Microbiol.* **30**, 911–921.

Fratamico, P. M. and C. G. Crawford (1999). Detection by commercial immunomagnetic particle-based assays. In: R. Robison, C. Batt and P. Patel (eds), *Encyclopedia of Food Microbiology*, Vol. 1, pp. 654–663. Academic Press, London.

Fratamico, P. M., L. K. Bagi and T. Pepe (2000). A multiplex polymerase chain reaction assay for rapid detection and identification of *Escherichia coli* O157:H7 in foods and bovine feces. *J. Food Prot.* **63**, 1032–1037.

Fratamico, P. M., J. L. Smith and R. L. Buchanan (2002). *Escherichia coli.* In: D. O. Cliver and H. P. Riemann (eds), *Foodborne Diseases*, 2nd edn, pp. 79–101. Academic Press, New York, NY.

Fratamico, P. M., C. E. Briggs, D. Needle *et al.* (2003). Sequence of the *Escherichia coli* O121 O-antigen gene cluster and detection of enterohemorrhagic *E. coli* O121 by PCR amplification of the *wzx* and *wzy* genes. *J. Clin. Microbiol.* **41**, 3379–3383.

Friedrich, A. W., M. Bielaszewska, W. L. Zhang *et al.* (2002). *Escherichia coli* harboring Shiga toxin 2 gene variants, frequency and association with clinical symptoms. *J. Infect. Dis.* **185**, 74–84.

Gaastra, W. and A.-M. Svennerholm (1996). Colonization factors of human enterotoxigenic *Escherichia coli* (ETEC). *Trends Microbiol.* **4**, 444–452.

Gagliardi, J. V. and J. S. Karns (2002). Persistence of *Escherichia coli* O157:H7 in soil and on plant roots. *Environ. Microbiol.* **4**, 89–96.

Glass, K. A., J. M. Loeffelholz, J. P. Ford and M. P. Doyle (1992). Fate of *Escherichia coli* O157:H7 as affected by pH or sodium chloride and in fermented, dry sausage. *Appl. Environ. Microbiol.* **58**, 2513–2516.

Gordillo, M. E., G. R. Reeve, J. Pappas *et al.* (1992). Molecular characterization of strains of enteroinvasive *Escherichia coli* O143, including isolates from a large outbreak in Houston, Texas. *J. Clin. Microbiol.* **30**, 889–893.

Grant, C. C. R., R. J. Messer and W. Cieplak (1994). Role of trypsin-like cleavage at arginine 192 in the enzymatic and cytotonic activities of *Escherichia coli* heat-labile enterotoxin. *Infect. Immun.* **62**, 4270–4278.

Grauke, L. J., I. T. Kudva, J. W. Yoon *et al.* (2002). Gastrointestinal tract location of *Escherichia coli* O157:H7 in ruminants. *Appl. Environ. Microbiol.* **68**, 2269–2277.

Greenberg, D. E., Z.-D. Jiang, R. Steffen *et al.* (2002). Markers of inflammation in bacterial diarrhea among travelers, with a focus on enteroaggregative *Escherichia coli* pathogenicity. *J. Infect. Dis.* **185**, 944–949.

Grewal, H. S., H. Valvatne, M. K. Bhan *et al.* (1997). A new putative fimbrial colonization factor, CS19, of human enterotoxigenic *Escherichia coli. Infect. Immun.* **65**, 507–513.

Griffin, P. M (1995). *Escherichia coli* O157:H7 and other enterohemorrhagic *Escherichia coli. In:* M. J. Blaser, P. D. Smith, J. I. Ravdin *et al.* (eds), *Infections of the Gastrointestinal Tract,* pp. 739–761. Raven Press, New York, NY.

Guerrant, R. L. and N. M. Thielman (1995). Types of *Escherichia coli* enteropathogens. *In:* M. J. Blaser, P. D. Smith, J. I. Ravdin *et al.* (eds), *Infections of the Gastrointestinal Tract*, pp. 687–690. Raven Press, New York, NY.

Gunzer, F., I. Hennig-Pauka, K. H. Waldmann *et al.* (2002). Gnotobiotic piglets develop thrombotic microangiopathy after oral infection with enterohemorrhagic *Escherichia coli. Am. J. Clin. Pathol.* **118**, 364–375.

Gyles, C. L (1992). *Escherichia coli* cytotoxins and enterotoxins. *Can. J. Microbiol.* **38**, 734–746.

Hale, T. L., P. Echeverria and J. P. Nataro (1997). Enteroinvasive *Escherichia coli.* In: M. Sussman (ed.), Escherichia coli, *Mechanisms of Virulence*, pp. 449–468. Cambridge University Press, Cambridge.

Harris, J. R., J. Mariano, J. G. Wells *et al.* (1985). Person-to-person transmission in an outbreak of enteroinvasive *Escherichia coli. Am. J. Epidemiol.* **122**, 245–252.

Hashimoto, H., K. Mizukoshi, M. Nishi *et al.* (1999). Epidemic of gastrointestinal tract infection including hemorrhagic colitis attributable to Shiga toxin-producing *Escherichia coli* O118:H2 at a junior high school in Japan. *Pediatrics* **103(1)**, E2.

Higgins, L. M., G. Frankel, G. Douce *et al.* (1999). *Citrobacter rodentium* infection in mice elicits a mucosal Th1 cytokine response and lesions similar to those in murine inflammatory bowel disease. *Infect. Immun.* **67**, 3031–3039.

Hobbs, B. C., B. Rowe, M. Kendall *et al.* (1976). *Escherichia coli* O27 in adult diarrhea. *J. Hygiene Camb.* **77**, 393–400.

Horne, C., B. A. Vallance, W. Deng and B. B. Finlay (2002). Current progress in enteropathogenic and enterohemorrhagic *Escherichia coli* vaccines. *Expert Rev. Vaccines* **1**, 483–493.

Houng, H S. H., O. Sethabutr and P. Echeverria (1997). A simple polymerase chain reaction technique to detect and differentiate *Shigella* and enteroinvasive *Escherichia coli* in human feces. *Diagn. Microbiol. Infect. Dis.* **28**, 19–25.

Huerta, M., I. Grotto, M. Gdalevaich *et al.* (2000). A waterborne outbreak of gastroenteritis in the Golan Heights due to enterotoxigenic *Escherichia coli. Infection* **28**, 267–271.

Huppertz, H.-I., S. Rutkowski, D. H. Busch *et al.* (1999). Bovine colostrum ameliorates diarrhea in infection with diarrheagenic *Escherichia coli*, Shiga toxin-producing *E. coli*, and *E. coli* expressing intimin and hemolysin. *J. Pediatr. Gastroenterol. Nutr.* **29**, 452–456.

Ibekwe, A. M. and C. M. Grieve (2003). Detection and quantification of *Escherichia coli* O157:H7 in environmental samples by real-time PCR. *J. Appl. Microbiol.* **94**, 421–431.

Itoh, Y., I. Nagano, M. Kunishima and T. Ezaki (1997). Laboratory investigation of enteroaggregative *Escherichia coli* O untypeable, H10 associated with a massive outbreak of gastrointestinal illness. *J. Clin. Microbiol.* **35**, 2546–2550.

Jallat, C., V. Livrelli, A. Darfeuille-Michaud *et al.* (1993). *Escherichia coli* strains involved in diarrhea in France, high prevalence and heterogeneity of diffusely adhering strains. *J. Clin. Microbiol.* **31**, 2031–2037.

Janka, A., M. Bielaszewska, U. Dobrindt *et al.* (2003). Cytolethal distending toxin gene cluster in enterohemorrhagic *Escherichia coli* O157:H$^-$ and O157:H7, characterization and evolutionary considerations. *Infect. Immun.* **71**, 3634–3638.

Jiang, X., J. Morgan and M. P. Doyle (2002). Fate of *Escherichia coli* O157:H7 in manure-amended soil. *Appl. Environ. Microbiol.* **68**, 2605–2609.

Judge, N. A., H. S. Mason and A. D. O'Brien (2004). Plant cell-based intimin vaccine given orally to mice primed with intimin reduces time of *Escherichia coli* shedding in feces. *Infect. Immun.* **72**, 168–175.

Juneja, V., B. Marmer and B. Eblen (1999). Predictive model for the combined effect of temperature, pH, sodium chloride, and sodium pyrophosphate on the heat resistance of *Escherichia coli* O157:H7. *J. Food Safety* **19**, 147–160.

Karch, H., M. Bielaszewska, M. Bitzan and H. Schmidt (1999). Epidemiology and diagnosis of Shiga toxin-producing *Escherichia coli* infections. *Diagn. Microbiol. Infect. Dis.* **34**, 229–243.

Karmali, M. A. (2003). The medical significance of Shiga toxin-producing *Escherichia coli* infections. *Methods Mol. Med.* **73**, 1–7.

Katz, J. M., X. Lu, S. A. Young and J. C. Galphin (1997). Adjuvant activity of heat-labile enterotoxin from enterotoxigenic *Escherichia coli* for oral administration of inactivated influenza virus vaccine. *J. Infect. Dis.* **175**, 352–363.

Keusch, G. T., D. W. K. Acheson, C. Marchant and J. McIver (1998). Toxoid-based active and passive immunization to prevent and/or modulate hemolytic-uremic syndrome due to Shiga toxin-producing *Escherichia coli.* In: J. B. Kaper and A. D. O'Brien (eds), Escherichia coli *O157:H7 and other Shiga toxin-producing* E. coli *Strains*, pp. 409–418. ASM Press, Washington, DC.

Klapproth, J.-M. A., I. C. A. Scaletsky, B. P. McNamara *et al.* (2000). A large toxin from pathogenic *Escherichia coli* strains that inhibits lymphocyte action. *Infect. Immun.* **68**, 2148–2155.

Knutton, S., R. K. Shaw, M. K. Bhan *et al.* (1992). Ability of enteroaggregative *Escherichia coli* strains to adhere in vitro to human intestinal mucosa. *Infect. Immun.* **60**, 2083–2091.

Kobayashi, M., T. Sasaki, N. Saito *et al.* (1999). Houseflies, not simple mechanical vectors of enterohemorrhagic *Escherichia coli* O157:H7. *Am. J. Trop. Med. Hyg.* **61**, 625–629.

Kolling, G. I. and K. R. Matthews (1999). Export of virulence genes and Shiga toxin by membrane vesicles of *Escherichia coli* O157:H7. *Appl. Environ. Microbiol.* **65**, 1843–1848.

Konadu, E., J. C. Parke, A. Donohue-Rolfe *et al.* (1998). Synthesis and immunologic properties of O-specific polysaccharide-protein conjugate vaccines for prevention and treatment of infections with *Escherichia coli* O157 and other causes of hemolytic-uremic syndrome. J. B. Kaper and A. D. O'Brien (eds), Escherichia coli *O157:H7 and other Shiga toxin-producing* E. coli *Strains*, pp. 419–424. ASM Press, Washington, DC.

Kudoh, Y., H. Zen-Yoji, S. Matsushita *et al.* (1977). Outbreaks of acute enteritis to heat-stable enterotoxin-producing strains of *Escherichia coli. Microbiol. Immunol.* **21**, 175–178.

Lan, R., B. Lumb, D. Ryan and P.R. Reeves (2001). Molecular evolution of large virulence plasmid in *Shigella* clones and enteroinvasive *Escherichia coli. Infect. Immun.* **69**, 6303–6309.

Lan, R. and P. R. Reeves (2002). *Escherichia coli* in disguise, molecular origins of *Shigella. Microbes Infect.* **4**, 1125–1132.

Law, D. and H. Chart (1998). Enteroaggregative *Escherichia coli. J. Appl. Microbiol.* **84**, 685–697.

LeBlanc, J. J (2003). Implication of virulence factors in *Escherichia coli* O157:H7 pathogenesis. *Crit. Rev. Microbiol.* **29**, 277–296.

Leung, P. H., J. S. Peiris, W. W. Ng *et al.* (2003). A newly discovered verotoxin variant, VT2g, produced by bovine verocytotoxigenic *Escherichia coli. Appl. Environ. Microbiol.* **69**, 7549–7553.

Levine, M. M. (1987). *Escherichia coli* that cause diarrhea, enterotoxigenic, enteropathogenic, enteroinvasive, enterohemorrhagic, and enteroadherent. *J. Infect. Dis.* **155**, 377–389.

Levine, M. M., E. S. Caplan, D. Waterman *et al.* (1977). Diarrhea caused by *Escherichia coli* that produce only heat-stable enterotoxin. *Infect. Immun.* **17**, 78–82.

Lin, J., M. P. Smith, K. C. Chapin *et al.* (1996). Mechanisms of acid resistance in enterohemorrhagic *Escherichia coli. Appl. Environ. Microbiol.* **62**, 3094–3100.

Line, J. E., A. R. Fain Jr, A. B. Moran *et al.* (1991). Lethality of heat to *Escherichia coli* O157:H7, D-value and z-value determinations in ground beef. *J. Food Prot.* **54**, 762–766.

Liu, H., H. Wang, Z. Shi *et al.* (2003). Identification of *Escherichia coli* O157:H7 with oligonucleotide arrays. *Bull. Environ. Contam. Toxicol.* **71**, 826–832.

López-González, V., P. S. Murano, R. E. Brennan and E. A. Murano (1999). Influence of various packaging conditions on survival of *Escherichia coli* O157:H7 to irradiation by electron beam versus gamma rays. *J. Food Prot.* **62**, 10–15.

López-Saucedo, C., J. F. Cerna, N. Villegas-Sepulveda *et al.* (2003). Single multiplex polymerase chain reaction to detect diverse loci associated with diarrheagenic *Escherichia coli. Emerg. Infect. Dis.* **9**, 127–131.

López-Vidal, Y., J. J. Calva, A. Trujillo *et al.* (1990). Enterotoxin and adhesions of enterotoxigenic *Escherichia coli*, are they risk factors for acute diarrhea in the community? *J.. Infect. Dis.* **162**, 442–447.

Lortie, L.-A., J. D. Dubreuil and J. Harel (1991). Characterization of *Escherichia coli* strains producing heat-stable enterotoxin b (STb) isolated from humans with diarrhea. *J. Clin. Microbiol.* **29**, 656–659.

Lumish, R. M., R. W. Ryder, D. C. Anderson *et al.* (1980). Heat-labile enterotoxigenic *Escherichia coli* induced diarrhea aboard a Miami-based cruise ship. *Am. J. Epidemiol.* **111**, 432–436.

Lyte, M., A. K. Erickson, B. P. Arulanandam *et al.* (1997). Norepinephrine-induced expression of the K99 pilus adhesin of enterotoxigenic *Escherichia coli. Biochem. Biophys. Res. Commun.* **232**, 682–686.

MacDonald, K. L., M. Eidson, C. Strohmeyer *et al.* (1985). A multistate outbreak of gastrointestinal illness caused by enterotoxigenic *Escherichia coli* in imported semisoft cheese. *J. Infect. Dis.* **151**, 716–720.

Mainil, J.G., E. Jacquemin and E. Oswald (2003). Prevalence and identity of *cdt*-related sequences in necrotoxigenic *Escherichia coli. Vet. Microbiol.* **94**, 159–165.

Makino, K., K. Ishii, T Yasunaga *et al.* (1998). Complete nucleotide sequences of 93-kb and 3.3-kb plasmids of an enterohemorrhagic *Escherichia coli* O157:H7 derived from Sakai outbreak. *DNA Res.* **5**, 1–9.

Mariani-Kurkdjian, P., E. Denamur, A. Milon *et al.* (1993). Identification of a clone of *Escherichia coli* O103:H2 as a potential agent of hemolytic-uremic syndrome in France. *J. Clin. Microbiol.* **31**, 296–301.

Mason, H. S., T. A. Haq, J. D. Clements and C. J. Arntzen (1998). Edible vaccine protects mice against *Escherichia coli* heat-labile enterotoxin (LT), Potatoes expressing a synthetic LT-B gene. *Vaccine* **16**, 1336–1343.

Matthews, K. R., P. A. Murdough and A. J. Bramley (1997). Invasion of bovine epithelial cells by verocytotoxin-producing *Escherichia coli* O157:H7. *J. Appl. Microbiol.* **82**, 197–203.

Matussek, A., L. Lauber, A. Bergau *et al.* (2003). Molecular and functional analysis of Shiga toxin-induced response patterns in human vascular endothelial cells. *Blood* **102**, 1323–1332.

McCarthy, T.A., N. L. Barrett, J. L. Hadler *et al.* (2001). Hemolytic-uremic syndrome and *Escherichia coli* O121 at a lake in Connecticut, 1999. *Pediatrics* **108(4)**, E59.

McGee P., D. J. Bolton, J. J. Sheridan *et al.* (2002). Survival of *Escherichia coli* O157:H7 in farm water, its role as a vector in the transmission of the organism within herds. *J. Appl. Microbiol.* **93**, 706–713.

McMaster, C., E. A. Roche, G. A. Willshaw *et al.* (2001). Verocytotoxin-producing *Escherichia coli* serotype O26:H11 outbreak in an Irish crèche. *Eur. J. Clin. Microbiol. Infect. Dis.* **20**, 430–432.

Mead, P. S. and P. M. Griffin (1998). *Escherichia coli* O157:H7. *Lancet* **352**, 1207–1212.

Mead, P. S., L. Slutsker, V. Dietz *et al.* (1999). Food-related illness and death in the United States. *Emerg. Infect. Dis.* **5**, 607–625.

Melton-Celsa, A. R. and A. D. O'Brien (2003). Animal models for STEC-mediated disease. *Methods Mol. Med.* **73**, 291–305.

Ménard, L.-P. and J. D. Dubreuil (2002). Enteroaggregative *Escherichia coli* heat-stable enterotoxin 1 (EAST1), a new toxin with an old twist. *Crit. Rev. Microbiol.* **28**, 43–60.

Meng, J., M. P. Doyle, T. Zhao and S. Zhao (2001). Enterohemorrhagic *Escherichia coli*. In: M. P. Doyle, L. R. Beuchat and T. J. Montville (eds), *Food Microbiology. Fundamentals and Frontiers*, 2nd edn, pp. 193–213. ASM Press, Washington, DC.

Merson, M. H., G. K. Morris, D. A. Sack *et al.* (1976). Traveler's diarrhea in Mexico. A prospective study of physicians and family members attending a congress. *J. Am. Med. Assoc.* **294**, 1299–1305.

Michino, H., K. Araki, S. Minami *et al.* (1999). Massive outbreak of *Escherichia coli* O157:H7 infection in school children in Sakai City, Japan, associated with consumption of white radish sprouts. *Am. J. Epidemiol.* **150**, 787–796.

Milon, A., E. Oswald and J. De Rycke (1999). Rabbit EPEC, a model for the study of enteropathogenic *Escherichia coli. Vet. Res.* **30**, 203–219.

Miqdady, M. S., Z-D. Jiang, J. P. Nataro and H. L. DuPont (2002). Detection of enteroaggregative *Escherichia coli* with formalin-preserved HEp-2 cells. *J. Clin. Microbiol.* **40**, 3066–3067.

Misselwitz, J., H. Karch, M. Bielazewska *et al.* (2003). Cluster of hemolytic-uremic syndrome caused by Shiga toxin-producing *Escherichia coli* O26:H11. *Pediatr. Infect. Dis. J.* **22**, 349–354.

Mitsuda, T., T. Muto, M. Yamada *et al.* (1998). Epidemiological study of a food-borne outbreak of enterotoxigenic *Escherichia coli* O25:NM by pulsed-field gel electrophoresis and randomly amplified polymorphic DNA analysis. *J. Clin. Microbiol.* **36**, 652–656.

Mizan, S., M. D. Lee, B. G. Harmon *et al.* (2002). Acquisition of antibiotic resistance plasmids by enterohemorrhagic *Escherichia coli* O157:H7 within rumen fluid. *J. Food Prot.* **65**, 1038–1040.

Mol, O. and B. Oudega (1996). Molecular and structural aspects of fimbriae biosynthesis and assembly in *Escherichia coli. FEMS Microbiol. Rev.* **19**, 25–52.

Mølbak, K., P. S. Mead and P. M. Griffin (2002). Antimicrobial therapy in patients with *Escherichia coli* O157:H7 infection. *J. Am. Med. Assoc.* **288**, 1014–1016.

Morabito, S., H. Karch, P. Mariani-Kurkdjian *et al.* (1998). Enteroaggregative, Shiga toxin-producing *Escherichia coli* O111:H2 associated with an outbreak of hemolytic-uremic syndrome. *J. Clin. Microbiol.* **36**, 840–842.

Morais, T. B., T. A. T. Gomes and D. M. Sigulem (1997). Enteroaggregative *Escherichia coli* in infant feeding bottles. *Lancet* **349**, 1448–1449.

Mulvey, G., D. J. Rafter and G. D. Armstrong (2002). Potential for using antibiotics combined with a Shiga toxin-absorbing agent for treating O157:H7 *Escherichia coli* infections. *Can. J. Chem.* **80**, 871–874.

Muniesa, M., J. Recktenwald, M. Bielaszewska *et al.* (2000). Characterization of a Shiga toxn 2e-converting bacteriophage from an *Escherichia coli* strain of human origin. *Infect. Immun.* **68**, 4850–4855.

Naimi, T. S., J. H. Wicklund, S. J. Olsen *et al.* (2003). Concurrent outbreaks of *Shigella sonnei* and enterotoxigenic *Escherichia coli* infections associated with parsley, implications for surveillance and control of foodborne illness. *J. Food Prot.* **66**, 535–541.

Nair, G. B. and Y. Takeda (1998). The heat-stable enterotoxins. *Microbial Pathogen.* **24**, 123–131.

Nascimento de Araújo, A. and L.G. Giugliano (2000). Human milk fractions inhibit the adherence of diffusely adherent *Escherichia coli* (DAEC) and enteroaggregative *E. coli* (EAEC) to HeLa cells. *FEMS Microbiol. Lett.* **184**, 91–94.

Nataro, J. P. and J. B. Kaper (1998). Diarrheagenic *Escherichia coli. Clin. Microbiol. Rev.* **11**, 142–201.

Nataro, J. P., Y. Deng, D. R. Maneval *et al.* (1992). Aggregative adherence fimbriae I of enteroaggregative *Escherichia coli* mediate adherence to HEp-2 cells and hemagglutination of human erythrocytes. *Infect. Immun.* **60**, 2297–2304.

Nataro, J. P, Y. Deng, S. Cookson *et al.* (1995). Heterogeneity of enteroaggregative *Escherichia coli* virulence demonstrated in volunteers. *J. Infect. Dis.* **171**, 465–468.

Nataro, J. P., S. Hicks, A. D. Philips *et al.* (1996). T84 cells in culture as a model for enteroaggregative *Escherichia coli* pathogenesis. *Infect. Immun.* **64**, 4761–4768.

Nataro, J. P., T. Steiner and R. L. Guerrant (1998). Enteroaggregative *Escherichia coli. Emerg. Infect. Dis.* **4**, 251–261.

Navarro-García, F., C. Eslava, J. M. Villaseca *et al.* (1998). In vitro effects of a high-molecular weight heat-labile enterotoxin from enteroaggregative *Escherichia coli. Infect. Immun.* **66**, 3149–3154.

Naylor, S. W., J. C. Low, T. E. Besser *et al.* (2003). Lymphoid follicle-dense mucosa at the terminal rectum is the principal site of colonization of enterohemorrhagic *Escherichia coli* O157:H7 in the bovine host. *Infect. Immun.* **71**, 1505–1512.

Neill, M. A., P. I. Tarr, D. N. Taylor and A. F. Trofa (1994). *Escherichia coli.* In: Y. H. Hui, J. R. Gorham, K. D. Murrell and D. O. Cliver (eds), *Foodborne Disease Handbook*, Vol. 1, pp. 169–213. Marcel Dekker, New York, NY.

Nirdnoy, W., O. Serichantalergs, A. Cravioto *et al.* (1997). Distribution of colonization factor antigens among enterotoxigenic *Escherichia coli* strains isolated from patients with diarrhea in Nepal, Indonesia, Peru and Thailand. *J. Clin. Microbiol.* **35**, 527–530.

Noller, A. C., M. C. McEllistrem, A. G. F. Pacheco *et al.* (2003). Multilocus variable-number tandem repeat analysis distinguishes outbreak and sporadic *Escherichia coli* O157:H7 isolates. *J. Clin. Microbiol.* **41**, 5389–5397.

O'Brien, A. D. and R. K. Holmes (1996). Protein toxins of *Escherichia coli* and *Salmonella*. In: F. C. Neidhardt (ed.), Escherichia coli *and* Salmonella, *Cellular and Molecular Biology*, 2nd edn, pp. 2788–2802. ASM Press, Washington, DC.

O'Donnell, J. M., L. Thornton, E. B. McNamara *et al.* (2002). Outbreak of Vero cytotoxin-producing *Escherichia coli* O157 in child day care facility. *Commun. Dis. Public Health* **5**, 54–58.

Oelschlaeger, T. A., T. J. Barrett and D. J. Kopecko (1994). Some structures and processes of human epithelial cells involved in uptake of enterohemorrhagic *Escherichia coli* O157:H7 strains. *Infect. Immun.* **62**, 5142–5150.

Ogden, I. D., N. F. Hepburn, M. MacRae *et al.* (2002). Long-term survival of *Escherichia coli* O157 on pasture following an outbreak associated with sheep at a scout camp. *Lett. Appl. Microbiol.* **34**, 100–104.

Okamoto, K. and M. Takahara (1990). Synthesis of *Escherichia coli* heat-stable enterotoxin STp as a pre-pro form and role of the pro sequence in secretion. *J. Bacteriol.* **172**, 5260–5265.

Okeke, I. N. and J. P. Nataro (2001). Enteroaggregative *Escherichia coli. Lancet Infect. Dis.* **1**, 304–313.

Okamoto, K., Y. Fujii, N. Akashi *et al.* (1993). Identification and characterization of heat-stable enterotoxin II-producing *Escherichia coli* from patients with diarrhea. *Microbiol. Immunol.* **37**, 411–414.

Okeke, I. N., H. Steinrück, K. J. Kanack *et al.* (2002). Antibiotic-resistant cell-detaching *Escherichia coli* strains from Nigerian children. *J. Clin. Microbiol.* **40**, 301–305.

Okeke, I.N., I.C.A. Scaletsky, E.H. Soars *et al.* (2004). Molecular epidemiology of the iron utilization genes of enteroaggregative *Escherichia coli. J. Clin. Microbiol.* **42**, 36–44.

O'Loughlin, E. V. and R. M. Robins-Browne (2001). Effect of Shiga toxin and Shiga-like toxins on eukaryotic cells. *Microbes Infect.* **3**, 493–507.

Olsen, S. J., G. Miller, T. Breuer *et al.* (2002). A waterborne outbreak of *Escherichia coli* O157:H7 infections and hemolytic uremic syndrome, implications for rural water systems. *Emerg. Infect. Dis.* **8**, 370–375.

Ørskov, F. (1984). Genus 1. *Escherichia.* In: N. R. Krieg (ed.), *Bergey's Manual of Systematic Bacteriology*, Vol. 1, pp. 420–423. Williams & Wilkins, Baltimore, MD.

Ørskov, F. and I. Ørskov (1992). *Escherichia coli* serotyping and disease in man and animals. *Can. J. Microbiol.* **38**, 699–704.

Pal, T., N. A. Al-Sweih, M. Herpay and T. D. Chugh (1997). Identification of enteroinvasive *Escherichia coli* and *Shigella* strains in pediatric patients by an IpaC-specific enzyme-linked immunosorbent assay. *J. Clin. Microbiol.* **35**, 1757–1760.

Palumbo, S. A., J. E. Call, F. J. Schultz and A. C. Williams (1995). Minimum and maximum temperatures for growth and verotoxin production by hemorrhagic strains of *Escherichia coli. J.. Food Prot.* **58**, 352–356.

Park, C. H., H. J. Kim and D. L. Hixon (2002). Importance of testing stool specimens for Shiga toxins. *J. Clin. Microbiol.* **40**, 3542–3543.

Parry, S. H. and D. M. Rooke (1985). Adhesins and colonization factors of *Escherichia coli.* In: M. Sussman (ed.), *The Virulence of* Escherichia coli, *Reviews and Methods*, pp. 79–155. Academic Press, New York, NY.

Paton, A. W., R. M. Ratcliff, R. M. Doyle *et al.* (1996). Molecular microbiological investigation of an outbreak of hemolytic-uremic syndrome caused by dry fermented sausage contaminated with Shiga-like toxin-producing *Escherichia coli. J. Clin. Microbiol.* **34**, 1622–1627.

Paton, A. W., M. C. Woodrow, R. M. Doyle *et al.* (1999). Molecular characterization of a Shiga toxigenic *Escherichia coli* O113:H21 strain lacking *eae* responsible for a cluster of cases of hemolytic-uremic syndrome. *J. Clin. Microbiol.* **37**, 3357–3361.

Paton, J. C., T. J. Rogers, R. Morona and A. W. Paton (2001). Oral administration of formaldehyde-killed recombinant bacteria expressing a mimic of the Shiga toxin receptor protects mice from fatal challenge with Shiga-toxigenic *Escherichia coli. Infect. Immun.* **69**, 1389–1393.

Perna, N. T., G. Plunkett, III, V. Burland *et al.* (2001). Genome sequence of enterohemorrhagic *Escherichia coli* O157:H7. *Nature* **409**, 529–533.

Pickering, L. K., R. L. Guerrant and T. G. Cleary (1995). Microorganisms responsible for neonatal diarrhea. In: J. S. Remington and J. O. Klein (ed.), *Infectious*

Diseases of the Fetus & Newborn Infant, 4th edn, pp. 1142–1222. W. B. Saunders Co., Philadelphia, PA.

Pierson, M. D. and L. M. Smoot (2001). Indicator microorganisms and microbiological criteria. In: M. P. Doyle, L. R. Beuchat and T. J. Montville (eds), *Food Microbiology, Fundamentals and Frontiers*, 2nd edn, pp. 71–87. ASM Press, Washington, DC.

Poitrineau, P., C. Forestier, M. Meyer *et al.* (1995). Retrospective case-control study of diffusely adhering *Escherichia coli* and clinical features in children with diarrhea. *J. Clin. Microbiol.* **33**, 1961–1962.

Potter, A. A., S. Klashinsky, Y. Li *et al.* (2004). Decreased shedding of *Escherichia coli* O157:H7 by cattle following vaccination with type III secreted proteins. *Vaccine* **22**, 362–369.

Pradel, N., S. Leroy-Setrin, B. Joly and V. Liverelli (2002). Genomic subtraction to identify and characterize sequences of Shiga toxin-producing *Escherichia coli* O91:H21. *Appl. Environ. Microbiol.* **68**, 2316–2325.

Pradel, N., C. Ye, V. Livrelli *et al.* (2003). Contribution of the twin arginine translocation system to the virulence of enterohemorrhagic *Escherichia coli* O157:H7. *Infect. Immun.* **71**, 4908–4916.

Pruimbom-Brees, I. M., T. W. Morgan, M. R. Ackermann *et al.* (2000). Cattle lack vascular receptors for *Escherichia coli* O157:H7 Shiga toxins. *Proc. Natl Acad. Sci. USA* **97**, 10 325–10 329.

Qadri, F., C. Wenneras, F. Ahmed *et al.* (2000). Safety and immunogenicity of an oral, inactivated enterotoxigenic *Escherichia coli* plus cholera toxin B subunit vaccine in Bangladeshi adults and children. *Vaccine* **18**, 2704–2712.

Rabinowitz, R. P. and M. S. Donnenberg (1996). *Escherichia coli.* In: L. J. Paradise, M. Bendinelli and H. Friedman (eds), *Enteric Infections and Immunity*, pp. 101–131. Plenum Press, New York, NY.

Reid, S. D., R. K. Selander and T. S. Whittam (1999). Sequence diversity of flagellin (*fliC*) alleles in pathogenic *Escherichia coli. J. Bacteriol.* **181**, 153–160.

Ricci, L. C., F. P. de Faria, P. S. de S. Porto *et al.* (1997). A new fimbrial putative colonization factor (PCF02) in human enterotoxigenic *Escherichia coli* isolated in Brazil. *Res. Microbiol.* **148**, 65–69.

Rich, C., S. Favre-Bronte, F. Sapena *et al.* (1999). Characterization of enteroaggregative *Escherichia coli* isolates. *FEMS Microbiol. Lett.* **173**, 55–61.

Riemann, H., and F. L. Bryan (eds) (1979). *Foodborne Infections and Intoxications*, 2nd edn. Academic Press, New York, NY.

Riordan, T., R. J. Gross, B. Rowe *et al.* (1985). An outbreak of food-borne enterotoxigenic *Escherichia coli* diarrhoea in England. *J. Infect.* **11**, 167–171.

Ritchie, J. M., C. M. Thorpe, A. B. Rogers and M. K. Waldor (2003). Critical roles for stx_2, *eae*, and *tir* in enterohemorrhagic *Escherichia coli*-induced diarrhea and intestinal inflammation in infant rabbits. *Infect. Immun.* **71**, 7129–7139.

Roels, T. H., M. E. Proctor, L. C. Robinson *et al.* (1998). Clinical features of infections due to *Escherichia coli* producing heat-stable toxin during an outbreak in Wisconsin, a rarely suspected cause of diarrhea in the United States. *Clin. Infect. Dis.* **26**, 898–902.

Russo, T. A. and J. R. Johnson (2003). Medical and economic impact of extraintestinal infections due to *Escherichia coli*, focus on an increasingly important endemic problem. *Microbes Infect*. **5**, 449–456.

Ryan, C. A., R. V. Tauxe, G. W. Hosek *et al*. (1986). *Escherichia coli* O157:H7 diarrhea in a nursing home, clinical, epidemiological, and pathological findings. *J. Infect. Dis*. **154**, 631–638.

Ryder, R. W., I. K. Wachsmuth, A. E. Buxton *et al*. (1976). Infantile diarrhea produced by heat-stable enterotoxigenic *Escherichia coli. N. Engl. J. Med*. **295**, 849–853.

Salyers, A. A. and D. D. Whitt (1994). *Bacterial Pathogenesis, A Molecular Approach*. ASM Press, Washington, DC.

Sanderson, M.W., T. E. Besser, J. M. Gay *et al*. (1999). Fecal *Escherichia coli* O157:H7 shedding patterns of orally inoculated calves. *Vet. Microbiol*. **69**, 199–205.

Sarantuya, J., J. Nishi, N. Wakimoto *et al*. (2004). Typical enteroaggregative *Escherichia coli* is the most prevalent pathotype among *E. coli* strains causing diarrhea in Mongolian children. *J. Clin. Microbiol*. **42**, 133–139.

Sargeant, J. M., M. W. Sanderson, R. A. Smith and D. D. Griffin (2003). *Escherichia coli* O157 in feedlot cattle feces and water in four major feeder-cattle states in the USA. *Prev. Vet. Med*. **61**, 127–135.

Savarino, S. J., A. Fasano, D. C. Robertson and M. M. Levine (1991). Enteroaggregative *Escherichia coli* elaborate a heat-stable enterotoxin demonstrable in an in vitro rabbit intestinal model. *J. Clin. Invest*. **87**, 1450–1455.

Savarino, S. J., A. Fasano, J. Watson *et al*. (1993). Enteroaggregative *Escherichia coli* heat-stable enterotoxin 1 represents another subfamily of *E. coli* heat-stable toxin. *Proc. Natl Acad. Sci. USA* **90**, 3093–3097.

Savarino, S. J., A. McVeigh, J. Watson *et al*. (1996). Enteraggregative *Escherichia coli* heat-stable enterotoxin is not restricted to enteroaggregative *E. coli. J. Infect. Dis*. **173**, 1019–1022.

Savarino, S. J., E. R. Hall, S. Bassily *et al*. (1999). Oral, inactivated whole cell enterotoxigenic *Escherichia coli* plus cholera toxin B subunit vaccine, results of the initial evaluation in children. *J. Infect. Dis*. **179**, 107–114.

Scaletsky, I. C. A., M. L. M. Silva and L. R. Trabulsi (1984). Distinctive patterns of adherence of enteropathogenic *Escherichia coli* to HeLa cells. *Infect. Immun*. **45**, 534–536.

Schmidt, H. (2001). Shiga-toxin-converting bacteriophages. *Res. Microbiol*. **152**, 687–695.

Schmidt, H., C. Knop, S. Franke *et al*. (1995). Development of PCR for screening of enteroaggregative *Escherichia coli. J. Clin. Microbiol*. **33**, 701–705.

Schmidt, H., J. Scheef, S. Morabito *et al*. (2000). A new Shiga toxin 2 variant (Stx2f) from *Escherichia coli* isolated from pigeons. *Appl. Environ. Microbiol*. **66**, 1205–1208.

Schroeder, C. M., J. Meng, S. Zhao *et al*. (2002a). Antimicrobial resistance of *Escherichia coli* O26, O103, O111, O128, and O145 from animals and humans. *Emerg. Infect. Dis*. **8**, 1409–1414.

Schroeder, C. M., C. Zhao, C. DebRoy *et al*. (2002b). Antimicrobial resistance of *Escherichia coli* O157 isolated from humans, cattle, swine, and food. *Appl. Environ. Microbiol*. **68**, 576–581.

Shaikh, N. and P. I. Tarr (2003). *Escherichia coli* O157:H7 Shiga toxin-encoding bacteriophages, integrations, excisions, truncations, and evolutionary implications. *J. Bacteriol.* **185**, 3596–3605.

Sharma, V. K. and E. A. Dean-Nystrom (2003). Detection of enterohemorrhagic *Escherichia coli* O157:H7 by using a multiplex real-time PCR assay for genes encoding intimin and Shiga toxins. *Vet. Microbiol.* **93**, 247–260.

Shin, K., Y. Yamauchi, S. Teraguchi *et al.* (1998). Antibacterial activity of bovine lactoferrin and its peptides against enterohemorrhagic *Escherichia coli* O157:H7. *Lett. Appl. Microbiol.* **26**, 407–411.

Smith, H. R. and T. Cheasty (1998). Diarrhoeal diseases due to *Escherichia coli* and *Aeromonas*. In: L. Collier, A. Balows and M. Sussman (eds), *Topley and Wilson's Microbiology and Microbial Infections*, pp.513–537 Arnold, London.

Smith, H. R., T. Cheasty and B. Rowe (1997). Enteraggregative *Escherichia coli* and outbreaks of gastroenteritis in UK. *Lancet* **350**, 814–815.

Smith, J. L (1988). Heat-stable enterotoxins, properties, detection and assay. *Devel. Indust. Microbiol.* **29**, 275–286.

Sperandio, V (2001). Genome sequence of *E. coli* O157:H7. *Trends Microbiol.* **9**, 159.

Sperandio, V., J. L. Mellies, W. Nguyen *et al.* (1999). Quorum sensing controls expression of the type III secretion gene transcription and protein secretion in enterohemorrhagic and enteropathogenic *Escherichia coli. Proc. Natl Acad. Sci. USA* **96**, 15 196–15 201.

Sperandio, V., A. G. Torres, J. A. Giròn, and J. B. Kaper (2001). Quorum sensing is a global regulatory mechanism in enterohemorrhagic *Escherichia coli* O157:H7. *J. Bacteriol.* **183**, 5187–5197.

Sperandio, V., A. G. Torres, B. Jarvis *et al.* (2003). Bacteria-host communication, the language of hormones. *Proc. Natl Acad. Sci. USA* **100**, 8951–8956.

Steiner, T. S., A. A. M. Lima, J. P. Nataro and R. L. Guerrant (1998). Enteroaggregative *Escherichia coli* produce intestinal inflammation and growth impairment and cause interleukin-8 release from intestinal epithelial cells. *J. Infect. Dis.* **177**, 88–96.

Steinsland, H., P.Valentiner-Branth, H. M. S. Grewal *et al.* (2003). Development and evaluation of genotypic assays for the detection and characterization of enterotoxigenic *Escherichia coli. Diagn. Microbiol. Infect. Dis.* **45**, 97–105.

Stephan, R., S. Ragettli and F. Untermann (2000). Prevalence and characteristics of verotoxin-producing *Escherichia coli* (VTEC) in stool samples from asymptomatic human carriers working in the meat processing industry in Switzerland. *J. Appl. Microbiol.* **88**, 335–341.

Su, C. and L. J. Brandt (1995). *Escherichia coli* O157:H7 infection in humans. *Ann. Intern. Med.* **123**, 698–714.

Sueyoshi, M. and M. Nakazawa (1994). Experimental infection of young chicks with attaching and effacing *Escherichia coli. Infect. Immun.* **62**, 4066–4071.

Sugiyama, K., K. Inoue and R. Sakazaki (2001). Mitomycin-supplemented washed blood agar for the isolation of Shiga toxin-producing *Escherichia coli* other than O157:H7. *Lett. Appl. Microbiol.* **33**, 193–195.

Sussman, M (1985). *Escherichia coli* in human and animal disease. In: M. Sussman (ed.), *The Virulence of* Escherichia coli, *Reviews and Methods*, pp. 7–45. Academic Press, New York, NY.

Swaminathan, B., T. J. Barrett, S. B. Hunter, R. V. Tauxe and the PulseNet Task Force (2001). PulseNet, the molecular subtyping network for foodborne bacterial disease surveillance, United States. *Emerg. Infect. Dis.* **7**, 382–389.

Takahashi, I., M. Marinaro, H. Kiyono *et al.* (1996). Mechanisms for mucosal immunogenicity and adjuvancy of *Escherichia coli* labile enterotoxin. *J. Infect. Dis.* **173**, 627–635.

Takeda, T., K. Yoshino, E. Adachi *et al.* (1999). *In vitro* assessment of a chemically synthesized Shiga toxin receptor analog attached to Chromosorb P (Synsorb Pk) as a specific absorbing agent of Shiga toxin 1 and 2. *Microbiol. Immunol.* **43**, 331–337.

Tana, S. Watarai, E. Isogai, and K. Oguma (2003). Induction of intestinal IgA and IgG antibodies preventing adhesion of verotoxin-producing *Escherichia coli* to Caco-2 cells by oral immunization with liposomes. *Lett. Appl. Microbiol.* **36**, 135–139.

Taneike, I., H.-M. Zhang, N. Wakisaka-Saito and T. Yamamoto (2002). Enterohemolysin operon of Shiga toxin-producing *Escherichia coli*, a virulence function of inflammatory cytokine production from human monocytes. *FEBS Lett.* **524**, 219–224.

Tarr, P. I., L. M. Schoening, Y.-L. Yea *et al.* (2000). Acquisition of the *rfb-gnd* cluster in evolution of *Escherichia coli* O55 and O157. *J. Bacteriol.* **182**, 6183–6191.

Tatsuno, I., M. Horie, H. Abe *et al.* (2001). *toxB* gene on pO157 of enterohemorrhagic *Escherichia coli* O157:H7 is required for full epithelial cell adherence phenotype. *Infect. Immun.* **69**, 6660–6669.

Tauschek, M., R. A. Strugnell and R. M. Robins-Browne (2002). Characterization and evidence of mobilization of the LEE pathogenicity island of rabbit-specific strains of enteropathogenic *Escherichia coli. Mol. Microbiol.* **44**, 1533–1550.

Taylor, D. E., M. Rooker, M. Keelan *et al.* (2002). Genomic variability of O islands encoding tellurite resistance in enterohemorrhagic *Escherichia coli* O157:H7 isolates. *J. Bacteriol.* **184**, 4690–4698.

Taylor, W. R., W. L. Schell, J. G. Wells *et al.* (1982). A foodborne outbreak of enterotoxigenic *Escherichia coli* diarrhea. *N. Engl. J. Med.* **306**, 1093–1095.

Tilden, J., W. Young, A. M. McNamara *et al.* (1996). A new route of transmission for *Escherichia coli*, infection from dry fermented salami. *Am. J. Public Health* **86**, 1142–1145.

Trachtman, H. and E. Christen (1999). Pathogenesis, treatment, and therapeutic trials in hemolytic uremic syndrome. *Curr. Opin. Pediatr.* **11**, 162–168.

Tsen, H.Y., L.Z. Jian and W.R. Chi (1998). Use of a multiplex PCR system for the simultaneous detection of heat labile toxin I and heat stable toxin II genes of enterotoxigenic *Escherichia coli* in skim milk and porcine stool. *J. Food Prot.* **61**, 141–145.

Tsuji, T., Y. H. Kamiya, Y. Kawamoto and Y. Asano (1997). Relationship between a low toxicity of the mutant A subunit of enterotoxigenic *Escherichia coli* enterotoxin and its strong adjuvant action. *Immunology* **90**, 176–182.

Tuttle, J., T. Gomez, M. P. Doyle *et al.* (1999). Lessons from a large outbreak of *Escherichia coli* O157:H7 infections, insights into the infectious dose and method of widespread contamination of hamburger patties. *Epidmiol. Infect.* **122**, 185–192.

Tzipori, S., J. Montanaro, R. M. Robins-Browne *et al.* (1992). Studies with enteroaggregative *Escherichia coli* in the gnotobiotic piglet gastroenteritis model. *Infect. Immun.* **60**, 5302–5306.

Tzipori, S., F. Gunzer, M. S. Donnenberg *et al.* (1995). The role of the *eaeA* gene in diarrhea and neurological complications in a gnotobiotic piglet model of enterohemorrhagic *Escherichia coli* infection. *Infect. Immun.* **63**, 3621–3627.

Vallance, B. A. and B. B. Finlay (2000). Exploitation of host cells by enteropathogenic *Escherichia coli. Proc. Natl Acad. Sci. USA* **97**, 8799–8806.

Vallence, B. A., C. Chan, M. L. Robertson and B. B. Finlay (2002). Enteropathogenic and enterohemorrhagic *Escherichia coli* infections, Emerging themes in pathogenesis and prevention. *Can. J. Gastroenterol.* **16**, 771–778.

Viboud, G.I., N. Binsztein and A.-M. Svennerholm (1993). Characterization of monoclonal antibodies against putative colonization factors of enterotoxigenic *Escherichia coli* and their use in an epidemiological study. *J. Clin. Microbiol.* **31**, 558–564.

Walmsley, A. M. and C. J. Arntzen (2000). Plants for delivery of edible vaccines. *Curr. Opin. Biotechnol.* **11**, 126–129.

Wang, G., C. G. Clark and F. G. Rodgers (2002). Detection in *Escherichia coli* of the genes encoding the major virulence factors, the genes defining the O157:H7 serotype, and the components of the type 2 Shiga toxin family by multiplex PCR. *J. Clin. Microbiol.* **40**, 3613–3619.

Wang, L., W. Qu and P.R. Reeves (2001a). Sequence analysis of four *Shigella boydii* O-antigen loci, implication for *Escherichia coli* and *Shigella* relationships. *Infect. Immun.* **69**, 6923–6930.

Wang, L., C. E. Briggs, D. Rothemund *et al.* (2001b). Sequence of the *E. coli* O104 antigen gene cluster and identification of O104 specific genes. *Gene* **270**, 231–236.

Wang, L., S. Huskic, A. Cisterne *et al.* (2002). The O-antigen gene cluster of *Escherichia coli* O55:H7 and identification of a new UDP-GlcNAc C4 epimerase gene. *J. Bacteriol.* **184**, 2620–2625.

Warburton, D. W., J. W. Austin, B. H. Harrison and G. Sanders (1998). Survival and recovery of *Escherichia coli* O157:H7 in inoculated bottled water. *J. Food Prot.* **61**, 948–952.

Werber, D., A. Fruth, A. Liesegang *et al.* (2002). A multistate outbreak of Shiga toxin-producing *Escherichia coli* O26:H11 infections in Germany, detected by molecular subtyping surveillance. *J. Infect. Dis.* **186**, 419–422.

Wilkerson, C. and N. van Kirk (2004). Antibiotic resistance and distribution of tetracycline resistance genes in *Escherichia coli* O157:H7 isolates from humans and bovines. *Antimicrob. Agents Chemother.* **48**, 1066–1067.

Willshaw, G. A., T. Cheasty and H. R. Smith (2000). *Escherichia coli.* In: B. M. Lund, T. C. Baird-Parker and G. W. Gould (eds), *The Microbiological Safety and Quality of Food*, Vol II, pp. 1136–1177. Aspen Publishers, Gaithersburg, MD.

Wood, L. V., W. H. Wolfe, G. Ruiz-Palacios *et al.* (1983). An outbreak of gastroenteritis due to a heat-labile enterotoxin-producing strain of *Escherichia coli. Infect. Immun.* **41**, 931–934.

Woods, J. B., C. K. Schmitt, S. C. Darnell *et al.* (2002). Ferrets as a model system for renal disease secondary to intestinal infection with *Escherichia coli* O157, H7 and other Shiga toxin-producing *E. coli. J. Infect. Dis.* **185**, 550–554.

World Health Organization (1998). Zoonotic non-O157 Shiga toxin-producing *Escherichia coli* (STEC). Report of a WHO Scientific Working Group Meeting, 23–26 June 1998, Berlin, Germany (available online at http, //www.who.int/emcdocuments/zoonoses/docs/whocsraph988.html/shigaindex.html).

Yamagami, S., M. Motoki, T. Kimura *et al.* (2001). Efficacy of postinfection treatment with anti-Shiga toxin (Stx) 2 humanized monoclonal antibody TMA-15 in mice lethally challenged with Stx-producing *Escherichia coli. J. Infect. Dis.* **184**, 738–742.

Yamamoto, T., and P. Echeverria (1996). Detection of the enteroaggregative *Escherichia coli* heat-stable enterotoxin 1 gene sequence in enterotoxigenic *E. coli* strains pathogenic to humans. *Infect. Immun.* **64**, 1441–1445.

Yamamoto, T. and M. Nakazawa (1997). Detection and sequences of enteroaggregative *Escherichia coli* heat-stable enterotoxin 1 gene in enterotoxigenic *E. coli* strains isolated from piglets and calves with diarrhea. *J. Clin. Microbiol.* **35**, 223–227.

Yamanaka, H., M. Kameyama, T. Baba *et al.* (1994). Maturation pathway of *Escherichia coli* heat-stable enterotoxin I, requirements of DsbA for disulfide bond formation. *J. Bacteriol.* **176**, 2906–2913.

Yamanaka, H., T. Nomura, Y. Fujii and K. Okamoto (1997). Extracellular secretion of *Escherichia coli* heat-stable enterotoxin I across the outer membrane. *J. Bacteriol.* **179**, 3383–3390.

Yamanaka, H., T. Nomura, Y. Fujii and K. Okamoto (1998). Need for TolC, an *Escherichia coli* outer membrane protein, in the secretion of heat-stable enterotoxin I across the outer membrane. *Microbial Pathogen.* **25**, 111–120.

Yaron, S., G. I. Kolling, L. Simon and K. R. Matthews (2000). Vesicle-mediated transfer of virulence genes from *Escherichia coli* O157:H7 to other enteric bacteria. *Appl. Environ. Microbiol.* **66**, 4414–4420.

Zhang, W., M. Bielaszewska, T. Kuczius and H. Karch (2002). Identification, characterization, and distribution of a Shiga toxin gene variant (Stx_{1c}) in *Escherichia coli* strains isolated from humans. *J. Clin. Microbiol.* **40**, 1441–1446.

Zhao, T. and M. P. Doyle (1994). Fate of enterohemorrhagic *Escherichia coli* O157:H7 in commercial mayonnaise. *J. Food Prot.* **57**, 780–783.

Zhao, T., M. P. Doyle and R. Besser (1993). Fate of enterohemorrhagic *Escherichia coli* O157:H7 in apple cider with and without preservatives. *Appl. Environ. Microbiol.* **59**, 2526–2530.

Campylobacter and related infections

7

Sean F. Altekruse and Guillermo I. Perez-Perez

1 *Campylobacter jejuni*

1.1 Historical aspects and current problems

The first description of a bacterium now belonging to the genus *Campylobacter* is attributed to Theodore Escherich, at the end of the nineteenth century (Escherich, 1886). At the beginning of the twentieth century a *Campylobacter* species, described as a related *Vibrio*, was recognized as causing abortions in sheep. Only after a suitable isolation medium was developed in the 1970s were two closely related pathogens, *C. jejuni* and *C. coli*, recognized to be common human enteric pathogens (DeKeyser *et al.*, 1972). *C. jejuni* accounts for approximately 90 % of human *Campylobacter* infections (Kramer *et al.*, 2000).

In 1999, the Centers for Disease Control and Prevention (CDC) estimated that there were 2.5 million cases of human campylobacteriosis in the United States each year (Mead *et al.*, 1999). While the incidence of campylobacteriosis declined by 27 % in FoodNet surveillance sites between 1996 and 2001, *Campylobacter* spp. remains one of the most common bacterial foodborne infectious diseases in the United States (CDC, 2002a). Post-infectious sequelae of infection include Guillain-Barré syndrome and reactive arthritis. Current issues for the prevention of this zoonosis include implementation of pathogen reduction measures along the food chain, zoonotic infections, and infections caused by fluoroquinolone resistant *Campylobacter* strains.

Foodborne Infections and Intoxications 3e
ISBN-10: 0-12-588365-X
ISBN-13: 978-012-588365-8

1.2 Characteristics of *C. jejuni*, classification and virulence factors

1.2.1 Classification

Campylobacter and other related genera (i.e. *Arcobacter, Helicobacter*) form DNA Superfamily IV (On, 2001). *C. jejuni* and *C. coli* are closely related to one another. Like most of the characterized species in this taxonomic family, they are recognized to be human or animal pathogens and have fastidious growth requirements reflecting dependence on a warm-blooded host for replication (On and Harrington, 2001).

1.2.2 Pathogenesis, survival in the environment

The sequencing of the *C. jejuni* genome (Parkhill *et al.*, 2000) is an important milestone for understanding *Campylobacter* pathogenesis. Already it is clear that the 1.5 million base-pair *C. jejuni* genome is small, with few homologues of the virulence determinants in other foodborne pathogens. Suspected virulence determinants of *C. jejuni* include motility, adherence, exotoxin production, iron regulation and cell invasion.

C. jejuni does not normally replicate outside the intestinal tract of warm-blooded animals (Nachamkin, 2003). The infectious dose is reported to be less than 1000 organisms (Black *et al.*, 1988). Other adaptations to an intestinal niche include a single polar flagellum and the cell's corkscrew shape (Figure 7.1). These traits facilitate motility in the viscous intestinal mucus. Requirements for growth in the laboratory (Nachamkin, 2003) also reflect this narrow ecologic niche – a micro-aerophilic nitrogen atmosphere with low oxygen (5–7 %) and high carbon dioxide tension (7–13 %). *C. jejuni* is unable to replicate at temperatures below the body temperature of warm-blooded animals (approximately 30°C), or at a pH < 4.9. The organism is also sensitive to desiccation and osmotic stress (e.g. NaCl concentrations above 2 %).

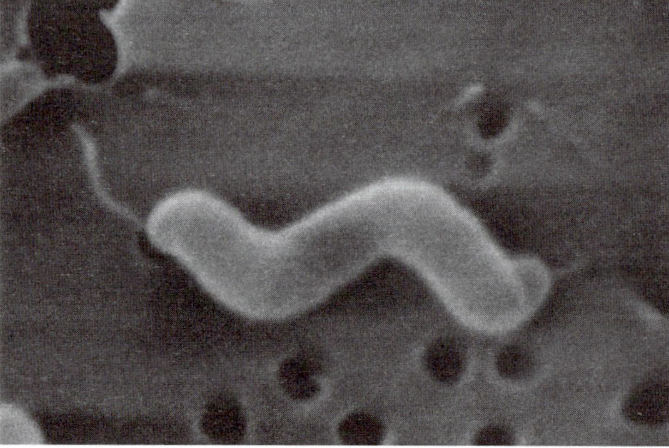

Figure 7.1 Scanning electron micrograph illustrating the single polar flagella and corkscrew shape of *C. jejuni*. These morphologic characteristics contribute to the characteristic 'darting' motility of *C. jejuni* in the viscous mucous layer of the intestinal lumen.

C. jejuni gradually die outside the host intestinal tract. In one study, 58 of 85 (68 %) *C. jejuni* strains could not be cultured from water after 3 weeks; however, a few strains were detected in unstirred water after 60 days (Talibart *et al.*, 2000). Environmental factors may facilitate *Campylobacter* survival under adverse conditions. Survival times are longer in nutrient-rich water than in de-ionized water (Thomas *et al.*, 1999). Similarly, biofilms are reported to facilitate the survival of *C. jejuni* in broiler houses (Trachoo *et al.*, 2002). Some researchers postulate that campylobacters can survive in water in a viable but non-cultivable form (Cappelier *et al.*, 2000; Thomas *et al.*, 2002); however, the role of this dormant stage in the *Campylobacter* life cycle is controversial (van de Giessen *et al.*, 1996a).

1.3 Nature of infection in man and animals

1.3.1 Man

1.3.1.1 The acute clinical illness

While *C. jejuni* and *C. coli* can exist as commensal organisms of domestic poultry and livestock, they are considered human pathogens. The clinical spectrum of human *Campylobacter* enteritis ranges from loose stools to dysentery. Self-limiting acute enteritis is the most common syndrome. Prodromal symptoms are common, and include headache, low fever, and myalgia lasting from a few hours to a few days. Symptoms of acute infection often begin with abdominal cramps followed by diarrhea and high fever, peaking during the first days of illness (Blaser, 1997). *C. jejuni*-specific serum antibodies confer immunity to symptomatic infection; however, the duration of protective immunity is not known (Blaser *et al.*, 1987; Walz *et al.*, 2001).

An estimated 100 fatal *C. jejuni* infections occur each year in the United States. These fatal infections occur most often in infants, in the elderly, or in immunosuppressed individuals (Mead *et al.*, 1999). Bacteremia is most often seen in patients with underlying disease (Pigrau *et al.*, 1997), and is a potentially fatal complication of HIV/AIDS (Manfredi *et al.*, 1999). Chronic diarrhea is also a complication of HIV-associated campylobacteriosis. HIV-positive individuals who develop campylobacteriosis have shorter survival times and higher rates of bacteremia and hospitalization than HIV-positive individuals without campylobacteriosis (Angulo and Swerdlow, 1995). This aspect of campylobacteriosis has major public health significance in developing nations (Coker *et al.*, 2002).

1.3.1.2 Sequelae to infection

Sequelae to infection include Guillain-Barré syndrome and reactive arthritis.

With several thousand cases each year, *Guillain-Barré syndrome* (GBS) is the most common etiology of acute flaccid paralysis in the United States (Nachamkin, 2002). Guillain-Barré syndrome (GBS) is an acute immune-mediated disorder of the peripheral nervous system. Leg weakness is often the presenting sign, followed by ascendant paralysis. After one year, 70 % of patients make complete neurological recovery, 22 % partially recover, 8 % remain unable to walk, and 2 % remain bedridden or require ventilation. Most cases of GBS are believed to follow an infectious disease. Approximately 40 % of GBS cases are thought to follow *Campylobacter* infection, and

GBS is estimated to occur in 1 in 1000 patients infected with *Campylobacter*. Although a diverse group of strains is associated with GBS (Dingle *et al.*, 2001), the syndrome is strongly linked to a few strains of *C. jejuni*, such as Penner serotypes HS:19 (Yuki *et al.*, 1997) and HS:41 (Prendergast *et al.*, 1998). *Campylobacter* strains contain sialic acid linkages to lipo-oligosaccharides resembling sialic acid moieties on the gangliosides of peripheral nerve tissues (Ho *et al.*, 1999). Patients with GBS develop antibodies against these gangliosides, resulting in autoimmune targeting of peripheral nerve sites. Complement-mediated damage (Hafer-Macko *et al.*, 1996) and blockage of neuro-transmission (Goodyear *et al.*, 1999) are suspected to play a role in GBS pathogenesis.

Since many individuals are exposed to *C. jejuni* strains that mimic gangliosides and only a few develop GBS, it is suspected that host factors contribute to GBS. In one study (Rees *et al.*, 1995) *Campylobacter*-related GBS was associated with major histo-compatibility antigen, HLA-DQB1*03; however, this association was not replicated in another well-designed study (Yuki *et al.*, 1997). Proposed treatments for GBS have not been fully evaluated in clinical trials, but include corticosteroid therapy, immunoglobulin therapy and plasmapheresis (Hughes, 2002).

Reactive arthritis, or Reiter's syndrome, is another sterile post-infectious sequela to acute gastrointestinal campylobacteriosis. Onset of reactive arthritis occurs 7–10 days after onset of diarrheal illness. The frequency of reactive arthritis as a sequela of campylobacteriosis has not been well described in the US. In Finland, 45 of 870 (7 %) patients with laboratory confirmed campylobacteriosis developed reactive arthritis (Hannu *et al.*, 2002). Arthritis was oligo- or polyarticular, and in most cases mild. In the Finnish study, 37 of the 45 patients (82 %) had *C. jejuni* and 8 (18 %) had *C. coli* infections. No cases of reactive arthritis occurred in children. In a Danish study, patients with joint pain had more severe gastrointestinal symptoms and longer duration of diarrhea than those without joint pain. Anti-*Campylobacter* antibody levels were similar in both patient groups. Antibiotic treatment did not prevent reactive arthritis (Locht and Krogfelt, 2002).

1.3.1.3 Treatment of acute campylobacteriosis

Supportive measures, particularly fluid and electrolyte replacement, are the principal therapies for most patients with campylobacteriosis. Severely dehydrated patients should receive rapid volume restoration with intravenous fluids. For most other patients, oral rehydration is indicated. Although *Campylobacter* infections are usually self-limiting, antibiotic therapy may be prudent for patients who have high fever, bloody diarrhea or more than eight stools in 24 hours, immunosuppressed patients, patients with bloodstream infections, and those whose symptoms worsen or persist for a week or more from the time of diagnosis. When indicated, antimicrobial therapy soon after the onset of symptoms can reduce the median duration of illness from approximately 10 days to 5 days. When treatment is delayed (e.g. until *C. jejuni* infection is confirmed by a medical laboratory), therapy may not be successful (Smith *et al.*, 1999). Ease of administration, lack of serious toxicity and a high degree of efficacy made erythromycin the historical drug of choice for *C. jejuni* infection; however, other antimicrobial agents, particularly the quinolones and newer macrolides (including azithromycin), are also widely used.

In 2005 the US Food and Drug Administration banned the use of fluoroquinolones in poultry in response to the emergence of fluoroquinolone resistant *Campylobacter* strains as a cause of human infections in the United States. The FDA partially attributed this trend to veterinary use of fluoroquinolones, concluding that use of fluoroquinolones in poultry compromises the clinical utility of fluoroquinolones in humans. An FDA risk assessment estimated that each year thousands of people infected with fluoroquinolone resistant *Campylobacter* strains are treated with a fluoroquinolone, resulting in prolonged duration of illness (Food and Drug Administration, 2000).

Fluoroquinolone resistant *Campylobacter* strains were not detected in the United States in 1986 – the year when this class of antimicrobials was first introduced for human use (Friedman *et al.*, 2001). Resistance rates increased to 5 % in the next few years. The proportion of *Campylobacter* isolates from humans that exhibited resistance to fluoroquinolones increased after 1995, around the time when fluoroquinolones were approved for the treatment of colibacillosis in poultry flocks. Since 1997, 14–18 % of *Campylobacter* strains isolated from humans in the United States have been resistant to ciprofloxacin (Marano *et al.*, 2000). A study by the Minnesota Department of Health suggested that the epidemiology of infection with fluoroquinolone resistant *Campylobacter* strains shifted, beginning in 1995 with the emergence of a domestic reservoir of fluoroquinolone resistant *C. jejuni*. In Minnesota the molecular subtypes of fluoroquinolone resistant *C. jejuni* strains isolated from humans who had not traveled outside the US matched the molecular subtypes of fluoroquinolone resistant *C. jejuni* isolated from locally purchased retail poultry products (Smith *et al.*, 1999). An increase in the frequency of infections with fluoroquinolone resistant strains was also observed by the National Antimicrobial Resistance Monitoring System (NARMS). Case-control studies conducted by NARMS showed that chicken consumption is an important risk factor for infection with domestically acquired fluoroquinolone resistant strains. Infections with such strains were also associated with longer duration of diarrhea and increased likelihood of hospitalization (CDC, 2002b).

A report by Engberg and colleagues documented the emergence of fluoroquinolone resistant *Campylobacter* strains as a cause of human infection in 10 developing nations during the 1990s in relation to the approval of this class of antimicrobial drugs for use in veterinary practice (Engberg *et al.*, 2001). Most fluoroquinolone resistance is caused by spontaneous point mutations in the DNA gyrase A subunit region that alter the fluoroquinolone-binding site (Hooper *et al.*, 1987). Strains with this mutation have elevated minimum inhibitory concentrations (MIC). This attribute confers selective advantage to the bacterium in the presence of fluoroquinolones (McDermott *et al.*, 2002).

1.3.1.4 Risk factors for human illness

- *Poultry consumption*. The initial epidemiologic studies of sporadic campylobacteriosis conducted in the United States (Harris *et al.*, 1986; Deming *et al.*, 1987; Hopkins and Scott, 1993) and western Europe (Norkrans and Svedhem, 1982; Oosterom *et al.*, 1984; Kapperud *et al.*, 1992) revealed robust associations with the handling

(Norkrans and Svedhem, 1982; Hopkins and Scott, 1993) or eating (Oosterom *et al.*, 1984; Deming *et al.*, 1987; Kapperud *et al.*, 1992) of poultry, particularly undercooked poultry (Hopkins *et al.*, 1984; Harris *et al.*, 1986). More recent epidemiologic studies in the United States (Effler *et al.*, 2001), the United Kingdom (Rodrigues *et al.*, 2001) and New Zealand (Eberhart-Phillips *et al.*, 1997) confirmed the association between human campylobacteriosis and poultry consumption, and added an additional nuance – an association between *Campylobacter* infection and eating commercially prepared poultry (Effler *et al.*, 2001). These associations are not unexpected, given that the majority of chicken in stores is contaminated with *C. jejuni* (Zhao *et al.*, 2001). Molecular subtyping studies demonstrate partial correspondence between poultry and human isolates (Smith *et al.*, 1999; Hanninen *et al.*, 2000). In Quebec, 20 % of genotypes from humans and poultry had matching pulsed field gel electrophoresis patterns (Nadeau *et al.*, 2002).

- *Commercially prepared foods*. Case-control studies in the United States and other developed nations indicate that eating chicken in restaurants is associated with increased risk of infection (Eberhart-Phillips *et al.*, 1997; Effler *et al.*, 2001; Rodrigues *et al.*, 2001). On occasions, other foods prepared in restaurants or commercial kitchens have been implicated in outbreaks of camplylobacteriosis, including tuna salad (Roels *et al.*, 1998), sweet potatoes (Winquist *et al.*, 2001), and lettuce (CDC, 1998). Cross-contamination during food preparation is suspected to be a contributory factor in such outbreaks, and it has been clearly shown that *C. jejuni* can survive on food contact surfaces and thereby cross-contaminate other foods (Cogan *et al.*, 1999).
- *Other food items*. In addition to poultry, several types of meat have been epidemiologically implicated as sources of human campylobacteriosis in developed nations. Some of these implicated food items include pork loins, barbequed foods (Studahl and Andersson, 2000), and liver pâté (Gillespie *et al.*, 2002).
- *Unpasteurized milk*. Drinking unpasteurized milk is the principal risk factor for outbreaks of campylobacteriosis. Between 1981 and 1990, 20 outbreaks of *Campylobacter* enteritis were reported in the United States (Wood *et al.*, 1992). Of these 20 outbreaks, 14 (70 %) occurred among children who drank unpasteurized milk on school field trips or other youth activities. Unlike sporadic *Campylobacter* infections; which peak during the summer and are also associated with activities such as eating chicken, eating at restaurants and international travel, milk-associated outbreaks have a bimodal seasonality, with peaks during the spring and fall corresponding with the peak seasons for youth activities such as school field trips. Despite regulatory efforts to address the hazard, unpasteurized milk-associated outbreaks continue to occur (CDC, 2002c). Recently, molecular typing studies have linked outbreak-associated infectious strains with unpasteurized milk from implicated dairies (Lehner *et al.*, 2000; CDC, 2002c).
- *Water*. One of the first case-control studies of campylobacteriosis, conducted in Colorado (Hopkins *et al.*, 1984), found an association with consumption of untreated surface water. More recently, a study conducted in England found that people with *C. coli* infection were more likely to report drinking bottled water than were those with *C. jejuni* infection (Gillespie *et al.*, 2002). Waterborne outbreaks of

campylobacteriosis typically involve lapses in community water sanitation (Sacks *et al.*, 1986; Melby *et al.*, 2000). The proportion of *Campylobacter* infections caused by contaminated water is likely to vary by region, and with economic development.

- *Zoonotic transmission.* Case-control studies identify contact with pet dogs and cats, and especially juvenile or diarrheic pets, as risk factors for *Campylobacter* infection, accounting for perhaps 5 % of human campylobacteriosis (Norkrans and Svedhem, 1982; Hopkins *et al.*, 1984; Deming *et al.*, 1987; Kapperud et al., 1992; Saeed *et al.*, 1993). The hazard of zoonotic campylobacteriosis may be greatest for young children, an age group with elevated rates of campylobacteriosis (Tauxe *et al.*, 1988). In one case report, a 3-week-old girl in a household with a recently acquired Labrador retriever puppy developed bloodstream *C. jejuni* infection. Amplified fragment-length polymorphism (AFLP) analysis confirmed that the human and canine isolates were genetically similar (Wolfs *et al.*, 2001). In an Australian case-control study, children less than 3 years of age who lived in a home with a pet puppy had a 17-fold increase in risk of campylobacteriosis compared to children with no puppy. Elevated risk of pediatric campylobacteriosis was also associated with pet chicken ownership (Tenkate and Stafford, 2001). Occupational risk factors for campylobacteriosis include farm residence, poultry-related occupations, and daily contact with chickens (Studahl and Andersson, 2000).
- *Foreign travel.* Foreign travel is a commonly reported risk factor for campylobacteriosis (Eberhart-Phillips *et al.*, 1997; Smith *et al.*, 1999; Rodrigues *et al.*, 2001). In nations where campylobacters are uncommon in chicken (i.e. some Scandinavian nations), international travel is the dominant source of human *Campylobacter* infections (Norkrans and Svedhem, 1982). In the United States it is estimated that between 20 and 25 % of *Campylobacter* infections are acquired during international travel (CDC, 2002a). Campylobacteriosis was the most frequently reported enteric bacterial infection in Austrian tourists returning from Southern Europe and Asia (Reinthaler *et al.*, 1998). In England, travel to South Africa was associated with *C. coli* infection (Gillespie *et al.*, 2002). Causal exposures (e.g. food, beverage, dining venue, antimicrobial usage, animal contact) for travel-associated infections remain to be determined.
- *Antimicrobial usage.* In a Hawaiian case-control study, antibiotic use in the month before onset of illness was associated with campylobacteriosis – a unique finding in studies to date (Effler *et al.*, 2001). One hypothesis for the observation is that antimicrobial usage lowers the infectious dose of drug resistant *C. jejuni* strains. Another potential explanation for the above finding is that the use of antibiotics may alter colonic flora, resulting in decreased resistance to infection even with antimicrobial susceptible *C. jejuni* strains.

1.3.2 Campylobacter ecology

1.3.2.1 Wildlife reservoirs

The evidence that wildlife is a reservoir for human *Campylobacter* infections is somewhat equivocal. To be a substantial source of human infections, feces from wildlife would need to enter the human food or water supply. Although some wild birds are colonized with *Campylobacter*, a Danish study of *C. jejuni* isolates indicated that the serotype distribution in wildlife was different from the distributions in broiler chickens and humans (Petersen *et al.*, 2001). *C. jejuni* contamination rates in wild bird-associated specimens vary markedly, between 0 and 50 % in one US study (Craven *et al.*, 2000). The finding of *Campylobacter* in wildlife may also indicate contact with food animals. In a study from Japan, 3 of 13 *C. jejuni* isolates from sparrows exhibited quinolone resistance, suggesting that sparrows may also acquire campylobacters from food animal populations (Chuma *et al.*, 2000).

1.3.2.2 Poultry

Although not all flocks become colonized, *C. jejuni* is introduced into many broiler flocks during the production cycle (Wedderkopp *et al.*, 2000; Denis *et al.*, 2001; Stern *et al.*, 2001). Risk factors for flock colonization include season, caretakers who work with other animals, and drinking-water sanitation (Denis *et al.*, 2001; Kapperud *et al.*, 1993; van de Giessen *et al.*, 1996b). Associations are also reported with type of air handling system and level of beetle infestation (Refregier-Petton *et al.*, 2001).

Infections spread rapidly within flocks after introduction. Colonization typically occurs by 3 to 4 weeks of age (Evans and Sayers, 2000; Shreeve *et al.*, 2000). While most campylobacters do not survive in cleaned and disinfected houses (Evans and Sayers, 2000), certain strains appear to persist in successive broiler flock rotations (Petersen and Wedderkopp, 2001). Recent studies suggest that the crop is among the most frequently infected organ of broilers entering processing plants, with overall crop carriage rates in the order of 60 % (Byrd *et al.*, 1998), similar to the frequency of contamination reported on broiler carcasses after processing (Zhao *et al.*, 2001).

1.3.2.3 Cattle, pigs, and sheep

Campylobacter species often inhabit the bovine intestinal tract, particularly of calves. In a Swiss study, the overall prevalence of *C. jejuni* in calves during the first 3 months of life on large cow-calf farms was 39 % (Busato *et al.*, 1999). In a Danish study, 20 of 24 cattle herds were positive and young animals had a higher prevalence than older animals (Nielsen, 2002). In 40 % of positive herds, all *C. jejuni* isolates had the identical serotype and PFGE type. *Campylobacter* prevalence in a multiple herd study of adult beef cattle in California was 5 % (Hoar *et al.*, 2001). The number of adult cows on the farm was positively associated with the proportion that tested positive.

In a study of sheep raised around Lancaster, England, *C. jejuni* accounted for 90 % of all campylobacters. Overall, 92 % of sheep were carriers at slaughter. Between one-quarter and one-half shed campylobacters while on pasture. The highest rate of shedding (100 %) coincided with stresses of lambing, weaning, and movement onto new pasture; and no shedding was detected among sheep fed on hay or silage (Jones *et al.*, 1999).

C. coli is the predominant *Campylobacter* species of swine. In a study of fecal specimens from healthy Belgian swine, 61 of 65 (94 %) *Campylobacter* isolates were *C. coli* (Van Looveren *et al.*, 2001). A Dutch study suggests that breeding management can eliminate most *C. coli* by breaking the chain of transmission from sows to piglets (Weijtens *et al.*, 2000).

1.4 *Campylobacter* in food and water

1.4.1 Food

Retail food surveillance programs in developed nations provide valuable data on food-borne hazards by type of retail meat and poultry product. In 2002, FoodNet sites in the United States began routine retail food surveys to compare genotypic and antimicrobial resistance patterns of campylobacters from human and food isolates. In an English study of nearly 500 retail specimens, chicken meat had the highest contamination rate (83 %), followed by lamb liver (73 %), pork liver (72 %) and beef liver (54 %). *C. jejuni* predominated in chicken meat (77 %), while *C. coli* predominated in pork liver (42 %) (Kramer *et al.*, 2000). In metropolitan Washington, DC, 130 of 184 (70 %) packages of chicken sold at retail outlets contained *C. jejuni* or *C. coli*, followed by 4 % of 172 turkey, and less than 2 % of pork and beef (Zhao *et al.*, 2001). In a study of more than 2000 lamb carcasses from six large processing plants, less than 1 % of carcasses were contaminated with *C. jejuni* or *C. coli* (Zhao *et al.*, 2001).

1.4.2 Milk and water

Surveys of bulk tank milk indicate that unpasteurized milk is a source of *C. jejuni*. In one study, approximately 10 % of unpasteurized milk specimens from dairy bulk tanks were contaminated with *C. jejuni* (Jayarao and Henning, 2001). Surface waters are often contaminated with campylobacters. In a Norwegian study, 32 of 60 water specimens from the Bo River contained campylobacters, and *C. coli* was detected more often than *C. jejuni* (Rosef *et al.*, 2001). In that study, fecal coliform counts were not a reliable indicator of low-level *Campylobacter* contamination.

1.5 Culturing

1.5.1 Specimen transport

Campylobacters are sensitive to stress and die outside their host, so stool specimens should be chilled if possible (not frozen) and submitted to a laboratory within 24 hours of collection. Storing specimens in deep, airtight containers minimizes exposure to oxygen, and desiccation. If a specimen cannot be processed within 24 hours or is likely to contain small numbers of organisms, a rectal swab placed in a specimen transport medium (e.g. Cary-Blair) should be used. Individual laboratories can provide guidance on specimen-handling procedures (Nachamkin, 2003).

1.5.2 Culture

Numerous procedures are available for recovering *C. jejuni* from clinical specimens. Direct plating is cost-effective for testing large numbers of specimens; however, testing sensitivity may be reduced. Pre-enrichment (raising the temperature from 36°C to 42°C over several hours), filtration, or both are used in some laboratories to improve recovery of stressed bacteria from specimens (e.g. stored foods or swabs exposed to oxygen). Isolation can be facilitated by use of a selective media containing a combination of antimicrobial agents such as cephalothin, oxygen quenching agents, and a low oxygen atmosphere (Nachamkin, 2003).

1.5.3 Polymerase chain reaction (PCR)

The polymerase chain reaction provides an important alternative to traditional microbiological culture techniques for detection (Denis *et al.*, 2001) and characterization (Chuma *et al.*, 2000; Wang *et al.*, 2002) of *Campylobacter* strains. In one study, when PCR ELISA was performed on samples of 48-hour enrichment cultures of foods, it was 99 % sensitive and 96 % specific for the detection of *C. jejuni* and *C. coli* compared to selective culture (Bolton *et al.*, 2002). Multiplex polymerase chain reaction assays can also be used to confirm the identity of a *Campylobacter* isolate among the five clinically most important species: *C. jejuni, C. coli, C. lari, C. upsaliensis* and *C. fetus* subsp. *fetus* (Wang *et al.*, 2002). Advantages of multiplex PCR over traditional biochemical tests for characterization of campylobacter strains include rapidity, ease of use, and high sensitivity and specificity. Conversely, selective culture is less expensive than PCR and provides an isolate for additional typing (Kulkarni *et al.*, 2002).

1.5.4 Typing schemes

The two most accepted *Campylobacter* serotyping schemes are the Penner scheme (Penner *et al.*, 1983), based on heat-stable antigens, and the Lior scheme, based on heat-labile antigens (Garcia *et al.*, 1985). Both techniques yield a high proportion of untypable strains, and are technically demanding and costly. These limitations have led to a development of alternative subtyping schemes, and genotypic approaches are increasingly used to characterize *Campylobacter* isolates (Wassenaar and Newell, 2000). Options include pulsed field gel electrophoresis (PFGE), *fla* typing, and amplified restriction fragment length polymorphism (AFLP) analysis. Some schemes (e.g. *fla* typing) have advantages for use in limited situations related to ease and adequacy of discriminatory power; others (e.g. AFLP) provide the reproducibility and stability needed for large epidemiologic and taxonomic studies.

1.6 Control of *Campylobacter* infection

1.6.1 On-farm controls

Efforts to reduce pathogen loads at the farm increase the likelihood that pathogen reduction steps at processing and in the kitchen will enhance the safety of foods of animal origin. For this reason, intervention to reduce broiler intestinal colonization is an active field of investigation. In one study, biosecurity measures reduced *Campylobacter* colonization rates in flocks by 50 % (Gibbens *et al.*, 2001). Reduction in colonization was associated with use of disinfectant footbaths, daily water disinfection, and the location of ventilation units. In another study, cleaning and disinfection, change of footwear at the entrance to broiler houses, and control of vermin significantly reduced *Campylobacter* infections of broiler flocks on two farms (van de Giessen *et al.*, 1996b).

Several studies have focused on interventions during broiler flock depopulation. In one study, lactic acid treatment of drinking water during the 8-hour pre-slaughter feed withdrawal period reduced carcass contamination by 15 % (Byrd *et al.*, 1998). Other studies indicate that when flocks are depopulated in batches, colonization rates increase (Wedderkopp *et al.*, 2000; Hald *et al.*, 2001). Recent studies suggest that the transport crates brought to the farm at the time of depopulation may expose birds to campylobacters (Slader *et al.*, 2002).

Competitive exclusion products have also been proposed to reduce broiler colonization. Various products containing defined poultry isolates of *C. jejuni* (Chen and Stern, 2001), *Lactobacillus* (Chang and Chen, 2000) and undefined cultures are reported to reduce colonization under experimental conditions (Hakkinen and Schneitz, 1999). Diet may alter the composition of intestinal mucus, thereby affecting the colonization potential of campylobacters (Fernandez *et al.*, 2000).

1.6.2 Processing controls

Carcass processing is a promising site for pathogen reduction efforts. The microbial quality of broiler carcasses has been associated with the abattoir where processing occurred (Wedderkopp *et al.*, 2000). Treatment of wash water is a potential processing control to reduce contamination, and the poor microbial quality of poultry wash water is thought to contribute to higher contamination rates of poultry than of red meat. The use of electrolyzed water for washing poultry carcasses reduced *C. jejuni* counts on chicken by 3 $\log_{10}$ units (Park *et al.*, 2002). Washing in 10 % oleic acid significantly reduced the number of *Campylobacter* that remained attached to poultry skin (Hinton and Ingram, 2000). Campylobacters are also very sensitive to active chlorine (Yang *et al.*, 2001). The chlorination of carcass wash water, an important component of the HACCP programs in many processing plants (Hulebak and Schlosser, 2002), may have contributed to the decline in human campylobacteriosis in the United States since the mid-1990s.

Post-processing interventions have also been investigated. Freezing poultry carcasses to –20°C resulted in a 2-$\log_{10}$ reduction in *Campylobacter* counts compared to refrigeration (Stern *et al*, 1985). Electron beam irradiation of poultry would virtually eliminate campylobacters from poultry products; however, some consumers report that the color and texture of chicken fillets are altered by irradiation (Lewis *et al.*, 2002). If these food quality considerations are successfully resolved, irradiation of poultry products may become among the most important technologies for the prevention of foodborne campylobacteriosis in the United States.

1.6.3 Food handling

Epidemiologic studies indicate that restaurants (Effler *et al.*, 2001) and home kitchens (Hopkins *et al.*, 1983) are both important venues for *C. jejuni* infection (Eberhart-Phillips *et al.*, 1997; Rodrigues *et al.*, 2001). Surveys indicate that safe food-handling skills could be improved in demographic groups of the US population, including males and young adults (Altekruse *et al.*, 1999). Kitchen sanitation guidelines should emphasize cleaning and disinfection of food contact services, hands and utensils following contact with raw meat and poultry. In addition, raw meat and poultry should be stored separately from foods that are served without subsequent cooking. Meat thermometers are recommended to measure the internal temperature of meat and poultry when it is cooking; poultry should be heated to an internal temperature of 82°C (180°F) to kill *Campylobacter*.

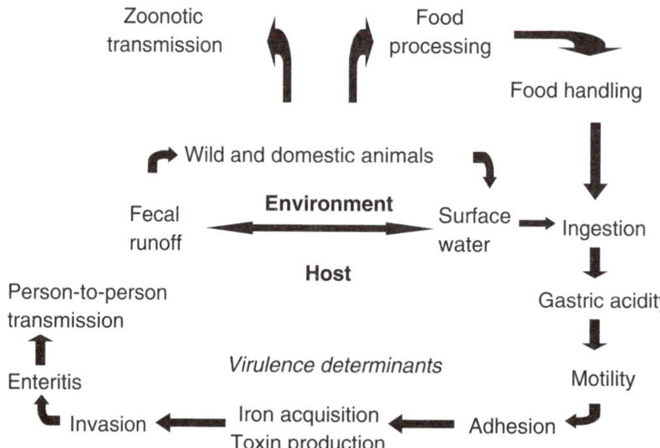

Figure 7.2 A model of the transmission cycle of *C. jejuni* depicting suspected environmental, food and animal sources, and putative virulence determinants for infection of susceptible hosts.

1.6.4 Zoonosis prevention

Handwashing after animal contact is a sensible step to prevent zoonotic campylobacteriosis in both household and occupational settings. Additional sanitary precautions are recommended with juvenile or diarrheic pets. It is particularly important to ensure that children wash their hands after animal contact (Friedman *et al.*, 1998). If sufficient attention is given to hygiene, many immunocompromised patients can safely enjoy animal companionship (Angulo *et al.*, 1994).

1.7 Conclusion

It is humbling that a bacterium as sensitive to physiologic stress as *C. jejuni* remains a common cause of foodborne infection in the new millennium. Well-defined hazards for the transmission of campylobacters exist in the environment and food chain (Figure 7.2). Because no single intervention will eliminate these hazards, a combination of prevention efforts is needed – on farms, in processing plants, in kitchens, and related to animal contact. When considering options for pathogen reduction, it is important to balance cost and effectiveness; however, cost should not be an excuse for inaction.

2 Related organisms

2.1 Other *Campylobacter* species

Campylobacter species other than *C. jejuni* and *C. coli* include *C. cervus, C. concisus, C. fetus* subspecies *fetus, C. hyointestinalis, C. lari, C. mucosalis, C. gracilis, C. rectus, C. showae, C. sputorum* and *C. upsaliensis*. Many are suspected to be human or animal pathogens (Cardarelli-Leite *et al.*, 1996). Many of these related species are also inhibited by the antibiotics in selective media used for routine isolation of *C. jejuni* and *C. coli*, and several species (e.g. *C. concisus, C. sputorum, C. cervus, C. rectus* and strains of *C. hyointestinalis*) require different micro-aerophilic incubation conditions than *C. jejuni* for growth. Furthermore, procedures for the accurate identification of *Campylobacter* organisms to the species level are time consuming and difficult. Thus the true prevalence of human infections with these other *Campylobacter* species is unknown.

2.1.1 *C. lari*

Campylobacter lari was first isolated from mammalian and avian species (Waldenstrom *et al.*, 2002). In 1984, the first case of human disease related to *C. lari* was reported – fatal bacteremia in an immunocompromised patient with multiple myeloma (Nachamkin *et al.*, 1984). Soon after, sporadic cases of enteric infection were also described (Tauxe *et al.*, 1985). Although *C. lari* bacteremia is most often reported in patients with underlying disease (Martinot *et al.*, 2001), cases of *C. lari* bacteremia in immune-competent individuals have also been described (Krause *et al.*, 2002; Werno *et al.*, 2002).

2.1.2 *C. fetus* subspecies *fetus*

Until recently *C. fetus* subspecies *fetus* was regarded as an animal pathogen, causing bovine and ovine abortion and sterility. Between 1980 and 1995, *C. fetus* was implicated in at least four reported outbreaks of human disease in North America; three were associated with foods – raw milk, a supplement containing raw calf liver, and cottage cheese (Blom *et al.*, 1995). In addition to being isolated from stools of patients with gastroenteritis, it is recognized to cause invasive infections and has been isolated from human blood, spinal fluid, abscesses, and cellulitis associated with bacteremia (Briedis *et al.*, 2002). Bacteremia is usually seen in patients with underlying disease, such as metastatic malignancy or HIV infection (Blom *et al.*, 1995).

2.1.3 *C. hyointestinalis*

Between 1979 and 1985, two of four laboratory confirmed cases of *C. hyointestinalis* that were reported to CDC were from stools of homosexual men (Edmonds *et al.*, 1987). Stool isolates were also obtained from an 8-month-old girl who lived on a farm with livestock, and a 79-year-old woman who had traveled to Egypt. A small outbreak among family members in Canada may have been associated with drinking raw milk (Salama *et al.*, 1992). *C. hyointestinalis*, with or without *C. mucosalis*, has been implicated as a causative agent of proliferative ileitis in swine (Boosinger *et al.*, 1985) and diarrhea of calves (Diker *et al.*, 1990).

2.1.4 *C. upsaliensis*

Since the first isolation of *C. upsaliensis* was reported in 1983, from the stools of healthy and diarrheic dogs (Sandstedt *et al.*, 1983), pet animals have continued to be suspected as a principal source of human infection (Bourke *et al.*, 1998). In Los Angeles, for example, *C. upsaliensis* was isolated from pet dogs in the households of two of six patients with *C. upsaliensis* infection (Labarca *et al.*, 2002).

Initially, *C. upsaliensis* infections were associated with extremes of age or with underlying disease. Of 11 human isolates reported by the CDC between 1980 and 1986, 8 originated from blood (Patton *et al.*, 1989). Blood isolates originated from two infants with fever and respiratory symptoms, a woman with an ectopic pregnancy, three elderly men with underlying diseases, and two immunocompromised adults. Of the three stool isolates, one originated from an immunocompromised patient with persistent diarrhea.

It is suspected that there is underreporting of enteric *C. upsaliensis* infections because the antibiotics that are used in selective media for isolation of *C. jejuni* (e.g. cephalothin) inhibit the growth of *C. upsaliensis* (Walmsley and Karmali, 1989). In a Swedish study, *C. upsaliensis* was the second most common *Campylobacter* species in children with diarrhea, after *C. jejuni* (Lindblom *et al.*, 1995); and in Los Angeles County in 1998, *C. upsaliensis* accounted for 4 % of fluoroquinolone resistant *Campylobacter* isolates from human patients (Labarca *et al.*, 2002). Regional differences in the prevalence of *C. upsaliensis* infection are suspected, with low prevalence of human infection reported in the United Kingdom (Aspinall *et al.*, 1996) and Denmark (Engberg *et al.*, 2000) compared to Sweden and Los Angeles.

2.2 *Arcobacter*

2.2.1 Historical aspects and current problems

The genus *Arcobacter* was proposed in 1991 to include several aerotolerant campylobacter-like organisms that grow at 15° to 30°C, which is lower than the temperature range used for incubation of *Campylobacter*. Arcobacters have been recovered from livestock and from humans with enteritis (Kiehlbauch *et al.*, 1991) as well as bacteremia (On *et al.*, 1995; Hsueh *et al.*, 1997). Two *Arcobacter* species are most suspected to cause human disease: *A. butzleri* and *A. cryaerophilus. A. butzleri* infection was reported in an outbreak of gastroenteritis in schoolchildren (Vandamme *et al.*, 1992), in two patients with chronic disease who developed severe diarrhea (Lerner *et al.*, 1994), and in a neonate with bacteremia (On *et al.*, 1995). *A. cryaerophilus* has been recovered from the blood of a uremic patient with pneumonia (Hsueh *et al.*, 1997). The burden of human illness remains uncertain. In a Danish study of 3267 patients with diarrhea, one stool specimen tested positive for *A. butzleri* and *A. cryaerophilus*, respectively (Engberg *et al.*, 2000).

2.2.2 Environmental sources

A. butzleri and *A. cryaerophilus* have been found in various environments, and a focus of research has been the development of protocols for their isolation (Ohlendorf and Murano, 2002). *A. butzleri*-associated diarrheal illness has been reported in non-human primates (Anderson *et al.*, 1993), and *A. cryaerophilus* and *A. butzleri* have each been recovered from late-term aborted porcine and equine fetuses (Wesley *et al.*, 1995). In a study conducted in Denmark, of 10 broiler carcasses obtained from supermarkets and 15 from abattoirs, all carcasses were confirmed to carry *Arcobacter butzleri* (Atabay *et al.*, 1998). Three supermarket and 10 abattoir carcasses also carried *A. cryaerophilus*, and two abattoir carcasses also carried *A. skirrowii*. In the Netherlands, *A. butzleri* was present in 53 of 220 (24 %) poultry meat specimens, as well as in some beef and pork meat specimens (de Boer *et al.*, 1996). In Germany, the pathogen was detected over several months in a drinking water reservoir (Jacob *et al.*, 1993). To the extent that *A. cryaerophilus* and *A. butzleri* are found to be human pathogens, the findings from environmental microbiological studies such as these may have significance for public health protection.

2.3 *Helicobacter pylori*

2.3.1 Historical aspects and current problems

Even though *Helicobacter pylori* organisms were cultured for the first time in the early 1980s, descriptions of their presence in the gastric mucosa of humans date from the beginning of the twentieth century. The organism was first named *Campylobacter*-like organism (CLO) because of its resemblance to the *Campylobacter* species. Then it was named *C. pyloridis*, later changed to *C. pylori*, and finally in 1989 it was placed in the new genus *Helicobacter*, and renamed *H. pylori* (Goodwin *et al*, 1989; Versalovic and Fox, 2003).

H. pylori colonizes approximately 50 % of the adult population in the world; however, clinical disease occurs in less than 10 % of those who are colonized (Torres *et al.*, 2000; Blaser and Berg, 2001). The diagnosis and treatment of the upper gastrointestinal disease have changed with the recognition of *H. pylori* as a pathogen of the gastric mucosa. Superficial gastritis is now considered the histopathological confirmatory sign of gastric mucosal infection. *H. pylori* infection is associated with chronic gastritis, duodenal and gastric ulcers, gastric carcinoma, and gastric mucosa-associated lymphoid tissue. The causal role of *H. pylori* in gastric cancer is almost universally accepted (Peek and Blaser, 2002).

2.3.2 Characteristics and virulence factors of *H. pylori*

H. pylori is a spiral, Gram-negative rod that grows well at 37°C in a micro-aerobic atmosphere (between 5 % and 10 % CO_2). The organism prefers enriched media such as blood agar, or media supplemented with 5 % newborn calf serum, starch, fetal bovine serum or others (Perez-Perez, 2000). Colonies may be observed after 48 hours of incubation, but in primary cultures from gastric biopsies the incubation should be prolonged for up to 7 days. The organism is oxidase and catalase positive, and has a potent urease activity that produces large amounts of NH_3 which allows the neutralization of the normal acid pH of the stomach. The genomes of representative *H. pylori* strains have been sequenced (Tomb *et al.* 1997; Alm *et al.*, 1999). The genome encodes for approximately 1500 proteins. Initial studies indicated that urease production and motility were essential for the first steps of the infection (Eaton *et al.*, 1991, 1996). Later studies showed that *H. pylori* binds tightly to epithelial cells using adhesins such as BabA and other members of the Hop protein family (Ilver *et al.*, 1998). Almost all *H. pylori* strains express a 95-kDa vacuolating cytotoxin (VacA). Since *in vitro* vacuolating activity is detected in only 60 % of the isolates, the pathogenic role of this toxin is debated. Although VacA is not essential for colonization, mutant strains of *H. pylori* that are VacA negative are outcompeted by wild- type bacteria in mouse models (Atherton *et al.*, 1997; Salama *et al.*, 2001). Another important virulence factor of *H. pylori* is the Cag pathogenicity island (cag-PAI), which is present in 60 % of all isolates (Censini *et al.*, 1996). The cag-PAI is a 37-kb genomic fragment containing 29 genes that most likely was acquired by horizontal gene transfer (Akopyants *et al.*, 1998). Some of the encoded components belong to a type-IV secretion apparatus that injects the 120-kDa protein CagA into host cells where the CagA protein is phosphorylated and induces a strong response of cytokine production by the host cells (Higashi *et al.*, 2002).

2.3.3 Epidemiology and transmission

H. pylori infection is present in every country where it has been investigated, but the prevalence varies from country to country, and even among different population groups in the same country. The presence of *H. pylori* is mainly associated with socioeconomic level and age (Malaty and Graham, 1994; Perez-Perez *et al.*, 2002). Prevalence among adults ranges from greater than 80 % in many developing countries to between 20 % and 40 % in developed countries.

It is generally accepted that the organism is acquired by the oral route, and that person-to-person transmission within the family during early childhood is important in the natural history of infection. Food and water are not established vehicles for transmission of the bacterium. There is also no conclusive evidence of animal reservoirs for the transmission of *H. pylori* to humans; although the organism has been found in non-human primates and other animals such as cats (Handt *et al.*, 1994). Once acquired, infection with *H. pylori* is usually chronic, and spontaneous elimination of the bacterium is relatively uncommon (Perez-Perez *et al.*, 2002).

2.3.4 Clinical outcome of *H. pylori* colonization in man

Disease expression as a consequence of *H. pylori* colonization is unpredictable, and is influenced by bacterial and host factors. As was mentioned before, histological gastritis is universally present in patients infected with *H. pylori*. However, the pattern and distribution of gastritis are associated with particular disease outcomes. Patients with antral-predominant gastritis are more likely to develop a duodenal ulcer, whereas patients with corpus predominant gastritis are more likely to have a gastric ulcer and possibly gastric carcinoma (Suerbaum and Michetti, 2002). The role of *H. pylori* in the development of peptic ulcer disease has been clearly demonstrated. Evidence of the role of *H. pylori* in the pathogenesis of peptic ulcer disease includes the recognition that antimicrobial therapy to eradicate infection dramatically reduces the recurrence rate of *H. pylori*-associated peptic ulcer disease (Marshall *et al.*, 1988). Some of the principal lines of evidence that chronic *H. pylori* infection increases the risk of gastric cancer were derived from large seroepidemiologic cases-control studies (Parsonnet *et al.*, 1991; Nomura *et al.*, 1995), studies using experimental animal models (Fox, 1998), and prospective clinical data from Japan (Uemura *et al.*, 2001). Another impressive line of evidence of the important causal role of *H. pylori* in the development of gastric cancer is the parallel decline in *H. pylori* infection rates and gastric cancer rates in most industrialized countries (Parsonnet, 1995). Eradication and reduced rates of acquisition of *H. pylori* are associated with several clinical advantages, including the cure of peptic ulcer disease and the decrease of gastric cancer; however, the decline of *H. pylori* prevalence has also been suggested as being the main reason for the increase of gastro-esophageal reflux disease (GERD) and esophageal cancer (Meining *et al.*, 1998; Blaser, 1999). Data from case-control and cohort studies have, for example, suggested that *H. pylori* infection protects against GERD (Chow *et al.*, 1998; Vicari *et al.*, 1998). This hypothesis remains controversial. The high prevalence of *H. pylori* and the interest in its potential role in the etiology of other chronic diseases have prompted epidemiologic studies that suggest that *H. pylori* may be a risk factor for extra-gastrointestinal chronic diseases, including coronary arteriosclerosis (Perez-Perez *et al.*, 1999) and pancreatic cancer (Stolzenberg-Solomon *et al.*, 2001).

2.3.5 Diagnostic methods for *H. pylori* infection

Methods for the diagnosis of *H. pylori* can be divided into two categories – invasive and non-invasive – based on whether endoscopic biopsies are performed. Diagnostic options vary depending on the clinical setting and purpose of the test. When endoscopy is clinically indicated, invasive tests are preferred. The diagnostic screening test of first choice is the urease test, because it is relatively simple to perform and less expensive than other diagnostic tests. A histologic diagnosis is definitive, and provides information on the stage of disease. A third option, bacterial culture, permits evaluation of antimicrobial susceptibility patterns with potential implications for therapy. The selection of an invasive diagnostic method is based largely on cost. Cultures to detect *H. pylori*, for example, are not routinely requested, but are recommended when antimicrobial therapy failures occur (Perez-Perez, 2000).

Non-invasive tests are useful for screening large numbers of patients, and have positive applications for epidemiological studies. The non-invasive diagnostic methods include the urea breath test, serological test, and stool antigen detection. The sensitivity of these methods is high, because in theory they are sampling the entire stomach while biopsy samples only a minuscule portion of it. Under certain circumstances, the sensitivity of serologic tests may be diminished. For example, serology is not the optimal test to follow eradication of infection, since the patient's antibody titers may persist after treatment (Cutler *et al.*, 1995; Perez-Perez *et al.*, 1997).

2.3.6 Treatment of *H. pylori* infection

Once eradication of *H. pylori* has been achieved, reinfection is a rare event. For this reason, the benefit of treatment is enormous. *H. pylori* eradication regimens using antibiotics alone have cure rates of less than 80 %. Most eradication regimens combine antibiotics with proton-pump inhibitors or other agents that inhibit gastric acid secretion (Lind *et al.*, 1999; Suerbaum and Michetti, 2002). The use of a proton-pump inhibitor in combination with two or three antimicrobial agents for 14 days is often effective in eradicating *H. pylori* infection; however, non-compliance is a concern with this regimen. The use of a combination of antimicrobial agents reduces the risk of selecting resistant *H. pylori* strains. Although most *H. pylori* are susceptible to amoxicillin and tetracycline, resistance to clarithromycin and metronidazole is common, particularly in developing countries (Torres *et al.*, 2001).

3 Summary

C. jejuni is a common bacterial cause of human gastroenteritis in the US and elsewhere. Sequelae of campylobacteriosis add to the burden of disease. Mishandled or improperly prepared poultry products are commonly implicated vehicles for transmission of *C. jejuni* infections to humans, and restaurants are increasingly recognized as being important venues for infection. Each link in the food chain from producer to consumer has a role in preventing illnesses caused by this pathogen. For organisms related to *C. jejuni* (e.g. *C. upsaliensis*, *Arcobacter butzleri*), a need exists to determine the reservoirs, the clinical syndromes, and the burden of the illnesses they cause, so as to better define prevention priorities and strategies. *H. pylori* colonizes approximately 50 % of the adult population worldwide. *H. pylori* infection is the principal cause of chronic gastritis, gastric ulcer disease and gastric cancer; however, these diseases only occur in a fraction of people who are colonized. The diagnosis and treatment of upper gastrointestinal diseases have changed with the recognition of *H. pylori* as a major gastrointestinal pathogen.

Bibliography

Akopyants, N. S., S. W. Clifton, D. Kersulyte *et al.* (1998). Analyses of the cag pathogenicity island of *Helicobacter pylori. Mol. Microbiol.* **28**, 37–53.

Alm, R. A., L. S. Ling, D. T. Moir *et al.* (1999). Genomic-sequence comparison of two unrelated isolates of the human gastric pathogen *Helicobacter pylori. Nature* **397**, 176–180.

Altekruse, S. F., S. Yang, B. B. Timbo and F. J. Angulo (1999). A multi-state survey of consumer food-handling and food-consumption practices. *Am. J. Prev. Med.* **16**, 216–221.

Anderson, K. F., J. A. Kiehlbauch, D. C. Anderson *et al.* (1993). *Arcobacter (Campylobacter) butzleri*-associated diarrheal illness in a nonhuman primate population. *Infect. Immun.* **61**, 2220–2223.

Angulo, F. J. and D. L. Swerdlow (1995). Bacterial enteric infections in persons infected with human immunodeficiency virus. *Clin. Infect. Dis.* **21** (Suppl. 1), S84–S93.

Angulo, F. J., C. A. Glaser, D. D. Juranek *et al.* (1994). Caring for pets of immunocompromised persons. *J. Am. Vet. Med. Assoc.* **205**, 1711–1718.

Aspinall, S. T., D. R. Wareing, P. G. Hayward and D. N. Hutchinson (1996). A comparison of a new campylobacter selective medium (CAT) with membrane filtration for the isolation of thermophilic campylobacters including *Campylobacter upsaliensis. J. Appl. Bacteriol.* **80**, 645–650.

Atabay, H. I., J. E. Corry and S. L. On (1998). Diversity and prevalence of *Arcobacter* spp. in broiler chickens. *J. Appl. Microbiol.* **84**, 1007–1016.

Atherton, J. C., R. M. Peek Jr, K. T. Tham *et al.* (1997). Clinical and pathological importance of heterogeneity in vacA, the vacuolating cytotoxin gene of *Helicobacter pylori. Gastroenterology* **112**, 92–99.

Black, R. E., M. M. Levine, M. L. Clements *et al.* (1988). Experimental *Campylobacter jejuni* infection in humans. *J. Infect. Dis.* **157**, 472–479.

Blaser, M. J. (1997). Epidemiologic and clinical features of *Campylobacter jejuni* infections. *J. Infect. Dis.* **176** (Suppl. 2), S103–S105.

Blaser, M. J. (1999). Hypothesis: the changing relationships of *Helicobacter pylori* and humans: implications for health and disease. *J. Infect. Dis.* **179**, 1523–1530.

Blaser, M. J. and D. E. Berg (2001). *Helicobacter pylori* genetic diversity and risk of human disease. *J. Clin. Invest.* **107**, 767–773.

Blaser, M. J., E. Sazie and L. P. Williams Jr (1987). The influence of immunity on raw milk-associated *Campylobacter* infection. *J. Am. Med. Assoc.* **257**, 43–46.

Blom, K., C. M. Patton, M. A. Nicholson and B. Swaminathan (1995). Identification of *Campylobacter fetus* by PCR-DNA probe method. *J. Clin. Microbiol.* **33**, 1360–1362.

Bolton, F. J., A. D. Sails, A. J. Fox *et al.* (2002). Detection of *Campylobacter jejuni* and *Campylobacter coli* in foods by enrichment culture and polymerase chain reaction enzyme-linked immunosorbent assay. *J. Food Prot.* **65**, 760–767.

Boosinger, T. R., H. L. Thacker and C. H. Armstrong (1985). *Campylobacter sputorum* subsp mucosalis and *Campylobacter hyointestinalis* infections in the intestine of gnotobiotic pigs. *Am. J. Vet. Res.* **46**, 2152–2156.

Bourke, B., V. L. Chan and P. Sherman (1998). *Campylobacter upsaliensis*: waiting in the wings. *Clin. Microbiol. Rev.* **11**, 440–449.

Briedis, D. J., A. Khamessan, R. W. McLaughlin *et al.* (2002). Isolation of *Campylobacter fetus* subsp. fetus from a patient with cellulitis. *J. Clin. Microbiol.* **40**, 4792–4796.

Busato, A., D. Hofer, T. Lentze *et al.* (1999). Prevalence and infection risks of zoonotic enteropathogenic bacteria in Swiss cow-calf farms. *Vet. Microbiol.* **69**, 251–263.

Byrd, J. A., D. E. Corrier, M. E. Hume *et al.* (1998). Effect of feed withdrawal on *Campylobacter* in the crops of market-age broiler chickens. *Avian Dis.* **42**, 802–806.

Cappelier, J. M., A. Rossero and M. Federighi (2000). Demonstration of a protein synthesis in starved *Campylobacter jejuni* cells. *Intl J. Food Microbiol.* **55**, 63–67.

Cardarelli-Leite, P., K. Blom, C. M. Patton *et al.* (1996). Rapid identification of *Campylobacter* species by restriction fragment length polymorphism analysis of a PCR-amplified fragment of the gene coding for 16S rRNA. *J. Clin. Microbiol.* **34**, 62–67.

Censini, S., C. Lange, Z. Xiang *et al.* (1996). cag, a pathogenicity island of *Helicobacter pylori*, encodes type I-specific and disease-associated virulence factors. *Proc. Natl. Acad. Sci. USA* **93**, 14 648–14 653.

CDC (Centers for Disease Control and Prevention) (1998). Outbreak of *Campylobacter* enteritis associated with cross-contamination of food – Oklahoma, 1996. *J. Am. Med. Assoc.* **279**, 1341.

CDC (Centers for Disease Control and Prevention) (2002a). Preliminary FoodNet data on the incidence of foodborne illnesses – selected sites, United States, 2001. *Morbid. Mortal. Weekly Rep.* **51**, 325–329.

CDC (Centers for Disease Control and Prevention) (2002b). Centers for Disease Control and Prevention, National Antimicrobial Resistance Monitoring System Worldwide website, http, //www.cdc.gov/narms/pus/camp.htm (accessed 30 January 2003).

CDC (Centers for Disease Control and Prevention) (2002c). Outbreak of *Campylobacter jejuni* infections associated with drinking unpasteurized milk procured through a cow-leasing program – Wisconsin, 2001. *Morbid. Mortal. Weekly Rep.* **51**, 548–549.

Chang, M. H. and T. C. Chen (2000). Reduction of *Campylobacter jejuni* in a simulated chicken digestive tract by Lactobacilli cultures. *J. Food Prot.* **63**, 1594–1597.

Chen, H. C. and N. J. Stern (2001). Competitive exclusion of heterologous *Campylobacter* spp. in chicks. *Appl. Environ. Microbiol.* **67**, 848–851.

Chow, W. H., M. J. Blaser, W. J. Blot *et al.* (1998). An inverse relation between cagA+ strains of *Helicobacter pylori* infection and risk of esophageal and gastric cardia adenocarcinoma. *Cancer Res.* **58**, 588–590.

Chuma, T., S. Hashimoto and K. Okamoto (2000). Detection of thermophilic *Campylobacter* from sparrows by multiplex PCR, the role of sparrows as a source of contamination of broilers with *Campylobacter. J. Vet. Med. Sci.* **62**, 1291–1295.

Cogan, T. A., S. F. Bloomfield and T. J. Humphrey (1999). The effectiveness of hygiene procedures for prevention of cross-contamination from chicken carcasses in the domestic kitchen. *Lett. Appl. Microbiol.* **29**, 354–358.

Coker, A. O., R. D. Isokpehi, B. N. Thomas *et al.* (2002). Human campylobacteriosis in developing countries. *Emerg. Infect. Dis.* **8**, 237–244.

Craven, S. E., N. J. Stern, E. Line *et al.* (2000). Determination of the incidence of Salmonella spp., *Campylobacter jejuni*, and *Clostridium perfringens* in wild birds near broiler chicken houses by sampling intestinal droppings. *Avian Dis.* **44**, 715–720.

Cutler, A. F., S. Havstad, C. K. Ma *et al.* (1995). Accuracy of invasive and noninvasive tests to diagnose *Helicobacter pylori* infection. *Gastroenterology* **109**, 136–141.

de Boer, E., J. J. Tilburg, D. L. Woodward *et al.* (1996). A selective medium for the isolation of *Arcobacter* from meats. *Lett. Appl. Microbiol.* **23**, 64–66.

DeKeyser, P., M. Gossuin-Detrain, J. P. Butzler and J. Sternon (1972). Acute enteritis due to related vibrio, first positive stool cultures. *J. Infect. Dis.* **125**, 390–392.

Deming, M. S., R. V. Tauxe, P. A. Blake *et al.* (1987). *Campylobacter* enteritis at a university, transmission from eating chicken and from cats. *Am. J. Epidemiol.* **126**, 526–534.

Denis, M., J. Refregier-Petton, M. J. Laisney *et al.* (2001). *Campylobacter* contamination in French chicken production from farm to consumers. Use of a PCR assay for detection and identification of *Campylobacter jejuni* and Camp. *coli. J. Appl. Microbiol.* **91**, 255–267.

Diker, K. S., S. Diker and M. B. Ozlem (1990). Bovine diarrhea associated with *Campylobacter hyointestinalis. Zentralbl. Veterinarmed.* [B] **37**, 158–160.

Dingle, K. E., B. N. Van Den, F. M. Colles *et al.* (2001). Sequence typing confirms that *Campylobacter jejuni* strains associated with Guillain-Barré and Miller-Fisher syndromes are of diverse genetic lineage, serotype, and flagella type. *J. Clin. Microbiol.* **39**, 3346–3349.

Eaton, K. A., C. L. Brooks, D. R. Morgan and S. Krakowka (1991). Essential role of urease in pathogenesis of gastritis induced by *Helicobacter pylori* in gnotobiotic piglets. *Infect. Immun.* **59**, 2470–2475.

Eaton, K. A., S. Suerbaum, C. Josenhans and S. Krakowka (1996). Colonization of gnotobiotic piglets by *Helicobacter pylori* deficient in two flagellin genes. *Infect. Immun.* **64**, 2445–2448.

Eberhart-Phillips, J., N. Walker, N. Garrett *et al.* (1997). Campylobacteriosis in New Zealand, results of a case-control study. *J. Epidemiol. Comm. Health* **51**, 686–691.

Edmonds, P., C. M. Patton, P. M. Griffin *et al.* (1987). *Campylobacter hyointestinalis* associated with human gastrointestinal disease in the United States. *J. Clin. Microbiol.* **25**, 685–691.

Effler, P., M. C. Ieong, A. Kimura *et al.* (2001). Sporadic *Campylobacter jejuni* infections in Hawaii, associations with prior antibiotic use and commercially prepared chicken. *J. Infect. Dis.* **183**, 1152–1155.

Engberg, J., S. L. On, C. S. Harrington and P. Gerner-Smidt (2000). Prevalence of *Campylobacter*, *Arcobacter*, *Helicobacter*, and *Sutterella* spp. in human fecal samples as estimated by a reevaluation of isolation methods for Campylobacters. *J. Clin. Microbiol.* **38**, 286–291.

Engberg, J., F. M. Aarestrup, D. E. Taylor *et al.* (2001). Quinolone and macrolide resistance in *Campylobacter jejuni* and *C. coli*, resistance mechanisms and trends in human isolates. *Emerg. Infect. Dis.* **7**, 24–34.

Escherich T. (1886). Über das Vorkommen von Vibrionen im Darmkanal und den Stuhlgängen der Säuglinge. *Münch. Med. Wochenschr.* **33**, 815–817, 833–835.

Evans, S. J. and A. R. Sayers (2000). A longitudinal study of campylobacter infection of broiler flocks in Great Britain. *Prev. Vet. Med.* **46**, 209–223.

Fernandez, F., R. Sharma, M. Hinton and M. R. Bedford (2000). Diet influences the colonisation of *Campylobacter jejuni* and distribution of mucin carbohydrates in the chick intestinal tract. *Cell Mol. Life Sci.* **57**, 1793–1801.

Food and Drug Administration (2000). FDA hastening withdrawal of fluoroquinolone approvals for poultry. *J. Am. Vet. Med. Assoc.* **217**, 1447, 1452.

Fox, J. G. (1998). Review article, *Helicobacter* species and *in vivo* models of gastrointestinal cancer. *Aliment. Pharmacol. Ther.* **12** (Suppl. 1), 37–60.

Friedman, C. R., C. Torigian, P. J. Shillam *et al.* (1998). An outbreak of salmonellosis among children attending a reptile exhibit at a zoo. *J. Pediatr.* **132**, 802–807.

Friedman C. R., J. Niemann, H. C. Wegener and R. V. Tauxe (2001). Epidemiology of *Campylobacter jejuni* infections in the United States and other industrialized nations. In: I. Nachamkin and M. J. Blaser (eds), *Campylobacter*, 2nd edn, pp. 121–138. American Society for Microbiology, Washington, DC.

Garcia, M. M., H. Lior, R. B. Stewart *et al.* (1985). Isolation, characterization, and serotyping of *Campylobacter jejuni* and *Campylobacter coli* from slaughter cattle. *Appl. Environ. Microbiol.* **49**, 667–672.

Gibbens, J. C., S. J. Pascoe, S. J. Evans *et al.* (2001). A trial of biosecurity as a means to control *Campylobacter* infection of broiler chickens. *Prev. Vet. Med.* **48**, 85–99.

Gillespie, I. A., S. J. O'Brien, J. A. Frost *et al.* (2002). A case–case comparison of *Campylobacter coli* and *Campylobacter jejuni* infection, a tool for generating hypotheses. *Emerg. Infect. Dis.* **8**, 937–942.

Goodwin, C. S., J. A. Armstrong, T. Chilvers *et al.* (1989). Transfer of *Campylobacter pylori* and *Campylobacter mustelae* to *Helicobacter* gen. nov. as *Helicobacter pylori* comb. nov. and *Helicobacter mustelae* comb. nov., respectively. *Intl J. Syst. Bacteriol.* **39**, 397–405.

Goodyear, C. S., G. M. O'Hanlon, J. J. Plomp *et al.* (1999). Monoclonal antibodies raised against Guillain-Barré syndrome-associated *Campylobacter jejuni* lipopolysaccharides react with neuronal gangliosides and paralyze muscle-nerve preparations. *J. Clin. Invest.* **104**, 697–708.

Hafer-Macko, C., S. T. Hsieh, C. Y. Li *et al.* (1996). Acute motor axonal neuropathy, an antibody-mediated attack on axolemma. *Ann. Neurol.* **40**, 635–644.

Hakkinen, M. and C. Schneitz (1999). Efficacy of a commercial competitive exclusion product against *Campylobacter jejuni. Br. Poult. Sci.* **40**, 619–621.

Hald, B., E. Rattenborg and M. Madsen (2001). Role of batch depletion of broiler houses on the occurrence of *Campylobacter* spp. in chicken flocks. *Lett. Appl. Microbiol.* **32**, 253–256.

Handt, L. K., J. G. Fox, F. E. Dewhirst *et al.* (1994). *Helicobacter pylori* isolated from the domestic cat, public health implications. *Infect. Immun.* **62**, 2367–2374.

Hanninen, M. L., P. Perko-Makela, A. Pitkala and H. Rautelin (2000). A three-year study of *Campylobacter jejuni* genotypes in humans with domestically acquired infections and in chicken samples from the Helsinki area. *J. Clin. Microbiol.* **38**, 1998–2000.

Hannu, T., L. Mattila, H. Rautelin *et al.* (2002). *Campylobacter*-triggered reactive arthritis, a population-based study. *Rheumatology (Oxford)* **41**, 312–318.

Harris, N. V., N. S. Weiss and C. M. Nolan (1986). The role of poultry and meats in the etiology of *Campylobacter jejuni/coli* enteritis. *Am. J. Public Health* **76**, 407–411.

Higashi, H., R. Tsutsumi, A. Fujita *et al.* (2002). Biological activity of the *Helicobacter pylori* virulence factor CagA is determined by variation in the tyrosine phosphorylation sites. *Proc. Natl. Acad. Sci. USA* **99**, 14 428–14 433.

Hinton, A. Jr and K. D. Ingram (2000). Use of oleic acid to reduce the population of the bacterial flora of poultry skin. *J. Food Prot.* **63**, 1282–1286.

Ho, T. W., H. J. Willison, I. Nachamkin *et al.* (1999). Anti-GD1a antibody is associated with axonal but not demyelinating forms of Guillain-Barré syndrome. *Ann. Neurol.* **45**, 168–173.

Hoar, B. R., E. R. Atwill, C. Elmi and T. B. Farver (2001). An examination of risk factors associated with beef cattle shedding pathogens of potential zoonotic concern. *Epidemiol. Infect.* **127**, 147–155.

Hooper, D. C., J. S. Wolfson, E. Y. Ng and M. N. Swartz (1987). Mechanisms of action of and resistance to ciprofloxacin. *Am. J. Med.* **82**, 12–20.

Hopkins, R. S. and A. S. Scott (1983). Handling raw chicken as a source for sporadic *Campylobacter jejuni* infections. *J. Infect. Dis.* **148**, 770.

Hopkins, R. S., R. Olmsted and G. R. Istre (1984). Endemic *Campylobacter jejuni* infection in Colorado, identified risk factors. *Am. J. Public Health* **74**, 249–250.

Hsueh, P. R., L. J. Teng, P. C. Yang *et al.* (1997). Bacteremia caused by *Arcobacter cryaerophilus. Br. J. Clin. Microbiol.* **35**, 489–491.

Hughes, R. A. (2002). Systematic reviews of treatment for inflammatory demyelinating neuropathy. *J. Anat.* **200**, 331–339.

Hulebak, K. L. and W. Schlosser (2002). Hazard analysis and critical control point (HACCP) history and conceptual overview. *Risk Analysis* **22**, 547–552.

Ilver, D., A. Arnqvist, J. Ogren *et al.* (1998). *Helicobacter pylori* adhesin binding fucosylated histo-blood group antigens revealed by retagging. *Science* **279**, 373–377.

Jacob, J., H. Lior and I. Feuerpfeil (1993). Isolation of *Arcobacter butzleri* from a drinking water reservoir in eastern Germany. *Zentralbl. Hyg. Umweltmed.* **193**, 557–562.

Jayarao, B. M. and D. R. Henning (2001). Prevalence of foodborne pathogens in bulk tank milk. *J. Dairy Sci.* **84**, 2157–2162.

Jones, K., S. Howard and J. S. Wallace (1999). Intermittent shedding of thermophilic campylobacters by sheep at pasture. *J. Appl. Microbiol.* **86**, 531–536.

Kapperud, G., E. Skjerve, N. H. Bean *et al.* (1992). Risk factors for sporadic *Campylobacter* infections, results of a case-control study in southeastern Norway. *J. Clin. Microbiol.* **30**, 3117–3121.

Kapperud, G., E. Skjerve, L. Vik *et al.* (1993). Epidemiological investigation of risk factors for campylobacter colonization in Norwegian broiler flocks. *Epidemiol. Infect.* **111**, 245–255.

Kiehlbauch, J. A., D. J. Brenner, M. A. Nicholson *et al.* (1991). *Campylobacter butzleri* sp. nov. isolated from humans and animals with diarrheal illness. *J. Clin. Microbiol.* **29**, 376–385.

Kramer, J. M., J. A. Frost, F. J. Bolton and D. R. Wareing (2000). *Campylobacter* contamination of raw meat and poultry at retail sale, identification of multiple types and comparison with isolates from human infection. *J. Food Prot.* **63**, 1654–1659.

Krause, R., S. Ramschak-Schwarzer, G. Gorkiewicz *et al.* (2002). Recurrent septicemia due to *Campylobacter fetus* and *Campylobacter lari* in an immunocompetent patient. *Infection* **30**, 171–174.

Kulkarni, S. P., S. Lever, J. M. Logan *et al.* (2002). Detection of campylobacter species, a comparison of culture and polymerase chain reaction based methods. *J. Clin. Pathol.* **55**, 749–753.

Labarca, J. A., J. Sturgeon, L. Borenstein *et al.* (2002). *Campylobacter upsaliensis*, another pathogen for consideration in the United States. *Clin. Infect. Dis.* **34**, E59–E60.

Lehner, A., C. Schneck, G. Feierl *et al.* (2000). Epidemiologic application of pulsed-field gel electrophoresis to an outbreak of *Campylobacter jejuni* in an Austrian youth centre. *Epidemiol. Infect.* **125**, 13–16.

Lerner, J., V. Brumberger and V. Preac-Mursic (1994). Severe diarrhea associated with *Arcobacter butzleri. Eur. J. Clin. Microbiol. Infect. Dis.* **13**, 660–662.

Lewis, S. J., A. Velasquez, S. L. Cuppett and S. R. McKee (2002). Effect of electron beam irradiation on poultry meat safety and quality. *Poult. Sci.* **81**, 896–903.

Lind, T., F. Megraud, P. Unge *et al.* (1999). The MACH2 study, role of omeprazole in eradication of *Helicobacter pylori* with 1-week triple therapies. *Gastroenterology* **116**, 248–253.

Lindblom, G. B., E. Sjogren, J. Hansson-Westerberg and B. Kaijser (1995). *Campylobacter upsaliensis*, *C. sputorum sputorum* and *C. concisus* as common causes of diarrhoea in Swedish children. *Scand. J. Infect. Dis.* **27**, 187–188.

Locht, H. and K. A. Krogfelt (2002). Comparison of rheumatological and gastrointestinal symptoms after infection with *Campylobacter jejuni/coli* and enterotoxigenic *Escherichia coli. Ann. Rheum. Dis.* **61**, 448–452.

Malaty, H. M. and D. Y. Graham (1994). Importance of childhood socioeconomic status on the current prevalence of *Helicobacter pylori* infection. *Gut* **35**, 742–745.

Manfredi, R., A. Nanetti, M. Ferri and F. Chiodo (1999). Fatal *Campylobacter jejuni* bacteraemia in patients with AIDS. *J. Med. Microbiol.* **48**, 601–603.

Marano, N. N., S. Rossiter, K. Stamey *et al.* (2000). The National Antimicrobial Resistance Monitoring System (NARMS) for enteric bacteria, 1996–1999, surveillance for action. *J. Am. Vet. Med. Assoc.* **217**, 1829–1830.

Marshall, B. J., C. S. Goodwin, J. R. Warren *et al.* (1988). Prospective double-blind trial of duodenal ulcer relapse after eradication of *Campylobacter pylori. Lancet* **2**, 1437–1442.

Martinot, M., B. Jaulhac, R. Moog *et al.* (2001). *Campylobacter lari* bacteremia. *Clin. Microbiol. Infect.* **7**, 96–97.

McDermott, P. F., S. M. Bodeis, L. L. English *et al.* (2002). Ciprofloxacin resistance in *Campylobacter jejuni* evolves rapidly in chickens treated with fluoroquinolones. *J. Infect. Dis.* **185**, 837–840.

Mead, P. S., L. Slutsker, V. Dietz *et al.* (1999). Food-related illness and death in the United States. *Emerg. Infect. Dis.* **5**, 607–625.

Meining, A., G. Kiel and M. Stolte (1998). Changes in *Helicobacter pylori*-induced gastritis in the antrum and corpus during and after 12 months of treatment with ranitidine and lansoprazole in patients with duodenal ulcer disease. *Aliment. Pharmacol. Ther.* **12**, 735–740.

Melby, K. K., J. G. Svendby, T. Eggebo *et al.* (2000). Outbreak of *Campylobacter* infection in a subartic community. *Eur. J. Clin. Microbiol. Infect. Dis.* **19**, 542–544.

Nachamkin, I. (2002). Chronic effects of *Campylobacter* infection. *Microbes Infect.* **4**, 399–403.

Nachamkin, I. (2003). *Campylobacter* and *Arcobacter*. In: P. R. Murray, E. J. Baron, M. A. Pfaller *et al* (eds), *Manual of Clinical Microbiology*, 8th edn, pp. 902–914. ASM Press, Washington, DC.

Nachamkin, I., C. Stowell, D. Skalina *et al.* (1984). *Campylobacter laridis* causing bacteremia in an immunosuppressed patient. *Ann. Intern. Med.* **101**, 55–57.

Nadeau, E., S. Messier and S. Quessy (2002). Prevalence and comparison of genetic profiles of *Campylobacter* strains isolated from poultry and sporadic cases of campylobacteriosis in humans. *J. Food Prot.* **65**, 73–78.

Nielsen, E. M. (2002). Occurrence and strain diversity of thermophilic campylobacters in cattle of different age groups in dairy herds. *Lett. Appl. Microbiol.* **35**, 85–89.

Nomura, A. M., G. N. Stemmermann and P. H. Chyou (1995). Gastric cancer among the Japanese in Hawaii. *Jpn J. Cancer Res.* **86**, 916–923.

Norkrans, G. and A. Svedhem (1982). Epidemiological aspects of *Campylobacter jejuni* enteritis. *J. Hyg. (Lond.)* **89**, 163–170.

Ohlendorf, D. S. and E. A. Murano (2002). Sensitivity of three methods used in the isolation of *Arcobacter* spp. in raw ground pork. *J. Food Prot.* **65**, 1784–1788.

On, S. L. (2001). Taxonomy of *Campylobacter*, *Arcobacter*, *Helicobacter* and related bacteria, current status, future prospects and immediate concerns. *Symp. Ser. Soc. Appl. Microbiol.*1S–15S, 2001.

On, S. L. and C. S. Harrington (2001). Evaluation of numerical analysis of PFGE-DNA profiles for differentiating *Campylobacter fetus* subspecies by comparison with phenotypic, PCR and 16S rDNA sequencing methods. *J. Appl. Microbiol.* **90**, 285–293.

On, S. L., A. Stacey and J. Smyth (1995). Isolation of *Arcobacter butzleri* from a neonate with bacteraemia. *J. Infect.* **31**, 225–227.

Oosterom, J., C. H. den Uyl, J. R. Banffer and J. Huisman (1984). Epidemiological investigations on *Campylobacter jejuni* in households with a primary infection. *J. Hyg. (Lond.)* **93**, 325–332.

Park, H., Y. C. Hung and R. E. Brackett (2002). Antimicrobial effect of electrolyzed water for inactivating *Campylobacter jejuni* during poultry washing. *Intl J. Food Microbiol.* **72**, 77–83.

Parkhill, J., B. W. Wren, K. Mungall *et al.* (2000). The genome sequence of the food-borne pathogen *Campylobacter jejuni* reveals hypervariable sequences. *Nature.* **403**, 665–668.

Parsonnet, J. (1995). The incidence of *Helicobacter pylori* infection. *Aliment. Pharmacol. Ther.* **9**(Suppl. 2), 45–51.

Parsonnet, J., G. D. Friedman, D. P. Vandersteen *et al.* (1991). *Helicobacter pylori* infection and the risk of gastric carcinoma. *N. Engl. J. Med.* **325**, 1127–1131.

Patton, C. M., N. Shaffer, P. Edmonds *et al.* (1989). Human disease associated with '*Campylobacter* upsaliensis' (catalase-negative or weakly positive *Campylobacter* species) in the United States. *J. Clin. Microbiol.* **27**, 66–73.

Peek, R. M. Jr and M. J. Blaser (2002). *Helicobacter pylori* and gastrointestinal tract adenocarcinomas. *Natl Rev. Cancer* **2**, 28–37.

Penner, J. L., J. N. Hennessy and R. V. Congi (1983). Serotyping of *Campylobacter jejuni* and *Campylobacter coli* on the basis of thermostable antigens. *Eur. J. Clin. Microbiol.* **2**, 378–383.

Perez-Perez, G. I. (2000). Accurate diagnosis of *Helicobacter pylori*. Culture, including transport. *Gastroenterol. Clin. North Am.* **29**, 879–884.

Perez-Perez, G. I., A. F. Cutler and M. J. Blaser (1997). Value of serology as a noninvasive method for evaluating the efficacy of treatment of *Helicobacter pylori* infection. *Clin. Infect. Dis.* **25**, 1038–1043.

Perez-Perez, G. I., R. M. Peek, A. J. Legath *et al.* (1999). The role of CagA status in gastric and extragastric complications of *Helicobacter pylori*. *J. Physiol. Pharmacol.* **50**, 833–845.

Perez-Perez, G. I., A. Salomaa, T. U. Kosunen *et al.* (2002). Evidence that cagA(+) *Helicobacter pylori* strains are disappearing more rapidly than cagA(−) strains. *Gut* **50**, 295–298.

Petersen, L. and A. Wedderkopp (2001). Evidence that certain clones of *Campylobacter jejuni* persist during successive broiler flock rotations. *Appl. Environ. Microbiol.* **67**, 2739–2745.

Petersen, L., E. M. Nielsen, J. Engberg *et al.* (2001). Comparison of genotypes and serotypes of *Campylobacter jejuni* isolated from Danish wild mammals and birds and from broiler flocks and humans. *Appl. Environ. Microbiol.* **67**, 3115–3121.

Pigrau, C., R. Bartolome, B. Almirante *et al.* (1997). Bacteremia due to *Campylobacter* species, clinical findings and antimicrobial susceptibility patterns. *Clin. Infect. Dis.* **25**, 1414–1420.

Prendergast, M. M., A. J. Lastovica and A. P. Moran (1998). Lipopolysaccharides from *Campylobacter jejuni* O, 41 strains associated with Guillain-Barré syndrome exhibit mimicry of GM1 ganglioside. *Infect. Immun.* **66**, 3649–3655.

Rees, J. H., R. W. Vaughan, E. Kondeatis and R. A. Hughes (1995). HLA-class II alleles in Guillain-Barré syndrome and Miller Fisher syndrome and their association with preceding *Campylobacter jejuni* infection. *J. Neuroimmunol.* **62**, 53–57.

Refregier-Petton, J., N. Rose, M. Denis and G. Salvat (2001). Risk factors for *Campylobacter* spp. contamination in French broiler-chicken flocks at the end of the rearing period. *Prev. Vet. Med.* **50**, 89–100.

Reinthaler, F. F., G. Feierl, D. Stunzner and E. Marth (1998). Diarrhea in returning Austrian tourists, epidemiology, etiology, and cost-analyses. *J. Travel. Med.* **5**, 65–72.

Rodrigues, L. C., J. M. Cowden, J. G. Wheeler *et al.* (2001). The study of infectious intestinal disease in England, risk factors for cases of infectious intestinal disease with *Campylobacter jejuni* infection. *Epidemiol. Infect.* **127**, 185–193.

Roels, T. H., B. Wickus, H. H. Bostrom *et al.* (1998). A foodborne outbreak of *Campylobacter jejuni* (O, 33) infection associated with tuna salad, a rare strain in an unusual vehicle. *Epidemiol. Infect.* **121**, 281–287.

Rosef, O., G. Rettedal and L. Lageide (2001). Thermophilic campylobacters in surface water, a potential risk of campylobacteriosis. *Intl J. Environ. Health Res.* **11**, 321–327.

Sacks, J. J., S. Lieb, L. M. Baldy *et al.* (1986). Epidemic campylobacteriosis associated with a community water supply. *Am. J. Public Health* **76**, 424–428.

Saeed, A. M., N. V. Harris and R. F. DiGiacomo (1993). The role of exposure to animals in the etiology of *Campylobacter jejuni/coli* enteritis. *Am. J. Epidemiol.* **137**, 108–114.

Salama, N. R., G. Otto, L. Tompkins and S. Falkow (2001). Vacuolating cytotoxin of *Helicobacter pylori* plays a role during colonization in a mouse model of infection. *Infect. Immun.* **69**, 730–736.

Salama, S. M., H. Tabor, M. Richter and D. E. Taylor (1992). Pulsed-field gel electrophoresis for epidemiologic studies of *Campylobacter hyointestinalis* isolates. *J. Clin. Microbiol.* **30**, 1982–1984.

Sandstedt, K., J. Ursing and M. Walder (1983). Thermotolerant *Campylobacter* with no or weak catalase activity isolated from dogs. *Curr. Microbiol.* **8**, 209–213.

Shreeve, J. E., M. Toszeghy, M. Pattison and D. G. Newell (2000). Sequential spread of *Campylobacter* infection in a multipen broiler house. *Avian Dis.* **44**, 983–988.

Slader, J., G. Domingue, F. Jorgensen *et al.* (2002). Impact of transport crate reuse and of catching and processing on *Campylobacter* and *Salmonella* contamination of broiler chickens. *Appl. Environ. Microbiol.* **68**, 713–719.

Smith, K. E., J. M. Besser, C. W. Hedberg *et al.* (1999). Quinolone-resistant *Campylobacter jejuni* infections in Minnesota, 1992–1998. Investigation Team. *N. Engl. J. Med.* **340**, 1525–1532.

Stern, N. J., P. J. Rothenberg and J. M. Stone (1985). Enumeration and reduction of *Campylobacter jejuni* in poultry and red meats. *J. Food Prot.* **48**, 606–610.

Stern, N. J., P. Fedorka-Cray, J. S. Bailey *et al.* (2001). Distribution of *Campylobacter* spp. in selected US poultry production and processing operations. *J. Food Prot.* **64**, 1705–1710.

Stolzenberg-Solomon, R. Z., M. J. Blaser, P. J. Limburg *et al.* (2001). *Helicobacter pylori* seropositivity as a risk factor for pancreatic cancer. *J. Natl Cancer Inst.* **93**, 937–941.

Studahl, A. and Y. Andersson (2000). Risk factors for indigenous campylobacter infection, a Swedish case-control study. *Epidemiol. Infect.* **125**, 269–275.

Suerbaum, S. and P. Michetti (2002). *Helicobacter pylori* infection. *N. Engl. J. Med.* **347**, 1175–1186.

Talibart, R., M. Denis, A. Castillo *et al.* (2000). Survival and recovery of viable but noncultivable forms of *Campylobacter* in aqueous microcosm. *Intl J. Food Microbiol.* **55**, 263–267.

Tauxe, R. V., C. M. Patton, P. Edmonds *et al.* (1985). Illness associated with *Campylobacter laridis*, a newly recognized *Campylobacter* species. *J. Clin. Microbiol.* **21**, 222–225.

Tauxe, R. V., N. Hargrett-Bean, C. M. Patton and I. K. Wachsmuth (1988). *Campylobacter* isolates in the United States, 1982–1986. *Morbid. Mortal. Weekly Rep. CDC Surveill. Summ.* **37**, 1–13.

Tenkate, T. D. and R. J. Stafford (2001). Risk factors for campylobacter infection in infants and young children, a matched case-control study. *Epidemiol. Infect.* **127**, 399–404.

Thomas, C., D. J. Hill and M. Mabey (1999). Evaluation of the effect of temperature and nutrients on the survival of *Campylobacter* spp. in water microcosms. *J. Appl. Microbiol.* **86**, 1024–1032.

Thomas, C., D. Hill and M. Mabey (2002). Culturability, injury and morphological dynamics of thermophilic *Campylobacter* spp. within a laboratory-based aquatic model system. *J. Appl. Microbiol.* **92**, 433–442.

Tomb, J. F., O. White, A. R. Kerlavage *et al.* (1997). The complete genome sequence of the gastric pathogen *Helicobacter pylori. Nature* **388**, 539–547.

Torres, J., G. Perez-Perez, K. J. Goodman *et al.* (2000). A comprehensive review of the natural history of *Helicobacter pylori* infection in children. *Arch. Med. Res.* **31**, 431–469.

Torres, J., M. Camorlinga-Ponce, G. Perez-Perez *et al.* (2001). Increasing multidrug resistance in *Helicobacter pylori* strains isolated from children and adults in Mexico. *J. Clin. Microbiol.* **39**, 2677–2680.

Trachoo, N., J. F. Frank and N. J. Stern (2002). Survival of *Campylobacter jejuni* in biofilms isolated from chicken houses. *J. Food Prot.* **65**, 1110–1116.

Uemura, N., S. Okamoto, S. Yamamoto *et al.* (2001). *Helicobacter pylori* infection and the development of gastric cancer. *N. Engl. J. Med.* **345**, 784–789.

Vandamme, P., P. Pugina, G. Benzi *et al.* (1992). Outbreak of recurrent abdominal cramps associated with *Arcobacter butzleri* in an Italian school. *J. Clin. Microbiol.* **30**, 2335–2337.

van de Giessen, A. W., C. J. Heuvelman, T. Abee and W. C. Hazeleger (1996a). Experimental studies on the infectivity of non-culturable forms of *Campylobacter* spp. in chicks and mice. *Epidemiol. Infect.* **117**, 463–470.

van de Giessen, A. W., B. P. Bloemberg, W. S. Ritmeester and J. J. Tilburg (1996b). Epidemiological study on risk factors and risk reducing measures for campylobacter infections in Dutch broiler flocks. *Epidemiol. Infect.* **117**, 245–250.

Van Looveren, M., G. Daube, L. De Zutter *et al.* (2001). Antimicrobial susceptibilities of *Campylobacter* strains isolated from food animals in Belgium. *J. Antimicrob. Chemother.* **48**, 235–240.

Versalovic, J. and J. G. Fox (2003). *Helicobacter*. In: P. R. Murray, E. J. Baron, M. A. Pfaller *et al.* (eds), *Manual of Clinical Microbiology*, 8th edn, pp. 915–928. ASM Press, Washington, DC.

Vicari, J. J., R. M. Peek, G. W. Falk *et al.* (1998). The seroprevalence of cagA-positive *Helicobacter pylori* strains in the spectrum of gastroesophageal reflux disease. *Gastroenterology* **115**, 50–57.

Waldenstrom, J., T. Broman, I. Carlsson *et al.* (2002). Prevalence of *Campylobacter jejuni*, *Campylobacter lari*, and *Campylobacter coli* in different ecological guilds and taxa of migrating birds. *Appl. Environ. Microbiol.* **68**, 5911–5917.

Walmsley, S. L. and M. A. Karmali (1989). Direct isolation of atypical thermophilic *Campylobacter* species from human feces on selective agar medium. *J. Clin. Microbiol.* **27**, 668–670.

Walz, S. E., S. Baqar, H. J. Beecham *et al.* (2001). Pre-exposure anti-*Campylobacter jejuni* immunoglobulin A levels associated with reduced risk of *Campylobacter* diarrhea in adults traveling to Thailand. *Am. J. Trop. Med. Hyg.* **65**, 652–656.

Wang, G., C. G. Clark, T. M. Taylor *et al.* (2002). Colony multiplex PCR assay for identification and differentiation of *Campylobacter jejuni*, C. *coli*, C. *lari*, C. *upsaliensis*, and C. *fetus* subsp. fetus. *J. Clin. Microbiol.* **40**, 4744–4747.

Wassenaar, T. M. and D. G. Newell (2000). Genotyping of *Campylobacter* spp. *Appl. Environ. Microbiol.* **66**, 1–9.

Wedderkopp, A., E. Rattenborg and M. Madsen (2000). National surveillance of *Campylobacter* in broilers at slaughter in Denmark in 1998. *Avian Dis.* **44**, 993–999.

Weijtens, M. J., H. A. Urlings and P. J. Van der Plas (2000). Establishing a campylobacter-free pig population through a top-down approach. *Lett. Appl. Microbiol.* **30**, 479–484.

Werno, A. M., J. D. Klena, G. M. Shaw and D. R. Murdoch (2002). Fatal case of *Campylobacter lari* prosthetic joint infection and bacteremia in an immunocompetent patient. *J. Clin. Microbiol.* **40**, 1053–1055.

Wesley, I. V., L. Schroeder-Tucker, A. L. Baetz *et al.* (1995). *Arcobacter*-specific and *Arcobacter butzleri*-specific 16S rRNA-based DNA probes. *J. Clin. Microbiol.* **33**, 1691–1698.

Winquist, A. G., A. Roome, R. Mshar *et al.* (2001). Outbreak of campylobacteriosis at a senior center. *J. Am. Geriatr. Soc.* **49**, 304–307.

Wolfs, T. F., B. Duim, S. P. Geelen *et al.* (2001). Neonatal sepsis by *Campylobacter jejuni*, genetically proven transmission from a household puppy. *Clin. Infect. Dis.* **32**, E97–E99.

Wood, R. C., K. L. MacDonald and M. T. Osterholm (1992). *Campylobacter* enteritis outbreaks associated with drinking raw milk during youth activities. A 10-year review of outbreaks in the United States. *J. Am. Med. Assoc.* **268**, 3228–3230.

Yang, H., Y. Li and M. G. Johnson (2001). Survival and death of *Salmonella* typhimurium and *Campylobacter jejuni* in processing water and on chicken skin during poultry scalding and chilling. *J. Food Prot.* **64**, 770–776.

Yuki, N., Y. Tagawa, F. Irie *et al.* (1997). Close association of Guillain-Barré syndrome with antibodies to minor monosialogangliosides GM1b and GM1 alpha. *J. Neuroimmunol.* **74**, 30–34.

Zhao, C., B. Ge, J. De Villena *et al.* (2001). Prevalence of *Campylobacter* spp., *Escherichia coli*, and *Salmonella* serovars in retail chicken, turkey, pork, and beef from the Greater Washington, DC, area. *Appl. Environ. Microbiol.* **67**, 5431–5436.

Yersinia infections

8

Truls Nesbakken

1 Historical aspects and contemporary problems

The genus *Yersinia* of the family *Enterobacteriaceae* includes three well-established pathogens (*Yersinia pestis, Yersinia pseudotuberculosis* and *Yersinia enterocolitica*) and several non-pathogens (Mollaret *et al.*, 1979). *Y. pestis* was isolated by Alexandre Yersin in 1894 (Yersin, 1894). The most important *Yersinia* infection, plague, caused by *Y. pestis*, is one of the most ancient recognized human diseases. Disease due to *Y. pseudotuberculosis* (first described in 1884) has been recognized since the beginning of the twentieth century, and *Y. enterocolitica* was first shown to be associated with human disease in 1939 (Mollaret, 1995).

The current interest in *Y. enterocolitica* started in 1958 following a number of epizootics among chinchillas and hares (Mollaret *et al.*, 1979; Hurvell, 1981), and after the establishment of a causal relationship with abscedizing lymphadenitis in man. The similarity between the human and animal isolates was established in 1963 (Knapp and Thal, 1963), and in 1964 the species name *Y. enterocolitica* was formally proposed by Frederiksen (1964). Over the past 30 years the bacterium has been found with increasing frequency as a cause of human disease, and from animals and inanimate sources.

Foodborne Infections and Intoxications 3e
ISBN-10: 0-12-588365-X
ISBN-13: 978-012-588365-8

Yersinia enterocolitica is an important cause of gastroenteritis in humans, especially in temperate countries (Mollaret *et al.*, 1979; World Health Organization, 1983, 1987). Evidence from large outbreaks of yersiniosis in the US, Canada, and Japan (Cover and Aber, 1982) and from epidemiological studies of sporadic cases has shown that *Y. enterocolitica* is a foodborne pathogen, and that in many cases pork is implicated as the source of infection (Morris and Feeley, 1976; Hurvell, 1981; Tauxe *et al.*, 1987; Lee *et al.*, 1991; Ostroff *et al.*, 1994). Due to the relative lack of information on *Y. pseudotuberculosis* as a foodborne pathogen (Schiemann, 1989), this bacterium will be considered less extensively than *Y. enterocolitica. Y. pseudotuberculosis* mainly causes epizootic disease, especially in rodents, with necrotizing granulomatous disease of liver, spleen and lymph nodes (Aleksic and Bockemühl, 1990; Aleksic *et al.*, 1995). In humans it may cause acute abdominal disease, septicemia, arthritis and erythema nodosum (Ahvonen, 1972a; Knapp, 1958).

2 Classification

The genus *Yersinia* was proposed in 1944 by Van Loghem (1944) for bacteria that were related to the genus *Pasteurella* and that were pathogenic. Thal (1954) drew attention to evidence relating *Yersinia* to the *Enterobacteriaceae*. A general numerical taxonomic study from 1958 placed *Yersinia* between *Klebsiella* and *Escherichia coli* (Sneath and Cowan, 1958). The allocation of *Yersinia* to the family *Enterobacteriaceae* was further supported by Frederiksen (1964).

3 Differentiation of *Y. enterocolitica* from other *Yersinia* species

A range of strains of *Yersinia* variants has been isolated from animals, water, and food (Mollaret *et al.*, 1979; Hurvell, 1981; Lee *et al.*, 1981). Many of these bacteria have characteristics that deviate considerably from the typical pattern shown by *Y. enterocolitica*, but can be classified as belonging to the genus *Yersinia* (Mollaret *et al.*, 1979). Such *Y. enterocolitica*-like bacteria have now been divided on a genetic basis into seven new species (Aleksic *et al.*, 1987; Bercovier *et al.*, 1980a, 1980b, 1984; Brenner *et al.*, 1980a, 1980b, Ursing *et al.*, 1980; Brenner, 1981; Wauters *et al.*, 1988a): *Yersinia frederiksenii*, *Yersinia kristensenii*, *Yersinia intermedia*, *Yersinia aldovae*, *Yersinia rohdei*, *Yersinia mollaretii* and *Yersinia bercovieri*.

Y. enterocolitica is a Gram-negative, oxidase-negative, catalase-positive, nitrate reductase-positive, facultative anaerobic rod (occasionally coccoid), 0.5–0.8 × 1–3 μm in size (Bercovier and Mollaret, 1984). It does not form a capsule or spores. It is non-motile at 35–37°C, but motile at 22–25°C with relatively few, peritrichous flagellae. Some human pathogenic strains of serovar O:3 are, however, non-motile at both temperatures. In addition, the bacterium is urease positive, H_2S negative, ferments mannitol, and produces acid (but not gas) from glucose.

3.1 Phenotypic characterization

3.1.1 Biotyping

Wauters *et al.* (1987) described a revised biotyping scheme that differentiates between pathogenic (biovars 1B, 2, 3, 4, 5) and non-pathogenic (only biovar 1A) variants. The proposed biovar 6 (Wauters *et al.*, 1987) is re-classified into two new species: *Y. bercovieri* and *Y. mollaretii* (Wauters *et al.*, 1988a).

3.1.2 Serotyping using O- and H-antigens

Y. enterocolitica can be divided into serovars using O-antigens. Seventy-six different O-factors have been described in both *Y. enterocolitica* and *Y. enterocolitica*-like bacteria (Wauters, 1981; Wauters *et al.*, 1991). A few strains, however, cannot be typed by this system, and the number of described antigen factors is therefore likely to increase in the future. Fifty-four H-factors have been recognized (Wauters *et al.*, 1991; S. Aleksic, personal communication, 1995), but H-antigen determination is rarely carried out and most studies are limited to the O-antigens.

3.1.3 Correlation between biovars and serovars and pathogenicity

Strains of biovar 1B belong to a small number of pathogenic serovars, the most frequent being O:8. Biovar 2 only includes two serovars, O:9 and O:5,27, which are pathogenic for man. Biovar 4/serovar, O:3, is the main pathogenic serovar for man.

3.1.4 Phage typing

Phage typing requires a battery of phages and indicator strains. Two European phage-typing systems are described, but are not used in many laboratories (Mollaret and Nicolle, 1965; Niléhn, 1969; Nicolle, 1973). A bacteriophage typing system that allows greater differentiation of American O:8 strains has also been described (Baker and Farmer, 1982).

3.2 Genotyping

Methods based on the characterization of the genotype include plasmid profile analysis, restriction enzyme analysis of plasmid and chromosomal DNA (DNA fingerprinting), pulsed field gel electrophoresis (PFGE) (Buchrieser *et al.*, 1994; Najdenski *et al.*, 1994; Saken *et al.*, 1994), and the use of DNA or RNA probes (Wachsmuth, 1985; Mayer, 1988; Tenover, 1988; Andersen and Saunders, 1990). Methods based on the characterization of the genotype, within some sero-/biovar combinations like serovar O:8/biovar 1B, often result in a number of different patterns. However, within serovar O:3/biovar 4, the diversity of patterns is limited (Nesbakken *et al.*, 1987). Though several pulsotypes are found among O:3/biovar 4 strains, most of the strains belong to one or two dominating pulsotypes (Buchrieser *et al.*, 1994; Najdenski *et al.*, 1994; Saken *et al.*, 1994; Asplund *et al.*, 1998; Frederiksson-Ahomaa *et al.*, 1999).

4 Virulence factors

4.1 The virulence plasmid

Human pathogenic strains of *Yersinia* spp. *enterocolitica* possess a special plasmid 40–50 MDa in size (Brubaker, 1979; Portnoy and Martinez, 1985). The presence of this plasmid is an essential, though not sufficient, prerequisite for the bacterium to be able to induce disease. The presence of this virulence plasmid has been associated with several properties, most of which are phenotypically expressed only at elevated growth temperatures of 35–37°C (Portnoy and Martinez, 1985). The list of such plasmid-mediated and temperature-regulated properties includes Ca^{2+}-dependent growth (Gemski *et al.*, 1980), production of V and W antigens (Perry and Brubaker, 1983), spontaneous autoagglutination (Laird and Cavanaugh, 1980), mannose-resistant haemagglutination (Kapperud *et al.*, 1987), serum resistance (Pai and DeStephano, 1982), binding of Congo red dye (Prpic *et al.*, 1985), hydrophobicity (Lachica *et al.*, 1984), mouse virulence (Ricciardi *et al.*, 1978; Nesbakken *et al.*, 1987), and production of a number of proteins (Portnoy *et al.*, 1984; Bölin *et al.*, 1985; Portnoy and Martinez, 1985), of which one is a true outer membrane protein (YadA, previously termed Yop1) (Michiels *et al.*, 1990). This true outer membrane protein forms a fibrillar matrix on the bacterial surface and mediates cellular attachment and entry (Bliska and Falkow, 1994). It also confers resistance to the bactericidal effect of normal human serum, and inhibition of the anti-invasive effect of interferon.

4.2 The chromosome

Elements encoded by the chromosome are also necessary for maximum virulence. The pathogenic yersiniae share at least two chromosomal loci, *inv* and *ail*, that play a role in their entry into eukaryotic cells (Miller *et al.*, 1988). The *inv* and *ail* gene products can be classified as adhesins since they mediate adherence to the eukaryotic surface. Unlike other previously characterized bacterial adhesins, they also mediate entry into a variety of mammalian cells. A high pathogenicity island in pathogenic species of *Yersinia* encodes genes for three yersiniabactin (Ybt) transport proteins, six Ybt biosynthetic enzymes, one transcriptional regulator (YbtA) and one protein of unknown function (YbtX) (Perry *et al.*, 2001).

5 Ability to survive and grow

Y. enterocolitica is a facultative organism able to multiply in both aerobic and anaerobic conditions. The ability of *Y. enterocolitica* to multiply at low temperatures is of considerable concern to food producers. Optimum temperature is 28–29°C, and the reported growth range is –2–42°C (Bercovier and Mollaret, 1984). The minimum pH for growth has been reported as being between 4.2 and 4.4 (Kendall and Gilbert, 1980). The ability to propagate at refrigeration temperature in vacuum-packed foods with a prolonged shelf-life (Hanna *et al.*, 1976, 1979) is of considerable significance in food hygiene. *Y. enterocolitica* may survive in frozen foods for long periods (Schiemann, 1989).

Y. enterocolitica is not able to grow at pH < 4.2 or > 9.0 (Kendall and Gilbert, 1980; Stern *et al.*, 1980a), or at salt concentrations greater than 7% (a_w < 0.945) (Stern *et al.*, 1980a). The organism does not survive pasteurization or normal cooking, boiling, baking, and frying temperatures. Heat treatment of milk and meat products at 60°C for 1–3 minutes effectively inactivates *Y. enterocolitica* (Lee *et al.*, 1981). D-values determined in scalding water were 96, 27 and 11 seconds at 58°C, 60°C and 62°C respectively (Sörqvist and Danielsson-Tham, 1990).

According to many reports, the ability of *Y. enterocolitica* to compete with other psychrotrophic organisms normally present in food may be poor (Stern *et al.*, 1980b; Fukushima and Gomyoda, 1986; Schiemann, 1989; Kleinlein and Untermann, 1990). In contrast, a number of studies have shown that *Y. enterocolitica* is able to multiply in foods kept under chilled storage and might even compete successfully (Hanna *et al.*, 1977; Stern *et al.*, 1980a; Grau, 1981; Lee *et al.*, 1981; Gill and Reichel, 1989; Brocklehurst and Lund, 1990; Lindberg and Borch, 1994; Borch and Arvidsson, 1996; Bredholt *et al.*, 1999).

6 Nature of the infection in man

6.1 Clinical symptoms of *Y. enterocolitica* infection

Gastroenteritis is by far the most common symptom of *Y. enterocolitica* infection (yersiniosis) in humans (Bottone, 1977; Mollaret *et al.*, 1979; Wormser and Keusch, 1981; Cover and Aber, 1989). The clinical picture is usually one of a self-limiting diarrhea associated with mild fever and abdominal pain (Wormser and Keusch, 1981). Nausea and vomiting occur, but less frequently. The portion of the gastrointestinal tract usually involved is the ileocaecal region (Sandler *et al.*, 1982). The colon may also be affected and the infection may simulate Crohn's disease, which has a different prognosis (Vantrappen *et al.*, 1977; Gleason and Patterson, 1982). Occasionally the infection is limited to the right fossa iliaca in the form of terminal ileitis or mesenteriel lymphadenitis, with symptoms that can be confused with those of acute appendicitis. In several studies of patients with the appendicitis-like syndrome, *Y. enterocolitica* has been found in up to 9% of patients (Niléhn and Sjöström, 1967; Ahvonen, 1972a; Jebsen *et al.*, 1976; Pai *et al.*, 1982; Samadi *et al.*, 1982; Attwood *et al.*, 1987; Megraud, 1987). Infections with serovars O:3 or O:9 are, in some patients, followed by reactive arthritis (Aho *et al.*, 1981), which is most common in patients possessing the tissue type HLA-B27. Often, although not always, the patient has shown prior gastrointestinal symptoms. Other complications seen with *Y. enterocolitica* infection are reactive skin complaints, erythema nodosum being the most common. Many such patients have no history of prior gastrointestinal involvement. Septicemia due to *Y. enterocolitica* is seen almost exclusively in individuals with underlying disease (Bottone, 1977), while those with cirrhosis and disorders associated with excess iron are particularly predisposed to infection and increased mortality.

Gastroenteritis dominates in children and young people, while various forms of reactive arthritis are most common in young adults, and most patients with skin manifestations are adult females (Wormser and Keusch, 1981). In Scandinavia there is a relatively high incidence of both reactive arthritis (10–30%) (Winblad, 1975) and erythema nodosum (30%) (Ahvonen, 1972b), caused by serovars O:3 and O:9.

6.2 Pathogenesis and immunity

Human infection due to *Y. enterocolitica* is most often acquired by the oral route. The minimal infectious dose required to cause disease is unknown. In one volunteer, ingestion of 3.5×10^9 organisms was sufficient to produce illness (Szita *et al.*, 1973). The incubation period is uncertain, but has been estimated as being between 2 and 11 days (Szita *et al.*, 1973; Ratnam *et al.*, 1982).

Enteric infection leads to proliferation of *Y. enterocolitica* in the lumen of the bowel and in the lymphoid tissue of the intestine. Adherence to and penetration into the epithelial cells of the intestinal mucosa are essential factors in the pathogenesis of *Y. enterocolitica* infection (Portnoy and Martinez, 1985; Cornelis *et al.*, 1987; Miller *et al.*, 1988; Bliska and Falkow, 1994). When the bacteria reach the lymphoid tissues in the terminal ileum, a massive multiplication and inflammatory response takes place in the Peyer's patches. Reactive arthritis and erythema nodosum appear to be delayed immunologic sequelae of the original intestinal infection.

7 Nature of infection or carrier state in animals

7.1 Food animals

7.1.1 Pigs

Newborn piglets are easily colonized and become long-term healthy carriers of *Y. enterocolitica* in the oral cavity and intestines (Schiemann, 1989). In one study (Skjerve *et al.*, 1998), an enzyme-linked immunosorbent assay (ELISA) (Nielsen *et al.*, 1996) was used to detect IgG antibodies against *Y. enterocolitica* O:3 in sera from 1605 slaughter pigs from 321 different herds. Positive titers were found in 869 (54.1%) of the samples. Healthy pigs are often carriers of strains of *Y. enterocolitica* that are pathogenic to humans, in particular strains of serovar O:3/biovar 4 and serovar O:9/biovar 2) (Hurvell, 1981; Schiemann, 1989). The organisms are present in the oral cavity (especially the tongue and tonsils), the submaxillary lymph nodes, the intestine and feces (Nesbakken *et al.*, 2003a, 2003b; Figure 8.1).

Strains of O:3 have been found frequently on the surface of freshly slaughtered pig carcasses, in frequencies up to 63.3% (Nesbakken and Kapperud, 1985; Nesbakken, 1988). This is probably the result of spread of the organism via feces and intestinal contents during slaughter and dressing operations.

Other pathogenic strains do not appear to be as closely associated with pigs, and may have a different ecology. In western Canada, O:8 and O:5,27 strains have

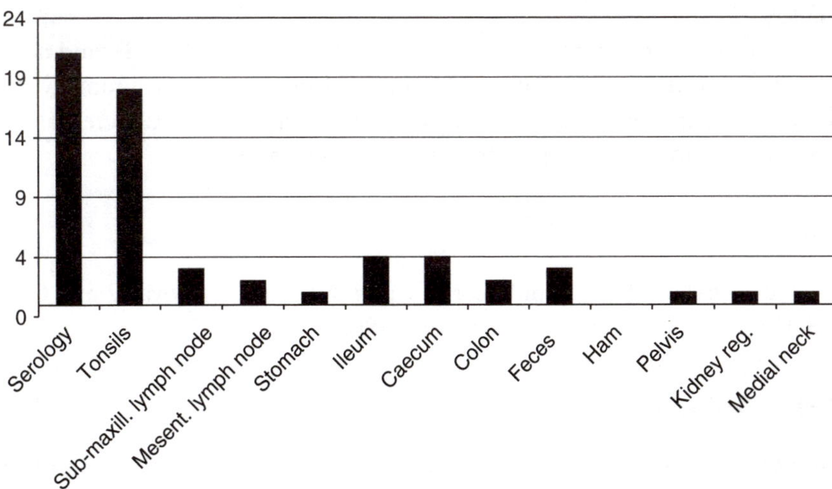

Figure 8.1 Antibodies against *Y. enterocolitica* O:3 in blood, and *Y. enterocolitica* O:3 in lymphoid tissues, intestinal contents and on pigs' carcasses ($n = 24$). Taken from: Nesbakken,T., Eckner, K., Høidal, H. K. and Røtterud, O.-J. (2003b), Occurrence of *Yersinia enterocolitica* in slaughter pigs and consequences for meat inspection, slaughtering, and dressing procedures. In: M. Skurnik, J. A. Bengoechea and K. Granfors (eds), *The Genus Yersinia. Entering the Functional Genomic Area. Advances in Experimental Medicine and Biology*, Vol. 529, pp. 303–308. Kluwer Academic/Plenum Publishers, New York, NY. Reproduced with kind permission of Kluwer Academic Publishers.

been found most commonly in humans, but only O:5,27 strains were found in the throats of slaughter-age pigs (Schiemann, 1989). In the US, O:5,27 strains were isolated from the cecal contents and feces of 2 out of 50 pigs at slaughter (Kotula and Sharar, 1993). Serovar O:8/biovar 1B, until recently considered to be the most common human pathogenic strain of *Y. enterocolitica* in the US (Ostroff, 1995) and in western Canada (Toma and Lafleur, 1981), has seldom been reported in pigs.

7.1.2 Cattle

Positive tests, in serological control programs for brucellosis in cattle, have in some cases proved to be cross-reactions against *Y. enterocolitica* serovar O:9 (Nielsen *et al.*, 1996; Wauters, 1981; Weynants *et al.*, 1996a). However, cattle are generally not considered to be carriers of human pathogenic *Y. enterocolitica*.

7.1.3 Sheep and goats

In Norway, Krogstad (1974) demonstrated outbreaks of *Y. enterocolitica* infection in goat herds in which serovar O:2/biovar 5 was implicated. He also described a case in which an animal attendant was infected by the same serovar. Biovar 5 has also been isolated from goats in New Zealand (Lanada, 1990). Enteritis in sheep and goats due to infection of *Y. enterocolitica* O:2,3, biovar 5 is also seen in Australia (Slee and Button, 1990). Serovar O:3 was isolated from the rectal contents in two (3.0%) of 66 lambs in New Zealand (Bullians, 1987).

7.1.4 Poultry

Stengel (1985) isolated *Y. enterocolitica* serovars O:3 (n = 3), O:9 (n = 3), and non-virulent *Y. enterocolitica* (n = 13) from 130 samples of poultry. This is probably the first time these virulent serovars have been isolated from poultry, and there was no obvious opportunity for cross-contamination from pigs or pork.

7.2 Deer

Surveys in New Zealand have found deer to carry both O:5,27/biovar 2 and O:9/biovar 2 (S. Fenwick, personal communication, 1996).

7.3 Other animals

Dogs, cats, and rodents, such as rats, may also occasionally be fecal carriers of O:3 and O:9 (Hurvell; 1981; Fukushima *et al.*, 1984; Fenwick *et al.*, 1994; Hayashidani *et al.*, 1995). The relatively intimate contact between man and pets suggests a potential reciprocal transmission, although such an epidemiological link has not been clearly confirmed (Nesbakken *et al.*, 1991a; Fenwick *et al.*, 1994; Ostroff *et al.*, 1994). O:3 has been isolated from 1 (0.9%) of 117 crows in Japan, and O:9 from 1 (0.7%) of 156 Japanese serows (Kato *et al.*, 1985).

It has been suggested that rodents may be reservoirs of O:8 and O:21 strains in North America (Schiemann, 1989), and evidence has been published that this is indeed the case in Japan (Hayashidani *et al.*, 1995). Serovars O:8 and O:21 are closely related in many ways (biochemical profile, H antigens and animal virulence). Serovar 21 ('O:Tacoma') has been isolated from wild rodent fleas in the western US (Quan *et al.*, 1974).

As is the case with human pathogenic variants, the animal pathogenic strains also belong to particular combinations of serovars and biovars (Mollaret *et al.*, 1979; Hurvell, 1981). Serovar O:1/biovar 3 was responsible for widespread outbreaks in chinchilla both in Europe and the US during the period 1958–1964 (Mollaret *et al.*, 1979; Hurvell, 1981). During the same period, epizootics were observed among hares along the French–Belgian frontier, caused by serovar O:2/biovar 5 (Mollaret *et al.*, 1979; Hurvell, 1981).

8 Foods most often associated with sporadic cases and outbreaks

8.1 Pork

In contrast to the frequent occurrence of the bacterium in pigs and on freshly slaughtered carcasses, pathogenic *Y. enterocolitica* have only exceptionally been found in pork products at the retail sale stage, with the exception of fresh tongues (Vidon, 1985; Nesbakken *et al.*, 1985; DeBoer *et al.*, 1986; Ternström and Molin, 1987; World Health Organization, 1987; Wauters *et al.*, 1988b; Schiemann, 1989; Delmas and Nesbakken *et al.*, 1991b).

Because of the high prevalence of *Y. enterocolitica* in pig herds, strict slaughter hygiene will remain an important means to reduce carcass contamination with *Y. enterocolitica* as well as other pathogenic microorganisms (Skjerve *et al.*, 1998). During lairage, pathogenic *Y. enterocolitica* may spread from infected to non-infected pigs (Fukushima *et al.*, 1990).

Technological solutions have already been found which allow removal of the rectum without soiling of the carcass. The sealing off of the rectum with a plastic bag immediately after it has been freed can significantly reduce the spread of *Y. enterocolitica* to pig carcasses (Nesbakken *et al.*, 1994). Meat inspection procedures concerning the head also seem to represent a cross-contamination risk: incision of the submaxillary lymph nodes is a compulsory procedure according to the EU regulations (European Commission, 1995). In a screening of 97 animals, 5.2% of samples from the submaxillary lymph nodes were positive; when sampling 24 of these lymph nodes in a follow-up study, 12.5% of the samples were positive (Nesbakken *et al.*, 2003b; Figure 8.1). The compulsory incision of submaxillary lymph nodes may result in the bacterium being transported from the medial neck region to other parts of the carcass by the knives and hands of the meat inspection personnel (Nesbakken, 1988; Nesbakken *et al.*, 2003a).

8.2 Milk and dairy products

Worldwide studies indicate that *Y. enterocolitica* is fairly common in raw milk (Lee *et al.*, 1981). *Y. enterocolitica* was also isolated from ice cream (Mollaret *et al.*, 1972) and pasteurized milk (Sarrouy, 1972; Zen-Yoji, 1973) as early as 1970. However, it is almost solely in connection with outbreaks caused by contaminated pasteurized milk (Tacket *et al.*, 1984; Greenwood and Hooper, 1990; Alsterlund *et al.*, 1995), reconstituted powdered milk (Morse *et al.*, 1984) and contaminated chocolate milk (Black *et al.*, 1978) that the pathogenic strains have been found.

8.3 Water

Shallow wells in particular, and also rivers and lakes, are susceptible to contamination by surface runoff from rain or snow melt. Such runoff may become fecally contaminated by wild or domestic animals, or by leakage from septic tanks or open latrines in the surrounding areas. Water is a significant reservoir of *Y. enterocolitica* (Lassen, 1972; Harvey *et al.*, 1976; Kapperud and Jonsson, 1978; Saari and Jansen, 1979; Langeland, 1983; Brennhovd, 1991). However, most isolates of *Y. enterocolitica* and *Y. enterocolitica*-like bacteria obtained from water are variants with no known pathogenic significance to man.

9 Sporadic cases

Y. enterocolitica has been isolated from humans in many countries of the world, but it seems to be found most frequently in cooler climates (North America; the western coast of South America; Europe; northern, central and eastern Asia; Australia; New Zealand;

and South Africa) (Mollaret *et al.*, 1979; World Health Organization, 1983, 1987; Aleksic and Bockemühl, 1990). The widespread nature of *Y. enterocolitica* has been well documented; by the mid-1970s Mollaret *et al.* (1979) had compiled reports of isolates from 35 countries on six continents. *Y. enterocolitica* infections are an important cause of gastroenteritis in the developed world, occurring particularly as sporadic cases in northern Europe (Black and Slome, 1988; Cover and Aber, 1989), where a clustering of cases during the autumn and winter has been reported (World Health Organization, 1983).

There are appreciable geographic differences in the distribution of the different phenotypes of *Y. enterocolitica* isolated from man (Mollaret *et al.*, 1979; Wauters, 1991). There is also a strong correlation between the serovars isolated from humans and pigs in the same geographical area (Esseveld and Goudzwaard, 1973; Pedersen, 1979; Wauters, 1979; Bercovier *et al.*, 1980a; Schiemann and Fleming, 1981). Serovar O:3 is widespread in Europe, Japan, Canada, Africa, and Latin America. Sometimes, but not always, phage typing makes it possible to distinguish between European, Canadian and Japanese strains (Mollaret *et al.*, 1979; Kapperud *et al.*, 1990b). Serovar O:3 seems to be responsible for more than 90% of the cases in Denmark, Norway, Sweden, and New Zealand, and as many as 78.8% of the cases in Belgium. Serovar O:9/biovar 2 is the second most common in Europe, but its distribution is uneven; while it still accounts for a relatively high percentage of the strains isolated in France, Belgium and the Netherlands, only a few strains have been isolated in Scandinavia (World Health Organization, 1983). Until recently, the most frequently reported serovars in the US were O:8 followed by O:5,27 (Mollaret *et al.*, 1979; Bisset *et al.*, 1990; Ostroff, 1995; World Health Organization, 1995). In recent years, serovar O:3 has been on the increase in the US; O:3 now accounts for the majority of sporadic *Y. enterocolitica* isolates in California (Bisset *et al.*, 1990). In 1989, the estimated cost of yersiniosis in the US was $138 million (World Health Organization, 1995). Principal foodborne infections, as estimated for 1997, are ranked by estimated number of cases caused by foodborne transmission each year in the United States. *Y. enterocolitica* is number 10 in the list (among the bacteria in the list, *Y. enterocolitica* is number 7) (Mead *et al.*, 1999). The appearance of strains of serovars O:3 and O:9 in Europe and Japan in the 1970s (Anon., 1976), and in North America by the end of the 1980s (Lee *et al.*, 1990, 1991), is an example of a global pandemic (Tauxe, 2002).

The first Japanese case of *Y. enterocolitica* O:8 infection was linked to consumption of imported raw pork (Ichinohe *et al.*, 1991), although O:8 infections from raw water have also occurred in Japan (Hayashidani *et al.*, 1995).

The incidence of *Y. enterocolitica* infection in patients with acute endemic enterocolitis ranges from 0% to 4%, depending on the geographic location, study method, and population (Kapperud and Slome, 1998). Only a few epidemiological studies have been performed to investigate the sources of sporadic human infections. A 1985 study of *Y. enterocolitica* in Belgium identified consumption of raw pork as a risk factor for disease (Tauxe *et al.*, 1987). The following variables were found to be independently related to an increased risk of yersiniosis in a case-control study conducted in Norway: drinking untreated water, general preference for meat to be prepared raw or rare, and frequency of consumption of pork and sausages (Ostroff *et al.*, 1994).

10 Outbreaks

In the United States, chocolate milk (Black *et al.*, 1978), pasteurized milk (Tacket *et al.*, 1984), soybean curd (tofu) (Tacket *et al.*, 1985) and bean sprouts (Aber *et al.*, 1982) have been implicated as sources in outbreaks of *Y. enterocolitica* infection. These outbreaks, all of which occurred before 1983, were caused by *Y. enterocolitica* serovars that have been infrequently associated with human disease (serovars O:13, O:18), or which no longer predominate in the US (serovar O:8). More recently, the preparation of raw pork intestines (chitterlings) was associated with an outbreak of *Y. enterocolitica* O:3 infections among black US infants in Georgia (Lee *et al.*, 1990); the organism was isolated from samples of the pork intestines. Also, in outbreaks in Buffalo, New York, between 1994 and 1996 (Kondracki *et al.*, 1996), chitterlings were the vehicle.

The milkborne outbreak in Sweden in 1988 (Alsterlund *et al.*, 1995) was probably caused by recontamination of pasteurized milk because of lack of chlorination of the water supply. In the multistate outbreak in 1982 (Tacket *et al.*, 1984), milk cartons were contaminated with mud from a pig farm (Aulisio *et al.*, 1982). In the case of the outbreak described by Greenwood and Hooper (1990), post-pasteurization contamination may have occurred from bottles. Previous studies have shown that milk-associated *Y. enterocolitica* outbreaks have been linked to the addition of ingredients after pasteurization (Black *et al.*, 1978; Morse *et al.*, 1984).

In 1981, an outbreak of infection due to *Y. enterocolitica* O:8 in Washington State occurred in association with the consumption of tofu packed in untreated spring water (Tacket *et al.*, 1985). The outbreak serovar was isolated from the spring water samples. Another outbreak caused by serovar O:8 was traced to ingestion of contaminated water used in manufacturing or preparation of food (Schiemann, 1989). Two other *Yersinia* outbreaks have been associated with well water. One occurred among members of a Pennsylvania girl-scout troop after they ate bean sprouts grown in contaminated well water (Aber *et al.*, 1982); the other was a familial outbreak of yersiniosis in Canada (Thompson and Gravel, 1986).

The epidemiology of yersiniosis in the US seems to have evolved into a pattern similar to the picture in Europe (Bottone *et al.*, 1987; Bisset *et al.*, 1990; Ostroff, 1995), where foodborne *Yersinia* outbreaks are rare, and where serovar 3 predominates (Mollaret *et al.*, 1979; Prentice *et al.*, 1991; Verhaegen *et al.*, 1991). Although yersiniosis appears to be more common in Europe than in the United States, only five foodborne outbreaks have been reported in Europe (Toivanen *et al.*, 1973; Olsovsky, *et al.*, 1975; Greenwood and Hooper, 1990; Alsterlund *et al.*, 1995; Swedish Institute for Infectious Disease Control, 1995).

In Japan, several outbreaks connected to schools (Zen-Yoji *et al.*, 1973; Maruyama, 1987) and communities (Asakawa *et al.*, 1973) have been reported. In all cases the vehicle was unknown. Often a few hundred persons were ill out of several hundred who were at risk. Serovar O:3 was the agent involved in all cases. In one outbreak in China, caused by serovar O:3 from pickled vegetables, 351 persons were ill (Anon., 1987).

11 Principles of detection

The analytical methods available today for the isolation of pathogenic *Y. enterocolitica* suffer from limitations such as insufficient selectivity and, in particular, inadequate differentiation between pathogenic and non-pathogenic strains.

11.1 Specific principles for isolation

A three-step method, based on a combination of cold enrichment in a non-selective medium with subsequent inoculation onto a highly selective medium, has been developed for the Nordic Committee on Food Analysis (1987).

Wauters *et al.* (1988b) developed a method for isolation of serovar O:3 from meat and meat products. The procedure is based on a 2- to 3-day selective enrichment period in irgasan-ticarcillin-potassium chlorate (ITC) enrichment broth at room temperature, and therefore saves time compared with the method described above (Figure 8.2).

Both *Y. enterocolitica* and *Y. pseudotuberculosis* seem to be more tolerant of alkaline conditions than do most other *Enterobacteriaceae*, and treatment of food enrichments with potassium hydroxide (KOH) may be used to selectively reduce the level of background flora (Aulisio *et al.*, 1980) (Figure 8.3). Elements of the methods from the Nordic Committee on Food Analysis (1987), Schiemann (1982) and Wauters *et al.* (1988b), and KOH treatment (Schiemann, 1983), are incorporated into the International Organization for Standardization (ISO) method (ISO 10273) (Figures 8.2, 8.3; International Organization for Standardization, 1994).

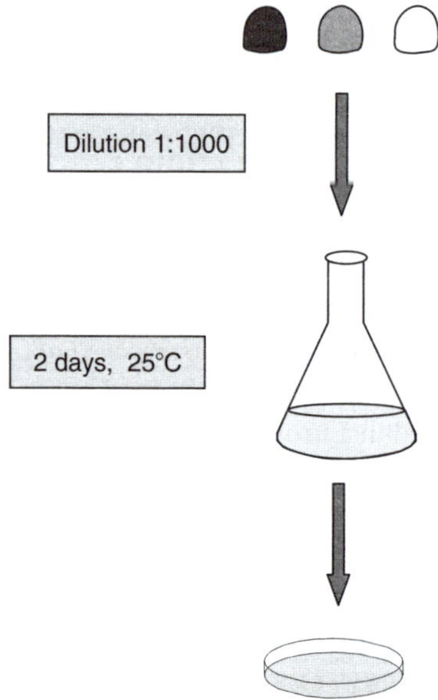

Figure 8.2 Method for recovery of *Y. enterocolitica* from foods according to the International Organization for Standardization (1994). This element of the method is recommended for serovar O:3 in particular. ITC, irgasan-ticarcillin-potassium chlorate enrichment broth; SSDC, Salmonella-Shigella + sodium deoxycholate, $CaCl_2$ agar.

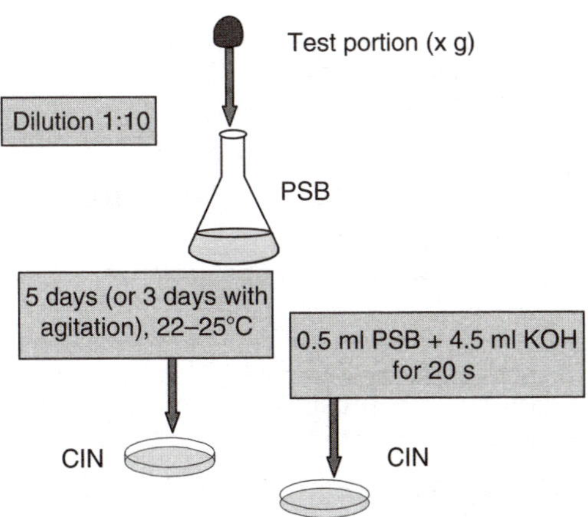

Figure 8.3 Method for recovery of *Y. enterocolitica* from foods according to the International Organization for Standardization (1994). This element of the method is recommended for all pathogenic serovars. PSB, phosphate-buffered saline, sorbitol, and bile salts; KOH, KOH and NaCl; CIN, cefsulodin-irgasan-novobiocin agar.

11.2 Detection by DNA colony hybridization

Genetic probes can also be used in DNA colony hybridization to demonstrate virulent *Y. enterocolitica* strains (Wachsmuth, 1985; Tenover, 1988; Kapperud *et al.*, 1990a). Isolation plus hybridization increased the detection rate from 16% to 38% for the method according to Wauters *et al.* (1988a), and from 10% to 48% for the Nordic method. The results of this investigation (Nesbakken *et al.*, 1991b) support the supposition that conventional culture methods lead to underestimation of virulent *Y. enterocolitica* in pork products.

11.3 Detection by polymerase chain reaction

Polymerase chain reaction (PCR) methods often use primers targeting the *virF* (Thisted-Lambertz *et al.*, 1996; Weynants *et al.*, 1996b) or *yadA* (Kapperud *et al.*, 1993) gene, but the *IcrE* gene (Viitanen *et al.*, 1991) and the *yopT* gene (Arnold *et al.*, 2001) from the virulence plasmid have also been used. *Y. enterocolitica* may lose the virulence plasmid during culture, subculture or storage (Blais and Philippe, 1995). Accordingly, PCR methods based on chromosomal virulence genes, often the *ail* gene, have been developed. Often a combination of genes from the virulence plasmid and the chromosome are used. A common gene combination in such a multiplex PCR assay is the *virF* and *ail* genes (Kaneko *et al.*, 1995; Nilsson *et al.*, 1998).

Rasmussen *et al.* (1995) detected *Y. enterocolitica* O:3 in fecal samples and tonsil swabs from pigs using IMS and PCR based on the *inv* gene. O:3 cells were detected after pre-enrichment, but direct detection needed further optimization of the sample preparation procedures. By combining *inv*, *virF* and *ail* genes in a multiplex PCR assay, Weynants *et al.* (1996b) could differentiate between *Y. pseudotuberculosis*, virulent *Y. enterocolitica* and *Y. enterocolitica* O:3.

Bibliography

Aber, R. C., M. A. Mc Carthy, R. Berman *et al.* (1982). An outbreak of *Yersinia enterocolitica* illness among members of a Brownie troop in Centre County, Pennsylvania. *Program and Abstracts of the 22nd Interscience Conference on Antimicrobial Agents and Chemotherapy. Miami Beach.* American Society for Microbiology, Washington, DC.

Aho, K., P. Ahvonen, O. Laitinen and M. Leirisalo (1981). Arthritis associated with *Yersinia enterocolitica* infection. In: E. J. Bottone (ed.), *Yersinia enterocolitica,* pp. 113–124. CRC Press, Inc., Boca Raton, FL.

Ahvonen, P. (1972a). Human yersiniosis in Finland. I. Bacteriology and serology. *Ann. Clin. Res.* **4**, 30–38.

Ahvonen, P (1972b). Human yersiniosis in Finland. II. Clinical features. *Ann. Clin. Res.* **4**, 39–48.

Aleksic, S. and J. Bockemühl (1990). Mikrobiologie und Epidemiologie der Yersiniosen. *Immun. Infekt.* **18**, 178–185.

Aleksic, S., A. G. Steigerwalt, J. Bockemühl *et al.* (1987). *Yersinia rohdei* sp. nov. isolated from human and dog feces and surface water. *Intl J. Syst. Bacteriol.* **37**, 327–332.

Aleksic, S., J. Bockemühl and H. H. Wuthe (1995). Epidemiology of *Y. pseudotuberculosis* in Germany, 1983–1993. *Contrib. Microbiol. Immunol.* **13**, 55–58.

Alsterlund, R., M.-L. Danielsson-Tham, T. Edén *et al.* (1995). *Yersinia enterocolitica*-utbrott på Bjärehalvön. Risker med kylda matvaror. *Svensk Vet. Tidsskr.* **47**, 257–260.

Andersen, J. K. and N. A. Saunders (1990). Epidemiological typing of *Yersinia enterocolitica* by analysis of restriction fragment length polymorphisms with a cloned ribosomal RNA gene. *J. Med. Microbiol.* **32**, 179–187.

Anonymous (1976). Worldwide spread of infections with *Yersinia enterocolitica. WHO Chronicle* **30**, 494–496.

Anonymous (1987). First report of an outbreak caused by *Yersinia enterocolitica* serotype O, 3 in Shengyang. *Chin. J. Epid.* **8**, 264–267.

Arnold, T., A. Hensel, R. Hagen *et al.* (2001). A highly specific one-step PCR-assay for the rapid discrimination of enteropathogenic *Yersinia enterocolitica* from pathogenic *Yersinia pseudotuberculosis* and *Yersinia pestis. Syst. Appl. Microbiol.* **24**, 285–289.

Asakawa, Y., S. Akahane, N. Kagata *et al.* (1973). Two community outbreaks of human infection with *Yersinia enterocolitica. J. Hygiene* **71**, 715–723.

Asplund, K., T. Johansson and A. Siitonen (1998). Evaluation of pulsed field gel electrophoresis of genomic restriction fragments in the discrimination of *yersinia enterocolitica* O:3. *Epidemiol. Infect.* **121**, 579–586.

Attwood, S. E. A., K. Mealy, M. T. Cafferkey *et al.* (1987). *Yersinia* infection and acute abdominal pain. *Lancet* **i**, 529–533.

Aulisio, C. C. G., I. J. Mehlman and A. C. Sanders (1980). Alkali method for rapid recovery of *Yersinia enterocolitica* and *Yersinia enterocolitica* from foods. *Appl. Environ. Microbiol.* **39**, 135–140.

Aulisio, C. C. G., J. M. Lanier and M. A. Chappel (1982). *Yersinia enterocolitica* and O:13 associated with outbreaks in three southern states. *J. Food. Prot.* **45**, 1263.

Baker, P. M. and J. J. Farmer III (1982). New bacteriophage typing system for *Yersinia enterocolitica, Yersinia kristensenii, Yersinia frederiksenii*, and *Yersinia intermedia*, correlation with serotyping, biotyping, and antibiotic susceptibility. *Clin. Microbiol.* **15**, 491–502.

Bercovier, H. and H. H. Mollaret (1984). Genus XIV. *Yersinia*. In: N. R. Krieg (ed.), *Bergey's Manual of Systematic Bacteriology*, Vol. 1, pp. 498–506. Williams & Wilkins, Baltimore, MD.

Bercovier, H., D. J. Brenner, J. Ursing *et al.* (1980a). Characterization of *Yersinia enterocolitica* sensu stricto. *Curr. Microbiol.* **4**, 201–206.

Bercovier, H., J. Ursing, D. J. Brenner *et al.* (1980b). *Yersinia kristensenii*, a new species of *Enterobacteriaceae* composed of sucrose negative strains (formerly called atypical *Yersinia enterocolitica* or *Yersinia enterocolitica*-like). *Curr. Microbiol.* **4**, 219–224.

Bercovier, H., A. G. Steigerwalt, A. Guiyoule *et al.* (1984). *Yersinia aldovae* (formerly *Yersinia enterocolitica*-like group X2), a new species of *Enterobacteriaceae* isolated from aquatic ecosystems. *Intl J. Syst. Bacteriol.* **34**, 166–172.

Bissett, M. L., C. Powers, S. L. Abbott and J. M. Janda (1990). Epidemiologic investigations of *Yersinia enterocolitica* and related species, sources, frequency, and serogroup distribution *J. Clin. Microbiol.* **28**, 910–912.

Black, R. E. and S. Slome (1988). *Yersinia enterocolitica. Infect. Dis. Clin. North Am.* **2**, 625–641.

Black, R. E., R. J. Jackson, T. Tsai *et al.* (1978). Epidemic *Yersinia enterocolitica* infection due to contaminated chocolate milk. *N. Engl. J. Med.* **298**, 76–79.

Blais, B. W. and L. M. Philippe (1995). Comparative analysis of *yadA* and *ail* polymerase chain reaction methods for virulent *Yersinia enterocolitica. Food Control* **6**, 211–214.

Bliska, J. B. and S. Falkow (1994). Interplay between determinants of cellular entry and cellular disruption in the enteropathogenic *Yersinia. Curr. Opin. Infect. Dis.* **7**, 323–328.

Bölin, I., D. A. Portnoy and H. Wolf-Watz (1985). Expression of the temperature-inducible outer membrane proteins of yersiniae. *Infect. Immun.* **48**, 234–240.

Borch, E. and B. Arvidsson (1996). Growth of *Yersinia enterocolitica* O, 3 in pork. In: *Food Associated Pathogens. Proceedings of the International Union of Food Science and Technology, Uppsala, Sweden*, pp. 202–203.

Bottone, E. J. (1977). *Yersinia enterocolitica*, a panoramic view of a charismatic microorganism. *Crit. Rev. Microbiol.* **5**, 211–241.

Bottone, E. J., C. R. Gullans and M. F. Sierra (1987). Disease spectrum of *Yersinia enterocolitica* serogroup O:3, the predominant cause of human infection in New York City. *Contrib. Microbiol. Immunol.* **9**, 56–60.

Bredholt, S., T. Nesbakken and A. Holck (1999). Protective cultures inhibit growth of *Listeria monocytogenes* and *Escherichia coli* O157:H7 in cooked, sliced vacuum- and gas-packaged meat. *Intl J. Food Microbiol.* **53**, 43–52.

Brenner, D. J. (1981). Classification of *Yersinia enterocolitica*. In: E. J. Bottone (ed.), *Yersinia enterocolitica*, pp. 1–8. CRC Press, Inc., Boca Raton, FL.

Brenner, D. J., H. Bercovier, J. Ursing *et al.* (1980a). *Yersinia intermedia*, a new species of *Enterobacteriaceae* composed of rhamnose-positive, melibiose-positive, raffinose-positive strains (formerly called *Yersinia enterocolitica* or *Yersinia enterocolitica*-like). *Curr. Microbiol.* **4**, 207–212.

Brenner, D. J., J. Ursing, H. Bercovier *et al.* (1980b). Deoxyribonucleic acid relatedness in *Yersinia enterocolitica* and *Yersinia enterocolitica*-like organisms. *Curr. Microbiol.* **4**, 195–200.

Brennhovd, O. (1991). Termotolerante *Campylobacter* spp. og *Yersinia* spp. i noen norske vannforekomster. Thesis, Norwegian College of Veterinary Medicine, Oslo.

Brocklehurst, T. F. and B. M. Lund (1990). The influence of pH, temperature and organic acids on the initiation of growth of *Yersinia enterocolitica. J. Appl. Bacteriol.* **69**, 390–397.

Brubaker, R. R. (1979). Expression of virulence in yersiniae. In: D. Schlessinger (ed.), *Microbiology – 1979*, pp. 168–171. American Society for Microbiology, Washington, DC.

Buchrieser, C., S. D. Weagant and C. W. Kaspar (1994). Molecular characterization of *Yersinia enterocolitica* by pulsed-field gel electrophoresis and hybridization of DNA fragments to *ail* and pYV probes. *Appl. Environ. Microbiol.* **60**, 4371–4379.

Bullians, J. A. (1987). *Yersinia* species infection of lambs and cull cows at an abattoir. *NZ Vet. J.* **35**, 65–67.

Cornelis, G., Y. Laroche, G. Balligand *et al.* (1987). *Yersinia enterocolitica*, a primary model for bacterial invasiveness. *Rev. Infect. Dis.* **9**, 64–87.

Cover, T. L. and R. C. Aber (1989). *Yersinia enterocolitica. N. Engl. J. Med.* **321**, 16–24.

DeBoer, E., W. M. Seldam and J. Oosterom (1986). Characterization of *Yersinia enterocolitica* and related species isolated from foods and porcine tonsils in the Netherlands. *Intl J. Food Microbiol.* **3**, 217–224.

Delmas, C. L. and D. J.-M. Vidon (1985). Isolation of *Yersinia enterocolitica* and related species from foods in France. *Appl. Environ. Microbiol.* **50**, 767–771.

Esseveld, H. and C. Goudzwaard (1973). On the epidemiology of *Y. enterocolitica* infections: pigs as the source of infections in man. *Contrib. Microbiol. Immunol.* **2**, 99–101.

European Commission (1995). Council Directive 64/433/EEC on Health Conditions for the Production and Marketing of Fresh Meat. EC, Brussels.

Fenwick, S. G., P. Madie and C. R. Wicks (1994). Duration of carriage and transmission of *Yersinia enterocolitica* biotype 4, serotype O:3 in dogs. *Epidemiol. Infect.* **113**, 471–477.

Frederiksen, W. (1964). A study of some *Yersinia pseudotuberculosis*-like bacteria ('*Bacterium enterocoliticum*' and '*Pasteurella* X'). In: *Proceedings of the XIV Scandinavian Congress of Pathology and Microbiology*, pp. 103–104. Universitetsforlagets Trykningssentral, Oslo.

Fredriksson-Ahomaa, M., T. Autio and H. Korkeala (1999). Efficient subtyping of *Yersinia enterocolitica* bioserotype 4/O:3 with pulsed-field gel electrophoresis. *Lett. Appl. Microbiol.* **29**, 308–312.

Fukushima, H. and M. Gomyoda (1986). Inhibition of *Yersinia enterocolitica* serotype O3 by natural microflora of pork. *Appl. Environ. Microbiol.* **51**, 990–994.

Fukushima, H., M. Tsubokura, K. Otsuki and Y. Kawaoka (1984). Biochemical heterogenity of serotype O3 strains of 700 *Yersinia* strains isolated from humans, other mammals, flies, animal feed, and river water. *Curr. Microbiol.* **11**, 149–154.

Fukushima, H., K. Maruyama, I. Omori *et al.* (1990). Contamination of pigs with *Yersinia* at the slaughterhouse. *Fleischwirtschaft* **70**, 1300–1302.

Gemski, P., J. R. Lazere and T. Casey (1980). Plasmid associated with pathogenicity and calcium dependency of *Yersinia enterocolitica. Infect. Immun.* **27**, 682–685.

Gill, C. and M. Reichel (1989). Growth of cold-tolerant pathogens *Yersinia enterocolitica, Aeromonas hydrophila* and *Listeria monocytogenes* on high-pH beef packaged under vacuum or carbon dioxide. *Food Microbiol.* **6**, 223–230.

Gleason, T. H. and S. D. Patterson (1982). The pathology of *Yersinia enterocolitica* ileocolitis. *Am. J. Surg. Pathol.* **6**, 347–355.

Grau, F. H. (1981). Role of pH, lactate, and anaerobiosis in controlling the growth of some fermentative gram-negative bacteria on beef. *Appl. Environ. Microbiol.* **42**, 1043–1050.

Greenwood, M. H. and W. L. Hooper (1990). Excretion of *Yersinia* spp. associated with consumption of pasteurized milk. *Epidemiol. Infect.* **104**, 345–350.

Hanna, M. O., D. L. Zink, Z. L. Carpenter and C. Vanderzant (1976). *Yersinia enterocolitica*-like organisms from vacuum packaged beef and lamb. *J. Food Sci.* **41**, 1254–1256.

Hanna, M. O., J. C. Stewart, D. L. Zink *et al.* (1977). Development of *Yersinia enterocolitica* on raw and cooked beef and pork at different temperatures. *J. Food Sci.* **42**, 1180–1184.

Hanna, M. O., J. C. Stewart, Z. L. Carpenter *et al.* (1979). Isolation and characteristics of *Yersinia enterocolitica*-like bacteria from meats. *Contrib. Microbiol. Immunol.* **5**, 234–242.

Harvey, S., M. J. Greenwood, M. J. Pickett and A. M. Robert (1976). Recovery of *Yersinia enterocolitica* from streams and lakes of California. *Appl. Environ. Microbiol.* **32**, 352–354.

Hayashidani, H., Y. Ohtomo, Y. Toyokawa *et al.* (1995). Potential sources of sporadic human infection with *Yersinia enterocolitica* serovar O:8 in Aomori prefecture. *Jpn. J. Clin. Microbiol.* **33**, 1253–1257.

Hurvell, B (1981). Zoonotic *Yersinia enterocolitica* infection, host range, clinical manifestations, and transmission between animals and man. In: E. J. Bottone (ed.), *Yersinia enterocolitica*, pp. 145–159. CRC Press, Inc., Boca Raton, FL.

Ichinohe, H., M. Yoshioka, H. Fukushima *et al.* (1991). First isolation of *Yersinia enterocolitica* serotype O8 in Japan. *J. Clin. Microbiol.* **29**, 846–847.

International Organization for Standardization (1994). Microbiology – general guidance for the detection of presumptive pathogenic *Yersinia enterocolitica* (ISO 10273). International Organization for Standardization, Geneva.

Jebsen, O. B., B. Korner, K. B. Lauritsen *et al.* (1976). *Yersinia enterocolitica* infection in patients with acute surgical abdominal disease. A prospective study. *Scand. J. Infect. Dis.* **8**, 189–194.

Kaneko, S., N. Ishizaki and Y. Kokubo (1995). Detection of pathogenic *Yersinia enterocolitica* and *Yersinia pseudotuberculosis* from pork using polymerase chain reaction. *Contrib. Microbiol. Immunol.* **13**, 153–155.

Kapperud, G. and B. Jonsson (1978). *Yersinia enterocolitica* et bactéries apparentées isolées à partir d'écosystèmes d'eau douce en Norvège. *Med. Mal. Infect.* **8**, 500–506.

Kapperud, G. and S. B. Slome (1998). *Yersinia enterocolitica* infections. In: A. Evans and P. F. Brachman (eds), *Bacterial Infections of Humans,* 3rd edn, pp. 859–873. Plenum Medical Book Company, New York, NY.

Kapperud, G., E. Namork, M. Skurnik and T. Nesbakken (1987). Plasmid-mediated surface fibrillae of *Yersinia pseudotuberculosis* and *Yersinia enterocolitica*, relationship to the outer membrane protein YOP1 and possible importance for pathogenesis. *Infect. Immun.* **55**, 2247–2254.

Kapperud, G., K. Dommarsnes, M. Skurnik and E. Hornes (1990a). A synthetic oligonucleotide probe and a cloned polynucleotide probe based on the *yopA* gene for detection and enumeration of virulent *Yersinia enterocolitica. Appl. Environ. Microbiol.* **56**, 17–23.

Kapperud, G., T. Nesbakken, S. Aleksic and H. H. Mollaret (1990b). Comparison of restriction endonuclease analysis and phenotypic typing methods for differentiation of *Yersinia enterocolitica* isolates. *J. Clin. Microbiol.* **28**, 1125–1131.

Kapperud, G., T. Vardund, E. Skjerve *et al.* (1993). Detection of pathogenic *Yersinia enterocolitica* in foods and water by immunomagnetic separation, nested polymerase chain reactions, and colorimetric detection of amplified DNA. *Appl. Environ. Microbiol.* **59**, 2938–2944.

Kato, Y., K. Ito, Y. Kubokura *et al.* (1985). Occurrence of *Yersinia enterocolitica* in wild-living birds and Japanese serows. *Appl. Environ. Microbiol.* **49**, 198–200.

Kendall, M. and R. J. Gilbert (1980). Survival and growth of *Yersinia enterocolitica* in media and in food. In: G. W. Gould and J. E. L. Corry (eds), *Microbial Growth and Survival in Extremes of Environment*, Society for Applied Bacteriology Technical Series, No 15, pp. 215–226. Academic Press, London.

Kleinlein, N. and F. Untermann (1990). Growth of pathogenic *Yersinia enterocolitica* strains in minced meat with and without protective gas with consideration of the competitive background flora. *Intl J. Food Microbiol.* **10**, 65–72.

Knapp, W (1958). Mesenteric adenitis due to *Pasteurella pseudotuberculosis* in young people. *N. Engl. J. Med.* **259**, 776–778.

Knapp, W. and E. Thal (1963. Untersuchungen über die kulturellbiochemischen, serologischen, tierexperimentellen und immunologischen Eigenschaften einer vorläufig '*Pasteurella* X' benannten Bakterienart. *Zbl. Bakteriol. Orig. A* **190**, 472–484.

Kondracki, S., G. Balzano, J. Schwartz *et al.* (1996). Recurring outbreaks of yersiniosis associated with pork chitterlings. *Abstracts of the 36th Interscience Conference on Antimicrobial Agents and Chemotherapy, New Orleans*, p. 259.

Kotula A. W. and A. K. Sharar (1993). Presence of *Yersinia enterocolitica* serotype O:5,27 in slaughter pigs. *J. Food Prot.* **56**, 215–218.

Krogstad, O. (1974). *Yersinia enterocolitica* infection in goat. A serological and bacteriological investigation. *Acta Vet. Scand.* **15**, 597–608.

Lachica, R. V., D. L. Zink and W. R. Ferris (1984). Association of fibril structure formation with cell surface properties of *Yersinia enterocolitica. Infect. Immun.* **46**, 272–275.

Laird, W. J. and D. C. Cavanaugh (1980). Correlation of autoagglutination and virulence of yersiniae. *J. Clin. Microbiol.* **11**, 430–432.

Lanada, E. B (1990). The epidemiology of *Yersinia* iInfections in goat flocks. Thesis, Massey University, Palmerston North, New Zealand.

Langeland, G. (1983). *Yersinia enterocolitica* and *Yersinia enterocolitica*-like bacteria in drinking water and sewage sludge. *Acta Pathol. Microbiol. Immunol. Scand. Sect. B* **91**, 179–185.

Lassen, J. (1972). *Yersinia enterocolitica* in drinking-water. *Scand. J. Infect. Dis.* **4**, 125–127.

Lee, L. A., A. R. Gerber, D. R. Lonsway *et al.* (1990). *Yersinia enterocolitica* O:3 infections in infants and children, associated with the household preparation of chitterlings. *N. Engl. J. Med.* **322**, 984–987.

Lee, L. A., J. Taylor, G. P. Carter *et al.* (1991). *Yersinia enterocolitica* O:3, an emerging cause of pediatric gastroenteritis in the United States. *J. Infect. Dis.* **163**, 660–663.

Lee, W. H., C. Vanderzant and N. Stern (1981). The occurrence of *Yersinia enterocolitica* in foods. In: E. J. Bottone (ed.), *Yersinia enterocolitica*, pp. 161–171. CRC Press, Inc., Boca Raton, FL.

Lindberg, C. W. and E. Borch (1994). Predicting the aerobic growth of *Yersinia enterocolitica* O:3 at different pH values, temperatures and *L*-lactate concentrations using conductance measurements. *Intl J. Food Microbiol.* **22**, 141–153.

Maruyama, T (1987). *Yersinia enterocolitica* infection in humans and isolation of the microorganism from pigs in Japan. *Contrib. Microbiol. Immunol.* **9**, 48–55.

Mayer, L. W. (1988). Use of plasmid profiles in epidemiologic surveillance of disease outbreaks and in tracing the transmission of antibiotic resistance. *Clin. Microbiol. Rev.* **1**, 228–243.

Mead, P. S., L. Slutsker, V. Dietz *et al.* (1999). Food-related illness and death in the United States. *Emerg. Infect. Dis.* **5**, 607–625.

Megraud, F. (1987). *Yersinia* infection and acute abdominal pain. *Lancet* **i**, 1147.

Michiels, T., P. Wattiau, R. Brasseur *et al.* (1990). Secretion of *Yop* proteins by yersiniae. *Infect. Immun.* **58**, 2840–2849.

Miller, V. L., B. B. Finlay and S. Falkow (1988). Factors essential for the penetration of mammalian cells by *Yersinia. Curr. Top. Microbiol. Immunol.* **138**, 15–39.

Mollaret, H. H. (1995). Fifteen centuries of yersiniosis. *Contrib. Microbiol. Immunol.* **13**, 1–4.

Mollaret, H. H. and P. Nicolle (1965). Sur la fréquence de la lysogénie dans l'espèce nouvelle *Yersinia enterocolitica. C. R. Acad. Sci.* **260**, 1027–1029.

Mollaret, H. H., P. Nicolle, J. Brault and R. Nicolas (1972). Importance actuelle des infections a *Yersinia enterocolitica. Bull. Acad. Natl Med. Paris* **156**, 704.

Mollaret, H. H., H. Bercovier and J. M. Alonso (1979). Summary of the data received at the WHO Reference Center for *Yersinia enterocolitica. Contrib. Microbiol.* **5**, 174–184.

Morris, G. K. and J. C. Feeley (1976). *Yersinia enterocolitica*, a review of its role in food hygiene. *Bull. WHO* **54**, 79–85.

Morse, D. L., M. Shayegani and R. J. Gallo (1984). Epidemiologic investigation of a *Yersinia* camp outbreak linked to a food handler. *Am. J. Publ. Health* **74**, 589–592.

Najdenski, H., I. Iteman and E. Carniel (1994). Efficient subtyping of pathogenic *Yersinia enterocolitica* strains by pulsed-field gel electrophoresis. *J. Clin. Microbiol.* **32**, 2913–2920.

Nesbakken, T. (1988). Enumeration of *Yersinia enterocolitica* O:3 from the porcine oral cavity, and its occurrence on cut surfaces of pig carcasses and the environment in a slaughterhouse. *Intl J. Food Microbiol.* **8**, 287–293.

Nesbakken, T. and G. Kapperud (1985). *Yersinia enterocolitica* and *Yersinia enterocolitica*-like bacteria in Norwegian slaughter pigs. *Intl J. Food Microbiol.* **1**, 301–309.

Nesbakken, T., B. Gondrosen and G. Kapperud (1985). Investigation of *Yersinia enterocolitica, Yersinia enterocolitica*-like bacteria, and thermotolerant campylobacters in Norwegian pork products. *Intl J. Food Microbiol.* **1**, 311–320.

Nesbakken, T., G. Kapperud, H. Sørum and K. Dommarsnes (1987). Structural variability of 40-50 Mdal virulence plasmids from *Yersinia enterocolitica*. Geographical and ecological distribution of plasmid variants. *Acta Path. Microbiol. Immunol. Scand. Sect. B.* **95**, 167–173.

Nesbakken, T., G. Kapperud, J. Lassen and E. Skjerve (1991a). *Yersinia enterocolitica* antibodies in slaughterhouse employees, veterinarians, and military recruits. Occupational exposure to pigs as a risk factor for yersiniosis. *Contrib. Microbiol. Immunol.* **12**, 32–39.

Nesbakken, T., G. Kapperud, K. Dommarsnes *et al.* (1991b). Comparative study of a DNA hybridization method and two isolation procedures for detection of *Yersinia enterocolitica* O:3 in naturally contaminated pork products. *Appl. Environ. Microbiol.* **57**, 389–394.

Nesbakken, T., E. Nerbrink, O. J. Røtterud and E. Borch (1994). Reduction of *Yersinia enterocolitica* and *Listeria* spp. on pig carcasses by enclosure of the rectum during slaughter. *Intl J. Food Microbiol.* **23**, 197–208.

Nesbakken, T., K. Eckner, H. K. Høidal and O.-J. Røtterud (2003a). Occurrence of *Yersinia enterocolitica* and *Campylobacter* spp. in slaughter pigs and consequences for meat inspection, slaughtering, and dressing procedures. *Intl J. Food Microbiol.* **80**, 231–240.

Nesbakken, T., K. Eckner, H. K. Høidal and O.-J. Røtterud (2003b). Occurrence of *Yersinia enterocolitica* in slaughter pigs and consequences for meat inspection, slaughtering, and dressing procedures. In: M. Skurnik, J. A. Bengoechea and K. Granfors (eds), *The Genus Yersinia. Entering The Functional Genomic Area. Advances in Experimental Medicine and Biology*, Vol. 529, pp. 303–308. Kluwer Academic/Plenum Publishers, New York, NY.

Nicolle, P. (1973). *Yersinia enterocolitica*. In: H. Rische (ed.), *Lysotypie und andere spezielle epidemiologische laboratoriums-Methoden*, pp. 377–387. VEB Gustav Fischer Verlag, Jena.

Nielsen, B., C. Heisel and A. Wingstrand (1996). Time course of the serological response to *Yersinia enterocolitica* O:3 in experimentally infected pigs. *Vet. Microbiol.* **48**, 293–303.

Niléhn, B. (1969). Studies on *Yersinia enterocolitica* with special reference to bacterial diagnosis and occurrence in human acute enteric disease. *Acta Pathol. Microbiol. Scand. Contrib. Microbiol. Immunol.* **206** (Suppl.), 1.

Niléhn, B. and B. Sjöström (1967). Studies on *Yersinia enterocolitica*, occurrence in various groups of acute abdominal disease. *Acta Pathol. Microbiol. Scand.* **71**, 612–628.

Nilsson, A., S. T. Lambertz, P. Stålhandske *et al.* (1998). Detection of *Yersinia enterocolitica* in food by PCR amplification. *Lett. Appl. Microbiol.* **26**, 140–141.

Nordic Committee on Food Analysis (1987). *Yersinia enterocolitica*. Detection in Food. Method no. 117, 2nd edn. Nordic Committee on Food Analysis, Esbo, Finland.

Olsovsky, Z., V. Olsakova, S. Chobot and V. Sviridov (1975). Mass occurrence of *Yersinia enterocolitica* in two establishments of collective care of children. *J. Hyg. Epidemiol. Immunol.* **19**, 22–29.

Ostroff, S. M. (1995). *Yersinia* as an emerging infection, epidemiologic aspects of yersiniosis. *Contrib. Microbiol. Immunol.* **13**, 5–10.

Ostroff, S. M., G. Kapperud, L. C. Hutwagner *et al.* (1994). Sources of sporadic *Yersinia enterocolitica* infections in Norway, a prospective case-control study. *Epidemiol. Infect.* **112**, 133–141.

Pai, C. H. and L. DeStephano (1982). Serum resistance associated with virulence in *Yersinia enterocolitica. Infect. Immun.* **35**, 605–611.

Pai, C. H., F. Gillis and M. I. Marks (1982). Infection due to *Yersinia enterocolitica* in children with abdominal pain. *J. Infect. Dis.* **146**, 705.

Pedersen, K. B. (1979). Occurrence of *Yersinia enterocolitica* in the throat of swine. *Contrib. Microbiol. Immunol.* **5**, 253–256.

Perry, R. D. and R. R. Brubaker (1983). Vwa^+ phenotype of *Yersinia enterocolitica. Infect. Immun.* **40**, 166–171.

Perry, R. D., S. W. Bearden and J. D. Fetherston (2001). Iron and heme acquisition and storage systems of *Yersinia pestis. Rec. Res. Devel. Microbiol.* **5**, 13–27.

Portnoy, D. A., H. Wolf-Watz, I. Bölin *et al.* (1984). Characterization of common virulence plasmids in *Yersinia* species and their role in the expression of outer membrane proteins. *Infect. Immun.* **43**, 108–114.

Portnoy, D. A. and R. J. Martinez (1985). Role of a plasmid in the pathogenicity of *Yersinia* species. *Curr. Topics Microbiol. Immunol.* **118**, 29–51.

Prentice, M. B., D. Cope and R. A. Swann (1991). The epidemiology of *Yersinia enterocolitica* infection in the British Isles 1983–1988. *Contrib. Microbiol. Immunol.* **12**, 17–25.

Prpic, J. K., R. M. Robins-Browne and R. B. Davey (1985). *In vitro* assessment of virulence in *Yersinia enterocolitica* and related species. *J. Clin. Microbiol.* **22**, 105–110.

Quan, T. J., J. L.Meek, K. R. Tsuchiya *et al.* (1974). Experimental pathogenicity of recent North American isolates of *Yersinia enterocolitica. J. Infect. Dis.* **129**, 341–344.

Rasmussen, H. N., O. F. Rasmussen, H. Christensen and J. E. Olsen (1995). Detection of *Yersinia enterocolitica* O:3 in fecal samples and tonsil swabs from pigs using IMS and PCR. *J. Appl. Bacteriol.* **78**, 563–568.

Ratnam, S., E. Mercer, B. Pico *et al.* (1982). A nosocomial outbreak of diarrheal disease due to *Yersinia enterocolitica* serotype O:5 biotype 1. *J. Infect. Dis.* **145**, 242–247.

Ricciardi, I. D., A. D. Pearson, W. G. Suckling and C. Klein (1978). Long-term fecal excretion and resistance induced in mice infected with *Yersinia enterocolitica. Infect. Immun.* **21**, 342–344.

Saari, T. N. and G. P. Jansen (1979). Waterborne *Yersinia enterocolitica* in the midwest United States. *Contrib. Microbiol. Immunol.* **5**, 185–196.

Saken, E., A. Roggenkamp, S. Aleksic and J. Heesemann (1994). Characterisation of pathogenic *Yersinia enterocolitica* serogroups by pulsed-field gel electrophoresis of genomic *NotI* restriction fragments. *J. Med. Microbiol.* **41**, 329–338.

Samadi, A. R., K. Wachsmuth, M. I. Huq *et al.* (1982). An attempt to detect *Yersinia enterocolitica* infection in Dacca, Bangladesh. *Trop. Geogr. Med.* **34**, 151–154.

Sandler, M., A. H. Girdwood, R. E. Kottler and I. N. Marks (1982). Terminal ileitis due to *Yersinia enterocolitica. S. Afr. Med. J.* **62**, 573–576.

Sarrouy, J. (1972). Isolement d'une *Yersinia enterocolitica* a partir du lait. *Med. Mal. Infect.* **2**, 67.

Schiemann, D. A. (1982). Development of a two-step enrichment procedure for recovery of *Yersinia enterocolitica* from food. *Appl. Environ. Microbiol.* **43**, 14–27.

Schiemann, D. A. (1983). Alkalotolerance of *Yersinia enterocolitica* as a basis for selective isolation from food enrichments. *Appl. Environ. Microbiol.* **46**, 22–27.

Schiemann, D. A. (1989). *Yersinia enterocolitica* and *Yersinia pseudotuberculosis*. In: M. P. Doyle (ed.), *Foodborne Bacterial Pathogens*, pp. 601–672. Marcel Dekker, New York, NY.

Schiemann, D. A. and C. A. Fleming (1981). *Yersinia enterocolitica* from throats of swine in eastern and western Canada. *Can. J. Microbiol.* **27**, 1326–1333.

Skjerve, E., B. Lium, B. Nielsen and T. Nesbakken (1998). Control of *Yersinia enterocolitica* in pigs at herd level. *Intl J. Food Microbiol.* **45**, 195–203.

Slee, K. J. and C. Button (1990). Enteritis in sheep and goats due to *Yersinia enterocolitica* infection. *Austr. Vet. J.* **67**, 396–398.

Sneath, P. H. A. and S. T. Cowan (1958). An electro-taxonomic survey of bacteria. *J. Gen. Microbiol.* **19**, 551.

Sörqvist, S. and M. L. Danielsson-Tham (1990). Survival of *Campylobacter, Salmonella* and *Yersinia* spp. in scalding water used at pig slaughter. *Fleischwirtschaft* **70**, 1460–1466.

Tacket, C. O., J. Ballard, N. Harris *et al.* (1985). An outbreak of *Yersinia enterocolitica* infections caused by contaminated tofu (soybean curd). *Am. J. Epidemiol.* **121**, 705–711.

Stengel, G. (1985). *Yersinia enterocolitica*. Vorkommen und Bedeutung in Lebensmitteln. *Fleischwirtschaft* **65**, 1490–1495.

Stern, N. J., M. D. Pierson and A. W. Kotula (1980a). Effects of pH and sodium chloride on *Yersinia enterocolitica* growth at room and refrigeration temperature. *J. Food Sci.* **45**, 64–67.

Stern, N. J., M. D. Pierson and A. W. Kotula (1980b). Growth and competitive nature of *Yersinia enterocolitica* in whole milk. *J. Food Sci.* **45**, 972–974.

Swedish Institute for Infectious Disease Control (1995). *Smittskyddsinstitutets Epidemiologiska Årsrapport 1994*. Stockholm.

Szita, M. I., M. Káli and B. Rédey (1973). Incidence of *Yersinia enterocolitica* infection in Hungary. *Contrib. Microbiol. Immunol.* **2**, 106–110.

Tacket, C. O., J. P. Narain, R. Sattin *et al.* (1984). A multistate outbreak of infections caused by *Yersinia enterocolitica* transmitted by pasteurized milk. *J. Am. Med. Assoc.* **251**, 483–486.

Tauxe, R. V. (2002). Emerging foodborne pathogens. *Intl J. Food Microbiol.* **78**, 31–42.

Tauxe, R. V., G. Wauters, V. Goossens *et al.* (1987). *Yersinia enterocolitica* infections and pork, the missing link. *Lancet* **i**, 1129–1132.

Tenover, F. C. (1988). Diagnostic deoxyribonucleic acid probes for infectious diseases. *Clin. Microbiol. Rev.* **1**, 82–101.

Ternström, A. and G. Molin (1987). Incidence of potential pathogens on raw pork, beef and chicken in Sweden, with special reference to *Erysipelothrix rhusiopathiae. J. Food Prot.* **50**, 141–146.

Thal, E. (1954). Untersuchungen über *Pasteurella pseudotuberculosis*. Thesis, University of Lund, Lund, Sweden.

Thisted Lambertz, S., A. Ballagi-Pordány, A. Nilsson *et al.* (1996). A comparison between a PCR method and a conventional culture method for detection of pathogenic *Yersinia enterocolitica* in foods. *J. Appl. Bacteriol.* **81**, 303–308.

Thompson, J. S. and M. J. Gravel (1986). Family outbreak of gastroenteritis due to *Yersinia enterocolitica* serotype O:3 from well water. *Can. J. Microbiol.* **32**, 700–701.

Toivanen, P., A. Toivanen, L. Olkkonen and S. Aantaa (1973). Hospital outbreak of *Yersinia enterocolitica* infection. *Lancet* **i**, 1801–1803.

Toma, S. and L. Lafleur (1981). *Yersinia enterocolitica* infections in Canada 1966 to August 1978. In: E. J. Bottone (ed.), *Yersinia enterocolitica*. pp. 183–191 CRC Press, Inc., Boca Raton, FL.

Ursing, J., D. J. Brenner, H. Bercovier *et al.* (1980). *Yersinia frederiksenii*, a new species of *Enterobacteriaceae* composed of rhamnose-positive strains (formerly called atypical *Yersinia enterocolitica* or *Yersinia enterocolitica*-like). *Curr. Microbiol.* **4**, 213–217.

Van Loghem, J. J. (1944). The classification of plague-bacillus. *Antonie van Leeuwenhoek* **10**, 15.

Vantrappen, G., H. O. Agg, E. Ponette *et al.* (1977). *Yersinia* enteritis and enterocolitis, gastroenterological aspects. *Gastroenterology* **72**, 220–227.

Verhaegen, J., L. Dancsa, P. Lemmens *et al.* (1991. *Yersinia enterocolitica* surveillance in Belgium (1979–1989). *Contrib. Microbiol. Immunol.* **12**, 11–16.

Viitanen, A. M, P. Arstila, R. Lahesmaa *et al.* (1991). Application of the polymerase chain reaction and immunofluorescence techniques to the detection of bacteria in *Yersinia*-triggered reactive arthritis. *Arthritis Rheum.* **34**, 89–96.

Wachsmuth, K. (1985). Genotypic approaches to the diagnosis of bacterial infections, plasmid analyses and gene probes. *Infect. Control.* **6**, 100–109.

Wauters, G. (1979). Carriage of *Yersinia enterocolitica* serotype 3 by pigs as a source of human infection. *Contrib. Microbiol. Immunol.* **5**, 249–252.

Wauters, G. (1981). Antigens of *Yersinia enterocolitica*. In: E. J. Bottone (ed.), *Yersinia enterocolitica*, pp. 41–53. CRC Press, Inc., Boca Raton, FL.

Wauters, G. (1991). Taxonomy, identification and epidemiology of *Yersinia enterocolitica* and related species. In: H. Grimme, E. Landi and S. Dumontet (eds), *I Problemi della Moderna Biologia, Ecologia Microbica, Analitica di Laboratorio, Biotecnologia*, Vol. 1, pp. 93–101. Atti del IV Convegno Internazionale, Sorrento.

Wauters, G., S. Aleksic, J. Charlier and G. Schulze (1991). Somatic and flagellar antigens of *Yersinia enterocolitica* and related species. *Contrib. Microbiol. Immunol.* **12**, 239–243.

Wauters, G., M. Janssens, A. G. Steigerwalt and D. J. Brenner (1988a). *Yersinia mollaretii* sp. nov., formerly called *Yersinia enterocolitica* biogroups 3A and 3B. *Intl J. Syst. Bacteriol.* **38**, 424–429.

Wauters, G., K. Kandolo and M. Janssens (1987). Revised biogrouping scheme of *Yersinia enterocolitica. Contrib. Microbiol. Immunol.* **9**, 14–21.

Wauters, G., M. Janssens, A. G. Steigerwalt and D. J. Brenner (1988a). *Yersinia mollaretii* sp. Nov., formerly called *Yersinia enterocolitica* biogroups 3A and 3B. *Intl. J. Syst. Bacteriol.*, **38**, 424–429.

Wauters, G., V. Goossens, M. Janssens and J. Vandepitte (1988b). New enrichment method for isolation of pathogenic *Yersinia enterocolitica* O:3 from pork. *Appl. Environ. Microbiol.* **54**, 851–854.

Weynants, V., A. Tibor, P. A. Denoel *et al.* (1996a). Infection of cattle with *Yersinia enterocolitica* O:9: a cause of the false positive serological reactions in bovine brucellosis diagnostic tests. *Vet. Microbiol.* **48**, 101–112.

Weynants, V., V. Jadot, P. A. Denoel *et al.* (1996b. Detection of *Yersinia enterocolitica* serogroup O:3 by a PCR method. *J. Clin. Microbiol.* **34**, 1224–1227.

Winblad, S. (1975). Arthritis associated with *Yersinia enterocolitica* infections. *Scand. J. Infect. Dis.* **7**, 191–195.

World Health Organization (1983). *Yersiniosis. Report on a WHO meeting, Paris 1981.* WHO Regional Office for Europe, Euro Reports and Studies 60, Copenhagen.

World Health Organization (1987). *Report of the Round Table Conference on Veterinary Public Health Aspects of Yersinia enterocolitica, Orvieto 1985*. WHO Collaborative Centre for Research and Training in Veterinary Public Health, Instituto Superiore di Sanità, Rome.

World Health Organization (1995). *Report of the WHO Consultation on Emerging Foodborne Diseases*. Berlin.

Wormser, G. P. and G. T. Keusch (1981). *Yersinia enterocolitica*, clinical observations. In: E. J. Bottone (ed.), *Yersinia enterocolitica*, pp. 83–93. CRC Press, Inc., Boca Raton, FL.

Yersin, A. (1894). La peste bubonique a Hong Kong. *Ann. Inst. Pasteur, Paris* **8**, 662–667.

Zen-Yoji, H., T. Maruyama, S. Sakai *et al.* (1973). An outbreak of enteritis due to *Yersinia enterocolitica* occurring at a junior high school. *Jpn. J. Microbiol.* **17**, 220–222.

Listeria monocytogenes infections

9

Franco Pagotto, Nathalie Corneau and Jeff Farber

1 Historical aspects and contemporary problems

In 1926, Murray and colleagues described symptoms in rabbits and guinea pigs caused by a Gram-positive bacillus organism (Murray *et al.*, 1926). They named this organism *Bacterium monocytogenes*, in reference to the mononuclear leucocytosis observed in the affected animals (Murray *et al.*, 1926). In 1927, Pirie was able to isolate the same organisms from gerbils and later suggested that the generic name *Listerella* (named after the surgeon Lord Lister) be changed to *Listeria*, reflecting proper nomenclature (Pirie, 1927, 1940). Following its early discovery, the disease caused by this organism (listeriosis) was rare. Only after a Canadian outbreak in the early 1980s where human illnesses were noted did the organism become a household name in the food industry (McLauchlin, 1993).

Listeriosis is a generic term for a variety of syndromes caused by *Listeria monocytogenes*, and is a problem to the food industry in part due to the ubiquitous nature of the organism. As such, this human pathogen has the potential to cause disease in all aspects of the farm-to-fork continuum. In farm animals such as sheep, abortion, encephalitis and/or septicemia can occur when listeriosis strikes (Gitter, 1989). However, the majority of human listeriosis infections arise via the consumption of contaminated food (McLauchlin and Low, 1994).

Foodborne Infections and Intoxications 3e
ISBN-10: 0-12-588365-X
ISBN-13: 978-012-588365-8

Today, many countries have developed policies that range from 'zero tolerance' to tolerable limits, depending on the food item. Available data would suggest that listeriosis, averaging 2–7 cases per million population, is neither increasing nor decreasing in developing or developed countries. However, some believe that listeriosis is likely to have a greater impact on society in the future. This is due to an expected increase in the number of susceptible individuals, including the elderly population, as well as an increase in the consumption of extended shelf-life and refrigerated foods. In this chapter, we attempt to summarize the current state of knowledge of this important foodborne pathogen.

2 Characteristics of *Listeria*

The ninth edition of *Bergey's Manual of Determinative Bacteriology* (Holt, 1994) groups the genus *Listeria* along with the genera *Brochothrix*, *Carnobacterium*, *Caryophanon*, *Erysipelothrix*, *Kurthia*, *Lactobacillus,* and *Renibacterium*. They have collectively been placed in Group 19, under the heading of 'regular, non-sporing, Gram-positive rods'. *Listeria* species are described as short, Gram-positive, non-spore-forming, facultatively anaerobic rods. They vary in size (0.4–0.5 × 0.5–2 μm), have rounded ends, and are not encapsulated. They are motile by means of a few peritrichous flagella, with motility manifesting itself at 20–25 °C. However, motility is absent at 37 °C. It has been shown that transcription of the flagellin-encoding gene from *L. monocytogenes* is elevated at 22 °C, but undetectable at 37 °C (Peel *et al.*, 1988). Within the genus *Listeria*, *L. innocua* has been shown to produce flagellin at 37 °C, indicating that significant differences exist at the species level (Kathariou *et al.*, 1995).

Listeria are able to grow at temperatures ranging from < 0 °C to 45 °C (Farber and Peterkin, 1991, 2000). The organisms exhibit fermentative activities on carbohydrates, producing lactate and no gas from glucose. *Listeria* spp. are catalase positive and oxidase negative. The G + C content of DNA is 36–42 mol %. At present there are seven species that belong to the genus *Listeria* that can be identified, based on a variety of characteristics (Table 9.1). DNA–DNA hybridization, multilocus enzyme electrophoresis (MEE) and rRNA restriction fragment length polymorphism techniques have shown high levels of similarity between *L. murrayi* and *L. grayi*. This has led to a proposition that they are in fact the same, and that *L. murrayi* be renamed as *L. grayi* (Rocourt *et al.*, 1992).

The spectrum of human infections due to *L. monocytogenes* mimics that observed in non-humans. In humans, perinatal infection presents itself as the following manifestations: abortion, stillbirth, neonatal sepsis, and meningitis (Schuchat *et al.*, 1991). In adults, meningitis and encephalitis are the most common clinical symptoms. However, other symptoms have been attributed to listeriosis (Table 9.2).

L. monocytogenes is the major pathogenic species in both animals and humans (Rocourt and Seeliger, 1985; Farber and Peterkin, 2002). However, in humans, occasional infections due to *L. ivanovii* (Cummins *et al.*, 1994) and *L. seeligeri* (Rocourt *et al.,* 1987) have been reported. So far, no cases of listeriosis have been attributed to *L. innocua*, *L. welshimeri* or *L.grayi*.

Table 9.1 Characteristics differentiating the species of the genus *Listeria*

Characteristic	*L. monocytogenes*	*L. innocua*	*L. seeligeri*	*L. welshimeri*	*L. ivanovii*	*L. grayi*	*L. murrayi*
Gram stain[a]	+[b]	+	+	+	+	+	+
β-hemolysis[a]	+[c]	–	+	–	+[d]	–	–
Mannitol (acid production)[a]	–	–	–	–	–	+	+
L-Rhamnose (acid production)[a]	+	d[b]	–	d	–	–	d
D-Xylose	–	–	+	+	+	–	–
CAMP-test (*S. aureus*)[a]	+[e]	–	+	–	–	–	–
CAMP-test (*R. equi*)	–	–	–	–	+	–	–
Acid production from:							
L-Arabinose	–	–	–	–	–	–	–
Dextrin	d	–			–	+	+
Galactose	d	–			d	+	+
Glycogen	–	–	–	–	–	–	–
Lactose	d	+			+	+	+
D-Lyxose	–	–	–			+	+
Melezitose	d	d			d	–	–
Melibiose	–	–	–	–	–	–	–
α-methyl-D-glucoside	+	+			+	+	+
α-Methyl-D-mannoside	+	+	–[f]	+	–		
Sorbitol	d	–			–	–	–
Soluble starch	–	–			–	+	+
Sucrose	–	d			d	–	–
Voges-Proskauer	+	+	+	+	+	+	+
Hydrolysis of:							
Cellulose	–	–	–	–	–	–	–
Hippurate	+	+			+	–	–
Starch	d	d			–	–	–
Lecithinase	d	d			+	–	–
Phosphatase	+	+	+	+	+	+	+
NO_3^- to NO_2^- reduction	–	–			–	–	+
Pathogenicity for mice	+	–	–	–	+	–	–

[a] Tests most often used in differentiating the listeriae.
[b] Standard symbols: +, ≥ 90 % positive; –, ≥ 90 % negative; d, 11–89 % of strains are positive.
[c] Not all strains of *L. monocytogenes* exhibit β-hemolysis – the type strain ATCC 15313 is non-hemolytic on horse, sheep, and bovine blood.
[d] A very wide zone or multiple zones of hemolysis are usually exhibited by *L. ivanovii* strains.
[e] Of 30 strains, ATCC 15313, the type strain, did not give a positive reaction.
[f] Of 10 strains tested, 1 gave a positive reaction.

Table 9.2 Clinical manifestations attributed to listeriosis

Group	Symptoms[a]
Pregnant women	Fever, chills, myalgia, flu-like syndromes, preterm delivery, preterm labor, abortions, stillbirth, amnionitis, intrauterine/cervical infections
Newborns	Sepsis, meningitis, granulomatosis, infantiseptica, neonatal infections, pneumonia
'Healthy' adults	Diarrhea, fever, gastroenteritis, fatigue, headache, general malaise, nausea, vomiting, meningitis, septicemia
Other clinical manifestations	Cutaneous listeriosis, conjunctivitis, convulsions, hepatitis, meningoencephalitis, osteomyelitis, upper respiratory tract symptoms

Adapted from Slutsker *et al.* (2000), Donnelly (2001), and Pagotto and Farber (2003).
[a] Invasive listeriosis has a median incubation period of approximately 30 days, while non-invasive listeriosis has an 18–20-hour incubation period.

DNA sequencing studies that included genes encoding for the flagellin (*fla*), listeriolysin (*hlyA*) and invasive-associated protein p60 (*iap*) have elucidated at least three evolutionary lines for *L. monocytogenes* (Rasmussen *et al.*, 1995). Other technologies, such as MEE, pulsed field gel electrophoresis (PFGE), randomly amplified polymorphic DNA (RAPD) and restriction fragment length polymorphism (RFLP), have been able to group serotypes 1/2a, 1/2c, 3a and 3c into a division I; serotypes 1/2b, 3b, 4b, 4d and 4e to a division II; and serotype 4a into a separate group (Bibb *et al.*, 1990; Mazurier and Wernars, 1992; Vines *et al.*, 1992; Brosch *et al.*, 1994; Rocourt, 1996). Serotype 4b can be divided into two further groups (Ericsson *et al.*, 1995; Zheng and Kathariou, 1995).

The pathogenesis of *L. monocytogenes* is one of the better understood amongst human foodborne pathogens (Dramsi and Cossart, 1998; Braun and Cossart, 2000; Kreft and Vázquez-Boland, 2001; Vázquez-Boland *et al.*, 2001; Cossart, 2002; Kathariou, 2002). Initially serving as a model for the study of induction of T-cell mediated immunity in the mouse model, *L. monocytogenes* is perhaps the best studied invasive/intracellular bacteria. Listeriosis as an infective process is exceptional, since the organism is able to cross the gastrointestinal, materno-fetal and blood–brain protective barriers. There are many virulence factors involved in the intracellular journey that *L. monocytogenes* must endure in order to survive the host's defense (immune) system (Figure 9.1).

The initial step for the development of listeriosis typically involves the ingestion of the organism, followed by its survival against the non-specific immune system defenses of the gastrointestinal tract. There, it can multiply in the intestines and be eliminated in the feces. After food contaminated with *L. monocytogenes* has been ingested, the organism is believed to invade the intestinal epithelium and/or Peyer's patches. From here, the bacteria can enter the draining lymph nodes and disseminate via the bloodstream to the liver and spleen. In most cases infections by *L. moncytogenes* are limited, with clinical symptoms appearing in the immunocompromised, the elderly, pregnant women, and neonates. However, it is equally important to note that listeriosis has one of the highest

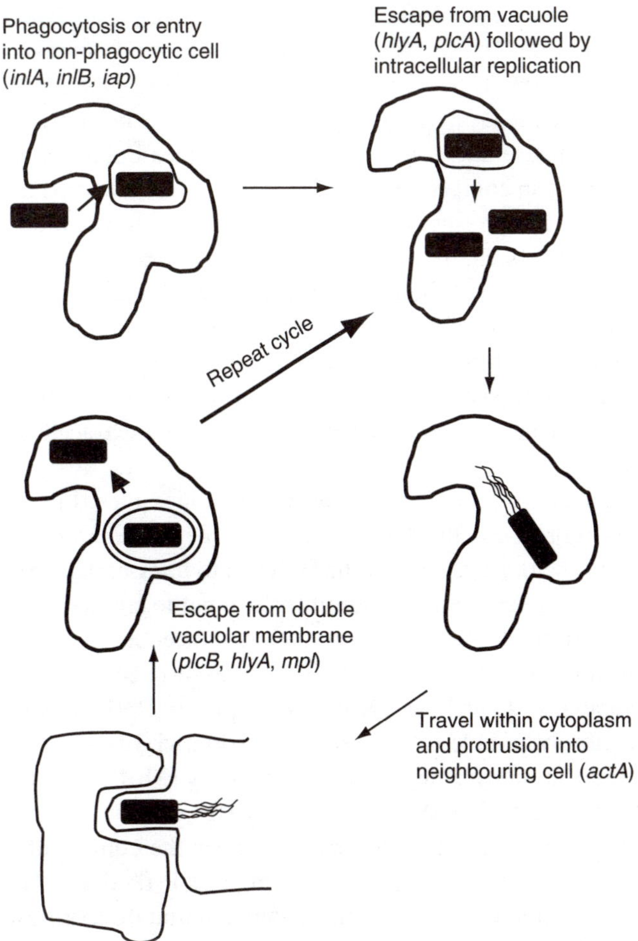

Figure 9.1 Diagramatic steps involved in the intracellular life cycle and cell-to cell spread of *L. monocytogenes*. Genes involved at each step are indicated. Modified from Pagotto and Farber, 2003.

case-fatality ratios of all foodborne bacterial infections. Once inside the host cell, *L. monocytogenes* is able to circumvent the immune response, as shown in Figure 9.1

L. monocytogenes is able to invade phagocytic and non-phagocytic cells, survive, and replicate inside them (Gaillard *et al.*, 1987; Campbell, 1994; Drevets *et al.*, 1995; Guzman *et al.*, 1995). In phagocytic cells, the organism is internalized within membrane-bound vacuoles (Figure 9.1). DeChastellier and Berche (1994) have estimated that approximately 10–15 % of *L. monocytogenes* are able to escape the vacuole in approximately 30 minutes through the action of the hemolysin protein, known as listeriolysin O (LLO), encoded by the gene *hlyA* (Gaillard *et al.*, 1986; Mengaud *et al.*, 1988; Portnoy *et al.*, 1988; Cossart *et al.*, 1989). In non-phagocytic cells, *L. monocytogenes* enters via the action of two internalin proteins, InlA and InlB, encoded by the *inlA* and *inlB* genes, respectively. InlA mediates entry into cells of epithelial nature (Gaillard *et al.*, 1991), whereas InlB is thought to be involved with entry into many types of cells, such as hepatocytes and fibroblasts (Dramsi *et al.*, 1995; Ireton *et al.*, 1996).

Once the bacterium has escaped the vacuole it enters the host cell cytoplasm, where growth and multiplication can occur. In the cytoplasm, the organism becomes surrounded by polymerized host cell actin (Tilney and Portnoy, 1989; Alvarez-Dominguez *et al.*, 1997; Sechi *et al.*, 1997). The actin is preferentially polymerized at the older pole of the bacterium, presumably after it has undergone cell division (Vázquez-Boland *et al.*, 2001). The ability to polymerize actin confers intracellular mobility to the bacterium. The resulting 'comet tail'-like structure pushes the bacterial cell into an adjacent host cell, where it again becomes encapsulated, this time in a double-membraned vacuole. A lecithinase is involved in the dissolution of these membranes, although the hemolysin may also contribute to this process. Intracellular growth and movement in the newly invaded cell are then repeated (Figure 9.1; Sheehan *et al.*, 1994).

Other virulence factors have been shown to be important in invasion and cell-to-cell spread of *L. monocytogenes*. These include two phospholipases, a major extracellular protein and a metalloprotease.

The *plcA* gene encodes a phospholipase C, while the *plcB* encodes a lecithinase, which is activated by a metalloprotease, product of the *mpl* gene (Sheehan *et al.*, 1994). The protein p60, encoded by the gene *iap*, has been shown to be essential in the invasion process. Believed to be a mureine hydrolase, its role in pathogenesis is still unclear (Wuenscher *et al.*, 1993). In addition, cells producing reduced levels of p60 show a 'rough' morphology and exhibit (depending on how the experiment is set up) reduced adherence, invasiveness, and virulence (Kohler *et al.*, 1990, 1991). Recent work suggests that there is no observable difference, comparing wild-type to *iap* mutants, of uptake by human monocyte-derived dendritic cells (MoDC) in the presence or absence of human plasma (Kolb-Maurer *et al.*, 2001).

Bacterial regulation of the virulence genes is currently under the control of a transcription factor, PrfA, encoded by the *prfA* gene (Figure 9.2). PrfA is the only bacterial virulence factor identified to date that plays a significant and direct role in the virulence of the organism. Belonging to the same transcriptional activator family as Crp-Fnr (Sheehan *et al.*, 1995), PrfA binds to a 'PrfA box' in the promoter region of the following virulence genes: *actA*, *inlA*, *hly*, *plcA*, *mpl*, as well as its own promoter region (i.e. *prfA*) (Vázquez-Boland *et al.*, 2001). PrfA contains a helix-turn-helix (HTH) motif in its C-terminal region that allows it to bind a 14 bp PrfA box (5′-TTAACANNTGTTAA-3′; where N can be any nucleotide) located in the promoter region of many virulence genes (Brehm *et al.*, 1996; Goebel *et al.*, 2000).

Non-bacterial factors that affect the expression of virulence genes or the activity of PrfA have been shown experimentally to include temperature (Dramsi *et al.*, 1993), growth in activated charcoal (Ripio *et al.*, 1996), growth in a defined medium (Milenbachs *et al.*, 1997), fermentable carbohydrates (Ripio *et al.*, 1996; Behari and Youngman, 1998; Brehm *et al.*, 1999;), low pH (Behari and Youngman, 1998), high osmolarity (Myers *et al.*, 1993), and iron (Conte *et al.*, 1996; 2000) (see Figure 9.2B).

A key question that remains unresolved with respect to virulence aspects of *L. monocytogenes* is why, despite its ubiquitous nature, most clinical and/or food isolates recovered belong to serovars 1/2a, 1/2b and 4b (there are at least 13 known serovars; see below). Actual data addressing this phenomenon in the form of animal studies or cell culture experiments are lacking. In 1979, Seeliger and Höhne described

A

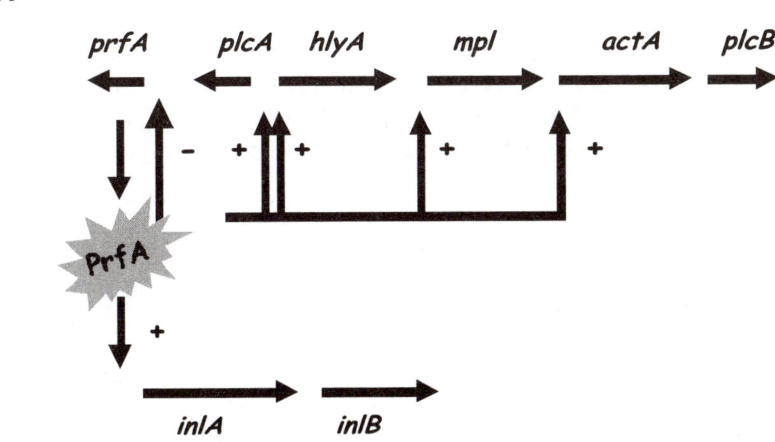

B

External source	Effect on PrfA activity or virulence gene expression
Fermentable sugars	Repression of PrfA-activated genes
Low pH	Repression of PrfA-activated genes
Miminal media	*prfA* transcription
Activated charcoal	*prfA* transcription
Iron (high)	*inlAB* transcription
Iron (low)	*actA* transcription
Osmolarity (high)	*Hly* transcription
Temperature (above 30°C)	*prfA* transcription

Figure 9.2 PrfA is the master regulator (A) of many virulence genes found on the chromosome of *Listeria monocytogenes*. +, positive effect on transcription; –, negative effect (autoregulation). Non-bacterial influences on virulence gene expression or PrfA activity is shown in B (adapted from Kreft and Vázquez-Boland, 2001).

a serotyping scheme where the letters a, b, c, etc. referred to the flagellar antigens. The numerical aspect (i.e. 1/2, 3, 4, etc.) described somatic antigens, being mostly teichoic acid-associated (Seeliger and Höhne, 1979). While much work has been done in the area of pathogenesis, it is important to remember that these studies have utilized serotype 1/2a (e.g. strains 10403S, EGD, NCTC7973, Mack) or serotype 1/2c (LO28). Serotype 4b is responsible for the majority of outbreaks, and it is predictable that significant differences may exist between serotype 4b and those serotypes currently being investigated. This is especially true based on several gel electrophoresis-based studies, which put the classification of *L. monocytogenes* into two genetic groups: one that includes 'a' and 'c' serotypes (1/2a, 1/2c, 3a, 3c) and another that includes the 'b' serotypes (1/2b, 3b, 4b) (Bibb *et al.*, 1990). Thus, while there exists a broad range of common virulence genes in all pathogenic strains of *L.* monocytogenes, serotype-specific genes may exist to explain the tropisms observed (e.g. the occurrence of serotype 4b in the majority of the large invasive outbreaks).

In the last few years, the genomes of *L. monocytogenes* EGD-e (serovar 1/2a), as well as *L. innocua* CLIP-11262 have become available (Glaser *et al.*, 2001; see also http://genolist.pasteur.fr/ListiList/, and www.tigr.org). The analyses of completely sequenced genomes will aid in resolving some of the issues that surround the pathogenesis of *L. monocytogenes*.

3 Nature of infection in man and animals

Listeriosis can be thought of as a zoonotic disease that has a minor yet important secondary host: humans. While most transmission of listeriosis occurs via contamination of ingested foods (Farber and Peterkin, 1991, 2000), transmission from infected animals to humans has been reported (Cain and McCann, 1986).

L. monocytogenes has been shown to be quite persistent in manufacturing plants. Nocera *et al.* (1990) investigated the genetic relatedness of isolates implicated in a Swiss epidemic of listeriosis that occurred over a 5-year period. Miettinen *et al.* (1999) also demonstrated that the same clone of *L. monocytogenes* persisted in an ice-cream plant for a period spanning 7 years. It would seem that the organism, through its ubiquitous nature, has ample opportunity to enter the food chain at all points of the farm-to-fork continuum. It is difficult to deny *Listeria* entry into the food chain, and as such it is a major cause of post-processing contamination (see Figure 9.3).

4 Prevalence of *Listeria* in foods, feeds and water

4.1 Animal feed

L. monocytogenes is highly adaptable and is capable of existing as a plant saprophyte. Its ability to multiply at low temperatures, within a wide pH range and in habitats having a low water activity, promotes the growth of this organism in the environment, and as such often results in contamination of feed and food (Skovgaard and Morgen, 1988). *L. monocytogenes* is sometimes present in very low numbers in grass or other forage crops used for ensilage (Donald *et al.*, 1995; Wilkinson, 1999). Storing harvested materials in silos or compacted into bales to achieve anaerobic conditions is usually sufficient to prevent growth of the organism (Wilkinson, 1999), but animal listeriosis can arise through the consumption of contaminated feedstuffs resulting from poor quality, improperly fermented silage having a pH of > 4.0 (Fernandez-Garayzabal *et al.*, 1992; Wiedmann *et al.*, 1994; Donald *et al.*, 1995; Ryser *et al.*, 1997). Sanaa *et al.* (1993) have shown that poor-quality silage can be a major source of contamination of raw milk by *L. monocytogenes*. Most cases of animal listeriosis have been associated with baled silage (Fenlon, 1988), usually via the introduction of oxygen in the wrapping step or through punctured membranes. Oxygen infiltration promotes the growth of *L. monocytogenes* by disturbing anaerobic fermentation.

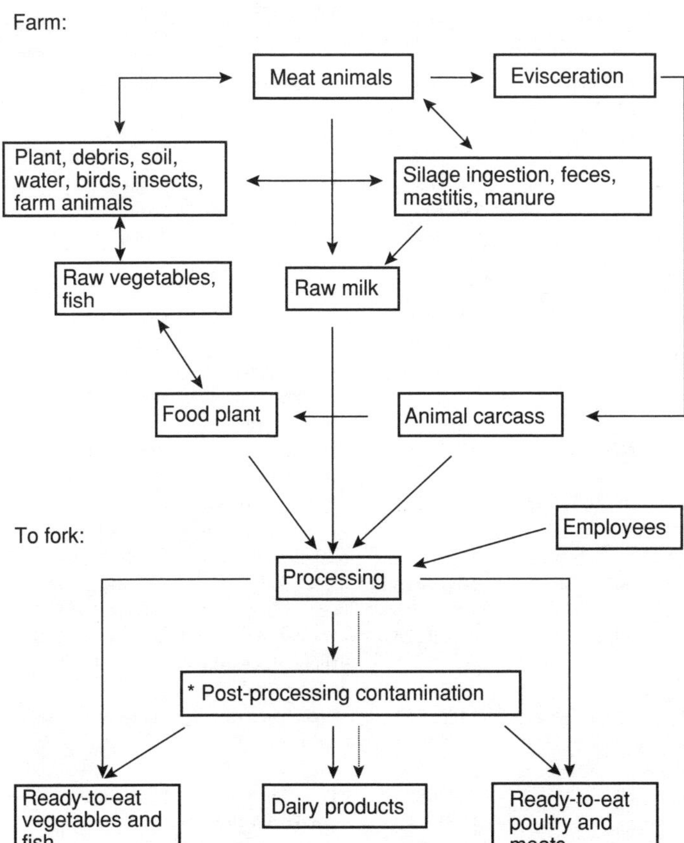

Figure 9.3 Various modes of entry of *Listeria monocygenes* in the farm-to-fork process. * indicates the major source of end-product contamination with *L. monocytogenes*.

Consequently, the outer layer of the bale and the layer of silage just below the top, in clamp silos, might contain viable and possibly large numbers of the pathogen (Wilkinson, 1999). While only part of the silage may be contaminated, it is difficult to separate this material from the uncontaminated crop silage.

4.2 Poultry and meat products

L. monocytogenes has been isolated from mammalian and avian species, including domestic chickens and turkeys (Gray and Killinger, 1966). Contamination of poultry and meat products by this environmentally ubiquitous bacterium can originate from agricultural ecosystems (Skovgaard and Morgen, 1988; Van Renterghem *et al.*, 1991), abattoirs (Lowry and Tiong, 1988; Loncarevic *et al.*, 1994; Ojeniyi *et al.*, 1996; Chasseignaux *et al.*, 2001), processing plants, or even at retail (Lawrence and Gilmour, 1995; Miettinen *et al.*, 2001). Because of the different methodologies being used to isolate *L. monocytogenes*, there is considerable variation in the incidence of *Listeria* in poultry and meat products. So far, a variety of meats (raw and ready-to-eat) have been found to be contaminated with *L. monocytogenes* (Table 9.3).

Table 9.3 Incidence of *Listeria monocytogenes* (Lm) in poultry and meat products

Country	Meat type	Incidence of *Listeria* spp. (%)	Incidence of Lm (%)	Comments	Reference
Australia	Retail vacuum-packaged processed meats	93/175 (53)	78/175 (44.6)	Obtained from 2 of 50 corned-beef packs contaminated with Lm	Grau and Vanderlinde, 1992
Belgium, France	Raw poultry products	ND	111/279 (39.8) 156/434 (35.9)	> 1 CFU/cm^2 or g	Uyttendaele *et al.*, 1999
Finland	Raw broiler pieces	ND	38/61 (62)	Broiler pieces bought from retail stores	Miettinen *et al.*, 2001
Norway	Chicken carcasses	ND	55/90 (61)		Rørvik and Yndestad, 1991
Spain	Retail chicken carcasses	95/100 (95)	32/100 (32)	Poultry carcasses purchased from 20 retail outlets	Capita *et al.*, 2001
UK	Raw chickens	ND	60/100 (60)		Pini and Gilbert, 1988
US	Fresh poultry carcasses	34/90 (38)	21/90 (23)	Broilers from three processing plants	Bailey *et al.*, 1989
US	Sliced ham and luncheon meats	ND	118/2287 (5.2)	Prevalence of Lm from 1990 through 1999	Levine *et al.*, 2001
US	Small cooked sausages	ND	243/6820 (3.6)	Prevalence of Lm from 1990 through 1999	Levine *et al.*, 2001
US	Cooked poultry products	ND	145/6836 (2.1)	Prevalence of Lm from 1990 through 1999	Levine *et al.*, 2001
US	Cooked, roast, corned beef	ND	163/5272 (3.1)	Prevalence of Lm from 1990 through 1999	Levine *et al.*, 2001

On the farm, *L. monocytogenes* seems to be predominantly associated with cattle excretion where a high percentage of animals (52 %) harbor the bacterium (Skovgaard and Morgen, 1988; Colburn *et al.*, 1990). Van Renterghem *et al.* (1991) found that almost 20 % of fresh pig and cattle feces were positive for *L. monocytogenes*; however, none of the manure samples obtained from storage tanks or manured soil samples contained the organism. These results indicate that *L. monocytogenes* may be incapable of surviving for long periods of time in soil or liquid manure. The authors concluded that soil and manure might not be good reservoirs for *Listeria*. However, the plant–soil rhizosphere is considered a more likely natural habitat for *L. monocytogenes* (Van Renterghem *et al.*, 1991).

L. monocytogenes is likely introduced into the abattoir through live animals (Ojeniyi *et al.*, 1996). Fecal contamination, combined with unhygienic practices, could be the cause of frequent contamination of raw meat products by *Listeria* (Skovgaard and Morgen, 1988; Hudson and Read, 1989). Contamination of broiler carcasses in abattoirs most likely occurs during or following the chilling step in the skin-removing machine (Miettinen *et al.*, 2001). Clouser *et al.* (1995) obtained similar results, and concluded that the contamination of turkey carcasses probably occurred after de-feathering, or during evisceration or chilling of the birds. Furthermore, Ojeniyi *et al.* (1996) found that in five out of seven poultry abattoirs the incidence of *L. monocytogenes* isolates increased after the spin chiller, indicating that this processing step may contribute to the spreading of the bacterium. Major sources of contamination in cattle and lamb are linked to hides and pelts, whereas in pigs, fecal contamination has been identified as the prime source of meat contamination (Adesiyun and Krishnan, 1995).

Chasseignaux *et al.* (2001) monitored the major points of contamination of meat products at two processing plants (poultry and pork) as a way of identifying the origin of *L. monocytogenes* in food. *L. monocytogenes* was isolated from poultry and pork abattoirs, respectively, in samples taken from the floors (28 % and 4 %), working tables (21 % and 37 %), boxes (30 % and 14 %), transport belts (6 % and 18 %), machines (14 % and 31 %), knives (50 % and 50 %) and water supply (100 % and 0 %). In this study, there was an increase in contamination in the poultry plant compared to the original contamination level of the raw meat, while the contamination of raw pork was similar to that of the finished products (Chasseignaux *et al.*, 2001). Uyttendaele *et al.* (1999) showed that processed poultry, cut without skin, harbored significantly more *L. monocytogenes* cells than non-processed poultry, supporting the notion that contamination occurs in the processing environment. Most *L. monocytogenes* cells are located on the food surface; however, this pathogen was found in the interior muscle core in 5 out of 110 samples of beef, pork and lamb roasts (Johnson *et al.*, 1990). This would indicate that the organisms were probably present in the muscle at the time of slaughter. Contamination of processing plants can also occur when a small number of healthy carriers (beef, pork or poultry) are introduced into the facility. Inadequate cleaning of the equipment, along with contaminated hands and gloves of workers, are other important sources of contamination in the final processing area (Clouser *et al.*, 1995).

The most frequent type of *L. monocytogenes* isolated from meats worldwide belongs to serotype 1 (Farber and Peterkin, 1991; Jay, 1996). Few serotyping studies have been performed on poultry samples, but it seems that isolates belonging to serotype 1 are also the most prevalent in this type of meat (Farber and Peterkin, 2000). Some isolates of *L. monocytogenes* have been found colonizing processing surfaces in plants, and, by persisting for long periods of time (months to years), these can become an important source of cross-contamination (Lawrence and Gilmour, 1995; Nesbakken *et al.*, 1996; Miettinen *et al.*, 2001). Lundén *et al.* (2002) reported the transfer of a persistent strain of *L. monocytogenes* contamination from one plant to two others after the relocation of a dicing machine. The incidence of *L. monocytogenes* in raw chicken can range from 12 % to 92 % (Hudson and Read, 1989; Johnson *et al.*, 1990; Farber and Peterkin, 1991; Loncarevic *et al.*, 1994), and from 2 % to 27 % in cooked and ready-to-eat poultry products (Gilbert *et al.*, 1989; Kerr *et al.*, 1990; Ribeiro and Burge 1992; Levine *et al.*, 2001). Recent work indicates that typical incidences would be in the range of 2–5 % (Levine *et al.*, 2001).

Introduction of *L. monocytogenes* in cooked products is often attributed to their contact with a soiled area, to cross-contamination between raw and cooked meat, or to inadequate cleaning and disinfection practices (Salvat *et al.*, 1995; Ojeniyi *et al.*, 1996). The high incidence of this pathogen in meat presents a potential risk for sporadic cases of listeriosis. To date, there have been several reported recalls and outbreaks in the US, Canada and the UK of meat products containing *L. monocytogenes*. The products recalled have included cocktail frankfurters, salami, pâté, cooked ham, hot dogs, chicken, farm sausages, ham sandwiches, turkey salad, cheeseburger, egg and ham salads, chicken salad sandwiches and many more. Because of the widespread distribution of *L. monocytogenes*, it is unlikely that this organism can be eradicated from meat products.

4.3 Eggs

Even though there is a high incidence of *L. monocytogenes* in raw poultry, the information available regarding the incidence of this organism in eggs is limited. *Listeria* spp. have been isolated from egg-wash water and from commercially available, broken liquid whole eggs (Farber *et al.*, 1992). Leasor and Foegeding (1989) found *Listeria* spp. in 15 of 42 (36 %) egg samples, with *L. monocytogenes* being found in 5 % of the samples. Moore and Madden (1993) found a higher incidence in blended raw eggs, as 27.2 % (47 out of 173 samples) were positive for *L. monocytogenes*. It has been shown by Sionkowski and Shelef (1990) that *L. monocytogenes* is able to grow in raw whole eggs at 5 °C. Moreover, Brackett and Beuchat (1992) concluded that home cooking of eggs by frying 'sunnyside-up' reduced the population of *L. monocytogenes* by only 0.4 log, while cooking scrambled eggs was a much more effective way of destroying the cells.

Canadian and US pasteurization standards for liquid whole eggs are based on the thermal properties of *Salmonella* spp. (US Department of Agriculture, 1969). Current minimum pasteurization standards are 3.5 minutes at 60 °C for unsalted, or 63 °C for salted, liquid whole eggs (Canadian Food Inspection Agency, 2001). This treatment allows a reduction of viable cells of 1.7–4.4 log in unsalted, liquid whole egg (Bartlett and Hawke, 1995). However, the pasteurization of liquid whole eggs containing 10 % NaCl only results in a 0.2–0.6-log reduction of *L. monocytogenes* (Bartlett and Hawke, 1995). Thus, if the initial count of *Listeria* is high, survival after current minimum pasteurization requirements could potentially occur (Bartlett and Hawke, 1995; Palumbo *et al.*, 1995). When tested in eggs under similar experimental conditions, *L. monocytogenes* and *L. innocua* were found to be as much as eight-fold more heat resistant than *Salmonella* spp. (McKenna, 1991; Palumbo *et al.*, 1995; Muriana *et al.*, 1996; Schuman and Sheldon, 1997).

4.4 Water

Listeria spp. are present in aqueous ecosystems such as river waters, sewage sludge, fresh water, sediment and bay water (Watkins and Sleath, 1981; Colburn *et al.*, 1990; Bernagozzi *et al.*, 1994; Arvanitidou *et al.*, 1997). Bernagozzi *et al.* (1994) confirmed the presence of *L. monocytogenes* in 40 % of the surface water sampled along the Idice River in Italy, whereas 5.9 % of the river samples in Greece were positive for this pathogen (Arvanitidou *et al.*, 1997). A study in Humboldt-Arcata Bay, California, carried out by Colburn *et al.* (1990), detected *Listeria* spp. in 81 % (30/37) of freshwater samples. The most predominant *Listeria* species found in freshwater by this group was *L. monocytogenes*, which was isolated from 62 % of the total samples (Colburn *et al.*, 1990). Sediments were collected at the same location as the surface water, and 17.4 % (8/46) harbored *L. monocytogenes* (Colburn *et al.*, 1990). The presence of domesticated animals in the surrounding area seems to have an influence on the incidence and predominance of the species recovered. When animals were nearby, *Listeria* spp. were recovered at a higher rate (Colburn *et al.*, 1990). In one study, *L. monocytogenes* was also frequently isolated from urban wastewater. Geuenich *et al.* (1985) found that 92.5 % of isolated listeriae were *L. monocytogenes*.

4.5 Milk and milk products

The occurrence of *L. monocytogenes* in raw milk has been linked with insufficient lighting of milking barns, inadequate sanitation of the exercise area, improper cow hygiene, and poor disinfection of towels between milkings (Sanaa *et al.*, 1993). Furthermore, this pathogen may contaminate milk as a consequence of mastitis, encephalitis, or abortion related to this bacterium (Gray and Killinger, 1966). Bovine mastitis, caused by *L. monocytogenes*, most often occurs in one-quarter of the udder of the infected cow (Jensen *et al.*, 1996). Jensen *et al.* (1996) proposed, as a conservative estimate, that *Listeria*-infected quarters could shed as many as 10 000 CFU/ml of milk. Such shedding would contribute to the contamination of the entire milk lot pooled in a tank. Through a 23-year study period in Denmark, 0.04 % of cows were found to have one-quarter of the udder infected, resulting in 1.2 % of the herd being positive for *L. monocytogenes* (Jensen *et al.*, 1996).

The incidence of *L. monocytogenes* in raw milk and dairy products is presented in Table 9.4. The level of the pathogen in raw milk can range from less than 1 CFU/ml to 62 CFU/ml (Fenlon *et al.*, 1995; O'Donnell, 1995). A higher incidence of *Listeria* spp. in raw milk during the indoor season, mostly spring, has been reported from different studies (Rea *et al.*, 1992; Fenlon *et al.*, 1995; Hassan *et al.*, 2000; Waak *et al.*, 2002). The presence of this organism in raw milk can lead to the contamination of processed products such as cheeses made from non-pasteurized milk. Loncarevic *et al.* (1995) found that 42 % of the cheeses made from raw milk were contaminated with *L. monocytogenes*, as compared with 2 % of the cheeses made from heat-treated milk. However, it should be noted that this value is quite elevated when compared to other reports.

L. monocytogenes can survive during the manufacture and ripening of certain cheeses and, if present initially, is likely to grow in high pH and a_w soft cheeses such as Camembert (Pearson and Marth, 1990). Throughout cheese ripening, the levels of *L. monocytogenes* have been shown to decrease steadily in artificially inoculated Cheddar or Colby (Pearson and Marth, 1990). Pasteurization is an efficient method to reduce or eliminate *Listeria* in milk. However, *L. monocytogenes* often occupies certain niches in the dairy processing plant, and as such can promote contamination of the final product (Charlton *et al.*, 1990; Canillac and Mourey, 1993; Pritchard *et al.*, 1995).

Table 9.4 Incidence of *L. monocytogenes* in milk and milk products

Country	Milk product	Incidence of Lm (%)	Comments	Reference
Brazil	Raw milk	21/220 (9.5)		Moura *et al.*, 1993
Canada	Raw milk	6/455 (1.3)	Bulk tank milk	Farber *et al.*, 1988
Czech Republic	Raw milk	2/111 (1.8)	Bulk tank milk	Schlegelova *et al.*, 2002
Denmark	Quarter milk samples	0.01-0.1	Mean of 0.4 %	Jensen *et al.*, 1996
European Countries	Red smear cheese	21/329 (6.4)		Rudolff and Scherer, 2001
Finland	Raw milk	1/59 (1.7)	Milked by hand	Husu *et al.*, 1990
Ireland	Raw milk	29/589 (4.9)	Bulk tank milk	Rea *et al.*, 1992
Italy	Soft cheese	2/121 (1.6)		Massa *et al.*, 1990
Korea	Raw milk	1/50 (2.0)		Ha *et al.*, 2002
Spain	Raw ovine milk	23/1052 (2.2)	From farm bulk tanks	Rodriguez *et al.*, 1994
Spain	Raw caprine milk	37/1445 (2.6)	From farm bulk tanks	Gaya *et al.*, 1996
Sweden	Raw milk	58/295 (19.6)	From dairy silos	Waak *et al.*, 2002
Switzerland	Dairy products	3722/76271 (4.9)	1990–1999 study	Pak *et al.*, 2002
UK	Soft cheese	2/45 (4.0)		Pini and Gilbert, 1988
US	Raw milk	12/292 (4.1)	From farm bulk tanks	Rohrbach *et al.*, 1992
US	Raw milk	6/131 (4.6)	Bulk tank milk	Jayarao and Henning, 2001

Rudolff and Scherer (2001) reported an incidence of *L. monocytogenes* of 8.0 % from red-smear cheese prepared from pasteurized milk and 4.8 % from redsmear cheese made from raw milk. No major variation was found regarding the incidence of this pathogen in red-smear cheese made from pasteurized and raw cows' milk (6.2 %), ewes' milk (5.9 %) and goats' milk (7.7 %) (Rudolff and Scherer, 2001).

4.6 Other foods

L. monocytogenes can be isolated from vegetables and seafood (Table 9.5). Vegetables can become contaminated via many routes that include soil, water, animal manure, decomposing flora, and sewage effluents (Farber and Peterkin, 2000). While many types of vegetables have been found to contain *Listeria*, it appears that potatoes and radishes are contaminated by *L. monocytogenes* on a regular basis (Heisick *et al.*, 1989; Sizmur and Walker, 1998). The predominant serovar found on vegetables appears to be serotype 1 (Farber and Peterkin, 2000).

Since listeriae are present in most aqueous ecosystems, water may be the original source of contamination for fish and seafood (Colburn *et al.*, 1990; Dillon and Patel, 1992). In fact, a study has shown that *L. monocytogenes* can be cultured from seawater for up to 3 weeks (Bremer *et al.*, 1998). Contamination of seafood can also take place in the processing establishment, smokehouse or slaughterhouse (Rørvik *et al.*, 1995). As in any other meat-processing plant, good hygiene practices and disinfection protocols are

Table 9.5 Incidence of *L. monocytogenes* in raw and processed vegetables and seafood

Product	Incidence (%)	Comments	Reference
Live shellfish	11/120 (9.2)	Shellfish collected from nine littoral	Monfort *et al.*, 1998
Fresh shrimp	8/74 (10.8)	No *Listeria* spp. found in oysters	Motes, 1991
Smoked salmon	3/33 (9.1)		Rørvik and Yndestad, 1991
Cooked blue crab	10/126 (7.9)	MPN enumeration estimates at < 100/g	Rawles *et al.*, 1995
Mussels		Outbreak due to contaminated product	Brett *et al.*, 1998
Vegetable salads	1/63 (1.6)	4 supermarkets, 14 fastfood chains and 13 family restaurants surveyed	Lin *et al.*, 1996
Potatoes, Radishes	28/132 (21.2) 19–132 (14.4 %)	*L. monocytogenes* also isolated from cucumbers (10.9 %), lettuce (1.1 %), mushrooms (12 %), and cabbage (2.2 %) purchased from supermarkets; broccoli, carrots, cauliflower and tomatoes were negative for *Listeria* spp. tested from supermarkets	Heisick *et al.*, 1989
Prepared salads	15/146 (10.2)	Purchased in restaurants and delicatessen shops	de Simon and Ferrer, 1998

essential to avoid product contamination. Although the incidence of *L. monocytogenes* in raw seafood can be fairly high (Table 9.5), the organism is usually found in very low numbers (< 100 CFU/g) (Farber, 1991; Eklund *et al.*, 1995; Rawles *et al.*, 1995).

5 Foods most often associated with human infections

It is now generally accepted that listeriosis is mostly transmitted through food, and that foods play a major role in humans becoming sick from infection. As a result of both sporadic cases and outbreaks, consumption of contaminated foods (and prevention thereof) has been a key area of focus in the research community. The most common sources of *L. monocytogenes* include raw and processed meat, dairy products, vegetables and seafood products (Farber and Peterkin, 1991). Also noteworthy are food production facilities and storage environments (Cox *et al.*, 1989). Refrigerated foods that may be consumed without prior cooking are of special concern because of the ability of *L. monocytogenes* to grow at refrigeration temperatures.

6 Principles of detection of *Listeria*

The ability of listeriae to grow at refrigeration temperatures has been used in various cold enrichments, and was probably the first method developed for isolation of the organism (Gray and Killinger, 1966). Microscopy was used to take advantage of the fact that cells grown on semi-transparent agars, when illuminated by oblique light, have a blue color with a center that is described as 'ground glass' (Gray, 1957). This technique became known as Henry illumination, and is still in use today.

There are several media currently used to select for the growth of listeriae while inhibiting competitors. The chemical compounds used in the selective broths and agars eventually allow for the isolation of listeriae. The most common broths and agars used are listed in Table 9.6.

Following incubation of the selective broths, samples are usually streaked onto selective agars, most often Oxford (Curtis *et al.*, 1989) or Palcam (Van Netten *et al.*, 1989). Palcam agar contains the same agents as Palcam broth, whereas Oxford agar contains aesculin, lithium chloride, cycloheximide, colistin, acriflavin, cefotetan and phosphomycin. Colonies of *L. monocytogenes* appear grey-green with a black sunken center and a black halo on a cherry-red background on Palcam agar, and black surrounded by with a sunken center and a black halo on Oxford agar.

Because listeriae are found throughout nature and in a wide variety of foods, at the moment there is not one universal protocol appropriate for all foodstuffs. The majority of identification methods for the detection and enumeration of *L. monocytogenes* are cultural in nature. Other methods include the use of DNA probes, amplification-based methods, antibody-based technologies, and other rapid methods (Table 9.7). Monoclonal antibodies have been used to serotype the genus *Listeria*, as have hemolysis of blood and *in vivo* mouse assays (Table 9.8).

Table 9.6 Common broths used to enrich selectively for listeriae

Method	Broth	Selective and/or screening agents/comments
US Food and Drug Administration Method (FDA)	Buffered enrichment broth (EB)	EB is Trypticase Soy Broth with yeast extract that contains the selective agents acriflavin, nalidixic acid, cycloheximide, and pyruvic acid
US Department of Agriculture Method (USDA)	UVM1	Primary enrichment broth (aesculin, nalidixic acid, acriflavin) which is subcultured into Fraser broth (aesculin, nalidixic acid, lithium chloride, acriflavin)
Netherlands Government Food Inspection Service (NGFIS)	Liquid Palcam	Selective agents include polymyxin B, acriflavin, lithium chloride, ceftazidime, aesculin, and mannitol
Health Canada Method (MFHPB-30)	Listeria enrichment broth (LEB)	Selective agents include aesculin, nalidixic acid, and acriflavin

Table 9.7 Some alternative methods for the identification and/or detection of *Listeria monocytogenes*[a]

Name	Type or format (target)	Company/Supplier
Accuprobe *L. monocytogenes*	DNA based (probe)	GenProbe Inc.
BAX® system	DNA based (PCR)	Qualicon
GENE-Trak®	DNA based (probe)	Gene-Trak Systems
PROBELIA®	DNA based (PCR)	Sanofi Diagnostics Pasteur
VIDAS™	Immunoassay based (enzyme linked fluorescent assay)	bioMerieux Vitek
Listeria VIA	Immunoassay based (ELISA)	Tecra Diagnostics
Reveal (Listeria)®	Immunoassay based (immunochromatography)	Neogen
ClearView™	Immunoassay based (immunochromatography)	Oxoid
Assurance EIA® (Listeria)	Immunoassay based (ELISA)	BioControl
EIAFOSS Listeria™	Immunoassay based (ELISA)	Foss Electric
API®	Biochemical	bioMerieux
Microbact™	Biochemical	Microgen
Malthus	Biochemical (conductance)	Malthus
Microlog™	Biochemical (carbon oxidation)	Biolog
MIS	Biochemical (fatty acid)	Microbial-ID

[a] This is only a partial list, and is not meant as an endorsement by the authors.

Table 9.8 Serology, hemolytic activity, and mouse virulence for *Listeria* species

Species	Serotype	Hemolysis of 7 % horse blood agar[a]	Mouse virulence
L. monocytogenes	1/2a, 1/2b, 1/2c, 3a, 3b, 3c, 4a, 4ab, 4b, 4b(x), 4c, 4d, 4e, 7	+	+
L. ivanovii	5	+	+
L. innocua	4ab, 6a, 6b, un[a]	–	–
L. welshimeri	6a, 6b	–	–
L. seeligeri	1/2b, 4c, 4d, 6b, un[a]	+	–

[a] Portion of colony is stabbed into the blood plate. +, positive reaction; –, negative reaction; un, undefined.

There are many protocols described that have as their basis the hybridization of genetic material (DNA). Many probes have been described and used in colony hybridization methods, as have polymerase chain reaction (PCR) protocols (Spencer and Ragout de Spencer, 2001). A possible disadvantage of these methods is that isolation of *L. monocytogenes* is not required. As such, there cannot be a database of isolates to do further studies, such as surveillance-based typing with PFGE. A further disadvantage of current non-cultural methods is that no information regarding the serotype is obtained. The advantages of the savings in time must therefore be viewed in light of the many disadvantages of non-cultural methods. The diagnosis of listeriosis depends on isolation of the organism from a normally sterile site, usually blood or cerebrospinal fluid. There are no characteristic clinical features to confirm the diagnosis without culture. On sheep blood agar, narrow zones of β-hemolysis can be found, but the colony may have to be moved to confirm the presence of hemolysis. The CAMP (Christie–Atkins–Munch–Petersen) test with *Staphylococcus aureus* for *L. monocytogenes* and *L. seeligeri* and *Rhodococcus equi* for *L. ivanovii* may help to distinguish the three species (Pagotto *et al.*, 2001).

7 Treatment and prevention of listeriosis

Because foodborne listeriosis outbreaks are often associated with high case-fatality rates, it is imperative that effective antibiotics are used for treatment. Interestingly enough, antibiotic resistance is not an issue at the moment. As such, current treatments for listeriosis involve a combination of ampicillin and gentamicin (MacGowan, 1990). Two to three weeks of therapy appears to be sufficient to prevent relapses. Most β-lactam antibiotics may be used. However, cephalosporins are not effective against *L. monocytogenes*, and should not be used for treatment of serious infections such as neonatal meningitis. Other chemotherapeutic agents, considered to be secondary treatments, include erythromycin, fluoroquinolones, and vancomycin. A recent study of 84 isolates of *L. monocytogenes* revealed that susceptibilities of clinical isolates to penicillin, ampicillin, erythromycin, tetracycline, and gentamicin remained unchanged (Safdar and Armstrong, 2003). Unexpectedly, a high prevalence of clindamycin resistance was observed (Safdar and Armstrong, 2003).

Listeriae are present in many 'ready-to-eat' or minimally processed foods, and therefore prevention of illness from foodborne sources is difficult. The introduction of HACCP (hazard analysis of critical control points) programs in the food industry has been identified as an important factor in diminishing the environmental contamination by *L. monocytogenes* in 'ready-to-eat' and minimally processed foods not subject to further processing. For immunocompromised patients, a number of recommendations have been made to decrease the risk of foodborne listeriosis. Dietary recommendations for preventing foodborne listeriosis are summarized in Table 9.9. Currently, no vaccines exist for humans.

Table 9.9 How to lower the risk of acquiring foodborne listeriosis

Group of interest	Food item	Recommendations
'Normal' individuals		
	Meat, fish, poultry	Cook all food of animal origin thoroughly (70°C for a minimum of 2 min)
	All foods	Separate uncooked from cooked
	Milk and milk products	Avoid unpasteurized milk
	Non-food items	Wash hands, knives, and kitchen utensils properly
	Fruits and vegetables	Wash all raw fruits and vegetables prior to consumption

Table 9.9 How to lower the risk of acquiring foodborne listeriosis—*continued*

Group of interest	Food item	Recommendations
High risk (elderly, pregnant, immunocompromised) individuals		
	Dairy products	Do not consume soft cheeses such as Brie, Camembert or Feta (hard cheese, cottage or processed cheese and yoghurt are fine)
	Leftovers	Thoroughly heat to 70 °C for a minimum of 2 min
	Hot dogs/wieners	Heat to 70 °C for a minimum of 2 min
	Deli meats	Avoid unless thoroughly heated
	Pâté	Should be avoided unless in-can pasteurization used

Adapted from Pagotto and Farber, 2003.

Bibliography

Adesiyun, A. A. and C. Krishnan (1995). Occurrence of *Yersinia enterocolitica* O:3, *Listeria monocytogenes* O:4 and thermophilic *Campylobacter* spp. in slaughter pigs and carcasses in Trinidad. *Food Microbiol.* **12**, 99–107.

Alvarez-Dominguez, C., R. Roberts and P. D. Stahl (1997). Internalized *Listeria monocytogenes* modulates intracellular trafficking and delays maturation of the phagosome. *J. Cell Sci.* **110** (Pt 6), 731–743.

Arvanitidou, M., A. Papa, T. C. Constantinidis *et al.* (1997). The occurrence of *Listeria* spp. and *Salmonella* spp. in surface waters. *Microbiol. Res.* **152**, 395–397.

Bailey, J. S., D. L. Fletcher and N. A. Cox (1989). Recovery and serotype distribution of *Listeria monocytogenes* from broiler chickens in the southeastern United States. *J. Food Prot.* **52**, 148–150.

Bartlett, F. M. and A. E. Hawke (1995). Heat resistance of *Listeria monocytogenes* Scott A and HAL957E1 in various liquid egg products. *J. Food Prot.* **58**, 1211–1214.

Behari, J., and P. Youngman (1998). Regulation of *hly* in *Listeria monocytogenes* by carbon sources and pH occurs through separate mechanisms mediated by PrfA. *Infect. Immun.* **66**, 3635–3642.

Bernagozzi, M., F. Bianucci, R. Sacchetti and P. Bisbini (1994). Study of the prevalence of *Listeria* spp. in surface water. *Zentralbl. Hyg. Umweltmed.* **196**, 237–244.

Bibb, W. F., B. G., Gellin, R. Weaver *et al.* (1990). Analysis of clinical and food-borne isolates of *Listeria monocytogenes* in the United States by multilocus enzyme electrophoresis and application of the method to epidemiological investigations. *Appl. Environ. Microbiol.* **56**, 2133–2141.

Brackett, R. E. and L. R. Beuchat (1992). Survival of *Listeria monocytogenes* on the surface of eggshells and during frying of whole and scrambled eggs. *J. Food Prot.* **55**, 862–865.

Braun, L. and P. Cossart (2000). Interactions between *Listeria monocytogenes* and host mammalian cells. *Microbes Infect.* **2**, 803–811.

Brehm, K., J. Kreft, M. T. Ripio and J. A. Vázquez-Boland (1996). Regulation of virulence gene expression in pathogenic *Listeria*. *Microbiology* **12**, 219–236.

Brehm, K., M. T., Ripio, J. Kreft and J. A. Vázquez-Boland (1999). The *bvr* locus of *Listeria monocytogenes* mediates virulence gene repression by β-glucosides. *J. Bacteriol.* **181**, 5024–5032.

Bremer, P. J., C. M. Osborne, R. A. Kemp and J. J. Smith (1998). Survival of *Listeria monocytogenes* in sea water and effect of exposure on thermal resistance. *J. Appl. Microbiol.* **85**, 545–553.

Brett, M. S., P. Short and J. McLauchlin (1998). A small outbreak of listeriosis associated with smoked mussels. *Intl J. Food Microbiol.* **43**, 223–229.

Brosch, R., J. Chen and J. B. Luchansky (1994). Pulsed-field fingerprinting of listeriae: identification of genomic divisions for *Listeria monocytogenes* and their correlation with serovar. *Appl. Environ. Microbiol.* **60**, 2584–2592.

Cain, D. B. and V. L. McCann (1986). An unusual case of cutaneous listeriosis. *J. Clin. Microbiol.* **23**, 976–977.

Campbell, P. A. (1994). Macrophage-*Listeria* interactions. *Immunol. Ser.* **60**, 313–328.

Canadian Food Inspection Agency (2001). Processed egg regulations (available at http://laws.justice.gc.ca/en/C-0.4/C.R.C.-c.290/index.html).

Canillac, N., and A. Mourey (1993). Sources of contamination by *Listeria* during the making of semi-soft surface-ripened cheese. *Sci. Aliments* **13**, 533–544.

Capita, R., C. Alonso-Calleja, B. Moreno and M. C. Garcia-Fernandez (2001). Occurrence of *Listeria* species in retail poultry meat and comparison of a cultural/immunoassay for their detection. *Intl J. Food Microbiol.* **65**, 75–82.

Charlton, B. R., H. Kinde and L. H. Jensen (1990). Environmental survey for *Listeria* species in California milk processing plants. *J. Food Prot.* **53**, 198–201.

Chasseignaux, E., M. T. Toquin, C. Ragimbeau *et al.* (2001). Molecular epidemiology of *Listeria monocytogenes* isolates collected from the environment, raw meat and raw products in two poultry- and pork-processing plants. *J. Appl. Microbiol.* **91**, 888–899.

Clouser, C. S., S. Doores, M. G. Mast and S. J. Knabel (1995). The role of defeathering in the contamination of turkey skin by *Salmonella* species and *Listeria monocytogenes*. *Poultry Sci.* **74**, 723–731.

Colburn, K. G., C. A. Kaysner, C. Abeyta Jr and M. M. Wekell (1990). *Listeria* species in a California coast estuarine environment. *Appl. Environ. Microbiol.* **56**, 2007–2011.

Conte, M. P., C. Longhi, M. Polidoro *et al.* (1996). Iron availability affects entry of *Listeria monocytogenes* into the enterocyte-like cell line Caco-2. *Infect. Immun.* **64**, 3935–3929.

Conte, M. P., C. Longhi, G. Petrone *et al.* (2000). Modulation of *actA* gene expression in *Listeria monocytogenes* by iron. *J. Med. Microbiol.* **49**, 681–683.

Cossart, P. (2002). Molecular and cellular basis of the infection by *Listeria monocytogenes*, an overview. *Intl J. Med. Microbiol.* **291**, 401–409.

Cossart, P., M. F. Vicente, J. Mengaud *et al.* (1989). Listeriolysin O is essential for virulence of *Listeria monocytogenes*, direct evidence obtained by gene complementation. *Infect. Immun.* **57**, 3629–3636.

Cox, L. J., T. Kleiss, J. L. Cordier *et al.* (1989). *Listeria* spp. in food processing, non-food and domestic environments. *Food Microbiol.* **6**, 49–61.

Cummins, A. J., A. K. Fielding and J. McLauchlin (1994). *Listeria ivanovii* infection in a patient with AIDS. *J. Infect.* **28**, 89–91.

Curtis, G. D. W., R. G. Mitchell, A. F. King and E. J. Griffin (1989). A selective differential medium for the isolation of *Listeria monocytogenes*. *Lett. Appl. Microbiol.* **8**, 95–98.

de Chastellier, C. and P. Berche (1994). Fate of *Listeria monocytogenes* in murine macrophages: evidence for simultaneous killing and survival of intracellular bacteria. *Infect. Immun.* **62**, 543–353.

de Simon, M. and M. D. Ferrer (1998). Initial numbers, serovars and phagevars of *Listeria monocytogenes* isolated in prepared foods in the city of Barcelona (Spain). *Intl J. Food Microbiol.* **44**, 141–144.

Dillon, R. and T. Patel (1992). *Listeria* in seafoods: a review. *J. Food Prot.* **55**, 1009–1015.

Donald, A. S., D. R. Fenlon and B. Seddon (1995). The relationship between ecophysiology, indigenous microflora and growth of *Listeria monocytogenes* in grass silage. *J. Appl. Bacteriol.* **79**, 141–148.

Donnelly, C. W. (2001). *Listeria monocytogenes*. In: R. G. Labbé and S. García (eds), *Guide to Foodborne Pathogens*, pp. 99–132. Wiley Interscience, New York, NY.

Dramsi, S. and P. Cossart (1998). Intracellular pathogens and the actin cytoskeleton. *Ann. Rev. Cell Dev. Biol.* **14**, 137–166.

Dramsi, S., C. Kocks, C. Forestier and P. Cossart (1993). Internalin-mediated invasion of epithelial cells by *Listeria monocytogenes* is regulated by the bacterial growth state, temperature and the pleiotropic activator prfA. *Mol. Microbiol.* **9**, 931–941.

Dramsi, S., I. Biswas, E. Maguin *et al.* (1995). Entry of *Listeria monocytogenes* into hepatocytes requires expression of inIB, a surface protein of the internalin multigene family. *Mol. Microbiol.* **16**, 251–261.

Drevets, D. A., R. T. Sawyer, T. A. Potter and P. A. Campbell (1995). *Listeria monocytogenes* infects human endothelial cells by two distinct mechanisms. *Infect. Immun.* **63**, 4268–4276.

Eklund, M. W., F. T. Poysky, R. N. Pardnjpye *et al.* (1995). Incidence and sources of *Listeria monocytogenes* in cold-smoked fishery products and processing plants. *J. Food Prot.* **58**, 502–508.

Ericsson, H., P. Stalhandske, M. L. Danielsson-Tham *et al.* (1995). Division of *Listeria monocytogenes* serovar 4b strains into two groups by PCR and restriction enzyme analysis. *Appl. Environ. Microbiol.* **61**, 3872–3874.

Farber, J. M. (1991). *Listeria monocytogenes* in fish products. *J. Food Prot.* **54**, 922–924, 934.

Farber, J. M. and P. I. Peterkin (1991). *Listeria monocytogenes*, a food-borne pathogen. *Microbiol. Rev.* **55**, 476–511.

Farber, J. M. and P. I. Peterkin (2000). *Listeria monocytogenes*. In: B. M. Lund, T. C. Parker and G. W. Gould (eds), *The Microbiological Safety and Quality of Food*, pp. 1178–1232. Aspen Publishers, Gaithersburg, MD.

Farber, J. M., G. W. Sanders and S. A. Malcolm (1988). The presence of *Listeria* spp. in raw milk in Ontario. *Can. J. Microbiol.* **34**, 95–100.

Farber, J. M., E. Daley and F. Coates (1992). Presence of *Listeria* spp. in whole eggs and wash water sample from Ontario and Quebec. *Food Res. Intl.* **25**, 143–145.

Fenlon, D. R. (1988). Listeriosis. In: J. M. Wilkinson (ed.), *Silage and Health*, pp. 7–18. Chalcombe Publications, Lincoln.

Fenlon, D. R., T. Stewart and W. Donachie (1995). The incidence, numbers and types of *Listeria monocytogenes* isolated from farm bulk tank milks. *Lett. Appl. Microbiol.* **20**, 57–60.

Fernandez-Garayzabal, J. F., M. Blanco, J. A. Vazquez-Boland *et al.* (1992). A direct plating method for monitoring the contamination of *Listeria monocytogenes* in silage. *Zentralbl. Veterinärmed.* **39**, 513–518.

Gaillard, J. L., P. Berche and P. Sansonetti (1986). Transposon mutagenesis as a tool to study the role of hemolysin in the virulence of *Listeria monocytogenes. Infect. Immun.* **52**, 50–55.

Gaillard, J. L., P. Berche, J. Mounier *et al.* (1987). *In vitro* model of penetration and intracellular growth of *Listeria monocytogenes* in the human enterocyte-like cell line Caco-2. *Infect. Immun.* **55**, 2822–2829.

Gaillard, J. L., P. Berche, C. Frehel *et al.* (1991). Entry of *Listeria monocytogenes* into cells is mediated by internalin, a repeat protein reminiscent of surface antigens from gram-positive cocci. *Cell* **65**, 1127–1141.

Gaya, P., C. Saralegui, M. Medina and M. Nunez (1996). Occurrence of *Listeria monocytogenes* and other *Listeria* spp. in raw caprine milk. *J. Dairy Sci.* **79**, 1936–1941.

Geuenich, H. H., H. E. Muller, A. Schretten-Brunner and H. P. Seeliger (1985). The occurrence of different *Listeria* species in municipal waste water. *Zentralbl. Bakteriol. Mikrobiol. Hyg. [B]*. **181**, 563–565.

Gilbert, R. J., K. L. Miller and D. Roberts (1989). *Listeria monocytogenes* and chilled foods. *Lancet* **18**, 383–384.

Gitter, M. (1989). Veterinary aspects of listeriosis. *PHLS Microbiol. Dig.* **6**, 38–42.

Glaser, P., L. Frangeul, C. Buchrieser *et al.* (2001). Comparative genomics of *Listeria* species. *Science* **26**, 849–852.

Goebel, W., J. Kreft and R. Böckmann (2000). Regulation of virulence genes in pathogenic *Listeria*. In: V. A. Fischetti, R. P. Novick, J. J. Ferretti *et al.* (eds), *Gram-positive Pathogens*, pp. 499–506. ASM Press, Washington, DC.

Grau, F. H. and P. B. Vanderlinde (1992). Occurrence, numbers, and growth of *Listeria monocytogenes* on some vacuum-packaged processed meats. *J. Food Prot.* **55**, 4–7.

Gray, M. L. (1957). A rapid method for the detection of colonies of *Listeria monocytogenes. Zentralbl. Bakteriol. Mikrobiol. Hyg.* **169**, 373–377.

Gray, M. L. and A. H. Killinger (1966). *Listeria monocytogenes* and listeric infections. *Bacteriol. Rev.* **30**, 309–382.

Guzman, C. A., M. Rohde, T. Chakraborty *et al.* (1995). Interaction of *Listeria monocytogenes* with mouse dendritic cells. *Infect. Immun.* **63**, 3665–3673.

Ha, K.-S., S.-J. Park, J.-H. Seo and D.-H. Chung (2002). Incidence and polymerase chain reaction assay of *Listeria monocytogenes* from raw milk in Gyeongnam province of Korea. *J. Food Prot.* **65**, 111–115.

Hassan, L., H. O. Mohammed, P. L. McDonough and R. N. Gonzalez (2000). A cross-sectional study on the prevalence of *Listeria monocytogenes* and *Salmonella* in New York dairy herds. *J. Dairy Sci.* **83**, 2441–2447.

Heisick, J. E., D. E. Wagner, M. L. Nierman and J. T. Peeler (1989). *Listeria* spp. found on fresh market produce. *Appl. Environ. Microbiol.* **55**, 1925–1927.

Holt, J. J. (1994). *Bergey's Manual of Determinative Bacteriology*, 9th edn. William and Wilkins, Baltimore, MD.

Hudson, W. R. and G. C. Read (1989). *Listeria* contamination at a poultry processing plant. *Lett. Appl. Microbiol.* **9**, 211–214.

Husu, J. R., J. T. Seppanen, S. K Sivela and A. L. Rauramaa (1990). Contamination of raw milk by *Listeria monocytogenes* on dairy farms. *Zentralbl. Veterinarmed.* **37**, 268–275.

Ireton, K., B. Payrastre, H. Chap *et al.* (1996). A role for phosphoinositide 3-kinase in bacterial invasion. *Science* **274**, 780–782.

Jay, J. M. (1996). Prevalence of *Listeria* spp. in meat and poultry products. *Food Control* **7**, 209–214.

Jayarao, B. M. and D. R. Henning (2001). Prevalence of foodborne pathogens in bulk tank milk. *J. Dairy Sci.* **84**, 2157–2162.

Jensen, N. E., F. M. Aarestrup, J. Jensen and H. C. Wegener (1996). *Listeria monocytogenes* in bovine mastitis. Possible implication for human health. *Intl J. Food Microbiol.* **32**, 209–216.

Johnson, J. L., M. P. Doyle and R. G. Cassens (1990). *Listeria monocytogenes* and other *Listeria* spp. in meat and meat products: a review. *J. Food Prot.* **53**, 81–91.

Kathariou, S. (2002). *Listeria monocytogenes* virulence and pathogenicity, a food safety perspective. *J. Food Prot.* **65**, 1811–1129.

Kathariou, S., R. Kanenaka, R. D. Allen *et al.* (1995). Repression of motility and flagellin production at 37 degrees C is stronger in *Listeria monocytogenes* than in the nonpathogenic species *Listeria innocua. Can. J. Microbiol.* **41**, 572–577.

Kerr, K. G., N. A. Rotowa, P. M Hawkey and R. W. Lacey (1990). Incidence of *Listeria* spp. in pre-cooked, chilled chicken products as determined by culture and enzyme-linked immunoassay (ELISA). *J. Food Prot.* **53**, 606–607.

Kohler, S., M. Leimeister-Wachter, T. Chakraborty *et al.* (1990). The gene coding for protein p60 of Listeria monocytogenes and its use as a specific probe for *Listeria monocytogenes. Infect. Immun.* **58**, 1943–1950.

Kohler, S., A. Bubert, M. Vogel and W. Goebel (1991). Expression of the iap gene coding for protein p60 of *Listeria monocytogenes* is controlled on the posttranscriptional level. *J. Bacteriol.* **173**, 4668–4674.

Kolb-Maurer, A., S. Pilgrim, E. Kampgen *et al.* (2001). Antibodies against listerial protein 60 act as an opsonin for phagocytosis of *Listeria monocytogenes* by human dendritic cells. *Infect. Immun.* **69**, 3100–3109.

Kreft, J. and J. A. Vázquez-Boland (2001). Regulation of virulence genes in *Listeria. Intl J. Med. Microbiol.* **291**, 145–157.

Lawrence, L. M. and A. Gilmour (1995). Characterization of *Listeria monocytogenes* isolated from poultry products and from the poultry-processing environment by random amplification of polymorphic DNA and multilocus enzyme electrophoresis. *Appl. Environ. Microbiol.* **61**, 2139–2144.

Leasor, S. B. and P. M. Foegeding (1989). *Listeria* species in commercially broken raw liquid whole egg. *J. Food Prot.* **52**, 777–780.

Levine, P., B. Rose, S. Green *et al.* (2001). Pathogen testing of ready-to-eat meat and poultry products collected at federally inspected establishments in the United States, 1990 to 1999. *J. Food Prot.* **64**, 1188–1193.

Lin, C.-M., S. Y. Fernando and C.-I. Wei (1996). Occurrence of *Listeria monocytogenes*, *Salmonella* spp., *Escherichia coli* and *E. coli* O157, H7 in vegetable salads. *Food Control* **7**, 135–140.

Loncarevic, S., A. Milanovic, F. Caklovica *et al.* (1994). Occurrence of *Listeria* species in an abattoir for cattle and pigs in Bosnia and Hercegovina. *Acta Vet. Scand.* **35**, 11–15.

Loncarevic, S., M. L. Danielsson-Tham and W. Tham (1995). Occurrence of *Listeria monocytogenes* in soft and semi-soft cheeses in retail outlets in Sweden. *Intl J. Food Microbiol.* **26**, 245–250.

Lowry, P. D. and I. Tiong (1988). The incidence of *Listeria monocytogenes* in meat and meat products-Factors affecting distribution. In: C. S. Chandler and R. F. Thornton (eds), *Proceedings of the 34th International Congress of Meat Science and Technology, Part B. Brisbane, Australia*, pp. 528–30.

Lundén, J. M., T. J. Autio and H. J. Korkeala (2002). Transfer of persistent *Listeria monocytogenes* contamination between food-processing plants associated with a dicing machine. *J. Food Prot.* **65**, 1129–1133.

MacGowan, A. P. (1990). Listeriosis – the therapeutic options. *J. Antimicrob. Chemother.* **26**, 721–722.

Massa, S., D. Cesaroni, G. Poda and L. D. Trovatelli (1990). The incidence of *Listeria* spp. in soft cheeses, butter and raw milk in the province of Bologna. *J. Appl. Bacteriol.* **68**, 153–156.

Mazurier, S. I. and K. Wernars (1992). Typing of *Listeria* strains by random amplification of polymorphic DNA. *Res. Microbiol.* **143**, 499–505.

McKenna, R. T. (1991). Thermal resistance of *Listeria monocytogenes* in raw liquid egg yolk. *J. Food. Prot.* **54**, 816.

McLauchlin, J. (1993). Listeriosis and *Listeria monocytogenes. Environ. Policy Pract.* **3**, 201–214.

McLauchlin, J. and J. C. Low (1994). Primary cutaneous listeriosis in adults: an occupational disease of veterinarians and farmers. *Vet. Record* **24**, 615–617.

Mengaud, J., M. F. Vicente, J. Chenevert *et al.* (1988). Expression in *Escherichia coli* and sequence analysis of the listeriolysin O determinant of *Listeria monocytogenes. Infect. Immun.* **56**, 766–772.

Miettinen, M. K., K. J. Björkroth and H. J. Korkela (1999). Characterization of *Listeria monocytogenes* from an ice cream plant by serotyping and pulsed-field gel electrophoresis. *Intl J. Food Microbiol.* **46**, 187–192.

Miettinen, M. K., L. Palmu, K. J. Bjorkroth and H. Korkeala (2001). Prevalence of *Listeria monocytogenes* in broilers at the abattoir, processing plant, and retail level. *J. Food Prot.* **64**, 994–999.

Milenbachs, A. A., D. P. Brown, M. Moors and P. Youngman (1997). Carbon-source regulation of virulence gene expression in *Listeria monocytogenes. Mol. Microbiol.* **23**, 1075–1085.

Monfort, P., J. Minet, J. Rocourt *et al.* (1998). Incidence of *Listeria* spp. in Breton live shellfish. *Lett. Appl. Microbiol.* **26**, 205–208.

Moore, J. and R. H. Madden (1993). Detection and incidence of *Listeria* species in blended raw egg. *J. Food Prot.* **56**, 652–654.

Motes, M. L. Jr (1991). Incidence of *Listeria* spp. in shrimp, oysters, and estuarine waters. *J. Food Prot.* **54**, 170–173.

Moura, S. M., M. T. Destro and B. D. Franco (1993). Incidence of *Listeria* species in raw and pasteurized milk produced in Sao Paulo, Brazil. *Intl J. Food Microbiol.* **19**, 229–237.

Muriana, P. M., H. Hou and R. K. Singh (1996). A flow-injection system for studying heat inactivation of *Listeria monocytogenes* and *Salmonella enteritidis* in liquid whole egg. *J. Food Prot.* **59**, 121–126.

Murray, E. G. D., R. A. Webb and M. B. R. Swann (1926). A disease of rabbits characterised by a large mononuclear leucocytosis, caused by a hitherto undescribed bacillus *Bacterium monocytogenes* (n.sp.). *J. Pathol. Bacteriol.* **29**, 407–439.

Myers, E. R., A. W. Dallmier and S. E. Martin (1993). Sodium chloride, potassium chloride, and virulence in *Listeria monocytogenes. Appl. Environ. Microbiol.* **59**, 2082–2086.

Nesbakken, T., G. Kapperud and D. A. Caugant (1996). Pathways of *Listeria monocytogenes* contamination in the meat processing industry. *Intl J. Food Microbiol.* **31**, 161–171.

Nocera, D., E. Bannerman, J. Rocourt *et al.* (1990). Characterization by DNA restriction endonuclease analysis of *Listeria monocytogenes* strains related to the Swiss epidemic of listeriosis. *J. Clin. Microbiol.* **28**, 2259–2263.

O'Donnell, E. T. (1995). The incidence of *Salmonella* and *Listeria* in raw milk from farm bulk tanks in England and Wales. *J. Soc. Dairy Technol.* **48**, 25–29.

Ojeniyi, B., H. C. Wegener, N. E. Jensen and M. Bisgaard (1996). *Listeria monocytogenes* in poultry and poultry products: epidemiological investigations in seven Danish abattoirs. *J. Appl. Bacteriol.* **80**, 395–401.

Pagotto, F. J. and J. M. Farber (2003). *Listeria*/listeriosis. In: B. Caballero, L. Trugo and P. Finglas (eds), *Encyclopedia of Food Sciences and Nutrition*, pp. 3582–3588. Academic Press, London.

Pagotto, F., E. Daley, J. Farber and D. Warburton (2001). MFHPB-30: isolation of *Listeria monocytogenes* from all food and environmental samples. Official methods for the microbiological analysis of foods. *Compendium of Analytical Methods*, Vol. 2, Health Protection Branch (available at http, //www.hc-sc.gc.ca/food-aliment/mh-dm/mhe-dme/compendium/volume_2/pdf/e_mfhpb30.pdf).

Pak, S. I., U. Spahr, T. Jemmi and M. D. Salman (2002). Risk factors for *L. monocytogenes* contamination of dairy products in Switzerland, 1990–1999. *Prev. Vet. Med.* **53**, 55–65.

Palumbo, M. S., S. M. Beers, S. Bhaduri and S. A. Palumbo (1995). Thermal resistance of *Salmonella* spp. and *Listeria monocytogenes* in liquid egg yolk and egg yolk products. *J. Food Prot.* **58**, 960–966.

Pearson, L. J. and E. H. Marth (1990). *Listeria monocytogenes* – threat to a safe food supply: a review. *J. Dairy Sci.* **73**, 912–928.

Peel, M., W. Donachie and A. Shaw (1988). Temperature-dependent expression of flagella of *Listeria monocytogenes* studied by electron microscopy, SDS-PAGE and western blotting. *J. Gen. Microbiol.* **134**, 2171–2178.

Pini, P. N. and R. J. Gilbert (1988). The occurrence in the U.K. of *Listeria* species in raw chickens and soft cheeses. *Intl J Food Microbiol.* **6**, 317–326.

Pirie, J. H. H. (1927). A new disease of veld rodents, 'Tiger River Disease'. *Publ. S. Afr. Inst. Med. Res.* **3**, 163–186.

Pirie, J. H. H. (1940). *Listeria*: change of name for a genus of bacteria. *Nature (Lond.)* **145**, 264.

Portnoy, D. A., P. S. Jacks and D. J. Hinrichs (1988). Role of hemolysin for the intracellular growth of *Listeria monocytogenes. J. Exp. Med.* **167**, 1459–1471.

Pritchard, T. J., K. J. Flanders and C. W. Donnelly (1995). Comparison of the incidence of *Listeria* on equipment versus environmental sites within dairy processing plants. *Intl J. Food Microbiol.* **26**, 375–384.

Rasmussen, O. F., P. Skouboe, L. Dons *et al.* (1995). *Listeria monocytogenes* exists in at least three evolutionary lines: evidence from flagellin, invasive associated protein and listeriolysin O genes. *Microbiology* **141**, 2053–2061.

Rawles, D., G. Flick, M. Pierson *et al.* (1995). *Listeria monocytogenes* occurrence and growth at refrigeration temperatures in fresh blue crab (*Callinectes sapidus*) meat. *J. Food Prot.* **58**, 1219–1221.

Rea, M. C., T. M. Cogan and S. Tobin (1992). Incidence of pathogenic bacteria in raw milk in Ireland. *J. Appl. Bacteriol.* **73**, 331–336.

Ribeiro, C. D. and S. H. Burge (1992). Developing microbiological guidelines for food: first results from cooked chicken portions. *Publ. Health Lab. Service Digest.* **9**, 100 102.

Ripio, M. T., G. Dominguez-Bernal, M. Suarez *et al.* (1996). Transcriptional activation of virulence genes in wild-type strains of *Listeria monocytogenes* in response to a change in the extracellular medium composition. *Res. Microbiol.* **147**, 371–384

Rocourt, J. (1996). Risk factors for listeriosis. *Food Control* **7**, 195–202.

Rocourt, J. and H. P. R. Seeliger (1985). Distribution des espèces du genre *Listeria. Zentralbl. Bakteriol. Mikrobiol. Hyg A* **259**, 317–330.

Rocourt, J., A. Schrettenbrunner, H. Hof and E. P. Espaze (1987). Une nouvelle espèce du genre *Listeria*: *Listeria seeligeri. Pathol. Biol.* **35**, 1075–1080.

Rocourt, J., P. Boerlin, F. Grimont *et al.* (1992). Assignment of *Listeria grayi* and *Listeria murrayi* to a single species, *Listeria grayi*, with a revised description of Listeria grayi. *Intl J. Syst. Bacteriol.* **42**, 171–174.

Rodriguez, J. L., P. Gaya, M. Medina and M. Nunez (1994). Incidence of *Listeria monocytogenes* and other *Listeria* spp. in raw ewes milk. *J. Food Prot.* **57**, 571–575.

Rohrbach, B. W., F. A. Draughon, P. M. Davidson and S. P. Oliver (1992). Prevalence of *Listeria monocytogenes*, *Campylobacter jejuni*, *Yersinia enterocolotica*, and

Salmonella in bulk tank milk: risk factors and risk of human exposure. *J. Food Prot.* **55**, 93–97.

Rørvik, L. M. and M. Yndestad (1991). *Listeria monocytogenes* in foods in Norway. *Intl J. Food Microbiol.* **13**, 97–104.

Rørvik, L. M., D. A. Caugant and M. Yndestad (1995). Contamination pattern of *Listeria monocytogenes* and other *Listeria* spp. in a salmon slaughterhouse and smoked salmon processing plant. *Intl J. Food Microbiol.* **25**, 19–27.

Rudolff, M. and S. Scherer (2001). High incidence of *Listeria monocytogenes* in European red smear cheese. *Intl J. Food Microbiol.* **63**, 91–98.

Ryser, E. T., S. M. Arimi and C. W. Donnelly (1997). Effects of pH on distribution of *Listeria* ribotypes in corn, hay, and grass silage. *Appl. Environ. Microbiol.* **63**, 3695–3697.

Safdar, A. and D. Armstrong (2003). Antimicrobial activities against 84 *Listeria monocytogenes* isolates from patients with systemic listeriosis at a comprehensive cancer center (1955–1997). *J. Clin. Microbiol.* **41**, 483–485.

Salvat, G., M. T. Toquin, Y. Michel and P. Colin (1995). Control of *Listeria monocytogenes* in the delicatessen industries, the lessons of a listeriosis outbreak in France. *Intl J. Food Microbiol.* **25**, 75–81.

Sanaa, M., B. Poutrel, J. L. Menard and F. Serieys (1993). Risk factors associated with contamination of raw milk by *Listeria monocytogenes* in dairy farms. *J. Dairy Sci.* **76**, 2891–2898.

Schlegelova, J., V. Babak, E. Klimova *et al.* (2002). Prevalence of and resistance to anti-microbial drugs in selected microbial species isolated from bulk milk samples. *J. Vet. Med. B* **49**, 216–225.

Schuchat, A., B. Swaminathan and C. V. Broome (1991). Epidemiology of human listeriosis. *Clin. Microbiol. Rev.* **4**, 169–183.

Schuman, J. D. and B. W. Sheldon (1997). Thermal resistance of *Salmonella* spp. and *Listeria monocytogenes* in liquid egg yolk and egg white. *J. Food Prot.* **60**, 634–638.

Sechi, A. S., J. Wehland and J. V. Small (1997). The isolated comet tail pseudopodium of *Listeria monocytogenes*: a tail of two actin filament populations, long and axial and short and random. *J. Cell Biol.* **137**, 155–167.

Seeliger, H. P. R. and K. Höhne (1979). Serotyping of *Listeria monocytogenes* and related species. In: T. Bergan and J. R. Norris (eds), *Methods in Microbiology*, Vol. 13, pp. 31–49. Academic Press, London.

Sheehan, B., C. Kocks, S. Dramsi *et al.* (1994). Molecular and genetic determinants of the *Listeria monocytogenes* infectious process. *Curr. Topics Microbiol. Immunol.* **192**, 187–216.

Sheehan, B., A. Klarsfeld, R. Ebright and P. Cossart (1995). Differential action of virulence gene expression by PrfA, the *Listeria monocytogenes* virulence regulator. *J. Bacteriol.* **177**, 6469–6476.

Sionkowski, P. J. and L. A. Shelef (1990). Viability of *Listeria monocytogenes* strain Brie-1 in avian egg. *J. Food Prot.* **53**, 15–17.

Sizmur, K. and C. W. Walker (1988). *Listeria* in prepackaged salads. *Lancet* **1**, 1167.

Skovgaard, N. and C. A. Morgen (1988). Detection of *Listeria* spp. in faeces from animals, in feeds, and in raw foods of animal origin. *Intl J. Food Microbiol.* **6**, 229–242.

Slutsker, L., M. C. Evans and A. Schuchat (2000). Listeriosis. In: W. M. Scheld, W. A. Craig and J. M. Hughes (eds), *Emerging Infections*, Vol. 4, pp. 83–106. ASM Press, Washington, DC.

Spencer, J. F. T. and A. L. Ragout de Spencer (2001). *Food Microbiology Protocols*. Humana Press, New Jersey, NJ.

Tilney, L. G. and D. A. Portnoy (1989). Actin filaments and the growth, movement, and spread of the intracellular bacterial parasite, *Listeria monocytogenes*. *J. Cell Biol.* **109**, 1597–1608.

US Department of Agriculture (1969). *Egg Pasteurization Manual*. ARS 94–48. Poultry Laboratory, Agriculture Research Services, US Department of Agriculture, Albany, CA.

Uyttendaele, M., P. De Troy and J. Debevere (1999). Incidence of *Salmonella*, *Campylobacter jejuni*, *Campylobacter coli*, and *Listeria monocytogenes* in poultry carcasses and different types of poultry products for sale on the Belgian retail market. *J. Food Prot.* **62**, 735–740.

Van Netten, P., I. Perales, A. van de Moosdijk *et al.* (1989). Liquid and solid selective differential media for the detection and enumeration of *L. monocytogenes* and other Listeria spp. *Intl J. Food Microbiol.* **8**, 299–316.

Van Renterghem, B., F. Huysman, R. Rygole and W. Verstraete (1991). Detection and prevalence of *Listeria monocytogenes* in the agricultural ecosystem. *J. Appl. Bacteriol.* **71**, 211–217.

Vazquez-Boland, J. A., M. Kuhn, P. Berche *et al.* (2001). *Listeria* pathogenesis and molecular virulence determinants. *Clin. Microbiol. Rev.* **14**, 584–640.

Vines, A., M. W. Reeves, S. Hunter and B. Swaminathan (1992). Restriction fragment length polymorphism in four virulence-associated genes of *Listeria monocytogenes*. *Res. Microbiol.* **143**, 281–294.

Waak, E., W. Tham and M. L. Danielsson-Tham (2002). Prevalence and fingerprinting of *Listeria monocytogenes* strains isolated from raw whole milk in farm bulk tanks and in dairy plant receiving tanks. *Appl. Environ. Microbiol.* **68**, 3366–3370.

Watkins, J. and K. P. Sleath (1981). Isolation and enumeration of *Listeria monocytogenes* from sewage, sewage sludge and river water. *J. Appl. Bacteriol.* **50**, 1–9.

Wiedmann, M., J. Czajka, N. Bsat *et al.* (1994). Diagnosis and epidemiological association of *Listeria monocytogenes* strains in two outbreaks of listerial encephalitis in small ruminants. *J. Clin. Microbiol.* **32**, 991–996.

Wilkinson, J. M. (1999). Silage and animal health. *Nat. Toxins* **7**, 221–232.

Wuenscher, M. D., S. Kohler, A. Bubert *et al.* (1993). The *iap* gene of *Listeria monocytogenes* is essential for cell viability, and its gene product, p60, has bacteriolytic activity. *J. Bacteriol.* **175**, 3491–3501.

Zheng, W. and S. Kathariou (1995). Differentiation of epidemic-associated strains of *Listeria monocytogenes* by restriction fragment length polymorphism in a gene region essential for growth at low temperatures (4 °C). *Appl. Environ. Microbiol.* **61**, 4310–4314.

Infections with other bacteria

10

Maha N. Hajmeer and Daniel Y. C. Fung

1 Introduction

Some foodborne disease agents are well recognized. Some are newly identified and are considered to be emerging. There are, however, several foodborne disease-causing agents that are well recognized, but are also considered to be emerging. These are microorganisms that either have recently become more common, due possibly to a number of newly reported outbreaks (e.g. *Salmonella* spp.), or have only recently had the role they play in the transmission of disease via food recognized (e.g. *Escherichia coli* O157:H7 and *Listeria monocytogenes*).

This chapter will focus on the following types of bacterial agents:

- Those that 'historically' have been important foodborne pathogens (i.e. micro-organisms that are known to cause threats via food; these might have been eradicated in certain parts of the world, but are still a threat or are making a comeback in other places)
- Those that are emerging foodborne pathogens (i.e. agents whose role in the transmission of disease via food has recently been, or is beginning to be, understood, or ones that are becoming more common)
- Those that are transmitted occasionally via food, but are known to pose a greater problem if transmitted by other routes (e.g. lesions, air droplets)
- Those that are questioned as foodborne pathogens (i.e. they are sometimes found in foods, may or may not be pathogenic, or might be pathogenic to certain susceptible populations – for example, young children, the elderly, the immunocompromised and the immunosuppressed).

Foodborne Infections and Intoxications 3e
ISBN-10: 0-12-588365-X
ISBN-13: 978-012-588365-8

For each bacterial agent discussed, attempts have been made to provide information relating to classification, history of occurrence and contemporary problems, pathogenesis, ability to survive and grow in the environment, reservoirs and transmission, nature of infection in man and in animals, the food vehicles most often involved, and control measures.

2 Synopses of individual bacteria

Many bacterial genera and species, besides those discussed in detail in other parts of this book, have been reported directly and indirectly to have been involved in foodborne infections and intoxications. This section presents synopses of these organisms and their potential roles as foodborne pathogens. For more detailed information about these organisms, it is suggested that readers consult the following works: *Bergey's Manual of Systematic Bacteriology*, first edition (Krieg and Holt, 1984; Sneath *et al.*, 1986; Staley *et al.*, 1989; Williams *et al.*, 1989); *Bergey's Manual of Systematic Bacteriology*, second edition (Boone *et al.*, 2001); *Modern Food Microbiology*, sixth edition (Jay, 2000); *Food Microbiology: Fundamentals and Frontiers*, second edition (Doyle *et al.*, 2001); *Compendium of Methods for the Microbiological Examination of Foods*, fourth edition (Downes and Ito, 2001); *Microbiology Control for Foods and Agricultural Products* (Bourgeois *et al.*, 1995); *Introduction of Food Microbiology*, third edition (Fung and Goetsch, 2000); *Microbiology: An Introduction* (Batzing, 2002); *Foodborne Pathogens: Hazards, Risk Analysis and Control* (Blackburn and McClure, 2002); and *Fundamental Food Microbiology* (Ray, 2004). Much of the earlier information on 'other potential pathogens', before 1980, was detailed in the second edition of *Foodborne Infections and Intoxications* (Riemann and Bryan, 1979).

In developing the synopses of bacteria in this section, the 'traditional' names of familiar bacteria are used. For more recent nomenclature of bacteria, readers are referred to *Bergey's Manual of Determinative Bacteriology*, ninth edition (Holt *et al.*, 1994). Here, the bacteria are listed alphabetically by genus.

2.1 *Aeromonas*

Aeromonas species, including *A. hydrophila*, *A. caviae*, *A. sorbia* and *A. veronii*, are aquatic bacteria that occur in sewage and in fresh and brackish water environments worldwide. Sometimes members of this species are mistaken for *Escherichia coli*. More is known about *A. hydrophila* spp. compared to the others.

A. hydrophila is a facultative anaerobic, Gram-negative, catalase-positive, oxidase-positive, fermentative and asporogenous microorganism. It is a straight rod, and the cells have rounded ends approaching a spherical shape. It can occur singly, in pairs or short chains, and is 0.3–1.0 μm in diameter and 1.0–3.5 μm in length. Usually it is motile by a single polar flagellum, although peritrichous flagella may be seen in young cultures. Biochemically, it is similar to *E. coli* and *Klebsiella*. The optimal and maximal temperatures for growth are 28°C and 42°C, respectively. Some strains can grow at 5°C, which is a temperature usually considered adequate to control growth of foodborne pathogens.

Some *Aeromonas* spp. can cause illness to fish, frogs (red leg disease) and humans (diarrhea or bacteremia). In 1891, *Aeromonas* spp. were recognized as colonizers and pathogens of cold-blooded animals such as fish (Ewing *et al.*, 1961). It was not until 1968 that they were identified as human pathogens (Von Graevenitz and Mensch, 1968). Since then, they have been associated with a number of human diseases (Holmberg *et al.*, 1986; Janda and Duffey, 1988).

A. hydrophila has been associated with foodborne infection, but the evidence is not clear. The organism is often found in normal and diarrheal human intestines. Certain strains of *A. hydrophila* are able to produce enterotoxins (Sanyul *et al.*, 1975; Stephen *et al.*, 1975; Ljungh *et al.*, 1977). Diseases caused by *A. hydrophila* include gastroenteritis (cholera- and dysentery-like illness) and extra-intestinal infections such as septicemia and meningitis in immunocompromised individuals or people with malignancies. In the illness known as *Aeromonas* diarrhea, the signs and symptoms include watery diarrhea, abdominal pain, chills, nausea and headache. The incubation period tends to be between 1 and 2 days.

A. caviae and *A. sorbia* may also cause similar infections. Infection in man may be acquired through open wounds, or by ingestion of food or water contaminated with the bacterium. Of the *Aeromonas* species, *A. punctata* is pathogenic to frogs and *A. salmonicida* to salmon and other fish (Bryan, 1979).

A. hydrophila is proposed as a cause of diarrhea in humans, especially young children. To date, the causation of diarrhea or transmission of bacterium via food or water has yet to be conclusively proven. This organism has been isolated from seafood (fish, shrimp, crabs, scallops, oysters), meats (red meats, pork, poultry) and milk. Contaminated water is another vehicle – even bottled mineral water.

Factors contributing to foodborne illness with *Aeromonas* include contamination of foods via sewage or surface water. It is therefore important to assure that sewage material does not come in contact with food and that the surface water does not become contaminated. Proper heating of food offers sufficient protection against this organism. Eating of undercooked or raw food, such as raw shellfish, is not recommended.

2.2 *Alcaligenes faecalis*

Alcaligenes faecalis is not considered to be a major foodborne pathogen, although occasional foodborne outbreaks have been reported, such as one large outbreak in India. A total of 270 people were affected, with typical symptoms including abdominal pain, diarrhea, headache and vomiting. Following the outbreak, *A. faecalis* was suggested as the etiologic agent.

A. faecalis is a Gram-negative, oxidase-positive and catalase-positive microorganism with a rod, coccal-rod, or even coccal shape. It is an obligate aerobic bacterium, motile with peritrichous flagella. Some strains are capable of anaerobic respiration in the presence of nitrate or nitrite. Usually it occurs singly, with no more than eight peritrichous flagella (occasionally up to 12), and has a size of $0.5–1.0 \times 0.5–2.6\ \mu m$. The optimum temperature range for growth is 20–37°C.

A. faecalis is an important food spoilage bacterium. It has also been isolated from diverse sources such as soil, water, medical specimens (blood, urine, feces, sputum, wounds, pleural fluid), nematodes and insects. Some species are common inhabitants of the intestinal tract of vertebrates.

2.3 *Arizona*

The genus *Arizona* is currently classified within the genus *Salmonella*, as the DNA of *Arizona* has been characterized as having close similarities to other salmonellae. In past times, the name *Arizona* was applied to a group of *Salmonella*-like organisms called *Paracolobactrum arizonae*. These are Gram-negative, motile, rod-shaped microorganisms. They are serologically similar to *Salmonella*, along with the atypical abilities to ferment lactose and liquefy gelatin. In later years, *Arizona* nomenclature included *Arizona arizonae*, *Arizona hinshawii* and *Salmonella arizonae* before it was absorbed into the genus *Salmonella*.

Arizona has been reported to cause infections in animals, including diarrheal and fatal infections in fowls (Edwards *et al.*, 1956). Important reservoirs for the microorganism are turkeys, chickens and reptiles (e.g. snakes). *Arizona* can cause diarrheal infections in man. Several serotypes of *Arizona* have been associated with foodborne outbreaks (Bryan, 1979), and these include O1,2:H1,2,5; O1,4:H1,2,5; O5:H1,2,5; O7:H1,2,6; O7:H1,7,8; O10:H1,2,5; and O10:H1,3,11. Food vehicles associated with infections included cream pie, chocolate éclairs, custard, eggs, chicken and turkey. *Arizona* infections are as severe as salmonellosis. Preventive measures include proper cooking of food, efficient hygienic practices to avoid cross-contamination, effective cooling and holding of food (especially poultry and egg products), and adequate cleaning and disinfection of utensils and food-contact equipment.

2.4 *Bacillus*

The genus *Bacillus*, which includes *B. anthracis* and *B. subtilis*, is very diverse. Species in this genus are spore-forming, aerobic or facultatively anaerobic, rod-shaped organisms that are ubiquitous in nature. The endospores are usually oval, but can sometimes be round or cylindrical. They can be found in soil, water and airborne dust. The rods are straight, with rounded or squared ends. They have dimensions of $0.5–2.5 \times 1.2–10\,\mu m$, and can be found in pairs or chains. *Bacillus* spp. are usually catalase positive, and the cells stain Gram-positive.

Differentiation of the species within the genus *Bacillus* can be difficult because of the large number of species and their many shared characteristics. Presently, the genus *Bacillus* is divided into six subgroups based on spore morphology. Among the six groups, four (*B. subtilis*, *B. anthracis, B. mycoides,* and *B. thuringiensis*) are closely related. Differentiation of the four closely related groups can be based on criteria such as colony morphology, motility, hemolysis, susceptibility to the antibiotic penicillin, formation of parasporal body (crystals in the sporangial cell that are proteinous), and virulence to mice. The vegetative cells of *Bacillus* spp. are relatively easily destroyed by most food-processing methods, but the spores are resistant to a variety of treatments – such as high and low temperature, chemicals, radiation, pressure, drying, etc. When members of this genus are in the vegetative state, they are motile by peritrichous flagella.

Although most *Bacillus* spp. are harmless, some are pathogenic to vertebrates or invertebrates. Two species (*B. cereus* and *B. anthracis*) are recognized as pathogens. *B. cereus*, which falls in the *B. subtilis* group, has been implicated in foodborne disease. *B. anthracis* is a pathogen of humans and animals. It can infect humans perorally, but this is considered an inefficient mode of transmission. Other routes for infection with this species are discussed below.

2.4.1 *Bacillus anthracis*

This organism has received international attention because of its use as an agent of bioterrorism. It causes anthrax in animals such as cows and sheep, and sometimes in humans. The vegetative cells and spores of this organism can be found in soil, water, air and plant materials. The spores persist in contaminated materials for a long time. The pathogen is probably not a major source of foodborne infection, although its presence in food is not desirable.

Human anthrax is commonly contracted through the skin via abrasions, cuts or animal bites, by direct contact with contaminated animal products, and by inhalation of endospores. Most cases occur in individuals who handle or come in contact with infected animals or animal products (e.g. veterinarians). Human anthrax is rarely contracted through ingestion, but it is not impossible. The majority (95%) of human cases of anthrax are associated with cutaneous infections. In the case of cutaneous infections, small lesions (known as black eschar) develop 1–7 days following the entry of spores. This develops to local necrosis, and may be followed by toxic septicemia and death, depending on the spread of the pathogen. Infection via inhalation or the gastrointestinal route tends to be rare. Inhalation anthrax occurs when the spores are taken up by the alveolar macrophages that carry the pathogen to the regional lymph nodes, causing necrotic hemorrhaging. This can eventually lead to death. Gastrointestinal anthrax, which is usually an outcome of eating contaminated animal products (e.g. contaminated meat), results in systemic symptoms that can also lead to death. Although gastrointestinal anthrax is rare, mortality may reach 50%.

In order for anthrax infections to occur, the pathogen has to produce a capsule and exotoxins. The capsule allows the organism to survive phagocytosis. Animals with high levels of phagocytic activity are more resistant to anthrax infection, as the capsule of this pathogen is poorly immunogenic. *B. anthracis* produces a three-part toxin (edema factor, protective antigen and lethal factor) mediated by plasmids of the cell. Human or animal survivors of an anthrax infection may develop specific immunity as a result of the production of antitoxin antibody. Some antibiotics, such as penicillin, tetracycline and erythromycin, may be used in the treatment of anthrax.

Proper animal-handling and sanitary procedures are important factors in reducing the incidence of anthrax infections. Examples include the use of appropriate protective equipment and clothing when handling infected animals or animal products, and the use of effective methods for decontamination of infected animal products and even whole animal carcasses (e.g. deep burial). Proper food preparation, cooking and storage practices are essential to prevent food poisoning (or gastrointestinal anthrax).

2.4.2 *Bacillus subtilis*

This organism, closely related to *B. cereus*, is a spore-forming bacterium that occurs as single rods and, rarely, in chains. It decomposes the pectin and polysaccharide of plant tissues, and may produce rots in potato tuber. The organism colonizes the developing root system of plants, and is able to compete well with other micro-organisms – especially plant-disease causing agents such as *Fusarium*, *Aspergillus*, *Alternaria* and *Rhizoctonia*. It has been used in fungicide products as a treatment for a number of commodities, including cotton, barley, corn and peanut seeds. Due to the ability of *B. subtilis* to form levan and dextran from sucrose, it can cause ropiness and slime in bread.

B. subtilis has been a topic of research as a foodborne pathogen in recent years. According to Bennett (2003), some strains of *B. subtilis* are capable of forming an emetic toxin. Foods implicated as vehicles of *B. subtilis* include sausage rolls, meat pasties, turkey stuffing, chicken stuffing, pizza, wholemeal bread, steak pie and pork sandwiches. In a food-poisoning case attributed to *B. subtilis* involving ingestion of an Oriental lamb dish, the level of *B. subtilis* was more than 6 log CFU/g, with an incubation period of 1 hour, and resulted in vomiting. Kramer and Gilbert (1989) reported that in United Kingdom from 1975 to 1986 there were 49 outbreaks involving 175 cases of food poisoning associated with *B. subtilis*. The vehicles included meat pies, rolls, and Indian and Chinese meat and seafood dishes with curry and rice, bread etc. The incubation time ranged from 10 minutes to 14 hours, with a median of 2.5 hours. The symptoms were vomiting (highest number) followed by diarrhea, abdominal pain, cramps, nausea, headaches and flushing/sweating, with duration of 1.5–8 hours. *Bacillus subtilis* levels in implicated foods ranged from 5 log to 9 log CFU/g, with a mean of 6.7 log CFU/g.

It is worth noting that there is a popular Japanese food, called *Natto*, which is made by fermenting cooked soybean with *Bacillus subtilis natto*. The level of *B. subtilis* in *Natto* exceeds 9 log CFU/g, with the formation of strings of polymer of glutamic acid. This breakfast dish is eaten daily by a large proportion of the Japanese population. Apparently, this strain is a non-toxin producing *B. subtilis*.

2.5 *Brucella*

Six species are currently recognized within the genus *Brucella* – *B. abortus* (found in cattle), *B. canis* (dogs), *B. melitensis* (goats), *B. neotomae*, *B. ovis* (sheep), and *B. suis* (swine). Although not very common, brucellosis is a worldwide problem. It has been reported in the Americas, Europe, Asia, and Africa. *Brucella* species cause disease in a number of animals, including cattle, goats, sheep, pigs and dogs. It can also cause infections in humans. *B. melitensis* and *B. suis* are more easily transmissible to people, especially via the oral route. The three species of *Brucella* of concern to human health are *B. abortus*, related to cattle; *B. suis*, related to swine; and *B. melitensis*, related to goats and sheep.

Brucella is an aerobic, small, Gram-negative, catalase- and oxidase-positive, non-motile, non-spore-forming short rod or coccobacillus that may become pleomorphic. It can be arranged singly, or in pairs, short chains or small groups. Its size ranges from

0.5 × 0.7 μm to 0.6 × 4.0 μm. Growth of the organism occurs at between 20°C and 40°C, with an optimum growth temperature of 37°C. The optimum pH is 6.6–7.4.

Transmission of *Brucella* is zoonotic, as the organism is carried and shed by animals. Livestock such as cattle (beef and dairy) seem to be a primary source of the pathogen, at least in the US. Some pets (e.g. dogs) are also a source. If proper treatment is applied to animals, infection is cleared within a few days. However, some body fluids (e.g. blood) may be infectious for weeks. Reinfection is always a possibility.

Brucella is an important human pathogen that causes undulant fever, especially in veterinarians and workers handling infected animals. Humans contract the illness when they eat or drink contaminated food, come in contact with infected animals, or inhale the organism. The organism can also get into the body through skin cuts and wounds. Most of the reported foodborne infection cases related to *Brucella* involved unpasteurized milk and milk products (e.g. cheese). Proper milk pasteurization will eliminate this problem. *Brucella* has been found to occur and survive in meat and meat products. Since meat products are usually cooked before consumption, the likelihood of contracting brucellosis from ingesting cooked meat is very low. Inhaling the organism and infections through the skin are more of a problem for individuals working with animals (e.g. veterinarians, hunters, workers in slaughterhouses and meat-packing facilities, laboratory personnel).

Person-to-person contact is not a common route for the spread of the infectious agent. Breastfeeding mothers can transmit the infections to their infants. Sexual transmission has also been reported.

Brucella causes a disease called brucellosis (Malta fever). The microorganism seems able to enter the body via the oral route, the respiratory tract or the skin, and eventually migrates to the blood and lymph vessels. The intracellular microorganism will multiply inside phagocytes and cause bacteremia (presence of bacteria in the blood). Symptoms include fever, sweating, chills, fatigue, headache, weight loss and endocarditis – some of which are similar to influenza symptoms. The severity of the symptoms depends on several factors, including the virulence of the pathogen, the dose, and the susceptibility of individuals. Mortality is estimated at less than 2%, and is mainly associated with endocarditis. The incubation period for brucellosis is 7–21 days.

Hygiene is an important preventative measure against *Brucella*. Pasteurization will destroy the *Brucella*, and therefore it is recommended that individuals avoid drinking unpasteurized milk and eating raw cheese. Treatment with antibiotics such as streptomycin, erythromycin, tetracycline, tetracycline plus gentamicin, and doxycycline seems to work against this pathogen. No vaccines are available for humans. Live vaccines are used for animals, and these may cause disease to humans.

2.6 *Citrobacter*

Citrobacter is a Gram-negative, catalase-positive, oxidase-negative straight rod (~1 μm in diameter and 2–6 μm in length). It occurs singly and in pairs, and follows the same general description as a member of the family *Enterobacteriaceae*. It is a facultatively anaerobic microorganism that is motile by peritrichous flagella. The optimal temperature for *Citrobacter* growth is 37°C.

Citrobacter has been isolated from feces of healthy humans and animals (birds, reptiles and amphibians), as well as from water, soil, sewage, clinical specimens (urine, throat, sputum, blood and wound swabs) and food (fresh meat and poultry), and is considered to be an opportunistic human pathogen. *Citrobacter* has been found in enteral nutrient solution or medical foods along with *Staphylococcus epidermidis*, *Corynebacterium* and *Acinetobacter* spp. The organism can cause diarrhea in patients, and may produce a heat-stable enterotoxin. Meat and milk have been implicated as sources of foodborne outbreaks, but the evidence is not conclusive. Tschape *et al.* (1995) reported an outbreak of severe gastroenteritis followed by hemolytic uremic syndrome (HUS), and thrombotic thrombocytopenic purpura was reported in a nursery school and kindergarten. The vehicle in this case was parsley, which had been used to make green butter that was added to sandwiches. Apparently the parsley came from an organic garden where pig manure was used instead of artificial fertilizers. Control measures against *Citrobacter* infections include effective hygienic practices, and proper handling, cooking and storage of food.

2.7 *Clostridium bifermentans*

Clostridium bifermentans falls within the genus *Clostridium*, which contains about 100 species. Members of the genus *Clostridium* are Gram-positive, catalase-negative, anaerobic, spore-forming, rod-shaped (0.3–2.0 × 1.5–20.0 μm), pleomorphic bacteria. They often occur in pairs or short chains with rounded or pointed ends, and are usually motile by peritrichous flagella. They form oval or spherical endospores.

Within the genus *Clostridium*, *C. botulinum* and *C. perfringens* are well known foodborne intoxication and infection agents (see Chapters 13 and 4, respectively). *C. bifermentans* is characterized by the ability to ferment both carbohydrates and amino acids – thus the name *bifermentans*. The organism has been isolated from soil, fresh water, marine sediments, human feces and a variety of human medical specimens, and from foods such as clams, cheese, canned tomatoes and vacuum-packed smoked fish. It was implicated in an outbreak of food poisoning involving about 75 people. The associated vehicle was meat-and-potato pie.

C. bifermentans can enter the human body through wounds in contact with soil or feces, along with *C. perfringens*, *C. ramosum*, *C. sporogenes*, etc., and may play a role in the development of gas gangrene. It can spoil canned food as a sulfide spoiler. The radiation D value of the spores of *C. bifermentans* is 1.4 kGy. The role of *C. bifermentans* in food poisoning is not clear.

2.8 *Corynebacterium diphtheriae*

The genus *Corynebacterium* belongs to the family *Mycobacteriaceae*. *Corynebacterium diphtheriae* is the etiologic agent of a life-threatening disease known as diphtheria. Diphtheria has occurred worldwide for centuries, and still exists today. Hippocrates provided the first clinical description of the disease in the fourth century BC. A diphtheria epidemic swept through Europe in the seventeenth century, when the disease was called *el garatillo* or 'the strangler' in Spain, and the 'gullet disease' in Italy. In the eighteenth century, the disease reached the American colonies, and whole families lost their lives. Occasional outbreaks of diphtheria still occur almost yearly.

C. diphtheriae is a Gram-positive, non-motile, catalase-positive microorganism. It is a straight or slightly curved rod, frequently swollen at one or both ends (Chinese character morphology). This distinctive characteristic of the cells, also known as a 'club-shaped' appearance, is the result of thin spots in the cell walls that lead to some Gram variability and ballooning of cells. Old cultures of *Corynebacterium* are easily decolorized, making them a Gram-variable culture. Older cells may look like metachromatic granules when stained, due to storage of inorganic phosphate.

C. diphtheriae is the organism that causes the severe disease, diphtheria, in humans – a toxigenic infection, usually of the upper respiratory tract. Humans are the principal reservoirs of *C. diphtheriae*. The disease is characterized by muscle weakness, pharyngitis, fever, edema or swelling (of the neck or area surrounding the skin lesion), and a pseudomembranous material covering the lesions. The toxin may eventually reach other target organs via the circulatory system, leading to paralysis and congestive heart failure. Treatment is available with antitoxin (barring hypersensitivity to horse serum) and some antibiotics such as erythromycin or penicillin.

Other than in the respiratory tract, the organism occurs on the skin of infected persons or normal carriers, and in wounds. It is spread by droplets or direct contact. The virulent strains will grow on mucous membranes or skin abrasions and start to produce toxins. All toxigenic strains are capable of producing the same disease-producing heat-labile polypeptide (62 000 Da) exotoxin. The human lethal dose is 0.1 μg/kg. Before artificial immunization, diphtheria was mainly a disease of small children. This is because when clinical and subclinical infections occurred in a population, there was an increase of production of antitoxin in that population; thus most people apart from children were immune to the toxin. However, with the introduction of artificial active immunization, antitoxin levels are adequate until adulthood. Young adults must receive a booster shot of toxoid to keep the antitoxin level high, since in developed countries toxigenic diphtheria bacilli are not sufficiently prevalent in the population to stimulate subclinical infection and sustain resistance. Levels of antitoxin decline with time, and many older persons have insufficient amounts of circulating antitoxin to protect them against diphtheria. The best ways to prevent diphtheria are to limit the distribution of toxigenic strains of diphtheria in the population and to maintain a high level of active immunization programs worldwide.

The role of food in the dissemination of *C. diphtheriae* is uncertain. The pathogen has been transmitted by raw milk. Milkborne outbreaks were recorded in the US before widespread practice of immunization and pasteurization of milk. No foodborne outbreak from *C. diphtheriae* has been reported in recent years in the US.

2.9 *Coxiella burnetii*

Coxiella burnetii belongs to the genus *Coxiella* in the Order *Rickettsiales*. Rickettsiae are small bacteria that are obligate intracellular parasites, except for Q fever-causing *C. burnetii*. They are transmitted to humans by arthropods. They are pleomorphic, small (600 × 300 nm in size), short rods or cocci, occurring singly, in pairs, short chains or filaments. They need to be stained with Giemsa's stain (blue color) or with Macchiavello's stain (red color, contrasting with the blue-staining cytoplasm). Human infection by *C. burnetii* is usually by inhalation of dust (airborne transmission). In 1999, Q fever became a notifiable disease in the US (CDC, 2005).

The disease caused by *C. burnetii* resembles influenza, non-bacterial pneumonia, hepatitis or encephalopathy. There is no rash or local lesion. Transmission to humans usually results from inhalation of dust contaminated with *C. burnetii* from dried feces, urine, or milk or from aerosols in slaughterhouses. Q fever, a zoonotic disease, is recognized around the world, and occurs mainly in persons associated with goats, sheep or dairy cattle, which are primary reservoirs for this pathogen. The organism is also common in some domesticated pets, wild animals, birds and ticks. The organism is shed in the feces, urine or milk of infected animals. Also, shedding of pathogen occurs in high numbers within the amniotic fluids and the placenta during the birthing process.

C. burnetii causes outbreaks in veterinary and medical centers, where large numbers of people are exposed to animals shedding the pathogen. About one-half of individuals infected with *C. burnetii* show signs of clinical illness. The infectious dose may be only a few organisms. The principal sign of infection is high fever that usually lasts for 1–2 weeks. Other signs and symptoms include headache, general fatigue, chills, sweats, nausea, vomiting and abdominal pain. The incubation period for Q fever is usually 2–3 weeks, and about 1–2% of individuals stricken with Q fever die of the disease. Most people recover without any treatment. If needed, treatment with antibiotics such as doxycycline (treatment of choice), quinolone, tetracyclines or chloramphenicol is available. People seriously infected with *C. burnetii* may, rarely, develop chronic Q fever. This is characterized by an infection that lasts for more than 6 months. According to the CDC, 65% of people infected with chronic Q fever may die of the disease (CDC, 2005).

Among rickettsial agents, *C. burnetii* is most resistant to drying. It is also resistant to heat and to a number of disinfectants. This hardy microorganism can therefore survive for long periods in the environment. The organism can survive pasteurization at 60°C for 30 minutes, and can survive for months in dried feces or milk due to the formation of endospore-like structures. For this reason the current milk pasteurization time/temperatures have been set at 63°C for 30 minutes or 72°C for 15 seconds to destroy *C. burnetii*, *Mycobacterium tuberculosis* and other milkborne pathogens.

C. burnetii is a highly infectious agent, and it is important that effective prevention and control measures be implemented in order to avoid the risk of infection. Examples of control and preventive measures include proper education of practitioners, workers and individuals who are at risk of occupational exposure to the pathogen, regarding the hazard, sources of infection and necessary precautions; appropriate disposal of infected animal materials such as placenta, aborted fetuses or fetal membranes; and pasteurization of milk and milk products.

2.10 *Edwardsiella*

The genus *Edwardsiella* belongs to the family *Enterobacteriaceae*. There are three species of *Edwardsiella* – *hoshinae*, *ictaluri* and *tarda*. Members of this genus are Gram-negative, catalase-positive, oxidase-positive, small straight rods. They are about 1 μm in diameter × 2–3 μm in length, and are motile by peritrichous flagella. Motility may be adversely impacted by temperature, as with *E. ictaluri*. This is motile at 25°C but non-motile at 37°C. *Edwardsiella* spp. are facultative anaerobic organisms, and need vitamins and amino acids to grow. They are relatively inactive in fermentation of carbohydrates compared with other members of *Enterobacteriaceae*. *Edwardsiella* is biochemically close to *E. coli*, *Salmonella* and *Shigella*.

Members of the genus *Edwardsiella* are frequently isolated from poikilothermic aquatic animals (fish, eels, frogs, turtles); domestic, zoo and marine animals; rats; and birds. They are pathogenic to eels and catfish. They can cause economic loss by developing 'red disease' in pond-cultured eels or emphysematous putrefactive disease of channel catfish. *E. hoshinae* has been isolated from animals and humans. *E. ictaluri* is a pathogen of catfish. *E. tarda*, also known as *E. anguillimortifera*, is usually found in aquatic animals. It is also found in the intestinal tract of reptiles (e.g. snakes) and marine animals (e.g. seals).

Members of *Edwardsiella* can also be opportunistic pathogens for humans. They have been associated with gastroenteritis and wound infections. In India they have been found in children with diarrhea. They have been isolated from urine, blood and fecal samples of humans. They may cause diarrhea in humans in association with the protozoon *Entamoeba histolytica*. At a minimum, *Edwardsiella* is undesirable in food and water due to the potential link to animal feces and other enteric pathogens.

2.11 *Enterococcus faecalis* and *Enterococcus faecium*

The genus *Enterococcus* is the new name for fecal *Streptococcus*. Schleifer and Kilpper-Bälz (1984) proposed transferring streptococci informally referred to as enterococci to a separate genus called *Enterococcus*; hence *Streptococcus faecalis* and *S. faecium* are now known as *Enterococcus faecalis* and *E. faecium*.

Members of the genus *Enterococcus* are Gram-positive, catalase-negative, non-motile, facultatively anaerobic microorganisms. The cells (0.6–2.0 × 0.6–2.5 μm) are spherical or ovoid, and occur in pairs and short chains. Their optimum growth temperature is 37°C, but they can grow at both 10°C and 45°C.

Enterococcus or enterococci as a group has received greater recognition in the area of water microbiology recently due to its greater ability to be an indicator for waterborne pathogens compared with fecal coliforms (National Research Council, 2004). Both *Enterococcus faecalis* and *Enterococcus faecium* belong to Lancefield's group D, and are collectively named *Enterococcus* or enterococci. The group-specific antigenic determinant is an intracellular glycerol teichoic acid associated with the cytoplasmic membrane, and is released in the conversion of cells to protoplasts. They belong to the homofermentative lactic-acid-bacteria group because of the formation of lactic acid as the major end product of glucose fermentation. *Enterococcus* is normally present in mammalian feces and a wide variety of environments. Its presence in foods may indicate direct or indirect contamination by fecal materials. From the point of clinical microbiology, enterococci cause 10% of urinary tract infections and 16% of nosocomial urinary tract infections. Enterococcal bacteriuria usually occurs in patients undergoing urologic manipulations. Intra-abdominal or pelvic wound infections are the next most commonly encountered infections. Bacteremia is the third most common type of infection, and enterococci are the third leading cause of nosocomial bacteremia (Murray *et al.*, 1995). Although the role of enterococci in direct foodborne infection is not clear, the ubiquitous nature of these organisms in the environment warrants further study of these organisms as agents of foodborne infection. *Enterococcus* spp. have been repeatedly proposed as causes of diarrheal illness in humans. However, in studies involving human volunteers, the microorganisms failed to cause illness.

The major differential characteristics between the two species are reduction of tetrazolium, decarboxylation of tyrosine, and fermentation of melezitose by *E. faecalis* and not by *E. faecium*. *E. faecalis* is generally non-motile, survives heating at 60°C for 30 minutes in neutral media, but is markedly affected by the pH of the medium and age of the cells. Enterococci have been isolated from feces of humans and homoiothermic and poikilothermic animals, insects and plants. They are common in many non-sterile foods. Human volunteer studies related to foods to show the pathogenicity of *Enterococcus* have produced mixed results (Stiles, 1989). They can be pathogenic agents in urinary tract infections and subacute endocarditis. Despite its potential pathogenic nature, *E. faecalis* has been involved in natural Swiss cheese fermentation, where it stimulated the growth of *Lactobacillus delbrueckii* subsp *bulgaricus*, *L. helveticus* and various gas-forming types.

E. faecium is similar to *E. faecalis* in morphology and in many biochemical reactions except those listed above. It is usually non-motile, but motile strains exist with differences in esterase and protease patterns on polyacrylamide gel. Similarly to *E. faecalis*, it also survives at 60°C for 30 minutes. Strict anaerobic strains isolated from bovine alimentary tract have been found to adapt to grow aerobically. *E. durans*, a species closely related to *E. faecium*, has been used as starter culture in Mozzarella cheese making (Kosikowski and Mistry, 1997).

2.12 *Erysipelothrix rhusiopathiae*

Erysipelothrix is a Gram positive, facultative to micro aerophilic anaerobic, catalase negative microorganism. It is a straight or slightly curved slender rod (0.2–0.4 × 0.8–2.5 μm), and can form filaments ≥ 60 μm long. It is non-motile, and does not form capsules or spores. The optimum temperature for growth of *Erysipelothrix* is 30–37°C. This organism is closely related to *Listeria*. It differs from *Listeria* in being non-motile, catalase negative and hydrogen sulfide positive, and contains L-lysine as the major diamino acid in its murein.

Erysipelothrix spp. are ubiquitous in nature. They are usually parasitic on mammals, birds and fish, but some strains of *Erysipelothrix* are pathogenic. The organism causes swine erysipelas – a specific acute, cutaneous, inflammatory disease characterized by hot, red, edematous, brawny and sharply defined eruptions. This bacterium may sometimes be pathogenic to man, causing erysipeloid, especially as an occupational disease through contact with infected animal products, including pork, poultry and fish. The organism survives salting, pickling and smoking of meat, and may survive in such meats for 1–3 months. It is killed by a temperature of 55°C for 15 minutes (Stiles, 1989). The organism was isolated from 4–54% of pork loins in

Sweden, and 34% of retail pork samples in Japan (Jay, 2000). No direct foodborne infection cases have been reported, but the potential exists for transmission of the organism from the meat plant environment to humans.

2.13 *Francisella tularensis*

Francisella tularensis is a Gram-negative, non-motile, obligate-aerobic, rod-shaped cell (0.2×0.2–$0.7\ \mu m$) that exhibits pleomorphism after active growth. It is oxidase negative and catabolizes carbohydrate slowly, with the production of acid but no gas. The organism is found on all continents except Australia and Antarctica. It is the agent of tularemia in man and animal. Human tularemia is an acute, febrile, granulomatous, infectious, zoonotic disease. The organism can be isolated from streams, rivers, lakes and ponds frequented by voles and water rats in the former Soviet Union (USSR), and beavers, muskrats and voles in North America. The organism has been isolated from wildlife such as rabbits, muskrats, water rats, squirrels, sheep, game birds, etc., and the bacteria can be transmitted to humans handling infected animal carcasses. It can also be transmitted through insect bite, ingestion of inadequately cooked meat or contaminated water, or inhalation of airborne organisms. Human-to-human transmission is extremely rare. The organism is destroyed by a temperature of 56–58°C for 10 minutes; thus proper cooking of meat is a preventive measure against this bacterium.

2.14 *Klebsiella*

Klebsiella is a member of the family *Enterobacteriaceae*, and exhibits the general bacteriological characteristics of this family. Its main differential characteristics from other members of the family *Enterobacteriaceae* are production of acid and gas at 37°C from lactose, and non-motility. *Klebsiella* belongs to the group of bacteria called coliforms, which includes *Escherichia*, *Enterobacter*, *Serratia* and *Citrobacter*. Coliforms share similar morphological and biochemical characteristics.

Klebsiella is a Gram-negative, catalase-positive and oxidase-negative bacterium. The cells are straight rods (0.3–1.0×0.6–$6.0\ \mu m$), are encapsulated, and occur singly, in pairs or in short chains. This microorganism can fix nitrogen in the environment. It can also cause pneumonia in humans. *Klebsiella* has been reported to cause nosocomial infections in urological, neonatal, intensive care and geriatric patients (Holt *et al.*, 1994). Generally, this microorganism is an opportunistic pathogen.

The three important species are *K. pneumoniae*, *K. oxytoca* and *K. ozaenae*. *K. pneumoniae* and *K. oxytoca* both have a large polysaccharide capsule. This capsule gives rise to large mucoid colonies, particularly if the microorganisms are cultured on a carbohydrate-rich medium. *K. pneumoniae* is the most important species, and is present in the respiratory tract and feces of approximately 5% of healthy humans. It causes about 3% of the total bacterial pneumonia in humans, and occasionally produces urinary tract infection and bacteremia with focal lesions in already weakened patients. It has been isolated from food such as fruits and vegetables, meat, milk and dairy products, salads, and drinking water; from the general environments of soil, dust, air and water; and from social environments, such as hospital and industrial settings. Frequent hand washing and proper sanitary practices are recommended for control of *Klebsiella*. Since the organism can produce heat-stable and heat-labile enterotoxins, causes diarrhea in humans and occurs in many food systems, it is reasonable to suspect that this organism can be an agent of foodborne infection and intoxication and should be monitored carefully. Fung and Niemiec (1977) developed an acriflavine violet-red bile agar for the isolation of *Klebsiella pneumonia*, where the bacterium forms a distinctive bright yellow colony on the agar. Chein and Fung (1980) confirmed this finding on the isolation and enumeration of *K. pneumoniae* from the environment and foods. Further studies are warranted to ascertain the role of *K. pneumoniae* as an agent of foodborne diseases, beside its importance as an agent of bacterial pneumonia.

2.15 *Leptospira*

Leptospira is a spiral-shaped microorganism, as its name implies. It is a flexible, helicoidal rod (0.1 × 6–12 μm), Gram-negative, highly motile and obligately aerobic bacterium. It has a multilayered outer membrane, helical-shaped peptidoglycans, and flagella located in the periplasmic space. The optimum temperature for *Leptospira* growth is around 28–30°C, with a generation time of 6–16 hours. It is oxidase, catalase and/or peroxidase positive. The bacterium is chemo-organotrophic, using fatty acids or fatty alcohols as energy and carbon sources. Some strains are parasitic and may be pathogenic for man and animals, while other strains are free-living in soil, freshwater or marine habitats. It is believed to be the most widespread zoonosis in the world. It is considered as an occupational risk for dairy and hog farmers; meat, poultry, and fish processors; and veterinarians working with animals and animal products.

The disease caused by *Leptospira*, known as leptospirosis or 'Weil's disease', was first described in 1886. Leptospirosis is often misdiagnosed as meningitis or hepatitis. Its severity can vary from subclinical to fatal. Treatment with antibiotics is ineffective after symptoms have persisted for more than 4 days. Foodborne infection results from ingestion of food or water contaminated with the organism. After 1–2 weeks, a febrile onset occurs with spirochetes present in the bloodstream. It then affects the liver and kidneys, producing hemorrhage and necrosis of tissue, resulting in dysfunction of the organs. Renal failure is one of the most common causes of death in leptospirosis. There are no known exo- or endotoxins produced by this microorganism. The disease is essentially an animal infection, with human infection being accidental. *Leptospira* remains viable in stagnant water for weeks. Drinking, swimming, bathing and food contamination may lead to human infection. Control measures include reducing the prevalence of *Leptospira* in domestic animals, preventing exposure to contaminated water, and reducing contamination by rodent control. Rodents are the primary reservoirs, and once infected they shed the microorganism for life. It is not certain whether *Leptospira* causes classical cases of foodborne infection.

2.16 *Mycobacterium bovis*

Mycobacteria are members of the family *Mycobacteriaceae* and part of the *Corynebacteria*, *Mycobacteria* and *Nocardia* (CMN) group. The group *Mycobacteria* contains a single genus – *Mycobacterium*. *M. bovis* belongs to the genus *Mycobacterium*, which is a large, heterogeneous group of acid- and alcohol-fast, non-spore-forming bacilli. This group of organisms grows very slowly; sometimes it takes weeks to obtain typical colonies. Recently, much work has been done in the use of the Polymerase Chain Reaction to identify this organism rapidly (within a few days) using genetic methods.

Members of the genus *Mycobacterium* are usually weakly Gram-positive, as the Gram-stain method does not readily stain them. They are aerobic, non-motile and non-spore-forming bacteria. The cells (0.2–0.7 × 1.0–10 μm) sometimes exhibit branching, filamentous or mycelium-like growth.

Members of the genus *Mycobacterium* occur in soil, water, food and animals. The most important species of *Mycobacterium* is *M. tuberculosis*, the agent of the disease tuberculosis (TB) worldwide. The World Health Organization estimates that 8 million new cases and 3 million deaths are directly attributable to this disease each year (Kochi, 1991). Incidentally, although the occurrence of TB seemed to have decreased in the past quarter of a century, recently there has been an increase in cases worldwide – including the US.

Mycobacterium bovis infects cattle and causes bovine tuberculosis. This is a contagious and debilitating disease in humans and animals. Bovine tuberculosis is presently rare in Northern Europe and North America, but is still a great threat in other places around the world. In 2002, California lost its TB-free status after three herds tested positive for the disease.

It is hard to detect *M. bovis* and to separate it from other *Mycobacterium* spp. Compared with *M. tuberculosis*, *M. bovis* causes illness in humans, guinea pigs, fowl and cattle, while *M. tuberculosis* will not infect cattle. The milk of tuberculous cows is the source of infection where bovine tuberculosis is not well controlled, and where milk is not pasteurized. Infections with *M. bovis* in cattle are long lasting. Infected cows can shed viable organisms in their milk. Upon consumption of infected milk (or dairy products made from it) bovine tuberculosis can occur in humans, but the duration is short and lesions of the lung heal spontaneously. Some symptoms of the disease include weakness, fatigue, loss of appetite and loss of weight. Additionally, infection of the vertebra by this bacterium has resulted in 'hunchbacks' in previous generations. Inhalation of aerosols containing the causative agent (TB bacteria) is one of the most common means of contracting the disease. Sometimes infected animals can exhale the bacteria, and therefore individuals in direct or prolonged contact with infected animals (e.g. veterinarians) are at risk.

The true incidence of bovine tuberculosis in humans is not known, and is probably under-estimated and under-reported. The best way to control bovine tuberculosis is by good sanitation on the farm and in food preparation. Pasteurization of milk will essentially eliminate the problem as far as bovine tuberculosis in humans is concerned.

2.17 *Mycobacteria* other than *bovis*

In addition to tubercle bacilli, *Mycobacterium tuberculosis* and *M. bovis*, there is a set of *Mycobacterium* collectively named 'atypical' mycobacteria, grouped according to speed of growth at various temperatures and the production of specific pigments. After proper preparation of a sputum specimen and the necessary growth substances, such as N-acetyl-L-cysteine, sodium hydroxide, trisodium citrate, phosphate buffer, bovine albumin fraction V, selective agents and Lowenstein-Jensen medium, the cultures are incubated in a carbon dioxide environment at 35–37°C, with inspection of the cultures weekly. When acid-fast organisms are present, *Mycobacterium* species are separated according to the following scheme (Jawetz *et al.*, 1987):

1. *Rapid growers* (less than 7 days)
   a. Positive arylsulfatase test (3 days) and growth on MacConkey agar
      i) nitrate reduction positive – *M. fortuitum*
      ii) nitrate reduction negative – *M. chelonei.*
   b. Negative arylsulfatase test – various non-pathogenic *Mycobacteriunm* spp.
2. *Slow growers* (more than 7 days)
   a. Non-pigmented growth
      i) niacin test positive, nitrate reduction positive – *M. tuberculosis*
      ii) niacin test negative, nitrate reduction variable – *Mycobacteriun* other than *M. tuberculosis*, e.g. *M. avium-intracellulare.*
   b. Pigmented growth
      i) pigmented when grown in light, non-pigmented in dark – photochromogen, e.g. *M. kansasii,*
      ii) pigmented when grown in light or dark – Scotochromogen, e.g. *M. scrofulaceum.*

The importance of these *Mycobacterium* spp. in foodborne infection is not well known. Improvement of dairy herd management, effective veterinarian services, advancement of sanitation on the farm as well as in milk processing facilities, adequate pasteurization of milk, and safety in handling milk in the distribution systems from the farm to the table will greatly eliminate the problem of *Mycobacterium* in human food supplies.

2.18 *Plesiomonas*

Plesiomonas shigelloides has been a suspect in cases of foodborne disease. The rate of occurrence of *P. shigelloides* infections in the US is not known. Infections with this type of microorganism are rarely reported, which may be in part because the disease is self-limiting in nature and people often do not seek medical attention. Most reported cases of gastroenteritis involve individuals with pre-existing health problems (e.g. cancer, sickle-cell anemia). A few outbreaks associated with gastrointestinal illness caused by *P. shigelloides* are reported in the literature (CDC, 1989, 1998a).

P. shigelloides is a Gram-negative, facultatively anaerobic, catalase-positive and fermentive organism. It is oxidase positive, and can be differentiated from bacteria in the family *Enterobacteriaceae* by this test, since *Enterobacteriaceae* are oxidase negative. The organism also resembles *Shigella*, but can be differentiated from *Shigella* by being motile. *P. shigelloides* has rounded, straight, rod-shaped cells (0.8–1.0 × 3.0 μm). It is motile by polar flagella. Its optimal growth temperature is 37°C. *Plesiomonas* ferments D-glucose and other carbohydrates with acid production but no gas. MacDonell and Colwell (1985) recommended that *Plesiomonas* be transferred to the genus *Proteus* (family *Enterobacteriaceae*), since its 5S rRNA is closely related to that of *Proteus mirabilis*. However, it has been noted by others that such a change would cause problems in the phenotypic definition of the genus *Proteus* (Holt *et al*., 1994). *P. shigelloides* has been isolated from fish, shellfish, aquatic animals, fresh water (e.g. rivers, ponds, streams), cattle, swine, goats, cats, dogs, monkeys, vultures and toads.

P. shigelloides is capable of producing many diseases, ranging from enteritis to meningitis. It can cause occasional opportunistic pathogen infections in humans (known as *Plesiomonas* enteritis or gastroenteritis). The signs and symptoms include diarrhea (usually for 1–2 days) with blood and mucus in stools, abdominal pain, headache, nausea, chills, fever and vomiting. *Plesiomonas* gastroenteritis is usually a mild, self-limiting disease. The duration of the illness may be 1–7 days. The infectious dose is presumed at one million cells and higher. The disease agent is suspected to be waterborne in most human infections, with toxigenic and invasive modes of attack. It may be present in contaminated water that is used for drinking or for rinsing foods to be eaten raw. Unsanitary water used for recreational purposes is a concern also. Eating contaminated, raw shellfish may lead to illness. All reported foods involved with cases of gastroenteritis were of aquatic origin (salted fish, crabs and oysters). The organism can be isolated from a variety of sources, including humans, birds, fish, reptiles and crustaceans. It can also be isolated from healthy people (0.2–3.2% of the population). The use of clean water is an important control measure against *Plesiomonas* gasteroenteritis. Additionally, proper cooking of food is recommended to destroy the microorganism. The true nature of *P. shigelloides* as a foodborne agent is not fully known, because the organism has not been well studied to date.

2.19 *Proteus*

Proteus is a member of the family *Enterobacteriaceae*. It is a Gram-negative, catalase-positive, oxidase-negative straight rod (0.4–0.8 × 1–3 μm) that is actively motile by peritrichous flagella. Most strains cause swarming on the surface of agar plates, making it difficult to isolate other colonies on the same plate. It deaminates phenylalanine and tryptophan, hydrolyzes urea, and produces hydrogen sulfide. *Proteus* is an important nosocomial pathogen, and can produce infections in humans after leaving the intestinal tract. The organisms are found in urinary tract infections, and can produce bacteremia, pneumonia, and focal lesions in patients receiving intravenous infusion. Due to hydrolysis of urea by urease, ammonia is produced; this causes urine to be alkaline and promotes stone formation in patients. *Proteus* spp. have been isolated from ham, eggs, and fresh meats and poultry. In eggs *P. vulgaris* and *P. intermedium* cause 'custard' rots, and in ham *Proteus* causes sourness. The D value (decimal reduction) of *Proteus* in fresh food is 0.2 kGy, making *Proteus* more resistant to radiation than *Vibrio, Yersinia enterocolitica, Pseudomonas putida, Campylobacter jejuni, Shigella* spp., *Aeromonas hydrophila* and *Bacillus cereus* (vegetative cells) with a range of 0.03–0.17 kGy, and more sensitive than *Escherichia coli, Staphylococcus aureus, Brucella abortus, Salmonella* spp., *Listeria monocytogenes, Lactobacillus* spp., *Streptococcus faecalis, Clostridium perfringens* (vegetative cells), *Moraxella phenylpyruvica, Bacillus cereus* (spores), *Clostridium sporogenes* (spores), *Clostridium botulinum* types A and B (spores), *Deinococcus radiodurans* and *Deinobacter* spp. with a range of 0.23–5.05 kGy (Doyle *et al.*, 2001). The role of *Proteus* in food infection and intoxication is not clear. Stiles (1989) cited some reports in Eastern Europe of foodborne illness due to *Proteus* because large numbers of *Proteus* cells were isolated from incriminated food and in stools of patients with gastroenteritis.

2.20 *Pseudomonas aeruginosa*

To date, over 140 *Pseudomonas* spp. have been described. These are common inhabitants of soil and water. *Pseudomonas aeruginosa* belongs to the bacterial family *Pseudomonadaceae*. It is a Gram-negative, catalase-positive, oxidase-positive or negative, straight or slightly curved rod ($0.5–1.0 \times 1.5–5.0\,\mu m$). It is motile by one or several polar flagella. *P. aeruginosa* is aerobic, with a strictly respiratory type of metabolism with oxygen as the terminal electron acceptor. This microorganism can grow at 42°C, but the optimum temperature is 37°C. It produces fluorescent (pyoverdin) and non-fluorescent (pyocyanin) pigments. *P. aeruginosa* forms two morphologically different types of colonies on agar. The large ones, usually observed in clinical materials, are smooth with flat edges and an elevated center – a 'fried egg' appearance; the small one is commonly isolated from natural sources, and has a rough and convex shape. This microorganism is resistant to high concentrations of salts and dyes, and to many of the commonly used antibiotics.

P. aeruginosa is widely distributed in nature, and is common in moist environments in hospitals. It can colonize healthy humans, and can cause disease in people with abnormal host defenses. It can be isolated from soil and water, and is commonly associated with spoilage of food such as eggs, cured meats, fish and milk. *P. aeruginosa* is pathogenic only when introduced into areas lacking normal defenses, such as tissue damage of mucous membranes and skin, severe burns, and intravenous or urinary catheters. It produces two extracellular protein toxins – Exoenzyme S and Exotoxin A. Toxin A is the most toxic and can lead to loss of host-cell protein-synthetic capability (a mechanism similar to that of diphtheria toxin). The bacterium attaches to and colonizes these damaged tissues and produces systemic disease. The infection of wounds and burns can give rise to blue-green pus; meningitis through lumbar puncture; urinary tract infection through catheters, instruments and irrigating solutions; and a variety of other diseases related to blood poisoning, the eyes, respiratory tract, etc. *P. aeruginosa* is highly resistant to many antibiotics. Treatment is by penicillins (ticarcillin, mezlocillin and pipaeracillin) in combination with an aminoglycoside such as gentamicin, tobramycin or amikacin.

P. aeruginosa is an opportunistic pathogen, and has been found in many food items such as eggs, cured meats, fish, milk, tomatoes, radishes, celery, cucumbers, onions, lettuce, etc. It is alleged to cause gastroenteritis in humans if ingested in large numbers. For healthy persons, the presence of this organism in food will not pose a threat. However, for immunocompromised, debilitated and burn-unit patients, this organism can result in fatal infections. One undocumented case indicated that burn-unit patients had a high mortality rate due to *P. aeruginosa* infection. It was determined that the organisms had infected the patients through consumption of contaminated raw salad. After the removal of all raw foods, with only cooked food being served to burn-unit patients, the mortality rate dropped dramatically (Komonis, personal communication, 1978). Transmission of *P. aeruginosa* via food and water is proposed, but unproven.

2.21 *Pseudomonas cocovenenans*

Pseudomonas cocovenenans is a typical *Pseudomonas* in terms of morphology and biochemical characteristics. The word 'cocovenenans' came from *Cocos*, the genus name for coconut, and *veneno*, which means poison. Thus, cocovenenans means coconut poisoning. *P. cocovenenans* produces a yellow, poisonous compound toxoflavin. This bacterium is listed among the *species incertae sedis* in *Bergey's Manual of Determinative Bacteriology*, Vol. 1 (Krieg and Holt, 1984).

In the making of *bongkrek* and *tempeh bongkrek* in Indonesia, a mixture of soybeans, peanut and coconut is fermented by *Rhizopus*, resulting in a tempeh-like product. Normally *Rhizopus* grows very fast and completes the fermentation in a few days. In unusual conditions the mold may grow abnormally slowly; then bacteria, including *P. cocovenenans*, and yeast may grow and spoil the product. *P. cocovenenans* produces toxoflavin and bongkrek acid, which are stable antibiotics. Bongkrek acid is very toxic, and can result in spasms, hypoglycemia and the death of patients. To prevent this from happening it is essential to ferment bongkrek properly and introduce the use of starter cultures such as *Rhizopus oligosporus* in tempeh fermentation. This foodborne intoxication is limited to the Far East and is not a concern in other parts of the world where tempeh and tempeh-like products are not eaten regularly in large quantities.

2.22 *Shigella*

An estimated 300 000 cases of shigellosis occur every year in the US (CDC, 1998b). Shigellosis or bacillary dysentery, the illness caused by *Shigella*, accounts for less than 10% of the reported outbreaks of foodborne illness in the US (FDA, 2005).

Shigella spp. include *S. boydii*, *S. flexneri*, *S. sonnei* and *S. dysenteriae*. These are Gram-negative, straight rod-shaped, non-motile and non-spore foming bacteria within the family *Enterobacteriaceae*. They are often confused with *Salmonella* in the bacteriological diagnostic process. *Shigella* spp. are non-motile and hydrogen-sulfide negative. They are facultatively anaerobic bacteria, and have both a respiratory and a fermentive type of metabolism. *Shigella* spp. are catalase positive and oxidase negative. Production of indole varies, and hydrogen sulfide is not produced. Some carbohydrates fermented by *Shigella* spp. include D-mannose, D-mannitol, maltose and trehalose.

Shigella is an intestinal pathogen of humans and other primates, such as chimpanzees and monkeys. It is a human-specific pathogen, and rarely occurs in animals. Water polluted with human feces tends to be a reservoir for this pathogen. In terms of foodborne infection *Shigella* is not as prevalent as *Salmonella*, but it is very important in waterborne diseases – especially in tropical and subtropical countries where sanitation conditions are poor, and particularly after natural disasters such as heavy rainfall, flooding and tsunamis. The organism is transmitted by water, food, humans and animals. The '4Fs' involved in the transmission of *Shigella* are food, fingers, feces and flies.

Shigella infections in humans are initiated when the pathogen is ingested via contaminated food or water (fecal–oral mode of transmission). Depending on the infective dose (which can be as few as a 10 cells, depending on the age and condition of the target individual), shigellosis can occur if the pathogen is able to attach to and penetrate the epithelial cells of the intestinal tract. Following successful attachment, the pathogen multiplies intracellularly, spreads to other adjacent epithelial cells, and in the process destroys the tissue. Virulent strains produce a toxin known as Shiga toxin. People inflicted with shigellosis may experience abdominal pain, diarrhea (with possible blood, pus or mucus in stools), cramps, vomiting, fever and tenesmus. Fatality may be as high as 10–15%. Some people may experience complications such as Reiter's disease, hemolytic uremic syndrome and reactive arthritis in the aftermath of shigellosis. The incubation period can be from 0.5–7 days (typically 1–2 days), with an onset of illness from 12–50 hours. Foods usually involved fall under the ready-to-eat category where the implicated food has been contaminated by an infected person. It has been associated with salads such as potato, tuna, shrimp, macaroni and chicken; poi; raw vegetables; milk and dairy commodities; poultry; and water.

Usually, 1–4 days after ingestion of the organisms there will be inflammation of the walls of the large intestine and ileum. Invasion of the blood is rare. Bloody stools will occur due to superficial ulceration. The lysed cell wall of *Shigella* will form endotoxins. *S. dysenteriae* produces an exotoxin that is a highly toxic neurotoxin. This toxin can be neutralized by homologous antibody. The mortality rate of shigellosis is higher than that of salmonellosis. Shigellosis can be prevented by good sanitation and hygiene, proper treatment of water, detection of carriers, prevention of infected persons from handling the food, prevention of contamination, and isolation of patients from the general public. Additionally, effective hygienic practices and proper refrigeration, cooking or reheating (if applicable) are useful. According to a CDC survey of the period 1993–1997 (CDC, 2000), there were 43 foodborne *Shigella* outbreaks involving 1555 cases with no deaths. During the same period there were 357 *Salmonella* outbreaks involving 32 610 cases and 13 deaths.

2.23 *Streptococcus moniliformis*

Streptococcus moniliformis is a Gram-positive, facultative-anaerobic, non-motile, catalase-, oxidase- and indole-negative, rod-shaped bacterium with rounded or pointed ends. It occurs singly or in the form of long wavy chains or filaments 10–150 μm in length. It may be highly pleomorphic.

Serum, ascitic fluid or blood is required for growth. *Streptococcus moniliformis* produces micro-colonies due to the formation of L-form cells. It causes a disease called rat-bite fever (or Haverhill fever). The disease is acquired from bites of infected rats, mice or cats, and is not communicable between humans. The organism is destroyed by heating at 56–58°C for 10 minutes. Under-pasteurized, contaminated milk is the only food associated with this disease; thus pasteurization of milk and heating food to 60°C for more than 10 minutes will eliminate this problem.

2.24 *Streptococcus pyogenes*

Streptococci are Gram-positive, catalase-negative, non-motile, non-spore-forming, facultatively anaerobic bacteria. They are spherical or ovoid in shape (0.5–2.0 μm), and occur in pairs or chains. They produce mainly lactic acid from carbohydrate fermentation, and have fastidious nutritional requirements (nutritionally rich media are needed for growth). The optimum temperature of growth is 37°C, with a range of 25–45°C.

The genus *Streptococcus* is classified into Groups A, B, C, D, F and G, based on a combination of antigenic, hemolytic and physiological characteristics. Both group A, which contains one species (*S. pyogenes*) with 40 antigenic types, and group D, which contains five species (*S. faecalis*, *S. faecium*, *S. durans*, *S. avium* and *S. bovis*), can be foodborne and cause illness in humans. Of the streptococci, *S. pyogenes* and *S. pneumonia* are considered as primary pathogens. *S. pyogenes* is classified as the leading cause of bacterial tonsillitis and pharyngitis. Treatment is available, and is usually with antibiotics. It is estimated that 5–15% of healthy individuals carry this bacterium. Other *Streptococcus* species can cause illness; however, they are considered to be opportunistic. Additionally, some are part of the normal flora.

S. pyogenes is a member of the lactic acid-bacteria group. *S. pyogenes* is beta hemolytic on blood-agar plates, and forms pus in patients. It belongs to Lancefield group A streptococci, based on the serology of the cell wall polysaccharides. The name 'Lancefield' is taken from Rebecca Lancefield, who originally established 18 groups.

S. pyogenes and related species can cause a variety of diseases in human. The first type is invasive, involving beta-hemolytic, group-A *S. pyogenes*. The portal of entry determines the principal clinical picture. This is a diffuse and rapidly spreading infection that involves the tissues and extends along lymphatic pathways with minimal local suppuration. In the lymphatic infection, the organisms can enter the bloodstream. Erysipelas is a condition that involves entry of the bacterium into the skin and causes massive brawny edema with a rapidly advancing margin of infection. Puerperal fever involves the bacterium's entry into the uterus after childbirth, resulting in what is essentially a septicemia originating in the infected wound, or endometritis. Sepsis or surgical scarlet fever is the infection of traumatic or surgical wounds with *Streptococcus*.

The second type of disease is attributable to local infection with beta-hemolytic Group A streptococci and their products. Streptococcal sore throat falls into this category. Virulent group A streptococci adhere to the pharyngeal epithelium by means of lipoteichoic acid covering surface pili. In infants and small children, the sore throat occurs as a subacute nasopharyngitis with a thin serous discharge and little fever. The infection can involve the middle ear, the mastoid and the meninges. The illness can persist for weeks. For older children, the disease is more acute and is characterized by intense nasopharyngitis, tonsillitis, and intense redness and edema of the mucus membranes, with purulent exudate. Scarlet fever rash can occur if the organisms produce erythrogenic toxin. Another disease is streptococcal pyoderma.

This is a local infection of the superficial skin layers called impetigo. It consists of superficial blisters that break down to leave eroded areas, covered with pus or crusts. It spreads by continuity, and is highly communicable in children.

The third type of disease is infective endocarditis, including acute endocarditis and subacute endocarditis. In the case of acute endocarditis, the bacteria may settle on normal or previously deformed heart valves, producing acute inflammation; rapid destruction of the valves frequently leads to fatal cardiac failures in days or weeks unless treated promptly. The subacute endocarditis often involves abnormal valves. Subacute endocarditis is most frequently due to members of the normal flora of the respiratory or intestinal tract that have accidentally reached the blood.

S. pyogenes has been implicated in meat and poultry due to contamination of workers through skin infection. Airborne spread is also a possibility. Before the time when pasteurization and refrigeration were used widely, milkborne outbreaks of scarlet fever and septic sore throat caused by Group A were frequent. Milk products, eggs, meats and salads were implicated in earlier outbreaks of *Streptococcus*. Milk products, eggs, meats, various salads (including shrimp), mousse desserts, rice dressing and sautéed vegetables in canned beef broth with added rice have been implicated more recently. Most current outbreaks have involved complex foods such as salads (particularly at salad bars), where infected food handlers contaminate the food vehicles. Similar to other foodborne outbreak situations, sanitation measures, cooking food properly, pasteurization of milk, and rapid chilling and holding of prepared foods at temperatures below 4°C will help to prevent these foodborne diseases. *Streptococcus* group A infection is an uncommon foodborne disease, even though it is one of the most important causes of upper respiratory tract and skin diseases. According to the CDC (2000), between 1993 and 1997 there was one foodborne outbreak of *Streptococcus* group A involving 122 cases and no deaths; there was also one outbreak of another *Streptococcus* group comprising 6 cases and no deaths. *Streptococcus* Group A may be a possible cause of foodborne infection, but is probably not an important one compared with other foodborne pathogens.

3 Conclusion

This review of less-known, potentially foodborne infection and intoxication bacteria provides some insight into the diversity of activities of microorganisms in our environment and food supply. Pasteur exclaimed, '*Messieurs, c'est les microbes que auront le dernier mot!*' (Gentlemen, it is the microbes who shall have the last word!) as the argument disproving the theory of spontaneous generation prompted us to ponder the power of the microbes in human affairs. New microorganisms can emerge and become major players in foodborne infection and intoxication in the world. Existing microbes may mutate and become more virulent and damaging to our health and food supply. We may yet discover more friendly organisms that might provide us with more efficient, better, less expensive, more nutritious foods for feeding the world. Thus the search goes on in understanding the activities of microbes in our food supplies and our environment. It will never end.

Bibliography

Batzing, B. L. (2002). *Microbiology: An Introduction*. Brooks/Cole Thomson Learning, Stanford, CT.

Bennett, R. W. (2003). Current perspective and rapid identification of Staphylococcal and *Bacillus* toxins. In: D. Y. C. Fung (ed.), *Handbook for 23rd Rapid Methods and Automation in Microbiology*, Sectin C No. 5, pp. 1–16. Kansas State University, Manhattan, KS.

Blackburn, C. M. and R. J. McClure (2002). *Foodborne Pathogens: Hazards Risk Analysis and Control*, Woodhead Publishing Ltd., Cambridge.

Boone, D. R., R. W. Castenholz and G. M. Garrity (2001). In: N. R. Krieg and J. G. Holt (eds), *Bergey's Manual of Systematic Bacteriology*, Vol. 1. Williams and Williams, Baltimore, MD.

Bourgeois, C. M., J. Y. Levean and D. Y. C. Fung (1995). *Microbiological Control for Foods and Agricultural Products*. VCH, New York, NY.

Bryan, F. L. (1979). Other bacteria. In: H. Riemann and F. L. Bryan (eds), *Foodborne Infections and Intoxications*, 2nd edn, pp. 211–297. Academic Press, New York, NY.

CDC (Centers for Disease Control and Prevention) (1989). Epidemiologic notes and reports aquarium-associated *Plesiomonas shigelloides* infection – Missouri. *Morbid. Mortal. Weekly Rep.* **38**, 617–619.

CDC (Centers for Disease Control and Prevention) (1998a). *Plesiomonas shigelloides* and *Salmonella* serotype Hartford infections associated with contaminated water supply – Livingston County, New York, 1996. *Morbid. Mortal. Weekly Rep.* **47**, 394–396.

CDC (Centers for Disease Control and Prevention) (1998b). Summary of notifiable diseases, United States 1997. *Morbid. Mortal. Weekly Rep.* **46**, 1–87.

CDC (Centers for Disease Control and Prevention) (2000). Surveillance for foodborne disease outbreaks, 1993–1997. *Morbid. Mortal. Weekly Rep.* **49**(SS-1), 11.

CDC (Centers for Disease Control and Prevention) (2005). Q fever (available at http://www.cdc.gov/ncidod/dvrd/qfever/ accessed January 2005).

Chein, S. P. K. and D. Y. C. Fung (1980). Acriflavine violet red bile agar for the isolation and enumeration of *Klebsiella pneumoniae. Food Microbiol.* **7**, 73–79.

Downes, F. P. and K. Ito (eds) (2001). *Compendium of Methods for the Microbiological Examination of Foods*, 5th edn. American Public Health Association, Washington, DC.

Doyle, M. P., L. R. Beuchat and T. J. Montville (2001). *Food Microbiology: Fundamentals and Frontiers*, 2nd edn. ASM Press, Washington, DC.

Edwards, P. R., A. C. McWhorter and M. A. Fife (1956). The *Arizona* group of enterobacteriaceae in animals and man; occurrence and distribution. *Bull. WHO* **14**, 511–528.

Ewing W. H., R. Hugh and J. G. Johnson (1961). *Studies on the* Aeromonas *Group*. US Department of Health, Education, and Welfare, Public Health Service, Communicable Disease Center, Atlanta, GA.

FDA (Food and Drug Administration) *Shigella* spp. (available at http://www.cfsan.fda.gov/~mow/chap19.html, posted January 1992; accessed January 2005).

Fung, D. Y. C. and S. J. Goetsch (2000). *Introduction to Food Microbiology*, 3rd edn. Kansas State University Press, Manhattan, KA.

Fung, D. Y. and M. Niemiec (1977). Acriflavine violet red bile agar for the isolation of *Klebsiella. Health Lab Sci.* **14**, 273–278.

Holmberg S. D., W. L. Schell, G. R. Fanning *et al.* (1986). *Aeromonas* intestinal infections in the United States. *Ann. Intern. Med.* **105**, 683–689.

Holt, J. G., N. R. Krieg, P. H. A. Sneath *et al.* (1994). *Bergey's Manual of Determinative Bacteriology*, 9th edn. Williams and Wilkins, Baltimore. MD.

Janda, J. M. and P. S. Duffey (1988). Mesophilic Aeromonads in human disease: current taxonomy, laboratory identification and infectious disease spectrum. *Rev. Infect. Dis.* **10**, 980–997.

Jawetz, E., J. L. Melnick and E. A. Adelberg (1987). *Review of Medical Microbiology*. Appleton and Lange, Norwalk, CT.

Jay, J. M. (2000). *Modern Food Microbiology*, 6th edn. Chapman and Hall, New York, NY.

Kochi, A. (1991). The global tuberculosis situation and the new control strategy of World Health Organization. *Tubercle* **71**, 1–6.

Kosikowski, F. V. and V. V. Mistry (1997). *Cheese and Fermented Milk Foods*, Vol. 1. F. V. Kosikowski, LLC, Westport, CT.

Kramer, J. M. and R. J. Gilbert (1989). *Bacillus cereus* and other *Bacillus* species. In: M. P. Doyle (ed.), *Foodborne Bacterial Pathogens*, pp. 22–70. Marcel Dekker, New York, NY.

Krieg, N. R. and J. G. Holt (1984). *Bergey's Manual of Systematic Bacteriology*, Vol. 1. Williams and Wilkins, Baltimore, MD.

Ljungh A., M. Popoff and T. Wadstrom (1977). *Aeromonas hydrophila* in acute diarrheal disease: detection of enterotoxin and biotyping of strains. *J. Clin. Microbiol.* **6**, 96–100.

MacDonell, M. T. and R. R. Colwell (1985). Phylogeny of the family Vibrionaceae and recommendation for two new genera: *Listonella* and *Shewanella. Syst. Appl. Microbiol.* **6**, 171 182.

Murray, P. R., E. J. Baron, M. A. Pfaller *et al.* (1995). *Manual of Clinical Microbiology*, 6th edn. ASM Press, Washington, DC.

National Research Council (2004). *Indicators for Waterborne Pathogens*. National Academies Press, Washington, DC.

Ray, B. (2004). *Fundamental Food Microbiology*, 3rd edn. CRC Press, Boca Raton, FL.

Riemann, H. and F. L. Bryan (eds) (1979). *Food-borne Infections and Intoxications*, 2nd edn. Academic Press, New York, NY.

Sanyul S. G., S. J., Singh and P. C. Sen (1975). Enteropathogenicity of *Aeromonas hydrophila* and *Plesiomonas shigelloides. J. Med. Microbiol.* **8**, 195–198.

Schleifer, K. H. and R. Kilpper-Bälz (1984). Transfer of *Streptococcus faecalis* and *Streptococcus faecium* to the genus *Enterococcus* nom Rev. as *Enterococcus faecalis* comb. Nov. and *Enterococcus faecium* comb. Nov. *Intl J. Sys. Bacteriol.* **34**, 31–34.

Sneath, P. H. A., N. S. Mair, M. E. Sharpe and J. G. Holt (1986). *Bergey's Manual of Systematic Bacteriology*, Vol. 2. Williams and Wilkins, Baltimore, MD.

Staley, J. T., M. P. Bryant, N. Pfennig and J. G. Holt (1989). *Bergey's Manual of Systematic Bacteriology*, Vol. 3. Williams and Wilkins, Baltimore, MD.

Stephen S., J. N. K. Rao, M. S. Kumar and R. Indrani (1975). Human infection with *Aeromonas* species: varied clinical manifestations. *Ann. Intern. Med.* **83**, 368–369.

Stiles, M. E. (1989). Less recognized or presumptive foodborne pathogenic bacteria. In: M. P. Doyle (ed.), *Foodborne Bacterial Pathogens,* pp. 674–733. Marcel Dekker, New York, NY.

Tschape, H., R. Prager, W. Streckel *et al.* (1995). Verotoxinogenic *Citrobacter freundii* associated with severe gastroenteritis and cases of haemolytic uraemic syndrome in a nursery school: green butter as the infection source. *Epidemiol. Infect.* **114**, 441–450.

Von Graevenitz, A. and A. H. Mensch (1968). The genus *Aeromonas* in human bacteriology: report of 30 cases and review of the literature. *N. Engl. J. Med.* **278**, 245–249.

Williams, S. T., M. E. Sharpe and J. G. Holt (1989). *Bergey's Manual of Systematic Bacteriology*, Vol. 4. Williams and Wilkins, Baltimore, MD.

Infections with viruses and prions

11

Dean O. Cliver, Suzanne M. Matsui, and Michael Casteel

1 Historical aspects and contemporary problems

Recognition of viruses as pre-eminent causes of foodborne disease, at least in the US and parts of Europe, has finally arrived. Between early lack of understanding of viruses as pathogens, followed by the continuing problems of diagnosing human viral infections and of detecting viruses in food and water, the impact of viruses has long been underestimated in this context. Even now, when diagnostic and detection methods have become much more adequate, relatively few diagnoses of infections with viruses transmitted via food and water are made, and still fewer tests of food and water samples for viral contamination. Nevertheless, a landmark paper from the CDC estimates that 67 % of foodborne illnesses in the US are caused by the Norwalk-like group of viruses, more recently renamed noroviruses (Mead *et al.*, 1999). A follow-up paper that focuses on the role of noroviruses estimates that half of foodborne-disease outbreaks in the US in 2000 were caused by noroviruses, but that these tended to comprise more cases per outbreak than those caused by bacteria

Foodborne Infections and Intoxications 3e
ISBN-10: 0-12-588365-X
ISBN-13: 978-012-588365-8

(Widdowson *et al.*, 2005). The role of other viruses is also recognized. Essentially no dissent from this estimate has been heard, but food safety microbiology remains remarkably unaltered. That is, many discussions – and even careers – said to be devoted to food safety microbiology address only bacteria. This chapter will attempt to promote some balance. Although studies in food and water virology have always been conducted in a number of countries, the field has grown greatly in recent times, especially in Europe. The 2003 PAHO/WHO List of Food Virologists (http://faculty.vetmed.ucdavis.edu/faculty/docliver/PAHO.pdf), which is surely incomplete, includes 73 scientists in 18 countries.

Outbreaks of poliomyelitis in the US after World War II were attributed to transmission via milk, at a time when the nature of the disease and of many viruses was poorly understood (Cliver, 1997). Poliomyelitis was later recognized as a human-specific disease spread by a fecal–oral cycle. Improved sanitation and pasteurization of milk apparently solved this problem before vaccines were developed to eradicate the disease. Later, a large outbreak of hepatitis A with an oyster vehicle was reported from Sweden (Roos, 1956; Gard, 1957). It has since been recognized that viruses shed in human feces are often present in water in which shellfish are grown or held, and that these mollusks are very efficient in collecting and retaining the viruses (Richards, 1985). Shellfish are still a pre-eminent vehicle for virus transmission, although many other foods have since been implicated. An exceptional incident involved transmission of West Nile virus from an infected mother to her infant via breast milk (CDC, 2002a).

The common general properties of most viruses transmitted via food and water are that they are shed in the feces of infected humans and are transmitted as particles too small to be visible with a light microscope. Most contain RNA (usually single-stranded, plus-sense) rather than DNA; these agents replicate without reverse transcription, so DNA plays no part in their infectious cycles. Because they multiply only in suitable living cells, they cannot multiply in contaminated food or water, but can only persist or be inactivated (lose infectivity). Viral infections are generally not treatable, so there is a disincentive to do laboratory testing to confirm a diagnosis, in that the findings will not affect the course of treatment; laboratory diagnoses are usually done in the US only if an outbreak is discerned or in the case of hepatitis A, which is a reportable disease. Diagnosis of viral illness is often based on detection of antiviral antibody in the host's blood, rather than on detection of the causal virus. This approach is not applicable to detecting viruses in foods, which is usually attempted only in the course of investigating an outbreak of human disease. Food testing after outbreaks is often complicated by the relatively long incubation periods of some viral diseases (whereby samples are no longer available by the time the diagnosis is known), and by the tendency to test at length for bacteria before viral testing is attempted. Nevertheless, methods are evolving and improving as the problem becomes better recognized. The application of cell cultures to the detection of viruses in food, water and environmental samples constituted a great stride, but in time it became evident that the noroviruses and the hepatitis A virus, which are the most important viruses transmitted via food (Koopmans and Duizer, 2004), either did not replicate or did not express themselves in cell cultures. The later advent of the reverse transcription-polymerase chain reaction (RT-PCR) technique enabled detection of the most important foodborne viruses, but introduced some other problems that will be addressed in this chapter. At the same time, it is important to note that not all viruses, including some of those that are most feared, are transmissible via food. Questions remain about others, inexplicably, such as transmission of rabies via the milk of rabid cows.

A relatively new problem is the transmissible spongiform encephalopathies (TSEs). Some of these (e.g., scrapie in sheep and goats) have been known for centuries, but transmission to humans was evidently not part of the picture. However, the apparent transmission of the agent of bovine spongiform encephalopathy (BSE, 'mad cow disease') to humans, giving rise to variant Creutzfeldt-Jakob Disease (vCJD), opened a new era in perception of foodborne disease risks. The apparent agents of the TSEs, called prions, are certainly not viruses, but they will be discussed in this chapter for want of a more appropriate disposition. Although very few people have been affected by vCJD to date, the social and political impact of this disease has been profound in the UK, where the outbreak began, and in most other developed countries.

The following sections will discuss the groups of viruses that are known to be transmitted via food and water, providing details of the nature of the agent and how it is transmitted. Where possible, means of control will be suggested.

2 Hepatitis A virus (HAV)

Epidemic hepatitis in human populations has been described throughout history, but it was not until the twentieth century that distinct forms of the disease were characterized and ascribed to specific infectious agents (Melnick, 1995). Studies involving humans and primates and retrospective epidemiological analyses of outbreaks and cases in the 1950s and 1960s revealed distinct forms of infectious hepatitis. One form of the disease was transmitted by the fecal–oral route with a relatively short incubation period, and a second form of the disease was transmitted parenterally; these diseases were later designated hepatitis A and B, respectively (MacCallum, 1947). In 1973, virus-like particles in the stools of human patients with hepatitis A were observed by immune electron microscopy (Feinstone *et al.*, 1973) and presumptively called hepatitis A virus (HAV). A major development occurred with the demonstration that HAV could be serially propagated in cultured cells following numerous (31) passages in marmosets (*Saguinus mystax*) (Provost and Hilleman, 1979). Molecular cloning (Ticehurst *et al.*, 1983) and complete sequencing (Najarian *et al.*, 1985) of HAV's genome were followed by the licensure of hepatitis A vaccines in the US in 1995 (Bell *et al.*, 1998). While there is no specific chemotherapy for hepatitis A, immune globulin derived from plasma may be administered during and after suspected outbreaks of HAV (Carl *et al.*, 1983) as a means of passive immunization.

Currently, hepatitis A is endemic in all regions of the world. Although the disease is less prevalent in North America, Japan, and other places with reasonably good sanitation practices, outbreaks and cases of hepatitis A associated with water, food, and household contact with an infected individual are a significant source of morbidity and occasional mortality in these and other countries. The true disease burden of hepatitis A may be greater than reported (Koff, 1995), as subclinical cases of infection with HAV in the US may not be brought to the attention of medical and public health officials and therefore escape reporting. Furthermore, passive reporting may not be relied upon for assessing the true incidence of disease. While methods are currently under development for the recovery of HAV in various foods, such methods are generally dependent upon detection using molecular enzymatic amplification of viral RNA instead of infectivity, and there is generally a lack of consensus or standardization among various laboratories. The occurrence of HAV in foods and its ability to survive during modern food production and processing procedures are not well characterized. Also problematic is that contamination of foods with HAV may not correlate with fecal coliform or *Escherichia coli* counts, which are the historically used indicators for assessment of the hygienic quality of water and food (see section 8.3). For these reasons, the risk of infection from foods contaminated with HAV remains poorly characterized.

2.1 Nature of agent

Hepatitis A virus is a member of the *Picornaviridae* family, and is the sole species in the *Hepatovirus* genus. The non-enveloped virions exhibit icosahedral symmetry and are approximately 27 nm in diameter (Feinstone *et al.*, 1973). Based on cloning and sequencing data, the positive-sense, single-stranded ribonucleic acid (RNA) of HAV is approximately 7500 nucleotides in length (Ticehurst *et al.*, 1983). HAV's genome consists of a 5′-untranslated region of about 735 nucleotides, which contains an internal ribosomal entry site, followed by structural and non-structural protein-encoding regions, and a 3′-untranslated region with a terminal poly(A) tract (Cohen *et al.*, 1987; Hollinger and Ticehurst, 1996). Various strains of HAV share up to 90 % similarity at the nucleotide sequence level and 98 % similarity at the amino acid level (Lemon *et al.*, 1987). While only one serotype has been identified (Lemon *et al.*, 1992), seven genotypes of HAV have been identified (four human – I, II, III and IV; three simian – V, VI and VII), based on differences in the VP1/2A region of HAV's genome (Robertson *et al.*, 1991, 1992).

In cultured kidney cells from African-green and rhesus monkeys, wild-type or clinical isolates of HAV tend to grow slowly and establish persistent infections, with little or no evidence of cytopathic effect (CPE) (Provost, 1984; Gust and Feinstone, 1988). Consequently, viral capsids in infected cells are detectable only by immunofluorescence or radioimmunoassay, or viral RNA is detected by hybridization or by amplification using the reverse transcription-polymerase chain reaction (RT-PCR). In contrast, rapidly replicating cytopathic variants of HAV have been identified, selected from cell culture adapted isolates of wild-type HAV, including derivation from a well-characterized isolate, HAV HM175 (Cromeans *et al.*, 1987, 1989; Lemon *et al.*, 1991). Several variants from the original HAV HM175, which was initially isolated from a family outbreak in Australia in 1976 (Gust *et al.*, 1985), cause rapid CPE in monkey kidney cells. Some of HM175's cytopathic variants (e.g. HM175/18f, HM175 clone 24A) (Cromeans *et al.*, 1989; Lemon *et al.*, 1997) have been used in laboratory disinfection and environmental persistence studies.

2.2 Environmental persistence

HAV is considered relatively stable in the environment (Brown and Stapleton, 2003), and is more heat-resistant than other picornaviruses (Siegl *et al.*, 1984). In general, HAV can survive from days to weeks in various liquid media (e.g. marine-, surface-, ground- and wastewater; phosphate buffered saline), on fomites and other surfaces, and in various foods. HAV is resistant to extremes of pH; in liquids it is stable at pH 3 but is inactivated by pH > 10 (Gust and Feinstone, 1988), and it can survive at pH 1 for several minutes (Scholz *et al.*, 1989). Most information on HAV's fate in environmental media and food has been generated in the laboratory, using well-characterized cytopathic variants of HAV HM175. In such studies, HAV HM175 is seeded or inoculated into food or other matrices, exposed to the test condition(s), recovered from the matrix, and enumerated by infectivity in cultured cells. Virus survival is expressed as the remaining percentage of the original inoculum's infectivity titer, or the results are reported as the percent or $\log_{10}$ inactivation of the original inoculum titer after a unit period. There are fewer published studies on the persistence or fate of wild-type HAV in the environment using primate infectivity as the endpoint determination. The validity of extrapolating survival data from a laboratory-adapted strain of HAV (such as HM175) as determined by cell culture to *in vivo* assays is questionable, as repeated serial passage of HAV HM175 results in partial attenuation, as demonstrated in chimpanzees (Feinstone *et al.*, 1983). While Grabow *et al.* (1983) concluded that low passage level lab strains of HAV are unlikely to be different in terms of inactivation from clinical isolates (Siegl *et al.*, 1981; Tratschin *et al.*, 1981), most laboratory survival studies using HAV HM175 do not report the passage level of the virus used in experiments.

The survival of HAV in liquids is prolonged by colder temperatures and shortened by heat. In phosphate buffered saline (PBS), HAV is instantly inactivated at 85 °C and is completely inactivated in 5 seconds at 80 °C or in 4 minutes at 70 °C (Siegl *et al.*, 1984). In contrast, HAV remains infectious in PBS for as long as 12 hours at 60 °C, and can survive in raw milk heated at 63 °C or 72 °C, 60 % in 30 minutes and 73 % in 15 seconds, respectively (Mariam and Cliver, 2000). In marine water at a temperature of 20 °C, HAV is inactivated by 99 % (2 $\log_{10}$) in 2 days (Callahan *et al.*, 1995). HAV can remain infectious in marine water for 60 days at 5 °C (Chung and Sobsey, 1993), in ground water for 17 months at 9 °C or 8 weeks at 25 °C (Sobsey *et al.*, 1986; Cromeans *et al.*, 1994), and in mineral water at 25 °C for 300 days (Biziagos *et al.*, 1988). Little or no information is available on the persistence of HAV in surface water (e.g. rivers, lakes), but infectious HAV persisted for 50 days in tap water (0 mg/l total chlorine) at 23 °C (Enriquez *et al.*, 1995). Further evidence for HAV's persistence in surface water used as a drinking water source is that outbreaks and cases of hepatitis A have been associated with the consumption of fecally contaminated drinking water obtained from lake or river sources (Lippy and Waltrip, 1984). In addition, HAV has been detected (using a combination of cell culture infectivity plus RT-PCR) in treated drinking water that had fecal coliform counts of 0/100 ml (Grabow *et al.*, 2001).

Several studies have also investigated the persistence of HAV in various foods and on other materials. The survival of HAV on environmental surfaces is favored in environments with low relative humidity (Mbithi *et al.*, 1991), and virus survival may be lengthened when the virus is embedded or adsorbed to food particles or components (i.e. sugars, fats, proteins), sediment, soil particles or fecal droplets, all of which may have a protective effect on the virus. Infectious HAV persists in cookies for 30 days at 21 °C (Sobsey *et al.*, 1991), in live Eastern oysters (*Crassostrea virginica*) for 3 weeks (Kingsley and Richards, 2003), in acid-marinated mussels for 4 weeks (Hewitt and Greening, 2004), and on lettuce for 9 days at 4 °C (Croci *et al.*, 2002). On an aluminum surface and in an environment with high relative humidity, infectious HAV is detectable in fecal suspensions after 60 days at 20 °C (Abad *et al.*, 1994). HAV in fecal suspensions is more resistant to heat than are virions in aqueous suspension (Gust and Feinstone, 1988), and HAV remains infectious for up to 1 month in dried feces (McCaustland *et al.*, 1982) and in septic tank effluent for 70 days at 5 °C (Deng and Cliver, 1995a). HAV remains completely infectious in marine water in the presence of sediment, but is inactivated by 99.98 % in marine water in the absence of sediment (Chung and Sobsey, 1993). The presence of extraneous material may also shield or stabilize the virus during thermal treatments. For example, HAV remains infectious in strawberry mashes heated at 85 °C and 90 °C for up to 5 and 3 minutes, respectively (Deboosere *et al.*, 2004). Similar phenomena have been observed for HAV in dairy products (Bidawid *et al.*, 2000a) and in shellfish tissue heated for 1 minute at 100 °C (Croci *et al.*, 1999a), for 3 minutes at 85 °C (Millard *et al.*, 1987) or for 19 minutes at 60 °C (Peterson *et al.*, 1978).

2.3 Reservoirs and transmission

Humans are the primary reservoir of HAV, maintained through serial propagation from infected to susceptible persons (Koff, 1995). The great apes and both New- and Old-World monkeys may also be either naturally or experimentally infected (Zachoval and Deinhardt, 1998), but these animals are considered insignificant in the spread of HAV to man. HAV is distributed in human populations worldwide (Catton and Locarnini, 1998), but varying epidemiologic patterns are observed (Papaevangelou, 1984). In the hyperendemic areas of Africa, Asia, and Latin America, where crowding and poor hygiene are commonplace, asymptomatic seroconversion in children is widespread, and nearly all adults are immune to HAV. An increase in the mean age of exposure to HAV is observed in people living in developing regions of the world with steadily improving sanitation and public health education programs, and in areas such as the US children generally remain unexposed to HAV and do not develop antibodies. Consequently, a large proportion of adolescents and adults in the US and other low-endemicity countries (Pebody *et al.*, 1998) are susceptible to infection with HAV, thus underscoring hepatitis A's importance as a foodborne disease.

Large amounts of HAV are shed in the feces of infected people, and in the absence of adequate personal hygiene or environmental sanitation, the fecal–oral route may transmit HAV. Fecal contamination of foods may occur in a food-service setting, in the home, in the field, or in the processing facility by the hands of HAV-infected people – either through direct manipulation or by contact with hand-contaminated surfaces or utensils.

In contrast, some foods may become contaminated by contact with feces-containing water or soil (see section 2.5). From a food safety perspective, the infected host is one of the most important elements to consider in controlling the foodborne transmission of HAV, as it is estimated that thousands of food handlers have hepatitis A in a given year (CDC, 2003a). In addition, the secondary attack rate of clinically apparent infection among household members with clinical symptoms is 20–50 % (Koff, 1998), and contact with a person suffering from hepatitis A is the most commonly identified risk factor in the spread of the disease in the US (Meyerhoff and Jacobs, 2001). Infectious HAV survives on hands for up to 4 hours, and HAV can be transferred from the fingers to inert surfaces (Mbithi *et al.*, 1992) and to food items such as lettuce (Bidawid *et al.*, 2000b). Hence, HAV-infected individuals who handle or share food or food-preparation objects should exercise particular care in washing their hands after defecation.

Handwashing and disinfection or sanitation practices in the food-preparation and food service settings are important barriers against the transmission of HAV and other human enteric viruses (Barker *et al.*, 2001), but compliance with such practices may be poor (Altekruse *et al.*, 1996; Jay *et al.*, 1999; Bermudez-Millan *et al.*, 2004) or incomplete (i.e. not washing hands for long enough) (Allwood *et al.*, 2004a; Anderson *et al.*, 2004). Food preparation personnel may not always be relied upon to remove themselves from work while symptomatic with enteric viral disease (Brockmann *et al.*, 1995); from 1973 to 1982, about 24 % of outbreaks of foodborne disease associated with food-service establishments were attributed to an infected or 'colonized' person handling the implicated food (Bryan, 1988). In addition, monitoring worker health for signs of hepatitis A may be of limited effectiveness (Appleton, 1990), as HAV is usually shed in feces at concentrations of 8–9 $\log_{10}$ viruses per gram from 10 days to 2 weeks before the onset of clinical illness, and fecal shedding may last for 2–4 weeks (Cliver, 1994). Vigorous handwashing procedures (using hospital handwashing agents) only reduce the levels of HAV and other enteric viruses on the hands by 1–2 $\log_{10}$ (Mbithi *et al.*, 1993), and HAV is generally resistant to many commercially available disinfectants used for disinfecting agri-food (Jean *et al.*, 2003) and other environmental (Mbithi *et al.*, 1990; Abad *et al.*, 1997a) surfaces. Hence, even small amounts of HAV-containing fecal material on surfaces or hands may still contain appreciable numbers of virions. These issues highlight the important fact that procedures such as handwashing and environmental disinfection may reduce, but not completely eliminate, levels of HAV and other viral contaminants on hands and on surfaces. A combination of steps or multiple barriers is the overall best approach to reduce the likelihood of HAV transmission via food. Such steps include heightened surveillance of food handler health, ongoing training and education using a multitude of teaching methods (Allwood *et al.*, 2004a), proper handwashing (CDC, 2003b), the use of gloves (Paulson, 1997), and adequate disinfection and sanitation practices.

2.4 Nature of infection in man and animals

The incubation period of hepatitis A ranges from 2 to 7 weeks, with an average duration of 28 days (Feinstone and Gust, 1997). Following exposure to HAV a prodromal or pre-icteric (i.e. before appearance of jaundice) phase occurs, lasting from several days to more than a week, consisting of anorexia, fever (usually < 103°F), fatigue, malaise and myalgia. Nausea, vomiting, and diarrhea may also be present during this time, and are more frequent in children than in adults. The prodromal period is followed by jaundice, the classical sign of which is dark urine, and pale or clay-colored stools. Collectively, these symptoms are called classical hepatitis A. The severity of hepatitis A is dependent on age; middle-aged and older people and those with chronic liver disease (e.g. cirrhosis) are more susceptible to severe HAV infections (Willner *et al.*, 1998). Children experience asymptomatic hepatitis A more frequently than adults, and over 90 % of infections in children < 5 years of age may be silent (Battegay *et al.*, 1995). Infection with HAV is usually not fatal in younger persons; the mortality rates for adolescents or adults aged 15–39 and ≥ 40 years are 0.3 % and 2.1 %, respectively. In addition to classical hepatitis A, a biphasic or relapsing form of hepatitis A occurs in 6–10 % of cases (Glikson *et al.*, 1992), during which time two or more episodes of the classical symptoms of hepatitis A are experienced. Cholestatic hepatitis occurs in approximately 10 % of adult patients, and is characterized by fever, severe itching of the skin and jaundice (Gordon *et al.*, 1984). Fulminant hepatitis, or the most severe form of hepatitis A, occurs in approximately 0.35 % of cases, and in some instances can require liver transplantation (O'Grady, 1992). It is characterized by increasingly severe jaundice and deteriorating liver function; as blood bypasses the liver, the patient may succumb to coma (Zachoval and Deinhardt, 1998). At-risk groups for fulminant hepatitis include those with chronic liver disease, including patients with hepatitis C (Vento *et al.*, 1998).

HAV is secreted from the liver with the bile, and virus particles are present in the feces of infected individuals 1–2 weeks before the onset of clinical symptoms. Fecal shedding of virus can last for 2–4 weeks (Cliver, 1994), during which time peak fecal titers may exceed 8 $\log_{10}$ infectious units or particles per milliliter or gram of feces (Purcell *et al.*, 1984). Virus has also been detected in the stools of children (Robertson *et al.*, 2000) and adult (Yotsuyanagi *et al.*, 1996) patients 10 and 12 weeks, respectively, after the appearance of symptoms. During the incubation period of the disease HAV levels increase in blood, and HAV is also occasionally found in urine, oropharyngeal (including saliva) secretions and semen (Koff, 1995), but little or no convincing evidence exists to support the notion that these materials play an appreciable role in the foodborne transmission of HAV, and the concentrations of virus in them is appreciably lower than in feces. Human and non-human primate feeding studies conducted from the 1940s to the 1970s (for examples and reviews, see Krugman *et al.*, 1962; Purcell and Dienstag, 1979; Zachoval and Deinhardt, 1998), which were performed using filtered stool and serum extracts from acutely ill individuals, proved that HAV was transmissible via the oral route. These studies provided some dose–response data for HAV, but the actual virus concentrations in these inoculums is unknown. However, Krugman *et al.* (1962) reported that 0.1 g of infected stool constituted the minimal human infectious dose, and a more recent study, using a strain of HAV that had been recovered from a patient involved in a foodborne outbreak, demonstrated that one oral infectious dose in non-human primates is the equivalent of 4.5 $\log_{10}$ intravenous doses for tamarins and chimpanzees (Purcell *et al.*, 2002). Currently, however, the minimal human oral infectious dose of wild-type HAV remains undefined (Lemon *et al.*, 1997).

2.5 Foods most often involved

The association of hepatitis A with the consumption of fecally contaminated foods is well-documented (Cliver, 1967, 1985; Fiore, 2004). While some foods, such as bivalve mollusks (shellfish) and fruits and vegetables (produce), may become contaminated by exposure to HAV-containing feces or water in the environment (see below), many instances of foodborne hepatitis A in the US and elsewhere have been traced to infected food handlers (infected people who might handle food during self-service in a buffet, in a cafeteria operation or perhaps in some other setting) or to cross-contamination with an environmental surface, container or utensil. In some scenarios, a single food handler may be responsible for preparing large batches of food potentially consumed by great numbers of people. A wide variety of food vehicles is likely to have played a role in this type of indirect contamination, and these include, but are not limited to, beverages such as orange juice (Eisenstein *et al.*, 1963), milk (Raška *et al.*, 1966), mai-tai punch (Philp *et al.*, 1973) and ice-slush beverages (Beller, 1992); cheeseburgers (Parkin *et al.*, 1983) and other sandwiches (Levy *et al.*, 1975; Meyers *et al.*, 1975); pastas (Leger *et al.*, 1975), breads (Warburton *et al.*, 1991) and iced pastries (Schoenbaum *et al.*, 1976); and caviar (Glerup *et al.*, 1994). Any food can serve as a vehicle (Cliver *et al.*, 1983a) for a fecal virus such as HAV.

The risk of infection from HAV may increase when a food type or group is eaten raw or partially cooked. Shellfish (oysters, clams, cockles and mussels) are representative of such a food group, having caused one of the largest outbreaks of foodborne enteric disease in history. In 1988, almost 300 000 cases of infectious hepatitis in Shanghai, China, were reported, and epidemiological evidence implicated the consumption of raw clams in 90 % of cases (Halliday *et al.*, 1991). Hepatitis A virus and HAV RNA were subsequently detected in the implicated mollusks (Hu, 1989). Starting with the recognition of an outbreak of hepatitis A associated with shellfish consumption in Sweden during the 1950s, shellfish currently comprise one of the most widely implicated foods associated with the transmission of HAV (Richards, 1985; Rippey, 1994; Lees, 2000; Potasman *et al.*, 2002), and several studies report the detection of HAV in naturally-contaminated shellfish (Desenclos *et al.*, 1991; Le Guyader *et al.*, 1993, 1994, 2000; Yang and Xu, 1993; Lees *et al.*, 1995; Chung *et al.*, 1996; Bosch *et al.*, 2001; Kingsley and Richards, 2001). It has been estimated that individuals consuming raw shellfish in the US have a 1 in 100 chance of becoming infected with an enteric virus (Rose and Sobsey, 1993), and several factors make shellfish a high-risk food for transmission of HAV. First, shellfish-growing beds are often located in close proximity to municipal wastewater outfalls in the near-shore marine or estuarine environment, and viral contamination of shellfish occurs because these animals obtain their food by filtering water from their immediate surroundings – often at a rate of hundreds to thousands of liters per week (Sobsey and Jaykus, 1991). Fecal contamination of shellfish growing areas may also occur because of recreational use or discharge of sewage from boats. HAV can survive for extended periods, both in marine water and in shellfish tissue (see section 2.2, above). Second, methods such as monitoring growing waters and shellfish tissue for the traditional indicators of fecal contamination (i.e. fecal coliforms or *E. coli*) may not always indicate the presence or absence of HAV and other enteric viruses, which have different rates of survival in the environment than do bacteria (see section 8). Shellfish-growing areas and shellfish tissue meeting bacterial indicator standards have still been associated with multi-focal or multi-state outbreaks of HAV (Desenclos *et al.*, 1991). Third, shellfish-cleansing methods such as depuration (in which shellfish are temporarily placed in tanks of clean water to remove microbial and other contaminants) and relaying (transferring the animals to cleaner natural waters) have been found to be less effective and slower for virus removal than for removal of enteric bacteria (Lees, 2000).

Fruits and vegetables are additional examples of a potentially higher-risk food group or type for human enteric viruses such as HAV. Produce items have become an increasingly popular food group with the growing recognition of their potential health benefits, and they are often eaten raw. Produce is grown and distributed worldwide, and the short shelf-life of many fruits and vegetables ensures a rapid and efficient distribution process, thus limiting conditions (time, temperature) that would contribute to viral inactivation. A wide variety of produce commodities has been linked epidemiologically to transmission of HAV as well as other enteric diseases, and evidence suggests that the incidence of produce-associated enteric disease has increased in recent years (Sivapalasingam *et al.*, 2004). For example, 86 058 cases of foodborne disease (from all causes and vehicles) were identified and reported to the Centers for Disease Control and Prevention from 1993 to 1997 (CDC, 2000a). Of those outbreaks and cases, fruits and vegetables were responsible for 12 357 cases of disease, compared to the 2448 cases reported from 1988 to 1992 (CDC, 1996). Multi-focal outbreaks of hepatitis A have been associated with strawberries, green onions, raspberries, blueberries, and lettuce (Ramsay and Upton, 1989; Rosenblum *et al.*, 1990; Niu *et al.*, 1992; Hutin *et al.*, 1999; Dentinger *et al.*, 2001; Calder *et al.*, 2003; CDC, 2003c). The mechanism of fecal contamination is less clear in these episodes, compared to occurrences involving infected food handlers, because produce commodities are extensively exposed to soil and water – two other environmental media that also can become fecally contaminated, especially when human fecal wastes are used as fertilizer or municipal wastewater as an irrigation source. To date, only one study has reported the occurrence of HAV in produce, which was lettuce grown in Costa Rica (Hernández *et al.*, 1997). However, human hands may also extensively handle produce during the farm-to-table continuum; many types of produce, such as strawberries and green onions, require hand harvesting and are frequently hand-sorted because of their delicate natures. In addition, produce that is destined for some sort of minimal processing (e.g. slicing, cutting, blending) is usually first rinsed in water to remove spoilage microbes and other extraneous material, and it has been postulated that such procedures may result in the cross-contamination of produce from a small amount of fecal or soil contamination (Beuchat and Ryu, 1997).

2.6 Principles of detection of HAV in food and environment

Despite the wide acceptance and use of bacterial indicators of fecal contamination in some foods and in food processing, there is convincing evidence that the traditional bacterial indicators such as the fecal coliforms and *E. coli* are poor predictors of human enteric viruses such as HAV (see section 8). Therefore, the direct detection of HAV and other pathogens in food and water remains an important objective, and is an important tool for use in environmental surveillance and occurrence studies, in retrospective epidemiological investigations following outbreaks, and for establishing and demonstrating the effectiveness of new or alternative virus reduction and control technologies. However, testing of food is difficult because viruses such as HAV may be present in low levels or unevenly distributed throughout a food sample, and the goal of detection is at minimum one infectious unit, if it is assumed that the minimal oral dose of HAV is low. In addition, since wild-type HAV and other viruses are obligate parasites (meaning that they do not ever reproduce outside of a host or host cell), and given the problems associated with *in vitro* cultivation, the standard and readily available enrichment or growth methods used to increase the assay sensitivity of many bacterial pathogens recovered from foods are not applicable for HAV. Hence relatively large samples or volumes of food (or multiple smaller amounts) must be examined, and detection methods dependent upon antigenic, morphologic or genotypic characteristics are used. However, techniques such as enzyme immunoassays and electron microscopy, with detection limits frequently exceeding 5–6 $\log_{10}$ virus particles per gram (Koopmans and Duizer, 2004), do not have the requisite sensitivity for pathogen detection in food. Consequently, nucleic acid amplification methods, particularly the polymerase chain reaction (PCR), have emerged as the detection methods of choice for HAV in foods and environmental samples.

Two general methods are used in the initial recovery of viruses such as HAV from food (Cliver *et al.* 1983a, 1983b, 1990). In the first approach, referred to as surface elution or extraction, an intact food sample is washed or rinsed in a liquid solution or solvent (i.e. the eluent) that is conducive for the recovery of virus. In a variation of this approach, the pH of the food–eluent mixture is lowered, thereby facilitating adsorption of the virus to food particles; following centrifugation the supernatant liquid is removed, and the virus-containing pellet is again subjected to elution (i.e. acid adsorption-elution). The second approach blends or homogenizes the food in a similar liquid medium, and is used when viral contaminants are suspected to be sequestered or internalized within a food. The formulation of an eluent generally relies upon the manipulation of the net charges of the virus particle and of the surface to which it is adsorbed; at neutral pH, an unenveloped virus capsid such as HAV's is presumed to carry an overall negative charge. Elution of viruses on surfaces, including foods, is possible by decreasing available adsorption sites on the surface of food particles by the introduction of other small organic molecules (proteins) that would adsorb to these sites instead. Techniques for eluting viruses, including bacteriophages, generally involve diminishing an eluent's ionic strength or increasing its soluble protein content. Examples of the most commonly used eluents for virus recovery include beef extract, amino acid (such as glycine) solutions or combinations of the two; quite often, investigators will employ alkaline forms of these solutions and include detergents or other reagents.

Following the initial recovery procedure, primary food eluates or homogenates are usually large in comparison to the amount that can be assayed using various detection procedures for HAV. For example, a food eluate may be several hundreds of milliliters, and sample volumes used in PCR are typically ≤ 0.01 ml. Also problematic is the fact that food eluates and homogenates contain substances inhibitory to molecular detection, including such things as particulate matter, salts, proteins, lipids, carbohydrates, and natural organic matter. Therefore, extensive concentration and purification procedures are typically required to optimize virus detection and to concentrate virions into an aqueous medium compatible with PCR detection. A number of techniques have been used successfully for HAV recovery, detection and confirmation in environmental and clinical samples that also are applicable for the concentration and purification of HAV from foods, including polyethylene glycol and other chemical precipitations, ultracentrifugation, centrifugal ultrafiltration, fluorocarbon extraction, nucleic acid (RNA) extraction and antigen/RNA immunocapture using coated beads or tubes. Detection and confirmation of HAV in concentrated and purified environmental samples includes the use of RT-PCR, nested PCR, real-time PCR, nucleic acid hybridization and nucleic acid sequencing. Such techniques have been previously used successfully to concentrate and purify HAV and other enteric viruses in eluates and extracts from adsorbent water filters, sewage solids and other material (Graff *et al.*, 1993; Prévot *et al.*, 1993; Deng *et al.*, 1994; Monceyron and Grinde, 1994; Schwab *et al.*, 1995, 1996; Shieh *et al.*, 1995, 1997; Gilgen *et al.*, 1997; Jothikumar *et al.*, 1998; Casas and Sunen, 2002; D'Souza and Jaykus, 2002; Li *et al.*, 2002; Morace *et al.*, 2002; Abd El Galil *et al.*, 2004).

The cytopathic variants of HAV (e.g. HM175 clone 24A) are the most commonly used strains in methods development and validation, where infection is quantified by the number of plaque-forming units (PFU) in monolayers of cultured monkey kidney (e.g. FRhK-4) cells (Cromeans *et al.*, 1989). These assays provide the environmental virologist with an ample supply of HAV for use in laboratory experiments assessing the efficacy of various recovery, concentration and purification techniques for HAV in seeded or artificially contaminated food samples. In addition, quantitative infectivity data are often used to standardize and interpret a RT-PCR result, which is semi-quantitative at best, and which by itself does not indicate the presence of an infectious virus in a sample. Using RT PCR as the detection method, several studies have reported methods for the recovery and molecular detection of HAV seeded in shellfish tissue or extracts (Goswami *et al.*, 1993, 2002; Atmar *et al.*, 1995; Jaykus *et al.*, 1996; Cromeans *et al.*, 1997; López-Sabater *et al.*, 1997; Dix and Jaykus, 1998; Croci *et al.*, 1999b; Legeay et al., 2000; Casas and Sunen, 2001; Kingsley and Richards, 2001; Mullendore *et al.*, 2001; DiPinto *et al.*, 2003; Romalde *et al.*, 2004), with most studies reporting the corresponding minimum amount of detectable infectious virus in artificially-contaminated shellfish tissue ranges from about 0.02 to 67 PFU/g of oyster, clam or mussel tissue. Some information is also available on the detection of HAV in other types of foods, including produce, hamburger and deli meats; detection limits are similar to values reported for shellfish, and include 0.5 PFU/g of various fresh and frozen berries and fresh vegetables (Dubois *et al.*, 2002); 20 PFU/g of lettuce and hamburger (Leggitt and Jaykus, 2000); 2.5–10 PFU per strawberry or lettuce piece (Bidawid *et al.*, 2000c); 1.7 PFU/g hamburger and 17 PFU/g of lettuce (Sair *et al.*, 2002); and 0.02–0.2 PFU/g of ham, turkey and roast beef (Schwab *et al.*, 2000).

The application of commonly used environmental virology methods for the recovery and detection of HAV in foods has made appreciable progress in recent years. Reverse transcription-PCR and other PCR-based techniques are theoretically able to detect one viral genome (or one virus particle), and have been used in actual practice to link infected food handlers, contaminated water or implicated food items to exposed individuals in outbreaks of HAV (De Serres *et al.*, 1999; Pina *et al.*, 2001; Sanchez *et al.*, 2002; Calder *et al.*, 2003; Chironna *et al.*, 2004). However, PCR-based detection techniques are associated with some significant limitations for the virological analysis of foods and environmental samples. The PCR is unable to indicate the presence of an infectious virus, and may in fact indicate the presence of an inactivated (Sobsey *et al.*, 1998), damaged or defective (Nuesch *et al.*, 1988) virus. Furthermore, the PCR is not quite as sensitive for detection of HAV in food and other environmental samples as its theoretical detection limit (one viral genome or intact particle). For example, while some of the aforementioned studies report the detection of HAV in foods in the vicinity of one infectious unit (PFU), previous studies reporting the ratio of HAV particles (as determined by RT-PCR or electron microscopy) to infectious units (as determined by *in vitro* or *in vivo* infectivity assays) range widely, from 1:1 to 1000:1 (Zhou *et al.*, 1991; Deng *et al.*, 1994; Polish *et al.*, 1999; Schwab *et al.*, 2000; Kingsley and Richards, 2001). Hence the corresponding minimum number of HAV particles detected in many food samples could actually have ranged upwards from thousands to tens of thousands of HAV particles per gram, which is far less than the theoretical sensitivity of the PCR. Additional limitations include the fact that PCR amplification methods may be jeopardized by contaminating nucleic acid sequences, either from prior procedures or from other positive samples. Therefore, interpretation of results from PCR-based methods may be inaccurate due to false-positive or false-negative results, leading to potential under- or overestimates of the occurrence of HAV in food, and the inability to assess the infectious potential of viral contaminants may result in an inaccurate quantification of risk to human health. For these reasons, HAV and other virus presence in foods based on PCR-based techniques must be interpreted cautiously when predicting human health risks. The continued investigation of concentration and purification methods should do much to address the inhibition issue. New approaches, including integrated cell culture-PCR (ICC-PCR) (Reynolds *et al.*, 2001; Jiang *et al.*, 2004), chemical treatment of samples so that inactivated viruses are not detected by PCR (Nuanualsuwan and Cliver, 2002) and use of molecular techniques to discriminate between infectious and non-infectious HAV (Bhattacharya *et al.*, 2004), may partially address the limitations of PCR in regard to infectivity determination.

3 Hepatitis E virus (HEV)

The development of diagnostic serological tests for HAV and hepatitis B virus (HBV) in the 1970s resulted in an improved understanding of the geographic distribution and circulation of these viruses in human populations. However, a large proportion of hepatitis cases worldwide remained unattributable to either HAV or HBV.

Retrospective analyses of sera from individuals involved in waterborne epidemics of acute hepatitis (e.g. Viswanathan, 1957) showed the presence of an enterically transmitted, non-A, non-B (ET-NANB) hepatitis (Khuroo, 1980; Wong *et al.*, 1980; Skidmore *et al.*, 1992). While ET-NANB hepatitis was similar to hepatitis A in terms of its fecal–oral route of transmission and failure to establish chronic disease or persistent infection, it appeared to have a longer incubation period and a lower rate of secondary cases. Other distinguishing characteristics of ET-NANB included severity of disease in pregnant women (Khuroo *et al.*, 1981) and a relatively higher clinical attack rate in young to middle-aged adults compared to hepatitis A. The virus now associated with ET-NANB was first visualized by immune electron microscopy in the 1980s (Bradley *et al.*, 1987), cloned, partially sequenced, and named hepatitis E virus (HEV) (Reyes *et al.*, 1990). The development and use of immunoassays to detect antibodies to HEV (anti-HEV) showed that the virus circulates in epidemic and endemic patterns in Asia, Africa and Latin America. While rare or silent in North America and other regions of the world with relatively high standards of sanitation, HEV has been isolated from some domesticated livestock and wild animals in these areas, and there is growing evidence that hepatitis E is a zoonotic disease.

A major contemporary problem involving HEV is the lack of productive and reliable *in vitro* infectivity assays or convenient small animal models. Consequently, little virus is available for studies addressing its persistence in environmental media, its inactivation by disinfectants or procedures used by the food and water industries, or the development of methods to recover the virus from foods. There is no hyperimmune globulin to HEV available for pre- or postexposure prophylaxis (Mast and Krawczynski, 1996) and, while clinical trials are underway in Nepal with a recombinant vaccine (Emerson and Purcell, 2003), an inactivated or attenuated vaccine is not yet available (Wang and Zhuang, 2004; Worm and Wirnsberger, 2004). Hence public health control of HEV remains problematic in many regions of the world, including those places suffering from civil conflict, warfare or natural disasters (Rey *et al.*, 1999; Chironna *et al.*, 2001; Sencan *et al.*, 2004). In such areas, where crowding and poor sanitation are prevalent, hepatitis E has been shown to infect hundreds to thousands of people (Emerson and Purcell, 2004). There are only a few reports of probable foodborne transmission of HEV, and its role as a zoonosis has yet to be determined. However, its similarity to HAV in terms of fecal–oral route of exposure and documented association with waterborne outbreaks suggests that HEV may become an increasingly recognized foodborne pathogen in many parts of the world.

3.1 Nature of agent

HEV is a non-enveloped, icosahedral particle about 32–34 nm in diameter (Bradley, 1992). HEV's genome contains a polyadenylated, linear, positive-sense RNA genome of approximately 7200 nucleotides (Reyes *et al.*, 1990), consisting of three separate but overlapping open reading frames (ORFs) with non-structural (ORF1) and structural genes (ORF2) located at the 5′ and 3′ ends, respectively (Tam and Bradley, 1998). The function of the protein expressed by the third ORF is currently unknown, but it may serve as an anchor site during the assembly of virus particles

(Zafrullah *et al.*, 1997). Although previously classified as a member of the *Caliciviridae* family, results from comparative nucleotide sequencing in the polymerase region show that HEV is distinct from caliciviruses and other hepatitis viruses (Berke and Matson, 2000). HEV has recently been classified by the International Committee on Taxonomy of Viruses as the protoype member of the *Hepevirus* genus of the family *Hepeviridae* (Kasorndorkbua *et al.*, 2004).

The amino acid sequence of HEV's ORF2 gene that codes for the capsid protein is well conserved among geographical isolates, and most strains of HEV belong to a single serotype (Emerson and Purcell, 2003). In contrast, four genotypes of HEV have been identified (Schlauder and Mushahwar, 2001) based on nucleotide divergence of the nucleotides in the ORF2 region (Worm *et al.*, 2002). An additional five genotypes may exist, based on partial sequence data from both ORF1 and ORF2 (Schlauder and Mushahwar, 2001). HEV isolates from patients in the US (Kwo *et al.*, 1997) and Europe (Schlauder *et al.*, 1999; Zanetti *et al.*, 1999; Pina *et al.*, 2000) form a group of isolates that are genetically divergent from strains from HEV-endemic areas (Schlauder and Mushahwar, 2001). In addition to isolation of the virus from humans, HEV has been detected in various wild and domestic animals, including swine (see section 3.3). Some swine and human HEV isolates from the same geographic area show a close genetic relationship to each other (Kwo *et al.*, 1997; Meng, X.-J. *et al.*, 1997; Schlauder *et al.*, 1998; Hsieh *et al.*, 1999; Pina *et al.*, 2000; Okamoto *et al.*, 2001; Nishizawa *et al.*, 2003; Banks *et al.*, 2004a), with some studies reporting > 99 % homology.

3.2 Environmental persistence

Attempts at infection of cultured mammalian cells have not yielded satisfactory production of either HEV antigen or virus particles (Schlauder and Dawson, 2003), although detection of negative-sense RNA and antibody neutralization have been demonstrated (Meng, J., *et al.*, 1997; Favorov and Margolis, 1999). Consequently, there is only a limited amount of information regarding HEV's fate in the environment. When visualized by immune electron microscopy, HEV capsids appear to disintegrate after exposure to high concentrations of salts, following centrifugation in sucrose gradients, after storage for 3–5 days at 4–8 °C, or by repeated freeze-thawing, suggesting that the virus is relatively labile compared to other enteric viruses (Bradley, 1992). In contrast, some investigators claim that HEV preparations stored at room temperature for over 1 week remain infectious to non-human primates (Panda and Nanda, 1997), but this may be due to a protective effect of fecal material on virions, such as seen with HAV (McCaustland *et al.*, 1982). HEV must survive in water and wastewater long enough to reach and then infect susceptible hosts, because it has been associated with waterborne outbreaks of hepatitis E. In addition, HEV RNA has been detected in wastewater (Jothikumar *et al.*, 1993; Pina *et al.*, 1998, 2000; Vaidya *et al.*, 2002; Clemente-Casares *et al.*, 2003), suggesting that virus capsids remain intact for extended periods in such matrices. Axiomatically, HEV's demonstrated transmission by drinking water and its ability to remain intact in wastewater suggest that the virus may be relatively stable in the environment. There are some epidemiologic data to suggest that boiling water may inactivate HEV (Velazquez *et al.*, 1990; Corwin *et al.*, 1995, 1999).

3.3 Reservoirs and transmission

Reservoirs of HEV may consist of infected humans and swine, maintained through serial transmission between infected hosts and susceptible individuals. Some other domesticated and wild animals may be included within this category, although the reservoir for HEV during inter-epidemic periods is not known conclusively (Aggarwal *et al.*, 2001). Hepatitis E is considered endemic in human populations in Southeast and Central Asia, India, Africa and Mexico, and sporadic cases have been reported in some Latin American countries (Focaccia *et al.*, 1995; Souto *et al.*, 1997; Goncales *et al.*, 2000) and Cuba (Lemos *et al.*, 2000). In these and other so-called 'non-industrialized' countries and regions, seroprevalence of anti-HEV (IgG-class antibody) in adults ranges from 7.2 % to 54.8 % (Smith, 2001). In some regions of Asia and Africa more than half the cases of acute viral hepatitis in humans are attributable to HEV, and clinical disease is most often seen in young to middle-aged adults 15–40 years of age (Arankalle *et al.*, 1995). Unlike HAV, where children typically acquire anti-HAV by the age of 10, clinical disease and anti-HEV are infrequently detected in children < 5 years of age in regions such as India, Bangladesh and Mexico (Aggarwal *et al.*, 1997; Alvarez-Munoz *et al.*, 1999; Favorov and Margolis, 1999; Khan *et al.*, 2000; Arankalle *et al.*, 2001a). In contrast, HEV has been reported to be a common cause of pediatric hepatitis in Cairo, Egypt (Hyams *et al.*, 1992; El-Zimaity *et al.*, 1993), and more than half (60 %) of Egyptian children ≤ 10 years old were found to be seropositive for HEV (Fix *et al.*, 2000).

Hepatitis E is a rare disease in North America, Europe and Japan, and most cases in these areas have been associated with travel to HEV-endemic regions (Ooi *et al.*, 1999; Schwartz *et al.*, 1999; Khuroo, 2003). However, cases have been identified in these countries (Schlauder *et al.*, 1998) in which travel had not occurred and in the absence of other identifiable risk factors (Mast *et al.*, 1997). Surveys of blood donors in the US have shown the presence of anti-HEV in 0.4–3 % of such cases (Mast *et al.*, 1997), and a single study reported a 21.3 % seroprevalence in blood donors from Baltimore, Maryland (Thomas *et al.*, 1997). In addition, the detection of HEV RNA in municipal wastewater from industrialized countries (Pina *et al.*, 1998, 2000; Clemente-Casares *et al.*, 2003) suggests that circulation of HEV in humans may be more prevalent than previously thought.

The reasons for the unexpectedly high seropositivity in countries where hepatitis E is uncommon and the differences in age-related epidemiologic patterns between regions with similar socioeconomic and sanitary conditions (i.e. India and Egypt) have not yet been clearly resolved (Emerson and Purcell, 2003). However, the epidemiology of HEV in humans may be partly explained by the presence of animal reservoirs in many regions of the world. HEV gene sequences and anti-HEV have been detected in swine and rats (Hsieh *et al.*, 1999; Kabrane-Lazizi *et al.*, 1999; Karetnyi *et al.*, 1999; Favorov *et al.*, 2000; Arankalle *et al.*, 2001b; Garkavenko *et al.*, 2001; van der Poel *et al.*, 2001; Wang *et al.*, 2002; Hirano *et al.*, 2003; Banks *et al.*, 2004b), in cattle in a low-endemicity region of the Ukraine (Favorov and Margolis, 1999) and in dogs and goats (Arankalle *et al.*, 2001b). In addition, a virus similar to HEV has been found in chickens (Haqshenas *et al.*, 2001). The genetic homology between swine and human isolates of HEV and the prevalence of anti-HEV in humans with occupational exposure (e.g. swine farm workers, veterinarians) (Karetnyi *et al.*, 1999; Drobeniuc *et al.*, 2001; Meng *et al.*, 2002) supports an early hypothesis that hepatitis E is a zoonotic disease (Balayan *et al.*, 1990).

Like HAV, HEV is transmitted between human hosts and between animal hosts by the fecal–oral route of exposure. Transmission of HEV by some foods has been reported (see section 3.5), but hepatitis E is most often associated with the consumption of fecally contaminated water in geographical and political regions with relatively poor sanitary conditions. Large waterborne epidemics associated with HEV have occurred in these and other areas, including an episode that involved more than 200 000 cases in the Xinjiang Region of China from 1986 to 1988 (Favorov and Margolis, 1999). In some of these occurrences of waterborne transmission, a seasonal epidemic pattern coinciding with periods of heavy rains has been described (Krawczynski and Bradley, 1989). Person-to-person transmission of HEV plays a diminished role as is evident in the low percentage of secondary cases observed during outbreaks (Naik *et al.*, 1992; Bradley *et al.*, 1997). There has been a single report of HEV isolation in flies and cockroaches (Tomar, 1998), but confirmation or replication of these results have not yet been reported by others.

3.4 Nature of infection in man and animals

Like hepatitis A virus, human infection with HEV does not appear to result in chronic disease and is not persistent, and mortality rates of those infected range from 0.5 % to 3 % (Mast and Krawczynski, 1996). In contrast, HEV-associated disease in pregnant women, especially those in their third trimester, may be severe (Khuroo *et al.*, 1981), with mortality rates ranging from 15 % to 25 % (Krawczynski, 1993; Mast and Krawczynski, 1996; Smith, 1999). HEV RNA is also detectable in the colostrum of nursing mothers (Chibber *et al.*, 2004). The incubation period of hepatitis E is somewhat longer than that of hepatitis A, with onset of disease usually occurring 2–9 weeks post-exposure (Krawcyznki, 1993). Symptoms are similar to those of hepatitis A and include jaundice, loss of appetite, nausea, vomiting, fever and abdominal pain (Worm *et al.*, 2002); the average length of illness may be as long as 100 days (Wald, 1995). Subclinical or inapparent forms of hepatitis E may exist (Aggarwal *et al.*, 2001), although convincing data supporting this assumption have yet to be obtained from epidemiological and serological studies. Shedding of HEV particles in the stool begins about 7 days before disease onset and can last as long as 50 days (Lemon, 1995). Fecal titers of HEV appear to be lower compared to other enteric viruses (Ticehurst *et al.*, 1992).

Experimental infection using human isolates of HEV has been accomplished in a human (Balayan *et al.*, 1983), in non-human primates (e.g. African green, squirrel, owl, cynomolgus and rhesus monkeys, and chimpanzees) (Kane *et al.*, 1984; Bradley *et al.*, 1987, 1988, 1997; Arankalle *et al.*, 1988; Ticehurst *et al.*, 1992; Nanda *et al.*, 1994; Tsarev *et al.*, 1995; Erker *et al.*, 1999a; McCaustland *et al.*, 2000; Aggarwal *et al.*, 2001), in swine and sheep (Balayan *et al.*, 1990; Meng *et al.*, 1998), and in rats (Maneerat *et al.*, 1996). Cynomolgus macaques are the most widely used animal model in experimental infections with HEV (Bradley *et al.*, 1987; Tsarev *et al.*, 1994), although swine are increasingly used (Meng *et al.*, 1998; Halbur *et al.*, 2001; Kasordorkbua *et al.*, 2004). Hepatitis in these and other non-human primates is often mild, but these animals shed virus in feces and their antibody response is similar to that of humans (Lemon, 1995; Panda and Nanda, 1997; Halbur *et al.*, 2001), in addition to exhibiting histopathologic changes similar to those in humans. The appearance of hepatitis E disease is dose-dependent in non-human primates, and may require doses ≥ 1000 times greater than required for infectivity alone (Tsarev *et al.*, 1994). For example, a dosage of 10–100 infectious units in macaques did not result in biochemical signs of disease, but did cause the animals to shed virus particles in stools (Aggarwal *et al.*, 2001). The minimal human oral infectious dose of HEV is unknown, but studies using swine HEV showed that the minimum amount required to infect other susceptible swine was 2 $\log_{10}$ genome equivalents (GEs; defined as the number of HEV genomes present in the highest serial 10-fold dilution positive by RT-PCR), and at least 6 $\log_{10}$ GEs was sufficient to cause disease in rhesus monkeys and a chimpanzee (Meng *et al.*, 1998). Studies in chimpanzees suggest that geographically distinct isolates of human HEV display different virulence properties (Bradley, 1995), and a human isolate of HEV caused more severe disease in experimentally infected swine than did swine HEV (Halbur *et al.*, 2001). It should be noted that many of the experimental studies in primates and swine were performed using an intravenous rather than oral route of exposure. This has important implications for attempts to place epidemiologic patterns and environmental transmission into context, because it is estimated that the infectious dose of HEV for cynomolgus macaques is 10 000-fold higher when inoculated intravenously compared to ingestion (Emerson and Purcell, 2003), and such phenomena also have been reported for experimental transmission of HEV in swine (Kasorndorkbua *et al.*, 2004).

3.5 Foods most often involved

Compared to HAV, there is relatively little evidence documenting the transmission of HEV by food vehicles. However, association of hepatitis E with the consumption of uncooked liver (Matsuda *et al.*, 2003) or partially cooked tissue of wild boar (Tamada *et al.*, 2004) and swine (Yazaki *et al.*, 2003), as well as uncooked deer meat (Tei *et al.*, 2003, 2004) in Japan have been reported. In addition, elevated anti-HEV has been associated with drinking unpasteurized cow's milk (Drobeniuc *et al.*, 2001), and the consumption of shellfish has been identified as a risk factor in the development of hepatitis E in Italy (Cacopardo *et al.*, 1997) and Hong Kong (Chau *et al.*, 2001).

3.6 Principles of detection in food and environment

Compared to HAV, there are relatively few reports on the recovery and detection of HEV in foods, although a few reports document the detection of HEV RNA or anti-HEV detection in swine (Kasorndorkbua *et al.*, 2002; Yazaki *et al.*, 2003), wild boar (Sonoda *et al.*, 2004), and deer tissue (Tei *et al.*, 2003; Sonoda *et al.*, 2004). Methods for virus recovery from infected tissue in these studies consisted simply of homogenization and suspension in a buffer followed by chemical extraction of viral RNA (e.g. using solutions of phenol-guanidine isothiocyanate). Such techniques are probably sufficient for recovery of virus from infected clinical specimens, as a target virus is expected to be present in relatively large amounts. However, environmental matrices such as foods and water differ, because the suspect virus is often expected to occur sporadically and at low concentrations, necessitating the analysis of larger sample sizes or volumes. These samples must often be followed by additional concentration and purification steps, as primary sample concentrates are often too large and are inhibitory for molecular assays such as PCR.

Techniques used for virus recovery and purification from environmental matrices and clinical samples have been reported for HEV, and various combinations of these methods could be applied to most foods. These methods include centrifugation, polyethylene glycol and chemical precipitation, pH adjustment, amino acid and beef extract elution, solvent extraction using fluorocarbons and alcohols, and RNA extraction for the recovery and purification of HEV from feces (McCaustland *et al.*, 1991; Aggarwal and McCaustland, 1998; Pina *et al.*, 2000), cell culture supernatants (Huang *et al.*, 1995); wastewater (Jothikumar *et al.*, 1993; Pina *et al.*, 1998, 2000; Clemente-Casares *et al.*, 2003); membrane filters (Grimm and Fout, 2002); and other media for recovery of viruses from surface and drinking water (Jothikumar *et al.*, 1995, 2000; Grimm and Fout, 2002). As with HAV, it is conceivable that such methods could also be applied toward the recovery of HEV in foods.

Numerous reports and published primer and probe sequences spanning the entire genome of HEV (Schlauder and Mushawar, 1994) are available and are used in RT-PCR (Turkoglu *et al.*, 1996; Chobe *et al.*, 1997; Huang *et al.*, 2002; Jameel *et al.*, 2002; Kasorndorkbua *et al.*, 2002), in nested (Choi *et al.*, 2004) and multiplex PCR formats (Jothikumar *et al.*, 2000; Singh and Naik, 2001), and in real-time quantitative PCR (Mansuy *et al.*, 2004; Orru *et al.*, 2004) assays for amplification of HEV RNA. Sequencing of the genome from various geographical isolates of HEV shows that most variation occurs in ORF1 (Huang *et al.*, 1992), while sequences within ORF2 are highly conserved amongst all isolates. Hence PCR primers, especially degenerate ones (Erker *et al.*, 1999b; Meng, J., *et al.*, 1997) derived from the capsid protein in ORF2 may be useful to amplify target RNA from all isolates of HEV, if broad-based detection of all strains of HEV in foods and other environmental matrices is the goal. In such cases, the use of 'touchdown' PCR (i.e. a gradual decrease of the annealing temperature that improves the chance of amplifying variants) may improve chances for detection of HEV (Schlauder and Dawson, 2003) in concentrated and purified environmental samples.

4 Noroviruses

Noroviruses are a leading cause of outbreaks of foodborne gastroenteritis, estimated to account for over 67 % of all cases of foodborne illness caused by a known agent (Mead *et al.*, 1999). In the US, the fraction of outbreaks reported to the CDC confirmed to be due to norovirus was less than 1 % in 1991 (Widdowson *et al.*, 2005). The development and application of reverse transcription (RT) PCR strategies to outbreak investigations have helped to better define the extent of disease due to noroviruses, as shown by the increase in confirmed norovirus outbreaks to 12 % in 2000. This, however, is likely a marked underestimate, given inconsistencies in reporting cases and testing of samples for viruses, as well as variable capabilities in state and local laboratories to test for viruses. The true disease burden remains to be determined.

4.1 Nature of agent

During the investigation of a 1968 outbreak of gastroenteritis at a school in Norwalk, Ohio, the viral agent responsible for the outbreak was discovered by Kapikian and colleagues in 1972, using immune electron microscopy (Kapikian *et al.*, 1972). This landmark study not only provided evidence linking a specific virus (now known as Norwalk virus) to human gastrointestinal disease, as long suspected (Reimann *et al.*, 1945), but it also led to the discovery of other viruses in human disease.

Even after the identification of the Norwalk virus and numerous other small, round, structured viruses (SRSVs) that were associated with similar episodes of gastroenteritis (reviewed in Green *et al.*, 2001) and preliminary biochemical characterization suggesting relatedness to caliciviruses (Greenberg *et al.*, 1981), these viruses proved challenging to study because they eluded adaptation to cell culture (Duizer *et al.*, 2004a) and only infected humans and higher-order primates (Wyatt *et al.*, 1978). Early studies relied heavily on human volunteers and reagents, such as 8FIIa, the 2 % stool filtrate that was derived from a secondary case in the 1968 Norwalk outbreak and served as the viral inoculum in the volunteer studies (Dolin *et al.*, 1971).

Cloning and sequencing of the Norwalk virus genome in the early 1990s (Jiang *et al.*, 1990, 1993; Matsui *et al.*, 1991) enabled detailed characterization of the viral genome and provided a means to identify closely related viruses that now comprise the genus *Norovirus* in the family *Caliciviridae*. Expression of the viral capsid led to the production of abundant virus-like particles (VLPs) that have advanced knowledge of viral structure and immunity and have provided reagents for viral detection (Jiang *et al.*, 1992, 2000). VLPs of norovirus strains representing three genogroups (see below), I, II and III, have been produced.

Noroviruses are now recognized as the cause of the vast majority of outbreaks of acute non-bacterial gastroenteritis in the US and abroad (Fankhauser *et al.*, 1998; Koopmans *et al.*, 2001). Although a diverse group genetically, the noroviruses share fundamental characteristics: a positive-sense, single-stranded RNA genome that is approximately 7.7 kb in length, with the non-structural proteins encoded in ORF1 at the 5′ end of the genome; the single, predominant, 60-kDa capsid protein (VP1) encoded in ORF-2; and a minor structural protein (VP2) encoded in ORF3 at the 3′ end (Glass, P. J. *et al.*, 2000). The viral genome is enclosed in an icosahedral capsid that has been studied thoroughly using the prototype Norwalk virus VLPs. Electron cryomicroscopy and X-ray crystallography studies indicate that the capsid is composed of 180 copies (90 dimers) of VP1, and measures 38 nm in diameter if protrusions found at the local and strict two-fold axes are included (Prasad *et al.*, 1994, 1999).

Phylogenetic analysis of the RNA-dependent RNA polymerase or VP1 shell domain sequences of scores of norovirus strains from human infection has allowed the classification of these strains primarily into two genogroups (GI and GII) that can be further subdivided into a number of genetic clusters (reviewed in Green *et al.*, 2001). Members of a cluster share 80 % or more VP1 sequence identity. Bovine noroviruses segregate into a separate genogroup (GIII). A fourth genogroup (GIV) has been proposed to include human viruses, such as Alphatron (Vinjé and Koopmans, 2000) and Fort Lauderdale (Fankhauser *et al.*, 2002), that are distinct from GI and GII viruses. A fifth genogroup (GV) has been proposed to accommodate the recently identified Murine Norovirus 1 (MNV-1) on the basis of its distinct capsid protein sequence (Karst *et al.*, 2003).

4.2 Virulence factors

The determinants of norovirus virulence have not been well studied, due to limitations imposed by the lack of an animal model and a cell culture system. Recently, however, a mouse norovirus, MNV-1, was discovered (Karst *et al.*, 2003), and it has been shown to replicate *in vitro* (Wobus *et al.*, 2004). These promising new developments have begun to provide some initial insights into viral virulence.

MNV-1, administered to mice by the oral route, has been shown to cause infection in the intestinal tract of mice and has been detected in liver and spleen as well (Karst *et al.*, 2003). The cellular transcription factor, STAT1 (signal transducer and activator of transcription 1) and interferon (IFN) receptors appear to be necessary for resistance to MNV-1 infection, suggesting an important role for innate immunity in mouse norovirus pathogenesis. STAT1-deficient mice were highly susceptible to serious infection, while mice that were deficient in both IFNαβ and IFNγ receptors were substantially more susceptible to lethal infection.

The ability of macrophage and dendritic cells to support MNV-1 replication *in vitro* also implies the importance of the innate immune system in infection (Wobus *et al.*, 2004). In addition, in the course of adaptation of the virus to cell culture, a virus with attenuated virulence *in vivo* (compared to the virulent parental strain) emerged in passage 2 (P2), while a virus that was avirulent for mice emerged after the third passage (P3). The intermediate phenotype of the P2 virus was associated with a mutation in the region predicted to encode the 3A-like non-structural polyprotein. The homologous 3A protein of poliovirus is thought to affect viral pathogenicity through modulation of cytokine secretion from cells (Dodd *et al.*, 2001). The P3 avirulent mutant was noted to harbor a critical mutation in a hypervariable domain of the capsid protein that has been postulated to form a receptor-binding site. This region also appears to be important in the pathogenicity of porcine enteric calicivirus of the genus *Sapovirus* (Guo *et al.*, 2001).

4.3 Environmental persistence

Gastroenteritis and hepatitis viruses are particularly well adapted to persist in the environment, and are not easily inactivated (Appleton, 2000). Viruses that enter the host through the gastrointestinal tract, or cause enteric infection, must effectively survive passage through the acidic milieu of the stomach. Noroviruses are known to remain infectious after exposure to an acidic environment of pH less than 3 (Dolin *et al.*,

1972). A recent study showed that norovirus RNA remained intact even after a 30-minute exposure to pH 2 at body temperature, while surrogate caliciviruses that do not cause enteric disease were more susceptible to this treatment (Duizer *et al.*, 2004b).

Noroviruses are stable over a wide range of temperatures. These viruses can remain infectious despite being frozen, refrigerated, or heated to 60 °C for 30 minutes. Ice made with contaminated water (Cannon *et al.*, 1991) or ice contaminated on the surface only (Khan *et al.*, 1994) has been found to cause outbreaks of norovirus gastroenteritis. In one outbreak in which contaminated oysters were implicated, people who ate what they believed to be thoroughly cooked oysters were as likely to develop gastroenteritis as those who ate raw oysters (McDonnell *et al.*, 1997).

These viruses can also be detected on a variety of inanimate surfaces. Contaminated surfaces present a cleaning challenge, and if they are not cleaned properly they could be a reservoir that prolongs an outbreak, most notably in closed environments such as hospitals, nursing homes and cruise ships. Although chlorine-based disinfectants are thought to be the most effective for these non-enveloped viruses, Norwalk virus appears to be more resistant to chlorine treatment than other gastrointestinal viruses such as poliovirus and rotavirus. Treatment with 10 mg chlorine per liter (or 5–6 mg of free chlorine per liter measured in the reaction vessel after 30 minutes) was needed to inactivate the Norwalk virus inoculum, compared to 3.75 mg chlorine per liter for the other viruses (Keswick *et al.*, 1985). In cases where there has been a fecal or emetic spill, it is recommended that the soiled surface be wiped clean first, using detergent and hot water, then treated with a chlorine-based disinfectant (Barker *et al.*, 2004). In this study, the use of hypochlorite in the final step at a concentration of 5000 ppm, found in most household disinfectants, was effective in eliminating norovirus from contaminated surfaces, as judged by the inability to detect viral RNA by RT-PCR. For surfaces that are not amenable to treatment with chlorine-based disinfectants (e.g. carpeting), cleaning with steam or detergent and hot water is recommended (Chadwick *et al.*, 2000).

4.4 Reservoirs and transmission

For noroviruses to cause infection, only a very low dose of virus is needed (Cubitt, 1989; Green *et al.*, 2001). Some estimates suggest that numbers as low as 10–100 viral particles are sufficient to cause infection (Glass, R. I. *et al.*, 2000).

Outbreaks of norovirus infection have occurred in a wide variety of settings, including hospitals and nursing homes, restaurants and catered events, schools and child-care centers, cruise ships, camps and other vacation settings. The viruses have been transmitted by ingestion of contaminated food (including oysters) or water, exposure to fomites and contaminated surfaces, as well as person-to-person contact. Person-to-person transmission was dramatically shown when a college football team that was afflicted with norovirus infection transmitted the illness to the opposing team during a game (Becker *et al.*, 2000).

Close contact, closed communities, poor sanitation and poor personal hygiene all facilitate viral transmission. To minimize nosocomial transmission, strict implementation of universal precautions and isolation, with provision of separate facilities for affected patients, are recommended (Chadwick *et al.*, 2000).

4.5 Nature of infection in man

Individuals of all ages may develop norovirus infection. Infection may occur throughout the year, but in temperate climates more cases occur in the winter. The incubation period is 24–48 hours. Norovirus gastroenteritis is characterized by the sudden onset of diarrhea and/or vomiting that is self-limiting, lasting 1–3 days. While the symptoms are usually mild, some patients experience marked and incapacitating diarrhea and/or vomiting during the acute phase. Fluid depletion and electrolyte disturbance, although uncommon, may develop, especially in very young and elderly patients. Resuscitation with oral or parenteral rehydration may be needed in such patients. In general, no chronic sequelae of illness develop after symptoms resolve, if patients are otherwise healthy. In immunocompromised patients, however, diarrhea and viral shedding may become chronic (Kaufman *et al.*, 2003; Nilsson *et al.*, 2003). Infected individuals have a high likelihood of passing infection to close contacts, and secondary cases are common.

Noroviruses cause a spectrum of illness, primarily affecting the gastrointestinal tract. In the first volunteer study, in which the stool filtrate 8FIIa was administered to three volunteers (Dolin *et al.*, 1971), two developed symptomatic illness: one had diarrhea without vomiting, and the other experienced numerous episodes of vomiting without diarrhea. The latter volunteer also had fever to 101°F (38.3 °C) and an elevated white blood cell count during the acute symptomatic phase of the illness, while the former only had a low-grade temperature elevation of less than 100°F (37.7 °C). Both volunteers reported associated symptoms of nausea, abdominal cramping, headache, malaise, myalgia and anorexia to varying degrees. Subsequent volunteer studies conducted with other noroviruses induced illness that was indistinguishable from that caused by Norwalk virus (Dolin *et al.*, 1975; Dolin, 1978). In an unusual norovirus outbreak involving British military troops in Afghanistan, four infected soldiers complained of extraintestinal symptoms, including neck stiffness, photophobia and confusion; one of them also developed disseminated intravascular coagulation (CDC, 2002b).

Vomiting has been observed more often in children, while diarrhea tends to be seen more often in adults (Gotz *et al.*, 2001). The significantly delayed gastric emptying that has been documented in ill volunteers may underlie the nausea and vomiting many infected individuals experience (Meeroff *et al.*, 1980). Droplets of vomitus may be an important means through which noroviruses are transmitted.

Intestinal biopsies taken from infected volunteers during the symptomatic phase of illness were notable for flattening of the villi, hypertrophy of intestinal crypts, and inflammation of the mucosa (Agus *et al.*, 1973; Schreiber *et al.*, 1973, 1974; Dolin *et al.*, 1975). Of note, volunteers who were given Hawaii virus, whether they developed symptoms or not, developed the characteristic intestinal changes transiently. Associated with these histological lesions were significantly decreased levels of brush-border enzymes and transient malabsorption of fat, D-xylose and lactose (Schreiber *et al.*, 1973). All of these changes resolved after recovery from illness.

Subclinical infections also occur in some individuals (Graham *et al.*, 1994). These persons may shed virus for up to 3 weeks after taking the viral inoculum and may develop a Norwalk virus-specific antibody response, but do not have diarrhea, vomiting or other clinical manifestations of illness.

Volunteer studies additionally demonstrated that acute gastroenteritis could be induced in 50 % or more of those given the inoculum, but some individuals appeared resistant. One study, in particular, showed that half of the volunteers were susceptible to developing symptomatic infection and developed symptomatic illness when challenged again with the same inoculum a year or two later. The other volunteers did not develop any symptoms either when first exposed to the inoculum or with subsequent challenge (Parrino *et al.*, 1977). The study suggested that long-term immunity was not established in susceptible individuals, and raised the possibility that susceptibility and resistance to Norwalk virus infection are genetically determined.

Recent studies have shown an association between an individual's blood ABO phenotype and the risk of developing symptomatic infection. Studies with the genogroup I prototype strain, Norwalk virus, have shown that blood group O volunteers were more likely to develop infection with viral challenge, while blood group B volunteers were less likely to become infected and less likely to develop symptomatic illness (Hutson *et al.*, 2002). Recombinant Norwalk virus VLPs were also found to recognize cell surface H-type (secretor-type) oligosaccharides, and could be internalized into cells expressing such cell surface receptors (Marionneau *et al.*, 2002). Furthermore, it was demonstrated that individuals who are homozygous-recessive for the $\alpha(1,2)$ fucosyltransferase (FUT2) gene that determines secretor phenotype (i. e. non-secretors) were resistant to infection (Lindesmith *et al.*, 2003). However, a fraction of the volunteers with functional FUT2 (secretors), who would be predicted to be susceptible to infection, were resistant. This suggested that other factors, possibly memory immune response, may contribute to an individual's susceptibility or resistance. It should be noted that the findings with Norwalk virus and its VLPs may not be directly extended to other strains of norovirus. Carbohydrate binding of VLPs has been shown to be strain-dependent (Harrington *et al.*, 2002; Huang *et al.*, 2003). VLPs of a strain of norovirus clustering with the predominant strains in circulation at present (GII.4) bound saliva and synthetic carbohydrates of all secretors, regardless of ABO type. In a recent study from the Netherlands, correlations with histo-blood group phenotype could be demonstrated with genogroup I but not genogroup II noroviruses (Rockx *et al.*, 2005). An earlier study of British troops in Afghanistan, who were afflicted with a severe, genogroup II norovirus infection that had atypical features, did find a trend correlating blood group and susceptibility to, and severity of, infection in a similar pattern to that observed with Norwalk virus, but the association was not statistically significant (Hennessy *et al.*, 2003). Clearly, more work is needed to understand fully the intricacies of susceptibility and resistance to infection for this diverse group of viruses.

4.6 Foods most often involved as vehicles

Any of a number of foods have served as vehicles for transmission of noroviruses, including oysters, sandwiches, salads, fresh raspberries and cake frosting. In general, food may be contaminated at the source, or contaminated during preparation. Sandwiches and salads, for example, require preparation by a food handler, without further heating, and may include fresh produce that could have been contaminated at the source (Widdowson *et al.*, 2005). These types of foods were implicated as vehicles of transmission in more than 56 % of foodborne outbreaks attributable to noroviruses in the US over the 10-year period from 1991 to 2000. Transmission by a food handler may be difficult to prove, since norovirus illness may be not be recognized or reported, given the mild or subclinical course in some individuals. In addition, viral shedding may occur for some time (up to 2 weeks by RT-PCR detection methods) after symptoms of gastroenteritis have resolved. It is not known whether virus shed beyond the acute phase of the illness remains infectious.

Contaminated oysters have been identified as the source of a number of large, multistate outbreaks of norovirus gastroenteritis, but oysters were not frequently found to be a cause of norovirus outbreaks over the 10-year period examined in a recent study (Widdowson *et al.*, 2005). As filter feeders, oysters and other bivalve mollusks can achieve higher viral concentrations in their tissues, up to 100-fold greater than found in the surrounding sea water (Appleton, 2000). They pose a further risk to the consumer as they are often eaten raw or, if eaten after light cooking, the internal temperature of the shellfish meat may not have reached the requisite 90 °C maintained for 1.5 minutes for proper heat inactivation of viruses (Millard *et al.*, 1987).

4.7 Principles of detection in food and environment

Development of methods to detect norovirus in food has lagged behind that for clinical samples, but is no less important. Many of the challenges encountered in detecting hepatitis A virus in food also apply to noroviruses (see section 2.6): traditionally used bacterial indicators do not adequately reflect levels of viral contamination; low levels and patchy areas of viral contamination within a sample necessitate examining large volumes; and the low infectious dose requires highly sensitive assays for detection. The lack of a cell culture system for human noroviruses further limits the ability to test samples for the presence of infectious virus.

Detection of noroviruses, whether in clinical, food or environmental samples, depends largely on RT-PCR. Methods such as direct electron microscopy or enzyme immunoassays require much higher concentrations of virus (~10^5–10^6 viral particles per ml) to yield a positive result (summarized in Glass, R. I., 2000) than those present in food and environmental samples (e.g. less than 10 infectious units per gram of food).

To apply RT-PCR strategies to food, the viruses must first be separated from the food matrix and concentrated. Treatment of the sample must also take into consideration elements in the food that may inhibit or interfere with reverse transcription or PCR. Viscous samples that are difficult to solubilize, or having to dilute a final RNA sample to perform RT-PCR, may indicate the presence of inhibitors, and the sample may need to undergo further treatment (Sair *et al.*, 2002). Plant food matrices, compared to animal meat, may also be more resistant to elimination of all inhibitors (Sair *et al.*, 2002). Finally, a common problem for all norovirus RT-PCR tests is selecting the primer set that will detect as many genetically diverse norovirus strains as possible (Vinjé *et al.*, 2003).

Shellfish, particularly oysters, have been the most extensively studied of the food vehicles. The approach has been to use RT-PCR and a potentially time-consuming confirmatory test (e.g. nucleic acid hybridization, sequencing) to verify the specificity of amplified product. Typically, several primer sets and several more probes are used to detect noroviruses of a specific genogroup (Le Guyader *et al.*, 1996, 2000). A recent, promising advance in the molecular detection of noroviruses in clinical samples has been the application of real-time RT-PCR, in which both amplification of viral nucleic acid in the relatively conserved 5′ end of ORF2 (encoding capsid protein VP1) and detection of virus-specific products by a reporter dye can be achieved in a quantitative fashion (Kageyama *et al.*, 2003, 2004; Hohne and Schreier, 2004). When this technique was applied to oysters contaminated with 8FIIa (genogroup I) or S35 (genogroup II.4) noroviruses or RNA from oyster samples previously collected (French coast, January 1995–October 2003) and processed, it worked well for genogroup II noroviruses (Loisy *et al.*, 2005). However, the assay for genogroup I noroviruses performed poorly in identifying shellfish that were contaminated with these viruses, and requires improvement. This study also encountered problems with inhibitors in the samples that precluded reliable quantification of results.

RT-PCR can be applied to the evaluation of potentially contaminated groundwater as well. Preparation of the sample, however, must take into account the large volumes of water that must be processed so that a reasonably representative sample concentrated to a suitably small volume can be derived for use in RT-PCR and the effective removal of organic and inorganic inhibitors of enzymatic amplification (Fout *et al.*, 2003). A filtration apparatus outfitted with a positively charged 1MDS filter is recommended for concentrating viruses from large volumes of groundwater. The viral particles on the filter are eluted in a two-stage process that uses a non-flocculating beef extract that in and of itself does not inhibit RT-PCR. Viruses from each eluate are concentrated using a celite (a diatomaceous earth product) method (Dahling and Wright, 1986), which yields approximately 90 % virus recovery (Dahling, 2002). As a first step to remove inhibitors, celite concentrates undergo ultracentrifugation through sucrose in tubes coated with bovine serum albumin (to optimize viral recovery). The viral pellets are resuspended and further treated with a solvent combination of dithiozone and 8-hydroxyquinoline, chelators of heavy metals, in chloroform. Final concentration of the aqueous phase (containing the viral particles), as well as removal of residual solvents and inhibitors less than 100 kDa, is achieved using Microcon-100 filters. Viral RNA can then be extracted for use in RT-PCR. The multiplex format, developed by Fout *et al.* (2003), is a promising technique for detecting a range of enteric viruses, including enterovirus, reovirus, rotavirus, HAV, and noroviruses in groundwater.

For public safety, the challenge remains to develop, refine and make widely available a reliable, rapid and inexpensive diagnostic test to detect the multitude of noroviruses that can potentially contaminate food, water and other environmental samples. Equally important are efforts to improve reporting of outbreaks, as well as sporadic cases, to enable more state laboratories to diagnose norovirus infection rapidly, and to educate those who harvest, prepare and serve food at all levels.

5 Rotaviruses

Although it is the leading cause of severe, acute, childhood diarrhea, group A rotavirus is not considered a major cause of foodborne gastroenteritis. The likelihood that infections caused by rotavirus are foodborne is estimated to be low, in the range of 1 % (Kapikian and Chanock, 1996). In other terms, about 39 000 cases and 500 hospitalizations are estimated to be attributable to foodborne rotavirus in the US annually (Mead *et al.*, 1999).

5.1 Nature of agent

The rotaviruses constitute a genus of the family *Reoviridae*, and are characterized by a genome consisting of 11 segments of double-stranded RNA enclosed in an icosahedral triple-layered capsid (reviewed in Estes, 2001). The innermost layer is composed of primarily of viral protein 2 (VP2), in association VP1 (viral polymerase) and VP3. The middle layer protein, VP6, bridges the space between the inner and outer layers and specifies viral subgroup. The outer capsid is composed of two important viral proteins that each specifies a virus's serotype: VP7, the glycoprotein that forms the smooth outer surface of the viral capsid and determines the G serotype, and VP4, the spike-like surface protein that is protease-sensitive and is the determinant of P serotype. During infection and morphogenesis, nascent double-layered viral particles bud into the endoplasmic reticulum and transiently acquire a lipid envelope. The mature particles released from the infected cells, however, are non-enveloped. Recent evidence suggests that VP7 may play a critical role in excluding lipids from the final infectious viral particle (Lopez *et al.*, 2005).

Rotaviruses were discovered by Bishop and co-workers in 1973 (Bishop *et al.*, 1973), and were named after the spoke-wheeled appearance (*rota* is Latin for wheel) evident by electron microscopy. The diameter of the virions was estimated to be 70 nm initially, but more recent viral reconstructions using the technique of electron cryomicroscopy estimates the diameter of the particle, including the VP4 spike protein, to be 100 nm (Yeager *et al.*, 1990; Prasad *et al.*, 2001).

Seven different serogroups of rotavirus (designated A–G) have been identified, according to VP6 serology (Kohli *et al.*, 1992), RNA genome electropherotype (Saif and Jiang, 1994) and group-specific PCR (Gouvea *et al.*, 1991). Three of these, groups A, B and C, are known to infect humans. Group A rotaviruses are the most common cause of severe diarrhea in young children throughout the world, causing high rates of morbidity and mortality, especially in developing countries. Group B rotaviruses have been reported to cause large waterborne outbreaks of diarrhea among adults, primarily in rural China. Antibodies to human group B rotavirus, however, have been detected in a number of countries throughout the world. In the US, it is estimated that 5 % of the population is seropositive. Group C rotaviruses have been sporadically associated with nausea, vomiting and diarrhea in children, and have been implicated in institutional outbreaks in Japan and the UK, some of which were thought to be foodborne. For an in-depth review of group B and group C rotaviruses, the reader is referred to Mackow (2002). The remainder of this section focuses on group A rotavirus.

5.2 Virulence factors

The pathogenesis of rotavirus diarrhea is likely multifactorial. Rotavirus infection has been localized to the mature enterocytes of the mid-to-tip portions of intestinal villi, causing cell death, villous atrophy and malabsorption. Whether these histological changes account entirely for the severe, watery diarrhea seen during acute infection with group A rotaviruses has been challenged by investigators who observed that the onset of profuse diarrhea occurred before blunting of villi or malabsorption developed in infected animals (reviewed in Estes, 2003). Conversely, in the rabbit model, no diarrhea was seen, despite the presence of histological lesions (Ciarlet *et al.*, 1998). Two other mechanisms have been proposed in recent years.

Rotavirus gene segment 10 that encodes a nonstructural glycoprotein, designated NSP4 (previously called NS28), was identified by reassortment studies as a virulence gene associated with the induction of diarrhea in a piglet model of infection (Hoshino *et al.*, 1995). Initially it was not clear how NSP4, a protein known to be involved in viral morphogenesis, could cause diarrhea. In 1996 another role for NSP4, that of a viral enterotoxin, was reported after it was found that NSP4 and a peptide of NSP4 (amino acid residues 114–135), administered intraperitoneally or intraluminally, caused diarrhea in neonatal suckling mice in a dose-dependent and age-dependent manner, without any associated histological lesions (Ball *et al.*, 1996). The complex functions of NSP4 in rotavirus infection intracellularly and as a secreted protein are summarized in reviews (Estes, 2003; Ramig, 2004).

Rotavirus also has been shown to stimulate intestinal secretion through the enteric nervous system (ENS) (Lundgren *et al.*, 2000). Drugs that have an inhibitory effect on the ENS decreased the host secretory response to rotavirus infection by about two-thirds. The precise mechanism by which ENS activation is triggered by rotavirus is not known at present.

5.3 Environmental persistence

Agricultural use of wastewater for irrigation and sewage sludge for fertilization may contribute to contamination of the environment with viruses and present risks of enteric virus infection when produce is eaten raw (reviewed in Richards, 2001). One foodborne outbreak among the crew of a British military ship in the Northern Arabian Gulf, linked to eating salad, is instructive in this regard (Gallimore *et al.*, 2004). Of note, five different enteric viruses – three noroviruses, one sapovirus (a separate genus of the *Caliciviridae*) and one rotavirus – were recovered from affected individuals (no mixed infections were identified), suggesting that the fresh components of the salad were contaminated during cultivation by wastewater or sludge containing several pathogens.

Rotavirus has been shown to be sensitive to inactivation by a number of chemical treatments (Harakeh and Butler, 1984; Lloyd-Evans *et al.*, 1986; Vaughn *et al.*, 1986, 1987; Chen and Vaughn, 1990), including chloramine-T, chlorhexidine gluconate (as found in Hibiclens), glutaraldehyde, hydrochloric acid, isopropyl alcohol, peracetic acid, povidone iodine, sodium-*o*-benzyl-*p*-chlorphenate, chlorine (at alkaline pH levels), chlorine dioxide, and ozone (at acidic and neutral pH). Combinations, such as *o*-phenylphenol and ethanol in a spray (e.g. Lysol disinfectant spray) and chlorine mixed with a phenolic compound, have also been shown to be effective viral disinfectants (Sattar *et al.*, 1994). Quaternary ammonium compounds have been variably effective (Lloyd-Evans *et al.*, 1986; Sattar *et al.*, 1994). Since handwashing with soap or common disinfectants does not completely inactivate rotavirus (Ansari *et al.*, 1991), using a waterless, alcohol-based hand-cleaning agent is recommended for sanitizing hands.

Disinfecting fresh produce for raw consumption may be more difficult than disinfecting inanimate surfaces. The complex topography and variable porosity of the surfaces of certain foods and the limits on the concentration or types of disinfecting chemicals that can be included in the wash may inhibit elimination of viral contaminants.

Rotavirus and other viruses are also less sensitive than bacteria to inactivation by penetrating ionizing radiation (e.g. gamma irradiation), as well as surface (direct contact) ultraviolet irradiation. The dose of gamma radiation required to achieve 90 % elimination of rotavirus from shellfish was 3.1 kGy or less (Mallet *et al.* 1991). It is not clear whether having 10 % of the initial viral load remaining in the food poses a significant or unacceptable risk of developing gastrointestinal infection. Ultraviolet light at 254 nm decreases the number of viruses on surfaces it contacts, but the ultraviolet dose required for rotavirus inactivation is three to four times that needed for *E. coli* (Chang *et al.*, 1985).

5.4 Reservoirs and transmission

Since many rotavirus infections across the age spectrum occur without symptoms, the affected individuals serve as a reservoir from which infection may be transmitted to others who are susceptible. Infected individuals shed large quantities of rotavirus particles in their stools. In addition, the dose of rotavirus required to induce infection may be as low as one infectious virus particle (i.e. 1 PFU) under optimal conditions, as shown in studies using animal models (Graham *et al.*, 1987). The virus is transmitted person-to-person through the fecal–oral route. Transmission by aerosol is suspected but not proven in humans (reviewed in Dennehy, 2000).

Rotavirus and other enteric viruses have been found in environmental waters that may serve as reservoirs for infections by these pathogens (Mehnert and Stewien, 1993; Pusch *et al.*, 2005). In settings where outbreaks of rotavirus occur (e. g. hospitals, nursing homes and day-care centers), fomites and environmental surfaces may serve as reservoirs for rotavirus (Sattar *et al.*, 1986). Viruses persisted best on non-porous surfaces at 4 °C in conditions of low humidity (~25 %). Porous surfaces, such as cotton-polyester fabric, also supported viral survival, much more so than paper products (including currency).

5.5 Nature of infection in man

The incubation period is 1–3 days, during which time viral shedding may commence. Fever, vomiting and watery diarrhea are the major symptoms in the first 2–3 days of illness, while diarrhea may continue for 5–8 days. Infected children shed in the order of 10^{11} viral particles in each gram of stool (Flewett, 1983). Patients may develop varying degrees of dehydration, usually within the first 3 days of illness. Although rotavirus infection is generally self-limited and symptoms are confined to the gastrointestinal tract, some immunocompromised patients may develop protracted diarrhea with shedding of viruses that have unusual gene rearrangements (Eiden *et al.*, 1985), and others may have infection spread to extraintestinal sites (Gilger *et al.*, 1992). The host and viral factors that promote systemic rotavirus infection are the focus of ongoing studies (reviewed in Ramig, 2004).

In the foodborne outbreak of rotavirus infection among college students (see below; CDC, 2000b), diarrhea and abdominal pain/discomfort were the symptoms most frequently reported (90 % or more) by those who submitted a complete list of symptoms (83 of 85 who had illness that met the case definition). About 80 % experienced loss of appetite, nausea and fatigue; 67 % had vomiting; and 50–60 % noted headache, chills, feverishness or low-grade fever, and myalgia. Illness lasted a median of 4 days (range of 1–8 days). Nine students (11 %) developed dehydration and were given intravenous fluids. In children, serotype G2 rotaviruses tend to induce more severe dehydration with diarrhea (Bern *et al.*, 1992). This outbreak was found to be caused by a serotype G2 rotavirus that has been implicated in other outbreaks of rotavirus gastroenteritis among older children and adults (Griffin *et al.*, 2002; Mikami *et al.*, 2004). The reason for the apparent predominance of G2 rotavirus in this segment of the population is not well understood. Older children and adults are generally expected to have immunity to rotavirus, developed after infection(s) early in childhood, but because G2 strains belong to a genetically distinct genogroup (DS-1) and are less common than the G1, G3 and G4 strains of the Wa genogroup, it is speculated that the levels of G2-specific neutralizing antibodies may be suboptimal in these individuals (Griffin *et al.*, 2002).

5.6 Foods most often involved as vehicles

Deli sandwiches served at a college campus dining hall were implicated as the source of an outbreak of rotavirus gastroenteritis among students and some food-service workers in late March–early April, 2000 (CDC, 2000b). Although none of the employees who assembled and served the sandwiches reported illness, one tested positive for the serotype P[4],G2 rotavirus in the stool specimen submitted. It was not possible to determine whether this employee was infected before the onset of the outbreak. At about the same time (April, 2000), a serotype G2 rotavirus was also implicated as the cause of an outbreak of infection among adults who dined at a restaurant in Japan (National Institute of Infectious Diseases, 2000).

Lettuce, contaminated at the time of harvest, was the suspected vehicle in outbreaks of diarrhea in Costa Rica (Hernández *et al.*, 1997). Rotavirus, detected by enzyme immunoassay and electron microscopy, was found in lettuce sold at farmers' markets during winter months, when the incidence of diarrhea was high. Hepatitis A virus and fecal coliforms and bacteriophages were also present in a number of the lettuce samples.

Waterborne outbreaks of rotavirus illness have also been reported. In one well-studied outbreak that had the features of norovirus outbreaks, failure of effective water treatment at four different levels culminated in community-wide cases of rotavirus gastroenteritis (Hopkins *et al.*, 1984).

5.7 Principles of detection in food and environment

It is well known that rotaviruses are frequently found in environmental samples, and as such may pose a risk of human infection. Of the enteric pathogens, human rotavirus has been deemed 'the most infectious' since ingestion of the virus results in infection of 10–15 % of those so exposed (Gerba *et al.*, 1996), and about 50% of those infected develop symptomatic illness (Rose *et al.*, 1995). It is therefore important to monitor environmental samples, such as shellfish-growing waters and drinking water, for the presence of rotavirus. Given the low infectious dose of rotavirus, detection assays must be extremely sensitive. Methods based on nucleic acid amplification (RT-PCR and NASBA – nucleic acid sequence-based amplification) have been developed to attain this goal (Barardi *et al.*, 1999; Jean *et al.*, 2002; Fout *et al.*, 2003; Parshionikar *et al.*, 2004; Brassard *et al.*, 2005). It should be kept in mind, however, that these techniques do not distinguish infectious from non-infectious virus (Richards, 1999). Pretreatment of samples prepared for RT-PCR, using proteinase K and RNase, has been shown to effectively discriminate between infectious feline calicivirus (FCV), a surrogate for norovirus infection, and FCV inactivated by ultraviolet light, hypochlorite or heat (Nuanualsuwan and Cliver, 2002). Alternatively, since rotavirus can be adapted to grow in cell culture (e.g. CaCo-2 cells), it may also be possible to select infectious rotavirus from environmental samples in this manner, then characterize it further by RT-PCR or immunological reactivity (Bosch *et al.*, 2004).

The behavior of rotavirus (and other enteric viruses) in the field is another important aspect that requires further study. For obvious reasons, it is not possible simply to release infectious virus into the environment for such monitoring studies. Fortunately, recombinant rotavirus-like particles (VLPs) – devoid of viral nucleic acid, but structurally similar to native virus – can be produced in milligram quantities and released in the environment as non-infectious surrogates to track rotavirus survival and response to virus inactivation practices (Caballero *et al.*, 2004; Loisy *et al.*, 2004). Initial studies have indicated that rotavirus VLPs composed of VP2 and VP6 (VLP2/6) are as stable as authentic viral particles in seawater (Loisy *et al.*, 2004). In a separate study, VLP2/6, fluorescent VLP2/6 (with green fluorescent protein (GFP) inserted at the amino terminus of VP2 to allow detection by flow cytometry) and a pseudovirus (VLP2/6 containing a heterologous RNA that can be amplified by RT-PCR to assess surrogate decay) were evaluated (Caballero *et al.*, 2004). The GFP-VLP and pseudovirus were more stable than intact viral particles, suggesting that they may provide a more conservative estimate of the behavior of the virus in the environment and may be superior to models based on bacteriophages or unrelated, non-enteric viruses.

6 Enteroviruses

6.1 Nature of agent

The enteroviruses are a genus in the family *Picornaviridae* and are therefore similar in many ways to HAV, described in section 2. The virions comprise a single, plus-sense strand of RNA, coated with a capsid made up of 60 replicates of each of 4 polypeptides (Racaniello, 2001). The overall diameter of the roughly spherical particle is approximately 28 nm.

The earliest of these agents recognized were the three serotypes of polioviruses, which cause paralytic poliomyelitis (Plotkin *et al.*, 1962). These infect humans and other primates. In an unsuccessful attempt to adapt the polioviruses to multiply in intracerebrally inoculated mice, the coxsackieviruses (named for the city of Coxsackie in the state of New York, US) were detected (Oberste *et al.*, 2003). When cell-culture methods for propagation of the polioviruses became available, similar agents that did not cause poliomyelitis were discovered and named enteric, cytopathogenic, human orphan (for want of association with a specific disease) viruses, or echoviruses (Hsiung, 1962; Wenner, 1962). Finally, some of the later-discovered human enteroviruses were simply called enteroviruses, as the classifications into the poliovirus, coxsackievirus and echovirus groups had become contentious. Enteroviruses are by no means limited to humans; they have counterparts that infect cattle (Cliver and Bohl, 1962), swine (Betts *et al.*, 1962), etc. However, none of the human enteroviruses is known with certainty to infect non-primates, and none of the enteroviruses of other animals has been shown to infect humans.

The genetic organization of the enteroviruses, typified by poliovirus 1, has been known for some time (Dewalt and Semler, 1989). The positive-sense RNA has a small protein initiator, designated VPg, attached to a 5′-nontranslated region comprising 743 of the total 7441 nucleotides. The central region, which is translated as a single polyprotein, ends with nucleotide 7370 and is followed by another nontranslated region ending in a poly-A tail. The polyprotein comprises three major segments: P1, which comprises the four capsid polypeptides, and P2 and P3, which code for other proteins (protease to cleave the polyprotein, RNA-dependent RNA polymerase, VPg, etc.) that are essential to replication of the virus by the host cell. P1 gives rise to capsid proteins VP0 (cleaved finally into VP4 and VP2), VP3 and VP1. Great progress in genetic analysis, etc., has been made using the polymerase chain reaction (PCR); since these are RNA viruses, reverse transcription (RT) is required to provide the DNA used in PCR.

Although the approximately 70 serotypes of human enteroviruses have traditionally been identified serologically, typing by genetic sequencing is now reported (Oberste *et al.*, 2003). A set of primers for a highly conserved region of the enterovirus genome enables detection of all of the human enteroviruses (Andreoletti *et al.*, 1996), and sequencing of the amplicon affords an apparently straightforward typing method. Because there is intratypic variability in the sequences of the 5′-nontranslated region, random fragment length polymorphism (RFLP) can analyze transcripts of this region for strain differentiation (e.g. in epidemiological investigations; Siafakas *et al.*, 2003).

6.2 Virulence factors

All of the human enteroviruses are presumed capable of causing disease in man, although pathogenesis has not been proven for every type. Perhaps most importantly, the three polioviruses, which are the most virulent of the enteroviruses, have all been attenuated for use as 'live-virus' vaccines (Sabin, 1965), showing that viral virulence is subject to genetic manipulation by humans. This outstanding accomplishment of many years ago contributed greatly to the eradication of poliomyelitis from much of the world; total eradication seems attainable. Now that knowledge of the genetics of these viruses is much more sophisticated, attenuation would be undertaken very differently if done today, but it might not be done at all. Although all of the enteroviruses infect the intestine and are shed in feces, it is noteworthy that they generally do not cause diarrhea. Pathogenesis most often involves the central nervous system (poliomyelitis, but also meningitis from coxsackieviruses and some echoviruses), but other tissues and organs are sometimes involved. Tissue tropisms are evidently related to the presence of receptors on cell membranes, to which the virus can attach. Obviously it matters whether the virus can gain access to the cells with receptors, as viruses have no means of transporting themselves through the body. Once the virus has attached to a cell receptor and been engulfed, the apparatus needed to replicate the virus is probably present in the cytoplasms of most cells, including cells not of primate origin. A cell in which virus is replicating may be stopped from producing messenger RNA on its own DNA templates, whereby death of the cell eventually results. If this process were to continue unchecked until all susceptible cells were killed, the host organism might well die, yet few infections with enteric viruses are lethal. This indicates the effectiveness of the host defense mechanisms that will be discussed below.

6.3 Environmental persistence

Like other viruses, enteroviruses cannot multiply outside the host, but can only persist or be inactivated. In the absence of human intervention, enteric viruses can persist for very long times at temperatures from 0 °C to 37 °C (Nuanualsuwan and Cliver, 2003). However, inactivation takes place at all temperatures above 0 °C at some rate (Salo and Cliver, 1976), and resident bacteria may hasten the process, apparently by specific enzymatic action (Cliver and Herrmann, 1972; Fujioka *et al.*, 1980; Toranzo *et al.*, 1983; Deng and Cliver, 1992, 1995b). Drying on surfaces inactivates many enteric viruses, although HAV, discussed earlier, is less affected. Because enteroviruses lack a lipid envelope, they are not inactivated by organic solvents; chemical inactivation is usually done with strong oxidants, such as chlorine, ozone, chlorine dioxide, etc. (Engelbrecht *et al.*, 1980; Jensen *et al.*, 1980; Young and Sharp, 1985; Tree *et al.*, 1997). Viruses in food or water are readily inactivated by heating and, in clear water or on food surfaces, are also susceptible to chlorination and to ultraviolet light (Gerba *et al.*, 2002; Nuanualsuwan and Cliver, 2002; Nuanualsuwan *et al.*, 2002). Freezing preserves enteroviruses. They are generally more susceptible to alkaline than acid conditions (alkalinity is rare in foods) (Grabow *et al.*, 1978;

Derbyshire and Brown, 1979) and acidity is tolerated for limited periods, after which inactivation will occur at a rate depending on temperature (Salo and Cliver, 1976).

6.4 Reservoirs and transmission

Although close analogs of human enteroviruses occur in other species, there is no indication of a non-human reservoir for any of the enteroviruses that infect humans. Therefore, transmission is from person to person – directly or indirectly. Virus shed in feces may be passed to other people via unwashed hands or other carriers of fecal residue, including fomites and mechanical vectors. Food sometimes becomes a vehicle of virus transmission as a result of fecal contamination. However, the bulk of human feces in developed countries is disposed via the water-carriage toilet and therefore occurs in sewage. The result is that the bulk of virus (not only enteroviruses) shed in feces passes into wastewater, in the custody of the system that will, hopefully, treat and disinfect the sewage before it impacts the lives of the general public. This outcome is likely if the sewers are properly maintained and operated, and if the sewage treatment facility is properly designed and operated. Although there are numerous opportunities for failure, the system works remarkably well most of the time – so that alternative means of feces disposal probably represent a greater risk than the systems that are currently in place. However, communities that lack the quantity of water needed for water-carriage waste disposal are obliged to use whatever means are at hand.

6.5 Nature of infection in man

Enteroviruses infect after being ingested. There are apparently some virus receptors on the tonsils, but best indications are that infection usually begins in the small intestine (usually the ileum), which means that the virus must withstand stomach acidity and the array of duodenal digestive agents, notably proteases, to reach the susceptible intestinal lining. Much study has been devoted to the interaction of enteroviruses with receptors on cells in culture, but relatively little to how viruses attach and infect the intestinal mucosa (Heinz *et al.*, 1987; Heinz and Cliver, 1988). In any case, it is clear that virus produced in the infected mucosa is transported via the lymph and blood for at least some period before establishing secondary infections in other organs and tissues. Periods of viremia are apparently brief in enterovirus infections, hence it has not been found necessary to screen donated blood for these agents, as is done, say, for hepatitis B virus. From the viremia stage, the race is on between infection and killing of cells in some functional organ and a protective response based either on antibody from pre-existing immunity or on interferon. Loss or impairment of the function of an infected organ is what gives rise to symptoms. Fever is common and may exert some antiviral effect, in that the polioviruses developed for use in attenuated vaccines were selected primarily on the basis of their inability to multiply at temperatures significantly above 37 °C. A host's first infection with a given virus is likely to be limited by production of interferon, whereas longer-term immunity results from the body's production of antibody that reacts with the viral protein coat.

The early systemic response is production of antibody of the IgM class, whereas long-term immunity is mediated by IgG-class antibody. Therefore, the presence of IgM-class antibody against a virus in a serum sample indicates that the donor is currently or has recently been infected by that virus. Induction of protective antibody production before the infection, by means of vaccination, has been the essence of the approach to eradicating poliomyelitis. How the antibody protects, or limits virus infection, is not entirely known. Pooled human immune serum globulin can be used for passive protection of at-risk people in some situations, assuming that the collective immunological experience of the donor population leads to a protective level of antibody against the virus in question. Shedding of virus from the intestine in feces often begins during the incubation period and may continue during convalescence. Virus in feces may be contained in packets comprising plasma membranes of the infected host cells, and may also be neutralized by IgA that is called coproantibody (i.e. antibody that occurs in feces). Neutralization of fecal virus by coproantibody may complicate detection of the virus, but it does not prevent infection if someone ingests the virus, in that the antibody is readily digested off the virus, restoring the virus to infectivity (Kostenbader and Cliver, 1986). It is likely that some coproantibody-neutralized virus is also restored to infectivity by the action of microbial enzymes in the environment. With the exception of paralytic poliomyelitis, most enterovirus illnesses leave no permanent sequelae; unfortunately, it is being observed that some of those who appeared to have recovered from poliomyelitis paralysis are subject to insidious relapses years later.

6.6 Foods most often involved as vehicles

As with some of the viruses discussed earlier, bivalve mollusks are clearly the most frequent vehicles. Although outbreaks of enteroviral illness associated with mollusks have not been described, the enteroviruses have been used frequently as models in studies of accumulation, depuration and extraction (Liu *et al.*, 1967; Landry *et al.*, 1982). Other candidate vehicles include any food that is touched by fecally soiled hands, and water that is contaminated with human waste. It seems likely that water has been a vehicle in some community outbreaks of poliomyelitis.

6.7 Principles of detection in food and environment

Most, but not all, of the human enteroviruses can replicate in cell cultures and produce cytopathic effects or plaques to denote their presence. Therefore, infection of cell cultures was the basic method of detection of viruses in food and water (Herrmann and Cliver, 1968; Cliver and Grindrod, 1969; Kostenbader and Cliver, 1977). The challenges in developing detection methods were in liquefaction of solid samples as necessary and concentration of the virus from liquid suspension. No single cell culture line or type is susceptible to all of the enteroviruses that can infect cell cultures (Richards and Cliver, 2001), so either multiple lines must be used or a single, broad-spectrum line (such as the BGM line) be trusted (Morris, 1985). Extraction and concentration methods are numerous (Kostenbader and Cliver, 1972;

Richards and Cliver, 2001); indeed, the enteroviruses have served principally as model agents for development of extraction methods that can be applied to the viruses that are more frequently foodborne but are not detectable in cell cultures. There is no analog of the enrichment methods so often used in detecting bacteria in foods, so concentration of virus from samples is critical (Cliver and Yeatman, 1965; Cliver, 1967; Herrmann and Cliver, 1968; Fattal *et al.*, 1976; Logan *et al.*, 1981; Hurst *et al.*, 1984; López-Sabater *et al.*, 1997; Katayama *et al.*, 2002). If cell cultures are to serve as the basis for viral detection, it is also necessary to eliminate bacterial contaminants by chemical treatment, filtration or antibiotics. Alternately, the human enteroviruses can be detected by RT-PCR, as can essentially all of the other enteric viruses (Chung *et al.*, 1996; López-Sabater *et al.*, 1997; Rosenfield and Jaykus, 1999; Casas and Sunen, 2001; Fout *et al.*, 2003); however, the RT-PCR test will give a positive result with inactivated virus, unless precautions are taken (Nuanualsuwan and Cliver, 2002). Fortuitously, a primer set has been devised that reacts with all of the human enteroviruses (Andreoletti *et al.*, 1996), so that a broad-spectrum test is available. It happens that bovine enteroviruses also produce an amplicon with these primers; further, bovine enteroviruses infect monkey kidney cell cultures as if they were of human origin, although human enteroviruses do not infect bovine cells as the bovine enteroviruses do. Clearly, a positive test result with either cell culture or PCR leads to a need to identify the agent that has been detected. The identity of a virus is ultimately based on reactions of its antigens with antibodies, but molecular methods of identification are now available for human enteroviruses (Oberste *et al.*, 2003).

7 Other viruses

7.1 Astroviruses

At least seven, possibly eight, serotypes of astroviruses are known to infect humans (Wilhelmi *et al.*, 2003). Other serotypes of astroviruses are infectious for animals, but apparently not for humans (Appleton, 1994; Glass *et al.*, 1996). *Astrovirus* is the only genus in the *Astroviridae*, a family said to be distinct from other viruses (Monroe *et al.*, 1993). The viral particles have typically been seen by electron microscopy as 28 nm in diameter, with a five- or six-pointed star pattern visible by negative staining; this may have been a preparation artifact, as it seems that the native virus is ~41 nm in diameter, with a layer of surface spikes (Risco *et al.*, 1995). The particles contain plus-sense, single-stranded RNA about 7500 nucleotides in length (Wilhelmi *et al.*, 2003).

The mode of pathogenesis is not firmly established, but the virus appears to infect mature enterocytes of the intestinal lining (Wilhelmi *et al.*, 2003). After a 3- to 4-day incubation period, the infected person may show diarrhea, fever, headache, malaise, nausea and occasional vomiting (Appleton, 1994). Diarrhea, with shedding of the virus, usually lasts 2–3 days, but may extend to 1–2 weeks. Astrovirus diarrhea most commonly affects young children, less commonly people > 65 years old, and still less commonly younger adults (Glass *et al.*, 1996). Transmission via food and water is apparently relatively uncommon, but astrovirus has been implicated in a foodborne outbreak involving over 4700 people in Japan (Oishi *et al.*, 1994) and has been detected in the feces of ill people who had eaten raw oysters (Yamashita *et al.*, 1991).

Although detection is most often by RT-PCR (Marx *et al.*, 1997), a method that entails inoculation of the CaCo-2 human cell line followed by RT-PCR allows determination of whether the virus is infectious when detected (Pintó *et al.*, 1996). By this method, artificially inoculated astrovirus was shown to persist in drinking water for perhaps 3 months, at diminishing levels, at room temperature (Abad *et al.*, 1997b). Astroviruses are also said to withstand pH 3, exposure to lipid solvents, and heating at 50 °C for 30 minutes (Appleton, 1994). The virus titer falls 3 $\log_{10}$ in 5 minutes and 6 $\log_{10}$ in 15 minutes at 60 °C. Procedures are available for recovery of astroviruses from shellfish (Legeay *et al.*, 2000).

7.2 Tickborne encephalitis viruses

The tickborne encephalitis viruses (TBE) are a widespread group that has important impacts on human health. The focus in the present section is on infections of dairy animals that lead to human infections via milk and milk products. Wider ranging reviews of the group, from which most of the information that follows derives, are available (Grešíková, 1972, 1994).

The TBE viruses belong to the genus *Flavivirus* in the family *Flaviviridae*. They are spherical, with diameters of 45–60 nm, and contain a single strand of plus-sense RNA. There are three structural proteins in the nucleocapsid, which is surrounded by a lipid bilayer. Procedures used to prepare the viruses for electron microscopy evidently produce a variety of distortions, as is true of many viruses with lipid envelopes. The genomes of two of the TBE agents, the Russian spring–summer encephalitis virus (Pletnev *et al.*, 1986) and the central European encephalitis virus (Mandl *et al.*, 1988), have been published.

Whether they infect via a tick bite or by ingestion with contaminated milk or milk product, these viruses produce encephalitis with various accompanying symptoms. The central European variety is said to be less severe than the Russian spring–summer variety, but all cause significant, prolonged debility. The TBE viruses that have been implicated in milkborne outbreaks are said to be more acid-resistant than most flaviviruses (Grešíková, 1994).

The TBE viruses are apparently not persistent in the environment. Rather, they persist by infecting small reservoir animals, such as rodents (field mice and voles) and insectivores (shrews and moles). The virus also multiplies in its tick vectors, producing persistent infections that can be transmitted sexually and transovarially among ticks, as well as to warm-blooded animals by bites.

The animal reservoirs of the TBE are small rodents and insectivores, as stated above. Ticks that become infected by biting these reservoir animals serve as biological vectors. Especially in central Europe, dairy animals are at risk of being bitten by the infected ticks (particularly *Ixodes ricinus*) and acquiring inapparent infections. Bites on the udder are said to be especially prevalent. All dairy species in the area (cattle, sheep and goats) have been shown experimentally to be susceptible; however, the terrain in the focal areas is probably more conducive to grazing by goats and sheep than by cattle. Shedding of virus in the milk of the infected animal begins about 2 days after inoculation, and may continue until day 7.

As stated above, central nervous system infections with these agents are often severe in humans. Vaccination is a more important mode of prevention than milk pasteurization, in that more infections are probably contracted from tick bites than from ingestion of virus-containing milk. However, people in urban areas, who are at limited risk of tick bite, are best protected by pasteurization of milk.

Outbreaks attributed to drinking raw goats' milk have been reported at least from the Rožňava natural focus in Slovakia (Grešíková, 1972), Moscow (Drozdov, 1959), Leningrad (Smorodintsev, 1958), and the Styrian region of Austria (Van Tongeren, 1955). The virus is apparently inactivated by pasteurization, but can persist for long periods in refrigerated milk products (Grešíková-Kohútová, 1959a, 1959b, 1959c). Fresh cheese made from sheep's milk was implicated in an outbreak in Slovakia (Grešíková *et al.*, 1975).

The TBE viruses are said to propagate in cells cultured from several mammalian and avian species, and to form plaques in some (Grešíková, 1994). However, recoveries of TBE viruses from milk and other environmental sources are typically expressed in mouse intracranial LD_{50} (Grešíková, 1972). It appears that isolation of these viruses from such sources is of principally research interest. As is true with viruses in general, means of prevention (for TBE, vaccination of people and dairy animals, and pasteurization of milk) do not require laboratory testing.

7.3 Coronaviruses

The coronaviruses have an RNA genome coated with protein and a lipid-containing envelope (Lai and Holmes, 2001). Their diameter is 80–120 nm. Most are transmitted by the respiratory route, and few infect humans. They are sometimes regarded as causes of diarrhea in humans, but their role is not well proven (Wilhelmi *et al.*, 2003). There has been one outbreak of ostensibly foodborne coronavirus disease recorded at a sports club in Scotland (Communicable Diseases and Environmental Health Scotland Weekly Report, 1991), and others certainly may have occurred. The sudden acute respiratory syndrome (SARS) coronavirus that dominated news in 2003 may have been foodborne on occasion, in association with wild animals such as civet cats that were eaten in China (Enserink, 2003; Martina *et al.*, 2003; Xu *et al.*, 2004), but this association has really not been proven (Bell *et al.*, 2004).

7.4 Adenoviruses

The adenoviruses contain double-stranded DNA and have a protein coat without lipid envelope (Shenk, 2001; Wilhelmi *et al.*, 2003). Their diameter is 60–90 nm. The majority of adenoviruses are transmitted by the respiratory route, and are not associated with food or water. However, serotypes 40 and 41 are known to be transmitted via a fecal–oral cycle and to cause gastroenteritis (Vizzi *et al.*, 1996). Even these serotypes are not clearly associated with foodborne or waterborne disease (Fleet *et al.*, 2000).

7.5 Reoviruses

These are the type genus of the family *Reoviridae*, which includes the rotaviruses discussed earlier (Nibert and Schiff, 2001). They have a segmented genome of double-stranded RNA and a double coat of protein that becomes infectious after the outer coat is removed by proteolytic action (Gibbs and Cliver, 1965). Their diameter is perhaps 80 nm. Although the reoviruses are widespread in the environment (Matsuura *et al.*, 1988; Pianetti *et al.*, 2000; Spinner and Di Giovanni, 2001) and may occur in food, they are not associated with overt illness in humans and are therefore not regarded as foodborne disease agents.

7.6 West Nile virus

The West Nile virus is a flavivirus, with an RNA genome and an enveloped, spherical capsid of 40–60 nm diameter (Lindenbach and Rice, 2001). It was first reported in the Middle East and arrived in North America in 1999; it is typically transmitted via mosquito bites and, less frequently, potentially by blood transfusion (Hinson and Tyor, 2001). However, there has been one instance reported in which a mother, having received two units of blood in conjunction with her delivery, apparently transmitted the virus to her infant via her breast milk (CDC, 2002c). Authentication of such events is extremely difficult.

7.7 Nipah virus

Nipah virus originally caused an outbreak of human illness among pig farmers in Malaysia; the contact was evidently occupational, and no transmission via food was described (Hinson and Tyor, 2001). More recently, this encephalitis virus has been shown to have a reservoir in fruit bats in Bangladesh (*Independent*, 2004). Incidence is highly seasonal, as it appears that only pregnant bats are infected, and the breeding season is winter. People who eat fruit or drink fruit-tree saps contaminated by the urine, feces or saliva of infected bats are subject to potentially fatal infection.

8 Surrogates

As has been shown in previous sections, testing food or water for viral contaminants is difficult, slow and costly. In that the majority of these viruses emanate from the human intestines, it is natural that indicators of fecal contamination have been considered as surrogates for the viruses themselves in test procedures. Various surrogates have been evaluated in this context; some show at least limited promise, depending on how they are to be applied. Surrogates may be used to indicate the probability that human virus is present in a sample of food or water, but also to assess the adequacy of a process step to remove or inactivate foodborne or waterborne virus. The requirements for these purposes are very different.

8.1 Indicators of probable virus contamination

Viruses will be present in a sample of food or water if fecal contamination has occurred, either directly or indirectly. Therefore, surrogates may be selected to facilitate testing food or water samples and to obviate testing for the viruses themselves. To serve this purpose, the surrogate should always be present if virus may be present, and ideally always absent if viral contamination has not occurred. Indicators of fecal contamination are legion, as will be discussed in more detail below. However, most human feces do not contain viruses, so indicators of fecal contamination *per se* are likely to overestimate viral contamination. Another problem is that most indicators of fecal contamination are not specific to human feces, so fecal contamination that is unlikely to introduce human viruses may give a positive test result.

8.2 Assessment of antiviral processes

Processes applied to food or water that are intended to enhance safety, or even palatability, may inactivate virus (Mariam and Cliver, 2000; Strazynski *et al.*, 2002). In such circumstances, a surrogate that is harmless may be used to determine whether virus would be removed (Caballero *et al.*, 2004) or inactivated by the process. Because the surrogate is likely to be inoculated into the food or water, specificity is not at issue. What is important is that the surrogate be at least as difficult to remove or inactivate as is the virus of concern.

8.3 Bacterial surrogates

The rationale for using bacteria as surrogates in this context is usually that certain bacteria would inevitably have been present in the feces in which the virus was shed. The prototype fecal bacterium is *Escherichia coli*, which is present in the feces of all warm-blooded animals. This is normal flora; the great majority of strains are nonpathogenic, but there are exceptions (see Chapter 6). Some salient taxonomic features are: Gram-negative short rods, non-spore-forming, aerobic or facultative. In recent times, rapid tests for *E. coli* have used 4-methylumbelliferyl-β-D-glucuronide (MUG), which, acted upon by *E. coli* β-glucuronidase, produces a fluorogen that fluoresces bright blue under UV excitation (Greenberg *et al.*, 1992, Section 9221 F). Over 90 % of *E. coli* strains are MUG-positive, but most strains of enterohemorrhagic *E. coli* (see Chapter 6) are MUG-negative.

Before the advent of the MUG test, identification of *E. coli* in food or water samples was laborious, so a broader group of bacteria, called 'coliforms' (resembling *E. coli*), was defined and sought. These are all Gram-negative short rods, non-spore-forming, aerobic or facultative, like *E. coli*; they produce acid and gas from lactose within 48 hours at 35 °C. The key enzyme in lactose metabolism is β-galactosidase, which hydrolyzes lactose into its constituent monosaccharides, glucose and galactose. A simple test for β-D-galactosidase yields a colored product from ortho-nitrophenyl-β-D-galactopyranoside or chlorophenol red-β-D-galactopyranoside, and is now the basis for a simplified coliform test (Greenberg *et al.*, 1992, Section 9223). The coliform group subsumes at least four genera and many species that are not necessarily of fecal origin, so the presence of coliforms has sanitary significance in drinking water or pasteurized milk but would usually not be associated with viral contamination. A more restrictive group is the thermotolerant (or fecal) coliforms, consisting largely of the genera *Escherichia* and *Klebsiella*. These produce acid from lactose within 48 hours at 44.5 °C. The association with fecal contamination is somewhat closer, but high levels of environmental *Klebsiella* have been recorded in some instances, so the association may be tenuous.

Perhaps most importantly, although *E. coli* is always of fecal origin, it is by no means exclusively of *human* fecal origin. Since virtually all viruses transmissible to humans via food and water are human-specific, the majority of *E. coli* detections would not be indicative of a human viral risk. Another concern is the ability of *E. coli* to multiply outside the host under some circumstances; this is a property that all viruses lack.

Other bacterial surrogates have been proposed in certain applications, including *Clostridium perfringens* (Payment *et al.*, 1988) and *Bacteroides fragilis. C. perfringens* offers greater environmental stability than *E. coli*, whereas *B. fragilis* is more closely associated with human fecal contamination and, as a strict anaerobe, is unlikely to multiply outside the host. Each of these has its place in certain situations, but both are more difficult to culture and detect from food or water than are *E. coli*. In fact, bacteriophages of *B. fragilis* (see below) are generally sought, rather than the host bacterium, to yield a more reproducible test.

8.4 Viral surrogates

Detecting viruses is never really easy, but the human enteroviruses (see section 5), including the vaccine polioviruses, are much easier to detect than the noroviruses (see section 3) or HAV (see section 2). Tests using broad-spectrum cell culture lines such as BGM have been available for years. Problems are that some laboratories that can now detect viruses by molecular methods lack a cell-culture capability, and that cell-culture testing is more costly and exacting than tests for bacterial surrogates. Additionally, an infected person seldom sheds more than one enteric virus at a time, so the presence of enteroviruses in food or water is more likely to indicate contamination by community sewage or some other pooled fecal source than direct fecal contamination by, say, a food worker. In certain contexts a positive test may result from the presence of *bovine* enterovirus, but other non-human viruses seldom express themselves in primate cultures such as BGM. Most importantly, infection of a cell culture shows that the sample source had not been disinfected, whereas molecular methods can give positive results with inactivated virus.

Surrogates for assessment of antiviral processes have long included agents such as vaccine polioviruses and the cell-culture-adapted strain of HAV. With the recognition that noroviruses were major causes of foodborne disease, the inability of these agents to replicate in cell culture led many laboratories to use a feline calicivirus as a surrogate (Slomka and Appleton, 1998). More recently, a non-human norovirus has been identified that is cytopathic in cell culture (Karst *et al.*, 2003; see section 4.2).

8.5 Bacteriophages

Another class of surrogates that has been proposed is bacteriophages – specifically, viruses that infect enteric bacteria. Three classes have been most studied: somatic coliphages, male-specific coliphages, and phages of *B. fragilis*. Each has its assets and liabilities, and none is an all-purpose surrogate.

8.5.1 Somatic coliphages

Somatic coliphages are viruses that infect *E. coli* via receptors on the cell wall. Although they infect *E. coli* by definition, some may have broader host ranges. A wide variety of phages meet the definition; some are near the sizes of picornaviruses and noroviruses, whereas others are much larger. There are standard tests for somatic coliphages in water, and some of these could be applied to foods. These tests use a standard strain of *E. coli* as host, may be incubated at ambient temperature in some instances, and can be read in 24 hours or less. Negative aspects are that the phages can replicate in *E. coli* in the environment (i.e. outside a host animal) and that they are generally no more host-specific than is *E. coli* itself.

8.5.2 Male-specific coliphages

Male-specific coliphages infect only 'male' strains of *E. coli* via the F-pilus. Because of special conditions regarding the genetics of the F-pilus and its expression, most male-specific coliphages evidently cannot multiply in the environment (Woody and Cliver, 1995, 1997). The host cells must be incubated at temperatures above 30 °C for the pilus to form, so ambient incubation of the test (an advantage for applications in developing countries) is not an option. Of four specificity groups, two are said to be most associated with human origin. A special host strain has been produced by inserting the gene for production of the *E. coli* f-pilus into a strain of *Salmonella* Typhimurium, to avoid confusion when somatic coliphages are present in environmental samples. This strain can be infected by somatic phages of *Salmonella*, but these are said to be much less prevalent than phages of *E. coli*.

8.5.3 Phages of *B. fragilis*

Phages of *B. fragilis* are supposed to represent an order of specificity comparable to that of their host bacterium. To the extent that *B. fragilis* is more typical than *E. coli* of human feces, the same is true of the phage. Surveys have been done of the relative prevalence of this group, and it looks useful in some situations. Inevitably, the need to incubate the host cultures anaerobically imposes some additional inconvenience compared to testing in *E. coli*.

8.6 Applications

A longstanding use of surrogate testing has been monitoring waters from which shellfish are harvested for the presence of fecal bacteria. However, this application began out of concern for typhoid (a bacterial disease caused by *Salmonella typhi*), so it was not primarily directed to predicting viral contamination of the shellfish.

More recently, association of fecal bacteria with viruses in shellfish has been more carefully assessed, and the results are not favorable. Shellfish beds are often subject to sewage contamination, whereby it might be expected that indicators would be present along with the viruses, but bacterial indicators have not met expectations. Surveys of enteroviruses (Shieh *et al.*, 2003; Dubois *et al.*, 2004) and of bacteriophages (Chung *et al.*, 1998; Muniain-Mujika *et al.*, 2000) in shellfish have been reported, but neither of these surrogates is yet used in any official test method for shellfish.

Surrogates (coliform bacteria) are commonly used in verifying the safety of drinking water and may be of some value in demonstrating that viruses are not present in this context. US communities that derive their drinking water from surface sources are expected to test the raw water periodically for viruses that express themselves in cell cultures; the presence of such viruses is presumably indicative of the presence of non-cytopathic viruses, such as noroviruses. Methods for detecting or quantifying bacteriophages in water have been standardized, and are used in various ways.

Where virus inactivation is being studied, the use of surrogates is attractive – especially if the concern is for viruses that do not kill cells in culture. Poliovirus has been a surrogate for other viruses for evaluating inactivation in food processing (Strazynski *et al.*, 2002). Both feline caliciviruses (Nuanualsuwan and Cliver, 2002; Nuanualsuwan *et al.*, 2002; Allwood *et al.*, 2004b; Hewitt and Greening, 2004) and male-specific coliphages (Mariam and Cliver, 2000; Nuanualsuwan *et al.*, 2002; Allwood *et al.*, 2003; Allwood *et al.*, 2004b; De Roda Husman *et al.*, 2004) have been used in inactivation studies, compared to viral pathogens.

Vaccine poliovirus shed by an infected infant was a surrogate in finger disinfection studies (Cliver and Kostenbader, 1984). Feline calicivirus has been the surrogate for norovirus used in studies of virus removal by handwashing (Bidawid *et al.*, 2004). Removal of viruses from beneath fingernails by various handwashing techniques has been studied using feline caliciviruses and non-pathogenic *E. coli* (Lin *et al.*, 2003). These same surrogates, along with F-specific coliphages, have been used in studies of virus removal from leafy salad vegetables (Allwood *et al.*, 2004b); and F-specific coliphages and *E. coli* have been evaluated as predictors of the presence of noroviruses on market produce (Allwood *et al.*, 2004c).

9 Prions

9.1 Nature of agent

Several transmissible spongiform encephalopathies (TSEs) are known (Johnson and Gibbs, 1998). Scrapie in sheep and goats has been recognized for at least 200 years. Several TSEs have been known to affect man, most notably Creutzfeldt-Jakob disease (CJD), but also Gerstmann-Straussler-Scheinker syndrome, fatal familial insomnia, and kuru. Classic CJD (cCJD) is known to occur in a familial (inherited) form (fCJD), an iatrogenic form (iCJD) that is transmitted by transplantation of tissue from an incubating case of CJD or a cadaver to a susceptible person or by contaminated surgical instruments, and a sporadic form (sCJD) that is seen in approximately one person per million worldwide, regardless of suspected risk factors. Kuru might be regarded as foodborne, in that it was evidently transmitted by ritual cannibalism in a small community in Papua New Guinea and was eradicated when the practice was prohibited.

TSEs that affected other, non-human species were transmissible mink encephalopathy (TME) and chronic wasting disease (CWD) of North American deer and elk. In *ca.* 1985, cattle in the UK began to be stricken with a disease that was named bovine spongiform encephalopathy (BSE, also 'mad cow disease'), followed by an outbreak of a similar disease in zoo ungulates and of feline spongiform encephalopathy (FSE) in cats and zoo felids, also in the UK. In 1994, a new TSE in humans, called new variant CJD (vCJD), was recognized in the UK (Will *et al.*, 1996). Since then BSE has occurred in several other countries (mostly in Europe), and vCJD has occurred in a few persons outside the UK, though some of these seem to have been in people who contracted the disease while living in the UK. FSE and vCJD were perceived as resulting from transmission of the agent of BSE to cats and humans, respectively. Homology of the BSE and vCJD agents has been established (Schonberger, 1998; Brown *et al.*, 2001a).

The causation of scrapie was already under study when BSE arose. The new disease proved similar, except that long observation had shown that scrapie (which has many different 'strains') was transmissible only among sheep and goats, whereas BSE was eventually seen to 'jump the species barrier'. There is still debate regarding the etiology of the TSEs, but a powerful consensus favors the abnormal 'prion' as the agent of disease. Prions are relatively low molecular weight peptides that appear to occur naturally in many animal species. Their normal function is essentially unknown (Erdtmann and Sivitz, 2004), but some are susceptible to refolding and aggregation that render them resistant to heat and to proteases, whereupon they accumulate in various tissues, depending on the host species.

Prions in the normal, nascently folded configuration may be called 'PrP^C', whereas the resistant form is called 'PrP^{Sc}'(for scrapie). This change appears to entail conversion of portions of the molecule from α-helices to β-strands (Zahn *et al.*, 2000; Wille *et al.*, 2002). In time, the accumulation of PrP^{Sc} in the central nervous system, particularly the brain, leads to structural and functional abnormalities that are ultimately fatal, as no treatment exists to date (Erdtmann and Sivitz, 2004). The amino acid sequence of a prion is genetically determined by the species, and to a lesser degree by the individual (i.e. there are inherited differences in the amino acid sequence of prions within a species). The folded configurations of PrP^C appear to be similar but not identical among species, as are the abnormal folded configurations of PrP^{Sc}, despite differences in amino acid sequences. Therefore, it is not clear why human PrP^C is apparently subject to convert to PrP^{Sc} under the influence of bovine PrP^{Sc}, but not of sheep PrP^{Sc}. The perception is that tertiary structural features of the PrP^C, yet to be defined, allow bovine PrP^{Sc} but not sheep PrP^{Sc} to cause this conversion to human PrP^{Sc}, and eventually cause vCJD. A further complication is that prions have two glycosylation sites each, so they may have zero, one or two glycosyl residues attached. A great many possible configurations of these residues have been demonstrated (Baldwin, 2001); glycosylation is evidently not required to convert PrP^{C} to PrP^{Sc}, but may play a role in *in vivo* pathogenesis. Conversion to PrP^{Sc} probably also involves participation by auxiliary proteins, some of which have been called 'chaperonins', that govern refolding (King *et al.*, 2002). The newly formed PrP^{Sc} apparently forms aggregates that resist dispersion and suspension in aqueous solvents.

PrP^{Sc} is extremely heat resistant, whereby it would be expected to persist in animal tissues after rendering and has been shown to persist on surgical instruments after rigorous cleaning and disinfection (Zobeley *et al.*, 1999; Flechsig *et al.*, 2001). Proteinase K completely digests PrP^{C} but only partly digests PrP^{Sc}, leaving a residue sometimes called PrP^{res} that is quite resistant to further digestion. Because available antibodies generally cannot distinguish PrP^{C} from PrP^{Sc} or PrP^{res}, tests for abnormal prions usually begin with a proteinase K digestion that eliminates the PrP^{C} from the sample, with some loss of PrP^{Sc} and thus of sensitivity. An alternate serological approach that does not rely on proteinase K digestion, called 'conformation-dependent immunoassay', appears to circumvent some of these problems and achieve greater sensitivity (Safar *et al.*, 2002). Although ovine and bovine PrP^{Sc} differ in amino acid sequence and in ability to convert human PrP^{C}, the same antibodies that are used to monitor for BSE will detect PrP^{Sc} from sheep scrapie, and apparently from CWD in cervids.

9.2 Virulence factors

Because PrP^{C} occurs generally, it may be supposed that the TSE threat is universal. However, TSEs have been reported in only a few species, and documented inter-species transmission is apparently rarer still. CJD has a familial form (fCJD) that is obviously inherited, as is fatal familial insomnia. However, no risk factors have been identified for sCJD. This indicates that the disease arises *de novo*, spontaneously, everywhere in the world, though with very low frequency (Erdtmann and Sivitz, 2004). It is noteworthy that all vCJD victims to date have been homozygous for methionine at codon 129 of their prion genes (Johnson and Gibbs, 1998; National CJD Surveillance Unit, 2002). Beyond this, no treatment is known to be effective against any TSE, and the outcome is invariably fatal, which suggests that there are no gradations of virulence. However, it is not known that clinical disease results from every infection. Studies in transgenic mice (carrying human, rather than mouse, prion genes) indicate that having valine rather then methionine at codon 129 does not prevent infection, but leads to a different type of abnormal prion formation when challenged by BSE or vCJD prions (Soldan, 2004). In addition to death from other causes during the long period of incubation, there may be infections that do not progress to cause clinical illness.

9.3 Environmental persistence

Because the PrP^{Sc} are extremely resistant to heat and are not easily digested by proteases, they are assumed to persist well in the environment outside the animal that produced them. It is suggested that the BSE epidemic resulted from changes in the rendering process in the UK, whereby temperatures were lower and organic solvents were no longer used to extract the last of the fat from the tissues. The presumably contaminated meat and bone meal (MBM) that resulted was stored and shipped without refrigeration, and the PrP^{Sc} apparently remained infectious for months or years. The PrP^{Sc} are not shed by infected cattle or humans, so the environment would be contaminated only as a result of decomposition of the cadaver. On the other hand, PrP^{Sc} from sheep, and especially cervids, are probably shed by live animals and may contaminate the environment. Specific data are lacking, but it appears that CWD, at least, is regularly transmitted via PrP^{Sc} persisting in soil (Williams and Miller, 2002). Much more needs to be learned about this.

9.4 Reservoirs and transmission

Each susceptible species is apparently a reservoir unto itself of its respective TSE, subject further to individual and strain variations in susceptibility. Incubation periods are a significant portion of the life expectancy of the host species, and PrP^{Sc} are harbored in various parts of the body during incubation, as well as in the clinical phase. Tissue distributions of PrP^{Sc} in one species are of little value in predicting tissue distributions in another, though all known TSEs culminate in accumulation of PrP^{Sc} in the brain. Within a species, tissue distributions differ over the course of the infection, which is of great importance if screening tests are to be applied. Natural transmission is most often peroral, and there seems to be a significant 'species barrier', even when the TSE in question is known to be transmissible from one species to another. For example, it appears that substantially less BSE PrP^{Sc} is required to infect a cow than a human, other things being equal.

Route of administration has been shown to be important in experiments, in that intracerebral inoculation is more efficient than parenteral inoculation, which in turn is more efficient than peroral administration (Erdtmann and Sivitz, 2004). Intracerebral inoculation has no natural counterpart; however, iCJD has been shown to result from corneal transplants and implants of cadaver dura mater. Transmission of scrapie and BSE among sheep by blood transfusion has been accomplished experimentally (Hunter *et al.*, 2002), but no human case of cCJD has yet been associated with transfusion of blood. There are two reports, at this writing, of vCJD deaths associated with transfusions from earlier vCJD victims, although this has not been verified through official channels (BBC News, 2003; Medical News Today, 2004). Vertical (*in utero*) transmission is also theoretically possible with most TSEs, but seems to be the exception, rather than the rule. In fact, vertical transmission has not been decisively demonstrated in most TSEs. In the absence of demonstrated alternatives, it appears that infection usually begins perorally.

If the PrP^{Sc} are not shed, transmission becomes possible only by ingestion of infected tissue. Prevention of this ingestion is the essence of how the BSE epidemic has been controlled in the UK: prohibiting feeding of ruminant remains (MBM or other, eventually from all mammals) to ruminants has essentially broken the cycle of transmission, although a few unexplained cases are still occurring in cattle born after the ban. Feeding of cattle-derived products from the UK to cattle in other countries is thought to have been the means of introducing BSE into those countries. In the UK presently, specified bovine offals may not be used as food, and animals over 30 months of age do not go into the food supply (this measure may soon be rescinded). Other European countries test animals 24 or 30 months old at slaughter, and eliminate any whose brain stem is found to contain PrP^{Sc}; in Japan, all cattle are tested at slaughter regardless of age, including veal calves. In the US, in the aftermath of the first BSE diagnosis in December 2003, expanded testing has been applied particularly to cattle over 30 months old that die on the farm, cannot walk, or arrive for slaughter showing possible neurological symptoms. Antemortem tests, particularly of lymphoid tissue (e.g. tonsils) or of excreta, have been actively sought; but none has yet been fully validated. Since various parts of the body may silently harbor PrP^{Sc} during the long incubation period, even postmortem testing is not absolutely reliable.

Restrictions on who may donate blood have been widely imposed, based on assumptions as to the risk that a potential donor may be incubating a case of vCJD (FDA, 2002). Risk assessments have been based principally on periods of time spent in the UK and in other European countries, on the assumption that this entailed possible acquisition of vCJD from eating BSE beef. Means of testing blood and blood products for PrP^{Sc} are being sought (Brown *et al.*, 2001b), as well as tests to be applied to tissues and organs considered for transplantation. Other concerns are bovine materials in nutritional supplements and other ingested products that are not regulated as food. In the US, which recently reported its first case of BSE, the brain and spinal cord of a USDA-inspected and passed carcass are food only if the animal is under 30 months of age at slaughter, but central nervous system tissue is not permitted in meat obtained by advanced meat recovery systems designed to harvest the last bits of meat from bones. PrP^{Sc} have also been shown to persist and retain their infectivity on surgical instruments that have been used on CJD patients and then on others; transmission has occurred despite cleaning, disinfection and prolonged storage of the instruments before reuse (Zobeley *et al.*, 1999; Flechsig *et al.*, 2001). Although it cannot be proven that BSE derived from scrapie, the BSE–vCJD connection has served as an incentive in many countries to undertake eradication of scrapie as an indirect public health measure.

9.5 Nature of infection in man

As stated above, vCJD infections to date apparently have resulted largely from ingestion of bovine tissues contaminated with BSE PrP^{Sc}. The exact tissues by which infection was transmitted have not been identified, nor has the portion of the human digestive tract that serves as the portal of entry. The frequent association of PrP^{Sc} with lymphoid tissue in various TSEs suggests that lymphoid tissue in the intestinal lining (e.g. Peyer's patches) may be the entry site. The Peyer's patches are also suspected of being the source of PrP^{Sc} shed from the body in CWD. Migration of the PrP^{Sc} from lymphoid tissue in the intestine to the brain has not been demonstrated; conjectures include migration along nerve trunks, via the blood or through the lymphoreticular system (Weissmann *et al.*, 2001). Association of PrP^{Sc} with tonsils, spleens and appendices has been recorded (Wadsworth *et al.*, 2001; Hilton *et al.*, 2002); whether these are primary or secondary sites of infection is not known. In any case, clinical symptoms are associated with significant accumulation of PrP^{Sc} in the brain.

Age at onset of vCJD is usually under 40, with the highest incidence in the 20s. One case has been recorded in a 74-year-old man, so susceptibility may be life-long; however, this is the only case to date to have begun beyond the age of 60. First indications of vCJD are often behavioral changes that lead to referral to a psychiatrist (Spencer *et al.*, 2002; World Health Organization, 2001). Ataxia and dementia follow, and sometimes myoclonus. Distinctive features have been identified in electroencephalograms and magnetic resonance imaging of patients. Postmortem, the brain shows spongiform degeneration, usually with accumulation of amyloid plaques. At this writing, approximately 157 people have died of vCJD, the vast majority in the UK. Peak incidence has been *ca.* 28 cases in 2000, the only year in which vCJD exceeded half the UK incidence of sCJD (Erdtmann and Sivitiz, 2004). The epidemic curve appears to have held steady or trended downward since then.

9.6 Foods most often involved as vehicles

Only bovine tissues ingested as food have been suspected of transmitting BSE to humans, resulting in vCJD. In earlier stages of infection, BSE prions are found in the terminal ileum; clinically ill animals have high concentrations of Pr^{Sc} in their brains and spinal cords, as well as the conjunctiva, but not in voluntary muscle. No PrP^{Sc} has yet been found in milk (see below). At present, the supposition is that people who contracted vCJD got it by eating central nervous system tissue from cattle in a late stage of incubation. The central nervous system tissue may have been included in comminuted meat products, since the majority of victims to date claimed or were thought not to have eaten bovine brain or spinal cord intentionally. Feline spongiform encephalopathy may have resulted from feeding bovine central nervous system tissues to cats, as well as use of these tissues in commercial cat foods. It is hoped that the elimination of specified risk materials from the food chain in the UK and other BSE countries has minimized the risk of transmission of vCJD by ingestion. Extensive testing of red meat (voluntary muscle) and milk from experimentally infected cattle has yet to reveal PrP^{Sc} in either. However, PrP^{Sc} has been demonstrated in the muscles of experimentally infected rodents (Bosque *et al.*, 2002).

9.7 Principles of detection in food and environment

Neither foods nor the environment are likely to be monitored for contamination with BSE prions, but experimental surveys have been made of bovine tissues and milk as a function of the stage of BSE incubation in the animal. Milk from experimentally infected cows has been fed to calves (to avoid the 'species barrier'; Brown, 2001) and injected intracerebrally into susceptible rodents (to use the most efficient route of administration); thus far, no TSE has resulted in the recipient animals. There are some problems in testing.

First, the PrP^{Sc} contains no nucleic acid, so amplification methods analogous to the polymerase chain reaction are not available, and sensitivities are relatively low. A first approach has been classical histopathology, in which haematoxylin and eosin-stained sections of the brain are shown to contain classical 'florid plaques' (Erdtmann and Sivitz, 2004). When antibody against PrP became available, it was possible to stain sections with fluorescent antibody; this produced an important gain in specificity, and perhaps in sensitivity (Bieschke *et al.*, 2000). Other applications of anti-PrP antibody have been in immunoblotting and enzyme-linked immunosorbent assay (ELISA) tests (Biffiger *et al.*, 2002), which vary in sensitivity as well as in specificity but are much more compatible with testing large numbers of samples. Immunohistochemistry affords a high-specificity test for PrP^{Sc} in neural tissue (Erdtmann and Sivitz, 2004). Non-serological tests have been based on inoculation of animals (usually rodents) and of cell cultures (Klohn *et al.*, 2003). Each of these has its advantages, but neither offers sensitivity competitive with the better serological methods.

Second, the PrP^{Sc} apparently does not accumulate in most bovine tissues during the preclinical phase of the infection, with the exceptions of terminal ileum and tonsils at certain stages. Obviously, more sensitive test methods might change this perception, if they enabled detection of lower levels of PrP^{Sc}.

Third, most of the antibodies prepared to date cannot distinguish between PrP^{Sc} and PrP^{C} (Safar *et al.*, 2000), so the specificity of the test depends on prior treatment with proteinase K to remove the PrP^{C} (Ingrosso *et al.*, 2002). Although PrP^{Sc} is considered resistant to proteinase K, some is lost in the treatment, which leads to a reduction in test sensitivity. Proposed approaches to improvement are the development of enhanced serologic tests (Safar *et al.*, 2002) or of non-antibody test reagents that specifically detect PrP^{Sc}, which would obviate the proteinase K treatment, or development of other specifically reactive molecules ('aptamers') that would transcend the limits of antibodies used for this purpose (Jayasena, 1999). Other detection methods, based on multi-spectral ultraviolet fluorescent spectroscopy (Rubenstein *et al.*, 1998), Fourier transform infrared spectroscopy (Schmitt *et al.*, 2002), proteomics, high performance liquid chromatography, etc., are also being explored (Erdtmann and Sivitz, 2004).

The limited sensitivities of available detection methods are a problem in studies of inactivation of PrP^{Sc}, in that reducing PrP^{Sc} levels below the limit of detection may not indicate a safe outcome of the treatment being tested. This has not generally been a problem in tests of environmental samples, which have been limited. However, the problem of how to clean and disinfect surgical instruments requires sensitive methods for detecting residual PrP^{Sc} on surfaces; the needed methods are not currently available. In this particular application, detection of any residual PrP, whether PrP^{Sc} or PrP^{C}, would be significant, so the test need not be specific for the infectious form. Equipment used by butchers to process animals that are later shown to have or are suspected of having TSEs is also in need of disinfection; regulatory authorities are facing a dilemma as to how to deal with this problem, although cross-contamination of meat is a less urgent threat than contamination of the human brain in neurosurgery.

10 Overview

The preceding sections have shown that viruses are very common (perhaps the most common) causes of foodborne disease, at least in the US. They are also transmitted via water, but they may be less pre-eminent in that vehicle. The majority of foodborne viral illnesses are caused by the noroviruses, which produce transient gastroenteritis and are very contagious. Unlike most other viral infections, norovirus infections evidently do not evoke a durable immune response; this suggests that development of an effective vaccine is unlikely. HAV and HEV cause many fewer individual illnesses, but they cause disabilities that last for weeks, sometimes with permanent sequelae. HEV causes significant mortality in pregnant women. Several other groups of viruses are transmissible via foods but are relatively rare, for various reasons. With the possible exception of HEV, most of the foodborne and waterborne viruses are transmitted by the fecal–oral route and are essentially specific to humans. Viruses cannot multiply in foods, but only persist or be inactivated. Their extremely small size and the difficulties of propagating them in the laboratory make detection of foodborne and waterborne viruses an especially challenging task.

Prions are an entirely different class of infectious agents. They are low molecular weight peptides whose normal function is unknown. Under various circumstances, prions that are in their normal configuration, acquired by folding as synthesis proceeds, become reconfigured in such a way that they cannot be cycled or turned over by the body in normal fashion. These refractory prions accumulate in various tissues; their deposition in the brain ultimately leads to neural degeneration and death. To date, the only prion disease known to be transmitted to humans via food is BSE ('mad cow disease'), which in humans causes new variant Creutzfeldt-Jakob disease (vCJD). This is apparently a true foodborne disease, but so rare that the extreme and costly measures directed against it may well cost more lives (through neglect of other foodborne hazards) than they save.

Bibliography

Abad, F. X., R. M. Pintó, and A. Bosch. 1994. Survival of enteric viruses on environmental fomites. *Appl. Environ. Microbiol.* **60**, 3704–3710.

Abad, F. X., R. M. Pintó, and A. Bosch (1997a). Disinfection of human enteric viruses on fomites. *FEMS Microbiol. Lett.* **156**, 107–111.

Abad, F. X., R. M. Pintó, C. Villena *et al.* (1997b). Astrovirus survival in drinking water. *Appl. Environ. Microbiol.* **63**, 3119–3122.

Abd El Galil, K. H., M. A. El Sokkary, S. M. Kheira *et al.* (2004). Combined immunomagnetic separation-molecular beacon-reverse transcription-PCR assay for detection of hepatitis A virus from environmental samples. *Appl. Environ. Microbiol.* **70**, 4371–4374.

Aggarwal, R. and K. A. McCaustland (1998). Hepatitis E virus RNA detection in serum and feces specimens with the use of microspin columns. *J. Virol. Methods* **74**, 209–213.

Aggarwal, R., H. Shahi, S. Naik *et al.* (1997). Evidence in favour of high infection rate with hepatitis E virus among young children in India [letter]. *J. Hepatol.* **26**, 1425–1426.

Aggarwal, R., S. Kamili, J. R. Spelbring and K. Krawczynski (2001). Experimental studies on subclinical hepatitis E virus infection in cynomolgus macaques. *J. Infect. Dis.* **184**, 1380–1385.

Agus, S. G., R. Dolin, R. G. Wyatt *et al.* (1973). Acute infectious nonbacterial gastroenteritis: Intestinal histopathology. Histologic and enzymatic alterations during illness produced by the Norwalk agent in man. *Ann. Intern. Med.* **79**, 18–25.

Allwood, P. B., Y. S. Malik, C. W. Hedberg and S. M. Goyal (2003). Survival of F-specific RNA coliphage, feline calicivirus, and *Escherichia coli* in water, a comparative study. *Appl. Environ. Microbiol.* **69**, 5707–5710.

Allwood, P. B., T. Jenkins, C. Paulus *et al.* (2004a). Hand washing compliance among retail food establishment workers in Minnesota. *J. Food Prot.* **67**, 2825–2828.

Allwood, P. B., Y. S. Malik, C. W. Hedberg and S. M. Goyal (2004b). Effect of temperature and sanitizers on the survival of feline calicivirus, *Escherichia coli*, and F-specific coliphage MS2 on leafy salad vegetables. *J. Food Prot.* **67**, 1451–1456.

Allwood, P. B., Y. S. Malik, S. Maherchandani *et al.* (2004c). Occurrence of *Escherichia coli*, noroviruses, and F-specific coliphages in fresh market-ready produce. *J. Food Prot.* **67**, 2387–2390.

Andreoletti, L., D. Hober, S. Belaich *et al.* (1996). Rapid detection of enterovirus in clinical specimens using PCR and microwell capture hybridization assay. *J. Virol. Methods* **62**, 1–10.

Altekruse, S. F., D. A. Street, S. B. Fein and A. S. Levy (1996). Consumer knowledge of foodborne microbial hazards and food-handling practices. *J. Food Prot.* **59**, 287–294.

Alvarez-Munoz, M. T., J. Torres, L. Damasio *et al.* (1999). Seroepidemiology of hepatitis E virus infection in Mexican subjects 1 to 29 years of age. *Arch. Med. Res.* **30**, 251–254.

Anderson, J. B., T. A. Shuster, K. E. Hansen *et al.* (2004). A camera's view of consumer food-handling behaviors. *J. Am. Diet. Assoc.* **104**, 186–191.

Ansari, S. A., V. S. Springthorpe and S. A. Sattar (1991). Survival and vehicular spread of human rotaviruses, possible relation to seasonality of outbreaks. *Rev. Infect. Dis.* **13**, 448–461.

Appleton, H. (1990). Foodborne illness, foodborne viruses. *Lancet* **336**, 362–364.

Appleton, H. (1994). Norwalk virus and the small round viruses causing foodborne gastroenteritis. In: Y. H. Hui, J. R. Gorham, K. D. Murrell and D. O. Cliver (eds), *Foodborne Disease Handbook*, Vol. 2, pp. 57–79. Marcel Dekker, New York, NY.

Appleton, H. (2000). Control of food-borne viruses. *Br. Med. Bull.* **56**, 172–183.

Arankalle, V. A., J. Ticehurst, M. A. Sreenivasan *et al.* (1988). Aetiological association of a virus-like particle with enterically transmitted non-A, non-B hepatitis. *Lancet* **1**, 550–554.

Arankalle, V.A., S. A. Tsarev, M. S. Chadha *et al.* (1995). Age-specific prevalence of antibodies to hepatitis A and E viruses in Pune, India, 1982 and 1992. *J. Infect. Dis.* **171**, 447–450.

Arankalle, V. A., M. S. Chadha, S. D. Chitambar *et al.* (2001a). Changing epidemiology of hepatitis A and hepatitis E in urban and rural India (1982–98). *J. Viral Hepatitis* **8**, 293–303.

Arankalle, V. A., M. V. Joshi, A. M. Kulkarni *et al.* (2001b). Prevalence of anti-hepatitis E virus antibodies in different Indian animal species. *J. Viral Hepatitis* **8**, 223–227.

Atmar, R. L., F. H. Neill, J. L. Romalde *et al.* (1995). Detection of Norwalk virus and hepatitis A virus in shellfish tissues with the PCR. *Appl. Environ. Microbiol.* **61**, 3014–3018.

Balayan, M. S., A. G. Andjaparidze, S. S. Savinskaya *et al.* (1983). Evidence for a virus in non-A/non-B hepatitis transmitted via the fecal oral route. *Intervirology* **20**, 23–31.

Balayan, M. S., R. K. Usmanov, N. A. Zamyatina *et al.* (1990). Experimental hepatitis E infection in domestic piglets. *J. Med. Virol.* **32**, 58–59.

Baldwin, M. A. (2001). Mass spectrometric analysis of prion proteins. *Adv. Protein Chem.* **57**, 29–54.

Ball, J. M., P. Tian, C. Q. Zeng *et al.* (1996). Age-dependent diarrhea induced by a rotaviral nonstructural glycoprotein. *Science* **272**, 101–104.

Banks, M., R. Bendall, S. Grierson *et al.* (2004a). Human and porcine hepatitis E virus strains, United Kingdom. *Emerg. Infect. Dis.* **10**, 953–955.

Banks, M., G. S. Heath, S. S. Grierson *et al.* (2004b). Evidence for the presence of hepatitis E virus in pigs in the United Kingdom. *Vet. Record* **154**, 223–227.

Barardi, C. R., H. Yip, K. R. Emsile *et al.* (1999). Flow cytometry and RT-PCR for rotavirus detection in artificially seeded oyster meat. *Intl J. Food Microbiol.* **49**, 9–18.

Barker, J., D. Stevens and S. F. Bloomfield (2001). Spread and prevention of some common viral infections in community facilities and domestic homes. *J. Appl. Microbiol.* **91**, 7–21.

Barker, J., I. B. Vipond and S. F. Bloomfield (2004). Effects of cleaning and disinfection in reducing the spread of Norovirus contamination via environmental surfaces. *J. Hosp. Infect.* **58**, 42–49.

Battegay, M., I. D. Gust and S. M. Feinstone (1995). Hepatitis A virus. In: G. L. Mandell, J. E. Bennett and R. Dolin (eds), *Mandell, Douglas, and Bennett's Principles and Practice of Infectious Diseases*, 4th edn, Vol. 2, pp.1636–1656. Churchill Livingstone, New York, NY.

BBC News (2003). Blood donor 'may have passed CJD' (available at http, //news.bbc.co.uk/1/hi/health/3327745.stm, accessed 20 December 2003).

Becker, K. M., C. L. Moe, K. L. Southwick and J. N. MacCormack (2000). Transmission of Norwalk virus during a football game. *N. Engl. J. Med.* **343**, 1223–1227.

Bell, B. P., C. N. Shapiro, M. J. Alter *et al.* (1998). The diverse patterns of hepatitis A epidemiology in the United States – implications for vaccine strategies. *J. Infect. Dis.* **178**, 1579–1584.

Bell, D., S. Roberton and P. R. Hunter (2004). Animal origins of SARS coronavirus, possible links with the international trade in small carnivores. *Phil. Trans. R. Soc. Lond. B Biol. Sci.* **359**, 1107–1114.

Beller, M. (1992). Hepatitis A outbreak in Anchorage, Alaska, traced to ice slush beverages. *West. J. Med.* **156**, 624–627.

Berke, T. and D. O. Matson (2000). Reclassification of the *Caliciviridae* into distinct genera and exclusion of hepatitis E virus from the family on the basis of comparative phylogenetic analysis. *Arch. Virol.* **145**, 1421–1436.

Bermudez-Millan, A., R. Perez-Escamilla, G. Damio *et al.* (2004). Food safety knowledge, attitudes, and behaviors among Puerto Rican caretakers living in Hartford, Connecticut. *J. Food Prot.* **67**, 512–516.

Bern, C., L. Unicomb, J. R. Gentsch *et al.* (1992). Rotavirus diarrhea in Bangladeshi children, correlation of disease severity with serotypes. *J. Clin. Microbiol.* **30**, 3234–3238.

Betts, A. O., P. H. Lamont and D. F. Kelly (1962). Porcine enteroviruses other than the virus of Teschen disease. *Ann. NY Acad. Sci.* **101**, 428–435.

Beuchat, L. R. and J.-H. Ryu (1997). Produce handling and processing practices. *Emerg. Infect. Dis.* **3**, 1–9.

Bhattacharya, S. S., M. Kulka, K. A. Lampel *et al.* (2004). Use of reverse transcription and PCR to discriminate between infectious and non-infectious hepatitis A virus. *J. Virol. Methods* **116**, 181–187.

Bidawid, S., J. M. Farber, S. A. Sattar, and S. Hayward (2000a). Heat inactivation of hepatitis A virus in dairy foods. *J. Food Prot.* **63**, 522–528.

Bidawid, S., J. M. Farber and S. A. Sattar (2000b). Contamination of foods by food handlers, experiments on hepatitis A virus transfer to food and its interruption. *Appl. Environ. Microbiol.* **66**, 2759–2763.

Bidawid, S., J. M. Farber and S. A. Sattar (2000c). Rapid concentration and detection of hepatitis A virus from lettuce and strawberries. *J. Virol. Methods* **88**, 175–185.

Bidawid, S., N. Malik, O. Adegbunrin *et al.* (2004). Norovirus cross-contamination during food handling and interruption of virus transfer by hand antisepsis, experiments with feline calicivirus as a surrogate. *J. Food Prot.* **67**, 103–09.

Bieschke, J., A. Giese, W. Schulz-Schaeffer *et al.* (2000). Ultrasensitive detection of pathological prion protein aggregates by dual-color scanning for intensely fluorescent targets. *Proc. Natl. Acad. Sci. USA* **97**, 5468–5473.

Biffiger, K., D. Zwald, L. Kaufmann *et al.* (2002). Validation of a luminescence immunoassay for the detection of PrP(Sc) in brain homogenate. *J. Virol. Methods* **101**, 79–84.

Bishop, R. F., G. P. Davidson, I. H. Holmes and B. J. Ruck (1973). Virus particles in epithelial cells of duodenal mucosa from children with acute nonbacterial gastroenteritis. *Lancet* **1**, 1281–1283.

Biziagos, E., J. Passagot, J. M. Crance and R. Deloince (1988). Long-term survival of hepatitis A virus and poliovirus type 1 in mineral water. *Appl. Environ. Microbiol.* **54**, 2705–2710.

Bosch, A., G. Sanchez, F. Le Guyader *et al.* (2001). Human enteric viruses in Coquina clams associated with a large hepatitis A outbreak. *Water Sci. Technol.* **43**, 61–65.

Bosch, A., R. M. Pinto, J. Comas and F. X. Abad (2004). Detection of infectious rotaviruses by flow cytometry. *Methods Mol. Biol.* **268**, 61–68.

Bosque, P. J., C. Ryou, G. Telling *et al.* (2002). Prions in skeletal muscle. *Proc. Natl. Acad. Sci. USA* **99**, 3812–3817.

Bradley, D. W. (1992). Hepatitis E, epidemiology, aetiology, and molecular biology. *Rev. Med. Virol.* **2**, 19–28.

Bradley, D. W. (1995). Hepatitis E virus, a brief review of the biology, molecular virology, and immunology of a novel virus. *J. Hepatol.* **22**, 140–145.

Bradley, D. W., K. Krawczynski, E. H. Cook Jr *et al.* (1987). Enterically transmitted non-A, non-B hepatitis, Serial passage of disease in cynomolgus macaques and tamarins and recovery of disease-associated 27- to 34 nm virus-like particles. *Proc. Natl. Acad. Sci. USA* **84**, 6277–6281.

Bradley, D. W., A. Andjaparidze, E. H. Cook Jr. *et al.* (1988). Aetological agent of enterically-transmitted non-A, non-B hepatitis. *J. Gen. Virol.* **69**, 731–738.

Bradley, D. W., K. Krawczynski and M. A. Purdy (1997). Epidemiology, natural history and experimental models. In: A. J. Zuckerman and H. C. Thomas (eds), *Viral Hepatitis – Scientific Basis and Clinical Management*, pp. 379–383. Churchill Livingstone, London.

Brassard, J., K. Seyer, A. Houde *et al.* (2005). Concentration and detection of hepatitis A virus and rotavirus in spring water samples by reverse transcription-PCR. *J. Virol. Methods* **123**, 163–169.

Brockmann, R. A., D. D. Lenaway and C. D. Humphrey (1995). Norwalk-like viral gastroenteritis, a large outbreak on a university campus. *J. Environ. Health* **57**, 19–22.

Brown, E. A. and J. T. Stapleton (2003). Hepatitis A virus. In: P. R. Murray, E. J. Baron, J. H. Jorgenson *et al.* (eds), *Manual of Clinical Microbiology*, 8th edn, pp.1452–1463. American Society for Microbiology Press, Washington, DC.

Brown, P (2001). Afterthoughts about bovine spongiform encephalopathy and variant Creutzfeldt-Jakob disease. *Emerg. Infect. Dis.* **7**, 598–600.

Brown, P., R. G. Will, R. Bradley *et al.* (2001a). Bovine spongiform encephalopathy and variant Creutzfeldt-Jakob disease, background, evolution, and current concerns. *Emerg. Infect. Dis.* **7**, 6–16.

Brown, P., L. Cervenakova and H. Diringer (2001b). Blood infectivity and the prospects for a diagnostic screening test in Creutzfeldt-Jakob disease. *J. Lab. Clin. Med.* **137**, 5–13.

Bryan, F. L. (1988). Risks of practices, procedures and processes that lead to outbreaks of foodborne diseases. *J. Food Prot.* **51**, 63–73.

Caballero, S., F. X. Abad, F. Loisy *et al.* (2004). Rotavirus virus-like particles as surrogates in environmental persistence and inactivation studies. *Appl. Environ. Microbiol.* **70**, 3904–3909.

Cacopardo, B., R. Russo, W. Preiser *et al.* (1997). Acute hepatitis E in Catania (eastern Sicily) 1980–1994. The role of hepatitis E virus. *Infection* **25**, 313–316.

Calder, L., G. Simmons, C. Thornley *et al.* (2003). An outbreak of hepatitis A associated with consumption of raw blueberries. *Epidemiol. Infect.* **131**, 745–751.

Callahan, K. M., D. J. Taylor and M. D. Sobsey (1995). Comparative survival of hepatitis A virus, poliovirus and indicator viruses in geographically diverse seawaters. *Water Sci. Technol.* **31**, 189–193.

Cannon, R. O., J. R. Poliner, R. B. Hirschhorn *et al.* (1991). A multistate outbreak of Norwalk virus gastroenteritis associated with consumption of commercial ice. *J. Infect. Dis.* **164**, 860–863.

Carl, M., D. P. Francis and J. E. Maynard (1983). Food-borne hepatitis A, recommendations for control. *J. Infect. Dis.* **148**, 1133–1135.

Casas, N. and E. Sunen (2001). Detection of enterovirus and hepatitis A virus RNA in mussels (*Mytilus* spp.) by reverse transcriptase-polymerase chain reaction. *J. Appl. Microbiol.* **90**, 89–95.

Casas, N. and E. Sunen (2002). Detection of enteroviruses, hepatitis A virus and rotaviruses in sewage by means of an immunomagnetic capture reverse transcription-PCR assay. *Microbiol. Res.* **157**, 169–175.

Catton, M. G. and S. A. Locarnini (1998). Epidemiology. In: A. J. Zuckerman and H. C. Thomas (eds), *Viral Hepatitis*, 2nd edn, pp. 29–41. Churchill Livingstone, London.

CDC (Centers for Disease Control and Prevention) (1996). Surveillance for foodborne disease outbreaks – United States, 1988–1992. *Morbid. Mortal. Weekly Rep.* **45**, 1–66.

CDC (Centers for Disease Control and Prevention) (2000a). Surveillance for foodborne disease outbreaks – United States, 1993–1997. *Morbid. Mortal. Weekly Rep.* **49**, 1–51.

CDC (Centers for Disease Control and Prevention) (2000b). Foodborne outbreak of Group A rotavirus gastroenteritis among college students – District of Columbia, March–April 2000. *Morbid. Mortal. Wkly. Rep.* **49**, 1131–1133.

CDC (Centers for Disease Control and Prevention) (2002a). Possible West Nile virus transmission to an infant through breast-feeding – Michigan, 2002. *Morbid. Mortal. Weekly Rep.* **51**, 877–878.

CDC (Centers for Disease Control and Prevention) (2002b). Outbreak of acute gastroenteritis associated with Norwalk-like viruses among British military personnel – Afghanistan, May 2002. *Morbid. Mortal. Weekly Rep.* **51**, 477–479.

CDC (Centers for Disease Control and Prevention) (2002c). From the Centers for Disease Control and Prevention. Possible West Nile virus transmission to an infant through breast-feeding – Michigan, 2002. *J. Am. Med. Assoc.* **288**, 1976–1977.

CDC (Centers for Disease Control and Prevention) (2003a). Foodborne transmission of hepatitis A – Massachusetts, 2001. *Morbid. Mortal. Weekly Rep.* **52**, 565–567.

CDC (Centers for Disease Control and Prevention) (2003b). CDC's hand washing guidelines. *Health Care Nutr. Focus* **20**, 1–7.

CDC (Centers for Disease Control and Prevention) (2003c). Hepatitis A outbreak associated with green onions at a restaurant – Monaca, Pennsylvania, 2003. *Morbid. Mortal. Weekly Rep.* **52**, 1155–1157.

Chadwick, P. R., G. Beards, D. Brown *et al.* (2000). Management of hospital outbreaks of gastro-enteritis due to small round structured viruses. *J. Hosp. Infect.* **45**, 1–10.

Chau, T. N., S. T. Lai, C. Tse *et al.* (2001). Parenteral and sexual transmission are not risk factors for acute hepatitis E virus infection in Hong Kong [letter]. *Am. J. Gastroenterol.* **96**, 3046–3047.

Chang, J. C. H., S. F. Ossoff, D. C. Lobe *et al.* (1985). UV inactivation of pathogenic and indicator microorganisms. *Appl. Environ. Microbiol.* **49**, 1361–1365.

Chen, Y. S. and J. M. Vaughn (1990). Inactivation of human and simian rotaviruses by chlorine dioxide. *Appl. Environ. Microbiol.* **56**, 1363–1366.

Chibber, R. M., M. A. Usmani and M. H. Al-Sibai (2004). Should HEV infected mothers breast feed? *Arch. Gynecol. Obstet.* **270**, 15–20.

Chironna, M., C. Germinario, P. L. Lopalco *et al.* (2001). Prevalence of hepatitis virus infections in Kosovar refugees. *Intl J. Infect. Dis.* **5**, 209–213.

Chironna, M., P. Lopalco, R. Prato *et al.* (2004). Outbreak of infection with hepatitis A virus (HAV) associated with a foodhandler and confirmed by sequence analysis reveals a new HAV genotype IB variant. *J. Clin. Microbiol.* **42**, 2825–2828.

Chobe, L. P., M. S. Chadha, K. Banerjee and V. A. Arankalle (1997). Detection of HEV RNA in faeces by RT-PCR during the epidemics of hepatitis E in India (1976–1995). *J. Viral Hepatitis* **4**, 129–133.

Choi, C., S. K. Ha and C. Chae (2004). Development of nested RT-PCR for the detection of swine hepatitis E virus in formalin-fixed, paraffin-embedded tissues and comparison with in situ hybridization. *J. Virol. Methods* **115**, 67–71.

Chung, H. and M. D. Sobsey (1993). Comparative survival of indicator viruses and enteric viruses in seawater and sediment. *Water Sci. Technol.* **27**, 425–428.

Chung, H., L. A. Jaykus and M. D. Sobsey (1996). Detection of human enteric viruses in oysters by in vivo and in vitro amplification of nucleic acids. *Appl. Environ. Microbiol.* **62**, 3772–3778.

Chung, H., L. A. Jaykus, G. Lovelace and M. D. Sobsey (1998). Bacteriophages and bacteria as indicators of enteric viruses in oysters and their harvest waters. *Water Sci. Technol.* **38**, 37–44.

Ciarlet, M., M. A. Gilger, C. Barone *et al.* (1998). Rotavirus disease, but not infection and development of intestinal histopathological lesions, is age-restricted in rabbits. *Virology* **251**, 343–360.

Clemente-Casares, P., S. Pina, M. Buti *et al.* (2003). Hepatitis E virus epidemiology in industrialized countries. *Emerg. Infect. Dis.* **9**, 448–454.

Cliver, D. O. (1967). Food-associated viruses. *Health Lab. Sci.* **4**, 213–221.

Cliver, D. O. (1985). Vehicular transmission of hepatitis A. *Publ. Health Rev.* **13**, 235–292.

Cliver, D. O (1994). Viral foodborne disease agents of concern. *J. Food Prot.* **57**, 176–178.

Cliver, D. O. (1997). Virus transmission via foods, an IFT Scientific Status Summary. *Food Technol.* **51**, 71–78.

Cliver, D. O. and E. H. Bohl (1962). Isolation of enteroviruses from a herd of dairy cattle. *J. Dairy Sci.* **45**, 921–925.

Cliver, D. O. and J. Grindrod (1969). Surveillance methods for viruses in foods. *J. Milk Food Technol.* **32**, 421–425.

Cliver, D. O. and J. E. Herrmann (1972). Proteolytic and microbial inactivation of enteroviruses. *Water Res.* **6**, 797–805.

Cliver, D. O. and K. D. Kostenbader Jr (1984). Disinfection of virus on hands for prevention of food-borne disease. *Intl J. Food Microbiol.* **1**, 75–87.

Cliver, D. O. and J. Yeatman (1965). Ultracentrifugation in the concentration and detection of enteroviruses. *Appl. Microbiol.* **13**, 387–392.

Cliver, D. O., R. D. Ellender and M. D. Sobsey (1983a). Methods for detecting viruses in foods, background and general principles. *J. Food Prot.* **46**, 248–259.

Cliver, D. O., R. D. Ellender, and M. D. Sobsey (1983b). Methods to detect viruses in foods, testing and interpretation of results. *J. Food Prot.* **46**, 345–357.

Cliver, D. O., R. D. Ellender, G. S. Fout *et al.* (1990). Foodborne viruses. In: C. Vanderzant and D. F. Splittstoesser (eds), *Compendium of Methods for the Microbiological Examination of Foods*, 3rd edn, pp. 763–787. American Public Health Association, Washington, DC.

Cohen, J. I., J. R. Ticehurst, R. H. Purcell *et al.* (1987). Complete nucleotide sequence of wild-type hepatitis A virus, comparison with different strains of hepatitis A virus and other picornaviruses. *J. Virol.* **61**, 50–59.

Communicable Diseases and Environmental Health Scotland Weekly Report (1991). Outbreak of diarrhoeal illness associated with coronovirus [sic] infection. *Comm. Dis. Environ. Health Scotland Weekly Rep.* **25**(46), 1.

Corwin, A., K. Jarot, I. Lubis *et al.* (1995). Two years' investigation of epidemic hepatitis E virus transmission in West Kalimantan (Borneo), Indonesia. *Trans. R. Soc. Trop. Med. Hyg.* **89**, 262–265.

Corwin, A. L., N. T. K. Tien, K. Bounlu *et al.* (1999). The unique riverine ecology of hepatitis E virus transmission in South-East Asia. *Trans. R. Soc. Trop. Med. Hyg.* **93**, 255–260.

Croci, L., M. Ciccozzi, D. De Medici *et al.* (1999a). Inactivation of hepatitis A virus in heat-treated mussels. *J. Appl. Microbiol.* **87**, 884–888.

Croci, L., D. De Medici, G. Morace *et al.* (1999b). Detection of hepatitis A virus in shellfish by nested reverse transcription-PCR. *Intl J. Food Microbiol.* **48**, 67–71.

Croci, L., D. De Medici, C. Scalfaro *et al.* (2002). The survival of hepatitis A virus in fresh produce. *Intl J. Food Microbiol.* **73**, 29–34.

Cromeans, T., M. D. Sobsey and H. A. Fields (1987). Development of a plaque assay for a cytopathic, rapidly replicating isolate of hepatitis A virus. *J. Med. Virol.* **22**, 45–56.

Cromeans, T., H. A. Fields and M. D. Sobsey (1989). Replication kinetics and cytopathic effect of hepatitis A virus. *J. Gen. Virol.* **70**, 2051–2062.

Cromeans, T., O. V. Nainan, H. A. Fields *et al.* (1994). Hepatitis A and E viruses. In: Y. H. Hui, J. R. Gorham, K. D. Murrell and D. O. Cliver (eds), *Foodborne Disease Handbook*, Vol. 2, pp. 1–56. Marcel Dekker, New York, NY.

Cromeans, T. L., O. V. Nainan and H. S. Margolis (1997). Detection of hepatitis A virus RNA in oyster meat. *Appl. Environ. Microbiol.* **63**, 2460–2463.

Cubitt, W. D. (1989). Diagnosis, occurrence and clinical significance of the human 'candidate' caliciviruses. *Progr. Med. Virol.* **36**, 103–119.

Dahling, D. R. (2002). An improved filter elution and cell culture assay procedure for evaluating public groundwater systems for cultural enteroviruses. *Water Environ. Res.* **74**, 564–568.

Dahling, D. R. and B. A. Wright (1986). Recovery of viruses from water by a modified flocculation procedure for second-step concentration. *Appl. Environ. Microbiol.* **51**, 1326–1331.

Deboosere, N., O. Legeay, Y. Caudrelier and M. Lange (2004). Modeling the effect of physical and chemical parameters on heat inactivation kinetics of hepatitis A virus in a fruit model system. *Intl J. Food Microbiol.* **93**, 73–85.

Deng, M. Y., and D. O. Cliver (1992). Inactivation of poliovirus type 1 in mixed human and swine wastes and by bacteria from swine manure. *Appl. Environ. Microbiol.* **58**, 2016–2021.

Deng, M. Y. and D. O. Cliver (1995a). Persistence of inoculated hepatitis A virus in mixed human and animal wastes. *Appl. Environ. Microbiol.* **61**, 87–91.

Deng, M. Y. and D. O. Cliver (1995b). Antiviral effects of bacteria isolated from manure. *Microbial Ecol.* **30**, 43–54.

Deng, M. Y., S. P. Day and D. O. Cliver (1994). Detection of hepatitis A virus in environmental samples by antigen-capture PCR. *Appl. Environ. Microbiol.* **60**, 1927–1933.

Dennehy, P. H. (2000). Transmission of rotavirus and other enteric pathogens in the home. *Pediatr. Infect. Dis. J.* **19**, S103–S105.

Dentinger C. M., W. A. Bower, O. V. Nainan *et al.* (2001). An outbreak of hepatitis A associated with green onions. *J. Infect. Dis.* **183**, 1273–1276.

Derbyshire, J. B. and E. G. Brown (1979). The inactivation of viruses in cattle and pig slurry by aeration or treatment with calcium hydroxide. *J. Hygiene (Lond.)* **82**, 293–299.

De Roda Husman, A. M., P. Bijkerk, W. Lodder *et al.* (2004). Calicivirus inactivation by nonionizing (253.7-nanometer-wavelength [UV]) and ionizing (gamma) radiation. *Appl. Environ. Microbiol.* **70**, 5089–5093.

Desenclos, J.-C., K. C. Klontz, M. H. Wilder *et al.* (1991). A multistate outbreak of hepatitis A caused by the consumption of raw oysters. *Am. J. Publ. Health* **81**, 1268–1272.

De Serres, G., T. L. Cromeans, B. Levesque *et al.* (1999). Molecular confirmation of hepatitis A virus from well-water, epidemiology and public health implications. *J. Infect. Dis.* **179**, 37–43.

Dewalt, P. G. and B. L. Semler (1989). Molecular biology and genetics of poliovirus protein processing. In: B. L. Semler and E. Ehrenfeld (eds), *Molecular Aspects of Picornavirus Infection and Detection*, pp. 73–93. American Society for Microbiology, Washington, DC.

Di Pinto, A., V. T. Forte, G. M. Tantillo *et al.* (2003). Detection of hepatitis A virus in shellfish (*Mytilus galloprovincialis*) with RT-PCR. *J. Food Prot.* **66**, 1681–1685.

Dix, A. B. and L.-A. Jaykus (1998). Virion concentration methods for the detection of human enteric viruses in extracts of hard-shelled clams. *J. Food Prot.* **61**, 458–465.

Dodd, D. A., T. H. Giddings Jr and K. Kirkegaard (2001). Poliovirus 3A protein limits interleukin-6 (IL-6), IL-8, and beta interferon secretion during viral infection. *J. Virol.* **75**, 8158–8165.

Dolin, R. (1978). Norwalk agent-like particles associated with gastroenteritis in human beings. *J. Am. Vet. Med. Assoc.* **173**, 615–619.

Dolin, R., N. R. Blacklow, H. DuPont *et al.* (1971). Transmission of acute infectious nonbacterial gastroenteritis to volunteers by oral administration of stool filtrates. *J. Infect. Dis.* **123**, 307–312.

Dolin, R., N. R. Blacklow, H. DuPont *et al.* (1972). Biological properties of Norwalk agent of acute infectious nonbacterial gastroenteritis. *Proc. Soc. Exp. Biol. Med.* **140**, 578–583.

Dolin, R., A. G. Levy, R. G. Wyatt *et al.* (1975). Viral gastroenteritis induced by the Hawaii agent. Jejunal histopathology and serologic response. *Am. J. Med.* **59**, 761–768.

Drobeniuc, J., M. O. Favorov, C. N. Shapiro *et al.* (2001). Hepatitis E virus antibody prevalence among persons who work with swine. *J. Infect. Dis.* **184**, 1594–1597.

Drozdov, S. G (1959). Role of domestic animals in epidemiology of diphasic milk fever [in Russian]. *Zh. Mikrobiol. Epidemiol. Immunobiol.* **30**, 102–108.

D'Souza, D. H. and L. A. Jaykus (2002). Zirconium hydroxide effectively immobilizes and concentrates human enteric viruses. *Lett. Appl. Microbiol.* **35**, 414–418.

Dubois, E., C. Agier, O. Traore *et al.* (2002). Modified concentration method for the detection of enteric viruses on fruits and vegetables by reverse transcriptase-polymerase chain reaction or cell culture. *J. Food Prot.* **65**, 1962–1969.

Dubois, E., G. Merle, C. Roquier *et al.* (2004). Diversity of enterovirus sequences detected in oysters by RT-heminested PCR. *Intl J. Food Microbiol.* **92**, 35–43.

Duizer, E., K. J. Schwab, F. H. Neill *et al.* (2004a). Laboratory efforts to cultivate noroviruses. *J. Gen. Virol.* **85**, 79–87.

Duizer, E., P. Bijkerk, B. Rockx *et al.* (2004b). Inactivation of caliciviruses. *Appl. Environ. Microbiol.* **70**, 4538–4543.

Eiden, J., G. A. Losonsky, J. Johnson and R. H. Yolken (1985). Rotavirus RNA variation during chronic infection of immunocompromised children. *Pediatr. Infect. Dis.* **4**, 632–637.

Eisenstein, A. B., R. D. Aach, W. Jacobsohn and A. Goldman (1963). An epidemic of infectious hepatitis in a general hospital, probable transmission by contaminated orange juice. *J. Am. Med. Assoc.* **185**, 171–174.

El-Zimaity, D. M. T., K. C. Hyams, I. Z. E. Imam *et al.* (1993). Acute sporadic hepatitis E in an Egyptian pediatric population. *Am. J. Trop. Med. Hyg.* **48**, 372–376.

Emerson, S. U. and R. H. Purcell (2003). Hepatitis E virus. *Rev. Med. Virol.* **13**, 145–154.

Emerson, S. U. and R. H. Purcell (2004). Running like water – the omnipresence of hepatitis E. *N. Engl. J. Med.* **351**, 2367–2368.

Engelbrecht, R. S., M. J. Weber, B. L. Salter and C. A. Schmidt (1980). Comparative inactivation of viruses by chlorine. *Appl. Environ. Microbiol.* **40**, 249–256.

Enriquez, C. E., C. J. Hurst and C. P. Gerba (1995). Survival of the enteric adenoviruses 40 and 41 in tap, sea, and waste water. *Water Res.* **29**, 2548–2553.

Enserink, M. (2003). Infectious diseases. Clues to the animal origins of SARS. *Science* **300**, 1351.

Erdtmann, R. and L. Sivitz (eds) (2004). *Advancing Prion Science, Guidance to the National Prion Research Program.* The National Academies Press, Washington, DC.

Erker, J. C., S. M. Desai, G. G. Schlauder *et al.* (1999a). A hepatitis E virus variant from the United States, molecular characterization and transmission in cynomolgus macaques. *J. Gen. Virol.* **80**, 681–690.

Erker, J. C., S. M. Desai and I. K. Mushawar (1999b). Rapid detection of hepatitis E virus RNA by reverse transcription-polymerase chain reaction using universal oligonucleotide primers. *J. Virol. Methods* **81**, 109–113.

Estes, M. K. (2001). Rotaviruses and their replication. In: D. M. Knipe, P. M. Howley, D. E. Griffin *et al.* (eds), *Fields Virology*, 4th edn, Ch. 54. Lippincott, Williams & Wilkins, Baltimore, MD.

Estes, M. K. (2003). The rotavirus NSP4 enterotoxin, current status and challenges. In: U. Desselberger and J. Gray (eds), *Viral Gastroenteritis. Perspectives in Medical Virology*, Vol. 9, pp. 207–224. Elsevier, London.

Fankhauser, R. L., J. S. Noel, S. S. Monroe *et al.* (1998). Molecular epidemiology of 'Norwalk-like viruses' in outbreaks of gastroenteritis in the United States. *J. Infect. Dis.* **178**, 1571–1578.

Fankhauser, R. L., S. S. Monroe, J. S. Noel *et al.* (2002). Epidemiologic and molecular trends of 'Norwalk-like viruses' associated with outbreaks of gastroenteritis in the United States. *J. Infect. Dis.* **186**, 1–7.

Fattal, B., E. Katzenelson, T. Hostovsky and H. Shuval (1976). Comparison of adsorption elution methods for concentration and detection of viruses in water. *Water Res.* **11**, 955–958.

Favorov, M. O., M. Y. Kosoy, S. A. Tsarev *et al.* (2000). Prevalence of antibody to hepatitis E virus among rodents in the United States. *J. Infect. Dis.* **181**, 449–455.

FDA (US Food and Drug Administration) (2002). Guidance for industry: revised preventive measures to reduce the possible risk of transmission of Cruetzfeldt-Jakob disease (CJD) and variant Creutzfeldt-Jakob Disease (VCJD) by blood and blood products (available at http://www.fda.gov/cber/gdlns/cjdvcjd.pdf>, accessed 20 December 2003).

Feinstone, S. M., R. J. Daemer, I. D. Gust and R. H. Purcell (1983). Live attenuated vaccine for hepatitis A. *Dev. Biol. Stand.* **54**, 429–432.

Feinstone, S. M., A. Z. Kapikian and R. H. Purcell (1973). Hepatitis A, detection by immune electron microscopy of a viruslike antigen associated with acute illness. *Science* **182**, 1026–1028.

Feinstone, S. M. and I. D. Gust (1997). Hepatitis A virus. In: D. D. Richman, R. J. Whitley and F. G. Hayden (eds), *Clinical Virology*, pp. 1049–1072. Churchill Livingstone, New York, NY.

Fiore, A. E. (2004). Hepatitis A transmitted by food. *Clin. Infect. Dis.* **38**, 705–715.

Fix, A.D., M. Abdel-Hamid, R. H. Purcell *et al.* (2000). Prevalence of antibodies to hepatitis E virus in rural Egyptian communities. *Am. J. Trop. Med. Hyg.* **62**, 519–523.

Flechsig, E., I. Hegyi, M. Enari *et al.* (2001). Transmission of scrapie by steel-surface-bound prions. *Mol. Med.* **7**, 679–684.

Fleet, G. H., P. Heiskanen, I. Reid and K. A. Buckle (2000). Foodborne viral illness – status in Australia. *Intl J. Food Microbiol.* **59**, 127–136.

Flewett, T. (1983). Rotavirus in the home and hospital nursery. *Br. Med. J.* **287**, 568–569.

Focaccia, R., H. Sette Jr and O. J. G. Conceicao (1995). Hepatitis E in Brazil [letter]. *Lancet* **346**, 1165.

Fout, G. S., B. C. Martinson, M. W. Moyer and D. R. Dahling (2003). A multiplex reverse transcription-PCR method for detection of human enteric viruses in groundwater. *Appl. Environ. Microbiol.* **69**, 3158–3164.

Fujioka, R. S., P. C. Loh and L. S. Lau (1980). Survival of human enteroviruses in the Hawaiian ocean environment, evidence for virus-inactivating microorganisms. *Appl. Environ. Microbiol.* **39**, 1105–1110.

Gallimore, C. I., C. Pipkin, H. Shrimpton *et al.* (2004). Detection of multiple enteric virus strains within a foodborne outbreak of gastroenteritis, an indication of the source of contamination. *Epidemiol. Infect.* **133**, 41–47.

Gard, S. (1957). Discussion. In: F. W. Hartman (ed.), *Hepatitis Frontiers*, pp. 241–243. Little, Brown, Boston, MA.

Garkavenko, O., A. Obriadina, J. Meng *et al.* (2001). Detection and characterization of swine hepatitis E virus in New Zealand. *J. Med. Virol.* **65**, 525–529.

Gerba, C. P., J. B. Rose, C. N. Haas and K. D. Crabtree (1996). Waterborne rotavirus, a risk assessment. *Water Res.* **30**, 2929–2940.

Gerba, C. P., D. M. Gramos and N. Nwachuku (2002). Comparative inactivation of enteroviruses and adenovirus 2 by UV Light. *Appl. Environ. Microbiol.* **68**, 5167–5169.

Gibbs, T. and D. O. Cliver (1965). Methods for detecting minimal contamination with reovirus. *Health Lab. Sci.* **57**, 81–88.

Gilgen, M., D. Germann, J. Luthy and P. Hubner (1997). Three-step isolation method for sensitive detection of enterovirus, rotavirus, hepatitis A virus, and small round structured viruses in water samples. *Intl J. Food Microbiol.* **37**, 189–199.

Gilger, M. A., D. O. Matson, M. E. Conner *et al.* (1992). Extraintestinal rotavirus infections in children with immunodeficiency. *J. Pediatr.* **120**, 912–917.

Glass, P. J., L. J. White, J. M. Ball *et al.* (2000). Norwalk virus open reading frame 3 encodes a minor structural protein. *J. Virol.* **74**, 6581–6591.

Glass, R. I., J. Noel, D. Mitchell *et al.* (1996). The changing epidemiology of astrovirus-associated gastroenteritis, a review. *Arch. Virol. Suppl.* **12**, 287–300.

Glass, R. I., J. Noel, T. Ando *et al.* (2000). The epidemiology of enteric caliciviruses from humans, a reassessment using new diagnostics. *J. Infect. Dis.* **181** (Suppl. 2), S254–S261.

Glerup, H., H. T. Sorensen, A. Flyvbjerg *et al.* (1994). A 'mini epidemic' of hepatitis A after eating Russian caviar. *J. Hepatol.* **21**, 479.

Glikson, M., E. Galun, R. Oren *et al.* (1992). Relapsing hepatitis A. A review of 14 cases and literature survey. *Medicine (Baltimore)* **71**, 14–23.

Goncales, N. S. L., J. R. R. Pinho, R. C. Moreira *et al.* (2000). Hepatitis E virus immunoglobulin G antibodies in different populations in Campinas, Brazil. *Clin. Diagn. Lab. Immunol.* **7**, 813–816.

Gordon, S. C., K. R. Reddy, L. Schiff and E. R. Schiff (1984). Prolonged intrahepatic cholestasis secondary to acute hepatitis A. *Ann. Intern. Med.* **101**, 635–637.

Goswami, B. B., W. H. Koch and T. A. Cebula (1993). Detection of hepatitis A virus in *Mercenaria mercenaria* by coupled reverse transcription and polymerase chain reacton. *Appl. Environ. Microbiol.* **59**, 2765–2770.

Goswami, B. B., M. Kulka, D. Ngo *et al.* (2002). A polymerase chain reaction-based method for the detection of hepatitis A virus in produce and shellfish. *J. Food Prot.* **65**, 393–402.

Gotz, H., K. Ekdahl, J. Lindback *et al.* (2001). Clinical spectrum and transmission characteristics of infection with Norwalk-like virus, findings from a large community outbreak in Sweden. *Clin. Infect. Dis.* **33**, 622–628.

Gouvea, V., J. R. Allen, R. I. Glass *et al.* (1991). Detection of group B and C rotaviruses by polymerase chain reaction. *J. Clin. Microbiol.* **29**, 519–523.

Grabow, W. O., I. G. Middendorff and N. C. Basson (1978). Role of lime treatment in the removal of bacteria, enteric viruses, and coliphages in a wastewater reclamation plant. *Appl. Environ. Microbiol.* **35**, 663–669.

Grabow, W. O. K., V. Gauss-Müller, O. W. Prozesky and F. Deinhardt (1983). Inactivation of hepatitis A virus and indicator organisms in water by free chlorine residuals. *Appl. Environ. Microbiol.* **46**, 619–624.

Grabow, W. O. K., M. B. Taylor and J. C. de Villers (2001). New methods for the detection of viruses, call for review of drinking water quality guidelines. *Water Sci. Technol.* **43**, 1–8.

Graff, J., J. Ticehurst and B. Flehmig (1993). Detection of hepatitis A virus in sewage sludge by antigen capture polymerase chain reaction. *Appl. Environ. Microbiol.* **59**, 3165–3170.

Graham, D. Y., G. R. Dufour and M. K. Estes (1987). Minimal infective dose of rotavirus. *Arch. Virol.* **92**, 261–271.

Graham, D. Y., S. Jiang, T. Tanaka *et al.* (1994). Norwalk virus infection of volunteers, new insights based on improved assays. *J. Infect. Dis.* **170**, 34–43.

Green, K. Y., R. M. Chanock and A. Z. Kapikian (2001). Human caliciviruses. In: D. M. Knipe and P. M. Howley (eds), *Fields Virology*, 4th edn, pp. 841–874.. Lippincott Williams & Wilkins, Philadelphia, PA.

Greenberg, A. E., L. S. Clesceri and A. D. Eaton (eds) (1992). *Standard Methods for the Examination of Water and Wastewater*, 18th edn. American Public Health Association, Washington, DC.

Greenberg, H. B., J. R. Valdesuso, A. R. Kalica *et al.* (1981). Proteins of Norwalk virus. *J. Virol.* **37**, 994–999.

Guo, M., J. Hayes, K. O. Cho *et al.* (2001). Comparative pathogenesis of tissue culture-adapted and wild-type Cowden porcine enteric calicivirus (PEC) in gnotobiotic pigs and induction of diarrhea by intravenous inoculation of wild-type PEC. *J. Virol.* **75**, 9239–9251.

Grešíková, M. (1972). Studies on the tick-borne arbovirus isolated in Central Europe. *Biologické Práce* **8**(2), 1–109.

Grešíková, M. (1994). Tickborne encephalitis. In: Y. H. Hui, J. R. Gorham, K. D. Murrell and D. O. Cliver (eds), *Foodborne Disease Handbook* (in three volumes), volume 2, pp. 113–135. Marcel Dekker, New York, NY.

Grešíková, M., M. Sekeyová, S. Stúpalová and S. Necas (1975). Sheep milk-borne epidemic of tick-borne encephalitis in Slovakia. *Intervirology* **5**, 57–61.

Grešíková-Kohútová, M. (1959a). The effect of heat on infectivity of the tick-borne encephalitis virus. *Acta Virol.* **3**, 215–221.

Grešíková-Kohútová, M. (1959b). Effect of pH on infectivity of the tick-borne encephalitis virus. *Acta Virol.* **3**, 159–167.

Grešíková-Kohútová, M. (1959c). Persistence of the virus of tick-borne encephalitis in milk & milk products (in Czech). *Česk. Epidemiol. Mikrobiol. Imunol.* **8**, 26–32.

Griffin, D. D., M. Fletcher, M. E. Levy *et al.* (2002). Outbreaks of adult gastroenteritis traced to a single genotype of rotavirus. *J. Infect. Dis.* **185**, 1502–1505.

Grimm, A. C. and G. S. Fout (2002). Development of a molecular method to identify hepatitis E virus in water. *J. Virol. Methods* **101**, 175–188.

Gust, I. D. and S. M. Feinstone (1988). *Hepatitis A*. CRC Press, Boca Raton, FL.

Gust, I. D., N. I. Lehmann, S. Crowe *et al.* (1985). The origin of the HM175 strain of hepatitis A virus. *J. Infect. Dis.* **151**, 365–367.

Halbur, P. G., C. Kasorndorkbua, C. Gilbert *et al.* (2001). Comparative pathogenesis of infection of pigs with hepatitis E viruses recovered from a pig and a human. *J. Clin. Microbiol.* **39**, 918–923.

Halliday, M. L., L. Y. Kang, T. K. Zhou *et al.* (1991). An epidemic of hepatitis A attributable to the ingestion of raw clams in Shanghai, China. *J. Infect. Dis.* **164**, 852–859.

Haqshenas, G., H. L. Shivaprasad, P. R. Woolcock *et al.* (2001). Genetic identification and characterization of a novel virus related to human hepatitis E virus from chickens with hepatitis-splenomegaly syndrome in the United States. *J. Gen. Virol.* **82**, 2449–2462.

Harakeh, M. and M. Butler (1984). Inactivation of human rotavirus SA11 and other enteric viruses in effluent by disinfectants. *J. Hygiene (Lond.)* **93**, 157–163.

Harrington, P. R., L. Lindesmith, B. Yount *et al.* (2002). Binding of Norwalk virus-like particles to ABH histo-blood group antigens is blocked by antisera from infected human volunteers or experimentally vaccinated mice. *J. Virol.* **76**, 12 335–12 343

Heinz, B. A. and D. O. Cliver (1988). Coxsackievirus-cell interactions that initiate infection in porcine ileal explants. *Arch. Virol.* **101**, 35–47.

Heinz, B. A., D. O. Cliver and B. Donohoe (1987). Enterovirus replication in porcine ileal explants. *J. Gen. Virol.* **68**, 2495–2499.

Hennessy, E. P., A. D. Green, M. P. Connor *et al.* (2003). Norwalk virus infection and disease is associated with ABO histo-blood group type. *J. Infect. Dis.* **188**, 176–177.

Hernández, F., R. Monge, C. Jimenez and L. Taylor (1997). Rotavirus and hepatitis A virus in market lettuce (*Latuca sativa*) in Costa Rica. *Intl J. Food Microbiol.* **37**, 221–223.

Herrmann, J. E. and D. O. Cliver (1968). Methods for detecting food-borne enteroviruses. *Appl. Microbiol.* **16**, 1564–1569.

Hewitt, J. and G. E. Greening (2004). Survival and persistence of norovirus, hepatitis A virus, and feline calicivirus in marinated mussels. *J. Food Prot.* **67**, 1743–1750.

Hilton, D. A., A. C. Ghani, L. Conyers *et al.* (2002). Accumulation of prion protein in tonsil and appendix, review of tissue samples. *Br. Med. J.* **325**, 633–634.

Hinson, V. K. and W. R. Tyor (2001). Update on viral encephalitis. *Curr. Opin. Neurol.* **14**, 369–374.

Hirano, M., X. Ding, T. C. Li *et al.* (2003). Evidence for widespread infection of hepatitis E virus among wild rats in Japan. *Hepatol. Res.* **27**, 1–5.

Hohne, M. and E. Schreier (2004). Detection and characterization of norovirus outbreaks in Germany, Application of a one-tube RT-PCR using a fluorogenic real-time detection system. *J. Med. Virol.* **72**, 312–319.

Hollinger, F. B. and J. Ticehurst (1996). Hepatitis A virus. In: B. N. Fields, D. M. Knipe and P. M. Howley (eds), *Fields Virology*, 3rd edn, pp. 735–782. Lippincott-Raven, Philadelphia, PA.

Hopkins, R. S., G. B. Gaspard, F. P. Williams Jr *et al.* (1984). A community waterborne gastroenteritis outbreak, evidence for rotavirus as the agent. *Am. J. Publ. Health* **74**, 263–265.

Hoshino, Y., L. J. Saif, S. Y. Kang *et al.* (1995). Identification of group A rotavirus genes associated with virulence of a porcine rotavirus, host range restriction of a human rotavirus in the gnotobiotic piglet model. *Virology* **209**, 274–280.

Hsieh, S.-Y., X.-J. Meng, Y.-H. Wu *et al.* (1999). Identity of a novel swine hepatitis E virus in Taiwan forming a monophyletic group with Taiwan isolates of human hepatitis E virus. *J. Clin. Microbiol.* **37**, 3828–3834.

Hsiung, G. D. (1962). Further studies on characterization and grouping of ECHO viruses. *Ann. NY Acad. Sci.* **101**, 413–422.

Hu, M. D. (1989). Isolation of HAV and HAV RNA from hairy clams in Qidong region by using cell culture and RNA hybridization technique. *Shanghai Med. J.* **12**, 72.

Huang, C.-C., D. Nguyen, J. Fernandez *et al.* (1992). Molecular cloning and sequencing of the Mexico isolate of hepatitis E virus (HEV). *Virology* **191**, 550–558.

Huang, F. F., G. Haqshenas, D. K. Guenette *et al.* (2002). Detection by reverse transcription-PCR and genetic characterization of field isolates of swine hepatitis E virus from pigs in different geographic regions of the United States. *J. Clin. Microbiol.* **40**, 1326–1332.

Huang, P., T. Farkas, S. Marionneau *et al.* (2003). Noroviruses bind to human ABO, Lewis, and secretor histo-blood group antigens, identification of 4 distinct strain-specific patterns. *J. Infect. Dis.* **188**, 19–31.

Huang, R., N. Nakazono, K. Ishii *et al.* (1995). Existing variations on the gene structure of hepatitis E virus strains from some regions of China. *J. Med. Virol.* **47**, 303–308.

Hunter, N., J. Foster, A. Chong *et al.* (2002). Transmission of prion diseases by blood transfusion. *J. Gen. Virol.* **83**, 2897–2905.

Hurst, C. J., D. R. Dahling, R. S. Safferman and T. Goyke (1984). Comparison of commercial beef extracts and similar materials for recovering viruses from environmental samples. *Can. J. Microbiol.* **30**, 1253–1263.

Hutin, Y. F., V. Pool, E. H. Kramer *et al.* (1999). A multistate, foodborne outbreak of hepatitis A. *N. Engl. J. Med.* **340**, 595–602.

Hutson, A. M., R. L. Atmar, D. Y. Graham and M. K. Estes (2002). Norwalk virus infection and disease is associated with ABO histo-blood group type. *J. Infect. Dis.* **185**, 1335–1337.

Hyams, K. C., M. C. McCarthy, M. Kaur *et al.* (1992). Acute sporadic hepatitis E in children living in Cairo, Egypt. *J. Med. Virol.* **37**, 274–277.

Independent, Bangladesh (2004), available at http://independent-bangladesh.com/news/nov/21/21112004ts.htm A1, accessed 30 December 2004.

Ingrosso, L., V. Vetrugno, F. Cardone and M. Pocchiari (2002). Molecular diagnostics of transmissible spongiform encephalopathies. *Trends Mol. Med.* **8**, 273–280.

Jameel, S., M. Zafrullah, Y. K. Chawla and J. B. Dilawari (2002). Reevaluation of a North India isolate of hepatitis E virus based on the full-length genomic sequence obtained following long RT-PCR. *Virus Res.* **86**, 53–58.

Jay, L. S., D. Comar and L. D. Govenlock (1999). A video study of Australian domestic food-handling practices. *J. Food Prot.* **62**, 1285–1296.

Jayasena, S. D. (1999). Aptamers, an emerging class of molecules that rival antibodies in diagnostics. *Clin. Chem.* **45**, 1628–1650.

Jaykus, L. A., R. De Leon and M. D. Sobsey (1996). A virion concentration method for detection of human enteric viruses in oysters by PCR and oligoprobe hybridization. *Appl. Envr. Microbiol.* **62**, 2074–2080.

Jean, J., B. Blais, A. Darveau and I. Fliss (2002). Simultaneous detection and identification of hepatitis A virus and rotavirus by multiplex nucleic acid sequence-based amplification (NASBA) and microtiter plate hybridization system. *J. Virol. Methods* **105**, 123–132.

Jean, J., J. F. Vachon, O. Moroni *et al.* (2003). Effectiveness of commercial disinfectants for inactivating hepatitis A virus on agri-food surfaces. *J. Food Prot.* **66**, 115–119.

Jensen, H., K. Thomas and D. G. Sharp (1980). Inactivation of coxsackieviruses B3 and B5 in water by chlorine. *Appl. Environ. Microbiol.* **40**, 633–640.

Jiang, X., D. Y. Graham, K. Wang and M. K. Estes (1990). Norwalk virus genome, cloning and characterization. *Science* **250**, 1580–1583.

Jiang, X., M. Wang, D. Y. Graham and M. K. Estes (1992). Expression, self-assembly, and antigenicity of the Norwalk virus capsid protein. J. Virol. **66**, 6527–6532.

Jiang, X., M. Wang, K. Wang and M. K. Estes (1993). Sequence and genome organization of Norwalk virus. *Virology* **195**, 51–61.

Jiang, X, N. Wilton, W. M. Zhong *et al.* (2000). Diagnosis of human caliciviruses by use of enzyme immunoassays. *J. Infect. Dis.* **181** (Suppl. 2), S349–S359.

Jiang, Y. J., G. Y. Liao, W. Zhao *et al.* (2004). Detection of infectious hepatitis A virus by integrated cell culture/strand-specific reverse transcriptase-polymerase chain reaction. *J. Appl. Microbiol.* **97**, 1105–1112.

Johnson, R. T. and C. J. Gibbs Jr (1998). Creutzfeldt-Jakob disease and related transmissible spongiform encephalopathies. *N. Engl. J. Med.* **339**, 1994–2004.

Jothikumar, N., K. Aparna, S. Kamatchiammal *et al.* (1993). Detection of hepatitis E virus in raw and treated wastewater with the polymerase chain reaction. *Appl. Environ. Microbiol.* **59**, 2558–2562.

Jothikumar, N., P. Khanna, R. Paulmurugan *et al.* (1995). A simple device for the concentration and detection of enterovirus, hepatitis E virus and rotavirus from water samples by reverse transcription-polymerase chain reaction. *J. Virol. Methods* **55**, 401–415.

Jothikumar, N., D. O. Cliver and T. W. Mariam (1998). Immunomagnetic capture PCR for rapid concentration and detection of hepatitis A virus from environmental samples. *Appl. Environ. Microbiol.* **64**, 504–508.

Jothikumar, N., R. Paulmurugan, P. Padmanabhan *et al.* (2000). Duplex RT-PCR for simultaneous detection of hepatitis A and hepatitis E virus isolated from drinking water samples. *J. Environ. Monit.* **2**, 587–590.

Kabrane-Lazizi, Y., J. B. Fine, J. Elm *et al.* (1999). Evidence for widespread infection of wild rats with hepatitis E virus in the United States. *Am. J. Trop. Med. Hyg.* **61**, 331–335.

Kageyama, T., S. Kojima, M. Shinohara *et al.* (2003). Broadly reactive and highly sensitive assay for Norwalk-like viruses based on real-time quantitative reverse transcription-PCR. *J. Clin. Microbiol.* **41**, 1548–1557.

Kageyama, T., M. Shinohara, K. Uchida *et al.* (2004). Coexistence of multiple genotypes, including newly identified genotypes, in outbreaks of gastroenteritis due to Norovirus in Japan. *J. Clin. Microbiol.* **42**, 2988–2995.

Kane, M. A., D. W. Bradley, S. M. Shresta *et al.* (1984). Epidemic non-A, non-B hepatitis in Nepal, recovery of a possible etiologic agent and transmission studies to marmosets. *J. Am. Med. Assoc.* **252**, 3140–3145.

Kapikian, A. Z. and R. M. Chanock (1996). Rotaviruses. In: B. N. Fields, D. M. Knipe and P. M. Howley (eds), *Fields Virology*, 3rd edn, pp. 1657–1708. Lippincott-Raven, Philadelphia, PA.

Kapikian, A. Z., R. G. Wyatt, R. Dolin *et al.* (1972). Visualization by immune electron microscopy of a 27-nm particle associated with acute infectious non-bacterial gastroenteritis. *J. Virol.* **10**, 1075–1081.

Karetnyi, Y. V., M. J. R. Gilchrist and S. J. Naides (1999). Hepatitis E virus infection prevalence among selected populations in Iowa. *J. Clin. Virol.* **14**, 51–55.

Karst, S. M., C. E. Wobus, M. Lay *et al.* (2003). STAT1-dependent innate immunity to a Norwalk-like virus. *Science* **299**, 1575–1578.

Kasorndorkbua, C., P. G. Halbur, P. J. Thomas *et al.* (2002). Use of a swine bioassay and a RT-PCR assay to assess the risk of transmission of swine hepatitis E virus in pigs. *J. Virol. Methods* **101**, 71–78.

Kasorndorkbua, C., D. K. Guenette, F. F. Huang *et al.* (2004). Routes of transmission of swine hepatitis E virus in pigs. *J. Clin. Microbiol.* **42**, 5047–5052.

Katayama, H., A. Shimasaki and S. Ohgaki (2002). Development of a virus concentration method and its application to detection of enterovirus and Norwalk virus from coastal seawater. *Appl. Environ. Microbiol.* **68**, 1033–1039.

Kaufman, S. S., N. K. Chatterjee, M. E. Fuschino *et al.* (2003). Calicivirus enteritis in an intestinal transplant recipient. *Am J. Transplant.* **3**, 764–768.

Keswick, B. H., T. K. Satterwhite, P. C. Johnson *et al.* (1985). Inactivation of Norwalk virus in drinking water by chlorine. *Appl. Environ. Microbiol.* **50**, 261–264.

Khan, A. S., C. L. Moe, R. I. Glass *et al.* (1994). Norwalk virus-associated gastroenteritis traced to ice consumption aboard a cruise ship in Hawaii, comparison and application of molecular method-based assays. *J. Clin. Microbiol.* **32**, 318–322.

Khan, W. I., R. Sultana, M. Rahman *et al.* (2000). Viral hepatitis, recent experiences from serological studies in Bangladesh. *Asian Pac. J. Allergy Immunol.* **18**, 99–103.

Khuroo, M. S. (1980). Study of an epidemic of non-A, non-B hepatitis. Possibility of another human hepatitis virus distinct from post-transfusion non-A, non-B type. *Am. J. Med.* **68**, 818–824.

Khuroo, M. S. (2003). Viral hepatitis in international travelers, risks and prevention. *Intl J. Antimicrob. Agents* **21**, 143–152.

Khuroo, M. S., M. R. Teli, S. Skidmore *et al.* (1981). Incidence and severity of viral hepatitis in pregnancy. *Am. J. Med.* **70**, 252–255.

King, J., C. Haase-Pettingell and D. Gossard (2002). Protein folding and misfolding. *Am. Scientist* **90**, 445–453.

Kingsley, D. H. and G. P. Richards (2001). Rapid and efficient extraction method for reverse-transcription-PCR detection of hepatitis A and Norwalk-like viruses in shellfish. *Appl. Environ. Microbiol.* **67**, 4152–4157.

Kingsley, D. H. and G. P. Richards (2003). Persistence of hepatitis A virus in oysters. *J. Food Prot.* **66**, 331–334.

Klohn, P. C., L. Stoltze, E. Flechsig *et al.* (2003). A quantitative, highly sensitive cell-based infectivity assay for mouse scrapie prions. *Proc. Natl. Acad. Sci. USA* **100**, 11 666–11 671.

Koff, R. S. (1995). Seroepidemiology of hepatitis A in the United States. *J. Infect. Dis.* **171**, S19–S23.

Koff, R. S. (1998). Hepatitis A. *Lancet* **351**, 1643–1649.

Kohli, E., L. Maurice, J. F. Vautherot *et al.* (1992). Localization of group specific epitopes as the major capsid protein of group A rotavirus. *J. Gen. Virol.* **73**, 907–914.

Koopmans, M. and E. Duizer (2004). Foodborne viruses, an emerging problem. *Intl J. Food Microbiol.* **90**, 23–41.

Koopmans, M., J. Vinjé, E. Duizer *et al.* (2001). Molecular epidemiology of human enteric caliciviruses in the Netherlands. *Novartis Found. Symp.* **238**, 197–214; discussion 214–218.

Kostenbader, K. D. Jr and D. O. Cliver (1972). Polyelectrolyte flocculation as an aid to recovery of enteroviruses from oysters. *Appl. Microbiol.* **24**, 540–543.

Kostenbader, K. D. Jr and D. O. Cliver (1977). Quest for viruses associated with our food supply. *J. Food Sci.* **42**, 1253–1257, 1268.

Kostenbader, K. D. Jr and D. O. Cliver (1986). Reactivation with pancreatin of coproantibody-neutralized virus in ground beef. *J. Food Prot.* **49**, 920–923.

Krawczynski, K. (1993). Hepatitis E. *Hepatology* **17**, 932–941.

Krawczynski, K. and D. W. Bradley (1989). Enterically transmitted non-A, non-B hepatitis, identification of virus associated antigen in experimentally infected cynomolgus macaques. *J. Infect. Dis.* **159**, 1042–1049.

Krugman, S., R. Ward and J. P. Giles (1962). The natural history of infectious hepatitis. *Am. J. Med.* **32**, 717–728.

Kwo, P. Y., G. G. Schlauder, H. A. Carpenter *et al.* (1997). Acute hepatitis E by a new isolate acquired in the United States. *Mayo Clin. Proc.* **72**, 1133–1136.

Lai, M. M. C. and K. V. Holmes (2001). Coronaviridae, the viruses and their replication. In: D. M. Knipe and P. M. Howley (eds), *Fundamental Virology*, 4th edn, pp. 641–663. Lippincott Williams & Wilkins, Philadelphia, PA.

Landry, E. F., J. M. Vaughn, T. J. Vicale and R. Mann (1982). Inefficient accumulation of low levels of monodispersed and feces-associated poliovirus in oysters. *Appl. Environ. Microbiol.* **44**, 1362–1369.

Lees, D. (2000). Viruses and bivalve shellfish. *Intl J. Food Microbiol.* **59**, 81–116.

Lees, D. N., K. Henshilwood and S. Butcher (1995). Development of a PCR-based method for the detection of enteroviruses and hepatitis A virus in molluscan shellfish and its application to polluted field samples. *Water Sci. Technol.* **31**, 457–464.

Legeay, O., Y. Caudrelier, C. Cordevant *et al.* (2000). Simplified procedure for detection of enteric pathogenic viruses in shellfish by RT-PCR. *J. Virol. Methods* **90**, 1–14.

Leger, R. T., M. K. Boyer, C. P. Pattison and J. F. Maynard (1975). Hepatitis A, report of a common-source outbreak with recovery of a possible etiologic agent. I. Epidemiologic studies. *J. Infect. Dis.* **131**, 163–166.

Leggitt, P. R. and L. A. Jaykus (2000). Detection methods for human enteric viruses in representative foods. *J. Food Prot.* **63**, 1738–1744.

Le Guyader, F., V. Apaire-Marchais, J. Brillet and S. Billaudel (1993). Use of genomic probes to detect hepatitis A virus and enterovirus RNAs in wild shellfish and relationship of viral contamination to bacterial contamination. *Appl. Environ. Microbiol.* **59**, 3963–3968.

Le Guyader, F., E. Dubois, D. Menard and M. Pommepuy (1994). Detection of hepatitis A virus, rotavirus, and enterovirus in naturally contaminated shellfish and sediment by reverse transcription-seminested PCR. *Appl. Environ. Microbiol.* **60**, 3665–3671.

Le Guyader, F., F. H. Neill, M. K. Estes *et al.* (1996). Detection and analysis of a small round-structured virus strain in oysters implicated in an outbreak of acute gastroenteritis. *Appl. Environ. Microbiol.* **62**, 4268–4272.

Le Guyader, F., L. Haugarreau, L. Miossec *et al.* (2000). Three-year study to assess human enteric viruses in shellfish. *Appl. Environ. Microbiol.* **66**, 3241–3248.

Lemon, S. M. (1995). Hepatitis E virus. In: G. L. Mandell, J. E. Bennett and R. Dolin (eds), *Mandell, Douglas, and Bennett's Principles and Practice of Infectious Diseases*, 4th edn, Vol. 2, pp. 1663–1666. Churchill Livingstone, New York, NY.

Lemon, S. M., S. F. Chao, R. W. Jansen *et al.* (1987). Genomic heterogeneity among human and nonhuman strains of hepatitis A virus. *J. Virol.* **61**, 735–742.

Lemon, S. M., P. C. Murphy, P. A. Shields *et al.* (1991). Antigen and genetic variation in cytopathic hepatitis A virus variants arising during persistent infection, evidence for genetic recombination. *J. Virol.* **65**, 2056–2065.

Lemon, S. M., R. W. Jansen and E. A. Brown (1992). Genetic, antigenic, and biological differences between strains of hepatitis A virus. *Vaccine* **10**, S40–S44.

Lemon, S. M., D. E. Schultz and D. R. Shaffer (1997). The molecular basis of attentuation of hepatitis A virus. In: T. J. Harrison and A. J. Zuckerman (eds), *The Molecular Medicine of Viral Hepatitis*, pp. 3–31. John Wiley and Sons, Chichester.

Lemos, G., S. Jameel, S. Panda *et al.* (2000). Hepatitis E virus in Cuba. *J. Clin. Virol.* **16**, 71–75.

Levy, B. S., R. E. Fontaine, C. A. Smith *et al.* (1975). A large food-borne outbreak of hepatitis A. Possible transmission via oropharyngeal secretions. *J. Am. Med. Assoc.* **234**, 289–294.

Li, J. W., X. W. Wang, C. Q. Yuan *et al.* (2002). Detection of enteroviruses and hepatitis A virus in water by consensus primer multiplex RT-PCR. *World J. Gastroenterol.* **8**, 699–702.

Lin, C. M., F. M. Wu, H. K. Kim *et al.* (2003). A comparison of hand washing techniques to remove *Escherichia coli* and caliciviruses under natural or artificial fingernails. *J. Food Prot.* **66**, 2296–2301.

Lindenbach, B. D. and C. M. Rice (2001). Flaviviridae, the viruses and their replication. In: D. M. Knipe and P. M. Howley (eds), *Fundamental Virology*, 4th edn, pp. 589–639. Lippincott Williams & Wilkins, Philadelphia, PA.

Lindesmith, L., C. Moe, S. Marionneau *et al.* (2003). Human susceptibility and resistance to Norwalk virus infection. *Nature Med.* **9**, 548–553.

Lippy, E. C. and S. C. Waltrip (1984). Waterborne disease outbreaks – 1946–1980, a thirty-five year perspective. *J. Am. Water Works Assoc.* **76**, 60–67.

Liu, O. C., H. R. Seraichekas and B. L. Murphy (1967). Viral depuration of the northern quahaug. *Appl. Microbiol.* **15**, 307–315.

Lloyd-Evans, N., V. S. Springthorpe and S. A. Sattar (1986). Chemical disinfection of human rotavirus-contaminated inanimate surfaces. *J. Hygiene (Lond.)* **97**, 163–173.

Logan, K. B., G. E. Scott, N. D. Seeley and S. B. Primrose (1981). A portable device for the rapid concentration of viruses from large volumes of natural freshwater. *J. Virol. Methods* **3**, 241–249.

Loisy, F., R. L. Atmar, J. Cohen *et al.* (2004). Rotavirus VLP2/6, a new tool for tracking rotavirus in the marine environment. *Res. Microbiol.* **155**, 575–578.

Loisy, F., R. L. Atmar, P. Guillon *et al.* (2005). Real-time RT-PCR for norovirus screening in shellfish. *J. Virol. Methods* **123**, 1–7.

Lopez, T., M. Camaco, M. Zayas *et al.* (2005). Silencing the morphogenesis of rotavirus. *J. Virol.* **79**, 184–192.

López-Sabater, E. I., M. Y. Deng and D. O. Cliver (1997). Magnetic immunoseparation PCR assay (MIPA) for detection of hepatitis A virus (HAV) in American oyster (*Crassostrea virginica*). *Lett. Appl. Microbiol.* **24**, 101–104.

Lundgren, O., A. Timar-Peregrin, K. Persson *et al.* (2000). Role of the enteric nervous system in the fluid and electrolyte secretion of rotavirus diarrhea. *Science* **287**, 491–495

Mackow, E. R. (2002). Human group B and C rotaviruses. In: M. J. Blaser, P. D. Smith, J. I. Ravdin *et al.* (eds), *Infections of the Gastrointestinal Tract*, 2nd edn, pp. 879–901. Lippincott Williams & Wilkins, Philadelphia, PA.

Mallett, J. C., L. E. Beghian, T. G. Metcalf and J. D. Kaylor (1991). Potential of irradiation technology for improving shellfish sanitation. *J. Food Safety* **11**, 231–245.

Mandl, C. W., F. X. Heinz and C. Kunz (1988). Sequence of the structural proteins of tick-borne encephalitis virus (western subtype) and comparative analysis with other flaviviruses. *Virology* **166**, 197–205.

Maneerat, Y., E. T. Clayson, K. S. A. Myint *et al.* (1996). Experimental infection of laboratory rats with the hepatitis E virus. *J. Med. Virol.* **48**, 121–128.

Mansuy, J.-M., J. M. Peron, F. Abravanel *et al.* (2004). Hepatitis E in the South West of France in individuals who have never visited an endemic area. *J. Med. Virol.* **74**, 419–424.

Mariam, T. W. and D. O. Cliver (2000). Small round coliphages as surrogates for human viruses in process assessment. *Dairy Food Environ. San.* **20**, 684–689.

Marionneau, S., N. Ruvoen, B. Le Moullac-Vaidye *et al.* (2002). Norwalk virus binds to histo-blood group antigens present on gastroduodenal epithelial cells of secretor individuals. *Gastroenterology* **122**, 1967–1977.

Martina, B. E., B. L. Haagmans, T. Kuiken *et al.* (2003). Virology, SARS virus infection of cats and ferrets. *Nature* **425**, 915.

Marx, F. E., M. B. Taylor and W. O. K. Grabow (1997). A comparison of two sets of primers for the RT-PCR detection of astroviruses in environmental samples. *Water SA (Pretoria)* **23**, 257–262.

Mast, E. E. and K. Krawczynski (1996). Hepatitis E, an overview. *Ann. Rev. Med.* **47**, 257–266.

Mast, E. E., I. K. Kuramoto, M. O. Favorov *et al.* (1997). Prevalence of and risk factors for antibody to hepatitis E virus seroreactivity among blood donors in Northern California. *J. Infect. Dis.* **176**, 34–40.

Matsuda, H., K. Okada, K. Takahashi and S. Mishiro (2003). Severe hepatitis E virus infection after ingestion of uncooked liver from a wild boar. *J. Infect. Dis.* **188**, 944.

Matsui, S. M., J. P. Kim, H. B. Greenberg *et al.* (1991). The isolation and characterization of a Norwalk virus-specific cDNA. *J. Clin. Invest.* **87**, 1456–1461.

Matsuura, K., M. Ishihura, T. Nakayama *et al.* (1988). Ecological studies on reovirus pollution of rivers in Toyama Prefecture. *Microbiol. Immunol.* **32**, 1221–1234.

Mbithi, J. N., V. S. Springthorpe and S. A. Sattar (1990). Chemical disinfection of hepatitis A virus on environmental surfaces. *Appl. Environ. Microbiol.* **56**, 3601–3604.

Mbithi, J. N., V. S. Springthorpe and S. A. Sattar (1991). Effect of relative humidity and air temperature on survival of hepatitis A virus on environmental surfaces. *Appl. Environ. Microbiol.* **57**, 1394–1399.

Mbithi, J. N., V. S. Springthorpe, J. R. Boulet and S. A. Sattar (1992). Survival of hepatitis A virus on human hands and its transfer on contact with animate and inanimate surfaces. *J. Clin. Microbiol.* **30**, 757–763.

Mbithi, J. N., V. S. Springthorpe and S. A. Sattar (1993). Comparative in vivo efficiencies of handwashing agents against hepatitis A virus (HM175) and poliovirus type 1 (Sabin). *Appl. Environ. Microbiol.* **59**, 3463–3469.

MacCallum, F. O. (1947). Homologous serum jaundice. *Lancet* **2**, 691–692.

McCaustland, K. A., W. W. Bond, D. W. Bradley *et al.* (1982). Survival of hepatitis A virus in feces after drying and storage for 1 month. *J. Clin. Microbiol.* **16**, 957–958.

McCaustland, K. A., S. Bi, M. A. Purdy and D. W. Bradley (1991). Application of two RNA extraction methods prior to amplification of hepatitis E virus nucleic acid by the polymerase chain reaction. *J. Virol. Methods* **35**, 331–342.

McCaustland, K. A., K. Krawczynski, J. W. Ebert *et al.* (2000). Hepatitis E virus infection in chimpanzees, a retrospective analysis. *Arch. Virol.* **145**, 1909–1918.

McDonnell, S., K. B. Kirkland, W. G. Hlady *et al.* (1997). Failure of cooking to prevent shellfish-associated viral gastroenteritis. *Arch. Intern. Med.* **157**, 111–116

Mead, P. S., L. Slutsker, V. Dietz *et al.* (1999). Food-related illness and death in the United States. *Emerg. Infect. Dis.* **5**, 607–625.

Medical News Today (2004). Concerns about second wave of 'mad cow' prion infection. http, //www.medicalnewstoday.com/medicalnews.php?newsid=15809 (3 November 2004).

Meeroff, J. C., D. S. Schreiber, J. S. Trier and N. R. Blacklow (1980). Abnormal gastric motor function in viral gastroenteritis. *Ann. Intern. Med.* **92**, 370–373.

Mehnert, D. U. and K. E. Stewien (1993). Detection and distribution of rotavirus in raw sewage and creeks in Sao Paulo, Brazil. *Appl. Environ. Microbiol.* **59**, 140–143.

Melnick, J. L. (1995). History and epidemiology of hepatitis A virus. *J. Infect. Dis.* **171**, S2–S8.

Meng, J., P. Dubreuil and J. Pillot (1997). A new PCR-based seroneutralization assay in cell culture for diagnosis of hepatitis E. J. Clin. *Microbiol. J. Clin. Microbiol.* **35**, 1373–1377.

Meng, X.-J., R. H. Purcell, P. G. Halbur *et al.* (1997). A novel virus in swine is closely related to the human hepatitis E virus. *Proc. Natl. Acad. Sci. USA* **94**, 9860–9865.

Meng, X.-J., P. G. Halbur, M. S. Shapiro *et al.* (1998). Genetic and experimental evidence for cross-species infection by swine hepatitis E virus. *J. Virol.* **72**, 9714–9721.

Meng, X.-J., B. Wiseman, F. Elvinger *et al.* (2002). Prevalence of antibodies to hepatitis E virus in veterinarians working with swine and in normal blood donors in the United States and other countries. *J. Clin. Microbiol.* **40**, 117–122.

Meyerhoff, A. S. and R. J. Jacobs (2001). Transmission of hepatitis A through household contact. *J. Viral Hepatitis* **8**, 454–458.

Meyers, J. D., F. J. Romm, W. S. Tihen and J. A. Bryan (1975). Food-borne hepatitis A in a general hospital. Epidemiologic study of an outbreak attributed to sandwiches. *J. Am. Med. Assoc.* **231**, 1049–1053.

Mikami, T., T. Nakagomi, R. Tsutsui *et al.* (2004). An outbreak of gastroenteritis during school trip caused by serotype G2 group A rotavirus. *J. Med. Virol.* **73**, 460–464.

Millard, J., H. Appleton and J. V. Parry (1987). Studies on heat inactivation of hepatitis A virus with special reference to shellfish, Part 1. Procedures for infection and recovery of virus from laboratory-maintained cockles. *Epidemiol. Infect.* **98**, 397–414.

Monceyron, C. and B. Grinde (1994). Detection of hepatitis A virus in clinical and environmental samples by immunomagnetic separation and PCR. *J. Virol. Methods* **46**, 157–166.

Monroe, S. S., B. Jiang, S. E. Stine *et al.* (1993). Subgenomic RNA sequence of human astrovirus supports classification of Astroviridae as a new family of RNA viruses. *J. Virol.* **67**, 3611–3614.

Morace, G., F. A. Aulicino, C. Angelozzi *et al.* (2002). Microbial quality of wastewater, detection of hepatitis A virus by reverse transcriptase-polymerase chain reaction. *J. Appl. Microbiol.* **92**, 828–836.

Morris, R. (1985). Detection of enteroviruses, an assessment of ten cell lines. *Water Sci. Technol.* **17**(10), 81–88.

Mullendore, J. L., M. D. Sobsey and Y.-S. C. Shieh (2001). Improved method for the recovery of hepatitis A virus from oysters. *J. Virol. Methods* **94**, 25–35.

Muniain-Mujika, I., R. Girones and F. Lucena (2000). Viral contamination of shellfish, evaluation of methods and analysis of bacteriophages and human viruses. *J. Virol. Methods* **89**, 109–118.

Naik, S. R., R. Aggarwal, P. N. Salunke and N. N. Mehrotra (1992). A large waterborne viral hepatitis E epidemic in Kanpur, India. *Bull. WHO* **70**, 597–604.

Najarian, R., D. Caput, W. Gee *et al.* (1985). Primary structure and gene organization of human hepatitis A virus. *Proc. Natl. Acad. Sci. USA* **82**, 2627–2631.

Nanda, S. K., S. K. Panda, H. Durgapal and S. Jameel (1994). Detection of the negative strand of hepatitis E virus RNA in the livers of experimentally infected rhesus monkeys, evidence for viral replication. *J. Med. Virol.* **42**, 237–240.

National CJD Surveillance Unit (2002). *Creutzfeldt-Jakob Disease Surveillance in the UK, Eleventh Annual Report* (available at http: //www.cjd.ed.ac.uk/eleventh/rep2002.htm, accessed 20 December 2003).

National Institute of Infectious Diseases (2000). An outbreak of group A rotavirus infection among adults from eating meals prepared at a restaurant, April 2000 – Shimane. *Infect. Agents Surv. Rep.* **2**, 145–146.

Nibert, M. L. and I. A. Schiff (2001). Reoviruses and their replication. In: D. M. Knipe and P. M. Howley (eds), *Fundamental Virology*, 4th edn, pp. 793–842. Lippincott Williams & Wilkins, Philadelphia, PA.

Nilsson, M., K. O. Hedlund, M. Thorhagen *et al.* (2003). Evolution of human calicivirus RNA in vivo, accumulation of mutations in the protruding P2 domain of the capsid leads to structural changes and possibly a new phenotype. *J. Virol.* **77**, 13 117–13 124.

Nishizawa, T., M. Takahashi, H. Mizuo *et al.* (2003). Characterization of Japanese swine and human hepatitis E virus isolates of genotype IV with 99 % identity over the entire genome. *J. Gen. Virol.* **84**, 1245–1251.

Niu, M. T., L. B. Polish, B. H. Robertson *et al.* (1992). Multistate outbreak of hepatitis A associated with frozen strawberries. *J. Infect. Dis.* **166**, 518–524.

Nuanualsuwan, S. and D. O. Cliver (2002). Pretreatment to avoid positive RT-PCR results with inactivated viruses. *J. Virol. Methods* **104**, 217–225.

Nuanualsuwan, S. and D. O. Cliver (2003). Capsid functions of inactivated human picornaviruses and feline calicivirus. *Appl. Environ. Microbiol.* **69**, 350–357.

Nuanualsuwan, S., T. W. Mariam, S. Himathongkham and D. O. Cliver (2002). Ultraviolet inactivation of feline calicivirus, human enteric viruses and coliphages. *Photochem. Photobiol.* **76**, 406–410.

Nuesch, J., S. Krech and G. Siegl (1988). Detection and characterization of subgenomic RNAs in hepatitis A virus particles. *Virology* **165**, 419–427.

Oberste, M. S., W. A. Nix, K. Maher and M. A. Pallansch (2003). Improved molecular identification of enteroviruses by RT-PCR and amplicon sequencing. *J. Clin. Virol.* **26**, 375–377.

O'Grady, J. (1992). Management of acute and fulminant hepatitis A. **Vaccine 10**, S21–S26.

Oishi, I., K. Yamazaki, T. Kimoto *et al.* (1994). A large outbreak of acute gastroenteritis associated with astrovirus among students and teachers in Osaka, Japan. *J. Infect. Dis.* **170**, 439–443.

Okamoto, H., M. Takahashi, T. Nishizawa *et al.* (2001). Analysis of the complete genome of indigenous swine hepatitis E virus isolated in Japan. *Biochem. Biophys. Res. Commun.* **289**, 926–936.

Ooi, W. W., J. M. Gawoski, P. O. Yarbough and G. A. Pankey (1999). Hepatitis E seroconversion in United States travelers abroad. *Am. J. Trop. Med. Hyg.* **61**, 822–824.

Orru, G., G. Masia, G. Orru *et al.* (2004). Detection and quantitation of hepatitis E virus in human faces by real-time quantitative PCR. *J. Virol. Methods* **118**, 77–82.

Panda, S. K. and S. K. Nanda (1997). Development of vaccines against hepatitis E virus infection. In: T. J. Harrison and A. J. Zuckerman (eds), *The Molecular Medicine of Viral Hepatitis*, pp. 45–62. John Wiley and Sons, Chichester.

Papaevangelou, G. J (1984). Global epidemiology of hepatitis A. In: R. J. Gerety (ed.), *Hepatitis A*, pp. 101–132. Academic Press, Orlando, FL.

Parkin, W. E., P. Marzinsky and M. R. Griffin (1983). Foodborne hepatitis A associated with cheeseburgers. *J. Med. Soc. NJ* **80**, 612–615.

Parrino, T. A., D. S. Schreiber, J. S. Trier *et al.* (1977). Clinical immunity in acute gastroenteritis caused by Norwalk agent. *N. Engl. J. Med.* **297**, 86–89.

Parshionikar, S. U., J. Cashdollar and G. S. Fout (2004). Development of homologous viral internal controls for use in RT-PCR assays of waterborne enteric viruses. *J. Virol. Methods* **121**, 39–48.

Paulson, D. S. (1997). Foodborne disease – controlling the problem. *J. Environ. Health* **59**, 15–19.

Payment, P., F. Gamache and G. Paquette (1988). Microbiological and virological analysis of water from two water filtration plants and their distribution systems. *Can. J. Microbiol.* **34**, 1304–1309.

Pebody, R. G., T. Leino, P. Ruutu *et al.* (1998). Foodborne outbreaks of hepatitis A in low endemic country, an emerging problem? *Epidemiol. Infect.* **120**, 55–59.

Peterson, D. A., L. G. Wolfe, E. P. Larkin and F. W. Deinhardt (1978). Thermal treatment and infectivity of hepatitis A virus in human feces. *J. Med. Virol.* **2**, 201–206.

Philp, J. R., T. P. Hamilton II, T. J. Albert *et al.* (1973). Infectious hepatitis outbreak with mai tai as the vehicle of transmission. *Am. J. Epidemiol.* **97**, 50–54.

Pianetti, A., W. Baffone, B. Citterio *et al.* (2000). Presence of enteroviruses and reoviruses in the waters of the Italian coast of the Adriatic Sea. *Epidemiol. Infect.* **125**, 455–462.

Pina, S., J. Jofre, S. U. Emerson *et al.* (1998). Characterization of a strain of infectious hepatitis E virus isolated from sewage in an area where hepatitis E is not endemic. *Appl. Environ. Microbiol.* **64**, 4485–4488.

Pina, S., M. Buti, M. Cotrina *et al.* (2000). HEV identified in serum from humans with acute hepatitis and in sewage of animal origin in Spain. *J. Hepatol.* **33**, 826–833.

Pina, S., M. Buti, R. Jardi *et al.* (2001). Genetic analysis of hepatitis A virus strains recovered from the environment and from patients with acute hepatitis. *J. Gen. Virol.* **82**, 2955–2963.

Pintó, R. M., F. X. Abad, R. Gajardo and A. Bosch (1996). Detection of infectious astroviruses in water. *Appl. Environ Microbiol.* **62**, 1811–1813.

Plager, H. (1962). The Coxsackie viruses. *Ann. NY Acad. Sci.* **101**, 390–397.

Pletnev, A. G., V. F. Yamshchikov and V. M. Blinov (1986). Tick-borne encephalitis virus genome. The nucleotide sequence coding for virion structural proteins. *FEBS Lett* **200**, 317–321.

Plotkin, S. A., R. I. Carp and A. F. Graham (1962). The polioviruses of man. *Ann. NY Acad. Sci.* **101**, 357–389.

Polish, L. B., B. H. Robertson, B. Khanna *et al.* (1999). Excretion of hepatitis A virus (HAV) in adults, comparison of immunologic and molecular detection methods and relationship between HAV positivity and infectivity in tamarins. *J. Clin. Microbiol.* **37**, 3615–3617.

Potasman, I., A. Paz and M. Odeh (2002). Infectious outbreaks associated with bivalve shellfish consumption, a worldwide perspective. *Clin. Infect. Dis.* **35**, 921–928.

Prasad, B. V., R. Rothnagel, X. Jiang and M. K. Estes (1994). Three-dimensional structure of baculovirus-expressed Norwalk virus capsids. *J. Virol.* **68**, 5117–5125.

Prasad, B. V., M. E. Hardy, T. Dokland *et al.* (1999). X-ray crystallographic structure of the Norwalk virus capsid. *Science* **286**, 287–290.

Prasad, B. V., S. Crawford, J. A. Lawton *et al.* (2001). Structural studies on gastroenteritis viruses. *Novartis Found. Symp.* **238**, 26–37; discussion 37–46.

Prevot, J. S., S. Dubrou and J. Marchel (1993). Detection of human hepatitis A virus in environmental water by an antigen-capture polymerase chain reaction. *Water Sci. Technol.* **27**, 227–233.

Provost, P. J. (1984). *In vitro* propagation of hepatitis A virus. *In* R. J. Gerety (eds) *Hepatitis A*, . pp. 245–261. Academic Press,. Orlando, FL.

Provost, P. J. and M. R. Hilleman (1979). Propagation of human hepatitis A virus in cell culture in vitro. *Proc. Soc. Exp. Biol. Med.* **160**, 213–221.

Purcell, R. H. and J. L. Dienstag (1979). Experimental hepatitis A virus infection. In: T. Oda (ed.), *Hepatitis Viruses, Proceedings of the International Symposium on Hepatitis Viruses held November 18–20, 1976, Tokyo*, pp. 3–12. University Park Press, Baltimore, MD.

Purcell, R. H., S. M. Feinstone and J. R. Ticehurst (1984). Hepatitis A virus. In: G. N. Vyas, J. L. Dienstag and J. H. Hoofnagle (eds), *Viral Hepatitis and Liver Disease*, pp. 9–22. Grune and Stratton, Orlando, FL.

Purcell, R. H., D. C. Wong and M. Shapiro (2002). Relative infectivity of hepatitis A virus by the oral and intravenous routes in 2 species of nonhuman primates. *J. Infect. Dis.* **185**, 1668–1671.

Pusch, D., D.-Y. Oh, S. Wolf *et al.* (2005). Detection of enteric viruses and bacterial indicators in German environmental waters. *Arch. Virol.*, published online 13 January 2005; ISSN, 0304-8608 (paper), 1432-8798 (online).

Racaniello, V. R. (2001). Picornaviridae, the viruses and their replication. In: D. M. Knipe and P. M. Howley (eds), *Fundamental Virology*, 4th edn, pp. 529–566. Lippincott Williams & Wilkins, Philadelphia, PA.

Ramig, R. F. (2004). Pathogenesis of intestinal and systemic rotavirus infection. *J. Virol.* **78**, 10 213–10 220.

Ramsay, C. N. and P. A. Upton (1989). Hepatitis A and frozen raspberries. *Lancet* **1**, 43–44.

Raška, K., J. Helcl, J. Ježek *et al.* (1966). A milk-borne infectious hepatitis epidemic. *J. Hyg. Epidemiol. Microbiol. Immunol.* **10**, 413–428.

Reimann, H. A., A. H. Prince and J. H. Hodges (1945). The cause of epidemic diarrhea, nausea and vomiting (viral dysentery?). *Proc. Soc. Exp. Biol. Med.* **59**, 8–9.

Rey, J. L., Q. Ramadani, J. L. Soares *et al.* (1999). Sero-epidemiological study of the hepatitis epidemic in Mitrovica in the aftermath of the war in Kosovo. *Bull. Soc. Pathol. Exot.* B3–7.

Reyes, G. R., M. A. Purdy, J. P. Kim *et al.* (1990). Isolation of a cDNA from the virus responsible for enterically transmitted non-A, non-B hepatitis. *Science* **247**, 1335–1339.

Reynolds, K. A., C. P. Gerba, M. Abbaszadegan and I. L. Pepper (2001). ICC/PCR detection of enteroviruses and hepatitis A virus in environmental samples. *Can. J. Microbiol.* **47**, 153–157.

Richards, G. P. (1985). Outbreaks of shellfish-associated enteric virus illness in the USA requisite for development of viral guidelines. *J. Food Prot.* **48**, 815–823.

Richards, G. P. (1999). Limitations of molecular biological techniques for assessing the virological safety of foods. *J. Food Prot.* **62**, 691–697.

Richards, G. P. (2001). Enteric virus contamination of foods through industrial practices, a primer on intervention strategies. *J. Ind. Microbiol. Biotechnol.* **27**, 117–125.

Richards, G. P. and D. O. Cliver (2001). Foodborne viruses. In: F. P. Downes and K. Ito (eds), *Compendium of Methods for the Microbiological Examination of Foods*, 4th edn, pp. 447–461. American Public Health Association, Washington, DC.

Rippey, S. R. (1994). Infectious diseases associated with molluscan shellfish consumption. *Clin. Microbiol. Rev.* **7**, 419–425.

Risco, C., J. L. Carrascosa, A. M. Pedregosa *et al.* (1995). Ultrastructure of human astrovirus serotype 2). *J. Gen. Virol.* **76**, 2075–2080.

Robertson, B. H., B. Khanna, O. V. Nainan and H. S. Margolis (1991). Epidemiologic patterns of wild-type hepatitis A virus determined by genetic variation. *J. Infect. Dis.* **163**, 286–292.

Robertson, B. H., R. W. Jansen, B. Khanna *et al.* (1992). Genetic relatedness of hepatitis A virus strains recovered from different geographic regions. *J. Gen. Virol.* **73**, 1365–1377.

Robertson, B. H., F. Averhoff, T. L. Cromeans *et al.* (2000). Genetic relatedness of hepatitis A virus isolates during a community-wide outbreak. *J. Med. Virol.* **62**, 144–150.

Rockx, B. H., H. Vennema, C. J. Hoebe *et al.* (2005). Association of histo-blood group antigens and susceptibility to norovirus infections. *J. Infect. Dis.* **191**, 749–754.

Romalde, J. L., C. Ribao, M. Luz Villarino and J. L. Barja (2004). Comparison of different primer sets for the RT-PCR detection of hepatitis A virus and astrovirus in mussel tissues. *Water Sci. Technol.* **50**, 131–136.

Roos, B. (1956). Hepatitepidemi, spridd genom ostron. *Svenska Läkartidn.* **53**, 989–1003.

Rose, J. B., C. N. Haas and C. P. Gerba (1995). Linking microbiological criteria for foods with quantitative risk assessment. *J. Food Safety* **15**, 121–132.

Rose, J. B. and M. D. Sobsey (1993). Quantitative risk assessment for viral contamination of shellfish and coastal waters. *J. Food Prot.* **56**, 1043–1050.

Rosenblum, L. S., I. R. Mirkin, D. T. Allen *et al.* (1990). A multifocal outbreak of hepatitis A traced to commercially distributed lettuce. *Am. J. Publ. Health* **80**, 1075–1079.

Rosenfield, S. I. and L. A. Jaykus (1999). A multiplex reverse transcription polymerase chain reaction method for the detection of foodborne viruses. *J. Food. Prot.* **62**, 1210–1214.

Rubenstein, R., P. C. Gray, C. M. Wehlburg *et al.* (1998). Detection and discrimination of PrP^{Sc} by multi-spectral ultraviolet fluorescence. *Biochem. Biophys. Res. Commun.* **246**, 100–106.

Sabin, A. B. (1965). Oral poliovirus vaccine. History of its development and prospects for eradication of poliomyelitis. *J. Am. Med. Assoc.* **194**, 872–876.

Safar, J. G., M. Scott, J. Monaghan *et al.* (2002). Measuring prions causing bovine spongiform encephalopathy or chronic wasting disease by immunoassays and transgenic mice. *Natl Biotechnol.* **20**, 1147–1150.

Saif, L. J. and B. Jiang (1994). Non group A rotaviruses of human and animals. In: R. F. Ramig (ed.), *Rotaviruses*, pp. 339–371. Springer-Verlag, New York, NY.

Sair, A. I., D. H. D'Souza, C. L. Moe and L. A. Jaykus (2002). Improved detection of human enteric viruses in foods by RT-PCR. *J. Virol. Methods* **100**, 57–69.

Salo, R. J. and D. O. Cliver (1976). Effect of acid pH, salts, and temperature on the infectivity and physical integrity of enteroviruses. *Arch. Virol.* **52**, 269–282.

Sanchez, G., R. M. Pinto, H. Vanaclocha and A. Bosch (2002). Molecular characterization of hepatitis A virus isolates from a transcontinental shellfish-borne outbreak. *J. Clin. Microbiol.* **40**, 4148–4155.

Sattar, S. A., N. Lloyd-Evans, V. S. Springthorpe and R. C. Nair (1986). Institutional outbreaks of rotavirus diarrhea, potential role of fomites and environmental surfaces as vehicles for virus transmission. *J. Hygiene (Lond.)* **96**, 277–289.

Sattar, S. A., H. Jacobsen, H. Rahman *et al.* (1994). Interruption of rotavirus spread through chemical disinfection. *Infect. Control Hosp. Epidemiol.* **15**, 751–756.

Schlauder, G. G. and G. J. Dawson (2003). Hepatitis E virus. In: P. R. Murray, E. J. Baron, J. H. Jorgenson *et al.* (eds), *Manual of Clinical Microbiology*, 8th edn, pp. 1495–1511. ASM Press, Washington, DC.

Schlauder, G. G. and I. K. Mushahwar (1994). Detection of hepatitis C and E viruses by the polymerase chain reaction. *J. Virol. Methods* **47**, 243–253.

Schlauder, G. G. and I. K. Mushahwar (2001). Genetic heterogeneity of hepatitis E virus. *J. Med. Virol.* **65**, 282–292.

Schlauder, G. G., G. J. Dawson, J. C. Erker *et al.* (1998). The sequence and phylogenetic analysis of a novel hepatitis E virus isolated from a patient with acute hepatitis reported in the United States. *J. Gen. Virol.* **79**, 447–456.

Schlauder, G. G., S. M. Desai, A. R. Zanetti *et al.* (1999). Novel hepatitis E virus (HEV) isolates from Europe, evidence for additional genotypes of HEV. *J. Med. Virol.* **57**, 243–251.

Schmitt, J., M. Beekes, A. Brauer *et al.* (2002). Identification of scrapie infection from blood serum by Fourier transform infrared spectroscopy. *Anal. Chem.* **74**, 3865–3868.

Schoenbaum, S. C., O. Baker and Z. Jerek (1976). Common-source epidemic of hepatitis due to glazed and iced pastries. *Am. J. Epidemiol.* **104**, 74–80.

Scholz, E., U. Heinricy and B. Flehmig (1989). Acid stability of hepatitis A virus. *J. Gen. Virol.* **70**, 2481–2485.

Schonberger, L. B. (1998). New variant Creutzfeldt-Jakob disease and bovine spongiform encephalopathy. *Infect. Dis. Clin. North Am.* **12**, 111–121.

Schreiber, D. S., N. R. Blacklow and J. S. Trier (1973). The mucosal lesion of the proximal small intestine in acute infectious nonbacterial gastroenteritis. *N. Engl. J. Med.* **288**, 1318–1323.

Schreiber, D. S., N. R. Blacklow and J. S. Trier (1974). The small intestinal lesion induced by Hawaii agent acute infectious nonbacterial gastroenteritis. *J. Infect. Dis.* **129**, 705–708.

Schwab, K. J., R. De Leon and M. D. Sobsey (1995). Concentration and purification of beef extract mock eluates from water samples for the detection of enteroviruses, hepatitis A virus, and norwalk virus by reverse transcription-PCR. *Appl. Environ. Microbiol.* **61**, 531–537.

Schwab, K. J., R. De Leon and M. D. Sobsey (1996). Immunoaffinity concentration and purification of waterborne enteric viruses for detection by reverse transcriptase PCR. *Appl. Environ. Microbiol.* **62**, 2086–2094.

Schwab, K. J., F. H. Neill, R. L. Fankhauser *et al.* (2000). Development of methods to detect 'Norwalk-like viruses' (NLVs) and hepatitis A virus in delicatessen foods, application to a food-borne NLV outbreak. *Appl. Environ. Microbiol.* **66**, 213–218.

Schwartz, E., N. P. Jenks, P. Van Damme and E. Galun (1999). Hepatitis E virus infection in travelers. *Clin. Infect. Dis.* **29**, 1312–1314.

Sencan, I., I. Sahin, D. Kaya *et al.* (2004). Assessment of HAV and HEV seroprevalence in children living in post-earthquake camps from Druzce, Turkey. *Eur. J. Epidemiol.* **19**, 461–465.

Shenk, T. E. (2001). Adenoviridae, the viruses and their replication. In: D. M. Knipe and P. M. Howley (ed.), *Fundamental Virology*, 4th edn, pp. 1053–1088. Lippincott Williams & Wilkins, Philadelphia, PA.

Shieh, Y. C., R. S. Baric, J. W. Woods and K. R. Calci (2003). Molecular surveillance of enterovirus and norwalk-like virus in oysters relocated to a municipal-sewage-impacted gulf estuary. *Appl. Environ. Microbiol.* **69**, 7130–7136.

Shieh, Y.-S. C., D. Wait, L. Tai and M. D. Sobsey (1995). Method to remove inhibitors in sewage and other fecal wastes for enterovirus detection by polymerase chain reaction. *J. Virol. Methods* **54**, 51–66.

Shieh, Y.-S. C., R. S. Baric and M. D. Sobsey (1997). Detection of low levels of enteric viruses in metropolitan and airplane sewage. *Appl. Environ. Microbiol.* **63**, 4401–4407.

Siafakas, N., P. Markoulatos, C. Vlachos *et al.* (2003). Molecular sub-grouping of enterovirus reference and wild type strains into distinct genetic clusters using a simple RFLP assay. *Mol. Cell. Probes* **17**, 113–123.

Siegl, G., G. G. Frosner, V. Gauss-Miller *et al.* (1981). The physicochemical properties of infectious hepatitis A virions. *J. Gen. Virol.* **57**, 331–341.

Siegl, G., M. Weitz and G. Kronauer (1984). Stability of hepatitis A virus. *Intervirology* **22**, 218–226.

Singh, V. K. and S. Naik (2001). Simultaneous amplification of DNA and RNA virus using multiplex PCR system. *Clin. Chim. Acta* **308**, 179–181.

Sivapalasingam, S., C. R. Friedman, L. Cohen and R. V. Tauxe (2004). Fresh produce, a growing cause of outbreaks of foodborne illness in the United States, 1973 through 1997. *J. Food Prot.* **67**, 2342–2353.

Skidmore, S. J., P. O. Yarbough, K. A. Gabor and G. R. Reyes (1992). Hepatitis E virus, the cause of a waterbourne hepatitis outbreak. *J. Med. Virol.* **37**, 58–60.

Slomka, M. J. and H. Appleton (1998). Feline calicivirus as a model system for heat inactivation studies of small round structured viruses in shellfish. *Epidemiol. Infect.* **121**, 401–407.

Smith, J. L. (1999). Foodborne infections during pregnancy. *J. Food Prot.* **62**, 818–829.

Smith, J. L. (2001). A review of hepatitis E virus. *J. Food Prot.* **64**, 572–586.

Smorodintsev, A. A. (1958). Tick-borne spring-summer encephalitis. *Perspect. Med. Virol.* **1**, 210–247.

Sobsey, M. D. and L. A. Jaykus (1991). Human enteric viruses and depuration of bivalve mollusks. In: W. S. Otwell, G. E. Rodrick and R. E. Martin (eds), *Molluscan Shellfish Depuration*, pp. 71–114.. CRC Press, Boston, MA.

Sobsey, M. D., P. A. Shields, F. H. [S.] Hauchman *et al.* (1986). Survival and transport of hepatitis A virus in soils, groundwater and wastewater. *Water Sci. Technol.* **18**, 97–106.

Sobsey, M. D., P. A. Shields and F. S. Hauchman (1991). Survival and persistence of hepatitis A virus in environmental samples. In: F. B. Hollinger, S. M. Lemon and H. S. Margolis (eds), *Viral Hepatitis and Liver Disease*, pp. 121–124. Williams and Wilkins, Baltimore, MD.

Sobsey, M. D., D. A. Battigelli, G.-A. Shin and S. Newland (1998). RT-PCR amplification detects inactivated viruses in water and wastewater. *Water Sci. Technol.* **38**, 91–94.

Soldan, C. (2004). Pathology and biochemistry of prion disease varies with genotype in transgenic mice. *Eurosurveillance* **8**(47) (available at http: //www.eurosurveillance.org/ew/2004/041118.asp 2).

Sonoda, H., M. Abe, T. Sugimoto *et al.* (2004). Prevalence of hepatitis E virus (HEV) infection in wild boars and deer and genetic identification of a genotype 3 HEV from a boar in Japan. *J. Clin. Microbiol.* **42**, 5471–5374.

Souto, F. J., C. J. Fontes, R. Parana and L. G. Lyra (1997). Short report, further evidence for hepatitis E in the Brazilian Amazon. *Am. J. Trop. Med. Hyg.* **57**, 149–150.

Spencer, M. D., R. S. Knight and R. G. Will (2002). First hundred cases of variant Creutzfeldt-Jakob disease, retrospective case note review of early psychiatric and neurological features. *Br. Med. J.* **324**, 1479–1482.

Spinner, M. L. and G. D. Di Giovanni (2001). Detection and identification of mammalian reoviruses in surface water by combined cell culture and reverse transcription-PCR. *Appl. Environ. Microbiol.* **67**, 3016–3020.

Strazynski, M., J. Kramer and B. Becker (2002). Thermal inactivation of poliovirus type 1 in water, milk and yoghurt. *Intl J. Food Microbiol.* **74**, 73–78.

Tam, A. W. and D. W. Bradley (1998). Structure and molecular virology. In: A. J. Zuckerman and H. C. Thomas (eds), *Viral Hepatitis*, 2nd edn, pp. 395–401. Churchill Livingstone, London.

Tamada, Y., K. Yano, H. Yatsuhashi *et al.* (2004). Consumption of wild boar linked to cases of hepatitis E [letter]. *J. Hepatol.* **40**, 869–873.

Tei, S., N. Kitajima, K. Takahashi and S. Mishiro (2003). Zoonotic transmission of hepatitis E virus from deer to human beings. *Lancet* **362**, 371–373.

Tei, S., N. Kitajima, S. Ohara *et al.* (2004). Consumption of uncooked deer meat as a risk factor for hepatitis E virus infection, an age- and sex-matched case-control study. *J. Med. Virol.* **74**, 67–70.

Thomas, D. L., G. H. Yarbough, D. Vlahov *et al.* (1997). Seroreactivity to hepatitis E virus in areas where the disease is not endemic. *J. Clin. Microbiol.* **35**, 1244–1247.

Ticehurst, J. R., V. R. Racaniello, B. M. Baroudy *et al.* (1983). Molecular cloning and characterization of hepatitis A virus cDNA. *Proc. Natl. Acad. Sci. USA* **80**, 5885–5889.

Ticehurst, J., L. L. Rhodes Jr, K. Krawczynski *et al.* (1992). Infection of owl monkeys (*Aotus trivirgatus*) and cynomolgus monkeys (*Macaca fascicularis*) with hepatitis E virus from Mexico. *J. Infect. Dis.* **165**, 835–845.

Tomar, B. S. (1998). Hepatitis E in India. *Zhonghua Min Guo Xiao Er Ke Yi Xue Hui Za Zhi.* **39**, 150–156.

Toranzo, A. E., J. L. Barja and F. M. Hetrick (1983). Mechanism of poliovirus inactivation by cell-free filtrates of marine bacteria. *Can. J. Microbiol.* **29**, 1481–1486.

Tratschin, J. D., G. Siegl, G. G. Frosner and F. Deinhardt (1981). Characterization and classification of virus particles associated with hepatitis A. III. Structural proteins. *J. Virol.* **38**, 151–156.

Tree, J. A., M. R. Adams and D. N. Lees (1997). Virus inactivation during disinfection of wastewater by chlorination and UV irradiation and the efficacy of F+ bacteriophage as a 'viral indicator'. *Water Sci. Technol.* **35**, 227–232.

Tsarev, S. A., T. S. Tsareva, S. U. Emerson *et al.* (1994). Infectivity titration of a prototype strain of hepatitis E virus in cynomolgus monkeys. *J. Med. Virol.* **43**, 161–177.

Tsarev, S. A., T. S. Tsareva, S. U. Emerson *et al.* (1995). Experimental hepatitis E in pregnant Rhesus monkeys, failure to transmit hepatitis E virus (HEV) to offspring and evidence of naturally acquired antibodies to HEV. *J. Infect. Dis.* **172**, 31–37.

Turkoglu, S., Y. Lazizi, H. Meng *et al.* (1996). Detection of hepatitis E virus RNA in stools and serum by reverse transcription-PCR. *J. Clin. Microbiol.* **34**, 1568–1571.

Vaidya, S. R., S. D. Chitambar and V. A. Arankalle (2002). Polymerase chain reaction-based prevalence of hepatitis A, hepatitis E and TT viruses in sewage from an endemic area. *J. Hepatol.* **37**, 131–136.

van der Poel, W. H. M., F. Verschoor, R. van der Heide *et al.* (2001). Hepatitis E virus sequences in swine related to sequences in humans, the Netherlands. *Emerg. Infect. Dis.* **7**, 970–976.

Van Tongeren, H. A. (1955). Encephalitis in Austria. IV. Excretion of virus by milk of the experimentally infected goat. *Arch. Gesamte Virusforsch.* **6**, 158–162.

Vaughn, J. M., Y. S. Chen and M. Z. Thomas (1986). Inactivation of human and simian rotaviruses by chlorine. *Appl. Environ. Microbiol.* **51**, 391–394.

Vaughn, J. M., Y. S. Chen, K. Lindburg and D. Morales (1987). Inactivation of human and simian rotaviruses by ozone. *Appl. Environ. Microbiol.* **53**, 2218–2221.

Velazquez, O., H. C. Stetler, C. Avila *et al.* (1990). Epidemic transmission of enterically transmitted non-A, non-B hepatitis in Mexico, 1986–1987). *J. Am. Med. Assoc.* **263**, 3281–3285.

Vento, S., T. Garofano, C. Renzini *et al.* (1998). Fulminant hepatitis associated with hepatitis A virus superinfection in patients with chronic hepatitis C. *N. Engl. J. Med.* **338**, 286–290.

Vinjé, J. and M. P. G. Koopmans (2000). Simultaneous detection and genotyping of 'Norwalk-like viruses' by oligonucleotide array in a reverse line blot hybridization format. *J. Clin. Microbiol.* **38**, 2595–2601.

Vinjé, J., H. Vennema, L. Maunula *et al.* (2003). International collaborative study to compare reverse transcriptase PCR assays for detection and genotyping of noroviruses. *J. Clin. Microbiol.* **41**, 1423–1433.

Viswanathan, R. (1957). Infectious hepatitis in Delhi (1955–56), a critical study, epidemiology. *Indian J. Med. Res.* **45**, 1–30.

Vizzi, E., D. Ferraro, A. Cascio *et al.* (1996). Detection of enteric adenoviruses 40 and 41 in stool specimens by monoclonal antibody-based enzyme immunoassays. *Res. Virol.* **147**, 333–339.

Wadsworth, J. D., S. Joiner, A. F. Hill *et al.* (2001). Tissue distribution of protease resistant prion protein in variant Creutzfeldt-Jakob disease using a highly sensitive immunoblotting assay. *Lancet* **358**, 171–180.

Wald, A. (1995). Hepatitis E. *Adv. Pediatr. Infect. Dis.* **10**, 157–166.

Wang, L. and H. Zhuang (2004). Hepatitis E, an overview and recent advances in vaccine research. *World J. Gastroenterol.* **10**, 2157–2162.

Wang, Y.-C., H. Zhang, N. Xia *et al.* (2002). Prevalence, isolation, and partial sequence analysis fo hepatitis E virus from domestic animals in China. *J. Med. Virol.* **67**, 516–521.

Warburton, A. R., T. G. Wreghitt, A. Rampling *et al.* (1991). Hepatitis A outbreak involving bread. *Epidemiol. Infect.* **106**, 199–202.

Weissmann, C., A. J. Raeber, F. Montrasio *et al.* (2001). Prions and the lymphoreticular system. *Philos. Trans. R. Soc. Lond. B. Biol. Sci.* **356**, 177–184.

Wenner, H. A. (1962). The ECHO viruses. *Ann. NY Acad. Sci.* **101**, 398–412.

Widdowson, M.-A., A. Sulka, S. N. Bulens *et al.* (2005). Norovirus and foodborne disease, United States, 1991–2000. *Emerg. Infect. Dis.* **11**, 95–102.

Wilhelmi, I., E. Roman and A. Sanchez-Fauquier (2003). Viruses causing gastroenteritis. *Clin. Microbiol. Infect.* **9**, 247–262.

Will, R. G., J. W. Ironside, M. Zeidler *et al.* (1996). A new variant of Creutzfeldt-Jakob disease in the UK. *Lancet* **347**, 921–925.

Wille, H., M. D. Michelitsch, V. Guenebaut *et al.* (2002). Structural studies of the scrapie prion protein by electron crystallography. *Proc. Natl. Acad. Sci. USA* **99**, 3563–3568.

Williams, E. S. and M. W. Miller (2002). Chronic wasting disease in deer and elk in North America. *Rev. Sci. Technol.* **21**, 305–316.

Willner, I. R., M. D. Uhl, S. C. Howard *et al.* (1998). Serious hepatitis A, an analysis of patients hospitalized during an urban epidemic in the United States. *Ann. Intern. Med.* **128**, 111–114.

Wobus, C. E., S. M. Karst, L. B. Thackray *et al.* (2004). Replication of Norovirus in cell culture reveals a tropism for dendritic cells and macrophages. *PLoS Biol.* **2**, e432.

Wong, D. C., R. H. Purcell, M. A. Sreenivasan *et al.* (1980). Epidemic and endemic hepatitis in India, evidence for non-A / non-B hepatitis virus etiology. *Lancet* **2**, 876–878.

Woody, M. A. and D. O. Cliver (1995). Effects of temperature and host cell growth phase on replication of F-specific RNA coliphage Q beta. *Appl. Environ. Microbiol.* **61**, 1520–1526.

Woody, M. A. and D. O. Cliver (1997). Replication of coliphage Q beta as affected by host cell number, nutrition, competition from insusceptible cells and non-FRNA coliphages. *J. Appl. Microbiol.* **82**, 431–440.

World Health Organization (2001). The revision of the surveillance case definition for variant Creutzfeldt-Jakob disease (vCJD). WHO/CDS/CSR/EPH/2001.5

Worm, H. C., and G. Wirnsberger (2004). Hepatitis E vaccines, progress and prospects. Drugs **64**, 1517–1531.

Worm, H. C., W. H. M. van der Poel and G. Brandstatter (2002). Hepatitis E, an overview. *Microbes Infect.* **4**, 657–666.

Wyatt, R. G., H. B. Greenberg, D. W. Dalgard *et al.* (1978). Experimental infection of chimpanzees with the Norwalk agent of epidemic viral gastroenteritis. *J. Med. Virol.* **2**, 89–96.

Xu, R. H., J. F. He, M. R. Evans *et al.* (2004). Epidemiologic clues to SARS origin in China. *Emerg. Infect. Dis.* **10**, 1030–1037.

Yamashita, T., S. Kobayashi, K. Sakae *et al.* (1991). Isolation of cytopathic small round viruses with BS-C-1 cells from patients with gastroenteritis. *J. Infect. Dis.* **164**, 954–957.

Yang, F. and X. Xu (1993). A new method of RNA preparation for detection of hepatitis A virus in environmental samples by the polymerase chain reaction. *J. Virol. Methods* **43**, 77–84.

Yazaki, Y., H. Mizuo, M. Takahashi *et al.* (2003). Sporadic acute or fulminant heptatitis E in Hokkaido, Japan, may be food-borne, as suggested by the presence of hepatitis E virus in pig liver as food. *J. Gen. Virol.* **84**, 2351–2357.

Yeager, M., K. A. Dryden, N. H. Olson *et al.* (1990). Three-dimensional structure of rhesus rotavirus by cryoelectron microscopy and image reconstruction. *J. Cell Biol.* **110**, 2133–2144.

Yotsuyanagi, H., K. Koike, K. Yasuda *et al.* (1996). Prolonged fecal excretion of hepatitis A virus in adult patients with hepatitis A as determined by polymerase chain reaction. *Hepatology* **24**, 10–13.

Young, D. C. and D. G. Sharp (1985). Virion conformational forms and the complex inactivation kinetics of echovirus by chlorine in water. *Appl. Environ. Microbiol.* **49**, 359–364.

Zachoval, R. and F. Deinhardt (1998). Natural history and experimental models. In: A. J. Zuckerman and H. C. Thomas (eds), *Viral Hepatitis*, 2nd edn, pp.43–57. Churchill Livingstone, London.

Zafrullah, M., M. H. Ozdener, S. K. Panda and S. Jameel (1997). The ORF3 protein of hepatitis E is a phosphoprotein that associates with the cytoskeleton. *J. Virol.* **71**, 9045–9053.

Zahn, R., A. Liu, T. Luhrs *et al.* (2000). NMR solution structure of the human prion protein. *Proc. Natl. Acad. Sci. USA* **97**, 145–150.

Zanetti, A., G. G. Schlauder, L. Romano *et al.* (1999). Identification of a novel variant of hepatitis E virus in Italy. *J. Med. Virol.* **57**, 356–360.

Zhou, Y.-J., M. K. Estes, X. Jiang and T. G. Metcalf (1991). Concentration and detection of hepatitis A virus and rotavirus from shellfish by hybridization tests. *Appl. Environ. Microbiol.* **57**, 2963–2968.

Zobeley, E., E. Flechsig, A. Cozzio *et al.* (1999). Infectivity of scrapie prions bound to a stainless steel surface. *Mol. Med.* **5**, 240–243.

Foodborne parasites

12

J. P. Dubey, K. D. Murrell, and J. H. Cross

1 Introduction

Although there are many parasites that are transmissible to human beings from animals, important parasites acquired by eating uncooked or undercooked meat or fish are discussed in this chapter. Parasites transmissible by ingestion of food contaminated with soil (*Toxocara, Bayliscaris* and others) or water (*Cryptosporidium, Cyclosplora, Isospora, Giardia* and others) are also discussed.

2 Meatborne protozoa and helminths

2.1 Toxoplasmosis

2.1.1 Etiologic agent

Toxoplasmosis is caused by the protozoon *Toxoplasma gondii.* It is an intracellular coccidian parasite that is classified in the Phylum *Apicomplexa*, Class *Sporozoasida*, Order *Eucoccidiorida*, Suborder *Eimeriorina* and Family *Toxoplasmatidae*. Felids including the domestic cat (*Felis catus*) and wild *Felidae* are definitive hosts, and various warm-blooded animals are intermediate hosts (Figure 12.1).

Like all coccidian parasites, oocysts are excreted only by the definitive hosts, cats. Unsporulated, non-infective oocysts (10 × 11 μm) are shed in feces of infected cats. Oocysts sporulate outside in cat feces within 1 or more days, depending on environmental conditions (Figure 12.2 D, E). Sporulated oocysts can survive in soil and elsewhere in the environment for many months, and they are resistant to freezing and

Foodborne Infections and Intoxications 3e
ISBN-10: 0-12-588365-X
ISBN-13: 978-012-588365-8

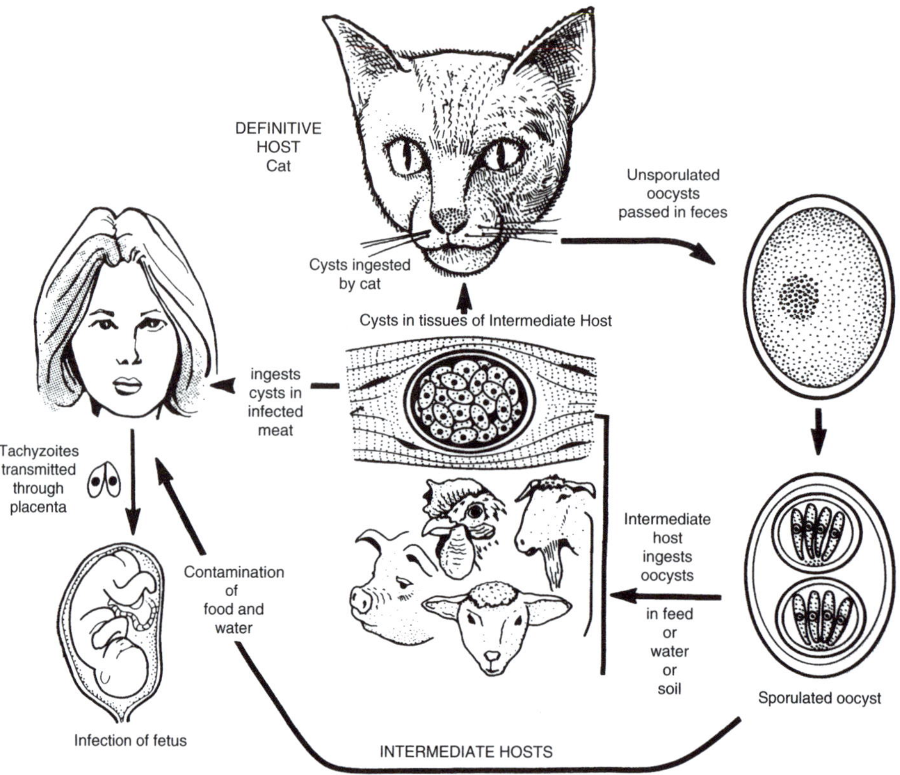

Figure 12.1 Lifecycle of *Toxoplasma gondii*.

drying. Each sporulated oocyst contains two sporocysts, in each of which there are four banana-shaped sporozoites ($8 \times 2\,\mu$m). Sporulated oocysts are infectious to virtually all warm-blooded hosts, including human beings.

Intermediate hosts, including human beings, can become infected by ingesting sporulated oocysts in food or water. Sporozoites are released from oocysts in the gut lumen and become tachyzoites (Greek *tachy*, speed; *zoite*, organism) within 12 hours of penetration into host cells. Tachyzoites ($6 \times 2\,\mu$m) multiply in virtually all cells of the body by division into two zoites (Figure 12.2A). Within 3–4 days, tachyzoites may become encysted in tissues and are called bradyzoites (*brady*, slow). Bradyzoites ($7 \times 2\,\mu$m) are enclosed in a thin, elastic wall, and the entire structure is called a tissue cyst (Figure 12.2B). Tissue cysts are formed in many locations, but particularly in the central nervous system (CNS) – both striated and smooth muscles (Figure 12.2C) – and in many edible organs (reviewed in Dubey *et al.*, 1998a). They are often elongated (up to 100 μm) in muscles and round (up to 70 μm) in the central nervous system. Tissue cysts may persist for the life of the host.

All hosts, including the definitive felid hosts, can become infected by ingesting tissue cysts. In the intermediate host, bradyzoites become tachyzoites within 18 hours of infection, and the tachyzoite–bradyzoite cycle is repeated. However, in the definitive host, the cat, bradyzoites give rise to a conventional coccidian cycle in the small-intestinal epithelium. This coccidian cycle consists of an asexual cycle (schizonts)

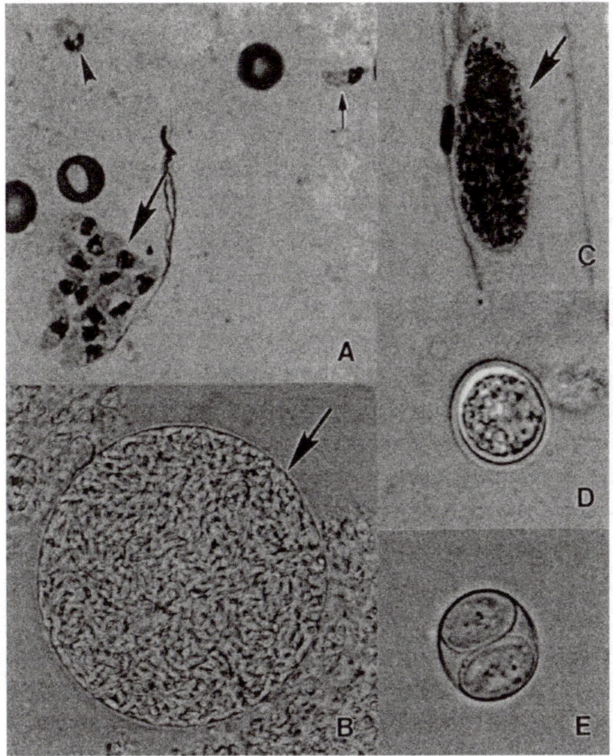

Figure 12.2 Stages of *Toxoplasma gondii*. A, D and E, × 1500; B and C, × 750. (A) Tachyzoites. Note group (large arrow), individual (small arrow) and dividing (arrowhead) organisms in smear of lung. Giemsa stain. (B) A tissue cyst from mouse brain homogenate. Note thin cyst wall (arrow) enclosing hundreds of bradyzoites. Unstained. (C) A tissue cyst in section of skeletal muscle. Note bradyzoites enclosed in a thin cyst wall (arrow). Periodic acid Schiff (PAS) hematoxylin stain. The bradyzoites stain red with PAS (they appear black in this photograph). (D) Unsporulated oocyst. Unstained. (E) Sporulated oocyst. Unstained.

followed by the sexual cycle (gamonts). The male parasite (microgamete) fertilizes the female (macrogamont) parasite, giving rise to an oocyst. The entire asexual and sexual cycle can be completed in the feline intestine within 3 days of ingestion of tissue cysts. The extraintestinal cycle of *T. gondii* in the cat is like that in the intermediate hosts.

The ingestion of tissue cysts or oocysts can cause parasitemia in the mother and lead to infection of the fetus. Congenital *T. gondii* infections are frequent in humans, sheep and goats.

2.1.2 Clinical signs and lesions

Clinical signs and symptoms vary with hosts, mode of infection, immune status, and the organ parasitized. Congenitally acquired toxoplasmosis is generally more severe than is postnatally acquired toxoplasmosis. Chorioretinitis, hydrocephalus, mental retardation and jaundice may all be seen together in severely affected infants, but chorioretinitis is the most common sequela of congenital infection in children (Figure 12.3).

Most infections in immunocompetent human beings are asymptomatic; however, toxoplasmosis can be fatal in immunocompromised people, such as those with

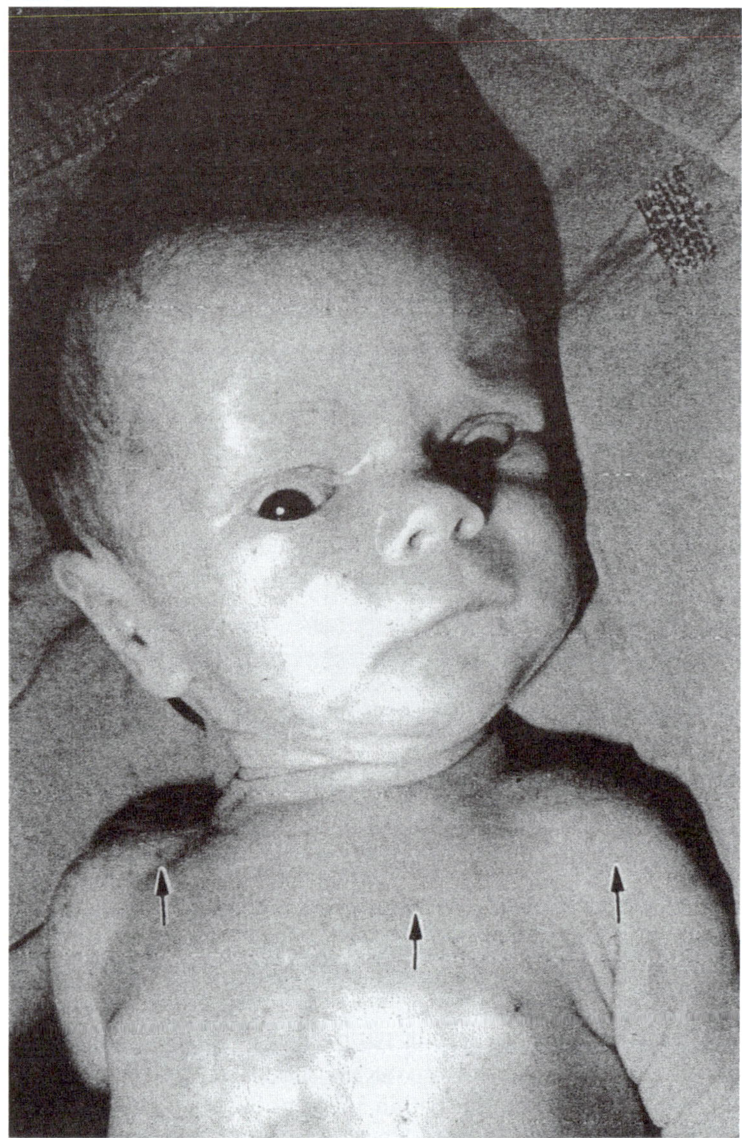

Figure 12.3 Child congenitally infected with *T. gondii*. This child had maculopapular skin rash (arrows), mild hydrocephalus and nystagmus. (courtesy of Dr George Desmonts, Paris, France).

acquired immunodeficiency syndrome (AIDS) or receiving immunotherapy for tumors or in connection with organ transplants. Whereas lymphadenopathy is the most common symptom in immunocompetent humans, encephalitis predominates in immunosuppressed patients. Toxoplasmosis is a leading cause of mortality in AIDS patients in whom clinical disease results from reactivation of latent infection (Figure 12.4). In the immunocompetent person, *T. gondii* rarely causes serious illness. Often, flu-like symptoms accompanied by lymphadenopathy may result. Fever, headache, fatigue, muscle and joint pains, a maculopapular rash, nausea and

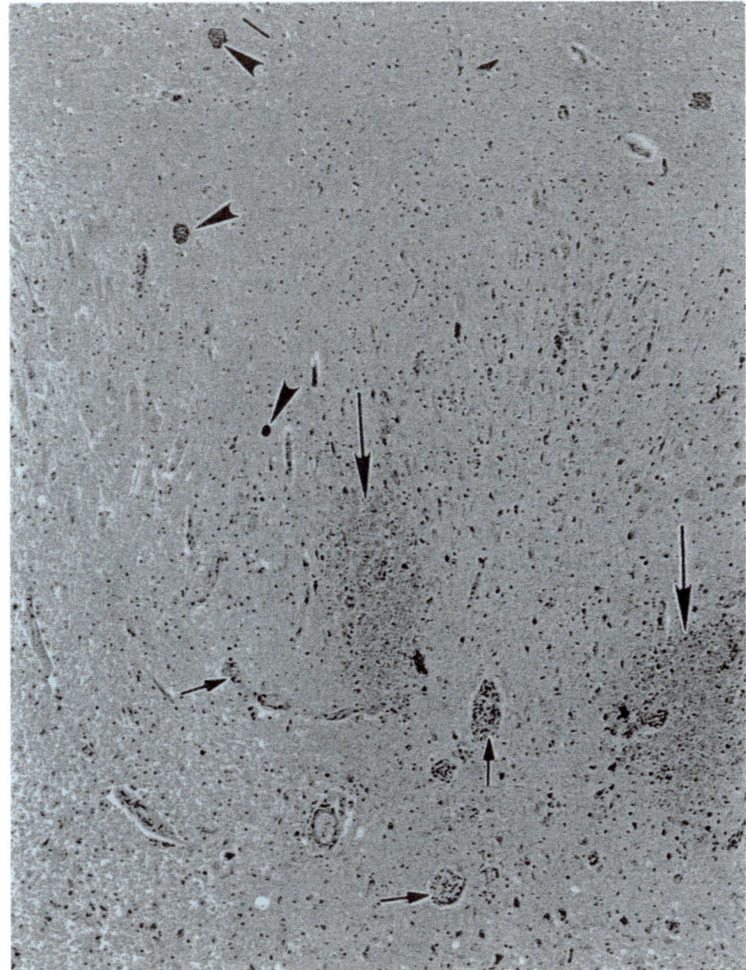

Figure 12.4 Section of brain of an AIDS patient with fatal toxoplasmosis. Note two areas of necrosis (large arrows), vasculitis (small arrows), and few tissue cysts (arrowheads). Numerous tachyzoites are present in necrotic lesions but are not visible at this magnification. Immunohistochemical stain with anti-*T. gondii* serum.

abdominal pain, and loss of vision may result (Dubey and Beattie, 1988; Choi *et al.*, 1997; Dubey, 1997a).

Toxoplasma gondii is a major cause of abortion in sheep and goats, and causes mortality in many other species of animals. In ewes infected during pregnancy, lambs may be mummified, macerated, aborted or stillborn, or may be born weak and die within a week of birth. Severe congenital toxoplasmosis has been reported in cats, dogs and pigs, but not in cattle or horses.

Clinical toxoplasmosis also occurs in adult animals, including dogs and cats. Clinical toxoplasmosis has not been documented in cattle or horses. Toxoplasmosis is a leading cause of mortality in Australian marsupials, especially in zoos.

2.1.3 Pathogenesis

Toxoplasma gondii causes necrosis by active multiplication in cells; it does not produce a toxin. The extent of lesions and associated clinical signs vary depending upon the organ parasitized – for example, even small lesions in eyes are debilitating. Early in the infection, *T. gondii* produces enteritis and mesenteric lymph node necrosis before lesions develop in other organs. Some hosts may die of enteritis. Pneumonia and non-suppurative encephalitis are other important lesions. Concurrent infections, stress and immunosuppressive conditions can aggravate toxoplasmosis. Certain species of animals, e.g. Australian marsupials and New World monkeys, are highly susceptible to infection.

2.1.4 Diagnosis

Clinical signs of toxoplasmosis are non-specific and unreliable for definitive diagnosis. Diagnosis is aided by biologic, serologic or histologic methods, or by combinations of these. *T. gondii* can be isolated from patients by inoculation of laboratory animals (generally mice) or tissue cultures with appropriate material. Secretions, excretions, body fluids, and tissues taken antemortem or tissues taken at necropsy are all possible specimens from which to attempt to isolate *T. gondii*. However, these procedures are too complicated for a routine use in most diagnostic laboratories and are performed only in specialized laboratories.

Finding of *T. gondii* antibodies aids diagnosis. There are numerous serologic tests for detecting humoral antibodies; details can be found in Dubey and Beattie (1988). A number of agglutination tests and enzyme-linked immunosorbent assays (ELISA) are commercially available, which have been modified to detect both IgG and IgM antibodies. The IgM antibodies appear and disappear sooner than do the IgG antibodies, and the difference in IgM and IgG antibody titers is helpful in determining time since infection occurred. IgG antibody titers may persist for life; therefore their presence establishes only that a host has been exposed to *T. gondii*. A 16-fold higher antibody titer in a serum sample, taken 2–4 weeks after the first sample was collected, more accurately indicates an acute acquired infection.

Diagnosis can be made by finding *T. gondii* in host tissue removed by biopsy or at necropsy. A rapid diagnosis may be made by making impression smears of lesions on glass slides. After drying for 10–30 minutes, smears should be fixed in methyl alcohol and stained with Giemsa or any of several other stains used routinely to stain blood smears. *Toxoplasma gondii* tachyzoites (Figure 12.2A) are crescent-shaped, are about $6 \times 2\,\mu$m, have a well-defined nucleus, and are often found in macrophages. Immunohistochemical staining, and detection of parasite DNA by various methods (e.g. polymerase chain reaction), can aid diagnosis. Additional details of diagnosis of toxoplasmosis, especially in children, can be found in Remington *et al.* (1995).

2.1.5 Treatment

Sulfadiazine and pyrimethamine are two drugs widely used for treatment of human toxoplasmosis. These drugs act on multiplying tachyzoites, and have little or no effect on tissue cysts. Spiramycin is used in some countries as a prophylactic to minimize transmission of the parasite from mother to the fetus.

2.1.6 Public health significance

Cats are the key hosts in the epidemiology of toxoplasmosis. They can excrete millions of oocysts in a gram of feces. Cats generally become infected by eating tissue cysts of *T. gondii* in birds and mammals. Cats excrete oocysts for only 1- to -2 week periods in their life.

Food animals become infected with *T. gondii* by ingesting food and water contaminated with sporulated oocysts. Tissue cysts can persist in tissues of live animals for many months (probably for life). *T. gondii* infections are more common in sheep, goats and pigs than in horses and cattle, and therefore ingestion of beef is of little importance in the epidemiology of toxoplasmosis. Human beings become infected by ingesting tissue cysts in undercooked meat, or oocysts in food and water contaminated with cat feces. Approximately 25% of the adult human beings in the US have antibodies to *T. gondii*, and it is estimated that twice that level of prevalence occurs in Central and South America and in France. Approximately 1 in 1000 children is born infected with *T. gondii* in the US, and there are enormous financial and emotional costs involved in raising congenitally infected children (Roberts *et al.*, 1994). Therefore, it is essential to prevent infections in human beings during pregnancy.

2.1.7 Prevention and control

To prevent human infection, hands should be washed thoroughly with soap and water after handling meat. All cutting boards, sink tops, knives and other materials that come in contact with uncooked meat should be washed with soap and water because the stages of *T. gondii* that occur in meat are killed by water. The meat of any animal should be cooked to 70°C before human or animal consumption, and tasting meat while cooking it or while seasoning homemade sausages should be avoided. Pregnant women should definitely avoid contact with cat feces or litter, with soil and with raw meat. Pet cats should be fed only dry, canned or cooked food. Cat litter should be changed every day, preferably not by a pregnant woman. Gloves should be worn while gardening. Vegetables should be washed thoroughly before eating because of the risk of contamination with cat feces.

Because most cats become infected by eating infected tissues, cats should never be fed uncooked meat, viscera or bones, and efforts should be made to keep cats indoors to prevent hunting. Trash cans should be covered to prevent scavenging. Freezing meat overnight in a domestic freezer (–8° to –12°C) can kill most *T. gondii* tissue cysts (Dubey, 1993). Cats should be spayed to control the feline population on farms. Dead animals should be removed promptly to prevent scavenging by cats. A vaccine for prevention of toxoplasmosis in human beings is not yet available, although a vaccine containing live but non-persistent tachyzoites is available in New Zealand and Europe to prevent abortion in sheep.

There is no inspection of meat for *T. gondii* infection for slaughtered animals in any country. Gamma irradiation (0.5 kGy) is effective in rendering *T. gondii* non-infective (Dubey and Thayer, 1994; Dubey *et al.*, 1998b).

2.2 Sarcocystosis

2.2.1 Etiologic agent

Sarcocystis spp. are coccidian parasites that belong to the Phylum *Apicomplexa*, Class *Sporozoasida*, Subclass *Coccidiasina*, Order *Eucoccidiorida* and Family *Sarcocystidae*. The name *Sarcocystis* (Greek *sarkos*, flesh; *kystis*, bladder) implies parasites in muscle, and refers to sarcocysts found in striated muscles of mammals, birds and poikilothermic animals. *Sarcocystis* species have an obligatory prey–predator two-host cycle. The asexual cycle develops only in the intermediate host, which in nature is often a prey animal. Sexual stages develop only in a carnivorous definitive host. A given host may be parasitized by more than one species of *Sarcocystis*, and intermediate and definitive hosts may vary for each species of *Sarcocystis* (Dubey *et al.*, 1989). The definitive host becomes infected by eating sarcocysts containing infective zoites (bradyzoites). The bradyzoites transform into male and female gamonts in the small intestine, and after fertilization oocysts are produced. The oocysts sporulate in the lamina propria and contain two sporocysts, each with four sporozoites. The oocyst wall is thin and often breaks, so that both sporocysts and oocysts are excreted in feces, usually 1 week after ingestion of sarcocysts. The asexual cycle occurs initially in vascular endothelium, later in cells in the bloodstream, and finally in muscles. Two or more asexual cycles (schizonts) are produced in blood vessels. After brief multiplication in leukocytes, merozoites enter muscles and produce sarcocysts.

Sarcocysts mature in about 1 to 2 months, and become infectious for the carnivore host. Sarcocysts of some species of *Sarcocystis* may become grossly visible – for example, *S. gigantea* of sheep.

2.2.2 Pathogenicity

Only some species of *Sarcocystis* are pathogenic (Dubey *et al.*, 1989). Generally, species using canids as definitive hosts are more pathogenic than those using felids. For example, of the three species in cattle, *S. cruzi* (for which the dog is the definitive host) is the most pathogenic, whereas *S. hirsuta* and *S. hominis* (which undergo sexual development in cats and primates, respectively) are only mildly pathogenic. Pathogenicity is manifested in the intermediate host. *Sarcocystis* generally does not cause illness in definitive hosts.

2.2.3 Sarcocystosis in humans

There are two known species of *Sarcocystis* for which humans serve as the definitive host, *S. hominis* and *S. suihominis*. Humans also serve as accidental intermediate hosts for several unidentified species of *Sarcocystis*. Symptoms in persons with intestinal sarcocystosis are different from those in persons with muscular sarcocystosis and vary with the species of *Sarcocystis* causing the infection.

2.2.3.1 *Intestinal sarcocystosis*

Sarcocystis hominis infection is acquired by ingesting uncooked beef containing *S. hominis* sarcocysts. *Sarcocystis hominis* is only mildly pathogenic for humans. A volunteer who ate raw beef from an experimentally infected calf developed nausea, stomach ache and diarrhea 3–6 hours after ingesting the beef; these symptoms lasted 24–36 hours. The volunteer excreted *S. hominis* sporocysts between 14 and 18 days after ingesting the beef. During the period of patency he had diarrhea and stomach ache. Somewhat similar but milder symptoms were experienced by other volunteers who ate uncooked, naturally infected beef (reviewed in Dubey, 1997b).

Sarcocystis suihominis, which is acquired by eating undercooked pork, is more pathogenic than *S. hominis*. Human volunteers developed hypersensitivity-like symptoms; nausea, vomiting, stomachache, diarrhea and dyspnea within 24 hours of ingestion of uncooked pork from naturally or experimentally infected pigs. Sporocysts were shed 11–13 days after ingesting the infected pork (reviewed in Dubey, 1997b).

2.2.3.2 *Muscular sarcocystosis*

Sarcocysts have been found in striated muscles of human beings, mostly as incidental findings. Judging from the published reports, sarcocysts in humans are rare. Most reported cases are from Asia and Southeast Asia (reviewed in Dubey, 1997b). The clinical significance of sarcocysts and the lifecycles of sarcocysts of humans are unknown.

2.2.4 Diagnosis

The antemortem diagnosis of muscular sarcocystosis can only be made by histological examination of muscle collected by biopsy. The finding of immature sarcocysts with metrocytes suggests recently acquired infection. The finding of mature sarcocysts only indicates past infection.

The diagnosis of intestinal sarcocystosis is easily made by fecal examination. As said earlier, sporocsts or oocysts are shed fully sporulated in feces, while those of *Isospora belli* are often shed unsporulated. It is not possible to distinguish one species of *Sarcocystis* from another by examination of sporocysts.

Before the discovery of the lifecycle of *Sarcocystis* and recognition of cattle and pigs as sources of human infection, *Sarcocystis* sporocysts in human feces were referred to as *Isospora hominis*. Because of structural similarities between *S. hominis* and *S. suihominis* sporocysts, it is not possible to distinguish these two species by microscopic examination. Therefore, surveys do not distinguish between these two species. Based on published reports, it appears that intestinal sarcocystosis is more common in Europe than in other continents (reviewed in Dubey, 1997b).

2.2.5 Epidemiology and control

Sarcocystis infection is common in many species of animals worldwide (Dubey *et al.*, 1989). A variety of conditions exist that permit such high prevalence: a host may harbor any of several species of *Sarcocystis*, many definitive hosts are involved in transmission, and large numbers of sporocysts may be shed. *Sarcocystis* oocysts and sporocysts develop in the lamina propria and are discharged over a period of many months and, as oocysts and sporocysts are resistant to freezing, they can overwinter on the pasture; *Sarcocystis* sporocysts and oocysts remain viable for many months in the environment. They may be spread by invertebrate transport hosts. There is little or no immunity to reshedding of sporocysts, and therefore each meal of infected meat can initiate a new round of production of sporocysts. The fact that *Sarcocystis* oocysts, unlike those of many other species of coccidia, are passed in feces in the infective form frees them from dependence on weather conditions for maturation and infectivity.

The poor hygiene practiced in underdeveloped countries during handling of meat from slaughter place to kitchen can be a source of *Sarcocystis* infection. In one survey in India, *S. suihominis* oocysts were found in feces of 14 of 20 children aged 3 to 12 years (Banerjee *et al.*, 1994), indicating that meat was consumed raw at least by some because *S. suihominis* can only be transmitted to humans by the consumption of raw pork. In another study, 3- to 5-year-old children from a slum area were found to consume meat scraps virtually raw, and many pigs from that area harbored *S. suihominis* sarcocysts (Solanki *et al.*, 1991). In European countries where consumption of raw or undercooked meat is relatively common, humans are likely to have intestinal sarcocystosis. In one survey, 60.7% of pigs had *S. suihominis* sarcocysts in their muscles.

2.2.6 Chemotherapy

There is no treatment known for *Sarcocystis* infection of humans.

2.2.7 Control

There is no vaccine to protect livestock or humans against sarcocystosis. Shedding of *Sarcocystis* oocysts and sporocysts in feces of the definitive hosts is the key factor in the spread of *Sarcocystis* infection. Therefore, to interrupt this cycle carnivores should be excluded from animal houses, and from feed, water and bedding for livestock. Uncooked meat or offal should never be fed to carnivores. As freezing can drastically reduce or eliminate infectious sarcocysts in meat, meat should be frozen if not cooked. Exposure to heat at 55°C for 20 minutes kills sarcocysts, so only limited cooking or heating is required to kill sporocysts. Dead livestock should be buried or incinerated. Dead animals should never be left in the field for vultures and carnivores to eat.

2.3 Trichinellosis

2.3.1 The etiologic agent

Trichinellosis (also called trichinosis or trichiniasis) results from infection by a parasitic nematode belonging to the genus *Trichinella*. The classic and most frequent cause of human trichinellosis is the species, *T. spiralis* (Table 12.1); it is normally derived from domestic pork. The pig–man epidemiologic pattern is usually referred to as the synanthropic or domestic cycle. The other species of *Trichinella* (Table 12.1) are morphologically similar, and can infect humans (Murrell *et al.*, 2000). All species of *Trichinella* have a direct lifecycle with complete development in a single host. There is a degree of host specificity, but the complete host range for each species is not fully known. The host capsule surrounding the infective larvae is a modified striated muscle structure called a nurse

Table 12.1 Biological and zoogeographical characteristics of the species of *Trichinella*

	Trichinella spp.				
Trait	***spiralis***	***nativa***	***nelsoni***	***britovi***	***pseudospiralis***
Infectivity for:					
Humans	High	High	High	Moderate	Moderate(?)
Swine	High	Low	Low	Low	Low
Rats	High	Low	Low	Low	Moderate
Mice	High	Low	Low	Low	Moderate
Chickens	No	No	No	No	Yes
Pathogenicity for humans	High	High	Low	Moderate	Low–Moderate
Resistance to freezing	Low	High	Low	Low	Low
Nurse cell development (in days)	16–37	20–30	34–60	24–42	No capsule forms around L_1
Distribution	Cosmopolitan	Arctic	Equatorial Africa	Temperate	Cosmopolitan
Major host	*Suidae*, *Rattus*	*Ursidae*, *Canidae*	*Hyaenidae*, *Felidae*	*Suidae*, *Canidae*	Mammals, birds
Diagnostic DNA probes available	Yes	Yes	Yes	Yes	Yes
Unique alloenzyme markers	6	2	4	1	12

See Pozio *et al.* (1992).

cell, which is digested away in the stomach when the infected muscle is ingested by the next host (Despommier, 1983). The free larvae (L_1) then move into the upper small intestine and invade the columnar epithelial intestinal cells. Within 30 hours the larvae undergo four molts to reach the mature adult stages, the males and females. After mating, the female begins shedding live newborn larvae (NBL), about 5 days post-infection; the NBL are early developmental forms of the L_1 stage. The persistence of adult worms in the intestine of humans may last for many weeks (Murrell and Bruschi, 1994). The NBL migrate throughout the body, via the blood and lymph circulatory system. Although the NBL may attempt to invade many different tissues, they are only successful if they can enter striated skeletal muscle cells. The NBL continue to grow and develop during the first 2 weeks of intracellular life, until they reach the fully developed L_1 infective stage. The longevity of the nurse cell–L_1 complex appears to vary by parasite and host species, but it generally persists for one to several years before calcification and death occur. The lifecycle is completed when the host's infected muscle is ingested by a suitable host. A host capsule does not develop around *T. pseudospiralis* muscle larvae.

2.3.2 Epidemiology and prevalence

The prevalence of swine trichinellosis and the incidence of human trichinellosis appear to be greater in developing countries such as China, Thailand, Mexico, Argentina, Bolivia and some Central European countries (Table 12.2). The more important features of this parasite's epidemiology are its obligatory transmission by

Table 12.2 Summary of incidence reports of human Trichinellosis for Europe, Africa, Asia and Latin America

Country	Years	Number of cases	Source
Argentina	1965–1995	1677	Pork
Austria	1988	12	Pork
Bulgaria	1993–1995	2335	Pork?
Canada/Alaska	1970–1985	424	Polar bear, walrus
Chile	1963–1983	1926	Pork
China	1982–1992	5872	Pork, dog, game
Czechoslovakia	1980	77	Wild boar
Egypt	1975–1985	2 outbreaks	Pork
France	1983–1986	1505	Horse, pork, wild boar
Germany	1980–1982	258	Pork, wild boar
Greenland	1948–1984	600	Polar bear, walrus
Italy	1984–1996	507	Horse, wild boar
Japan	1974–1981	87	Bear
Kenya	1959–1963	40	Wild 'pig'
Lebanon	1981	100	Pork
Lithuania	1992–1995	1221	Pork, wild boar
Mexico	1963–1983	83	Pork
Poland	1983–1996	1427	Pork, wild boar
Russia	1993–1994	1720	Wild animals, pork?
Senegal	1967	9	Warthog
Spain	1981–1996	117	Pork, wild boar
Tanzania	1977	11	Wild 'pig'
Thailand	1962–1988	3000	Pork
US	1970–1995	1875	Pork, bear, cougar
Yugoslavia (former)	1983–1985	1734	Pork

ingestion of meat and its existence in two normally separate ecological systems, the sylvatic and the domestic (Murrell *et al.*, 1987; Campbell, 1991). In certain circumstances the two biotopes are linked through man's activities, resulting in the exposure of humans to *Trichinella* species normally confined to sylvatic animals. The species still most frequently associated with human infection is *T. spiralis*, the type that is normally found in domestic pigs. The domestic cycle of *T. spiralis* involves a complex set of potential routes (Murrell and Bruschi, 1994). Transmission on a farm may result from predation on or scavenging on other animals (e.g. rodents), hog cannibalism, and the feeding of uncooked meat scraps. Increasingly, the importance of wild animal reservoirs is being recognized (Pozio, 2000). Examples include recent human infections attributed to *T. pseudospiralis* in New Zealand in 1994 (Andrews *et al.*, 1995), apparently derived from the consumption of unidentified meat, and more recently in Thailand, where 59 people were involved after consumption of raw meat from a feral pig (Jongwutiwes *et al.*, 1998).

The worldwide incidence of human trichinellosis has declined substantially over the past few decades, but outbreaks are still frequent, especially in developing regions (Table 12.2). In developed countries, the epidemiology of human trichinellosis is typified by urban common-source outbreaks. In the US, the largest human outbreaks have occurred among ethnic groups with preferences for raw or only partially cooked pork. Infected meat is typically purchased from local supermarkets, butchers' shops or other commercial outlets. However, in recent years nearly a third or more of human infections in the US have been derived from wild animal meat (Table 12.3). In Europe, where the safeguard of pork inspection is mandatory, most recent outbreaks have resulted from infected horsemeat or wild boar. The resurgence of trichinellosis in Central Europe appears to result from increased transmission from both pork and wild game (Dupouy-Camet *et al.*, 1994). In Latin America and Asia, however, domestic pork appears to be the chief source of infection.

2.3.3 Pathology and clinical aspects

The ingestion of 500 or more larvae by a human risks clinical disease. As will be discussed below, there is evidence that most infections with *T. spiralis* are unnoticed or confused with some other illness (e.g. influenza) (Murrell and Bruschi, 1994; Capo and Despommier, 1996). In heavy infections, illness is reflected in gastrointestinal signs such as nausea, abdominal pain and diarrhea. Coinciding with muscle invasion by NBL is acute muscular pain, facial edema, fever and eosinophilia. Cardiomyopathy is not uncommon; it results from unsuccessful invasion of larvae in cardiac muscle. The chief factor in the acute muscle phase of this disease is the host's immune response to invasion; hence immunosuppressants are often administered in life-threatening cases. The host's intestinal immune response contributes, along with drugs, to the expulsion of adult parasites from the intestine. The symptoms in humans can differ according to which species is involved (Table 12.1), and hence treatment may be dependent upon proper identification of the infecting species.

2.3.4 Diagnosis

Guidelines for the diagnosis of infection have been recommended by the International Commission on Trichinellosis. Diagnosis may be made by either a direct or an indirect demonstration of infection (Murrell and Bruschi, 1994; Capo and Despommier, 1996).

Table 12.3 Cases of trichinellosis by reported food source, USA

	Reporting period			
	1975–1981		1982–1986	
Reported food source	**No. of cases**	**% of total**	**No. of cases**	**% of total**
Pork products:				
Domestic pig	740	69.4	172	59.9
Wild pig	9	0.8	225	7.7
Subtotal	749		194	
Non-pork products:				
Bear meat	64	6.0	42	14.6
Other wild animal	68	6.4	2	4.2
Ground beef[a]	67	6.3	12	4.2
Subtotal	199		56	
Unknown	118	11.1	37	12.9
Total	1066		787	

[a] Mixed with pork during processing.
From Bailey and Schantz (1990).

2.3.4.1 Enteral phase

Adult *Trichinella* can sometimes be recovered from the intestinal mucosa at post-mortem examination and could theoretically (but not practicably) be recovered from a living patient by duodenal aspiration or biopsy.

2.3.4.2 Dissemination stage

Newborn larvae (NBL) are transported via the bloodstream and are disseminated to various parts of the body. Experimentally, they can be filtered from the blood by means of a 3-μm pore filter, but it is unlikely that this method will be useful in routine diagnosis because of the relatively small number of NBL in the blood at any given time. However, a PCR test for circulatory NBL has been reported (Uparanukraw and Morakote, 1997).

2.3.4.3 Muscle stage

This stage offers the best chance for direct demonstration of the organism. The larvae may be found by examination of a biopsy specimen taken from a superficial skeletal muscle. The procedure should not be undertaken if a firm diagnosis can be made by other means. Larvae that have reached the musculature within the first 3 weeks of infection are more readily detected by the compression and histologic techniques; otherwise, digestion of muscle tissue is the preferred method (Murrell and Bruschi, 1994).

2.3.4.4 Indirect demonstration

A negative biopsy does not exclude the possibility of infection. Circulating antibody can be detected even in lightly infected patients 3–4 weeks after infection, and as early as 2 weeks in heavily infected individuals. A variety of serologic tests can be used, but the enzyme-linked immunosorbent assay (ELISA) has proved most useful (see Murrell and Bruschi, 1994). The ELISA is more sensitive, and can also detect a rise and fall in antibody levels. However, some problems in specificity have been observed when these tests are used with crude antigen extracts; therefore, considerable effort has been made recently to produce more refined antigens. Stichosome antigens derived either from the larva's excretions and secretions during *in vitro* culture, or by somatic extraction, have proved superior to crude extracts. Recently, N-glycan antigen from *T. spiralis* larvae has been synthesized and has proved to be very promising as a diagnostic antigen (Reason *et al.*, 1994). Skin tests can also be performed, but these lack specificity and are further handicapped by the persistence of reactivity in patients as long as 10 years after initial infection.

2.3.5 Prevention and control

2.3.5.1 Pre-slaughter prevention

The establishment and maintenance of *Trichinella*-free pig herds is based on the requirement that pigs be prevented from eating uncooked meat, whether meat scraps, rodents, other wild animals or dead pig carcasses (Murrell, 1995). Because *Trichinella* is transmitted only through ingestion of meat, interruption of its transmission can be targeted at this unique pathway. However, the farm management level required to accomplish this is frequently difficult to achieve because of entrenched cultural, socioeconomic and historical conditions. Regardless, the elements of an effective preventive pig management program are:

- Strict adherence to garbage feeding regulations, particularly cooking requirements for waste material (212°F, 100°C for 30 minutes)
- Stringent rodent control
- Preventing exposure of pigs to dead animal carcasses, including other pigs
- Prompt and proper disposal of dead pig and other animal carcasses (e.g. burial, incineration or rendering); this minimizes infection risk for commensal wild animals
- Construction of effective barriers between pigs, wild animals and even domestic pets.

The importance of wild animals in the epidemiology and control of synanthropic (domestic) trichinellosis cannot be overly emphasized (Pozio *et al.*, 1996). There is strong, but indirect, evidence that feral dogs and cats and farm-associated wild animals may serve as important reservoirs of infection and can reintroduce *Trichinella* into a pig herd unless great care is taken to prevent exposure (Murrell *et al.*, 1987).

Public health and veterinary services authorities have sufficient means to establish and maintain continuous surveillance programs. Such programs allow identification of infected herds and make it possible to attempt eradication. National-level surveillance can be carried out using slaughter inspection data coupled with trace-backs to farms of origin. These are standard procedures in most European countries. In many developed countries (e.g. the European Union and Chile), inspection at slaughter is mandatory. In some countries (e.g. the US) national prevalence studies are periodically conducted using serological and/or tissue digestion tests to monitor herd status and pinpoint 'hot spots'.

2.3.5.2 Post-slaughter control

Meat inspection has proved to be a very successful strategy for controlling trichinellosis. Government-organized and -supervised inspection has been a crucial factor in the dramatic lowering of the incidence of this zoonosis in many countries (e.g. the European Union). Inspection currently is performed by either of two direct methods: microscopical examination or muscle sample digestion (Gamble and Murrell, 1987). The trichinoscope method utilizes small pieces of the cura of the diaphragm, which are compressed between two glass plates and examined under low microscopical power. This method is expensive and labor-intensive, and is not standardized worldwide (i.e. regarding the number and sizes of pieces to be examined). The practical limit of sensitivity is about three trichinae larvae per gram of muscle. In the US, where 60–68% of infected animals harbor less than one larva, the microscopical method has serious drawbacks for an eradication strategy (Schad *et al.*, 1985). The digestion method is rapidly supplanting the trichinoscope procedure in most countries where inspection for trichinae is practiced (e.g. the European Union). This method, introduced about 15 years ago, involves the artificial digestion (pepsin–HCl) of diaphragm tissue pooled into batches in order to reduce the number of samples and time required for examination. The digestion method offers considerable economic advantages because of lower labor demands, especially in abattoirs with high throughputs; the cost may be one-tenth that of trichinelloscopy. In the US and the European Union, all horsemeat for human consumption is inspected by the pooled digestion procedure. More recently, immunological tests have been developed for use in abattoir testing. An ELISA test using stichosome antigens has been tested under modern high-volume slaughterhouse conditions and found to be highly sensitive and practical (Oliver *et al.*, 1989). These results have prompted the US Department of Agriculture to promulgate a regulation permitting the use of immunological tests for slaughterhouse inspection. This regulation (9 CFR. Ch. III, 318.10f) allows approval of any test that detects at least 98% of swine bearing trichinae at larval densities equal to or less than one trichina per diaphragm pillar muscle at a 95% confidence interval. As improvements continue to be made in instrumentation and procedures, alternative replacement of digestion test in large volume abattoirs may be expected.

In countries where meat inspection for trichinae is not mandatory, other strategies for reducing consumer risk are followed. In the US, for example, consumers are advised on proper meat-handling procedures (e.g. cooking, freezing, curing) for killing any trichinae present. Cooking to an internal temperature of 60°C for at least 1 minute is advised. Consumers are also urged to freeze pork at either −15°C for 20 days, −23°C for 10 days, or −30°C for 6 days if the meat is less than 15 cm thick. These temperatures may not be adequate, however, for wild game meat infected with species such as *T. nativa.* The process for curing pork must be one that has been demonstrated to be effective (FSIS, 1990). Commercial production of ready-to-eat pork products is carried out under scrutiny of regulatory agencies to ensure food safety.

A role for food irradiation in the treatment of meat to reduce the risk of foodborne microorganisms and parasites is receiving serious consideration. Gamma irradiation is highly effective for the devitalization of *T. spiralis* larvae in pork (Brake *et al.*, 1985). The Food and Drug Administration in the US has approved low-dose irradiation (up to 1.0 kGy) for the treatment of pork to control trichinellosis.

Because wild game is an important source of infection for humans and pigs, all such meat should be considered as suspect and should only be consumed either after inspection by the trichinoscope or digestion method or after thorough cooking. Many countries (e.g. Russia, Germany) have instituted mandatory inspection of wild game, especially for wild boars. Others provide educational programs for hunters and consumers of game foods.

2.4 Taeniasis/cysticercosis

2.4.1 Etiologic agents

The terms *cysticercosis* and *taeniasis* refer to infections with larval and adult tapeworms, respectively. The important feature of this particular zoonosis is that the larvae are meatborne (beef or pork) and the adult stages develop only in the intestines of humans (obligate host). There are two species, *Taenia saginata* ('beef tapeworm') and *T. solium* ('pork tapeworm').

The adult tapeworm stage of *T. saginata* and *T. solium* resides in the human small intestine and is composed of a chain (strobila) of segments (proglottids) which contain both male and female reproductive systems (Murrell *et al.*, 1986). As the segments mature and fill with eggs, they detach and pass out of the anus, either free or in the fecal bolus. The life span of an adult tapeworm may be as long as 30–40 years. The number of eggs shed per day from a host may be very high (500 000–1 million) which leads to high environmental contamination. These eggs contain an infective stage (oncosphere), which matures in the environment. It is probably impossible to distinguish the two species by morphology of the eggs; however, it is possible now to do this with PCR technology (see below). Symptoms vary in their intensity, and some people never realize that they have a tapeworm. Most people, however, experience symptoms which '...may include nervousness, insomnia, anorexia, loss of weight, abdominal pain, and digestive disturbances' (Roberts and Murrell, 1993). Occasionally the appendix, uterus or biliary tract is invaded, and serious disorders can occur.

In the US, diagnostic laboratories have diagnosed *Taenia* tapeworms in 0.056% of stool specimens, over half of which were reported from western states (Schantz and McAuley, 1991). This is probably a conservative estimate because many cases of taeniasis, both *T. solium* and *T. saginata*, are asymptomatic. An annual average of 1104 tapeworm cases was documented by the Centers for Disease Control and Prevention (CDC) between 1978 and 1981. Medical treatment to eliminate the tapeworm generally requires a couple of visits to a physician, at least one laboratory test, and drug therapy (Roberts and Murrell, 1993). The medical cost plus the income loss for these mild cases is estimated at $238 per case; annual costs, then, for 1104 cases in the US total US $263 228.

Around 45 million people have been estimated to harbor a *T. saginata* tapeworm – 11 million in Europe, 15 million in Asia, 18 million in Africa and 1 million in South America (Roberts *et al.*, 1994). In Laos, 12% are infected with either *T. saginata* or *T. solium*. *T. solium* tapeworm is a significant public health problem in Latin America, where infection rates range from 0.3% in Chile to 1.13% in Guatemala.

Biological and epidemiological studies in Southeast Asia have demonstrated the presence of a subspecies of *T. saginata* ('Taiwan Taenia') (Fan, 1991). This Asian form is most notable for its use of pigs rather than cattle as an intermediate host.

When *T. saginata* eggs are ingested by cattle, the oncosphere stage is released in the intestine; it then penetrates the gut and migrates throughout the body via the circulatory system. Oncospheres that invade skeletal muscle or heart muscle develop to the cysticercus stage, a fluid-filled cyst or small bladder. When beef that is either raw or improperly cooked is eaten by humans, the larval cyst is freed and attaches by means of a small head (scolex) with suckers to the intestinal wall. Over the span of a couple of months, the tapeworm develops and begins to shed eggs, completing the lifecycle (Murrell *et al.*, 1986).

The development of *T. solium* is similar in its intermediate host (pigs or humans), except that the cysticerci are distributed throughout the liver, brain, central nervous system (neurocysticercosis), skeletal muscle and myocardium. Neurocysticercosis is considered to be one of the most common infections of the human central nervous system (CNS). This disease is increasingly recognized as a public health problem, especially in developing countries (Flisser, 1988). Although it is uncommon in the US, Canada and Western Europe, its prevalence in Latin America, China and Africa appears to be relatively high (Flisser, 1988; Cao *et al.*, 1996). Neurocysticercosis is reported from some regions to be a major cause of epilepsy (Schantz, 1991; Garcia *et al.*, 1993).

2.4.2 Epidemiology

Taeniid eggs are capable of surviving for long periods in the environment, and are resistant to moderate desiccation, to disinfectants and to low temperature (4–5°C). The eggs may remain infective for 4–6 months, depending on the moisture content of the microenvironment (Snyder and Murrell, 1986). For example, the longevity may be 71 days in liquid manure, 33 days in river water, and up to 150 days on pasture.

Feedlot outbreaks of bovine cysticercosis are usually traced to human carriers who have handled cattle or feed with contaminated hands, from the passage out of the anus of egg-bearing segments (proglottids) into the environment while working in the lot, or by indiscriminate defecation in and around feed storage facilities (Murrell *et al.*, 1986).

Sewage sludge and effluent should be regarded as high-risk factors in the epidemiology of this zoonosis, especially when used on agricultural lands. This infection source has emerged as a particular problem in Europe in recent years (Barbier *et al.*, 1990; Murrell, 1995). This has been exacerbated by the introduction of various chemicals and detergents in households and industry which affect the efficiency of the sewage treatment process. Mechanical transmission by birds that have fed on sewage sludge has also been implicated (see Murrell, 1995).

Epidemiologic investigations on porcine cysticercosis have been less extensive than those for bovine infections. However, studies in Mexico indicate that the chief risk factors are primitive husbandry conditions that permit pigs access to human feces (in the absence of indoor toilets), or the open access to outdoor latrines (Sarti Gutierrez *et al.*, 1992). Similar risk factors have been reported in China (Cao *et al.*, 1996).

The increased immigration of people into the US, especially from endemic areas such as Central and South America, has yielded an increasing incidence of neurocysticercoses (Schantz, 1991). Because of improved diagnosis and the increased immigration from endemic countries, the number of neurocysticercosis cases diagnosed in cities such as Los Angeles has increased four-fold since the late 1970s (Schantz, 1991). Locally acquired and travel acquired cases are also being reported more frequently (Sorvillo *et al.*, 1992). Similarly, the employment of domestic workers from endemic regions is also a major risk factor (Schantz *et al.*, 1992).

Recently, *T. solium* cysticercosis in black bears was reported from California (Theis *et al.*, 1996). It is believed that infected humans were the source of the infection in bears. This reflects not only the potential hazards of a human incursion into the wild, but also the risk for game meat transmission.

Swine and bovine cysticercosis are also significant economic problems in certain regions due to condemnation of infected carcasses at slaughter. In Mexico, more than US $43 million was lost in 1980 – the equivalent of 68% of the country's total investment in pig production (Flisser, 1988). In the US and Europe, for example, the infection rate for *T. saginata* in beef is generally less than 0.03%; therefore the inspection cost to find one infected carcass is quite large. In Africa and Latin America, however, the prevalence of *T. saginata* is relatively high and the resulting commodity losses from condemnation or treatment to destroy cysticerci are heavy. In Africa as a whole, the cost is about US $1.5–2.0 billion/year (see Murrell, 1991). The epizootic nature of bovine cysticercosis (sporadic outbreaks) makes it difficult to assess directly the economic impact of bovine cysticercosis in countries with low prevalences.

2.4.3 Clinical cysticercosis

Most cases of human cysticercosis (*T. solium*) are asymptomatic and are not recognized by either the individual or a physician. Symptomatic infections may be characterized as disseminated, ocular or neurological. Disseminated infections may localize in the viscera, muscles, connective tissue and bone; subcutaneous cysticerci may present a nodular appearance (Flisser, 1988). These localizations are often asymptomatic, but may produce pain and muscular weakness. Only about 3% of infections involve the eye. Central nervous system involvement may include invasion of the cerebral parachynus subarachnoid space, ventricles and spinal cord. The larval cysts may persist for years; cysts that die often calcify. Symptoms of infection may include partial paralysis, dementia, encephalitis, headache, meningitis, epileptic seizures and stroke. These manifestations of infection are determined by the numbers and locations of the cysts, and the host's inflammatory response to them (Flisser, 1988).

2.4.4 Diagnosis

Detection of the adult worm (intestinal) infection is based primarily on identification of proglottids or eggs in the feces. Occasionally, the species of tapeworm is determined on the basis of morphology of recovered proglottids or scolex. Recently, a species-specific DNA probe for *T. solium* and *T. saginata* eggs, in a dot blot assay configuration, has been developed (Chapman *et al.*, 1995). Serological diagnosis of humans with *T. solium* cysticercus infection is now quite reliable, with the introduction of an enzyme-linked immunotransfer blot (EITB) using purified worm glycoprotein antigen; this test has proven to be 98% sensitive in parasitologically proven cases, and is 100% specific (Tsang and Wilson, 1995). However, clinicians recommend that diagnosis of human cysticercosis be based on a proper interpretation of the patient's symptoms together with radiological and immunological data (Del Brutto *et al.*, 1996). These criteria, utilizing CT or MRI examinations, are presented in Table 12.4. The reliability of even these criteria, however, will rest on continuing evaluation in population-based studies.

Table 12.4 Diagnostic criteria for human cysticercosis

Absolute criteria:

1. Histologic demonstration of the parasite from biopsy of a subcutaneous nodule or brain lesion
2. Direct visualization of the parasite by fundoscopic examination
3. Evidence of cystic lesions showing the scolex on CT or MRI

Major criteria:

1. Evidence of lesions suggestive of neurocysticercosis on neuroimaging studies[a]
2. Positive immunologic tests for the detection of anticysticercal antibodies[b]
3. Plain X-ray films showing multiple 'cigar-shaped' calcifications in thigh and calf muscles.

Minor criteria:

1. Presence of subcutaneous nodules (without histological confirmation)
2. Evidence of punctuate soft-tissue or intracranial calcifications on plain X-ray films
3. Presence of clinical manifestations suggestive of neurocysticercosis[c]
4. Disappearance of intracranial lesions after a trial with anticystecereal drugs

Table 12.4 Diagnostic criteria for human cysticercosis—*continued*

Epidemiologic criteria:

1. Individuals coming from or living in an area where cysticercosis is endemic
2. History of frequent travel to cysticercosis-endemic areas
3. Evidence of a household contact with *T. solium* infection

[a] CT or MRI showing cystic lesions, ring-enhancing lesions, parenchymal brain calcifications, hydrocephalus, and abnormal enhancement of the leptomeninges. Myelograms showing multiple filling defects in the column of contract medium.
[b] Serum imunoblot, and CSF ELISA.
[c] Epilepsy, focal neurologic signs, intracranial hypertension and dementia.
From Del Brutto *et al.* (1996)

2.4.5 Prevention and control

New guidelines have been published for the surveillance, prevention and control of taeniasis and cysticercosis have recently been published by international agencies (Murrell, 2005). Reliance on abattoir inspection has several weaknesses. In some countries, clandestine marketing of swine and pork bypasses inspection (The Cysticercosis Working Group in Peru, 1993). Further, the accuracy of the meat inspection procedure may be less than 50% for light to moderate infection (Dewhirst *et al.*, 1967). Consequently, considerable research is underway to develop more rapid and sensitive detection technologies. Immunodiagnostic tests are of particular interest, and several tests are under development (Hayunga *et al.*, 1991; Brandt *et al.*, 1992; Draelants *et al.*, 1995).

Regardless of the success of efforts to improve the detection of infected animals at slaughter, effective control will always be best achieved by the minimization of risk factors. Governments must take steps to ensure that all beef and pork is marketed through channels that allow for meat inspection. On-farm management must ensure that animals are protected from human feces. Whenever possible, farm workers should be examined for parasites and treated if warranted. The use of sewage sludge and effluents for agricultural purposes should receive more careful scrutiny. For example, cattle should be screened from direct contact with streams carrying effluent and untreated wastewater from plants (Ilsoe *et al.*, 1990). Legislation may also be needed to regulate the agricultural use of sewage sludge, especially on grazing and pasture land (Barbier *et al.*, 1990). Research is urgently needed on the effects of composting and alternative chemical treatment in the production of safe sewage sludges and effluents for agricultural use.

In countries where inspection is not mandatory, it is generally recommended that consumers cook their meat to at least 60°C. The control of porcine cysticercosis in developing countries is especially urgent. The obstacles to change are poverty, tradition and vested interests (Schantz *et al.*, 1993). The WHO and the Pan American Health Organization have developed two alternative strategies for the control of human *T. solium* infections (Pawlowski, 1990; Schantz etal., 1993): comprehensive long-term intervention, and short-term intervention based on mass treatment of adult worm (intestinal) infections in existing transmission foci. The comprehensive long-term program includes appropriate legislation, modernization of swine production systems, improvement of meat-inspection efficiency and coverage, provision of adequate sanitary facilities, and adoption of measures to identify and treat human tapeworm carriers. As pointed out by Schantz *et al.* (1993), several of these features will be difficult to achieve because of social, political and economic realities in endemic areas. Therefore, the short-term program of identifying foci and treating all diagnosed or suspected human cases has been developed. The prospects for eradication of *T. solium* are considered good because of the vulnerability of the lifecycle to improved sanitation, the introduction of confined (intensive) swine housing, and rigorous meat inspection (Schantz *et al.*, 1993).

Other strategies for control of porcine cysticercosis include vaccines for pigs. Homologous antigens from *T. solium* cysticerci are able to induce a high level of protective immunity (71%) against a challenge infection (Nascimento *et al.*, 1995). Current efforts are aimed at identifying the protective antigens; whether a practicable vaccine can be produced will depend upon a complex set of issues (efficacy, cost/effectiveness, compatibility with other control efforts, etc.). Recently, a promising protovaccine for bovine cysticercosis was announced (Lightowlers *et al.*, 1996).

3 Fishborne helminths

3.1 Nematodes

3.1.1 *Capillaria philipiensis*

Capillaria philippiensis is a trichurid nematode first reported from humans in the Philippines. The parasite is small (males 1.5 × 3.9 mm and females 2.3 × 5.3 mm), and the adult worms have rows of cells (stichocytes) that surround the esophagus at the anterior half of the worm. The worms reside in the small intestines, and eggs produced by the females pass in the feces and reach freshwater. The eggs are eaten by freshwater fish and hatch in the fish intestine, and the larvae develop into the infective stage in 3 weeks. When the fish is eaten by a definitive host, the larvae develop into adults in 2 weeks, and the first generation of female worms produce larvae. These larvae mature in 2 weeks, and the females produce eggs which pass in the feces. There are always a few adult female worms that remain and produce larvae that develop into adults. This is a process of auto-infection to maintain the infection and increase the population. Auto-infection is an integral part of the lifecycle.

Man is presently the only confirmed definitive host, but fish-eating birds are probably the natural host that has been demonstrated experimentally, and one bird was found naturally infected. Many species of small freshwater fish in the Philippines and Thailand are intermediate hosts; *Hypseleotria bipartite* in the Philippines and *Puntius gonionotus* in Thailand are two species (Cross and Bhaibulaya, 1983). Infections in nature are probably maintained by fish-eating migratory birds that disseminate infections along the migratory flyways. Human infections are reported mostly from the Phillippines and Thailand, with sporadic reports from Japan, Korea, Taiwan, Egypt, India, Iran, Italy and Spain (Cross, 1992). As the numbers of parasites increase in humans, the symptoms of diarrhea, abdominal pain and borborygmus become more severe. The loss of potassium, protein and other essential elements leads to cachexia. If treatment is not initiated in time, the patient will die. Treatment is replacement of potassium, and administration of an antidiarrheal drug and an anthelminthic. The recommended drugs are mebendazole 400 mg/day in divided doses for 20 days or albendazole 400 mg/day for 10 days. The diagnosis is based on detecting *C. philippinensis* eggs, larvae and adults in the feces.

3.1.2 *Gnathostoma spp.*

Several species of *Gnathostoma* larvae are acquired from eating a variety of animals. Adult worms are found in tumors in the stomach wall of fish-eating mammals. Eggs passed by female worms pass in animal feces, reach water, and embryonate. The larva hatches from the egg and is eaten by a freshwater copepod, in which it develops into the second stage. When the infected copepod is eaten by a bird, fish, frog, turtle or mammal, the larva enters the tissue and develops into the third stage. When the second intermediate host is eaten by a paratenic host, the larva enters the tissue but does not develop further. It will, however, migrate in the tissue and cause disease. When the second intermediate host or paratenic host is eaten by the final host, the larva will penetrate the intestinal wall, migrate to the liver and other organs and eventually back to the peritoneal cavity, and penetrate the stomach wall, where it provokes a tumor. Cats and dogs are definitive hosts for *G. spinigerum*, and pigs are the definitive host for *G. hispidium*. Intermediate hosts are fish, frogs, snakes, chickens, ducks, rats, etc. The major source of infection in Thailand is a snake-headed fish, *Ophicephalus* sp., which is eaten raw or fermented. Japanese eat raw fish as *shashimi* and become infected, and in Mexico fish prepared as *ceviche* is a source of infection (Burke and Cross, 1970).

Gnathostoma larvae are short and stumpy, with a globose head armed with rows of hooklets. There are also spines halfway down the body. They measure 11–25 mm for males and 25–54 mm for females.

Humans are unnatural hosts, and the parasite never matures in humans. The worm migrates through any organ. In the subcutaneous tissue it develops migratory tracks, causing necrosis and hemorrhage. Toxic products cause transient swellings, pain and edema. The larvae may enter the eye, causing retinal damage, and in the central nervous system invasion leads to encephalitis, myelitis, radiculitis,and subarachnoid hemorrhage.

The diagnosis can based on a history of eating uncooked fish and on symptoms. Serologic tests are available (Noppartana *et al.*, 1991). A definitive diagnosis is made only when the larvae are recovered from surgical specimens, urine or vaginal discharge. Larvae may emerge from the subcutaneous tissue following long-term treatment with albendazole (Kravivichian *et al.*, 1992).

3.1.3 *Anasakis simplex*

Anisakiasis is infection with the larval stages of anisakid nematodes. *Anisakis simplex* and *Pseudoterranova decipiens* are the species most commonly involved. These are parasites of marine mammals found in the stomachs of cetaceans and pinnepeds. Female worms pass eggs in the feces, and the eggs remain on the ocean floor until a larva develops. The larva hatches from the egg and is eaten by a small crustacean (euphausid). The third-stage larva develops in the crustacean, and when it is eaten by squid or marine fish the larva penetrates the gut and passes into the peritoneal cavity or into musculature. When eaten by the mammalian definitive host, the larva is released from the fish or squid tissue and enters the animal's stomach.

Whales, dolphins and porpoises are definitive hosts for *A. simplex*, and seals, sea lions and walruses are the hosts for *P. decipiens*. Mackerel, herring, cod, salmon, and squid are second intermediate hosts for *A. simplex*; and cod, halibut, flatfish and red snapper are the second intermediate hosts for *P. decipiens*.

The larvae are the pathogenic stage of the parasite to humans. The larvae are 20–50 mm long and 0.3–0.22 mm wide. They are yellowish brown in color; the mouth has three lips and a boring tooth. The digestive tract of the worm has a postesophageal expansion called a ventriculus. The excretory pore is anterior, and the mucron is at the end of the posterior end. Cross sections of the worms reveal Y-shaped lateral cords.

When humans eat infected marine fish or squid raw, prepared as *shashimi*, *sushi* or *ceviche*, the larvae penetrate the tissue of the gastrointestinal tract and cause an eosinophilic granuloma. *Anisakis simplex* has been reported in the stomachs of over 1000 Japanese (Ishikura *et al*., 1993). *Pseudoterranova decipiens* is reported from humans in Northern Japan and along the California coast. In these areas the sea lion populations have increased to high levels along with the intermediate host. *Pseudoterranova decipiens* infections usually cause 'tickle throat'.

The diagnosis is made by the recovery of the parasite surgically following the diagnosis of an acute abdominal pain. However, the larvae are now being recovered by the use of fiber-optic gastroscopy. Immunodiagnostic methods are available, but are not readily available or completely reliable.

3.2 Trematodes

3.2.1 *Clonorchis sinensis*

The Chinese liver fluke is the most important member of the opisthorchiid flukes. It is widespread in Asia, being reported from China, Japan, Korea, Taiwan and Vietnam. The hermaphroditic adult worms reside in the distal tributaries of the bile passages. Eggs produced by the worms pass down the bile ducts to the intestines and out with the feces. The eggs must reach water, and are eaten by the first intermediate snail host (*Parafossaurulus, Bythnia* and *Alocinma* spp.). The larval miracidium is released from the egg and enters the snail tissue to go through polyembryony, producing sporocysts, rediae and cercariae. The cercariae leave the snail and search for cyprinid fish, the second intermediate hosts. The cercariae enter the skin of the fish and encyst as metacercariae. When the fish is eaten raw, the metacercariae excyst in the intestine of the definitive host and migrate through the Ampulla of Vater to the bile radicles, where they mature into adults.

The adult worms measure 10–25 mm long and 3–5 mm wide, and are lanceolate, flat and pinkish in color. There are oral and ventral suckers. Stained preparations reveal large, branching testes in the posterior portion of the worm.

The parasites in the biliary tract provoke hyperplasia of the epithelium, leading to fibrous development of the ducts. Several thousand worms may be involved. Fever and chill may develop, and the liver may become large and tender. Long-term infections may lead to carcinoma of the bile ducts.

Humans and animals acquire the infection by eating raw, pickled or smoked carp fish; *Pseudorasbora* or *C. tenopharyngodon* spp. are the most common sources of infection. There are over 113 species of freshwater fish recorded as second intermediate hosts for it (World Health Organization, 1995).

3.2.2 *Opisthorchis viverrini*

Opisthorchis viverrini is a liver fluke found in Thailand and the surrounding countries in Southeast Asia, and *O. feliensis* is reported from Eastern Europe and Siberia. The lifecycles of these flukes are similar to that of the Chinese liver fluke, but the intermediate hosts are different. *Bithynia* spp. are snail hosts for *O. viverrini*, and *Codiella* spp. for *O. feliensis*. There are many fish intermediate hosts for these species, but the important ones for *Puntius* spp. are *O. viverrini* and *Abrranis* spp. for *0. feliensis*. Infections are acquired by eating improperly cooked, preserved or fermented fish. Diseases associated with these species are similar to those caused by *C. sinensis*. The diagnosis of the liver fluke infection is based on the detection of eggs in feces or duodenal aspirates, or by immunological methods.

3.2.3 Other heterophyid flukes

Metagonimus yokogawi and *Heterophyes heterophyes* are two species of flukes acquired by eating raw freshwater fish. They are tiny intestinal parasites that pass eggs in the feces. The eggs hatch in the water, releasing a miracidium. *Pirenella* spp. of snails provide the first intermediate host for *H. heterophyes*, and *Semisulcospira* spp. are the snail hosts for *M. yokogawai*. Cercariae produced by polyembryony in the snail search for a fish second intermediate host: *Mugil* spp. for *H. heterophyes*, and salmonid species for *M. yokogawi*. *H. heterophyes* is endemic in Egypt and China, and *M. yokogawi* in Japan and Korea.

Both flukes are very small, 1–2 mm in length and 0.3–0.7 mm in width. Attachment of the fluke may cause small ulcers in the mucosa, leading to abdominal pain, diarrhea and lethargy. The severity depends on the number of worms involved. There are reports of eggs of these worms in ectopic locations such as the brain and heart. Infections are often reported in sports fisherman who eat the fish raw shortly after being caught in the cool mountain streams in Northern Japan. The diagnosis is based on finding eggs in the feces.

3.2.4 Echinostomes

There are many species of *Echinostoma* that are acquired by eating a variety of aquatic animal life, especially freshwater fish in Asia. They are generally parasites of other animals, and humans are accidental hosts. The flukes reside in the intestines, and eggs pass in the feces and reach water. The miracidium hatches from the egg and enters a susceptible snail, *Gyraulus* spp. Cercariae produce may enter other snails or other aquatic animal life; *Echinostoma lindoensis*, *E. ilocanum* and *E. misyanum* are endemic in certain parts of Indonesia, the Philippines and Malaysia. The parasites have a characteristic collar of spines around the oral sucker, are spindle-shaped, and measure 4–7 mm in length by 1.0–1.35 mm in width. Little disease is associated with infection, except for occasional colic and diarrhea. The diagnosis is based on the presence of eggs in the feces (Cross and Basaca-Sevilla, 1986).

3.3 Cestodes

3.3.1 *Diphyllobothrium latum*

There are several species of fish tapeworms, but the most important is *D. latum*, which is widespread in the temperate and subarctic regions of the northern hemisphere. Fish-eating mammals are definitive hosts. The worm resides in the small intestine and produces eggs, which pass in the feces into water. A ciliated larva (coricidium) develops in the egg and, when released from the egg, swims in the water until ingested by a copepod. A procercoid larva develops in the copepod, and when eaten by a fish second intermediate host the larva develops into a plerocercoid larva or sparganum in the fish tissue. When the fish is eaten uncooked, the plerocercoid larva attaches to the intestinal mucosa and develops into an adult tapeworm. The worm is the largest parasite to infect humans, and ranges in length from 2–15 mm with a maximum width of the gravid proglottids of 20 mm. The scolex is 2 mm long and 1 mm wide, and has a dorsal ventral groove or bothrium. The mature proglottid has a rosette-shaped uterus that fills with eggs.

There is little disease associated with fish tapeworm infections. The worms may compete for vitamin B12 with the host and cause megaloblastic anemia. This occurs more often in Finland, where patients experience fatigue, weakness, diarrhea, the desire to eat salt, epigastric pain, and fever (Cross, 1994). The diagnosis is made by finding eggs in the stools.

A variety of fish serve as second intermediate hosts, including pike, perch, turbot, salmon and trout. The fish are usually poorly cooked, or pickled or smoked. Japanese eat the fish as *shashimi* or *sushi*. Other definitive hosts include dog, fox, bear, mink, seals and sea-lions.

At times humans may acquire infections with the spargana of *Spirometra* spp. This is a diphyllobothrid tapeworm of felines and canines. The plerocercoid larva is a larval migrant that causes migratory transient swelling.

4 Protozoa and helminths disseminated in fecally contaminated food and water

4.1 Protozoa

4.1.1 Isosporosis

Isospora belli is the cause of intestinal coccidiosis in humans. It belongs to the family *Eimeriidae* (Lindsay *et al.*, 1997). Most recorded cases have occurred in the tropics, rather than in the temperate zone. Infection is now seen more frequently in compromised patients, particularly those with AIDS (Dubey, 1993, 1997b).

Isospora belli oocysts are elongate and ellipsoidal, and are 20 × 19 μm. Sporulated oocysts contain two ellipsoidal sporocysts without a Stieda body. Each sporocyst is 9–14 by 7–12 μm and contains four crescent-shaped sporozoites and a residual body. Sporulation occurs within 5 days, both within the host and in the external environment; thus both unsporulated and sporulated oocysts may be shed in feces.

Infection occurs by the ingestion of food contaminated by oocysts. Merogony and gametogony occur in the upper small intestinal epithelial cells, from the level of the crypts to the tips of the villi. The number of generations of merogony is unknown. In AIDS patients the parasite may be disseminated to extra-intestinal organs, including the mesenteric and mediastinal lymph nodes, the spleen and the liver. Single zooites surrounded by a capsule (cyst wall) have a prominent refractile or crystalloid body, indicating that the encysted organisms are sporozoites (Lindsay *et al.*, 1997). Organisms with a cyst wall are found only in extra-intestinal organs. *Isospora belli* can cause severe symptoms with an acute onset, particular in AIDS patients. Infection has been reported to cause fever, cholecystitis, persistent diarrhea, weight loss and even death (Dubey, 1993).

Diagnosis can be established by finding characteristic bell-shaped oocysts in the feces or coccidian stages in intestinal biopsy material. Affected intestinal portions may have a flat mucosa similar to that found in sprue. The stools during infection are fatty and at times very watery. Sulfonamides are considered effective against coccidiosis (St Georgiev, 1993).

4.1.2 Cyclosporosis

Cyclospora cayetanensis is the cause of disease in humans. *Cyclospora cayetanensis* oocysts are approximately 8 μm in diameter and contain two ovoid 4 × 6 μm sporocysts (Ortega *et al.*, 1994). Each sporocyst has two sporozoites. Thus, there are a total of four sporozoites in a sporulated oocyst.

Unsporulated oocysts are excreted in feces. Sporulation occurs outside the body. Other stages in the lifecycle are not known. Outbreaks of cyclosporosis in humans have been associated with ingestion of fruits, salads and herbs contaminated with oocysts (Herwaldt *et al.*, 1997).

Both immunocompetent and immunosuppressed patients of all ages may have diarrhea, fever, fatigue and abdominal cramps. Infection has been reported from several countries (Goodgame, 1996; Herwaldt *et al.*, 1997). Diagnosis can be made by fecal examination. *Cyclospora* oocysts are approximately 8 μm, remarkably uniform, and contain a sporont (inner mass) that occupies most of the oocyst. They are acid-fast and need to be distinguished from *Cryptosporidium* sporocysts; *Cyclospora cayetanensis* oocysts have a much thicker oocyst wall, and their contents are more granular than are those of cryptosporidial oocysts.

Treatment with sulfamethoxazole and trimethopium is cosidered effective in relieving symptoms (Goodgame, 1996). Irradiation is one proposed method of killing coccidian oocysts on fruits and vegetables (Dubey *et al.*, 1998a).

4.1.3 Cryptosporidiosis

Cryptosporidiosis in humans is caused by minute coccidia of the genus *Cryptosporidium*, which comprises perhaps 14 species and is in taxonomic flux

(Fayer, 2004). Two species known to infect humans are *C. hominis*, which is human-specific, and *C. parvum*, which infects many species and appears to have a reservoir in cattle. Members differ from other coccidians in location and structure; cryptosporidial species are located at the surface of a cell, mostly in the microvillous border of enterocytes. Cryptosporidial microgametes are bullet-like (without flagella) and are thus distinct from all other coccidia.

Cryptosporidium parvum oocysts are approximately 5 μm in diameter and contain four sporozoites. After ingestion of food or water contaminated with sporulated oocysts, sporozoites excyst and penetrate the surface of the microvillus borders of host cells. Asexual development leads to first- and second-generation meronts. Merozoites released from second-generation meronts form male (micro) and female (macro) gamonts. After fertilization, oocysts are formed. Oocysts sporulate in the host, and sporulated oocysts are excreted in feces. As many as 79 species of mammals are considered a host for *C. parvum* (Fayer *et al.*, 1997).

Cryptosporidial infection is common, but disease is rare in immunocompetent humans. Infections in immunosuppressed (such as AIDS) patients can be fatal. The most characteristic symptoms are profuse watery diarrhea, cramps, abdominal pain, vomiting and low-grade fever. Rarely, cryposporidiosis may involve extra-intestinal organs, including the gall bladder, lungs, eyes and vagina (Fayer *et al.*, 1997).

Diagnosis is made by fecal examination, including direct examination of fecal floats, staining of fecal smears using acid-fast and other stains, and staining with anti-cryptosporidial antibodies (Current, 1997).There is no specific therapy. Removal of oocysts during filtration and sedimentation of municipal water is an engineering problem. Boiling of water kills all coccidian oocysts.

4.1.4 Giardiasis

Giardia lamblia is the cause of giardiasis in humans. It is a bilaterally symmetrical flagellated protozoa. Trophozoites and cysts are the two stages in its simple fecal–oral cycle. Trophozoites are pear-shaped, 10–20 μm long and 5–15 μm wide. They contain two nuclei and have eight (four lateral, two ventral and two caudal) flagella (Garcia, 1997). The dorsal surface is convex, while the ventral surface is usually concave and has a sucking disc at the broader end. Cysts are round to ellipsoidal, and are 8–19 μm by 11–14 μm in size. There are four nuclei and no flagella. Trophozoites attach to the intestinal epithelium by the adhesive disc. Trophozoites divide into two by longitudinal binary fission. Cysts and trophozoites may be passed in feces. Trophozoites do not survive for a long period outside the host, whereas cysts can survive outside the host.

Humans become infected by ingesting food or water contaminated with cysts. The exact mechanism of pathogenesis is unknown because the parasite is extracellular. Giardiasis can cause a severe intestinal disorder, resulting in diarrhea, nausea, steatorrhea and weight loss. Diagnosis is made by fecal examination. To find trophozoites, smears of freshly passed stool should be made in isotonic saline (not water). Flotation in zinc sulfate is useful for detecting cysts. The discharge of *Giardia* in the stool is not always regular; therefore, multiple stool examinations may be necessary. Immunodiagnostic methods, including an ELISA and a fluorescent antibody test to detect *Giardia* in stools, have been described.

Quinacrine and metronidazole are two drugs recommended for the treatment of giardiasis (Garcia, 1997). Because of the potential for wild and domestic animals as reservoirs of infection, good personal hygiene is necessary to prevent infection with *Giardia*. Boiling will kill *Giardia* in water.

4.2 Helminths

4.2.1 Visceral larva migrans

There are several nematodes transmitted via ingestion of contaminated soil (Crompton, 1997). Nematodes (e.g. *Trichuris, Enterobius*) that complete full development in humans will not be discussed further. Visceral larva migrans (VLM) usually results from the migration of nematode larvae in the accidental human host. Although several nematodes may cause VLM, the most common causes of VLM in humans are *Toxocara canis* and *Bayliascaris procyonis* (Bowman, 1995; Despommier, 1997). Dogs and other canids are the definitive hosts for *T. canis*, whereas raccoons are the definitive hosts for *B. procyonis*. Adult worms live in the stomach and small intestines, and pass unembryonated eggs in the feces. First-stage larvae develop inside the egg and undergo a molt to form second-stage larvae. *Toxocara* and *Bayliascaris* eggs are sticky, and survive in the environment for many years. Humans become infected by ingesting embryonated eggs in contaminated water or food, or by ingesting contaminated soil. In accidental human hosts, second-stage larvae can migrate extensively in many organs but do not mature. Even one or a few larvae in the eye or brain can cause extensive damage, and symptoms vary depending on the organs involved. Other nematodes (e.g. *Toxocara cati* from cats, *Ascaris suum* from pigs, *Ancylostoma* from dogs and cats) can also cause VLM or cutaneous larval migrans (Bowman, 1995; Crompton, 1997).

In the definitive host (dog), *T. canis* larvae return to the gut and mature into adult worms. Ascarids can lay a large number of eggs for a long period. *Toxocara canis* is transmitted transplacentally and via milk to pups. Three-week-old pups can shed *T. canis* eggs in feces, and these are a health hazard for humans, especially children. *Bayliascaris procyonis* is only transmitted fecally, and eggs shed by raccoons (e.g. in chimneys and garages, and near hiking trails) are a source of infection, especially for children.

Diagnosis of VLM in humans is difficult because of imprecise symptoms. Eosinophilia can arouse suspicion of helminth infection. Serologic tests can aid diagnosis. Mebendazole and albendazole have been used to treat VLM in humans. Dogs should be dewormed regularly to reduce contamination with eggs in the environment.

Bibliography

Andrews, J. R. H., C. Bandi, E. Pozio *et al.* (1995). Identification of *Trichinella pseudospiralis* from a human case using random amplified polymorphic DNA. *Am. J. Trop. Med. Hyg.* **53**, 185–188.

Bailey, T. M. and P. M. Schantz (1990). Trends in the incidence and transmission patterns of trichinosis in humans in the United States: comparisons of the periods 1975–1982 and 1982–1986. *Rev. Infect. Dis.* **12**, 5–11.

Banerjee, P. S., B. B. Bhatia, and B. A. Pandit (1994). *Sarcocystis suihominis* infection in human beings in India. *J. Vet. Parasitol.* **8**, 57–58.

Barbier, D., D. Perrine, C. Duhamel *et al.* (1990). Parasitic hazards with sewage sludge applied to land. *Appl. Environ. Microbiol.* **56**, 1420–1422.

Bowman, M. S. (1995). *Georgis' Parasitology for Veterinarians*, 6th edn. W B Saunders, Philadelphia, PA.

Brake, R. J., K. D. Murrell, E. E. Ray *et al.* (1985). Destruction of *Trichinella spiralis* by flow dose irradiation of infected pork. *Food Safety* **7**, 127–134.

Brandt, J. R. A., S. Geerts, R. De Deken *et al.* (1992). A monoclonal antibody-based ELISA for the detection of circulating excretory-secretory antigens in *Taenia saginata* cysticercosis. *Intl J. Parasitol.* **22**, 471–477.

Burke, J. M. and J. H. Cross (1997). Gnathostomiasis. In: D. H. Connor, F. W. Chandler, D. A. Schwartz *et al.* (eds), *Pathology of Infectious Diseases*, pp. 1431–1435. Appleton and Lange, Stamford, CT.

Campbell, W. C. (1991). *Trichinella* in Africa and the *nelsoni* affair. In: C. N. C. Macpherson and P. S. Craig (eds), *Parasite Helminths and Zoonoses in Africa*, pp. 83–100. Unwin Human, London.

Cao, W. C., C. P. B. Vander Ploeg, C. L. Gao *et al.* (1996). Seroprevalence and risk factors of human cysticercosis in a community of Shandong, China. *SE Asian J. Trop. Med. Publ. Health* **27**, 279–285.

Capo, V. and D. D. Despommier (1996). Clinical aspects of infection with *Trichinella spp. Clin. Microbiol. Rev.* **9,** 47–54.

Chapman, A., V. Vallejo, K. G. Mossier *et al.* (1995). Isolation and characterization of species-specific DNA probes from *Taenia solium* and *Taenia saginata* and their use in egg detection assay. *J. Clin. Microbiol.* **33**, 1238–1288.

Choi, W. Y., H. W. Nam, N.H. Kwak *et al.* (1997). Foodborne outbreaks of human toxoplasmosis. *J. Infect. Dis.* **175**, 1280–1282.

Crompton, D. W. T. (1997). Gastrointestinal nematodes – *Ascaris*, hookworm, *Trichuris*, and *Enterobius*. In: F. E. G. Cox, J. P. Kreier and D. Wakelin (eds), *Topley and Wilson's Microbiology and Microbial Infections*, 9th edn, Vol. V, pp. 561–584. Arnold, London.

Cross, J. H. (1992). Intestinal capillariasis. *Clin. Microbiol. Rev.* **5**, 120–129.

Cross, J. H. (1994). Fish and invertebrate-borne helminthes. In: Y. H. Hui, J. R. Gorham, K. D. Murrell and D. O. Cliver (eds), *Foodborne Disease Handbook*, Vol. 2, pp. 279–329. Marcel Dekker, New York, NY.

Cross, J. H. and V. Basaca-Sevilla (1986). Studies on *Echinostoma ilocanum* in the Philippines, *SE Asian J. Trop. Med. Publ. Health* **17**, 23–27.

Cross, J. H. and M. Bhaibulaya (1983). Intestinal capillariasis in the Philippines and Thailand. In: N. Croll and J. H. Cross (eds), Human *Ecology and Infectious Diseases*, pp. 103–136. Academic Press, New York, NY.

Current, W. L. (1997). Cryptosporidiosis. In: F. E. G. Cox, J. P. Kreier and J. H. Cross (eds), *Topley and Wilson's Microbiology and Microbial Infections*, 9th edn, Vol. V, pp. 329–347. Arnold, London.

Del Brutto, O. H., N. H. Wadia, M. Dumas *et al.* (1996). Proposal of diagnostic criteria for human cysticercosis and neurocysticercosis. *J. Neurol. Sci.* **142**, 1–6.

Despommier, D. D. (1983). Biology. In W. C. Campbell (ed.), Trichinella *and Trichinellosis*, pp. 75–151. Plenum Press, New York, NY.

Despommier, D. D. (1997). *Trichinella* and *Toxocara*. In: F. E. G. Cox, J. P. Kreier and D. Wakelin (eds), *Topley and Wilson's Microbiology and Microbial Infections*, 9th edn, Vol. V, pp. 596–607. Arnold, London.

Dewhirst, L. W., J. D. Cramer and J. J. Sheldon (1967). An analysis of current inspection procedures for detecting bovine cysticercosis. *J. Am. Vet. Med. Assoc.* **150**, 412–417.

Draelants, E., E. Hofkens, E. Harding *et al.* (1995). Development of a dot-enzyme immunoassay for the detection of circulating antigen in cattle infected with *Taenia saginata. Res. Vet. Sci.* **58**, 99–100.

Dubey, J. P. (1993). *Toxoplasma, Neospora, Sarcocystis*, and other tissue cyst-forming coccidia of humans and animal. In J. P. Kreier (ed.), *Parasitic Protozoa*, Vol. VI, pp. 1–158. Academic Press, New York, NY.

Dubey, J. P. (1997a). Toxoplasmosis. In: F. E. G. Cox, J. P. Kreier and D. Wakelin (eds), *Topley and Wilson's Microbiology and Microbial Infections*, 9th edn, Vol. V, pp. 303–318. Arnold, London.

Dubey, J. P. (1997b). *Sarcocystis, Isospora*, and *Cyclospora*. In: F. E. G. Cox, J. P. Kreier and D. Wakelin (eds), *Topley and Wilson's Microbiology and Microbial Infections*, 9th edn, Vol. V, pp. 319–327. Arnold, London.

Dubey, J. P. and C. P. Beattie (1988). *Toxoplasmosis of Animals and Man*, pp. 1–220. CRC Press, Boca Raton, FL.

Dubey, J. P., and D. W. Thayer. (1994). Killing of different strains of *Toxoplasma gondii* tissue cysts by irradiation under defined conditions. *J. Parasitol.* **80**, 764–767.

Dubey, J. P., C. A. Speer, and R. Fayer (1989). *Sarcocystosis of Animals and Man*, p. 1–215. CRC Press, Boca Raton, FL.

Dubey, J. P., D. S. Lindsay and C. A. Speer (1998a). Structure of *Toxoplasma gondii* tachyzoites, bradyzoites, sporozoites, and biology and development of tissue cysts. *Clin. Microbiol. Rev.* **11**, 267–299.

Dubey, J. P., D. W. Thayer, C. A. Speer and S. K. Shen (1998b). Effect of gamma irradiation on unsporulated and sporulated Toxoplasma gondii oocysts. *Intl J. Parasitol.* **28**, 369–375.

Dupouy-Camet, J., C. L. Soule and T. Ancelle (1994). Recent news on trichinellosis, another outbreak due to horsemeat consumption in France in 1993. *Parasite* **1**, 99–103.

Fan, P. C. (1991). Asian *Taenia saginata*, species or strain. *SE Asian J. Trop. Med. Publ. Health*, **22**, 245–250.

Fayer, R. (2004). Waterborne zoonotic protozoa. In: J. A. Cotruvo, A. Dufour, G. Rees *et al.* (eds), *Waterborne Zoonoses, Identification, Causes and Control*, pp. 255–282. IWA (International Water Association) Publishing, London.

Fayer, R., C. A. Speer and J. P. Dubey (1997). The general biology of *Cryptosporidium*. In: R. Fayer (ed.), *Cryptosporidium and Cryptosporidiosis,* pp. 1–41. CRC Press, Boca Raton, FL.

Flisser, A. (1988). Neurocysticercosis in Mexico. *Parasitol. Today* **4**, 131–137.

Food Safety and Inspection Service (1990). Approval of tests for trichinosis in pork. *Title 9 Code of Federal Regulations*, Ch. III (1-1-90 edition), p. 220. US Dept. of Agriculture, Washington, DC.

Gamble, H. R. and K. D. Murrell (1987). Swine trichinellosis, diagnosis and control. *Agri-Practice* **8**, 12–15.

Garcia, H. H., R. Gilman, M. Martinez, V. C. W. Tsang *et al.* (1993). Cysticercosis as a major cause of epilepsy in Peru. *Lancet* **341**, 197–200.

Garcia, L. S. (1997). Giardiasis. In: F. E. G. Cox, J. P. Kreier and J. H. Cross (eds), *Topley and Wilson's Microbiology and Microbial Infections*, 9th edn, Vol. V, pp.193–214. Arnold, London.

Goodgame, R. W. (1996). Understanding intestinal spore-forming protozoa, cryptosporidia, microsporidia, isospora, and cyclospora. *Ann. Intern. Med.* **124**, 429–441.

Hayunga, E. G., M. P. Sumner, M. L. Rhoads *et al.* (1991). Development of a serologic assay for cysticercosis, using an antigen isolated from *Taenia spp* cyst fluid. *Am. J. Vet. Res.* **52**, 462–470.

Herwaldt, B. L., M. L. Ackers and The Cyclospora Working Group (1997). An outbreak in 1996 of cyclosporiasis associated with imported raspberries. *N. Engl. J. Med.* **336**, 1548–1556.

Ilsoe, B., N. C. Kyvsgaard, P. Nansen and S. A. Henriksen (1990). Bovine cysticercosis in Denmark. *Acta Vet. Scand.* **31**, 159–168.

Ishikura, H., K. Kikuchi, K. Nagasawa *et al.* (1993). *Anisakidae and anisakidosis. Prog. Clin. Parasitol.* **3**, 43–102.

Jongwutiwes, S., N. Chantachum, P. Kraivichain *et al.* (1998). First outbreak of human trichinellosis caused by *Trichinella pseudospiralis. Clin. Infect. Dis.* **26**, 111–115.

Kravivichian, P., M. Kulkumthorn, P. Yingyourd *et al.* (1992). Albendazole for the treatment of human gnathostomiasis. *Trans. R. Soc. Trop. Med. Hyg.* **86**, 418–421.

Lightowlers, M. W., R. Rolfe and C. G. Gavci (1996). *Taenia saginata*, vaccination against cysticercosis in cattle with recombinant oncosphere antigens. *Exp. Parasit.* **84**, 330–338.

Lindsay, D. S., J. P. Dubey and B. L. Blagburn (1997). Biology of *Isospora* spp. from humans, nonhuman primates, and domestic animals. *Clin. Microbiol. Rev.* **10**, 19–34.

Murrell, K. D. (1991). Economic losses from food-borne parasitic zoonoses. *SE Asian J. Trop. Med. Publ. Health* **22**, 377–380.

Murrell, K. D. (1995). Foodborne parasites. *Intl J. Environ. Health Res.* **5**, 63–85.

Murrell, K.D. (ed) (2005). *WHO/FAO/OIE Guidelines for surveillance, prevention and control of taeniosis/oysticercosis.* International Office for Epizootiology, Paris.

Murrell, K. D. and F. Bruschi (1994). Clinical trichinellosis. *Prog. Clin. Parasitol.* **4**, 117–150.

Murrell, K. D., R. Fayer and J. P. Dubey (1986). Parasitic organisms. *Adv. Meat Res.* **2**, 311–377.

Murrell, K.D., R. J. Lictenfels, D.S. Zarlenga and E. Pozio (2000). The systematics of the genus *Trichinella* with a key to species. *Vet. Parasitol.* **93**, 293–307.

Murrell, K. D., F. Stringfellow, J. P. Dame *et al.* (1987). *Trichinella spiralis* in an agricultural ecosystem. II. Evidence for natural transmission of *Trichinella spiralis* from domestic swine to wildlife. *J. Parasitol.* **73**, 103–109.

Nascimento, E., J. O. Costa, M. P. Guimarães and C. A. P. Tavares (1995). Effective immune protection of pigs against cysticercosis. *Vet. Immunol. Immunopathol.* **45**, 127–137.

Nopparatana, C., P. Setasuban, W. Cahicumpa and P. Tapchhaisi (1991). Purification of Gnathostoma spinigerum speific antigen and immunodiagnosis of human gnathostomiasis. *Intl J. Parasitol.* **21**, 677–687.

Oliver, D. G., P. Singh, D. E. Allison *et al.* (1989). Field evaluation of an enzyme immunoassay for the detection of trichinellosis in hogs in a high volume North Carolina abattoir. *Proceedings of the 7th International Trichinellosis Conference, Alicante, Spain. Consejo Superiore de Investigaciones Cientificas Press, Madrid*, pp. 439–444.

Ortega, Y. R., R. H. Gilman and C. R. Sterling (1994). A new coccidian parasite (Apicomplexa, *Eimeriidae*) from humans. *J. Parasitol.* **80**, 625–629.

Pawlowski, Z. S. (1990). Perspectives on the control of *Taenia solium. Parasitol. Today* **6**, 371–373.

Pozio, E., E. La Rosa, K. D. Murrell and R. Lichtenfels (1992). Taxonomic revision of the genus *Trichinella. J. Parasitol.* **78**, 654–659.

Pozio, E. (2000). Factors affecting the flow among domestic, synanthropic and sylvatic cycles of *Trichinella. Vet. Parasitol.* **93**, 241–262.

Pozio, E.,G. A. La Rosa, F. A. Serrano *et al.* (1996). Environmental and human influence on the ecology of *Trichinella spiralis* and *Trichinella britovi* in Western Europe. *Parasitology* **113**, 527–533.

Reason, A. J., L. A. Ellis, J. A. Appelton *et al.* (1994). Novel Tyrelose-containing tri- and tertiary N-glycans in the immunodominant antigens of the intracellular parasite *Trichinella spiralis. Glycobiology* **4**, 593–603.

Remington, J. S., R. McLeod and G. Desmonts (1995). Toxoplasmosis. In: J. S. Remington and J. O. Klein (eds), *Infectious Diseases of the Fetus and Newborn Infant*, 4th edn, pp. 140–267. W. B. Saunders, Philadelphia, PA.

Roberts, T. and K. D. Murrell (1993). Economic losses caused by foodborne parasitic diseases. *Proceedings of the Symposium on Cost–Benefit Aspects of Food Irradiation Processing. IAEA/FAO/WHO. Aix-En-Provence, 1–5 March*, pp. 51–75. IAEA-SM-328/66, Vienna.

Roberts, T., K. D. Murrell and S. Marks (1994). Economic losses caused by foodborne parasites. *Parasitol. Today* **10**, 419–423.

Sarti Gutierrez, E., P. M. Schantz, J. Aguilera and A. Lopez (1992). Epidemiologic observations on porcine cysticercosis in a rural community of Michoacan State, Mexico. *Vet. Parasitol.* **41**, 195–201.

Schad, G. A., D. A. Leiby, C. H. Duffy and K. D. Murrell (1985). Swine trichinosis in New England slaughterhouses. *Am. J. Vet. Res.* **46**, 2008–2010.

Schantz, P. M. (1991). Parasitic zoonoses in perspective. *Intl J. Parasitol.* **21**, 161–170.

Schantz, P. M. and J. McAuley (1991). Current status of foodborne parasitic zoonoses in the United States. *SE Asian J. Trop. Med. Publ. Health* **22**, 65–71.

Schantz, P. M., A. C. Moore, J. L. Munoz *et al.* (1992). Neurocysticercosis in an Orthodox Jewish community in New York City. *N. Engl. J. Med.,* **327**, 692–695.

Schantz, P. M., M. Cruz, E. Sarti and Z. Pawlowski (1993). Potential eradicability of taeniasis and cysticercosis. *Bull. PAHO* **27**, 397–403.

Snyder, G. R. and K. D. Murrell (1986). Bovine cysticercosis. In: G. T. Woods (ed.), *Practices in Veterinary Public Health and Preventive Medicine in the United States*, pp. 161–187. Iowa State University Press, Ames, IA.

Solanki, P. K., H. O. P. Shrivastava and H. L. Shah (1991). Prevalence of *Sarcocystis* in naturally infected pigs in Madhya-Pradesh with an epidemiological explanation for the higher prevalence of *Sarcocystis suihominis. Indian J. An. Sci.* **61**, 820–821.

Sorovillo, F. J., S. H. Waterman, F. O. Richards and P. M. Schantz (1992). Cysticercosis surveillance, locally acquired and travel-related infection and detection of intestinal tapeworm carriers in Los Angeles County. *Am. J. Trop. Med. Hyg*. **47**, 365–371.

St Georgiev, V. (1993). Opportunistic infections, treatment and developmental therapeutics of cryptosporidiosis and isosporiasis. *Drug Develop. Res*. **28**, 445–459.

The Cysticercosis Working Group in Peru (1993). The marketing of cysticercotic pigs in the Sierra of Peru. *Bull. WHO* **71**, 223–228.

Theis, J. H., M. Cleary, M. Syrvanen *et al.* (1996). DNA-confirmed *Taenia solium* cysticercosis in black bears (*Ursus americanus)* from California. *Am. J. Trop. Med. Hyg*. **55**, 456–458.

Tsang, V. C. W. and M. Wilson (1995). *Taenia solium* cysticercosis, an unrecognized but serious public health problem. *Parasitol. Today* **11**, 124–126.

Uparanukraw, P. and N. Morakote (1997). Detection of circulating *Trichinella spiralis* larvae by polymerase chain reaction. *Parasitol. Res*. **83**, 52–56.

World Health Organization (1983). Guidelines for surveillance, prevention and control of Taeniasis/cysticercosis. WHO Document UPH/83.49.

World Health Organization (1995). Control of foodborne trematode infections. WHO Tech. Rep. Ser. 849.

Foodborne Intoxications

Clostridium botulinum

13

Nina G. Parkinson and Keith A. Ito

1 Introduction

The spore-forming bacterium *Clostridium botulinum* and its neurotoxin have been studied extensively. The organism's distribution around the world, the heat resistance of the spore, the toxin and conditions under which it is produced, its toxicity, the illnesses attributed to the toxin, the modes of transmission and the treatment have been well documented. Thermal processes for commercially processed low-acid, shelf-stable foods have been successful in minimizing cases of foodborne botulism. As the demand by consumers for 'fresh' packaged and minimally processed foods has increased, the microorganisms' ability to grow and produce toxin in 'new' environments has posed new challenges. In recent years it has become apparent that, in addition to foodborne botulism cases, there are three other types of illnesses affecting humans associated with this bacterium and its toxin: infant botulism, wound botulism and adult infectious botulism. The toxin is also being used for medicinal purposes and could be used for germ warfare. For these reasons, the bacterium and its toxin continue to be investigated throughout the world.

Foodborne Infections and Intoxications 3e
ISBN-10: 0-12-588365-X
ISBN-13: 978-012-588365-8

2 Historical aspects and contemporary problems

Although the illness now known as botulism has been around for hundreds of years, it was not until the early nineteenth century that it was first documented and studied. Between 1815 and 1828, in Württemberg, Germany, Justinus Kerner noted similarities in symptoms in over 230 patients. Kerner discovered that all had consumed different types of sausage that had been prepared in different ways. He noted that air pockets left in the sausage prevented it from becoming toxic. He determined that only the boiled or smoked sausages were toxic (Smith, 1997). Kerner attributed the disease to a toxic fatty acid developing within the sausage over a period of time. Because of his early findings, the illness is sometimes referred to as 'Kerner's disease'.

In 1896, Emile Pierre Marie van Ermengem, a Professor of Bacteriology at the University of Ghent, investigated an illness outbreak in Ellezelles, Belgium. Several members of a musical band ate a raw salted ham and developed symptoms similar to those described by Kerner. Van Ermengem isolated a spore-forming anaerobic bacterium, whose culture filtrates, when injected into various species of laboratory animals, produced characteristic and often fatal paralysis. He contended that a neurotoxic metabolite was produced within the ham in which this irregularly distributed bacillus had grown. Van Ermengem suggested the name *Bacillus botulinus*, because of similarities in symptoms to the sausage studies by Kerner (*botulus* is Latin for sausage). His studies yielded all the essential facts about foodborne botulism (Smith, 1997):

- It is not an infection, but an intoxication
- The toxin is produced by a specific bacterium
- The toxin is ingested with the food and is not inactivated by the digestive processes
- The organism is relatively resistant to mild chemical agents, but it is resistant to heat
- The toxin is not produced if the concentration of sodium chloride is sufficiently high
- Not all species of animals are susceptible.

In the early twentieth century, much work was dedicated to understanding the conditions of growth and toxin production in foods. Initially, the belief was that only meat was associated with cases of botulism. Later it was realized that vegetables and seafood could also be vehicles for botulinum poisoning. The basic knowledge that *C. botulinum* is unable to grow in acid conditions and that the spore is extremely heat resistant have led to regulations in the US and other countries regarding thermal processing of canned foods.

The Centers for Disease Control and Prevention (CDC) in Atlanta have documented 724 cases of verified foodborne botulism in American adults from 1973 through 1996 (Shapiro *et al.*, 1998). The majority were associated with prepared home-canned vegetables. In addition to regulations, better community education concerning home-canning procedures and precautions has minimized botulinum food poisoning in the US.

3 Characteristics of *Clostridium botulinum*

Clostridium botulinum is Gram-positive, straight to slightly curved, rod-shaped, motile by peritrichous flagella, and forms oval and subterminal spores, which usually swell the cell (Cato *et al.*, 1986). The various types are differentiated based upon the antigenically specific toxins that they produce. The toxins are serologically distinct, but do not necessarily differ toxicologically.

3.1 Classification

Historically, as each new serologically distinct type of toxin was discovered it was added alphabetically to the list of existing types. Thus, there are currently seven known types, A–G, with type C having a C1 and a C2 toxin. It is now known that some strains produce a mixture of two toxin types. A definition based on toxin type has resulted in a metabolically diverse group of organisms. In 1970, Holdeman and Brooks proposed dividing the types known at that time into three physiological groups based upon the ability to metabolize certain substrates. They suggested that Group I contains the proteolytic types A, B and F, while Group II contains types C and D and Group III contains the non-proteolytic types B, E and F. Lee and Riemann (1970), based on DNA binding studies, confirmed that the types could be so grouped. A fourth group was added with the discovery of type G. Suen *et al.* (1988a) determined that the type G organisms were related to each other but were not related to the Group I organisms. They suggested changing the name to *Clostridium argentinense* as a first step to changing the nomenclature for classifying the group.

Smith (1997) suggested that the strains be placed into four groups based upon the proteolytic activities of the groups. Currently, it is accepted that the four groups consist of Group I, the proteolytic types A, B and F; Group II, the non-proteolytic types B, F and E; Group III, the types C and D; and Group IV, the type G (Cato *et al.*, 1986; Gibson and Eyles, 1989; Hatheway, 1993). Smith (1997) acknowledged that *C. botulinum* was not a single species but was rather a mixture of different groups. He felt that, although taxonomically it made sense to divide the species, practically it did not. Collins and East (1998), after analyzing 16S rRNA gene sequencing information, concluded that the strains within a group were highly related but that the relationship between groups was not that close. Using this information along with information on the toxin, DNA–DNA pairing studies and non-toxic, non-hemagglutinating proteins, they concluded that the groups should represent four different species.

Group I contains the proteolytic type A, B and F strains. They digest meat in cooked meat medium and casein in milk medium and liquefy gelatin in gelatin medium. They ferment glucose but not mannose or sucrose. The optimal temperature for growth of this group is 35–40 °C, and the minimum temperature for growth is 10 °C. Fermentation by-products from growth in peptone yeast extract glucose broth include acetic, isobutyric, isovaleric and butyric acids, along with a small amount of other acids and alcohols (Holdeman and Brooks, 1970). In trypticase soy broth, hydrocinnamic acid and hydrogen gas are produced. On egg-yolk agar, colonies have a mother-of-pearl layer due to the action of lipase on the fat in the medium. Some type B strains may not show this reaction because they lack the lipase. Growth is inhibited by 10 % sodium chloride. Although plasmids have been found, the toxicity does not appear to be plasmid-directed (Weickert *et al.*, 1986).

Group II includes the non-proteolytic strains of type B and F and all the type E strains. They do not digest casein or meat but they do liquefy gelatine. They produce lipase, which is detected on egg-yolk agar. Glucose is fermented, as are mannose and sucrose. When they are grown in peptone yeast extract glucose broth, butyric, propionic and acetic acids (Holdeman and Brooks, 1970) are produced, as is hydrogen gas. The optimal growth temperature is 18–25 °C, and the minimum growth temperature is 3.3 °C. Growth is inhibited by 5 % sodium chloride. Plasmids have been found in this group, but they do not appear to be associated with the production of toxicity. The plasmids may be related to the production of boticin, an inhibitor of the organism's growth (Scott and Duncan, 1978). There are some strains of organisms that are non-toxigenic and culturally resemble type E (Kautter *et al.*, 1966). There is some cross-reactivity of the type E toxin with the type F proteolytic and/or non-proteolytic toxin (Eklund *et al.*, 1967; Yang and Sugiyama, 1975).

Group III consists of types C and D. These organisms are variably non-proteolytic or proteolytic. They produce lipase, which is detected on egg-yolk agar. Gelatin is liquefied. Milk and meat are slowly digested by some of the strains. Glucose and mannose are fermented, while sucrose is not. When they are grown in peptone yeast extract glucose broth, acetic and butyric acids are produced (Holdeman and Brooks, 1970). The optimal temperature for growth is 40 °C; they grow very slightly at 15 °C and not at all at 10 °C (Segner *et al.*, 1971). Growth is inhibited by 3 % sodium chloride. DNA–DNA homology studies (Nakamura *et al.*, 1983) showed that there were two different groups of type C, and that these groups exhibited different abilities to ferment melibiose and galactose. One group fermented these sugars, while the other did not. Plasmids have been found in this group. Bacteriophage is involved in the toxin production of this group (Inoue and Iida, 1970) and also in the conversion from one type to the other (Eklund and Poysky, 1974; Fujii *et al.*, 1988).

Group IV consists of only type G. These organisms are proteolytic. The strains in this group do not produce lipase. They slowly digest meat, while gelatin and milk are rapidly digested. Glucose, mannose and sucrose are not fermented. In peptone yeast extract broth, acetic, butyric and isovaleric acids are among the acids produced. The optimal temperature for growth is 37 °C. Growth is inhibited by 6.5 % sodium chloride. A plasmid is carried by type G strains, and it is involved in the production of the toxin (Eklund *et al.*, 1988).

There have been several cases of botulism caused by established species of *Clostridium* besides *Clostridium botulinum*. The two clostridial species involved are *C. butyricum* and *C. baratii*. The *C. butyricum* has been associated with type E outbreaks and the *C. baratii* with type F outbreaks. *C. butyricum* is a Gram-positive, straight rod, motile by peritrichous flagella, and forms oval central to subterminal spores that do not swell the cell (Cato *et al.*, 1986). The strains are non-proteolytic and do not liquefy gelatin. They ferment glucose, mannose and sucrose. The optimal growth temperature is 30–37 °C, and growth can occur at 10 °C. Growth is inhibited by 6.5 % sodium chloride. Fermentation by-products from growth in peptone yeast extract glucose broth include butyric, formic and acetic acids. It is lipase negative and lecithinase negative. The phenotypic characteristics of an isolate of *C. butyricum* that produced type E toxin (McCrosky *et al.*, 1986) indicated that it would be identified as *C. butyricum. C. baratii* is a Gram-positive, non-motile rod with round to oval spores, subterminal to terminal, which swell the cell. The strains are non-proteolytic and do not liquefy gelatin. They ferment glucose, mannose and sucrose. The optimal growth temperature is 30–45 °C. Growth is inhibited by 6.5 % sodium chloride. Fermentation by-products from growth in peptone yeast extract glucose broth are butyric, acetic and lactic acids. Hydrogen gas is produced. It is lipase negative and lecithinase positive. The phenotypic characteristic of an isolate of *C. baratii* that produced type F toxin (Hall *et al.*, 1985) indicated that it had the characteristics of *C. baratii*.

Suen *et al.* (1988b) did DNA hybridization studies and concluded that *C. baratii and C. butyricum* were indeed the species to which the neurotoxigenic isolates belonged. Thus it is apparent that in addition to *C. botulinum* there are other clostridial species in which botulinal toxins can be found. The genes for the toxin in the case of *C. butyricum* strains are located on the chromosome (Zhou *et al.*, 1993; Wang *et al.*, 2000). Wang *et al.* (2000) studied 13 isolates of toxin-producing *C. butyricum* and found three clusters based on random-amplified polymorphic DNA assay, pulsed-field gel electrophoresis and Southern-blot hybridization. These clusters also coincided with three fermentation patterns of arabinose and inulin: arabinose is fermented and inulin is not; neither is fermented; or inulin is fermented and arabinose is not.

3.2 Factors affecting germination, outgrowth and sporulation

Spore germination is often thought of as the changing of the dormant spore into a multiplying vegetative cell. Many of the early studies used this criterion to evaluate conditions for germination of *C. botulinum* spores. It is now known that germination involves several steps, including activation, changes in the spore's resistance and refractivity, and finally the emergence of vegetative growth. There are a number of factors (Roberts and Hobbs, 1968), including activation temperature, pH, incubation temperature and anaerobiosis, that could affect germination.

Activation is usually accomplished with sub-lethal heating of the spore. This heating is quite varied, and various temperature–time combinations have been used. Usually a higher temperature is used for the activation of the proteolytic strains (70–80 °C for 5–20 minutes) than for the non-proteolytic strains (60–75 °C for 10–15 minutes). It has been noted (Montville, 1981) that heat activation is affected by many factors, including the amount of heat treatment, the conditions for determining the germination when a criterion is outgrowth, and strain variation. Montville (1981) found that heat activation of types A and B was not required for maximum colony formation when plated on botulinum-assay medium, but was required when plated on reinforced-clostridial medium. Rowley and Feeherry (1970) noted that maximum activation of type A occurred when heating for 60 minutes at 80 °C, as opposed to heating for 10 minutes at 80 °C. Ward and Carroll (1966) found no difference in germination results of type E when no heat of activation was compared with heating at 60 °C for 15 minutes or at

75 °C for 10 minutes. On a psychrotolerant, non-toxigenic, non-proteolytic strain of *C. botulinum*, it has been observed (Evans *et al.*, 1997) that heat shocking has a greater effect on stored spores than on fresh spores. They also noted that the temperature at which spores formed and the length and conditions of storage affected the ability of the spore to germinate. A heat shock of 10 minutes at 65 °C was found to stimulate germination of type E (Ando, 1971), but was varied in its effectiveness based on the germination compounds involved.

Germination has been found to occur under a variety of conditions, and has been found to be affected by a number of different compounds. Riemann (1953), using heat-activated type A spores in a casein-hydrolyzate yeast-extract medium, found that the addition of sodium hydrogencarbonate (sodium bicarbonate, $NaHCO_3$) had a beneficial effect on spore germination. Wynne *et al.* (1954) placed heat-activated type A and B spores in 0.02 M phosphate-buffered glucose (2 %) medium and found that addition of glucose caused germination of the spores. Treadwell *et al.* (1958), using type A spores, found that $NaHCO_3$ aided germination. They also found that soluble starch and dipotassium phosphate had no noticeable effect on germination. Germination occurred in vitamin-free casamino acid. Rowley and Feeherry (1970) found that type A spores in a chemically defined medium could be germinated in 8 mM L-cysteine. Increasing the concentration of the L-cysteine increased the rate of germination, but not the extent of the germination. They noted that the presence of $NaHCO_3$ enhanced germination. They also reported that L-alanine was an effective germinant. With L-cysteine, they found that the optimum pH for germination was between pH 6.5 and 7.5 and the optimal temperature was 37 °C. It was found (Billon *et al.*, 1997) that when the germination of individual spores of type A was followed, temperature affected the germination lag. As temperature increased, the lag was shorter and the standard deviation of the germination lag was smaller in the population studied. Ando (1973) found that type A spores germinated in a phosphate-buffered bicarbonate solution with added lactate. He found that germination occurred under both aerobic and anaerobic conditions. He also found (Ando, 1974) that alpha-hydroxy acids, glycolic acid, and L- and D-lactic acid aided germination of type A when added to an L-alanine germination medium. Foegeding and Busta (1983) used two different crops of type A, and found that the concentration of L-alanine needed for germination after hypochlorite treatment was different for each crop. They also noted that the addition of calcium lactate promoted germination with suboptimal concentrations of L-alanine. Smoot and Pierson (1981) found that potassium sorbate was a competitive inhibitor of L-alanine, and L-cysteine induced germination of type A spores. The inhibition is pH dependent, and can be overcome when the spores are removed from the presence of sorbate. EDTA has also been found to inhibit the germination of type A spores (Winarno *et al.*, 1971). The inhibition is pH-dependent and can be overcome with the addition of $MgCl_2$ or $CaCl_2$. Essential oils have been found (Chaibi *et al.*, 1997) to inhibit germination of type A spores. The essential oils from artemisia, cedar, eucalyptus, orange, grapefruit, savage carrots and vervain inhibited germination by L-alanine by inhibiting the commitment to germinate.

Ward and Carroll (1966) used type E spores in chemically defined medium to determine that differences in germination occurred between newly harvested and stored spore suspensions. They found, by using single amino-acid solution supplementation of a chemically defined medium, that glycine and L-cysteine were the best germinants, and phenylalanine or leucine and alanine showed some delays. The omission of p-aminobenzoic acid was the only vitamin that appeared to affect germination. Ando (1971) studied germination of type E spores in a chemically defined medium. He found L-alanine germination was assisted by the presence of L- and D-lactate, inosine, or glucose. Germination with L-alanine was assisted by bicarbonate and had an optimum pH of 7.1–7.3. He also noted that L-cysteine and L-serine were effective germinants. Moreover, he found (Ando, 1974) that in an L-alanine germination system the addition of alpha-hydroxy acids aided germination, as did glycolic acid and L- and D-lactic acids. Ando and Iida (1970) found that germination of type E could occur in air. Germination occurred with an Eh of +198 to +414 mV. They also determined that the rate of germination increased with aging of the spores. Ando *et al.* (1975) found that glycine at high levels inhibited germination of type A, B and E spores. It has been noted that the presence of lysozyme is important in the recovery of heated spores (Alderton *et al.*, 1974). It was postulated that, in type E spores, lysozyme enhanced the germination process (Sebald and Ionesco, 1972). Peck *et al.* (1992) used non-proteolytic type B and type E spores, and determined that lysozyme replaced the germination system in heated spores.

The growth of types A and E has been studied in chemically defined media. Ward and Carroll (1966) found that for type E, growth was not dependent upon any single amino acid. The omission of aspartic acid, tryptophan or valine was the most detrimental to growth. They found that there was no absolute vitamin requirement, and that the presence of nucleic acids stimulated growth, as did the presence of ribose. The addition of lactic or citric acid enabled the formation of toxin. Strasdine and Melville (1968), using type E, also found that the omission of amino acids or vitamins did not prevent growth. They noted that the exception to this was the omission of valine, which inhibited spore outgrowth. They found that the carbon source affected growth and sporulation. Growth occurred with galactose, but no sporulation occurred. Sporulation and growth occurred with maltose or glucose. Hawirko *et al.* (1979) determined that type E, in a chemically defined medium, needed isoleucine and/or valine for growth to occur. Smoot and Pierson (1979) reported, for type A in trypticase soy broth, that they observed no difference in growth or toxin production at pH 7 of Eh values of −60 mV to −145 mV. Ando and Iida (1970) found that growth of type E occurred from an Eh of −28 mV to −350 mV. Mager *et al.* (1954) found that nine amino acids were essential for growth, as were biotin, para-aminobenzoic acid and thiamine. Kindler *et al.* (1956) found that CO_2 was needed for the growth of types A and B. They also noted that the chemically defined medium that supported the growth of types A and B did not support the growth of types C, D and E. Several time-of-growth models (Baker and Genigeorgis, 1990; Whiting and Call, 1993; Whiting and Strobaugh, 1998) have been developed to assist in the determination of time of growth and the effects of environmental factors such as acidity, temperature, spore-inoculum size and sodium chloride concentration on the growth.

Sporulation usually occurs in media that employ meat or vegetable infusions containing particulate matter (Roberts, 1967). However, sporulation has been achieved in aparticulate media for type A using enzymatic hydrolyzates of casein and animal tissues (Tsuji and Perkins, 1962). Perkins and Tsuji (1962) devised a synthetic medium for the sporulation of type A spores. They determined that arginine played an important role in sporulation. Day and Costilow (1964) found that sporulation did not occur without an exogenous source of energy. Spore maturation occurred in the presence of L-alanine and L-proline, L-isoleucine and L-proline, or L-alanine and L-arginine. They found that arginine or citrulline alone would not support the maturation process. There does not appear to be any protein or nucleic acid synthesis in progress. Hawirko *et al.* (1979) devised a synthetic medium for the sporulation of type E. The medium consisted of 18 amino acids, vitamins, minerals, nucleic acid components, glucose, acetate and bicarbonate. The degree of sporulation was affected by the levels of purines and pyrimidines, acetate and glucose, and the presence of ornithine, proline, serine, threonine and valine. The presence of poly-beta-hydroxybutyrate has been suggested (Emeruwa and Hawirko, 1973) as an endogenous energy source for the sporulation of type E spores. Strasdine (1972) has indicated that an exogenous source of energy is not needed for type E spores, as they can accumulate, intracellularly, a starch-like glucan which reaches maximum concentrations about the time of forespore formation.

3.3 Conditions for toxin production

Studies involving formation of toxin usually involve factors that affect the growth of the organism. Indication of growth, particularly in foods, is usually measured by the presence of toxin. It has been observed that cell autolysis is related to the amount of toxin produced (Boroff, 1955) in the culture medium. Autolysis, however, is not the only mechanism by which toxin is released into culture medium, since the presence of extracellular toxin has been observed in young cells in the absence of autolysis (Bonventre and Kempe, 1960). Investigators have suggested that simple diffusion could account for the presence of the toxin in this situation.

It is difficult to separate the nutritional needs for growth and the needs for toxin formation. It does appear that a carbohydrate source such as glucose is needed for toxin formation (Bonventre and Kemp, 1959a), whether in a still culture or in an agitating fermentor (Siegel and Metzger, 1979). For type E, excess tryptophan decreased toxin formation in minimal growth medium (Leyer and Johnson, 1990). Toxin is formed at pH values that allow growth. At pH values > 7, the toxin has been found to be unstable and the biological activity of the toxin is reduced (Bonventre and Kemp, 1959b). The investigators also noted that, for the type A strain tested, toxicity disappeared when the culture was held at 48 °C. Although toxin is formed at temperatures that allow the growth of the organism, the toxin is relatively unstable at temperatures above 30 °C. Pedersen (1955) found that for type E, toxin titer fell at temperatures above 37 °C. Toxin formation does not appear to be influenced by water activity (Baird-Parker and Freame, 1967). Baird-Parker (1971) noted that when *C. botulinum* grows in a food, toxin is likely to be produced. The amount of toxin produced and the stability of the toxin depend upon factors such as those previously mentioned.

4 Characteristics of the toxin

The botulinum toxins of the various types have been found to be similar in molecular size, with a molecular mass of about 130–170 kDa. The size variation reported among a specific type is often due to the procedures used to make the determination. The toxin molecule is generally considered to have a molecular weight of about 150 kDa. The toxin molecule consists of a heavy chain of about 100 kDa, and a light chain of about 50 kDa. These chains are joined by a disulfide bridge. The type E molecule is a single-chain molecule until it becomes activated with trypsin (DasGupta and Sugiyama, 1972), and then the typical dichain is observed. All the toxins appear to be formed as relatively low-toxicity molecules, which require a molecular change to increase the toxicity.

The toxin is naturally present as a complex in culture fluids (Sugiyama, 1980). The complexes vary in size depending upon the procedure used to grow the organism (Sakaguchi *et al.*, 1979). The complexes are the toxin associated with non-toxic proteins, and may have hemagglutinating activity. The complexing proteins appear to provide a protective function to the toxin itself during oral ingestion (Ohishi *et al.*, 1977). A molecular model was developed for the passage of type A toxin and the complex through the gastrointestinal tract (Chen *et al.*, 1998). For type C toxin, the hemagglutinin part of the complex seemed to have a critical role in the binding and absorption of the toxin (Fujinaga *et al.*, 1997). The cluster of genes involved in the complex for all the types has been mapped (Bhandari *et al.*, 1997; East and Collins, 1994; Fujii *et al.*, 1993; Hauser *et al.*, 1994; Henderson *et al.*, 1996; Kubota *et al.*, 1998; Marvaud *et al.*, 1998), which has aided investigation of the various functions of the parts of the complex.

The genes for the type G toxin complex are found on a plasmid (Zhou *et al.*, 1995) and not on chromosomal DNA. The neurotoxin genes for types C and D are found in bacteriophages (Eklund *et al.*, 1987). The genes of the other types are found on the chromosome (Binz *et al.*, 1990; Hutson *et al.*, 1994; Cordoba *et al.*, 1995; Elmore *et al.*, 1995).

5 Nature of the intoxication in man and animals

There are four types of illnesses attributed to *C. botulinum* in humans (foodborne botulism, wound botulism, infant botulism, and child or adult infectious botulism), which are distinguished in the manner they are contracted. The disease known as foodborne botulism is considered an intoxication because it results from the ingestion of the botulism toxin. The other three illnesses can result from the ingestion of the cell or spore, or contamination of a wound, and are considered infections.

In cases of foodborne botulism, the microorganism and its spore are usually harmless; it is only the toxin that is of concern for healthy adults. However, the ingestion of spores can be lethal to babies and to individuals with certain health concerns, as is seen in cases of infant botulism and child or adult infectious botulism. Likewise, toxin is produced in wound botulism if the spore is present and the conditions are favorable, resulting in illness. All three types of botulism are seen in the US (Table 13.1).

Table 13.1 Recorded botulism cases in the US, 1973–1996 (Shapiro *et al.*, 1998)

Type	Median (per year)	Range (per year)	Total (all years)
Foodborne	24	8–86	724
Infant	71	0–99	1444
Wound	3	0–25	103
Undetermined	Not available	Not available	39

Foodborne botulism was the first type of these illnesses to be identified. The numbers of cases and fatalities have decreased dramatically since it was first identified and controls established. The CDC (Centers for Disease Control and Prevention) (1998) reports that in the US, for the 50-year period between 1899 and 1949, 1281 cases of botulism were recorded, with a case-fatality rate of approximately 60 %. During the 46 years between 1950 and 1996, an additional 1087 cases were recorded, with a case-fatality rate of 15.5 %. The number of cases per outbreak remained constant (2.5–2.6 cases/outbreak). Wound botulism, which was the next to be identified, is relatively uncommon, although a large number of cases have been noted in needle-drug users since 1994. Infant botulism was identified in 1976, but cases likely occurred before that. Currently this is the most common type of botulism in the US, and treatment and prevention programs are still being studied. Finally, the most recently identified form of the illness, an extension of infant botulism (adult infectious botulism) that has been seen in children and adults with specific health restrictions, is also being studied.

The diagnosis of botulism is based on compatible clinical findings, history of exposure to suspect foods, and supportive ancillary testing to rule out other causes of neurologic dysfunction that mimic botulism. Intoxication symptoms can be misleading, and may be misdiagnosed because of similarity to other illnesses. Since actual incidents are few, many doctors may not be familiar with symptoms and treatment may be delayed. Some of the symptoms may be transient or mild; and the illness may be misdiagnosed as myasthenia gravis, stroke, Guillain Barré syndrome, bacterial or chemical food poisoning, tick paralysis, chemical intoxication, mushroom poisoning, medication reactions, poliomyelitis, diphtheria or psychiatric illness. Clinicians must know how to recognize, diagnose and treat this rare but potentially lethal disease (CDC, 1998; Shapiro *et al.*, 1998).

Collecting stool and serum samples early during the course of illness increases the likelihood of obtaining positive results, but laboratory results may not be reported until many hours or days after the specimens are received. The administration of antitoxin is the only specific therapy available for botulism, and evidence suggests that it is effective only if done very early in the course of neurologic dysfunction (Sugiyama, 1980). Hence the diagnosis of this illness cannot await the results of studies that may be long delayed and be confirmatory only in some cases. The diagnosis should be made on the basis of the case history and physical findings (CDC, 1998). Laboratory confirmation of suspected cases is performed at the CDC and some state laboratories (CDC, 1998; Shapiro *et al.*, 1998).

CDC (1998) reports that the clinical syndrome of botulism, whether foodborne, infant, wound or intestinal colonization, is dominated by the neurologic symptoms and signs resulting from a toxin-induced blockade of the voluntary motor and autonomic cholinergic junctions, and is quite similar for each cause and toxin type. Onset usually occurs 18–36 hours after exposure, with a range of 6 hours to 10 days (Shapiro *et al.*, 1998). The ingestion of other bacteria or their toxins in the improperly preserved food, or changes in bowel motility, are likely to account for the abdominal pain, nausea, vomiting and diarrhea that often precede or accompany the neurologic symptoms of foodborne botulism. Dryness of the mouth, inability to focus to a near point (prompting the patient to complain of 'blurred vision') and diplopia are usually the earliest neurologic complaints. Fever is usually absent (Reed, 1994). Symptoms from any toxin type may range from subtle motor weakness or cranial nerve palsies to rapid respiratory arrest (Reed, 1994; Shapiro *et al.*, 1998). If the disease is mild, no other symptoms may develop and the initial symptoms will gradually resolve. The person with mild botulism may not need medical attention. In more severe cases, however, these initial symptoms may be followed by dysphonia, dysarthria, dysphagia and peripheral-muscle weakness (Griffin *et al.*, 1997). Symmetric descending paralysis is characteristic of botulism; paralysis begins with the cranial nerves, then affects the upper extremities, followed by the respiratory muscles and finally the lower extremities in a proximal-to-distal pattern (Shapiro *et al.*, 1998). If illness is severe, respiratory muscles are involved, leading to ventilatory failure and death unless supportive care is provided. Electron microscopic evidence suggests that clinical recovery correlates with the formation of new presynaptic end plates and neuromuscular junctions. Before mechanical ventilation and intensive supportive care, up to 60 % of patients died; since the 1950s, however, the mortality rate from botulism has steadily decreased. Recovery follows the regeneration of new neuromuscular connections. A 2- to 8-week duration of ventilatory support is common, although patients have required ventilatory support for up to 7 months before the return of muscular function (CDC, 1998).

5.1 Foodborne botulism

Foodborne botulism is caused by ingestion of preformed toxin produced in food by *C. botulinum*, in which spores that are present in the food germinate, reproduce and produce toxin in the anaerobic environment of the food (Shapiro *et al.*, 1998).

The clinical syndrome of foodborne botulism is dominated by neurologic symptoms and signs resulting from a toxin-induced blockade of the voluntary motor and autonomic cholinergic junctions. Although the syndrome is similar for each toxin type, type A toxin has been associated with more severe disease and a higher fatality rate than has type B or type E toxin (Chia *et al.*, 1986).

Death now occurs in 5–10 % of cases of foodborne botulism; early deaths result from a failure to recognize the severity of disease, whereas deaths after 2 weeks result from complications of long-term mechanical ventilatory management (Griffin *et al.*, 1997).

5.2 Wound botulism

Wound botulism was first recognized in 1943 in either deep lacerations or compound fractures, but it is currently emerging among chronic drug abusers. It occurs when *C. botulinum* cells or spores alone or with other microorganisms infect a deep wound and produce toxin due to the anaerobic conditions (CDC, 1998; Shapiro *et al.*, 1998). The clinical manifestations are similar to those seen in foodborne botulism, except that gastrointestinal symptoms are absent and the median incubation period is longer (7 days; range 4–14 days). The case-fatality rate for wound botulism is approximately 15 %. In the US, 80 % of cases are caused by toxin type A and 20 % by type B (Shapiro *et al.*, 1998).

In the US, particularly the western US, more than 40 cases were noted in the early 1990s; in most cases the illnesses were attributed to subcutaneous injection or 'skin popping' of 'black tar' heroin (CDC, 1995). In Europe, similar cases involving injecting drug users have been suspected since the late 1990s. Five cases were reported in Switzerland in 1998 and 1999, three cases in Norway in 1997, and two cases in the UK in 2000. Three cases of wound botulism had been reported in Italy between 1979 and 1999, all of them associated with injuries sustained in building and agricultural occupations (Kuusi *et al.*, 1999; Athwal and Gale, 2000; Khan and Chay, 2000).

5.3 Infant botulism

Infant botulism was recognized as one of the forms of botulism in 1976 (Midura and Arnon, 1976; Pickett *et al.*, 1976). However, it is not a new disease, and in fact a misclassified case has been shown to have existed as early as 1931 (Arnon *et al.*, 1979a). Since its recognition in 1976, infant botulism has become, at least in the US, the most common form of botulism. The CDC (1998) noted that 1442 cases were reported in the US between 1976 and 1996. In contrast, 444 foodborne botulism outbreaks involving 1087 cases were reported from 1950 to 1996. Recent studies of cases in other countries seem to indicate that it occurs in countries with cases of foodborne botulism. In Europe, a total of 9 cases were reported from 1994 to 1999 – 3 in Spain, 2 each in Italy and Germany, and one each in the UK and Denmark (Therre, 1999). Canada reported 13 cases of infant botulism between 1990 and 1999 (Austin and Slinger, 2000).

Unlike foodborne botulism, which is caused by the ingestion of preformed toxin, infant botulism is caused by the ingestion of spores of *Clostridium botulinum*. A *C. baratii* strain that produces type F botulinum toxin (Hall *et al.*, 1985) and a *C. butyricum* strain producing a type E botulinum toxin (Aureli *et al.*, 1986) indicate that other potential carriers of the botulinum toxin could be involved in this disease. The ingested spores germinate, the colon is colonized, and botulinum neurotoxin is produced. The toxin is absorbed and carried by the bloodstream to peripheral cholinergic synapses. The toxin is bound there and prevents the release of acetylcholine. The toxin's action results in flaccid paralysis and hypotonia. Constipation also occurs due to slower gut motility. It has been demonstrated in infant mice that there is a limited age of susceptibility to germination of *C. botulinum* spores in the gut (Sugiyama and Mills, 1978). It also appears that the natural intestinal flora contribute to the prevention of *C. botulinum* multiplication (Moberg and Sugiyama, 1979).

Infant botulism occurs over a restricted age range. Most of the cases occur in patients who are between 1 and 6 months of age, and all the known cases to date have occurred in children 1 year of age or younger (Arnon, 1998). Since spores must be ingested before the illness is observed, food risk factors have been evaluated. Honey has been identified (Arnon *et al.*, 1979b) as an avoidable risk, and the recommendation made not to feed honey to infants. Because of the ubiquity of the organism, it has been found in the dust and soil around homes involved in incidences of infant botulism. However, in most cases there has been no identification of source of organism associated with the diagnosed illness. Thus, even with the minimization of honey as an infant food, cases of infant botulism still occur.

The clinical manifestations of infant botulism can vary widely. Usually constipation is noticed first, with listlessness and poor feeding. Weakness and hypotonia are the usual early symptoms. The first sign of illness is found in the cranial nerve; there is difficulty in controlling the head, feeble crying, and general weakness. The swallow reflex is impaired. It may be possible to detect toxin in the serum, but feces collection is usually the definitive manner to identify the organism and toxin.

There is treatment available for this disease. A botulism immune globulin (Arnon, 1993) has been developed and tested. Treatment also includes good supportive care. Meticulous attention to the feeding and breathing requirements of the infant is paramount. The chance of recovery for infants with gradual onset of the illness is good, with hospitalization and care. The existing equine botulinum antitoxin cannot not be used in infants because of its unacceptably high incidence of serious side-effects (anaphylaxis, serum sickness) when used in adults with foodborne botulism (Arnon, 1993), and a specific treatment for infant botulism has been studied using human-derived botulinum antitoxin, formally known as botulism immune globulin (BIG).

Studies were conducted in the late 1970s in which autopsy specimens from 280 California infants who died from a variety of causes were tested for botulism. In 10 infants, all of whom died suddenly and unexpectedly, *C. botulinum* organisms were isolated. Of the 10 deaths, 9 had been ascribed to Sudden Infant Death Syndrome (SIDS). This work and a similarity of age distribution between children with infant botulism and SIDS has led some researchers to suspect that there may be a relationship between the two conditions (Arnon, 1980).

5.4 Adult infectious botulism

There have been several instances in adults where botulism has been diagnosed as a result of *in vivo* multiplication of the organism and the subsequent production of toxin. This occurs in wound botulism, but has also been seen when the multiplication results from the colonization of the intestinal tract. This is similar to what happens in infant botulism; however, in the adult there has usually been surgery involving alteration of the gastrointestinal tract before the botulism symptoms appear. Minervin (1966) provided evidence from observations of botulism cases in Russia of the potential for the toxico-infectious nature of botulism. Chia *et al.* (1986) described an incident of foodborne intestinal infection from the consumption of a food containing *Clostridium botulinum* organisms but no toxin. McCrosky and Hatheway (1988) investigated four cases in which no preformed toxin was detected in foods, but toxin was detected in clinical samples over a prolonged time period and conditions had been altered in the gastrointestinal tract. Griffin *et al.* (1997) reported a case in which an adult with Crohn's disease developed botulism. Although initially Guillain-Barré syndrome was suspected, botulism was diagnosed due to identification of the toxin in stool and serum specimens.

Typical symptoms of botulism – weakness of the musculature, diplopia and constipation – are usually noted. Because of the rareness of intestinal colonization in adults, the symptoms are often misdiagnosed and associated with other diseases. However, in individuals with altered gastrointestinal tracts, botulism should be considered if the symptoms warrant.

6 Therapeutic use of the toxin

In the 1980s, research was conducted using the paralyzing ability of the botulinum toxin to limit muscle contractions in illnesses where such contractions are a disadvantage. Initially it was licensed as Oculinum to treat two eye conditions – blepharospasm and strabismus – characterized by excessive muscle contractions. It is now marketed under the trade name Botox (Vangelova, 1995).

Small doses of the toxin are injected into the affected muscles. As happens with botulism, the toxin binds to the nerve endings, blocking the release of the chemical acetylcholine, which would otherwise signal the muscle to contract. The toxin thus paralyzes or weakens the injected muscle but leaves other muscles unaffected. The injections block extra contractions (of the muscle) but leave enough strength for normal use (Vangelova, 1995). Purified botulinum toxin is currently used to treat other medical conditions, such as torticollis (contractions of the neck and shoulder muscles), oromandibular dystonia (clenching of the jaw muscles) and spasmodic dysphonia (which results in speech that it is difficult to understand) (Shapiro *et al.*, 1998).

Several factors explain why botulinum toxin has proved to be such a valuable drug. First, the toxin is highly selective in acting on cholinergic cells – a characteristic that diminishes the chances of adverse side effects. Any adverse effects that do occur are usually caused by toxin diffusion and action on nearby cells. Second, the toxin has a long duration of action. Clinically significant responses can last from several months to more than a year, after which time the toxin must be re-administered. Finally, the dose of toxin can be individually titrated for each patient to ensure maximal benefits. No other form of therapy can match the combination of these three qualities and the toxin's impressive clinical efficacy (Simpson, 1996).

7 Other potential uses of the toxin

The potential for intentional poisoning has come into clearer focus in recent years. As many as 17 countries are suspected to include or be developing biological agents in their offensive weapons programs. Botulinum toxin often is one of these agents because it is relatively easy to produce and is highly lethal in small quantities (Shapiro *et al.*, 1998).

After the 1991 Gulf War, Iraq revealed that it had produced 19 000 l of concentrated botulinum toxin, of which approximately 10 000 l were loaded into military weapons. The 19 000 have been not fully accounted for, and constitute approximately three times the amount needed to kill the entire current human population by inhalation (Arnon *et al.*, 2001).

8 Animal botulism

Botulism in wild and domesticated animals, birds and aquaculture fish occurs worldwide. Documentation of the incidence in most wild animals is difficult to obtain. However, the presence in wild birds is well documented (Jensen and Price, 1987). In some instances there are control procedures, such as vaccination, which work well. In others, there are no practical control procedures available.

Botulism in mink fur farms has been known to exist since its first reporting in 1938 (Hall and Stiles). The disease has been reported in the US, Canada, Europe and Japan. Although not a common problem, its occurrence has caused a high mortality. This is probably due to the tendency to feed animals with large quantities of meat that is not fit for human consumption (Smith and Sugiyama, 1988). The disease is usually caused by type C strains. The availability of a type C toxoid has enabled farm mink to be successfully vaccinated against the disease (Burger *et al.*, 1963).

Botulism in dogs has been reported in the US (Richmond *et al.*, 1978), Europe (Claessens *et al.*, 1983; Cornelissen *et al.*, 1985), Australia (Farrow *et al.*, 1983), New Zealand (Wallace and McDowell, 1986), South America (Fain Binda *et al.*, 1998) and Africa (Doutre, 1983). The type reported is usually C, although type D has also been reported. Dogs suffering from this illness usually have signs of lower motor neuron dysfunction, which may range from limb weakness to flaccid paralysis. Their bark is weak and constipation may occur. Toxin can be found in the serum; it may also be found in the feces. Farrow *et al.* (1983) found toxin and spores 114 days after ingestion of the suspect food, and suspected that colonization of the gut had occurred.

Botulism in pond-raised fish was reported in trout in 1974 (Huss and Eskildsen, 1974). Eklund *et al.* (1982, 1984) reported incidents involving salmon from 1979 to 1984. They concluded that dead fish, which were decomposing on the bottoms of the ponds, were the likely source of the toxin. Type E toxin is the cause because of the sensitivity of fish to this toxin type. Fish with this illness are unable to swim against the current. They have progressive inability to control their swimming behavior, swimming on one side or another, and sinking or rising to the surface. Once symptoms develop, death usually occurs. Toxin can be demonstrated in the intestines of sick fish.

Botulism in cattle has been observed in a number of countries, including the US (Divers *et al.*, 1986; Wilson *et al.*, 1995; Galey *et al.*, 2000), Canada (Bienvenu and Morin, 1990; Wobeser *et al.*, 1997), Australia (Main and Gregory, 1996), Europe (Haagsma and Ter Laak, 1979; Muller, 1981; Notermans *et al.*, 1981; Popoff *et al.*, 1984; McLoughlin *et al.*, 1988; Hogg *et al.*, 1990; Chiers *et al.*, 1998),

South America (Doberreiner *et al.*, 1992; Ortolani *et al.*, 1997) and Africa (Theiler *et al.*, 1927). The disease is caused by the ingestion of preformed toxin. The usual toxin type detected is C or D. Type B is frequently found, while type A is rarely encountered. A progressive paralysis of the animal occurs, leading to difficulty in rising. A characteristic of the disease is the loss of lingual tone – when the tongue is pulled out of the mouth, it is not immediately retracted but hangs at the side of the mouth. The chance of recovery from the disease is good if the animal can stand; however, once cattle are down their recovery is problematic. Toxin can be demonstrated in the serum if the serum is taken in the early stages of the disease. Toxin may also be present in rumen contents or feed material. Circulating antibodies may be detected in recovered animals (Gregory *et al.*, 1996; Main and Gregory, 1996). The type of toxin found is often dependent on the source of the toxin. In areas where the land is phosphorus deficient, cattle obtain the phosphorus by consuming bones and the carcasses of small animals. These small animals may be contaminated with *C. botulinum* type C or D spores, which grow in the dead animals; these are then consumed by the cattle, which become affected by the toxin. Forage that is contaminated with a dead animal produces type C or D illness (Haagsma and TerLaak, 1979; Galey *et al.*, 2000). Poultry litter has been found to produce type C illness (Hogg *et al.*, 1990; Ortolani *et al.*, 1997). Type B illness has been found in cattle that have been fed brewer's malt (Notermans *et al.*, 1981), rye silage (Divers *et al.*, 1986), beet pulp (Chiers *et al.*, 1998), and hay that has been wrapped in plastic (Wilson *et al.*, 1995). Whitlock and Williams (1999) have summarized the infection in cattle, including symptoms, diagnosis, clinical signs and therapy. In Australia, toxoid shots against types C and D (Gregory *et al.*, 1996) are given to immunize cattle. However, cattle are not routinely vaccinated against botulism.

Avian botulism occurs worldwide, and is usually caused by type C. Type E has been reported to be found in gulls (Kaufmann *et al.*, 1967). Botulism occurs in wild (Giltner and Couch, 1930; Jensen, 1981) as well as domesticated birds (Dickson, 1917; Harrigan, 1980). Jensen and Price (1987) indicated that over 100 species of wild birds are affected by botulism. Toxin can be demonstrated in the sera. Fecal and intestinal samples are often useful for testing for the presence of the organism, since it can be found normally in these. For birds of flight, difficulty in flying is an early symptom of the illness, followed by the inability to walk and difficulty in holding the head up. As the disease progresses, muscle tone is lost and the typical limber-neck posture occurs. The chance of recovery from the illness is good in the early stages of the illness. Recovery is more problematic as the severity of the disease increases. Toxoid treatment is possible, and is effective; however, it might not be cost-effective if the population to be treated is large. Control procedures usually involve management of the habitat to ensure that it is free of dead birds and small animals, preventing standing water (which can become anaerobic), and removal of other sources of the organism and its potential to grow. A very good compilation of information on avian botulism can be found in *Avian Botulism: An International Perspective* (Eklund and Dowell, 1987).

9 Distribution of spores in food and the environment

C. botulinum is found in various environments, including soil, decaying vegetation, streams, lakes and coastal waters. Surveys testing for the presence of *C. botulinum* in the environment were started in the US, Europe and Canada in the early 1920s by K. F. Meyer and B. J. Dubovsky (Dubovsky and Meyer, 1922; Meyer and Dubovsky, 1922a, 1922b, 1922c). These surveys were carried out in response to numerous cases of foodborne botulism following consumption of vegetables and meats. Much of this work was done before all of the types were identified and their characteristics fully known; therefore, the data focus primarily on types A and B. The subsequent discovery of *C. botulinum* type E in the 1960s generated new investigations of its distribution (Huss, 1980; Dodds, 1993).

Data show that different types of *C. botulinum* are more prevalent in certain parts of the world, which has resulted in a close association between the types of botulism outbreaks and the occurrence of spores in the environment (Dodds, 1993):

- Types A and B are generally the cause of outbreaks in more temperate and warmer zones (with one type often predominating over the other).
- Type A spores are found predominantly in soils of the Western US, Brazil, Argentina and China, and the most frequently implicated foods are vegetables. They favor neutral to alkaline soils (average pH of 7.5) with low organic content, hence their virtual absence from the Eastern US and Europe, where the soil is heavily farmed.
- Distribution of type B spores seems more uniform, but those found in the soils of Eastern US are proteolytic, and cases of foodborne illness usually involve vegetables. Those found in the UK and northern Europe are typically non-proteolytic, and the vehicle is most often a meat product. Type B spores favor a slightly acidic soil (pH 6.25) and a higher organic content.
- Type E is common in colder regions of the northern hemisphere and is mostly associated with fish and marine mammals. It is most prevalent in aquatic environments of the northern sub-arctic and temperate zones. High prevalence has been found in marine sediments in the coastal waters of Canada, Alaska, Greenland, Japan, Scandinavia and Russia. Freshwater sediments in various parts of the world, such as the US, Japan, Sweden and Finland, have also been found to be contaminated with type E (Huss, 1980). Type E grows well in carrion of fish, marine mammals and invertebrates, and these organisms also carry spores in their intestines. Water currents are significant in influencing the distribution of spores through the marine environment.
- Types C and D mainly cause botulism in non-human species, including cattle and birds. These types have been found in marshes inhabited by birds in localized areas around the world, including the Gulf Coast and northwest in the US; in sediments in the Netherlands and Czechoslovakia; in Bangladesh, South Africa, Indonesia and Java; and in specific regions of Japan, Thailand and New Zealand (Dodds, 1993).

- Type F is found in specific locations throughout the world. In the US it has been detected in sediment and soil from the Pacific Northwest, along with salmon from some rivers in Oregon and Washington. Isolates have been detected in soil from Switzerland, Paraguay and Argentina. It has also been found in certain locations in the UK, Taiwan, Japan, Brazil, the USSR, Indonesia and Java.
- The first survey reporting a substantial number of *C. botulinum* type G was carried out in Switzerland. This type was first detected in soil in Argentina. Type G is difficult to detect in mixed cultures.

In the US, during the period 1971–1989, the CDC reported a total of 272 outbreaks (597 cases) of foodborne botulism, of which 252 were typed. Of these, 61 % were type A, 21 % were type B, 17 % were type E, and 1 % were type F. Although type F is able to poison humans, it is rarely implicated in outbreaks. The data can be subdivided into those from the eastern and western US: in the east, the incidence of types A and B was roughly equal. In the west, the number of type A outbreaks was eight times greater than that of those caused by type B (this is consistent with the predominance of type A in the soils of the western US). The highest number of cases resulted from eating fruits and vegetables, with the majority due to vegetables. There were equivalent numbers from meat and fish, with most of the outbreaks associated with fish occurring in Alaska. In the US, 92 % of the outbreaks were from foods that had been canned or processed in the home (Hauschild, 1993).

Worldwide, an average of 450 outbreaks of foodborne botulism was reported annually from 1951 to 1990. The highest rate of outbreaks was reported in Poland, with an average of 325 (448 cases) per year between 1984 and 1987. Most of the cases were caused by type B, and 83 % of cases were associated with meat products, 75 % of which were prepared in the home. China reported an average of 38 annual outbreaks (168 cases) between 1958 and 1983; most were type A botulism, and the vehicles were fruits and vegetables. Between 1978 and 1989 France reported 123 outbreaks, with almost all of the cases attributed to type B toxin, and 89 % of the contributing food was meat products. In Italy type B is also dominant, but fruits and vegetables are implicated in 77 % of the cases. In Spain, between 1969 and 1988, fruits and vegetables were implicated in 60 % of botulism cases, and 38 % were attributed to meat products (Hauschild, 1993).

In countries where type E predominates in environmental surveys, it is also the type responsible for food-poisoning outbreaks. For example, in Denmark between 1984 and 1989, all 11 food poisoning incidents were attributed to type E. However, in all cases the type of food involved was meat. In Iran between 1972 and 1974, 314 cases were reported, and 97 % were type E, with fish or fish products as the main source. However, fleshy fish were only responsible for 10 % of the cases, while fish eggs (*ashbal*) accounted for 90 % of the cases. Preparation of *ashbal* includes salt curing of the eggs for some months; it is then consumed without further preparation (similar to caviar). The rate of detection of type E spores in the Caspian Sea was very high. In Japan, where a large number of fish and fish products are consumed, 96 % of the 97 outbreaks (479 cases) that occurred between 1951 and 1987 were type E, and 99 % were attributable to fish or fish products; 98 % of these incidents involved domestically prepared food (Hauschild, 1993).

10 Foods associated with botulism

In addition to the geographic location from which a food originates, several other factors can influence the likelihood that a food may be associated with a foodborne botulism outbreak. These include preparation practices, holding temperatures and conditions. For shelf-stable foods, it is critical that the spores be destroyed in a thermal process adequately designed for the particular food. If the process is inadequate and other conditions are favorable, including absence of oxygen, moderate temperatures, and pH above 4.6, the spore can grow and produce toxin. For this reason the US and other countries regulate commercially prepared, shelf-stable foods to comply with certain measures in order to control *C. botulinum* spores. The rate of outbreaks from commercially processed, shelf-stable foods has dropped significantly since these regulations have been in existence. However, these controls are hard to enforce with home-preserved foods, and the majority of cases worldwide are attributed to these foods.

Acidification and temperature control are popular ways to control the growth of *C. botulinum*. As the demand increases for convenience foods, including refrigerated items, and new technologies are introduced to the food-processing industry, food scientists are continuing to study the circumstances under which *C. botulinim* can grow and produce toxin.

Improperly home-prepared foods have been implicated in several cases of foodborne botulism over the past few years. One case of foodborne botulism resulted from improperly pickled eggs in Illinois in 1997 (CDC, 2000). Home-prepared bamboo shoots in Thailand in 1998 (CDC, 1999) were implicated in 13 cases and 2 deaths. In Europe, mushrooms packed in oil were found to be the source of 2 cases and 1 death in the UK; they were packed in Italy and transported to the UK (Brusin and Salmasco, 1998). Likewise, a case of foodborne botulism identified in Austria in 2001 apparently resulted from home-canned vegetables brought from Hungary (Allerberger, 2001).

Restaurant- and delicatessen-associated outbreaks accounted for 42 % of foodborne botulism cases, but only 4 % of the outbreaks, in the US from 1976 to 1994. In 1983, 28 people developed botulism symptoms after eating sautéed onions served on patty-melt sandwiches in Illinois. Apparently the onions were prepared daily and held in a pan with a large volume of melted margarine on a warm stove (below 60 °C), and were not reheated before serving (CDC, 1984). A more recent example in the US involved skordalia (a dip made with potato, oil and spices), which was found to be the source of 30 botulism cases in Texas in 1994. Potatoes baked in aluminum foil were left at room temperature for 18 hours before they were used in the dip. Investigators believed that the foil may have provided the necessary anaerobic environment for spore germination (Angulo *et al.*, 1998). In Argentina, *matambre* (a refrigerated meat roll) was the source of 9 cases of foodborne botulism in 1998 in a food-service operation. The meat roll had been cooked and sealed in heat-shrunk plastic wrap and stored in refrigerators that did not cool it adequately (Villar *et al.*, 1999).

Commercially-prepared foods are still occasionally implicated in cases of foodborne botulism around the world. Salted or fermented ungutted white fish, known by different names, have been the source of outbreaks in several separate incidents since the 1980s (CDC, 1985, 1987, 1992). Although the mechanism of contamination of the fish has not been established, *C. botulinum* spores are found in the intestinal contents of fish and the fish in this case were not eviscerated; so this was likely the source (CDC, 1987). In the UK in 1989, a batch of hazelnut yogurt was the source of 27 cases and 1 death. The yogurt was made with canned hazelnut puree that was prepared with aspartame instead of sugar, causing an increase in water activity and ultimately an inadequate process, allowing the survival of *C. botulinum* spores and production of toxin (Brett, 1999).

There are some areas of the world that have unusually high numbers of cases of foodborne botulism, most likely due to regional practices. Among them is Poland, where a large portion of the cases are caused by meat preserved in *weck* jars (glass jars with rubber seals and a device used to create a vacuum) (Galazka and Przbylska, 1999). A second area with high incidence of foodborne botulism is the Pacific Northwest. From 1950 to 2000, Alaska recorded 226 cases of foodborne botulism from 114 outbreaks. All the patients were Alaska Natives, and all the cases with known causes were associated with eating fermented foods. In traditional fermentation, food is kept in a grass-lined hole in the ground or a wooden barrel sunk into the ground, or is placed in a shady area above ground for several weeks to months. Since the 1970s, however, plastic or glass containers have been used and fermentation has been done above ground or indoors. The anaerobic condition of sealed containers and warmer temperatures make fermentation more rapid and production of botulism toxin more likely. These non-traditional methods were associated with increased botulism rates in Alaska between 1970 and 1989 (CDC, 2001).

11 Detection of *Clostridium botulinum*

The need to detect the presence of *C. botulinum* occurs when an illness having the symptoms associated with botulism is noted. Thus the primary means of determining the presence of the organism is to determine whether a sample is toxic or not. The sample may be from a food that was ingested and caused the illness, a duplicate sample from the same production lot, an environmental sample, or a clinical sample such as blood, vomitus or feces. There is no single culture medium, incubation temperature or sample preparation procedure that is used to isolate all the types. The procedures to determine the presence and the toxin-producing type involve an enrichment, isolation and verification procedure (Kautter and Solomon, 1977; Solomon and Lilly, 1998; Solomon *et al.*, 2001).

Enrichment is accomplished by placing 1–2 ml of the sample in about 15 ml of enrichment medium. The medium used is usually cooked-meat medium (CMM) or liver broth (LB) and trypticase peptone glucose yeast extract (TPGY) medium, from which air has been exhausted in flowing steam and which has been cooled before inoculation. These media are used because the usual cause of illness in humans is organisms from Groups I, II and IV. The Group I organisms grow well in CMM (or on LB), while the Group II organisms grow well on TPGY and the Group IV organisms can grow on both, as do the non-botulinum clostridia that produce botulinum toxin. A medium using TPGY with 1 mg of trypsin/ml was found to be effective (Lilly *et al.*, 1971) as an enrichment medium for type E. A cooked medium combined with fluid thioglycollate broth (Quagliaro, 1977) was found useful, as growth of proteolytic strains produced a black ring at the surface of the broth while non-proteolytic strains produced a white opaque ring. If, on microscopic examination of the sample, spores are observed, heat shocking of the samples will aid in the isolation procedure by reducing competitive organisms. However, a heat shock of 10 minutes at 80 °C, usually used for Group I isolation, will prevent the isolation of Group II organisms. A milder heat shock, such as 10 minutes at 60 °C, is necessary to recover organisms from this group, but even this may not enable recovery. Thus, non-heat shocked cultures should also be incubated. Anaerobiosis is obtained in the CMM with the meat particles and in the TPGY with the thioglycollate. However, a mineral oil or vaspar overlay is sometimes used, as is an anaerobic jar, to maintain anaerobiosis. The CMM should be incubated at 30–37 °C. The TPGY should be incubated at 20–30 °C. Isolation of Group IV organisms may require incubation at 37 °C. If only a single incubation temperature is available, 30 °C is best. However, incubation should last for at least 5–7 days to allow for growth and toxin formation. After incubation, each culture should be examined for gas formation, turbidity, odor and digestion of the meat particles in the CMM. The cultures should be examined microscopically, either under phase contrast or as a stained smear, for the presence of typical clostridial cells, particularly the tennis racket-shaped spores. If growth of typical cells is observed, the culture should be toxin-tested and the type of toxin determined.

Isolation is assisted if good sporulation has occurred in the enrichment medium. A small amount of material (1–2 ml) is saved from the enrichment culture. This sample is either heat shocked or treated with ethanol (Johnston *et al.*, 1964) to destroy vegetative cells. To treat with ethanol, 1 ml of sample is mixed with an equal volume of filter-sterilized absolute ethanol and held at room temperature for 1 hour. Heat shock at 80 °C for 10 minutes is useful for Group I organisms. Heating of Group II organisms may deter isolation. The treated suspension is streaked on blood agar, liver veal egg yolk agar (LVEY) or anaerobic egg yolk agar (EYA) plates; the plates should be dry and incubated anaerobically at 30–35 °C. Blood agar provides a good growth medium, but does not have any differential or selective properties. Growth on egg yolk agar produces colonies that exhibit a surface iridescence when they are examined under oblique light (McClung and Toabe, 1947). This is due to the lipase activity that is exhibited by some strains. Additionally, some of the strains have a lecithinase activity, which causes a yellow precipitate to form. Other media have been used in the isolation procedure. Hauschild and Hilsheimer (1977) used Wynne agar, supplemented with 0.4 % egg yolk, in a pour-plate method. They found the agar needed to be covered with a layer of agar containing 0.01 % dithiothreitol. Spencer (1969) found types A and B to be inhibited by 200 μg/ml but not 50 μg/ml of neomycin in blood agar. Type E was more sensitive, being inhibited at 25 μg/ml. Dezfulian and Dowell (1980) found that types A and B were resistant to a high level of cycloserine, sufamethazole and trimethoprin. Dezfulian *et al.* (1981) developed *C. botulinum* isolation medium (CBI), based on this information, and found it useful for the isolation of *C. botulinum* from fecal samples. Glasby and Hatheway (1985) found it to be effective on isolates from infant botulism cases. Swenson *et al.* (1980) tested 224 strains of *C. botulinum* against 13 different antimicrobial agents, and found 90 % of the strains to be susceptible to all of the agents except gentamicin, nalidixic acid and sufamthoxazole-trimethoprin. Silas *et al.* (1985) developed a peptone glucose yeast extract medium for types A, B and proteolytic F, containing cycloserine, sulfamethoxazole and trimethoprim.

The identification of *Clostridium botulinum* has been evaluated using the bioMérieux rapid ID 32 A test kit (Brett, 1998). It was found that some strains of *C. botulinum* might not be correctly identified down to the species level using this test procedure. Chromatographic techniques have been developed for the identification of *C. botulinum*. Detection of butyric acid and other short-chained fatty acids (Mayhew and Gorbach, 1975) and the determination of cellular fatty acids (Ghanem *et al.*, 1991) have been found useful in identifying *C. botulinum*. Pyrolysis gas chromatography (Gutteridge *et al.*, 1980) has been used to discriminate between established groups. Techniques using genetic materials have been utilized to identify *C. botulinum*. ELISA has been used for types F (Ferreira *et al.*, 1990) and E (Dezfulian, 1993). Pulsed field gel electrophoresis has been used on type A (Lin and Johnson, 1995) and Group II strains (Hielm *et al.*, 1998). It has been found useful in studying these organisms at the molecular level, and is a useful tool for epidemiological studies. Polymerase chain reaction (PCR) has been used to identify types A, B, E and F (Hielm *et al.*, 1996; Aranda *et al.*, 1997; Cordoba *et al.*, 1998). Ribotyping has also been found suitable for use in identification (Hielm *et al.*, 1999).

12 Detection of botulinum toxin

The verification of the presence of *C. botulinum* toxin occurs generally via a mouse test (Kautter and Solomon, 1977; Solomon and Lilly, 1998; Solomon *et al.*, 2001). The type of toxin produced is also determined with a mouse test. The sample is prepared for injection into mice. For food samples, a clear liquid may be tested directly, but a food with a solid matrix may need to be mixed with gelatin-phosphate buffer in order to extract the toxin. Liquid–solid materials should be centrifuged to remove the particulate matter. For clinical samples, sera may be injected directly. Feces and other samples may need centrifugation in order to obtain a clear sample for testing, and the sample may need to be filtered through a 0.45 μm filter to avoid non-specific death of the mice. If non-specific death is suspected, the use of antibiotics such as chloramphenicol and oxytetracycline is suggested (Segner and Schmidt, 1968). The sample may need to be trypsinized (Duff *et al.*, 1956), particularly if the presence of non-proteolytic types is suspected. To trypsinize, 1 g of trypsin (Difco, 1:250) is dissolved in 20 ml of sterile distilled water and 0.2 ml of the trypsin solution is added to 1.8 ml of the sample. The mix is incubated at 37 °C for 1 hour, with occasional gentle agitation.

Swiss-Webster mice, 16–24 g, should be injected intraperitoneally with 0.5 ml of the prepared sample. A sample that has been boiled for 10 minutes and cooled should also be injected. This is a control, as the botulinal toxin will be destroyed by heating. The mice should be observed for 48 hours for symptoms of botulism and death. Death without appropriate symptoms is not evidence of the botulinum toxin; symptoms include ruffling of fur, weakness of limbs and labored breathing. Death immediately after injection usually indicates an injection injury or a reaction to a toxic material other than the botulinum toxin.

Toxin type is determined with monovalent antitoxins. The mice are injected with the specific antitoxin about 1 hour before injection with the suspect sample. Death will occur, except with the appropriately protected mice. However, it should be noted that there is some cross-reactivity with some types, particularly types C and D (Jansen *et al.*, 1976) and types E and F (Solomon and Lilly, 1998). The cross-reactivity can be verified by retesting with combinations of monovalent antitoxin.

A number of rapid tests have been developed to ascertain the presence of toxin. A hemagglutination inhibition test was described by Johnson *et al.* (1966). They used types A and B, and toxin-coupled sheep red blood cells, or bentonite with known amounts of the respective antitoxins. The sensitivity was not as good as the mouse test. A gel diffusion test (Vermilyea *et al.*, 1968) was used with types A, B and E. Some cross-reaction was noted between types A and B. The test was not as sensitive as the mouse test. Thomas (1991) described an enzyme-linked immunosorbent assay (ELISA) to detect type C and D toxins; he utilized polyclonal antibodies to give a semi-purified toxic complex of the toxin. Some cross-reaction was noted against *C. novyi* type A. The test was not as sensitive as the mouse test. Ogert *et al.* (1992) described a biosensor that incorporated optical fibers and an argon-ion laser to measure fluorescence from an antibody sandwich-based immunoassay. Toxin could be detected at levels as low as 5 ng/ml. An ELISA test using an enzyme-linked coagulation assay (ELCA) against types A, B and E toxins was found to be effective (Doellgast *et al.*, 1993, 1994). The venom from a Russell viper was used to make a coagulation-activating enzyme, which was conjugated to affinity-purified horse antibodies specific to the toxin type. This test was found to be as sensitive as a mouse bioassay. ELISA was combined with antimicrobial susceptibility testing to identify types A and B (Dezfulian and Bartlett, 1994), and identification was obtained in 48 hour. An ELISA–ELCA test was used (Roman *et al.*, 1994) to determine type E in fish-challenge studies, and the authors found that the test could be used as an alternative to the mouse bioassay procedure. ELISA has also been used in inoculated meat and/or chicken systems (Huhtanen *et al.*, 1992), green beans and mushroom samples (Rodriguez and Dezfulian, 1997), and found to be useful as a means of rapidly screening food samples for the presence of type A and B toxin.

The polymerase chain reaction (PCR) has been used to detect genes encoding proteolytic and non-proteolytic type B toxins (Szabo *et al.*, 1992). A radio-labeled oligonucleotide probe was used. A two-primer set was used to amplify a 1.5-kbp fragment corresponding to the light chain of the toxin. Type A was detected using a two-primer set and an oligonucleotide detection probe against type A neurotoxin gene (Fach *et al.*, 1993). PCR procedures have been optimized (Szabo *et al.*, 1993) and used for the detection of type F (Ferreira *et al.*, 1994) and of the presence of an unexpressed type B gene in a toxigenic type A (Franciosa *et al.*, 1994), and to detect types A, B, E, F and G in foods (Szabo *et al.*, 1994; Fach *et al.*, 1995; Sciacchitano and Hirshfield, 1996). Takeshi *et al.* (1996) used a method to detect the neurotoxin genes of types A–F by using oligonucleotide-primer sets of about 300 bp corresponding to the light chains of the toxin. Kimura *et al.* (2001) used a nuclease assay (TaqMan assay) as a detection system for PCR products to quantify type E by amplifying a 280-bp sequence of the toxin. The procedure worked well to assay for toxin in fish. Hyytiä *et al.* (1999) evaluated a random amplified polymorphic DNA analysis against a repetitive-element, sequence-based PCR (rep-PCR) for their abilities to characterize Group I and Group II strains. They found the discriminating power of rep-PCR to be inferior to that of RAPD.

Bibliography

Alderton, G., J. K. Chen and K. A. Ito (1974). Effect of lysozyme on the recovery of heated Clostridium spores. *Appl. Microbiol.* **27**, 613–615.

Allerberger, F. (2001). An isolated case of foodborne botulism, Austria 2001. *Eurosurv. Weekly*. **5** 1–2 (available online at http://www.eurosurveillance.org/ew/2001/010412.asp).

Ando, Y. (1971). The germination requirements of spores of *Clostridium botulinum* type E. *Jpn J. Microbiol.* **15**, 515–25.

Ando, Y. (1973). Studies on germination of spores of clostridial species capable of causing food poisoning. I. Factors affecting the germination of spores of *Clostridium botulinum* type A on a chemically defined medium. *J. Food Hyg. Soc. Jpn* **14**, 457.

Ando, Y. (1974). Alpha-hydroxy acids as co-germinants for some clostridial spores. *Jpn J. Microbiol.* **18**, 100–102.

Ando, Y. and H. Iida (1970). Factors affecting the germination of spores of *Clostridium botulinum* type E. *Jpn J. Microbiol.* **14**, 361–370.

Ando, Y., K. Kameyama and T. Karashimada (1975). Studies on germination of spores of clostridial species capable of causing food poisoning (IX) Effect of glycine on the growth from spores of *Clostridium botulinum*. *J. Food Hyg. Soc. Jpn* **16**, 258.

Angulo, F. J., J. Getz, J. P. Taylor *et al.* (1998). A large outbreak of botulism: the hazardous baked potato. *J. Infect. Dis.* **178**, 172–177.

Aranda, E., M. M. Rodriguez, M. A. Asensio and J. J. Cordoba (1997). Detection of *Clostridium botulinum* types A, D, E, and F in foods by PCR and DNA probe. *Lett. Appl. Microbiol.* **25**, 186–190.

Arnon, S. S. (1980). Infant botulism. *Ann. Rev. Med.* **31**, 541–560.

Arnon, S. S. (1993). Clinical trial of human botulism immune globulin. In: B. R. DasGupta (ed.), *Botulinum and Tetanus Neurotoxins: Neurotransmission and Biomedical Aspects*, pp. 477–482. Plenum Press, New York, NY.

Arnon, S. S. (1998). Infant botulism. In: D. Feigen and J. D. Cherry (eds), *Textbook of Pediatric Infectious Disease*, 4th edn, pp. 1570–1577. W.B. Saunders, Philadelphia, PA.

Arnon, S. S., S. B. Werner, H. K. Faber and W. H. Farr (1979a). Infant botulism in 1931: discovery of a misclassified case. *Am. J. Dis. Child.* **133**, 580–582.

Arnon, S. S., T. F. Midura, K. Damus *et al.* (1979b). Honey and other environmental risk factors for infant botulism. *J. Pediatr.* **94**, 331–336.

Arnon, S. S., R. Schechter, T. V. Inglesby *et al.* (2001). Botulinum toxin as a biological weapon, medical and public health management. *J. Am. Med. Assoc.* **285**, 1059–1070.

Athwal, B. and A. Gale (2000). Wound botulism in an injecting drug user in London. *Eurosurv. Weekly* **20**, 1–3 (available online at http://www.eurosurveillance.org/ew/2000/000518.asp 3).

Aureli, P., L. Fenicia, B. Pasolini, M. Gianfranceschi, L. M. McCroskey and C. L. Hatheway (1986). Two cases of type E infant botulism in Italy caused by neurotoxigenic *Clostridium butyricum*. J. Infect. Dis. **54**, 207–211.

Austin, J. and R. Slinger (2000). Foodborne and Infant Botulism in Canada — January 1990 to December 1999. Canadian Association for Clinical Microbiology and Infectious Disease. [Online.] http://www.cacmid.ca/abstracts/a24.html.

Baird-Parker, A. C. (1971). Factors affecting the production of bacterial food poisoning toxins. *J. Appl. Bacteriol.* **34**, 181–197.

Baird-Parker, A. C. and F. Freame (1967). Combined effect of water activity, pH and temperature on the growth of *Clostridim botulinum* from spore and vegetative cell inocula. *J. Appl. Bacteriol.* **30**, 420–429.

Baker, D. A. and C. Genigeorgis (1990). Predicting the safe storage of fresh fish under modified atmospheres with respect to *Clostridium botulinum* toxigenesis by modeling length of the lag phase of growth. *J. Food Prot.* **53**, 131–140.

Bhandari, M., K. D. Campbell, M. D. Collins and A. K. East (1997). Molecular characterization of the clusters of genes encoding the botulinum neurotoxin complex in *Clostridium botulinum* (*Clostridium argentinense*) type G and nonproteolytic *Clostridium botulinum* type B. *Curr. Microbiol.* **35**, 207–214.

Bienvenue, J.-G. and M. Morin (1990). Poultry litter associated botulism (type C). *Cattle Can. Vet. J.* **31**, 711.

Billon, C. M.-P., C. J. McKirgan, P. J. McClure and C. Adair (1997). The effect of temperature on the germination of single spores of *Clostridium botulinum* 62A. *J. Appl. Microbiol.* **82**, 48–56.

Binz, T., H. Kurazono, M. Wille *et al.* (1990). The complete sequence of botulinum neurotoxin Type A and comparison with other clostridial neurotoxins. *J. Biol. Chem.* **265**, 9153.

Bonventre, P. F. and L. L. Kempe (1959a). Physiology of toxin production by *Clostridium botulinum* types A and B. II. Effect of carbohydrate source on growth, autolysis and toxin production. *Appl. Microbiol.* **7**, 372–374.

Bonventre, P. F. and L. L. Kempe (1959b). Physiology of toxin production by *Clostridium botulinum* types A and B. III. Effect of pH and temperature during incubation on growth, autolysis and toxin production. *Appl. Microbiol.* **7**, 374–377.

Bonventre, P. F. and L. L. Kempe (1960). Physiology of toxin production by *Clostridium botulinum* types A and B. IV. Activation of the toxin. *J. Bacteriol.* **79**, 24–32.

Boroff, D. A. (1955). Study of toxins of *Clostridium botulinum*. III. Relation of autolysis to toxin production. *J. Bacteriol.* **70**, 363–367.

Brett, M. M. (1998). Evaluation of the use of the bioMerieux Rapid ID 32A for the identification of *Clostridium botulinum*. *Lett. Appl. Microbiol.* **26**, 81–84.

Brett, M. (1999). Botulism in the United Kingdom. *Eurosurv. Monthly* **4**, 9–11 (available online at http://www.eurosurveillance.org/em/v04n01/0401-224.asp).

Brusin, S. and S. Salmasco (1998). Botulism associated with home-preserved mushrooms. *Eurosurv. Weekly* **2**, 1–2 (available online at http://www.eurosurv.org/1998/980430.html bot).

Burger, D., J. R. Gorham and R. K. Farrell (1963). Mink disease: the immunization of young mink against botulism. *Natl Fur News* **35**, 19.

Cato, E. P., W. L. George and S. M. Finegold (1986). Section 13: Genus Clostridium. In: P. H. Sneath, N. S. Mair, M. E. Sharpe and J. G. Holt (eds), *Bergey's Manual of Systematic Bacteriology*, p. 1141. Williams and Wilkens, Baltimore, MD.

CDC (Centers for Disease Control and Prevention) (1984). Foodborne botulism – Illinois. *Morbid. Mortal. Weekly Rep.* **33**, 22–23.

CDC (Centers for Disease Control and Prevention) (1985). Botulism associated with commercially distributed kapchunka – New York City. *Morbid. Mortal. Weekly Rep.* **34**, 546–547.

CDC (Centers for Disease Control and Prevention) (1987). International outbreak of type E botulism associated with ungutted, salted whitefish. *Morbid. Mortal. Weekly Rep.* **36**, 812–813.

CDC (Centers for Disease Control and Prevention) (1992). Outbreak of type E botulism associated with an uneviscerated, salt-cured fish product. *Morbid. Mortal. Weekly Rep.* **41**, 521–522.

CDC (Centers for Disease Control and Prevention) (1995). Wound botulism – California, 1995. *Morbid. Mortal. Weekly Rep.* **44**, 889–892.

CDC (Centers for Disease Control and Prevention) (1998). Botulism in the United States, 1989–1996. In: *Handbook for Epidemiologists, Clinicians, and Laboratory Workers*. Centers for Disease Control and Prevention, Atlanta, GA.**69

CDC (Centers for Disease Control and Prevention) (1999). Foodborne botulism associated with home-canned bamboo shoots – Thailand, 1998. *Morbid. Mortal. Weekly Rep.* **48**, 437–439.

CDC (Centers for Disease Control and Prevention) (2000). Foodborne botulism from eating home-pickled eggs – Illinois, 1997. *Morbid. Mortal. Weekly Rep.* **49**, 778–780.

CDC (Centers for Disease Control and Prevention) (2001). Botulism outbreak associated with eating fermented food – Alaska. *Morbid. Mortal. Weekly Rep.* **32**, 680–682.

Chaibi, A., L. H. Ababouch, K. Belasri *et al.* (1997). Inhibition of germination and vegetative growth of *Bacillus cereus* T and *Clostridium botulinum* 62A spores by essential oils. *Food Microbiol.* **14**, 161–174.

Chen, F., G. M. Kuziemko and R. C. Stevens (1998). Biophysical characterization of the stability of the 150-kilodalton botulinum toxin, the nontoxic component and the 900-kilodalton botulinum toxin complex species. *Infect. Immun.* **66**, 2420–2425.

Chia J. K., J. B. Clark, C. A. Ryan and M. Pollack (1986). Botulism in an adult associated with foodborne intestinal infection with *Clostridum botulinum. N. Engl. J. Med.* **315**, 239–241.

Chiers, K., F. Haesebrouck and L. Devriese (1998). An outbreak of botulism in cattle caused by *Clostridium botulinum* type B. *Vlaams Diergeneeskd. Tijdschr.* **67**, 296–299.

Claessens, J., H. Thoonen, J. Hoorens and D. Maenhout (1983). Een Uitbraak van Botulisme Bij de Hond. *Vlaams Diergeneeskd. Tijdschr.* **52**, 202–208.

Collins, M. D. and A. K. East (1998). Phylogeny and taxonomy of the food-borne pathogen *Clostridium botulinum* and its neurotoxins. *J. Appl. Microbiol.* **84**, 5–17.

Cordoba, J. J., M. D. Collins and A. K. East (1995). Studies on the genes encoding botulinum neurotoxin type A of *Clostridium botulinum* from a variety of sources. *Syst. Appl. Microbiol.* **18**, 13–22.

Cordoba, M. G., J. J. Cordoba and R. Jordano (1998). Identification of *Staphylococcus aureus, Clostridium perfringens* and *Clostridium botulinum* in foods using nucleic acid techniques. *Alimentation* **35**, 25–30.

Cornelissen, J. M. M., J. Haagsma and J. J. van Nes (1985). Type C botulism in 5 dogs. *J. Am. Anim. Hosp. Assoc.* **21**, 401–404.

DasGupta, B. R. and H. Sugiyama (1972. A common subunit structure in *Clostridium botulinum* type A, B and E toxins. *Biochem. Biophys. Res. Commun.* **48**, 108–112.

Day, L. E. and R. N. Costilow (1964). Physiology of the sporulation process in *Clostridium botulinum.* II. Maturation of forespores. *J. Bacteriol.* **88**, 695–701.

Dezfulian, M. (1993). A simple procedure for identification of *Clostridium botulinum* colonies. *World J. Microbiol. Biotechnol.* **9**, 125–127.

Dezfulian, M. and J. G. Bartlett (1994). Rapid identification of *Clostridium botulinum* colonies by in vitro toxicity and antimicrobial susceptibility testing. *World J. Microbiol. Biotechnol.* **10**, 27–29.

Dezfulian, M. and V. R. Dowell Jr (1980). Cultural and physiological characteristics and antimicrobial susceptibility of *Clostridium botulinum* isolates from food-borne and infant botulism cases. *J. Clin. Microbiol.* **11**, 604–609.

Dezfulian, M., L. M. McCrosky, C. L. Hatheway and V. R. Dowell Jr (1981). Selective medium for isolation of *Clostridium botulinum* from human feces. *J. Clin. Microbiol.* **13**, 526–531.

Dickson, E. C. (1917). Botulism, a case of limberneck in chickens. *J. Am. Vet. Med. Assoc.* **3**, 612–613.

Divers, T. J., R. C. Bartholomew, J. B. Messick *et al.* (1986). *Clostridium botulinum* type B toxicosis in a herd of cattle and a group of mules. *J. Am. Vet. Med. Assoc.* **188**, 382–386.

Dobereiner, J., C. H. Tokarnia, J. Langenegger and I. S. Dutra (1992). Epizootic botulism of cattle in Brazil. *Dtsche. Tierarztliche Wochenschr.* **99**, 188–190.

Dodds, K. L. (1993). *Clostridium botulinum* in the environment. In: A. H. W. Hauschild and K. L. Dodds (eds), Clostridium botulinum *Ecology and Control in Foods*, pp. 21–51. Marcel Dekker, New York, NY.

Doellgast, G. J., M. X. Triscott, G. A. Beard *et al.* (1993). Sensitive enzyme-linked immunosorbent assay for detection of *Clostridium botulinum* neurotoxins A, B, and E using signal amplification via enzyme-linked coagulation assay. *J. Clin. Microbiol.* **31**, 2402–2409.

Doellgast, G. J., M. X. Triscott, G. A. Beard and J. D. Bottoms (1994). Enzyme-linked immunosorbent assay-enzyme-linked coagulation assay for detection of antibodies to *Clostridium botulinum* neurotoxins A, B, and E and solution-phase complexes. *J. Clin. Microbiol.* **32**, 851–853.

Doutre, M. P. (1983). Seconde observation de botulisme de type D chez le chien au Senegal. *Rev. Elev. Med. Vet. Pays Trop.* **36**, 131–132.

Dubovsky, B. J. and K. F. Meyer (1922). The distribution of the spores of *B. botulinus* in the Territory of Alaska and the Dominion of Canada. *V. J. Infect. Dis.* **31**, 595–599.

Duff, J. T., G. G. Wright and A. Yarinsky (1956). Activation of *Clostridium botulinum* type E toxin by trypsin. *J. Bacteriol.* **72**, 455–460.

East, A. K. and M. D. Collins (1994). Conserved structure of genes encoding components of botulinum neurotoxin complex M and the sequence of the gene coding for the nontoxic component in nonproteolytic *Clostridium botulinum* type F. *Curr. Microbiol.* **29**, 69–77.

Eklund, M. W. and V. R. Dowell Jr (1987). *Avian Botulism: An International Perspective*. Charles C. Thomas, Springfield, IL.

Eklund, M. W. and F. T. Poysky (1974). Interconversion of type C and D strains of *Clostridium botulinum* by specific bacteriophages. *Appl. Microbiol.* **27**, 251–258.

Eklund, M. W., F. T. Poysky and D. I. Wieler (1967). Characteristics of *Clostridium botulinum* type F isolated from the Pacific Coast of the United States. *Appl. Microbiol.* **15**, 1316–1323.

Eklund, M. W., M. E. Peterson, F. T. Poysky *et al.* (1982). Botulism in juvenile coho salmon (*Onocorynchus kisutch*) in the United States. *Aquaculture* **27**, 1–11.

Eklund, M. W., F. T. Poysky, M. E. Peterson *et al.* (1984). Type E botulism in salmonids and conditions contributing to outbreaks. *Aquaculture* **41**, 293–309.

Eklund, M. W., E. Poysky, K. Oguma *et al.* (1987). Relationship of bacteriophages to toxin and haem-agglutinin production and its significance in avian botulism outbreaks. In: M. W. Eklund and V. R. Dowell Jr (eds), *Avian Botulism: An International Perspective*. Charles C. Thomas, Springfield, IL.

Eklund, M. W., F. T. Poysky, L. M. Mseitif and M. S. Strom (1988). Evidence for plasmid-mediated toxin and bacteriocin production in *Clostridium botulinum* type G. *Appl. Environ. Microbiol.* **54**, 1405–1408.

Elmore, M. J., R. A. Hutrson, M. D. Collins *et al.* (1995). Nucleotide sequence of the gene coding for proteolytic (Group I) *Clostridium botulinum* type F neurotoxin: genealogical comparison with other clostridial neurotoxins. *Syst. Appl. Microbiol.* **18**, 23–31.

Emeruwa, A. C. and R. Z. Hawirko (1973). Poly-beta-hydroxybutyrate metabolism during growth and sporulation of *Clostridium botulinum*. *J. Bacteriol.* **116**, 989–993.

Evans, R. I., N. J. Russell, G. W. Gould and P. J. McClure (1997). The germinability of spores of a psychrotolerant, non-proteolytic strain of *Clostridium botulinum* is influenced by their formation and storage temperature. *J. Appl. Microbiol.* **83**, 273–280.

Fach, P., D. Hauser, J. P. Guillou and M. R. Popoff (1993). Polymerase chain reaction for the rapid identification of *Clostridium botulinum* type A strains and detection in food samples. *J. Appl. Bacteriol.* **75**, 234–239.

Fach, P., M. Gilbert, R. Griffais *et al.* (1995). PCR and gene probe identification of botulinal neurotoxin A, B, E, F and G producing *Clostridium spp.* and evaluation in food samples. *Appl. Environ. Microbiol.* **61**, 389–392.

Fain Binda, J. C., R. A. Fernandez, R. E. Blotta *et al.* (1998). Natural botulism in dogs due to ingestion of botulinum toxin C. *Vet. Argent.* **15**, 216–221.

Farrow, B. R. H., W. G. Murrell, M. L. Revington *et al.* (1983). Type C botulism in young dogs. *Aust. Vet. J.* **60**, 374–377.

Ferreira, J. L., H. K. Hamdy, S. G. McCay and F. A. Zapatka (1990). Monoclonal antibody to type F *Clostridium botulinum* toxin. *Appl. Environ. Microbiol.* **56**, 808–811.

Ferreira, J. L., M. K. Hamdy, S. G. McCay *et al.* (1994). Detection of *Clostridium botulinum* type F using the polymerase chain reaction. *Mol. Cell. Probes* **8**, 365–373.

Foegeding, P. M. and F. F. Busta (1983). Differing L-alanine germination requirements of hypochlorite-treated *Clostridium botulinum* spores from two crops. *Appl. Environ. Microbiol.* **45**, 1415–1417.

Franciosa, G., J. L. Ferreira and C. L. Hatheway (1994). Detection of type A, B and E botulism neurotoxin genes in *Clostridium botulinum* and other *Clostridium* species by PCR: evidence of unexpressed type B toxin genes in type A toxigenic organisms. *J. Clin. Microbiol.* **32**, 1911–1917.

Fujii, N., K. Oguma, N. Yokosawa *et al.* (1988). Characterization of bacteriophage nucleic acids obtained from *Clostridium botulinum* types C and D. *Appl. Environ. Microbiol.* **54**, 69–73.

Fujii, N., K. Kimura, N. Yokosawa *et al.* (1993). The complete nucleotide sequence of the gene encoding the nontoxic component of *Clostridium botulinum* type E progenitor toxin. *J. Gen. Microbiol.* **139**, 79–86.

Fujinaga, Y., K. Inoue, S. Watanabe *et al.* (1997). The haemagglutinin of *Clostridium botulinum* type C progenitor plays an essential role in binding of toxin to the epithelial cells of guinea pig small intestine, leading to the efficient absorption of the toxin. *Microbiology* **143**, 3841–3847.

Galazka, A. and A. Przbylska (1999). Surveillance of foodborne botulism in Poland: 1960–1998. *Eurosurv. Monthly* **4**, 69–72 (available online at http://www.eurosurveillance.org/em/v04n06/0406-223.asp).

Galey, F. D., R. Terra, R. Walker *et al.* (2000). Type C botulism in dairy cattle from feed contaminated with a dead cat. *J. Vet. Diagn. Invest.* **12**, 204–209.

Ghanem, F. M., A. C. Ridpath, W. E. C. Moore and L. V. H. Moore (1991). Identification of *Clostridium botulinum, Clostridium argentinense*, and related organisms by cellular fatty acid analysis. *J. Clin. Microbiol.* **29**, 1114–1124.

Gibson, A. M. and M. J. Eyles (1988). Changing perceptions of foodborne botulism. *CSIRO Food Res. Q.* **49**, 46–59.

Giltner, L. T. and J. F. Couch (1930). Western duck sickness and botulism. *Science* **72**, 660.

Glasby, C. and C. L. Hatheway (1985). Isolation and enumeration of *Clostridium botulinum* by direct inoculation of infant fecal specimens on egg yolk agar and *Clostridium botulinum* isolation media. *J. Clin. Microbiol.* **21**, 264–266.

Gregory, A. R., T. M. Ellis, T. F. Jubb *et al.* (1996). Use of enzyme-linked immunoassays for antibody to types C and D botulinum toxins for investigations of botulism in cattle. *Aust. Vet. J.* **73**, 55–61.

Griffin, P. M., C. L. Hatheway, R. B. Rosenbaum and R. Sokolow (1997). Endogenous antibody production to botulinum toxin in an adult with intestinal colonization botulism and underlying Crohn's Disease. *J. Infect. Dis.* **175**, 633–637.

Gutteridge, C. S., B. M. Mackey and J. R. Norris (1980). A pyrolysis gas-liquid chromatogaphy study of *Clostridium botulinum* and related organisms. *J. Appl. Bacteriol.* **49**, 165–174.

Haagsma, J. and E. A. Ter Laak (1979). First case of type D botulism in cattle in the Netherlands. *Tijdschr. Diergeneeskd.* **104**, 609–613.

Hall, I. C. and G. W. Stiles (1938). An outbreak of botulism in captive mink on a fur farm in Colorado. *J. Bact.* **36**, 282.

Hall, J. D., L. M. McCroskey, B. J. Pincomb and C. L. Hatheway (1985). Isolation of an organism resembling *Clostridium barati* which produces type F botulinal toxin from an infant with botulism. *J. Clin. Microbiol.* **21**, 654–655.

Harrigan, K. E. (1980). Botulism in broiler chickens. *Aust. Vet. J.* **56**, 603–605.

Hatheway, C. L. (1993). *Clostridium botulinum* and other clostridia that produce botulinum neurotoxin. In: A. H. W. Hauschild and K. L. Dodds (eds), Clostridium botulinum, *Ecology and Control in Foods*, pp. 3–20. Marcel Dekker, New York, NY.

Hauschild, A. H. W. (1993). Epidemiology of human foodborne botulism. In. A. H. W. Hauschild and K. L. Dodds (eds), Clostridium botulinum *Ecology and Control in Foods*, pp. 69–104. Marcel Dekker, New York, NY.

Hauschild, A. H. W. and R. Hilsheimer (1977). Enumeration of *Clostridium botulinum* spores in meats by a pour plate procedure. *Can. J. Microbiol.* **23**, 829–832.

Hauser, D., M. W. Eklund, P. Boquet and M. R. Popoff (1994). Organization of the botulinum neurotoxin C1 gene and its associated non-toxic protein genes in *Clostridium botulinum* C 468. *J. Mol. Gen. Genet.* **243**, 631–640.

Hawirko, R. Z., C. A. Naccarato, R. P. W. Lee and P. Y. Maeba (1979). Outgrowth and sporulation studies on *Clostridium botulinum* type E: influence of isoleucine. *Can. J. Microbiol.* **25**, 522–527.

Henderson, I., S. M. Whelan, T. O. Davis and N. P. Minton (1996). Genetic characterisation of the botulinum toxin complex of *Clostridium botulinum* strain NCTC 2916. *FEMS Microbiol. Lett.* **140**, 151–158.

Hielm, S., E. Hyytiä, J. Ridell and H. Korkeala (1996). Detection of *Clostridium botulinum* in fish and environmental samples using polymerase chain reaction. *Intl J. Food Microbiol.* **31**, 357–365.

Hielm, S., J. Björkroth, E. Hyytiä and H. Korkeala (1998). Genomic analysis of *Clostridium botulinum* Group II by pulsed-field gel electrophoresis. *Appl. Environ. Microbiol.* **64**, 703–708.

Hielm, S., J. Björkroth, E. Hyytiä and H. Korkeala (1999). Ribotyping as an identification tool for *Clostridium botulinum* strains causing human botulism. *Intl J. Food Microbiol.* **47**, 121–131.

Hogg, R. A., V. J. White and G. R. Smith (1990). Suspected botulism in cattle associated with poultry litter. *Vet. Rec.* **126**, 476–479.

Holdeman. L. V. and J. B. Brooks (1970). Variation among strains of *Clostridium botulinum* and related clostridia. In: *Proceedings of the First US Japan Conference on Toxic Microorganisms*, pp. 278–286. UJNR Joint Panels on Toxic Microorganisms and the US Department of the Interior, Washington, DC.

Huhtanen, C. N., R. C. Whiting, A. J. Miller and J. E. Call (1992). Qualitative correlation of the mouse neurotoxin and enzyme-linked immunoassay for determining *Clostridium botulinum* types A and B toxins. *J. Food Safety* **12**, 119–127.

Huss, H. H. (1980). Distribution of *Clostridium botulinum. Appl. Environ. Microbiol.* **39**, 764–769.

Huss, H. H. and U. Eskildsen (1974). Botulism in farmed trout caused by *Clostridium botulinum* type E. *Vet. Med.* **26**, 733–738.

Hutson, R. A., M. D. Collins, A. K. East and D. E. Thompson (1994). Nucleotide sequence of the gene coding for non-proteolytic *Clostridium botulinum* type B neurotoxin: comparison with other clostridial neurotoxins. *Curr. Microbiol.* **28**, 101–110.

Hyytiä, E., J. Björkroth, S. Hielm and H. Korkeala (1999). Characterization of *Clostridium botulinum* Groups I and II by randomly amplified polymorphic DNA analysis and repetitive element sequence-based PCR. *Intl J. Food Microbiol.* **48**, 179–189.

Inoue, K. and H. Iida (1970). Conversion of toxigenicity in *Clostridium botulinum* type C. *Jpn J. Microbiol.* **14**, 87–89.

Jansen, B. C., P. C. Knoetze and F. Visser (1976). The antibody reponse of cattle to *Clostridium botulinum* types C and D toxoids. *Onderstepoort J. Vet. Res.* **43**, 165–174.

Jensen, W. I. (1981). Evaluation of coproexamination as a diagnostic test for avian botulism. *J. Wildl. Dis.* **17**, 171–176.

Jensen, W. I. and J. I. Price (1987). The global importance of type C botulism in wild birds. In: M. W. Eklund and V. R. Dowell Jr (eds), *Avian Botulism: An International Perspective*, pp. 33–54. Charles C. Thomas, Springfield, IL.

Johnson, H. M., K. Brenner, R. Angelotti and H. E. Hall (1966). Serological studies on types A, B and E botulinal toxins by passive hemagglutination and bentonite flocculation. *J. Bacteriol.* **91**, 967–974.

Johnston, R., S. Harmon and D. Kautter (1964). Method to facilitate the isolation of *Clostridium botulinum* type E. *J. Bacteriol.* **88**, 1521–1522.

Kaufmann, O. W., H. M. Solomon and R. H. Monheimer (1967). Experimentally induced type E botulism in gulls hatched under natural field conditions and reared in captivity. In: M. Ingram and T. A. Roberts (eds), *Botulism 1966*, pp. 400–406. Chapman and Hall Ltd., London.

Kautter, D. A. and H. M. Solomon (1977). Microbiological methods: collaborative study of a method for the detection of *Clostridium botulinum* and its toxins in foods. *J. AOAC* **60**, 541–545.

Kautter, D. A., S. M. Harmon, R. K. Lynt Jr and T. Lilly Jr (1966). Antagonistic effect on *Clostridium botulinum* type E by organisms resembling it. *Appl. Microbiol.* **14**, 616–622.

Khan, M. and S. Chay (2000). Wound botulism in injecting drug user: second case in England. *Eurosurv. Weekly* **25**, 2–3 (available online at http://www.eurosurv.org/2000/000622.htm 3).

Kimura, B., S. Kawasaki, H. Nakano and T. Fujii (2001). Rapid, quantitative PCR monitoring of growth of *Clostridium botulinum* type E in modified-atmosphere-packaged fish. *Appl. Environ. Microbiol.* **67**, 206–216.

Kindler. S. H., J. Mager and N. Grossowicz (1956). Nutritional studies with the *Clostridium botulinum* group. *J. Gen. Microbiol.* **15**, 86–393.

Kubota, T., N. Yonekura, Y. Hariya *et al.* (1998). Gene arrangement in the upstream region of *Clostridium botulinum* type E and *Clostridium butyricum* BL6340 progenitor toxin genes is different from that of other types. *FEMS Microbiol. Lett.* **158**, 215–221.

Kuusi, M., V. Hasseltvedt and P. Aavitsland (1999). Botulism in Norway. *Eurosurv. Monthly* **4**, 11–12 (available online at http://www.eurosurveillance.org/em/v04n01/0401-225.asp).

Lee, W. H. and H. Riemann (1970). Correlation of toxic and non-toxic strains of *Clostridium botulinum* by DNA composition and homology. *J. Gen. Microbiol.* **60**, 117–123.

Leyer, G. J. and E. A. Johnson (1990). Repression of toxin production by tryptophan in *Clostridium botulinum* type E. *Arch. Microbiol.* **154**, 443–447.

Lilly, T. Jr, S. M. Harmon, D. A. Kautter *et al.* (1971). An improved medium for detection of *Clostridium botulinum* type E. *J. Milk Food Technol.* **34**, 492–497.

Lin, W.-J. and E. A. Johnson (1995). Genome analysis of *Clostridium botulinum* type A by pulsed-field gel electrophoresis. *Appl. Environ. Microbiol.* **61**, 4441–4447.

Mager, J., S. H. Kindler and N. Grossowicz (1954). Nutritional studies with *Clostridium parabotulinum* type A. *J. Gen. Microbiol.* **10**, 130–141.

Main, D. C. and A. R. Gregory (1996). Serological diagnosis of botulism in dairy cattle. *Aust., Vet. J.* **73**, 77–78.

Marvaud, J. C., M. Gilbert, K. Inoue *et al.* (1998). botR/A is a positive regulator of botulinum neurotoxin and associated non-toxin protein genes in *Clostridium botulinum* A. *Mol. Microbiol.* **29**, 1009–1018.

Mayhew, J. W. and S. L. Gorbach (1975). Rapid gas chromatographic technique for presumptive detection of *Clostridium botulinum* in contaminated food. *Appl. Microbiol.* **29**, 297–299.

McClung, L. S. and R. Toabe (1947). The egg yolk plate reaction for the presumptive diagnosis of *Clostridium sporogenes* and certain species of the gangrene and botulinum groups. *J. Bacteriol.* **53**, 139–147.

McCroskey, L. M. and C. L. Hatheway (1988). Laboratory findings in four cases of adult botulism suggest colonization of the intestinal tract. *J. Clin. Microbiol.* **26**, 1052–1054.

McCroskey, L. M., C. L. Hatheway, L. Fenicia *et al.* (1986). Characterization of an organism that produces type E botulinal toxin but which resembles *Clostridium butyricum* from the feces of an infant with type E botulism. *J. Clin. Microbiol.* **23**, 201–202.

McLoughlin, M. F., S. G. McIlroy and S. D. Neill (1988). A major outbreak of botulism in cattle being fed ensiled poultry litter. *Vet. Rec.* **122**, 579–581.

Meyer, K. F. and B. J. Dubovsky (1922a). The distribution of the spores of *B. botulinus* in California. II. *J. Infect. Dis.* **31**, 541–555.

Meyer, K. F. and B. J. Dubovsky (1922b). The distribution of the spores of *B. botulinus* in the United States. IV. *J. Infect. Dis.* **31**, 558–594.

Meyer, K. F. and B. J. Dubovsky (1922c). The occurrence of the spores of *B. botulinus* in Belgium, Denmark, England, The Netherlands and Switzerland. VI. *J. Infect. Dis.* **31**, 600–609.

Midura, T. F. and S. S. Arnon (1976). Infant botulism: identification of *Clostridium botulinum* and its toxin in faeces. *Lancet* **2**, 934–936.

Minervin, S. M. (1966). On the parenteral-enteral method of administering serum in cases of botulism. In: M. Ingram and T. A. Roberts (eds), *Proceedings of the 5th International Symposium on Food Microbiology*, pp. 336–345. Chapman and Hall Ltd, London.

Moberg, L. J. and H. Sugiyama (1979). Microbial ecologic basis of infant botulism as studied with germfree mice. *Infect. Immun.* **25**, 653–657.

Montville, T. J. (1981). Effect of plating medium on heat activation requirement of *Clostridium botulinum* spores. *Appl. Environ. Microbiol.* **42**, 734–736.

Muller, J. (1981). The incidence and distribution of bovine botulism in Denmark 1959–1978 and its relation to 'contagious bulbar paralysis in cattle'. *Nord. Vet. Med.* **33**, 33–41.

Nakamura, S., I. Kimura, K. Yamakawa and S. Nishida (1983). Taxonomic relationships among *Clostridium novyi* types A and B, *Clostridium haemolyticum* and *Clostridium botulinum* type C. *J. Gen. Microbiol.* **129**, 1473–1479.

Notermans, S., J. Dufrenne and J. Oosterom (1981). Persistence of *Clostridium botulinum* type B on a cattle farm after an outbreak of botulism. *Appl. Environ. Microbiol.* **41**, 179–183.

Ogert, R. A., J. E. Brown, B. R. Singh *et al.* (1992). Detection of *Clostridium botulinum* toxin A using fiber optic-based biosensor. *Anal. Biochem.* **205**, 306–312.

Ohishi, I., S. Sugii and G. Sakaguchi (1977). Oral toxicities of *Clostridium botulinum* toxins in response to molecular size. *Infect. Immun.* **16**, 107–109.

Ortolani, E. L., L. A. B. Rito, C. S. Mori *et al.* (1997). Botulism outbreak associated with poultry litter consumption in three Brazilian cattle herds. *Vet. Hum. Toxicol.* **39**, 89–92.

Peck, M. W., D. A. Fairbarn and B. M. Lund (1992). Factors affecting growth from heat-treated spores of non-proteolytic *Clostridium botulinum*. *Lett. Appl. Microbiol.* **15**, 152–155.

Pederson, H. O. (1955). On type E botulism. *J. Appl. Bact.* **18**, 619.

Perkins, W. E. and K. Tsuji (1962). Sporulation of *Clostridium botulinum*. II. Effect of arginine and its degradation products on sporulation in a synthetic medium. *J. Bacteriol.* **84**, 86–94.

Pickett, J., B. Berg, E. Chaplin and M. A. Brunstetter-Shafer (1976). Syndrome of botulism in infancy: clinical and electrophysiologic study. *N. Engl. J. Med.* **295**, 770–772.

Popoff, M. R., R. Rose, J. C. Legardinier *et al.* (1984). Identification de foyers de botulisme bovin de type D en France. *Rev. Med. Vet.* **135**, 103–107.

Quagliaro, D. A. (1977). An improved cooked meat medium for the detection of *Clostridium botulinum*. *J. AOAC* **60**, 563–569.

Reed, G. H. (1994). Foodborne illness (Part 7) *Clostridium botulinum. Dairy Food Environ. Sanit.* **14**, 268–269.

Richmond, R. N., C. Hatheway and A. F. Kaufmann (1978). Type C botulism in a dog. *J. Am. Vet. Med. Assoc.* **173**, 202–203.

Riemann, H. (1953). Germination of anaerobic spores. *Estratto dagli Atti del VI Congr. Int. di Microbiolgia, Roma* **4**, 150–155.

Roberts, T. A. (1967). Sporulation of mesophilic clostridia. *J. Appl. Bacteriol.* **30**, 430–443.

Roberts, T. A. and Hobbs, G. (1968). Low temperature growth characteristics of clostridia. *J. Appl. Bacteriol.* **31**, 75–88.

Rodriguez, A. and M. Dezfulian (1997). Rapid identification of *Clostridium botulinum* and botulinal toxin in food. *Folia Microbiol.* **42**, 149–151.

Roman, M. G., J. Y. Humber, P. A. Hall *et al.* (1994). Amplified immunoassay ELISA–ELCA for measuring *Clostridium botulinum* type E neurotoxin in fish fillets. *J. Food Prot.* **57**, 985–990.

Rowley, D. B. and F. Feeherry (1970). Conditions affecting germination of *Clostridium botulinum* 62A spores in chemically defined medium. *J. Bacteriol.* **104**, 1151–1157.

Sakaguchi, G., I. Ohishi and S. Sugii (1979). Pathogenetic significance of the molecular sizes of *Clostridium botulinum* toxins. *Abstr. Gen. Meet. Am. Soc. Microbiol.* **79**, 217.

Sciacchitano, C. J. and I. N. Hirshfield (1996). Molecular detection of *Clostridium botulinum* type E neurotoxin gene in smoked fish by polymerase chain reaction and capillary electrophoresis. *J. AOAC Intl* **79**, 861–865.

Scott, V. N. and C. L. Duncan (1978). Cryptic plasmids in *Clostridium botulinum* and *C. botulinum*-like organisms. *FEMS Microbiol. Lett.* **4**, 55–58.

Sebald, M. and H. Ionesco (1972). Germination IzP-dependante des spores de *Clostridium botulinum* type E. C. *R. Acad. Sci. Paris* t.**275 Serie D**, 2175–2177.

Segner, W. P. and C. F. Schmidt (1968). Nonspecific toxicities in the mouse assay test for botulinum toxin. *Appl. Microbiol.* **16**, 1105–1109.

Segner, W. P., C. F. Schmidt and J. K. Boltz (1971). Enrichment, isolation and cultural characteristics of marine strains of *Clostridium botulinum* type C. *Appl. Microbiol.* **22**, 1017–1024.

Shapiro, R. L., C. L. Hatheway and D. L. Swerdlow (1998). Botulism in the United States: a clinical and epidemiologic review. *Ann. Intern. Med.* **129**, 221–228.

Siegel, L. S. and J. F. Metzger (1979). Toxin production by *Clostridium botulinum* type A under various fermentation conditions. *Appl. Environ Microbiol.* **38**, 606–611.

Silas, J. C., J. A. Carpenter, M. K. Hamdy and M. A. Harrison (1985). Selective and differential medium for detecting *Clostridium botulinum. Appl. Environ. Microbiol.* **50**, 1110–1111.

Simpson, L. L. (1996). Botulinum toxin: a deadly poison sheds its negative image. *Ann. Intern. Med.* **125**, 616–617.

Smith, L. D. S. (1997). *Botulism: The Organism, its Toxin, the Disease*. Charles C. Thomas, Springfield, IL.

Smith, L. D. S. and H. Sugiyama (1988). Botulism in animals and fish. In: L. D. S. Smith (ed.), *Botulism: The Organism, its Toxin, the Disease*, 2nd edn, p. 139. Charles C. Thomas, Springfield, IL.

Smoot, L. A. and M. D. Pierson (1979. Effect of oxidation–reduction potential on the outgrowth and chemical inhibition of *Clostridium botulinum* 10755A spores. *J. Food Sci.* **44**, 700–704.

Smoot, L. A. and M. D. Pierson (1981). Mechanisms of sorbate inhibition of *Bacillus cereus* T and *Clostridium botulinum* 62A spore germination. *Appl. Environ. Microbiol.* **42**, 477–483.

Solomon, H. M. and T. Lilly Jr (1998). *Clostridium botulinum*. In: *FDA Bacteriological Analytical Manual*, 8th edn, Ch. 17. AOAC International, Gaithersburg, MD.

Solomon, H. M., E. A. Johnson, D. T. Bernard *et al.* (2001). *Clostridium botulinum* and its toxins. In: F. P. Downes and K. Ito (eds), *Compendium of Methods for the Microbiological Examination of Foods*, 4th edn, pp. 317–324. APHA, Washington, DC.

Spencer, R. (1969). Neomycin-containing media in the isolation of *Clostridium botulinum* and food poisoning strains of *Clostridium welchii*. *J. Appl. Bacteriol.* **32**, 170–174.

Strasdine, G. A. (1972). The role of intracellular glucan in endogenous fermentation and spore maturation in *Clostridium botulinum* type E. *Can. J. Mirobiol.* **18**, 211–217.

Strasdine, G. A. and J. Melville (1968). Growth and spore production of *Clostridium botulinum* type E in chemically defined media. *J. Fish. Res. Board Can.* **25**, 547–553.

Suen, J. C., C. L. Hatheway, A. G. Steigerwalt and D. J. Brenner (1988a). *Clostridium argentinense sp.nov.*: a genetically homogeneous group composed of all strains of *Clostridium botulinum* toxin type G and some nontoxigenic strains previously identified as *Clostridium subterminale* or *Clostridium hastiforme*. *J. Syst. Bacteriol.* **38**, 375–381.

Suen, J. C., C. L. Hatheway, A. G. Steigerwealt and D. J. Brenner (1988b). Genetic confirmation of identities of neurotoxigenic *Clostridium baratii* and *Clostridium butyricum* implicated as agents of infant botulism. *J. Clin. Microbiol.* **26**, 2191–2192.

Sugiyama, H. (1980). *Clostridim botulinum* neurotoxin. *Microbiol. Rev.* **44**, 419–448.

Sugiyama, H. and D. C. Mills (1978). Intraintestinal toxin in infant mice challenged intragastrically with *Clostridum botulinum* spores. *Infect. Immun.* **21**, 59–63.

Swenson, J. M., C. Thornsberry, L. M. McCrosky *et al.* (1980). Susceptibility of *Clostridium botulinum* to thirteen antimicrobial agents. *Antimicrob. Agents Chemother.* **18**, 13–19.

Szabo, E. A., J. M. Pemberton and P. M. Desmarchelier (1992). Specific detection of *Clostridium botulinum* type B by using the polymerase chain reaction. *Appl. Environ. Microbiol.* **58**, 418–420.

Szabo, E. A., J. M. Pemberton and P. M. Desmarchelier (1993). Detection of the genes encoding botulinum neurotoxin types A to E by the polymerase chain reaction. *Appl. Environ. Microbiol.* **59**, 3011–3020.

Szabo, E. A., J. M. Pemberton, A. M. Gibson *et al.* (1994). Polymerase chain reaction for detection of *Clostridium botulinum* types A, B and E in food, soil and infant feces. *J. Appl. Microbiol.* **76**, 539–545.

Takeshi, K., Y. Fujinaga, K. Inoue *et al.* (1996). Simple method for detection of *Clostridium botulinum* type A to F neurotoxin genes by polymerase chain reaction. *Microbiol. Immunol.* **40**, 5–11.

Theiler, A., P. R. Viljoen, H. H. Green *et al.* (1927). *Lamsiekte (Parabotulism) in Cattle in South Africa.* 11th and 12th Report of the Director of Veterinary Education and Res. Dep. Agric. Union of South Africa.

Therre, H. (1999). Botulism in the European Union. *Eurosurv. Monthly* **4**, 2–7 (available online at http://www.eurosurveillance.org/em/v04n01/0401-222.asp).

Thomas, R. J. (1991). Detection of *Clostridium botulinum* type C and D toxin by ELISA. *Aust. Vet. J.* **68**, 111–113.

Treadwell, P. E., G. J. Jann and A. J. Salle (1958). Studies on factors affecting the rapid germination of spores of *Clostridium botulinum. J. Bacteriol.* **76**, 549–556.

Tsuji, K. and W. E. Perkins (1962). Sporulation of *Clostridium botulinum*. I. Selection of an aparticulate sporulation medium. *J. Bacteriol.* **84**, 81–85.

Vangelova, L. (1995). Botulinum toxin: a poison that can heal. *FDA Consumer Magazine* **29**, 16–19.

Vermilyea, B. L., H. W. Walker and J. C. Ayres (1968). Detection of botulinal toxins by immunodiffusion. *Appl. Microbiol.* **16**, 21–24.

Villar, R. G., R. L. Shapiro, S. Busto *et al.* (1999). Outbreak of type A botulism and development of a botulism surveillance and antitoxin release system in Argentina. *J. Am. Med. Assoc.* **281**, 1334–1340.

Wallace, V. and D. M. McDowell (1986). Botulism in a dog: first confirmed case in New Zealand. *NZ Vet. J.* **34**, 149–150.

Wang, X., T. Maegawa, T. Karasawa *et al.* (2000). Genetic analysis of type E botulinum toxin producing *Clostridium butyricum* strains. *Appl. Environ. Microbiol.* **66**, 4992–4997.

Ward, B. Q. and B. J. Carroll (1966). Spore germination and vegetative growth of *Clostridium botulinum* type E in synthetic media. *Can. J. Microbiol.* **12**, 1145–1156.

Weikert, M. J., G. H. Chambliss and H. Sugiyama (1986). Production of toxin by *Clostridium botulinum* type A strains cured of plasmids. *Appl. Environ. Microbiol.* **51**, 52–56.

Whiting, R. C. and J. E. Call (1993). Time of growth model for proteolytic *Clostridium botulinum. Food Microbiol.* **10**, 295–301.

Whiting, R. C. and T. P. Strobaugh (1998). Expansion of the time-to-turbidity model for proteolytic *Clostridium botulinum* to include spore numbers. *Food Microbiol.* **15**, 449–453.

Whitlock, R. H. and J. M. Williams (1999). Botulism toxicosis of cattle. *Proc. Annu. Conf./Conv. Bovine Pract.*, **32**, 45–53.

Wilson, R. B., M. T. Boley and B. Corwin (1995). Presumptive botulism in cattle associated with plastic-packaged hay. *J. Vet. Diagn. Invest.* **7**, 165–169.

Winarno, F. G., C. R. Stumbo and K. M. Hayes (1971). Effect of EDTA on the germination of and outgrowth from spores of *Clostridium botulinum* 62A. *J. Food Sci.* **36**, 781–785.

Wobeser, G., K. Baptiste, E. G. Clark and A. W. Deyo (1997). Type C botulism in cattle in association with a botulism die-off in waterfowl in Saskatchewan. *Can. Vet. J.* **38**, 782.

Wynne, E. S., D. A. Mehl and W. R. Schmieding (1954). Germination of *Clostridium* spores in buffered glucose. *J. Bacteriol.* **67**, 435–437.

Yang, K. H. and H. Sugiyama (1975). Purification and properties of *Clostridium botulinum* type F toxin. *Appl. Microbiol.* **29**, 598–603.

Zhou, Y., H. Sugiyama and E. A. Johnson (1993). Transfer of neurotoxigenicity from *Clostridium butyricum* to a nontoxigenic *Clostridium botulinum* type E-like strain. *Appl. Environ. Microbiol.* **59**, 3825–3831.

Zhou, Y., H. Sugiyama, H. Nakano and E. A. Johnson (1995). The genes for the *Clostridium botulinum* type G toxin complex are on a plasmid. *Infect. Immun.* **63**, 2087–2091.

Staphylococcal intoxications

14

Merlin S. Bergdoll[1] *and Amy C. Lee Wong*

1 Historical aspects and contemporary problems

1.1 Historical aspects

Among the toxic substances produced by the staphylococci are the enterotoxins, the causative agents of staphylococcal food poisoning. Ingestion of any of the enterotoxins by humans usually results in intestinal disturbances, involving vomiting and diarrhea within a few hours of the ingestion. The illness can be serious, but usually lasts only a few hours with no sequela.

There is no record of when illnesses similar to staphylococcal food poisoning were first observed, but it is likely that humans have been afflicted with this illness as long as they have been consuming foods in which staphylococci could grow. There are records of illnesses of this type as early as 1830, although the organisms themselves were not recognized until 1878 and 1880 by Koch and Pasteur, respectively. Although Ogston is credited with applying the name '*Staphylococcus*' to these organisms in 1881 because of the grapelike clusters of cocci he observed in cultures, it was Rosenbach who in 1884 obtained pure cultures of the microorganisms on solid media and accepted the name *Staphylococcus*.

Dack (1956), in his book *Food Poisoning*, relates several descriptions of foodborne illnesses similar to staphylococcal food poisoning. A number of food items were

[1] deceased

Foodborne Infections and Intoxications 3e
ISBN-10: 0-12-588365-X
ISBN-13: 978-012-588365-8

involved – sausage, rabbit pie, 'pork brawn', milk, ice cream, and of course cheese, where Vaughan and Sternberg first associated micrococci with the illnesses in 1884. Each of these investigators independently examined cheese that had been implicated in food-poisoning outbreaks in Michigan. Sternberg stated: 'It seems not improbable that the poisonous principle is a ptomaine developed in the cheese as a result of the vital activity of the above mentioned *Micrococcus*, or some other microorganisms which had preceded it, and had perhaps been killed by its own poisonous products.' In 1894 Denys concluded that the illness of a family who had consumed meat from a cow dead of 'vitullary fever' was due to the presence of pyogenic staphylococci, and in 1907 Owen recovered staphylococci from dried beef implicated in a foodborne illness characteristic of staphylococcal food poisoning (Dack, 1956).

In 1914, Barber was the first investigator actually to relate staphylococcal food poisoning to a toxic substance produced by the staphylococci (Barber, 1914). He discovered that milk from a mastitic cow caused illness when left unrefrigerated, and showed that the illness was due to growth in the milk of the staphylococci isolated from the mastitis. The significance of this excellent report was not recognized and, as a result, this type of food poisoning was ascribed for the most part to other bacterial agents. For example, an outbreak involving 2000 soldiers in the German army during World War I was attributed to *Proteus vulgaris* even though cocci were present in large numbers.

In 1929, Dack rediscovered the role of staphylococci in food poisoning with his classical work on two Christmas cakes that were responsible for the illness of 11 people (Dack *et al.*, 1930). These three-layer sponge cakes with thick cream fillings were baked possibly 1 day before delivery and eaten 2 days later. They were presumably refrigerated at the bakery but not after delivery. Dack and his associates showed, with the aid of human volunteers, that the sponge cake substance was responsible for the illness. Staphylococci isolated from this part of the cake produced a substance that caused typical food poisoning symptoms in human volunteers. In essence, this was the beginning of the research on staphylococcal food poisoning.

From 1930 until about 1948 a number of investigators studied this type of food poisoning from various angles, with Dack and Dolman independently carrying out much of the work – particularly during the early part of the period. Beginning around 1948, Casman (of the US Food and Drug Laboratories) and Surgalla and Bergdoll (in the Food Research Institute at the University of Chicago), began intensive studies of this subject. Casman continued his studies until his retirement in 1969. These studies were continued by Bennett, who had worked with Casman before his retirement. Surgalla worked with Bergdoll until 1953, when he left the project; Bergdoll continued the studies at the Food Research Institute until his retirement in 1988. During the middle part of this period, Sugiyama made a number of valuable contributions to the work at the Food Research Institute. Many other investigators have made valuable contributions to the field, although the interest in this area has diminished in the developed countries as other foodborne agents have been identified. However, Bergdoll concentrated his efforts on aiding developing countries in pursuing research in this field, particularly in Brazil where staphylococcal food poisoning is quite common.

1.2 Contemporary problems

Once the etiology of staphylococcal food poisoning had been established, it soon became recognized as a major cause of foodborne illness in the US and in many other countries. In the 1960s and 1970s it was recognized as one of the leading foodborne diseases in the US; however, by the late 1980s very few cases of this type of food poisoning were being reported to the Centers for Disease Control and Prevention (CDC) (Table 14.1; CDC, 1981a, 1981b, 1983a, 1983b; MacDonald and Griffin, 1986; Bean *et al.*, 1990, 1996; Olsen *et al.*, 2000). This was also true in England (Wieneke, 1993). It is suggested that outbreaks involving a few people, such as a family, were still occurring, but because of their relative mildness and short duration, these were not being reported. At the height of this disease in the US a number of large outbreaks were occurring, with only an occasional one reported since that time.

The decrease in the number of outbreaks in developed countries is probably due to greater care taken in preparing and handling foods that are to be served to large groups of individuals. For example, in Japan many outbreaks were due to rice balls, which were made by hand. Today, very few if any outbreaks involve rice balls, primarily because they are now made by machine, which eliminates the major source of contamination – the food handler. However, the situation is different in developing countries, such as Brazil. When Bergdoll started working with the food microbiologists in Brazil in 1976, this type of food poisoning was not being recognized even

Table 14.1 Staphylococcal food poisoning outbreaks reported in the US, 1978–1997

Year	Number	Total[a]	%	Cases	Total[a]	%
1978	23	105	22.0	1318	4466	29.5
1979	34	119	28.6	2391	6806	35.1
1980	27	136	19.9	944	6891	13.7
1981	44	185	23.8	2934	8208	35.7
1982	28	151	18.5	669	5501	12.2
1983	14	127	11.0	1257	7082	17.7
1984	11	128	8.6	1153	7307	15.8
1985	14	143	9.8	421	22 132	1.9
1986	7	119	5.9	248	4855	5.1
1987	1	83	1.2	100	8928	1.1
1988	8	139	5.8	245	7156	3.4
1989	14	171	8.2	524	6557	8.0
1990	13	196	6.6	372	9002	4.1
1991	9	173	5.2	331	6335	5.2
1992	6	117	5.1	206	4156	5.1
1993	7	135	1.4	335	10 402	2
1994	13	148	2.0	421	5487	2.6
1995	6	155	1.0	66	10 017	0.4
1996	7	112	1.5	178	14 219	0.8
1997	9	105	1.8	393	3696	3.3

Compiled from Annual Summaries of Foodborne Outbreaks (CDC).

[a] Number of reported and confirmed bacterial foodborne disease outbreaks and number of bacterial foodborne disease cases.

though there was interest in conducting research in this area. It was not until some 10 years later that the first staphylococcal food-poisoning outbreak was investigated and reported (Noleto and Tibana, 1987). In this case the staphylococci isolated (in large numbers) from the implicated food produced enterotoxin – adequate circumstantial evidence to conclude that the outbreak was due to staphylococcal food poisoning. A second group reported an outbreak in 1988 in which cheese was implicated (Sabioni *et al.*, 1988). Bergdoll had been working with both these groups and directed the PhD research of the leaders of both research groups. Apparently outbreaks were widespread in Brazil, but were not being investigated. For example, many outbreaks were occurring in Belo Horizonte and the state of Minas Gerais, but were not being investigated because no one in the Public Health Laboratory had any experience in this field. Finally Bergdoll was asked to assist in examining the staphylococci isolated from foods implicated in 18 outbreaks for enterotoxin production (do Carmo and Bergdoll, 1990). Bergdoll was able to give the researchers at São Paulo assistance in examining the food, implicated in six outbreaks, for the presence of enterotoxin (Cerqueira-Campos *et al.*, 1993). For the most part complete investigation of the outbreaks was not being performed. In one outbreak that Bergdoll was instrumental in investigating, the food handler was shown to be responsible for contaminating the implicated food (Pereira *et al.*, 1994). As recently as 1996, several hundred individuals became ill with staphylococcal food poisoning after consuming food at a political rally.

In some countries, obtaining information about the importance of staphylococcal food poisoning is impeded because, aside from the fact that other problems may be of more immediate concern, it is difficult to determine how important the staphylococci are in the digestive disturbances common to these countries. Although the type of foods consumed in any given country may affect the incidence of this type of food poisoning, the populations of most countries consume foods that can support the growth of staphylococci and the production of enterotoxin.

2 Characteristics of *Staphylococcus aureus*

2.1 Classification of *Staphylococcus*

Taxonomically the staphylococci have been placed in the Family *Micrococcaceae*. Baird-Parker (1963) proposed a system of classification of the micrococci and staphylococci based on certain physiological and biochemical tests. He divided the Family *Micrococcaceae* into Group I (*Staphylococcus* Rosenbach emend. Evans) and Group II (*Micrococcus* Cohn emend. Evans). These groups were then divided into subgroups on the basis of pigment production, coagulase and phosphatase reactions, acetoin production, and formation of acid from glucose (both aerobically and anaerobically) and other sugars. Six subgroups were recognized within the genus *Staphylococcus* and seven within the genus *Micrococcus*. Hajék and Marsálek (1971) further divided the pathogenic staphylococci found in Baird-Parker's subgroup I into six biotypes with characteristic biochemical and biological properties.

Staphylococcus can be differentiated from the other three members in the family, *Micrococcus, Stomatococcus*, and *Planococcus*, on the basis of the guanine plus cytosine (G + C) content of the DNA, cell wall composition, and the ability to grow and ferment glucose anaerobically. Only three species of *Staphylococcus* (*S. aureus, S. epidermidis* and *S. saprophyticus*) were included in the genus in 1974 (Buchanan and Gibbons, 1974). They were differentiated primarily on the basis of the ability to produce coagulase, ferment mannitol (both aerobically and anaerobically) and produce heat-stable endonuclease, and by the cell wall composition (Baird-Parker, 1974).

Kloos and Schleifer (1975) outlined a simplified scheme for the routine identification of human *Staphylococcus* species. They divided these into 11 species on the basis of coagulase activity, hemolysis, nitrate reduction, and aerobic acid production from several sugars. Since then the number of species and sub-species had increased to 32 as of 1994 (Holt *et al.*, 1994). This increase included the elevation of two of the *S. aureus* biotypes to species status, biotype E (from dogs) to *S. intermedius* and biotype F (from swine) to *S. hyicus*. An additional coagulase-positive species, *S. delphini* from dolphins, has been added.

2.2 Characteristics of *S. aureus*

For many years *S. aureus* was the only recognized species that produced coagulase, with the species further subdivided by biotyping to determine the animal source of the species. Identification of *S. aureus* was relatively easy, because any organism that produced coagulase was automatically classified as *S. aureus*. Addition of thermonuclease (TNase) production (Chesbro and Auborn, 1967) to the characteristics of *S. aureus* aided in the identification because the production of either coagulase or TNase was adequate to classify an organism as *S. aureus*. Although upgrading of the biotypes to species was helpful in identifying the major source of a particular coagulase-positive species, additional tests were necessary to identify *S. aureus*. Each biotype was predominately associated with particular animals, with humans being the major source of *S. aureus*, although *S. aureus* can be isolated from a wide range of animals. However, it is seldom that the other coagulase-positive species can be isolated from humans. Most of the staphylococcal species recognized to date do not produce coagulase, with the exception of *S. aureus, S. delphini, S. hyicus, S. intermedius* and *S. schleiferi* subspecies *coagulans*. All of the coagulase-positive species produce TNase, with the exception of *S. delphini*, but only *S. aureus* ferments mannitol both aerobically and anaerobically, and produces protein A and acetoin (Table 14.2). The fact that not all strains in any one species are positive for all of the characteristics of that species complicates the classification of *S. aureus*. For example, not all *S. aureus* strains ferment mannitol or are positive for acetoin (Bennett *et al.*, 1986; Roberson *et al.*, 1992), which could result in their being classified as another coagulase-positive species if additional testing were not performed. Although *S. aureus* is isolated from many animal species, it is the only coagulase-positive species normally isolated from humans, particularly from the nose and throat. *S. intermedius* can be transmitted from dogs to the skin, but seldom to the nose or throat (Talan *et al.*, 1989).

Table 14.2 Characterization of *Staphylococcus* species

Property	*S. aureus*	*S. intermedius*	*S. hyicus*	*S. delphini*
Pigment	+	–	–	–
Coagulase	+	+	±	+
TNase	+	+	+	–
Mannitol (aerobic)	+	±	–	+
Mannitol (anaerobic)	+	–	–	–
Hemolysins	+	±	–	+
Clumping factor	+	±	–	–
Acetoin production	+	–	–	–

S. aureus is the most important species involved in staphylococcal food poisoning, even though *S. intermedius* strains can produce enterotoxins (De la Fuente *et al.*, 1986; Hirooka *et al.*, 1988). Only one outbreak has been associated with *S. intermedius* (Khambaty *et al.*, 1994) since the identification of this species. In this outbreak the source of the staphylococci, whether animal or human, was not identified.

2.2 Hosts and reservoirs

The staphylococci are ubiquitous in nature, with humans and animals as the primary reservoirs. They are present in the nasal passages and throat, in the hair, and on the skin of probably 50 % or more of healthy individuals. The prevalence is usually higher in individuals associated with hospital environments because many infections and diseases are caused by the staphylococci. These organisms are associated with sore throats and colds, and are found in abundance in postnasal drip following colds. Staphylococci can be isolated from animals, with the bovine being the most important because of the involvement of staphylococci in mastitis. Although animals and humans are the major source, staphylococci also can be found in the air, dust, water, and human and animal wastes.

2.3 Ability to survive and grow in the environment

The staphylococci can survive indefinitely in the nasal passages and throats of humans and animals. From these sources they can be transferred to meat and other foods. Essentially all meats can be contaminated with staphylococci; however, the organisms may persist on raw meats but grow very poorly. In foods that provide a satisfactory medium they can grow to sufficient numbers to produce enterotoxin if the foods are not refrigerated. These organisms can be transferred to equipment; if the equipment is then not adequately cleansed before use, the organisms can be transferred to foods. A common source of contamination of dairy products is cows' udders, particularly in animals with staphylococcal mastitis. The organisms are destroyed when the milk is pasteurized, but any enterotoxin in the milk will not be inactivated (Read and Bradshaw, 1966; Evenson *et al.*, 1988).

3 The toxins

3.1 Conditions for enterotoxin production

The same factors that affect growth of the organism in general also affect the production of enterotoxin. The methods used for production of the enterotoxins depends to a large extent on the purpose of the production – for example, whether attempting to determine the enterotoxigenicity of a strain or conducting purification studies. In either case it is desirable to produce as much enterotoxin (per milliliter) as possible, but for purification the overall quantity needed is an important factor (Kato *et al.*, 1966).

3.1.1 Enterotoxigenic staphylococci

The staphylococci produce 12 related but serologically distinct enterotoxins – A (SEA) and B (SEB) (Casman *et al.* 1963), C (SEC) (Bergdoll *et al.*, 1965), D (SED) (Casman *et al.*, 1967), E (SEE) (Bergdoll *et al.*, 1971), G (SEG) (Munson *et al.*, 1998), H (SEH) (Ren *et al.*, 1994; Su and Wong, 1995), I (SEI) (Munson *et al.*, 1998), J (SEJ) (Zhang *et al.*, 1998), K (SEK) (Jarraud *et al.*, 2001; Orwin *et al.*, 2001), L (SEL) and M (SEM) (Jarraud *et al.*, 2001) – and possibly some unidentified ones. There is no enterotoxin F (SEF) because toxic shock syndrome toxin was misidentified as SEF when it was first isolated. Two SEKs were described independently by two different groups at about the same time (Jarraud *et al.*, 2001; Orwin *et al.* 2001); however they are different proteins based on their deduced amino acid sequences. Enterotoxin C has several members that are very closely related and react with the same antibodies (Reiser *et al.*, 1984; Bohach and Schlievert, 1989; Couch and Betley, 1989). These are labeled C_1, C_2 and C_3, in addition to others that are labeled C_{bovine}, C_{ovine}, and C_{canine} (Edwards *et al.*, 1997). Only an antibody to SEC_2 is needed to identify all of the SECs (Reiser *et al.*, 1984).

Enterotoxins A–E and SEH were identified by purification of the protein and their specific reaction to antibodies developed to them. It is important for purification purposes to identify staphylococcal strains that produce a maximum amount of enterotoxin. The SECs and SEB are normally produced in relatively large amounts (over 100 μg/ml), so there is no particular problem in identifying strains for production of these toxins for purification (Robbins *et al.*, 1974). This is not true for the other enterotoxins, as they are produced at only a few micrograms per milliliter, with SED produced at less than 1 μg/ml. To improve the production of SEA, several mutations were generated to increase the amount produced (Friedman and Howard, 1971). This also was done for an SED producer, but still the mutant produces only a few micrograms per milliliter (Chang and Bergdoll, 1979).

3.1.2 Media

The media developed in the Food Research Institute for the production of enterotoxins employed a pancreatic digest of casein from Mead Johnson and Co. (Evansville, Indiana), which had been used by Segalove (1947), supplemented with thiamin, niacin, and glucose (Surgalla *et al.*, 1951; Kato *et al.*, 1966) (Table 14.3). The use of pancreatic digests of casein became standard for the large-scale production of

Table 14.3 Effect of medium on production of SEB[a]	
Medium[b]	Enterotoxin (μg/ml)
PHP (Amigen), 2%	200
NZA-A, 2%	40
NZA-A, 4%	110
NZA-A, 4%, 0.1% K_2HPO_4	210
PHP, 1% + NZA-A, 1%	190
PHP, 3% + NAK, 3%	480

[a] Incubation at 37 °C for 18 hours on gyrotory shaker (280 rpm), 50 ml in 250 ml Erlenmeyer flasks (Kato *et al.*, 1966).
[b] Supplemented with 10 μg/ml niacin and 0.5 μg/ml thiamin and adjusted to pH 6.0.

enterotoxin, and these are still in use today; however, because of the discontinuance of production of Amigen, protein hydrolysate powder (PHP) and N-Z Amine NAK (NAK) (Humko-Sheffield Chemical Co., Lyndhurst, New Jersey), less suitable pancreatic digests of casein are used. One such product is N-Z Amine A (NZA-A) (Humko-Sheffield Chemical Co.), but it is necessary to supplement it with yeast extract (Metzger *et al.*, 1973). The addition of yeast extract has made it unnecessary to add niacin and thiamin to the media. Metzger *et al.* (1973) obtained excellent production of both SEA and SEB with 4 % NZA-A supplemented with 1 % yeast extract and 0.2 % glucose. Over 100 μg/ml of SEA and over 400 μg/ml of SEB were produced with this medium. Carpenter and Silverman (1974) also obtained good results with 4 % NZA-A supplemented with 0.4 % yeast extract and 0.1 % K_2HPO_4. This product is normally sold in bulk, but can be obtained in small amounts at an increased price. Other pancreatic digests of casein, such as tryptone, available in small quantities (Difco Laboratories, Detroit, MI), have proved useful for enterotoxin production when supplemented with yeast extract (unpublished data). These results show that various enzymatic hydrolysates of casein can be used to produce enterotoxin when they are supplemented, particularly with yeast extract. What is used will be determined to a large extent by the casein hydrolysate available.

Glucose is frequently added to the medium used for enterotoxin production, but in most of the experiments conducted in the Food Research Institute enterotoxin production was not increased when it was present. Metzger *et al.* (1973) reported a 40 % reduction in enterotoxin B production in the fermenter at a constant pH of 7 when glucose was eliminated from the medium.

Brain heart infusion (BHI) broth was promoted by Casman and Bennett (1963) for enterotoxin production, particularly for small volumes. It is primarily useful in testing staphylococcal strains for enterotoxigenicity because of the larger amount of enterotoxin that can be produced (Table 14.4). There appeared to be a variation in results among sources of the BHI for production of SEB (Reiser and Weiss, 1969), as well as within lots from the same source (Casman and Bennett, 1963). Because this medium is more readily available than casein digests, it is being used on a wider scale than other media for the production of small volumes with high concentrations of

Table 14.4 Enterotoxin production methods (μg/ml)[a]

Toxin	Shake flask	Semi-solid agar			Sac culture			Membrane-over-agar		
	1	1	2	3	1	2	3	1	2	3
SEA	2.6	4	7	8	14	21	26	19–27[b]	21	32
SEB	26.0	11	12	11	100	260	220	15–35	1	3
SEC	72.0	20	36	0	360	250	400	143–185	110	8
SED	0.5	1–2	1	1	4	4	5	4	5	4
SEE	4.1	10	4	5	16	25	28	16–32	25	28

[a] 1, 3% PHP + 3% NAK medium; 2, BHI(1X); 3, BHI(2X) (Robbins *et al.*, 1974).
[b] The higher values were obtained with new Spectra/Por 1 molecular porous dialysis membrane with a molecular weight cut-off of 6000–8000; the lower values were with 1-year-old membrane.

enterotoxin. Relatively large amounts can be obtained without supplementation with yeast extract; however, the addition of yeast extract does increase enterotoxin production (Bergdoll, unpublished data).

3.1.3 Temperature

The temperature range for enterotoxin production varies with the medium, but in general the minimum temperature for production is 10 °C and the maximum is 45 °C. The temperature for optimum production of enterotoxin is 35–40 °C, with 37 °C being the temperature generally used. There are variations with the different enterotoxins and with different media. Scheusner *et al.* (1973) showed that enterotoxins A–D were produced in BHI over a range of temperatures (19–39 °C). All but SEB were produced at 45 °C, but SEB was the only one produced at 13 °C; however, the SED-producing strain did not grow at this temperature. Vandenbosch *et al.* (1973) obtained maximal production of SEB (strain S-6) and SEC (strain FRI-137) in PHP-NAK medium at 40 °C with somewhat lesser amounts at 35 °C and 42 °C. Small amounts of toxin were produced at 15 °C and 45 °C, with none at 10 °C and 50 °C. Thota *et al.* (1973) obtained maximal SEE production at 40 °C. Pereira *et al.* (1982) produced maximum amounts of SEA and SEB in 4 % NAK medium by the sac culture method (Donnelly *et al.*, 1967) at 39.4 °C.

Several investigators have shown that enterotoxin production can occur in certain foods at temperatures as low as 10 °C (Genigeorgis *et al.*, 1969; Tatini, 1973) and as high as 40 °C (Fung, 1972). Normally growth is much slower at the lower temperatures, and since enterotoxin production is related to growth, a much longer period at these temperatures would be required before enterotoxin might be detectable.

3.1.4 pH values

Most strains of staphylococci will grow at pH values between 4.5 and 9.3, with the optimum being 7.0–7.5; however, the conditions for enterotoxin production are more restricted than for growth. For example, enterotoxin production is limited to pH 5.15–9.0 (Scheusner *et al.*, 1973), without any attempt to control the pH during fermentation. With the use of a fermenter, Metzger *et al.* (1973) produced the greatest amount of SEB when the pH was controlled at pH 7.0 (580 μg/ml), with much less enterotoxin produced at pH values of 6.0 (268 μg/ml) and 8.0 (32 μg/ml), after 10 hours of incubation using a 4 % N-Z Amine A medium containing 1 % yeast extract and 0.2 % glucose. Carpenter and Silverman (1974) also found that the greatest amount of SEB was produced in a fermenter when the pH was held constant at 7.0. Metzger *et al.* (1973) obtained more SEB production when the pH was controlled than when it was not controlled (starting pH, 6.5) (580 μg/ml vs 440 μg/ml), although Carpenter and Silverman (1974) obtained no more SEB at a constant pH of 7.0 than when the pH was not controlled. Jarvis *et al.* (1973) obtained higher levels of SEA when the pH was controlled at 6.5 than under any other conditions tested.

Most experiments have been done by adjusting the medium to a specific pH value with no attempt to control the pH during incubation. Experiments at the Food Research Institute in which the starting pH was adjusted to 5.0–8.0 at 0.5 pH-unit intervals resulted in the maximum amount of enterotoxin being produced at initial pH values of 6.0 and 6.5. It has become normal practice to adjust the pH of the medium to 6.5 before fermentation. It would appear that controlling the pH is not generally necessary to obtain high enterotoxin production.

3.1.5 Methods of production

Several different methods have been used for the production of enterotoxin, with the choice usually depending on the purpose of production, the equipment available, and the ease of operation. Semisolid media (Casman and Bennett, 1963) and solid media (membrane-over-agar) (Hallander, 1965; Robbins *et al.*, 1974) in Petri dishes are used for the production of small volumes of concentrated enterotoxin to identify enterotoxigenic strains (Table 14.4). More enterotoxin can be produced by these methods than when shake flasks are employed – up to 10 times as much for SEA, SED and SEE in the membrane-over-agar method. The increase for SEB and SEC is less than for the other enterotoxins (up to two-fold), but is of less concern because strains producing these enterotoxins normally produce much larger quantities than is the case with SEA, SED and SEE. Another method that is used in testing strains for enterotoxin production is the sac-culture technique of Donnelly *et al.* (1967) (Table 14.4). In this method, medium is placed in a dialysis tube that is positioned in the bottom of an Erlenmeyer flask to which is added about 20 ml of inoculum. Incubation is carried out for 24 hours, by shaking at a moderate speed, in order for the inoculum to make good contact with the sac. This method yields the largest amounts of SEB and SEC of any of the methods, and equivalent amounts of SEA, SED and SEE to those obtained with the membrane-over-agar method (Robbins *et al.*, 1974). Other investigators have found this method superior to alternative methods for obtaining concentrated solutions of enterotoxin (Simkovicova and Gilbert, 1971; Untermann, 1972). The membrane-over-agar method was used in the Food Research Institute because of the relative simplicity of the method and adequate production of all enterotoxins for classification of strains as enterotoxigenic. One problem with the membrane-over-agar method is that not all types of membranes, such as cellophane used by Hallander (1965), give satisfactory results. The material found to be most satisfactory is Spectra/Por 1® molecularporous dialysis membrane with a molecular weight cut-off of 6000–8000 (Spectrum, Houston, TX, USA) (Robbins *et al.*, 1974), which should be obtained fresh and stored in the refrigerator.

The production of enterotoxin for purification is normally carried out using much larger volumes of medium than can be obtained with the methods described above. Normally, large-scale production of enterotoxin is accomplished in Erlenmeyer or similar-type flasks with some method of shaking, or in fermenters. For the former method, flasks of different sizes and shapes have been used as well as different methods of shaking – such as reciprocal and gyrotory – with satisfactory results (Table 14.3) (Kato *et al.*, 1966). The method used in the Food Research Institute to produce relatively large quantities of enterotoxin was the incubation of 400–600 ml of inoculated media in 2-liter Erlenmeyer flasks for 18–24 h at 37 °C on a gyrotory shaker at 280 rpm (Kato *et al.*, 1966). This method can also be used to produce small volumes of more concentrated toxin by incubating 15 ml of medium in 250-ml Erlenmeyer flasks. The sac-culture method has been used for the production of higher concentrations of enterotoxin for purification, particularly for SED, by using larger sacs in larger Erlenmeyer flasks.

Production of enterotoxin in volumes greater than 10–15 l is best accomplished by deep culture aeration. Fermenters of various sizes, 0.5–2000 l, have been used for the production of enterotoxin. Metzger *et al.* (1973) reported high production of SEB (up to 600 μg/ml) in a 50-l volume of medium in a fermenter (Fermentation Design, Bethlehem, Pennsylvania) and up to 150 μg/ml of SEA (personal communication). Carpenter and Silverman (1974) produced equivalent amounts of SEB in 0.5-l of medium in a l-l fermenter. Jarvis *et al.* (1973) were able to produce more SEA in 2 l of medium in a fermenter than in shake flasks, but somewhat less SEB and much less SEC in the fermenter. The aeration rate appeared to affect SEC production: the higher the rate, the less enterotoxin was produced. Carpenter and Silverman (1974) found that more SEB was produced at 10–50 % dissolved oxygen (DO) levels than at zero or 100 % levels. In the larger fermenters, such as that used by Metzger *et al.* (1973), it was impossible to raise the DO level above zero during active fermentation, but this appeared to have little effect on toxin production.

3.1.6 Characteristics of the enterotoxins

3.1.6.1 Physicochemical characteristics

The enterotoxins are simple proteins that are hygroscopic and easily soluble in water and salt solutions, and have relatively low molecular weights of 25 000–29 000 Da. They are basic proteins with isoelectric points (pI) of 7.0–8.6, with the exception of SEG and SEH, which have pIs of 5.6 and 5.7 respectively. Different pIs for the same enterotoxin may be observed, which is due to the loss of amide groups (Chang and Dickie, 1971). The maximum absorbance of the SEs is 277 nm; the absorbance is higher than for normal proteins, primarily because of the high content of tyrosine.

One characteristic of the enterotoxins is the presence of two cysteine residues near the center of the molecule that are joined by a disulfide bond, forming what is referred to as the cystine loop (Bergdoll *et al.*, 1974). The exceptions are SEI, one of the SEKs (described by Orwin *et al.*), and SEM, which lack a second cysteine normally present in the center of the enterotoxin molecule. The sequence and number of residues in the loop are not the same for the different enterotoxins. The loops for SEA and SEE are identical, and those for the different SECs are identical (Figure 14.1). The significance of the loops is not known. It is assumed that they stabilize the molecular structure, although the cystine molecule can be reduced and the -SH 's substituted so the –S–S–bond cannot be reestablished without neutralizing the emetic reaction (Dalidowicz *et al.*, 1966). Two of the enterotoxins, SEB (Spero *et al.*, 1973) and SEC_1 (Spero *et al.*, 1976), can be nicked by trypsin in the cystine loop, again without neutralizing the emetic reaction. The compactness of the enterotoxin molecule is demonstrated by the fact that the SEB and SEC molecules can be split into two fragments by mild tryptic treatment without the two parts separating or the loss of the emetic reaction (Spero *et al.*, 1973, 1976). Directly downstream from

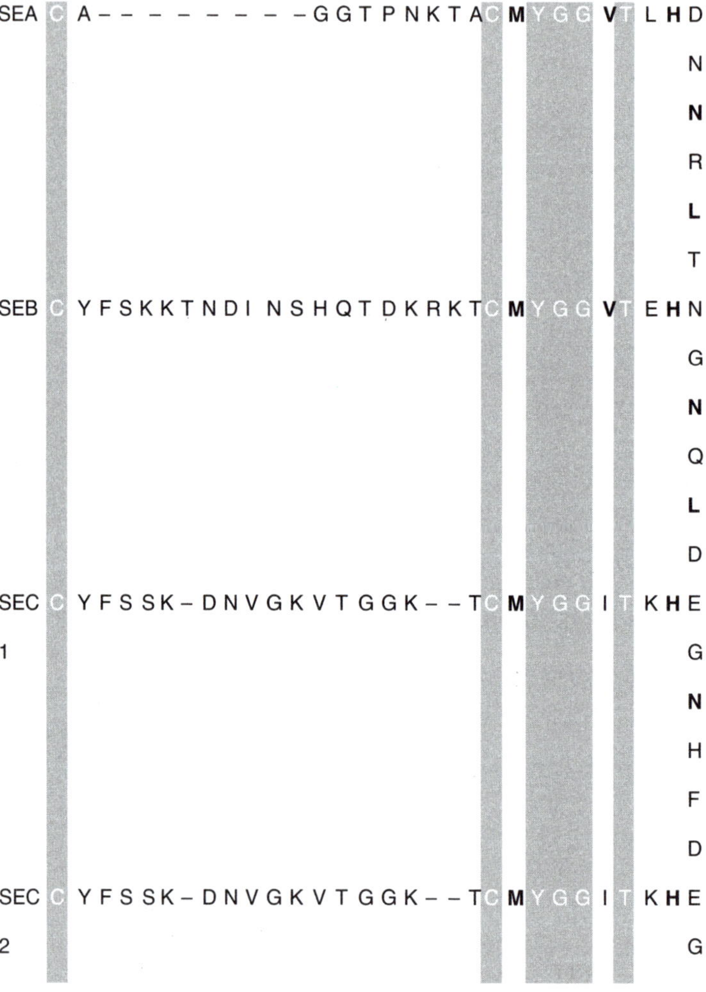

Figure 14.1 Amino acid sequence alignment in the cystine loop region. Residues conserved in all the SEs with the cystine loop are shaded. Residues conserved in the majority of them are in bold.

the cystine loop a number of amino acid residues are conserved among all the enterotoxins (Figure 14.1). It has been suggested that this region is involved in the emetic activity of the enterotoxin molecule (Bergdoll, 1989).

There is significant nucleotide and amino acid sequence homology among the enterotoxins. The SECs are the most similar (96–98 % amino acid identity), with only a three or four residue difference between them (Couch and Betley, 1989; Hovde *et al.*, 1990); antibodies to each of the SECs react with all SECs (Reiser *et al.*, 1984). SEB is most similar to the SECs; some antibodies produced against SEB cross-react with the SECs (Lee *at al.*, 1980). SEA and SEE have a high degree of homology

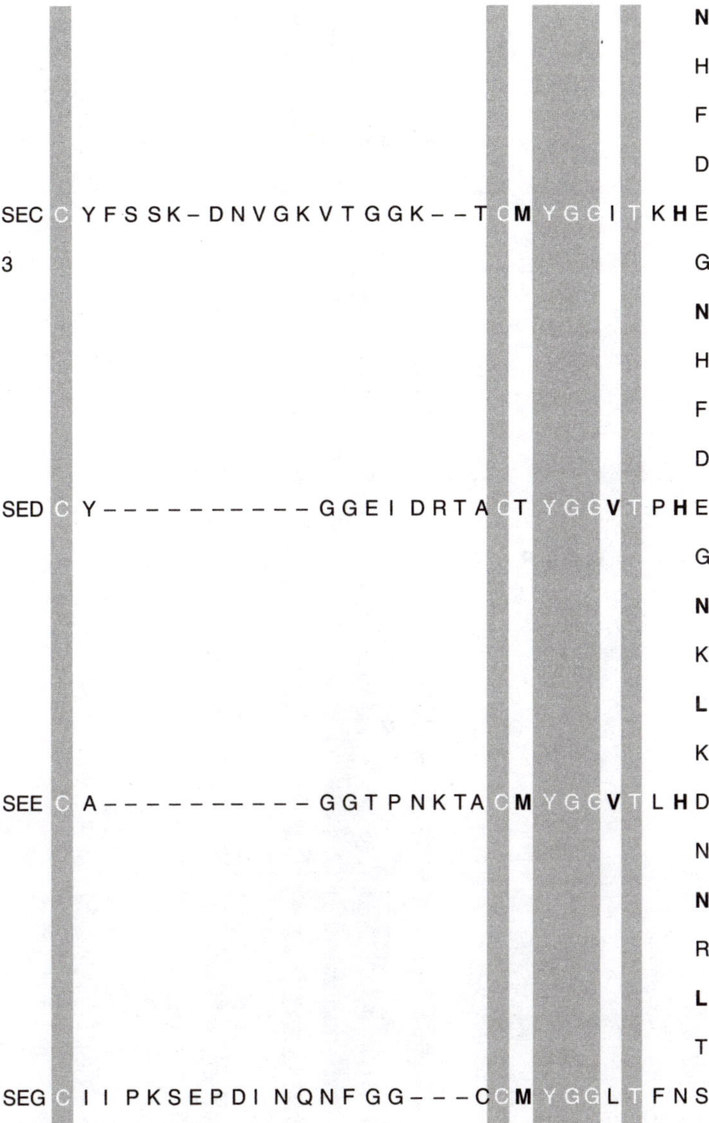

Figure 14.1—*Cont'd*

(82 %) (Betley and Mekalanos, 1988; Couch *et al.*, 1988); they have common antigenic sites in addition to specific sites (Lee *et al.*, 1978), and are also closely related to SED and SEJ (Zhang *et al.*, 1998). SEG shares 39 % and 38 % deduced amino acid sequence identity with SEB and SEC, respectively (Munson *et al.*, 1998).

Biological characteristics

The enterotoxins are immunosuppressive (Smith and Johnson, 1975) and mitogenic (Peavy *et al*, 1970), and can stimulate production of interferon (Archer *et al.*, 1979), interleukin-1 and -2 (Johnson and Magazine, 1988), and tumor necrosis factor

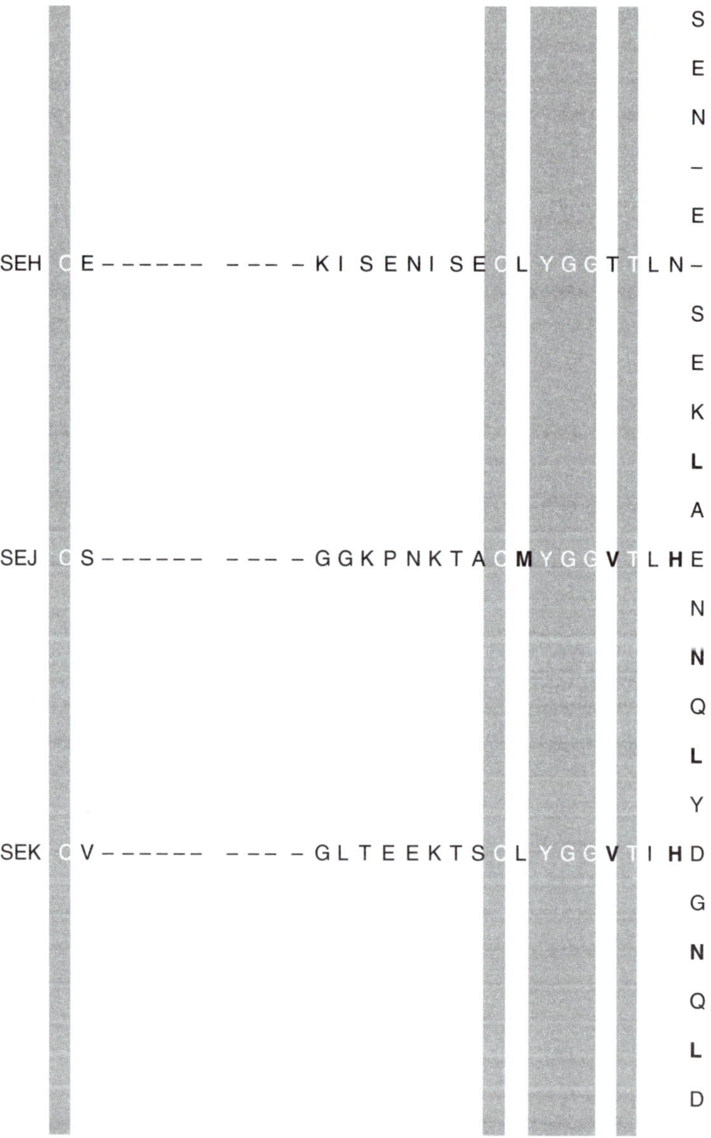

Figure 14.1—*Cont'd*

(Das and Langone, 1989). Other activities were observed when the toxins were injected intravenously into rhesus monkeys (Beisel, 1972).

All the enterotoxins have been shown to have superantigenic activity (Dinges *et al.*, 2000). The enterotoxins have been labeled superantigens because they can activate as many as 10 % of the mouse's T-cell repertoire, whereas conventional antigens stimulate less than 1 % of all T cells (Hoffman, 1990). Although they are considered to be T-cell mitogens, the amount required for stimulation is several magnitudes lower than those of conventional T-cell mitogens, such as phytohemagglutinin and concanavalin A (White *et al.*, 1989). They require antigen-presenting cells bearing major histocompatibility complex (MHC) class II molecules to stimulate T cells, but do not require preprocessing to peptides as conventional antigens do (Marrack and Kappler, 1990). The cystine loop must be intact for T-cell activation, as nicking the loop by mild tryptic action negates SEB's activity (Grossman *et al.*, 1990). Also, the superantigens attach to the outer face of the MHC molecule rather than in the groove as the conventional antigens do (Marrack and Kappler, 1990). The T-cell receptor (TCR) is a heterodimer composed of α and β chains, which include variable portions Vα, Jα, Vβ, Dβ and Jβ. The T lymphocytes stimulated by the staphylococcal toxins are those that contain particular Vβ elements, regardless of other cell surface proteins. Thus, they can stimulate both $CD4^+$ and $CD8^+$ cells (Misfeldt, 1990). Stimulation of T cells can be blocked by antibodies to MHC class II molecules and antibodies to the appropriate Vβ element.

3.1.7 Stability of the enterotoxins

The staphylococcal enterotoxins are more stable in many respects than most proteins. In the active state, they are resistant to proteolytic enzymes such as trypsin, chymotrypsin, rennin and papain. Although pepsin can digest the enterotoxins at pH values of 2.0 and below (Bergdoll, 1970), this acidic level does not exist in the stomach under normal conditions, particularly in the presence of food. This makes it possible for the enterotoxins to pass through the stomach to the intestinal tract, where they stimulate the emetic and diarrheal actions.

The enterotoxins are not easily inactivated by heat. Initial studies revealed that their emetic activity was not completely destroyed after solutions of crude enterotoxin were

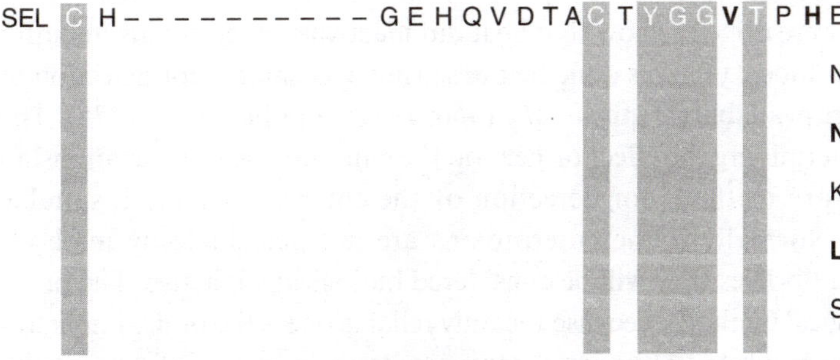

Figure 14.1—*Cont'd*

boiled for 30 minutes and fed to human volunteers (Jordan *et al.*, 1931) or injected intravenously into monkeys (Davison *et al.*, 1938). Ordinarily the effect of heat on the enterotoxins is not of much concern, because the majority of staphylococcal food-poisoning outbreaks are due to human contamination of foods that are not heat treated after their preparation. If the foods were heated within an hour after preparation, any staphylococci would be destroyed before sufficient growth took place to produce enterotoxin.

Milk in the US has not been involved in food poisoning because it is kept cold until it can be pasteurized. The one food-poisoning outbreak that did occur from milk happened because the milk was inadvertently held at a warm temperature for several hours because of a problem with the pasteurizer (Evenson *et al.*, 1988). Soo *et al.* (1974) reported that little loss of SEA and SED occurred in milk, skim milk or cream during pasteurization – that is, 72 °C for 15 seconds. Read and Bradshaw (1966) concluded that SEB in milk was affected very little by pasteurization. They also concluded that spray-drying processes used for milk would not inactivate the enterotoxin; spray-dried milk has been involved in several staphylococcal food-poisoning outbreaks. Cheese has been involved in food poisoning, but this was before pasteurizing milk was the norm.

It is unusual for commercially processed foods to be implicated in staphylococcal food poisoning, but a number of cases occurred in England from corned beef canned in Argentina, Brazil and Malta (Anonymous, 1979a, 1979b, 1979c). However, this was not due to the failure of the processing procedures, but to leaky seams or improperly sealed cans that allowed staphylococci to enter the cans during the cooling process. Sufficient growth of the staphylococci in the can occurred to produce enterotoxin before the corned beef was consumed. Several outbreaks in the US from mushrooms canned in China resulted in their importation being stopped (Anonymous, 1989). It was never satisfactorily determined how the mushrooms were contaminated, except that apparently some cans leaked. Although it was reported by the Food and Drug Administration laboratories that enterotoxin was present in mushrooms from several Chinese canneries (Anonymous, 1990), other scientists were unable to detect enterotoxin in the Chinese mushrooms they examined. How the mushrooms could have become contaminated before they were processed, or if they actually were, was never determined. One study indicated that staphylococci could grow during transportation in sealed plastic bags (Hardt-English *et al.*, 1990); however, this was not confirmed (Brunner and Wong, 1992).

Although studies were done to show that heat did inactivate enterotoxins in buffer and different types of foods, this was done before sensitive techniques for detection of the enterotoxins were available (Denny *et al.*, 1966, 1971; Humber *et al.*, 1975). The major problem in determining the effect of heat on the enterotoxins is the relationship between the laboratory methods for detection of the enterotoxins and a suitable biological method. Normally if the enterotoxins are not detectable by methods employing specific antibodies they will be considered biologically inactive. The problem is with the biological methods, because the only reliable one is the oral administration of enterotoxin to monkeys. Although it compares to the ingestion of enterotoxin by humans, humans are several times more sensitive. The other biological method frequently used is the intravenous injection of cats or kittens (Casman and Bennett, 1963; Denny and Bohrer, 1963; Hammon, 1941), but intravenous injection does not compare to oral administration. It is necessary to treat the sample with trypsin or pancreatin to destroy any interfering substances that can produce reactions similar to those of enterotoxin. The use of this method to test the biological activity of heated enterotoxin is inconclusive, especially if the heated sample is not treated with a proteolytic enzyme, as apparently was the case in the results reported by Bennett and Berry (1987). The heated enterotoxin would probably be destroyed in the stomach by pepsin, as was the case in the studies by Schwabe *et al.* (1990) when the heated enterotoxin was administered to monkeys. The fact that SEA heated at 121 °C lost its mitogenicity more rapidly than its serological activity indicated that biological activity may be lost before serological activity (Stelma *et al.*, 1980). The studies of Bennett and Berry (1987) and Schwabe *et al.* (1990) showed that enterotoxins were inactivated, as determined by ELISA methods, by heating procedures normally used in commercial processing of foods.

Attention has been given to the possible renaturation of heated enterotoxin by treatment with urea. It is known that urea can partially unfold the enterotoxins, with refolding occurring when the urea is removed (Avena and Bergdoll, 1967); however, heat denaturation of proteins is more complicated than a mere unfolding of the molecule, and it has not been shown that treatment with urea can restore the molecule to its original structure. Only relatively small amounts of enterotoxin were detectable after urea extraction of heat-treated enterotoxin (Brunner and Wong, 1992; Bennett, 1994), whereas Akhtar *et al.* (1996) were unable to show any increase in recovery after urea treatment of heated enterotoxin.

All the information available indicates that enterotoxin in food is not easily inactivated by heat, and that the larger the amount present, the more heat is required to reduce the quantity to below detection levels. However, in general, the higher the temperature the more rapidly the enterotoxin is denatured, with the times and temperatures used in normal processing of canned food sufficient to destroy the quantity of enterotoxin usually present in foods involved in food-poisoning outbreaks (< 0.5–$10\,\mu g/100$ g food). However, the processing procedures should not be depended upon to eliminate any hazards that may arise from mishandling the food at any stage during the preparation of the food for canning. No matter what the circumstances, there is no justification for allowing organisms to grow in foods with the possible production of deleterious substances, even though no one may be affected.

The effect of gamma irradiation on SEB and SEA has been reported. A dose of 50 kGy (cobalt-60 source) was required to reduce the concentration of SEB in 0.04-M Veronal buffer (pH 7.2) from 31 μg/ml to less than 0.7 μg/ml (Read and Bradshaw, 1967). In milk, a dose of 200 kGy was needed to reduce the concentration from 30 μg/ml to less than 0.5 μg/ml. These authors concluded that irradiation processes used for pasteurization or sterilization of foods would not inactivate the enterotoxin.

A dose of 8 kGy was insufficient to inactivate all of 111.1 ng/ml SEA (27–34 % remained) in lean minced-beef slurries, although SEA was denatured in gelatin phosphate buffer (Rose *et al.*, 1988; Modi *et al.*, 1990). The more concentrated the beef slurries, the less SEA was denatured.

4 Nature of the intoxication in man and animals

4.1 Symptoms

The symptoms of staphylococcal food poisoning are quite characteristic for this illness in comparison to other foodborne diseases, with one possible exception. In 1974–1975 in England, symptoms similar to staphylococcal food poisoning were noted for outbreaks due to *Bacillus cereus* in fried rice. The onset of symptoms is quite rapid, usually 1–6 hours after the ingestion of food containing enterotoxin, with the average time in a large outbreak being 2–3 hours. Occasionally, symptoms will occur in less than 1 hour or later than 6 hours. The development of symptoms is determined by the amount of toxin consumed and the sensitivity of the individuals involved. The amount of toxin consumed in any given outbreak varies with the amount of food eaten and the distribution of the toxin within the food, which can vary considerably.

The most common symptoms are nausea, vomiting, retching, abdominal cramping and diarrhea. Vomiting is the symptom most frequently observed. In one typical outbreak involving a relatively large number of people, of 48 interviewed, 22 reported nausea, 33 cramps, 37 vomiting, and 34 diarrhea. Of these, 33 had both vomiting and diarrhea, 9 vomited without having diarrhea, and 6 had diarrhea without vomiting (Dennison, 1936). Only five had either nausea or cramps or both symptoms, but without vomiting or diarrhea. Three became ill in less than 2 hours, 39 in 2–5 hours, and 6 after 5 hours, with 1 in 7 hours and 1 in 8 hours.

In severe cases, headache, muscular cramping and marked prostration may occur. In the latter instances, fever may develop or the temperature may drop slightly. Normally there is no change in blood pressure, but in one severe case a dramatic drop in blood pressure was noted – from 120/80 to 60/40. In most cases recovery is rapid, occurring in a few hours to 1 day; the more severe the symptoms, the longer the recovery period. The mortality rate is extremely low but occasionally deaths do occur, usually involving the elderly or very young.

The variation in severity of staphylococcal food poisoning symptoms experienced by individuals suggests development of resistance to previous exposure to the enterotoxin, but there is no evidence to support this. One unreported attempt to check individuals involved in an outbreak for antibodies was unsuccessful.

4.2 Emetic dose

The amount of SE required to produce food poisoning in humans is difficult to determine. Reliable results from the examination of food implicated in food-poisoning outbreaks are difficult to obtain because normally the enterotoxin is not uniformly distributed in the food and it is impossible to know how much food any one individual consumed. A food-poisoning outbreak among school children following the consumption of chocolate milk provided the opportunity to obtain an estimate of the amount of enterotoxin required to cause illness. The quantity of milk consumed by a majority of the children who became ill was half a pint (~240 ml). The analysis of

12 half-pint cartons produced an average of 144 ng SEA/half-pint (94–184 ng SEA/half-pint). Although the SEA was not distributed uniformly in the milk, the minimum amount of SEA required to result in illness in sensitive individuals was not more than 184 ng (Evenson *et al.*, 1988) (Table 14.5).

The emetic dose for rhesus monkeys is somewhat variable for the different SEs; SEA appears to be the most potent. Monkeys were the animals used at the Food Research Institute as they most closely resemble the human in their response to the SEs. The least amount of SEA that was observed to produce an emetic reaction when given intragastrically was 5 μg/3 kg monkey, with SEB, SEC and SEE requiring 10 μg and SED requiring 20 μg. Much smaller amounts were needed to produce an emetic action when the SE was given intravenously; 20 ng/kg monkey for SEA and SEC_3 (Bergdoll, 1989). The minimum amounts required for the other SEs were not determined.

4.3 Diagnosis

Any foodborne illness with the symptoms outlined here, particularly if it involves more than one person, is suspected of being staphylococcal food poisoning. A list of foods consumed at the previous meal or meals is needed to aid in the diagnosis, as there are certain foods that support the growth of staphylococci and are frequently involved in this type of illness. Any suspected food should be examined for the presence of staphylococci; if large numbers are present, it can be concluded with some degree of certainty that the illness is staphylococcal food poisoning. Additional information that can remove any doubt is whether the staphylococci are enterotoxigenic and/or whether enterotoxin can be detected in the suspected food. Although the latter is definite proof of the cause, often an insufficient quantity of food (10 g can be used, but larger amounts are better) is available for examination. The presence of enterotoxigenic staphylococci in the food is reasonable assurance that these organisms were the cause of the illness.

4.4 Incidence

The true incidence of staphylococcal food poisoning in the US, as well as in other countries, is unknown, because this illness is not a reportable disease in most countries.

Table 14.5 Sea analysis in 2% chocolate milk

Sample	ng per ml	ng per 1/2 pint (~240 ml)
1	0.40	94
2	0.73	172
3	0.48	113
4	0.63	149
5	0.78	184
6	0.65	153
Average	0.61	144

From Evenson *et al.* (1988).

The numbers of confirmed outbreaks reported in the US to the CDC from 1977 to 1997 are given in Table 14.1. Many additional outbreaks of food poisoning reported were probably due to staphylococcal food poisoning but not classified as such because insufficient information was given – such as the staphylococcal count. Over 40 outbreaks were reported in some years, with staphylococcal food poisoning being the leading bacterial foodborne disease. After 1982 no more than 14 confirmed outbreaks were reported per year, and less than 10 have been confirmed for several years.

Todd (1996) reviewed the worldwide surveillance of foodborne diseases between 1985 and 1989, looking at 17 countries. The highest incidence of staphylococcal foodborne disease per 10^7 population was 58.5 in Cuba, 9–15 in Hungary, Finland, Japan and Israel, and 0.4 in the United States. The largest number of outbreaks was 128 per year in Japan, compared to only 9.4 per year in the US. Denmark and the Netherlands reported no outbreaks.

In a later report of food poisoning in Korea and Japan (1971–1990; Lee *et al.* 1996), 14.9 % of outbreaks in Korea were due to staphylococcal food poisoning and 24.6 % of outbreaks in Japan were due to staphylococcal food poisoning. In Taiwan, 169 of 555 (30.5 %) foodborne disease outbreaks from 1986 to 1995 were due to staphylococcal food poisoning (Pan *et al.* 1997). The average number of outbreaks from 1988 to 1993 was 21.5 per year, with only 13 and 12 outbreaks reported in 1994 and 1995, respectively.

4.5 Prevention and control

The staphylococci are ubiquitous organisms that cannot be eliminated from our environment. At least 30–50 % of individuals carry these organisms in their nasal passages or throats, or on their hands. Any time a food is exposed to human handling, there is the possibility that the food will be contaminated with staphylococci. Not all of these may be enterotoxin producers, but 30–50 % may well be. Heating of the food after handling will normally assure against food poisoning unless the food has been held unrefrigerated for several hours before the heating; if enterotoxin has formed in the food, the heating might not be sufficient to destroy it. In many cases foods are not processed further after handling, and unless proper care is taken the organisms may grow and produce enterotoxin. The course of action recommended is to keep susceptible foods refrigerated at all times except when being prepared and while being served. Refrigeration should be carried out in such a manner as to facilitate quick cooling of the entire food mass. Most food-poisoning outbreaks could be prevented if this simple precaution were taken. To illustrate its importance, in one food-poisoning outbreak, cream-filled coffee cake that was kept unrefrigerated by one outlet caused a number of illnesses, while the same cake kept refrigerated by another outlet resulted in no illnesses. If it is impossible to guarantee that a susceptible food will be kept refrigerated, special care should be taken in its preparation to avoid contamination if at all possible. Bryan (1968, 1976a) has reviewed the subject and listed helpful hints for those responsible for handling foods.

5 Prevalence of *S. aureus* in foods

Most meat is contaminated with staphylococci, but normally this is not of concern because the organisms do not usually multiply rapidly on raw meat and they are destroyed in the cooking process. Staphylococci that are present on the meat before processing are rarely involved in food-poisoning outbreaks. In the case of ham involvement in food poisoning, the causative organism is a result of post-processing contamination. One exception is fermented sausage – particularly sausages prepared without sufficient heating to destroy the staphylococci. It has been demonstrated by laboratory experiments that sufficient growth of the staphylococci can occur and produce enterotoxin before the fermenting organisms produce enough acid to inhibit the staphylococcal growth (Bergdoll, unreported results). In fact, outbreaks from this type of fermented sausage (Genoa) did occur, due primarily to inadequate fermentation conditions. The contamination of raw meats is probably from multiple sources, including human handlers. Examination of the fermented sausage revealed the presence of several different strains of staphylococci.

Several surveys have been conducted on different types of foods in the marketplace in different countries. Normally foods were found to be contaminated with staphylococci, particularly if they had been handled by humans. However, staphylococci would not grow sufficiently on many of the foods to produce enterotoxin, and cooking before consumption would destroy the organisms. In a majority of staphylococcal food-poisoning outbreaks, the foods were contaminated during their preparation for eating and mishandled after the preparation. One example is the cream-filled coffee cake mentioned above.

6 Foods most often associated with Staphylococcal food poisoning

Any food that provides a good medium for the growth of staphylococci may be involved in this type of foodborne illness. In the US, pork – particularly baked ham – is the food most frequently involved in outbreaks; poultry, salads (meat, potato, etc.) and cream-filled bakery goods are responsible for many of the remaining outbreaks. The frequency of the involvement of baked ham may be for a number of reasons. It is a common food item for picnics, some of which involve large numbers of people. In the latter cases, particularly, it is very difficult to keep the ham properly refrigerated until the food is consumed. Refrigeration in shallow pans is necessary to prevent the growth of any staphylococci that may be present. The warmer summertime temperatures complicate the situation. Another factor is that some people believe that cooked food is safe and do not realize the necessity for adequate refrigeration. Cream-filled bakery goods can pose a particular problem in the summer if they are not stored under adequate refrigeration from the time of preparation until they are consumed. This is much less of a problem in the US now than it was a few years ago (Bryan, 1976b).

The foods involved in other countries vary with the diet as well as the local conditions. In Japan, rice balls are a common item taken on picnics and outings, and once were the major item involved in staphylococcal food poisoning. Rice is an excellent growth medium for staphylococci. The balls used to be prepared by hand and were not usually refrigerated, but now they are made by machine to avoid human handling. In some of the European countries, such as Poland and the former Czechoslovakia, ice cream made by small producers is a cause of this type of foodborne illness. In Brazil, the two foods involved most frequently are cream-filled cake and a white cheese frequently produced on the farm or in small establishments.

7 Principles of detection of *S. aureus*

No specific test may be useful in every case to isolate the staphylococci from the wide variety of foods in which they are found. As a result, attempts have been made to find a combination of selective and enrichment media that will support the growth of the staphylococci and at the same time suppress the growth of other microflora present that tend to overgrow the staphylococci. A three-tube isolation procedure using trypticase soy broth with 10 % sodium chloride and 1 % sodium pyruvate was accepted as the official method for recovery of the largest numbers of coagulase-positive staphylococci from the widest variety of foods (Lancette and Lanier, 1987); however, thermally stressed cells of *S. aureus* are unable to grow in the medium. As a result, food samples likely to contain a small population of injured cells were incubated in double-strength trypticase soy broth before the addition of 13 % NaCl and spread-plating on Baird-Parker agar (Lancette and Tatini, 1992). For detecting small numbers of *S. aureus* in raw food ingredients and non-processed foods expected to contain large numbers of competing organisms, incubation is in trypticase soy broth containing 10 % NaCl and 1 % sodium pyruvate before transferring to Baird-Parker agar plates. For detecting relatively large numbers of staphylococci, the food extract is plated directly on Baird-Parker agar.

Typical colonies of *S. aureus* on Baird-Parker agar are circular, smooth, convex, moist, ~1.5 mm in diameter on uncrowded plates, gray-black to jet-black, smooth with entire margins and off-white edges, and may show an opaque zone with a clear halo extending beyond it. Normally those colonies that appear to be *S. aureus* will be counted, and one or more of each type tested for coagulase and TNase production. However, the upgrading of biotypes E and F to *S. intermedius* and *S. hyicus* complicates the species classification because all three species can be coagulase- and TNase-positive. If at least one test is positive and the food being examined is from a food-poisoning outbreak, the staphylococci are probably *S. aureus* from human contamination. An additional positive anaerobic mannitol fermentation test will confirm *S. aureus*. The number of colonies on the triplicate plates represented by the *S. aureus* positive colonies is multiplied by the dilution factor, and the result reported as the number of *S. aureus* per gram of food.

Agglutination kits employing the clumping factor, protein A, and specific antigens of *S. aureus* are available for identification of *S. aureus* strains. However, these kits are designed primarily for use in the clinical field where large numbers of staphylococci are being examined and where large numbers of coagulase-negative species are also encountered (Personne *et al.*, 1997; Wilkerson *et al.*, 1997). The clumping factor test is not satisfactory because *S. intermedius* and some coagulase-negative species can be positive for this factor.

An alternative method has been proposed by Roberson *et al.* (1992) in which P agar supplemented with acriflavin and the β-galactosidase test are used. Of the coagulase-positive species, only *S. aureus* will grow on the supplemented P agar and is negative with the β-galactosidase test. *S. intermedius* does not grow on the modified P agar and is 100 % positive with the β-galactosidase test, whereas *S. hyicus* is negative by both tests. This method is useful if the staphylococci being tested are from sources other than clinical.

Another method that has been proposed to identify *S. aureus* from non-clinical sources employs an immunoenzymatic assay using a monoclonal antibody prepared against endo-β-acetyl-glucosaminidase – an enzyme produced by all isolates of this species. Comparison of this method with six kits available for identification of *S. aureus* has shown it to be specific for *S. aureus*, whereas the kits were positive for *S. intermedius*, *S. schleiferi*, and *S. lugdunensis* (Guardati *et al.*, 1993).

8 Principles of detection of the enterotoxins

8.1 Introduction

It was not possible to develop specific methods for the detection of the enterotoxins before Bergdoll *et al.* (1959a) identified and purified the first enterotoxin. Until that time, the only means of detecting the presence of the enterotoxins was by the use of animals that gave emetic reactions to the toxin, either intragastrically or intravenously. Fortunately, at the time Surgalla and Bergdoll began their research to identify the enterotoxin, immunological methods were being developed for the specific detection of individual proteins. These investigators were able to show that specific antibodies could be produced to the enterotoxin when the emetic reaction in monkeys was neutralized by antisera produced against the crude toxin (Bergdoll *et al.*, 1959b; Surgalla *et al.*, 1954). Subsequently, all laboratory methods for the enterotoxins have been based on the use of specific antibodies to each of the enterotoxins for their detection, because it is almost impossible to detect individual proteins by chemical methods.

8.2 Biological methods

Before the first enterotoxin was purified, many types of animals (such as pigs, dogs, cats and kittens, and monkeys) were tested in the search for an inexpensive specific test method. All of these animals, with the exception of monkeys, were relatively insensitive to the enterotoxins, unless the toxin was injected intraperitoneally or intravenously. Emesis is the most readily observable reaction to enterotoxin; hence animals without a vomiting mechanism, such as rodents, were of little value as test subjects.

The feeding of young monkeys (Surgalla *et al.*, 1953) – preferably rhesus although cynomologus monkeys have been used successfully – provides the most reliable bioassay for enterotoxins because, of the biologically active substances produced by the staphylococci, only enterotoxins cause emesis when administered by the oral route. Assays are performed by administering solutions of the enterotoxins (up to 50 ml) to monkeys (2–3 kg) by catheter. The animals are observed for 5 hours for emesis. A response in at least two animals is accepted as a positive reaction. Other investigators have used the production of diarrhea as well as emesis as a criterion for positive reactions (Schantz *et al.*, 1965), as this occurs in humans suffering from staphylococcal food poisoning about as frequently as emesis. The sensitivity of the monkey-feeding test was increased by feeding 20-fold concentrates of bacterial culture supernatants. This made it possible to show that strains once thought to be non-enterotoxigenic did produce enterotoxin. The cost of monkeys, the expense of their upkeep and the difficulty in getting approval for the use of animals in research has limited their use for routine testing. One additional drawback is that the animals become resistant to the enterotoxin after several feedings

The intravenous injection of cats and kittens (Hammon, 1941) also proved useful for the detection of the enterotoxins. When materials other than the purified enterotoxins are injected intravenously, it is necessary to inactivate any interfering substances by treatment with trypsin (Denny and Bohrer, 1963) or pancreatin (Casman and Bennett, 1963). Cats are not as reliable as monkeys because they are subject to non-specific reactions. More recent studies have suggested that the Asian house shrew (*Suncus murinus*) could be used as an animal model. Hu *et al.* (1999) showed that the 50 % emetic dose of SEA by oral and intraperitoneal administration was 32 μg and 3 μg per kg body weight, respectively.

Because antibodies are specific for each enterotoxin, it is necessary to continue the use of animal testing until each new enterotoxin has been purified and antibodies produced against it. Animal testing is also necessary for assessing the effect of various treatments, such as heat, on the enterotoxins.

8.3 Immunological methods

The most specific and sensitive tests for the enterotoxins are based on their reactions with specific antibodies. The first tests developed were based on the reaction of the enterotoxin with the specific antibodies in gels to give a precipitin reaction. These were the only laboratory methods available until radioimmunoassay (RIA) was applied, and later the enzyme-linked immunosorbent assay (ELISA) and the reversed passive latex agglutination (RPLA) method were developed. The gel-diffusion methods have been used primarily for the detection of enterotoxin production by staphylococcal strains, although the RPLA method is used for testing strains for low production of enterotoxin. The RIA method was used for testing for enterotoxin in foods until the ELISA and RPLA were available.

8.4 Detection of enterotoxigenic strains

8.4.1 Gel-diffusion methods

Many types of gel reactions have been used in the detection of the enterotoxins, the most common ones being some form of either the Ouchterlony gel plate or the microslide (Ouchterlony, 1949, 1953; Crowle, 1958). A modification of the Ouchterlony gel-plate test used in the Food Research Institute for detection of enterotoxin-producing staphylococcal strains, and recommended to others, is the optimum sensitivity plate (OSP) method (Robbins *et al.*, 1974). It is easy to use and, in conjunction with production of the enterotoxins by the membrane-over-agar method (Robbins *et al.*, 1974) or the sac-culture method (Donnelly *et al.*, 1967), is adequate in sensitivity to detect most enterotoxigenic staphylococci. The normal sensitivity is 0.5 μg/ml, but this can be increased to 0.1 μg/ml by a five-fold concentration of the staphylococcal culture supernatant fluids.

The microslide method is the most sensitive of the gel-diffusion methods (0.05–0.1 μg/ml), but care is needed in preparing the slides (Casman *et al.*, 1969); even so, the results are often difficult to interpret. For example, Gibbs *et al.* (1978) reported that a high percentage of staphylococcal strains isolated from poultry produced SEA as determined by the microslide method. These strains were examined in the Bergdoll laboratory and found to be negative for SEA. Careful examination of the microslides produced in the Gibbs *et al.* laboratory showed a line of impurity that was incorrectly observed to form a line of identity with the SEA control precipitin line. Many things can go wrong with this method, and experience is very important in using it successfully (Casman *et al.*, 1969). A sensitivity of 30 ng/ml has been reported for this method, but this is exceptional as most operators are unable to achieve a sensitivity of greater than 100 ng/ml.

Although monoclonal antibodies have been developed for the enterotoxins (Thompson *et al.*, 1985), they cannot be used in gels because their reactions with the enterotoxins do not result in the formation of precipitates, even when a mixture of monoclonals is used.

8.4.2 The RPLA method

Through the application of the RPLA method to the detection of enterotoxin production by staphylococci it was possible to detect low-producing strains (approximately 10–20 ng/ml) that were not detectable by the OSP gel-diffusion method (Igarashi *et al.*, 1986; Bergdoll, 1990; Table 14.6). Some staphylococcal strains that were implicated in food poisoning by the monkey feeding test but negative by OSP were positive by ELISA (Table 14.7; Kokan and Bergdoll, 1987). The production of 10–20 ng of enterotoxin/ml is probably of significance, because only 100–200 ng of enterotoxin A was shown to be necessary to produce food poisoning (Evenson *et al.*, 1988), with the amount present in the vehicle (2 % chocolate milk) being 0.40–0.78 ng/ml (Table 14.5). The OSP method can be used as a preliminary method for the testing of strains, as only a small percentage of strains are low enterotoxin producers.

Table 14.6 Staphylococcal strain testing by RPLA method

Strain[a]	Enterotoxins detected	
	RPLA	OSP
188	SEA, SED	SEA, SED
228	SEA, SED	–
311	SEA	SEA
365	SEC	–
452	SEA	SEA
581	SEC	SEC
609	SEA, SED	–
754	SEA, SEB	SEA, SEB
802	SEB	SEB
887	SEA	–
896	SEA, SEB	SEA, –
965	SEA, SED	SEA, SED

From Bergdoll (1990).
[a] Strains were received from Dr James K. Todd, The Children's Hospital, Denver, CO.

Table 14.7 *S. aureus* strains negative for enterotoxin by OSP[a]

Source	Monkey positive	ELISA positive
Food poisoning	38	10
Foods	25	4
Fish (raw)	11	6
Nares (human)	26	4
Nares (horse)	4	1
Miscellaneous	6	1

[a] OSP, optimum sensitivity plate (Kokan and Bergdoll, 1987).

8.4.3 Polymerase chain reaction (PCR) and DNA hybridization

Since DNA sequence information is available for all the described SEs, DNA probes (Notermans *et al.*, 1988; Ewald *et al.*, 1990; Neill *et al.*, 1990; Jaulhac *et al.*, 1992) and PCR (Johnson *et al.*, 1991; Wilson *et al.*, 1991, 1994; Tsen and Chen, 1992; McLauchlin *et al.*, 2000) are commonly employed for detection of the SE genes. Several reports have described the development of multiplex PCR for the detection of multiple SE genes (Becker *et al.*, 1998; Mehrotra *et al.*, 2000; Martin *et al.*, 2003). Sharma *et al.* (2000) were able to detect five enterotoxin genes, *sea–see*, in a single PCR reaction in 3–4 hours. PCR is highly sensitive and specific, and allows the detection of enterotoxigenic staphylococci in a relatively short time with little sample preparation. One advantage of the method is that dead enterotoxigenic cells can also be detected, which is important in the analysis of heated foods.

8.5 Detection in foods

The detection of enterotoxin in foods requires methods that are sensitive to less than 1 ng/g of food. The quantity of enterotoxin present in foods involved in food-poisoning outbreaks may vary from less than 1 ng/g to greater than 50 ng/g. Although little difficulty is usually encountered in detecting the enterotoxin in foods involved in food-poisoning outbreaks, outbreaks do occur in which the amount of enterotoxin is less than 1 ng/g – such as the case of the 2 % chocolate milk. In such instances, the enterotoxin can be detected only by the most sensitive methods. Another situation in which it is essential to use a very sensitive method is in determining the safety of a food for consumption, where it is necessary to use the most sensitive methods available in order to show that no enterotoxin is present.

8.5.1 The ELISA method

ELISA methods were applied to the detection of the enterotoxins in foods soon after they were originally developed for the detection of other proteins. The most common type of ELISA is the sandwich method, in which the antibody is reacted with the unknown sample before the antibody–enterotoxin complex is treated with the enzyme–antibody conjugate (Saunders and Bartlett, 1977). This format is preferred because the amount of enzyme, and thus the color developed from the enzyme–substrate reaction, is directly proportional to the amount of enterotoxin present in the sample. This eliminates the need for highly purified enterotoxins, as crude or only partially purified enterotoxin is needed for preparation of a standard curve.

The majority of users of ELISA methods employ microtiter plates or strips to which the antibodies are attached. The large number of wells in a microtiter plate provide for analyzing several samples at one time. A plate reader is useful for recording the results, which adds to the expense of the method, although in many cases the test can be read visually. One commercial kit that uses microtiter strips is the RIDASCREEN®, developed in Germany by Kraatz-Wadsack *et al.* (1991). The polyclonal antibodies were prepared in sheep in the Institute of Milk Hygiene, University of Munich, Germany. A collaborative study conducted in Canada indicated that the specificity, sensitivity, repeatability and reproducibility met food safety criteria (Park *et al.*, 1996a).

An alternate procedure has been developed; this is the use of polystyrene balls to which the antibodies are attached (Stiffler-Rosenberg and Fey, 1978; Freed *et al.*, 1982; Fey and Pfister, 1983). The ball method is more cumbersome because each ball must be handled separately. The main advantage is that a relatively large volume of the unknown sample can be used, thus increasing the amount of enterotoxin adsorbed per sample. This makes possible the use of 1-ml volumes of substrate so that the color developed can be read in a simple colorimeter, an instrument that most laboratories have available. The sensitivities of the ELISA methods are between 0.5 and 1.0 ng/g of food. One kit that employs the ELISA ball method was developed in Switzerland by Fey and Pfister (1983) and produced by Diagnostische Laboratorien, Bern, Switzerland. Those who used it have found it to be a very good method for detecting enterotoxin in foods (Wieneke and Gilbert, 1987). Although the test could be completed in a single day, the recommendation was that the antibody-coated balls be shaken with 20 ml of food extract overnight to obtain the highest sensitivity. The method could be used quantitatively, although this is not necessary in checking foods for the presence of enterotoxins. This method was found to be superior when compared to other methods (Wieneke and Gilbert, 1987; Wienecke, 1991). Unfortunately this kit is no longer available.

A third method is based on a fluorimetric reaction with a sensitivity of less than 1 ng/ml (Armstrong *et al.*, 1993). This method is produced and marketed by bioMérieux Vitek, Inc. (Hazelwood, Missouri) as VIDAS® (Vitek ImmunoDiagnostic Assay System) S.E.T. An automated detection system is used with this assay.

A dip-stick method was developed by IGEN, Inc., with monoclonal antibodies developed at the Food Research Institute, and licensed to Transia, Transia-Diffchamb, Lyon, France. The antibodies to each of the enterotoxins were adsorbed onto nitrocellulose paper attached to wells in plastic sticks. There was some problem with the SEC antibodies, as only one monoclonal antibody for SEC was available and the second antibody was a polyclonal antibody. One report indicated that false-positive reactions were obtained for SEC in some food samples (Wieneke, 1991). This method was very easy to apply, as only one test was needed to check for all of the enterotoxins and it could be done in 1 day. It did not prove to be as sensitive as the other methods (Wieneke, 1991).

Many investigators have developed ELISA methods for the detection of the staphylococcal enterotoxins with slight differences in procedures. One difference involved the use of biotinylated antibodies and avidin-alkaline phosphatase instead of coupling the enzyme directly to the antibodies (Hahn *et al.*, 1986; Edwin, 1989). The biotinylated antibodies apparently were more stable than the enzyme antibodies. A sensitivity of 0.1 ng/ml was possible.

8.5.2 The RPLA method

An RPLA kit also is available commercially for use in the detection of enterotoxins in foods. It is produced by Denka Seiken Co. Ltd., Niigata, Japan. The method is adequately sensitive for the detection of enterotoxin in most foods that are implicated in food-poisoning outbreaks (Igarashi *et al.*, 1985); however, it may be inadequate for detection of the small amounts of enterotoxin that are sometimes present. This was indicated when Wieneke and Gilbert (1987) compared the RPLA method to ELISA methods for the detection of enterotoxin in foods. The RPLA method was adequate in all but two tests; in one case the sensitivity was inadequate, and in the other case the food extract gave a non-specific agglutination.

8.5.3 Screening methods

In some instances it may not be necessary to determine the type of enterotoxin if enterotoxin is present; for example, in examining the marketability of suspect foods. The food would not be marketable if any enterotoxin were present. For this purpose, including all of the enterotoxins in one test saves time. However, this would not save time in examining foods implicated in food-poisoning outbreaks, as identification of the type of enterotoxin is valuable in tracing the source of the contamination.

Two kits that include all of the enterotoxins in one test are available commercially. One kit utilizes small tubes coated with monoclonal antibodies to all of the enterotoxins (Transia, Transia-Diffchamb, Lyon, France). The method is sensitive to 0.2 ng/ml of extract; a bright blue color is developed with this concentration of enterotoxins. Although this is a screening test, if the enterotoxins were present at these low levels it would be difficult to analyze for individual enterotoxins because the other methods are less sensitive. If more than one enterotoxin were present, it would be necessary to concentrate the extract. An international collaborative study using this method has been reported in which acceptable results were obtained by the collaborators and were validated by the French Normalization Agency for identification of staphylococcal enterotoxins in foods and culture fluids (Lapeyre *et al.*, 1996).

The second kit, TECRA®, is produced by Bioenterprises Pty Ltd., Roseville, New South Wales, Australia. It utilizes microtiter plates and has a sensitivity of at least 1 ng of SEA/g of ham (Park *et al.*, 1996b). Although the sensitivity is adequate for most foods involved in food-poisoning outbreaks, it is doubtful that the SEA in the chocolate milk outbreak would have been detectable. This is critical, particularly in the case of testing suspect foods whose marketability is contingent on the absence of enterotoxin. Two collaborative studies have been conducted utilizing the TECRA kit, one by Bennett and McClure. (1994) in the US for the purpose of establishing it as an official method, and one by Park *et al.* (1996b) in Canada. Unfortunately, the minimum amount of enterotoxin included in any of the foods was 4 ng/g of food in the Bennett and McClure study and 1 ng/g of food in the Park *et al.* study. Although all collaborators were successful in detecting these minimum amounts, they were not tested for detecting 0.5 ng/g – the minimum amount that should be detectable by any acceptable method, as was pointed out by Bergdoll (1994). Despite this shortcoming, the Bennett and McClure collaborative study was accepted by the Official Methods Committee as an official method for detection of enterotoxin in foods.

9 Summary

The staphylococci, isolated in the 1800s as specific organisms, were identified as the cause of human infections that included food poisoning. It was not until the 1900s that the cause of food poisoning was identified as a toxin produced by the staphylococci growing in foods. This toxin was given the name enterotoxin because of its effect on the intestinal tract. The staphylococci produce several enterotoxins (SEA, B, C, D, E, G, H, I, J, K, L, M) that are related proteins, with varying degrees of sequence homology. Although the staphylococci produce other toxic substances that may be involved in human infections, the emphasis of this chapter is on the enterotoxins and their involvement in food poisoning. Staphylococcal food poisoning occurs worldwide, but has declined in the developed countries, such as the US and England. Many cases may not be reported because the illness is relatively mild and short-lived, with no sequelae. The enterotoxins are low molecular weight, single-chain proteins, with a cystine loop in the center of the molecule. They are identified by specific antibodies, which are the basis of the detection methods. The gel-diffusion methods in which a precipitin reaction occurs between the enterotoxin and its specific antibody are used for detecting enterotoxin production by staphylococcal strains. The majority of the methods used for detection of enterotoxins in foods are some form of the enzyme-linked immunosorbent assay (ELISA), with a sensitivity of less than 1 ng/g of food.

Bibliography

Akhtar, M., C. E. Park and K. Rayman, K. (1996). Effect of urea treatment on recovery of staphylococcal enterotoxin A from heat-processed foods. *Appl. Environ. Microbiol.* **62**, 3274–3276.

Anonymous (1979a). *Corned Beef.* Commun. Dis. Rep. No. 23, Communicable Disease Surveillance Centre, London.

Anonymous (1979b). *Canned Corned Beef from Malta.* Commun. Dis. Rep. No. 38, Communicable Disease Surveillance Centre, London.

Anonymous (1979c). *Corned Beef – Argentina Establishment 1.* Commun. Dis. Rep. No. 48, Communicable Disease Surveillance Centre, London.

Anonymous (1989). Multiple outbreaks of staphylococcal food poisoning caused by canned mushrooms. *Morbid. Mortal. Weekly Rep.* **38**, 417–418.

Anonymous (1990). Big problem with mushrooms. *Food Ind. Rep.*, **23 July**, 2, 6–7.

Archer, D. L., B. G. Smith, J. T. Ulrich and H. M. Johnson (1979). Immune interferon induction by T-cell mitogens involves different T-cell subpopulations. *Cell. Immunol.* **48**, 420–426.

Armstrong, T., C. Gravens, F. Hoag *et al.* (1993). Automated ELISA detection of staphylococcal enterotoxins by the VIDAS[tm] system. *Abstr. Annu. Meet, Am. Soc. Microbiol.*, **56**, 341.

Avena, R. M. and M. S. Bergdoll (1967). Purification and some physicochemical properties of enterotoxin C, *Staphylococcus aureus* strain 361. *Biochemistry* **6**, 1474–1480.

Baird-Parker, A. C. (1963). A classification of micrococci and staphylococci based on physiological and biochemical tests. *J. Gen. Microbiol.* **30**, 409–426.

Baird-Parker, A. C. (1974). The basis for the present classification of staphylococci and micrococci. *Ann. NY Acad. Sci.* **236**, 6–13.

Barber, M. A. (1914). Milk poisoning due to a type of *Staphylococcus albus* occurring in the udder of a healthy cow. *Philippine J. Sci., Sect. B. (Trop. Med.)* **9**, 515–519.

Bean, N. H., P. M. Griffin, J. S. Goulding and C. B. Ivey (1990). Foodborne disease outbreaks, 5-year summary, 1983–1987. *Morbid. Mortal. Weekly Rep. CDC Surv. Summ.* **39**(SS-1), 15–57.

Bean, N. H., J. S. Goulding, C. Lao and F. J. Angulo (1996). Surveillance for foodborne-disease outbreaks – United States, 1988–1992. *Morbid. Mortal. Weekly Rep. CDC Surv. Summ.* **45**(SS-5), 1–66.

Becker, K., R. Roth and G. Peters (1998). Rapid and specific detection of toxigenic *Staphylococcus aureus*: use of two multiplex PCR enzyme immunoassays for amplification and hybridization of staphylococcal enterotoxin genes, exfoliative toxin genes, and toxic shock syndrome toxin 1 gene. *J. Clin. Microbiol.* **36**, 2548–2553.

Beisel, W. R. (1972). Pathophysiology of staphylococcal enterotoxin type B, (SEB) toxemia after intravenous administration to monkeys. *Toxicon* **10**, 433–440.

Bennett, R. W. (1994). Urea renaturation and identification of staphylococcal enterotoxin. In: R. C. Spencer, E. P. Wright and S. W. B. Newsom (eds), *Rapid Methods and Automation in Microbiology and Immunology*, pp. 401–411. Intercept Ltd, Andover.

Bennett, R. W. and M. R. Berry Jr (1987). Serological activity and in vivo toxicity of *Staphylococcus aureus* enterotoxins A and D in selected canned foods. *J. Food Sci.* **52**, 416–418.

Bennett, R. W. and F. D. McClure (1994). Visual screening with enzyme immunoassay for staphylococcal enterotoxins in foods: collaborative study. *J. AOAC Intl* **77**, 357–364.

Bennett, R. W., M. Yeterian, W. Smith *et al.* (1986). *Staphylococcus aureus* identification characteristics and enterotoxigenicity. *J. Food Sci.* **51**, 1337–1339.

Bergdoll, M. S. (1970). Enterotoxins. In: T. C. Montie, S. Cadis and S. J. Ajl (eds), *Microbial Toxins*, pp. 265–326. Academic Press, New York, NY.

Bergdoll, M. S. (1989). *Staphylococcus aureus.* In: M. P. Doyle (ed.), *Foodborne Bacterial Pathogens*, pp. 463–523. Marcel Dekker, New York, NY.

Bergdoll, M. S. (1990). Analytical methods for *Staphylococcus aureus. Intl J. Food Microbiol.* **10**, 91–100.

Bergdoll, M. S. (1994). The AOAC official methods for detection of staphylococcal enterotoxins in food. *J. AOAC Intl* **77**, 31A.

Bergdoll, M. S., H. Sugiyama and G. M. Dack (1959a). Staphylococcal enterotoxin. I. Purification. *Arch. Biochem. Biophys.* **85**, 62–69.

Bergdoll, M. S., M. J. Surgalla and G. M. Dack (1959b). Staphylococcal enterotoxin. Identification of a specific precipitating antibody with enterotoxin-neutralizing property. *J. Immunol.* **83**, 334–338.

Bergdoll, M. S., C. R. Borja and R. M. Avena (1965). Identification of a new enterotoxin as enterotoxin C. *J. Bacteriol.* **90**, 1481–1485.

Bergdoll, M. S., C. R. Borja, R. N. Robbins and K. F. Weiss (1971). Identification of enterotoxin E. *Infect. Immun.* **4**, 593–595.

Bergdoll, M. S., I.-Y. Huang and E. J. Schantz (1974). Chemistry of the staphylococcal enterotoxins. *Agric. Food Chem.* **22**, 9–13.

Betley, M. J. and J. J. Mekalanos (1988). Nucleotide sequence of the type A staphylococcal enterotoxin gene. *J. Bacteriol.* **170**, 34–41.

Bohach, G. A. and P. M. Schlievert (1989). Conservation of the biologically active portions of staphylococcal enterotoxins C_1 and C_2. *Infect. Immun.* **57**, 2249–2252.

Brunner, K. G. and A. C. L. Wong (1992). *Staphylococcus aureus* growth and enterotoxin production in mushrooms. *J. Food Sci.* **57**, 700–703.

Bryan, F. L. (1968). What the santitarian should know about staphylococci and salmonellae in non-dairy products. I. Staphylococci. *J. Milk Food Technol.* **31**, 110–116.

Bryan, F. L. (1976a). *Staphylococcus aureus.* In: M. P. Defigueiredo and D. F. Splittstoesser (eds), *Food Microbiology: Public Health and Spoilage Aspects*, pp. 12–128. AVI Publishing Company, Westport, CT.

Bryan, F. L. (1976b). Public health aspects of cream-filled pastries. A review. *J. Milk Food Technol.* **39**, 289–296.

Buchanan, R. E. and N. E. Gibbons (eds) (1974). *Bergey's Manual of Determinative Bacteriology*, 8th edn. Williams and Wilkins, Baltimore, MD.

Carpenter, D. F., and G. J. Silverman (1974). Staphylococcal enterotoxin B and nuclease production under controlled dissolved oxygen conditions. *Appl. Microbiol.* **28**, 628–637.

Casman, E. P. and R. W. Bennett (1963). Culture medium for the production of staphylococcal enterotoxin A. *J. Bacteriol.* **86**, 18–23.

Casman, E. P., M. S. Bergdoll and J. Robison (1963). Designation of staphylococcal enterotoxins. *J. Bacteriol.* **85**, 715–716.

Casman, E. P., R. W. Bennett, A. E. Dorsey and J. A. Issa (1967). Identification of a fourth staphylococcal enterotoxin, enterotoxin D. *J. Bacteriol.* **94**, 1875–1882.

Casman, E. P., R. W. Bennett, A. E. Dorsey and J. E. Stone (1969). The micro-slide gel diffusion test for the detection and assay of staphylococcal enterotoxins. *Health Lab. Sci.* **6**, 185–198.

CDC (Centers for Disease Control and Prevention) (1981a). *Foodborne and Waterborne Disease Outbreaks Annual Summary 1978*. Revised edition, reissued February 1981. CDC, Atlanta, GA.

CDC (Centers for Disease Control and Prevention) (1981b). *Foodborne and Waterborne Disease Outbreaks Annual Summary 1979*, issued April 1981. CDC, Atlanta, GA.

CDC (Centers for Disease Control and Prevention) (1983a). *Foodborne and Waterborne Disease Outbreaks Annual Summary 1980*, issued February 1983. CDC, Atlanta, GA.

CDC (Centers for Disease Control and Prevention) (1983b). *Foodborne and Waterborne Disease Outbreaks Annual Summary 1981*, issued June 1983. CDC, Atlanta, GA.

Cerqueira-Campos, M. L., S. M. P. Furlanetto, S. T. Iaria and M. S. Bergdoll (1993). Staphylococcal food poisoning outbreaks in Sao Paulo (Brazil). *Rev. Microbiol.* **24**, 261–264.

Chang, H.-C. and M. S. Bergdoll (1979). Purification and some physicochemical properties of staphylococcal enterotoxin D. *Biochemistry* **18**, 1937–1942.

Chang, P.-C. and N. Dickie (1971). Fractionation of staphylococcal enterotoxin B by isoelectric focusing. *Biochim. Biophys. Acta* **236**, 367–375.

Chesbro, W. R. and K. Auborn (1967). Enzymatic detection of the growth of *Staphylococcus aureus* in foods. *Appl. Microbiol.* **15**, 1150–1159.

Couch, J. L. and M. J. Betley (1989). Nucleotide sequence of the type C_3 staphylococcal enterotoxin gene suggests that intergenic recombination causes antigenic variation. *J. Bacteriol.* **171**, 4507–4510.

Couch, J. L., M. T. Soltis and M. J. Betley (1988). Cloning and nucleotide sequence of the type E staphylococcal enterotoxin gene. *J. Bacteriol.* **170**, 2954–2960.

Crowle, A. J. (1958). A simplified micro double-diffusion agar precipitation technique. *J. Lab. Clin. Med.* **52**, 784–787.

Dack, G. M. (1956). *Food Poisoning*, pp. 109–116. University of Chicago Press, Chicago, IL.

Dack, G. M., W. E. Cary, O. Woolpert and H. Wiggers (1930). An outbreak of food poisoning proved to be due to a yellow hemolytic staphylococcus. *J. Prevent. Med.* **4**, 167–175.

Dalidowicz, J. E., S. J. Silverman, E. J. Schantz *et al.* (1966). Chemical and biological properties of reduced and alkylated staphylococcal enterotoxin B. *Biochemistry* **5**, 2375–2381.

Das, C. and J. J. Langone (1989). Dissociation between murine spleen cell mitogenic activity of enterotoxin contaminants and anti-tumor activity of staphylococcal protein A. *J. Immunol.* **142**, 2943–2948.

Davison, E., G. M. Dack and W. E. Cary (1938). Attempts to assay the enterotoxic substance produced by staphylococci by parenteral injection of monkeys and kittens. *J. Infect. Dis.* **62**, 219–223.

De la Fuente, R., J. Almazan, L. Fuentes *et al.* (1986). Enterotoxigenicity of *S. aureus* and *S. intermedius* strains isolated from dogs. *Proc. 2nd World Congr. Foodborne Infect. Intox.* **1**, 362–364.

Dennison, G. A. (1936). Epidemiology and symptomatology of staphylococcus food poisoning. *Am. J. Publ. Health* **26**, 1168–1175.

Denny, C. B. and C. W. Bohrer (1963). Improved cat test for enterotoxin. *J. Bacteriol.* **86**, 347–348.

Denny, C. B., P. L. Tan and C. W. Bohrer (1966). Heat inactivation of staphylococcal enterotoxin A. *J. Food Sci.* **31**, 762–767.

Denny, C. B., J. Y. Humber and C. W. Bohrer (1971). Effect of toxin concentration on the heat inactivation of staphylococcal enterotoxin A in beef bouillon and in phosphate buffer. *Appl. Microbiol.* **21**, 1046–1066.

Dinges, M. M., P. M. Orwin and P. M. Schlievert (2000). Exotoxins of *Staphylococcus aureus. Clin. Microbiol. Rev.* **13**, 16–34.

do Carmo, L. S. and M. S. Bergdoll (1990). Staphylococcal food poisoning in Belo Horizonte (Brazil). *Rev. Microbiol.* **21**, 320–323.

Donnelly, S. R., J. E. Leslie, L. A. Black and K. H. Lewis (1967). Serological identification of enterotoxigenic staphylococci from cheese. *Appl. Microbiol.* **15**, 1382–1387.

Edwards, V. M., J. R. Deringer, S. C. Callantine *et al.* (1997). Characterization of the canine type C enterotoxin produced by *Staphylococcus intermedius* pyoderma isolates. *Infect. Immun.* **65**, 2346–2352.

Edwin, C. (1989). Quantitative determination of staphylococcal enterotoxin A by an enzyme-linked immunosorbent assay using a combination of polyclonal and monoclonal antibodies and biotin-streptavidin interaction. *J. Clin. Microbiol.* **27**, 1496–1501.

Evenson, M. L., M. W. Hinds, R. S. Bernstein *et al.* (1988). Estimation of human dose of staphylococcal enterotoxin A from a large outbreak of staphylococcal food poisoning involving chocolate milk. *Intl J. Food Microbiol.* **7**, 311–316.

Ewald, S., C. J. Heuvelman and S. Notermans (1990). The use of DNA probes for confirming enterotoxin production by *Staphylococcus aureus* and micrococci. *Intl J. Food Microbiol.* **11**, 251–258.

Fey, H. and H. Pfister (1983). A diagnostic kit for the detection of staphylococcal enterotoxins (SET) A, B, C and D (SEA, SEB, SEC, SED). In: S. Avrameas,

P. Druet, R. Masseyeff and G. Feldman (eds), *Immunoenzymatic Techniques*, pp. 345–348. Elsevier/North-Holland Publishing Co., Amsterdam.

Freed, R. C., M. L. Evenson, R. F. Reiser and M. S. Bergdoll (1982). Enzyme-linked immunosorbent assay for detection of staphylococcal enterotoxins in foods. *Appl. Environ. Microbiol.* **44**, 1349–1355.

Friedman, M. E. and M. Howard (1971). Induction of mutants to *Staphylococcus aureus* 100 with increased ability to produce enterotoxin A. *J. Bacteriol.* **106**, 289–291.

Fung, D. Y. C. (1972). Experimental production of enterotoxin B in fish protein concentrate. *J. Milk Food Technol.* **35**, 577–581.

Genigeorgis, C., H. Riemann and W. W. Sadler (1969). Production of enterotoxin B in cured meats. *J. Food Sci.* **34**, 62–68.

Gibbs, P. A., J. T. Patterson and J. Harvey (1978). Biochemical characteristics and enterotoxigenicity of *Staphylococcus aureus* strains isolated from poultry. *J. Appl. Bacteriol.* **44**, 57–74.

Grossman, D., R. G. Cook, J. T. Sparrow *et al.* (1990). Dissociation of the stimulatory activities of staphylococcal enterotoxins for T cells and monocytes. *J. Exp. Med.* **172**, 1831–1841.

Guardati, M. C., C. A. Guzman, G. Piatti and C. Pruzzo (1993). Rapid methods for identification of *Staphylococcus aureus* when both human and animal staphylococci are tested: comparison with new immunoenzymatic assay. *J. Clin. Microbiol.* **32**, 1606–1608.

Hahn, I. F., P. Pickenhahn, W. Lenz and H. Brandis (1986). An avidin-biotin ELISA for the detection of staphylococcal enterotoxins A and B. *J. Immunol. Methods* **92**, 25–29.

Hajék, V. and E. Marsálek (1971). A study of staphylococci isolated from the upper respiratory tract of different animal species. IV. Physiological properties of *Staphylococcus aureus* strains of hare origin. *Zbl. Bakt., I. Abt. Orig.* **216**, 168–174.

Hallander, H. O. (1965). Production of large quantities of enterotoxin B and other staphylococcal toxins on solid media. *Acta Pathol. Microbiol. Scand.* **63**, 299–305.

Hammon, W. M. (1941). *Staphylococcus* enterotoxin, improved cat test, chemical and immunological studies. *Am. J. Publ. Health* **31**, 1191–1198.

Hardt-English, P. G. York, R. Stier and P. Cocotas (1990). Staphylococcal food poisoning outbreaks caused by canned mushrooms from China. *Food Technol.* **44**(12), 74, 76–77.

Hirooka, E. Y., E. E. Müller, J. C. Freitas *et al.* (1988). Enterotoxigenicity of *Staphylococcus intermedius* of canine origin. *Intl J. Food Microbiol.* **7**, 185–191.

Hoffman, C (1990). 'Superantigens' may shed light on immune puzzle. *Science* **248**, 695–686.

Holt, J. G., N. R. Krieg, P. H. A. Sneath *et al.* (1994). Characteristics differentiating the species and subspecies of the genus *Staphylococcus*. *Bergey's Manual of Determinative Bacteriology*, 9th edn, pp 544–551. Williams and Wilkins, Baltimore, MD.

Hovde, C. J., S. P. Hackett and G. A. Bohach (1990). Nucleotide sequence of the staphylococcal enterotoxin C_3 gene: sequence comparison of all three type C staphylococcal enterotoxins. *J. Mol. Gen. Genet.* **220**, 329–333.

Hu, D.-L., K. Omoe, H. Shimura *et al.* (1999). Emesis in the shrew mouse (*Suncus murinus*) induced by peroral and intraperitoneal administration of staphylococcal enterotoxin A. *J. Food Prot.* **62**, 1350–1353.

Humber, J. Y., C. B. Denny and C. W. Bohrer (1975). Influence of pH on the heat inactivation of staphylococcal enterotoxin A as determined by monkey feeding and serological assay. *Appl. Microbiol.* **30**, 755–758.

Igarashi, H., M. Shingaki, H. Fujikawa *et al.* (1985). Detection of staphylococcal enterotoxins in food poisoning outbreaks by reversed passive latex agglutination. In: J. Jeljaszewicz (ed.), *The Staphylococci*, pp. 255–257. Gustav Fischer Verlag, Stuttgart.

Igarashi, H., H. Fujikawa, M. Shingaki and M. S. Bergdoll (1986). Latex agglutination test for staphylococcal toxic shock syndrome toxin 1. *J. Clin. Microbiol.* **23**, 509–512.

Jarraud, S., M. A. Peyrat, A. Lim *et al.* (2001). *egc*, A highly prevalent operon of enterotoxin gene, forms a putative nursery of superantigens in *Staphylococcus aureus. J. Immunol.* **166**, 669–677.

Jarvis, A. W., R. C. Lawrence and G. G. Pritchard (1973). Production of staphylococcal enterotoxins A, B, and C under conditions of controlled pH and aeration. *Infect. Immun.* **7**, 847–854.

Jaulhac, B., M. Bes, N. Bornstein *et al.* (1992). Synthetic DNA probes for detection of genes for enterotoxins A, B, C, D, and E for TSST-1 in staphylococcal strains. *J. Appl. Bacteriol.* **72**, 386–392.

Johnson, H. W. and H. I. Magazine (1988). Potent mitogenic activity of staphylococcal enterotoxin A requires induction of interleukin 2. *Intl Arch. Allergy Appl. Immunol.* **87**, 87–90.

Johnson, W. M., S. D. Tyler, E. P. Ewan *et al.* (1991). Detection of genes for enterotoxins, exfoliative toxins, and toxic shock syndrome toxin 1 in *Staphylococcus aureus* by the polymerase chain reaction. *J. Clin. Microbiol.* **29**, 426–430.

Jordan, E. O., G. M. Dack and O. Woolpert (1931). The effect of heat, storage and chlorination on the toxicity of staphylococcus filtrates. *J. Prev. Med.* **5**, 383–386.

Kato, E., M. Khan, L. Kujovich and M. S. Bergdoll (1966). Production of enterotoxin A. *Appl. Microbiol.* **14**, 966–972.

Khambaty, F. M., R. W. Bennett and D. B. Shah (1994). Application of pulsed-field gel electrophoresis to the epidemiological characterization of *Staphylococcus intermedius* implicated in a food-related outbreak. *Epidemiol. Infect.* **113**, 75–81.

Kloos, W. E. and K. H. Schleifer (1975). Simplified scheme for routine identification of human staphylococci. *Clin. Microbiol.* **1**, 82–88.

Kokan, N. P. and M. S. Bergdoll (1987). Detection of low-enterotoxin-producing *Staphylococcus aureus* strains. *Appl. Environ. Microbiol.* **53**, 2675–2676.

Kraatz-Wadsack, G., W. Ahrens and J. von Stuckrad (1991). ELISA screening of staphylococcal enterotoxins by means of a specially developed test kit. *Intl J. Food Microbiol.* **13**, 265–272.

Lancette, G. A. and J. Lanier (1987). Most probable number method for isolation and enumeration of *Staphylococcus aureus* in foods: collaborative study. *J. AOAC* **70**, 35–38.

Lancette, G. A. and S. R. Tatini (1992). *Staphylococcus aureus*. In: C. Vanderzant and D. F. Splittstoesser (eds), *Compendium of Methods for the Microbiological Examination of Foods*, 3rd edn, pp. 533–550. American Public Health Association, Inc., Washington, DC.

Lapeyre, C., M. N. Desolan and X. Drouet (1996). Immunoenzymatic detection of staphylococcal enterotoxins: international interlaboratory study. *J. AOAC Intl* **79**, 1095–1101.

Lee, A. C.-M., R. N. Robbins and M. S. Bergdoll (1978). Isolation of specific and common antibodies to staphylococcal enterotoxins A and E by affinity chromatography. *Infect. Immun.* **21**, 387–391.

Lee, A. C.-M., R. N. Robbins, R. F. Reiser and M. S. Bergdoll (1980). Isolation of specific and common antibodies to staphylococcal enterotoxins B, C_1, and C_2. *Infect. Immun.* **27**, 431–434.

Lee, W.-C., T. Sakai, M.-J. Lee *et al.* (1996). An epidemiological study of food poisoning in Korea and Japan. *Intl J. Food Microbiol.* **29**, 141–148.

MacDonald, K. L. and Griffin, P. M. (1986). Foodborne disease outbreaks annual summary, 1982. Morbid. *Mortal. Weekly Rep. CDC Surv. Summ.* **35**(SS-1), 7–16.

Marrack, P. and J. Kappler (1990). The staphylococcal enterotoxins and their relatives. *Science* **248**, 705–711.

Martin, M. C., M. A. González-Hevia and M. C. Mendoza (2003). Usefulness of a two-step procedure for detection and identification of enterotoxigenic staphylococci of bacterial isolates and food samples. *Food Microbiol.* **20**, 605–610.

McLauchlin, J., G. L. Narayanan, V. Mithani and G. O'Neill (2000). The detection of enterotoxins and toxic shock syndrome toxin genes in *Staphylococcus aureus* by polymerase chain reaction. *J. Food Prot.* **63**, 479–488.

Mehrotra, M., G. Wang and W. M. Johnson (2000). Multiplex PCR for detection of genes for *Staphylococcus aureus* enterotoxins, exfoliative toxins, toxic shock syndrome toxin 1, and methicillin resistance. *J. Clin. Microbiol.* **38**, 1032–1035.

Metzger, J. F., A. D. Johnson, W. S. Collins II and V. McGann (1973). *Staphylococcus aureus* enterotoxin B release (excretion) under controlled conditions of fermentation. *Appl. Microbiol.* **25**, 770–773.

Misfeldt, M. L. (1990). Superantigens. *Infect. Immun.* **58**, 2409–2413.

Modi, N. K., S. A. Rose and H. S. Tranter (1990). The effects of irradiation and temperature on the immunological activity of staphylococcal enterotoxin A. *Intl J. Food Microbiol.* **11**, 85–92.

Munson, S. H., M. T. Tremaine, M. J. Betley and R. A. Welch (1998). Identification and characterization of staphylococcal enterotoxin types G and I from *Staphylococcus aureus. Infect. Immun.* **66**, 3337–3348.

Neill, R. J., G. R. Fanning, F. Delahoz *et al.* (1990). Oligonucleotide probes for detection and differentiation of *Staphylococcus aureus* strains containing genes for enterotoxins A, B, and C and toxic shock syndrome toxin 1. *J. Clin. Microbiol.* **28**, 1514–1518.

Noleto, A. L. S. and A. Tibana (1987). Outbreak of staphylococcal enterotoxin B food poisoning. *Rev. Microbiol.* **18**, 144–145.

Notermans, S., K. J. Heuvelman and K. Wernars (1988). Synthetic enterotoxin B DNA probes for detection of enterotoxigenic *Staphylococcus aureus* strains. *Appl. Environ. Microbiol.* **54**, 531–533.

Olsen, S. J., L. C. MacKinnon, J. S. Goulding *et al.* (2000). Surveillance for food-borne-disease outbreaks – United States, 1993–1997. *Morbid. Mortal. Weekly Rep.* **49**(SS-1), 1–61.

Orwin, P. M., D. M. Leung, H. L. Donahue *et al.* (2001). Biochemical and biological properties of staphylococcal enterotoxin K. *Infect. Immun.* **69**, 360–366.

Ouchterlony, O. (1949). Antigen-antibody reactions in gels. *Acta Pathol. Microbiol. Scand.* **26**, 507–515.

Ouchterlony, O. (1953). Antigen-antibody reactions in gels. IV. Types of reactions in coordinated systems of diffusion. *Acta Pathol. Microbiol. Scand.* **29**, 231–240.

Owen, R. (1907). The bacteriology of meat poisoning. *Physician Surg.* **29**, 289–298.

Pan, T. M., T. K. Wang, C. L. Lee *et al.* (1997). Food-borne disease outbreaks due to bacteria in Taiwan, 1986–1995. *J. Clin. Microbiol.* **35**, 1260–1262.

Park, C. E., D. Warburton and P. J. Laffey (1996a). A collaborative study on the detection of staphylococcal enterotoxins in foods by an enzyme immunoassay kit (RIDASCREEN). *Intl J. Food Microbiol.* **29**, 281–295.

Park, C. E., D. Warburton and P. J. Laffey (1996b). A collaborative study on the detection of staphylococcal enterotoxins in foods with the enzyme immunoassay kit (TECRA). *J. Food Prot.* **59**, 390–397.

Peavy, D. L., W. H. Adler and R. T. Smith (1970). The mitogenic effects of endotoxin and staphylococcal enterotoxin B on mouse spleen cells and human peripheral lymphocytes. *J. Immunol.* **106**, 1453–1458.

Pereira, J. L., S. P. Salzberg and M. S. Bergdoll (1982). Effect of temperature, pH and sodium chloride concentration on production of enterotoxins A and B. *J. Food Prot.* **45**, 1306–1309.

Pereira, M. L., L. S. do Carmo, E. J. dos Santos and M. S. Bergdoll (1994). Staphylococcal food poisoning from cream-filled cake in a metropolitan area of South-Eastern Brazil. *Rev. Saude Publica* **28**, 406–409.

Personne, P., M. Bes, G. Lina *et al.* (1997). Comparative performances of six agglutination kits assessed by using typical and atypical strains of *Staphylococcaus aureus. J. Clin. Microbiol.* **35**, 1138–1140.

Read, R. B. and J. G. Bradshaw (1966). Staphylococcal enterotoxin B thermal inactivation in milk. *J. Dairy Sci.* **49**, 202–203.

Read, R. B. and J. G. Bradshaw (1967). Gamma irradiation of staphylococcal enterotoxin B. *Appl. Microbiol.* **15**, 603–605.

Reiser, R. F., R. N. Robbins, A. L. Noleto *et al.* (1984). Identification, purification, and some physicochemical properties of staphylococcal enterotoxin C_3. *Infect. Immun.* **45**, 625–630.

Reiser, R. F. and K. F. Weiss (1969). Production of staphylococcal enterotoxins A, B, and C in various media. *Appl. Microbiol.* **18**, 1041–1043.

Ren, K., J. D. Bannan, V. Pancholi *et al.* (1994). Characterization and biological properties of a new staphylococcal exotoxin. *J. Exp. Med.* **180**, 1675–1683.

Robbins, R., S. Gould and M. Bergdoll (1974). Detecting the enterotoxigenicity of *Staphylococcus aureus* strains. *Appl. Microbiol.* **28**, 946–950.

Roberson, J. R., L. K. Fox, D. D. Hancock and T. E. Besser (1992). Evaluation of methods for differentiation of coagulase-positive staphylococci. *J. Clin. Microbiol.* **30**, 3217–3219.

Rose, S. A., N. K. Modi, H. S. Tranter *et al.* (1988). Studies on the irradiation of toxins *Clostridium botulinum* and *Staphylococcus aureus. J. Appl. Bacteriol.* **65**, 223–229.

Sabioni, J. G., E. Y. Hirooka and M. L. R. de Souza (1988). Intoxiçao alimentar por queijo Minas contaminado com *Staphylococcus aureus. Rev. Saude Publica* **22**, 458–461.

Saunders, G. C. and M. L. Bartlett (1977). Double-antibody solid-phase enzyme immunoassay for the detection of staphylococcal enterotoxin A. *Appl. Environ. Microbiol.* **34**, 518–522.

Schantz, E. J., W. G. Roessler, J. Wagman *et al.* (1965). Purification of staphylococcal enterotoxin B. *Biochemistry* **4**, 1011–1016.

Scheusner, D. L., L. L. Hood and L. G. Harmon (1973). Effect of temperature and pH on growth and enterotoxin production by *Staphylococcus aureus. J. Milk Food Technol.* **36**, 249–252.

Schwabe, M., S. Notermans, R. Boot *et al.* (1990). Inactivation of staphylococcal enterotoxins by heat and reactivaton by high pH treatment. *Intl J. Food Microbiol.* **10**, 33–42.

Segalove, M. (1947). The effect of penicillin on growth and toxin production by enterotoxic staphylococci. *J. Infect. Dis.* **81**, 228–243.

Sharma, N. K., C. E. Rees and C. E. Dodd (2000). Development of a single-reaction multiplex PCR toxin typing assay for *Staphylococcus aureus* strains. *Appl. Environ. Microbiol.* **66**, 1347–1353.

Simkovicova, M. and R. J. Gilbert (1971). Serological detection of enterotoxin from food poisoning strains of *Staphylococcus aureus*. *J. Med. Microbiol.* **4**, 19–30.

Smith, B. G. and H. M. Johnson (1975). The effect of staphylococcal enterotoxins on the primary in vitro immune response. *J. Immunol.* **115**, 575–578.

Soo, H. M., S. R. Tatini and R. W. Bennett (1974). Thermal inactivation of staphylococcal enterotoxins A and D in certain foods. *Abstr. Annu. Meet. Am. Soc. Microbiol.*, 14.

Spero, L., J. R. Warren and J. F. Metzger (1973). Effect of single peptide bond scission by trypsin on the structure and activity staphylococcal enterotoxin B. *J. Biol. Chem.* **248**, 7289–7294.

Spero, L., B. Y. Griffin, J. L. Middlebrook and J. F. Metzger (1976). Effect of single and double peptide bond scission by trypsin on the structure and activity of staphylococcal enterotoxin C. *J. Biol. Chem.* **251**, 5580–5588.

Stelma, G. N. Jr, J. G. Bradshaw, P. E. Kaufman and D. L. Archer (1980). Thermal inactivation of mitogenic and serological activities of staphylococcal enterotoxin A at 121 °C. *IRCS Med. Sci.: Microbiol., Parasitol., Infect. Dis.* **8**, 629.

Stiffler-Rosenberg, G. and H. Fey (1978). Simple assay for staphylococcal enterotoxins A, B, and C: modification of enzyme-linked immunosorbent assay. *J. Clin. Microbiol.* **8**, 473–479.

Su, Y.-C. and A. C. L. Wong (1995). Identification and purification of a new staphylococcal enterotoxin, H. *Appl. Environ. Microbiol.* **61**, 1438–1443.

Surgalla, M. J., J. L. Kadavy, M. S. Bergdoll and G. M. Dack (1951). Staphylococcal enterotoxin: production methods. *J. Infect. Dis.* **89**, 180–184.

Surgalla, M. J., M. S. Bergdoll and G. M. Dack (1953). Some observations on the assay of staphylococcal enterotoxin by the monkey-feeding test. *J. Lab. Clin. Med.* **41**, 782–788.

Surgalla, M. J., M. S. Bergdoll and G. M. Dack (1954). Staphylococcal enterotoxin. Neutralization by rabbit antiserum. *J. Immunol.* **72**, 398–403.

Talan, D. A., D. Staatz, A. Staatz and G. D. Overturf (1989). Frequency of *Staphylococcus intermedius* as human nasopharyngeal flora. *J. Clin. Microbiol.* **27**, 2393.

Tatini, S. R. (1973). Influence of food environments on growth of *Staphylococcus aureus* and production of various enterotoxins. *J. Milk Food Technol.* **36**, 559–563.

Thompson, N. E., M. S. Bergdoll, R. F. Myer *et al.* (1985). Monoclonal antibodies to the enterotoxins and to the toxic shock syndrome toxin produced by *Staphylococcus aureus*. In: A. J. L. Macario and E. C. de Macario (eds), *Monoclonal Antibodies against Bacteria*, pp. 23–59. Academic Press, Inc., New York, NY.

Thota, F. H., S. R. Tatini and R. W. Bennett (1973). Effect of temperature, pH, and NaCl on production of staphylococcal enterotoxins E and F. *Abstr. Annu. Meet. Am. Soc. Microbiol.*, 1.

Todd, E. C. D. (1996). Worldwide surveillance of foodborne disease: the need to improve. *J. Food Prot.* **59**, 82–92.

Tsen, H.-Y. and T.-R. Chen (1992). Use of the polymerase chain reaction for specific detection of type A, D, and E enterotoxigenic *Staphylococcus aureus* in foods. *Appl. Microbiol. Biotechnol.* **37**, 685–690.

Untermann, F. (1972). Cultivation methods for the production of staphylococcal enterotoxins on a bigger scale. *Zbl. Bakt. Hyg., I. Abt. Orig. A.* **219**, 426–434.

Vandenbosch, L. L., D. Y. C. Fung and M. Widomski (1973). Optimum temperature for enterotoxin production by *Staphylococcus aureus* S-6 and 137 in liquid medium. *Appl. Microbiol.* **25**, 498–500.

White, J., A. Herman, A. M. Pullen *et al.* (1989). The Vβ-specific superantigen staphylococcal enterotoxin B: stimulation of mature T cells and clonal deletion in neonatal mice. *Cell* **56**, 27–35.

Wieneke, A. A. (1991). Comparison of four kits for the detection of staphylococcal enterotoxin in foods from outbreaks of food poisoning. *Intl J. Food Microbiol.* **14**, 305–312.

Wieneke, A. A. (1993). Staphylococcal food poisoning in the United Kingdom, 1969–1990. *Epidemiol. Infect.* **110**, 519–531.

Wieneke, A. A. and R. J. Gilbert (1987). Comparison of four methods for the detection of staphylococcal enterotoxin in foods from outbreaks of food poisoning. *Intl J. Food Microbiol.* **4**, 135–143.

Wilkerson, M., S. McAllister, J. M. Miller *et al.* (1997). Comparison of five agglutination tests for identification of *Staphylococcus aureus*. *J. Clin. Microbiol.* **35**, 48–151.

Wilson, I. G., J. E. Cooper and A. Gilmour (1991). Detection of enterotoxigenic *Staphylococcus aureus* in dried skim milk: use of the Polymerase Chain Reaction for the amplification and detection of staphylococcal enterotoxin genes *ent* B and *sec*+ and thermonuclease gene *nuc. Appl. Environ. Microbiol.* **57**, 1793–1798.

Wilson, I. G., J. E. Cooper and A. Gilmour (1994). Some factors inhibiting amplification of the *Staphylococcus aureus* enterotoxin C_1 (*sec*+) by PCR. *Intl J. Food Microbiol.* **22**, 55–62.

Zhang, S., J. J. Iandolo and G. C. Stewart (1998). The enterotoxin D plasmid of *Staphylococcus aureus* encodes a second enterotoxin determinant (*sej*). *FEMS Microbiol. Lett.* **168**, 227–233.

Bacillus cereus gastroenteritis

15

Heidi Schraft and Mansel W. Griffiths

1 Introduction

Bacillus cereus is very widespread, but the foodborne diseases it causes are relatively mild and may occur at a much higher frequency than indicated by reports. Still, the acute signs and symptoms of these diseases can be quite violent, making the victim incapable of operating complicated machinery.

The incidence of foodborne diseases caused by *B. cereus* seems to be much higher in Canada and especially in Northern Europe than in England, Wales, Japan and the USA. The reason for this is unknown, but part of the difference may due to the frequency with which diagnostic laboratories test for *B. cereus*; in the USA no causative agent is found in about 50 % of investigated foodborne disease outbreaks, and a number of these may well have been caused by *B. cereus*.

2 Historical aspects and contemporary problems

Bacillus species other than *B. anthracis* had long been suspected of causing foodborne diseases when Steinar Hauge in 1955 demonstrated that *B. cereus* growing in vanilla pudding caused a diarrheal syndrome. Even after that it took several years to convince all food safety experts, because initial experiments with volunteers failed to

Foodborne Infections and Intoxications 3e
ISBN-10: 0-12-588365-X
ISBN-13: 978-012-588365-8

produce disease. The emetic syndrome caused by *B. cereus* growing in fried rice was discovered about 15 years later.

The toxins responsible for the two syndromes have been identified, and specific tests for their identification developed. It has also been demonstrated that *B. cereus* spores may have a heat resistance comparable to that of *C. botulinum* spores, and that vegetative growth may occur over a broad range of temperatures, including refrigerator temperature. This raises questions about how to ensure that prepacked, ready-to-eat food is safe to eat.

3 Characteristics of *Bacillus cereus*

Bacillus cereus are Gram-positive, catalase-positive, endospore-forming rods. The genus *Bacillus* is very heterogeneous; its 34 different species are divided into 6 groups (Logan, 1994) based on spore morphology (Table 15.1). Earlier work classified *Bacillus cereus* into one of three groups, the *B. subtilis* group, along with *B. anthracis*, *B. thuringiensis* and *B. mycoides* (Priest, 1993). All four species are very closely related, and recent data for rRNA homology suggest that they be classified into one single species (Ash *et al*., 1991; Lechner *et al*., 1998). *Bacillus anthracis*, the causative agent of anthrax in mammals, is easily distinguished from the other three members of the group based on its susceptibility to penicillin and the absence of hemolysis on sheep blood agar (Table 15.2). The remaining three species are identified based on

Table 15.1 Classification of the Genus *Bacillus*

Bacillus **Group according to**			**Morphology of cells and spores**	**Examples of members**
Parry *et al*. (1983), Priest (1993)	**Logan (1994)**			
I	I	*B. cereus* group	Large cells, ellipsoidal spores that do not swell sporangium	*B. cereus*, *B. anthracis*, *B. thuringiensis*, *B. mycoides*
	II	*B. subtilis* group	Small cells, ellipsoidal spores that do not swell sporangium	*B. subtilis*, *B. amyloliquefaciens*, *B. licheniformis*, *B. pumilus*
II	III	*B. circulans* group	Ellipsoidal spores that swell sporangium	*B. circulans*, *B. macerans*, *B. polymyxa*
	IV	*B. coagulans* group	Thermophilic; ellipsoidal spores that swell sporangium	*B. coagulans*, *B. stearothermophilus*
III	V	*B. sphaericus* group	Spherical spores that swell sporangium	*B. sphaericus*
IV	VI	*B. firmus* group	Similar to Logan's group I, but 16S rDNA sequences not closely related to group I	*B. firmus*

Table 15.2 Characteristics of *B. anthracis, B. cereus, B. mycoides* and *B. thuringiensis*

Species	Colony appearance	Motility	Hemolysis	Penicillin susceptibility	Crystalline parasporal inclusion	Virulence in mice
B. anthracis	White	–	–	+	–	+
B. cereus	White	+	+	–	–	–
B. mycoides	Rhizoid	–	–	–	–	–
B. thuringiensis	White/gray	+	+	–	+	–

motility and formation of crystalline parasporal inclusion bodies (Kramer and Gilbert, 1989; Drobniewski, 1993). However, these properties are encoded on plasmids and may be lost during subculturing of isolates. Therefore, many authors combine *B. cereus*, *B. mycoides* and *B. thuringiensis* into the '*B. cereus* group', and treat all isolates as one entity.

Two additional members of the *B. cereus* group have been proposed recently. Nakamura (1998) characterized a genetically distinct group of *B. mycoides* by measuring DNA relatedness and phenotypic properties, including fatty acid composition, and suggested the introduction of a new species, named *B. pseudomycoides*. Lechner *et al.* (1998) described a sub-group comprising psychrotolerant, but not mesophilic, *B. cereus* strains. Members of this new group, named *B. weihenstephanensis*, grow at 4–7 °C, but not at 43 °C. Introduction of this new species is based on the fact that many foodborne *B. cereus* are capable of growing at refrigeration temperatures and thus pose a special challenge to food microbiologists.

With the exception of *B. anthracis*, which has been suppressed in most areas of the developed world, *B. cereus* and the closely related species are widely distributed in the environment. They are frequently isolated from soil and plants. Dust, soil and raw vegetables can be considered as primary contamination sources for foods.

4 Characteristics of *Bacillus cereus* food poisoning

Two distinct types of illness have been attributed to the consumption of foods contaminated with *B. cereus*:

- The diarrheal syndrome has an incubation time of 4–16 hours, and is manifested as abdominal pain and diarrhea that usually subsides within 12–24 hours
- The emetic syndrome has an incubation time of 1–5 hours, causing nausea and vomiting that lasts for 6–24 hours.

In some *B. cereus* outbreaks, there appears to be a clear overlap of the diarrheal and emetic syndromes (Kramer and Gilbert, 1989). Each syndrome has striking similarities to another foodborne disease – *Clostridium perfringens* toxico-infection and *Staphylococcus aureus* intoxication, respectively (Table 15.3).

Table 15.3 Comparison of symptoms in food poisoning caused by *B. cereus*, *C. perfringens* and *S. aureus*

Pathogen	Incubation period (h)	Predominant symptom	Duration of illness (h)	Type of disease
C. perfringens	8–16	Diarrhea	12–24	Toxico-infection
B. cereus, diarrheal	8–16	Diarrhea	12–24	Toxico-infection
B. cereus, emetic	1–5	Vomiting	12–24	Intoxication
S. aureus	1–5	Vomiting	12–24	Intoxication

4.1 *Bacillus cereus* diarrheal syndrome

4.1.1 The disease

The diarrheal illness caused by *B. cereus* has an incubation time of approximately 4–16 hours, and is manifested by abdominal pain and diarrhea that usually subsides within 12–24 hours. Nausea is sometimes observed, but vomiting is rare. In most cases, the disease is self-limiting and no treatment is necessary. In severe cases, fluid replacement therapy may be indicated. The infective dose is reported as being 10^5–10^7 cells. The characteristics of the syndrome are shown in Table 15.4.

This type of *B. cereus* food poisoning was first described in detail by Hauge (1955), who evaluated an outbreak of gastroenteritis in a hospital that was traced back to vanilla pudding contaminated with high numbers (up to 10^8 CFU/ml) of *B. cereus*. Since that first publication, reports on the incidence of *B. cereus* diarrheal disease have increased worldwide.

4.1.2 Pathogenesis of the diarrheal syndrome

Goepfert *et al.* (1972) and Spira and Goepfert (1972) were the first to relate the mechanism of pathogenicity of *B. cereus* to a possible enterotoxin. Their findings were based on results from the ligated rabbit ileal-loop test (LRIL), in which fluid accumulation in the loop is measured after injection of *B. cereus* cultures or culture supernatants. This test, along with the vascular permeability reaction (VPR), an

Table 15.4 Characteristics of *B. cereus* diarrheal and emetic syndromes

Syndrome	Incubation period (h)	Duration of illness (h)	Infective dose	Symptoms	Foods commonly implicated
Diarrheal	8–16	12–24	10^3–10^7 cells ingested	Abdominal pain, watery diarrhea, occasional nausea	Meat products, soups, milk and milk products, vegetables, puddings and sauces
Emetic	1–5	12–24	10^5–10^8 cells/g of food	Nausea, vomiting, occasional diarrhea	Rice, pasta, noodles, pastries

assay that scores the increase of permeability in veins of rabbits or guinea pigs after intradermal injection of bacterial culture supernatants, is still used as a reference test to evaluate the toxicity of toxin extracts. In addition to these biological assays, cell culture tests using CHO, HEp-2 or McCoy cell lines are now frequently used to test for toxicity of enterotoxin preparations (Buchanan and Schultz, 1992; Jackson, 1993). It was also observed in early experiments that hemolysis and enterotoxin activity are correlated to some degree.

Initial attempts to purify the enterotoxin of *B. cereus* revealed a three-component protein complex that exhibited positive results in LRIL and VPR assays. These early findings were supported by results of Wong and co-workers at the University of Wisconsin, who described a three-component enterotoxin, hemolysin BL (HBL), consisting of three proteins termed B, L_1 and L_2 (Beecher and Wong, 1994, 1997; Ryan *et al.*, 1997). All three components are apparently required to produce maximal fluid accumulation in the LRIL assay. The HBL enterotoxin is very heterogeneous, and some *B. cereus* strains may produce more than one set of the three HBL components (Schoeni and Wong, 1999; Beecher and Wong, 2000). Further research is required to determine the significance of the HBL homologues in pathogenicity.

In contrast, Granum's research group in Norway (Lund and Granum, 1996) has purified and characterized a non-hemolytic enterotoxin (NHE). This toxin is also composed of three protein components which are different from the HBL subunits. Sequencing and gene expression assays show that both HBL and NHE are transcribed from one operon, with maximum enterotoxin activity produced during late exponential or early stationary growth (Granum *et al.*, 1999). Testing for the presence of HBL and NHE in three different *B. cereus* strains indicated that any enterotoxigenic *B. cereus* isolate may produce both toxins, or only one (Lund and Granum, 1997). Agata and co-workers described yet another protein, *B. cereus* enterotoxin T (BceT), which showed positive reactions in VPR and LRIL tests. The protein and the encoding DNA sequence are not related to HBL or NHE (Agata *et al.*, 1996). However, additional research indicated that this protein is not likely to cause diarrheal disease (Choma and Granum, 2002). A fourth possible *B. cereus* enterotoxin (EntFM) has been cloned and sequenced by Asano *et al.* (1997), but information on its biological activity has not been published.

A cytotoxic protein (CytK) implicated in necrotic enteritis has been isolated from a *B. cereus* strain involved in a severe outbreak (three deaths) of *B. cereus* diarrheal disease. The protein was very similar to β-barrel channel-forming toxins found in *S. aureus* and *C. perfringens* (Lund *et al.*, 2000). The *cytK* gene was detected in 73 % of *B. cereus* isolates involved in diarrheal disease, but in only 37 % of isolates from food not implicated in disease (Guinebretiere *et al.*, 2002).

Information about the currently known enterotoxins is summarized in Table 15.5. *B. thuringiensis*, closely related to *B. cereus* and still widely used as insecticide, has been reported to contain *B. cereus* enterotoxins (Jackson *et al.*, 1995; Rusul and Yaacob, 1995; Perani *et al.*, 1998; Rivera *et al.*, 2000). Enterotoxin genes have also been found in mosquito-larvicidal *B. sphaericus* (Yuan *et al.*, 2002). Such findings should be of concern to regulators approving the use of *Bacillus* species as biological pesticides.

Table 15.5 Properties of *B. cereus* toxins

Toxin	Characteristics	Mol. Mass	References
Enterotoxins	Heat-labile protein(s), inactivated by trypsin, pepsin and pronase, not stable at pH < 4		Kramer and Gilbert (1989)
Hemolysin BL (HBL)	3-component protein		Beecher and Wong (1994)
	B-complex, encoded by *hblA* gene	38.0 kDa	
	L_1-complex, encoded by *hblC* gene	38.5 kDa	
	L_2-complex, encoded by *hblD* gene	43.2 kDa	
Non-hemolytic enterotoxin (NHE)	3-component protein		Lund and Granum (1996), Granum *et al.* (1999)
	Protein NheA, encoded by *NheA* gene	41.0 kDa	
	Protein NheB, encoded by *NheB* gene	39.8 kDa	
	Protein NheC, encoded by *NheC* gene	36.5 kDa	
B. cereus enterotoxin (bceT)	Single protein, encoded by *bceT* gene	41 kDa	Agata *et al.* (1996)
B. cereus enterotoxin (entFM)	Single protein, encoded by *entFM* gene	45 kDa	Asano *et al.* (1997), Shinagawa (1990)
B. cereus cytotoxin (CytK)	Single protein, encoded by *cytK* gene	34 kDa	Lund *et al.* (2000)
Emetic toxin	Heat stable, not inactivated by proteolytic enzymes, stable at pH 2		Kramer and Gilbert (1989)
Cereulide	Cyclic dodecadepsipeptide, coding gene(s) unknown	1.2 kDa	Agata *et al.* (1995)

It has long been thought that the *B. cereus* diarrheal syndrome was a classical intoxication, due to the ingestion of toxin produced during growth of *B. cereus* in the food. However, it has been postulated that the disease is caused by ingested *B. cereus* cells that grow and produce enterotoxin within the intestinal tract of the patient – i.e. by a toxico-infection (Granum, 1994; Granum and Lund, 1997). This claim is based on three observations:

- Most strains produce enterotoxin in significant amounts only after reaching cell concentrations of 10^7/ml, while the infectious dose in many foodborne outbreaks has been found to be 10^3–10^4/ml
- The activity of enterotoxin exposed to a pH of 3.1 for 20 minutes is reduced by 80 %, and the toxin is completely destroyed within 20 minutes of further exposure to trypsin and chymotrypsin
- The incubation time of 12–24 hours is too long for simple enterotoxin action, but several *B. cereus* isolates are able to grow well under anaerobic conditions at 37 °C and produce significant levels of enterotoxin within 6 hours.

However, the above studies were conducted using enterotoxin produced in brain heart infusion broth, and it should be noted that enterotoxin produced and ingested in a food matrix may be protected from damage through low pH and enzymes.

It may well be possible that both modes of pathogenesis exist, especially in light of the existence of several diarrheagenic enterotoxins. In particular, a toxico-infection might be occurring in a newly observed, more severe form of *B. cereus* diarrheagenic syndrome. Granum (1997) has described two outbreaks where people were suffering from more severe symptoms than those normally associated with the diarrheagenic syndrome. In the first outbreak, 17 of 24 people were affected after eating stews and 3 of the patients were hospitalized, 1 for 3 weeks. The infective dose observed in this outbreak was 10^4–10^5 cells, somewhat lower than that usually associated with *B. cereus*. The time to onset of symptoms for these patients was more than 24 hours – also longer than commonly observed for *B. cereus*. The second outbreak affected competitors at a skiing event in Norway, where 152 out of 252 people became ill after drinking milk. The illness was restricted to young skiers between the ages of 16 and 19, while the older coaches and officials did not exhibit symptoms. Again, the incubation period was greater than 24 hours in some cases, and the patients were ill for periods of between 2 and several days. Granum and co-workers postulated that this more severe form of *B. cereus* gastroenteritis results from adhesion of spores to the epithelial cells followed by enterotoxin production within the intestinal tract. The longer incubation time seen in these cases may be due to the time required for germination of the spores.

Work by Andersson *et al.* (1998a) may support this new virulence mechanism: the authors have shown that spores of *B. cereus* strains with high hydrophobicity adhere significantly better to CaCo-2 cells than do spores with low hydrophobicity. After adhesion, the spores were able to germinate and produce enterotoxins.

Only limited studies have been carried out on the mode of action of the enterotoxin. It reverses absorption of fluid, Na^+ and Cl^- by epithelial cells, and causes malabsorption of glucose and amino acids, as well as necrosis and mucosal damage. It has been suggested that the effects on fluid absorption are due to stimulation of adenylate cyclase (Kramer and Gilbert, 1989).

4.2 *Bacillus cereus* emetic syndrome

4.2.1 The disease

A second type of *B. cereus* gastroenteritis, the emetic syndrome, was identified in the 1970s, associated with the consumption of fried rice (Kramer and Gilbert, 1989). The emetic illness has an incubation time of 1–5 hours, and is manifested by nausea and vomiting that lasts for 6–24 hours. Diarrhea is observed only occasionally. The symptoms are usually self-limiting, and treatment is seldom necessary. To transmit this type of *B. cereus* food poisoning, the food involved will typically contain 10^5–10^8 cells/g (Table 15.4). Emetic *B. cereus* may also cause more severe episodes of disease: a 17-year-old boy exhibited *B. cereus* emetic syndrome and died of liver failure. His father suffered hyperbilirubinemia and rhabdomyolysis, but he recovered (Mahler *et al.*, 1997).

4.2.2 Pathogenesis of the emetic syndrome

Early experiments, relying on monkey-feeding trials, identified the cause of the emetic syndrome as a toxin because cell-free supernatants produced the same symptoms as cultures in which the cells remained (Kramer and Gilbert, 1989). The toxin was found to be a peptide with a molecular weight of around 10 000 Da, stable at pH 2–11, and able to withstand heating at 121 °C for 90 minutes, as well as treatment with trypsin and pepsin (Table 15.5). The toxin is not antigenic. However, investigations on the mechanism of pathogenicity of the emetic syndrome have been limited because the nature of the toxin remained unknown until 1995, when Agata *et al.* isolated the emetic toxin and named it cereulide. It is a dodecapepsipeptide, consisting of a ring structure, which is closely related to the potassium ionophore valinomycin. Its characterization and the development of a pathogenicity test based on vacuolation of HEp-2 cells have allowed further elucidation of the pathogenicity of the emetic syndrome. It is now known that cereulide stimulates the vagus afferent by binding to 5-HT_3 receptors, and induces swelling of mitochondria with toxic effects due to potassium-ionophoretic properties (Sakurai *et al.*, 1994; Agata *et al.*, 1995; Mikkola *et al.*, 1999). The toxin also inhibits natural killer cells *in vitro* (Paananen *et al.*, 2002). A rapid bioassay which tests the toxicity of cereulide to boar spermatozoa (Andersson *et al.*, 1998b) and a chemical assay which is based on high performance liquid chromatography and mass spectroscopy (HPLC-MS; Haeggblom *et al.*, 2002) are now available for further clarification of pathogenicity and incidence of the *B. cereus* emetic syndrome.

4.3 Incidence and transmission through food

The mild and transient character of the *B. cereus* foodborne diseases likely results in the disease being vastly under reported. Reports on foodborne disease outbreaks caused by *B. cereus* are summarized in Table 15.6. None of the reports listed

Table 15.6 Reports of foodborne disease outbreaks caused by *B. cereus*

Country	Period	Number of outbreaks with identified causative agent	Prevalence of *B. cereus* (%)	Total *B. cereus*	Reference
Canada	Mean from 2–5 years	100.2/year	14/year	14	Todd, 1996
Denmark	1985–1992	211	14	6.6	WHO, 1992
Finland	1988–1992	101	12	11.9	WHO, 1992
Germany	1989–1992	350	2	0.57	WHO, 1992
Iceland	1985–1993	27	7	25.9	WHO, 1992
Japan	Mean from 2–5 years	236.5/year	10.5/year	4.4	Todd, 1996
Netherlands	1985–1991	191	39	20.4	WHO, 1992
Netherlands	1991–1994	217	42	19.0	Simone *et al.*, 1997

Table 15.6 Reports of foodborne disease outbreaks caused by *B. cereus*—cont'd

Country	Period	Number of outbreaks with identified causative agent	Prevalence of *B. cereus* (%)	Total *B. cereus*	Reference
Norway	1988–1993	NA	NA	33	Granum and Lund, 1997
Spain	1985–1992	4367	11	0.3	WHO, 1992
Sweden	1990–1992	60	1	1.7	WHO, 1992
UK: England and Wales	1989–1993	1101	68	6.2	WHO, 1992
UK: Scotland	1989–1993	1435	17	1.2	WHO, 1992
USA	1988–1992	1001	21	2.1	Bean *et al.* 1996

distinguishes between emetic and diarrheal syndromes, but it is more likely that the outbreaks reflect the diarrheal syndrome. The incidence of *B. cereus* foodborne diseases seems to be higher in the countries of Northern Europe (20–33 %) and in Canada (14 %). England and Wales, Japan and the USA (1.2–4.4 %) have a clearly lower incidence. The reasons for these differences are not known. Foods most often implicated as vehicles for transmission of the diarrheal syndrome include meat and meat dishes, vegetables, cream, spices, poultry and eggs. The emetic syndrome is most frequently associated with starchy foods, especially rice dishes (Kramer and Gilbert, 1989).

5 Presence, growth and survival of *B. cereus* in foods

5.1 Presence of *B. cereus* in foods

Members of the *B. cereus* group are ubiquitously distributed in the environment, mainly because of their spore-forming capabilities. Thus *B. cereus* can easily contaminate various types of foods, especially products of plant origin. The organism is also frequently isolated from milk and dairy products, meat and meat products, pasteurized liquid egg, rice, ready-to-eat vegetables, and spices. Based on the ubiquitous distribution of the organism, it is virtually impossible to obtain raw products that are free from *B. cereus* spores.

The appearance of psychrotrophic strains in the dairy industry has added a new dimension to *B. cereus* surveillance in food. Studies indicate that both raw and pasteurized milk will harbor psychrotrophic *B. cereus*, with a prevalence of 9–37 % in raw milk and 2–35 % in pasteurized milk (te Giffel *et al.*, 1995). With an average generation time of 17 hours at 6 °C, *B. cereus* may produce enterotoxin during extended storage at slightly unfavorable temperatures (Griffiths, 1990; Griffiths and Philips, 1990). For example, Odumeru *et al.* (1997) found *B. cereus* enterotoxin in 38 % of pasteurized milk samples stored at 10 °C until their expiry date. Feijoo *et al.* (1997) reported rapid growth of *B. cereus* in cream, reaching population levels high enough for toxin production within 11 hours at 23 °C. Furthermore, reports by Becker *et al.* (1994) and Rowan and Anderson (1998) indicate a relatively high prevalence of enterotoxigenic *B. cereus* in milk-based infant formulas, and Hatakka (1998) found *B. cereus* to be the most common pathogen in hot meals served by airlines between 1991 and 1994. The latter two findings are rather alarming, as two commodities with traditionally strict microbiological standards (baby food and airline meals) seem to be quite frequently contaminated with enterotoxigenic *B. cereus*.

5.2 Growth and survival

An excellent overview of factors affecting the growth and survival of *B. cereus* can be found in the monograph *Microorganisms in Foods 5*, published by the International Committee on Microbiological Specifications of Foods (ICMSF) in 1996. The most important aspects are summarized below.

Under ideal conditions, *B. cereus* has an optimum growth temperature of 30–40 °C, with growth possible between 4° and 55 °C. Strains that grow at 7 °C or below are identified as psychrotrophic and will not grow above 43 °C. Mesophilic *B. cereus* usually grow at temperatures of 15–50°or 55 °C. The optimum pH for *B. cereus* growth has been reported as 6.0–7.0, with generation times of approximately 23 minutes at 30 °C. The minimum pH for growth is 5.0 and the maximum is 8.8. In the presence of NaCl as a humectant, *B. cereus* will not grow at a_w of 0.93. However, when glycerol was used as a humectant, growth was possible at a_w of 0.93, but not 0.92. Only few reports are available regarding the effects of organic acids and chemical preservatives on *B. cereus*. In rice filling held at 23 °C, growth was inhibited by 0.26 % sorbic acid and by 0.39 % potassium sorbate. Butylated hydroxyanisole (BHA) at concentrations of 0.1–0.5 % inhibited *B. cereus* growth at 32 °C for over 24 hours.

5.3 Heat resistance and germination of spores

Effective destruction of *B. cereus* spores is the ultimate goal for safe foods with an extended shelf-life. A $D_{121.1}$ value of 0.03–2.35 minutes (z = 7.9–9.9 °C) has been reported for spores suspended in phosphate buffer at pH 7.0. For spores suspended in milk, the D_{95} value ranged from 1.8 to 19.1 minutes. A similar wide range (2.7–15.3 minutes) was observed for spores in milk-based infant formula. In general, spores of psychrotrophic *B. cereus* are less heat resistant than are spores of mesophilic strains (Notermans and Batt, 1998).

Germination of spores requires the presence of purine ribosides and glycine or a neutral L-amino acid. L-alanine is the most effective amino acid at stimulating germination. Germination temperatures vary greatly between strains, and are strongly influenced by the substratum. In laboratory media, germination has been observed at –1–59 °C; in cooked rice, germination temperatures ranged from 5–50 °C (Kramer and Gilbert, 1989).

6 Isolation and identification

6.1 Cultural methods

Both emetic and diarrheagenic *B. cereus* have a relatively high infective dose, and detection of the organism is achieved primarily by direct plating onto selective agar media. Two selective media are widely used for the isolation and presumptive identification of *B. cereus*: mannitol–egg yolk–polymyxin (MYP) agar, and polymyxin–pyruvate–egg yolk–mannitol–bromothymol blue agar (PEMBA). Both rely on the presence of lecithinase (phospholipase C) and the absence of mannitol fermentation in *B. cereus*. Presumptive colonies are surrounded by an egg-yolk precipitate and have a violet-red background on MYP and a turquoise to peacock blue color on PEMBA. Both media contain polymyxin to inhibit the growth of competitive organisms (van Netten and Kramer, 1992). A medium that requires no polymyxin, VRM, has been described. Tris and resazurin are included in VRM to inhibit growth of *B. megaterium* and many other competitive bacteria present in foods (DeVasconellos and Rabinovitch, 1995). Peng *et al.* (2001) developed the BCM® *B. cereus*/*B. thuringiensis* chromogenic agar plating medium, which is based on the detection of the PI-PLC enzyme present in *B. cereus*.

If enrichment is required for detection of low numbers in foods, a nutrient rich broth such as brain heart infusion broth, or trypticase soy broth supplemented with polymyxin (1000 IU/l), is recommended (van Netten and Kramer, 1992).

A method for the detection of *B. cereus* spores in milk was developed by Christiansson *et al.* (1997). The procedure includes a heat treatment at 72 °C, and incubation with trypsin and Triton X-100 at 55 °C. The samples are filtered through a membrane filter to concentrate the spores. The filter is then placed onto solid agar medium to allow for growth of the spores.

Presumptive colonies isolated from selective media are characterized by various confirmatory tests: *B. cereus* and closely related species will be Gram-positive, catalase-positive rods with spores that do not swell the sporangium. They will produce acid from glucose anaerobically, and will show a positive reaction for nitrate reduction, Voges-Proskauer, L-tyrosine degradation, and growth in 0.001 % lysozyme (Harmon *et al.*, 1992). A biochemical *Bacillus* identification system, the API 50 CHB test, is available from bioMérieux. The test determines the ability of an isolate to assimilate 49 carbohydrates, and has become a reference method for the biochemical identification of *Bacillus* species.

6.2 Non-cultural detection methods

With the advent of rapid microbiological methods for the detection of bacteria in foods, several culture-independent approaches have been proposed.

6.2.1 Polymerase chain reaction

The polymerase chain reaction (PCR) has been used by several authors for the detection of *B. cereus*. One group of these assays aims at detecting *B. cereus per se*, while other assays target specific sub-groups of *B. cereus*, such as psychrotrophic or enterotoxigenic strains.

Schraft and Griffiths (1995) reported a sensitive test for the detection of *B. cereus* in milk, which is based on primers targeting the phospholipase gene of *B. cereus*. This gene encodes the positive egg-yolk reaction used for the presumptive identification of the organism on selective plating media. The authors used a DNA purification procedure that allowed the detection of less than 1 CFU/ml milk without a pre-enrichment step. Damgaard *et al.* (1996) combined magnetic capture hybridization (MCH) with a nested PCR assay targeting the phospholipase gene of *B. thuringiensis* to detect *B. cereus* and *B. thuringiensis* in naturally contaminated rhizosphere samples. Although no lower detection limits were reported, this MCH-PCR assay might be useful for the detection of *B. cereus* in foods. A PCR-based assay for the detection of all members from the *B. cereus* group, which targets 16s rDNA, has been developed by Hansen *et al.* (2001).

Yamada *et al.* (1999) have cloned and sequenced the gyrase B gene (*gyrB*) of *B. cereus* and designed a primer set for a PCR assay. The assay was evaluated with experimentally inoculated rice homogenates. A detection limit of < 1 CFU/g could be achieved when a 15-hour pre-enrichment and two-step filtration were applied before the actual PCR amplification.

PCR protocols have been described that would allow the selective detection of psychrotrophic *B. cereus*. The primers used in the assay target the psychrotrophic and mesophilic rDNA signature sequences (von Stetten *et al.*, 1998) and gene sequences for the major *B. cereus* cold shock proteins cspF and cspA (Francis *et al.*, 1998). However, it has been shown that *B. cereus* without these signature sequences can grow at low temperatures (Stenfors and Granum, 2001).

Several PCR assays have been described for the detection of enterotoxigenic *B. cereus*. These tests are based on the gene sequence of the toxin to be detected. Since there are several different proteins or protein complexes apparently capable of causing the *B. cereus* diarrheal syndrome (HBL, NHE, bceTt, entFM and CytK; see sections 2 and 4), it would be most prudent simultaneously to target all potential enterotoxin genes: a protocol for the detection of the HBL and bceT toxin complex was reported by Mäntynen and Lindström (1998). The DNA sequences for NHE and for CytK have been determined, and sequences for PCR primers published (Granum *et al.*, 1999, Guinebretiere *et al.*, 2002).

To date, the gene(s) responsible for the production of the emetic toxin (cereulide) have not been identified. Therefore, no PCR test for emetic *B. cereus* is available. However, there is evidence that emetic and non-emetic strains can be differentiated by randomly amplified polymorphic DNA typing (RAPD; Schraft *et al.*, 1996a).

6.2.2 Antibodies against spores and vegetative cells

Koo *et al.* (1998) have engineered a monoclonal single-chain antibody that is reported to be specific for *B. cereus* T spores. The antibody was fused with a streptavidin molecule, and the fusion protein expressed in *Escherichia coli*. This antibody–streptavidin protein was then used in a magnetic bead-based immunoassay to concentrate *B. cereus* spores from liquid samples. When samples were inoculated with 5×10^4 *B. cereus* spores, the assay was capable of removing more than 90 % of the spores from phosphate buffer and 37 % from whole milk. Charni *et al.* (2000) reported the production of monoclonal antibodies that reacted only with *B. cereus* vegetative cells, but not with spores.

6.3 Typing methods

Fast and accurate identification methods for *B. cereus* are also important in the food industry. Odumeru *et al.* (1999) have evaluated two automated microbial identification systems for their ability to identify, accurately and reproducibly, 40 *B. cereus* isolates. The API 50 CHB biochemical test was used as reference test. The fatty-acid based Microbial Identification System (MIS; MIDI Inc., Newark, DE) had a slightly better (55 %) sensitivity (i.e. proportion of reference positive strains correctly identified) than the biochemical VITEC system (42.5 %; bioMérieux Vitek, Hazelwood, MO). When the definition of a correct identification was expanded to include closely related members of the *B. cereus* group, e.g. *B. mycoides* and *B. thuringiensis*, the sensitivity of the systems increased to 82.5 % and 67.5 % respectively. The specificity, i.e. proportion of reference negative strains not identified as *B. cereus*, was very good (97.5 %) for both systems.

When *B. cereus* is repeatedly isolated in finished products or suspected of causing a foodborne disease outbreak, methods for tracing possible contamination sources throughout the production chain are indispensable. Traditionally, biochemical profiles, serology, phage typing and fatty-acid analysis have been used for typing. Phage typing and fatty-acid profiling have been most applied in more recent studies. Väisänen *et al.* (1991) found phage typing to be useful in excluding packaging materials as potential source of *B. cereus* in dairy products, and Ahmed *et al.* (1995) described a 12-phage typing scheme as a valuable epidemiological typing tool for the identification of sources of *B. cereus* food poisoning. Since phage typing requires continuous phage propagation, it has not found widespread application in the food industry.

Fatty-acid analysis can now be performed with a fully automated system, the MIS system described above for identification. Work by Schraft *et al.* (1996b) and Pirttijarvi *et al.* (1998) has shown that whole-cell fatty-acid analysis can be used to trace and identify *B. cereus* strains colonizing the processing environment of a given dairy plant.

In addition to phenotypic typing, genotypic methods are frequently used for epidemiological typing of *B. cereus*. Randomly amplified polymorphic DNA (RAPD) is one such technique which was applied by Ronimus *et al.* (1997) to type over 2000 mesophilic and thermophilic *Bacillus* species isolated from an industrial setting. The authors were able to trace individual species from the feedstock to the finished product in a food-processing plant. RAPD typing also allowed the separation of mesophilic from psychrotrophic *B. cereus* isolates (Lechner *et al.*, 1998). Amplified fragment length polymorphism (AFLP) is a genotyping method with significant higher reproducibility than RAPD. Ripabelli *et al.* (2000) evaluated AFLP for epidemiological typing of *B. cereus* with 21 cultures from 7 different outbreaks. All isolates could be typed with high reproducibility, and a unique AFLP pattern was generated for isolates of each outbreak. The introduction of a rapid DNA preparation method for *B. cereus* will facilitate large-scale typing of *B. cereus* with either RAPD or AFLP (Nilsson *et al.*, 1998).

An automated ribotyping instrument, the RiboPrinter, has been developed by Qualicon Inc. Andersson *et al.* (1999) compared RAPD typing with automated ribotyping, and found the latter to be a fast and reliable method that was only slightly less discriminatory than the more labor-intensive RAPD technique.

6.4 Detection of toxins

Two commercial kits are available for the detection of *B. cereus* diarrheagenic toxins. Both kits are immunoassays, but they detect different antigens. The reverse passive latex agglutination (RPLA) enterotoxin assay produced by Oxoid reacts with a 43-kDa protein, the subunit L_2, of the tripartite enterotoxin complex (HBL) described by Beecher and Wong (1994). The second assay uses the enzyme-linked immunosorbent assay format (ELISA). This kit is marketed under the name *Bacillus* Diarrhoeal Enterotoxin (BDE) Visual Immunoassay kit (TECRA). It will detect a 41-kDa protein (NheA) that is part of the non-hemolytic enterotoxin complex (NHE) described by Lund and Granum (1996). Any enterotoxigenic *B. cereus* strain may produce either HBL or NHE, or both. Thus it is not surprising that early comparisons of the two commercial test kits have triggered contradictory reports on their accuracy in detecting the enterotoxin. Based on today's knowledge about *B. cereus* enterotoxins, any isolate that reacts positively with one or both of the commercial kits should be considered enterotoxigenic (Granum and Lund, 1997). Final confirmation of enterotoxigenicity of an isolate can be obtained by cytotoxicity tests using Vero cells, CHO cells or human embryonic lung cells.

The bioassay using boar spermatozoa and the HPLC-MS procedure described in section 2 are the only tests reported for the detection of the emetic toxin Cereulide.

7 Outbreaks caused by *Bacillus* spp. other than *B. cereus*

Bacillus species that do not belong to the *B. cereus* group have been reported to cause foodborne disease (Kramer and Gilbert, 1989). *B. subtilis* has caused vomiting within a few hours of ingesting the incriminated food (meat, seafood with rice, and pastry). High numbers of *B. subtilis* (10^7 CFU/g) were isolated from the patients' vomitus. *B. licheniformis* has caused diarrheal illness with an incubation time of about 8 hours. High numbers (10^7 CFU/g) of *B. licheniformis* were isolated from the implicated foods (meat, poultry or vegetable pies, and stews).

Production of toxins by non-*B. cereus* isolates has been confirmed. The *B. cereus* enterotoxin RPLA assay (specific for the L_2 subunit of the HBL enterotoxin) was positive for isolates of *B. licheniformis, B. subtilis, B. circulans* and *B. megaterium*, indicating that these *Bacillus* species may potentially cause foodborne diarrheagenic disease (Rowan *et al.*, 2001). Inhibition of boar sperm motility, an indicator for *B. cereus* emetic toxin, was demonstrated for approximately 50 % of *B. licheniformis* isolated from food (Salkinoja-Salonen *et al.*, 1999).

8 Treatment and prevention

The symptoms of *B. cereus* infection are usually mild and self-limiting, and would not normally require treatment.

Because *B. cereus* can almost invariably be isolated from foods and can survive extended storage in dried food products, it is not practicable to eliminate low numbers of spores from foods. Control against food poisoning should be directed at preventing the germination of spores and minimizing the growth of vegetative cells. To accomplish this, foods should be rapidly and efficiently cooled to less than 7 °C or maintained above 60 °C, and thoroughly reheated before serving.

Bibliography

Agata, N., M. Otha, M. Mori and M. Isobe (1995). A novel dodecadepsipeptide, cereulide, is an emetic toxin of *Bacillus cereus. FEMS Microbiol. Lett.* **129**, 17–20.

Agata, N., M. Otha, Y. Arakawa and M. Mori (1996). The *bceT* gene of *Bacillus cereus* encodes an enterotoxic protein. *Microbiology* **141**, 983–988.

Ahmed, R., P. Sankar-Mistry, S. Jackson *et al.* (1995). *Bacillus cereus* phage typing as an epidemiological tool in outbreaks of food poisoning. *J. Clin. Microbiol.* **33**, 636–640.

Andersson, A., P. E. Granum and U. Ronner (1998a). The adhesion of *Bacillus cereus* spores to epithelial cells might be an additional virulence mechanism. *Intl J. Food Microbiol.* **39**, 93–99.

Andersson, M. A., R. Mikkola, J. Helin *et al.* (1998b). A novel sensitive bioassay for detection of *Bacillus cereus* emetic toxin and related depsipeptide ionophores. *Appl. Environ. Microbiol.* **64**, 1338–1343.

Andersson, A., B. Svensson, A. Christiansson and U. Ronner (1999). Comparison between automatic ribotyping and random amplified polymorphic DNA analysis of *Bacillus cereus* isolates from the dairy industry. *Intl J. Food Microbiol.* **47**, 147–151.

Asano, S., Y. Nukumizu, H. Bando *et al.* (1997). Cloning of novel enterotoxin genes from *Bacillus cereus* and *Bacillus thuringiensis. Appl. Environ. Microbiol.* **63**, 1054–1057.

Ash, C., J. A. E. Farrow, M. Dorsch *et al.* (1991). Comparative analysis of *Bacillus anthracis, Bacillus cereus* and related species on the basis of reverse transcriptase sequencing of 16S rRNA. *Intl J. Syst. Bacteriol.* **41**, 343–346.

Bean, N. H., J. S. Goulding, L. Christopher and F. J. Angulo (1996). Surveillance of foodborne disease outbreaks – United States, 1988–1992. *Morbid. Mortal. Weekly Rep.* **45**, 1–67.

Becker, H. G. Schaller, W. von Wiese and G. Terplan (1994). *Bacillus cereus* in infant foods and dried milk products. *Intl J. Food Microbiol.* **23**, 1–15.

Beecher, D. J. and A. C. L. Wong (1994). Improved purification and characterisation of hemolysin BL, a hemolytic dermonecrotic vascular permeability factor from *Bacillus cereus. Infect. Immun.* **62**, 980–986.

Beecher, D. J. and A. C. L. Wong (1997). Tripartite hemolysin BL from *Bacillus cereus*. Hemolytic analysis of component interaction and model for its characteristic paradoxical zone phenomenon. *J. Biol. Chem.* **272**, 233–239.

Beecher, D. J. and A. C. L. Wong (2000). Tripartite haemolysin BL: isolation and characterization of two distinct homologous sets of components from a single *Bacillus cereus* isolate. *Microbiology* **146**, 1371–1380.

Buchanan, R. L. and F. J. Schultz (1992). Evaluation of the Oxoid BCERLPA kit for the detection of *Bacillus cereus* diarrheal enterotoxin as compared to cell culture cytotonicity. *J. Food Prot.* **55**, 440–443.

Charni, N., C. Perissol, J. Le Petit and N. Rugani (2000). Production and characterization of monoclonal antibodies against vegetative cells of *Bacillus cereus. Appl. Environ. Microbiol.* **66**, 2278–2281.

Choma, C. and P. E. Granum (2002). The enterotoxin T (BcET) from *Bacillus cereus* can probably not contribute to food poisoning. *FEMS Microbiol. Lett.* **217**, 115–119.

Christiansson, A., K. Ekelund and H. Ogura (1997). Membrane filtration method for enumeration and isolation of spores of *Bacillus cereus* from milk. *Intl Dairy J.* **7**, 743–748.

Damgaard, P. H., C. S. Jacobsen and J. Sørensen (1996). Development and application of a primer set for specific detection of *Bacillus thuringiensis* and *Bacillus cereus* in soil using magnetic capture hybridization and PCR amplification. *Syst. Appl. Microbiol.* **19**, 436–441.

DeVasconellos, F. J. M. and L. Rabinovitch (1995). A new formula for an alternative culture medium, without antibiotics, for isolation and presumptive quantification of *Bacillus cereus* in foods. *J. Food Prot*. **58**, 235–238.

Drobniewski, F. A. (1993). *Bacillus cereus* and related species. *Clin. Microbiol. Rev*. **6**, 324–338.

Feijoo, S. C., L. N. Cotton, C. E. Watson and J. H. Martin (1997). Effect of storage temperatures and ingredients on growth of *Bacillus cereus* in coffee creamers. *J. Dairy Sci.* **80**, 1546–1553.

Francis, K. P., R. Mayr, F. von Stetten *et al.* (1998). Discrimination of psychrotrophic and mesophilic strains of the *Bacillus cereus* group by PCR targeting major cold shock protein genes. *Appl. Environ. Microbiol*. **64**, 3525–3529.

Goepfert, J. M., W. M. Spira and H. U. Kim (1972). *Bacillus cereus*: food poisoning organism. A review. *J. Milk Food Technol*. **35**, 213–227.

Granum, P. E. (1994). *Bacillus cereus* and its toxins. *J. Appl. Bacteriol. Symp*. **Suppl. 76**, 61S–66S.

Granum, P. E. (1997). *Bacillus cereus*. In: M. P. Doyle, L. R. Beuchat and T. J. Montville (eds), *Food Microbiology: Fundamentals and Frontiers*, pp. 327–336. ASM Press, Washington, DC.

Granum, P. E. and T. Lund (1997). *Bacillus cereus* and its food poisoning toxins. *FEMS Microbiol. Lett*. **157**, 223–228.

Granum, P. E., K. O'Sullivan and T. Lund (1999). The sequence of the non-hemolytic enterotoxin operon from *Bacillus cereus*. *FEMS Microbiol. Lett*. **177**, 225–229.

Griffiths, M. W. (1990). Toxin production by psychrotrophic *Bacillus* spp. present in milk. *J. Food Prot*. **53**, 790–792.

Griffiths, M. W. and J. D. Phillips (1990). Incidence, source and some properties of psychrotrophic *Bacillus* spp. found in raw and pasteurized milk. *J. Soc. Dairy Technol.* **43**, 62–66.

Guinebretiere, M. H., V. Broussolle and C. The-Nguyen (2002). Enterotoxigenic profiles of food-poisoning and food-borne *Bacillus cereus* strains. *J. Clin. Microbiol.* **40**, 3053–3056.

Haeggblom, M. M., C. Apetroaie, M. A. Andersson and M. S. Salkinoja-Salonen (2002). Quantitative analysis of cereulide, the emetic toxin of *Bacillus cereus*, produced under various conditions. *Appl. Environ. Microbiol.* **68**, 2479–2483.

Hansen, B. M., T. D. Leser and N. B. Hendriksen (2001). Polymerase chain reaction assay for the detection of *Bacillus cereus* group cells. *FEMS Microbiol. Lett.* **202**, 209–213.

Harmon, S. M., D. A. Kautter, D. A. Golden and E. J. Rhodehamel (1992). *Bacillus cereus*. In: *FDA Bacteriological Analytical Manual*, 7th edn, pp. 191–207. AOAC International, Arlington, VA.

Hattakka, M. (1998). Microbiological quality of hot meals served by airlines. *J. Food Prot.* **61**, 1052–1056.

Hauge, S. (1955). Food poisoning caused by aerobic spore forming bacilli. *J. Appl. Bacteriol.* **18**, 591–595.

ICMSF (International Commission on Microbiological Specifications for Foods) (1996). *Microorganisms in Foods 5: Microbiological Specifications of Food Pathogens*. Chapman and Hall, London.

Jackson, S. G. (1993). Rapid screening test for enterotoxin-producing *Bacillus cereus*. *J. Clin. Microbiol.* **31**, 972–974.

Jackson, S. G., R. B. Goodbrand, R. Ahmed and S. Kasataya (1995). *Bacillus cereus* and *Bacillus thuringiensis* isolated in a gastroenteritis outbreak investigation. *Lett. Appl. Microbiol.* **32**, 103–105.

Koo, K., P. M. Foegeding and H. E. Swaisgood (1998). Development of a streptavidin conjugated single-chain antibody that binds *Bacillus cereus* spores. *Appl. Environ. Microbiol.* **64**, 2497–2502.

Kramer, J. M. and R. J. Gilbert (1989). *Bacillus cereus* and other *Bacillus* species. In: M. P. Doyle (ed.), *Foodborne Bacterial Pathogens*, pp. 21–70. Marcel Dekker, New York, NY.

Lechner, S., R. Mayr, K. P. Francis *et al.* (1998). *Bacillus weihenstephanensis* sp. nov. is a new psychrotolerant species of the *Bacillus cereus* group. *Intl J. Syst. Bacteriol.* **48**, 1373–1382.

Logan, N. A. (1994). *Bacterial Systematics*. Blackwell Scientific Publications, Osney Mead.

Lund, T., M. De Buyser and P. E. Granum (2000). A new cytotoxin from *Bacillus cereus* that may cause necrotic enteritis. *Mol. Microbiol.* **38**, 254–261.

Lund, T. and P. E. Granum (1996). Characterisation of a non-hemolytic enterotoxin complex from *Bacillus cereus* isolated after a foodborne outbreak. *FEMS Microbiol. Lett.* **141**, 151–156.

Lund, T. and P. E. Granum (1997). Comparison of biological effect of two different enterotoxin complexes isolated from three different strains of *Bacillus cereus*. *Microbiology* **143**, 3329–3339.

Mahler, H., A. Pasi, J. M. Kramer *et al.* (1997). Fulminant liver failure in association with the emetic toxin of *Bacillus cereus. N. Engl. J. Med.* **336**, 1143–1148.

Mäntynen, V. and K. Lindström (1998). A rapid PCR-based test for enterotoxic *Bacillus cereus. Appl. Environ. Microbiol.* **64**, 1634–1639.

Mikkola, R., N. E. L. Saris, P. A. Grigoriev *et al.* (1999). Ionophoretic properties and mitochondrial effects of cereulide – the emetic toxin of *Bacillus cereus. Eur. J. Biochem.* **263**, 112–117.

Nakamura, L. K. (1998). *Bacillus pseudomycoides* sp. nov. *Intl J. Syst. Bacteriol.* **48**, 1031–1035.

Nilsson, J., B. Svensson, K. Ekelund and A. Christiansson (1998). A RAPD-PCR method for large-scale typing of *Bacillus cereus. Lett. Appl. Microbiol.* **27**, 168–172.

Notermans, S. and C. A. Batt (1998). A risk assessment approach for food-borne *Bacillus cereus* and its toxins. *J. Appl. Microbiol.* **84** (Suppl.), 51S–61S.

Odumeru, J. A., A. K. Toner, C. A. Muckle *et al.* (1997). Detection of *Bacillus cereus* diarrheal enterotoxin in raw and pasteurized milk. *J. Food Prot.* **60**, 1391–1393.

Odumeru, J. A., M. Steele, L. Fruhner *et al.* (1999). Evaluation of accuracy and repeatability of identification of food borne pathogens by automated bacterial identification systems. *J. Clin. Microbiol.* **37**, 944–949.

Paananen, A., R. Mikkola, T. Sareneva *et al.* (2002). Inhibition of natural killer cell activity by cereulide, an emetic toxin from *Bacillus cereus. Immunology* **129**, 420–428.

Parry, J. M., P. C. B. Turnbull and J. R. Gibson (1983). *A Colour Atlas of Bacillus Species* (Series Volume 19). Wolfe Medical Atlases, London.

Peng, H., V. Ford, E. W. Framton *et al.* (2001). Isolation and enumeration of *Bacillus cereus* from foods on a novel chromogenic plating medium. *Food Microbiol.* **18**, 231–238.

Perani, M., A. H. Bishop and A. Vaid (1998). Prevalence of beta-exotoxin, diarrhoeal toxin and specific delta-endotoxin in natural isolates of *Bacillus thuringiensis. FEMS Microbiol. Lett.* **160**, 55–60.

Pirttijarvi, T. S. M., L. M. Ahonen, L. M. Maunuksela and M. S. Salkinoja Salonen (1998). *Bacillus cereus* in a whey process. *Intl J. Food Microbiol.* **44**, 31–41.

Priest, F. G. (1993). Systematics and ecology of *Bacillus*. In: A. L. Sonenshein, J. A. Hoch and R. Losick (eds), Bacillus subtilis *and Other Gram-positive Bacteria*, pp. 3–16. American Society for Microbiology, Washington, DC.

Ripabelli, G., J. McLauchlin, V. Mithani and E. J. Threlfall (2000). Epidemiological typing of *Bacillus cereus* by amplified fragment length polymorphism. *Lett. Appl. Microbiol.* **30**, 358–363.

Rivera, A. M. G., P. E. Granum and F. G. Priest (2000). Common occurrence of enterotoxin genes and enterotoxicity in *Bacillus thuringiensis. FEMS Microbiol. Lett.* **190**, 151–155.

Ronimus, R. S., L. E. Parker and H. W. Morgan (1997). The utilization of RAPD-PCR for identifying thermophilic and mesophilic *Bacillus species. FEMS Microbiol. Lett.* **147**, 75–79.

Rowan, N. J. and J. G. Anderson (1998). Diarrhoeal enterotoxin production by psychrotrophic *Bacillus cereus* present in reconstituted milk-based infant formulae (MIF). *Lett. Appl. Microbiol.* **26**, 161–165.

Rowan, N. J., K. Deans, J. G. Anderson *et al.* (2001). Putative virulence factor expression by clinical and food isolates of *Bacillus* spp. after growth in reconstituted infant milk formulae. *Appl. Environ. Microbiol.* **67**, 3873–3881.

Rusul, G. and N. H. Yaacob (1995). Prevalence of *Bacillus cereus* in selected foods and detection of enterotoxin using TECRA-VIA and BCET-RPLA. *Intl J. Food Microbiol.* **25**, 131–139.

Ryan, P. A., J. D. MacMillan and B. A. Zilinskas (1997). Molecular cloning and characterization of the genes encoding the L-1 and L-2 components of hemolysin BL from *Bacillus cereus. J. Bacteriol.* **179**, 2551–2556.

Sakurai, N., K. Kolke, Y. Irie and H. Hayashi (1994). The rice culture filtrate of *Bacillus cereus* isolated from emetic type food poisoning causes mitochondrial swelling in a HEp-2 cell. *Microbiol. Immunol.* **38**, 337–343.

Salkinoja-Salonen, M. S., R. Vuorio, M. A. Andersson *et al.* (1999). Toxigenic strains of *Bacillus licheniformis* related to food poisoning. *Appl. Environ. Microbiol.* **65**, 4637–4645.

Schoeni, J. L. and A. C. L. Wong (1999). Heterogeneity observed in the components of hemolysin BL, an enterotoxin produced by *Bacillus cereus. Intl J. Food Microbiol.* **53**, 159–167.

Schraft, H. and M. W. Griffiths (1995). Specific oligonucleotide primers for the detection of lecithinase positive *Bacillus* spp. by PCR. *Appl. Environ. Microbiol.* **61**, 98–102.

Schraft, H., S. Miller-Piper, A. Buechin and M. W. Griffiths (1996a). Identification of DNA sequences specific to emetic *B. cereus. Proceedings of the 96th General Meeting of the American Society of Microbiologists, New Orleans, May 19–23*, Abstract No. P-86.

Schraft, H., M. Steele, B. McNab *et al.* (1996b). Epidemiological typing of *Bacillus* spp. isolated from food. *Appl. Environ. Microbiol.* **27**, 4229–4232.

Shinagawa, K. (1990). Purification and characterization of *Bacillus cereus* enterotoxin and its application to diagnosis. In: A. E. Pohland, V. R. Dowell Jr and J. L. Richard (eds), *Microbial Toxins in Foods and Feeds*, pp. 181–193.. Plenum Press, New York, NY.

Simone, E., M. Goosen, S. H. Notermans and M. W. Borgdorff (1997). Investigations of foodborne diseases by food inspection services in the Netherlands. *J. Food Prot.* **60**, 442–446.

Spira, W. M. and J. M. Goepfert (1972). *Bacillus cereus* induced fluid accumulation in rabbit ileal loops. *Appl. Microbiol.* **24**, 341–348.

Stenfors, L. and P. E. Granum (2001). Psychrotolerant species from the *Bacillus cereus* group are not necessarily *Bacillus weihenstephanensis. FEMS Microbiol. Lett.* **197**, 223–228.

te Giffel, M. C., R. R. Beumer, B. A. Slaghuis and F. M. Rombouts (1995). Occurrence and characterization of (psychrotrophic) *Bacillus cereus* on farms in the Netherlands. *Neth. Milk Dairy J.* **49**, 125–138.

Todd, E. C. D. (1996). Worldwide surveillance of foodborne disease, the need to improve. *J. Food Prot.* **59**, 82–92.

Väisänen, O. M., N. J. Mwaisumo and M. S. Salkinoja-Salonen (1991). Differentiation of dairy strains of the *Bacillus cereus* group by phage typing, minimum growth temperature, and fatty acid analysis. *J. Appl. Bacteriol.* **70**, 315–324.

van Netten, P. and J. M. Kramer (1992). Media for the detection and enumeration of *Bacillus cereus* in foods: a review. *Intl J. Food Microbiol.* **17**, 85–99.

von Stetten, F., K. P. Francis, S. Lechner *et al.* (1998). Rapid discrimination of psychrotolerant and mesophilic strains of the *Bacillus cereus* group by PCR targeting 16s rDNA. *J. Microbiol. Methods* **34**, 99–106.

WHO Surveillance Programme (1992). *Fifth Report of WHO Surveillance Programme for Control of Foodborne Infections and Intoxications in Europe, 1985–1989*. Institute for Veterinary Medicine, Berlin.

Yamada, S., E. Ohashi, N. Agata and K. Venkatesvaran (1999). Cloning and nucleotide sequence analysis of gyrB of *Bacillus cereus, B. thuringiensis, B. mycoides*, and *B. anthracis* and their application to the detection of *B. cereus* in rice. *Appl. Environ. Microbiol.* **65**, 1483–1490.

Yuan, Z. M., B. M. Hansen, L. Andrup and J. Eilenberg (2002). Detection of enterotoxin genes in mosquito-larvicidal *Bacillus* species. *Curr. Microbiol.* **45**, 221–225.

Mycotoxins and alimentary mycotoxicoses

16

Fun S. Chu

1 Introduction

Mycotoxin is a convenient generic term describing the toxic substances formed during the growth of fungi. 'Myco' means fungal (mold), and 'toxin' represents poison. In contrast to the bacterial toxins, which are mainly proteins with antigenic properties, the mycotoxins encompass a considerable variety of low molecular weight compounds with diverse chemical structures and biological activities. Like most microbial secondary metabolites, the functions of mycotoxins for the fungi themselves are still not clearly defined. In considering the effect of mycotoxins on the animal's body, it is important to distinguish between mycotoxicosis and mycosis. Mycotoxicosis is used, in general, to describe the action of mycotoxin(s), and is frequently mediated through a number of organs, notably the liver, kidney and lungs, and the nervous, endocrine and immune systems. On the other hand, mycosis refers to a generalized invasion of living tissue(s) by growing fungi. Mycotoxins and mycotoxicoses are an especially significant problem for human and animal health, because under certain conditions crops and foodstuffs can provide a favorable medium for fungus growth and toxin production. Because of the relative stability of mycotoxins to heat and other

Foodborne Infections and Intoxications 3e
ISBN-10: 0-12-588365-X
ISBN-13: 978-012-588365-8

treatments, they may remain in foods and feeds for a long period. Mycotoxins have caused great economic losses because not only is there an outright loss of crops and animals when a severe outbreak occurs; there are also a number of unseen losses, such as declines in production of milk and eggs. Toxin residue problems introduce effects on animal product quality, increased susceptibility to infection, refusal to eat by animals, manifestations of impaired nutritional status, decline in reproductive success, and the costs of controlling the problems (Vasanthi and Bhat, 1998; CAST, 2003; Chu and Bhatnagar, 2004).

The mycotoxin problem is actually an old one. Ergotism and mushroom poisoning, for example, have been known for centuries. Outbreaks of other types of toxicoses associated with the ingestion of moldy foods and feeds by humans and animals have also been recorded in the last century (Wilson and Hayes, 1973). A well documented example is the outbreak of a disease called alimentary toxic aleukia (ATA) that resulted in more than 5000 deaths in humans in the Orenberg district of the USSR during World War II (Joffe, 1974). The cause of this outbreak was later determined to be trichothecene mycotoxins produced by fungi growing on grain allowed to stand in the field during winter. Since the discovery in the early 1960s of aflatoxins, highly potent carcinogens produced by *Aspergillus flavus* and *A. parasiticus*, research has focused new attention on mycotoxins (Goldblatt, 1969; Busby and Wogan, 1981; Cole and Cox, 1981; Eaton and Groopman, 1994). Developments in the last four decades have disclosed many new fungal poisons that are attracting attention because of their association with foods and animal feeds, and because of their diverse toxic effects (Ciegler *et al.*, 1971; Rodricks *et al.*, 1977; Willie and Morehouse, 1977; Mirocha *et al.*, 1979; Cole and Cox, 1981; Shank, 1981; Hamilton, 1982; Richard and Thurston, 1986; Steyn and Vleggaar, 1986; Natori *et al.*, 1989; Sharma and Salunkhe, 1991; Bhatnagar *et al.*, 1992; Coulombe, 1993; Hui *et al.*, 1994; Miller and Treholm, 1994; Miller, 1995; Smith *et al.*, 1995; Steyn, 1995; D'Mello and MacDonald, 1997; Chu, 1998, 2000, 2002, 2003; Moss, 1998; Scudamore and Livesey, 1998; Sinha and Bhatnagar, 1998; Vasanthi and Bhat, 1998; D'Mello *et al.*, 1999; DeVries *et al.*, 2002; CAST, 2003). It is beyond the scope of this chapter to review all literature on this subject. Instead, the discussion will be focused on selected mycotoxins that most frequently contaminate our foods, their toxic effects and potential hazards to humans and animals, and their modes of action, as well as possible preventive measures.

2 Mycotoxins produced by toxigenic fungi

Invasion of fungi and production of mycotoxins in commodities can occur under favorable conditions in the field (pre-harvest), at harvest, and during processing, transportation and storage. Fungi that are frequently found in the field include *Aspergillus flavus*, *Alternaria longipes*, *Alternaria alternata*, *Claviceps purpurea*, *Fusarium moniliforme*, *F. graminearum*, and a number of other *Fusarium* spp. Species most likely introduced at harvest include *F. sporotrichioides*, *Stachybotrys atra*,

Cladosporium sp., *Myrothecium verrucaria, Trichothecium roseum*, as well as *A. alternata*. Most penicillia are storage fungi. These include *Penicillium citrinum, P. cyclopium, P. citreoviride, P. islandicum, P. rubrum, P. viridicatum, P. urticae, P. verruculosum, P. palitans, P. puberulum, P. expansum* and *P. roqueforti*, all of which are capable of producing mycotoxins in grains and foods. Other toxicogenic storage fungi are: *A. parasiticus, A. flavus, A. versicolor, A. ochraceus, A. clavatus, A. fumigatus, A. rubrum, A. chevallieri, F. moniliforme, F. tricinctum, F. nivale* and several other *Fusarium* spp. Thus, most of the mycotoxin-producing fungi belong to three genera, namely *Aspergillus, Fusarium* and *Penicillium* (Hesseltine, 1976). However, not all species in these genera are toxicogenic. Genetics, environmental and nutritional factors, time of incubation, etc., greatly affect the formation of mycotoxins. In the field, weather conditions, plant stress, invertebrate vectors, species and spore load of infective fungi, variations within plant and fungal species, and microbial competition all play important roles in the formation of mycotoxins. During storage and transportation, water activity (a_w), temperature, crop damage, time, blending with moldy components, and a number of chemical factors – such as aeration (O_2, CO_2 levels), types of grains, pH, lack of specific nutrients, presence of inhibitors, minerals, and chemical treatment – are important. In general, mold growth in the grains or foods is necessary before subsequent onset of toxin production, and optimal conditions for toxin formation generally have a narrower window than those for mold growth. For example, the optimal temperatures and a_w for the growth of *A. flavus* and *A. parasiticus* are around 35–37 °C (range 6–54 °C) and 0.95 (range 0.78–1.0), respectively; for aflatoxin production they are 28–33 °C and 0.90–0.95 (0.83–0.97), respectively (Lacey, 1989). The production of fumonisin by *F. moniliforme* on maize is another good example. Fumonisin levels only decreased threefold when a_w was lowered from 1.0 to 0.95, with no change in fungal growth. However, a 300-fold reduction in fumonisin production was found when a_w was lowered from 1.0 to 0.90, with only a 20 % decrease in fungal growth (Cahagenier *et al.*, 1995).

3 Natural occurrence and toxic effects of selected mycotoxins

Some of the most frequent naturally occurring mycotoxins, the toxin-producing fungi, and their major toxic effects in human and animals are shown in Table 16.1. It is apparent that whereas some mycotoxins, e.g. aflatoxin (AF), are produced only by a few species of fungi within a genus, others, e.g. ochratoxin A (OA), are produced by fungi across several genera. A number of commodities can be contaminated with different types of mycotoxins. In general, most mycotoxins affect specific organs, but due to diverse structures, the toxic effects of trichothecenes (TCTCs) affect many organs. Among the mycotoxins, AF, OA, fumonisin (Fm), deoxynivalenol (DON) and several other TCTCs, and some others are the most frequent contaminants in foods and feeds. Only these mycotoxins will be discussed in some detail.

Table 16.1 Natural occurrence of selected common mycotoxins

Mycotoxins	Major producing fungi	Typical substrate in nature
Aflatoxin B_1(AF) and other aflatoxins	*Aspergillus flavus*, *A. parasiticus*	Peanuts, corn, cottonseed, cereals, figs, most tree nuts, milk, sorghum, walnuts
Alternaria (AAL)[a] mycotoxins	*Alternaria alternata*	Cereal grains, tomato, animal feeds
Citrinin (CT)	*Penicillium citrinum*	Barley, corn, rice, walnuts
Cyclochlorotine (CC)	*P. islandicum*	Rice
Cyclopiazonic acid (CPA)	*A. flavus*, *P. cyclopium*	Peanuts, corn, cheese
Deoxynivalenol (DON)	*Fusarium graminearum*	Wheat, corn
Fumonisins (FM)	*F. moniliforme*	Corn, sorghum, rice
Luteoskyrin (LT)	*P. islandicum*, *P. rugulosum*	Rice, sorghum
Moniliformin (MN)	*F. moniliforme*	Corn
Ochratoxin A (OA)	*A. ochraceus*, *P. viridicatum*, *P. verrucosum*, etc.	Barley, beans, cereals, coffee, feeds, maize, oats, rice, rye, wheat
Patulin (PT)	*P. patulum*, *P. urticae*, *A. clavatus*, etc.	Apple, apple juice, beans, wheat
Penicillic acid (PA)	*P. puberulum*, *A. ochraceus*, etc.	Barley, corn
Penitrem A (PNT)	*P. palitans*	Feedstuffs, corn
Roquefortine (RQF)	*P. roqueforti*	Cheese
Rubratoxin B (RB)	*P. rubrum*, *P. purpurogenum*	Corn, soybeans
Sterigmatocystin (ST)	*A. versicolor*, *A. nidulans*, etc.	Corn, grains, cheese
T-2 toxin	*F. sporotrichioides*	Corn, feeds, hay
Zearalenone (ZE)	*F. graminearum*, etc.	Cereals, corn, feeds, rice
12-13, Epoxy-trichothecenes (TCTC) other than T-2 & DON	*F. nivale*, etc.	Corn, feeds, hay, peanuts, rice

From Chu (2002).

[a] Letters in the parenthesis are the abbreviations for the mycotoxin's name used throughout the text. The optimal temperatures for the production of mycotoxin are generally between 24°C and 28°C, except for T-2 toxin, which is generally produced maximally at 15°C.

3.1 Aflatoxin and aflatoxicoses

3.1.1 General considerations

Aflatoxins have chemical structures containing dihydrofuranofuran and tetrahydrofuran fused with a substituted coumarin. At least 16 different structurally related toxins have been found (Goldblatt, 1969). The toxins are primarily produced by *A. flavus* and *A. parasiticus* in a number of important agricultural commodities in the field and during storage. In addition, some *A. nominus* and *A. tamarii* strains are also AF producers (Goto *et al.*, 1996). In view of the significantly different genetics and morphology for the AF-producing isolates of *A. tamarii* from those of non-AF producers, a species name of *A. pseudotamarii* sp. nov. was given (Ito *et al.*, 2001). Because the four major toxins were originally isolated from fungal cultures of *A. flavus*, the first few letters of the fungus name (**A** and **fla**) were used to coin the toxin name. These toxins fluoresce either blue or green under UV light, and this distinguishes the **B** or **G** types of toxins. AFB1 is most toxic in this group, and is one of the most potent naturally occurring carcinogens; other AFs are less toxic (B1 > G1 > B2 > G2). Because AFB1 frequently contaminates several major commodities, extensive studies have been done on its toxicity, and biological and biochemical effects (Busby and Wogan, 1979, 1981; Eaton and Groopman, 1994; CAST, 2003). Consumption of AFB1-contaminated feed by dairy cows results in the shedding of AFM1 in milk. AFM1, a hydroxylated metabolite of AFB1, is about 10 times less toxic than AFB1, but its presence in milk is of concern for human health (Cullen *et al.*, 1987; Van Egmond, 1989a; Galvano *et al.*, 1996).

The main target organ of AF is the liver. Typical symptoms for aflatoxicoses in animals include proliferation of the bile duct, centrilobular necrosis and fatty infiltration of the liver, and hepatomas in addition to generalized hepatic lesions. The susceptibility of animals to AFB1 varies considerably with species, with a decrease in sensitivity (LD_{50}: mg/kg) in the following order: rabbits (0.3), ducklings (0.34), mink (0.5–0.6), cats (0.55), pigs (6–7-kg size, 0.6), trout (0.8), dogs (1.0), guinea pigs (1.4–2.0), sheep (2.0), monkeys (2.2), chickens (6.3), rats (5.5 for male, 17.9 female), mice (9.0) and hamsters (10.2). In addition to these acute (hepatotoxic) effects, carcinogenic effects of AFB1 are of the most concern (Eaton and Groopman, 1994). For example, a 10 % incidence of hepatocarcinoma was observed in male Fisher rats fed a diet containing only 1 μg AFB1/kg over a period of 2 years (Newberne and Rogers, 1981). Rats, rainbow trout, monkeys and ducks are most susceptible, but mice are relatively resistant to AFB1. Because of their presence in foods and strong evidence of their association with human carcinogenesis, aflatoxins are still a serious threat to human health even after more than 40 years of research (Eaton and Groopman, 1994; Wogan, 2000; CAST, 2003). Although the liver is the main target, AFB1 also affects other organs and tissues (Heinonen *et al.*, 1996; Massey, 1996; Kelly *et al.*, 1997), including the respiratory system, to certain degree.

3.1.2 Biosynthesis of aflatoxin

Although the biosynthetic pathway for AFB was postulated more than a decade ago, the major genes and enzymes involved in the biosynthesis have only very recently been identified and characterized. Starting with polyketide precursors such as acetate, at least 23 enzymatic steps have been identified (Dutton, 1988; Bhatnagar *et al.*, 1992, 1995; Bennett *et al.*, 1994; Trail *et al.*, 1995; Townsend, 1997). A simplified pathway, including the genes (*italicized*), enzymes (underlined), or both, involved in the formation of the products, is summarized as follows: Acetate → polyketide (*pks*/polyketide synthase; fas-1, fas-2/fatty acid synthase 1 and 2) → norsolorinic acid → averatin (*nor-1*/*reductase*) → averufanin (*avnA*/monooxygenase) → averufin → versiconal hemiacetal acetate (*avf1*) → versiconal (esterase) → versicolorin B (*vbs*/verB synthase) → versicolorin A → sterigmatocystin (*ver-1*/dehydrogenase) (ST) → *O*-methylsterigmatocystin (*omtA*/O-methyl-ST-transferase) → AFB_1 (*cyp450*/oxidoreductase). The genes for almost all the key enzymes and a regulatory gene (*aflR*), which encodes Gal4-type 47-kD protein, AFLR, that has been shown to be required for transcriptional expression of structural genes, have been cloned and characterized (Feng *et al.*, 1992; Payne *et al.*, 1993; Woloshuk *et al.*, 1994; Bhatnagar *et al.*, 1995; Chang *et al.*, 1995a, 1995b, 1996; Feng and Leonard, 1995; Trail *et al.*, 1995; Yu *et al.*, 1995, 1996; Cary *et al.*, 1996; Prieto and Woloshuk, 1997; Townsend, 1997; Payne and Brown, 1998). A defect in aflR expression in the koji mold *A. sojae* causes turn-off of the expression of AF genes that produce AF (Matsushima *et al.*, 2001). Investigation of the DNA sequences and restriction enzyme mapping of cosmid and phage libraries of *A. flavus* and *A. parasiticus* have further shown that these genes are clustered within an approximately 75-kb region of the fungal genome (Trail *et al.*, 1995; Yu *et al.*, 1995; Brown *et al.*, 1996; Keller and Hohn, 1997; Townsend, 1997; Klich *et al.*, 2000; Sidhu, 2002). G-protein signaling has recently been found to be involved in regulation of the biosynthesis of AF and sterigmatocystin (ST) (Tag *et al.*, 2000; Calvo *et al.*, 2002). The finding of the involvement of the same G-protein signal pathway for both ST production and sporulation of the fungus suggests that the production of fungal secondary metabolites and sporulation are closely related (Calvo *et al.*, 2002).

3.1.3 Aflatoxin in human foods

Aflatoxins have been found in corn, peanuts and peanut products, cotton seeds, peppers, rice, pistachios, tree nuts (Brazil nuts, almonds, pecans), pumpkin seeds, sunflower seeds and other oil seeds, copra, spices, dried fruits (figs, raisins) and yams (Vrabcheva, 2000; Bassa *et al.*, 2001; Reddy *et al.*, 2001; CAST, 2003). Among these products, frequent contamination with high levels of AF in peanuts, corn and cottonseed, mostly due to infestation with mold in the field, are of the most concern (Widstrom, 1996). Soybeans, beans, pulses, cassava, grain sorghum, millet, wheat, oats, barley and rice are resistant or only moderately susceptible to AF contamination in the field (Jelinek *et al.*, 1989; Wood, 1992; CAST, 2003). It should be reiterated that resistance to aflatoxin contamination in the field does not guarantee that the commodities are free of AF contamination during storage. Inadequate storage conditions, such as high moisture and warm temperatures (25–30 °C), can create conditions favorable for the growth of fungus and production of AF. High levels of AFB1 have been reported in some lots of rice, cassava, figs, spices, pecans and other nuts (Jelinek *et al.*, 1989; Wood, 1992; Rustom, 1997).

The potential hazard of AFs to human health has led to worldwide monitoring programs for the toxin in various commodities, as well as regulatory actions by nearly all countries. Levels varying from zero tolerance to 50 ppb have been set for total AFs. Most countries, including the US, have a regulatory level around 20 ppb in foods. However, considerably lower limits for AF in foods (2.0 ppb for AFB and 4.0 ppb for total AFs) were established by the European Economic Community in 1999 (Van Egmond, 2002). For AFM1 in dairy products, a level between zero tolerance and 0.5 ppb has been used (Van Egmond, 1989a, 1989b, 2002). To avoid contamination of milk and other dairy products with AFM, rigorous programs regulating AFB1 in feed have also been established. Most governments set a lower tolerance level for AFs in the feed for dairy cows. In the US the limit level of AFs in feed for dairy cows is 20 ppb; but for other animals going into the food chain the limits are between 100 and 300 ppb.

3.1.4 Aflatoxin and human carcinogenesis

Whereas AFB1 has been found to be a potent carcinogen in many animal species (Busby and Wogan, 1981; Wogan, 1992; Smela *et al.*, 2001), the role of AF in carcinogenesis in humans is complicated by hepatitis B virus (HBV) infections in humans

(Peers *et al.*, 1987; Hsieh, 1989; Wild *et al.*, 1992). Epidemiological studies have shown a strong positive correlation between AF levels in the diet and primary hepatocellular carcinoma (PHC) incidence in some parts of the world, including certain regions of the People's Republic of China, Kenya, Mozambique, the Philippines, Swaziland, Thailand, and the Transkei of South Africa (Zhu *et al.*, 1987; Wogan, 1992; Eaton and Groopman, 1994; Wild and Hall, 2000). Aflatoxin–DNA and AF–albumin adducts, as well as several AF metabolites, mainly AFM1, have been detected in serum, milk and urine of humans in these regions (Groopman *et al.*, 1994). However, the prevalence of HBV infection is also correlated to liver cancer incidence in these regions. Since multiple factors are considered to be important in carcinogenesis (Harris and Sun, 1986; Wogan, 2000), environmental contaminants such as AFs and other mycotoxins may, either in combination with HBV or independently, be important etiological factors. Several recent studies indicated that both heavy exposure to AF and high HBV infection are important factors (Chen *et al.*, 1996a, 1996b), and enhancement of mutation of the *p53* gene suggest the synergistic effect of these two risk factors for PHC in humans (Jackson *et al.*, 2001; Smela *et al.*, 2001).

3.2 Ochratoxin and ochratoxicoses

3.2.1 General considerations

Ochratoxins, a group of dihydroisocoumarin-containing mycotoxins, are produced by a number of fungi in the genera *Aspergillus* and *Penicillium*, including *A. sulphureus, A. sclerotiorum, A. melleus, A. ochraceus, A. awamori, P. viridicatum, P. palitans, P. commune, P. variabile, P. purpurescens, P. cyclopium* and *P. chrysogenum*. *Aspergillus ochraceus* and *P. viridicatum*, two species that were first reported as ochratoxin A (OA) producers, occur frequently in nature. Because of its distinct chemotype, *P. viridicatum* has been reclassified as *P. verrucosum* (Chu, 1974; Kuiper-Goodman and Scott, 1989; Pohland *et al.*, 1992). Other fungi, such as *Petromyces alliceus* (isolated from onion), *A. citricus* and *A. fonsecaeus* (of the *A. niger* group), have also been found to produce OA (Abarca *et al.*, 1994; Ono *et al.*, 1995). The discovery of the capability of *A. niger* for OA production led to the suggestion that all the *A. niger* should be screened for OA before its use in industrial fermentation (Schuster *et al.*, 2002). Because most OA producers are storage fungi, pre-harvest fungal infection and OA production are not serious problems; the toxins are generally produced in grains during storage in temperate regions. Although most OA producers can grow in a range of 4–37 °C and at an a_w as low as 0.78, optimal conditions for toxin production are narrower, with temperature of 24–25 °C and a_w values > 0.97 (minimum a_w for OA production is about 0.85). Worldwide, OA occurs primarily in cereal grains (barley, oats, rye, corn, wheat) and mixed feeds, and levels higher than 1 ppm have been reported. OA has been found in other commodities, including beans, coffee, fruit juices, nuts, olives, cheese, fish, pork, milk powder, peppers, wine, beer and bread (Pohland *et al.*, 1992; Van Egmond and Speijers, 1994; Studer-Rohr *et al.*, 1995; Filali *et al.*, 2001; Pietri *et al.*, 2001; Thirumala *et al.*, 2001; Petzinger and Weidenbach, 2002). The presence of OA residues in animal products is of concern because it binds tightly to serum albumin and has a long half-life in animal tissues and body fluids. Thus, OA can be carried through the food chain; and the natural occurrence of OA in kidneys, blood serum, blood sausage and other sausage made from pork has been reported.

3.2.2 Toxic effects

OA, the most toxic member (LD_{50} ~20–25 mg/kg in rats) and also most commonly found toxin in this group, has been found to be a potent nephrotoxin that causes kidney damage, including degeneration of the proximal tubule, in many animal species (Chu, 1974; Kuiper-Goodman and Scott, 1989; Fink-Gremmels *et al.*, 1995; Simon, 1996). Liver necrosis and enteritis were also observed in these animals. Other than acute toxic effects, OA also acts as an immunosuppressor (Boorman, 1988; Boorman *et al.*, 1992; Simon, 1996) and a teratogen in test animals (Hayes, 1981). Although OA has not been shown to be mutagenic in early studies, mutagenicity of OA was demonstrated using cell lines with stable human cytochrome P-450 enzymes (deGroene *et al.*, 1996). A weak genotoxic effect has also been demonstrated in several systems (Dirheimer, 1996; Ehrlich *et al.*, 2002). Ochratoxin A is considered to be a weak nephro-carcinogen because a high level of toxin and an extended period of exposure are necessary to induce the tumors. A dose-related induction of renal tubular cell tumors was found in Fisher rats (F344/N), with a significant increase in renal tubular cell tumors at levels of 70 and 210 μg OA/kg body weight per day, but cancer incidence was not significantly different from the control at a lower level of 21 μg OA/kg per day (Boorman, 1988; Schlatter *et al.*, 1996). Studies show that OA also causes liver cancer in rats.

3.2.3 OA and human health

Although the role of OA in human pathogenesis at present is still speculative, it has long been considered to be associated with the nephropathy of people residing in certain Scandinavian and Balkan regions, and more recently in Tunisia (Krogh, 1976; Petkova-Bocharova and Castegnaro, 1985; Maaroufi *et al.*, 1996), where exposure may be endemic. The pathological lesions of nephropathy in humans are similar to those observed in endemic porcine nephropathy, which is caused by the consumption of feeds contaminated with OA (Krogh, 1976). Since a high proportion of the patients suffering from endemic nephropathy in the Balkan regions develop tumors of the renal pelvis and ureter, the possible involvement of nephropathy in tumor development was suggested (Petkova-Bocharova and Castegnaro, 1985; Petkova-Bocharova *et al.*, 1988). Ochratoxin A has been found in human serum in Tunisia and in a number European countries, including Bulgaria, Poland, the former Yugoslavia, and Germany. In certain areas of the Balkans and Tunisia, OA levels in the food and in human serum in endemic regions are higher than in the non-endemic areas. OA also has been found in human milk and kidneys in some endemic regions.

Human exposure to OA could occur through consumption of OA-containing cereals or foods of animal origin in which the animals were fed with OA-contaminated feeds. Reports from a workshop on the impact of OA on human health (see a series of papers published in *Food Additives and Contaminants*, **14** (Suppl. 1–3), 1996) show that the mean daily intake of OA by humans in European Union member countries was 1.8 ng of OA/kg body weight (range 0.7–4.7) as calculated from data on OA levels in foods, and 0.9 ng/kg (range 0.2–2.4) when calculated from the levels of OA in human blood. The main dietary sources (55 %) were cereals and cereal products (0.2 to 1.6 μg OA/kg). Coffee (mean level 0.8 μg/kg; Stegen *et al.*, 1997), beer, pig meat, blood products and pulses also contribute to the OA intake in humans. These findings re-emphasize the possible involvement of OA in human carcinogenesis and the health hazard of OA exposure in humans. Among 77 countries which have regulations for different mycotoxins, 8 have specific regulations for OA, with limits ranging from 1 to 20 μg/kg in different foods.

3.3 Fumonisin

3.3.1 General considerations

Fumonisins (Fm) are a group of toxic metabolites produced primarily by *Fusarium verticillioides* (previously *F. moniliforme*), one of the most common fungi colonizing corn throughout the world (Riley and Richard, 1992; Riley *et al.*, 1993a; Scott, 1993; Dutton, 1996; Jackson *et al.*, 1996; Shier, 2000; ApSimon, 2001; Marasas, 2001; Rheeder *et al.*, 2002). More than 11 structurally related Fms (B1, B2, B3, B4, C1, C4, A2, A2, etc.) have been found since the discovery of FmB1 (diester of propane-1, 2, 3-tricarboxylic acid of 2 amino-12, 16-dimethyl-3, 5, 10, 14, 15-pentahydroxyicosane) in 1987 (Gelderblom *et al.*, 1988, 1992a, 1992b; Riley and Richard, 1992; Shier, 1992; Nelson *et al.*, 1993; Norred, 1993; Riley *et al.*, 1993a; Scott, 1993; Marasas, 1995; Jackson *et al.*, 1996). Several hydrolyzed derivatives of Fms, resulting from removal of all or one of the tricarballylic acid and other ester groups, have also been found in nature (Jackson *et al.*, 1996; Seo *et al.*, 1996). Instead of an amine group at the C-2 position, several new fumonisins that contain an N-linked 3-hydroxypyridine moiety at this position have also been identified (Musser *et al.*, 1996). *F. proliferatum*, another common naturally occurring species, also produces Fms. Although *F. anthophilum*, *F. napiforme* and *F. nygamai* are capable of producing Fms, they are not commonly isolated from food and feed. Other related fusaria, including *F. subglutinans*, *F. annulatum*, *F. succisae* and *F. beomiforme*, are not Fm producers (Nelson *et al.*, 1992). In addition to fusaria, *Alternaria alternata* f. sp. *lycopersici* produces a group of host-specific toxins, named AAL toxins, which have a chemical structure similar to the Fms and induce similar toxic effects (Abbas *et al.*, 1995). Production of FmB by *A. alternata* f. sp. *lycopersici* has been found for some cultures (Chen *et al.*, 1992; Abbas and Riley, 1996). Likewise, some Fm-producing isolates also produce small amounts of AAL toxins (Mirocha *et al.*, 1992). Extensive studies have been carried out in the last decade, and data on different aspects of the most recent studies were reported in the 2001 May issue of *Environmental Health Perspectives* (Vol. 109, supplement 2).

The biosynthetic pathway for Fms is not completely understood. Labeling studies indicate that the polyketide synthase, alanine or other amino acid, and methionine are involved. The 20-carbon chain backbone of Fms resembles those for fatty acids and linear polyketides (Desjardins *et al.*, 1996; Proctor *et al.*, 1999; Proctor, 2000; ApSimon, 2001). At least nine loci (designated as *Fum* 1 through *Fum* 9) have been identified by genetic analyses of *Gibberella fuijikuroi*, the sexual stage of *F. moniliforme* (*verticillioides*). Detailed genetic analyses of the fungi have uncovered 15 genes that are associated with fumonisin production, and these genes are clustered together. A PKS gene (*Fum5*) was identified to be involved in formation of the polyketide. Gene *Fum2* and *Fum3* are associated with the interconversion of different fumonisins (Desjardins *et al.*, 1996; Xu and Leslie, 1996; Proctor *et al.*, 1999). The four latest identified *F. verticillioides* genes (*Fum* 6–9) are adjacent to *Fum* 5 in the cluster (Seo *et al.*, 2001).

Fumonisins are most frequently found in corn, corn-based foods and other grains (such as sorghum and rice). The level of contamination varies considerably with different regions and year, ranging from 0 to > 100 ppm, generally below 1 ppm. FmB1 is most commonly found in the naturally contaminated samples; FmB2 generally accounts for one-third or less of the total. Although production of the toxin generally occurs in the field, continued production of toxin during post-harvest storage also contributes to the overall levels. In the laboratory, the highest yield was achieved in corn culture and less in rice culture; peanuts and soybeans were found to be poor substrates for toxin production. The toxin is very stable to heat as well as being resistant to other treatments. Acid hydrolysis causes the loss of tricarballylic acid, but the hydrolyzed products may still have toxic effects. Transmission of FmB to the edible portion of meat or egg is unlikely to occur because of the rapid excretion of the toxin in animals. No significant amount of FmB1 is transmitted to milk (Hammer *et al.*, 1996; Richard *et al.*, 1996).

3.3.2 Toxicological effects of FmB1

Fumonisin B1 is primarily a hepatotoxin and carcinogen in rats (Riley and Richard, 1992; Riley *et al.*, 1993a, 1994; Jackson *et al.*, 1996; Shier, 2000). Earlier studies showed that feeding culture material from *F. moniliforme* to rats resulted in cirrhosis and hepatic nodules, adenofibrosis, hepatocellular carcinoma-ductular carcinoma, and cholangiocarcinoma (Gelderblom *et al.*, 1988). Later studies indicated that the kidney is also a target organ (Norred and Voss, 1994; Badria *et al.*, 1996; Bucci and Howard, 1996), and tubular nephrosis was found both in rats and in horses in a field case associated with equine leukoencephalomalacia (ELEM). Results from these earlier observations were confirmed with the studies of purified Fms. Although FmB1 was originally found to be a potent cancer promoter (Gelderblom *et al.*, 1988), subsequent studies showed that it is also a carcinogen (Gelderblom *et al.*, 1991, 1992b, 1993, 1994), and it has been classified as a class II carcinogen (International Agency for Research on Cancer, 1993). The effective dose of FmB1 for cancer initiation in rat liver depends on both the level and the duration of exposure. For tumor formation, there appears to be a balance between the compensatory cell proliferation due to the hepatotoxicity of FmB1 and the inhibitory effect on the subsequent hepatocyte cell proliferation. All of the three major Fms, i.e. FmB1, FmB2 and FmB3, show cancer initiation and promoting activities in rats. In cell culture systems, FmB1 was found to be a mitogen, cytotoxic to the cells (Gelderblom *et al.*, 1993, 1994) and able to alter the expression of genes associated with the cell cycle. It induces *in vivo* genotoxicity, but has no mutagenic effect in the *Salmonella* system.

Fumonisin B1 was identified as an etiological agent responsible for ELEM in horses and other Equidae (donkeys and ponies), and for porcine pulmonary edema

(PPE) in pigs (Norred and Voss, 1994; Marasas, 1995; Jackson *et al.*, 1996). ELEM was originally found to be a seasonal disease occurring in late fall and early spring. The disease primarily causes neurotoxic effects in animals, including uncoordinated movements and apparent blindness shown by violent blundering into stalls and walls. Liver damage has also been reported. Death sometimes occurs with no nervous symptoms. Clinically, the disease is characterized by edema in the cerebrum of the brain. The levels of FmB1 and FmB2 in feeds associated with confirmed cases of ELEM ranged from 1.3 to 27 ppm. In pigs, PPE occurs only at high FmB levels (175 ppm), while liver damage occurs at much lower concentrations with a NOAEL of < 12 ppm. Cardiovascular function is altered by FmB1. The FmB1-induced PPE is caused by left-sided heart failure and not by altered endothelial permeability (Gumprecht *et al.*, 2001; Haschek *et al.*, 2001). In cattle, renal injury, hepatic lesions and alteration of sphingolipid in various organs was observed. A high level of Fm is also needed to cause disease in poultry. Similar to AAL toxin, Fms are also toxic to some plants: jimsonweed, black nightshade, duckweed, and tomatoes with the asc/asc genotype are most susceptible (Abbas *et al.*, 1995; Abbas and Riley, 1996).

3.3.3 Impact on human and animal health

While Fms are commonly detected in corn-based foods and feeds, the impact of low levels of Fms in human foods is not clear (Norred and Voss, 1994; Marasas, 1995). Although several reports have indicated a possible role of FmBl in the etiology of human esophageal cancer in the regions of South Africa, China and northeastern Italy, where *Fusarium* species are common contaminants, more data are necessary to sustain this hypothesis (Groves *et al.*, 1999). The co-occurrence of Fms with carcinogenic mycotoxins such as AFB (da-Silva *et al.*, 2000; Vargas *et al.*, 2001) or nitrosamines may play an important role in carcinogenesis in humans. Current data suggest that they may have more impact on the health of farm animals than on humans. To protect animal and human health, guidelines for Fm levels in feeds to be used for various animals were established by the FDA in 2000. The FmB1 levels are limited to 5, 20, 60 100, 30 and 10 ppm in corn and corn by-products to be used for horse and rabbit, catfish and swine, ruminants and mink, poultry, breeding stock (ruminant, poultry and mink), and others (dogs and cats), respectively. The FmB1 levels are limited to 1, 10, 30, 50, 15 and 5 ppm in the total ration for the above animal species, respectively (e.g. 1 ppm for horse and rabbit) (Verardi and Rosner, 1995; Park and Troxell, 2002; Van Egmond, 2002). Although FmB1 has been classified as a type 2 carcinogen, only Switzerland regulates the Fm level (1.0 ppm) in foods.

3.4 Selected important trichothecene (TCTC) mycotoxins and related mycotoxicoses

3.4.1 General considerations

Trichothecenes are a group of naturally occurring toxic tetracyclic sesquiterpenoids produced by many species of fungi in the genera *Fusarium* (most frequently), *Myrothecium*, *Trichoderma*, *Trichothecium*, *Cephalosporium*, *Verticimonosporium* and *Stachybotrys*. The term 'TCTC' is derived from trichothecin, the first compound to be isolated in this group. All the mycotoxins in this group contain a common 12-13 epoxytrichothecene (six-membered oxygen-containing ring) skeleton and an olefinic bond with different side chain substitutions. Based on the presence of a macrocyclic ester or ester–ether bridge between C-4 and C-15, TCTCs are generally classified as macrocyclic (type C) or non-macrocyclic. The non-macrocyclic TCTCs are further divided into two types: type A TCTCs, including T-2 toxin, HT-2 toxin, neosolaniol (NESO), diacetoxyscirpenol (DAS) and T-2 tetraol (T-4ol), which contain a hydrogen- or ester-type side chain at the C-8 position; and type B TCTCs, including DON, nivalenol (NIV) and fusarenon-x (FS-x, or 4-acetyl-niv), which contain a ketone. Type C group TCTCs, including roridins and verrucarins, contain a macrocyclic ring. In addition to fungi, extracts from a Brazilian shrub, *Baccharis megapotamica*, also contain macrocyclic TCTCs. Several books and a number of review articles dealing with TCTC mycotoxins have been published (Ueno, 1983, 1986; Marasas *et al.*, 1986; Beasley, 1989; Chelkowski, 1989; Chu, 1991a, 1997; Miller and Treholm, 1994; Jarvis *et al.*, 1995; D'Mello *et al.*, 1999).

Similarly to other sesquiterpenes, TCTCs are biosynthesized via the mevalonate pathway. The TCTC skeleton is formed by cyclization of farnesyl pyrophosphate via the intermediate trichodiene by an enzyme trichodiene synthase. The isovaleroxy side chain in T-2 toxin is derived from leucine. Several key enzymes in the biosynthetic pathway have been identified. At least six genes (*Tri1* to *Tri6*) involved in the biosynthesis of TCTC have been cloned, and these genes are clustered together on a chromosome. Genes *Tri3*, *Tri4*, and *Tri5*, which encode a transacetylase (15-O-acetyltransferase), a cytochrome P-450 monooxygenease, and trichodiene synthase, respectively, are contained within a 9-kb region, while *Tri5* and *Tri6* (a regulatory gene) are in a 5.7-kb region (Desjardins *et al.*, 1993; Hohn *et al.*, 1993; Proctor *et al.*, 1995a; Keller and Hohn, 1997; Proctor, 2000).

The structural diversity of TCTC mycotoxins results in different toxic effects in animals and humans. Unlike AFB1 and OA, where primary effects and clinical manifestations are well defined in the liver and kidney, TCTC mycotoxicoses are difficult to distinguish because they affect many organs, including the gastrointestinal tract, and the hematopoietic, nervous, immune, hepatobiliary and cardiovascular systems. Ingestion of foods and feeds that contain TCTC mycotoxins causes many types of mycotoxicoses in humans and animals, including moldy corn toxicosis, scabby wheat toxicosis (or red-mold, or akakbi-byo disease, or scabby barley poisoning), feed refusal and emetic syndrome (swine), fusaritoxicoses, hemorrhagic syndrome and alimentary toxic aleukia (ATA). More than 100 TCTCs have been identified in the laboratory (Chu, 1997); only a few selected toxins that have been found in foods are described here.

3.4.2 T-2 toxin and other type-A TCTCs

T-2 toxin, a highly toxic type A TCTC, was isolated in the middle 1960s. It was originally found to be produced by a strain of *F. tricinctum* isolated from moldy corn, and later found to be produced by *F. sporotrichioides* (major), *F. poae*, *F. sulphureum*, *F. acuminatum* and *F. sambucinum*. Unlike most mycotoxins, which are usually synthesized at around 25 °C, the optimal temperature for T-2 toxin production is around 15 °C. Higher temperatures (20–25 °C) are needed for the production of related metabolites, such as H-T2 toxin (Ueno, 1983; Park *et al.*, 1996). Although T-2 toxin occurs naturally in cereal grains, including barley, corn, corn stalk, oats, wheat and mixed feeds, contamination with T-2 toxin is less frequent than with deoxynivalenol (DON). However, T-2 toxin (LD_{50} in mice: 2–4 mg/kg) is much more toxic to animals, perhaps also to humans, than DON (LD_{50} in mice: 50–70 mg/kg).

Almost all the major TCTCs, including T-2 toxin, are cytotoxic and cause hemorrhage, edema, and necrosis of skin tissues. Inflammatory reactions near the nose and mouth of animals are similar to some lesions found in humans suffering from ATA disease. The severity of lesions is also related to chemical structure. Macrocyclic toxins such as verrucarin A are most active, followed by group A toxins (T-2 toxin), and the least active are the type B toxins, such as NIV. Neurologic dysfunctions, including emesis, tachycardia, diarrhea, refusal of feed/anorexia, and depression, were also observed. T-2 toxin and some TCTCs also induce major GI lesions including perioral dermatitis, stomatitis, esophagitis, gastritis, radiomimetic lesions, and sometimes hemorrhage in the intestines. However, the major lesion of T-2 toxin is its devastating effect on the hematopoietic system in many mammals, including humans. Typically, there is a marked initial increase in the number of circulating white blood cells, especially lymphocytes, followed by a rapid decrease to 10–75 % of normal values. Platelet counts are also reduced. There is also extensive cellular damage in the bone marrow, intestines, spleen and lymph nodes, and in severe cases there is complete atrophy of the bone marrow and marked alteration of plasma coagulation factors. T-2 toxin and related TCTCs are the most potent immunosuppressants of the known mycotoxins, and cause significant lesions in the lymph nodes, spleen, thymus and bursa of Fabricius. The heart and pancreas are other target organs for T-2 toxin intoxication. Although urinary and hepatobiliary lesions have been observed for T-2 toxin and DAS, these effects are considered to be secondary.

3.4.3 Deoxynivalenol (DON)

Deoxynivalenol is a major type B TCTC mycotoxin produced primarily by *F. graminearum* and other related fungi such as *F. culmorum* and *F. crookwellense*. Because DON causes feed refusal and emesis in swine, the name 'vomitoxin' is also used. Although DON is considerably less toxic than most other TCTC mycotoxins, the level of contamination of this toxin in corn and wheat is generally high, usually above 1 ppm and sometimes greater than 20 ppm. Contamination of DON in other commodities, including barley, oats, sorghum, rye, safflower seeds and mixed feeds, has also been reported. DON has been found in cereal grains in many countries, including Australia, Canada, China, Finland, France, Germany, Hungary, India, Italy, Japan, South Africa, the UK, the US and many others.

With wet and cold weather during maturation, grains are especially susceptible to *F. graminearum* infection, which causes so-called 'scabby wheat' and simultaneously produces the toxin. The optimal temperature for DON production is about 24 °C. Outbreaks of DON in winter wheat in the US, Finland and Canada usually occur when continental chilly and damp weather favoring the fungal infection is followed by a humid summer favorable for toxin production. For other crops, such as corn and rice, a continental humid warm summer is more favorable. Depending on the weather conditions, the infestation of *F. graminearum* in wheat and corn and subsequent production of toxins in the field varies considerably from year to year as well as by regions. Thus, the levels of DON in these commodities are difficult to predict.

Toxicologically, DON induces anorexia and emesis in both humans and animals. Swine are most sensitive to feed contaminated with DON. Because of the frequent occurrence of high levels of DON in wheat and corn, its stability, and reported food-poisoning outbreaks in humans, contamination of cereals with DON is a major concern of both the government and the food and feed industries. Contamination of DON in wheat and corn may be associated with other toxic effects because other fusarium toxins, including zearalenone and other TCTCs, may also be present. Other type B TCTCs, such as nivalenol and acetylated DON, which are more toxic than DON to test animals, occur naturally in some parts of the world.

3.4.4 The impact of TCTC on human and animal health

Owing to their their diverse toxic effects, and also because of frequent contamination by toxins or toxigenic fungi in foods and feeds, TCTCs are potentially hazardous to human and animal health. However, among the many types of TCTC mycotoxicoses mentioned earlier, only ATA and scabby wheat toxicosis have been demonstrated in human populations. The former, ATA, was attributed to the human consumption of overwintered cereal grains colonized by *F. sporotrichioides* and *F. poae*; it caused the deaths of hundreds of people in the USSR between 1942 and 1947. Later studies indicated that T-2 toxin and related TCTCs were the primary cause. The signs and symptoms of ATA disease, which include skin inflammation, vomiting, damage to hematopoietic tissues, leukocytosis and leukopenia, are common in humans and animals, including cats, cattle, guinea pigs, poultry, monkeys and swine.

DON has been found to be primarily responsible for outbreaks of scabby wheat toxicosis in humans that are quite common in several countries (Ueno, 1983, 1986), but these toxicoses rarely cause death (Luo, 1988). For example, between 1961 and 1985, 35 outbreaks involving 7818 cases were found to be caused by consumption of foods made from either scabby wheat or moldy corn meal in China. In a well documented study, 362 persons in a commune were involved. The symptoms, which occurred between 15 minutes and 1 hour after eating foods made from flour from moldy corn, included nausea (90 %), emesis (61 %), and headache and drowsiness (78 %); 5–6 % had abdominal pain, diarrhea and a low fever. People generally recovered 2–4 days after ingestion of the foods. Analysis of the leftover moldy corn revealed that the samples had 0.34–93.8 ppm of DON; no T-2 toxin or nivalenol (NIV) was found. Similar cases have been reported for people consuming scabby wheat flour.

The widespread natural occurrence of DON in wheat in Canada and the US in the late 1970s and early 1980s alerted the general public to the potential hazard of this mycotoxin (Anonymous, 1983; Trenholm *et al.*, 1988; Rotter *et al.*, 1996; CAST, 2003). Although DON is not as toxic as other TCTCs, the level of contamination in wheat and corn is high and intoxication of humans by DON occurs more often. The tolerable daily intakes of DON for adults and infants have been estimated to be 3 μg/kg and 1.5 μg/kg body weight, respectively. Consequently, a tolerance level of

1 ppm for DON in grains for human consumption has been set by a number of countries, including the US (Van Egmond, 1989b). In Canada, the guideline for DON in the uncleaned soft wheat used for non-staple foods is 2 ppm, but 1 ppm for infant foods (Rotter *et al.*, 1996). Although inproper storage may lead to the production of some TCTC mycotoxins, infestation of fusaria in wheat and corn in the field is of most concern regarding the DON problem.

TCTCs may also be involved in the so-called 'sick building syndrome' in humans. *Stachybotrys atra*, an indoor mold, was isolated from a badly water-damaged Chicago suburban home where the occupants complained about headaches, sore throats, hair loss, flu symptoms, diarrhea, fatigue, dermatitis and general malaise. Several TCTCs (verrucarins B and J, stratoxin H, trichoverrins A and B) were found in the *S. atra*-contaminated materials of this home. T-2 toxin, DAS, roridine A and T-2 tetraol were isolated from the dust samples from the air ventilation systems of three urban Montreal office buildings in another suspected case of sick building syndrome. Likewise, *S. chartarum* has also been associated with pulmonary hemorrhage cases in the Cleveland, Ohio, area (Hendry and Cole, 1993; Nikulin *et al.*, 1994; Jarvis et al., 1995; Dearborn *et al.*, 1999; Gravesen *et al.*, 1999; Johanning *et al.*, 1999; Mahmoudi and Gershwin, 2000; Vesper *et al.*, 2000; Assouline-Dayan *et al.*, 2002; Nielsen *et al.*, 2002). A remediation program involving removal of all contaminated wallboard, paneling and carpeting in the water-damaged areas of the home, as well as spraying a sodium hypochlorite solution to all surfaces during remediation, appeared to be effective. Air samples taken from post-remediation buildings showed no detectable levels of *S. chartarum* or related toxicity (Price and Ahearn, 1999). Several studies showed that other molds and mycotoxins are present in mold-damaged buildings. Thus it is important to identify both the mold and mycotoxins for the cause of sick building syndrome (Tuomi *et al.*, 2000).

3.5 Other selected mycotoxins

In addition to the mycotoxins discussed above, a number of other mycotoxins also occur naturally. The impacts of some of these mycotoxins on human and animal health are discussed in the following sections.

3.5.1 Other mycotoxins produced by *Aspergillus*

Sterigmatocystin (ST) is a naturally occurring hepatotoxic and carcinogenic mycotoxin produced by fungi in the *Aspergillus*, *Bipolaris* and *Chaetomium* genera and *Penicillium luteum*. Structurally related to AFB1, ST is known to be a precursor of AFB1 (Betina, 1989; Chu, 1991a; CAST, 2003), and rapid progress in understanding the biosynthesis of AFB1 has been made through the studies on the biochemistry and genetics of biosynthesis of ST (Brown *et al.*, 1996; Calvo *et al.*, 2002; Udwary *et al.*, 2002). ST is also a carcinogen, mutagen and genotoxin, but its carcinogenicity is 10–100 times less than that of AFB1 in test animals (Van der Watt, 1977; Mori and Kawai, 1989). ST occurs naturally in cereal grains such as barley, rice and corn, in coffee beans, and in foods such as cheese (Jelinek *et al.*, 1989). It has also been found in pickled foodstuffs in Linxian, a region of the People's Republic of China with a high esophageal cancer incidence (Sun *et al.*, 1988), and in foods from Mozambique, where high liver cancer incidence has been reported (Van der Watt, 1977). Toxigenic fungi have been isolated from patients with esophageal cancer, and these strains are capable of producing ST in many commodities. The role of ST in human carcinogenesis appears to be indirect and inclusive.

Aspergillus terreus and several other fungi (*A. flavus* and *A. fumigatus*) have been found to produce the tremorgenic toxins territrem A, B and C (Ling, 1995), aflatrem and fumitremorgin. *A. terreus*, *A. fumigatus* and *Trichoderma viride* also produce gliotoxin, an epipolythiopiperazine-3,6-diones-sulfur containing piperazines antibiotic, which may have immunosuppressive effects in animals (Waring and Beaver, 1996). In addition, *A. flavus*, *A. wentii*, *A. oryzae* and *P. atraovenetum* are capable of producing nitropropionic acid, a mycotoxin causing apnea, convulsions, congestion in the lungs and subcutaneous vessels, and liver damage in test animals. This toxin was also identified as an etiological agent for 'deteriorated sugarcane poisoning', a fatal form of food poisoning that occurred in China. However, the fungi involved in the contamination of the sugar cane and nitropropionic acid production were *Arthrinium sacchari*, *Arth. saccharicola* and *Arth. phaeospermum* (Liu *et al.*, 1988). An acceptable daily intake of 25 μg of NPA/kg per day (or 1.75 mg/day for a 70-kg human) in humans was recently established (Burdock *et al.*, 2001).

3.5.2 Other mycotoxins produced by *Penicillium*

Other than OA, penicillia produce many mycotoxins with diversified toxic effects. Cyclochlorotine, luteoskyrin (LS) and rugulosin (RS) have long been considered to be possibly involved in yellow rice disease during World War II (Ciegler *et al.*, 1971; Chu, 1977; Cole and Cox, 1981). They are hepatotoxins, and also produce hepatomas in test animals. However, incidents of food contamination with these toxins have not been well documented. Several other mycotoxins, including patulin (PT), penicillic acid (PA), citrinin (CT), cyclopiazonic acid (CPA), citreoviridin and xanthomegnin, which are produced primarily by several species of penicillia, have attracted some attention because of their frequent occurrence in foods (Wilson and Hayes, 1973; CAST, 2003).

PT and PA are produced by many species in the genera *Aspergillus* and *Penicillium*. A heat-resistant fungus, *Byssochlamys nivea* (Tournas, 1994), frequently found in foods also produces PT. Both toxins are hepatotoxic and teratogenic. PT is frequently found in damaged apples, apple juice and apple cider, and sometimes in other fruit juices and feed. PA has been detected in 'blue eye corn' and in meat. Due to its highly reactive double bonds, which readily react with sulfhydryl groups in foods, patulin is not very stable in foods containing these groups (Scott, 1975). Nevertheless, PT is considered a health hazard to humans. At least 10 countries have regulatory limits, most commonly at a level of 50 μg/kg, for PT in various foods and juices. Frequently associated with the natural occurrence of OA is CT, also a nephrotoxin, produced by *P. citrinum* and several other penicillia and aspergilli (Cole and Cox, 1981). Recent discovery of the capability of production of CT by *Monocuus ruber* and *M. purpureus*, two molds frequently used in the preparation of certain oriental foods, has emphasized the potential health hazard of CT to human health (Pastrana *et al.*, 1996).

One of the mycotoxins closely associated with the natural occurrence of AF in peanuts is CPA (Riley and Goeger, 1991; Fernandez *et al.*, 2001; CAST, 2003), which causes hyperesthesia and convulsions as well as liver, spleen, pancreas, kidney, salivary gland and myocardial damage. The toxin was originally found to be produced by *P. cyclopium*; but a number of other penicillia (*P. crustosum*, *P. griseofulvin*, *P. puberulum*, *P. camemberti*) and aspergilli (*A. versicolor*, *A. flavus*, *A. tamarii* but not *A. parasiticus*) also produce CPA (Goto *et al.*, 1996; Huang *et al.*, 1994). Natural occurrence of CPA in corn, peanuts and cheese has been reported. *Penicillium rubrum* and *P. purpurogenum* produce two highly toxic hepatotoxins (LD_{50}, 3.0 mg/kg mice, IP) called rubratoxins A (minor) and B (major), which are complex nonadrides fused with anhydrides and lactone rings (Wilson and Hayes, 1973). Rubratoxin B has shown to have synergistic effects with AFB1 (CAST, 2003).

In addition to the above hepatotoxins and nephrotoxins, penicillia produce many mycotoxins with strong pharmacological effects on neurosystems (Plumlee and Galey, 1994; Steyn, 1995). For example, *P. crustosum* and *P. cyclopium* produce tremorgenic indoloditerpenes called penitrem A–F. Penitrem A, the major toxin in this group, causes tremorgenic effects in mice (Yamaguchi *et al.*, 1993). Roquefortines A–C (C is most toxic), which are produced by *P. roqueforti* and several other penicillia, have neurotoxic effects in animals, and have been found in cheese. Tremorgens in the paspalitrem group (paspalicine, paspalinine, paspalitrem A and B, paspaline and paxilline) are produced by *Claviceps paspali* and some penicillia. Janthitrems are produced by *P. janthinellum* (Wilkins *et al.*, 1992; Penn *et al.*, 1993). Seventeen genes involved in the biosynthesis of indole-diterpene have been found to be clustered within a 50-kb region of chromosome in the fungus (Young *et al.*, 2001).

3.5.3 Other mycotoxins produced by *Fusarium*

Other than TCTCs and Fms, some fusaria can also produce other mycotoxins. Zearalenone [6-(10-hydroxy-6-oxo-trans-1-undecenyl)-beta-resorcyclic acid-u-lactone] is a mycotoxin produced by the scabby wheat fungus, *F. graminearum* (roseum), which also produces DON (CAST, 2003). Zearalenone (ZE), also called F-2, is a phytoestrogen causing hyperestrogenic effects and reproductive problems in animals, especially swine. Natural contamination with ZE primarily occurs in cereal grains such as corn and wheat. Contamination of feed with this mycotoxin, sometimes together with DON, may result in a large economic loss in the swine industry. *F. moniliforme* also produces several other mycotoxins, including fusarins A–F (Wiebe and Bjeldanes, 1981; Gelderblom *et al.*, 1984; Lu and Jeffrey, 1993), moniliformin (Burmeister *et al.*, 1979; Marasas *et al.*, 1986), fusarioic C (McBrien *et al.*, 1996), fusaric acid (Bacon *et al.*, 1996), fusaproliferin (Ritieni *et al.*, 1995) and beauvericin (Gupta *et al.*, 1991; Plattner and Nelson, 1994) in addition to Fms. Although the impact of these mycotoxins on human health is still not known, fusarin C (FC) has been identified as a potent mutagen. Moniliformin, which causes cardiomyopathy in test animals, may be involved in Keshan disease in humans in regions where dietary selenium deficiency is also a problem (Liu, 1996). In addition to *F. moniliforme* and *F. subglutinans*, several other fusaria, including *F. graminearum*, have also been identified as FC producers (Farber and Sanders, 1986). Fusarochromanones are a group of mycotoxins produced by some

Fusarium equiseti isolates (Lee *et al.*, 1985; Xie and Mirocha, 1995). These mycotoxins cause tibial dyschondroplasia in broiler chicks, turkeys and ducks, and are considered to be the possible cause of Kashin-Beck disease in China. *F. moniliforme* was also found most effective, among a group of several fungi tested, in reducing nitrates to form potent carcinogenic nitrosamines (Ji *et al.*, 1986). These observations further suggest that the contamination of foods with this fungus could be one of the etiological factors involved in human carcinogenesis in certain regions of the world (Li *et al.*, 1980).

3.5.4 Mycotoxins produced by other species

Other than the fungi in the *Aspergillus*, *Penicillium* and *Fusarium* genera, mycotoxins are also produced by *Alternaria*, a plant pathogen and another genus of fungi commonly found in our environment. The toxins can be produced in both pre-harvest and post-harvest commodities by these fungi. *Alternaria alternata* and other *Alternaria* species are capable of producing dibenzo-*a*-pyrone types of mycotoxins: alternariol, alternariol monomethyl ether (AME), altenuene, isoaltenuene and altenuisol, tetramic acid metabolites tenuazonic acid (TzA) and related compounds, and perylene derivatives altertoxins (ATX) I, II (also called stemphyltoxin II), III, and stemphyltoxin (Chelkowski and Visconti, 1992). Although most of those compounds are relatively non-toxic, AME has been shown to be positive in the Ames test at relatively high concentrations (Woody and Chu, 1992). TzA is a protein synthesis inhibitor, and is capable of chelating metal ions and forming nitrosamines. This mycotoxin is also produced by *Phoma sorghina* and *Pyricularia oryzae*, and may be related to 'Onyalai', a hematological disorder in humans living south of the Sahara in Africa. Although no extensive survey has been conducted to determine the occurrence of these mycotoxins in human foods, limited studies indicate that the incidence of some of these mycotoxins, such as AME and TzA, may be high in apple and tomato products (Jelinek *et al.*, 1989). As mentioned earlier, the discovery of the structural and functional similarity between fumonisins and AAL toxin further shows the importance of mycotoxins produced by fungi in the *Alternaria* family (Mirocha *et al.*, 1992).

Sporidesmines, a group of hepatotoxins discovered in the 1960s, are also worthy of mention. These mycotoxins, causing facial eczema in animals, are produced by *Pithomyces chartarum* and *Sporidesmium chartarum*, and are very important economically to the sheep industry. Slaframine, an interesting mycotoxin produced by *Rhizoctonia leguminicola* (in infested legume forage crops), causes excessive salivation or slobbering in ruminants as a result of blocking acetylcholine receptor sites (CAST, 2003).

4 Intoxication from naturally occurring toxic fungal metabolites

Two classical examples related to the mycotoxins are ergotism and mushroom poisons. Because these intoxications are a result of ingestion of a fungal body containing toxic metabolites, these two types of poisons are sometimes excluded from the modern discussion of mycotoxins. Nevertheless, ergots and poisonous mushrooms can still be unintentionally introduced into the food chain today. Some phytoalexins elaborated by plants as a result of alteration of their metabolism due to either fungal infection or other damages have also been found toxic to humans and animals.

4.1 Ergotism

Ergotism is a human disease that results from consumption of the ergot body, which is the sclerotium of the fungus, in rye or other grains infected by a parasitic fungus in the *Claviceps* genus – e.g. *C. purpurea* and *C. paspali* (Flieger *et al.*, 1997). Pharmacologically active alkaloids produced by the fungus in the ergot body are the major cause for human intoxication. Two types of ergotisms have been documented. In the convulsive type, the affected persons have general convulsions, a tingling sensation of muscles (e.g. their foot may 'go to sleep'), and sometimes the entire body is racked by spasms. Epidemics occurred between 1581 and 1928 in European and other countries. Although it has been suggested that the consumption of ergots that cause convulsive ergotism may have played a role in 'Salem Witchcraft' incidents, this is still a controversial issue. In the gangrenous type, the affected parts became swollen and inflamed with violent, burning pains – hence the 'Fire of St Anthony'. In general, the affected area first became numb, then turned black, shrank, and finally became mummified and dry. Outbreaks occurred from the Middle Ages to the nineteenth century. In some areas of France, grain contained as much as 25 % ergots, and patients died as a result of consumption of about 100 g of ergot over a few days. Between 1770 and 1771, about 8000 people died in one district alone in France. In general, 2 % ergots in the grain is sufficient to cause an epidemic. European and most other countries have a regulatory limit of 0.1–0.2 % ergots in flour. Biochemically, ergotisms are due to the intoxication by ergoline alkaloids that are produced by the fungus present in the sclerotia of *C. purpurea*. The most active components are amides of D-lysergic acid, including both cyclic-type peptides and non-peptide amides of ergot alkaloids. These alkaloids, causing smooth muscle contraction and blocking neurohormones, have both vasoconstriction and vasodilation effects, and also affect the central nervous system. Thus, they have some therapeutic uses.

4.2 Mushroom poisons (MP)

Whereas it has been known for centuries that consumption of poisonous mushrooms can cause severe injury and even death in humans, accidental ingestion of these poisonous mushrooms, generally called toadstools, still occurs in our modern day. The term 'toadstools', however, does not necessarily represent all poisonous mushrooms. As a result of the harvest and consumption of wild mushrooms, a number of cases of MP with different degrees of severity, including death, are reported yearly. Even after extensive research and an increased understanding of the chemical nature as well as toxicology of some of these poisons, no effective antidote is available for most mushroom poisons (Zahl, 1965; Hatfield and Brady, 1975; Bresinsky and Besl, 1990). Depending on their toxic effects and major target organs, mushroom poisons are generally classified into the groups described below.

4.2.1 Poisonous mushrooms causing severe gastrointestinal lesions and liver and kidney damage

This type of mushroom poisoning is primarily caused by potent toxic cyclopeptides present in several species of *Amanita*. About 90 % of these cases are due to the ingestion of *Amanita phalloides* and *A. verna*. The former, generally called 'death cap' or 'deadly *Agaricus*' is about 10–15 cm tall, with smooth, deep olive green and readily extending streaked caps up to 12 cm in diameter, white lamella and a pale greenish stem. The major toxins involved in this type of mushroom poisoning are amanitins and phalloidins (Wieland and Faulstich, 1978, 1979). In addition to these two species, *A. virosa*, which is commonly called 'white-destroying angel', also contains a series of actin-binding cyclic peptides called virotoxins. Other species implicated in this type of mushroom poisoning are *A. tenuifolia, A. bisporigera, A. ocreata* (North America) and *Galerina marginata*.

4.2.1.1 Amanitins, phalloidins and virotoxins

Although investigations into the agents causing mushroom poisoning began in the nineteenth century, much of the chemistry was studied in the twentieth century by Professor Weiland's group in Germany. A number of toxic cyclic peptides were identified and characterized.

The amanitins, sometimes called amanita toxins, are slow-acting toxins, with mortality occurring within 15 hours of eating toxic mushroom; the toxic dose or LD_{50} is 0.3 mg/kg in mice. The toxins have bicylic octapeptide skeletons containing hydroxylproline, alanine, glycine, isoleucine, aspartic acid and tryptophan. A sulfoxide substituted tryptophan serves as the nucleus and forms a trans annular bridge with other amino acids. Variations in some amino acids lead to different variants of amanitins. Cleavage of sulfoxide and peptide bond or removal of γ-OH changes the conformation of the molecules and decreases toxicity (Faulstich, 1980).

Phallotoxins are bicyclic heptapeptides with one amino acid less than the amanitins. They contain a thioether linkage rather than sulfoxide with one D-threonine instead of phenolic OH. Monocyclic or acid-cleaved toxins are non-toxic (Munekata, 1981). The phalloidins, which have a LD_{50} of 2–2.5 mg/kg in mice, are fast-acting toxins. At a lethal dose, death usually occurs within 1–2 hours.

The toxin content in toxic mushrooms varies with the regions where they are collected. In Europe, toxic *A. phalloides*, accounting for about 20–50 % of fatalities, contains about 10 mg of phalloidin, 8 mg of α-amanitin, 5 mg of β-amanitin and 1.5 mg γ-amanitin in 100 g of fresh tissue or 5 g of dry tissue in some species. Similar total amounts of toxins are present in this species found in the US, but β-amanitin is the predominant toxin. Other species, such as *A. bisporigera* in North America, have higher toxin levels – e.g.15 mg α-amanitin in 100 g of sample. The lethal dose of α-amanitin in humans is about 0.1 mg/kg, and phallotoxins are 10 to 20 times less potent. Thus, *50 g of a poisonous mushroom is sufficient to cause death in humans*.

Typical symptoms of amanita poisoning generally involve three phases. The first phase usually occurs within 10–14 hours of ingestion of the mushroom, with lesions in the gastrointestinal tract (especially the duodenum), and violent emesis and diarrhea with blood and mucus. The second phase is a transient remission period, and this is followed by a third phase, the acute phase. The patients suffer severe liver damage with swelling and tenderness, icterus, and increased GOT and glutamate-pyruvate transaminase activity. Lesser damage occurs in the kidney, with proteins, casts and erythrocytes observed in the urine.

The toxins act exclusively in the liver, causing blebs on hepatocytes *in vivo* and inhibiting the translocation of microfilaments in the hepatocytes by disturbing the structure or dynamics of liver cells. This may be caused by a fatal decrease of actin monomers. Biochemically, the amanita toxins inhibit type II RNA polymerase in the liver, through formation of a 1:1 complex with the enzyme ($K = 3.6 \times 10^{-9}$). Phalloidins depolymerize the 1:1 DNA:actin complex by formation of a 1:1 complex with actin protomer ($K = 3.6 \times 10^{-8}$). Both amanitin and phalloidin have also been found to increase cytosolic Ca^{++} and alter phosphoinositide turnover, which results in activation of phosporylase *a* and glycogenolysis. Phalloidin is more potent than amanitin (Kawaji *et al.*, 1992).

Although some antidotes, such as alpha-lipoic acid or thiotic acid and cytochrome C, have been suggested, most of them are ineffective. Early diagnosis of the symptoms is essential before the acute phase, which generally requires liver transplant, is developed. For the analysis of toxins, methods such as TLC (staining with cinnamaldehyde), HPLC and immunoassays are available, but no effective rapid field tests have been developed.

4.2.1.2 Cortinarius toxins

The largest group of wild mushrooms in Europe is in the genus *Cortinarius*, members of which were generally considered edible. However, an outbreak of mushroom poisoning (*C. orellanus*) with more than 100 identified cases and 10 % mortality in Poland in 1957 changed the situation. Several species, including *C. orellanus* (Poland), *C. speciosissimus, C. gentilis* (Scotland, Finland), C. *smithiana* and *C. splendens* (France), have been identified as toxic (Tebbett and Caddy, 1984). The toxins in these mushrooms are not as potent as those in the *Amanita* genus. Two major types of toxins are involved in the intoxication. Orellanine, a bipyridine (2,2'-bipyridine-3,3', 4,4'-tetrol-1,1-dioxide), has an oral LD_{50} of 33 mg/kg (Prast *et al.*, 1988). Orelline, which is produced from orellanine during analysis, has been found in the plasma and kidney tissues of intoxicated patients. Another group of toxins consists of cortinarins (cortinarin A, B and C), which are cyclic peptides containing nine amino acids with a thioether linkage between tryptophan and other amino acids. This linkage ($S–CH_2$) is broken by reduction of cortinarin A with Reney Ni to form cortinarin B (Faulstich *et al.*, 1980). Different amounts of these toxins have been found in toxic *C. violaceus, C. orellanus, C. orellanoides, C. speciosissimus* and *C. croceololius* as analyzed by TLC or HPLC. Both orellanine and cortinarin A and B have been implicated in outbreaks.

Toxicologically, cortinarins are slow acting nephrotoxins as compared to the amanitins and phalloidins, and have an unusual latent period (2–30 days). The symptoms include intensive thirst, nausea, gastrointestinal disturbance, vomiting, persistent headache and feeling of coldness, and diarrhea. These symptoms are followed by acute renal insufficiency, which in many cases becomes chronic, requiring treatment by dialysis or kidney transplantation. The toxins impair renal function, causing oliguria and sometimes anuria. Autopsies show severe renal lesions characteristic of toxic interstitial nephritis. The cortinarins are structually related to vasopressin, and may have a similar mode of action.

4.2.1.3 *Other types of mushroom poison affecting the gastrointestinal tract*

There are several other poisonous mushrooms which cause severe gastroenteritis, and some of them also provoke allergenic effects. One of the mushrooms most frequently involved is *Chlorophyllum molybdites* (also known as *Lepiota morgani, Chlorophyllum esculentum* or *Lepiota molybdites*). This mushroom, sometimes mistaken for edible species such as *Agaricus bisporus*, is generally regarded as a relatively benign gastrointestinal irritant (Stenklyft and Augenstein, 1990; Lehmann and Khazan, 1992). In two foci of *Lepiota* mushroom poisoning that involved 10 people in Spain in 1989 (Ramirez *et al.*, 1993), 5 people suffered intestinal symptoms and others incurred liver damage; 3 developed fulminant hepatitis and 2 subsequently died of acute respiratory distress. Other mushrooms involved are summarized by Hatfield and Brady (1975) and Bresinsky and Besl (1990).

4.2.2 Poisonous mushrooms affecting the nervous system

A number of mushrooms, including *Amanita muscaria, A. pantherina*, and species in the *Inocybe* and *Clitocybe* genera, contain pharmacologically active compounds affecting the nervous system. Two classic compounds in this group are ibotenic acid and muscarine. Several new compounds have also been found in the mushrooms collected in cases of poisoning.

4.2.2.1 *Ibotenic acid (IBA) and related compounds*

Ibotenic acid, muscimol, muscazone and related isoxazole derivatives are neurotoxins present in *A. muscaria* and *A. pantherina*. Symptoms including dizziness, ataxia, elevated mood and psychic stimulation are observed in humans 90 minutes after ingestion of a dose of 10 mg of muscimol. More toxic effects, including muscular incoordination, confusion, inarticulation and visual disturbance, have been observed 45 minutes after a dose of 15 mg of this compound. The hallucinogen, muscimol, is formed through metabolic activation (decarboxylation) of IBA. Structurally related to γ-aminobutyric acid, these compounds act as inhibitors of sympathetic transmitters in the CNS, acting on the γ-aminobutyric acid and glutamic acid receptors of neurons and affecting permeability to ions and membrane depolarization potential. Both IBA and muscimol have been reported to resemble LSD generally in their influence on the levels of neurotransmitters in rats.

4.2.2.2 *Muscarine (MUS)*

A. muscaria, all species of *Inocybe* and *Clitocybe* are the only genera of fungi containing MUS and related quaternary ammonium compounds. Intoxication reflects its potent cholinergic activity at postganglionic parasympathetic synapses. The symptoms, usually observed between 15 minutes and 2 hours after ingestion, include headache, vomiting, and constriction of the pharynx. Salivation, lacrimation and diffuse perspiration may be followed by severe vomiting and diarrhea, with irregular pulse and asthmatic breath. Mortality rate is about 5 %; death is generally caused by heart or respiratory failure. The LD_{50} of MUS in mouse via i.v. is about 0.2 mg/kg, and a decrease in blood pressure and reduction in heart rate have been observed at 0.01 mg/kg. Toxic effects were observed in monkey at an i.p. dose of 0.5 mg per monkey, but no poisoning signs were observed from 2 mg via the oral route because the toxin was not absorbed. The effect of MUS is due to blockage of postganglionic parasympathetic sites because of its structural similarity with acetylcholine. Unlike other types of mushroom poisoning, a treatment, atropine, is available for MUS poisoning.

Because of the poor absorption of the quaternary compounds in the intestinal tract and owing to the fact that the amount of MUS (0.0002–0.0003 % of fresh weight) in *A. muscaria* and *A. pantherina* is very small, the role of MUS in contributing to the overall intoxication in humans by these two groups of mushrooms is questionable. The isoxazoles, which are found in *A. muscaria* and *A. pantherina* at levels of 0.17–1.0 % and 0.02–0.53 % (dry weight), respectively, may play a more significant role. Symptoms of central nervous system depression, ataxia, waxing and waning obtundation, hallucination, intermittent hysteria or hyperkinetic behavior were observed 30–180 minutes after children (1–6 yr age) ingested *A. muscaria* or *A. pantherina* in a number of cases that occurred between 1979 and 1989 in the State of Washington in the US (Benjamin, 1992). A recent study has shown that both heating and drying cause a decrease in IBA and an increase in MUS, although both remained stable when mushrooms were stored under dry or salted conditions. Boiling and soaking of the mushroom in water leached out most of the active compounds.

4.2.2.3 *Other active agents*

Acromelic acids (Acro) A and B, isolated from poisonous mushroom *Clitocybe acromelalga*, are extremely potent neuroexcitants. Acromelic acid induced selective neuron damage in rat spinal cord. Injection of 5 mg of Acro/kg caused frequent cramps and generalized convulsions in rats (Kwak *et al.*, 1991). In addition, L-stizolobiac and L-stizolobinic acids, which are the precursors for acromelic acid (Fushiya *et al.*, 1992), and the neurological lathyrogens β-cyano-L-alanine and N-(γ-L-glutamyl) β-cyano-L-alanine, were also isolated from this mushroom (Fushiya *et al.*, 1993). Neurotoxic triterpene glycosides, hebevinosides XII, XIII, XIV, were found in the poisonous *Hebeloma vinosophyllum* (Fujimoto et *al.*, 1991).

4.2.3 Hemolytic poisonous mushrooms

Several large molecular weight proteins affecting the hemopoetic system have been reported in some mushrooms. Phallolysin types A and B were present in *A. phalloides* (LD_{50} in mice 0.2–0.7 mg/kg). These are high molecular weight thermal- and alcohol-labile hemolysins. Rubescenslysine, a protein-type hemolysin and cytolysin, was isolated from the edible mushroom *A. rubescens*. The toxin causes lysis of red cells and has a LD_{50} of 0.15–0.31 mg/kg in mice by i.v. injection.

4.2.4 Carcinogenic and mutagenic poisonous mushrooms

During the last two decades, extensive studies have investigated the possible carcinogenicity of both edible and toxic mushrooms.

4.2.4.1 *Hydrazine and related compounds*

A number of mushrooms contain hydrazine-related compounds, which are metabolically converted to active hydrazine, and can cause tumors in test animals (Toth, 1979, 1982, 1995). *Agaricus bisporus* and *Cortinellus shitake* contain agaritine, which is first metabolized to 4-OH-methylphenylhydrazine and then to 4-methylphenylhydrazine. Hydrazine, N?-acetyl-4-hydroxymethylphenylhydrazine, and 4-methylphenylhydrazine have been found to induce lung and blood vessel tumors in mice. A carcinogenic diazonium compound, 4-(hydroxymethyl)-benzenediazonium, has also been isolated from *A. bisporus*. This compound can directly interact with DNA and deoxyribonucleosides (Hiramoto *et al*., 1995). The false morel, *Gyromitra esculenta*, contains gyromitrin, which can be converted to N-methyl-N-formylhydrazine (MFH) and then to the N-methyhydrazine (MH) *in vivo* (Michelot and Toth, 1991). Intoxication with this type of poisonous mushroom usually occurs within 48 hours of consumption of the raw or poorly cooked mushrooms. Symptoms include vomiting and diarrhea, followed by jaundice, convulsions, and coma in severe cases. Frequent consumption of the mushroom can cause hepatitis, renal damage and neurological disorders. In experimental animals, MFH causes tumors of the lung, liver, blood vessels, gall bladder and bile duct of mice, and of the lung, Kupffer's cells, gall bladder and bile duct of hamsters. Methylhydrazine causes tumors in mouse lungs and in hamster Kupffer's cells and cecum. Mechanistically, the carcinogenicity of these compounds is due to its ability to methylate guanine in the DNA of target tissues to form O^6-methylguanine (Bergman and Hellenas, 1992).

4.2.4.2 *Other mutagens and active compounds*

Sterner *et al*. (1982) found that 48 species of mushrooms produced mutagens that are positive in the Ames test. Compounds in these mushrooms include isovelleral from *Lactarius* sp., β-nitroaminoalanine in *A. silvaticus*, and coprine (an aldehyde dehydrogenase inhibitor) from inky cap *Coprinus astramentarine*. Velleral, isovelleral and structurally related merulidial and marasmic acid are also positive in the Ames test. In another survey of the mutagenicity of 400 edible mushrooms with the Ames test, Swiss scientists (Gruter *et al*., 1991) found that 13 of 35 wild or commercial mushrooms were positive. One-fifth of the dried mushrooms were also positive. Extracts from *Lactarius necator* were most active; commercial samples, including *A. bisporus*, *Cantharellus cibarius*, *Stropjaria rugosoannulata* and dried *Auricularia* were also positive. Fruiting bodies of the edible mushrooms *Lactarius deliciosus, L. deterrimus* and *L. sanguifluus* contained sesquiterpenoids with a guaiane skeleton that had weak mutagenicity in the Ames assay (Anke *et al*., 1989). Aqueous extracts of nine wild and two cultivated species of Spanish edible mushrooms, including *Pleurotus ostreatus*, *P. eryngii*, *Agrocybe cylindracea* and *Marasminus oreades*, were positive in the Ames test (Morales *et al*., 1990).

4.2.4.3 *Public health and environmental implications of mutagenic substances in mushrooms*

Although recent studies have shown that some mushrooms have carcinogenic effects in test animals, and some show mutagenicity and genotoxicity in different testing systems, the significance of the presence of these compounds in these mushrooms to public health is still not clear. Based on the amount of the carcinogenic compounds present in *A. bisporus*, Toth and Gannett (1993) estimated that it would be necessary to consume > 1 lb (454 g) of raw mushrooms per day for 50 years to induce cancer in humans. Because compounds such as gyromitrin and its homologues are water soluble and heat sensititive, detoxification of these compounds could easily be carried out heating. The national FDA of Sweden (Larsson and Eriksson, 1989) recommends boiling mushrooms in a large amount of water for two 5-minute periods, with a change of water between boilings. As much as 90 % of the MH was removed after such treatments.

4.3 Toxic metabolites of fungal damaged foods of plant origin

Phytoalexins are a group of compounds produced in plants as a result of physical or biological damage that alters the metabolism of the host plant. Some of these compounds are toxic to humans and animals. One example is the fungal infected sweet potato (Wilson and Hayes, 1973). Before World War II, Japanese investigators found that sweet potatoes infected by black rot mold, *Ceratocystis fimbriata (Ceratostomella fimbriata)*, were toxic to animals. A number of compounds, including ipomeamarone, ipomeanine, β-furoic acid and batatic acid, which contained a furan moiety with a side chain attached at the 3 or β-position, were isolated. Ipomeamarone (IPM-one) gives a bitter taste to the sweet potato, and is most abundant. It is considered to be a hepatotoxin, but may also be toxic to the lungs. Ipomeanol and IPM-one are produced by *F. solani javanicum* (in the US) (Sharma and Salunkhe, 1991). Although these metabolites are considered to have diverse toxic effects in farm animals, their impact on humans is not known. Another group of compounds that may play a role in human carcinogenesis is that of nitroso-compounds produced by some commonly occurring fungi. A number of nitrosamines have been identified in foods, and it has been postulated that the presence of these nitrosamines in foods in regions of China with a high rate of esophageal cancer may play a role in human carcinogenesis (Li *et al.*, 1980).

5 Modes of action of mycotoxins

Due to the diversity of their chemical structures, mycotoxins may exhibit a number of biological effects, including both acute and chronic toxic effects as well as carcinogenic, mutagenic, genotoxic and teratogenic effects in both prokaryotic and eukaryotic systems. As shown in Table 16.2, the toxic effects of most mycotoxins are very organ-specific and their modes of action vary considerably. Once mycotoxins have been ingested, the toxins may either react directly with the target organs to exert toxic effects or be metabolized to an active intermediate (activated) and then exert their toxic effects. The toxins or their metabolites may also be detoxified and then excreted from the body. As will be shown below, the interaction of mycotoxins, either activated or as intact molecules, with cellular macromolecules plays a dominant role in exerting their toxic actions (Busby and Wogan, 1979; Kiessling, 1986;

Table 16.2 Toxicological effects of selected commonly occurring mycotoxins

Major mycotoxins	Toxicological effect[a]	Mode of action
Aflatoxin B1(AF) and other AFs	A, C, H, I, M, T	Form DNA adducts and Mutate p53 tumor suppressor gene
Alternaria (AAL) Mycotoxins[b]	A, Hr, M	Inhibition of ceramide synthase
Citrinin (CT)	C(?) Nh, M	Inhibition of protein synthesis
Cyclochlorotine (CC)	A, C, H	Binding with DNA
Cyclopiazonic acid (CPA)	Cv, I, M, Nr	Alteration of Ca^{++}hemostasis & inhibition of Ca^{++}-dependent ATPase
Deoxynivalenol (DON)	I, Nr	Inhibition of protein synthesis
Fumonisins (Fm)	A, C, Cv, H, Nr, R	Disruption of sphingolipid metabolism by inhibiting ceramide synthase
Luteoskyrin (LS)	A, C, H, M	Binding with DNA
Moniliformin (MN)	Cv, Nr	Binding with pyruvate dehydrogenase and affecting TCA cycle enzymes
Ochratoxin A (OA)	A, I, Nh, T	Inhibition of protein synthesis and several enzymes; enhancement of lipid peroxidation; impairment of Ca^{++} and cAMP homeostasis
Patulin (PT)	C(?), Nr, D, T	Binding with -SH groups of protein and inhibition of -SH enzymes
Penicillic acid (PA)	C(?), Nr, M	Inhibition of -SH enzymes
Rubratoxin B (RB)	A, H, T	Inhibition of protein synthesis
Sterigmatocystin (ST)	H, C, M	Same as aflatoxin
T-2	A, ATA, Cv, D, I, T	Inhibition of protein synthesis, binding with peptidyl transferase, membrane proteins & other enzymes
Tenuazonic acid (Tz)	Hr	Inhibition of protein synthesis
Tremorgenic toxins	Nr	Binding and inhibition of acetylcholinesterase
Zearalenone (ZE)	G, M	Binding with estrogen receptor
12-13, Epoxy trichothecenes (TCTC) other than T-2 & DON	A, D, I, Nr	Same as T-2 toxin

From Chu (2003)

[a] A, apoptosis; ATA, alimentary toxic aleukeria, C, carcinogenic; C(?), carcinogenic effect is still questionable; Cv, cardiovascular lesion; D, dermatoxin; G, genitotoxin and estrogenic effects; H, hepatotoxic; Hr, hemorrhagic; I, immuno-modulation effects, M, mutagenic, Nh, nephrotoxin; Nr, neurotoxins; R, respiratory; T, teratogenic. Source: CAST, 2003.

[b] Letters in parentheses are the abbreviations for the mycotoxin's name used throughout the text.

Chu, 1991a, 1998, 2003; Eaton and Gallagher, 1994). Modulation of the immune system also plays an important role for overall toxic effects; mycotoxins can be either immunosuppressive (most often) or immunostimulatory (Pestka and Bondy, 1990; Bondy and Pestka, 2000). Recent studies on mycotoxin-induced apoptosis further revealed their mode of action at the cellular level. For some mycotoxins, e.g. FmB, the kinetics of apoptosis, cellular proliferation and necrosis plays a dominant role in its toxic effects.

5.1 Role of metabolism on the toxicity and mode of action of mycotoxins

Because metabolism plays a key role in the activation and detoxification of mycotoxins, the biological effects of mycotoxins are greatly affected by the metabolic activities in different animals under various conditions (Busby and Wogan, 1981; Chu, 1977, 1991a; Wogan, 1992; Coulombe, 1993; Eaton and Gallagher, 1994; Eaton and Groopman, 1994; McLean and Dutton, 1995; Neal, 1995; Massey, 1996; Hussein and Brasel, 2001). For example, AFB1 is metabolically activated before formation of AFB–DNA adducts to exert its carcinogenic effect. However, the activated AFB1 can also conjugate with proteins and interact with glutathione and then be excreted from the body. Thus, glutathione S-transferase serves as a key enzyme in the detoxification process for AFB1 (Guengerich *et al.*, 1996). In addition, other types of mixed function oxygenases (cytochrome P450) are also capable of metabolizing AFB1 to various hydroxylated metabolites, including AFM1 and AFQ1, and demethylase converts AFB1 to AFP1 (Forrester *et al.*, 1990; Raney *et al.*, 1992a, 1992b, 1992c). The cytosolic steroid reductase can reversibly convert AFB1 to aflatoxicol, which serves as a reservoir for AFB1 (Busby and Wogan, 1981; Hsieh, 1989). Formation of AF dialdehyde by aldo-keto reductases and subsequent reaction with proteins also plays a role against AFB1 toxicity (Guengerich *et al.*, 2001; Neal, 1995). The impact of epoxide hydrolase on the detoxification of AFB was recently reiterated through the cloning of the enzyme in yeasts (Kelly *et al.*, 2002).

For other mycotoxins, metabolism is usually related to detoxification. For example, metabolic deacetylation and de-epoxidation of TCTCs form hydroxylated derivatives and de-epoxide-TCTCs that are less toxic (Beasley, 1989; Yagen and Bialer, 1993). Hydrolysis of OA by proteolytic enzymes leads to the non-toxic metabolites. Hydroxylation of T-2 at the C-3' position and OA at the C-4 position, by the microsomal enzymes, has also been demonstrated. A lactonohydrolase was found to be responsible for detoxification of ZE (Takahashi-Ando *et al.*, 2002). Most hydroxylated mycotoxins may be excreted directly or form conjugates of glucuronide or sulfate and then be excreted from the body. Thus, various factors affect the kinetics of formation of adducts and detoxification, greatly affecting the toxicity of mycotoxins. A number of factors, including sex and species of the animal (genetics), environmental factors, nutritional status and mycotoxin synergism, etc., can either directly or indirectly modulate toxic effects.

5.2 Interactions of mycotoxins with macromolecules

5.2.1 Non-covalent interaction of mycotoxins with macromolecules

Mycotoxins and their metabolites may either react directly with macromolecules non-covalently or form a covalent bond(s) with macromolecules. Both types of binding may be important in toxicity. Non-covalent interactions usually occur between mycotoxins and enzymes, hormone receptors or other macromolecules. These interactions are reversible, and generally involve competition with binding sites of bioactive macromolecules. For example, OA interacts strongly with serum albumin, which leads to a long half-life for the toxin in animal and human bodies. Both AFB1 and AFG1 also bind with serum albumin, but their affinities with albumin are not as high as that of OA. OA also has high affinity with some enzymes and proteins, and the binding to the renal proteins is highly specific to a new organic anion transporter or more likely, a cytosolic binding component of unknown function present in humans, rats, mice and possibly pigs (Heussner *et al.*, 2002). Zearalenone and aflatoxicol interact with estrogenic receptors and steroid receptors, respectively. Although early studies demonstrated that both AFB1 and AFG1 have high affinity with nucleic acids and other macromolecules, including lysine-rich histone and chromatin, the role of their interaction was not clear. Nevertheless, NMR analysis of the binding of AFB1 and AFG1 with double-stranded $d(ATGCAT)_2$ or $d(GCATGC)_2$ and B-DNA and gel shifting analysis of the binding both mycotoxins with plasmid pBR322 revealed that intercalation of aflatoxins between the base pair was involved in the interaction and that the association of AFG1 with the nucleotides was weaker than that of AFB1. Intercalation of stable epoxide-AFB1 with these nucleotides has also been demonstrated. In the presence of specific inhibitor to occupy the intercalation site, the adduct yield at the specific site was found to be greatly diminished (Kobertz *et al.*, 1997). These data suggest that intercalation plays a significant role in subsequent covalent attachment of the carcinogen at the N7 position of guanine in DNA (Gopalakrishnan *et al.*, 1990; Raney *et al.*, 1990). A three-component complex, i.e. DNA–Mg^{++}–luteoskyrin, was isolated, but the interaction of the toxin with RNA polymerase was considered to be more important for the overall inhibition of RNA polymerase activity. Non-covalent binding of TCTC mycotoxins with ribosomes, and specifically with peptidyl transferase, is a key step in the inhibition of protein synthesis. The interaction of TCTC and OA with some immunomodulators and receptors in suppressor cells is considered to be one of the mechanisms by which these mycotoxins cause immunosuppression (Holt *et al.*, 1988). A phosphorylation and Ca^{++} dependent interaction of fumonisin with GTP-binding proteins has been demonstrated (Ho *et al.*, 1996). Both beauvericin and fusaproliferin are capable of forming complexes with model oligonucleotides; but the former forms a more stable complex than the later; no complex has been formed with FmB1 (Pocsfalvi *et al.*, 1997, 2000).

5.2.2 Formation of mycotoxin–DNA adducts

The mode of action of AFB1 is one of the best examples demonstrating the importance of activation of a mycotoxin for its interaction with macromolecules (Iyer *et al.*, 1994). It has been known for some time that the double bond in the dihydrofuran ring of the AF molecule is important for carcinogenic effects and that metabolism plays an important role in the action of aflatoxins (Chu, 1977, 1991a; Lin *et al.*, 1977; Busby and Wogan, 1981; Wogan, 1992; Eaton and Gallagher, 1994; Eaton and Groopman, 1994; McLean and Dutton, 1995; Neal, 1995; Massey, 1996; Wang and Groopman, 1999). Considerable evidence has accumulated over the years demonstrating that AFB1 must first be activated by mixed-function oxidases, specifically cytochrome P450 1A2 and 3A4, to a putative short-lived AFB-8,9-exo-epoxide before exerting its carcinogenic effects. The activated intermediate can be converted to other hydroxylated metabolites, or conjugated to glutathione or glucouronic acid etc. and then excreted, or bound to DNA, RNA and protein to exert its toxic, carcinogenic and mutagenic effects. Adduct formation of this epoxide with DNA occurs through nucleophilic attack primarily at the N-7 guanine position of DNA. This reaction causes mutations through GC to TA transversion. The AFB1-N7 guanine adduct is very unstable, with only the open ring form stable enough to be considered biologically important (Essigmann *et al.*, 1983), and analysis of this adduct in human urine permits quantitative evaluation of adduct formation after exposure to AFB1 (Garner, 1989; Groopman *et al.*, 1992, 1993, 1994, 1995, 1996; Groopman and Kensler, 1993; Chang *et al.*, 1994; Kensler and Groopman, 1996; Wang and Groopman, 1999). Extensive studies on the mechanism and factors affecting the formation of AFB1–DNA adducts have led to the following observations:

- Formation of AFB1–DNA adducts in animals are not only organ-specific and dose-related, but are also quantitatively correlated to the susceptibility of animals to the carcinogenic effects of AFB1 regarding carcinogenesis (Choy, 1993)
- Formation of AFB1–DNA adducts is also correlated to mutagenic (frame shift) and genotoxic effects of AFB1, including sister chromatid exchanges, recombination, chromosomal aberrations, and clastogenic responses
- The AFB-8,9-*exo*-epoxide predominantly binds to the N-7 position of guanine at G:C-rich regions of DNA
- The predominant mutation induced by the activated AFB-epoxides in *E. coli*, SOS response and in mammalian cell systems appears to be the GC → TA transversion.

Substantial evidence shows that activated AFB1 induces mutations in oncogenes (Beer and Pitot, 1989), including the *ras* gene, in experimental animals by binding to DNA primarily at codon 12 (McMahon *et al.*, 1990; Wogan, 1992; Shen and Ong, 1996). With the increasing evidence of the involvement of tumor suppressor genes in human carcinogenesis, the role of AFB1 in inactivating the tumor suppressor gene p53 has been intensively studied. Aflatoxin-induced G:C mutations, both G to T and G to A, have been implicated in the inactivation of human p53 tumor suppressor gene, and a high frequency of mutations has been found at codon 249 of p53 (Aguilar *et al.*, 1993; Soini *et al.*, 1996). Prevalence of codon 249 mutations, i.e. AGG to AGT (arg to ser) was found to be around 50 % in human hepatocellular carcinoma (Lehman *et al.*, 1994; Harris, 1996; Wild and Kleihues, 1996). In contrast, the prevalence of such mutations is low in the non-HPC patients. Although more data are needed to further support such a role, the identification of mutations/inactivation of p53 at this site has been suggested as a biomarker for AFB-induced liver cancers in humans. Some epidemiological data indicate that the presence of hepatitis B virus in humans enhances mutations in the p53 gene, and further suggest a synergistic effect between these two risk factors.

Sterigmatocystin acts in a mechanistically similar way to AFB1 in carcinogenesis through the formation of DNA adducts (Essigmann *et al.*, 1979; Raney *et al.*, 1992a) and mutation of the p53 suppressor gene (Wang and Groopman, 1999). Although OA–DNA adducts have been found in the kidney, liver and several other organs of mice and rats, neither the adduct structures nor their toxic effects have been defined (Pfohl-Leszkowicz *et al.*, 1993; Grosse *et al.*, 1995). Administration of tritiated OA to rats showed no covalent binding of OA to DNA of kidney and liver, although OA has been found to bind to several proteins and enzymes non-covalently. Formation of a reactive intermediate such as those for AFB1 is less likely to be involved in the action of OA (Arlt *et al.*, 2001; Gautier *et al.*, 2001; Zepnik *et al.*, 2001; Gross-Steinmeyer *et al.*, 2002).

5.2.3 Covalent interactions of mycotoxins with enzymes/proteins

Several mycotoxins have a reactive group (or groups), and thus readily react with proteins and enzymes, either specifically or non-specifically. PT and PA, for example, have been shown to react covalently with –SH and also possibly $-NH_2$ groups of proteins through addition or substitution reactions with their α, β-unsaturated double bonds conjugated to the lactone ring. In the model systems, PT forms the same type of adducts with glutathione and N-acetyl-L-cysteine (NAC). However, free cysteine formed markedly different adducts, including mixed thiol/amino type adducts involving the alpha–NH_2 group (Fliege and Metzler, 2000a, 2000b). Although the inhibitory effects of PT and PA on certain –SH enzymes, such as lactate and alcohol dehydrogenase, aldolase, and several bacterial enzymes, were thought to be due to the blocking of –SH and $-NH_2$ groups at the active sites of the enzymes, inhibition of lactate dehydrogenase by both PT and PA and penicillic acid is reversible (Chu, 1997). PT–cysteine adduct(s) still have a teratogenic effect. PT and PA are not the only mycotoxins which bind with – SH group of proteins or enzymes (Scott, 1975); some epipolythiopiperazines (e.g. gliotoxin) have also been found to bind with –SH enzymes covalently (Waring and Beaver, 1996; Chu, 1977). Binding of membrane proteins with TCTC mycotoxins may be due to their interactions with the –SH groups in the proteins. Both T-2 toxin and fusarenon X interact with a number of '–SH enzymes', thus causing a reduction in enzyme activity (Ueno, 1983, 1986). Alteration of the barrier function of the intestinal epithelium by PT was also attributed to its binding to the –SH groups of cellular proteins because such effect was protected by glutathione (Mahfoud *et al.*, 2002)). Although it has been shown that an activated AFB-epoxide reacts with the nucleophiles, its role in covalent interactions with specific enzymes is less clear. Several aflatoxin metabolites have been shown to interact with enzymes and proteins. Both AFB2a and AFG2a, which are very reactive in interacting with the NH_2 groups of proteins to form Schiff bases, have been shown to be very effective inhibitors of DNase *in vitro*. The activated AFB1 and AFG1 also react with albumin to form albumin adducts through the same Schiff-base mechanism, followed by Amadori rearrangement and subsequent condensation with another aldehyde group (Skipper and Tannenbaum, 1990). The presence of this adduct in human serum has been used as one of the indexes for human exposure to AFs (Sabbioni and Wild, 1991; Shebar *et al.*, 1993).

5.3 Inhibition of specific enzymes and protein synthesis

5.3.1 Inhibition of specific enzymes

Mycotoxins inhibit many enzyme systems both *in vivo* and *in vitro*, but most of these effects are secondary (Kiessling, 1986; Chu, 2002). For example, deoxyribonuclease and RNA polymerase activity are inhibited by AFs, ST, PT, LS, TCTC and α-amanitin in both *in vitro* and *in vivo* experiments. Ochratoxin A inhibits carboxypeptidase A, renal phosphoenolpyruvate carboxykinase (PEPCK), phenylalanine-tRNA synthetase and phenylalanine hydroxylase activity (Chu, 1974, 1977, 2002). Both PT and PA inhibit dehydrogenase through their interactions with the SH group at the enzyme's active center. Several mycotoxins, including territrem B and slaframine, are acetylcholine esterase inhibitors (Chen *et al.*, 1999). The inhibition of formation of certain metabolites of caffeine and testosterone, and its inhibitory effect of mutagenesis of AFB, was attributed to its inhibitory effect to some cytochrome P-450 enzymes (Kuilman-Wahls *et al.*, 2002). Moniliformin binds strongly with pyruvate dehydrogenase and subsequently affects the enzymes in the TCA cycle. Strong binding of MN and aldose reductase has also been observed (Deruiter *et al.*, 1993).

The inhibition of ceramide synthase (sphinganine/sphingosine N-acyltransferase) by Fm and AAL toxins is considered to be a primary effect because complex sphingolipids and ceramides are heavily involved in cellular regulation, including cell differentiation, mitogenesis and apoptosis in multiple organs (such as brain, lung, liver and kidney) of the susceptible animals, and in different cell systems (Wang *et al.*, 1991; Riley *et al.*, 1993b, 1994, 1999, 2001; Schroeder *et al.*, 1994; Spiegel and Merrill, 1996; Garren *et al.*, 2001; Merrill *et al.*, 2001). The primary amino group of Fms is essential for its inhibitory and toxic effects (Norred *et al.*, 2001). The target organs are especially sensitive to the sphingolipid dysregulation. Alteration of sphingolipid metabolism results in a dramatic increase in the free sphingolipid bases, i.e. sphinganine (Sa) and sphingosine (So) in tissues, and a decrease in complex sphingolipid levels. Significant increases in the Sa/So ratio were found in the serum of pigs fed a diet containing as little as 5 ppm total FmB. Thus, testing Sa/So ratios in serum and urine has been suggested as an early biomarker of FmB exposure in animals and humans (Merrill *et al.*, 1993, 1996). FmB1 was also found to inhibit cytochrome P-450 enzyme selectively, with the most significant effect on CYP2C11, which was considered to be due to the suppression of protein kinase resulting from the inhibition of sphingolipid biosynthesis (Spotti *et al.*, 2000).

5.3.2 Inhibition of protein synthesis

A number of mycotoxins, including OA, CT, PR-toxin, TzA and TCTCs, have been found to inhibit protein synthesis, but most of these effects are considered secondary. However, in the case of TCTC mycotoxins, the inhibition of protein synthesis is one of the earlier events in the manifestation of their toxic effects (Feinberg and McLaughlin, 1989). OA inhibits protein synthesis by acting as a competitive inhibitor for phenylalanine-tRNA synthetase, and its inhibitory effect on renal PEPCK is due to inhibition of *de novo* synthesis of this enzyme because renal levels of mRNA coding this enzyme are decreased by OA (Meisner and Krogh, 1986). Even AFs are protein synthesis inhibitors, but this effect is secondary because aflatoxins primary act at the transcriptional level. The most potent protein inhibitory mycotoxins are TCTCs, which act at different steps in the translation process (Kiessling, 1986; Beasley, 1989). Inhibitory effects vary considerably with the chemical structure of the side chain. In general, T-2, HT-2, NIV, FS-x, DAS, verrucarin A and roridin A affect at the initiation step, whereas verrucarol, trichothecin and crotocin affect elongation. Inhibition of the elongation step by DAS and FS-x has also been demonstrated. Some of the lesser known TCTCs, such as trichodermol and trichodermone, affect the termination step (Feinberg and McLaughlin, 1989). In contrast, ZE stimulates protein synthesis by mimicking hormonal action.

5.4 Mechanism of modulation of immune systems

Although many mycotoxins affect immune systems in some animals or *in vitro* systems, only AFB, OA and TCTC mycotoxins modulate immune processes that may affect human and animal health. Depending on the nutritional status, toxin dose and other factors, these mycotoxins can be either immunosuppressive (most often) or immunostimulatory (Pestka and Bondy, 1990; Pestka, 1994; Bondy and Pestka, 2000). Because TCTC mycotoxins such as T-2 toxin and DAS exert major toxic effects on the bone marrow, lymph nodes, thymus and spleen in mammals, modulation of the immune system by this group of mycotoxins has been most extensively studied. Animals receiving OA showed general immunotoxic signs, including lymphocytopenia and depletion of lymphoid cells, especially in the thymus, bursa of Fabricius and Peyer's patches. The mechanisms of action of these three major groups of mycotoxins on humoral and cell-mediated immunity (CMI) are briefly discussed below. In general, TCTCs and OA affect both humoral immunity and CMI, while AFB primarily affects CMI.

5.4.1 Effect on humoral immune response

TCTCs have been shown to have a marked effect on humoral immunity. Administration of T-2 toxin to animals results in reduction of their resistance to infection, and decreased antibody formation. Using plaque-forming cells (PFC) as an indicator, an inhibition of anti-sheep red blood cell response by low doses of T-2 toxin and DAS was demonstrated. Stimulation of antibody response to T lymphocyte-independent immunogens such as polyvinylpyrrolidone and dinitrophenyl-Ficoll was demonstrated in spleen cells of mice exposed to T-2 toxin or DAS. Thus, T-2 toxin or DAS may selectively affect subpopulations of T-suppressor cells or their precursors; the suppression of antibody synthesis might be due either to impairment of antibody-forming or to T-helper cell activities. B lymphocyte precursor cells were also found to be sensitive targets of T-2 toxin (Holladay *et al.*, 1995). Similarly, serum IgM, IgA and complement C3 levels but not IgG levels were decreased when calves were administered a sublethal dose of T-2 toxin. Immunoglobulin G levels decreased substantially and IgM levels decreased slightly when monkeys were fed T-2 toxin for 4–5 weeks.

Immunosuppressive effects have also been observed for other TCTCs, including the macrocyclic types. However, DON has been found to have both immunostimulatory and immunosuppressive effects in experimental animals. Mice fed a diet containing more than 10 ppm of DON experienced thymic atrophy and other structural changes, decrease in antibody formation, and suppression of B and T cell proliferation. Serum and saliva IgA was significantly increased in mice fed high levels (25 ppm) of DON, whereas serum IgM and IgG decreased. The increased IgA production is related to IgA-mediated nephropathy in mice, thus it was postulated that DON might be one of the etiologic agents in IgA nephropathy, which is the most common glomerulonephritis in humans worldwide (Pestka and Bondy, 1990; Greene *et al.*, 1994). Likewise, an elevated serum IgA level as well as deposition of IgA in the glomerular mesangium have also been observed in experimental mice fed with low levels of NIV, and some of the pathological changes in mice resembled those in human IgA nephropathy (Hinoshita *et al.*, 1997). Depressed humoral immune responses, including lower serum IgM and IgG levels and suppressed antibody responses to sheep red blood cells and other antigens in mice, have been observed in animals fed or injected with OA. The immunosuppressive effects of OA are prevented by phenylalanine both *in vitro* and *in vivo*.

5.4.2 Cell-mediated immunity

Regarding CMI, both T-2 toxin and DAS inhibit proliferation of mitogen-stimulated human lymphocytes, especially T-cell populations. This effect is concentration dependent, and low concentrations of T-2 toxin were stimulatory. Structure–activity studies showed that T-2, HT-2, and 3'-OH T-2 toxins were most effective in inhibition, while 3'-OH HT-2, T-2 triol and T-2 tetraol toxins were 50–100 times less effective. The macrocyclic TCTC roridin and verrucarin A were 75–100 times more potent than T-2 toxin. The differences in the immunotoxic effects of various TCTCs were attributed to differences in both uptake and metabolism of toxins by the cells. Gliotoxin, which inhibits phagocytosis by macrophage as well as proliferation of T-cells, has been found to be a potent immunomodulating agent, and it specifically inhibits transcription factor of NF-kappa B (Pahl *et al.*, 1996).

DON and other TCTCs at concentrations required for partial or maximal protein synthesis inhibition have been shown to hyperinduce cytokines in T-helper cells and stimulate interleukin-1 and 2 (IL-1, 2) production in spleen cell cultures (Bondy and Pestka, 2000; Wong *et al.*, 2001). Induction of expression of mRNA of ILs-2, 4, 5 and 6 in a T-cell model EL4.IL-2 by DON was found at levels required for partial or maximal protein synthesis inhibition. A single oral gavage with DON is sufficient to induce IL-1 beta, IL-2, IL-6 and TNF α mRNA levels in Peyer's patches and spleen (Dong *et al.*, 1994; Warner *et al.*, 1994; Ouyang *et al.*, 1996;). The TCTC-induced cytokine superinduction could lead to the terminal differentiation of immunoglobulin-secreting cells via T-cell-mediated polyclonal differentiation of B-cells or their precursors. The Peyer's patch may be particularly sensitive to such dysregulation. Levels of IL-2, IL-3 and IFNγ mRNA from Con-A-activated splenocytes were higher in T-2 toxin treated CD-1 mice. DON-mediated anorexic and growth effects were largely independent of TNF-α, and its dysregulation of IgA production was partially dependent on the interaction of TNF-α with TNFR1 (Pestka and Zhou, 2002). Since the protein levels of all three cytokines were also increased, the T-2 toxin increased the translational/post-translation efficiency of these cytokines (Dugyala and Sharma, 1997). The DON-mediated inhibition of nuclear protein binding to NRE-A, an IL-2 promoter negative regulatory element, in EL-4 cells, was also attributed to the modification at the post-translation level. Such modification may be requisite for DON's down-regulatory effects (Yang and Pestka, 2002). Aflatoxin B1 preferentially affects macrophage functions, and was found to be also acting through regulation of cytokine levels (Dugyala and Sharma, 1996). OA, CPA, ZE and alpha-zearalenol also stimulated cytokine production with increases in IL-2 and IL-5 in the EL4.IL-2 cell system at certain toxin levels, but PT and T-2 toxin were inhibitory in this system.

Suppressed delayed hypersensitivity (DH), including cutaneous DH to phytohemagglutinin, was demonstrated in AFB- and OA-treated animals. An apparent increase in DH by T-2 toxin was demonstrated when the toxin was administered a few days after the mice were sensitized with sheep red blood cells. Pretreatment of animals with either T-2 toxin or DAS was also found to prolong skin graft survival time.

5.5 Induction of apoptosis

5.5.1 Induction of apoptosis by mycotoxins

In addition to necrosis, which is characterized by a typical morphological change in cellular membrane, and cellular swelling that causes cellular death, mycotoxins and many naturally occurring toxicants also cause apoptosis (APT), which is a regular programmed cell death. Different from necrosis, APT is characterized morphologically by cell shrinkage, membrane blebbing and chromatin condensations. Fragmentation of DNA and nuclear condensation in cells are used to detect the early phase of APT. Induction of APT by AFB1, beauvericin (Logrieco *et al.*, 1998), OA (Seegers *et al.*, 1994), several TCTCs (T-2 toxin, roridin A, NIV, DON, etc.), CT and luteoskyrin (LS) (Ueno *et al.*, 1995), rubratoxin B (RB) (Nagashima, 1996), FmB1, and AAL toxin-TA, gliotoxin (GT) (Sutton *et al.*, 1995) and wortmannin (WM), has been observed in a number of cell lines as well as in the cells of target organs after animals received the toxin. The concentration causing such induction varies with different toxins tested. Some mycotoxins may induce APT in one system but not in others. For example, several TCTCs, including roridin A, T-2 toxin, NIV and DON, induce APT in HL-60 at significant low levels, with minimum effective dose ranging from 0.001 to 0.2 μg/ml. In the same system, the medium level for AFB1, OA, CT and LT to induce APT was in the 1–20 μg/ml range, and PA at a level of 100 μg/ml. However, no induction of APT was observed for cytochalasin A, ST, rugulosin, cyclochlorotine, kojic acid, RB, Fm, WM (Ueno *et al.*, 1995) and fusaric acid (Voss *et al.*, 1999).

The effect of different TCTCs on APT varied with the system tested. In the RAW 264.7 murine macrophage and U937 human leukemic cells, satratoxin G, roridin and verrucarin A were most potent, followed by T-2 toxin, satratoxin F and H; NIV and DON were least (Yang *et al.*, 2000). T-2 toxin, satratoxin G, roridin and DAS were also found to be most potent in inducing APT in the HL-60 system (Nagase *et al.*, 2001). Both the acetyl group at the C-4 position and the isovaleryl or 3?-hydroxyisovaleryl group at the C-8 position of the T-2 molecule are important for inducing APT in the thymus (Islam *et al.*, 1998). Depending on the lymphocyte set, tissue source and glucocorticoid induction, DON can either inhibit or enhance apoptosis in murine T, B and IgA cells. Prostaglandin and glucocorticoid were involved in the DON-induced APT (Wong *et al.*, 2001). The hypothalamic–pituitary–adrenal axis plays a key role in LPS potentiating DON-induced lymphocyte APT (Islam *et al.*, 2002). An increase in nitric oxide and prostaglandin E-2 production was found when a murine macrophage cell line was grown in the presence of FmB1 (Meli *et al.*, 2000).

The effect of FmB and AAL toxin on APT in many cell lines has been extensively studied in recent years (Lim *et al.*, 1996; Tolleson *et al.*, 1996a, 1996b; Wang *et al.*, 1996a, 1996b). The kidney and liver of animals or cells of tomato plants treated with FmB or TA (one of the major AAL toxins) also showed typical apoptotic characters, with the degree of injury being related to dose and time of exposure. For example, with the FmB-treated keratinocytes, nucleosomal DNA fragments were found in the media 2–3 days after exposure to the toxin, and increased DNA strand breaks were detected in attached keratinocytes. Morphological changes, with typical features of APT, were also observed in the FmB1-treated keratinocytes and HET-1A cells. In feeding studies, induction of APT and hepatocellular and bile duct hyperplasia were observed in both male and female rats. Apoptosis was also observed in the kidneys; higher FmB1 levels were needed for female than for male rats. Addition of FmB1 or AAL toxin to African green monkey kidney cells (CV-1) induced formation of DNA ladders, compaction of nuclear DNA, and the subsequent appearance of apoptotic bodies. In plant cells, DNA ladders were observed during cell death in toxin-treated tomato protoplasts and leaflets. Coincident with the appearance of DNA ladders, there was a progressive delineation of fragmented DNA into distinct bodies during the death of toxin-treated tomato protoplasts. The intensity of DNA ladders was enhanced by Ca^{++} but inhibited by Zn^{++}. DNA fragmentation was also found in the dying cells of toxin-treated onion root caps and tomato leaf tracheal elements, and apoptotic-like bodies were present in sloughing root cap cells. In general, the fundamental elements of APT, as characterized in animals, are conserved in plants.

5.5.2 Mechanism of apoptotic effect induced by mycotoxins

Similar to other toxicants, modification of signal transduction pathway by mycotoxins plays a key role for APT indution. Inhibition of phosphatidylinositol 3-kinase by WM was considered to be the cause of apoptotic cell death in a number of cell systems (Padmore *et al.*, 1996; Yao and Copper, 1996). Gliotoxin-induced apoptosis in spleen cells was found to be mediated by cAMP (Sutton *et al.*, 1995). Regulation of cellular Ca^{++} ions play an important role in the induction of APT by several mycotoxins, including T-2 (Yoshino *et al.*, 1996), RB (Nagashima, 1996), FmB and TA, one of the major AAL toxins, GT (Liu *et al.*, 2000), sporidesmin and related epipolythiodioxopiperazines (Waring and Beaver, 1996). The mycotoxin-induced APT and Ca^{++} regulation was inhibited by Zn^{++} in several studies. Whereas p53 was found not to be involved in the RB-induced APT signal transduction (Nagashima, 1996), exposure of cells to ST resulted in failure of G1 arrest. Thus, the carcinogenic effects of ST were considered to be mediated by failure of the p53-mediated G1 checkpoint (Xie *et al.*, 2000). Ochratoxin A interacts with specific cell types of distinct members of the mitogen-activated protein kinase (MAPK) family, e.g. MAPK 1 and 2, at concentrations where no acute toxic effect can be observed. Induction of APT via the c-jun amino–terminal–kinase (JNK) pathway was suggested as the possible cause for the OA-induced changes in renal function, as well as part of its teratogenic action. Activation of caspase 3 was also involved in APT by OA (Schwerdt *et al.*, 1999; Gekle *et al.*, 2000). Contrary to other mycotoxins, ZE functioned as an anti-apoptotic agent by increasing the survival of MCF-7 cell cultures undergoing APT. ZE stimulates cytokine production and the growth of estrogen receptor-positive human breast carcinoma cells. The mitogen-activated protein kinase signaling cascade through the ras/Erk pathway is required for ZE's effects on cell-cycle progression in MCF-7 cells (Ahamed *et al.*, 2001).

Much effort in elucidating the sites involved in induction APT by mycotoxins in the signal transduction pathway was focused on FmB1 and TCTCs. Data shown in Table 16.3 clearly indicate that MAPKs play a significant role in the induction of APT and toxic effects by various TCTCs. The involvement of tumor necrosis factors

Table 16.3 Apoptotic effects induced by selected mycotoxins

Mycotoxins Cell line/species	Apoptotic effects	Reference
Fusarenon-X (FS-X)		
Human promyelotic leukemia HL-60	Releases cytochrome C and activates caspases −3, −8, and −9	1
Satratoxin and other TCTCs		
U937 human leukemia cells and murine macrophage RAW 264.7	Activation of MAPKs, extracelluar signal-regulated protein kinase (ERK), p38 MAPK, and stress-activated protein kinase /c-JNK	2
T-2 toxin & other TCTCs		
HL-60	Activates caspase-3 and 9 and DFF/CAD	3, 4
	Elevates Ca^{++} signal by the activation of endonuclease and protease	5, 6
Dorsal skin of hypotrichotic WBN/ILA-Ht rats	Depresses basal cell proliferation and elevates TGF-beta q mRA level	7
Thymocytes/mice	Independent of Fas/Fas	8
Fumonisin B1		
Mouse/liver and kidney	Increases expression of TNFα, interleukin (IL)-1β and TNFα signaling molecules, TNF receptor 55 and receptor protein with no alteration of Fas related signal	9–12
	Increases expression of IL-1α and IL-1 receptor antagonist (IL-1Ra) and caspase 8	
	Increases expression of oncogenic transcription factors B-Myc, c-Myc, Max and Mad	
	Increases apoptotic genes Bcl-2, Bax and Bad	
Renal epithelial LLC-PK1	Involves serine palmitoyltransferase and induces expression of calmodulin	13, 14

Table 16.3 Apoptotic effects induced by selected mycotoxins—*continued*

Mycotoxins Cell line/species	Apoptotic effects	Reference
Tomato suspension	Involves Le-pirin gene which encoding a protein 53% identical with human protein PIRIN that interacts with oncogene Bc1-3	15
CV-1 and MEF	Involves the TNF signal transduction pathway and cleavage of caspase; the tumor suppressor gene p53 was not involved	16
CV-1	Repressed specific isoforms of protein kinase C and cyclin-dependent kinase 2 (CDK2) Induced expression of CDK inhibitors, p21Waf1/Cip1, p27Kip1 and p57Kip2 The baculovirus IAP gene (blocks TNF induced APT) was involved Inhibition of interleukin converting enzymes proteases or caspases by the baculovirus gene p35 The tumour suppressor gene p53 and Bcl-2 were not involved	17, 18
Simian virus 40 transformed CV-1	Resistant to the antiproliferative or apoptotic effects of FB1 Sp1 or Sp1-related proteins mediate FB1induced activation p21 promoter	19

Abbreviations: CV-1, African green monkey kidney fibroblasts; MEF, mouse embryo fiberblasts, TNF, tumor necrosis factor

References: 1, Miura *et al.*, 2002; 2, Yang, G. H. *et al.*, 2000; 3, Yshino *et al.*, 1996; 4, Nagase *et al.*, 2001; 5, Shinozuka *et al.*, 1997; 6, Yoshino *et al.*, 1996; 7, Albarenque *et al.*, 2000; 8, Alam *et al.*, 2000; 9, Bhandari and Sharma, 2002a; 10, Bhandari and Sharma, 2002b; 11, Lemmer *et al.*, 1999; 12, Dugyala *et al.*, 1998; 13, Reiley *et al.*, 1999; 14, Kim *et al.*, 2001; 15, Orzaez *et al.*, 2001; 16, Jones *et al.*, 2001; 17, Zhang *et al.*, 2001; 18, Ciacci and Jones, 1999; 19, Zhang *et al.*, 1999.

(TNFs) for the FmB1-induced APT cannot be overemphasized (He *et al.*, 2001; Sharma *et al.*, 2001; Zhang *et al.*, 2001). The induction of APT by AFB1 and LS was attributed to their inhibitory effect on RNA synthesis as a result of their binding with DNA and the subsequent inhibitory effect of DNA-dependent RNA polymerase (Ueno *et al.*, 1995). Although suppression of protein synthesis was considered to be the mechanism for the induction of APT by TCTC, OA and several other mycotoxins (Ueno *et al.*, 1995), this effect was considered to be more important for the TCTC-induced APT (Yang *et al.*, 2000). The levels of satratoxin G and other TCTCs inhibiting protein synthesis correlated well with those for APT induction as well as activation of MAPKs (Shinozuka *et al.*, 1997; Yang *et al.*, 2000; Nagase *et al.*, 2001). Considerable evidence has shown that there is a close correlation between the inhibition of ceramide synthase with alteration of signaling and APT induction by FmB1 (Schmelz *et al.*, 1998; Tolleson *et al.*, 1999; He *et al.*, 2001; Bhandari and Sharma,

2002a, 2002b). Because complex sphingolipids and ceramide play a key role in the regulation of cell growth, the basic mechanism for FmB1-induced APT was attributed to its inhibitory effect on the enzyme (Merrill *et al.*, 1996, 1997, 2001). Inhibition of CER synthase by FmB1 leads to the accumulation of the sphingoid bases and CER, and subsequently promotes cell death. Because the cellular response depends on the factors and agents that protect/induce APT, the balance between the rate of APT and proliferation was considered to be important for the tumorigenesis induced by FmB1 (Riley *et al.*, 2001). Since FmB1 is neither genotoxic in bacterial mutagenesis screens nor affects unscheduled DNA synthesis, the apoptotic necrosis, atrophy and consequent regeneration together with its ability to alter gene expression in the signal transduction pathways may play an essential role in its carcinogenic effects (Dragan *et al.*, 2001; Jones *et al.*, 2001; Zhang *et al.*, 2001).

5.6 Other mechanisms

As well as those described above, several other mechanisms have been postulated for the mode of action of mycotoxins. Alteration of membrane structures by TCTCs and several other mycotoxins may directly affect cellular components (Bunner and Morris, 1988). Modulation of signal translocation of protein kinase (PK) by mycotoxins has been suggested as another mechanism. Because sphingolipids are natural inhibitors of the Ca^{++} activated-protein kinase C (PKC), it has been suggested that FmB may affect the PKC-regulated functions in the signal translocation process (Huang *et al.*, 1995). A direct interaction of the diacylglycerol binding site of PKC to phorbol esters, which are well-known tumor promotors, by FmB1 has been observed (Yeung *et al.*, 1996). FmB induced PKC translocation via a direct interaction at the diacyglycerol site that also binds phorbol esters, a group of well-known tumor promoters (Yeung *et al.*, 1996). CPA alters Ca^{++} hemostasis and induces charge alteration in plasma membranes and mitochondria; the reversible inhibition of reticulum Ca^{++}-dependent ATPase has been considered as the primary cause (Petr *et al.*, 1999). Paxilline, a reversible, non competitive inhibitor of the cerebellar inosital 1, 4, 5-triphosphate (InsP -3) receptor, inhibits the InsP-3-induced Ca^{++} release (Longland *et al.*, 2000) and the enzyme formation (Bilmen *et al.*, 2002). OA also may impair cellular Ca^{++} cAMP homeostasis. In the renal epithelial cells, OA interferes with hormonal Ca^{++} signaling and leads to altered cell proliferation (Benesic *et al.*, 2000). Secalonic acid D, a cleft palate-inducing mycotoxin produced by *P. oxalicum* and other fungi, reduces palatal cAMP levels. It inhibits the binding of the cAMP response elements to the binding protein and alters phosphorylation, and leads to altered expression of genes involved in cell proliferation – an event critical for normal palate development (Umesh *et al.*, 2000; Balasubramanian *et al.*, 2001).

Studies on the impact of oxidative DNA damage on carcinogenesis and mutagenesis have generated renewed interest in the role of formation of free radicals in the toxic effect of mycotoxins. Specific biomarkers, such as the formation of 8-hydroxydeoxyguanosine (8-OHdG) as a result of the interaction of DNA with the reactive oxygen species, have been used to identify such interaction. Luteoskyrin and several related anthraquinone-type mycotoxins are capable of inducing free radicals leading to the formation of 8-OHdG (Akuzawa *et al.*, 1992; Ueno *et al.*, 1993). Oxidative DNA damage has been found to be involved in the genotoxicity of AFB (Shen and Ong, 1996; Shen *et al.*, 1996), and lipid peroxidation was considered as a possible secondary mechanism for ST (Sivakumar *et al.*, 2001). Enhancement of lipid peroxidation is considered one of the manifestations of cellular damage in OA toxicity (Rahimtula *et al.*, 1988; Hoehler *et al.*, 1996a) and the induction of free radicals in a bacterial model system was found to be regulated by Ca^{++} ions (Hoehler *et al.*, 1996b). Superoxide dismutase and catalase have been shown to have some protective effect for OA-induced nephrotoxicity in rats (Baudrimont *et al.*, 1994a). OA-induced lipid peroxidation in the Vero cell was also found to be decreased in the presence of these two enzymes, but the effects of aspartame and piroxcam were less (Baudrimont *et al.*, 1994b). OA-evoked oxidative stress has been shown as an increase of alpha-tocopherol plasma levels and the expression of the oxidative haem oxygenase-1, specifically in the kidney (Gautier *et al.*, 2001).

6 Preventive measures

6.1 Controlling mycotoxin problems through prevention of toxin formation

Owing to the widespread nature of fungi in the environment and the high stability of the toxins, mycotoxin problems are difficult to control. The most effective approach is to prevent the formation of the toxins in the field as well as during storage (Sinha and Bhatnagar, 1998; Chu, 2002; CAST, 2003; Chu and Bhatnager, 2003). Some measures, including use of resistant varieties (most effective, but not all are successful), crop rotation, use of earlier harvest varieties, avoidance of overwintering in the field, minimizing bird and insect damage and mechanical damage, cleaning and drying grain quickly to less than 10–13 % moisture, increased sanitation in the field, and use of mold inhibitors, have been suggested (Northold and Bullerman, 1982; Magan and Lacey, 1988; Mills, 1989; Birzele-Barbara *et al.*, 2000). After harvest, crops should be stored in a clean area to avoid insect and rodent infestation, and moisture should be reduced by regular aeration to prevent mold contamination and growth. Rigorous quality control programs, including analysis of mycotoxins, examination of broken kernels and removal of suspected contaminated feed, should be established in the milling facility. Some of these practices have been implemented at different levels in farm practice and agricultural industries (Trenholm *et al.*, 1988, CAST, 2003).

Extensive research on the control of aflatoxin formation has been conducted in the last decade. Progress has been made in trying to breed and identify germplasm lines resistant to AF, DON, and Fms, with some success (Abbas *et al.*, 2002; Windham and Williams, 2002). Efforts have been made to clone the genes that are involved in the formation or inhibition of formation of mycotoxins. Some promising research results have been obtained for some mycotoxins within the last 10 years (Cleveland and Bhatnagar, 1992; Linz and Pestka, 1992; Bhatnagar *et al.*, 1993). For example, we now know that the genes involved in the biosynthesis of AFB, TCTC and Fm are located in clusters, and genes that regulate the formation of AFB and TCTC have also been found. Transformants of strains prepared by the deletion of key genes involved in the formation of AFB, TCTC and Fms have been obtained, and those transformants have lost the ability to produce these mycotoxins (Bhatnagar *et al.* 1995; Proctor *et al.*, 1995b; Kale *et al.*, 1996; Desjardins and Hohn, 1997; Keller and Hohn, 1997; Bacon *et al.*, 2001; Duvick, 2001). Such information is now gradually being transferred to practical applications. Genetically modified crops resistant to insect infestation (e.g. Bt-corn) have shown some effect in minimizing fungal propagation and AF, DON and Fms formation (Dowd, 2001; Bakan *et al.*, 2002; Magg *et al.*, 2002; Schaafsma *et al.*, 2002). Naturally occurring *Aspergillus* strains non-toxic to peanut, corn and cotton have resulted in a great reduction of AFB formation (Brown *et al.*, 1991; Dorner *et al.*, 1992; Bhatnagar *et al.*, 1992, 1993; Cotty and Bhatnagar, 1994; Horn and Dorner, 2002). Transfer of the control genes to naturally occurring fungi (wild type) would prevent/minimize toxin formation. Although application of non-toxic *Aspergillus* strains to the peanut, corn and cotton fields has shown promising results in reducing AFB formation in the field, it should be pointed out that non-aflatoxigenic strains should be rigorously tested for their potential of production of other mycotoxins before wide application. For non-aflatoxigenic *Aspergillus*, its ability for production of CPA should be determined. Because the process for the mycotoxin contamination is very complex (Payne, 1998), a combination of approaches will be required to eliminate or even control the pre-harvest toxin contamination problem (Bhatnagar *et al.*, 1995), and conditions for using these approaches should be rigorously evaluated.

In post-harvest storage, some antifungal agents, including organic acids (sorbic, propionic, benzoic), antibiotics (neatamycin or pimaricin, nisin), phenolic antioxidants (butylated hydroxyanisole and butylated hydroxytoluene) and fumigants (such as methylbromide, chlorine) have been found to be partially effective in inhibiting the growth and toxin production of some fungi. Some spices and essential oils also appear to have some antifungal effects (Doyle *et al.*, 1982; Ray and Bullerman, 1982; Park, 1993a, 1993b). Current research is aimed at identifying and characterizing naturally occurring antifungal agents in commodities such as corn (Huang *et al.*, 1997; Neucere and Cleveland, 1997; Woloshuk *et al.*, 1997; Gembeh *et al.*, 2001) and other crops (Burrow *et al.*, 1997). To enhance their natural resistance, the crops can be genetically modified through enhancing the genes that produce such agents. Aflastatin A, a novel inhibitor of aflatoxin production, has been isolated from a *Streptomyces* species (Sakuda *et al.*, 1996; Ono *et al.*, 1997).

6.2 Avoiding human exposure through rigorous monitoring programs

While it is impossible to remove mycotoxins completely from foods and feeds, effective measures to decrease the risk of exposure depend on a rigorous program of monitoring mycotoxins in foods and feeds. Consequently, governments of many countries have set limits for permissible levels or tolerance levels for a number of mycotoxins in foods and feeds (Van Egmond, 1989a, 1996, 2002; Stoloff *et al.*, 1991; Verardi and

Rosner, 1995; FAO/WHO, 1997; Sinha and Bhatnagar, 1998; Trucksess and Pohland, 2001; Park and Troxell, 2002). Monitoring of mycotoxins in animal feeds is also important because it not only provides a healthier diet for animals, but may also indirectly prevent any mycotoxin residue carryover into animal products intended for human consumption. These data, together with toxicological and exposure assessments, can then be used to make risk assessments for different mycotoxins and to determine the levels that are unlikely to be hazardous to humans (Kuiper-Goodman, 1990, 1995). Official methods of analysis for many mycotoxins have been established (Horwitz, 2000). However, because of the diverse chemical structures of mycotoxins, the presence of trace amounts of toxins in very complicated matrices that interfere with analysis, and the uneven distribution of the toxins in the sample, analysis of mycotoxins is a difficult task (Chu, 1991c, 1995; Richard *et al.*, 1993). In general, the samples are first ground to fine particles, then extracted with appropriate solvent systems, followed by a clean-up before they are subjected to separation and quantitation and confirmation protocols. Thus, it is not uncommon that the analytical error can amount to 20–30 % (Horwitz *et al.*, 1993). To obtain reliable analytical data, an adequate sampling program and an accurate analytical method are both important (Park and Pohland, 1989; Whitaker *et al.*, 1994, 1995; Van Dolah and Richard, 1999; Trucksess and Pohland, 2001, 2002; DeVries *et al.*, 2002).

To minimize the errors that might be introduced at each of the above steps, investigations into the development of new sensitive, specific and simple methods for mycotoxin analysis have been conducted since aflatoxin was first discovered in the early 1960s. Such studies have led to many improved and innovative analytical methods for mycotoxin analysis in the last few years (Chu, 1991b, 1995, 2001; Trucksess and Pohland, 2001; DeVries *et al.*, 2002). Simplified and efficient clean-up procedures, such as solid-phase extraction cartridges, have been established. Better analytical quality control has been established by using 'Certified Reference Materials' (Boenke, 1997). More sensitive TLC, HPLC and GC techniques are now available. Sensitive and versatile high-resolution MS and GC/tandem MS/MS are coming to the market. The MS methods have also been incorporated into HPLC systems. New chemical methods, including capillary electrophoresis, fluorescence polarization immunoassay and biosensors are emerging, and have gained application for mycotoxin analysis (Maragos, 1997; Chu, 2000; Maragos *et al.*, 2001; Maragos and Plattner, 2002; Nasir and Jolley, 2002). Recent progress in nanotechnology may lead to new innovative approaches for monitoring of mycotoxins in grains as well as new diagnostic tools for mycotoxicosis.

After a number of years of research, immunoassays have gained wide acceptance as analytical tools for mycotoxins. Antibodies against almost all of the mycotoxins are now available (Chu, 1986, 1991c, 1996; Pestka, 1988, 1994; Chu, 2000) and new immunoassay systems are emerging (Maragos, 1997; Maragos and and Plattner, 2002). Whereas quantitative immunoassays are still primarily used in the control laboratories, immunoscreening methods have been widely accepted as a simple approach to screening for mycotoxins in several commodities. Many immunoscreening kits, which require less than 15 minutes to complete, are commercially available. These assays eliminate many of the above steps; a sample after extraction usually proceeds directly to analysis (Chu, 1996; Trucksess and Wood, 1997; Trucksess and Pohland, 2001). The immunoaffinity method has also become popular as a clean-up method in conjunction with other chemical methods. Several immunoassay protocols have been adopted as first action by the AOAC. Automation of mycotoxin analysis has become a reality by combining these newer techniques. Instead of using antibodies, molecularly imprinted polymers are being used for solid-phase extraction of mycotoxins such as OA (Jodlbauer *et al.*, 2002). This new approach could provide a more stable affinity-based reagent for mycotoxin analysis. Rather than analysis of toxin, PCR methods, based on the primers of key enzymes involved in the biosynthesis of mycotoxins, have been introduced for the determination of toxicogenic fungi present in foods (Birzele *et al.*, 2000; Edwards *et al.*, 2001; Schnerr *et al.*, 2002). Detailed protocols for mycotoxin analysis can be seen in several of the most recent reviews and books (Chu, 1991b, 2001; Gilbert, 1999; Van Dolah and Richard, 1999; DeVries *et al.*, 2002) and the most recent edition of *AOAC* (Horwitz, 2000). The progress of research dealing with mycotoxin analysis can be seen from the numerous papers cited by the 'General Referee on Mycotoxins' in the Annual Report on Mycotoxins, which generally appears in every year's January/February issue of the *Journal of the Association of Official Analytical Chemists International* (Trucksess, 2003).

6.3 Removal of mycotoxins from commodities through detoxification and other physical and chemical means

Detoxification may be desirable in controlling mycotoxins, but it is very difficult and in some cases not economically feasible. The toxicity and nutritional values of the detoxified commodities should be tested extensively, and the cost of such treatment should be rigorously analyzed. Although several chemical detoxification methods, including acids, alkalis, aldehydes and oxidizing agents, and gases like chlorine, sulfur dioxide, $NaNO_2$, ozone and ammonia, have been tested for several mycotoxins, especially AF, TCTC and Fm, only the ammoniation process is an effective and practical method for detoxification of AF (Jorgensen and Price, 1981; Cole, 1989; Park, 1993a, 1993b; Park and Liang, 1993; Piva *et al.*, 1995; Rustom, 1997; Lemke *et al.*, 2001). After rigorous investigations on the safety of an ammoniated commodity that contained AF, a recent report showed that ammoniated peanut meals may still have different effects *in vivo* when incorporated into animal diets even when AF levels are reduced to acceptable levels (Neal *et al.*, 2001). Solvent extractions have been shown to be effective, but are not economically feasible. Physical methods such as thermal inactivation, photochemical or gamma irradiation have been tested, but have limited efficiency (Sinha and Bhatnager, 1998).

Physical screening and subsequent removal of damaged kernels by air blowing, washing with water or use of specific gravity methods have shown some effect for some mycotoxins, including DON, FmB and AFB1 (Trenholm *et al.*, 1988, 1992). For example, Fms in corn screens were 10 times higher than in the regular kernels. In rice, Fm is concentrated in the husks. Thus, it is possible to decrease the Fm levels in the contaminated grains through physical removal of damaged grains. Likewise, AFB1- and DON-affected grains are lighter and can be separated out.

6.4 Removal of mycotoxins during food processing

While cooking generally does not destroy mycotoxins, some mycotoxins can be detoxified or removed by certain food-processing procedures. For example, extrusion cooking appears to be effective for detoxifying DON but not AFB (Cazzaniga *et al.*, 2001). Roasting has had some effect in reducing levels of OA in coffee (Van der Stegen *et al.*, 2001). FmB1 can form Schiff's bases with reducing sugars such as fructose under certain conditions (Murphy *et al.*, 1995) and lose its hepato-carcinogenicity (Lu *et al.*, 1997; Liu *et al.*, 2001), but the hydrolyzed FmB1 was found to be still toxic (Voss *et al.*, 1996). Loss of FmB1 occurs during extrusion and baking of corn-based foods with sugars (Castelo *et al.*, 2001), and nixtamalization (alkaline cooking) and rinsing in the preparation of tortilla chips and masa (Dombrink *et al.*, 2000; Saunders *et al.*, 2001; Voss *et al.*, 2001). Four major products of the glucose and FmB interaction have been identified (Lu *et al.*, 2002). PT can be removed from apple juice by treatment with certain types of active carbons (Leggott *et al.*, 2001). Fermentations, including such as *Flavobacterium* and food-grade *Lactobacillus*, and enzymatic digestion can also adsorb and break down mycotoxins (Bhatnager *et al.*, 1991; El-Nezami *et al.*, 2002a, 2002b). The effect of food processing on various mycotoxins has been recently reviewed by several authors in an ACS symposium (DeVries *et al.*, 2002).

6.5 Dietary modifications

Dietary modification greatly affects the absorption, distribution and metabolism of mycotoxins, and this can subsequently affect their toxicity (Galvano *et al.*, 2001; Atroshi *et al.*, 2002). For example, the carcinogenic effect of AFB1 is affected by nutritional factors, dietary additives (Newberne, 1987) and anti-carcinogenic substances (Whitty and Bjeldanes, 1987; Dashwood *et al.*, 1989). Diets containing chemoprotective agents and antioxidants such as ascorbic acid, BHA (Wattenberg, 1986), BHT (Williams *et al.*, 1986; Klein *et al.*, 2002), ethoxyquin (Cabral and Neal, 1983), oltipraz (Buetler *et al.*, 1996), penta-acetyl geniposide (Lin *et al.*, 2000), Kolaviron biflavonoids (Nwankwo *et al.*, 2000) and even green tea (Qin *et al.*, 1997) have also been found to inhibit carcinogenesis caused by AFB1 in test animals. Dietary administration of the naturally occurring chemopreventive agents ellagic acid, coumarin or α-angelicalactone caused an increase in glutamate–cysteine ligase activity, a key enzyme for the synthesis of glutathione (Shepherd *et al.*, 2000). The mechanism of such protective effects is due to shifting of metabolism to a detoxification route by formation of AFB1-glutathione conjugate rather than formation of AFB1–DNA adducts (Jhee *et al.*, 1989). Among 77 naturally occurring compounds present in fruits, vegetables and spices tested, anthraquinones, coumarins and flavone-type flavonoids were found to be potential inhibitors for the formation of aflatoxin B-1-8,9-epoxide (Lee *et al.*, 2001).

Ebselen possesses a potent protective effect against aflatoxin B-1-induced cytotoxicity; the effect may be due to its strong capability in inhibiting intracellular reactive oxygen species formation and preventing oxidative damage (Yang *et al.*, 2000a, 2000b). Likewise, the toxic effect of OA on test animals was minimized when antioxidants such as vitamin C and E were added to the diet (Bose and Sinha, 1994; Hoehler and Marquardt, 1996). The apoptosis induced by OA was modified by antioxidants (Atroshi *et al.*, 2000). Ascorbic acid also gives some protective effect for AFB (Netke *et al.*, 1997). Fumonisin-induced DNA damage and apoptosis were reduced or inhibited by vitamin E (Atroshi *et al.*, 1999; Mobio *et al.*, 2000). Vitamin E was also found to prevent the cytotoxicity induced by FmB. The FmB1-induced neural tube defects were prevented by folic acid (Sadler *et al.*, 2002). The cytotoxicity of T-2 toxin was inhibited by some antioxidants (Shokri *et al*, 2000). Diets containing fish oil may impair early immunopathogenesis in DON-induced immunoglobulin A nephropathy (Pestka *et al.*, 2002). L-phenylalanine was found to have some protective effect for the toxic effects of OA because it diminishes OA's inhibitory effect on some of the enzymes discussed earlier. Aspartame, which is partially effective in decreasing the nephrotoxic and genotoxic effects of OA, competes with OA for binding to serum albumin (Creppy *et al.*, 1996).

Most mycotoxins have a high affinity for hydrated sodium calcium aluminasilicate (HSCAS or NovaSil) and related products. Diets containing NovaSil and related absorbers have been found to be effective in preventing absorption of AFB1 and several other mycotoxins in test animals, thus decreasing their toxicity (Colvin *et al.*, 1989; Harvey *et al.*, 1989; Beaver *et al.*, 1990; Kubena *et al.*, 1993; Smith *et al.*, 1994; Huwig *et al.*, 2001; Philips *et al.*, 2002). Several other adsorbents, such as zeolite, bentonite and superactive charcoal, have been found to be effective in decreasing the toxicity of other mycotoxins such as T-2 toxin.

7 Concluding remarks

Mycotoxins are a group of naturally occurring, low molecular weight fungal secondary metabolites that frequently contaminate agricultural commodities, foods and feeds. Numerous molds can produce mycotoxins, both pre-harvest and post-harvest, but not all are toxigenic. The major mycotoxins that we are currently most concerned about are aflatoxins, ochratoxins, fumonisins, and some trichothecene mycotoxins such as DON. Whereas natural occurrence of toxic fungi and mycotoxins in foods and feeds has been reported and outbreaks of mycotoxicoses have been documented, there are still some mycotoxicoses that have not been well characterized. Production of mycotoxins is controlled by the genetics of the mold, by the substrate, and by environmental conditions such as water activity (a_w) or moisture and relative humidity, temperature, time, atmosphere and microbial interactions. Although some mycotoxins are known to cause certain mycotoxicoses, others are not. Many mycotoxins are highly toxic to animals and probably to humans.

Mycotoxins can cause both acute and chronic effects in prokaryotic and eukaryotic systems, including humans, and their toxicities vary considerably with toxin and animal species. Most of their effects are organ-specific, but some mycotoxins may affect many organs. Induction of cancers through initiation and promotion processing by some mycotoxins in animals and possibly in humans is one of major concerns regarding their chronic effects. Modulation of immune systems by some mycotoxins is another concern. Mycotoxins also induce apoptosis, which has been shown to play a significant role in their toxic effects. Interaction with macromolecules, through either non-covalent or covalent bindings, or both, is the basis of their mode of action. For some mycotoxins, such as aflatoxins, metabolic activation is necessary before their binding with specific macromolecules; but for others, metabolic activation is not necessary. Nevertheless, metabolism plays a key role in modulating toxicity because metabolism can lead to either activated metabolites or non-toxic metabolites for subsequent conjugation and excretion. Thus, factors affecting the metabolism of mycotoxins greatly affect the toxicity. Recent studies on the mechanisms of mycotoxin-induced apoptosis have enhanced our understanding of the mode of action of mycotoxins.

Almost all of the mycotoxins are very stable; only limited detoxification methods are currently available, and some of these are not economically feasible. Because the toxins are formed both pre-harvest and post-harvest, it is very difficult to control toxin formation. Nevertheless, good farm management, the availability of resistant crops and the application of bio-control agents could minimize the problems. Genetically modified crops resistant to insect infestation, and application of non-toxic fungi in the field as competitive agents, have shown some success in reducing mycotoxin formation. With an increase in understanding of the biosynthesis of mycotoxins, new genetically modified crops or bio-control tools could be made available in the future. Dietary modification can also decrease the risk to some degree. Because most of these control measures are not very effective, rigorous programs for preventing human and animal exposure to contaminated foods and feed have been established, and effective methods for monitoring the toxin levels in the foods have been developed. Although risk assessment has been carried out for some mycotoxins, more epidemiological data for human exposure are needed for establishing toxicological parameters and the safe dose in humans. Mycotoxin problems may coexist with us for a long time, and there are many mycotoxicoses that need to be further studied. It is unrealistic to expect completely to eliminate the problem; only through multiple approaches and good farm management can we minimize the problem and enhance human and animal health.

Acknowledgments
The author thanks Professor Dean O. Cliver for providing excellent suggestions, and Ms. Ellen Doyle and Barbara Cochrane for their help in preparing the manuscript.

8 Abbreviations

AAL	*A. alternata* f. sp. *lycopersici*
Acro	Acromelic acid
AF	aflatoxin
AFB1	aflatoxin B1
AFG1	aflatoxin G1
AFM1	aflatoxin M1

AFP1	aflatoxin P1
AFQ1	aflatoxin Q1
AM	*Alternaria* mycotoxins
AME	altemariol monomethyl ether
AOAC	Association of Official Analytical Chemists
APT	apoptosis
ATA	alimentary toxic aleukia
ATX	altertoxin
BHA	butylated hydroxyanisole
BHT	butylated hydroxytoluene
CDK	cyclin-dependent kinase
CER	ceramide
CMI	cell-mediated immunity
CPA	cyclopiazonic acid
CT	citrinin
CV-1	African green monkey kidney fibroblasts
DAS	diacetoxyscirpenol
DON	deoxynivalenol
ELEM	equine leukoencephalomalacia
FC	fusarin C
FDA	US Food and Drug Administration
Fm	fumonisin
FmB1	fumonisin B1
FmB2	fumonisin B2
FmB3	fumonisin B3
FS-x	fusarenon-x
GC	gas chromatography
GOT	glutamate-oxaloacetate transaminase
GT	gliotoxin
HBV	hepatitis B virus
HSCAS	hydrated sodium calcium aluminasilicate
HPLC	high-performance liquid chromatography
IgG	immunoglobulin G
IgM	immunoglobulin M
IL	interleukin
InsP-3	inositol-l,4,5-triphosphate
IP	intraperitoneal
IPM-one	ipomeamarone
JNK	c-jun amino-terminal-kinase
LD_{50}	median lethal dose
LS	luteoskyrin
MAPK	mitogen activated protein kinase
MEF	mouse embryo fiberblasts
MFH	N-methyl-N-formylhydrazine
MH	N-methyhydrazine

MN	Moniliformin
MP	Mushroom poisons
MS	mass spectroscopy
MUS	Muscarine
NAC	N-acetyl-L-cysteine
NESO	neosolaniol
NIV	nivalenol
NPA	nitropropionic acid
OA	ochratoxin A
OhdG	hydroxydeoxyguanosine
PA	penicillic acid
PEPCK	renal phosphoenolpyruvate carboxykinase
PFC	plaque-forming cells
PG	prostaglandin
PHC	primary hepatocellular carcinoma
PK	protein kinase
PKS	polyketide synthase
PPE	porcine pulmonary edema
PT	patulin
RA	rubratoxin A
RB	rubratoxin B
RS	rugulosin
Sa	sphinganine
SBS	sick building syndrome
So	sphingosine
SP	deteriorated sugarcane poisoning
ST	sterigmatocystin
TCTCs	trichothecenes
TLC	thin-layer chromatography
TNF	tumor necrosis factor
TzA	tenuazonic acid
T-4ol	T-2 tetraol
WM	wortmannin
ZE	zearalenone

Bibliography

Abarca, M. L., M. R. Bragulat, G. Castellá and F. J. Cabañes (1994). Ochratoxin A production by strains of *Aspergillus niger* var. *niger*. *Appl. Environ. Microbiol.* **60**, 2650–2652.

Abbas, H. K. and R. T. Riley (1996). The presence and phytotoxicity of fumonisins and AAL-toxin in *Alternaria alternata*. *Toxicon* **34**, 133–136.

Abbas, H. K., T. Tanaka and W. T. Shier (1995). Biological activities of synthetic analogues of *Alternaria alternata* toxin (AAL) and fumonisin in plant and mammalian cells. *Phytochemistry* **40**, 1681–1689.

Abbas, H. K., W. P. Williams, G. L. Windham *et al.* (2002). Aflatoxin and fumonisin contamination of commercial corn (*Zea mays*) hybrids in Mississippi. *J. Agric. Food Chem.* **50**, 5246–5254.

Aguilar, F., S. P. Hussain and P. Cerutti (1993). Aflatoxin B_1 induces the transversion of G to T in codon 249 of the p53 tumor suppressor gene in human hepatocytes. *Proc. Natl Acad. Sci. USA* **90**, 8586–8590.

Ahamed, S., J. S. Foster, A. Bukovsky and J. Wimalasena (2001). Signal transduction through the ras/Erk pathway is essential for the mycoestrogen zearalenone-induced cell cycle progression in MCF7 cells. *Mol. Carcinogen.* **30**, 88–98.

Akuzawa, S., H. Yamaguchi, T. Masuda and I. Ueno (1992). Radical-mediated modification of deoxyguanine and deoxyribose by luteoskyrin and related anthraquinones. *Mutation Res.* **266**, 63–69.

Alam, M. M., M. Nagase, T. Yoshizawa and N. Sakato (2000). Thymocyte apoptosis by T-2 toxin in vivo in mice is independent of Fas/Fas ligand system. *Biosci. Biotechnol. Biochem.* **64**, 210–213.

Albarenque, S. M., J. Shinozuka, K. Suzuki *et al.* (2000). Kinetics and distribution of transforming growth factor (TGF) beta 1 mRNA in the dorsal skin of hypotrichotic WBN/ILAHt rats following topical application of T2 toxin. *Exp. Toxicol. Pathol.* **52**, 297–301.

Anke, H., O. Bergendorff and O. Sterner (1989). Assays of the biological activities of guaiane sesquiterpenoids isolated from the fruit bodies of edible *Lactarius* species. *Food Chem. Toxicol.* **27**, 393–397.

Anonymous (1983). *Protection against Trichothecenes.* Reported by the Committee on Protection Against Mycotoxins, National Research Council, National Academy of Sciences, National Academy Press, Washington, DC.

ApSimon, J. W. (2001). Structure, synthesis and biosynthesis of fumonisin B1 and related compounds. *Environ. Health Perspect.* **109** (Suppl. 2), 245–249.

Arlt, V. M., L. A. Pfohl, J. P. Cosyns and H. H. Schmeiser (2001). Analyses of DNA adducts formed by ochratoxin A and aristolochic acid in patients with Chinese herbs nephropathy. *Mutation Res.* **494**, 143–150.

Assouline-Dayan, Y., A. Leong, Y. Shoenfeld and M. E. Gershwin (2002). Studies of sick building syndrome. IV. Mycotoxicosis. *J. Asthma* **39**, 191–201.

Atroshi, F., A. Rizzo, I. Biese *et al.* (1999). Fumonisin B-1-induced DNA damage in rat liver and spleen. Effects of pretreatment with coenzyme Q(10), L-carnitine, alpha-tocopherol and selenium. *Pharmacol. Res.* **40**, 459–467.

Atroshi, F., I. Biese, H. Saloniemi *et al.* (2000). Significance of apoptosis and its relationship to antioxidants after ochratoxin a administration in mice. *J. Pharm. Pharmaceut. Sci.* **3**, 281–291.

Atroshi, F., A. Rizzo, T. Westermarck and T. Ali-Vehmas (2002). Antioxidant nutrients and mycotoxins. *Toxicology* **180**, 151–167.

Bacon, C. W., J. K. Porter, W. P. Norred and J. F. Leslie (1996). Production of fusaric acid by *Fusarium* species. *Appl. Environ. Microbiol.* **62**, 4039–4043.

Bacon, C. W., I. E. Yates, D. M. Hinton and F. Meredith (2001). Biological control of *Fusarium moniliforme* in maize. *Environ. Health Perspect.* **109** (Suppl. 2), 325–332.

Badria, F. A., S. N. Li and W. T. Shier (1996). Fumonisins as potential causes of kidney disease. *J. Toxicol.–Toxin Rev.* **15**, 273–292.

Bakan, B., D. Melcion, D. Richard-Molard and B. Cahagnier (2002). Fungal growth and Fusarium mycotoxin content in isogenic traditional maize and genetically modified maize grown in France and Spain. *J. Agric. Food Chem.* **50**, 728–731.

Balasubramanian, G., U. Hanumegowda and C. S. Reddy (2001). Secalonic acid D alters the nature of and inhibits the binding of the transcription factors to the phorbol 12-O-tetradecanoate-13 acetate-response element in the developing murine secondary palate. *Toxicol. Appl. Pharmacol.* **169**, 142–150.

Bassa, M., C. Mestres, D. Champiat *et al.* (2001). First report of aflatoxin in dried yam chips in Benin. *Plant Dis.* **85**, 1032.

Baudrimont, I., A.-M. Betbeder, A. Gharbi *et al.* (1994a). Effect of superoxide dismutase and catalase on the nephrotoxicity induced by subchronical administration of ochratoxin A in rats. *Toxicology* **89**, 101–111.

Baudrimont, I., R. Ahouandjivo and E. E. Creppy (1994b). Prevention lipid peroxidation induced by ochratoxin A in Vero cells in culture by several agents. *Chem. Biol. Interact.* **104**, 29–40.

Beasley, V. R. (1989). *Trichothecene Mycotoxicosis, Pathophysiologic Effects*, Vols I and II. CRC Press, Inc. Boca Raton, FL.

Beaver, R. W., D. M. Wilson, M. A. James *et al.* (1990). Distribution of aflatoxins in tissues of growing pigs fed an aflatoxin contaminated diet amended with a high affinity aluminosilicate sorbent. *Vet. Hum. Toxicol.* **32**, 16–18.

Beer, D. G. and H. C. Pitot (1989). Proto-oncogene activation during chemically induced hepatocarcinogensis in rodents. *Mutation Res.* **220**, 1–10.

Benesic, A., S. Mildenberger and M. Gekle (2000). Nephritogenic ochratoxin A interferes with hormonal signalling in immortalized human kidney epithelial cells. *Pflugers Arch.* **439**, 278–287.

Benjamin, D. R. (1992). Mushroom poisoning in infants and children: the *Amanita pantherina/muscaria* group. *Clin. Toxicol.* **30**, 13–22.

Bennett, J. W., D. Bhatnagar and P. K. Chang (1994). The molecular genetics of aflatoxin biosynthesis. *FEMS-Symp. Madison, Wisconsin*, **69**, 51–58.

Bergman, K. and K. E. Hellenas (1992). Methylation of rat and mouse DNA by the mushroom poison gyromitrin and its metabolite monomethylhydrazine. *Cancer Lett.* **61**, 165–170.

Betina, V. (1989). *Mycotoxins: Chemical, Biological and Environmental Aspects.* Elsevier, Amsterdam.

Bhandari, N. and R. P. Sharma (2002a). Fumonisin B1-induced alterations in cytokine expression and apoptosis signaling genes in mouse liver and kidney after an acute exposure. *Toxicology* **172**, 81–92.

Bhandari, N. and R. P. Sharma (2002b). Modulation of selected cell signaling genes in mouse liver by fumonisin B1. *Chem. Biol. Interact.* **139**, 317–331.

Bhatnagar, D., E. B. Lillehoj and J. W. Bennett (1991). Biological detoxification of mycotoxin. In: J. E. Smith and R. S. Henderson (eds), *Mycotoxins and Animal Foods*, p. 815. CRC Press, Boca Raton, FL.

Bhatnagar, D., E. B Lillehoj and D. K. Arora (1992). *Handbook of Applied Mycology, Vol. V. Mycotoxins in Ecological Systems*. Marcel Dekker, New York.

Bhatnagar, D., P. J. Cotty and T. E. Cleveland (1993). Pre-harvest aflatoxin contamination. Molecular strategies for its control. In: A. M. Spanier *et al.* (eds), *Food Flavor and Safety. Molecular Analysis and Design. ACS Symp. Ser.* **528**, 272–292.

Bhatnagar, D., G. Payne, J. E. Linz and T. E. Cleveland (1995). Molecular biology to eliminate aflatoxins. *Intl News Fats Oils Rel. Materials* **6**, 262–271.

Bilmen, J. G., L. L. Wootton and F. Michelangeli (2002). The mechanism of inhibition of the sarco/endoplasmic reticulum Ca^{2+} ATPase by paxilline. *Arch. Biochem. Biophys.* **406**, 55–64.

Birzele, B., A. Prange and J. Kraemer (2000). Deoxynivalenol and ochratoxin A in German wheat and changes of level in relation to storage parameters. *Food Addit. Contam.* **17**, 1027–1035.

Boenke, A. (1997). The food and feed chains–a possible strategy for the production of certified reference materials (CRMs) in the area of mycotoxins. *Food Chem.* **60**, 255–262.

Bondy, G. S. and J. J. Pestka (2000). Immunomodulation by fungal toxins. *J. Toxicol. Environ. Health B Crit. Rev*. **3**, 109–143.

Boorman, G. (1988). NTP Technical report on the toxicology and carcinogenesis studies of ochratoxin A (CAS no. 303-47-9) in F344/N rats (gavage studies), NIH publication No. 89–2813). US Department of Health and Human Services, NIH, Bethesda, MD.

Boorman, G. A., M. R. McDonald, S. Imoto and R. Persing (1992). Renal lesions induced by ochratoxin A exposure in the F344 rat. *Toxicol. Pathol* **20**, 236–245.

Bose, S. and S. P. Sinha (1994). Modulation of ochratoxin–produced genotoxicity in mice by vitamin C. *Food Chem. Toxicol*. **32**, 533–537.

Bresinsky, A. and H. Besl (1990). *A Colour Atlas of Poisonous Fungi* (trans. N. G. Bisset). Wolfe Publishing Ltd, London.

Brown, D. W., J. H. Yu, H. S. Kelkar *et al.* (1996). Twenty-five coregulated transcripts define a sterigmatocystin gene cluster in *Aspergillus nidulans*. *Proc. Natl Acad. Sci. USA* **93**, 1418–1422.

Brown, R. L., P. J. Cotty and T. E. Cleveland (1991). Reduction in aflatoxin content of maize by atoxigenic strains of *Aspergillus flavus*. *J. Food Prot*. **54**, 623–626.

Bucci, T. J. and P. C. Howard (1996). Effect of fumonisin mycotoxins in animals. *J. Toxicol.–Toxin Rev*. **15**, 293–302.

Buetler, T. M., T. K. Bammler, J. D. Hayes and D. L. Eaton (1996). Oltipraz–mediated changes in aflatoxin B1 biotransformation in rat liver: implication for human chemo intervention. *Cancer Res*. **56**, 2306–2313.

Bunner, D. L. and E. R. Morris (1988). Alteration of multiple cell membrane functions in L–6 myoblasts by T-2 toxin, an important mechanism of action. *Toxicol. Appl. Pharmacol*. **92**, 113–121.

Burdock, G. A., I. G. Carabin and M. G. Soni (2001). Safety assessment of beta nitropropionic acid: a monograph in support of an acceptable daily intake in humans. *Food Chem.* **75**, 1–27.

Burmeister, H. R., A. Ciegler and R. F. Vesonder (1979). Moniliformin, a metabolite of *Fusarium moniliforme* NRRL 6322: purification and toxicity. *Appl. Environ. Microbiol.* **37**, 11–13.

Burow, G. B., T. C. Nesbitt, J. Dunlap and N. P. Keller (1997). Seed lipoxygenase products modulate *Aspergillus* mycotoxin biosynthesis. *Mol. Plant–Microbe Interact.* **10**, 380–387.

Busby, W. F. Jr and G. N. Wogan (1979). Food-borne mycotoxins and alimentary mycotoxicoses. In: H. Riemann and F. L. Bryan (eds), *Food-borne Infections and Intoxications*, 2nd edn, pp. 519–610. Academic Press, New York, NY.

Busby, W. F. Jr and G. N. Wogan (1981). Aflatoxins. In: R. C. Shank (ed.), *Mycotoxins and N-nitrosocompounds: Environmental Risks*, Vol. II, pp. 3–45. CRC Press, Boca Raton, FL.

Cabral, J. R. P. and G. E. Neal (1983). The inhibitory effects of ethoxyquin on the carcinogenic action of aflatoxin B1 in rats. *Cancer Lett.* **19**, 125–132.

Cahagnier, B., D. Melcion and D. Richardmolard (1995). Growth of *Fusarium moniliforme* and its biosynthesis of fumonisin B1 on maize grain as a function of different water activities. *Lett. Appl. Microbiol* **20**, 247–251.

Calvo, A. M., R. A. Wilson, J. W. Bok and N. P. Keller (2002). Relationship between secondary metabolism and fungal development. *Microbiol. Molecular Biol.* **66**, 447.

Cary, J. W., M. Wright, D. Bhatnagar *et al.* (1996). Molecular characterization of an *Aspergillus parasiticus* dehydrogenase gene, *nor*A, located on the aflatoxin biosynthesis gene cluster. *Appl. Environ. Microbiol.* **62**, 360–366.

CAST (2003) (J. L. Richard and G. A. Payne, eds). *Mycotoxins: Risks in Plant, Animal. and Humans.* Council for Agricultural Science and Technology (CAST) Task Force Report No. 139, Ames, IA.

Castelo, M. M., L. S. Jackson, M. A. Hanna *et al.* (2001). Loss of fumonisin B1 in extruded and baked corn-based foods with sugars. *J. Food Sci.* **66**, 416–421.

Cazzaniga, D., J. C. Basilico, R. J. Gonzalez *et al.* (2001). Mycotoxins inactivation by extrusion cooking of corn flour. *Lett. Appl. Microbiol.* **33**, 144–147.

Chang, L. W., S. M. T. Hsia, P.-C. Chan and L.-L. Hsieh (1994). Macromolecular adducts: biomarkers for toxicity and carcinogenesis. *Ann. Rev. Pharmacol. Toxicol.* **34**, 41–67.

Chang, P.-K., J. W. Cary, J. J. Yu *et al.* (1995a). The *Aspergillus parasiticus* polyketide synthase gene pksA, a homolog of *Aspergillus nidulans* wA, is required for aflatoxin B-1 biosynthesis. *Mol. Gen. Genetics* **248**, 270–277.

Chang, P.-K., K. C. Ehrlich, J. Yu *et al.* (1995b). Increased expression of *Aspergillus parasiticus aflR*, encoding a sequence-specific DNA-binding protein, relieves nitrate inhibition of aflatoxin biosynthesis. *Appl. Environ. Microbiol.* **61**, 2372–2377.

Chang, P.-K., K. C. Ehrlich, J. E. Linz *et al.* (1996). Characterization of the *Aspergillus parasiticus nia*D and *nii*A gene cluster. *Curr. Genetics* **30**, 68–75.

Chelkowski, J. (ed) (1989). *Fusarium mycotoxins, taxonomy and pathogenicity.* Elsevier, New York.

Chelkowski, J. and A. Visconti (eds) (1992). *Alternaria – Metabolites, Biology and Plant Diseases*. Elsevier, Amsterdam.

Chen, C. J., L. Y. Wang, S. N. Lu *et al.* (1996a). Elevated aflatoxin exposure and increased risk of hepatocellular carcinoma. *Hepatology* **24**, 38–42.

Chen, C. J., M. W. Yu, Y. F. Liaw *et al.* (1996b). Chronic hepatitis B carriers with null genotypes of glutathione S-transferase M1 and T1 polymorphisms who are exposed to aflatoxin are at increased risk of hepatocellular carcinoma. *Am. J. Hum. Genetics* **59**, 128–134.

Chen, J., C. J. Mirocha, W. Xie *et al.* (1992). Production of the mycotoxin fumonisin B1 by *Alternaria alternata f. sp. lycopersici. Appl. Environ. Microbiol.* **58**, 3928–3931.

Chen, J. W., Y. L. Luo, M. J. Hwang *et al.* (1999). Territrem B, a tremorgenic mycotoxin that inhibits acetylcholinesterase with a noncovalent yet irreversible binding mechanism. *J. Biol. Chem.* **274**, 34 916–34 923.

Choy, W. N. (1993). A review of the dose-response induction of DNA adducts by aflatoxin B_1 and its implications to quantitative cancer-risk assessment. *Mutat. Res.* **296**, 181–198.

Chu, F. S. (1974). Studies on ochratoxins. *Crit. Rev. Toxicol.* **2**, 499–524.

Chu, F. S. (1977). Mode of action of mycotoxins and related compounds. *Adv. Appl. Microbiol.* **22**, 83–143.

Chu, F. S. (1986). Immunoassays for mycotoxins. In: R. J. Cole (ed.), *Modern Methods in the Analysis and Structural Elucidation of Mycotoxins*, pp 207–237. Academic Press, New York, NY.

Chu, F. S. (1991a). Mycotoxins: food contamination, mechanism, carcinogenic potential and preventive measures. *Mutat. Res.* **259**, 291–306.

Chu, F. S. (1991b). Detection and determination of mycotoxin. In: R. P. Sharma and D. K. Salunkhe (eds), *Mycotoxins and Phytoalexins in Human and Animal Health*, pp. 33–79. Telford Press, Inc., Caldwell, NJ.

Chu, F. S. (1991c). Development and use of immunoassays in detection of the ecologically important mycotoxins. In: D. Bhatnagar, E. B. Lillehoj and D. K. Arora (eds), *Handbook of Applied Mycology: Mycotoxins*, Vol. V, pp. 87–136. Marcel Dekker, New York, NY.

Chu, F. S. (1995). Mycotoxin analysis. In: I. J. Jeon and W. G. Ikins (eds), *Analyzing Food for Nutrition Labeling and Hazardous Contaminants,* pp. 283–332. Marcel Dekker, New York, NY.

Chu, F. S. (1996). Recent studies on immunoassays for mycotoxins. In: R. C. Beier and L. H. Stanker (eds), *Immunoassays for Residue Analysis: Food Safety*, pp. 294–313. ACS Symposium Series book no. 621, American Chemical Society, Washington, DC

Chu, F. S. (1997). Trichothecene mycotoxicosis. In: R. Dulbecco (ed.), *Encyclopedia of Human Biology*, 2nd edn, Vol. 8, pp. 511–522. Academic Press, New York, NY.

Chu, F. S. (1998). Mycotoxins – occurrence and toxic effects. In: *Encyclopedia of Food and Nutrition – Food Contaminants*, 2nd edn, pp. 858–869. Academic Press, New York, NY.

Chu, F. S. (2000). Immunoassays for mycotoxins. In: Y. H. Hui, R. A. Smith and D. G. Spoerke Jr (eds), *Foodborne Disease Handbook*, 2nd edn, Vol. 3, pp. 683–713. Marcel Dekker, New York, NY.

Chu, F. S. (2002). Mycotoxins. In: D. O. Cliver and H. Riemann (eds), *Foodborne Diseases*, 2nd edn, pp. 271–303. Academic Press, New York, NY.

Chu, F. S. (2003). Mycotoxins/toxicology. *Encyclopedia of Food Science and Nutrition*, 2nd edn, pp. 4096–4108. Elsevier Science Ltd, London.

Chu, F. S. and D. Bhatnagar (2004). Mycotoxins. In: D. K. Arora (ed.), *Fungal Biotechnology in Agricultural, Food and Environmental Applications*. Marcel Dekker, NewYork, NY, pp. 325–342.

Ciacci, Z. J. R. and C. Jones (1999). Fumonisin B1, a mycotoxin contaminant of cereal grains and inducer of apoptosis via the tumour necrosis factor pathway and caspase activation. *Food Chem. Toxicol.* **37**, 703–712.

Ciegler, A., S. Kadis and S. J. Ajl (1971). *Microbial Toxins, Vol. 6 & 7, Fungal Toxins*. Academic Press, NewYork, NY.

Cleveland, T. E. and D. Bhatnagar (1992). Molecular strategies for reducing aflatoxin levels in crops before harvest. In: D. Bhatnagar and T. E. Cleveland (eds), *Molecular Approaches to Improving Food Quality and Safety*, pp. 205–228. Van Nostrand Reinhold, New York, NY.

Cole, R. J. (1989). Technology of aflatoxin decontamination. In: S. Natori, K. Hashimoto and Y. Ueno (eds), *Mycotoxins and Phycotoxins 88*, pp. 177–184. Elsevier, Amsterdam.

Cole, R. J. and R. H. Cox (1981). *Handbook of Toxic Fungal Metabolites*. Academic Press, New York, NY.

Colvin, B. M., L. T. Sangster, K. D. Haydon *et al.* (1989). Effect of high affinity aluminosilicate sorbent on prevention of aflatoxicosis in growing pigs. *Vet. Hum. Toxicol.* **31**, 46–48.

Cotty, P. J. and D. Bhatnagar (1994). Variability among atoxigenic *Aspergillus flavus* strains in ability to prevent aflatoxin contamination and production of aflatoxin biosynthetic pathway enzymes. *Appl. Environ. Microbiol.* **60**, 2248–2251.

Coulombe, R.A. Jr (1993). Biological action of mycotoxins. *J. Dairy Sci.* **76**, 880–891.

Creppy, E. E., I. Baudrimont, A. Belmadani and A. M. Betbeder (1996). Aspartame as a preventive agent of chronic toxic effects of ochratoxin A in experimental animals. *J. Toxicol.–Toxin Rev.* **15**, 207–221.

Cullen, J. M., B. H. Ruebner, L. S. Hsieh *et al.* (1987). Carcinogenicity of dietary aflatoxin M1 in male Fischer rats compared to aflatoxin B1. *Cancer Res.* **47**, 1913–1917.

Dashwood, R. H., A. Arbogast, T. Fong *et al.* (1989). Quantitative inter-relationships between aflatoxin B1 carcinogen dose, indole-3-carbinol anti-carcinogen dose, target organ DNA adduction and final tumor response. *Carcinogenesis* **10**, 175–181.

Da-Silva, J. B., C. R. Pozzi, M. A. B. Mallozzi *et al.* (2000). Mycoflora and occurrence of aflatoxin B1 and fumonisin B1 during storage of Brazilian sorghum. *J. Agric. Food Chem.* **48**, 4352–4356.

Dearborn, D. G., I. Yike, W. G. Sorenson *et al.* (1999). Overview of investigations into pulmonary hemorrhage among infants in Cleveland, *Ohio. Environ. Health Perspect.* **107** (Suppl. 3), 495–499.

deGroene, E. M., A. Jahn, G. J. Horbach and J. Fink-Gremmels (1996). Mutagenicity and genotoxicity of the mycotoxin ochratoxin A. Environ. *Toxicol. Pharmacol.* **1**, 21–26.

Deruiter, J., J. M. Jacyno, H. G. Cutler and R. A. Davis (1993). Studies on aldose reductase inhibitors from fungi. 2. Moniliformin and small ring analogues. *J. Enzyme Inhibit.* **7**, 249–256.

Desjardins, A. E. and T. M. Hohn (1997). Mycotoxins in plant pathogenesis. *Mol. Plant-Microb. Interact.* **10**, 147–152.

Desjardins, A. E., T. M. Hohn and S. P. McCormick (1993). Trichothecene biosynthesis in *Fusarium* species: chemistry, genetics and significance. *Microbiol. Rev.* **57**, 595–604

Desjardins, A. E., R. D. Plattner and R. H. Proctor (1996). Linkage among genes responsible for fumonisin biosynthesis in *Gibberella fujikuroi* mating population A. *Appl. Environ. Microbiol.* **62**, 2571–2576.

DeVries, J. W., M. W. Trucksess and L. S. Jackson (eds) (2002). *Mycotoxins and Food Safety*. Kluwer Academic/Plenum Publishers, New York, NY.

Dirheimer, G. (1996). Mechanistic approaches to ochratoxin toxicity. *Food Addit. Contam.* **13** (Suppl. S), 45–48.

D'Mello, J. P. F. and A. C. E. MacDonald (1997). Mycotoxins. *Anim. Feed Sci. Technol.* **69**, 155–166.

D'Mello, J. P. F., C. M. Placinta and A. C. E. MacDonald (1999). *Fusarium* mycotoxins: a review of global implications for animal health, welfare and productivity. *Anim. Feed Sci. Technol.* **80**, 183–205.

Dombrink. K. M. A., T. J. Dvorak, M. Barron and L. W. Rooney (2000). Effect of nixtamalization (alkaline cooking) on fumonisin-contaminated corn for production of masa and tortillas. *J. Agric. Food Chem.* **48**, 5781–5786.

Dong, W., J. I. Azcona-Olivera, K. H. Brooks *et al.* (1994). Elevated gene expression and production of interleukins 2, 4, 5 and 6 during exposure to vomitoxin (deoxynivalenol) and cycloheximide in the EL-4 thymoma. *Toxicol. Appl. Pharmacol.* **127**, 282–290.

Dorner, J. W., R. J. Cole and P. D. Blankenship (1992). Use of a biocompetitive agent to control pre-harvest aflatoxin in drought stressed peanuts. *J. Food Prot.* **55**, 888–892.

Dowd, P. F. (2001). Biotic and abiotic factors limiting efficacy of Bt corn in indirectly reducing mycotoxin levels in commercial fields. *J. Econ. Entomol.* **94**, 1067–1074.

Doyle, M. P., R. S. Applebaum, R. E. Brackett and E. H. Marth (1982). Physical, chemical and biological degradation of mycotoxins in foods and agricultural commodities. *J. Food Prot.* **45**, 964–971.

Dragan Y. P., W. R. Bidlack, S. M. Cohen *et al.* (2001). Implications of apoptosis for toxicity, carcinogenicity and risk assessment: Fumonisin B-1 as an example. *Toxicol. Sci.* **61**, 6–17.

Dugyala, R. R. and R. P. Sharma (1996). The effect of aflatoxin B1 on cytokines mRNA and corresponding protein levels in peritoneal macrophages and splenic lymphocytes. *Intl J. Immunopharmacol.* **18**, 599–608.

Dugyala, R. R. and R. P. Sharma (1997). Alteration of major cytokines produced by mitogen-activated macrophages and splencytes in T-2 toxin treated male CD-1 mice. *Environ. Toxicol. Pharmacol.* **3**, 73–81.

Dugyala, R. R., R. P. Sharma, M. Tsunoda and R. T. Riley (1998). Tumor necrosis factor-alpha as a contributor in fumonisin B1 toxicity. *J. Pharmacol. Exp. Ther.* **285**, 317–324.

Dutton, M. F. (1988). Enzymes and aflatoxin biosynthesis. *Microbiol. Rev.* **52**, 274–295.

Dutton, M. F. (1996). Fumonisins, mycotoxins of increasing importance: their nature and their effects. *Pharmacol. Ther.* **70**, 137–161.

Duvick, J. (2001). Prospects for reducing fumonisin contamination of maize through genetic modification. *Environ. Health Perspect.* **109** (Suppl. 2), 337–342.

Eaton, D. L. and E. P. Gallagher (1994). Mechanisms of aflatoxin carcinogenesis. *Ann. Rev. Pharmacol. Toxicol.* **34**, 135–172.

Eaton, D. L. and J. D. Groopman (1994). *The Toxicology of Aflatoxins – Human Health, Veterinary and Agricultural Significance*. Academic Press, New York, NY.

Edwards, S. G., S. R. Pirgozliev, M.C. Hare and P. Jenkinson (2001). Quantification of trichothecene-producing *Fusarium* species in harvested grain by competitive PCR to determine efficacies of fungicides against fusarium head blight of winter wheat. *Appl. Environ. Microbiol.* **67**, 1575–1580.

Ehrlich, V., F. Darroudi, M. Uhl *et al.* (2002). Genotoxic effects of ochratoxin A in human-derived hepatoma (HepG2) cells. *Food Chem. Toxicol.* **40**, 1085–1090.

El-Nezami, H. S., A. Chrevatidis, S. Auriola *et al.* (2002a). Removal of common *Fusarium* toxins in vitro by strains of *Lactobacillus* and *Propionibacterium. Food Addit. Contam.* **19**, 680–686.

El-Nezami, H. S., N. Polychronaki, S. Salminen and H. Mykkanen (2002b). Binding rather than metabolism may explain the interaction of two food-grade *Lactobacillus* strains with zearalenone and its derivative alpha-zearalenol. *Appl. Environ. Microbiol.* **68**, 3545–3549.

Essigmann, J. M., L. J. Barker, K. W. Fowler *et al.* (1979). Sterigmatocystin-DNA interactions, Identification of a major adduct formed after metabolic activation in vitro. *Proc. Natl Acad. Sci. USA* **76**, 179–183.

Essigmann, J. M., C. L. Green, R. G. Croy *et al.* (1983). Interaction of aflatoxin B1 and alkylating agents with DNA: structural and functional studies. *Cold Spring Harbor Symp. Quant. Biol.* **47**, 327–337.

FAO/WHO (1997). Worldwide regulations for mycotoxins – 1995. A compendium. FAO Food and Nutrition Paper No. 64. FAO of UN, Rome.

Farber, J. M. and G. W. Sanders (1986). Production of fusarin C by *Fusarium* spp. *J. Agric. Food Chem.* **34**, 963–966.

Faulstich, H. (1980). The amatoxins. In: F. Halm (ed.), *Progress in Molecular and Subcellular Biology*, pp. 88–134. Springer Verlag, New York, NY.

Faulstich, H., A. Buka, H. Bodenmuller and T. Wieland. (1980). Virotoxins. *Biochemistry* **19**, 3334–3343.

Feinberg, B. and S. S. McLaughlin (1989). Biochemical mechanism of action of trichothecene mycotoxins. In: V. Beasley (ed.), *Trichothecene Mycotoxicosis, Pathophysiologic Effects*, Vol. I, pp. 27–35. CRC Press, Inc., Boca Raton, FL.

Feng, G. H. and T. J. Leonard (1995). Characterization of the polyketide synthase gene (pksL1) required for aflatoxin biosynthesis in *Aspergillus parasiticus. J. Bacteriol.* **117**, 6246–6254.

Feng, G. H., F. S. Chu and T. J. Leonard (1992). Molecular cloning of genes related to aflatoxin biosynthesis by differential screening. *Appl. Environ. Microbiol.* **58**, 455–460.

Fernandez, P. V., A. Patriarca, O. Locani and G Vaamonde (2001). Natural co occurrence of aflatoxin and cyclopiazonic acid in peanuts grown in Argentina. *Food Addit. Contam.* **18**, 1017–1020.

Filali, A., L. Ouammi, A. M. Betbeder *et al.* (2001). Ochratoxin A in beverages from Morocco: a preliminary survey. *Food Addit. Contam.* **18**, 565–568.

Fink-Gremmels, J., A. Jahn and M. J. Blom (1995). Toxicity and metabolism of ochratoxin A. *Nat. Toxins* **3**, 214–220.

Fliege, R. and M. Metzler (2000a). Electrophilic properties of patulin. Adduct structures and reaction pathways with 4-bromothiophenol and other model nucleophiles. *Chem. Res. Toxicol.* **13**, 363–372.

Fliege, R. and M. Metzler (2000b). Electrophilic properties of patulin. N-acetylcysteine and glutathione adducts. *Chem. Res. Toxicol.* **13**, 373–381.

Flieger, M., M. Wurst and R. Shelby (1997). Ergot alkaloids – sources, structures and analytical methods. *Folia Microbiol.* **42**, 3–30.

Forrester, L. M., G. E. Neal, D. J. Judah *et al.* (1990). Evidence for involvement of multiple forms of cytochrome P-450 in aflatoxin B_1 metabolism in human liver. *Proc. Natl Acad. Sci. USA* **87**, 8306–8310.

Fujimoto, H., K. Maeda and M. Yamazaki (1991). New toxic metabolites from a mushroom, *Hebeloma vinosophyllum*. III. Isolation and structures of three new glycosides, hebevinosides XXII, XIII and XIV and productivity of the hebevinosides at three growth stages of the mushroom. *Chem. Pharm. Bull.* **39**, 1958–1961.

Fushiya, S., S. Sato and S. Nozoe (1992). L-Stizolobic acid and L-stizolobinic acid from *Clitocybe acromelaga*, precursors of acromelic acids. *Phytochemistry* **31**, 2337–2339.

Fushiya, S., S. Sato, G. Kusano and S. Nozoe (1993). β-Cyano-L-alanine and N-(γ-glutamy)-β-cyano-L-alanine, neurotoxic constituents of *Clitocybe acromelaga. Phytochemistry* **33**, 53–55.

Galvano, F., V. Galofaro and G. Galvano (1996). Occurrence and stability of aflatoxin M1 in milk and milk products: a worldwide review. *J. Food Prot.* **59**, 1079–1090.

Galvano, F., A. Piva, A. Ritieni and G. Galvano (2001). Dietary strategies to counteract the effects of mycotoxins: a review. *J. Food. Prot.* **64**, 120–131.

Garner, R. C. (1989). Monitoring aflatoxin exposure at a macromolecular level in human with immunological methods. In: S. Natori, K. Hashimoto and Y. Ueno (eds), *Mycotoxins and Phycotoxins 88*, pp. 29–35. Elsevier, Amsterdam.

Garren, L., D. Galendo, C. P. Wild and M. Castegnaro (2001). The induction and persistence of altered sphingolipid biosynthesis in rats treated with fumonisin B1. *Food Addit. Contam.* **18**, 850–856.

Gautier, J. C., D. Holzhaeuser, J. Markovic *et al.* (2001). Oxidative damage and stress response from ochratoxin A exposure in rats. *Free Rad. Biol. Med.* **30**, 1089–1098.

Gekle, M., G. Schwerdt, R. Freudinger *et al.* (2000). Ochratoxin A induces JNK activation and apoptosis in MDCKC7 cells at nanomolar concentrations. *J. Pharmacol. Exp. Ther.* **293**, 837–844.

Gelderblom, W. C. A., W. F. O. Marasas, P. S. Steyn *et al.* (1984). Structure elucidation of fusarin C, a mutagen produced by *Fusarium moniliforme. J. Chem. Soc. Chem. Comm.* **1984**, 122–124.

Gelderblom. W. C. A., K. Jaskiewicz, W. F. O. Marasas *et al.* (1988). Fumonisins – novel mycotoxins with cancer-promoting activity produced by *Fusarium moniliforme. Appl. Environ. Microbiol.* **54**, 1806–1811.

Gelderblom, W. C. A., N. P. J. Kriek, W. F. O. Marasas and P. G. Thiel (1991). Toxicity and carcinogenicity of the *Fusarium moniliforme* metabolite, fumonisin B_1, in rats. *Carcinogenesis* **12**, 1247–1251.

Gelderblom.W. C., W. F. O. Marasas, R. Vleggaar *et al.* (1992a). Fumonisins: isolation, chemical characterization and biological effects. *Mycopathologia* **117**, 11–16.

Gelderblom, W. C., A. E. Semple, W. F. O. Marasas and E. Farber (1992b). The cancer-initiating potential of the fumonisin B mycotoxins. *Carcinogenesis* **13**, 433–437.

Gelderblom, W.C.A., M. E. Cawood, S. D. Snyman and W. F. O. Marasas (1993). Structure–activity relationships of fumonisins in short-term carcinogenesis and cytotoxicity assays. *Food Chem. Toxicol.* **31**, 407–414.

Gelderblom, W. C. A., M. E. Cawood, S. D. Snyman and W. F. O. Marasas (1994). Fumonisin B_1 dosimetry in relation to cancer initiation in rat liver. *Carcinogenesis* **15**, 209–214.

Gembeh, S.V., R. L. Brown, C. Grimm and T. E. Cleveland (2001). Identification of chemical components of corn kernel pericarp wax associated with resistance to *Aspergillus flavus* infection and aflatoxin production. *J. Agric. Food Chem.* **49**, 4635–4641.

Gilbert, J (1999). Overview of mycotoxin methods, present status and future needs. *Nat. Toxins* **7**, 347–352.

Goldblatt, L. A. (1969). *Aflatoxin – Scientific Background, Control and Implications.* Academic Press, New York, NY.

Gopalakrishnan, S., T. M. Harris and M. P. Stone (1990). Intercalation of aflatoxin-B1 in 2 oligodeoxynucleotide adducts – comparative H-1 NMR analysis of D(ATCAFBGAT). D(ATCGAT) and D(ATAFBGCAT)2). *Biochemistry* **29**, 10 438–10 448.

Goto, T., D. T. Wicklow and Y. Ito (1996). Aflatoxin and cyclopiazonic acid production by a sclerotium-producing *Aspergillus tamarii* strain. *Appl. Environ. Microbiol.* **62**, 4036–4038.

Gravesen, S., P. A. Nielsen, R. Iversen and K. F. Nielsen (1999). Microfungal contamination of damp buildings – examples of risk constructions and risk materia. *Environ. Health Perspect.* **107** (Suppl. 3), 505–508.

Greene, D. M., J. I. Azcona-Olivera and J. J. Pestka (1994). Vomitoxin (deoxynivalenol)-induced IgA nephropathy in the B6C3F1 mouse, dose response and male predilection. *Toxicology* **92**, 245–260.

Groopman, J. D. and T. W. Kensler (1993). Molecular biomarkers for human carcinogen exposure. *Chem. Res. Toxicol.* **6**, 764–770.

Groopman, J. D., J. Zhu, P. R. Donahue *et al.* (1992). Molecular dosimetry of urinary aflatoxin-DNA adducts in people living in Guangxi Autonomous Region, People's Republic of China. *Cancer Res.* **52**, 45–52.

Groopman, J. D., C. P. Wild, J. Hasler *et al.* (1993). Molecular epidemiology of aflatoxin exposures: validation of aflatoxin-N7-guanine levels in urine as a biomarker in experimental rat models and humans. *Environ. Health Perspect.* **99**, 107–113.

Groopman, J. D., G. N. Wogan, B. D. Roebuck and T. W. Kensler (1994). Molecular biomarkers for aflatoxins and their application to human cancer Prevention. *Cancer Res.* **54** (Suppl.), S1907–S1911.

Groopman, J. D., T. W. Kensler and J. M. Links (1995). Molecular epidemiology and human risk monitoring. *Toxicol. Lett.* **82**, 763–769.

Groopman, J. D., J. S. Wang and P. Scholl (1996). Molecular biomarkers for aflatoxins: from adducts to gene mutations to human liver cancer. *Can. J. Physiol. Pharmacol.* **74**, 203–209.

Grosse, Y., I. Baudrimont, M. Castegnaro *et al.* (1995). Formation of ochratoxin A metabolites and DNA-adducts in monkey kidney cells. *Chem. Biol. Interact.* **95**, 175–187.

Gross-Steinmeyer, K., J. Weymann, H. G. Hege and M. Metzler (2002). Metabolism and lack of DNA reactivity of the mycotoxin ochratoxin A in cultured rat and human primary hepatocytes. *J. Agric. Food Chem.* **50**, 938–945.

Groves F. D., L. Zhang, Y. S. Chang *et al.* (1999). *Fusarium* mycotoxins in corn and corn products in a high risk area for gastric cancer in Shandong Province, China. *J. AOAC Intl* **82**, 657–662.

Gruter, A., U. Friederich and F. E. Wurgler (1991). The mutagenicity of edible mushrooms in a histidine-independent bacterial test system. *Food Chem. Toxicol.* **219**, 159–165.

Guengerich, F. P., W. W. Johnson, Y. F. Ueng *et al.* (1996). Involvement of cytochrome P450, glutathione S-transferase and epoxide hydrolase in the metabolism of aflatoxin B-1 and relevance to risk of human liver cancer. *Environ. Health Perspect.* **104**, 557–562.

Guengerich, F. P., H. Cai, M. McMahon *et al.* (2001). Reduction of aflatoxin B1 dialdehyde by rat and human aldo keto reductases. *Chem. Res. Toxicol.* **14**, 727 737.

Gumprecht, L. A., G. W. Smith, P. C. Constable and W. M. Haschek (2001). Species and organ specificity of fumonisin-induced endothelial alterations: potential role in porcine pulmonary edema. *Toxicology* **160**, 71–79.

Gupta, S., S. B. Krasnoff, N. L. Underwood *et al.* (1991). Isolation of beauvericin as an insect toxin from *Fusarium semitectum* and *Fusarium moniliforme* var. *subglutinans. Mycopathologia* **115**, 185–189.

Hamilton, P. B (1982). Mycotoxins and farm animals. *Refush Vet.* **39**, 17–45.

Hammer, P., A. Bluthgen and H. G. Walte (1996). Carry-over of fumonisin B1 into milk of lactating cows. *Milchwissenschaft*, **51**, 691–695.

Harris, C. C. (1996). The 1995 Walter Hubert lecture – Molecular epidemiology of human cancer: insights from the mutational analysis of the p53 tumour-suppressor gene. *Br. J. Cancer* **73**, 261–269.

Harris, C. C. and T. T. Sun (1986). Interactive effects of chemical carcinogens and hepatitis B virus in the pathogenesis of hepatocellular carcinoma, *Cancer Surv.* **6**, 765–780.

Harvey, R. B., L. F. Kubena, T. D. Philips *et al.* (1989). Prevention of aflatoxicosis by addition of hydrated sodium calcium aluminosilicate to the diets of growing barrows. *Am. J. Vet. Res.* **50**, 416–420.

Haschek, W. M., L. A. Gumprecht, G. Smith *et al.* (2001). Fumonisin toxicosis in swine: an overview of porcine pulmonary edema and current perspectives. *Environ. Health Perspect.* **109** (Suppl. 2), 251–257.

Hatfield, G. M. and L. R. Brady (1975). Toxins of higher fungi. *Lloydia* **38**, 36–55.

Hayes, A. W. (1981). *Mycotoxin Teratogenicity and Mutagenicity*. CRC Press, Inc. Boca Raton, FL.

He, Q., R. T. Riley and R. P. Sharma (2001). Fumonisin induced tumor necrosis factor alpha expression in a porcine kidney cell line is independent of sphingoid base accumulation induced by ceramide synthase inhibition. *Toxicol. Appl. Pharmacol.* **174**, 69–77.

Heinonen, J. T., R. Fisher, K. Brendel and D. L. Eaton (1996). Determination of aflatoxin B-1 biotransformation and binding to hepatic macromolecules in human precision liver slices. *Toxicol. Appl. Pharmacol.* **136**, 1–7.

Hendry, K. M. and E. C. Cole (1993). A review of mycotoxins in indoor air. *J. Toxicol. Environ. Health* **38**, 183–198.

Hesseltine, C. W. (1976). Conditions leading to mycotoxin contamination of foods and feeds. *Adv. Chem. Soc.* **146**, 1–22.

Heussner, A. H., E. O'Brien and D. R. Dietrich (2002). Species- and sex-specific variations in binding of ochratoxin A by renal proteins in vitro. *Exp. Toxicol. Pathol.* **54**, 151–159.

Hinoshita, F., Y. Suzuki, K. Yokoyama *et al.* (1997). Experimental IgA nephropathy induced by a low-dose environmental mycotoxin, nivalenol. *Nephron* **75**, 469–478.

Hiramoto, K. M., M. Kaku, T. Kato and K. Kikugawa (1995). DNA strand breaking by the carbon-centered radical generated from 4-(hydroxymethyl)benzenediazonium salt, a carcinogen in mushroom *Agaricus bisporus. Chem. Biol. Interact.* **94**, 21–36.

Ho, A. K., R. Peng, A. A. Ho *et al.* (1996). Interactions of fumonisins and sphingoid bases with GTP-binding proteins. *Biochem. Arch.* **12**, 249–260.

Hoehler, D. and R. R. Marquardt (1996). Influence of vitamins E and C on the toxic effects of ochratoxin A and T-2 toxin in chicks. *Poultry Sci.* **75**, 1508–1515.

Hoehler, D., R. R. Marquardt and A. A. Frohlich (1996a). Lipid peroxidation as one mode of action in ochratoxins A toxicity in rats and chicken. *Can. J. Anim. Sci.* **77**, 287–292.

Hoehler, D., R. R. Marquardt, A. R. McIntosh and H. Xiao (1996b). Free radical generation as induced by ochratoxin A and its analogs in bacteria (*Bacillus brevis). J. Biol. Chem.* **271**, 27 388–27 394.

Hohn, T. M., S. P. McCormick and A. E. Desjardins (1993). Evidence for a gene cluster involving trichothecene-pathway biosynthetic gene in *Fusarium sporotrichioides. Curr. Genetics* **24**, 291–295.

Holladay, S. D., B. J. Smith and M. I. Luster (1995). B lymphocyte precursor cells represent sensitive targets of T 2 mycotoxin exposure. *Toxicol. Appl. Pharmacol.* **131**, 309–315.

Holt, P. S., D. E. Corrier and J. R. DeLoach (1988). Immunosuppressive and enhancing effect of T-2 toxin on murine lymphocyte activation and interleukin-2 production. *Immunopharmacol. Immunotoxicol.* **10**, 365–385.

Horn, B.W. and J. W. Dorner (2002). Effect of competition and adverse culture conditions on aflatoxin production by *Aspergillus flavus* through successive generations. *Mycologia* **94**, 741–751.

Horwitz, W. (ed.) (2000). Mycotoxins. Official *Methods of Analysis of AOAC International*, 17th edn, Ch. 49. AOAC International, Gaithersburg, MD.

Horwitz, W., R. Albert and S. Nesheim (1993). Reliability of mycotoxin assays – an update. *J. AOAC Intl* **76**, 461–491.

Hsieh, D. P. H. (1989). Potential human health hazards of mycotoxins. In: S. Natori, K. Hashimoto and Y. Ueno (eds), *Mycotoxins and Phycotoxins 88*, pp. 69–80. Elsevier, Amsterdam.

Huang, C., M. Dickman, G. Henderson and C. Jones (1995). Repression of protein kinase C and stimulation of cyclic AMP response elements by fumonisin, a fungal encoded toxin which is a carcinogen. *Cancer Res.* **55**, 1655–1659.

Huang, X., J. W. Dorner and F. S. Chu (1994). Production of aflatoxin and cyclopiazonic acid by various aspergilli: an ELISA analysis. *Mycotoxin Res.* **10**, 101–106.

Huang, Z., D. G. White and G. A. Payne (1997). Corn seed protein inhibitory to *Aspergillus flavus* and aflatoxin biosynthesis. *Pathopathology* **87**, 622–627.

Hui, Y. H., J. R. Gorham, K. D. Murrell and D. O. Cliver (eds) (1994). Fungal diseases. In: *Foodborne Disease Handbook*, Vol. 2, pp. 463–668. Marcel Dekker, New York, NY.

Hussein, H. and J. M. Brasel (2001). Toxicity, metabolism and impact of mycotoxins on humans and animals. *Toxicology* **167**, 101–134.

Huwig, A., S. Freimund, O. Kappeli and H. Dutler (2001). Mycotoxin detoxification of animal feed by different adsorbents. *Toxicology Lett.* **122**, 179–188.

International Agency for Research on Cancer (IARC) (1993). Some naturally occurring substances: food items and constituents, heterocyclic amines and mycotoxins. *IARC Monographs on the Evaluation of Carcinogenic Risk to Humans*, Vol. 56. IARC, Lyon.

Islam, Z., M. Nagase, A. Ota *et al.* (1998). Structure function relationship of T-2 toxin and its metabolites in inducing thymic apoptosis in vivo in mice. *Biosci. Biotechnol. Biochem.* **62**, 1492–1497.

Islam, Z.,Y. S. Moon, H. R. Zhou *et al.* (2002). Endotoxin potentiation of trichothecene-induced lymphocyte apoptosis is mediated by up-regulation of glucocorticoids. *Toxicol. Appl. Pharmacol.* **180**, 43–55.

Ito, Y., S. W. Peterson, D. T. Wicklow and T. Goto (2001). *Aspergillus pseudotamarii*, a new aflatoxin producing species in *Aspergillus* section Flavi. *Mycol. Res.* **105**, 233–239.

Iyer, R. S., B. F. Coles, K. D. Raney *et al.* (1994). DNA adduction by the potent carcinogen aflatoxin B1: mechanistic studies. *J. Am. Chem. Soc.* **116**, 1603–1609.

Jackson, L., J. W. DeVries and L. B. Bullerman (eds) (1996). *Fumonisins in Food.* Plenum, New York, NY.

Jackson, P. E., G. S. Qian, M. D. Friesen *et al.* (2001). Specific p53 mutations detected in plasma and tumors of hepatocellular carcinoma patients by electrospray ionization mass spectrometry. *Cancer Res.* **61**, 33–35.

Jarvis, B. B., J. Salemmeand and A. Morais (1995). *Stachybotrys* toxins.1. *Nat. Toxins* **3**, 10–16.

Jelinek, C. F., A. E. Pohland and G. E. Wood (1989). Worldwide occurrence of mycotoxins in foods and feeds – an update. *J. AOAC* **72**, 223–230.

Jhee, E. C., L. L. Ho, K. Tsuji *et al.* (1989). Effect of butylated hydroxyanisole pretreatment on aflatoxin B1-DNA binding and aflatoxin B1-glutathione conjugation in isolated hepatocytes from rats. *Cancer Res.* **49**, 1357–1360.

Ji, C., M. H. Li, J. L. Li and S. J. Lu (1986). Synthesis of nitrosomethyl isoamylamine from isoamylamine and sodium nitrite by fungi. *Carcinogenesis* **7**, 301–303.

Jodlbauer, J., N. M. Maier and W. Lindner (2002). Towards ochratoxin A selective molecularly imprinted polymers for solid–phase extraction. *J. Chromatogr A* **945**, 45–63.

Joffe, A. Z. (1974). Toxicity of *Fusarium poae* and *F. sporotrichioides* and its relation to alimentary toxic aleukia. In: I. F. H. Purchase (ed.), *Mycotoxins*, pp. 229–262. Elsevier, Amsterdam.

Johanning, E., P. Landsbergis, M. Gareis *et al.* (1999). Clinical experience and results of a sentinel health investigation related to indoor fungal exposure. *Environ. Health Perspect.* **107** (Suppl. 3), 489–494.

Jones, C., Z. J. R. Ciacci, Y. Zhang *et al.* (2001). Analysis of fumonisin B1induced apoptosis. *Environ. Health Perspect.* **109** (Suppl. 2), 315–320.

Jorgensen, K. and R. L. Price (1981). Atmospheric pressure–ambient temperature reduction of aflatoxin B1 in ammoniated cottonseed. *J. Agric. Food Chem.* **29**, 555–558.

Kale, S. P., J. W. Cary, D. Bhatnagar and J. W. Bennett (1996). Characterization of experimentally induced, nonaflatoxigenic variant strains of *Aspergillus parasiticus*. *Appl. Environ. Microbiol.* **62**, 3399–3404.

Kawaji, A., K. Yamauchi, S. Fujii *et al.* (1992). Effects of mushroom toxins on glycogenolysis: comparison of toxicity of phalloidin, α-amanitin and DL-propargyl-flycine in isolated rat hepatocytes. *J. Pharmacobiol. Dyn.* **15**, 107–112.

Keller, N. P. and T. M. Hohn (1997). Metabolic pathway gene clusters in filamentous fungi. *Fungal Genet. Biol.* **21**, 17–29.

Kelly, E. J., K. E. Erickson, C. Sengstag and D. L. Eaton (2002). Expression of human microsomal epoxide hydrolase in *Saccharomyces cerevisiae* reveals a functional role in aflatoxin B-1 detoxification. *Toxicol. Sci.* **65**, 35–42.

Kelly, J. D., D. L. Eaton, F. P. Guengerich and R. A. Coulombe (1997). Aflatoxin B-1 activation in human lung. *Toxicol. Appl. Pharmacol.* **144**, 88–95.

Kensler, T. W. and J. D. Groopman (1996). Carcinogen-DNA and protein adducts – biomarkers for cohort selection and modifiable endpoints in chemoprevention trials. *J. Cell. Biochem.* **25**(Suppl.), 85–91.

Kiessling, K.-H. (1986). Biochemical mechanism of action of mycotoxins. *Pure Appl. Chem.* **58**, 327–338.

Kim, M. S., D. Y. Lee, T. Wang and J. J. Schroeder (2001). Fumonisin B1 induces apoptosis in LLCPK1 renal epithelial cells via a sphinganine and calmodulin-dependent pathway. *Toxicol. Appl. Pharmacol.* **176**, 118–126.

Klein, P. J., T. R. Van Vleet, J. O. Hall and R. A. Coulombe (2002). Dietary butylated hydroxytoluene protects against aflatoxicosis in turkeys. *Toxicol. Appl. Pharmacol.* **182**, 11–19.

Klich, M. A., E. J. Mullaney, C. B. Daly and J. W. Cary (2000). Molecular and physiological aspects of aflatoxin and sterigmatocystin biosynthesis by *Aspergillus tamarii* and *A. ochraceoroseus. Appl. Microbiol. Biotechnol.* **53**, 605–609.

Kobertz, W. R., D. Wang, G. N. Wogan and J. M. Essigmann (1997). An intercalation inhibitor altering the target selectivity of DNA damaging agents-synthesis of site-specific aflatoxin B-1 adducts in a P53 mutational hotspot. *Proc. Natl Acad. Sci. USA* **94**, 9579–9584.

Krogh, P (1976). Epidemiology of mycotoxic porcine nephropathy. *Nord. Vet. Med.* **28**, 452–458.

Kubena, L. F., R. B. Harvey, W. E. Huff *et al.* (1993). Efficacy of a hydrated sodium calcium aluminosilicate to reduce the toxicity of aflatoxin and diacetoxyscirpenol. *Poultry Sci.* **72**, 51–59.

Kuilman-Wahls, M. E. M., M. S. Vilar, L. de Nijs-Tjon *et al.* (2002). Cyclopiazonic acid inhibits mutagenic action of aflatoxin B-1. *Environ. Toxicol. Pharmacol.* **11**(3–4 Special Issue), 207–212.

Kuiper-Goodman, T. (1990). Uncertainties in the risk assessment of three mycotoxins: aflatoxin, ochratoxin and zearalenone. *Can. J. Physiol. Pharmacol.* **68**, 1018–1024.

Kuiper-Goodman, T. (1995). Mycotoxins: risk assessment and legislation. *Toxicol. Lett.* **82**, 853–859.

Kuiper-Goodman, T. and P. M. Scott (1989). Risk assessment of the mycotoxin ochratoxin A. *Biomed. Environ. Sci.* **2**, 179–248.

Kwak, S., H. Aizawa, M. Ishida and H. Shinozaki (1991). Systemic administration of acromelic acid induces selective neuron damage in the rat spinal cord. *Life Sci.* **49**, PL91–PL96.

Lacey, J. (1989). Water availability and the occurrence of toxicogenic fungi and mycotoxins in stored products. *Proc. Jpn Assoc. Mycotoxicol.* Special issue **1**, 186–189.

Larsson, B. K. and A. T. Eriksson (1989). The analysis and occurrence of hydrazine toxins in fresh and processed false morel, *Gyromitra esculenta. Z. Lebensm. Unters. Forsch.* **189**, 438–442.

Lee, S. E., B. C. Campbell, R. J. Molyneux *et al.* (2001). Inhibitory effects of naturally occurring compounds on aflatoxin B-1 biotransformation. *J. Agric. Food Chem.* **49**, 5171–5177.

Lee, Y. W., C. J. Mirocha, D. J. Shroeder and M. M. Walser (1985). TDP-1, a toxic component causing tibial dyschondroplasia in broiler chickens and trichothecenes from *Fusarium roseum Graminearum. Appl. Environ. Microbiol.* **50**, 102–107.

Leggott, N. L., G. S. Shephard, S. Stockenstrom *et al.* (2001). The reduction of patulin in apple juice by three different types of activated carbon. *Food Addit. Contam.* **18**, 825–829.

Lehman, T. A., M. Greenblatt, W. P. Bebbett and C. C. Harris (1994). Mutational spectrum of the p53 tumor suppressor gene: clues to cancer etiology and molecular pathogenesis. *Drug Metab. Rev*. **26**, 221–236.

Lehmann, P. F. and U. Khazan (1992). Mushroom poisoning by *Chlorophyllum molybdites* in the Midwest United States. Case and a review of the syndrome. *Mycopathologia* **118**, 3–13.

Lemke, S. L., S. E. Ottinger, C. L. Ake, K. Mayura and T. D. Philips (2001). Deamination of fumonisin B1 and biological assessment of reaction product toxicity. *Chem. Res. Toxicol*. **14**, 11–15, 191.

Lemmer, E. R., H. P. de la Motte, N. Omori *et al*. (1999). Histopathology and gene expression changes in rat liver during feeding of fumonisin B1, a carcinogenic mycotoxin produced by *Fusarium moniliforme. Carcinogenesis* **20**, 817–824.

Li, M. H., P. Li and P. J. Li (1980). Research on esophageal cancer in China. *Adv. Cancer Res*. **33**, 174–249.

Lim, C. W., H. M. Parker, R. F. Vesonder and W. M. Haschek (1996). Intravenous fumonisin B1 induces cell proliferation and apoptosis in the rat. *Nat. Toxins* **4**, 34–41.

Lin, J., J. A. Miller and E. C. Miller (1977). 2,3-dihydro-2-(guan-7-yl)-3-hydroxy-aflatoxin B1, a major acid hydrolysis product of aflatoxin B1-DNA or -ribosomal RNA adducts form in hepatic microsome mediated reactions and in rat liver in vivo. *Cancer Res*. **37**, 4430–4438.

Lin, Y. L., J. D. Hsu, F. P. Chou *et al*. (2000). Suppressive effect of penta-acetyl geniposide on the development of gamma-glutamyl transpeptidase foci-induced by aflatoxin B-1 in rats. *Chem. Biol. Interact*. **128**, 115–126.

Ling, K. H. (1995). Territrems, tremorgenic mycotoxins isolated from *Aspergillus terreus. J. Toxicol. Toxin Rev*. **13**, 243–252.

Linz, J. E. and J. J. Pestka (1992). Mycotoxin, molecular strategies for control. In: D. B. Finkelstein and C. Ball (eds), Aspergillus, *Biology and Industrial Applications*, pp. 217–231. Butterworth-Heinemann, Stoneham, MA.

Liu, H., S. Jackman, H. Driscoll and B. Larsen (2000). Immunologic effects of gliotoxin in rats: mechanisms for prevention of autoimmune diabetes mellitus. *Ann. Clin. Lab. Sci*. **30**, 366–378.

Liu, H., Y. Lu, J. S. Haynes *et al*. (2001). Reaction of fumonisin with glucose prevents promotion of hepatocarcinogenesis in female F344/N rats while maintaining normal hepatic sphinganine/sphingosine ratios. *J. Agric. Food Chem*. **49**, 4113–4121.

Liu, X. J. (1996). Studies on moniliformin and the etiology of Keshan disease. *J. Hyg.Res*. **25**, 151–156.

Liu, X. J., X. Y. Luo and W. J. Hu (1988). *Arthrinium* spp. and the etiology of deteriorated sugarcane poisoning. In: S. Natori, K. Hashimoto and Y. Ueno (eds), *Mycotoxins and Phycotoxins 88*, pp. 109–118. Elsevier, Amsterdam.

Logrieco, A., A. Moretti, G. Castella *et al*. (1998). Beauvericin production by *Fusarium* species. *Appl. Environ. Microbiol*. **64**, 3084–3088.

Longland, C. L., J. L. Dyer and F. Michelangeli (2000). The mycotoxin paxilline inhibits the cerebellar inositol 1,4,5-trisphosphate receptor. *Eur. J. Pharmocol*. **408**, 219–225

Lu, F. X. and A. M. Jeffrey (1993). Isolation, structural identification and characterization of a mutagen from *Fusarium moniliforme. Chem. Res. Toxicol.* **6**, 91–96.

Lu, Y., L. Clifford, C. C. Hauck *et al.* (2002). Characterization of fumonisin B-1-glucose reaction kinetics and products. *J. Agric. Food Chem.* **50**, 4726–4733.

Lu, Z., W. R. Dantzer, E. C. Hopmans *et al.* (1997). Reaction with fructose detoxifies fumonisin B1 while stimulating liver-associated natural killer cell activity in rats. *J. Agric. Food Chem.* **45**, 803–809.

Luo, X. Y. (1988). *Fusarium* toxin contamination of cereals in China. *Proc. Jpn Assoc. Mycotoxicol.* Special issue No. 1, 97–98.

Maaroufi, K., A. Achour, A. Zakhama *et al.* (1996). Human nephropathy related to ochratoxin A in Tunisia. *J. Toxicol. Toxin Rev.* **15**, 223–237.

Magan, N. and J. Lacey (1988). Ecological determinants of mould growth in stored grain. *J. Food Microbiol.* **7**, 245–256.

Magg, T., A. E. Melchinger, D. Klein and M. Bohn (2002). Relationship between European corn borer resistance and concentration of mycotoxins produced by *Fusarium* spp. in grains of transgenic Bt maize hybrids, their isogenic counterparts and commercial varieties. *Plant Breeding* **121**, 146–154.

Mahfoud, R., M. Maresca, N. Garmy and J. Fantini (2002). The mycotoxin patulin alters the barrier function of the intestinal epithelium: mechanism of action of the toxin and protective effects of glutathione. *Toxicol. Appl. Pharmacol.* **181**, 209–218.

Mahmoudi, M. and M. E. Gershwin (2000). Sick building syndrome. III. *Stachybotrys chartarum. J. Asthma* **37**, 191–198.

Maragos, C. M. (1997). Measurement of mycotoxins in foods with fiber-optic immunosensor. *J. Clin. Ligand Assay* **20**, 136–140.

Maragos, C. M. and R. D. Plattner (2002). Rapid fluorescence polarization immunoassay for the mycotoxin deoxynivalenol in wheat. *J. Agric. Food Chem.* **50**, 1827–1832.

Maragos, C. M., M. E. Jolley, R. D. Plattner and M. S. Nasir (2001). Fluorescence polarization as a means for determination of fumonisin in maize. *J. Agric. Food Chem.* **49**, 596–602.

Marasas, W. F. O. (1995). Fumonisins: their implications for human and animal health. *Nat. Toxins* **3**, 193–198.

Marasas, W. F. O. (2001). Discovery and occurrence of the fumonisins: a historical perspective. *Environ. Health Perspect.* **109** (Suppl. 2), 239–243.

Marasas, W. F. O., P. E. Nelson and T. A. Toussoun (1986). *Toxigenic* Fusarium *Species: Identity and Mycotoxicology*. Pennsylvania State University Press, University Park, PA.

Massey, T. E. (1996). The 1995 Pharmacological Society of Canada Merck Frosst Award – Cellular and molecular targets in pulmonary chemical carcinogenesis – studies with aflatoxin B-1. *Can. J. Physiol. Pharmacol.* **74**, 621–628.

Matsushima, K., K. Yashiro, Y. Hanya *et al.* (2001). Absence of aflatoxin biosynthesis in koji mold (*Aspergillus sojae*). *Appl. Microbiol. Biotechnol.* **55**, 771–776.

McBrien, K. D., Q. Gao, S. Huang *et al.* (1996). Fusaricide, a new cytotoxic N-hydroxypridone from *Fusarium* sp. *J. Nat. Prod.* **59**, 1151–1153.

McLean, M. and M. F. Dutton (1995). Cellular interactions and metabolism of aflatoxin: an update. *Pharmacol. Ther.* **65**, 163–192.

McMahon, G., E. F. Davis, J. Huber *et al.* (1990). Characterization of c-Ki-ras and N-ras oncogenes in aflatoxin B1-induced rat liver tumors. *Proc. Natl Acad. Sci. USA* **87**, 1104–1108.

Meisner, H. and P. Krogh (1986). Phosphoenolpyruvate carboxykinase as a selective indicator of ochratoxin A induced nephropathy. *Dev. Toxicol. Environ. Sci.* **14**, 199–206.

Meli, R., M. C. Ferrante, G. M. Raso *et al.* (2000). Effect of fumonisin B-1 on inducible nitric oxide synthase and cyclooxygenase-2 in LPS-stimulated J774A.1 cells. *Life Sci.* **67**, 2845–2853.

Merrill, A. H. Jr, G. van-Echten, E. Wang and K. Sandhoff (1993). Fumonisin B1 inhibits sphingosine (sphinganine) N-acyltransferase and de novo sphingolipid biosynthesis in cultured neurons in situ. *J. Biol. Chem.* **268**, 27 299–27 306.

Merrill, A. H., D. C. Liotta and R. T. Riley (1996). Fumonisins: fungal toxins that shed light on sphingolipid function. *Trends Cell Biol.* **6**, 218–223.

Merrill, A. H. Jr, E. M. Schmelz, D. L. Dillehay *et al.* (1997). Sphingolipids – the enigmatic lipid class: biochemistry, physiology and pathophysiology. *Toxicol. Appl. Pharmacol.* **142**, 208–225.

Merrill, A. H. Jr, M. C. Sullards, E. Wang *et al.* (2001). Sphingolipid metabolism: roles in signal transduction and disruption by fumonisins. *Environ. Health Perspect.* **109** (Suppl. 2), 283–289.

Michelot, D. and B. Toth (1991). Poisoning by *Gyrometra esculenta* – a review. *J. Appl. Toxicol.* **11**, 235–243.

Miller, J. D. (1995). Fungi and mycotoxins in grain: implications for stored product research. *J. Stored Product Res.* **31**, 1–16.

Miller, J. D. and H. L. Treholm (1994). *Mycotoxins in Grain: Compounds Other Than Aflatoxin.* Eagan Press, St Paul, MN.

Mills, J. T. (1989). Ecology of mycotoxicogenic *Fusarium* species on cereal seeds. *J. Food Prot.* **52**, 737–742.

Mirocha, C. J., S. V. Pathre and C. M. Christensen (1979). Mycotoxins. *Adv. Cereal Sci. Technol.* **3**, 159–225.

Mirocha, C. J., D. G. Gilchrist, W. T. Shier *et al.* (1992). AAL toxins, fumonisins (biology and chemistry) and host-specificity concepts. *Mycopathologia* **117**, 47–56.

Miura, K., L. Aminova and Y. Murayama (2002). Fusarenon-X induced apoptosis in HL-60 cells depends on caspase activation and cytochrome c release. *Toxicology* **172**, 103–112.

Mobio, T. A., I. Baudrimont, A. Sanni *et al.* (2000). Prevention by vitamin E of DNA fragmentation and apoptosis induced by fumonisin B-1 in c6 glioma cells. *Arch. Toxicol.* **74**, 112–119

Morales, P. E., E. Bermudez, P. E. Hernandez and B. Sanz (1990). The mutagenicity of some Spanish edible mushrooms in the Ames test. *Food Chem.* **38**, 279–288.

Mori, H. and K. Kawai (1989). Genotoxicity in rodent hepatocytes and carcinogenicity of mycotoxins and related chemicals. In: S. Natori, K. Hashimoto and Y. Ueno (eds), *Mycotoxins and Phycotoxins 88*, pp. 81–90. Elsevier, Amsterdam.

Moss, M. O. (1998). Recent studies of mycotoxins. *Appl. Microbiol. Symp. Suppl.* **84**, 62s–76s.

Munekata, E. (1981). New advances in phallotoxin chemistry. In: A. Eberle, R. Geiger and T. Wieland (eds), *Perspectives in Peptide Chemistry – 1981*, pp.129–140. S. Karger, New York, NY.

Murphy, P. A., E. C. Hopmans, K. Miller and S. Henddrick (1995). Can fumonisin B1 in foods be detoxified? In: W. R. Bidlack and S. T. Omaye (eds), *Natural Protectants Against Natural Toxicants*, Vol. 1, pp. 105–117. Technomic Publishing Co., Lancaster, PA.

Musser, S. M., M. L. Gay, E. P. Mazzola and R. D. Plattner (1996). Identification of a new series of fumonisins containing 3-hydroxypyridine. *J. Pat. Prod.* **59**, 970–972.

Nagase, M., M. M. Alam, A. Tsushima *et al.* (2001). Apoptosis induction by T-2 toxin: activation of caspase-9, caspase-3 and DFF-40/CAD through cytosolic release of cytochrome c in HL-60 cells. *Biosci. Biotechnol. Biochem.* **65**, 1741–1747.

Nagashima, H. (1996). Cytotoxic effects of rubratoxin B on cultured cells. *Mycotoxins* **42**, 57–61.

Nasir, M. S. and M. E. Jolley (2002). Development of a fluorescence polarization assay for the determination of aflatoxins in grains. *J. Agric. Food Chem.* **50**, 3116–3121.

Natori, S., K. Hashimoto and Y. Ueno (eds) (1989). *Mycotoxins and Phycotoxins 88*. Elsevier, Amsterdam.

Neal, G. E. (1995). Genetic implications in the metabolism and toxicity of mycotoxins. *Toxicol. Lett.* **82**, 861–867.

Neal, G. E., D. J. Judah, P. Carthew *et al.* (2001). Differences detected in vivo between samples of aflatoxin-contaminated peanut meal, following decontamination by two ammonia-based processes. *Food Addit. Contam.* **18**, 137–149.

Nelson, P. E., R. D. Plattner, D. D. Shackelford and A. E. Desjardins (1992). Fumonisin B_1 production by *Fusarium* species other than *F. moniliforme* in section *Liseola* and by some related species. *Appl. Environ. Microbiol.* **58**, 984–989.

Nelson, P. E., A. E. Desjardins and R. D. Plattner (1993). Fumonisins, mycotoxins produced by *Fusarium* species: biology, chemistry and significance. *Ann. Rev. Phytopathol.* **31**, 233–252.

Netke, S. P., M. W. Roomi, C. Tsao and A. Niedzwiecki (1997). Ascorbic acid protects guinea pigs from acute aflatoxin toxicity. *Toxicol. Appl. Pharmacol.* **143**, 429–435.

Neucere, J. N. and T. E. Cleveland (1997). Antifungal and electrophoretic properties of isolated protein fractions from corn kernels. *J. Agric. Food Chem.* **45**, 299–301.

Newberne, P. M. (1987). Interaction of nutrients and other factors with mycotoxins. In: P. Krogh (ed.), *Mycotoxins in Food*, pp.177–216. Academic Press, New York, NY.

Newberne, P. M. and A. E. Rogers (1981). Animal toxicity of major environmental mycotoxins. In: R. C. Shank (ed.), *Mycotoxins and Nitroso compounds, Environmental Risks*, Vol. I, pp. 51–106. CRC Press, Inc., Boca Raton, FL.

Nielsen, K. F., K. Huttunen, A. Hyvarinen *et al.* (2002). Metabolite profiles of *Stachybotrys* isolates from water-damaged buildings and their induction of inflammatory mediators and cytotoxicity in macrophages. *Mycopathologia* **154**, 201–205.

Nikulin, M., A.-L. Pasanen, S. Berg and E.-L. Hintikka (1994). *Stachybotrys atra* growth and toxin production in some building materials and fodder under different relative humidities. *Appl. Environ. Microbiol.* **60**, 3421–3424.

Norred, W. P. (1993). Fumonisins – mycotoxins produced by *Fusarium moniliforme. J. Toxicol. Environ. Health.* **38**, 309–328.

Norred, W. P. and K. A. Voss (1994). Toxicity and role of fumonisins in animal diseases and human esophageal cancer. *J. Food Prot.* **57**, 522–527.

Norred, W. P., R. T. Riley, F. I. Meredith *et al.* (2001). Instability of N-acetylated fumonisin B1 (FA1) and the impact on inhibition of ceramide synthase in rat liver slices. *Food Chem. Toxicol.* **39**, 1071–1078.

Northolt, M. D. and L. B. Bullerman (1982). Prevention of mold growth and toxin production through control of environmental conditions. *J. Food Prot.* **45**, 519–526.

Nwankwo, J. O., J. G. Tahnteng and G. O. Emerole (2000). Inhibition of aflatoxin B1 genotoxicity in human liver-derived HepG2 cells by kolaviron biflavonoids and molecular mechanisms of action. *Eur. J. Cancer Prevent.* **9**, 351–361.

Ono, H., A. Kataoka, M. Koakutsu *et al.* (1995). Ochratoxin producibility by strain of *Aspergillus niger* group stored in IFO culture collection. *Mycotoxins* **41**, 47–51.

Ono, M., S. Sakuda, A. Suzuki and A. Isogai (1997). Aflastatin a, a novel inhibitor of aflatoxin production by aflatoxigenic fungi. *J. Antibiotics* **50**, 111–118.

Orzaez, D., A. J. deJong and E. J. Woltering (2001). A tomato homologue of the human protein PIRIN is induced during programmed cell death. *Plant Molec. Biol.* **46**, 459–468.

Ouyang, Y. L., J. I. Azcona-Olivera, J. Murtha and J. J. Pestka (1996). Vomitoxin-mediated IL-2, IL-4 and IL-5 superinduction in murine CD4(+) T cells stimulated with phorbol ester and calcium ionophore: relation to kinetics of proliferation. *Toxicol. Appl. Pharmacol.* **138**, 324–334.

Padmore, L., G. K. Radda and K. A. Knox (1996). Wortmannin-mediated inhibition of phosphatidylinositol 3-kinase activity triggers apoptosis in normal and neoplastic B lymphocytes which are in cell cycle. *Internatl Immunol.* **84**, 585–594.

Pahl, H. L., B. Krauss, K. SchulzeOsthoff *et al.* (1996). The immunosuppressive fungal metabolite gliotoxin specifically inhibits transcription factor NF-kappa B. *J. Exp. Med.* **183**, 1829–1840.

Park, D. L. (1993a). Controlling aflatoxin in food and feed. *Food Technol.* **47**, 92–96.

Park, D. L (1993b). Perspectives on mycotoxin decontamination procedures. *Food Addit. Contam.* **10**, 49–60.

Park, D. L. and B. Liang (1993). Perspectives on aflatoxin control for human food and animal feed. *Trends Food Sci. Technol.* **4**, 334–342.

Park, D. L. and A. E. Pohland (1989). Sampling and sample preparation for detection and quantitation of natural toxicants in food and feed. *J. Assoc. Anal. Chem.* **72**, 399–404.

Park, D. L. and T. C. Troxell (2002). US perspective on mycotoxin regulatory issues. In: J. W.DeVries, M. W. Trucksess and L. S. Jackson (eds), *Mycotoxins and Food Safety*, p. 277. Kluwer Academic/Plenum Publishers, New York, NY.

Park, J., E. B. Smalley and F. S. Chu (1996). Natural occurrence of *Fusarium* mycotoxins of the 1992 corn crop in the field. *Appl. Environ. Microbiol.* **62**, 1642–1648.

Pastrana, L., M. O. Loret, P. J. Blanc and G. Goma (1996). Production of citrinin by *Monascus ruber* submerged culture in chemical defined media. *Acta Biotechnol. Environ. Microbiol.* **16**, 315–319.

Payne, G. A. (1998). Process of contamination by aflatoxin-producing fungi and their impact on crops. In: K. K. Sinha and D. Bhatnager (eds), *Mycotoxin in Agriculture and Food Safety*, pp. 278–306. Marcel Dekker, New York. NY.

Payne, G. A. and M. P. Brown (1998). Genetics and physiology of aflatoxin biosynthesis. *Ann. Rev. Phytopathol.* **36**, 329–362.

Payne, G. A., G. J. Nystrom, D. Bhatnagar *et al.* (1993). Cloning of the afl-2 gene involved in aflatoxin biosynthesis for *Aspergillus flavus. Appl. Environ. Microbiol.* **59**, 156–162.

Peers, F., X. Bosch, J. Kaldor *et al.* (1987). Aflatoxin exposure, hepatitis B virus infection and liver cancer in Swaziland. *Intl J. Cancer* **39**, 545–553.

Penn, J., R. Swift, L. J. Wigley *et al.* (1993). Janthitrems B and C, two principal indole-diterpenoids produced by *Penicillium janthinellum. Phytochemistry* **32**, 1431–1434.

Pestka, J. J. (1988). Enhanced surveillance of foodborne mycotoxins by immunoassay. *J. Assoc. Anal. Chem.* **71**, 1075–1081.

Pestka, J. J (1994). Application of immunology to the analysis and toxicity assessment of mycotoxins. *Food Agric. Immunol.* **6**, 219–234.

Pestka, J. J. and G. S. Bondy (1990). Alteration of immune function following dietary mycotoxin exposure. *Can. J. Physiol. Pharmacol.* **68**, 1009–1016.

Pestka, J. J. and H. R. Zhou (2002). Effects of tumor necrosis factor type 1 and 2 receptor deficiencies on anorexia, growth and IgA dysregulation in mice exposed to the trichothecene vomitoxin. *Food Chem. Toxicol.* **40**, 1623–1631.

Pestka, J. J., H. R. Zhou, Q. S. Jia and A. M. Timmer (2002). Dietary fish oil suppresses experimental immunoglobulin A nephropathy in mice. *J. Nutri.* **132**, 261–269.

Petkova–Bocharova, T. and M. Castegnaro (1985). Ochratoxin A contamination of cereals in an area of high incidence of Balkan endemic nephropathy in Bulgaria. *Food Addit. Contam.* **2**, 267–270.

Petkova-Bocharova, T., I. N. Chernozemsky and M. Castegnaro (1988). Ochratoxin A in human blood in relation to Balkan endemic nephropathy and urinary system tumors in Bulgaria. *Food Addit. Contam.* **5**, 299–301.

Petr, J., J. Rozinek, Z. Vanourkova and F. Jilek (1999). Cyclopiazonic acid, an inhibitor of calcium-dependent ATPases, induces exit from metaphase I arrest in growing pig oocytes. *Reprod. Fertil. Develop.* **11**, 235–246.

Petzinger, E. and A. Weidenbach (2002). Mycotoxins in the food chain: the role of ochratoxins *Livestock Prod. Sci.* **76**, 245–250.

Pfohl-Leszkowicz, A., Y. Gross, A. Kane *et al.* (1993). Differential DNA adduct formation and disappearance in three mouse tissues after treatment with the mycotoxin ochratoxin A. *Mutat. Res.* **289**, 265–273.

Philips T. D., S. L Lemke and P. G. Grant (2002). Characterization of clay-based enterosorbents for the prevention of aflatoxicosis. In: J. W. DeVries, M. W. Trucksess and L. S. Jackson (eds), *Mycotoxins and Food Safety*, p. 157. Kluwer Academic/Plenum Publishers, New York, NY.

Pietri, A., T. Bertuzzi, L. Pallaroni and G. Piva (2001). Occurrence of ochratoxin A in Italian wines. *Food Addit. Contam.* **18**, 647–654.

Piva, G., F. Galvano, A. Pietri and A. Piva (1995). Detoxification methods of aflatoxins. A review. *Nutr. Res.* **15**, 767–776.

Plattner, R. D. and P. E. Nelson (1994). Production of beauvericin by a strain of *Fusarium proliferatum* isolated from corn fodder for swine. *Appl. Environ. Microbiol.* **60**, 3894–3896.

Plumlee, K. H. and F. D. Galey (1994). Neurotoxic mycotoxins: a review of fungal toxins that cause neurological disease in large animals. *J. Vet. Intern. Med.* **8**, 49–54.

Pocsfalvi, G., G. Dilanda, P. Ferranti *et al.* (1997). Observation of non-covalent interactions between beauvericin and oligonucleotides using electrospray ionization mass spectrometry. *Rapid Comm. Mass Spect.* **11**, 265–272.

Pocsfalvi, G., A. Ritieni, G. Randazzo *et al.* (2000). Interaction of Fusarium mycotoxins, fusaproliferin and fumonisin B1, with DNA studied by electrospray ionization mass spectrometry. *J. Agric. Food Chem.* **48**, 5795–5801.

Pohland, A. E., S. Nesheim and L. Friedman (1992). Ochratoxin A: a review. *Pure Appl. Chem.* **64**, 1029–1046.

Prast, H., E. R. Werner, W. Pfaller and M. Moser (1988). Toxic properties of the mushroom *Cortinarius orellanus*: I. Chemical characterization of the main toxin of *Cortinarius orellanus* (Fries) and *Cortinarius speciosissimus* (Kuelhn and Romagn) and acute toxicity in mice. *Arch. Toxicol.* **62**, 81–88.

Price, D. L. and D. G. Ahearn (1999). Sanitation of wallboard colonized with *Stachybotrys chartarum. Curr. Microbiol.* **39**, 21–26.

Prieto, R. and C. P. Woloshuk (1997). *ord1*, an oxidoreductase gene responsible for conversion of o–methylsterigmatocystin to aflatoxin in *Aspergillus flavus. Appl. Environ. Microbiol.* **63**, 1661–1666.

Proctor, R. H. (2000). Fusarium toxins: trichothecenes and fumonisin. In: J. W. Cary, J. E. Linz and D. Bhatnagar (eds), *Microbial Foodborne Diseases: Mechanism of Pathogenesis and Toxin Synthesis*, pp. 366–381. Technomic Publishing, Co. Inc., Lancaster, PA.

Proctor, R. H.., T. M. Hohn, S. P. McCormick and A. E. Desjardins (1995a). *Tri6* encodes an unusual zinc finger protein involved in regulation of trichothecene biosynthesis in *Fusarium sporotrichioides. Appl. Environ. Microbiol.* **61**, 1923–1930.

Proctor, R. H., T. M. Hohn and S. P. McCormick (1995b). Reduced virulence of *Gibberella zeae* caused by disruption of a trichothecene toxin biosynthetic gene. *Mol. Plant Microbe Interact.* **8**, 593–601.

Proctor, R. H., A. E. Desjardins and R. D. Plattner (1999). Biosynthetic and genetic relationships of B-series fumonisins produced by *Gibberella fujikuroi* mating population A. *Nat. Toxins* **7**, 251–258.

Qin, G., P. Gopalan-Kirczky, J. Su *et al.* (1997). Inhibition of aflatoxin B1-induced initiation of hepatocarcinogenesis in the rat by green tea. *Cancer Lett.* **112**, 149–154.

Rahimtula, A. D., J. C. Bereziat, V. Bussacchini-Griot and H. Bartsch (1988). Lipid peroxidation as a possible cause of ochratoxin toxicity. *Biochem. Pharmacol.* **37**, 4469–4477.

Ramirez, P. P., P. Parrilla, F. Sanchez Bueno *et al.* (1993). Fulminant hepatic failure after *Lepiota* mushroom poisoning. *J. Hepatol* **19**, 51–54.

Raney, K. D., S. Gopalakrishnan, S. Byrd *et al.* (1990). Alteration of the aflatoxin cyclopentenone ring to a delta-lactone reduces intercalation with DNA and decreases formation of guanine-N7 adducts by aflatoxin epoxides. *Chem. Res. Toxicol.* **3**, 254–261.

Raney, K. D., T. Shimada, D.-H. Kim *et al.* (1992a). Oxidation of aflatoxins and sterigmatocystin by human liver microsomes: significance of aflatoxin Q_1 as a detoxication product of aflatoxin B_1. *Chem. Res. Toxicol.* **5**, 202–210.

Raney, K. D., B. Coles, F. P. Guengerich and T. M. Harris (1992b). The *endo*-8, 9-epoxide of aflatoxin B_1: a new metabolite. *Chem. Res. Toxicol.* **5**, 333–335.

Raney, K. D., D. J. Meyer, B. Ketterer *et al.* (1992c). Glutathione conjugation of aflatoxin B_1 *exo*- and *endo*-epoxides by rat and human glutathione *S*-transferases. *Chem. Res. Toxicol.* **5**, 470–478.

Ray, L. L. and L. B. Bullerman (1982). Preventing growth of potentially toxic molds using antifungal agents. *J. Food Prot.* **45**, 953–963.

Reddy, S.V., D. K. Mayi, M. U. Reddy *et al.* (2001). Aflatoxins B1 in different grades of chillies (*Capsicum annum* L.) in India as determined by indirect competitive ELISA. *Food Addit. Contam.* **18**, 553–558.

Rheeder, J. P., W. F. O. Marasas and H. F. Vismer (2002). Production of fumonisin analogs by *Fusarium* species. *Appl. Environ. Microbiol.* **68**, 2101–2105.

Richard, J. L. and J. R. Thurston (1986). *Diagnosis of Mycotoxicoses*. Martinus Nijhoff Publishers, Dordrecht.

Richard, J. L., G. A. Bennett, P. F. Ross and P. E. Nelson (1993). Analysis of naturally occurring mycotoxins in feedstuffs and food. *J. Anim. Sci.* **71**, 2563–2574.

Richard, J. L., G. Meerdink, C. M. Maragos *et al.* (1996). Absence of detectable fumonisins in the milk of cows fed *Fusarium proliferatum* (Matsushima) Nirenberg culture material. *Mycopathologia* **133**, 123–126.

Riley, R. T. and D. E. Goeger (1991). Cyclopiazonic acid: speculation on its function in fungi. In: D. Bhatnagar, E. B. Lillehoj and D. K. Arora (eds), *Handbook of Applied Mycology*, Vol. V, pp. 385–402. Marcel Dekker, New York, NY.

Riley, R. T. and J. L. Richard (eds) (1992). Fumonisins: a current perspective and view to the future. *Mycopathologia* **117**, 1–124.

Riley, R. T., W. P. Norred and C. W. Bacon (1993a). Fungal toxins in foods: recent concerns. *Ann. Rev. Nutr.* **13**, 167–189.

Riley, R. T., N.-H. An, J. L. Showker *et al.* (1993b). Alteration of tissue and serine sphinganine to sphingosine ratio: an early biomarker of exposure to fumonisin-containing feeds in pigs. *Toxicol. Appl. Pharmacol.* **118**, 105–112.

Riley, R. T., K. A. Voss, H.-S. Yoo *et al.* (1994). Mechanism of fumonisin toxicity and carcinogenesis. *J. Food Prot.* **57**, 528–535.

Riley, R. T., K. A. Voss, W. P. Norred *et al.* (1999). Serine palmitoyltransferase inhibition reverses antiproliferative effects of ceramide synthase inhibition in cultured renal cells and suppresses free sphingoid base accumulation in kidney of BALBc mice. *Environ. Toxicol. Pharmacol.* **7**, 109–118.

Riley, R. T., E. Enongene, K. A. Voss *et al.* (2001). Sphingolipid perturbations as mechanisms for fumonisin carcinogenesis. *Environ. Health Perspect.* **109** (Suppl. 2), 301–308.

Ritieni, A., V. Fogliano, G. Randazzo *et al.* (1995). Isolation and characterization of fusaproliferin, a new toxic metabolite from *Fusarium proliferatum. Nat. Toxins* **3**, 17–20.

Rodricks, J. V., C. W. Hesseltine and M. A. Mehlman (1977). *Mycotoxins. Proceedings of the 1st International Conference on Mycotoxins in Human and Animal Health.* Pathotox Publ. Inc., Park Forest S., IL.

Rotter, B. A., D. B. Prelusky and J. J. Pestka (1996). Toxicology of deoxynivalenol (Vomitoxin). *J. Toxicol. Environ. Health* **48**, 1–34.

Rustom, I. Y. S. (1997). Aflatoxin in food and feed – occurrence, legislation and inactivation by physical methods [review]. *Food Chem.* **59**, 57–67.

Sabbioni, G. and C. P. Wild (1991). Identification of an aflatoxin G_1-serum albumin adduct and its relevance to the measurement of human exposure to aflatoxins. *Carcinogenesis* **12**, 97–103.

Sadler, T. W., A. H. Merrill, V. L. Stevens *et al.* (2002). Prevention of fumonisin B1-induced neural tube defects by folic acid. *Teratology* **66**, 169–176.

Sakuda, S., M. Ono, K. Furihata *et al.* (1996). Aflastatin A, a novel inhibitor of aflatoxin production of *Aspergillus parasiticus*, from *Streptomyces. J. Am. Chem. Soc.* **118**, 7855–7856.

Saunders, D. S., F. I. Meredith and K. A. Voss (2001). Control of fumonisin: effects of processing. *Environ. Health Perspect.* **109** (Suppl. 2), 333–336.

Schaafsma, A. W., D. C. Hooker, T. S. Baute and L. Illincic-Tamburic (2002). Effect of Bt-corn hybrids on deoxynivalenol content in grain at harvest. *Plant Dis.* **86**, 1123–1126.

Schlatter, C., J. Studerroh and T. Rasonyi (1996). Carcinogenicity and kinetic aspects of ochratoxin A. *Food Addit. Contam.* **13** (Suppl S), 43–44.

Schmelz, M., K. M. A. Dombrink, P. C. Roberts *et al.* (1998). Induction of apoptosis by fumonisin B1 in HT-29 cells is mediated by the accumulation of endogenous free sphingoid bases. *Toxicol. Appl. Pharmacol.* **148**, 252–260.

Schnerr, H., R. F. Vogel and L. Niessen (2002). Correlation between DNA of trichothecene-producing Fusarium species and deoxynivalenol concentrations in wheat-samples. *Lett. Appl. Microbiol.* **35**, 121–125.

Schroeder, J. J., H. M. Crane, J. Xia *et al.* (1994). Disruption of sphingolipid metabolism and stimulation of DNA syntheis by fumonisin B1: a molecular mechanism for carcinogenesis associated with *Fusarium moniliforme.* J. Biol. Chem. **269**, 3475–3481.

Schuster, E., N. Dunn-Coleman, J. C. Frisvad and P. W. M. van Dijck (2002). On the safety of *Aspergillus niger* – a review. *Appl. Microbiol. Biotechnol.* **59**, 426–435.

Schwerdt, G., R. Freudinger, S. Mildenberger *et al.* (1999). The nephrotoxin ochratoxin A induces apoptosis in cultured human proximal tubule cells. *Cell Biol. Toxicol.* **15**, 405–415.

Scott, P. M. (1975). Patulin. In: I. F. H. Purchase (ed.), *Mycotoxins*, pp. 383–403. Elsevier Scientific Publishing Co., New York.

Scott, P. M. (1993). Fumonisins. *Intl J. Food Microbiol.* **18**, 257–270.

Scudamore, K. A. and C. T. Livesey (1998). Occurrence and significance of mycotoxins in forage crops and silage: a review. *J. Sci. Food Agric.* **77**, 1–17.

Seegers, J. C., M.-L. Lottering and P. J. Garlinski (1994). The mycotoxin ochratoxin A causes apoptosis-associated DNA degradation in human lymphocytes. *Med. Sci. Res.* **22**, 417–419.

Seo, J. A., J. C. Kim and Y. W. Lee (1996). Isolation and characterization of two new type C fumonisins produced by *Fusarium oxysporum. J. Nat. Prod.* **59**, 1003–1005.

Seo, J. A., R. H. Proctor and R. D. Plattner (2001). Characterization of four clustered and coregulated genes associated with fumonisin biosynthesis in *Fusarium verticillioides. Fungal Genet. Biol.* **34**, 155–165.

Shank, R. C. (1981). *Mycotoxins and Nitroso Compounds: Environmental Risks*, Vols I and II. CRC Press, Inc., Boca Raton, FL.

Sharma, R. P. and D. K. Salunkhe (1991). *Mycotoxin and Phytoalexins in Human and Animal Health.* CRC Press, Inc., Boca Raton, FL.

Sharma, R. P., N. Bhandari, Q. He *et al.* (2001). Decreased fumonisin hepatotoxicity in mice with a targeted deletion of tumor necrosis factor receptor 1. *Toxicology* **159**, 69–79.

Shebar, F. Z., J. D. Groopman, G.-S. Qian and G. N. Wogan (1993). Quantitative analysis of aflatoxin–albumin adducts. *Carcinogenesis* **14**, 1203–1208.

Shen, H. M. and C. N. Ong (1996). Mutations of the p53 tumor suppressor gene and ras oncogenes in aflatoxin hepatocarcinogenesis. *Mutat. Res. Rev. Genet. Toxicol.* **366**, 23–44.

Shen, H. M., C. Y. Shi, Y. Shen and C. N. Ong (1996). Detection of elevated reactive oxygen species level in cultured rat hepatocytes treated with aflatoxin B-1. *Free Rad. Biol. Med.* **21**, 139–146.

Shepherd, A. G., M. M. Manson, H. W. L. Ball and L. I. McLellan (2000). Regulation of rat glutamate-cysteine ligase (gamma-glutamylcysteine synthetase) subunits by chemopreventive agents and in aflatoxin B-1-induced preneoplasia. *Carcinogenesis* **21**, 1827–1834.

Shier, W. T (1992). Sphingosine analogs: an emerging new class of toxins that includes the fumonisins. *J. Toxicol. Toxin Rev.* **11**, 241–257.

Shier, W. T (2000). The fumonisin paradox: a review of research on oral bioavailability of fumonisin B-1, a mycotoxin produced by *Fusarium moniliforme. J. Toxicol. Toxin Rev.* **19**, 161–187.

Shinozuka, J., L. Guanmin, K. Uetsuka *et al.* (1997). Process of the development of T-2 toxin-induced apoptosis in the lymphoid organs of mice. *Exp. Anim. Tokyo.* **46**, 117–126.

Shokri, F., M. Heidari, S. Gharagozloo and M. Ghazi-Khansari (2000). In vitro inhibitory effects of antioxidants on cytotoxicity of T-2 toxin. *Toxicology* **146**, 171–176.

Sidhu, G. S. (2002). Mycotoxin genetics and gene clusters. *Eur. J. Plant Pathol.* **108**, 705–711.

Simon, P (1996). Ochratoxin and kidney disease in the human. *J. Toxicol. Toxin Rev.* **15**, 239–249.

Sinha, K. K. and D. Bhatnagar (1998). *Mycotoxin in Agriculture and Food Safety*. Marcel Dekker, New York, NY.

Sivakumar, V., J. Thanislass, S. Niranjali and H. Devaraj (2001). Lipid peroxidation as a possible secondary mechanism of sterigmatocystin toxicity. *Human Exp. Toxicol* **20**, 398–403.

Skipper, P. L. and S. R. Tannenbaum (1990). Protein adducts in the molecular dosimetry of chemical carcinogens. *Carcinogenesis* **11**, 507–518.

Smela, M. E., S. S. Currier, E. A. Bailey and J. M. Essigmann (2001). The chemistry and biology of aflatoxin B-1 from mutational spectrometry to carcinogenesis. *Carcinogenesis* **22**, 535–545.

Smith, E. E., T. D. Phillips, J. A. Ellis *et al.* (1994). Dietary hydrated sodium calcium aluminosilicate reduction of aflatoxin M_1 residue in dairy goat milk and effects on milk production and components. *J. Anim. Sci.* **72**, 677–682.

Smith, J. E., G. Soloms, C. Lewis and [many] others (1995). Mycotoxins and toxic plant components. EC, WHO/IPCS, FAO and ILSI Europe Joint Workshop. *Nat. Toxins* **3**, 181–331.

Soini, Y., S. C. Chia, W. P. Bennett *et al.* (1996). An aflatoxin-associated mutational hotspot at codon 249 in the p53 tumor suppressor gene occurs in hepatocellular carcinomas from Mexico. *Carcinogenesis* **17**, 1007–1012.

Spiegel, S. and A. H. Merrill Jr (1996). Sphingolipid metabolism and cell growth regulation. *FASEB J.* **10**, 1388–1397.

Spotti, M., R. F. M. Maas, C. M. de Nijs and J. Fink-Gremmels (2000). Effect of fumonisin B-1 on rat hepatic p450 system. *Environ. Toxicol. Pharmacol.* **8**, 197–204.

Stegen, G., U. v. d. Jorissen, A. Pittet *et al.* (1997). Screening of European coffee final products for occurrence of ochratoxin A (OTA). *Food Addit. Contam.* **14**, 211–216.

Stenklyft, P. H. and W. L. Augenstein (1990). *Chlorophyllum molybdites* – severe mushroom poisoning in a child. *J. Toxicol. Clin. Toxicol.* **28**, 159–168.

Sterner, O., R. Bergman, E. Kesler *et al.* (1982). Mutagens in larger fungi. I. Forty-eight species screened for mutagenic activity in the Salmonella/microsome assay. *Mutat. Res.* **101**, 269–281.

Steyn, P. S. (1995). Mycotoxins, general view, chemistry and structure. *Toxicol. Lett.* **82**, 843–851.

Steyn, P. S. and R. Vleggaar (1986). *Mycotoxins and Phycotoxins*. Elsevier, Amsterdam.

Stoloff, L., H. Van Egmond and D. L. Park (1991). Rationales for the establishment of limits and regulations for mycotoxins. *Food Addit. Contam.* **8**, 213–222.

Studer-Rohr, I., D. R. Dietrich, J. Schlatter and C. Schlatter (1995). The occurrence of ochratoxin A in coffee. *Food Chem. Toxicol.* **33**, 341–355.

Sun, J. L, D. S. Wang, K. Q. Wang and H. J. Hong (1988). Studies of sterigmatocystin (ST) produced by *Aspergillus versicolor* in human gastric juice *in vitro. Proc. Jpn Assoc. Mycotoxicol.*, special issue 1, 127–130. Association of Mycotoxicology, Dept of Toxicology and Microbial Chemistry, Tokyo University of Science, Japan.

Sutton, P., J. Beaver and P. Waring (1995). Evidence that gliotoxin enhances lymphocyte activation and induces apoptosis by effects on cyclic AMP levels. *Biochem. Pharmacol.* **50**, 2009–2014.

Tag, A., J. Hicks, G. Garifullina *et al.* (2000). G-protein signalling mediates differential production of toxic secondary metabolites. *Molecular Microbiol.* **38**, 658–665.

Takahashi-Ando, N., M. Kimura, H. Kakeya *et al.* (2002). A novel lactonohydrolase responsible for the detoxification of zearalenone, enzyme purification and gene cloning. *Biochem. J.* **365**, 1–6.

Tebbett, I. R. and B. Caddy (1984). Mushroom toxins of the Genus *Cortinarius. Experientia* **40**, 441–446.

Thirumala, D. K., M. A. Mayo, G. Reddy *et al.* (2001). Occurrence of ochratoxin A in black pepper, coriander, ginger and turmeric in India. *Food Addit. Contam.* **18**, 830–835.

Tolleson, W. H., W. B. Melchior, S. M. Morris *et al.* (1996a). Apoptotic and antiproliferative effects of fumonisin B-1 in human keratinocytes, fibroblasts, esophageal epithelial cells and hepatoma cells. *Carcinogenesis* **17**, 239–249.

Tolleson, W. H., K. L. Dooley, W. G. Sheldon *et al.* (1996b). The mycotoxin fumonisin induces apoptosis in cultured human liver cells and in livers and kidney of rats. In: L. Jackson, J. W. DeVries and L. B. Bullerman (eds), *Fumonisins in Foods*, pp. 237–250. Plenum Press, New York, NY.

Tolleson, W. H., L. H. Couch, W. B. Melchior Jr *et al.* (1999). Fumonisin B1 induces apoptosis in cultured human keratinocytes through sphinganine accumulation and ceramide depletion. *Intl J. Oncol.* **14**, 833–843.

Toth, B. (1979). Mushroom hydrazines: occurrence, metabolism, carcinogenesis and environmental implications. In E. C. Miller, J. A. Miller, I. Hirono *et al.* (ed.), *Naturally Occurring Carcinogens – Mutagens and Modulators of Carcinogenesis*, pp. 57–65. Japanese Science Society Press, Tokyo/University Park Press.

Toth, B. (1982). Tumorigenicity of minute dose levels of N-methyl-N-formylhydrazine of *Gyrometra esculenta. Mycopathologia* **78**, 11.

Toth, B. (1995). Mushroom toxins and cancer [review]. *Intl J. Oncol.* **6**, 137–145.

Toth, B. and P. Gannett (1993). *Agaricus bisporus*: an assessment of its carcinogenic potency. *Mycopathologia* **124**, 73–77.

Tournas, V. (1994). Heat-resistant fungi of importance to the food and beverage industry. *Crit. Rev. Microbiol.* **20**, 243–263.

Townsend, C. A. (1997). Progress towards a biosynthetic rationale of the aflatoxin pathway. *Pure Appl. Chem.* **58**, 227–238.

Trail, F., N. Mahanti and J. Linz (1995). Molecular biology of aflatoxin biosynthesis. *Microbiology* **141**, 755–765.

Trenholm, H. L., D. B. Prelusky, J. C. Young and J. D. Miller (1988). *Reducing Mycotoxins in Animal Feeds*. Agriculture Canada Publication 1827E. Agriculture Canada, Ottawa, Canada.

Trenholm, H. L., L. L. Charmley, D. B. Prelusky and R. M. Warner (1992). Washing procedures using water or sodium carbonate solutions for the decontamination of three cereals contaminated with deoxynivalenol and zearalenone. *J. Agric. Food Chem.* **40**, 2147–2151.

Trucksess, M. W. (2003). Mycotoxins. *J. AOAC Intl* **86**, 129–138.

Trucksess. M. W. and A. E. Pohland (eds) (2001). *Mycotoxin Protocols*. (Vol. 157 of *Methods in Molecular Biology*). Humana Press, Inc., Totowa, NJ.

Trucksess, M. W. and A. E. Pohland (2002). Methods and method evaluation for mycotoxins. *Mol. Biotechnol*, **22**, 287–292.

Trucksess, M. W. and G. E. Wood (1997). Immunochemical methods for mycotoxins in foods. *Food Test. Anal.* **3**, 24–27.

Tuomi, T., K. Reijula, T. Johnsson *et al.* (2000). Mycotoxins in crude building materials from water-damaged buildings. *Appl. Environ. Microbiol.* **66**, 1899–1904.

Udwary, D. W., L. K. Casillas and C. A.Townsend (2002). Synthesis of 11-hydroxyl O-methylsterigmatocystin and the role of a cytochrome P-450 in the final step of aflatoxin biosynthesis. *J. Am. Chem. Soc.* **124**, 5294–5303.

Ueno, I., M. Hoshino, T. Maitani *et al.* (1993). Luteoskyrin, an anthraquinoid hepatotoxin and ascorbic acid generate hydroxyl radical in vitro in the presence of a trace amount of ferrous iron. *Free Rad. Res. Comm.* **19** (Suppl.), S95–S100.

Ueno, Y. (1983). *Trichothecenes: Chemical, Biological and Toxicological Aspects*. Elsevier, Amsterdam.

Ueno, Y. (1986). Toxicology of microbial toxins. *Pure Appl. Chem.* **58**, 339–350.

Ueno, Y., K. Umemori, E. Niimi *et al.* (1995). Induction of apoptosis by T-2 toxin and other natural toxins in HL-60 human promyelotic leukemia cells. *Nat. Toxins* **3**, 129–137.

Umesh, H., B. Ganesh and C. S. Reddy (2000). Secalonic acid D alters the expression and phosphorylation of the transcription factors and their binding to cAMP response element in developing murine secondary palate. *J. Craniofac. Genet. Devel. Biol.* **20**, 173–182.

Van der Stegen, G. H. D., P. J. M. Essens and J. van der Lijn (2001). Effect of roasting conditions on reduction of ochratoxin A in coffee. *J. Agric. Food Chem.* **49**, 4713–4715.

Van der Watt, J. J. (1977). Sterigmatocystin. In: I. F. H. Purchase (ed.), *Mycotoxins*, pp. 368–382. Elsevier Scientific Publishing Co., New York, NY.

Van Dolah, F. M. and J. L. Richard (1999). Advances in detection methods for fungal and algal toxins. *Nat. Toxins* **7**, 343–345.

Van Egmond, H. P. (1989a). Current situation on regulations for mycotoxins: overview of tolerances and status of standard methods of sampling and analysis. *Food Addit. Contam.* **6**, 139–188.

Van Egmond, H. P. (1989b). *Mycotoxins in Dairy Products*. Elsevier, London.

Van Egmond, H. P. (1996). Analytical methodology and regulations for ochratoxin A. *Food Addit. Contam.* **13** (Suppl. S), 11–13.

Van Egmond H. P. (2002). Worldwide regulation for mycotoxins. In: J. W. DeVries, M. W. Trucksess and L. S. Jackson (eds), *Mycotoxins and Food Safety*, p. 257. Kluwer/Academic/Plenum Publishers, New York.

Van Egmond, H. P. and G. J. A. Speijers (1994). Survey of data on the incidence and levels of ochratoxin A in food and animal feed worldwide. *Nat. Toxins* **3**, 125–144.

Vargas, E. A., R. A. Preis, L. Castro and C. M. G. Silva (2001). Co-occurrence of aflatoxins B1, B2, G1, G2, zearalenone and fumonisin B1 in Brazilian corn. *Food Addit. Contam.* **18**, 981–986.

Vasanthi, S. and R.V. Bhat (1998). Mycotoxins in foods – occurrence, health economic significance food control measures. *Indian J. Med. Res.* **108**, 212–224.

Verardi, G. and H. Rosner (1995). Some reflections on establishing a community legislation on mycotoxins. *Nat. Toxins* **3**, 337–340.

Vesper, S., D. G. Dearborn, I. Yike *et al.* (2000). Evaluation of *Stachybotrys chartarum* in the house of an infant with pulmonary hemorrhage quantitative assessment before, during and after re-mediation. *J Urban Health* **77**, 68–85.

Voss, K. A., C. W. Bacon, F. I. Meredith and W. P. Norred (1996). Comparative subchronic toxicity studies of nixtamalized and water-extracted *Fusarium moniliforme* culture material. *Food Chem. Toxicol.* **34**, 623–632.

Voss, K. A., S. M. Poling, F. I. Meredith *et al.* (2001). Fate of fumonisins during the production of fried tortilla chips. *J. Agric. Food Chem.* **49**, 3120–3126.

Voss, K. A., J. K. Porter, C. W. Bacon *et al.* (1999). Fusaric acid and modification of the sub-chronic toxicity to rats of fumonisins in *F. moniliforme* culture material. *Food Chem. Toxicol.* **37**, 853–861.

Vrabcheva, T. M. (2000). Mycotoxins in spices. *Voprosy Pitaniya* **69**, 40–43.

Wang, E., W. P. Norred, C. W. Bacon *et al.* (1991). Inhibition of sphingolipid biosynthesis by fumonisins. *J. Biol. Chem.* **226**, 14 486–14 490.

Wang, H., J. Li, R. M. Bostock and D. G. Gilchrist (1996a). Apoptosis: a functional paradigm for programmed plant cell death induced by a host selective phytotoxin and invoked during development. *Plant Cell* **8**, 375–391.

Wang, H., C. Jones, J. Ciacci-Zanella *et al.* (1996b). Fumonisins and *Alternaria alternata lycopersici* toxins: sphinganine analog mycotoxins induce apoptosis in monkey kidney cells. *Proc. Natl Acad. Sci. USA* **93**, 3461–3465.

Wang, J. S. and J. D. Groopman (1999). DNA damage by mycotoxins. *Mutat. Res.* **424**, 167–181.

Waring, P. and J. Beaver (1996). Gliotoxin and related epipolythiodioxopiperazines. *Gen. Pharmac.* **27**, 1311–1314.

Warner, R. L., K. Brooks and J. J. Pestka (1994). In vitro effects of vomitoxin (deoxynivalenol) on T-cell interleukin production and IgA secretion. *Food Chem. Toxicol.* **32**, 617–625.

Wattenberg, L. W. (1986). Protective effects of 2(3)-tert-butyl-4-hydroxanisole on chemical carcinogensis. *Food Chem. Toxicol.* **24**, 1099–1102.

Whitaker, T. B., F. E. Dowell, W. M. Hagler Jr *et al.* (1994). Variability associated with sampling, sample preparation and chemical testing for aflatoxin in farmers' stock peanuts. *J. AOAC Intl* **77**, 107–116.

Whitaker, T. B., J. Springer, P. R. Defize *et al.* (1995). Evaluation of sampling plans used in the United States, United Kingdom and The Netherlands to test raw shelled peanuts for aflatoxin. *J. AOAC Intl* **78**, 1010–1018.

Whitty, J. P. and L. F. Bjeldanes (1987). The effects of dietary cabbage on xenobiotic-metabolizing enzymes and the binding of aflatoxin B1 to hepatic DNA in rats. *Food Chem. Toxicol.* **25**, 581–587.

Widstrom, N. W. (1996). The aflatoxin problem with corn grain. *Adv. Agron.* **56**, 219–280.

Wiebe, L. A. and L. F. Bjeldanes (1981). Fusarin C, a mutagen from *Fusarium moniliforme* grown on corn. *J. Food Sci.* **46**, 1424–1246.

Wieland, T. and H. Faulstich (1978). Amatoxins, phallotoxins, phallolysin and antamanide, the biologically active components of poisonous *Amanita* mushroom. CRC *Crit. Rev. Biochem.* **5**, 185–260.

Wieland, T. and H. Faulstich (1979). The phalloidin story. In: Y. A. Ovchinnikov and M. N. Kolosov (eds), *Frontiers in Bioorganic Chemistry and Molecular Biology*, pp. 97–112. Elsevier/North–Holland Biomedical Press, New York, NY.

Wild, C. P. and A. J. Hall (2000). Primary prevention of hepatocellular carcinoma in developing countries. *Mutat. Res.* (*Rev. Mutat. Res.*) **462**, 381–393.

Wild, C. P. and P. Kleihues (1996). Etiology of cancer in human and animals. *Exp. Toxicol. Pathol.* **48**, 95–100.

Wild, C. P., S. M. Shrestha, W. A. Anwar and R. Montesano (1992). Field studies of aflatoxin exposure, metabolism and induction of genetic alterations in relation to HBV infection and hepatocellular carcinoma in The Gambia and Thailand. *Toxicol. Lett.* **64/65**, 455–461.

Wilkins, A. L., C. O. Miles, R. M. Ede *et al.* (1992). Structure elucidation of janthitrem B, a tremorgenic metabolite of *Penicillium janthinellum* and relative configuration of the A and B rings of janthitrems B, E and F. *J. Agric. Food Chem.* **40**, 1307–1309.

Williams, G. M., T. Tanaka and Y. Maeura (1986). Dose-related inhibition of aflatoxin B1 induced hepatocarcinogenesis by the phenolic antioxidants, butylated hydroxyanisole and butylated hydroxtoluene. *Carcinogenesis* **7**, 1043–1050.

Willie, R. D. and L. G. Morehouse (1977). *Mycotoxic Fungi, Mycotoxins and Mycotoxicoses*, Vols 1–3. Marcel Dekker, New York, NY.

Wilson, B. J. and A. W. Hayes (1973). Microbial toxins. In: *Toxicants Occurring Naturally in Foods*, pp. 372–423. NRC, National Academy of Science, Washington, DC.

Windham, G. L. and W. P. Williams (2002). Evaluation of corn inbreds and advanced breeding lines for resistance to aflatoxin contamination in the field. *Plant Dis.* **86**, 232–234.

Wogan, G. N. (1992). Aflatoxins as risk factors for hepatocellular carcinoma in humans. *Cancer Res.* **52** (Suppl.), 2114–2118.

Wogan, G. N. (2000). Impacts of chemicals on liver cancer risk. *Semin. Cancer Biol.* **10**, 201–210.

Woloshuk, C. P., K. R. Foutz, J. F. Brewer *et al.* (1994). Molecular characterization of *aflR*, a regulatory locus for aflatoxin biosynthesis. *Appl. Environ. Microbiol.* **60**, 2408–2414.

Woloshuk, C. P., J. R. Cavaletto and T. E. Cleveland (1997). Inducers of aflatoxin biosyntheisis from clonized maize kernels are generated by an amylase activity from *Aspergillus flavus. Phytopathology* **87**, 164–169.

Wong, S. S., R. C. Schwartz and J. J. Pestka (2001). Superinduction of TNF-alpha and IL-6 in macrophages by vomitoxin (deoxynivalenol) modulated by mRNA stabilization. *Toxicology* **161**, 139–149.

Wood, G. E. (1992). Mycotoxins in foods and feeds in the United States. *J. Anim. Sci.* **70**, 3941–3949.

Woody, M. A. and F. S. Chu (1992). Toxicology of *Alternaria* mycotoxins. In: J. Chelkowski (ed.), Alternaria – *Metabolites, Biology and Plant Ciseases*, pp. 409–434. Elsevier, Amsterdam.

Xie, W. and C. J. Mirocha (1995). Isolation and structure identification of two new derivatives of the mycotoxin fusarochromenone produced by *Fusarium equiseti*. *J. Nat. Prod.* **58**, 124–127.

Xie, T. X., J. Misumi, K. Aoki *et al.* (2000). Absence of p53-mediated G1 arrest with induction of MDM2 in sterigmatocystin-treated cells. *Intl J. Oncol.* **17**, 737–742.

Xu, J.-R. and J. F. Leslie (1996). A genetic map of *Gibberella fujikuroi* mating population A (*Fusarium moniliforme*). *Genetics* **143**, 175–189.

Yagen, B. and M. Bialer (1993). Metabolism and pharmacokinetics of T-2 toxin and related trichothecenes. *Drug Metab. Rev.* **25**, 281–323.

Yamaguchi, T., K. Nozawa, T. Hosoe *et al.* (1993). Indoloditerpenes related to tremorgenic mycotoxins, penitrems, from *Penicillium crustosum*. *Phytochemistry* **32**, 1177–1181.

Yang, C. F., J. Liu, H. M. Shen and C. N. Ong (2000a). Protective effect of ebselen on aflatoxin B-1-induced cytotoxicity in primary rat hepatocytes. *Pharmacol. Toxicol.* **86**, 156–161.

Yang, C. F., J. Liu, S. Wasser *et al.* (2000b). Inhibition of ebselen on aflatoxin B-1-induced hepatocarcinogenesis in Fischer 344 rats. *Carcinogenesis* **21**, 2237–2243.

Yang, G. H., B. B. Jarvis, Y. J. Chung and J. J. Pestka (2000). Apoptosis induction by the satratoxins and other trichothecene mycotoxins: relationship to ERK, p38 MAPK and SAPK/JNK activation. *Toxicol. Appl. Pharmacol.* **164**, 149–160.

Yang, G. H. and J. J. Pestka (2002). Vomitoxin (deoxynivalenol)-mediated inhibition of nuclear protein binding to NRE-A, an IL-2 promoter negative regulatory element, in E-4 cells. *Toxicology* **172**, 169–179.

Yao, R. and G. M. Cooper (1996). Growth factor-dependent survival of rodent fibroblasts requires phosphatidylinositol 3-kinase but is independent of pp70(S6K) activity. *Oncogene* **132**, 343–351.

Yeung, J. M., H. Y. Wang and D. B. Prelusky (1996). Fumonisin B-1 induces protein kinase c translocation via direct interaction with diacylglycerol binding site. *Toxicol. Appl. Pharmacol.* **141**, 178–184.

Yoshino, N., M. Takizawa, H. Akiba *et al.* (1996). Transient elevation of intracellular calcium ion levels as an early event in T-2 toxin-induced apopotosis in human promyelotic cell line HL-60. *Nat. Toxins* **4**, 234–241.

Young, C., L. McMillan, E. Telfer and B. Scott (2001). Molecular cloning and genetic analysis of an indole–diterpene gene cluster from *Penicillium paxilli*. *Mol. Microbiol.* **39**, 754–764.

Yu, J. H., R. A. E. Butchko, M. Fernandes *et al.* (1996). Conservation of structure and function of the aflatoxin regulatory gene *aflR* from *Aspergillus nidulans* and *A. flavus*. *Curr. Genetics* **29**, 549–555.

Yu, J. J., P. K. Chang, J. W. Cary *et al.* (1995). Comparative mapping of aflatoxin pathway gene clusters in *Aspergillus parasiticus* and *Aspergillus flavus. Appl. Environ. Microbiol.* **61**, 2365–2371.

Zahl, P. A. (1965). Bizarre world of fungi. *Natl Geographic* **128**, 502–527.

Zepnik, H., A. Paehler, U. Schauer and W. Dekant (2001). Ochratoxin A-induced tumor formation, Is there a role of reactive ochratoxin A metabolites? *Toxicol. Sci.* **59**, 59–67.

Zhang, Y., M. B. Dickman and C. Jones (1999). The mycotoxin fumonisin B1 transcriptionally activates the p21 promoter through a cis-acting element containing two Sp1 binding sites. *J. Biol. Chem.* **274**, 12 367–12 371.

Zhang, Y., C. Jones and M. B. Dickman (2001). Identification of differentially expressed genes following treatment of monkey kidney cells with the mycotoxin fumonisin B1. *Food Chem. Toxicol.* **39**, 45–53.

Zhu, J. Q., L. S. Zhang, X. Hu *et al.* (1987). Correlation of dietary aflatoxin B_1 levels with excretion of aflatoxin M_1 in human urine. *Cancer Res.* **47**, 1848–1852.

Miscellaneous natural intoxicants

17

Eric A. Johnson and Edward J. Schantz[1]

1 Introduction

Foodborne illness mediated by natural toxins of microbial, plant or animal sources is an important global public health problem. Foodborne illness has been defined by the World Health Organization (WHO) as 'a disease of infectious or toxic nature caused by, or thought to be caused by, the consumption of food or water' (WHO, 1997). Globally, the WHO has estimated that approximately 1.5 billion cases of foodborne illness and more than 3 million deaths occur in children under 5 years of age, and a significant proportion of these result from consumption of food contaminated with pathogenic microorganisms or natural intoxicants (WHO, 1997). These estimates of foodborne illlnesses are probably 100–300 times less than the actual occurrence (Bryan *et al.*, 1997; Lund *et al.*, 2000). The annual incidence of foodborne illness in industrialized countries has been estimated to affect 5–10 % of the population annually, and in many developing countries the incidence is probably considerably higher due to underlying morbidity, unsanitary conditions, inadequate food processing and handling, and a variety of other reasons (see Johnson, 2003).

Acute foodborne illnesses and intoxications are commonly classified into two main categories: (1) infections of the gastrointestinal tract by microbial pathogens, and (2) poisonings or intoxications resulting from consumption of preformed toxins or toxin precursors in foods. This classification, however, is overly simplistic, and does not take into account related aspects including chronic disease syndromes that

[1] deceased

Foodborne Infections and Intoxications 3e
ISBN-10: 0-12-588365-X
ISBN-13: 978-012-588365-8

can develop following acute foodborne infections, and other factors inherent in the nature and chemical complexity of foods. In natural intoxications, consumption of the toxin alone or its subsequent metabolism is responsible for the illness. Although believed to be of diminished importance compared to gastrointestinal infections, exposure to oral toxins also has a large impact on the health and mortality of humans and animals (CAST, 1994; WHO, 1997; Dabrowski and Sikorski, 2005).

In the US and in many other parts of the world, the large majority of foodborne illnesses are caused by viral and bacterial infections (Bean and Griffin, 1990; Bean *et al.*, 1996; Mead *et al.*, 1999; CDC, 2000; Taylor and Hefle, 2002), while only a relatively small subset (5–10 %) is documented to result from the consumption of natural intoxicants. This relatively low frequency of chemical intoxications compared to pathogens reflects in part the greater emphasis and resources that have been allocated to the study of the epidemiology and virulence of microbial pathogens. Although foodborne illnesses caused by natural intoxicants are an important public health problem, particularly in developing and poorer countries, this area of research has not received the attention given to foodborne diseases caused by microbial pathogens. This is probably due to the relatively low incidence of intoxications in industrialized countries and consequently limited resources for such research, as well as to the difficulties in studying foodborne intoxications by chemicals. In most cases of foodborne disease caused by pathogens, the causative microbial agents can be cultured from foods on artificial media, thus enabling investigation by classic means such as proof of Koch's postulates (see, for example, Johnson, 2003). In contrast, intoxications involve non-replicating chemical substances in the diet, and their detection often requires sensitive and sophisticated chemical methods. The involvement of natural intoxicants in dietary disease can be very difficult to demonstrate. Certain chemical intoxications also result from metabolic transformation in the host. Most food intoxications from natural compounds occur in poorer countries, and resources may not be available for adequate surveillance and epidemiological studies. Toxins from fish, shellfish and plants are the most common cause of chemical intoxications. Poisoning by natural intoxicants has also taken on a new dimension with the increasing cultivation of transgenic plants and other genetically engineered foods (Engeseth, 2001; Stewart, 2003; Toke, 2004), and the potential for biowarfare using plant toxins such as ricin (Khan *et al.*, 2001; Franz and Zajtchuk, 2002).

This chapter focuses on intoxications with an emphasis on naturally occurring organic toxins mainly of plant, algal, microbial and animal origins. Several treatises have extensively covered intoxications caused by inorganic compounds such as lead and mercury, man-made industrial chemicals, and pollutants including PCBs, food additives, herbicides and many other classes of compounds (Hayes, 2001; Klaassen, 2001; Kotsonis *et al.*, 2001; D'Mello, 2003), and these are not covered here. This chapter also does not include allergic responses, food intolerance, metabolic food reactions that occur due to excess intake and metabolism, and methods for analysis of food intoxicants. Excellent reviews of these latter topics are available elsewhere (FDA, 1995; Downes and Ito, 2001; Hui *et al.*, 2001a, 2001b; Kotsonis *et al.*, 2001; Taylor and Hefle, 2002).

1.1 Background and historical aspects of natural foodborne intoxications

Foodborne illnesses resulting from consumption of natural toxicants are distinct from microbial infections in several ways (Concon, 1988; Kotsonis *et al.*, 2001). Naturally occurring toxicants are products of the biosynthesis and metabolism of plants, algae, animals and microorganisms (National Academy of Sciences, 1973; Watson, 1998; Coulombe, 2000; Park *et al.*, 2000; Hui *et al.*, 2001a, 2001b; Dabrowski and Sikorski, 2005). Certain intoxicants are present in these biological materials constitutively, while others are formed in response to infections in plants or other metabolic processes. Improper food production practices and processing can affect the levels of toxicants (Rahman, 1999). Natural intoxicants that have been documented to cause food poisoning are produced by various species of bacteria, fungi, plants, insects and animals (Lund *et al.*, 2000; Klaassen, 2001; Reddy and Hayes, 2001). Unlike pathogens as etiologic agents, natural toxins are not able to reproduce in foods or in the gastrointestinal tract of humans and animals. In certain cases, the formation of toxicants such as nitrite can be mediated by microbial transformation of precursor molecules in foods or in the gut. In this chapter, intoxications due to natural toxicants are considered to be distinct from allergenic and anaphylactic responses mediated by food substances. The toxic response is not immune-mediated, although its toxic mechanism may involve the release of chemical mediators. As defined by Kotsonis *et al.* (2001), food toxicity (poisoning) is 'A term used to imply an adverse effect caused by the direct action of a food or food additive on the host recipient without the involvement of immune mechanisms.'

Through the centuries, humans probably learned to avoid consuming natural products that caused adverse reactions, and there is a rich historical record regarding human awareness and avoidance of specific foods (McNeil, 1976; Concon, 1988; Borzelleca, 2001). The ancient recognition of poisons in foods has been recorded in writings and illustrations from Egyptian, Chinese, Hindu, Roman, Arab, Greek and other civilizations (Borzelleca, 2001). Recognition of associations between food and chemical intoxications came about long before an understanding of toxicology, and some of the seminal events have been traced through history into the modern era (McNeil, 1976; Borzelleca, 2001).

The recognition and observations of poisons and poisoners was followed by eras of experimental, mechanistic and analytical toxicology (Borzelleca, 2001). Chemists and pharmacologists demonstrated that compounds in plants, insects, animals and macroscopic fungi could be poisonous when ingested. Knowledge of the microbial causes of foodborne disease began when Pasteur and Koch founded the science of microbiology, allowing microbiologists to isolate, characterize and systematically describe microorganisms associated with spoiled or poisonous foods (Brock, 1961; Tannahill, 1973; McNeil, 1976). In contrast, understanding of food poisoning by natural intoxicants came about through causal associations between consumption and illness, and from advances in analytical methods to identify the compounds and in pharmacology to understand the basis for their poisonous action in animal models. The age of safety, evaluation, quantification and prognostication followed (Borzelleca, 2001).

Landmark legislation occurred in the US with the 1906 Pure Food and Drugs Act and its successor, the 1938 Federal Food, Drug, and Cosmetic Act, to provide a safe and wholesome food supply (Hutt and Hutt, 1984; Middlekauf and Shubik, 1989; Miller and Taylor, 1989). Assurance of the safety and wholesomeness of foods is an important discipline fulfilled by legislators, industry and researchers. The US Food and Drug Administration published the *Redbooks I* and *II*, and the Organization for Economic Cooperation and Development (OECD) issued similar information, providing guidelines for sound science and data resources for determining safe exposure limits for consumers. The WHO Joint Expert Committee on Food Additives (JECFA) and other international organizations have applied sound toxicological thinking for establishing the safety of food chemicals and toxicants. The fields of surveillance and epidemiology have demonstrated the enormous impact that foodborne disease has on morbidity, mortality and economic losses throughout the world. In developing countries, foodborne disease is among the leading causes of morbidity and mortality in humans and animals, and has been a leading factor impeding technological progress (Miller and Taylor, 1989).

Food products can become contaminated by toxicants through a variety of means. Fish and shellfish can absorb toxins through feeding on toxic algae or bacteria. Certain other fish poisonings result from bacterial growth under non-refrigerated storage conditions. Some plants inherently produce toxins to high levels, while others form toxicants in response to microbial infections. Toxins can also accumulate during harvesting and processing of food commodities, such as by the mixing of toxic plants with edible crops or the formation of toxicants during processing. Paradoxically, a major source of food toxicants has resulted from advances in food technology and agricultural practices (Richards and Hefle, 2003). For example, enhanced agricultural productivity through the use of fertilizers, insecticides, pesticides, microbiocides, growth stimulants and antibiotics can sometimes lead to toxic levels of contaminants in a food through alterations in specific as well as global ecological systems. In the classic book *Silent Spring*, Rachel Carson lamented the presumed excessive use of pesticides with disregard for the ecological system, and its potential effect on global vitality.

Despite the importance of chemical intoxications, the study of foodborne illnesses has centered on viral, bacterial and fungal pathogens. Natural poisons and hazards in foods have focused mainly on pathogenic microorganisms and fungal toxins (National Research Council, 1985). Currently, the CDC limits chemical etiologies to certain seafood toxins, and most other countries also do not have thorough surveillance and preventive programs for natural toxicants. Persons suffering from chemical intoxications are generally referred to 'poison centers' (Spoerke, 2001a, 2001b), and systematic reporting programs for foodborne illness caused by most natural intoxicants have not been adequately implemented (National Research Council, 1985). The International Programme on Chemical Safety (IPCS) was established in 1980 to implement activities related to chemical safety (see www.who.int/ipcs/en/). This has established a world directory of poison centers (YellowTox). Although the IPCS collects and maintains data regarding some natural intoxicants, its emphasis has mainly been on inorganic minerals and synthetic chemicals.

1.2 Major principles of acute toxicology

What are natural toxins or poisons? According to Borzelleca (2001):

> 'A poison is any substance (chemical, physical or biological) that is harmful or destructive to a biological (living) system. A poison derived from a natural source is a toxin, and the study of toxins is toxicology.'

It is a well known paradigm that all chemicals can be toxic depending on the dose:

> 'What is there that is not a poison?
> All things are poison and nothing [is]
> without poison. Solely the dose
> determines the thing that is a poison.'

(Paracelsus, 1493–1541, cited in Klaassen, 2001.)

Although this paradigm is fundamentally true, it is impractical to consider natural food toxicants in this manner (Johnson and Pariza, 1989; Taylor and Hefle, 2002). The vast majority of components in foods are either present in too low a concentration or have a very low intrinsic toxicity and do not present a hazard under normal conditions of food production and consumption (Kotsonis *et al.*, 2001; Taylor and Hefle, 2002). The hazard to human health from natural toxicants results from their potency and capacity to cause illness on exposure. The potency of a toxin is often expressed as the median lethal dose (LD_{50}), a concept introduced by Trevan for the standardization of digitalis extracts, insulin, and diphtheria toxin (Trevan, 1927; DiPasquale and Hayes, 2001). The LD_{50} represents the dose that causes a toxic response (i.e. lethality) in 50 % of a population of test animals in a designated period of time (Johnson and Pariza, 1989; Hayes, 2001; Klaassen, 2001). The LD_{50} is chosen to quantify virulence or toxicity because of the nature of the dose–response relationship (Concon, 1988; Johnson and Pariza, 1989). In a typical sigmoid lethality curve, the rate of change in mortality (slope of the curve) as a function of dose reaches a maximum at the point of about 50 % survival. Curves with steeper slopes give a more accurate estimate of toxin concentration or infectious dose. The sigmoid shape of the LD_{50} curve results primarily from the chance distributions of lethal events in any given animal (Concon, 1988; Johnson and Pariza, 1989). Various factors influence the LD_{50} curve, including the type, strain, health and heterogeneity of the animal population; the route of administration; and the rate of metabolic detoxification. For these reasons, the LD_{50} determination is usually combined with other methods for determining the level of natural intoxicant required for illness in experimental animals. Furthermore, LD_{50} determines acute toxicity, and some compounds with low acute toxicity may have carcinogenic or teratogenic effects at doses that do not produce evidence of acute toxicity (Eaton and Klaassen, 2001). Compounds also may be toxic only when in combination with other substances, or may manifest toxicity on repeated exposure. Nonetheless, the dose–response curve using animal models is an essential and central tool for evaluating toxicity.

Natural toxins and poisons show tremendous variation in LD_{50} (see examples in Table 17.1). For a large population the lethal responses to a given dose of a

Table 17.1 Potencies of selected natural toxins

Estimated minimum lethal dose (μg/kg)	Toxin	Source or nature
0.00003	Botulinum neurotoxin type A	Bacterium: *Clostridium botulinum*
0.00010	Tetanus neurotoxin	Bacterium: *Clostridium tetani*
0.020	Ricin	Plant: castor bean, *Ricinus communis*
0.15	Palytoxin	Zoanthid: *Palythoa* spp.
0.20	Crotalus toxin	Rattlesnake: *Crotalus atrox*
0.30	Diphtheria toxin	Bacterium: *Corynebacterium diphtheriae*
0.30	Cobra neurotoxin	Snake: *Naja naja*
2.7	Kokoi venom	Frog: *Phyllobates bicolor*
8	Tarichatoxin	Newt: *Taricha torosa*
8	Tetrodotoxin	Fish: *Sphoeroides rubripes*
3.4–9	Saxitoxin	Dinoflagellate: *Gonyaulax catenella*
390	Bufotoxin	Toad: *Bufo vulgaris*
500	Curare	Plant: *Chondodendron tomentosum*
500	Strychnine	Plant: *Strychnos nux-vomica*
1100	Muscarin	Mushroom: *Amanita muscarina*
1500	Samandarin	Salamander: *Salamandra maculosa*
3000	Diisopropylfluorophosphate	Synthetic nerve gas
10 000	Sodium cyanide	Inorganic poison

From: Mosher *et al.*, 1964.
Minimal lethal dose refers to mouse except in the cases of ricin, where it refers to guinea pig, and bufotoxin and muscarin, where it refers to cat. In cat, administration was intravenous; in all other cases it was intraperitoneal. Since the survival times are variable and the experiments are not direct comparisons, these values are approximate and indicative only of relative toxicity by the designated route of administration.

compound may follow a normal distribution pattern (Gaussian), which presents difficulty in predicting the response of a given individual in the population because the response may be unique and fall at any place on the normal distribution curve, or outside it (Concon, 1988). Related to the LD_{50} are the No Effect Dose (NED), analogous to the No Observable Adverse Effective Level (NOAEL) and No Effective Level (NEL). For toxicants, various compounds can share the same LD_{50} but have substantially different NED values (Concon, 1988). For many compounds, the determination of the LD_{50} and NED is difficult owing to several factors, such as the time for the toxic response to occur, the variations among test animals and between species, interaction with other compounds, metabolism, the intestinal barrier and other factors (Concon, 1988).

An interesting aspect of toxicity requirements is the number of molecules needed to cause intoxication. It was proposed that toxic biological activity cannot occur for any single substance in a single cell below 10 000 molecules (Hutchinson, 1964; Dinman, 1972). However, many target-specific substances may show a lower threshold number in terms of total body cells. For example, when considered in terms of total body cells, it has been estimated that 2 μg or 8×10^{12} molecules of botulinum neurotoxin (molecular mass = 150 000 Da) is sufficient to produce lethality in an adult human (Lamanna, 1959). In reference to total body cells, this toxic dose is less than the 10 000-molecule threshold, but is exceeded when considered in terms of the total number of motor neurons. Thus, the 10 000-molecule threshold appears reasonable even for the most poisonous substance known.

The toxicity of natural poisons is also related to the frequency of exposure, since some compounds are more effective toxicants with repeated exposures. Repeated exposures may also have cumulative effects. The route of exposure is also an important variable, since (with few exceptions) compounds are least toxic by the oral route (Concon, 1988). Exceptions are those compounds that are activated during passage through the intestinal tract (Concon, 1988). Dietary factors and endogenous resistance, including detoxifying gut bacteria, will influence toxicity. Lastly, most toxicants show different binding affinities to plasma proteins (Concon, 1988), and their displacement from plasma proteins by compounds of greater affinity will increase the potency and effective dose (Concon, 1988). Other physiologic mechanisms of poisoning (such as organ specificity, systemic and organ detoxification, stability and clearance) are important, but are beyond the scope of this chapter; there are excellent treatises on this (see, for example, Kotsonis *et al.*, 2001).

1.3 Recognition, surveillance and epidemiology of foodborne intoxications

The recognition of a toxin as an etiologic agent of foodborne disease usually results first from associative epidemiological evidence, in which the occurrence of an illness in a human epidemic is examined and found to correlate with the consumption of a food (Evans and Brachman, 1992). The epidemiological association is ideally established by demonstration of suspected toxins from clinical samples and the causative food. In the case of miscellaneous chemical intoxications, metabolic reactions occurring in a food or by microorganisms may transform the naturally occurring form of the chemical to the toxicant. In practice, foodborne disease outbreaks caused by natural intoxicants are diagnosed by first examining the onset time of illness and the symptoms, and then isolating the toxin from the food and clinical samples (e.g. vomitus, feces, blood, organs) of the victim(s). The successful epidemiological investigation coupled with the etiological diagnosis can also help to facilitate both short-term and long-term control measures.

2 Bacterial toxins

2.1 Bongkrekic toxin and toxoflavin

Bongkrekic (BK) toxin can be produced during certain food fermentations, particularly during production of bongkrek, an Indonesian food produced by fermentation with the fungus *Rhizopus oligosporus* of coconut presscake or coconut milk wrapped in banana leaves. In certain fermentations the bacterium *Burkholderia cocovenenans* (formerly *Pseudomonas cocovenenans*) grows and produces two toxins: toxoflavin and bongkrekic acid (Van Veen, 1967; Garcia *et al.*, 1999; Jiao *et al.*, 2003). Consumption of tempe bongkrek has led to numerous human fatalities (Van Veen, 1967; Concon,

1988; Garcia *et al.*, 1999). The high mortality of bongkrek poisoning led Van Veen and Mertens to conduct extensive research on the nature of the toxicity in the 1920s and 1930s (Van Veen, 1967), and the two toxins, bongkrekic acid (BK) and toxoflavin, were found to be produced by the bacterium *B. cocovenenans* during the fermentation. *B. cocovenenans* has been isolated from foods including corn flour, edible fungi and soil (Hu *et al.*, 1984; Jiao *et al.*, 2003). The production of toxin is influenced by pH, fatty acids, and other intrinsic and extrinsic factors (Garcia *et al.*, 1999). The production of BK can be prevented by using oxalis leaves for wrapping the presscake, which promotes a rapid drop in pH and thus prevention of growth and toxin formation by *B. cocovenenans*. BK is heat-stable and survives food processing.

The structures and toxicities of toxoflavin (oral LD_{50} ~ 8 mg per kg bodyweight) and bongkrekic acid have been elucidated (Van Veen, 1967). Toxoflavin has the empirical chemical formula $C_7H_7N_5O_2$ and contains two six-membered rings, a pyrimidine and a triazine system, each with one methyl group. When presented orally to a 1–2 kg monkey, 1–1.5 mg of bongkrekic acid was fatal (Van Veen, 1967). BK shows strong antibiotic activity, particularly against molds and yeasts. BK ($C_{28}H_{50}O_7$) is a branched, unsaturated tricarboxylic acid. Since BK is much more toxic than toxoflavin, it is probably most responsible for illnesses and fatalities in humans. Following consumption of food containing BK, the onset of symptoms occurs within a few hours; the symptoms include abdominal pains, dizziness, excessive sweating, malaise, and eventually coma and death within 24 hours (Hu *et al.*, 1984). The mechanism of toxicity appears to involve interaction with the electron transport system and a decrease in carbohydrate metabolism (Van Veen, 1967).

Bongkrek poisoning and associated illnesses and deaths have also been reported in the People's Republic of China following the consumption of fermented corn flour or deteriorated *Tremella faciformis* (white fungi), with a fatality rate of more than 40 % during the 1970s (Meng *et al.*, 1988; Jiao *et al.*, 2003). Production of BK in this fermentation is prevented by keeping the pH above neutral, which discourages the formation of the toxin. The toxin responsible was reported to be BK produced by *Flavobacterium farinofermentans* (Meng *et al.*, 1988). *Flavobacterium farinofermentans* was later shown to be identical to *Pseudomonas cocovenenans* (Jiao *et al.*, 2003). In 1995, *P. cocovenenans* was transferred to the genus *Burkholderia* as *B. cocovenenans* (Jiao *et al.*, 2003).

2.2 *Bacillus* toxins

Bacillus cereus has long been known to produce two toxins – a diarrheal protein toxin and an emetic toxin (Granum and Baird-Parker, 2000) – whose structure remained elusive for several years (see Chapter 15). The prototype toxin is now known to comprise a cyclic peptide [D-*O*-Leu-D-Ala-L-*O*-Val-L-Val]$_3$ with a molecular mass of 1.2 kDa (Granum and Baird-Parker, 2000). The quantity of cereulide to cause an emetic reaction in monkeys is approximately 12–32 μg per kg, corresponding to 10^5–10^8 cells per gram food (Granum and Baird-Parker, 2000). The emetic toxin has remarkable heat stability, retaining toxicity after treatment for 90 minutes at 121 °C. Cereulide binds to 5-HT3 receptors, and the binding stimulates the vagus afferent nerve. In common with many preformed foodborne toxins, the onset of symptoms is rapid (1–5 hours) and the duration of the vomiting illness is relatively short (6–24 hours) (Granum and Baird-Parker, 2000).

Subsequent studies have shown that various *Bacillus* spp., including *B. licheniformis, B. pumilis, B. sphaericus, B. brevis* and possibly other species, also produce cereulide or other cyclic toxic peptides (Granum and Baird-Parker, 2000). It is becoming apparent that there is variation in structures depending on the species, and further characterization is required. Conditions influencing cereulide production in foods have been investigated (Granum and Baird-Parker, 2000).

2.3 *Clostridium* toxins

The genus *Clostridium* produces more protein toxins than other genera of bacteria (van Heyningen, 1950; Johnson, 1999). Most of the clostridial protein toxins causing foodborne illness are well-known, and include botulinum neurotoxin (see Chapter 13) and *C. perfringens* enterotoxin (see Chapter 4). Many species of *Clostridium* are common flora in the human and animal gastrointestinal tracts (Finegold *et al.*, 2002a). Recent evidence suggests that clostridia may produce toxins that are absorbed in the gastrointestinal tract and cause disease. For example, an association of *Clostridium bolteae* and other *Clostridium* spp. with late-onset autism has been suggested by microbiological and antimicrobial studies (Finegold *et al.*, 2002b). This is a very intriguing hypothesis – that neurotoxigenic clostridia in the gut could contribute to CNS-related behavioral diseases – but further research to evaluate this hypothesis is required. Nonetheless, these results suggest that relations between gut microbial flora and human disease may be more prevalent than previously realized.

2.4 Intoxications caused by bacterial formation of *N*-nitroso compounds (NOCs)

Although there are several dietary sources of NOCs (Henderson and Raskin, 1972; Keating *et al.*, 1973; Fassett, 1973; Bryan, 1982; Kotsonis *et al.*, 2001; Mensinga *et al.*, 2003), poisoning caused by nitrite (nitrite poisoning; methemoglobinemia) is usually related to consumption of foods high in nitrates and their subsequent reduction in the food or by the intestinal microbial flora. NOCs such as nitrosamines and nitrosamides can also be formed during drying or cooking of foods, or by migration from food contact materials (Henderson and Raskin, 1972; Fassett, 1973; Walley and Flanagan, 1987; Kotsonis *et al.*, 2001, Mensinga *et al.*, 2003). The environmental formation of nitrites in foods generally results from reduction of nitrogen precursors, particularly nitrate, and intoxications due to bacterial reduction of nitrate to nitrite have been reported in various foods. Certain plants, such as beets, broccoli, celery, brussels sprouts, corn, carrots, radish, rhubarb, spinach, turnip greens and others, may contain high levels of nitrite, depending on inherent and environmental factors (Keating *et al.*, 1973; Walley and Flanagan, 1987; Kotsonis *et al.*, 2001). Excessive fertilization of crops with nitrogen fertilizers, use of certain insecticides and herbicides, and other cultivation practices can increase plant nitrate levels. Various other factors also have an impact on the level of nitrate in foods, including which part of the plant is utilized as food; the environmental conditions, such as drought and harvest conditions; and the physiological state of the plant, including nutrient deficiencies and stage of maturity.

Nitrite poisoning has also resulted from *in vivo* bacterial activity in the oral cavity and stomach. Many species of bacteria, including pseudomonads, *Enterobacteriaceae*, staphylococci, clostridia and others, readily reduce nitrate to nitrite, which can result in methemoglobinemia. Storage of foods at cold temperatures will reduce microbial metabolism and diminish the reduction of nitrate to nitrite. Several outbreaks of methemoglobinemia have occurred from inadvertent addition of excess levels of curing salts to meats, fish and cheeses (Fassett, 1973; Kotsonis *et al.*, 2001, Mensinga *et al.*, 2003). The use of well-water containing high nitrate levels has caused methemoglobinemia in home-dialysis patients (Carlson and Shapiro, 1970; Kotsonis *et al.*, 2001). Methemoglobinemia initially manifests as darkened blood and a slate-gray cyanosis, which may occur mainly in the lips and mucous membranes in mild cases. Other symptoms may include nausea, vomiting, headache, shortness of breath and, occasionally, death. The cyanotic patient should receive respiratory support and airway management. In severe cases, adjunctive therapies may be required.

2.5 Bacterial endotoxins

Bacterial endotoxins are heat-stable lipopolysaccharides (LPS) that are associated with the outer membrane of Gram-negative bacteria. In itself, LPS is not considered to be a potent oral toxin. When injected into an animal, endotoxin rapidly causes shock and sepsis, and is often accompanied by severe diarrhea (Danner and Natanson, 1995; Klaassen, 2001). The absorption of endotoxin from the bowel is a major contributor to lethality in hemorrhagic shock. The importance of endotoxin does not appear to be limited to currently recognized infections. Certain Gram-negative bacteria are commonly found in the gut, and those causing enterotoxic lesions could facilitate the entry of endotoxin, leading to increased morbidity or death, depending on the exposure and host defense. Although not usually considered a foodborne oral toxin, endotoxin could be absorbed into circulation through intestinal lesions and cause intoxication to a number of organs (Danner and Natanson, 1995; Klaassen, 2001).

2.6 Miscellaneous bacterial toxins

A number of species of bacteria have occasionally been associated with foodborne disease through the production of toxic compounds. In particular, enterococci, group B streptococci and a variety of species of *Enterobacteriaceae* have been presumptively associated with foodborne illness (Bryan, 1979; Stiles, 2000). Streptococci and enterococci are known to produce a variety of protein toxins; these organisms occur commonly in foods and are also found in the gastric tracts of humans and animals, and it is possible that the toxins cause foodborne intoxications. Organisms transmitted as zoonoses have also been implicated in foodborne diseases. Conclusive evidence for illness association would ideally involve the solving of Koch's postulates (see, for example, Bryan, 1979; Stiles, 2000; Johnson, 2003), but there can be limitations to meeting the criteria, and molecular techniques are increasingly being used to identify new toxigenic organisms. Infrequent microbial infections and intoxications are discussed elsewhere in this book (see Chapter 10).

3 Seafood toxins

With the exception of fish and shellfish, most animals do not produce toxins that are known to cause illness on consumption – apart from the formation of prions by certain livestock. On the other hand, many fish and shellfish produce oral toxins that can cause human disease, and these are discussed in this section (reviewed in Halstead, 1967; Ahmed, 1991; Falconer, 1993; Yasumoto and Murata, 1993; Anderson, 2000; Fleming *et al.*, 2001; Llewellyn, 2001; Johnson and Schantz, 2002; Backer *et al.*, 2005).

Finfish and shellfish are nutritious food sources that contribute to a healthy and delicious human diet (Brown *et al.*, 1999; McGinn, 1999). Seafood consumption has increased in many countries during the past two decades. Currently, it is estimated that people throughout the world receive about 6 % of their total protein and 16 % of their animal protein from fish. Although seafoods have positive nutritious and health attributes, they can also serve as vehicles for a variety of foodborne illnesses (Halstead, 1967; Ahmed, 1991; Falconer, 1993; Yasumoto and Murata, 1993; Lipp and Rose, 1997; Anderson, 2000; Fleming *et al.*, 2001, Llewellyn, 2001; Johnson and Schantz, 2002; Backer *et al.*, 2005). They have been associated with the transmission of bacterial and viral gastroenteritis, and have also supported outbreaks of bacterial intoxications – including botulism and staphylococcal poisoning (see Chapters 13 and 14 of this book, respectively).

Seafoods are intriguing in also transmitting diseases mediated by non-protein, heat-stable, small molecular-weight toxins produced mainly by microalgae and bacteria (Ahmed, 1991; Yasumoto and Murata, 1993; Yasumoto and Murata, 1993; Plumley, 1997; Okada and Niwa, 1998; Llewellyn, 2001; Backer *et al.*, 2005). Seafood toxins cause a variety of illnesses of humans and animals in many areas of the world (Meyer *et al.*, 1928; Sommer and Meyer, 1937; Halstead, 1967; Ahmed, 1991; Falconer, 1993; Anderson, 2000; Fleming *et al.*, 2001; Johnson and Schantz, 2002; Backer *et al.*, 2005). There has also been concern that toxins are formed during intensive aquaculture of fish and shellfish (Jensen and Greenlees, 1997). Several of these illnesses – such as paralytic shellfish poisoning (PSP), puffer fish poisoning (PFP) and neurotoxic shellfish poisoning (NSP) – have been known for centuries, whereas others – such as amnesic shellfish poisoning (ASP), diarrhetic shellfish poisoning (DSP) and azaspiracid shellfish poisoning (AZP) – have been recognized more recently. Scombroid fish poisoning is one of the more common seafood diseases in many parts of the world, and is caused by histamine formed by bacteria in poorly refrigerated fish. Ciguatera is also a relatively common form of seafood poisoning that is transmitted in certain species of fish in tropical waters which are contaminated with ciguatoxin. Toxins produced by *Pfiesteria* and cyanobacteria are also recognized waterborne causes of diseases in fish, and could potentially cause human foodborne illnesses. Contaminated shellfish, including mussels, clams, cockles, oysters, scallops and other varieties feeding on toxic microalgae, are a main cause of seafood intoxications. Most toxic finfish and shellfish accumulate the toxins from water, are not visibly spoiled and cannot be distinguished from non-toxic seafoods on harvest. For some toxin-mediated illnesses, a single clam or mussel contains enough poison to kill a human, but without noticeable health or organoleptic effects to the shellfish (Fleming *et al.*, 2001; Johnson and Schantz, 2002; Backer *et al.*, 2005).

Preventive measures mainly rely on sampling algal blooms and foods, quantitative detection of the causative toxins, and warning the seafood industry and consumer (Halstead, 1967; Ahmed, 1991; Price *et al.*, 1991; Falconer, 1993; Smayda and Shimizu, 1993; Anderson, 2000; Fleming *et al.*, 2001, Johnson and Schantz, 2002; Backer *et al.*, 2005). In the US, Canada and certain other countries, governmental monitoring of coastal waters for toxic algae and toxins in seafoods and cautionary warnings have reduced the risk of consumption of poisonous seafood. Algal blooms and associated shellfish contamination have become more common during the past two decades, and it is anticipated that seafood illnesses will correspondingly increase in their incidence. Currently, the Centers for Disease Control and Prevention (CDC) conducts surveillance and reports the incidence of ciguatera, scombroid fish poisoning, and paralytic shellfish poisoning. Other seafood intoxications are prospective risks that may warrant enhanced surveillance and reporting. Mortality for the majority of seafood diseases is generally low, and treatment is mainly supportive (Rosen *et al.*, 1988; CDC, 2001; Fleming *et al.*, 2001; Johnson and Schantz, 2002; Backer *et al.*, 2005).

3.1 Overview of the causes of seafood intoxications

The main recognized human intoxications from fish and shellfish include ciguatera fish poisoning (CFP), paralytic shellfish poisoning (PSP), puffer fish poisoning (PFP), diarrhetic shellfish poisoning (DSP), neurotoxic shellfish poisoning (NSP), amnesic shellfish poisoning (ASP), scombroid fish intoxication (CFP) and certain other rare intoxications (Halstead, 1967; Ahmed, 1991; Falconer, 1993; Anderson, 2000; Fleming *et al.*, 2001; Hui *et al.*, 2001a; Johnson and Schantz, 2002; Backer *et al.*, 2005; Table 17.2). Except for scombroid toxin (histamine), which is produced by bacterial spoilage of improperly refrigerated fish, most of the toxins are produced by marine unicellular algae or phytoplankton. Of the more than 5000 known species of phytoplankton, about 60–80 are known to produce harmful toxins. Fewer than 25 toxic species were recognized only a decade ago. Occasionally, the algae grow to large numbers and form 'blooms' that are visible as patches near the water surface (Smayda and Shimizu, 1993; Anderson, 2000). 'Red tide' is a common name for a harmful algal bloom (HAB) in which the red pigments of the algal species give the ocean patch its characteristic color. Blooms vary in color depending on the algae, and appear as red, brown or green. Early records and folklore indicate that toxic algal blooms have occurred for hundreds of years, but their actual incidence and the associated causative algae were not accurately identified until relatively recently. Reports of toxic algal blooms are increasing worldwide in frequency, magnitude and

Table 17.2 Seafood intoxications (adapted from Johnson and Schantz, 2002)

Syndrome	Geographic areas	Source of toxin	Major toxin	Onset time; duration	Major symptoms[a]	Foods involved	Treatment	Prevention
Paralytic shellfish poisoning (PSP)	Worldwide	Toxic dino-flagellates: *Alexandrium* spp., *Gymnodinium catenatum*, *Pyrodinium bahamse*	Saxitoxin	5–30 min; occasionally few hours–few days	n, n, d, p, r	Mussels, clams, bay scallops, some fin fish	Supportive (respiratory)	Seafood surveillance; quarantine of seafood region; rapid reporting
Puffer fish poisoning	Pacific regions near Japan and China; rare in US	Puffer fish: poison in liver, gonads, and roe; possibly produced by bacteria	Tetrodotoxin	Similar to PSP	n, v, d, p, r, bp	Puffer or globe fish	Supportive (respiratory)	Regulated food source; preparation; rapid reporting
Ciguatera	Tropical areas around world; in US, mainly near Florida	Toxic dino-flagellates: *Gambierdiscus toxicus*, *Prorocentrum* spp., Ostreopsis spp., *Coolia monotis*, *Thecadinium* spp., *Amphidinium carterae*	Ciguatoxin	Hours; months–years	n, v, d, t, p	Edible tropical fish, commonly barracuda, kahala, snapper, grouper	Supportive	Seafood surveillance; quarantine of region; rapid reporting
Diarrhetic shellfish poisoning (DSP)	Mainly in Europe, Japan; rare cases in Chile, Southeast Asia, New Zealand	Toxic dino-flagellates: *Dinophysis* sp *Procentrum* spp.p.	Okadaic acid	Hours; days	d, n, v	Mussels, clams, scallops	Supportive	Seafood and water surveillance; quarantine of seafood, region; rapid reporting

Symptoms: a, allergic-like; b, bronchoconstriction; bp, decrease in blood pressure; d, diarrhea; n, nausea; p, paresthesias; r, respiratory distress; t, reversal of temperature sensation; v, vomiting.

(*Continued*)

Table 17.2 Seafood intoxications (adapted from Johnson and Schantz, 2002)—*continued*

Syndrome	Geographic areas	Source of toxin	Major toxin	Onset time; duration	Major symptoms[a]	Foods involved	Treatment	Prevention
Neurotoxic shellfish poisoning (NSP)	Gulf of Mexico, South Atlantic Bight, New Zealand	*Gymnodinium breve*; possibly other species	Brevetoxin	30 min–few hours; few hours	n, v, d, b, t, p	Bay scallops, clams, oysters, quahogs, cohinas	Supportive	Seafood and water surveillance; quarantine of seafood, region; rapid reporting
Amnesic shellfish poisoning (ASP)	NE Canada; rare or affects animals in NW US, Europe, Japan, Australia, New Zealand	Diatom: *Pseudonitschia* spp.	Domoic acid	Hours; months–years	n, v, d, p, r	Mussels, clams, crabs, scallops, anchovies	Supportive (respiratory)	Seafood surveillance; quarantine of seafood; rapid reporting
Scombroid fish poisoning	Worldwide	Bacterial decomposition of fish at elevated temperatures (> 5°C)	Histamine	10–90 min; hours	a, d, h, v	Various fish: common in mahimahi, tuna, bluefish, mackerel, skipjack	Supportive; antihista-mine	Regulated food handling; keep temperature < 5°C

Symptoms: a, allergic-like; b, bronchoconstriction; bp, decrease in blood pressure; d, diarrhea; n, nausea; p, paresthesias; r, respiratory distress; t, reversal of temperature sensation; v, vomiting.

geographical extent (Smayda and Shimizu, 1993; Anderson, 2000). They have a marked detrimental effect on fisheries, aquaculture and human health. The ecological factors contributing to this increased occurrence of toxic algal blooms are not completely understood, but recognized contributing factors include the increased availability of nutrients through pollution and oceanic currents and upwelling. Changes in climate have been proposed to contribute to blooms (Smayda and Shimizu, 1993; Anderson, 2000; Fleming *et al.*, 2001; Johnson and Schantz, 2002; Backer *et al.*, 2005).

Microalgal toxins are produced as secondary metabolites and are transferred through the food web, where they accumulate in shellfish and finfish. Since the toxins accumulate in the seafood through the food chain, usually the toxic seafoods appear unspoiled and seemingly harmless. Most of the toxins are tasteless and colorless at poisonous levels. This mode of toxin contamination combined with the heat-stability of the toxins presents considerable difficulty in the prevention of seafood intoxications. Currently, prevention of seafood intoxications depends mainly on surveillance and detection of toxins in the commodities at the point of harvest (Fleming *et al.*, 2001; Johnson and Schantz, 2002; Backer *et al.*, 2005).

3.2 Incidence of seafood intoxications

Globally, about 60 000 seafood intoxications and at least 100 deaths are reported worldwide each year. Like most other foodborne illnesses, this is certainly an underestimation by at least 100–300-fold, since many cases of seafood intoxications are mild, are not reported or are misdiagnosed. The most common intoxications worldwide are SFP, CFP, PSP, NTP and ASP. Globally, PSP is probably the most widespread geographically of seafood intoxications, while most seafood intoxications are clustered in geographic locations near to the area of harvest. Only CFP, SFP and PSP are reported in the *Morbidity and Mortality Weekly Report* by the CDC in the category of 'Chemical Poisonings'. In the US, SFP and CFP are responsible for more than 80 % of seafood intoxications. In the latest published statistics from the CDC for chemical poisonings, this category accounted for ~14 % of the outbreaks and ~2 % of the total foodborne illness cases for the period 1988–1992. In the US, seafood ranked third on the list of products that caused foodborne disease between 1983 and 1992 (Fleming *et al.*, 2001; Johnson and Schantz, 2002; Backer *et al.*, 2005).

3.3 Ciguatera fish poisoning (CFP)

Ciguatera fish poisoning (CFP) is one of the most common seafood illnesses, and is caused by eating finfish from tropical reef and island habitats that have accumulated ciguatera toxins (CTXs) from epibenthic dinoflagellates in the food chain. Approximately 20 000 cases have been estimated to occur worldwide annually. The CDC generally reports 10–20 outbreaks per year, with 50–100 cases in the US. The outbreaks usually occur in Hawaii, Puerto Rico, the Virgin Islands and Florida, but can occur in other regions to which the the fish are shipped.

As many as 400 species of fish have been implicated in ciguatera poisoning in the Caribbean and Pacific regions, and the fish most commonly involved are amberjack, snapper, grouper, barracuda, goatfish, and reef fish in the *Carrangidae*. Ciguatoxins reach particularly high concentrations in large predatory reef fishes, and these fish (such as barracuda) are frequently sought by sport fisherman on reefs of Hawaii, Guam and other South Pacific Islands. Various dinoflagellates are known to produce CTXs, including *Gambierdiscus toxicus, Procentrum* spp., *Ostreopsis* spp., *Coolia monotis, Thecadinium* spp. and *Amphinidium carterae*.

Ciguatera poisoning can involve gastrointestinal, neurological and cardiovascular symptoms. Gastrointestinal symptoms include diarrhea, abdominal pain, nausea and vomiting; their onset is a few hours after ingestion of the fish, and they last for only a few hours. Neurological symptoms usually begin 12–18 hours after consumption, and vary in severity. Neurological signs include reversal of temperature sensation (for example, ice cream tastes hot, hot coffee tastes cold); muscle aches; dizziness; tingling and numbness of the lips, tongue and digits; a metallic taste; dryness of the mouth; anxiety; sweating; and dilated eyes, blurred vision and temporary blindness. Paralysis and death have been documented, but these are rare. There is considerable variation of symptoms and recovery time in individual patients. Recovery may require weeks, months or even years, and the chronic effects of CFP have not been elucidated. Intravenous administration of mannitol can help to relieve acute symptoms, and amitryptiline or tocainide have been suggested for the treatment of chronic symptoms. The most common treatment is supportive, with attention to respiratory and cardiovascular functions. The fatality rate for CFP overall is less than 1 %, but has ranged from 0–12 % in various fish outbreaks (Fleming *et al.*, 2001; Johnson and Schantz, 2002; Backer *et al.*, 2005).

Several CTXs have been isolated from toxic fish and algae. They consist of a family of lipophilic, brevitoxin-type polyether compounds, and the prototype compound – gambiertoxin-4 – was isolated and characterized from *Gambierdiscus toxicus*. The total synthesis of brevetoxin A and ciguatoxin CTX3C have been accomplished (Hirami *et al.*, 2001). A family of CTXs has been found, since different algal species produce variant toxin structures, and the toxins may be modified by animal or human metabolism. The structures of at least 8 'native' CTXs have been determined, and 11 CTXs formed by oxidative metabolism have been detected. The toxic mechanism of CTXs involves the binding to and opening of sodium and calcium channels in excitable membranes. Like most other seafood toxins, CTXs are commonly detected by mouse bioassay. Immunoassays are available, including 'dipstick' tests and commercialized kits (Backer *et al.*, 2005).

Fish containing toxic levels of CTXs usually do not appear spoiled. Prevention of intoxications depends on surveillance and detection of toxins in fish and algae from endemic areas, and rapid reporting and treatment of cluster outbreaks. For the majority of US consumers, the illness is contracted from fish imported from endemic areas (Fleming *et al.*, 2001; Johnson and Schantz, 2002; Backer *et al.*, 2005).

3.4 Neurotoxic shellfish poisoning (NSP)

NSP was noticed centuries ago by Spanish explorers and Tampa Bay Indians to cause massive fish kills during certain seasons, when coastal waters became red in color. NSP occasionally accumulates in oysters, clams and mussels from the Gulf of

Mexico and the Atlantic coast and southern US states. Shellfish poisonings of humans were reported in the late 1800s and in 1946. NSP is usually confined to these regions, but a bloom was spread by the Gulf Stream leading to an outbreak in North Carolina, and outbreaks have occurred in New Zealand.

The symptoms of NSP mimic those of ciguatera, in which gastrointestinal and neurologic symptoms predominate. The onset of symptoms occurs 30 minutes to 3 hours after ingestion of the fish, and include nausea and vomiting; diarrhea; numbness and tingling in the mouth, arms and legs; incoordination; bronchorestriction; and paresthesias. The illness usually subsides within 2 days. The symptoms are usually less severe than in ciguatera, and no deaths have been reported, but the illness is still debilitating. Unlike ciguatera, which can persist for weeks, NSP generally subsides within a few days. Treatment is supportive, and there is no antidote. NSP blooms can become aerosolized in the surf and cause respiratory and asthma-like problems to people on the beach who breathe them. NSP appears to be rare throughout the world, with documented outbreaks mainly in the US and New Zealand. Algae related to *Gymnodinium breve* have been detected in Spain and Japan, and it is possible that intoxications could occur from shellfish harvested from these regions.

Gymnodinium breve scavenged from toxic blooms was found to produce polyether brevetoxins with structures related to certain ciguatoxins. Brevetoxin causes opening of the sodium channels in nerves and other tissues. Brevetoxin is detected by mouse bioassay or by ELISA tests. Prevention of poisoning from shellfish depends on surveillance of waters for toxic algae and rapid reporting. Coastal waters have been monitored for *G. breve* cell counts, and this has successfully prevented illnesses (Fleming *et al.*, 2001; Johnson and Schantz, 2002; Backer *et al.*, 2005).

3.5 Paralytic shellfish poisoning (PSP)

PSP is a serious and life-threatening intoxication that occurs by eating shellfish contaminated with saxitoxin (STX) and related toxins (Meyer *et al.*, 1928; Sommer and Meyer, 1937). PSP has a wider worldwide geographic distribution than other seafood intoxications caused by microalgal toxins. PSP was first reported in 1793, after five members of Captain George Vancouver's ship crew became ill and one sailor died after eating mussels from Poison Cove in central British Columbia (Kao, 1993). PSP from toxic mussels and clams was also recognized on the Pacific coast in the 1700s by Native Americans, who associated the poisoning with red (brownish-red) tides and accompanying bioluminescence. PSP occurs through ingestion of toxic bivalve mollusks (mainly mussels, clams, oysters and scallops) that have fed on toxic dinoflagellates including *Alexandrium* spp., *Gymnodinium catenatum* and *Pyrodinium bahamense* (Kao and Levinson, 1986; Ahmed, 1991). In the United States PSP has a wider geographical distribution than other dinoflagellate poisonings, and occurs in the Pacific Northwest Coast and Alaska, and in New England from Massachusetts to Maine. Toxic algal blooms of *Alexandrium* spp. and other PSP-producing microalgal species in northern California and other cold temperate regions are seasonal, occur mainly during the spring, and may be sustained through the summer in upwelling waters. Owing to current testing and control procedures, outbreaks are rare in commercial shellfish harvested from coastal regions. Most PSP outbreaks involve recreational collectors of bivalves, often from quarantined areas. The incidence of PSP appears to have increased since the 1970s (Fleming *et al.*, 2001; Johnson and Schantz, 2002; Backer *et al.*, 2005).

The symptoms of PSP generally begin within minutes after eating toxic shellfish, and initially affect the peripheral nervous system. The first signs of intoxication are a prickly or tingling feeling in the lips, tongue and fingertips, followed by numbness in the extremities and face. The intoxication continues with an ataxic gait and muscular incoordination followed by ascending paralysis. Death from respiratory failure may occur within 2–24 hours, depending on the quantity of toxin consumed (2–4 mg is considered to be the lethal dose for a human; Kao and Levinson, 1986). If a sufferer survives the first 24 hours, the prognosis for complete recovery is good and no chronic effects of the poisoning generally occur. There is no effective antidote; poisoned individuals should receive artificial respiration and supportive medical care as soon as these can be administered. Emergency treatment and first aid for victims of PSP have been described. In particular, attention should be given to cardiopulmonary resuscitation (CPR) and respiratory ability, and the victim should rapidly be transported to a hospital emergency facility (Rosen *et al.*, 1988).

STX was the first toxin recognized in shellfish, and it has been extensively characterized (Kao and Levinson, 1986). The nature of the poison responsible for PSP was elusive until it was discovered in the 1930s that culture supernatants lethal to mice were produced by phytoplankton and attributed to the genus *Gonyaulax*. A lethal substance was extracted from dinoflagellates harvested from blooms and from toxic shellfish. PSP or toxic mussel poison, now called saxitoxin (STX), was purified and identified by E. J. Schantz and colleagues (Schantz, 1992). Toxic extracts were prepared from large quantities of harvested poisonous California mussels (*Mytilus californius*) and butter clams (*Saxidomas giganteus*) from Alaska. The toxic substances were purified and found to consist of tetrahydropurine derivatives. Good-quality crystals of STX were obtained and the three-dimensional structure was resolved. The availability of purified toxin allowed the elucidation of the pharmacological mechanism of saxitoxin, and it was demonstrated to bind selectively and with high affinity to sodium channels of excitable membranes and to block completely the inward flux of sodium, in a manner very similar to tetrodotoxin. These toxins have become important neurobiological tools because of their selective and high-affinity blockade of the voltage-gated sodium channels of excitable membranes of neurons and skeletal muscle. Like most other marine microalgal toxins, PSP exists as a family of related compounds, called saxitoxins, neosaxitoxins or gonyautoxins, and more than 20 distinct structures have been elucidated (Kao and Levinson, 1986; Hui *et al.*, 2001a). Current taxonomic understanding is that PSP is produced by various dinoflagellate species of the genus *Alexandrium*, *Gymnodinium catenatum* and *Pyrodinium bahamense*, although reports have indicated that certain bacteria, including *Moraxella* sp., can produce low quantities of STX or inactive precursors and derivatives in culture.

The prevention of PSP occurs primarily by proactive monitoring of coastal algal blooms and seafoods for the presence of saxitoxin, and rapidly alerting the shellfish industry and public of a health hazard from eating clams, mussels and certain other shellfish from a designated region. Early investigators in California were instrumental in instigating a prevention program in the 1920s, which consisted mainly of posting warning placards on the beaches with instructions to not eat clams, mussels and certain other shellfish during the high season. The mouse bioassay is currently used by governmental personnel in the US and Canada to detect PSP and to determine whether dangerous levels are present in shellfish (Kao and Levinson, 1986; Hui *et al.*, 2001a). High-performance liquid chromatography (HPLC) is used as an alternative to the mouse bioassay, and capillary electrophoresis and ELISA methods are also being developed (Backer *et al.*, 2005). Industry personnel or recreational consumers of shellfish who plan to gather shellfish should contact their local, state or national health authorities to obtain information regarding the safety of these foods. There are no uniform tolerance levels for PSP, but most countries apply a level of 0.8 mg saxitoxin (equivalent to 400 mouse units) per kg mussel meat. If it is assumed that a person consumes 100 g of mussels, this level would afford a safety factor of two to four for an adult, and a minimum safety factor of seven to nine for severe intoxication or death (Kao and Levinson, 1986).

3.6 Puffer fish poisoning (PFP)

Puffer fish poisoning has traditionally been associated with eating certain species of fish belonging to the Tetraodontiformes. These fish are commonly referred to as *fugu*, pufferfish, globefish or swellfish, because they can inflate themselves. It has been recognized for centuries that eating these fish can result in a paralytic poisoning. Puffer fish poisoning most commonly occurs in countries that consume *fugu* as a delicacy, such as China and Japan. Puffer fish poisoning can be fatal, and it has been estimated that about 1800 Japanese have died in the past 40 years following consumption of PFP-tainted and improperly prepared *fugu*. The toxicity of puffer fish varies according to its source, the variety and species of fish, and whether it is wild-caught and grown or kept alive in aquaculture facilities.

The symptoms of pufferfish poisoning are similar to those of PSP, including an initial tingling and prickling sensation of the lips, tongue and fingers within a few minutes of eating poisonous fish. Nausea, vomiting and gastrointestinal pain may follow in some cases. Depending on the quantity of toxin consumed, the pupillary and corneal reflexes are lost and respiratory distress ensues. No antidote is currently available; treatment is supportive, with particular attention to maintaining respiration.

Puffer fish poison was first isolated in 1909 and named tetrodotoxin (TTX). The structure of TTX and its derivatives was reported from Japan and the US in 1964. TTX is an amino perhydroquinazoline compound with a molecular weight of about 400, depending on the form. The chemical structure is distinct from STX, although the symptoms are analogous. It is one of the most poisonous non-protein substances known; the lethal dose is about 0.2 μg for a mouse, 4 μg for a 1-kg rabbit and 1–4 mg for a human. Like STX, it has a highly specific action on sodium channels within excitable membranes.

Tetrodotoxin was long assumed to be produced by the *fugu*; but its detection in certain newts, frogs, marine snails, octopuses, squids, crabs, starfish and other creatures has indicated that it is formed lower in the food chain, possibly by bacteria including species of *Alteromonas, Vibrio*, and other bacterial genera. These and possibly other bacteria produce various forms of TTXs that vary in potency, including non-toxigenic precursors or derivatives (Fleming *et al.*, 2001; Johnson and Schantz, 2002; Backer *et al.*, 2005).

3.7 Amnesic shellfish poisoning (domoic acid)

Amnesic shellfish poisoning (ASP) is a life-threatening shellfish intoxication. It results from eating shellfish contaminated with domoic acid, which is produced by diatoms in the species *Pseudo-nitzschia*. It is a newly recognized seafood intoxication that was first described in 1987 from persons who ate poisonous blue mussels from Prince Edward Island, Canada (Quilliam and Wright, 1989). Until this outbreak, the diatom *Pseudo-nitzschia* was not thought to produce toxins poisonous to humans or animals. Most of the persons in the Canadian outbreak experienced gastroenteritis including vomiting (75 %), diarrhea (42 %) and abdominal cramps (49 %), while some older persons with underlying chronic diseases developed neurological symptoms including memory loss, confusion, disorientation, seizure, coma or cranial nerve palsies within 48 hours. Certain patients experienced short-term memory loss for at least 5 years, and some patients were unable to recognize family members or perform simple tasks. Interestingly, memory loss was more common in patients greater than 70 years of age than in the young. Evidence suggests that persons with impaired renal function may be at greater risk of domoic-acid neurotoxicity because of an impaired ability to inactivate and excrete the toxin. Of the 107 persons affected in the Canadian outbreak, 3 patients died within 3 weeks of eating the mussels. Currently there is no medical treatment for ASP other than supportive care. Medications have been administered to control seizures and potentially reduce the extent of brain lesions.

Investigators were unable to find infectious levels of pathogenic bacteria or viruses, or toxic levels of substances such as heavy metals or organophosphorous pesticides, in the poisonous Canadian mussels. Using the mouse assay designed to detect saxitoxin, it was found that mussel extracts caused death of the mice, usually within 15–45 minutes of intraperitoneal injection. However, the symptoms were distinct from those of PSP, with a unique scratching syndrome of the shoulders and hind leg, followed by convulsions and death. Upon examination of the mussels, it was found that the digestive glands of poisonous animals contained green phytoplankton. Investigators were able to purify domoic acid, and show that it caused similar unique symptoms in mice. The toxic mussels contained up to 900 mg of domoic acid per kg of tissue. Domoic acid was initially characterized in 1958 by Japanese researchers, and its ingestion in algae has been responsible for deaths of pelicans and cormorants in Monterey Bay, California. Although several algae can produce domoic acid, it was found that a bloom of the pennate diatom *Nitzschia pungens* f. *multiseries* that was occurring at the time of the outbreak contained domoic acid. Isolates of this *Nitzschia* strain produced domoic acid in culture, demonstrating that the diatom was the causative agent and not just a vehicle for the toxin. Domoic acid was produced as a secondary metabolite in axenic cultures of the alga *N. pungens* f. *multiseries*. Evidence indicated that domoic acid was produced by a limited number of species in the genera *Nitzschia, Digenea, Vidalia, Amansia* and *Chondriaarmata*. Anderson (2000) indicated that species of the genus *Pseudo-nitzschia* is the current taxon primarily responsible for domoic acid production.

Domoic acid is water-soluble, with a molecular weight of 311 Da. It contains a glutamate-like moiety, and is an analogue of kainic acid. Kainic acid binds to certain CNS receptors and stimulates the release of glutamate in the manner of excitotoxins (Llewellyn, 2001; Backer *et al.*, 2005). Evidence suggests that domoic acid affects calcium transport and stimulates a calcium-dependent process that regulates release of glutamate from presynaptic nerve endings. Domoic acid has been shown to be excitotoxic in a number of animal models, including rodents and primates, and induces seizures at high doses. Domoic acid produces a loss of neurons in various brain regions, particularly the hippocampus, in rodents. On autopsy, brain tissue from the victims of the 1987 Canadian outbreak had lesions in several regions, including the hippocampus, amygdala, thalamus and cerebral cortex (Todd, 1993). In mice, the IP LD_{50} was estimated to be 3.6 mg domoic acid per kg body weight. Human intoxication in ASP cases occurred after ingestion of an estimated 1–5 mg of domoic acid per kg of body weight. Domoic acid has also been shown to cause deaths among various marine animals, including birds and sea mammals such as sea lions and humpback whales.

Initially the IP mouse assay for PSP detection was used to survey seafood for domoic acid. However, this assay is not consistently sensitive enough to detect the toxin at the Canadian regulatory level of 20 μg/g tissue. A non-destructive extraction method, combined with reverse-phase chromatographic separation and UV detection at 242 nm, was employed and adopted as an official first action by the AOAC in 1990. The detection limit of this method is about 1 μg domoic acid per gram of tissue. Other biochemical methods have also been investigated for sensitive and accurate detection of domoic acid. The current Canadian and UK guideline for the limit of PSP in seafood is 20 mg per kg edible meat.

Although toxic *Pseudo-nitzschia* species occur in oceans worldwide, only two outbreaks of ASP affecting humans or animals have been reported. The first was the Canadian outbreak in 1987, followed by a large outbreak in seabirds in September 1991 in Monterey, California. High levels of domoic acid were found in *Pseudo-nitzschia australis* harvested from the area. Since anchovies are a major food source for seabirds in the area, it is possible that the intoxication could be transmitted in herbivorous finfish such as anchovies. An ASP outbreak affected 24 people who consumed razor clams and became ill with gastrointestinal symptoms; 2 people developed mild neurological symptoms. Surveys showed that razor clams and Dungeness crabs in Washington and Oregon contained domoic acid.

Monitoring of phytoplankton blooms for domoic acid, in response to the 1987 outbreak, has contributed to the prevention of ASP. It has been recommended that shellfish from suspect regions be tested and those that contain ≥ 20 mg/kg not be harvested for human consumption. If shellfish are found with levels above 5 mg/kg but below 20 mg/kg, then the harvest area is monitored closely. Domoic acid is permitted in the US in bivalve shellfish and cooked crab viscera at levels of 20 mg/kg and 30 mg/kg tissue, respectively. In 1988 the US started the National Shellfish Sanitation

Program, which is a cooperative program designed to reduce risks from the consumption of toxic shellfish. The program includes contingency plans in case of an outbreak; the certification of harvesters, processors and distributors; and tracking of the shipping of shellfish. The most extensive shellfish monitoring is directed at PSP, but domoic acid is also monitored to a lesser extent. As of 1998, monitoring programs in Canada and the US have found domoic acid in seafood products in Washington, California, Oregon, Alaska, the Bay of Fundy, British Columbia and Prince Edward Island. Domoic acid has also been detected at low concentrations in regions of Australia, Europe, Japan and New Zealand. In addition to monitoring for domoic acid, depuration has been considered as a detoxification method for shellfish. However, depuration rates vary greatly depending on the animal species and type of toxin, and overall this is not a consistent method of detoxification.

3.8 Diarrhetic shellfish poisoning (DSP)

Diarrhetic shellfish poisoning (DSP) was observed in the 1960s and 1970s, caused by toxic mussels, scallops or clams. Consumption of toxic shellfish causes gastrointestinal disturbances and diarrhea. DSP occurs mainly in Japan and northern Europe, but also in various other regions of the world – there have been outbreaks in South America, South Africa, southeastern Asia, and New Zealand. DSP is caused by toxins produced by *Dinophysis* spp. DSP is not usually fatal, but shellfish may become toxic in the presence of dinoflagellates at low cell densities (≥ 200 cells/ml). DSP is characterized by gastrointestinal symptoms such as severe diarrhea, nausea, vomiting, abdominal cramps and chills, which start 30 minutes to a few hours after eating toxic shellfish. Complete recovery usually occurs within 3 days. In more severe cases, hospitalization and administration of an electrolyte solution may be necessary.

Various toxins are produced by *Dinophysis* spp., including okadaic acid, pectenotoxins and yessotoxin. The parenteral toxicity of yessotoxins has been estimated to be approximately one order of magnitude greater than that of the oral dose. It has been estimated that symptoms from OA or DTX consumption begin at about 40–50 μg for an adult. Certain of the *Dinophysis* toxins appear to have mutagenic, cancer-inducing, hepatotoxic or immunogenic properties; but the chronic toxic effects in humans are not known. The mouse bioassay is most commonly used to detect the presence of DSP toxins, although HPLC combined with mass spectroscopy, immunoassays and cytotoxicity based assays has also been evaluated. Limits for diarrheal shellfish toxins (usually undetectable by mouse assay) have been proposed in Japan and certain European countries. The European Union applies a tolerance level of 0.16 OA equivalents per kg of mussel meat, which generally provides a safety factor of ≥ 2 before symptoms are noticed.

3.9 *Pfiesteria* intoxications

Pfiesteria heterotrophic dinoflagellates were proposed in the early 1990s to cause massive fish kills of millions to billions of finfish and shellfish in the coastal waters of the mid-Atlantic and southeastern US (Morris, 1999; Berry *et al.*, 2002; Collier and

Burke, 2002). The predatory organism, mainly the species *P. piscicida*, had also been reported from the Mediterranean Sea, the Gulf of Mexico and the western Atlantic. The organism exists in several life stages. It remains in river and coastal bottoms for years as cysts and, when induced by unknown factors in fish feces, the cysts bloom into a motile form of the organism. These then swarm to the upper waters and produce very potent toxins, resulting in a 'feeding frenzy'. After this, the organism transforms to an ameba state that feeds on microorganisms and fish remains, followed by reformation of cysts, which settle in the coastal sediment to complete the cycle.

Recently, considerable controversy has arisen regarding whether the organism dubbed the 'cell from hell' (Morris, 1999) is predatory to fish, and whether it also produces toxins that can cause illness in humans. Despite several documented fish kills, human illness due to environmental exposure has not been shown definitively (Morris, 1999, 2001; Berry *et al.*, 2002; Collier and Burke, 2002). It had been reported (Glasgow *et al.*, 1995) that researchers were intoxicated by chronic exposure to *P. piscicida* in aquarium water. Three workers reported symptoms of asthenia, skin lesions, emotional lability and memory dysfunction. Other symptoms included gastrointestinal pain, nausea, headache, spatial disorientation, impaired concentration and severe loss of short-term memory (Collier and Burke, 2002). However, neurologic examinations were normal. The speculation that humans could contract disease from *P. piscicida* was also questioned, since 254 crabbers who work in infested waters have not been known to contract the disease. After much debate, it has been concluded that exposure to the alga may cause human illness under the right conditions (Collier and Burke, 2002). This would not be surprising, since other dinoflagellates cause diseases that are mainly mediated by toxins. Despite preliminary evidence presented by Burkholder and colleagues (Glasgow *et al.*, 1995), the production of toxins has not been shown definitively for *P. piscicida* (Berry *et al.*, 2002; Collier and Burke, 2002). Experiments are also in progress to determine whether the toxic effects in humans are due to *Pfiesteria* solely, or to *Pfiesteria* together with associated microorganisms. Currently, *Pfiesteria* should be considered as a cause of human illness from contaminated waters, as well as an occupational and laboratory hazard. There have been no definitive reports of foodborne illness, but these may be forthcoming with effective investigation and diagnosis.

Recommendations have been suggested for the closing and reopening of waters affected by *Pfiesteria* or *Pfiesteria*-like events. Closure is recommended when a significant fish kill is reported and fish are found that contain sores and lesions consistent with the toxic activity of *Pfiesteria*, or when a significant number of fish exhibit erratic behavior that cannot be attributed to other factors such as low oxygen levels in the water. Waters may be reopened for recreational and commercial activities when these signs have not been apparent for 14 days.

3.10 Cyanobacterial intoxications

Unlike dinoflagellates and diatoms, which cause human food poisoning from marine finfish and shellfish, cyanobacteria (sometimes called blue-green algae) have caused severe animal illnesses from consumption of drinking water (Falconer, 1993; Chorus,

2001). The vast majority of illnesses due to cyanobacteria are waterborne and affect animals. The main toxic genera of prokaryotic cyanobacteria are the filamentous species *Anabena, Aphanizomenom, Nodularia* and *Oscillatoria*, and the unicellular species *Microcystis*. Like the marine eukaryotic microalgae, they form blooms under appropriate conditions in freshwater. Blooms usually occur in the summer and autumn during warm days, and are promoted by nutrient availability (especially nitrogen and phosphorous) that often derives from water runoff containing fertilizers, or from livestock or human waste. Toxic blooms occur in many lakes, ponds and rivers throughout the world. The primary toxicoses include gastrointestinal disturbances, acute hepatotoxicosis, neurotoxicoses, respiratory distress and allergic reactions.

Most of the poisonings by cyanobacteria involve acute hepatoxocosis and death mediated by microcystins and nodularin, which are heat-stable, small peptides (Shimizu, 2003). Certain cyanobacteria also produce neurotoxicoses due to the alkaloidal anatoxins and anaphatoxins. These toxins can cause death within minutes to a few hours, depending on the animal species and the quantity of toxin consumed. Cyanobacteria also produce a cholinesterase inhibitor called anatoxin-α(s), which has an organophosphate structure. Certain cyanobacteria have been reported to produce paralytic shellfish-like toxins, including saxitoxin and neosaxitoxin.

Cyanobacterial toxins have sporadically caused human intoxications from drinking water (Chorus, 2001). In most cases, water treatment systems are adequate to remove cyanobacteria by coagulation and filtration, and microcystins can be adsorbed by charcoal filters and are degraded by chlorine. However, extracellular cyanobacterial toxins may survive water treatment and are resistant to boiling. Cyanobacterial toxins in human drinking water have been documented to cause hepatic toxicity, gastroenteritis, contact dermatitis and allergic responses, and neuronal and brain damage. Some cyanobacteria, e.g. *Spirulina*, have been marketed as health foods. Although studies in rodents have shown no toxicity, it is important that *Spirulina* and other cyanobacteria food supplements are produced under hygienic conditions and do not contain cells or toxins from toxic species.

Diagnostic procedures for human and animal illnesses include association of a bloom of a toxigenic cyanobacterial species with consumption of water, the presence of characteristic symptoms in the animal or human, microscopic identification of the toxic species of cyanobacteria in the suspect water, and verification of the presence of toxin in the water by chemical and bioassays. Procedures for prevention of cyanobacterial intoxications from water rely on monitoring programs and quarantine measures when toxic algae reach a certain concentration in the bloom water. Methods with increased sensitivity compared to microscopic identification are being developed to detect toxin-producing cyanobacteria rapidly. Chemicals, particularly copper sulfate, have been added to lakes to kill cyanobacteria. The reduction of agricultural runoff and animal or human fecal contamination of water will also reduce bloom formation. Obviously, if a bloom occurs, animals and humans should avoid consumption of the water.

Although there are no documented outbreaks of food poisoning caused by cyanobacterial toxins, shellfish can filter cyanobacteria from water and may accumulate their toxins. Mussels (*Mytilus edulis*) fed *Microcystis* accumulated microcystins that persisted for several days after transfer to freshwater. Fresh fruits and vegetables washed with contaminated water could also potentially acquire these toxins. Since scenarios exist by which food could transmit cyanobacterial toxins, it may be important to monitor water from suspect sources that will have contact with foods. The UK and the State of Oregon have set limits of 1 ppm for microcystins in drinking water and dietary supplements. It is anticipated that cyanobacterial poisonings of humans will persist until we can prevent the blooms.

3.11 Scombroid (histamine) fish poisoning

Scombroid poisoning is probably the most prevalent of the seafood-transmitted illnesses worldwide (Taylor, 1986; Fleming *et al.*, 2001; Johnson and Schantz, 2002; Backer *et al.*, 2005). Scombroid poisonings have been commonly reported in Japan, Canada, the US, the UK and other countries that have a high dietary intake of fish. Scombroid poisoning symptoms mimic those of an Ig E-mediated food allergy, with flushing of the face, neck and upper arms, nausea, vomiting, diarrhea, abdominal pain, headache, dizziness, blurred vision, faintness, itching, rash, hives, and an oral burning sensation in the mouth. Hypotension, tachycardia, palpitations, respiratory distress and shock may occur in severe cases. The symptoms of scombroid illness usually occur within 10–90 minutes of eating contaminated fish. Individuals exposed to scombroid poison will usually experience only a few of these symptoms. The duration of the illness is usually less than 12 hours. Diagnosis of scombroid poisoning can generally be made by the short onset time, the non-specific yet characteristic symptoms, and a history of consumption of fish. The diagnosis can be confirmed by detection of histamine in the spoiled fish (scombroid poisoning has been diagnosed by measurement of plasma histamine). Corticosteroids and H_1 and H_2 antihistamines can be used to treat the symptoms.

Due to the variety of symptoms and their similarity to allergic responses, the illness is frequently misdiagnosed and is often confused with an allergic reaction. Many of the symptoms of allergic reactions mimic those apparent in scombroid illness, since histamine is a primary mediator of allergic disease. Normally treatment is unnecessary, as the vast majority of cases are mild and self-limiting, but antihistamine therapy can provide relief and rapid recovery. Hydration and electrolyte replacement may also be beneficial. Scombroid poisoning can be severe in persons with a history of allergic disease or with pre-existing cardiac or respiratory conditions, or in people being treated with certain drugs such as isoniazid or monoamine oxidase inhibitors. Antihistamines should be administered only under close medical supervision in these special situations.

Nearly all cases of scombroid poisoning have been associated with marine fish, particularly of the scombroid (dark flesh) variety, such as tuna, bonito and mackerel. Non-scombroid fish and shellfish have also been implicated in scombroid poisoning, including mahi-mahi, swordfish, salmon, dolphin, marlin, sardines, bluefish, amberjack, anchovy and abalone. Other foods have also transmitted scombroid poisoning, including Swiss cheese and some fermented foods and extracts. Scombroid poisoning is caused by certain bacterial species that grow in fish stored at inappropriate elevated temperatures, where the bacteria decarboxylate histidine to histamine. Histamine is heat-stable and withstands cooking. Since orally-administered histamine generally does not elicit symptoms, it is believed that potentiators such as the diamines putrescine and cadaverine promote the illness.

Several species of bacteria produce histamine through the action of the enzyme histidine decarboxylase. Bacterial species associated with scombroid poisoning include *Morganella morganii, Klebsiella pneumoniae, Vibrio* sp., *Enterobacter aerogenes, Clostridium perfringens, Hafnia alvei, Lactobacillus buchneri* and *Lactobacillus delbrueckii*. Other enteric *Enterobacteriaceae*, clostridia and vibrios have been associated with scombroid poisoning, but *M. morganii* and *K. pneumoniae* are the most common species implicated. These organisms are not frequently associated with living fish, and must contaminate the fish during handling and storage. Since the organisms forming scombroid toxin are not psychrophiles, temperatures above 15 °C are generally required to permit adequate growth and histamine formation. Histamine production on skipjack tuna was optimal at 30 °C, but once a large population of bacteria has been formed the enzyme histidine decarboxylase can stay active even under refrigeration conditions. Most of the histamine is produced near the intestines and then diffuses into the flesh.

The standard analytical method for detection of histamine and other biogenic amines is high-performance liquid chromatography, although other methods (including radioimmunoassay kits) are commercially available. The generally accepted toxic level of histamine in fish is 100 mg/100 g of flesh, the amounts in ingested fish actually causing illness have not been accurately defined. Histamine can be used to judge the freshness of certain raw fish. The FDA considers 20 mg of histamine per 100 g of gless, or 200 ppm, indicative of spoilage in tuna, and 50 mg/100g (500 ppm) an indication of a hazard. Since other finfish and shellfish intoxications show similar signs to scombroid poisoning, the final diagnosis may depend on detection of the toxins in the foods (Fleming *et al.*, 2001; Johnson and Schantz, 2002; Backer *et al.*, 2005).

The most important contributing factor to scombroid poisoning is improper refrigeration of the harvested fish, allowing bacterial proliferation. Fish should be chilled as rapidly as possible, and be brought below 15 °C and preferably below 10 °C within 4 hours; lengthier storage of fish should be at 0 °C (32 °F) or below, using ice, brine or mechanical refrigeration. Maintaining sanitary conditions during handling, processing and distribution will help to prevent bacterial contamination. Histamine is heat-stable and will withstand cooking. Improved reporting of scombroid incidences to public health agencies will increase awareness of the disease and its prevention.

3.12 Other finfish and shellfish toxins

Various substances have been implicated in toxic fish kills and potentially caused human disease. Food poisoning from eating parrot fish has been reported in Japan, and the causative toxin was identified as palytoxin (PTX). PTXs occur in various marine organisms such as seaweeds and crabs, and they also appear to be synthesized by microalgae. Tetramine occurs in the salivary gland of a few whelk species, and has occasionally caused human intoxications. New and emerging toxins posing hazards in seafoods have been proposed, including pinnatoxins, azaspiracids,, gymmodimine and spirolides (Backer *et al.*, 2005). Azaspiracids are polyether toxins produced by the dinoflagellate *Protoperindinium*, formerly thought to be benign. Other substances have been suggested to cause seafood illnesses or toxic blooms with resulting fish kills, including unique hemagglutinins, and reactive oxygen metabolites such as superoxide anions and hydroxyl radicals. Sardine poisoning associated with high mortality (~40 %) and hallucinatory fish poisoning have been described, but the causative toxins are not known.

3.13 Treatment and prevention of seafood intoxications

The structures of seafood toxins, as well as the signs, symptoms and pharmacologic and therapeutic treatments, have been published in several reviews (Fleming *et al.*, 2001; Johnson and Schantz, 2002; Backer *et al.*, 2005). The Centers for Disease Control and Prevention website (www.cdc.gov) also contains valuable information on the incidence, symptoms and treatment of seafood illnesses. Therapy for most shellfish intoxications depends on rapid supportive care, with particular attention given to cardiopulmonary sufficiency, respiratory distress and shock. Antidotes are not available for most of the shellfish toxins, although certain low molecular-weight compounds and monoclonal antibodies have been proposed to alleviate symptoms for some seafood intoxications. Scombroid poisoning can be treated with corticosteroids and H_1 and H_2 antihistamines, but physicians should refer to patient medication status and authoritative guidelines before administration of these.

Most seafood health risks originate in the environment, primarily from harmful algal blooms, and prevention depends on control at harvest. With few exceptions, risks cannot be detected by organoleptic inspection. Surveillance and sensitive detection of the causative algae and toxins by inspection and sampling provide the cornerstone for prevention of seafood intoxications caused by algal toxins. In contrast, prevention of scombroid poisoning requires prompt refrigeration and maintaining the temperature of the fish near to 0 °C. Reducing the incidence of seafood intoxications will require co-ordinated efforts of regulatory agencies and the seafood industries. A comprehensive surveillance and identification program of the etiologic agents and toxins responsible for seafood intoxications can provide valuable information for handling, processing and instituting programs such as hazard analysis critical control points (HACCP) to identify research needs and prevention strategies (Todd, 1997; Williams and Zorn, 1997; National Research Council, 1985).

3.14 Safety precautions for handling toxic seafoods and algae

Working with toxic seafoods, toxin-producing algae in culture, and extracts or purified toxins requires care and adequate safety precautions. Protective clothing, including lab coats, face and eye protection and impervious gloves, as well as air handling requirements, are recommended to prevent exposure to toxins or aerosols. Chlorine can be used to kill the organisms, but spills of toxins may require additional chemical treatment. The US Army has developed procedures for the chemical inactivation of various toxins. Brevetoxins, microcystins, tetrodotoxins, saxitoxins and palytoxins can be inactivated by 30-minute exposure to 2.5 % NaOCl or, more effectively, by 2.5 % NaOCl + 0.35 N NaOH. Algal seafood toxins are resistant to autoclaving at 121 °C or 10 minutes exposure to dry heat at 200°F, and chemical decontamination is usually required.

4 Plant toxins

4.1 Overview and importance of plant toxins

In most countries of the world, foods of plant origin supply most (~70 %) of the protein consumed by humans. Many food plants produce specific natural toxicants (Liener, 1980; Liener and Kadade, 1980; Hui *et al.*, 2001a; Norton, 2001; Panter, 2005). These toxicants can produce acute or chronic illness, or developmental perturbations (Table 17.3). The public perception of plant toxicants, relative to other

Table 17.3 Plant toxicants (adapted from Pariza, 1996)

Toxin	Chemical nature	Main food sources	Major toxicity symptoms
Protease inhibitors	Proteins (4000–24 000 kDa)	Beans (soy, kidney, mung, lima, navy); chick-peas; peas; potatoes (sweet, white); cereals	Impaired growth and food utilization; pancreatic hypertrophy
Hemagglutinins	Proteins (10 000 –124 000 kDa)	Beans (castor, soy, kidney, black, yellow, jack), lentils, peas	Impaired growth and food utilization; agglutination of erythrocytes *in vitro*; mitogenic activity to cell cultures *in vitro*
Saponins	Glycosides	Soybeans, sugar beets, peanuts, spinach, asparagus	Hemolysis of erythrocytes *in vitro*
Glucosinolates	Thioglycosides	Cabbage and related species, turnips; rutabaga, radish; rapeseed; mustard	Hyperthyroidism and thyroid enlargement
Cyanogens	Cyanogenic glycosides	Peas and beans; pulses; linseed; flax; fruit kernels, cassava	HCN poisoning
Gossypol	Gossypol pigments	Cottonseed	Liver damage; hemorrhage; edema
Lathyrogens	(β-aminopropionitrile and derivatives	Chick pea; vetch	Neurolathyrism (CNS damage)
Cycasin	Methylazoxymethanol	Nuts of *Cycas* genus	Cancer of liver and other organs
Favism	Vicine and convicine (pyrimidine-β-glucosides)	Fava beans	Acute hemolytic anemia
Phytoalexins	Simple furans (ipomeamarone)	Sweet potatoes	Pulmonary edema; liver and kidney damage

Table 17.3 Plant toxicants (adapted from Pariza, 1996)—*cont'd*

Toxin	Chemical nature	Main food sources	Major toxicity symptoms
Pyrrolizidine alkaloids	Dihydropyrroles	Families *Compositae* and *Boraginaceae*; herbal teas; sassafras; black pepper	Liver and lung damage; carcinogenesis
	Benzofurans (psoralins)	Celery, parsnips	Skin photosensitivity
	Acetylkenic furans (wyerone)	Broad beans	
	Isoflavonoids (pisatin and phaseollin)	Peas, French beans	Cell lysis *in vitro*
α-amantin	Bicyclic octapeptides	*Amanita phalloides* mushrooms	Salivation; vomiting; convulsions; death
Attractyloside	Steroidal glycoside	Thistle (*Atractylis gummifera*)	Depletion of glycogen

sources of foodborne illness, is rudimentary compared to other sources, such as salmonellosis. It is often perceived by the public that certain plants are toxic and should not be consumed, but that other plants are nutritious and non-toxic. Although many herbal medicines are perceived as not causing toxic effects because they derive from natural substances, several over-the-counter herbal preparations can have severe toxicity. The toxicity of herbal medicines is covered in some excellent reviews (Schilter *et al.*, 2003; Zhou *et al.*, 2004).

Populations in poor countries may be more susceptible to plant toxicants; this derives from the necessity to rely on plants as main sources of nutrition and is often related to poor cultivation practices and quality aspects of the foods. Environmental occurrences such as drought and flood, as well as war and civil unrest, can increase the dependence on poor-quality plants as foods. For example, about 400 million people in Africa rely on the root crop cassava (*Manihot esculenta*) for subsistence, but this plant contains natural toxins and can cause severe disease. Resources are often not available to remove potent toxins from cassava. The slowly developing and chronic diseases of many plants are also often not considered as hazards in the food supply. Moreover, livestock losses from toxic plants can be substantial (Hui *et al.*, 2001a; Panter, 2005).

The ability of plants to cause toxicity and illness depends on many factors, such as the disease state of the plant (many plants produce toxicants in response to bacterial or fungal infections), its maturity, the environmental conditions and soil characteristics, and processing (Hui *et al.*, 2001a; Wittstock and Gershenzon, 2002). As with other foodborne intoxications, risk increases in the very young or elderly, in those with underlying diseases or immunodeficiency, and in those suffering from malnutrition. Most plant toxicity occurs through preformed toxicants, while certain illnesses occur through metabolism, or by postharvest treatments and food processing (Rahmann, 1999). β-Carbolines such as norharman and harman, are formed during the cooking of foods from a reaction between tryptophan and aldehyde components with subsequent oxidation. Recently, acrylamide has been considered as an important toxicant, which is formed during cooking of certain foods – particularly fried foods such as potato snacks (Friedman, 2003).

Many diverse chemical classes of toxicants occur in plants, including glucosinolates, pyrrolizidine alkaloids, amino acid analogues, psoralens, lectins and others (Hui *et al.*, 2001a; Panter, 2005). Certain food plants also may produce hormone-disrupting toxicants, and many are well known to produce allergens. Importantly, particular plant toxicants can elicit long-term or delayed disease syndromes, but, despite their importance, the compounds and the effects of the toxicants have been poorly characterized.

4.2 Pyrrolizidine alkaloids

Alkaloids comprise a diverse family of compounds produced by various fungi and plants (Smith and Culvenor, 1981; Mattocks, 1986; Rizk, 1991). Most alkaloids are basic nitrogenous compounds and characteristically contain a heterocyclic ring system. They are mainly derived from amino acids, but terpenoids also serve as precursors for certain classes. Several have beneficial actions as medicines, but some of them are toxic to humans and animals by the oral route. Alkaloids are among the most important plant toxicants for humans. They can elicit acute illnesses, but their ingestion can also cause or trigger delayed or long-term effects.

Pyrrolizidine alkaloids (PAs) are among the most important food toxicants for humans and animals (Smith and Culvenor, 1981; Mattocks, 1986; Rizk, 1991). Consumption of toxic levels of pyrrolizidine alkaloids has been responsible for massive outbreaks of diseases in livestock, causing severe economic damage. PAs have also caused serious illnesses and deaths in humans, particularly in less-developed countries. The intoxication can also be contracted by ingestion of herbal teas and other preparations (Hui *et al.*, 2001a; Panter, 2005).

PAs have been found in flowering plants, but also are predominant in several groups of food plants (Smith and Culvenor, 1981; Mattocks, 1986; Rizk, 1991). They are primarily found in the plant families *Compositae* (*Asteraceae*), *Boraginaceae* and *Leguminosae*, but also have been detected in *Apocyanacae, Ranunculacae* and *Scrophulariacae*. In certain flowering plants, PAs have been detected at high levels (~3–5 %). They are most prevalent in buds, flowers and young leaves, compared to older leaves and stems. High levels have also been found in seeds of *Crotalaria*, and toxic levels in seeds of the grain for human consumption. Oral exposure in humans usually occurs through inadvertent contamination of foodstuffs or through consumption of herbal preparations. PAs, particularly comfrey, are consumed by humans in certain herbal medicines in the US and Europe. Since the PAs probably cause chronic effects, their poisoning of humans has been underestimated and the disease syndromes poorly characterized. Consumption of seeds of *Heliotropium* have caused liver pathology in persons in Afghanistan, and 28 patients died and 67 persons became ill in India following consumption of grain contaminated with seeds of *Crotolaria*. Contamination of milk from goats fed ragwort (*Senecio jacobaea*) has also led to human intoxications. Honey produced from *S. jacobaea* and *Echium platagineum* has been shown occasionally to contain high levels of PAs. Honey produced in hives in areas of prolific ragwort growth can accumulate high levels of PAs (Hui *et al.*, 2001a; Panter, 2005).

The chemistry and structure of PAs have been extensively reviewed (cited in Hui *et al.*, 2001a; Panter, 2005). PAs are activated by cytochrome P-450s primarily of the 3A4 family. Certain cytochrome P450s convert PAs to less toxic forms, and animals such as sheep that possess this pathway suffer less severe toxic effects than animals that convert PAs to the toxic pyrrole form (such as rats and horses). The ingestion of PAs is well known to cause damage to the liver. PAs also affect other organs, particularly the lungs, but may cause chronic effects in humans and grazing animals after long-term consumption. Although PAs are well-recognized to cause impairment of human health, the control of intoxications has been difficult to achieve. The plant sources grow prolifically in many areas of the world. It has been suggested that the primary food source posing the highest risk of large-scale poisoning is cereals that become contaminated with PAs (cited in Hui *et al.*, 2001a; Panter, 2005). PA consumption in herbal medicines also could pose a considerable risk for human intoxications and long-term harmful effects.

4.3 Psoralens

Psoralens comprise a group of phototoxic fucocoumarins present in many plant families, including the *Apiaceae* (formerly *Umbelliferae*; e.g. celery, parsnips), *Leguminosae* (certain legumes), *Rutaceae* (bergamot, limes, cloves) and *Moraceae*. When photoactivated by sunlight, compounds in this family can be mutagenic. Certain psoralens, such as xanthotoxin, methoxsalen and beghapten, have been studied for their potential by photoactivation to prevent skin diseases such as psoriasis and fungal infection.

Coumarin occurs widely in vegetables such as cabbage, radish and spinach, and in plants used as flavoring agents, such as lavender and sweet woodruff. Coumarin is widely found in herbal tea. At high concentrations, coumarin causes liver damage in test animals, and its former use as a food additive has been banned by the FDA. Coumarin is also a strong anticoagulant and is frequently used in rodent baits and in certain human medicines as a blood-thinning agent (Hui *et al.*, 2001a; Panter, 2005).

4.4 Cyanogenic glycosides

Cyanogenic glycosides have been detected in more than 2000 plant species, as well as in certain bacteria, fungi, and even members of the animal kingdom (Hui *et al.*, 2001a; Reddy and Hayes, 2001; Panter, 2005). Cyanogenic glycosides can release highly toxic hydrocyanic acid (HCN). Although found in many sources, they are mainly of concern in the seeds (kernels) of stone fruits including apples, apricots, cherries, peaches, pears, plums and quinces, and in almonds, sorghum, lima beans, cassava, corn, yams, chickpeas, cashews and kirsch. Most poisonings are associated with the consumption of cassava in Africa, Asia and Latin America (Reddy and Hayes, 2001). Although more than 20 cyanogenic glycosides have been identified, poisonings are typically associated with amygdalin, dhurrin, linamarin and lotaustralin. Cyanogenic lipids have also been detected in plants (Reddy and Hayes, 2001).

Accumulation of cyanogenic glycosides in plants is enhanced during stress conditions such as drought and frost, and availability of toxic precursors is stimulated by physical disruption – such as the trampling of plants during harvest, and food processing procedures including chopping. During ingestion, enzymes such as β-glucosidase, hydroxynitrile lyase and other enzymes found in the gastrointestinal tract of humans or in their gut flora release toxic HCN. Although HCN is quite toxic, the levels that occur in foods are relatively low and reports of poisoning are infrequent. The lethal dose has been estimated to be 0.5–3.5 mg HCN per kg body weight. Poisoning by cyanogenic glycosides is much more important in livestock, generally through the consumption of large quantities of forage sorghums, arrow grass and wild cherries. Cases of cyanosis from HCN poisoning have occurred following ingestion of lima beans, cassava and bitter almonds. Due to the high level of consumption of cassava in Africa and South America, cyanogenic glycosides have presented a substantial health risk in these areas. Processing of cassava, such as soaking, boiling, sun-drying and fermentation, can eliminate most of the cyanide. Cyanogenic glycosides have also been suspected to contribute to birth defects, diabetes, endemic goiter and 'konzo' (an upper myelopathic motorneuron disease endemic to East Africa). A number of glycosides occur in various plant species, but their risk as food toxicants is uncertain.

4.5 Allyl isothiocyanates

Allyl isothiocyanates are mainly responsible for the pungent flavor of certain foods including mustard and horseradish, where they are present at 50–150 ppm (Coulombe, 2000; Hui *et al.*, 2001a; Panter, 2005). They are also present at much lower levels in broccoli, cabbage and cassava. Isothiocyanates occur in cruciferous vegetables as glucosinolate conjugates, and cyanide can be released during digestion. Isothiocyanates have been implicated in causing hyperthyroidsm (goiter), particularly in geographical regions like India and Africa, where the consumption of minimally processed foods occurs together with iodine deficiency (Coulombe, 2000). As with cyanogenic glycosides, simple processing (chopping, rinsing, milling) can reduce the level of isothiocyanates.

4.6 Glycoalkaloids

Glycoalkaloids occur mainly in potatoes (*Solanum tuberosum*), particularly in green potatoes. They are also found in low levels in tomatoes. High levels of these glycoalkaloids can cause gastroenteritis, and can also impart a bitter taste to potatoes. They are of concern in new varieties of potatoes, including those derived by biotechnology. When injured, exposed to light or sprouted, potatoes can accumulate the glycoalkaloids α-solanine and α-chaconine. These compounds are potent inhibitors of acetylcholinesterase. Poisoning symptoms include gastric pain, weakness, nausea, vomiting and difficulty in breathing. Poisonings have occurred in animals fed damaged potatoes, greens or trim (Coulombe, 2000; Hui *et al.*, 2001a; Panter, 2005).

4.7 Hydrazines and other toxins in edible mushrooms

Commonly cultivated mushrooms, including *Agaricus bisporus*, shitake (*Cortinellus shitake*) and the false morel (*Gyromitra esculenta*), contain substantial quantities of hydrazines (up to 500 ppm). Certain hydrazines have been implicated as liver toxins and animal carcinogens. Shitake and false morel mushrooms also contain substantial levels of agaritine and gyromitrin, which can be transformed to carcinogens. Many wild mushroom species are known to contain other toxicants that can cause a variety of foodborne syndromes (Lovenberg, 1973; Reddy and Hayes, 2001).

4.8 Caffeic acid and chlorogenic acid

Caffeic acid, quinic acid derivative and chlorogenic acid phenolic compounds, are prevalent in a wide variety of plants and vegetables. Chlorogenic acid is hydrolyzed to caffeic acid and quinic acid in the gastrointestinal tract. Caffeic acid is metabolically transformed in humans to o-methylated derivatives, including ferulic, dihyroferulic and vanillic acids, as well as meta-hydroxyphenyl derivatives, which are excreted in the urine (Coulombe, 2000; Hui *et al.*, 2001a; Panter, 2005). Caffeic acid and its derivatives are present at relatively high levels in certain spices and seasonings (thyme, basil, dill, anise, caraway, rosemary, sage, tarragon and marjoram), vegetables (lettuce, potatoes, radishes and celery), and fruits (grapes, berries and tomatoes). Coffee also contains significant quantities of chlorogenic acids. Caffeic acid is an inhibitor of 5-lipoxygenase, which is a central enzyme for the biosynthesis of eicosanoids including leukotrienes and thromboxanes. At high doses caffeic acid and chlorogenic acid have also been demonstrated to be carcinogenic, and their presence in the diet at significant levels may also enhance carcinogenesis by other compounds.

4.9 Toxicants in spices

Many spices contain a vast and interesting array of secondary metabolite compounds, some of which can be toxicants (Coulombe, 2000). Examples of potentially toxic spice components are saffrole, myristicin, β-asarone and isosafrole, which have analogous structures to certain carcinogens such as alkyl benzenes. Capsaicin is a prominent flavoring agent in red and yellow chili peppers, and can cause irritation to the eyes and mucous membranes. It also affects neurotransmission and causes depletion of substance P, which is involved in pain mediation. Other components of spices that can have toxic effects include glycyrrhyzin, a saponin-like glycoside present in licorice. Licorice root extract was used as an expectorant in ancient times, and has been suspected of causing hypertension and other metabolic disorders. Certain other classes of compounds, such as terpenoids (δ-limonene), can be toxic to animals at high levels. The potential of spices for causing adverse chronic effects is evident, but it would be expected that prudent use would not affect normal human health.

4.10 Biologically active amines

Certain vaso- and psychoactive amines, including tyramine, octopamine, dopamine, epinephrine, norepinephrine, histamine, serotonin and others, are present in foods, particularly fermented products such as cheeses, yeast products, beer, wine and pickled herring. They also occur in coffee, chicken liver, broad beans, chocolate, pineapple, banana, plantain and avocado (Reddy and Hayes, 2001). Amines with vasoconstrictive properties are present in a variety of foods (Lovenberg, 1973; Reddy and Hayes, 2001). Pressor amines including tyramine and tryptamine, as well as certain related compounds (serotonin, adrenaline, noradrenaline, dopamine), affect central nervous system activity. Low levels of these compounds (10 mg) can cause severe hypertensive crisis in individuals treated with monoamine oxidase (MAO) inhibitors for depression and other mood disorders (Lovenberg, 1973; Reddy and Hayes, 2001).

4.11 Protease inhibitors

Plants contain inhibitors of difference classes of enzymes, including proteases, lipases, amylases and others (Kassell, 1970; Liener and Kadade, 1980). However, protease inhibitors are responsible for most toxic effects. The main protease inhibitor from soybean, commonly called Kunitz inhibitor, is capable of inhibiting trypsin and certain other proteases (Hui *et al.*, 2001a; Panter, 2005). Ingestion of raw soybean decreases digestion of protein, causing increased secretion of pancreatic enzymes and reduction in body mass gain. Other protease inhibitors from food sources, including beans, peas, egg white, cereal grains, alfalfa and potatoes, have also shown toxic effects in humans and animals. The main effects in animals include pancreatic hypertrophy, adenomas, and nodular hyperplasia associated with growth depression (Hui *et al.*, 2001a; Panter, 2005).

4.12 Lectins (phytohemagglutinins) and glutens

Lectins are high molecular weight (100–150 kDa), heat-labile proteins that have been detected in numerous edible plant species, particularly those belonging to the leguminoseae (beans, peas, etc.). Lectins have also been found in food animals, including crustaceans, mollusks, fish and even mammals (Reddy and Hayes, 2001). Their consumption can result in growth reduction in animals and humans, presumably due to their effects on the intestinal mucosa and disruption of nutrient transport. Necrosis of intestinal epithelia has also been observed. Systemic lectin exposure has caused fatalities due to liver damage. One of the most toxic lectins is ricin, present in the castor bean (see section 4.18 below). Ingestion can cause severe necrosis and eventually death from organ damage.

Gluten enteropathy (celiac sprue) has been associated with the consumption of wheat-germ agglutinin contaminating gluten in cereal foods. A strong genetic component contributes to the susceptibility to celiac sprue, but diet and probably environmental factors also participate in the etiology. Necrosis and loss of jejunal villi are characteristic of celiac sprue. The syndrome in children has been associated with vomiting, a bloated abdomen, behavioral changes (including irritability and restlessness), speech impairment, impaired growth, chronic diarrhea, and myopathy characterized by weakness and fatigue. Celiac sprue has been estimated to have a prevalence of 0.1 % in certain areas of Europe (Troncone *et al.*, 1996). The majority of patients respond within weeks to a gluten-free diet. Susceptible individuals are advised to avoid food products prepared from wheat, rye, barley and oats. Interestingly, a gluten-free diet in certain children has been associated with a reduction in neuropsychological phenomena such as autism (Dohan, 1976). The mechanism of the neuropsychological manifestations is unclear, but one possibility is that neuroactive peptides produced during the digestion of food proteins cross the gut barriers that have been necrotized. This hypothesis has previously been proposed as contributing to schizophrenia (Dohan, 1976), and it has been reported that a gluten-free diet may reduce the symptoms of schizophrenia. However, increased permeability of the small intestine was not observed in schizophrenic patients (Lambert *et al.*, 1989). Also, a link between celiac disease and childhood autism could not be demonstrated (Black *et al.*, 2002). Nonetheless, these findings raise the intriguing possibility of a connection between diet, gut flora, and behavioral diseases.

4.13 Phytates

Phytates (hexaphosphate esters of myo-inositol) bind di- and tri-valent metals and can lead to mineral deficiencies in human and animal diets. Such deficiencies primarily occur in developing countries where cereals are consumed as major or exclusive source of proteins. Treatment of foods and feeds with phytase results in phosphate utilization and reduction of environmental pollution from phosphates. The sequestration of metals by phytates can be alleviated by supplementing diets with minerals and vitamin D (Kotsonis *et al.*, 2001; Reddy and Hayes, 2001).

4.14 Estrogens

More than 200 species of plants contain estrogenic isoflavonoids (e.g. genistein) or their glycosides (e.g. genistin) (Reddy and Hayes, 2001). Coumestans and lignans are other important sources of plant estrogens. Phytoestrogens have been known to cause infertility in animals grazing in forages including subterranean clover and alfalfa. Genistin in soybeans has been associated with most human toxic effects and has reportedly interfered with steroid metabolism in infants fed soy-based formulas. In women, changes in menstrual cycles have been reported to occur owing to soy consumption. It is currently unclear whether consumption of phytoestrogens may lead to longer-term effects in humans (Reddy and Hayes, 2001).

4.15 Canavanine and other amino acid analogues

Canavanine is an arginine analogue (2-amino-4-(guanidinooxy)-butyric acid) that is widespread in seeds of *Leguminoseae*. Alfalfa sprouts (*Medicago sativa*) and Jackbean (*Canavalia ensiformis*) contain high levels (up to 15 000 ppm) of canavanine (Coulombe, 2000; Hui *et al.*, 2001a; Panter, 2005). Since canavanine is an analogue of arginine, it can be incorporated into cellular proteins and partially disrupt function. It is suspected of causing autoimmune disorders such as lupus erythematosus (Coulombe, 2000).

Amino acid derivatives that disrupt essential metabolic processes, including neurotransmission and bone development in humans and animals, are formed by various plants. For example, consumption of β-*N*-oxalyamino-L-alanine (BOAA), which is found in legumes such as grass-pea, can trigger lathyrism, a form of spastic paraparesis characterized by muscle weakness, increased muscle tone, and hyper-reflexia in the lower limbs (Spencer and Berman, 2003). Long-term consumption can lead to permanent inability to move the legs. Lathyrism is primarily a problem in regions such as areas of India, where edible material other than grass pea is scarce. The amino acid analogue β-(γ-L-glutamyl)-aminopropionitrile (BAPN) inhibits lysyl oxidase, which is important in collagen and bone formation. Feeding of BAPN to rodents results in joint and skeletal deformities.

Other disease syndromes have been observed in humans and animals that consume plants containing non-protein amino acids. Vomiting sickness, hypoglycemia and hepatic encephalopathy have been observed in people in West Africa who consume the fruit from the ackee tree, which produces hypoglycin (γ-glutamyl dipeptide). The ackee tree was imported to the Caribbean islands, and ingestion of unripe fruit has caused severe disease in poorly nourished people there.

4.16 Miscellaneous flavonoids

Flavonoids are widespread in plant-derived foods, including fruits and fruit juices, certain vegetables, tea, cocoa, red wine, dill, soybeans and others (Coulombe, 2000). These have been suspected as being carcinogenic. Rutin can be metabolized by intestinal bacteria to form quercetin, which has anticarcinogenic properties.

4.17 Other plant toxicants

Other plant compounds considered to be intoxicants in human and animal diets, including saponins (steroidal compounds and certain terpenoids), lipids (erucic acid, phytanic acid, cyclopropene fatty acids), anti-vitamin factors and mushroom compounds (psilocybin, coprine and others), have also been suspected to cause growth depression, psychoabnormality or toxic effects, and have been discussed in more comprehensive reviews (see, for example, Reddy and Hayes, 2001). The large number of plant-derived compounds showing potential adverse effects on humans and animals illustrates the paucity of knowledge in this area and the need for further research to evaluate toxicity and other effects.

4.18 Plant toxicants of potential risk in bioterrorism

There has been a heightened awareness of the potential for toxins to be used as bioterrorist agents in foods or by other means of dispersal (Khan *et al.*, 2001; Franz and Zajtchuk, 2002). Among the various pathogens and toxins, certain plant toxins have been considered as significant threat agents. In particular, ricin and other closely related plant toxins such as abrin have been considered as poisoning agents. Ricin is a potent protein cytotoxin that is derived from castor beans. Ricin can be aerosolized, and poisoning can occur through inhalation or ingestion. The symptoms of ricin poisoning depend on the route of administration and the quantity ingested. Following oral ingestion, initial symptoms usually occur with 6 hours. Gastrointestinal symptoms include severe abdominal pain, vomiting, diarrhea, and ulcerations and hemorrhages of the gastric and small-intestinal mucosa (as detected by endoscopy). The toxin also affects liver function, and tests are abnormal. Hallucinations, seizures, and blood in the urine are also characteristic signs, and death has been documented. People who know or suspect that they have been poisoned by ricin should seek immediate medical care. Currently there is no antidote other than passive immunotherapy, but improved vaccines and therapeutics are being developed.

5 Other foodborne toxicants

5.1 Insect- and mite-derived toxins in foods

Insect infestation of foods is clearly undesirable, and certain insects can produce chemical toxicants that can result in foodborne illnesses. Flour beetles (*Tribolium* spp.) produce benzoquinones that are carcinogenic in animals and possibly in humans (Wirtz *et al.*, 1978; Taylor and Hefle, 2002). No acute cases of human illness have been reported from flours infested with *Tribolium* spp., but the long-term effects of consumption are unknown. Human illnesses have been reported from foods contaminated with dust mites, which produce allergens that can cause allergic symptoms in humans on ingestion (Taylor and Hefle, 2002).

5.2 Intoxication of unknown etiology – bovine paraplegic syndrome

Bovine paraplegic syndrome (BPS) was first described in Venezuela in the mid-1950s, and has recently spread alarmingly in the cattle-growing areas of Venezuela and Paraguay (Sevcik *et al.*, 1993). Since the animals affected are generally in seemingly good condition, the disease is also called *enfermedad de las bonitas* ('disease of the pretty ones') by the farmers. It is mainly pregnant or lactating cows that are affected, and the mortality rate has been estimated to range from 5 % to 25 % in the animals at risk. The disease is characterized by ventral or sternal decubitis, and animals are unable to stand when stimulated. The diagnosis is dependent on eliminating other possible causes, including botulism, paralytic rabies, and blood parasites. All cows die within a few days, and there is no known treatment. The clinical etiology has not been established. It has been proposed that ruminal bacteria produced a heat- and acid-stable toxin (Sevcik *et al.*, 1993). The toxin blocked the sodium current in giant squid axons. Subsequent research has suggested that saxitoxin is produced by bacteria (the *Enterobacter asburiae, E. cloacae* and *Klebsiella pneumoniae* in the rumen) and that the characteristic sodium channel-blocking toxin is responsible for BPS (Sevcik *et al.*, 2003). However, other investigators have proposed that BPS is caused by toxins produced by *Clostridium perfringens* type D (Muller *et al.*, 1998) and *Lactobacillus vitulinum*, as well as several Gram-negative rumen isolates (Domínguez-Bello *et al.*, 1993). A possible role of bovine immunodeficiency virus has been indicated by immunochemical, virological and seroprevalence studies (Walder *et al.*, 1995). Thus the etiology remains unclear, although saxitoxin has also been isolated from bacteria and cyanobacteria (Mahmood and Carmichael, 1986; Carmichael *et al.*, 1997; Gallacher and Smith, 1999). The production of saxitoxin during cynanobacterial blooms has been implicated in the death of animals drinking water from ponds undergoing blooms (Schantz and Johnson, 1992).

5.3 Intoxications from genetically modified foods

Plant and animal genomes have long been modified by traditional breeding practices, but, with the development of biotechnology, specific new genes have been incorporated into foods and feed substances. Depending on the commodity and country of production, substantial percentages of crops and animals contain genes introduced by biotechnolology (Stewart, 2003; Toke, 2004). For example, ~25 % of corn cultivated in 1999 in the US contained an anti-pest gene from *Bacillus thuringiensis* (Stewart, 2003). Considerable controversy and debate has been voiced regarding the potential for toxicants generated by the insertion of genes into plants using biotechnology.

The safety of genetically modified crops and other foods has been evaluated by a number of methods and regulated under international laws. It is beyond the scope of this chapter to discuss the safety assessments and regulations, and the reader is referred to authoritative reviews (Kotsonis *et al.*, 2001; Taylor and Hefle, 2002).

5.4 Dietary supplements

The toxicology and safety of dietary supplements has also been a subject of considerable controversy. Although many consumers utilize dietary supplements, this class of foods has unique safety qualifications. Aspects of supplements' safety and toxicology are covered in recent authoritative reviews (Kotsonis *et al.*, 2001; Schilter *et al.*, 2003).

6 Laboratory practices, conclusions, and perspectives

Laboratory considerations regarding the handling of natural toxins have been reviewed and regulations implemented (Wannemacher, 1989; CDC, 1999; Fleming and Hunt, 2000), and these considerations have assumed increased importance with the implementation of Select Agent Regulations in the US Patriot Act. Intoxications by natural toxicants are illnesses resulting from exposure (usually by the oral route) to toxins produced by microorganisms, plants and, occasionally, animals. Comprehensive safety programs for laboratories working with toxins have been described (Wannemacher, 1989; CDC, 1999; Fleming and Hunt, 2000; Malizio *et al.*, 2000).

The provision of a nutritious and safe food supply is an essential goal of society to ensure the health and survival of humankind throughout the world. Epidemiological evidence has indicated that certain food-associated bacteria and natural toxicants are the major causes of illness and mortality transmitted by foods. Although progress has been made in the identification and understanding of the toxicology of miscellaneous intoxicants present in microorganisms, plants and animals, the body of knowledge in this research area is meager compared to our understanding of foodborne diseases caused by microbial and fungal pathogens. Natural intoxicants comprise a vast array of compounds with a myriad of structures and modes of action. Unlike many microbial pathogens, their involvement in animal and human foodborne disease cannot be determined by classical methods of identifying disease, such as solving of the famous Koch's postulates. Since natural intoxicants do not reproduce on their own, but depend on the host for production, they are often present in minute quantities in foodstuffs. Identification often depends on association of a plant or animal with disease, followed by sophisticated chemical tests to identify candidate compounds. Subsequently, *in vitro* and animal models are employed to evaluate toxicity. For many of these compounds minimum acceptable levels have not been established, and in certain cases it will not be possible to establish these. Since natural foodborne toxicants can be as deleterious as synthetic toxicants or even microbial pathogens, US and worldwide resources should be allocated to evaluate more thoroughly their impact on animal and human health. Natural foodborne intoxicants have an enormous medical and economic impact on societies, particularly in developing countries. Further surveillance, together with the identification and evaluation of health effects, could lead to a quantitative risk assessment. The education of producers, processors and consumers would have a tremendous benefit in improving the world's food supply.

Although microbial food safety is a major public health issue of increasing importance, many public health authorities in certain countries throughout the world do not fully appreciate its importance for human health and economic development (WHO, 1997). National and international programs to enhance food safety are considered a low priority in many countries, partly because resources are not available to develop food safety programs. Many countries have not developed legislation and the public health infrastructure to control foodborne disease. Although consumers are integral to the prevention of foodborne disease, many are unaware of their importance in enhancing food safety and do not receive adequate education to prevent illnesses within the home or at community events. As emphasized by the WHO (1997), strategies for decreasing the incidence of foodborne disease, enhancing human well-being and facilitating technological developments will require a shared responsibility among governments, industry, scholarly institutions and consumers to accomplish these goals.

Bibliography

Ahmed, F. E. (ed.) (1991). *Seafood Safety*. Committee on Evaluation of the Safety of Fishery Products. Food and Nutrition Board, Institute of Medicine, National Academy of Sciences Press, Washington, DC.

Anderson, D. M. (2000). Harmful algal web page, available at http://www.redtide.whoi.edu/hab/. National Office for Marine Biotoxins and harmful algal blooms. Woods Hole Oceanographic Institution, Woods Hole, MA.

Backer, L. C., H. Schurz-Rogers, L. E. Fleming *et al.* (2005). Marine phycotoxins in seafood. In: W. M. Dabrowski and Z. E. Sikorski (eds), *Toxins in Food*, pp. 155–189. CRC Press, Boca Raton, FL.

Bean, N. H. and P. M. Griffin (1990). Foodborne disease outbreaks in the United States, 1973–1987: pathogens, vehicles, and trends. *J. Food Prot.* **53**, 804–817.

Bean, N. H., J. S. Goulding, C. Lao and F. J. Angulo (1996). Surveillance for foodborne-disease outbreaks – United States, 1988–1992. *Morbid. Mortal. Weekly Rep.* **45**, 1–66.

Berry, J. P., K. S. Reece, K. S. Rein *et al.* (2002). Are *Pfiesteria* species toxigenic? Evidence against production of by *Pfiesteria shumwayae. Proc. Natl Acad. Sci. USA* **99**, 10 970–10 975.

Black, C., J. A. Kaye and H. Jick (2002). Relation of childhood gastrointestinal disorders to autism: nested case-control study using data from the UK General Practice Research Database. *Br. Med.* J. **325**, 419–421.

Borzelleca, J. F. (2001). The art, science, and the seduction of toxicology: an evolutionary development. In: A. W. Hayes (ed.), *Principles and Methods of Toxicology*, 4th edn, pp. 1–21. Taylor & Francis, Philadelphia, PA.

Brock, T. D. (1961). *Milestones in Microbiology*. Prentice Hall, Englewood Cliffs, NJ.

Brown, L. R., C. Flavin and H. French (eds) (1999). *State of the World.* The Worldwatch Institute, W. W. Norton & Co., New York, NY.

Bryan, F. L. (1979). Infections and intoxications caused by other bacteria. In: H. Riemann and F. L. Bryan (eds), *Foodborne Infections and Intoxications*, 2nd edn, pp. 211–297. Academic Press, Orlando, FL.

Bryan, F. L. (1982). *Diseases Transmitted by Foods. A Classification and Summary*, 2nd edn. Centers for Disease Control, Atlanta, GA.

Bryan, F. L., J. J. Guzewich and E. C. D. Todd (1997). Surveillance of foodborne disease. II. Summary and presentation of descriptive data and epidemiologic patterns; their value and limitations. *J. Food Prot.* **60**, 567–578.

Carlson, D. J. and F. L. Shapiro (1970). Methemoglobinemia from well water nitrates: a complication of home dialysis. *Ann. Intern. Med.* **74**, 757–759.

Carmichael, W. W., W. R. Evans, Q. Q. Yin, P. Bell and E. Moczydlowski (1997). Evidence for paralytic shellfish poisons in the freshwater cyanobacterium *Lyngbya wollei* (Farlow ex Gomont) comb. nov. *Appl. Environ. Microbiol.* **63**, 3104–3410.

CAST (Council for Agricultural Science and Technology) (1994). *Foodborne Pathogens: Risks and Consequences.* Task Force Report No. 122. Council for Agricultural Science and Technology, Ames, IA.

CDC (Centers for Disease Control and Prevention) (1999). *Biosafety in Microbiological and Biomedical Laboratories (BMBL)*, 4th edn. US Department of Health and Human Services, US. Government Printing Office, Washington, DC.

CDC (Centers for Disease Control and Prevention) (2000). Surveillance of foodborne disease outbreaks, 1993–1997. *Morbid. Mortal. Weekly Rep.* **49** (SS-1), 1–82.

CDC (Centers for Disease Control and Prevention) (2001). Diagnosis and management of foodborne illnesses: a primer for physicians. *Morbid. Mortal. Weekly Rep.* **50** (RR-2), 1–69.

Chorus, I. (ed.) (2001). *Cyanotoxins: Occurence, Causes, Consequences*. Springer Verlag, New York, NY.

Collier, D. N. and W. A. Burke (2002). *Pfiesteria* complex organisms and human illness. *Southern Med.* J. **95**, 720–726.

Concon, J. M. (1988). *Food Toxicology. Part A. Principles and Concepts*. Marcel Dekker, New York, NY.

Coulombe, R. A. Jr (2000). Natural toxins and chempreventives in plants. In: W. Helferich and C. K. Winter (eds), *Food Toxicology*, pp. 137–161. CRC Press, Boca Raton, FL.

Dabrowski, W. M. and Z. E. Sikorski (2005). *Toxins in Food*. CRC Press, Boca Raton, FL.

Danner, R. L. and C. Natanson (1995). Endotoxin: a mediator of and potential therapeutic target for septic shock. In: J. Moss, B. Iglewski, M. Vaughan and A. T. Tu (eds), *Bacterial Toxins and Virulence Factors in Disease. Handbook of Natural Toxins*, Vol. 8, pp. 589–617. Marcel Dekker, New York, NY.

Dinman, B. D. (1972). 'Non-concept' of 'no-threshold': chemicals in the environment. Stochastic determination impoose a lower limit on the dose–response relationship between cells and chemicals. *Science* **175**, 495–497.

DiPasquale, L. C. and A. W. Hayes (2001). Acute toxicity and eye irritancy. In: A. W. Hayes (ed.), *Principles and Methods of Toxicology*, 4th edn. Taylor & Francis, London.

D'Mello, J. P. F. (ed.) (2003). *Food Safety. Contaminants and Toxins*, CABI Publishing, Wallingford.

Dohan, F. C. (1976). Is celiac disease a clue to the pathogenesis of schizophrenia? *Mental Hygiene* **53**, 525–539.

Domíngues-Bello, M. G., M. Lovera, C. Sevcik and J. C. Brito (1993). Characterization of ruminal bacteria producing a toxin associated with bovine paraplegic syndrome. *Toxicon* **31**, 1595–1600.

Downes, F. P. and K. Ito (eds) (2001). *Compendium of Methods for the Examination of Foods*, 45th edn. American Public Health Association, Washington, DC.

Eaton, D. L., and C. D. Klaassen (2001). Principles of toxicology. In: A. W. Hayes (ed.), *Principles and Methods of Toxicology*, 4th edn, pp. 11–34. Taylor & Francis, London.

Engeseth, N. J. (2001). Safety of genetically engineered foods. In: W. Helferich and C. K. Winter (eds), *Food Toxicology*, pp. 77–92. CRC Press, Boca Raton, FL.

Evans, A. S. and P. S. Brachman (eds) (1992). *Bacterial Infections of Humans: Epidemiology and Control*, 2nd edn. Plenum Medical Book Company, New York, NY.

Falconer, I. R. (ed.) (1993). *Algal Toxins in Seafood and Drinking Water*. Academic Press Ltd, London.

Fassett, D. W. (1973). Nitrates and nitrites. In: Committee on Food Protection (eds), *Toxicants Occurring Naturally in Foods*, 2nd edn, pp. 7–25. National Academy of Sciences Press, Washington, DC.

FDA (Food and Drug Administration) (1995). *Food and Drug Administration/ Bacteriological Analytical Manual* (FDA/BAM), 8th edn. AOAC International, Gaithersburg, MD. Available at http://www.cfsan.fda.gov/~ebam/bam-toc.html.

Finegold, S.M., Y. Song and C. Liu (2002a). Taxonomy – general comments and update on taxonomy of clostridia and anaerobic cocci. *Anaerobe* **8**, 283–285.

Finegold, S.M., D. Molitoris, Y. Song *et al.* (2002b). Gastrointestinal studies in late-onset autism. *Clin. Infect. Dis.* **35** (Suppl. 1), S6–S16.

Fleming, D. O., and D. L. Hunt (eds) (2000). *Biological Safety. Principles and Practices*, 3rd edn. ASM Press, Washington, DC.

Fleming, L. E., D. Katz, J. A. Bean and R. Hammond (2001). Epidemiology of seafood poisoning. In: Y. H. Hui, D. Kitts and P. S. Stanfield (eds), *Foodborne Disease Handbook*, 2nd edn, Vol. 4, *Seafood and Environmental Toxins*, pp. 287–310. Marcel Dekker, New York, NY.

Franz, D. R. and R. Zajtchuk (2002). Biological terrorism: understanding the threat, preparation, and medical response. *Disease-A-Month* **48**, 493–564.

Friedman, M. (2003). Chemistry, biochemistry, and safety of acrylamide. A review. *J. Agric. Food Chem.* **51**, 4504–4526.

Gallacher, S. and E. A. Smith (1999). Bacteria and paralytic shellfish toxins. *Protist* **150**, 245–255.

Garcia, R. A., J. H. Hotchkiss and K. H. Steinkraus (1999). The effect of lipids on bongkreic (Bongkrek) acid toxin production by *Burkholderia cocovenenans* in coconut media. *Food Addit. Contam.* **16**, 63–69.

Glasgow, H. B. Jr, J. M. Burkholder, D. E. Schmechel *et al.* (1995). Insiduous effects of a toxic estuarine dinoflagellate on fish survival and human health. *J. Toxicol. Environ. Health* **46**, 501–522.

Granum, P. E. and T. C. Baird-Parker (2000). *Bacillus* species. In: B. M. Lund, T. C. Baird-Parker and G. W. Gould (eds), *The Microbiological Safety and Quality of Food*, Vol. 2, pp. 1029–1039. Aspen Publishers, Gaithersburg, MD.

Halstead, B. W. (1967). *Poisonous and Venomous Marine Animals of the World.* US Government Printing Office, Washington, DC.

Hayes, A. W. (ed.) (2001). *Principles and Methods of Toxicology*, 4th edn. Taylor & Francis, London.

Henderson, W. R. and N. H. Raskin (1972). Hot dog headache: individual susceptibility to nitrite. *Lancet* **2**, 1162–1163.

Hu, W. J., G. S. Zhang, F. S. Chu *et al.* (1984). Purification and partial characterization of flavotoxin A. *Appl. Environ. Microbiol.* **48**, 690–693.

Hui, Y. H., R. A. Smith and D. G. Spoerke Jr (eds) (2001a). *Foodborne Disease Handbook*, 2nd edn, Vol. 3. *Plant Toxicants*. Marcel Dekker, New York, NY.

Hui, Y. H., D. Kitts and P. S. Stanfield (eds) (2001b). *Foodborne Disease Handbook*, 2nd edn, Vol. 4, *Seafood and Environmental Toxins*. Marcel Dekker, New York, NY.

Hutchinson, G. E. (1964). The influence of the environment. *Proc. Natl Acad. Sci. USA* **51**, 930.

Hutt, P. B. and P. B. Hutt II (1984). A history of government regulation and misbranding of food. *Food Drug Cosmet. Law* J. **39**, 2–73.

Jensen, G. L. and K. J. Greenlees (1997). Public health issues in aquaculture. *Rev. Sci. Tech. Off. Int. Epiz*. **16**, 641–651.

Jiao, Z., Y. Kawamura, N. Mishima *et al.* (2003). Need to differentiate lethal toxin-producing strains of *Burkholderia gladioli*, which cause severe food poisoning: description of *B. gladioli* pathovar cocovenenans and an emended description of *B. gladioli. Microbiol. Immunol*. **47**, 915–925.

Johnson, E. A. (1999). Clostridial toxins as therapeutic agents: benefits of nature's most toxic proteins. *Ann. Rev. Microbiol*. **53**, 551–575.

Johnson, E. A. (2003). Bacterial pathogens and toxins in foodborne disease. In: D'Mello, J. P. F. (ed.), *Food Safety. Contaminants and Toxins*, pp. 25–45. CABI Publishing, Wallingford.

Johnson, E. A. and M. W. Pariza (1989). Microbiological principles for the safety of foods. In: R. D. Middlekauf and P. Shubik (eds), *International Food Regulation Handbook. Policy, Science, Law*, pp. 135–174. Marcel Dekker, New York, NY.

Johnson, E. A. and E. J. Schantz (2002). Seafood toxins. In: D. O. Cliver and H. P. Riemann (eds), *Foodborne Diseases*, 2nd edn, pp. 211–230. Academic Press, San Diego, CA.

Kao, C. Y. (1993). Paralytic shellfish poisoning. In: I. R. Falconer (ed.), *Algal Toxins in Seafood and Drinking Water*, pp. 75–86. Academic Press Ltd, London.

Kao, C. Y. and S. R. Levinson (eds) (1986). Tetrodotoxin, saxitoxin, and the molecular biology of the sodium channel. *Annals of the New York Academy of Sciences*, Vol. 479. New York Academy of Sciences, New York, NY.

Kassell, B. (1970). Inhibitors of proteolytic enzymes. *Methods Enzymol*. **19**, 839–906.

Keating, J. P., M. E. Lell, A. W. Straus *et al.* (1973). Infantile methemoglobinemia caused by carrot juice. *N. Engl. J. Med*. **288**, 824–826.

Khan A. S., D. Swerdlow and D. D. Juranek (2001). Precautions against biological and chemical terrorism directed at food and water supplies. *Publ. Health Rep*. **116**, 3–14.

Klaassen, C. D. (ed.) (2001). *Casarett & Doull's Toxicology: The Basic Science of Poisons*, 6th edn. McGraw Hill, New York, NY.

Kotsonis, F. N., G. A. Burdock and W. G. Flamm (2001). *Food Toxicology*. In: C. D. Klaassen (ed.), *Casarett and Doull's Toxicology: The Basic Science of Poisons*, 6th edn, pp. 1049–1088. McGraw-Hill, New York, NY.

Lamanna, C. (1959). The most poisonous poison. *Science* **130**, 763–772.

Lambert, M, T. I. Bjarnason, J. Connelly *et al.* (1989). Small intestine permeability in schizophrenia. *Br. J. Psychiatr*. **155**, 619–622.

Liener, I. E. (ed.) (1980). *Toxic Constituents of Plant Foodstuffs*, 2nd edn. Academic Press, New York, NY.

Liener, I. E. and M. L. Kadade (1980). Protease inhibitors. In: I. E. Liener (ed.), *Toxic Constituents of Plant Foodstuffs*, 2nd edn, pp. 7–71. Academic Press, New York, NY.

Lipp, E. K. and J. B. Rose (1997). The role of seafood in foodborne diseases in the United States of America. *Rev. Sci. Tech. Int. Epiz*. **16**, 620–640.

Llewellyn, L. E. (2001). Shellfish chemical poisoning. In: Y. H. Hui, D. Kitts and P. S. Stanfield (eds), *Foodborne Disease Handbook*, Vol. 4, *Seafood and Environmental Toxins*, pp. 77–108. Marcel Dekker, New York, NY.

Lovenberg, W. (1973). Some vaso- and psychoactive substances in food. In: Committee on Food Protection (eds), *Toxicants Occurring Naturally in Foods*, 2nd edn, pp. 170–178. National Academy of Sciences Press, Washington, DC.

Lund, B. M., T. C. Baird-Parker and G. W. Gould (eds) (2000). *The Microbiological Safety and Quality of Food*, Vol. 2. Aspen Publishers, Gaithersburg, MD.

Mahmood, N. A. and W. W. Carmichael (1986). Paralytic shellfish poisons produced by the freshwater cyanobacterium *Anabena flos-aquae* NRC-525-17. *Toxicon* **25**, 1221–1227.

Malizio, C. J., M. C. Goodnough and E. A. Johnson (2000). Purification of botulinum type A neurotoxin. *Methods Mol. Biol.* **145**, 27–39.

Mattocks, A. R. (1986). *Chemistry and Toxicology of Pyrrolizidine Alkaloids*. Academic Press, London.

McGinn, A. P. (1999). Charting a new course for the oceans. In: L. R. Brown, C. Flavin and H. French (eds), *State of the World*, Millennial edn, pp. 78–95. The Worldwatch Institute, W. W. Norton & Company, New York, NY.

McNeil, W. H. (1976). *Plagues and Peoples*. Doubleday Publishing, Garden City, NY.

Mead, P. S., L. Slutsker, V. Dietz *et al.* (1999). Food-related illness and death in the United States. *Emerg. Infect. Dis.* **5**, 607–625.

Meng, Z., Z. Li, J. Jin *et al.* (1988). Studies on fermented corn flour poisoning in rural areas of China. 1. Epidemiology, clinical manifestation, and pathology. *Biomed. Environ. Sci.* **1**, 101–104.

Mensinga, T. T., G. J. Speijers and J. Meulenbelt (2003). Health implications of exposure to environmental nitrogenous compounds. *Toxicol. Rev.* **22**, 41–51.

Meyer, K. F., H. Sommer and P. Schoenholz (1928). Mussel poisoning. *J. Prevent. Med.* **2**, 195–216.

Middlekauf, R. D. and P. Shubik (eds) (1989). *International Food Regulation Handbook. Policy, Science, Law*. Marcel Dekker, New York, NY.

Miller, S. A. and M. R. Taylor (1989). Historical developments of food regulation. In: R. R. Middelkauf and R. Shubik (eds), *International Food Regulation Handbook. Policy, Science, Law*, pp. 7–25. Marcel Dekker, New York, NY.

Morris, J. G. Jr (1999). *Pfiesteria*, 'the cell from hell,' and other toxic algal nightmares. *Clin. Infect. Dis.* **28**, 1191–1198.

Morris, J. G. Jr (2001). Human health effects and *Pfiesteria* exposure: a synthesis of available clinical data. *Environ. Health Perspect.* **109** (Suppl. 5), 787–790.

Mosher, H. S., F. A. Fuhrmann, H. D. Buchwald and H. G. Fischer (1964). Tarichatoxin-tetrodotoxin: a potent neurotoxin. *Science* **144**, 1100–1110.

Muller, W., B. A. Zucker, A. R. Sanches *et al.* (1998). Short communication: bovine paraplegic syndrome (Mal de Aquidaban) in Paraguay caused by *C. perfringens* toxovar D. *Berl. Münch. Tierärztl. Wochenschr.* **111**, 214–216.

National Academy of Sciences, Committee on Food Protection (1973). *Toxicants Occurring Naturally in Foods*, 2nd edn. National Academy of Sciences, Washington, DC.

National Research Council (1985). *An Evaluation of the Role of Microbiological Criteria for Foods and Ingredients.* National Academy Press, Washington, DC.

Norton, S. (2001). Toxic effect of plants. In: C. D. Klaassen (ed.), *Casarett and Doull's Toxicology. The Basic Science of Poisons*, 6th edn, pp. 965–976. McGraw-Hill, New York, NY.

Okada, K. and M. Niwa (1998). Marine toxins implicated in food poisoning. *J. Toxicol. Toxin Rev.* **17**, 373–384.

Panter, K. E. (2005). Natural toxins of plant origin. In: W. M. Dabrowski and Z. E. Sikorski (eds), *Toxins in Food*, pp. 11–63. CRC Press, Boca Raton, FL.

Pariza, M. W. (1996). Toxic substances. In: O. R. Fennema (ed.), *Food Chemistry*, 3rd edn, pp. 825–840. Marcel Dekker, New York, NY.

Park, D. L., C. E. Ayala, S. E. Guzman-Perez *et al.* (2000). Microbial toxins in foods: algal, fungal, and bacterial. In: W. Helferich and C. K. Winter (eds), *Food Toxicology*, pp. 93–135. CRC Press, Boca Raton, FL.

Plumley, F. G. (1997). Marine algal toxins: biochemistry, genetics, and molecular biology. *Limnol. Oceanogr.* **42**, 1252–1264.

Price, D., W. Kizer and H. K. Hansgen (1991). California's paralytic shellfish prevention program, 1927–89. *J. Shellfish Res.* **10**, 119–145.

Quilliam, M. A., and J. L. Wright (1989). The amnesic shellfish poisoning mystery. *Anal. Chem.* **61**, 1053A–1060A.

Rahman, M. S. (ed.) (1999). *Handbook of Food Preservation.* Marcel Dekker, New York, NY.

Reddy, C. S. and A. W. Hayes (2001). Food-borne toxicants. In: A. W. Hayes (ed.), *Principles and Methods of Toxicology*, 4th edn, pp. 491–530. Taylor & Francis, London.

Richards, H. and S. Hefle (2003). Plant biotechnology and food safety evaluation. In: C. N. Steward Jr (ed.), Transgenic Plants. Current Innovations and Future Trends, pp. 217–232. Horizon Scientific Press, Wymondham.

Rizk, A.-F. M. (1991). *Naturally Occurring Pyrrolizidine Alkaloids.* CRC Press, Boca Raton, FL.

Rosen, P., F. Baker, R. Barkin *et al.* (1988). *Emergency Medicine.* Mosby, St Louis, MI.

Schantz, E. J., and E. A. Johnson (1992). Properties and use of botulinum toxin and other microbial neurotoxins in medicine. *Microbiol. Rev.* **56**, 80–99.

Schilter, B., C. Andersson, R. Anton *et al.* (2003). Guidance for the safety assessment of botanical preparations for use in food and food supplements. *Food Chem. Toxicol.* **41**, 1625–1649.

Sevcik, C., J. C. Brito, F. D'Suze *et al.* (1993). Toxinology of a bovine paraplegic syndrome. *Toxicon* **12**, 1581–1594.

Sevcik, C., J. Noriega and G. D'Suze (2003). Identification of *Enterobacter* bacteria as saxitoxin producers in cattle's rumen and surface water from Venezuelan Savannahs. *Toxicon* **42**, 359–366.

Shimizu, Y. (2003). Microalgal metabolites. *Curr. Opin. Microbiol.* **6**, 236–243.

Smayda, T. J. and Y. Shimuzu (eds) (1993). *Toxic Phytoplankton Blooms in the Sea.* Elsevier, New York, NY.

Smith, L. W. and C. C. J. Culvenor (1981). Plant sources of pyrrolizidine alkaloids. *J. Nat. Prod.* **44**, 129–152.

Sommer, H. and K. F. Meyer (1937). Paralytic shellfish poisoning. *Arch. Pathol.* **24**, 560–598.

Spencer, P. S. and F. Berman (2003). Plant toxins and human health. In: J. P. F. D'Mello (ed.), *Food Safety. Contaminants and Toxins*, pp. 1–23. CABI Publishing, Wallingford.

Spoerke, D. G. Jr (2001a). US poison centers for exposures to plant and mushroom toxins. In: Y. H. Hui, R. A. Smith and D. G. Spoerke Jr (eds), *Foodborne Disease Handbook*, 2nd edn, Vol. 3, *Plant Toxicants*, pp. 1–36. Marcel Dekker, New York, NY.

Spoerke, D. G. Jr (2001b). Seafood and environmental toxin exposures: the role of poison centers. In: Y. H. Hui, R. A. Smith and D. G. Spoerke Jr (eds), *Foodborne Disease Handbook*, Vol. 4, *Seafood and Environmental Toxins*, pp. 1–21. Marcel Dekker, New York, NY.

Stewart, C. N. Jr (2003). *Transgenic Plants. Current Innovations and Future Trends*. Horizon Scientific Press, Wymondham.

Stiles, M. E. 2000. Less recognized and suspected foodborne bacterial pathogens. In: B. M. Lund, T. C. Baird-Parker and G. W. Gould (eds), *The Microbiological Safety and Quality of Food*, Vol. 2, pp. 1394–1419. Aspen Publishers, Gaithersburg, MD.

Tannahill, R. (1973). *Food in History*. Stein and Day, New York, NY.

Taylor, S. L. (1986). Histamine food poisoning: toxicology and clinical aspects. *CRC Crit. Rev. Toxicol.* **17**, 91–128.

Taylor, S. L. and S. L. Hefle (2002). Naturally occurring toxicants in foods. In: D. O. Cliver and H. P. Riemann (eds), *Foodborne Diseases*, 2nd edn, pp. 193–210. Academic Press, London.

Todd, E. C. D. (1993). Domoic acid and amnesic shellfish poisoning – a review. *J. Food Prot.* **56**, 69–83.

Todd, E. C. D. (1997). Seafood-associated diseases and control in Canada. *Rev. Sci. Tech. Int. Epiz.* **16**, 661–672.

Toke, D. (2004). *The Politics of GM Food. A Comparative Study of the UK, USA, and EU. Routledge*. Taylor & Francis, London.

Trevan, J. W. (1927). The error of determination of toxicity. *Proc. R. Soc. Lond.* **101B**, 483–514.

Troncone, R., L. Greco and S. Auricchio (1996). Gluten-sensitive enteropathy. *Pediatr. Clin. N. Am.* **43**, 355–357.

van Heyningen, W. E. (1950). *Bacterial Toxins*. Blackwell Scientific Publications, Oxford.

Van Veen, A. G. (1967). The bongkrek toxins. In: R. I. Mateles and G. N.Wogan (eds), *Biochemistry of some Foodborne Microbial Toxins*, pp. 43–50. MIT Press, Cambridge, MA.

Walder, R., Z. Kalvatchev, G. J. Tobin *et al.* (1995). Possible role of bovine immunodeficiency virus in bovine paraplegic syndrome: evidence from immunochemical virological and seroprevalence studies. *Res. Virol.* **146**, 313–323.

Walley, T. and M. Flanagan (1987). Nitrite-induced methemoglobinemia. *Postgrad. Med.* J. **63**, 643–644.

Wannemacher, R. W. (1989). Procedures for inactivation and safety containment of toxins. *Proceedings of the Symposium on Agents of Biological Origin*, US Army Research, Development and Engineering Center, Aberdeen Proving Ground, MD.

Watson, D. (1998). *Natural Toxicants in Foods*. Sheffield Academic Press, Sheffield.

WHO (World Health Organization) (1997). Food safety and foodborne diseases. *World Health Stats Q* **50** (No.1/2).

Williams, R. A. and R. A. Zorn (1997). Hazard analysis and critical control point systems applied to public health risks: the example of seafood. *Rev. Sci. Tech. Int. Epiz.* **16**, 349–358.

Wirtz, R. A., S. L. Taylor and H. G. Semey (1978). Concentrations of substituted p-benzoquinones and 1-pentadecene in the flour beetles *Tribolium consfusum* J. Du Val and *Tribolium castaneum* (Herbst). *Comp. Biochem. Physiol.* **61B**, 25–28.

Wittstock, U. and J. Gershenzon (2002). Constitutive plant constituents and their role in defense against herbivores and pathogens. *Curr. Opin. Plant Biol.* **5**, 1–8.

Yasumoto, T. and M. Murata (1993). Marine toxins. *Chem. Rev.* **93**, 1897–1909.

Zhou, S., H. L. Koh, Y. Gao *et al.* (2004). Herbal bioactivation: the good, the bad and the ugly. *Life Sci.* **74**, 935–968.

PART IV

Prevention of Foodborne Disease

Effects of food processing on disease agents

18

Roberto A. Buffo and Richard A. Holley

1 Introduction

Food-processing technology has evolved substantially since its industrial birth in the 1800s with the initial development of pasteurization and other thermal processes for packaged foods. However, it was not until many years later, in the 1930s, that safe canning processes were developed. In the 1940s, processes for frozen foods were introduced. In the 1950s, vacuum-packed and modified-atmosphere packaged foods became available, and the technology revolutionized the meat industry, changing it from carcass distribution to vacuum-packed and boxed cuts, which allowed a five-fold shelf-life extension. Modified atmospheres were successful in storage of fresh fruits and vegetables, and later for the packaging of cut produce. In the 1980s, aseptic food processing and packaging became a commercial reality, along with food irradiation (gamma and e-beam). The 1980s and 1990s witnessed the application of a variety of new thermal processes (microwave and radiofrequency treatments, and ohmic- and inductive-heat

Foodborne Infections and Intoxications 3e
ISBN-10: 0-12-588365-X
ISBN-13: 978-012-588365-8

technologies). Also, a number of non-thermal processes came into commercial use, with some still under study (pulsed X-rays, high-pressure processing, pulsed electric fields, pulsed light, UV light, magnetic fields and ultrasound). Advantages and limitations of these new processing technologies were recently outlined (Davidson and Harrison, 2002), and some new information is included in this chapter.

In this chapter the focus is on foodborne pathogens, although, for clarification, effects of processing on spoilage organisms are mentioned as necessary. The emergence of previously unrecognized organisms as significant contributors to foodborne illness in humans raises questions with respect to their origin and whether modern manufacturing practices associated with mass processing have provided unique opportunities for their survival. These organisms include *Campylobacter*, *Escherichia coli* O157:H7 and *Listeria monocytogenes*, and the protozoan *Cyclospora*.

The food-processing industry has become less reliant upon final product inspection and more dependent upon control of adequate processes to ensure the safety of food (HACCP). Accurate estimates of bacterial susceptibility to lethal steps used in food processing are essential, as the goal is to eliminate foodborne pathogens during processing. Better refrigeration of cooked cured meats and pasteurization of dairy products have reduced incidents of foodborne illness caused by *Staphylococcus aureus*; other organisms contaminating product at packaging have replaced this threat. Refrigeration of perishable products, such as minimally processed products and ready-to-eat prepared products, may select for the psychrotroph *Listeria*, at the same time the consolidation of industrial operations into larger plants with greater geographic distribution of product with an extended shelf-life has had an influence upon the size of foodborne illness outbreaks.

During the 1970s and the 1980s shelf-stable foods were developed using a combination of inhibitory treatments (salt, nitrite, temperature), which interacted to inhibit pathogens at lower individual levels than needed if each treatment was used alone. The system has worked well and is generally accepted. However, these multiple treatments may have contributed to the development of microbial resistance through a generalized stress response (GSR), which may contribute to salmonellosis.

Many challenges remain to produce safe and nutritious food with an adequate shelf-life. Low-salt, -sugar or -fat formulations may change water activity (a_w) or pH, and may provide new niches in foods for pathogens. Processes used for the manufacture of dry sausage and aged Cheddar cheese from thermized (unpasteurized) milk have failed to eliminate *E. coli* O157:H7 and *S.* Enteritidis, respectively. Many vegetables and sprouts cannot be freed from pathogens once contaminated. Alternative approaches are needed to eliminate such threats.

2 High-temperature preservation

2.1 History and present practices

Credit for discovering the value of heat as a preservative agent goes to the French chef, distiller and confectioner Nicholas Appert; he held the view that the cause of food spoilage was contact with air, and that the success of his technique was due to the exclusion of air from the product. This view persisted for another

50 years, sometimes with disastrous consequences, until Louis Pasteur established the relationship between microbial activity and putrefaction (Adams and Moss, 1995).

The use of high temperatures for processing of foods is the gold standard against which all other food-preservation technologies are evaluated. Thermal exposures (time–temperature combinations) have been developed that result in pathogen destruction with minimal changes in the functional characteristics of the food. High temperatures are used either to pasteurize or to sterilize foods, and each has commercial applications (Jay, 1996).

2.1.1 Pasteurization

Pasteurization (e.g. pasteurization of milk) implies the destruction of disease-producing, non-spore-forming organisms. The following equivalent heat treatments will eliminate organisms such as *Mycobacterium tuberculosis* and *Coxiella burnetti*: 63 °C for 30 minutes (low temperature, long time, LTLT); 72 °C for 15 seconds (high temperature, short time, HTST); 89 °C for 1 second; 90 °C for 0.5 seconds; 94 °C for 0.1 seconds; and 100 °C for 0.01 seconds (Jay, 1996).

Packaged products treated with mild heating combined with chilled storage are known as REPFEDs (i.e. refrigerated processed foods of extended durability) or cook-chill products, including '*sous vide*' meals (i.e. foods mildly heated within vacuum packs). For these chilled pasteurized foods specific attention is given to the control of non-proteolytic *Clostridium botulinum*, because of its ability to grow at chilled temperatures as low as 3 °C. Typically, an additional hurdle is necessary to ensure safety (Notermans *et al.*, 1990; Mossel and Struijk, 1991; Graham *et al.*, 1997). The hurdle concept is discussed later in this chapter.

2.1.2 Sterilization

Sterilization results when there is destruction of all viable organisms, as measured by an appropriate technique. The concept of commercial sterility has been applied to retorted or canned food when no viable organisms can be detected by the usual methods, or when the number of survivors is so low as to be non-significant under normal conditions. The term *appertization* (after Nicholas Appert) has been suggested for use as a replacement for the term 'commercially sterile', on the grounds that sterility is not a relative concept (Jay, 1996; Adams and Moss, 2000).

A major technological development in heat processing has been the use of ultra-high temperature, short-time treatments (UHT-ST), which involve rapid heating to temperatures of about 140 °C, holding for several seconds, and then rapidly cooling, to produce foods that are safe and shelf-stable at ambient temperatures. These processes provide opportunities for use of a variety of flexible packaging materials, and a large range of products is available (Lewis, 1993). UHT-ST treatment permits continuous operation, and is applied before packaging (Jelen, 1982). It is widely used for fruit juices, dairy products, creams and sauces and, more recently, for low-acid soups, including particulate-containing products (Rose, 1995).

2.2 Thermal destruction

Thermal processes are neither uniform nor instantaneous; a series of fundamental concepts have been developed to describe them (Jay, 1996):

- The Thermal Death Time (TDT) is the time needed to kill a given number of organisms at a specified temperature. Of lesser importance is the Thermal Death Point (TDP), defined as the temperature needed to kill a given number of microorganisms in a fixed time – usually 10 minutes.
- The *D* value (Decimal Reduction Time) is the time required to destroy 90 % of the organisms. Mathematically, the *D* value is represented by the slope of the logarithm of survivors plotted against time (see Tables 18.1 and 18.2; Adams and Moss, 1995).
- The *z* value is the number of degrees (Celsius or Fahrenheit) required to change the *D* value by a factor of ten; *z* values thus provide information on the destruction rate at different temperatures, allowing for the calculation of equivalent thermal processes at different temperatures.
- The *F* value is the equivalent time at 250 °F (121 °C) of all temperatures to which food is exposed during a heat process. The integrated lethal value of heat received at the coolest point in a container (generally close to the center) during processing is designated as F_0 and represents the capacity of a heat process (expressed as minutes at 250 °F or 121 °C) to reduce the number of *C. botulinum* spores.
- The 12-*D* concept refers to a process that reduces the probability of survival of proteolytic *Clostridium botulinum* spores by a factor of 10^{12}.

These thermal destruction parameters assume that the effects of heat on microorganisms are constant and unaffected by the heating rate in the sub-lethal temperature

Table 18.1 Heat resistance of vegetative bacteria

	D value (min. at temperature)				*z* value (°C)
Species	**70°C**	**65°C**	**60°C**	**55°C**	
Campylobacter jejuni				1.1	
Escherichia coli		0.1		4.0	
Lactobacillus plantarum		4.7–8.1			
Listeria monocytogenes		5.0–8.3			
Pseudomonas aeruginosa				1.9	
Pseudomonas fluorescens				1.0–2.0	
Salmonella spp.		0.02–0.25			
Salmonella Senftenberg	440[a]	0.3–2.0	10.8[b]	18.0[a]	
	2.6–6.1[c]				6.0[b]
					6.8–13.0[c]
Salmonella Typhimurium	816[a]	0.056			
Staphylococcus aureus		0.2–2.0	4.5–7.8		4.5

From Adams and Moss, 1995; Farkas, 2002.
[a] In milk chocolate
[b] In skim milk
[c] In heart infusion broth.

Table 18.2 Heat resistance of bacterial endospores

Species	D value (min at temperature)				z value (°C)
	121°C	110°C	100°C	80°C	
Bacillus cereus			5.0		
Bacillus coagulans	0.01–0.1				
Bacillus stearothermophilus	4.0–4.5		3000		7.0
Bacillus subtilis			11.0		7.0
Clostridium botulinum types A & B	0.1–0.2				10.0
Clostridium botulinum type E		> 1 s		0.1–0.3	
Clostridium butyricum			0.1–0.5		
Clostridium perfringens			0.3–20.0		10.0–30.0
Clostridium sporogenes	0.1–1.5				9.0–13.0
Clostridium thermosaccharolyticum	3.0–4.0				12.0–18.0
Desulfotomaculum nigrificans	2.0–3.0				

From Stumbo *et al.*, 1983; Farkas, 2001.

range. However, the rate of temperature elevation affects the subsequent isothermal death of several vegetative bacteria (Mackey and Derrick, 1986).

The assumption that the reduction in log numbers of survivors proceeds in a linear manner with time has served the food industry and regulatory agencies well. For bacterial spores, a good fit is generally observed. However, with milder thermal processes that use low processing temperatures significant deviations from linearity have been observed, including 'shoulder' and/or 'tailing' of the curves; therefore under these circumstances, *D* and *z* values cannot be accurately determined. Various explanations for these phenomena have been offered (Han, 1975; Cerf, 1977; Perkin *et al.*, 1980; Gould, 1989). In recent years mathematical models have included normal distribution of heat sensitivity and logistic functions, which are better able to predict thermal inactivation at relatively low heating rates (Cole *et al.*, 1993; Stephens *et al.*, 1994).

2.3 Heat resistance

2.3.1 Mesophilic organisms

Damage to DNA is the most probable key lethal event occurring when vegetative cells or spores are subjected to heat. Preventing spore germination is also important. Deviations from linearity in thermal death kinetics of vegetative cells indicate that there is a multiplicity of target sites, including the cytoplasmic membrane, key enzymes, RNA and the ribosome. This damage is cumulative rather than instantly lethal (Adams and Moss, 2000).

Heat resistance of microorganisms is related to their optimum growth temperatures. Psychrophilic and psychrotrophic organisms are the most heat sensitive, followed by mesophiles and thermophiles. Gram-positive bacteria tend to be more heat resistant than Gram-negatives, with cocci being more resistant than rods. Yeasts and molds tend to be fairly heat sensitive, and both yeast ascospores and asexual spores of molds are only slightly more heat resistant than their vegetative counterparts (Jay, 1996).

The heat resistance of bacterial endospores is primarily related to their ability to maintain a very low water content in the central DNA-containing protoplast (Adams and Moss, 1995; Jay, 1996).

Exposure of vegetative cells to temperatures that induce a heat shock improves resistance to higher temperatures (Mackey and Derrick, 1986). Synthesis of a family of proteins called heat-shock proteins (HSP), in response not only to heat shock but also to chemical or physical stresses, occurs in most organisms (Schlesinger *et al.*, 1982; Auffray *et al.*, 1995; Boutibonnes *et al.*, 1995). Many of the HSP inducers produce protein damage (Hightower, 1991), and it has been proposed that the signal for HSP induction is protein denaturation (Parsell and Sauer, 1989). For foods heated relatively slowly, the possibility of an increase in heat resistance of vegetative cells during heating must be taken into account to establish safe heat treatments (Farkas, 2002).

2.3.2 Thermophilic organisms

Thermophilic organisms have unique physiological characteristics that confer thermal tolerance.

The enzymes of thermophiles can be divided into three groups according to heat sensitivity (Jay, 1996). The base composition of rRNA has been shown to affect thermal stability. Pace and Campbell (1967) found that the microbial G-C content of rRNA tended to increase and the A-U content to decrease with increasing maximal growth temperatures. The increased G-C content makes for a more stable structure through more extensive hydrogen bonding. However, there seems to be no differences between thermophiles and mesophiles regarding the structure of DNA and mRNA.

Thermophilic growth is associated with a predominance of saturated lipids (Marr and Ingraham, 1962). Brock (1967) stated that thermophilic growth is linked to the nature of cellular membranes, much more so than to the properties of specific macromolecules or cytoplasmic organelles, and lethal injury may be primarily due to the melting of lipid constituents of the cell membrane, with resulting leakage of essential constituents and subsequent death (Jay, 1996)

2.4 Bacterial growth in canned foods

Bacterial growth in canned foods may result from under-processing permitting the survival of bacterial spores, or from contamination of the can content with spore-formers and non-spore-formers due to leakage through the seams (Jay, 1996). The extent of heat processing required depends largely on acidity: the more acidic a product, the milder the process needed (Adams and Moss, 2000). Thus, canned foods are classified based on acidity values (Adams and Moss, 1995; Jay, 1996).

2.4.1 Low-acid foods (pH > 4.6)

Examples are meat and marine products, milk, vegetables (e.g. corn and lima beans), and mixtures of meat and vegetables. Toxin production by surviving proteolytic *C. botulinum* may occur; these products must therefore undergo cooking sufficient to ensure 12 $\log_{10}$ reductions of the spores of this pathogen.

2.4.2 Acid foods (pH 3.7–4.0 to 4.6)

These foods are mostly fruits, such as pears, tomatoes and figs, often spoiled by thermophilic organisms. *Clostridium botulinum* cannot grow in acid foods.

2.4.3 High-acid foods (pH < 3.7–4.0)

This group includes fruits and brined or fermented vegetable products (e.g. grapefruit, rhubarb, sauerkraut and pickles). They are generally spoiled by non-spore-forming mesophiles, yeasts, molds and/or lactic acid bacteria.

Cans must be cooled rapidly after processing to prevent spoilage by thermophiles. (Adams and Moss, 2000).

Leakage is the most common cause of microbial spoilage in canned foods. During processing, cans are subjected to physical stress. The negative pressure created inside the can during the rapid cooling stage may lead to microorganisms in the cooling water being sucked inside through any small defect in the seam. To prevent this contamination, the outside of cans must remain clean, and chlorinated water has to be used to cool them (Adams and Moss, 2000).

2.5 Examples of heat resistance of pathogens

Veeramuthu *et al.* (1998) studied the thermal inactivation of two pathogens, *E. coli* O157:H7 and *Salmonella* Senftenberg, in ground turkey thigh meat, and examined the thermal denaturation of endogenous enzymes for their potential use as time–temperature indicators. The study was based on the premise that the USDA Food Safety and Inspection Service may amend cooking regulations to require that any thermal process used for poultry products be sufficient to cause a 7-*D* reduction in *Salmonella* spp. Bacteria counts were determined and muscle extracts assayed for residual enzyme activity or protein denaturation. *S.* Senftenberg had higher *D* values at all temperatures. The *z* values of *E. coli* were 6.0–5.7 °C; those of *S.* Senftenberg were 5.6–5.4 °C. Lactate dehydrogenase (LDH) was the most heat-stable enzyme at 64 °C. LDH, triose phosphate isomerase (TPI), acid phosphatase and immunoglobulin G had 10-fold activity reduction values of 3.8, 5.8, 6.3 and 8.6 °C, respectively. Temperature dependence of TPI was most similar to that of *S.* Senftenberg, suggesting it might be used to monitor adequacy of processing if a performance standard based on this pathogen is implemented.

D values and *z* values were determined for *L. monocytogenes* Scott A in raw ground pork by Ollinger-Snyder *et al.* (1995). *D* values were 108.81, 9.80 and 1.14 minutes at 50 °C, 55 °C and 60 °C, respectively, when heating in ground pork without soy hulls, and 113.64, 10.19 and 1.70 minutes, respectively, when heating in ground pork with soy hulls added; *z* values were 5.45 °C and 5.05 °C in ground pork prepared with and without soy hulls, respectively. Assuming that ground pork naturally contains *ca.* 10^2 *L. monocytogenes* cells per gram, and that safety can be assured with a 4-*D Listeria* cook (i.e. reducing the population by four orders of magnitude), this study indicated that ground pork should be heated to an internal temperature of 60 °C for at least 4.6 minutes without soy hulls and for at least 6.8 minutes with added soy hulls.

In beverage manufacture, cider and apple juice may be stored for a short time at ambient temperature before pasteurization. Storage time and temperature may affect the subsequent thermotolerance of bacteria in these beverages. Ingham and Uljas (1998) examined thermotolerance of two *E. coli* O157:H7 strains at 61 °C after storage in pH 3.4 apple cider or apple juice. Both strains exhibited biphasic survivor curves. Strain ATCC 43894 was consistently more thermotolerant than strain ATCC 43889, with 33–153 % greater *D* values derived from the linear portion of each survivor curve. Prior storage at 21 °C for 2 or 6 hours hastened thermal destruction of both strains in apple cider, but the effect was not statistically significant. However, when apple juice was tested, prior storage at 21 °C for 2 hours significantly decreased the thermotolerance of strain ATCC 43889, but not that of strain ATCC 43894. Storage at 21 °C for 6 hours in apple juice caused a decrease of 2.1 and 0.5 $\log_{10}$ CFU/ml in populations of strain ATCC 43889 and strain ATCC 43894, respectively. Experiments with filtered apple cider showed that the presence of filterable pulp enhanced the thermotolerance of both strains. These authors concluded that short-term (≤ 6 hours), room-temperature storage of pH 3.4 filtered apple cider or apple juice may enhance lethality of subsequent pasteurization.

Grijspeerdt and Herman (2003) determined a series of inactivation curves for *Salmonella* Enteritidis in boiled eggs using different conditions of time and temperature. The inactivation curves consistently showed an initial slow decline in bacterial numbers at lower temperatures, after which a very rapid inactivation took place. Results suggested that *S.* Enteritidis is more resistant to a slower heating process.

The safety of REPFEDs with respect to non-proteolytic *C. botulinum* is a central issue. Lindström *et al.* (2001) evaluated 'mild' and 'enhanced' heat treatments in relation to survival of type B spores in *sous vide* processed ground beef and pork cubes. At a reference temperature of 85 °C and a *z* value of 7.0 °C, the 'mild' treatment ran for 0–2 minutes and the 'enhanced' treatment for 67–515 minutes. Two concentrations of nisin were tested for the ability to inhibit growth and toxin production by non-proteolytic *C. botulinum* in the same products. A total of 96 samples were heat processed and analyzed for the presence of the pathogen, and for botulinum toxin after storage for 14–28 days at 4° and 8 °C. Predictably, after 'mild' processing all the samples of both products showed botulinal growth, and one ground beef sample became toxic at 8 °C. The 'enhanced' heat process resulted in growth but not toxin production by *C. botulinum* in one ground beef sample in 21 days at 8 °C. No growth was detected in the pork cube samples. The 'enhanced' processing of both products resulted in higher sensory quality than that corresponding to the 'mild' one. Nisin did not inhibit growth of *C. botulinum* in either product; growth was detected at 4 °C and 8 °C, and ground beef became toxic within 21–28 days at 8 °C. Aerobic and lactic acid bacterial counts were reduced by the addition of nisin at 4 °C. The study demonstrated that the mild processing temperatures commonly utilized in *sous vide* technology do not eliminate non-proteolytic *C. botulinum* type B spores. The heat treatment needs to be carefully evaluated for each product to ensure product safety in relation to non-proteolytic *C. botulinum*.

3 Low-temperature preservation

3.1 History and terminology

As early as the eleventh century BC the Chinese had developed ice houses as a means of storing ice through the summer months, and by the nineteenth century the cutting and transporting of natural ice had become a significant industry in Europe and the Americas in areas blessed with a cold climate (Adams and Moss, 1995). Mechanical methods of refrigeration and ice making were first patented in the 1830s. Shipment of chilled meat from North America to Europe was started in the 1850s, and by the 1890s the technology had been refined to the extent that shipping chilled and frozen meat from North and South America and Australia to Europe was a large and profitable enterprise. Presently, a highly sophisticated cold chain allows the distribution of all kinds of food products around the world (Adams and Moss, 1995).

There are three distinct temperature ranges used for low-temperature storage of foods. Chilling temperatures are those between the usual refrigerator (5–7 °C) temperatures and the slightly higher temperatures (10–15 °C) that are suitable for the storage of certain vegetables and fruits such as cucumbers, potatoes and limes. Refrigeration temperatures are those between 0 °C and 7 °C. Freezing temperatures are those at or below –18 °C. Growth of microorganisms is generally prevented at freezing temperatures, although some of them can and do grow slightly below 0 °C but at an extremely slow rate (Jay, 1996).

Bacterial strains able to grow at or below 7 °C are more common among Gram-negative than Gram-positive genera. Growth at temperatures below 0 °C is more likely for yeasts and molds than for bacteria. Among pathogenic bacteria, *Listeria*, *Yersinia* and non-proteolytic *C. botulinum* have the lowest minimum growth temperatures. *Salmonella* spp. have higher minimal growth temperatures than *Staphylococcus aureus* (Jay, 1996).

The range of chilled and refrigerated foods available has increased in recent years, as traditional products such as fresh meat, fresh fish and dairy products have been joined by a huge variety of new items, including complete meals, prepared salads and delicatessen foods, dairy desserts and the like. The development and almost general availability of an efficient cold chain from manufacturing to consumption has played a key role in satisfying this market need (Adams and Moss, 2000).

3.2 Freezing, chilling, and frozen storage

Freezing should not be regarded as a means of destroying foodborne microorganisms. The extent to which microorganisms lose viability upon freezing differs from strain to strain, and depends on the rate of freezing, the nature and composition of the food in question, and the length of time of frozen storage (Georgala and Hurst, 1963).

Quick freezing is done by lowering the temperature of foods to about –20 °C within 30 minutes. Small intracellular ice crystals are formed. Microorganisms undergo a rapid thermal shock (with no time for low temperature adaptation or blocking of suppression of metabolic activity), and there is only a brief exposure to adverse concentrations of solutes. Slow freezing takes place when the desired temperature (≤ –4 °C) is achieved within 3–72 hours. This is the type of freezing that occurs in the home freezer. Microorganisms are exposed for much longer to increased concentrations of solutes. There is no thermal shock effect, and this may allow for gradual adaptation to increasing concentrations of solutes and 'injury recovery' among survivors. It is important to emphasize that if the transition 0 to –4 °C is traversed quickly then small ice crystals are formed, whereas if the transition occurs slowly large ice crystals are formed; the latter scenario causes a higher lethality than the former (Jay, 1996; Adams and Moss, 2000). Ingram (1951) made three key observations related to the response of microorganisms freezing: (1) a sudden mortality occurs immediately on freezing; (2) the cells surviving immediately after freezing die gradually when stored in the frozen state; and (3) the decline in numbers is relatively rapid at temperatures just below freezing, especially about –2 °C, but less so at lower temperatures, and especially below –20 °C.

Jay (1996) has summarized the metabolic events that occur when cells freeze. During frozen storage, the death of survivors is fast initially and then slows gradually until the survival level stabilizes. The death rates tend to be lower than during the freezing process; survival seems to be related to the unfrozen, very concentrated residual solution formed by freezing. The concentration and composition of this residual solution may change during the course of storage, and the size of the ice crystals may increase, especially at fluctuating storage temperatures. In general, there is less loss of viability during frozen storage when the storage temperature is static rather than fluctuating. The lower the temperature of frozen storage, the slower the death rate of survivors. Gram-positive microorganisms survive frozen storage better than Gram-negative ones (International Commission on Microbiological Specifications for Foods, 1980).

Though psychrotrophs can grow in chilled foods, they do so relatively slowly. For example, the generation time for a *Pseudomonas* species isolated from fish was about 6.7 hours at 5 °C compared with 26.6 hours at 0 °C (Adams and Moss, 2000). Gill (1995) found that with fresh meat stored at 0 °C, the attainable shelf-life was only 70 % of that achievable at –2 °C (just at the freezing point of meat). This decreased to 50 %, 30 % and 15 % at 2 °C, 5 °C and 10 °C, respectively. Since chilling is not a bactericidal process, the use of raw materials of good microbiological quality and hygienic handling are key requirements for the production of safe chilled foods. Mesophiles that survive cooling, albeit in an injured state, can persist in the food for extended periods, and may recover and resume growth if the temperature becomes favorable at a later time. Thus, chilling will prevent an increase in the pathogenic risk from mesophiles, but will not assure their elimination (Adams and Moss, 2000). Psychrotrophic organisms can be found among yeasts, molds, and Gram-negative and Gram-positive bacteria. However, they all share the property of being inactivated at moderate heating temperatures, possibly because of excessive membrane fluidity at higher temperatures.

3.3 Thawing

The fate of microorganisms in food that has been frozen is partially determined by their ability to survive subsequent thawing. Three key factors should be considered (Fennema *et al.*, 1973).

The time–temperature combination during thawing is potentially detrimental to quality and safety. During thawing, the temperature rises rapidly to near the melting point and remains there throughout the course of thawing (which can be lengthy), thus allowing considerable opportunity for chemical reactions, recrystallization, and microbial growth if the process is slow enough.

Most frozen-food processors advise against the refreezing of foods once they have been thawed. Although this is mostly related to textural, flavor and nutritional qualities, the microbiology of thawed frozen foods is important. Some textural changes associated with freezing would seem to allow the invasion of surface microorganisms into deeper tissues, thus facilitating microbial growth. Thawing causes the release of enzymes such as nucleases, phosphatases, glycosidases and others, which may degrade macromolecules and generate simpler compounds that are more readily utilized by the microbial flora.

3.4 Cold resistance

Psychrophiles and psychrotrophs possess physiological mechanisms related to their ability to grow at low temperatures.

3.4.1 Cold-adapted enzymes and slower metabolic rates

The cold-adapted enzymes are more thermolabile than their mesophilic counterparts, so that at quite moderate temperatures (40–50 °C) they become too flexible, lose catalytic efficiency and eventually denature.

3.4.2 Lipid composition and membrane functionality

The usual lipid content of most bacteria is between 2 % and 5 %, most or all of which is located in the cell membrane. Even though different cold-adapted bacteria may have similar low-temperature growth ability, they will almost certainly have quite different fatty acyl composition. Differences will be influenced by phylogenetic distinctions and different metabolic capabilities, and by specific protein–lipid interactions. For example, *Salmonella* adapt to temperature almost entirely by changing lipid unsaturation, whereas *Listeria*, which contain predominantly branched fatty acyl chains, modify the anteiso/iso-branched ratio and the acyl chain length. *Bacillus* spp. use a combination of unsaturation and changes in branching pattern (Russell, 2002).

A second aspect of membrane structure is the need to preserve the bilayer (lamellar) structure and prevent the formation of non-bilayer phases such as hexagonal arrangements, which destroy the selective permeability properties of the membrane (Russell, 1989).

3.4.3 Cold shock and cold adaptation

Cold shock occurs when many cells of mesophilic bacteria die upon the sudden chilling of a suspension of viable cells grown at mesophilic temperatures. It is generally a property of Gram-negative but not of Gram-positive bacteria. However, spoilage organisms within the genera *Lactobacillus* and *Pseudomonas* and pathogens within the genera *Escherichia*, *Listeria*, *Salmonella* and *Yersinia* exhibit this response as well

(Jay, 1996). According to Rose (1968), cold shock results from the sudden release of cell constituents following cold damage to membrane lipids, with the consequent development of holes in the membrane.

Cold acclimation depends on the ability to synthesize stress proteins, usually called cold-shock proteins (CSP). Table 18.3 summarizes the major functions of CSP related to blocking deleterious cold-shock effects (Russell, 2002). The importance of the ribosome in sensing temperature changes is evident, as is the fact that the cellular function most sensitive to cold shock is the initiation of translation (Graumann and Marahiel, 1996, 1997). The regulation of CSP synthesis occurs at several levels, both transcriptional and translational, involving protein and mRNA stabilities. Significantly, compared to mesophiles, in a cold-shock event involving psychrotrophs there is no concomitant suppression of the expression of the 'housekeeping' genes that encode enzymes of central metabolic pathways. Thus, growth lag times are likely to be shorter or non-existent. Moreover, the number of CSP and the extent of their synthesis depend on the depth of the cold shock (Gounot and Russell, 1999; Hébraud and Potier, 2000). In addition, changes in lipid composition and synthesis of CSP are functionally linked, because balanced growth requires coordination of intracellular and extracellular events as well as those occurring within the membrane matrix (Hoch and Silhavy, 1995).

Table 18.3 Cold-shock response characteristics

Effect of low temperature	Cellular response
Block in initiation of protein synthesis	Synthesis of CSP to stabilize interaction of mRNA with 30S ribosomal subunits
Disruption of ribosome structure	Synthesis of CSP to stabilize protein–protein and protein–rRNA interactions within the ribosome
Formation of secondary structures	Synthesis of RNA chaperones to maintain mRNA in linear form
Increased negative supercoiling of DNA	Induction of DNA-unwinding enzymes and stabilizing histone-like proteins

From Russell, 2002.

4 Multiple intervention technology

4.1 Introduction

An essential task in the food industry is to maintain the quality and safety of foods during extended periods of storage. All foods deteriorate in quality at some specific rate following harvest, slaughter or manufacture. This quality deterioration is caused by a wide range of reactions, including physical (e.g. moisture migration to and from the environment or between the components of a composite food), chemical (e.g. rancidity due to oxidation), enzymatic (e.g. lipolysis leading to rancidity) and microbiological (spoilage and/or growth of pathogenic microorganisms) reactions.

While the aim of effective food preservation is to control all forms of quality deterioration, the overriding priority is to minimize the potential for the occurrence and growth of food spoilage and pathogenic microorganisms (Leistner and Gould, 2002).

Thus, preservation technologies are essentially aimed at inactivation of microorganisms or the delay or prevention of microbial growth. These preservative factors have been called 'hurdles'. This a less than fortunate term, because it may imply that microorganisms, in order to prosper, must overcome one hurdle at a time – thus giving the impression that the effect of several hurdles is the sum of the individual effects. However, this is not the case; because of interactions (or synergisms), the total effect may be more like the product of individual effects. An example is a factorial experiment on the preservation of canned cured meats where significant interactions were demonstrated for nitrite, sodium chloride and heat treatment (Riemann, 1963).The hurdle concept was originally introduced by Leistner (1978), and has since then become generally accepted; the term will therefore also be used in this text. There are six fundamental hurdles: temperature (high or low), water activity (a_w), acidity (pH), oxidation–reduction potential (E_h), preservatives (e.g. nitrites, sorbates, sulfite), and competitive microorganisms (e.g. lactic acid bacteria) (Leistner, 2000). However, the full list of possible hurdles is much more extensive, including natural antimicrobials, vacuum and modified atmosphere packaging, and alternative processing techniques, which are covered in later sections of this chapter.

In industrialized countries, the hurdle technology approach is currently of most interest for minimally processed foods that are mildly heated or fermented, and for establishing the microbial stability and safety of foods being developed in response to consumer demands for 'healthier' foods with less fat, sugar or salt. In developing countries, hurdle technology is applied to produce foods that remain stable, safe and tasty when stored at ambient temperature. Impressive success has been achieved in Latin America with the development of minimally processed, high-moisture fruit products.

4.2 Basic concepts

Food preservation implies generating an environment that inhibits the growth or causes the death of microorganisms in food. The fundamental biological concepts related to hurdle technology are as follows.

4.2.1 Homeostasis

Homeostasis is the stability of reactions in the cytoplasm of organisms. If hurdles in foods significantly disturb the homeostasis of microorganisms, they will not multiply – that is, they will remain in the lag phase and may even die before homeostasis is re-established. Thus, food preservation is achieved by disturbing the microbial homeostasis either temporarily or permanently (Leistner, 2000).

4.2.2 Metabolic exhaustion

Microorganisms respond to stress by activating repair mechanisms to re-establish homeostasis and overcome the hostile challenge. By doing this, they may use up their energy resources and die as they become metabolically exhausted. Essentially, an 'autosterilization' phenomenon occurs (Leistner, 1995a).

4.2.3 Stress reactions

Some bacteria become more resistant or even more virulent under stress as a result of generating shock proteins. This may interfere with safe food preservation, but the activation of genes for the synthesis of these 'novel' proteins has an energy cost. Thus, if different stresses are received simultaneously, microorganisms may become metabolically exhausted at an accelerated rate (Leistner, 1995a).

4.2.4 Multi-target preservation

This concept refers to the goal of an effective but mild food preservation approach, based on the fact that – as mentioned earlier – different hurdles in a food might not have just an additive antimicrobial effect but may also act synergistically (Leistner, 1978, 1995b). A synergistic effect could be achieved if the applied hurdles affected different targets within the microbial cell (e.g. the membrane, DNA, enzyme systems, pH). If so, restoration of homeostasis and activation of stress-shock proteins becomes more difficult (Leistner, 1995b). This means that it is more effective to employ several hurdles of small intensities than one inhibitory factor of larger intensity (Leistner, 1994). This approach may also minimize losses of product quality.

4.3 Survey of fundamental hurdles

Of the six fundamental hurdles, four are covered here: pH, a_w, E_h and competitive microorganisms. Temperature was discussed in the previous section on temperature preservation (high and low), whereas preservatives will be covered in the following sections on antimicrobial agents (natural and synthetic).

4.3.1 Acidity

Most microorganisms grow best at pH values around 7.0 (range 6.6–7.5), few grow below pH 4.0. Bacteria, especially pathogenic bacteria, tend to be more sensitive than molds and yeasts to low pH. Such foods as fruits, soft drinks, vinegar and wines have pH levels below that at which bacteria normally grow. Thus, the excellent keeping quality of these products is essentially due to pH. Not accidentally, fruits undergo primarily mold and yeast spoilage, because these organisms are able to grow at pH < 3.5. On the other hand, most meats and seafood have a final ultimate pH of about 5.6 and higher, which makes them susceptible to bacteria as well as mold and yeast spoilage. Vegetables tend to have higher pH values than fruits, and as such they are also prone to bacterial spoilage (Jay, 1996).

Some foods are characterized by inherent acidity. Others owe their low pH to the action of certain microorganisms, which is called biological acidity; this type of acidity is found in fermented dairy products, sauerkraut and pickles. Meats are more highly buffered than vegetables, owing to their higher protein content (Jay, 1996).

An adverse pH affects at least two essential characteristics of the metabolism of a microbial cell: the functioning of the enzymatic systems and the transport of nutrients into the cell. The cytoplasmic membrane is relatively impermeable to H^+ and OH^- ions in order to maintain a constant internal pH. Key compounds such as DNA and ATP require neutrality. Thus, when placed in an acid environment the cell must either keep H^+ from entering or expel H^+ ions as they enter in order to avoid a fatal alteration in its homeostasis. With respect to the transport of nutrients, the bacterial cell tends to have a negative charge. Therefore, non-ionized nutrients can enter the cell, whereas ionized ones cannot. At a neutral or alkaline pH organic acids do not enter, but in an acidic environment these compounds are non-ionized and can enter the negatively charged cell. In addition, the ionic character of ionizable groups is affected on either side of neutrality, resulting in increasing denaturation of membrane and transport enzymes (Jay, 1996).

4.3.2 Water activity

Water activity is defined as the ratio of the water vapor pressure over food to the vapor pressure over pure water at the same temperature. In mathematical terms, $a_w = p/p_0$, where p is the vapor pressure of the solution and p_0 the vapor pressure of the solvent (water). This concept is related to relative humidity (RH), as $RH = 100 \times a_w$. Most fresh foods have a_w values > 0.99. In general, bacteria require higher a_w values than fungi, and Gram-negative bacteria require higher a_w values than Gram-positive bacteria. Most spoilage bacteria do not grow below $a_w = 0.91$, whereas spoilage molds can grow at levels as low as 0.80. The lowest reported value for bacteria of any type is 0.75 for halophilic ('salt-loving') organisms. Xerophilic ('dry-loving') molds and osmophilic ('barophilic') yeasts can grow at a_w values of 0.65 and 0.61, respectively (Jay, 1996).

Three important relationships have been identified among a_w, temperature and nutrition (Morris, 1962): at any temperature, the ability of microorganisms to grow is reduced as the a_w is lowered; and the a_w range allowing growth is greatest at the optimum growth temperature and with optimum nutrients.

Curing by the addition of salt, conservation by the addition of sugar, the addition of other solutes (such as glycerol) and the removal of water from food by drying, as well as the immobilization of water by freezing, all lead to a reduction in water activity (Leistner and Gould, 2002). The general effect of lowering a_w below the optimum is an increase of the lag phase of growth and a decrease of the growth rate and size of the final microbial population. These effects result from adverse influences of lowered water availability on all metabolic activities, because all cellular chemical reactions take place in an aqueous environment. Lowering the water activity causes osmotic stress in microorganisms, and microorganisms respond by intracellular accumulation of compatible solutes. Bacteria can accumulate substances such as glutamate, glutamine, proline, alanine, sucrose and trehalose; whereas fungi produce polyhydric alcohols such as glycerol and arabitol (Jay, 1996). Halophilic bacteria are able to operate under low a_w conditions by virtue of their ability to accumulate potassium chloride (Brown, 1964).

The so-called intermediate moisture foods (IMF) rely on a lowered a_w as the main factor for microbial stability. IMF are characterized by moisture contents of around 15–50 % and a_w values of between 0.60 and 0.85, which protects against growth of pathogenic bacteria. Traditionally, the lowered a_w values are achieved by withdrawal of water by desorption or adsorption, and/or the addition of permissible additives such as salts and sugars. More recently other ingredients (called humectants) have been utilized, including glycerol, sorbitol and glycols (Jay, 1996). To control water by adsorption, food is first dried (often freeze-dried) and then subjected to controlled rehumidification until the desired composition is achieved. By desorption, the food is placed in a solution of higher osmotic pressure so that the desired a_w is reached at equilibrium (Robson, 1976). Although identical a_w values may be achieved by these two methods, IMF produced by adsorption are more inhibitory to microorganisms than those produced by desorption (Sloan *et al.*, 1976).

Gram-negative bacteria will not proliferate within the a_w range of IMF products. This is also true for most Gram-positive bacteria, with the exception of cocci, some spore-formers and lactobacilli. Molds and yeasts have the ability to grow in the IMF a_w range. Ultimately, antimicrobial activity results from interaction among the lowered a_w, the pH, the E_h, added preservatives (e.g., sorbates, benzoates and even some of the humectants), a competitive microflora, low storage temperatures, and the heat process applied during manufacturing (Jay, 1996).

4.3.3 Redox potential

The oxidation–reduction (or redox) potential (E_h) of a substrate is defined as the ease with which it gains or loses electrons. When an element or compound gains electrons it becomes reduced, whereas when it loses them it becomes oxidized. Oxidation also results from the addition of oxygen to the substrate. A substance that readily gives up electrons (i.e. that is readily oxidized) is a good reducing agent. A substance that readily takes up electrons (i.e. that is readily reduced) is a good oxidizing agent. When electrons are transferred from one compound to another, a potential difference is created between them. The more highly oxidized a substance is, the more positive will be its electrical potential; the more highly reduced a substance is, the more negative will be its electrical potential. When the concentration of oxidant and reductant is equal, a zero potential exists (Jay, 1996).

Aerobic microorganisms (e.g. *Bacillus*) require positive E_h values for growth, whereas anaerobic microorganisms (e.g. *Clostridium*) require negative E_h values. Some aerobic bacteria actually grow better under slightly reduced conditions; these organisms are called microaerophiles (e.g. *Lactobacillus* and *Campylobacter*). Other bacteria are able to grow under either aerobic or anaerobic conditions, and are called facultative anaerobes (e.g. *Staphylococcus* and *Salmonella*). Most molds encountered in food are aerobic, while most yeasts are facultative anaerobes (Jay, 1996).

Microorganisms affect the E_h of their environments during growth, just as they do pH. This is particularly true for aerobes, which can lower the E_h of their environment. As aerobes grow the oxygen is depleted, resulting in the lowering of E_h. However, growth is not slowed due to the ability of the microbial cell to make use of O_2-donating or H_2-accepting substances in the medium. The result is that the medium becomes poorer in oxidizing and richer in reducing substances (Morris, 1962). Microorganisms can reduce the E_h of their surrounding medium by producing metabolic by-products such as H_2S, which has the capacity to lower E_h to –300 mV. Because H_2S reacts readily with O_2, it will accumulate only in anaerobic environments (Jay, 1996).

Most plant and animal foods have a low E_h in their interiors because of the presence of reducing substances, such as reducing sugars and ascorbic acid in plants and sulfhydryl groups in meats. As long as the cells respire and remain active, they tend to balance the E_h system at a low level.Thus fresh fruits, vegetables and meats will have aerobic conditions at and near the surface. Plant foods, especially plant juices, tend to have E_h values of 300–400 mV, and aerobic bacteria and molds are the common cause of spoilage of products of this type. Solid meats can support the growth of aerobic slime-forming bacteria on the surface while simultaneously permitting growth of anaerobic bacteria in the interior (Genigeorgis and Riemann, 1979; Jay, 1996).

Food-processing operations can change the E_h of foods by destroying or altering reducing and oxidizing substances and/or allowing oxygen to diffuse into the tissues. As E_h changes as a result of biochemical changes in food and/or microbial metabolism, new species of microorganisms may succeed those that initiated growth.

4.3.4 Competitive microorganisms

In general, microbial interference is a phenomenon of non-specific inhibition or destruction of one species of microorganisms by others in the same environment, due to competition for nutrients, competition for attachment/adhesion sites, rendering the environment unfavorable, or a combination of these (Jay, 1995). It is known that the background biota needs to be numerically larger than the organism to be inhibited, and that the interfering biota are generally not homogeneous; however, the specific roles that individual species play are often unclear (Jay, 1996).

An example of interference is the inhibition or killing of a spectrum of food spoilage and/or pathogenic organisms by mixed cultures of lactic acid bacteria. Bacteriocins, pH depression, organic acids, hydrogen peroxide, diacetyl and possibly other products effect the inhibition (Jay, 1996). These compounds and their corresponding mechanisms of action are discussed in section 8.

Microorganisms that can be added to a food product to effect preservation have been designated protective cultures by Holzapfel *et al.* (1995). Lactic acid bacteria represent the largest and most important group in this category, and have the following properties: they present no health risks, they provide beneficial effects beyond the interference phenomenon, and they have no negative impact on sensory properties.

4.4 Examples of application

Kanatt *et al.* (2002) developed a number of ready-to-use, shelf-stable, intermediate-moisture (IM) spiced mutton and spiced chicken products, with a combination of reduced moisture, vacuum packaging and irradiation. The a_w of the products was reduced to about 0.80 either by grilling or by hot-air drying. These IM products were vacuum-packed and subjected to gamma radiation at 0–10 kGy. There was a dose-dependent radiation reduction in total viable counts and in numbers of *Staphylococcus* spp. IM meat products that did not undergo radiation showed visible mold growth within 2 months; products irradiated at 10 kGy showed an absence of viable microorganisms and retained high sensory acceptability for up to 9 months at ambient temperatures.

Marin *et al.* (2002) developed methods to prevent fungal growth of common contaminants on bakery products, including the genera *Eurotium*, *Aspergillus* and *Penicillium*. A factorial design was used to test logarithmically increasing levels of calcium propionate, potassium sorbate and sodium benzoate (0.003 %, 0.03 % and 0.3 %), pH (4.5, 6 and 7.5) and a_w (0.80, 0.85, 0.90 and 0.95). Potassium sorbate at 0.3 % was found to be the most suitable preservative in combination with the common levels of pH and a_w in Spanish bakery products. Sub-optimal concentrations led to an enhancement of fungal growth. None of the preservatives had a significant inhibitory effect at neutral pH.

Uyttendaele *et al.* (2001) studied the growth and survival of *E. coli* O157:H7 exposed to a combination of sub-optimal factors in a red meat medium (beef gravy). Hurdles were temperature (22 °C, 7 °C and –18 °C), pH (4.5, 5.4 and 7.0) and lactic acid. Prolonged survival was noted as the imposed stress became more severe, and as multiple growth factors became sub-optimal. At 7 °C or –18 °C, survival was prolonged at the more acid pH; and at a pH of 4.5, greater survival was observed at 7 °C than at 22 °C. The addition of lactic acid instead of HCl to reduce pH to 4.5 resulted in a more rapid decrease of the pathogen. However, high survival was observed in beef gravy, pH 5.4 at –18 °C (simulation of frozen meat), with an unsatisfactory reduction from log 3.0 to log 1.3 after 43 days, and in beef gravy at pH 4.5 with 5 % NaCl and 7 °C (simulation of a fermented dried meat product kept refrigerated). There was less than 1 log reduction in 43 days. These results suggest that preservatives may inhibit multiplication but induce prolonged survival of *E. coli* O157:H7.

Terebiznik *et al.* (2002) investigated the effect of nisin combined with pulsed electric fields (PEF) and a_w reduction by NaCl on the inactivation of *E. coli* in a simulated milk ultrafiltrate medium. The reduction of a_w from 0.99 to 0.95 by NaCl, without any other hurdle, did not affect *E. coli* viability. A reduction in PEF effectiveness occurred when the NaCl concentration was increased because of an increase in conductance, which reduced the pulse decay time. In cells subjected to PEF, nisin activity was decreased – probably due to the non-specific binding of nisin to cellular debris or the emergence of new binding sites in or on cells. However, the lethal effect of nisin was re-established and further enhanced when a_w was reduced to 0.95. A synergistic effect was evident when low-intensity PEF was applied. Decreasing a_w to 0.95 and applying PEF at 5 kV/cm (a non-lethal intensity when no other hurdle is used) with the further addition of nisin (1200 IU/ml) resulted in a 5-log reduction of the bacterial population.

Application of hurdle technology has largely been based on good judgment as well as trial and error. Although this empirical approach has proved successful in practical food design, food engineers question the lack of a mathematically and statistically sound base (Leistner and Gould, 2002). McMeekin *et al.* (2000) emphasized that a very sharp cut-off often occurs between conditions permitting growth and those preventing growth, allowing combinations of factors that deliver long-term stability and safety to be defined precisely and modelled (see also section 11, on predictive bacteriology).

5 Fermentation and safety

5.1 History and principles

Fermentation is one of the oldest methods of food processing. Foods such as bread, beer, wine and cheese originated in ancient times; and the principles involved in their manufacturing have hardly changed (Nout, 2001). There are many benefits associated with fermentation. It can preserve food (i.e. increase shelf-life), improve digestibility, enrich food nutritionally, and enhance taste and flavor. Furthermore, fermentation has the potential to enhance food safety by removing toxic components and controlling pathogens. Thus, it makes a significant contribution to human nutrition, particularly in developing countries where economic problems are a major barrier to ensuring food safety (e.g. the absence of freezing and/or refrigeration facilities) (Motarjemi *et al.*, 2001).

The microbes involved in fermentation include molds (mycelial fungi), yeasts (unicellular fungi) and bacteria. Microbial enzymes break down carbohydrates, lipids, proteins and other food components, thus improving food digestion and increasing bioavailability. Substances of microbial origin found in fermented foods include organic acids, alcohols, aldehydes, esters and many others. The presence of living microbial cells such as in non-pasteurized yogurt may have advantageous effects on the intestinal microflora and, indirectly, on human health (Nout, 2001).

In the strict sense, fermentation refers to a form of anaerobic energy metabolism. Nevertheless, in the context of fermented foods microbial growth and metabolism can also take place under aerobic conditions. Mold-related fermentations provide a typical example. There is a huge variety of fermented foods worldwide. Those of plant origin are derived from a variety of raw materials of different chemical composition and biophysical properties: cereals and potatoes having high starch contents; legumes and oil seeds rich in proteins; fruits containing high concentrations of reducing sugars; and green vegetables, carrots, tomatoes, olives and cucumbers of high moisture content. Most fermentative processes of vegetables and cereals are due to the action of lactic acid bacteria, often in combination with yeasts; other bacteria such as *Bacillus* spp., or mycelial fungi such as *Rhizopus* and *Aspergillus* spp., are equally important in the fermentation of legumes and oil seeds. Foods of animal origin are highly perishable products, and fermentation has long been an effective method for prolonging their shelf-life. Lactic acid bacteria are the main organisms carrying out the fermentation of animal products, although halotolerant bacteria and yeasts are also involved in fermenting fish and seafood (Nout, 2001).

Fermentations can be distinguished according to the physical state under which they take place. In liquid fermentations, the microorganisms are suspended while a mixing device is used to ensure homogeneity. Control of temperature and levels of dissolved oxygen can be achieved with immersion coolers or heaters, and aeration. In solid-state fermentations, even though the particulate matter contains sufficient water to allow microbial growth, water is not in the continuous phase; gas (air) is, but it is a poor heat conductor, and solid-state fermentations tend to develop gradients not only of temperature but also of gas composition. Controls of homogeneity and mixing systems are thus more complex than in liquid fermentations. There are also intermediate situations, such as shredded vegetables or olives, whose fermentations are carried out in brine; these are called immersed liquid fermentations (Nout, 2001).

5.2 Lactic acid bacteria

Lactic acid bacteria predominate in the majority of fermented foods. Their growth and metabolism inhibit the normal spoilage flora and bacterial pathogens through either bacteriostatic or bactericidal action. With toxigenic pathogens, the bacteriostatic action can effectively ensure safety (assuming that initial pathogen numbers are below those necessary to produce illness). With infectious pathogens, bacteriostatic action can be insufficient because the infectious dose of some is very low. In this case, bactericidal action is indispensable,and complete elimination of risk will also depend on other factors such as the type of pathogen considered and its initial numbers and physiological state (Adams, 2001).

Lactic acid bacteria form a group of Gram-positive, non-spore-forming, fermentative anaerobes that are often aerotolerant. They produce most of their cellular energy as a result of sugar fermentation. In the case of hexoses, this can proceed by one of two pathways (Axelsson, 1998): *homofermenters* ferment hexoses by the Embden–Meyerhof–Parmas glycolytic pathway, yielding almost exclusively lactic acid; and *heterofermenters* produce less acid overall as a mixture of lactic acid, acetic acid, ethanol and carbon dioxide, using the 6-phosphogluconate/phosphoketolase pathway.

Antimicrobial activity by lactic acid bacteria derives from low pH, organic acids, bacteriocins, carbon dioxide, hydrogen peroxide, diacetyl, ethanol and reuterin, as well as nutrient depletion and overcrowding. The specific modes of action of these antimicrobial agents are described in section 7 of this chapter. The predominant actions are the production of organic acids and pH reduction, with the others contributing to the aggregate effect, particularly by ensuring the successful early dominance of lactic acid bacteria (Adams, 2001).

5.3 Control of microbial hazards in fermented foods

5.3.1 Bacteria

Factors to be considered are decontamination of raw materials; prevention of product contamination by cleaning, disinfection and zoning (mainly segregation of raw from finished product); controlled fermentation and ripening processes (primarily related to temperatures); and whether the product will be consumed without terminal heating (Beumer, 2001).

Prevention of contamination of raw materials with pathogenic microorganisms is a primary safety measure. However, total removal of pathogens often becomes a hopeless task, considering that pathogens are frequently present in soil as well as in surface water and the gut of animals (and are therefore also found on the surfaces of fruits and vegetables or embedded in plant tissue). To minimize risks, pretreatment of raw materials against pathogens is helpful. For example, washing in clean water can eliminate up to 90 % of pathogens; cleaning and disinfection of surfaces and equipment and proper hygiene practices by plant personnel are also important. Another possibility is heat treatment (e.g. pasteurization), typically used for milk. If pasteurization is not possible, a post-process pasteurization step may be necessary to eliminate pathogens in the final product – for example in fermented sausages – although this may have serious effects on sensory quality (Beumer, 2001).

Zoning or segregation of process steps helps to contain contamination and should be part of an HACCP plan. As applied to fermentation processes, there should be separate areas for the storage of the raw materials, the preparation of raw materials (i.e. washing, cutting and adding ingredients), the fermentation process *per se*, the filling of packages, and the storage of final products. A typical example of an area requiring strict hygiene is the room where starter cultures are prepared for the lactic fermentation of milk, where potential dangers are contamination and the infection of cultures with bacteriophages (Beumer, 2001). In meat processing, the use of the same area of the plant for slicing raw fermented sausages and dry-cured ham containing high levels of viable lactic acid bacteria can result in contamination of cooked, vacuum-packed, cured meat products, and accelerate their spoilage. Every effort should therefore be made to ensure that separate slicing areas for these different product types (raw and cooked) are established.

The following are the major pathogenic bacteria and their association with fermented foods.

5.3.1.1 Aeromonas

Aeromonas are found in water (including sewage) and in food products that have been in contact with contaminated waters, such as seafood and vegetables (see Chapter 10). *Aeromonas hydrophila* is the most important pathogen (Roberts *et al.*, 1996).

In general, *Aeromonas* do not seem to be a serious risk in well-produced fermented foods. For example, the addition of *Aeromonas* to skim milk during lactic fermentation and to yogurt resulted in a sharp decrease in pathogen numbers (Aytac and Ozbas, 1994; Ozbas and Aytac, 1996). Spanish fermented sausages (longaniza and chorizo) contained *Aeromonas* in numbers between 1.0 and 4.5 $\log_{10}$ CFU/g; the hygienic states of the factories significantly influenced their incidence and numbers. However, aeromonads were rapidly inactivated during the early stages of manufacture regardless of initial contamination (Encinas *et al.*, 1999).

5.3.1.2 Campylobacter species

Campylobacter species, mainly *Campylobacter jejuni*, *Campylobacter coli* and *Campylobacter lari*, cause severe gastroenteritis (Roberts *et al.*, 1996; see Chapter 7).

Although the infective dose of *Campylobacter* is relatively low compared to that of other pathogenic microorganisms, *Campylobacter* transmission by fermented foods is regarded as not significant, since the organisisms do not grow below 30 °C and are sensitive to freezing, drying, low pH and sodium chloride. This has been confirmed by artificial contamination of fermented products such as yogurt and salami, where the numbers of the pathogen decreased rapidly (Northolt, 1983; Morioka *et al.*, 1996).

5.3.1.3 *Vibrios*

Among vibrios, *Vibrio cholerae* and *Vibrio parahaemolyticus* are the most important pathogenic species. These are found mostly in coastal marine waters, and are therefore associated with shellfish and other marine animals (Roberts *et al.*, 1996; see Chapter 5). Vibrios are not acid tolerant, and are inhibited by lactic acid bacteria isolated from fermented fish products (Ostergaard *et al.*, 1998). This has also been observed in pickles and squid shiokara that were artificially contaminated with *V. parahaemolyticus* (Yu and Chou, 1987; Wu *et al.*, 1999).

5.3.1.4 *Bacillus cereus*

Bacillus cereus causes both foodborne infection characterized by diarrhea and intoxication characterized by vomiting, after ingestion of the toxin cereulide. As a spore-former, *B. cereus* is ubiquitous and may be found in cereals and spices (Roberts *et al.*, 1996; see Chapter 15).

The behavior of *B. cereus* has been studied in fermented products such as tempeh, sauce-based fermented salads, and fish sausage. There was initial growth in all cases, but in products where pH decreased due to lactic acid bacteria activity, inhibition correlated directly with the degree of acidity. In tempeh, where no lactic acid fermentation is involved, *B. cereus* numbers reached 10^8 CFU/g, although soaking of the soybeans and subsequent acidification below pH 4.5 stopped its growth (Nout *et al.*, 1989; Aryanta *et al.*, 1991; Bonestroo *et al.*, 1993).

5.3.1.5 *Clostridium botulinum*

Botulism is a neuroparalytic syndrome caused by the botulinum neurotoxin. The organism responsible, *Clostridium botulinum*, is ubiquitous, and spores are widely distributed in the soil, shores and bottom deposits of lakes and coastal waters, and in the intestinal tracts of fish and animals (Roberts *et al.*, 1996; see Chapter 13).

Most outbreaks of foodborne botulism have been related to inadequate processing of vegetables, fish and meats (e.g. home processing). Nitrites inhibit this pathogen and are a safety factor in the production of cured and fermented meats (Beumer, 2001).

5.3.1.6 *Staphylococcus aureus*

Staphylococcus aureus produce several heat-stable enterotoxins. *S. aureus* is present on the skin and mucous membranes of warm-blooded animals, including humans. Contamination of cooked or ready-to-eat food products by a colonized person followed by storage at ambient temperatures is often implicated in outbreaks (Roberts *et al.*, 1996; see Chapter 14).

Fermented sausages and raw milk cheeses have been associated with *S. aureus* outbreaks, but in general the organism is regarded as a poor competitor, and its growth in fermented foods is typically associated with a failure of the normal fermenting flora (Bacus, 1986; Johnson *et al.*, 1990; Nychas and Arkoudelos, 1990). González-Fandos *et al.* (1999) investigated survival and toxigenesis of *S. aureus* in Spanish-type dry sausages (chorizo and salchichón). *Lactobacillus curvatus* in combination with dextrose and relatively low fermentation temperatures (< 20 °C) was an effective anti-staphylococcal agent during fermentation. Portocarrero *et al.* (2002) inoculated

S. aureus in fresh hams that were then cured, equilibrated to ≥ 2.5 % NaCl, cold- or non-smoked, and aged. Their results indicated that higher salt contents and lower a_w values played a decisive role in controlling the growth and toxin production of the pathogen.

5.3.1.7 *Listeria monocytogenes*

Listeria monocytogenes is ubiquitous, and is a human and animal pathogen. Approximately 30 % of cases of listeriosis are perinatal, with 20–25 % of those cases being fatal for the newborn. Most other cases occur in immuno-compromised persons, with a death rate varying from 30 % to 50 % (Roberts *et al.*, 1996; see Chapter 9).

Listeria is not particularly acid-tolerant, but in fermented products where a mold-ripening step is involved, the rise in pH can allow surviving cells to resume growth (Beumer, 2001). Products such as home-made sausages and mold-ripened cheeses have been associated with listeriosis (Farber *et al.*, 1993; Nissen and Holck, 1998).

5.3.1.8 *Enterobacteriaceae*

Enterobacteriaceae are typical components of the fecal flora of animals, which is the reason for the use of some members as hygiene indicators for processed foods (Beumer, 2001).

Salmonella is often found on raw meat. Widely distributed in the environment, salmonellae are present in the gut of infected animals and humans, and are shed in the feces (see Chapter 3).

Meat, milk, poultry, and eggs are the main vehicles for *Salmonella* transmission, and fermented foods derived from these raw materials, including salami and cheeses, have been occasionally associated with *Salmonella* outbreaks (Leyer and Johnson, 1992; Beumer, 2001). Sauer *et al.* (1997) studied a number of cases of human salmonellosis caused by *S.* Typhimurium associated with Lebanon bologna. The pathogen might have survived the fermentation process used by the manufacturer due to its high numbers in the raw meat (> 10^4 CFU/g). Stricter process controls in the manufacture of semi-dry fermented sausages were suggested. Inhot *et al.* (1998) inoculated a six-strain cocktail of *S.* Typhimurium into pepperoni batter. Fermentation and drying resulted in about a 3.0-$\log_{10}$ reduction in numbers of the pathogen, and subsequent vacuum storage at ambient temperature was more lethal than refrigerated storage.

Shigella is not a natural inhabitant of the environment (see Chapter 10). Person-to-person transmission due to poor personal hygiene and the consumption of contaminated water or foods washed with contaminated water cause infection (Roberts *et al.*, 1996). The rapid decrease in *Shigella* numbers at pH < 5.0 indicates that this microorganism is in fact a minor risk in fermented foods (Nout *et al.*, 1989; Kunene *et al.*, 1999).

Yersinia enterocolitica present in raw milk, seafood and raw pork has been associated with foodborne infections (see Chapter 8). Survival of *Yersinia* has been tested in fermenting milk and yogurt. Although there was some growth during the first few hours of the fermentation process, numbers fell below detection levels after completion of fermentation and a 4-day storage period (Ozbas and Aytac, 1996; Bodnaruk and Draughon, 1998).

E. coli O157:H7, because of its acid resistance, can survive the fermentation process. In recent years there have been a number of outbreaks involving fermented foods where this pathogen has been found responsible (Beumer, 2001; see Chapter 6).

The fate of various *E. coli* O157:H7 strains during the fermentation and storage of diluted cultured milk drink fermented with *Lactobacillus casei* spp. *casei* and *Lactobacillus delbrueckii* ssp. *bulgaricus* was investigated by Chang *et al.* (2000). All strains of the pathogen grew rapidly in skim milk and reached a maximum population of *ca.* 8.0–9.0 $\log_{10}$ CFU/ml after 24 hours. However, populations declined as cultivation proceeded further. Viable cells of *E. coli* O157:H7 were reduced to non-detectable levels in the non-sugar-added cultured drink. Sugar extended the survival period according to the pathogen strain and the amount of sugar added to the system.

Dineen *et al.* (1998) studied the persistence of *E. coli* O157:H7 as a post-pasteurization contaminant in fermented dairy products, its ability to compete against commercial starter culture in fermentation systems, and its survival in the yogurt production process. These authors concluded that *E. coli* O157:H7 is a serious potential health hazard in the case of post-processing entry into fermented dairy products, and that commercial starter cultures differ widely in their ability to reduce *E. coli* O157:H7 numbers in fermentation systems. The pathogen, inoculated at 10^5 CFU/ml in the starting milk, did not survive the yogurt manufacturing process after curd formation.

Faith *et al.* (1998) prepared beef jerky batter with fat contents of about 5 and 20 % and inoculated it with a five-strain mix of *E. coli* O157:H7 at 10^8 CFU/g. Pathogen numbers were determined in both the raw batter and the strips formed from it after a number of drying processes of different lengths and temperatures. There was no direct correlation between the moisture-to-protein ratio of the final product and the viability of the pathogen. However, higher fat contents, longer drying times and lower drying temperatures increased the viability of *E. coli* O157:H7.

Studies to determine the fate of *E. coli* O157:H7 during the production and storage of fermented dry sausage were conducted by Glass *et al.* (1992). A commercial sausage batter inoculated with 4.8×10^4 *E. coli* O157:H7 per g was fermented to pH 4.8 and dried until the moisture-to-protein ratio was $\leq 1.9:1$. The sausage chubs were then vacuum-packed and stored at 4 °C for 2 months. The organism survived but did not grow during fermentation, drying or subsequent storage, and had decreased by about 2 $\log_{10}$ CFU/g by the end of the storage period. The importance of using beef containing low populations of or no *E. coli* O157:H7 in sausage batter was stressed, because when initially present at 10^4 CFU/g, this organism could survive fermentation, drying and storage of sausage regardless of whether a starter culture was used.

5.3.2 Fungi

A number of species of fungi can produce relatively low molecular weight metabolites that are toxic to humans and domesticated animals and are called mycotoxins (see Chapter 16). Their biosynthesis has been associated with pre-harvest production by fungi that are obligate endophytes of plants, plant pathogens, or members of the flora responsible for the decay of plant materials. However, the highest concentrations of these toxic metabolites are produced by fungi growing on post-harvest commodities that are stored under inappropriate conditions. Considering that a number of fermented foods involve a mold-ripening stage, and that some of these molds are known to be toxigenic, the safety implications with regard to fermented foods are evident (Moss, 2001).

5.3.2.1 Aflatoxins

Aflatoxins are a family of complex heterocyclic metabolites regarded as acutely toxic, carcinogenic and immunosuppressive. They have been found in a wide range of tropical and subtropical products, such as figs, pistachio and Brazil nuts, spices, peanuts and maize. The most important of these commodities used as a raw material for fermented foods is maize, but there have also been reports of low concentrations of aflatoxins in rice and wheat.

Although aflatoxins are produced by a small number of species within the genus *Aspergillus* (e.g. *Aspergillus flavus* and *Aspergillus parasiticus*), they are widespread and have several routes of contamination (Moss, 2001). Direct contamination occurs through mold spoilage of stored products or pre-harvest growth by endophytic association of an aflatoxin producer with plants such as maize or groundnuts, followed by some kind of stress on the growing crop (primarily drought); this is followed by passage through the food chain into animal products such as milk, following the consumption of contaminated feed by farm animals.

Aflatoxins in milk can resist cold storage, heat treatment or spray drying; and they are not completely degraded by the fermentation processes used in the manufacture of cheese, cream or butter (Yousef and Marth, 1989). However, they may be detoxified by the action of *Lactococcus lactis*, which is able to degrade them into harmless compounds (El-Nezami *et al.*, 1998). Alternatively, aflatoxins can be removed from the gastrointestinal tract by probiotics, using strains of *Lactobacillus* and *Propionibacterium* (Ahokas *et al.*, 1998); however, the most effective strategies for limiting human exposure to aflatoxins are to avoid contamination in the first place or to use a chemical process (e.g. ammoniation) to degrade them irreversibly in animal feeds (Riley and Norred, 1999).

5.3.2.2 Ochratoxin A

Ochratoxin A is produced by *Penicillium verrucosum* and by a number of *Aspergillus* species, especially *Aspergillus ochraceus*, in temperate climates. It is most common in cereals such as barley, oats, rye and wheat, but has also been found in maize, coffee, cocoa, dried vine fruits, wine and beer (Moss, 1996; Pittet, 1998). Ochratoxin A is relatively thermostable and is not destroyed by most food processes. It is nephrotoxic and possibly carcinogenic (De Groene *et al.*, 1996).

Like aflatoxins, ochratoxin A can also pass through the food chain and may be found in meat products, especially pork, but does not seem to be secreted effectively in cow's milk (Krogh, 1987; Valenta and Goll, 1996). There have been several reports of its presence in meat products such as kidneys, liver and even sausages, due to transfer from animal feeds. However, its presence in moldy meat products such as smoked pork or sausages is much more serious (Kuiper-Goodman and Scott, 1989).

5.3.2.3 Patulin

Patulin is produced by a number of species of *Penicillium*, *Aspergillus* and *Byssochlamys*, but in the context of human foods, the most important species is *Penicillium expansum*. This mold is especially associated with a soft rot of apples, and the natural occurrence of patulin in commercial apple juice was reported as long ago as 1972. It is toxic to mammals, forming sarcomas (Preita *et al.*, 1994; Moss, 2001).

It is known that patulin disappears during fermentation of apple juice to cider using the yeast *Saccharomyces cerevisiae*; but most importantly, a patulin-free cider depends on the quality of the apple juice used in its manufacture (Harwig *et al.*, 1973; Moss, 2001).

5.3.2.4 Fumonisins

Fumonisins are produced by *Fusarium moniliforme* and related species within *Fusarium*. They are associated with esophageal carcinoma in humans and mostly present in maize and maize products (e.g. polenta, corn flakes and popcorn). In fact, fumonisins are remarkably widespread in corn-based products and can occur at high concentrations (Rheeder *et al.*, 1992; Pittet, 1998; Moss, 2001).

Fumonisins are relatively stable to elevated temperatures, and survive a range of cooking, baking and frying processes (Jackson *et al.*, 1997). They can also survive the alkaline process used in the manufacture of tortillas in Central and South America (Scott and Lawrence, 1996).

5.3.3 Viruses

Viruses differ significantly from bacteria because of the fact that they are obligate intracellular parasites and must replicate exclusively within an appropriate living host cell (see Chapter 11). Thus, the source of all viruses is an infected being shedding infectious particles in its immediate environment. Transmission to another host can be direct, as in person-to-person spread (e.g. through aerosols created by vomiting), or indirect, involving some other agent as a carrier for the virus (Carter and Adams, 2001).

Foods are contaminated with viruses as a result of the distribution of fecal derived and/or vomit-derived viruses through the environment, eventually contaminating the food or water of another potential host. This process can occur directly, as in the case of an infected food handler passing the pathogen to food immediately before consumption, or it could be the result of a long distribution process, moving virus from fecal material to supplying waters. Effectively water, either directly ingested or used as an ingredient or washing agent during food processing, is the chief vehicle for disseminating enteric viruses. It can carry viruses to plants in the fields through irrigation, and to the most significant vehicle of foodborne viral diseases, molluskan shellfish. Viruses can survive well in water, assisted by high protein, calcium and magnesium contents. These conditions are typical of sewage and sewage-contaminated waters (Carter and Adams, 2001).

There are few data related to the survival of viruses in fermented foods, which complicates the assessment of controlling foodborne viral diseases through fermentation. Nevertheless, some general inferences can be drawn (Carter and Adams, 2001). The type of food on which the fermented product is based is an important factor. Shellfish are obvious high-risk materials because they often live in sewage-contaminated estuarine waters and concentrate available viruses in their tissues. Thus, fermented foods prepared from organisms that consume shellfish, such as octopus or scavenging crabs, may be indirectly contaminated. Vegetable produce may be contaminated on the surface (through handling, washing or spraying with contaminated water), or deep within the tissues (e.g. resulting from the uptake of viruses contaminating irrigation waters). Surface contamination can be removed by careful peeling.

5.3.4 Parasites

Typically, parasitic infections are acquired by eating food products that are either raw or incompletely cooked, or poorly preserved (see Chapter 12). Most of those infections are preventable if proper processing is applied to destroy the pathogens. However, many infections are commonly associated with eating habits that have been in practice for generations (Taylor, 2001). There are few reports of parasite-related diseases following the ingestion of fermented foods. One example is the infection with *Giardia*, a protozoan parasite, occurring as the result of consumption of contaminated cheese dip. Fermentation alone and the physicochemical conditions associated with it may not be sufficient to prevent the transmission of foodborne parasites. Thus, potentially infected material should be avoided wherever possible or, alternatively, subjected to freezing or some form of heat treatment (Taylor, 2001).

6 Food packaging

6.1 Introduction

Modified atmosphere for food packaging (MAP) includes vacuum packaging, gas flushing, and naturally respiring products, and involves the use of special permeable films, and controlled-atmosphere packaging (CAP). In CAP the product is continually exposed to a constant mixture of gases, while in gas-flushing or gas-packaging the particular gas mixture desired is flushed only once at the time of packaging into an evacuated or non-evacuated environment surrounding the food (Farber, 1991).

Oxygen, nitrogen and carbon dioxide are the three main gases used. Oxygen (O_2) will stimulate the growth of aerobic bacteria and can inhibit the growth of strictly anaerobic bacteria, although there is a very wide variation in the sensitivity of anaerobes to oxygen. Oxygen is very important in MAP meats to maintain myoglobin in its oxygenated form, oxymyoglobin, which is the form that most consumers associate with fresh red meat. Nitrogen (N_2) is an inert, tasteless gas that displays little or no antimicrobial activity. Because of its low solubility in water, the presence of N_2 in a MAP food can the prevent pack collapse that can occur when high concentrations of carbon dioxide are used. In addition, nitrogen, by displacing oxygen in the pack, can delay rancidity and also inhibit the growth of aerobic microorganisms.

Carbon dioxide (CO_2) is both water- and lipid-soluble, and is mainly responsible for the bacteriostatic effect on microorganisms in MA environments. Not only does CO_2 have biostatic activity; it is also known to have an inhibitory effect on product respiration. Although the specific way in which CO_2 exerts its bacteriostatic effect is unknown, the overall effect on microorganisms is an extension of the lag phase of growth and a decrease in growth rate during the logarithmic phase (Farber, 1991). CO_2 acts on a bacterial cell by alteration of cell membrane function, including nutrient uptake and absorption, by the inhibition of enzymes. Penetration of bacterial membranes may lead to intracellular pH changes; and changes in the physicochemical properties of proteins (Daniels *et al.*, 1985; Dixon and Kell, 1989). The inhibitory effects of CO_2 on microorganisms in a culture medium or food depend on many factors, including its partial pressure and concentration, the volume of headspace, the water activity, the type of microorganism, the microbial growth phase and the growth medium. The storage temperature of a CO_2-MAP product should be kept as low as possible, because the solubility of CO_2 decreases dramatically with increasing temperature. Thus, improper temperature control will usually eliminate the beneficial effects of an elevated CO_2 concentration (Farber, 1991).

CAP may be done by the introduction at packaging of a sufficient volume of gas in the package in such a way that the concentration of gas in the headspace does not change during storage. An example is the use of 100 % CO_2 in master packs of fresh pork or beef, where the volume ratio of CO_2 to the total volume of meat in the package is > 2 : 1. During storage, the gas volume decreases as CO_2 dissolves in the meat, but there is sufficient CO_2 present to maintain the CO_2 concentration at 100 % in the main package headspace. At appropriate storage temperatures, this approach has yielded shelf-lives of > 10 weeks for pork and beef (Jeyamkondan *et al.*, 2000). While such systems have been successfully used for intercontinental transport of fresh beef, more common systems use CO_2, O_2 and N_2 alone or combined. Nitrogen oxide, sulfur dioxide and even carbon monoxide are other gases that have potential (Farber, 1991).

6.2 The use of MAP in food products

6.2.1 Red meats

Carbon dioxide, or combinations of CO_2 with O_2, can effectively extend the shelf-life of meats. The closer the temperature to 0 °C, the higher the CO_2 concentration and the lower the number of bacteria, the longer the shelf-life extension is (Clark and Lentz, 1969). When dealing with MAP of red meats, there are four areas of concern: the control of bacterial pathogens and spoilage microorganisms; the maintenance of meat color; the control of weight loss; and the development of meat tenderness (Farber, 1991).

Consumers relate the red color of meat to its freshness, attributing color changes to bacterial spoilage or to old animals. Color deterioration has been identified as the major limiting factor for the marketability of fresh red meats (Shay and Egan, 1987), and is known as 'loss of bloom' (Cross *et al.*, 1986). The brown color is due to myoglobin oxidation to form metmyoglobin, which signals approaching staleness; however, under vacuum or anoxic atmospheres, the red color is changed due to lack of oxygen. This has no correlation with freshness or bacterial spoilage, but rather with the transition from oxymyoglobin to deoxymyoglobin (i.e. the reduction of the heme moiety of oxymyoglobin, yielding a purplish red color). Such meats will normally 'bloom' to the desirable red color after 30 minutes exposure to air. Nevertheless, until consumers are educated regarding the color of meat in vacuum and MA packages, these types of packaging should not be used for the retail-ready sale of meats (Jeyamkondan *et al.*, 2000).

Pseudomonas, Moraxella, Psychrobacter and *Acinetobacter* species are the main spoilage bacteria of aerobically stored, chilled, fresh red meats, and are generally inhibited by concentrations of 20 % CO_2 or greater. Gram-positive microorganisms, such as *Lactobacillus* spp. and *Brochothrix thermosphacta*, are usually resistant to inhibition by CO_2. Thus, a shift from an initial Gram-negative aerobic spoilage flora to a predominantly Gram-positive facultative anaerobic microflora dominated by *Lactobacillus* spp. occurs in meat during MA storage or MAP (Newton and Gill, 1978; Holley *et al.*, 2003). Lactic acid bacteria are capable of continuously pumping out CO_2 from inside the cells to the environment, thereby maintaining metabolic balance. Because this process is rather energy consuming, the growth rate of lactic acid bacteria is fairly low (Jeyamkondan *et al.*, 2000) and the by-products of the metabolism of lactobacilli are inoffensive compared to the typical proteolytic spoilage odors produced by *Pseudomonas* spp. Depending on initial levels of sanitation, facultative Gram-negative organisms such as *Aeromonas* or *Shewanella* may be problematic in MAP-stored fresh meats (Newton and Gill, 1978; Holley *et al.*, 2004).

Potential pathogens in meat stored under anaerobic conditions and at low temperatures are *Yersinia enterocolitica, Listeria monocytogenes* and *Aeromonas hydrophila*, which can grow at low temperatures. However, when 100 % CO_2 and a storage temperature of −1.5 °C are simultaneously applied according to the hurdle concept, none of these pathogens can grow (Farber, 1991).

A study on the behavior of *L. monocytogenes* and *L. innocua* in raw minced beef packaged under MA was carried out by Franco-Abuín *et al.* (1997). Three gas atmospheres were tested with various CO_2 concentrations: 100 % CO_2; 65 % CO_2, 25 % O_2, 10 % N_2; and 20 % CO_2 with 80 % O_2. The 100 % CO_2 atmosphere was the most effective for the inhibition of growth of both species. Microbial inhibition was influenced by pH, but low pH values were not the most important factor in the inhibition of *Listeria*; instead, it was the direct effect of CO_2. Water activity values did not change during storage, and none of the gas mixtures was bactericidal.

Tsigarida *et al.* (2000) studied the effect of aerobic, MAP (40 % CO_2, 30 % N_2, 30 % O_2) and vacuum packaging (VP) on the growth and survival of *L. monocytogenes* on sterile and naturally contaminated beef fillets in relation to film permeability and oregano essential oil. The dominant microorganisms and the effect of endogenous flora on the growth and survival of *L. monocytogenes* were dependent on the type of packaging film. The pathogen increased whenever *Pseudomonas* sp. dominated, that is, aerobic storage and MAP/VP in O_2 high-permeability film, suggesting that this spoilage group might be able to enhance the growth of *Listeria. B. thermosphacta* constituted the major proportion of the total microflora in MAP/VP. Within the O_2 low-permeability film, no growth of *L. monocytogenes* was detected. The addition of 0.8 % (v/w) oregano essential oil resulted in an initial reduction of 2–3 log cycles of the majority of the bacterial population. Lactic acid bacteria and *L. monocytogenes* showed the most extensive decrease in all gaseous environments, but there was limited growth of *L. monocytogenes* in MAP/VP regardless of film O_2 permeability.

The growth and virulence of pathogenic *Yersinia enterocolitica* were investigated by Bodnaruk and Draughon (1998) on high (pH > 6.0) and normal (pH < 5.8) pH pork packaged in MA and stored at 4 °C. MAs used in the study were vacuum packaging and saturated CO_2. Pork was packaged in a high gas-barrier packaging film and examined over a 30-day period. Numbers of *Y. enterocolitica* on the lean surface of high pH pork slices increased by about 2.7 log CFU/cm^2 when vacuum packaged and stored at 4 °C for 30 days. Storage of inoculated normal-pH pork in 100 % CO_2 resulted in *Y. enterocolitica* remaining in the lag phase over the storage period. Virulence of the pathogen was maintained in 25–35 % of isolates following storage for 30 days at 4 °C in vacuum- and CO_2-packaged meats, and was not affected by pH.

6.2.2 Poultry

The organisms most often associated with foodborne diseases involving poultry include *Salmonella* spp., *Staphylococcus aureus*, *Clostridium perfringens*, *Bacillus cereus* and *Campylobacter*. *Campylobacter jejuni* and *Salmonella* spp. (which may be able to survive in a MAP product) and *L. monocytogenes* and *A. hydrophila* may, because of the extended storage lives of the MAP products, have additional time to grow to potentially high numbers (Farber, 1991). Although *C. perfringens* may be able to survive better in some MA as compared to air, it would not be able to grow at the chill temperatures commonly used for MAP products. Thus, it would not be much of a health hazard in a MAP product unless the product was temperature abused (Labbe, 1989). On the contrary, *Campylobacter* is a cause for constant concern because, although not a psychrotroph, the infective dose is low.

To determine the role of packaging and storage conditions on the survival of *Campylobacter jejuni* on chicken, the virulent strain *C. jejuni* 81116 was inoculated onto chicken skin pieces and stored at different temperatures and under various packaging conditions (air, N_2, CO_2, and VP). *C. jejuni* remained viable at –20 °C and –70 °C. The pathogen could also withstand repeated freeze–thaw cycles. The importance of packing *C. jejuni*-free chicken was stressed (Lee *et al.*, 1998).

6.2.3 Seafood and fish

Seafood and fish, unlike other muscle foods, are very susceptible to both microbiological and chemical deterioration. The major spoilage organisms found on spoiled seafood and fish include *Pseudomonas*, *Moraxella*, *Acinetobacter*, *Flavobacterium*, *Photobacterium*, *Aeromonas*, *Shewanella*, *Carnobacterium* and *Cytophaga* species.

L. monocytogenes has been identified as a significant contaminant in smoked fish, particularly salmon. For example, Domínguez *et al.* (2001) analyzed about 170 samples of smoked fish and 182 samples of pâté for sale in retail outlets and supermarkets in the nine provinces of Castille and León (Spain) for the prevalence of this pathogen and other *Listeria* spp. *L. monocytogenes* was isolated from 22.3 % of the smoked fish samples and 5.4 % of the pâté samples. Of these, 52 % of the former but only 10 % of the latter contained *L. monocytogenes* at > 100 CFU/g. According to Rørvik (2000), raw salmon is not an important source of *L. monocytogenes*, and contamination of smoked salmon occurs during processing and storage. Thus, the main issue for producers is to prevent colonization of the processing environment and spread of the bacteria to products. This should be achieved by the systematic implementation of Good Manufacturing Practices, together with HACCP.

From a microbiological safety standpoint, the organism of greatest concern when dealing with MAP seafood and fish products is the non-proteolytic *C. botulinum* type E. The restricted growth of the normal spoilage bacteria by MAP may actually enhance the proliferation of *C. botulinum*. In fact, *C. botulinum* type E is a natural seafood and fish contaminant that can grow at temperatures as low as 3.3 °C (Farber, 1991). Particularly critical are seafood and fish products that are eaten without terminal heating – for example, cooked peeled shrimp and smoked salmon. The health risks associated with eating MAP smoked fish products may be high because of the general contamination of fish with botulinal spores that can survive smoking, fewer competing microorganisms, and consumption without further cooking. However, spoilage and/or safety problems may be regarded as the result of insufficient salting, gross temperature abuse or heavy initial contamination with botulinal spores (Hauschild, 1989).

Cai *et al.* (1997) investigated the production of toxins by *C. botulinum* type E in MA-packaged channel catfish. Samples of catfish were inoculated with 3–4 log spores/g of a mixed pool of four strains of the pathogen, and packaged with an oxygen-barrier bag with a MA of 80 % CO_2, 20 % N_2, or in a master bag with the same MA. Packaged fish were stored at either 4 °C and sampled at intervals over 30 days or at 10 °C and sampled at intervals over 12 days. Under abusive storage conditions of 10 °C, *C. botulinum* toxin was detected on fish from each package type by day 6. At 4 °C, toxin production was detected by day 18. No toxin was found in the master bags held continuously at 4 °C. Spoilage preceded toxin production for samples stored at 4 °C for each type of packaging, whereas at 10 °C spoilage and toxin production coincided.

The growth and toxigenesis of *C. botulinum* type E in vacuum-packaged, unprocessed, raw, pickled or cold-smoked rainbow trout stored at slightly abusive temperatures were studied by Hyytia *et al.* (1999). In unprocessed fish there was only a 2-log increase in type E cell numbers at the time the toxicity first occurred after 2 weeks of storage at 8 °C. Neither growth nor toxin production was observed in raw pickled fish with a NaCl concentration of 6.7 % (w/v) during 6 weeks of storage at 6 °C. In cold-smoked fish with a NaCl level of 3.2 % (w/v), toxic samples were detected after 3 and 4 weeks of storage at 8 °C and 4 °C, respectively, with no increase in type E cell count.

Challenge studies were carried out by Lyver *et al.* (1998) to evaluate the safety of raw and cooked seafood nuggets inoculated with 10^3 CFU/g of *L. monocytogenes*. Nuggets were packaged in air or 100 % CO_2, with and without an oxygen-absorbent agent, and stored at 4 °C or 12 °C. Headspace O_2 decreased to < 1 % (v/v) in most samples, while headspace CO_2 ranged between 1 % and 100 %, depending on the packaging conditions and storage temperature. Most products maintained an acceptable appearance throughout storage, but nuggets stored at 12 °C developed sharp, acidic odors by day 28. *Bacillus* spp. and lactic acid bacteria numbers increased to 10^2 and 10^7 CFU/g respectively in raw nuggets, while only *Bacillus* spp. reached 10^4 CFU/g in cooked nuggets by 28 days. With the exception of nuggets packaged in 100 % CO_2 with or without an absorbent, numbers of *L. monocytogenes* increased to approximately 10^7 CFU/g in the uncooked product stored at both 4 °C and 12 °C after 28 days.

6.2.4 Vegetables

With the increased popularity of fresh produce in Europe and North America over the last 25 years, there has been a parallel, significant increase in the number of incidents and cases of foodborne illness from fresh fruit, juice, and vegetable consumption. Vehicles have included tomatoes, strawberries, raspberries, lettuce, sprouts (alfalfa and other species and varieties), parsley, basil, apple cider, orange juice, cantaloupe, watermelon and salads (potato, pea, garden fruit). Organisms implicated have included viruses (hepatitis A), protozoan parasites (*Cyclospora*, *Cryptosporidium* and *Giardia*), and bacteria (*Salmonella* spp., *E. coli* O157:H7, other enterotoxigenic *E. coli*, *Shigella*, *C. botulinum*, and *Listeria*). The change in frequency of illness from these sources has been so substantial that there are now almost as many outbreaks of foodborne illness from these products as there are from meat products. Many of these incidents arise from the use of poor-quality water during irrigation or washing after harvest, and from the growth of pathogens on cut produce surfaces when offered for retail sale without appropriate refrigeration. Of continuing concern is the growth of pathogens on produce with an extended shelf-life under MAP (Guan and Holley, 2003).

Demand for fresh, convenient, minimally processed vegetables has led to an increase in the quantity and variety of products available to the consumer. MAP, in combination with refrigeration, is increasingly being employed to ensure the quality and shelf-life of ready-to-use vegetables. The nature of these products and the storage conditions have presented microorganisms with new vehicles. Psychrotrophic pathogens and those capable of maintaining infectious potential are of particular concern. *L. monocytogenes*, *A. hydrophila*, *C. botulinum*, *Salmonella* spp. and *Shigella* spp. are the main ones with respect to vegetables (Francis *et al.*, 1999).

The effect of initial head spaces of air and 5 % CO_2, 95 % N_2 on the microflora of tomato salad (i.e. lactic acid bacteria, pseudomonads and yeasts) was studied at 4 °C and 10 °C by Drosinos *et al.* (2000). Lactic acid bacteria were the predominant organisms in all samples. The pH dropped during storage, particularly at 10 °C. The concentration of different organic acids, such as lactic, acetic, formic and propionic acids, increased in all samples stored under MAP conditions at both temperatures. The spoilage of tomatoes stored under 5 % CO_2, 95 % N_2 was delayed, as indicated by changes in texture, color and odor compared with those samples stored in air. When the salad was inoculated with *S.* Enteriditis, the pathogen survived at both temperatures but did not grow regardless of the packaging system used.

Francis and O'Beirne (1998) used a solid-surface model system to study the effects of gas atmospheres encountered in MAP of minimally processed lettuce on the survival and growth of *L. monocytogenes* and competing microorganisms. The effects of increasing CO_2 levels from 5 % to 20 %, of 3 % O_2, and of 80–95 % N_2 were determined. CO_2 concentrations of 5–10 % had no inhibitory effect on pure cultures of *L. monocytogenes*. Growth and inhibitory activities against *Listeria* or *Enterobacter cloacae* and *E. agglomerans* were inversely related to the concentration of CO_2. In contrast, the growth and anti-listerial activities of *Leuconostoc citreum* increased with elevated CO_2 concentrations. In the low O_2 atmosphere, *L. monocytogenes* grew considerably better in the presence of indigenous microflora of lettuce than when in pure culture. These results indicated that the gas atmospheres present within MA packages of minimally processed vegetables might affect the interactions between the pathogen and the natural competitive microflora sufficiently to enhance *L. monocytogenes* growth.

González-Fandos *et al.* (2001) studied the potential of *L. monocytogenes* to grow in mushrooms packaged in perforated and non-perforated polyvinylchloride (PVC) films and stored at 4 °C or 10 °C. CO_2 and O_2 contents inside the package, aerobic mesophiles, psychrotrophs, *Pseudomonas* spp., fecal coliforms, anaerobic spores and *L. monocytogenes* were determined. Mushrooms packaged in non-perforated film and stored at 4 °C had the most desirable quality parameters (texture, development stage, and absence of fungi). *L. monocytogenes* was able to grow at 4 °C and 10 °C in inoculated mushrooms packaged in perforated and non-perforated films, between 1 and 2 log units during the first 48 hours. After 10 days of storage, populations of *L. monocytogenes* were higher in mushrooms packaged in non-perforated films stored at 10 °C. It was concluded that MAP followed by storage at 4 °C or 10 °C extended shelf-life by maintaining an acceptable appearance, but allowed the growth and survival of *L. monocytogenes*.

The survival of *C. botulinum* in MA-packaged, fresh, whole ginseng roots was investigated by Macura *et al.* (2001). Ginseng roots were packaged in medium-barrier O_2 transmission films in air. Anaerobic conditions developed during 2 °C, 10 °C and 21 °C storage, but most rapidly at the elevated temperatures. *C. botulinum* challenge tests were performed for 10 °C and 21 °C samples. At 10 °C, the botulism toxin was recorded after 14 weeks of storage, before all product was spoiled and rendered unfit for human consumption. At 21 °C, the product spoiled before it came toxic. The authors recommended that commercial production of MAP fresh ginseng should not be contemplated until the safety of the packaged product can be ensured.

The survival and growth of *E. coli* O157:H7 and *L. monocytogenes* during storage at 4 °C and 8 °C on ready-to-use (RTU) packaged vegetables (lettuce, rutabaga, dry coleslaw mix, soybean sprouts) were studied by Francis and O'Beirne (2001). The vegetables were sealed with oriented polypropylene packaging film, and MA was developed in packs during storage due to produce respiration. Survival and growth patterns were dependent on vegetable type, package atmosphere, storage temperature and bacterial strain. *E. coli* O157:H7 generally survived and grew better than *L. monocytogenes*. Storage at 4 °C enabled survival of both pathogens on all products throughout the storage period.

During the last 20 years, an ever-increasing demand for RTU vegetables has led to a continuous growth in the quantity and diversity of products available to the consumer. Mild preservation technology, particularly refrigeration and MAP, is increasingly being relied upon to ensure the safety and quality of RTU vegetables. Such technology has resulted in the increased significance of psychrotrophic and facultative anaerobic pathogens, including *L. monocytogenes*, *A. hydrophila*, and *C. botulinum*. The emergence of such organisms, combined with continual product evolution, presents numerous questions with regard to the microbial safety of these products (Francis *et al.*, 1999). The importance of strict temperature control from process to consumption is assumed by default. Refrigerated temperatures must be maintained during transportation, distribution, storage and handling in supermarkets and by consumers. It is essential that contamination of produce be minimized through the use of good agricultural and strict hygiene practices, and that HACCP programs specific for the pathogen of concern be applied at all stages of production (Francis and O'Beirne, 2001).

6.3 Bacteriocins in packaging films

Examples of bacteriocin-coated surfaces include polyethylene films coated with nisin/methyl cellulose, nisin-coated poultry, and adsorption of nisin onto polyethylene, ethylene vinyl acetate, polypropylene, polyamide, polyester, acrylics and polyvinyl chloride (Appendini and Hotchkins, 2002). Bower *et al.* (1995) demonstrated that nisin adsorbed onto silanized silica surfaces inhibited the growth of *L. monocytogenes*. Nisin films were exposed to medium containing *L. monocytogenes*, and the contacting surfaces were evaluated at 4-hour intervals for 12 hours. Cells on surfaces that had been in contact with a high concentration of nisin (40000 IU/ml) exhibited no signs of growth, and many of them displayed evidence of cellular deterioration. Surfaces contacted with a 10-fold lower concentration of nisin (4000 IU/ml) had a smaller degree of inhibition. In contrast, surfaces contacted with films coated with heat-inactivated nisin allowed growth of *L. monocytogenes*.

7 Control of microorganisms by chemical antimicrobials

7.1 Introduction

Food antimicrobial agents are 'chemical compounds in foods that retard microbial growth or kill microorganisms, thereby resisting deterioration in safety and quality'. Most of the agents are only bacteriostatic or fungistatic, not bactericidal or fungicidal, and they will not preserve food indefinitely. Depending upon storage conditions, the food product eventually spoils or becomes hazardous. Thus, food antimicrobials are typically used in combination with other food preservation procedures, based on the hurdle concept discussed in section 4 (Davidson, 2001).

Food antimicrobials are sometimes called preservatives. However, this term often includes antioxidants and antibrowning agents in addition to antimicrobials. *Antimicrobials* is therefore a more specific term (Davidson and Harrison, 2002). In this chapter, the term 'preservative' is restricted to chemical preservatives obtained by synthesis. Classification of antimicrobials is often arbitrary, and here we have differentiated between chemical and natural antimicrobials. The first group comprises those agents that are either inorganic or produced by synthetic means, whereas the second group includes compounds that occur naturally and can be extracted from natural products. Nevertheless, some natural antimicrobials are now produced commercially by chemical synthesis. Cleaners and sanitizers used to remove dirt and to eliminate pathogenic and spoiling organisms from food facilities and equipment are also discussed under chemical antimicrobials.

7.2 Chemical preservatives

7.2.1 Nitrites

Sodium nitrite ($NaNO_2$) and potassium nitrite (KNO_2) have a specialized use in cured meat products. In fact, nitrites have other functions in cured meat in addition to serving as antimicrobial agents. As nitric oxide, they react with the meat pigment myoglobin to form the characteristic cured meat color, nitrosomyoglobin. They also contribute to flavor and texture and serve as antioxidants (Davidson, 2001). At one time, sodium nitrate ($NaNO_3$) and potassium nitrate (KNO_3) were used extensively; later, it was discovered that nitrates are converted into nitrites and that the latter are the effective antimicrobial agents (Gould, 2000). Nevertheless, nitrates are added to fermented dry sausage batter to serve as a reservoir for nitrite formation by bacterial reduction during the lengthy curing process.

The primary use of nitrites is to inhibit *Clostridium botulinum* growth and toxin production in cured meats. Essentially, nitrites inhibit outgrowth of the germinated spore, as only very high, unusable nitrite concentrations would significantly inhibit spore germination itself (Duncan and Foster, 1968). Nitrites are more inhibitory under anaerobic conditions. Ascorbate and isoascorbate enhance the antibotulinal action, probably by acting as reducing agents (Roberts *et al.*, 1991). In addition, these compounds are important as inhibitors of nitrosoamine formation. Nitrosoamines are carcinogenic agents resulting from the reaction of nitrites with secondary or tertiary amines (Tompkin, 1993).

Nitrites have effects on microorganisms other than *C. botulinum*. They have proven to have inhibitory effects against *C. perfringens*, *E. coli* O157:H7, *Listeria monocytogenes*, *Achromobacter*, *Enterobacter*, *Flavobacterium*, *Micrococcus* and *Pseudomonas* (Tarr, 1941; Gibson and Roberts, 1986a, 1986b; Pelroy *et al.*, 1994; Tsai and Chou, 1996). Certain strains of *Salmonella*, *Bacillus* and *Clostridium* are resistant (Perigo and Roberts, 1968; Rice and Pierson, 1982).

The antimicrobial action of nitrites on *C. botulinum* was elucidated by Woods *et al.* (1981) and Woods and Wood (1982), who showed that nitrites cause a reduction in intracellular ATP and excretion of pyruvate, thus inhibiting oxidative phosphorylation. A number of enzymes have been identified as being inhibited by nitrites, namely ferredoxin, oxido-reductase and pyruvate decarboxylase (Carpenter *et al.*, 1987; McMindes and Siedler, 1988; Tompkin, 1993). The mechanism of inhibition of non-spore-forming microorganisms may be different from that of spore-formers. Rowe *et al.* (1979) reported that nitrites are capable of blocking active transport, oxygen uptake and oxidative phosphorylation by oxidizing ferrous iron of the electron carrier, cytochrome oxidase.

7.2.2 Sulfites

Salts of sulfur dioxide include potassium sulfite and sodium sulfite (K_2SO_3 and Na_2SO_3), potassium bisulfite and sodium bisulfite ($KHSO_3$ and $NaHSO_3$), and potassium metabisulfite and sodium metabisulfite ($K_2S_2O_5$ and $Na_2S_2O_5$). As antimicrobials, sulfites are used primarily in fruit and vegetable products to control three groups of microorganisms: spoilage and fermentative yeasts and molds, acetic acid bacteria, and malolactic bacteria.

In addition, they act as antioxidants and clarifiers, and inhibit enzymatic and non-enzymatic browning of foods (Ough, 1993). In some countries, sulfites are used in fresh meats and meat products. For example, they are effective in delaying the growth of molds, yeasts and salmonellae in sausages during storage at refrigerated or room temperature (Banks and Board, 1982).

The most important factor influencing the antimicrobial activity of sulfites is pH. As pH decreases, the proportion of SO_2zH_2O increases and the bisulfite (HSO_3^-) ion concentration decreases; the latter has no antimicrobial activity (King *et al.*, 1981; Gould and Russell, 1991). Because of their extreme reactivity, it is difficult to pinpoint the exact antimicrobial mechanism for sulfites.

7.2.3 Phosphates

Phosphates have many important uses in food processing, including pH stabilization, acidification, alkalinization, sequestration or precipitation of metals, formation of complexes with organic polyelectrolytes (e.g. protein, pectin and starch), deflocculation, dispersion, peptization, emulsification, nutrient supplementation, anticaking, and leavening (Ellinger, 1972). Some phosphate salts, including sodium acid pyrophosphate (SAPP), tetrasodium pyrophosphate (TSPP), sodium tripolyphosphate (STPP), sodium hexametaphosphate (SHMP) and trisodium phosphate (TSP), have variable levels of antimicrobial activity in foods (Shelef and Seiter, 1993).

Gram-positive bacteria are generally more susceptible to phosphates than are Gram-negatives. Bacteria inhibited by phosphates include *L. monocytogenes*, *Staphylococcus aureus*, *Bacillus subtilis* and *Clostridium sporogenes* (Kelch and Bühlmann, 1958; Jen and Shelef, 1986; Zaika and Kim, 1993; Lee *et al.*, 1994). Post *et al.* (1968) preserved cherries against the fungal growth of *Penicillium*, *Rhizopus* and *Botrytis* using STPP. TSP is used as a sanitizer in chill water for raw poultry, particularly against *Salmonella* spp. (Davidson, 2001).

Several mechanisms have been suggested for bacterial inhibition by polyphosphates. Their ability to chelate metal ions appears to play an important role in their antimicrobial activity (Sofos, 1986). Maier *et al.* (1999) demonstrated that inhibition of *Bacillus cereus* by sodium polyphosphates is related to the chelation of divalent cations (Mg^{++} and Ca^{++}), which inhibits cell division by blocking cell septation. Knabel *et al.* (1991) reported that antimicrobial activity of polyphosphates is reduced at lower pH, owing to protonation of the chelating sites. They concluded that polyphosphates inhibited Gram-positive bacteria and fungi by removal of essential cations from binding sites on their cell walls.

7.2.4 Parabens

Parabens are alkyl esters of *p*-hydroxybenzoic acid. Esterification of the carboxyl group of benzoic acid allows the molecule to remain undissociated up to pH 8.5, giving the parabens an effective range of pH from 3.0 to 8.0. In most countries, the methyl, propyl and heptyl parabens are allowed for direct addition to foods as antimicrobial agents, whereas the ethyl and butyl esters are also approved in some countries. Generally speaking, the antimicrobial activity of parabens is inversely proportional to the chain length of the alkyl component. As the alkyl chain length increases, inhibitory activity generally increases. Increasing activity with decreasing polarity is more evident against Gram-positive than Gram-negative bacteria. Parabens are considered to be more active against yeasts and molds than against bacteria (Aalto *et al.*, 1953; Davidson, 2001).

Parabens act primarily on the cytoplasmic membrane, causing structural damage that results in leakage of metabolites. This effect is proportional to the alkyl chain length. It has also been reported that parabens inhibit nutrient uptake through the membrane by blocking the membrane transport system (Judis, 1963; Furr and Russell, 1972; Freese *et al.*, 1973).

7.2.5 Phenolic antioxidants

These compounds are used in foods primarily to delay oxidation of unsaturated lipids by interrupting the free radical chain mechanism of hydroxyperoxide formation during the autoxidation process (Davidson, 2001). However, they have also been shown to possess antimicrobial activity against a wide range of microorganisms – not only bacteria, yeasts and molds, but also viruses and protozoa. The most important members of this group are butylated hydroxyanisole (BHA), butylated hydroxytoluene (BHT) and t-butylhydroxyquinoline (TBHQ) (Jay, 1996).

In general, phenolic antioxidants are more effective against Gram-positive than Gram-negative bacteria. TBHQ has been reported as an extremely effective inhibitor of *L. monocytogenes* and *S. aureus*. BHA is a more effective antifungal agent than BHT or TBHQ, acting against *Aspergillus flavus*, *Aspergillus parasiticus*, *Saccharomyces cerevisiae*, *Geotrichum* and *Penicillium* spp. Their mechanism of action is similar to that of parabens, as they are chemically related (Davidson and Doan, 1993).

7.3 Cleaners

7.3.1 Cleaning media

Water is the standard cleaning medium. Air is an alternative for the removal of packaging material, dust and other debris where water is not acceptable, and solvents may be used for the removal of lubricants and other similar petroleum-derived products. The major functions of water as a cleaning medium have been reviewed by Marriot (1999).

7.3.2 Classification of cleaning compounds

The essential functions of a cleaning compound are to 'lower the surface tension of water so that soils may be dislodged and loosened, and to suspend soil particles for subsequent flushing away'. To complete the cleaning process, a sanitizer is applied to destroy residual microorganisms that are exposed as a result of the cleaning process. Most cleaning compounds that are used in the food industry are blends – in other words, ingredients are combined to produce a single product with specific characteristics that performs a given function for one or more cleaning applications (Marriott, 1999).

7.3.2.1 Soaps

Soaps are created by the reaction of an alkali with a fatty acid, and include either alkaline salts or carboxylic acids (Marriott, 1999).

7.3.2.2 *Alkaline compounds*

Alkaline compounds are divided into three simple classes: strong, heavy-duty, and mild. Strong alkaline cleaners have strong dissolving power and are very corrosive. They are used to remove heavy soils, but have little effect on mineral deposits. Examples are sodium hydroxide (caustic soda) and silicates having a high N_2O: SiO_2 ratio. Heavy-duty alkaline cleaners have moderate dissolving power and are slightly corrosive. Examples include sodium metasilicate (a good buffering agent), sodium hexametaphosphate, sodium pyrophosphate, sodium carbonate and tri-sodium phosphate (a good soil emulsifier). They are excellent for removing fats, but have no value for mineral deposit control. Mild alkaline cleaners are used for manual cleaning of lightly soiled areas. Examples are sodium bicarbonate, sodium sesquicarbonate, tetrasodium pyrophosphate, phosphate water conditioners (sequesters) and alkyl aryl sulfonates (surfactants). They have good water-softening capabilities, but are of no value for mineral deposit control (Marriott, 1999).

7.3.2.3 *Acid compounds*

Acid compounds are used for removing encrusted surface materials and dissolving mineral scale deposits. Organic acids such as citric, tartaric and gluconic are also excellent water softeners, rinse easily, and are mostly harmless. Inorganic acids are less used because they are corrosive and/or irritating to the skin. Two classes of acid cleaning compounds are distinguished: strong and mild. Strong acid cleaners are corrosive to concrete, metals and fabrics. When heated, they produce toxic gases. Examples are hydrochloric, hydrofluoric, sulfamic, sulfuric and phosphoric acids. They are mostly used to remove the encrusted surface matter and mineral scale frequently found on steam-producing equipment, boilers and some processing equipment. Mild acid cleaners are mildly corrosive. Examples are levulinic, hydroxyacetic, acetic, and gluconic acids, used primarily as water softeners (Marriott, 1999).

7.3.2.4 *Active-chlorine cleaners*

Active-chlorine cleaners containing active chlorine, such as sodium or potassium hypochlorite, are effective in the removal of carbohydrate and/or proteinaceous soils because they aggressively attack such materials and chemically modify them. Thus, active-chlorine products are especially valuable for cleaning surfaces containing starch and/or protein. Hypochlorites should be applied soon after they are made up, as they lack stability during storage (Wyman, 1996).

7.3.2.5 *Synthetic detergents*

Synthetic detergents are as good emulsifiers as soaps. They lower surface tension, promote wetting of particles, and deflocculate and suspend soil particles. Wetting agents are divided into three major categories: cationic, anionic and non-ionic. Anionic compounds are the most commonly used wetting agents because of their compatibility with alkaline cleaning agents and their good wettability properties. Examples are sulfated alcohols, olefins, oils, monoglycerides and amides, alkyl-aryl polyether sulfates, alkyl sulfonates and heterocyclic sulfonates. Because of their neutrality, non-ionic agents are effective under both acidic and alkaline conditions.

They are not affected by water hardness, but they tend to foam. Examples are thioethics, pluronics, amine fatty acid condensates, alkyl-aryl polyether alcohols and ethylene oxide fatty alcohol condensates. In general, synthetic agents are non-corrosive, non-irritating and easily rinsed from equipment and other surfaces (Anonymous, 1976).

7.3.2.6 Enzyme-based cleaners

Enzyme-based cleaners are proteases and work best under alkaline conditions. They are not as effective on all types of soils as are chlorine compounds (Marriott, 1999).

7.3.2.7 Cleaning auxiliaries

Auxiliary compounds protect sensitive surfaces or improve the cleaning properties of a compound. There are two main categories: sequestrants and surfactants. Sequestrants can chelate by complexing with magnesium and calcium ions. This action effectively reduces water hardness. Sequestrants consist of polyphosphates or organic amine derivatives, the latter being considered as generally more effective agents. Commercially, most organic agents are salts of ethylene-diamine-tetracetic acid (EDTA). Surfactants are surface-active agents that function to facilitate the transport of cleaning and sanitizing compounds over the surface to be cleaned. They are classified according to their net charge – that is, anionic (e.g. linear alkyl-benzene sulfonates), cationic (e.g. quaternary ammonium salts) and non-ionic (ethylene oxide derivatives, alkanol-amides and amine oxides). Important physicochemical characteristics of surfactants were summarized by Marriott (1999).

7.4 Sanitizers

Sanitizers, as defined by the US Environmental Protection Agency, are 'pesticide products that are intended to disinfect or sanitize, reducing or mitigating growth or development of microbiological organisms including bacteria, fungi, or viruses on inanimate surfaces in the household, institutional, and/or commercial environments' (CFR, 2001). Soil that remains on food-processing equipment after use is typically contaminated with microorganisms nourished by the nutrients of the soil deposits. A sanitary environment is obtained by thoroughly removing soil deposits with an appropriate cleaning solution and subsequently applying a sanitizer to destroy residual microorganisms.

7.4.1 Thermal sanitation

The two major sources for thermal sterilization are steam and hot water, but in general the process is rather inefficient because of high energy costs (Marriott, 1999).

7.4.2 Radiation sanitation

Ultraviolet (UV) light or high-energy cathode or gamma rays will destroy microorganisms. However, the methods of sanitizing have been restricted to certain commodities and are not really useful in food plants and food service operations because of limited total effectiveness, equipment cost and operator safety issues. In the case of UV, light rays must actually strike the microorganisms, and the latter can be protected by soil itself. In addition, light can be absorbed by dust, thin films of grease and opaque or turbid solutions, thus reducing killing power (Marriott, 1999). UV systems have potential, but require cleaned surfaces for effective use.

7.4.3 Chemical sanitation

The ideal sanitizer possesses important properties: broad-spectrum microbial destruction; resistance to factors such as soil load, detergent and soap residues, and water hardness; non-toxic and non-irritating properties; water solubility; acceptable odor or no odor; stability in solutions; ease of use; ready availability; low cost; and ease of measurement for preparation of solutions. The efficacy of chemical sanitizers depends on a number of physicochemical factors, as discussed by LeChevallier *et al.* (1988) and Marriott (1999).

7.4.4 Classification of sanitizers

7.4.4.1 Chlorine compounds

The main chemicals in this category are liquid chlorine, hypochlorites, inorganic chloramines and organic chloramines. When liquid chlorine (Cl_2) and hypochlorites are mixed with water, they hydrolyze to form hypochlorous acid (HOCl). This compound dissociates to form H^+ and OCl^-. Chlorine compounds are more effective antimicrobial agents at a low pH; the hypochlorite ion, which is not as effective as a bactericide, predominates at higher pH values. Another chlorine compound, chlorine dioxide (ClO_2), does not hydrolyze in aqueous solutions. ClO_2 is particularly suitable for sewage treatment because it is hardly affected by pH or organic matter (Marriott, 1999).

The mode of action of chlorine as an antimicrobial agent has not been fully determined. It appears to kill through inhibition of the glucose oxidation metabolic pathway and by oxidation of sulfhydryl groups of certain enzymes. However, other modes of action have also been proposed, including the disruption of protein synthesis, deleterious reactions with nucleic acids, inhibition of oxygen uptake, and oxidative phosphorylation coupled with leakage of some macromolecules, and promotion of chromosomal aberrations. Chlorine impairs membrane function, especially transport of extracellular nutrients. Chlorine is also known for stimulating spore germination and subsequently inactivating the germinated spore. Vegetative cells are more easily destroyed than are *Clostridium* spores, which in turn are killed more easily than *Bacillus* spores. Chlorine concentrations $\leq$ 50 ppm lack antimicrobial activity against *L. monocytogenes*, but higher concentrations effectively destroy this pathogen. This lethal effect is enhanced by increasing temperature; up to 52 °C, the reaction rate doubles for each 10 °C increase (Kulikoosky *et al.*, 1975; Camper and McFetters, 1979; Marriott, 1999; Meinhold, 1991).

7.4.4.2 Bromine and iodine compounds

Bromines have been used primarily for water treatment, either alone or in combination with other compounds. There is a synergistic effect when bromine and chlorine compounds are combined (Marriott, 1999).

The major iodine compounds used for sanitizing are iodophors (formed by complexing elemental iodine with surfactants such as nonyl phenolethylene oxide), alcohol–iodine solutions, and aqueous iodine solutions. Iodophors are the most important agents: they are used in water treatment, disinfection of equipment and surfaces, and as a skin antiseptic (Marriott, 1999). The mode of antibacterial action of iodine compounds is not fully understood. It appears that diatomic iodine (I_2) is the major agent, able to disrupt bonds that hold cell proteins together and to inhibit protein synthesis, but free elemental iodine and hypoiodous acid (HOI) have also been identified as possessing antimicrobial properties. The iodophor complex releases an intermediate triciodide ion that rapidly converts to diatomic iodine and hypoiodous acid at low pH (Anonymous, 1996).

Iodine-type sanitizers are somewhat more stable in the presence of organic matter than are the chlorine compounds. They are as active in the deactivation of vegetative cells as chlorine agents, but not as effective in spore inactivation. In a concentrated form, their shelf-life is long. However, in solution, iodine may be lost due to sublimation – especially at temperatures above 50 °C. Because they are acidic, iodine compounds are not affected by hard water and will prevent accumulation of minerals if used regularly. Iodine compounds cost more than chlorines and may cause off-flavors in some products. In addition, they tend to be very sensitive to pH changes (Marriott, 1999).

7.4.4.3 Quaternary ammonium compounds

These sanitizers, commonly called the 'quats', are ammonium compounds in which four organic groups are linked to a positively charged nitrogen atom. Typically, the organic radical is the cation, and either chloride or bromide the anion. Examples are alkyl-dimethyl-benzyl ammonium chloride and methyl-dodecylbenzyl-trimethyl ammonium chloride. Their mechanism of germicidal action is not fully understood, but it may be that the surface-active quat surrounds and covers the cell outer membrane, causing enzyme inhibition and leakage of internal constituents. Quats are essentially bacteriostatic, not bactericidal. They are used most frequently on floors, walls, furnishings and equipment. Being good penetrants, they have value for porous surfaces (Frank and Chmielewski, 1997; Marriott, 1999). The major advantages of quaternary ammonium compounds have been summarized (Anonymous, 1997).

7.4.4.4 Acid sanitizers

Acid sanitizers are frequently associated with cleaners. The acid neutralizes excess alkalinity that remains from the cleaning compound, prevents formation of alkaline deposits, and sanitizes. They act by penetrating and disrupting cell membranes and acidifying the cytoplasm. These compounds are especially effective on stainless steel surfaces or where contact time may be extended, and have a high antimicrobial activity against psychrotrophic organisms, yeasts and viruses (Marriott, 1999).

Organic acids, such as acetic, peroxyacetic, lactic, propionic and formic, are most frequently used. A pH value ≤ 3.0 is ideal for their performance. They can be applied by clean-in-place (CIP) methods, by spray or foaming. All cleaning compounds need to be thoroughly rinsed away before these sanitizers are applied, as they can lose all of their effectiveness in the presence of cationic surfactants (Marriott, 1999).

Carboxylic acids as sanitizers are assorted mixtures of free fatty acids, sulfonated fatty acids and other organic acids. They have a broad range of bactericidal activity, and are non-corrosive, cost-effective, and stable in dilutions both in the presence of organic matter and at high temperatures. They are negatively affected by cationic surfactants, so rinsing of previously used detergents is essential (Anonymous, 1997).

Increased interest in peroxyacetic acid has developed for CIP sanitizing systems. This sanitizer, which provides a rapid, broad-spectrum kill, works on the oxidation principle through reaction with the components of cell membranes. Because it is effective against yeasts (e.g. *Candida* and *Saccharomyces*) and molds (e.g., *Penicillium, Aspergillus, Mucor* and *Geotrichum*) it has gained acceptance in the soft drink and brewery industries. Its efficacy against *Listeria* and *Salmonella* justifies its application in dairy plants. It reduces pitting of equipment surfaces by being less corrosive than halogen compounds; it is also biodegradable (Anonymous, 1997).

Acid anionic sanitizers are formulated by combining an anionic surfactant with an acid sanitizer. Thus, an acidified rinse is combined with the sanitizing step. The advantages of these combinations have been discussed (Anonymous, 1997).

7.4.4.5 Miscellaneous

Hydrogen peroxide (H_2O_2) in 3–6 % aqueous solutions has been found to be effective against biofilms. As a sanitizer, H_2O_2 may be used on all types of surfaces: equipment, floors and drains, walls, steel-mesh gloves, belts, and other areas where contamination exists. It is effective against *L. monocytogenes* when applied to latex gloves (McCarthy, 1996). Ozone (O_3) has been evaluated as a chlorine substitute. Like ClO_2, O_3 is unstable and should be generated as needed at the site of application. Because it oxidizes rapidly, it has low environmental impact (Marriott, 1999). Glutaraldehyde has been used to control the growth of common Gram-negative and Gram-positive bacteria, as well as species of yeast and filamentous fungi found in conveyer lubricants used in the food industry (Marriott, 1999).

8 Control of microorganisms by natural antimicrobials

Consumer perception that use of synthesized food antimicrobials may be associated with toxicological problems has generated interest in the food industry regarding the use of naturally occurring compounds. Organic acids that are routinely produced in large quantities through chemical synthesis are also found naturally in many food products. The extraction of these and other antimicrobials from natural sources, however, can be complex, inefficient and expensive. Yet, synthetic agents may be considered less desirable than naturally occurring antimicrobials by a segment of consumers. Thus, interest in and the incentive for the development and use of naturally occurring antimicrobials in foods have increased because of the growing interest in so-called natural foods.

Numerous naturally occurring antimicrobial agents are present in animal and plant tissues, where they probably evolved as part of their hosts' defense mechanisms against invasion by microorganisms. Natural antimicrobials can be derived from barks, stems, leaves, flowers and fruits of plants; various animal tissues; and even from other microorganisms. Noted sources of natural antimicrobials are herbs, spices, fruits, milk, eggs, and lactic acid bacteria used in food fermentation. However, the selection, manufacture and commercial application of a proper antimicrobial are challenging due to the complexity of food, the variety of factors influencing preservation, and the complex chemical and sensory properties of natural antimicrobials themselves. Food preservation can be enhanced by interactions among multiple antimicrobial factors, which can yield additive or synergistic effects. These combined factors may include natural product composition, microbial flora, pH, water activity (a_w), added chemicals, packaging, and processing and storage temperatures (Sofos *et al.*, 1998).

For the successful application of a naturally occurring antimicrobial to a food, there is a need to determine its efficacy and the antimicrobial spectrum of the compound. The antimicrobial selected should not contribute to the development of resistant strains, nor alter the food in such a way that growth of another pathogen is possible. To be useful as a natural antimicrobial, a compound must show functionality in the targeted food system. Many antimicrobials act together and therefore might be most appropriately evaluated in combination. Success of application testing may be determined by increased shelf-life and reduced pathogen viability (Sofos *et al.*, 1998).

The exact mechanisms through which antimicrobials affect microbial growth are complex and difficult to elucidate fully, but knowledge of the antimicrobial mechanism of a compound will allow selection of combinations of antimicrobials with different mechanisms that could be optimally utilized against microorganisms in a food product (Sofos *et al.*, 1998).

An important aspect of any compound selected for use as a food preservative is its toxicological properties. A naturally occurring antimicrobial to be used in food must not be toxic either by animal testing or by its continuous consumption in a food over a long period. In addition to the absence of toxicity, the antimicrobial must be nonallergenic and be able to be metabolized and excreted so as not to lead to residue build-up. It should not react either to make important nutrients unavailable to humans or to destroy those nutrients, and should not interfere with the proliferation of desirable microorganisms, such as lactic acid bacteria.

Chemical and physical properties of the antimicrobial agent should be compatible with the composition and properties of the food to be preserved. Important properties to be considered include chemical reactivity, solubility, dissociation constant (pKa) and influence on product quality. The potential impact on the sensory characteristics is of significance. Many naturally occurring antimicrobials must be used at high concentrations to achieve antimicrobial activity. Obviously, compounds that negatively affect flavor or odor are unacceptable. For example, some spice extracts show antimicrobial activity, but only at concentrations that would cause the food to be rejected by most consumers. It is also unacceptable for a food antimicrobial to mask spoilage. Ultimately, the greatest roadblock to the use of naturally occurring antimicrobials is economics (Sofos *et al.*, 1998). Antimicrobials can be applied to food in various ways; the method used is dictated by what the existing processing and packaging procedures are (Sofos *et al.*, 1998).

8.1 Traditional antimicrobials

Antimicrobials that may be classified as traditional with long or frequent use in processed foods include sugars, common salt and wood smoke (Sofos and Busta, 1992).

8.1.1 Natural sugars

Natural sugars such as sucrose, fructose, glucose, syrups, and various corn and other products (which generally are useful in foods as sweeteners, flavorings and fermentable materials) can also exert antimicrobial activity through decrease of a_w. Sugars also have an indirect activity by serving as substrates in food fermentations leading to the formation of acids, alcohols and other antimicrobial agents (Foegeding and Busta, 1991).

Direct microbial inhibition requires sugar concentrations exceeding 40–50 %; for example, a 50 % sucrose concentration decreases a_w to 0.935, enough to inhibit growth of *Clostridium botulinum* (Sofos and Busta, 1992). Foods preserved with high sugar concentrations include jams, jellies, preserves, syrups, fruit juice concentrates, sweetened condensed milk and candies. Some yeasts and molds have the ability to grow in the presence of high sugar concentrations. Small concentrations of sugar act as substrates and support the growth of various microorganisms, including spoilage and pathogenic agents, as well as those useful in the production of fermented foods (Sofos *et al.*, 1998).

8.1.2 Common salt

Common salt (sodium chloride) has been used as a flavoring or a preservative in foods since ancient times. Foods treated with salt include meat, fish, cheese, butter, margarine and brined vegetables. Curing of meat with impure salt led to the discovery of nitrates as additives for the curing meat products (Sofos and Busta, 1980). In recent years, because of the potential link between sodium consumption and the development of hypertension, there has been a trend to decrease salt concentrations in processed foods. Potential partial replacers are potassium chloride and phosphates. However, if certain levels of potassium chloride are exceeded, bitter flavors may result. Phosphates also may contribute to undesirable textural and flavor effects. In any case, complete elimination of salt from certain foods may be impossible because of its important contribution to taste and to technological properties such as protein extraction and texture development (Sofos, 1986).

The amount of salt necessary to decrease a_w to 0.935, which inhibits growth of *C. botulinum*, is 10 % in the water phase of a product. However, even lower levels are important because they act synergistically with other antimicrobials such as acids, nitrite, sorbate and benzoate (Sofos, 1984). *Listeria monocytogenes* and *Staphylococcus aureus* can survive and even proliferate in environments exceeding 5 % salt concentration (Sofos, 1993).

8.1.3 Smoking

Smoking of foods is an ancient practice that is still in use. Wood smoke contributes to flavor but also incorporates antimicrobial components in a product, including phenolics, formaldehyde, acetic acid and creosote. However, their antimicrobial activity in today's mildly smoked foods is probably weak. The use of application of natural wood smoke to foods has not only decreased, but is also being replaced by liquid smoke flavorings isolated from natural wood-smoke condensates. These preparations are actually preferred because they are easy to apply uniformly, the concentration used can be controlled, pollution from crude tar products can be decreased, and polycyclic aromatic hydrocarbons can be removed (Sofos *et al.*, 1988).

8.2 Organic acids

Most bacteria prefer pH values near neutrality, whereas yeasts and molds are more tolerant of lower pH values. Increasing the acidity (i.e. lowering the pH) is an effective way of limiting microbial growth. This can be achieved through acidulant addition or natural fermentation resulting in production of acids by desirable microorganisms (Sofos and Busta, 1992).

Acids inhibit microbial growth by lowering pH or through the antimicrobial activity of undissociated molecules or anions. As pH decreases, the antimicrobial activity of short-chain organic acids increases more than that of long-chain acids. As pH decreases and approaches the pKa of a short-chain organic acid, the undissociated shorter molecule is able to enter the microbial cell, where it dissociates, acidifies the cytoplasm and interferes with chemical transport across the cell membrane and/or with enzymatic activity (Banwart, 1989). As the cell tries to maintain cytoplasmic homeostasis, protons are pumped out, disrupting the proton motive force and consequently interfering with oxidative phosphorylation and nutrient transport systems (Dillon and Cook, 1994).

The antimicrobial activity of organic acids is enhanced when they occur in mixtures; their spectrum of activity is also increased. Shorter-chain organic acids inhibit or inactivate both Gram-positive and Gram-negative bacteria, whereas longer-chain organic acids are effective primarily against Gram-positive bacteria because they cannot penetrate the outer membrane of Gram-negative bacteria. In addition, monounsaturated fatty acids are generally less inhibitory to microorganisms than are saturated fatty acids (Kabara, 1978). Acidic environments not only limit microbial growth but also enhance the destruction of microorganisms by heat; thus, decreased pasteurization or sterilization times are possible (Doores, 1993). Increased acidity also enhances the antimicrobial activity of other hurdles. For example, the antimicrobial activity of preservatives such as nitrites and sorbates increases as product pH decreases (Sofos and Busta, 1980).

8.2.1 Acetic acid

Acetic acid is present in vinegar, and is one of the oldest preservatives in use. A natural process for the production of acetic acid may involve an alcoholic fermentation of sugars naturally present in grapes, malts, grains and other plant materials, followed by aerobic oxidation of ethanol to yield acetic acid. Thus it is found in fermenting plant materials and some dairy products (Sofos and Busta, 1992). Acetic acid can be used as a general preservative because it is soluble in water, readily available, of low cost and low toxicity, and inhibits a wide spectrum of bacteria, including *Salmonella*, *S. aureus*, *L. monocytogenes* and the spore-formers *Bacillus* and *Clostridium* (Banwart, 1989; Doores, 1993). In the form of 1.5–2.5 % solutions, it can be applied in the decontamination of food animal carcasses (Dickens and Whittemore, 1994), and has also been tested as a vapor at low concentrations for control of fruit decay by post-harvest fungi (Sholberg and Gaunce, 1995).

Acetic acid derivatives such as sodium acetate, calcium diacetate and dehydroacetic acid are also used as antimicrobial agents. All these compounds are generally regarded as safe (GRAS). Sodium acetate has exhibited inhibitory activity against *L. monocytogenes* in catfish fillets and sausages (Chang *et al.*, 1995; Wederquist *et al.*, 1995; Rong-Yu *et al.*, 1996).

8.2.2 Benzoic acid

Benzoic acid is a natural constituent of cranberries, prunes, plums, apples, strawberries, cinnamon and ripe olives. In addition, it may be found in some yogurts as a by-product of microbial growth. Although pure benzoic acid is available commercially, sodium benzoate is more useful because of its higher water solubility (Chipley, 1993). As with other organic acids, the antimicrobial activity is due mostly to the undissociated form, even though the dissociated form has also shown activity (Eklund, 1988).

Benzoic acid is more effective against yeasts than against bacteria or molds, but the activity is highly dependent on food, pH, and a_w. In general, antimicrobial activity increases as the pH value of the food decreases near to its pKa of 4.19, and maximal antimicrobial activity occurs at pH values of 2.5–4.0 (Sofos, 1994). Pathogenic bacteria inhibited by benzoate include *Vibrio parahaemolyticus*, *S. aureus*, *L. monocytogenes*, *Bacillus cereus*, *Escherichia coli*, *Micrococcus*, *Pseudomonas* and *Streptococcus*. Susceptible yeasts include species in the genera *Candida*, *Rhodotorula*, *Debaryomyces* and *Saccharomyces*. Among the fungi affected are *Alternaria*, *Aspergillus*, *Mucor*, *Penicillium* and *Rhizopus*. Concentrations effective in inhibiting microorganisms are in the range of 0.05–0.1 % for yeasts and molds, and 0.01–0.2 % for bacteria (Chipley, 1993; Nassar *et al.*, 1995). Benzoate can be metabolized by some bacteria such as *Enterobacter*, *Pseudomonas*, *Corynebacterium glutamicum* and certain thermophilic bacilli. Osmotolerant species of yeasts such as *Zygosaccharomyces bailii*, as well as bacteria including *E. coli* and *Gluconobacter oxydans*, can acquire resistance to benzoic acid (Chipley, 1993).

Because of its low cost, among other advantages, benzoate is probably the most commonly used antimicrobial in commercial food preservation. When there are concerns regarding flavor defects, benzoate may be used at lower levels in combination with other preservatives such as sorbate or parabens. In general, benzoates are used extensively in the preservation of foods with pH < 4.5, including fruit products, beverage and bakery products, fruit juices and drinks, salads and salad dressings, pickles, sauerkraut, preserves, jams and jellies, and margarine. Typical usage levels range from 0.05 % to 0.1 % (Chipley, 1993; Sofos, 1994). Benzoic acid and its derivatives are GRAS substances and well tolerated by humans (Sofos and Busta, 1992; Chipley, 1993; Sofos, 1994).

8.2.3 Lactic acid

Lactic acid is formed by bacteria including *Lactobacillus*, *Lactococcus*, *Streptococcus*, *Pediococcus*, *Carnobacterium* and *Leuconostoc*, as well as certain molds. Heterofermentative organisms, e.g. *Leuconostoc*, also produce other end-products such as acetic acid, ethanol, carbon dioxide and diacetyl, as well as bacteriocins. Homo- and heterofermentative fermentations occur in cheese, sauerkraut, pickles and fermented meat. The lactic acid formed in these products through microbial degradation of sugars decreases the pH to levels unfavorable for growth of spoilage or pathogenic bacteria (Sofos and Busta, 1992).

Microorganisms inhibited by lactic acid and lactates include *C. botulinum*, *C. perfringens*, *C. sporogenes*, *E. coli*, *L. monocytogenes*, *Salmonella*, *Serratia liquefaciens*, *S. aureus*, *Yersinia enterocolitica*, *Aeromonas hydrophila* and *Enterobacter cloacae*. It is not effective against yeasts and molds (Maas *et al.*, 1989; Harmayani *et al.*, 1991; Shelef and Yang, 1991; Doores, 1993; Houtsma *et al.*, 1994; Meng and Genigeorgis, 1994; Miller and Acuff, 1994; Pelroy *et al.*, 1994). Overall, the antimicrobial activity of lactic acid ranges from good to poor depending on the substrate (Banwart, 1989). Used at a 3 % level in combination with potassium sorbate at 5 %, it extended the shelf-life of fresh, vacuum-packed poultry meat stored under refrigeration (Kolsarici and Candogan, 1995). The sodium and potassium salts, i.e. lactates, have been used relatively extensively in recent years as sensory potentiators, flavorings and antimicrobial agents in foods such as processed meats and poultry products (Shelef, 1994; Wederquist *et al.*, 1994, 1995). Lactic acid and its derivatives are highly soluble in water and classified as GRAS (Sofos *et al.*, 1998).

8.2.4 Propionic acid

Propionic acid is an oily liquid with a strong, pungent, rancid odor. It is produced by bacteria belonging to the genus *Propionibacterium*, as used in the ripening of Swiss-type cheeses. This natural acid acts as a flavoring agent and mold inhibitor in Swiss cheeses. In addition, propionic acid is formed by bacteria in the gastrointestinal tract of ruminant animals. In the pure form, it is miscible in water, alcohol, ether and chloroform, and it is somewhat corrosive. Its sodium and calcium salts, available commercially as white, free-flowing powders, are used as food preservatives (Sofos and Busta, 1992; Doores, 1993; Sofos, 1994). Propionic acid and its salts are classified as GRAS. Because they are metabolized as a fatty acid, propionates in concentrations as high as 1 % (as in Swiss-type cheeses) have no adverse health effects. Levels used in food preservation are in the range of 0.1–0.4 % (Sofos *et al.*, 1998).

The most common use of propionates is as mold and rope (e.g. *Bacillus subtilis*) inhibitors in bread and other bakery products. They are inexpensive, and do not interfere with the yeasts involved in leavening. Propionates are better mold inhibitors than benzoates, and they are used as such in cheeses, cheese products, fruits, vegetables, jams, jellies, preserves, malt extracts and tobacco (Robach, 1980; Sofos and Busta, 1992; Sofos, 1994). Microorganisms inhibited by propionates also include *E. coli*, *Salmonella*, *S. aureus*, *L. monocytogenes*, *Proteus vulgaris*, *Lactobacillus plantarum*, *Pseudomonas* spp., *Sarcina lutea*, *Aspergillus* spp., *Torula* spp and *Saccharomyces ellipsoideus* (Doores, 1993).

8.2.5 Sorbic acid

Sorbic acid is another naturally occurring antimicrobial substance available commercially as a synthetic compound. Sorbates (potassium, sodium and calcium salts) are agents effective against a variety of yeasts, molds and bacteria at concentrations of 0.1–0.3 % (Sofos, 1989). Sorbic acid and its salts are GRAS. The antimicrobial effect of sorbates increases at lower pH values.

Sorbates inhibit yeasts in fermented vegetables, fruit juices, wines, dried fruits, cheeses, fish and meats. Specific products in which sorbates are applied for inhibition of yeasts are carbonated beverages, salad dressings, syrups, tomato products, jams, candies, jellies and chocolate syrups. Sorbates are also effective inhibitors of many molds. Mold inhibition by sorbates is important in cheeses as well as in butter, fruits and fruit juices, grains, breads, cakes, smoked fish and dry sausages. In addition, they inhibit mycotoxin formation. Species of bacteria inhibited by sorbates include *Bacillus*, *Campylobacter*, *Clostridium*, *Enterobacter*, *Escherichia*, *Lactobacillus*, *Listeria*, *Mycobacterium*, *Salmonella*, *Staphylococcus*, *Vibrio* and *Yersinia* (Sofos and Busta, 1993).

Strains of molds belonging to the genera *Penicillium*, *Aspergillus* and *Mucor* are able to decompose sorbates in cheese and fruit products; lactic acid bacteria have also been reported to degrade sorbates in wine and fermented vegetables (Liewen and Marth, 1985; Sofos and Busta, 1993).

8.3 Lipid antimicrobials

Fatty acids and their soaps have been used since antiquity for cleansing and disinfecting. Fatty acids and their polyhydric alcohol esters are considered to be of low toxicity and have been used as emulsifiers in foods since the early 1900s (Kabara, 1993). In general, fats and oils from animals and vegetables are known for their ability to inhibit microorganisms by their fatty acids or oxidation products, mainly peroxides, and through associated antioxidative phenolic compounds formed by plants (Dallyn, 1994).

8.3.1 Fatty acids

Fatty acids, as components of natural fats, are primarily even-numbered straight-chain molecules, often with one or more double or triple bonds. The fatty acid compositions of natural fats differ considerably depending on origin. Plant oils often contain an abundance of medium-chain fatty acids, whereas oils from marine animals and plants are more abundant in unsaturated fatty acids; animal fats, on the other hand, are more saturated (Beuchat and Golden, 1989). Short-chain fatty acids show inhibitory activity at relatively high concentrations (1–3 %) against bacteria and fungi (Doores, 1993). Medium-chain saturated fatty acids (C_8–C_{14}) and their potassium and sodium salts are inhibitory mainly towards Gram-positive bacteria and yeasts. Inhibition generally occurs in media with concentrations from 0.0005 % to 0.005 %, but higher concentrations are usually required in foods. Lauric acid is the most inhibitory fatty acid against Gram-positive organisms, whereas capric acid is most active against yeasts (Kabara, 1993). The antimicrobial effect of saturated fatty acids is generally caused by the undissociated form of the molecule, and activity in foods is therefore controlled by pH. Activity is usually highest in acidic (pH ≤ 4.6) and mildly acidic (pH ≅5.0) foods (Tsuchido and Takano, 1988).

Branched-chain and hydroxylated fatty acids possess slightly less antimicrobial activity than their straight-chain counterparts. Fatty acids with chain lengths greater than C_{14} are not sufficiently soluble in the suspending solution for adequate cell contact, so their activity is lower. Similarly, introduction of hydrophobic groups, such as phenyl rings, decreases inhibitory activity due to low solubility, whereas increasing polarity with a hydroxyl or an amine group restores this activity. The presence of unsaturated linkages can markedly increase the antimicrobial activity of fatty acids. The magnitude of inhibition is influenced by degree and position of unsaturation. Thus the most active monounsaturated fatty acid is palmitoleic, whereas linoleic is the most active polyunsaturated fatty acid. The *cis* forms of fatty acids are active whereas the *trans* isomers are inactive, probably because steric hindrance of straight-chain acids prevents contact with cell membranes (Kabara, 1993).

All the above facts support the notion that the plasma membrane is the likely site of action. In addition to direct interaction of the undissociated molecule with the cell membrane, unsaturated fatty acids may exert some antimicrobial activity by autoxidation, which results in the formation of peroxides and other active oxygen metabolites. This oxygen-dependent activity would be expected to be relatively pH-independent under physiological conditions, and to be most effective against anaerobes and other organisms lacking enzymatic defenses against active oxygen metabolites (Sofos *et al.*, 1998).

Unsaturated fatty acids are well known to exert antimicrobial activity against Gram-positive bacteria and yeasts, but most Gram-negative bacteria are resistant. Oleic, linoleic and linolenic acids are inhibitory at 0.005–0.02 % against Gram-positive cocci, lactobacilli, corynebacteria, and spore-forming *Bacillus* and *Clostridium* (Nieman, 1954).

8.3.2 Fatty acid esters

Fatty acid esters, derived from the esterification of fatty acids with polyhydric alcohols or sugars, have shown considerable antimicrobial and emulsifying activities in foods (Shibasaki, 1982). In general, monoacylglycerols – that is, monoglycerides formed through the reaction of medium-chain fatty acids with glycerol – have more potent antimicrobial properties and show a wider spectrum of activity than do free fatty acids. Antimicrobial activity has been demonstrated for esters formed between fatty acids and a variety of polyhydric alcohols or compounds possessing hydroxyl groups, e.g. sugars and peptides. The requirement for antimicrobial activity seems to be that a hydrophilic group be attached to the lipid component (Conley and Kabara, 1973).

Monoacylglycerols are especially active against Gram-positive bacteria and certain fungi, and have little activity against Gram-negative bacteria. Several studies have indicated that many Gram-positive bacteria, including *B. subtilis*, *B. cereus*, *S. aureus*, *S. epidermidis*, streptococci groups A and D, *Micrococcus* spp., *Pneumococcus* spp and *Corynebacterium* spp. are sensitive to low concentrations of monolaurylglycerol in microbiological culture media (Conley and Kabara, 1973). Monolaurylglycerol prevents or delays growth and toxin formation by *S. aureus* and *Streptococcus* spp.

(Schlievert *et al.*, 1992). Antimycotic activity of this compound was demonstrated against *A. niger*, *Penicillium citrinum*, *Saccharomyces cerevisiae*, *Candida utilis*, *C. albicans*, *Cladosporium* spp and *Alternaria* spp. (Kato and Shibashaki, 1975; Kabara *et al.*, 1977; Marshall and Bullerman, 1986). Monoacylglycerols have also been demonstrated to inhibit *L. monocytogenes* in foods (Wang and Johnson, 1997).

8.3.3 Sucrose esters

Sucrose esters are heat-stable and, because they can withstand autoclaving, are useful in stabilizing canned or other heat-treated foods, especially because they seem active against spore-formers, including *Bacillus stearothermophilus*, *B. coagulans*, *Desulfotomaculum nigrificans* and several *Clostridium spp*. (Sofos *et al.*, 1998). They are also active against molds (Marshall and Bullerman, 1986).

8.3.4 Lipopeptides

Lipopeptides, which are derived from natural components by means of the condensation of peptides or amino acids and fatty acids, would be expected to have excellent antimicrobial properties. Sorboyl-tryptophan, sorboyl-D-alanine, myristoyl-D-aspartic acid and glycyl-D-alanine strongly inhibited *C. botulinum* when combined with 0.006 % sodium nitrite (Paquet and Rayman, 1987). Several lipopeptides, including polymyxin and lipopeptide antibiotics, have potent antimicrobial activity and could be effective food preservatives.

Polymyxins consist of a fatty acid moiety covalently linked to a cyclic peptide. In contrast to many other lipophilic antimicrobials, polymyxins have a strong effect on Gram-negative bacteria, causing direct membrane damage through a detergent-like action but at much lower concentrations than ordinary detergents. Membrane damage can be recognized by leakage of solutes, including nucleotides and inorganic ions, or by penetration of normally excluded molecules into the cell. Thus, polymyxins are unique among related lytic agents in being bactericidal in the absence of cell growth (Davis, 1990). They are useful in the control of *Salmonella* infections in poultry (Goodnough and Johnson, 1991). However, even though many of the lipopeptides are synthesized by food-related bacteria such as *Bacillus* spp., they are classified as antibiotics and it is unlikely that they will be used in food (Davis, 1990).

8.4 Plant substances

Compounds exhibiting various levels of antimicrobial activity are present naturally in plant stems, leaves, barks, flowers and fruits. Information on the antimicrobial activity of plant substances and extracts has been available since the nineteenth century, but interest in naturally occurring antimicrobials declined during the first half of the twentieth century, possibly because of the development of highly effective synthetic antimicrobials (Delaquis and Mazza, 1995). Compounds responsible for some of the flavors and aromas of foods are also inhibitory to microorganisms. In many instances, however, the concentrations of spices and herbs necessary for inhibiting microorganisms exceed those resulting from normal usage levels in foods (Beuchat, 1994).

Plants have developed mechanisms for defense against invasion by bacterial, fungal or insect and animal predators. Compounds involved in plant defense mechanisms may be classified as pre-infectional or post-infectional (Walker, 1994).

8.4.1 Prohibitins

Prohibitins include phenolic compounds, flavonols, glucosides, glycosides, alkaloids, dienes, lactones, polyacetylenes and protein-like compounds.

8.4.2 Inhibitins

Inhibitins are mostly phenolic or flavonoid in nature. The latter's activity is usually associated with the action of the enzyme diphenol oxidase. Many plants contain phenolic compounds in the form of hydrolyzable tannins that can denature proteins. These compounds cause an astringent taste and antimicrobial activity in plant extracts (Walker, 1994).

8.4.3 Post-inhibitins

Post-inhibitins are stored as inactive precursors and activated when needed to fight invasion. Activation is catalyzed by hydrolases or oxidases released by the host plant or the invading agent. Examples are the sulfoxides of onion and garlic. In garlic, the precursor alliin is degraded by the enzyme alliinase to yield allicin. Many plants contain cyanogenic glycosides hydrolyzed by specific α-glycosidases to release HCN, a microbial inhibitor in plants such as sorghum and lima beans, which is also toxic to animals (Walker, 1994).

Isothiocyanates are an important group of post-inhibitins derived from the glycosides glucosinolates, stored in cell vacuoles of plants in the family *Cruciferae* (cabbage, brussels sprouts, cauliflower, broccoli, rutabagas, mustard and rapeseed). When plants are injured, glucosinolates are hydrolyzed rapidly by the enzyme myrosinase to produce thiocyanates, isothiocyanates, nitriles, and glucose. Isothiocyanates are believed to exert inhibitory activity against molds, yeasts, and bacteria (Delaquis and Mazza, 1995).

8.4.4 Phytoalexins

Phytoalexins are primarily formed due to the activity of the diphenol oxidase enzymes (e.g. catecholases) acting on phenolic compounds to yield products of increased antimicrobial activity. These enzymes are present in almost all plants, where they oxidize dihydroxyphenols to form quinines, which are quite reactive and toxic. When the plant tissue and its membranes are damaged the enzyme is released and, as it comes in contact with the substrate, it forms the quinone inhibitors. Quinines can react with themselves or with proteins and amino acids of the plant or of the invading agent to form dark, highly oxidized melanoidins that are inhibitory to microorganisms. Common substrates for phenol oxidases include chlorogenic acid, catechin, epicatechin and DOPA (dihydroxyphenylalanine). In addition, plant defense mechanisms are enhanced by the action of hydrolytic enzymes on phenolic compounds resulting in the formation of diphenol oxidase substrates – that is, aglycones. The mechanism of action of phytoalexins, which are broad-spectrum antimicrobial agents, is not understood fully, although evidence suggests that they alter the properties of microbial plasma membranes (Walker, 1994).

Pisatin from peas was the first recognized phytoalexin, but the total number of compounds isolated exceeds 200 from more than 20 botanical families. Shredded carrots and carrot juice have a lethal effect on *L. monocytogenes* (Beuchat and Brackett, 1990). Extracts of carrot roots inhibit differentiation and aflatoxin formation by *Aspergillus parasiticus* (Batt *et al.*, 1980).

8.4.5 Garlic and onion

Plants in the *Allium* genus, namely garlic, onion and leek, are probably the most widely consumed foods with substantial antimicrobial activity, mainly related to the compound allicin. As aforementioned, intact tissues of *Allium* species do not contain allicin but do contain the precursor alliin. When bulb tissue is disrupted, alliin undergoes hydrolysis to yield allicin, pyruvate and ammonia by the action of the phosphopyridoxal enzyme alliinase. The mechanism of antimicrobial activity is inhibition of sulfhydryl-based enzyme activity, including alkaline phosphatase, invertase, urease and papain (Wills, 1956).

The spectrum of activity of *Allium* extracts is broad, including important pathogens such as *S. aureus*, *Bacillus* spp., *C. botulinum* type A (but not types B and E), *E. coli*, *Salmonella* spp and *Shigella* spp. Also, a large number of yeasts and molds are susceptible (Sofos *et al.*, 1998). *Allium* extracts are also able to inhibit mycotoxin formation (Mabrouk and El-Shayeb, 1981).

8.4.6 Spices and herbs

In addition to contributing to the sensory quality of foods, many spices and herbs also exhibit antimicrobial activity, mostly due to phenolic compounds. Although in many instances concentrations of these compounds necessary for inhibiting growth or various metabolic activities in microorganisms exceed those normally used in foods, the preservative effects of such seasoning agents should not be discounted. Among the spices with the greatest antimicrobial activity are cinnamon, cloves and allspice. Their active principles are often present in the essential oil or in the extracted, isolated and concentrated natural oil. The most important compounds are cinnamic aldehyde (present in cinnamon), eugenol (a major constituent in clove oil and present in considerable amounts in the essential oil of allspice), and thymol (present in the essential oils of thyme, oregano, savory, sage and several other herbs) (Sofos *et al.*, 1998).

These compounds exhibit a wide antimicrobial spectrum, including important pathogens such as *S.* Typhimurium, *S. aureus*, *E. coli* and *V. parahaemolyticus* (Deans and Richie, 1987), and inhibit *Aspergillus* growth and mycotoxin production (Bullerman *et al.*, 1977). The mode of action of phenolic compounds in spices and herbs against microorganisms has not been defined (Sofos *et al.*, 1998).

8.4.7 Plant pigments

Compounds responsible for the color of plant tissues have, in many instances, antimicrobial properties. Anthocyanins, present in almost all higher plants and predominantly in flowers and fruits, consist of an aglycone portion (i.e. anthocyanidin) esterified with one or more sugars. Although these pigments are better known for their food-coloring capabilities, they also inhibit bacteria, including *E. coli* and *S. aureus* (Powers *et al.*, 1960) and yeasts (Marwan and Nagel, 1986). The mechanism of antimicrobial activity is not fully understood, although it is generally accepted that their chelating ability may explain in part a proven anti-enzymatic action (Somaatmadja *et al.*, 1964).

The term 'annatto' refers to a series of preparations containing the carotenoid-type pigments *cis*-bixin and nor-bixin. Essentially non-toxic, annatto is commonly classified as a natural colorant and is used to impart distinctive flavor and color to foods. Annatto preparations are, in fact, the primary colorant for dairy foods such as cheese and butter. Annatto plant extracts show antimicrobial activity against Gram-positive bacteria, such as *B. cereus*, *C. perfringens*, *S. aureus* and *L. monocytogenes*, but have no activity against Gram-negative bacteria or yeasts (Galindo-Cuspinera *et al.*, 2003).

8.4.8 Other phenolic compounds

Several phenolic compounds, including oleuropein and its aglycone, are found in olives and olive oil. The hydrolysis products of oleuropein are complex alcohols, elenolic acid and the oleuropein aglycones. Oleuropein itself is not antimicrobial, but the elenolic acid and the aglycones are. Microorganisms inhibited by these compounds include *Lactobacillus* spp., *Leuconostoc mesenteroides*, *S. aureus* and fungi (Fleming *et al.*, 1973).

Hydroxy-cinnamic and cinnamic acids are phenolic compounds present in plant parts used as spices (Beuchat and Golden, 1989). Compounds of this chemical family with antimicrobial activity include caffeic, chlorogenic, *p*-coumaric, ferulic and quinic acids. Depending on the botanical species, hydroxy-cinnamic acids may be present at concentrations sufficient to retard microbial invasion and delay rotting of fruits and vegetables. Gram-positive and Gram-negative bacteria, molds and yeasts are sensitive to these compounds (Davidson and Branen, 1981). Caffeic, ferulic and *p*-coumaric acids, for example, inhibit *E. coli*, *S. aureus* and *B. cereus* (Herald and Davidson, 1983), and *S. cerevisiae* (Baranowski *et al.*, 1980). Caffeic and coumaric acids inhibited aflatoxin production (Paster *et al.*, 1988).

Tannins and tannic acid are present in the barks, rinds and other structural tissues of plants, and are known to possess antimicrobial activity against *S.* Enteritidis, *L. monocytogenes*, *E. coli*, *S. aureus*, *A. hydrophila* and *S. faecalis* (Chung and Murdock, 1991). The antimicrobial effect of red and white wines is proportional to the amount of flavonoid tannin present (Singleton and Esau, 1969).

8.4.9 Hops

Flowers of the hop vine are used to impart bitter flavors and other desirable properties to beer. Resins, commonly termed α-acids (represented by humulone and its derivatives), and β-acids (represented by lupulone and its derivatives) are the major compounds responsible for this flavor. Most of them also inhibit microbial growth. Gram-positive bacteria and some fungi are most sensitive (Mizobuchi and Sato, 1985). Although hops have antimicrobial activity, they contribute little to the microbial stability of beer (Richards and Macrae, 1964).

8.4.10 Coffee, tea, kola and cocoa

Caffeine is present in coffee and cocoa beans, tea and kola nuts. It is antimycotic as well as antibacterial. Inhibition of growth of several mycotoxigenic *Aspergillus* and *Penicillium* spp. at concentrations of caffeine as low as 0.0001 % has been documented, although the mechanisms by which caffeine inhibits polyketide mycotoxin synthesis is unknown (Buchanan *et al.*, 1983). Bacteria affected by caffeine include *S. aureus*, *Salmonella* spp., *E. coli*, *S. faecalis*, *B. cereus* and *L. monocytogenes* (Vanos and Bindschedler, 1985; Pearson and Marth, 1990). Tea and cocoa contain theophylline and theobromine, also regarded as antimicrobial agents. Growth of *S.* Typhimurium, *E. coli*, *S. aureus* and *B. cereus* was found to be inhibited in a 2 % total solids instant tea infusion, whereas *Lactobacillus plantarum* was inhibited at 10 % total solids (Vanos *et al.*, 1987).

8.4.11 Carbohydrates

Perhaps the most widely occurring and abundant group of indirect natural antimicrobial compounds is the simple sugars in foods of plant origin. Sucrose, fructose and other sugars decrease the a_w of foods and in sufficient concentration inhibit microbial spoilage. Carbohydrates also serve as substrates for microorganisms involved in fermentations, resulting in generation of antimicrobial components. Other carbohydrates are converted to toxic metabolites, thereby killing or inhibiting their producers (Scott, 1988).

8.5 Antimicrobial polypeptides

Among the natural substances considered to be safe alternatives to synthetic chemical preservatives in food processing are various enzymes and other peptides. Some polypeptides are being used, while others seem promising for food preservation. Polypeptides present in animal or plant tissue that are important in defense against pathogenic microorganisms are often classified as functioning by oxygen-dependent or -independent mechanisms (Sofos *et al.*, 1998).

The oxygen-dependent defenses involve enzymes and metabolic processes generating toxic metabolites of oxygen, such as hydrogen peroxide, superoxide ions, hydroxyl radicals and halides. These enzymes include peroxidases and other oxygen-metabolizing enzymes. Another group of enzymes related to oxygen is the oxidases (e.g. glucose oxidase and catalase), which can deplete oxygen and therefore inhibit aerobic microorganisms. Oxygen-independent means of inhibiting or killing microorganisms involve peptides and proteins that react with the cell surface and disrupt the structural surface layers or membranes of microorganisms. This group of polypeptides includes cationic peptides and proteins, bacteriocins, lysozyme and other lytic enzymes, and hydrolases including lipases and proteases. These polypeptides often function in combination with each other or with other groups of antimicrobials (Sofos *et al.*, 1998).

8.5.1 Lytic enzymes

Lysozymes act by cleaving glycoside bonds in the structural peptidoglycan of bacterial cell walls; by their activity lysozymes leave a punctured cell wall, which may lead to cell lysis in hypotonic media (Tranter, 1994). Lysozymes have desirable properties as food preservatives; it is economically feasible to obtain them from sources such as egg white in good yield and purity by ion-exchange resins. They have very low or no toxicity when consumed orally at levels exceeding 0.2 %, even after extended periods of consumption. Because they are odorless with a slightly sweet taste and, even more importantly, because the quantities used in foods are very low (0.002–0.04 %), they do not interfere with the sensory quality of products (Johnson, 1994).

Egg-white lysozyme remains stable in a number of food-processing operations. It can withstand boiling for 1–2 minutes at acidic pH values in solution, but it is denatured upon boiling at high pH values. It can be frozen, and is stable to spray drying (Tranter, 1994). This enzyme is active primarily against specific groups of Gram-positive bacteria, mostly non-pathogenic *Clostridium*, *Bacillus*, *Bifidobacterium*, *Corynebacterium* and *Streptococcus*, and the pathogenic *L. monocytogenes* and *C. botulinum*. Because of its specificity, it does not interfere with food and beverage fermentations carried out by beneficial lactic acid bacteria and yeasts (Carini *et al.*, 1985; Hughey and Johnson, 1989).

Lysozyme is used to prevent gas formation or 'late blowing' by *Clostridium tyrobutyricum* and other gas-producing *Clostridium* spp. during ripening of certain cheeses, such as Provolone, Grana, Padano, Emmental and Gouda (Carini *et al.*, 1985). Japanese investigators have studied lysozyme as a preservative in foods such as fresh vegetables, tofu, sausages, fish cakes and seafood (Tranter, 1994). Applications include coating fresh fruits, vegetables, meat and fish surfaces, as well as addition to wine and sake. Lysozyme extends the shelf-life of sake and mirin (Japanese wines) and prevents the malolactic fermentation in Western wines (Pitotti *et al.*, 1991). Using the Maillard reaction, lysozyme–dextran conjugates have been prepared. They show improved emulsifying properties and are able to inhibit both Gram-negative (*V. parahaemolyticus*, *E. coli*, *A. hydrophila* and *Klebsiella pneumoniae*) and Gram-positive bacteria (*B. cereus* and *S. aureus*) (Nakamura *et al.*, 1991).

In addition to lysozyme, a number of other enzymes are able specifically to cleave carbohydrate or peptide linkages in the cell walls of bacteria and fungi. These enzymes could be useful in the control of bacteria on foods but not as therapeutic agents, due to their interference with immune reactions (Leive and Davis, 1980).

8.5.2 Oxidases and peroxidases

Oxidases are enzymes oxidizing organic substrates, with molecular oxygen, generating hydrogen peroxide. Xanthine oxidase is an example of a naturally occurring enzyme in milk that generates hydrogen peroxide (Sofos *et al.*, 1998).

Glucose oxidase catalyzes a reaction between glucose and oxygen, and yields gluconic acid or D-glucono-δ-lactone and hydrogen peroxide. This enzyme has been approved to remove glucose from food and as an antioxidant. For example, it is used to remove glucose from egg white, to remove oxygen from beverages and headspace of packages, and to prevent Maillard browning reactions (Frank, 1992). It is effective against *S. infantis, S. aureus*, *C. perfringens* and *B. cereus*, although *C. jejuni*, *L. monocytogenes* and *Y. enterocolitica* have shown resistance (Tiina and Sandholm, 1989).

Peroxidases are widespread in nature and oxidize molecules at the expense of hydrogen peroxide. There are many peroxidases with different redox potentials and substrate specificities. Often, the rate-limiting substrate for peroxidase activity in foods is the availability of hydrogen peroxide, which is reactive and must be provided continuously (Ekstrand, 1994).

Lactoperoxidase, a product of the mammary glands and the most abundant enzyme in bovine milk, is also produced in salivary glands and is present in the saliva of mammals. It oxidizes thiocyanate or halogens, thereby producing toxic metabolites. Its combination with hydrogen peroxide, thiocyanate and/or iodide leads to a potent antibacterial system known as the *lactoperoxidase system*, which is based upon the inactivation of bacterial metabolic enzymes due to their oxidation by hypothiocyanate or hypoiodite. The enzyme alone has been documented to be responsible for some of the natural antibacterial activity present in cow's milk. The antimicrobial activity of the lactoperoxidase system is both species- and strain-specific. The system is bactericidal to Gram-negative bacteria (e.g. spoilage psychrotrophs) but is usually only bacteriostatic or temporarily inhibitory to Gram-positive organisms (e.g. *L. monocytogenes*, *B. cereus* and *Streptococcus uberis*) (Zajak *et al.*, 1981; Reiter and Harnulv, 1984; Marshall *et al.*, 1986; Siragusa and Johnson, 1989).

The lactoperoxidase/thiocyanate/hydrogen peroxide system has been utilized industrially, and has been suggested as a preservative in several systems, including the temporary preservation of raw milk in developing countries where refrigeration is unavailable. This system can inactivate *E. coli* (including the pathogenic O157:H7), *S.* Typhimurium, *Y. enterocolitica*, and *Pseudomonas aeruginosa* in milk and infant milk formula (Ekstrand, 1994). Lactoperoxidase has also been used to prevent spoilage in other dairy-based products, such as soft ice cream and pastry cream. Its activity against Gram-negative pathogens suggests that it could be beneficial for the preservation of other foods, particularly those of animal origin that may contain enteric pathogens (Johnson, 1994). An interesting potential application of lactoperoxidase is as a probiotic. Calves fed milk supplemented with the lactoperoxidase system components showed better weight gain than calves receiving unsupplemented milk (Ekstrand, 1994).

8.5.3 Transferrins

Growth and survival of many bacterial and fungal pathogens, namely *Staphylococcus*, *Clostridium*, *Listeria*, *Mycobacterium*, *Salmonella*, *Escherichia*, *Pseudomonas*, *Yersinia*, *Vibrio* and *Aeromonas*, depend on the availability of iron ions (Weinberg, 1978). In contrast, lactic acid bacteria have a metabolism based on manganese ions instead of iron. It is therefore possible, by means of iron sequestration, to favor the growth of beneficial lactic acid bacteria over that of either pathogens or spoilage microorganisms in foods. This can be accomplished through chelation by iron-binding polypeptides, especially the transferrins and related proteins (Bruyneel *et al.*, 1990). Unlike antimicrobial enzymes such as lysozyme and lactoperoxidase that act catalytically, transferrins must be present in stoichiometric excess of the quantity of iron ions available; therefore, they are useful only in foods such as milk or egg albumen that have low iron content (Sofos *et al.*, 1998).

About 13 % of egg-white protein is the iron binding ovotransferrin, also known as conalbumin. This compound is believed to be the key factor responsible for microbial inhibition in egg albumen. Its activity is enhanced in albumen's alkaline pH (8.5–9.5), but microbial growth resumes when the albumen is saturated with iron ions (Tranter, 1994).

Lactoferrin is a transferrin present in the milk of mammals as well as in tears, saliva and mucosal secretions (Masson *et al.*, 1969). The physiological functions of lactoferrin may include protection of young animals from enteropathogenic bacteria, protection of the non-lactating mammary gland against mastitis, the physiological transport and supply of iron ions, and immunoregulation (Ekstrand, 1994). In addition to metal chelation, lactoferrin damages the outer membrane of Gram-negative bacteria and sensitizes them to lysozyme, with which lactoferrin forms a complex (Ellison and Giehl, 1991). Microorganisms sensitive to this agent include *Vibrio cholerae*, *Streptococcus mutans*, *E. coli*, *P. aeruginosa* and the yeast *Candida albicans* (Arnold *et al.*, 1980), whereas *L. monocytogenes*, *S.* Typhimurium, *S. aureus*, *Pseudomonas fluorescens* and *Shigella sonnei* show resistance (Payne *et al.*, 1990).

Digestion of bovine lactoferrin with gastric pepsin yields a hydrolysate with antibacterial activity greater than that of native lactoferrin. This peptide, referred to as lactoferricin, is able to inactivate a broad range of Gram-positive and Gram-negative bacteria, including *E. coli*, *L. monocytogenes*, *S. aureus*, *C. perfringens*, *C. jejuni*, *Y. enterocolitica*, *Corynebacterium diphteriae*, *P. aeruginosa*, *Proteus vulgaris*, *K. pneumoniae*, *S.* Enteritidis and *S. mutans*. Resistant organisms include *P. fluorescens*, *Bifidobacterium bifidum* and *Enterococcus faecalis*. Lactoferricin is lethal at concentrations ranging from 0.00003 % to 0.0015 %, depending on the microorganism (Wakabayashi *et al.*, 1992; Hoek *et al.*, 1997).

8.5.4 Bacteriocins

Bacteriocins are small peptides, produced by bacteria, that possess antibiotic properties. Bacteriocins are normally not termed antibiotics in order to avoid confusion with and concerns about therapeutic antibiotics (Cleveland *et al.*, 2001). They differ from most therapeutic antibiotics in being proteinaceous and generally possessing a narrow specificity of action against strains of the same or closely related species (Tagg *et al.*, 1976). Bacteriocins are rapidly hydrolyzed by proteases in the human digestive tract (Joerger *et al.*, 2000). The effectiveness of bacteriocins as antimicrobial agents in foods can become limited for various reasons, and cost remains an issue impeding their broader use as food additives. Hence, not only do searches continue for new and more effective bacteriocins, but efforts are also being made to improve existing bacteriocins to address both biological and economic concerns (Chen and Hoover, 2003).

Most of the bacteriocins from lactic acid bacteria are cationic, hydrophobic or amphiphilic molecules composed of 20–60 amino acid residues (Nes and Holo, 2000). They are commonly classified into three groups (classes) that also include bacteriocins from other Gram-positive bacteria (Nes *et al.*, 1996).

Class I includes small peptides containing the unusual amino acids lanthionine, α-methyllanthionine, dehydroalanine and dehydrobutyrine. They are also called lantibiotics (from lanthionine-containing antibiotics). This class is further subdivided into type A and type B lantibiotics, according to chemical structures and antimicrobial activities (Guder *et al.*, 2000). Type A lantibiotics are elongated peptides with a net positive charge that exert their activity through the formation of pores in bacterial membranes. Type B are smaller globular peptides and have a negative or no net charge; antimicrobial activity is related to the inhibition of specific enzymes.

Class II includes small, heat-stable, non-lanthionine-containing peptides. These peptides are divided into three subgroups. Class IIa includes pediocin-like peptides with remarkable anti-*Listeria* activity (Ennahar *et al.*, 2000); class IIb are dipeptides, and class IIc contains the remaining peptides of the class.

Class III is not well characterized, and it includes large, heat-labile proteins that are of lesser interest to food scientists.

Nisin remains the commercially most important bacteriocin. It has led in popularity because of its relatively long history of safe use and its documented effectiveness against important Gram-positive foodborne pathogens and spoilage agents. In fact, it is the only purified bacteriocin approved for food use in the US, and has been successfully used for several decades as a food preservative in more than 50 countries (Chen and Hoover, 2003). Produced by *Lactococcus lactis*, nisin is a 34-amino acid peptide, categorized as a class I bacteriocin and a type A lantibiotic. At least six different forms have been discovered and characterized (designated as A–E and Z), with nisin A being the most active type (van Kraijj *et al.*, 1999).

Nisin usually has no effect on Gram-negative bacteria, yeasts and molds, although Gram-negative bacteria can be sensitized to nisin by permeabilization of the outer membrane layer through sublethal heating, freezing and chelating agents (Delves-Broughton *et al.*, 1996). Normally only Gram-positive bacteria are affected; and these types include lactic acid bacteria, *Listeria*, *Staphylococcus* and *Mycobacterium*, and the spore-forming *Bacillus* and *Clostridium*. The spores of bacilli and clostridia are actually more sensitive to nisin than are vegetative cells, although the antagonism is sporostatic, not sporicidal, thus requiring the continued presence of nisin to inhibit outgrowth of the spores. Heat damage of spores substantially increases their sensitivity to nisin, so that nisin is effective against spores in low-acid, heat-processed foods, resulting in its use as a processing aid in canned vegetables. The mechanism of its sporostatic action is distinct from its bactericidal effect on the cytoplasmic membrane of vegetative cells (Morris *et al.*, 1984).

Purified nisin has been evaluated for toxicological effects and found to be harmless or at least to have very low toxicity using rat and guinea pig models (Shtenberg and Ignatev, 1970). Examples of marketed food products that can legally be amended with nisin are canned soups (Australia), ice for storing fish (Bulgaria), baked goods and mayonnaise (Czech Republic), and milk shakes (Spain). The majority of products approved for its use are dairy products (especially cheeses) and canned goods. In the US, use of nisin-producing starter cultures has never been regulated as lactococci are considered GRAS (Chikindas and Montville, 2002).

Nisin (100 IU/ml) may control *L. monocytogenes* in ricotta-type cheeses for 8 weeks or more, according to the cheese type (Davies *et al.*, 1997). In a study using vacuum-packed cold-smoked rainbow trout, the inhibition of the above pathogen by nisin or sodium lactate, or their combination, was determined. Both antimicrobial agents were capable of inhibiting the pathogen growth, but the combination of the two compounds was even more effective (Nykänen *et al.*, 2000).

8.6 Miscellaneous antimicrobial agents

8.6.1 Ethanol

The value of ethanol as a naturally occurring antimicrobial has been recognized since the first alcoholic fermentations of fruits to produce wines, which occurred several thousand years ago. For ethanol to disinfect, water must be present. At 95 % ethanol most vegetative cells are resistant, whereas at 60–75 % ethanol most microorganisms are destroyed in less than a minute. Low concentrations are rarely biocidal alone, but may exert inhibitory effects. At 8–11 % (v/v), ethanol prevents growth of most molds and bacteria; a 15–18 % concentration is required to prevent growth of most yeasts. Gram-negative bacteria are more susceptible to ethanol than are Gram-positive bacteria, whereas bacterial spores are generally resistant. Various environmental factors influence the activity of ethanol as an antimicrobial. Increased sugar concentration, decreased temperature and decreased a_w increase its effectiveness; but the presence of organic matter decreases activity (Shelef and Seiter, 1993).

Because ethanol is amphiphilic, the primary site of activity is thought to be the cytoplasmic membrane. It is also thought that ethanol may have a direct effect on membranes or membrane-bound enzymes, or an indirect effect due to impairment of membrane biosynthesis. Dissolution of ethanol in the cell membrane increases fluidity of the lipid and decreases the gel-to-liquid crystalline phase transition temperature of lipids. This results in disruption of membrane organization, leakage of ions, leakage of low molecular weight solutes, and even leakage of macromolecules (Seiler and Russell, 1991).

Ethanol has GRAS status as a food additive. It is primarily used as a solvent for flavor and color compounds. Evaluation of its antimicrobial activity is actually quite recent. Most studies have focused on the antimycotic activity of the compound. Increased shelf-life of bakery products such as bread or pizza crust was obtained by the addition of ethanol and sterile water as a dip or a spray. This research first demonstrated that ethanol is a vapor phase inhibitor; thus techniques such as vacuum packaging with ethanol, adding sachets or strips impregnated with ethanol, or encapsulation techniques are possible methods for incorporating ethanol as a mold inhibitor (Seiler and Russell, 1991).

8.6.2 Natamycin

Natamycin was first isolated in 1955 from a culture of *Streptomyces natalensis*, a microorganism found in soil from Natal, South Africa. It is a polyene macrolide antibiotic (Brik, 1981). Natamycin is active against nearly all molds and yeasts, but has no effect on bacteria or viruses. Natamycin blocks mycotoxin production in genera such as *Aspergillus* and *Penicillium* (Gourama and Bullerman, 1988). The mode of action of polyene macrolides involves binding of ergosterol and other sterol groups in fungal cell membranes.

8.6.3 Reuterin

A number of low molecular weight compounds with antimicrobial activity have been isolated and identified from culture filtrates of lactic acid bacteria. The most studied to date is reuterin (β-hydroxy-propionaldehyde), produced by the heterofermenter *Lactobacillus reuteri*. Reuterin exists in three forms – the aldehyde, its hydrate, and a cyclic dimer – and has very broad spectrum activity against bacteria, fungi, protozoa and viruses. The addition of reuterin to foods such as minced beef, milk and cottage cheese has been shown to control coliform growth and to inactivate *L. monocytogenes* and *E. coli* O157:H7 (El-Ziney and Debevere, 1998; Muthukumarasamy *et al.*, 2003).

8.6.4 Diacetyl

Diacetyl is produced by the citrate-fermenting lactic acid bacteria *Leuconostoc cremoris* and *Lactococcus lactis* subsp. *lactis* var. *diacetylactis*. The compound produces a buttery flavor in fermented dairy products and in foods to which it is added for flavor. Acceptable sensory levels in dairy products range from 0.0001 % to 0.0007 %. The effective antimicrobial concentration is higher than its organoleptic detection threshold, limiting its use as a natural preservative. Its volatility also limits its usefulness.

8.6.5 Hydrogen peroxide

Hydrogen peroxide itself does not have antimicrobial activity. Rather, it produces powerful reaction products, such as singlet or superoxide oxygen, that are highly toxic to living organisms. Another possible mechanism is through the oxidation of sulfhydryl groups and double bonds in proteins and lipids. The compound is active against bacteria, molds, yeasts and viruses, and is particularly effective against anaerobes and facultative anaerobes because many lack catalase. In general, Gram-negative bacteria are more susceptible than Gram-positive ones (Cords and Dychdala, 1993). Organisms affected by hydrogen peroxide include *E. coli*, *E. aerogenes* and *L. monocytogenes*. Others, such as *Lactobacillus bulgaricus*, *L. lactis* and *Bacillus megaterium*, showed resistance. The susceptibility of spore-forming organisms (i.e. *Bacillus* and *Clostridium* species) depends on concentration, a_w and temperature (Dominguez *et al.*, 1987).

In the US, hydrogen peroxide is allowed as a direct additive at 0.05 % to pasteurized milk for making certain types of cheese and as an antimicrobial in whey (0.04 %) and starch (0.15 %). Catalase is added to products to remove residual hydrogen peroxide. It is also allowed for sterilization of polymeric food-packaging surfaces, which are used for aseptically packaged foods and for sanitizing of food contact surfaces (Code of Federal Regulations, 1992).

9 Alternative innovative technologies

Newer physical food preservation methods can be divided into thermal and non-thermal procedures. Special attention in research and development has been directed toward the non-thermal technologies that have the ability to inactivate microorganisms at ambient or near-ambient temperatures. In fact, alternative technologies for inactivating microorganisms without relying on heat are not new concepts, but their use as food preservation treatments has received considerable attention recently in response to consumer demands for more 'fresh' and 'natural' products (Ross *et al.*, 2003).

9.1 Ohmic heating

Ohmic heating is a thermal method that minimizes energy input and thus reduces thermal damage to food. If an electric current passes through a conductive medium, in this case the food, the medium warms up as a result of ionic movement. Essentially, ohmic heating utilizes the effect of the electrical resistance within a conductive liquid or solid material, allowing a direct conversion of electric energy into heat. It is therefore evident that its applicability is limited to foods with sufficient conductivity. In processing plants, the product is continuously pumped through a column equipped with several electrodes (Butz and Tauscher, 2002).

Ohmic heating is presently used for pasteurization and sterilization of liquid and particulate foods, especially ready-to-serve meals, fruits, vegetables, meat, poultry and fish, and is an alternative to sterilization of foods by means of conventional heat exchangers or autoclaves. There are several other potential applications, including blanching, evaporation, dehydration, fermentation and extraction (Butz and Tauscher, 2002).

Lethality within food particles undergoing ohmic heating was investigated by Kim *et al.* (1996). Meatballs containing spores of *Bacillus stearothermophilus* and precursors of chemical markers were thermally processed in a starch solution with 30–40 % solids content using a 5-kW ohmic system. Higher temperature and microbiological lethality were observed at the center of the meatballs rather than near the surface. The time–temperature history of the ohmically processed meatballs equivalent to 1.06 minutes at 133 °C corresponded to an F_0-value of 16.8, which is at least five times greater than that needed for a 12-log reduction in *Clostridium botulinum*.

9.2 Microwaves

The primary advantage of using microwaves, when compared to conventional electric oven heating, is time-saving. Within the food industry there are some unique applications associated with the singular heating properties of microwaves, including tempering, drying pasta and cooking bacon. Microwave technology has the potential and flexibility to be adapted for vacuum and freeze drying, pasteurizing, sterilizing, baking, roasting and blanching. A number of factors exert a significant influence in achieving uniform heating with microwave energy, namely the moisture and ionic content of foods, the specific heat of various food constituents, and the product density, shape and volume (Heddleson and Doores, 1994).

Microwave energy is a form of non-ionizing radiation that falls several orders of magnitude short of the energy necessary to break the weakest of chemical bonds (i.e. 1.2×10^{-5} eV for the quantum energy of microwaves versus 5.2 eV for the breakage of hydrogen bonds) (Rosen, 1972). Two major microwave constituents can be distinguished: a magnetic field and an electric field, oriented perpendicularly to one another. The latter is primarily responsible for heating, as it promotes the rotation of polar molecules and consequently yields heat generated by molecular friction (Curnutte, 1980).

There are two main schools of thought concerning the means by which microwaves injure or kill bacteria. On one hand, a number of studies have concluded that microwaves can reduce bacterial numbers entirely by heat. That includes irreversible heat-denaturation of enzymes, proteins, nucleic acids or other cellular constituents, resulting in cellular death as well as leakage of metabolites and/or cofactors crucial to cellular function through membranes damaged by heat. On the other hand, there are studies that suggest a non-thermal mechanism of lethality, an effect exclusive to microwave technology, apparently related to RNA damage (Chipley, 1980; Khalil and Villota, 1989).

Numerous studies have examined how microwave heating affects numbers of microorganisms present in various foods, often comparing microwave heating to similar time–temperature treatments performed in conventional ovens. The majority of these studies have focused on important pathogens, such as *Salmonella* spp., *L. monocytogenes*, *E. coli*, *S. aureus*, *Streptococcus faecalis* and *Clostridium perfringens* (Heddleson and Doores, 1994).

Heddleson *et al.* (1991) performed the first study using microwave heating to relate the composition of liquid model food systems to temperatures achieved and amounts of destruction of *Salmonella* spp. It was found that of the various food components examined, only NaCl significantly influenced temperatures and inactivation rates. The presence of salt caused large temperature gradients within small volumes, and the non-uniform heating contributed to a greater survival of the pathogen. Later, Heddleson *et al.* (1994) examined processing variables influencing the destruction of *Salmonella* spp. Heating medium volume, container shape and covering containers did not significantly alter the rate of inactivation at 60 °C. Increased post-heating holding times of ≥ 2 minutes increased bacterial destruction. Microwave heating with ovens of low power (*ca.* 450 W) was less effective than heating with units of high power (*ca.* 700 W).

Schnepf and Barbeau (1989) conducted a study to compare the effectiveness of microwave, convection-microwave and conventional electric ovens in eliminating *S.* Typhimurium from roasting chickens. The microwave oven proved to be the least efficient in killing the pathogen, while convection-microwave and the conventional oven proved to be of nearly equal efficiency. Minimum internal temperatures recommended by the US Department of Agriculture (USDA) (71 °C), the Food and Drug Administration (FDA) (74 °C) and the American Home Economics Association (85 °C) were used as the bases for selecting temperatures for examination. The researchers found that even at 85 °C, microwave heating did not eliminate *Salmonella*. It appears that *Salmonella* are not effectively killed by this thermal treatment when inoculated on the chicken surface, and therefore it may not be wise to assume that minimum internal temperatures can be used to recommend safe cooking practices for microwave ovens. Since microwaves penetrate within the food matrix and essentially exert a steam-cooking effect, the coolest temperatures may be found on the surface of solid food masses due to evaporation. This is the opposite effect to the one taking place in conventional electric ovens, in which conduction transfers heat from the surface to the inner mass, thus providing the rationale for current cooking recommendations based on minimum internal temperatures (Heddleson and Doores, 1994). In fact, previous researchers (Chen *et al.*, 1973; Lindsay *et al.*, 1986) reached the common conclusion that a minimum internal temperature is a poor criterion for determining safety standards in foods heated with microwave ovens.

In a later study, Reis-Tassinari and Landgraf (1997) evaluated the destruction of *S.* Typhimurium during reheating of foods in two different types of microwave ovens: a conventional 750-W and a 700-W unit with preset controls. Heating times in the conventional microwave oven were established at 50 seconds for baby food and 75 seconds for mashed potatoes and beef stroganoff samples, while for the preset oven time periods were determined by a built-in temperature sensor. The percentage of food samples positive for the pathogen after treatment in the conventional oven was 47.8 %, whereas in the microwave with preset controls it was 93.3 %. The results therefore suggested that reheating contaminated foods in microwave ovens might not be adequate to destroy *S.* Typhimurium and to assure food safety.

Galuska *et al.* (1988) conducted one of the first studies examining the destruction of *L. monocytogenes* by microwave heating. These authors examined the thermotolerance of *Listeria* suspended in non-fat dry milk heated by microwaves, and calculated *D*-values at five temperatures between 60 °C and 82.2 °C. They compared the results with those found by heating in a water bath. *D*-values were lower in the water bath-heated sample than in the microwave-heated sample. Microwave treatment accomplished a 4.5 $\log_{10}$ cycle reduction in viable cell numbers within 15 seconds at 71.1 °C. At conventional pasteurization processing temperatures, microwaves were as effective as conventional heating in destroying *L. monocytogenes*.

The effect of different microwave power levels (240, 400, 560 and 800 W) on the survival of *L. monocytogenes* in inoculated shrimp was investigated by Gundavarapu *et al.* (1995). *D*-values were determined using constant-temperature water baths to establish heat resistance of the pathogen in shrimp. Shrimp were inoculated with *ca.* 5×10^5 CFU/g of a five-strain mixture of *L. monocytogenes*. Shrimp samples were then cooked in the microwave oven at the different power levels using cooking times predicted by a mathematical model, as well as 20 % longer times than those obtained from the model. No viable *L. monocytogenes* was detected in uninoculated shrimp after microwave cooking at the lowest power treatment, but at least one replication of inoculated shrimp tested positive for the presence of *Listeria*. No viable pathogens were detected in shrimp cooked at 120 % of predicted times.

Dahl *et al.* (1981) studied the survival of *S. aureus* in model cook/chill foodservice systems similar to those used in hospitals for food preparation. *S. aureus* was surface-inoculated onto beef loaves, potatoes and canned green beans, which were heated to times and temperatures recommended by the HACCP plan of a specific facility. It was found that microwave heating to mean end temperatures of 74–77 °C did not eliminate *S. aureus* from food samples. Furthermore, the destruction kinetics of *S. aureus* could not be predicted consistently, and time and temperature were poor parameters upon which to base microwave thermal processes. *S. aureus* was inoculated into beef-soy loaves and assayed for recovery of viable cells during various stages of food handling in a hospital chill food-service system by Bunch *et al.* (1977). Heating the loaves at 121 °C in a conventional oven to an internal temperature of 60 °C substantially reduced the inoculum level at the center of the loaf. Chilling the loaves at 5 °C for 24–72 hours, followed by microwave reheating to an internal temperature of 80 °C, resulted in the elimination of *S. aureus* from the samples. However, it was noted that preformed toxin, if present, would not be inactivated by the microwave reheating treatment.

The microbiological quality of scrambled eggs and roast beef prepared in a hospital foodservice system and reheated in a microwave oven for later consumption was investigated by Cremer and Chipley (1980a, 1980b). The average time from cooking to microwave reheating was 26.3 hours for the scrambled eggs, which were then reheated to an average internal temperature of 67.1 °C. Nevertheless, the range of final temperatures was rather wide: 35–92 °C. At the time of microwave reheating the natural flora of the eggs reached a total plate count of 30 CFU/g of *Bacillus* spp., *Clostridium sporogenes*, *Staphylococcus epidermidis*, *E. coli* and *Enterobacter aerogenes*. This low count was attributed to the use of pasteurized, *Salmonella*-free frozen whole eggs as a starting material. Both coliforms and staphylococci were recovered after microwave heating, indicating inadequate thermal processing as revealed by the wide range of final internal temperatures. Roast beef was cooked in a conventional oven, stored for approximately 45 hours and then reheated by microwave energy, achieving a final average temperature of 68 °C, with a 43–93 °C range. Between the time of initial roasting and microwave reheating, bacterial numbers increased 3- to 11-fold from an initial level of ≤ 200 CFU/g, because of the lengthy storage time at temperatures favoring growth (7.5 °C was the mean internal temperature). Organisms present in the cooked meat included *C. sporogenes*, *C. perfringens*, *Bacillus* spp., *S. aureus* and *S. epidermidis*. Although microwave reheating lowered their numbers, it was unable to eliminate them. Expectedly, *Clostridium* and *Staphylococcus* were more heat resistant than coliforms. This study concluded that contamination from food handlers should be minimized, because the non-uniform temperatures found after microwave reheating were not sufficient to reduce the number of pathogens to acceptable levels.

Destruction of *E. coli* was studied by Koutchma and Ramaswamy (2000) in a continuous-flow microwave heating system in combination with low concentrations of hydrogen peroxide. The antimicrobial was added separately to the cell suspension at selected concentrations and immediately subjected to microwave heating, or the cell suspension was treated with hydrogen peroxide for 10 minutes at room temperature before microwave heating. Synergistic effects of microwave heating and hydrogen peroxide treatment on *E. coli* destruction were observed, and the interaction reached a maximum with exposure to hydrogen peroxide at 0.075 g/100 g, and microwave heating set at 60 °C.

In conclusion, microwave heating may lead to increased risk of foodborne illness due to poor uniformity of temperature within products and accelerated heating profiles. Factors that influence microwave technology include many of those that are significant in conventional processes, such as the mass and shape of foodstuffs, and their specific heat and thermal conductivity. There are, however, other factors unique to microwave heating due to the nature of the electric field involved in causing molecular friction; primarily moisture and salt contents of foods. It is therefore imperative that thermal processes involving microwave ovens take into account these additional unique factors (Heddleson and Doores, 1994).

9.3 Superheated steam

Superheated steam is defined as steam heated to a temperature higher than the boiling point corresponding to its pressure. It cannot exist in contact with water, nor contain water, and resembles a perfect gas (Dictionary.com, 2003). Typically, superheated steam is obtained by drastically dropping the pressure of saturated steam without changing the temperature, after which energy is applied to heat the steam to the desire temperature (Tang and Cenkowski, 2001).

Compared with hot-air drying, superheated-steam drying provides a number of advantages (Tang and Cenkowski, 2000). Due to its combination of high temperature and low moisture, superheated steam can be utilized to inactivate spoilage and pathogenic microorganisms in foods sensitive to moisture, such as ground spices and flour. Superheated steam is particularly effective for inactivating bacterial spores. Holley *et al.* (2003) found that treating oat groats with superheated steam at 145 °C and 15 psi for 10 minutes reduced the population of *B. stearothermophilus* spores from 6.3 to 3.6 $\log_{10}$.

9.4 High electric field pulses

High-intensity pulsed electric field (PEF) processing involves the application of pulses of high voltage (typically 20–80 kV/cm) to foods placed between two electrodes. PEF may be applied in the form of exponential decaying, square wave, bipolar or oscillatory pulses, and at ambient, sub-ambient or slightly above-ambient temperature for less than a second. Energy loss due to heating of foods is minimized, reducing the detrimental changes of the sensory and physical properties of foods (Barbosa-Cánovas *et al.*, 1999).

Microbial inactivation by PEF has been explained by a number of theories, among which electrical breakdown and electroporation predominate. Electric high-voltage impulses generate a trans-membrane potential across the cell membrane. If the difference between outer and inner membrane potential rises above a critical value of about 1 V, polarization and ultimately breakdown of the membrane are induced. At sufficiently high field-strength (above 10 kV/cm) and duration of the pulses (usually between nano- and micro-seconds), vegetative microorganisms in liquid media are inactivated due to irreversible membrane destruction. PEF effectiveness depends on process factors (electric field intensity, pulse width, treatment time and temperature, and pulse wave shapes), microbial entity factors (type, concentration and growth stage), and media factors (pH, antimicrobials and ionic compounds, conductivity and medium ionic strength). PEF differs from traditional thermal treatments in that the latter greatly affect the cell organelles, whereas the former does not rupture them. This is evidently related to a better final product from a textural and sensory viewpoint (Grahl and Maerkl, 1996; Pothakamury *et al.*, 1997; Butz and Tauscher, 2002).

Microbial inactivation by PEF is a function of growth phase and medium temperature. Logarithmic phase cells are more sensitive to electric fields than are stationary phase cells. Hülsheger *et al.* (1983) showed that a 4-hour culture of *E. coli* was more sensitive to PEF than was a 30-hour culture. Inactivation increases with an increase in temperature of the medium. With exponential decay pulses, the population of *E. coli* was reduced by 2 log cycles after 20 pulses at 40 °C and 50 pulses at 30 °C. On the other hand, when square waves were used, the population of *E. coli* was reduced by 2 log cycles after 10 pulses at 33 °C and 60 pulses at 7 °C (Pothakamury *et al.*, 1997). Inactivation also increased with a decrease in the ionic strength of the medium. After 30 pulses at 40 kV/cm, the population of *E. coli* decreased by 2 log cycles with a decrease in the ionic strength and a decrease in pH (Vega *et al.*, 1996). The presence of sodium or potassium in the treatment medium did not affect microbial inactivation, whereas calcium and magnesium induced a protective mechanism against PEF (Hülsheger *et al.*, 1981).

PEF reductions of 6–9 log cycles were obtained in skim milk inoculated with *E. coli* (Qin *et al.*, 1996). Furthermore, the *E. coli* O157:H7 population in apple juice was reduced by 5 logs using PEF for a maximum treatment time of 172 μs at temperatures lower than 35 °C (Evrendilek *et al.*, 1999). Similar results (i.e. 4–5 log reduction cycles) were obtained for *Staphylococcus aureus*, *Lactobacillus delbrueckii* and *Bacillus subtilis* in skim and raw milk (Pothakamury *et al.*, 1995; Qin *et al.*, 1995, 1998); however in yogurt the reduction was lower (2 log cycles) for *Streptococcus thermophilus*, *Lactobacillus bulgaricus* and *S. cerevisiae* (Dunn and Pearlman, 1987). *Mycobacterium paratuberculosis*, *L. monocytogenes* and *S.* Typhimurium have also been inactivated using PEF (Simpson *et al.*, 1999; Rowan *et al.*, 2001; Liang *et al.*, 2002), whereas *Corynebacterium* spp. and *Xanthomonas* spp. have shown resistance (Raso *et al.*, 1999).

PEF alone appears to have very little effect on bacterial spores (Pagán *et al.*, 1998), although some studies have reported successful destruction in saline solutions (Dantzer *et al.*, 1999). PEF does not induce germination of spores, but if germination is initiated by other methods, the resulting vegetative cells will become sensitive to an electric field (Barbosa-Cánovas *et al.*, 1998).

Pasteurization of milk was undertaken by Smith *et al.* (2002) using PEF alone and combined with nisin and lysozyme, added singly or together. A 7-log reduction was achieved through a combination of PEF treatment (80 kV/cm, 50 pulses), mild heat (52 °C) and the addition of both natural antimicrobials (38 IU/ml and 1638 IU/ml of nisin and lysozyme, respectively).

Similarly, using heat, acidity, antimicrobials (nisin or lysozyme) and PEF, Hodgins *et al.* (2002) pasteurized orange juice. Optimal conditions consisting of 20 pulses of an electric field of 80 kV/cm, pH 3.5 and a temperature of 44 °C with 100 IU/ml of nisin resulted in greater than a 6-log reduction in the microbial population. Following treatment, there was 97.5 % retention of vitamin C. Studying the inactivation of *S.* Typhimurium, Liang *et al.* (2002) also concluded that the bactericidal effects of nisin/lysozyme mixtures on PEF-treated cells were more pronounced than addition of either antimicrobial alone.

Application of PEF is restricted to food products that can withstand high electric fields – that is, that have low electrical conductivity and do not contain or form bubbles. It is expected that PEF will be developed into a reliable technology for reducing the pasteurization temperature of liquid foods (e.g. juices, milk, liquid whole egg). However, as existing PEF systems and experimental conditions are diverse, much more research is needed to assess properly the effect of critical processing factors on pathogens of concern, kinetics of inactivation, absence of potential health risks and process impact on food components. The economic feasibility of PEF compared to traditional pasteurization systems is also an issue (Butz and Tauscher, 2002).

9.5 Oscillating magnetic fields

The region in which a magnetic body is capable of magnetizing the particles present is called the magnetic field. When the susceptibility to magnetization is equal in all dimensions, the particle possesses isotropic susceptibility. On the other hand, when the susceptibility to magnetization is unequal along each dimension, the particle possesses anisotropic susceptibility. Isolated carbon atoms exhibit isotropic susceptibility, whereas two carbon atoms bonded by single, double or triple bonds exhibit anisotropic susceptibility (Barbosa-Cánovas *et al.*, 1998).

Magnetic fields are differentiated as static (SMF) or oscillating (OMF). An SMF exhibits constant field intensity with time, and the direction of the field remains the same. An OMF, applied in the form of pulses, reverses the charge for each pulse and the intensity of each pulse decreases with time to about 10 % of the initial intensity. Magnetic fields can also be classified as homogeneous or heterogeneous. In a homogeneous magnetic field the field intensity is uniform in the area enclosed by the magnetic coil, whereas in a heterogeneous one the field intensity is not uniform as it decreases at greater distances from the center of the coil. In addition, heterogeneous fields exert an accelerating force on the particles – an effect that is not present in homogeneous fields. Magnetic fields are usually generated by supplying current to electric coils. Magnetic flux is measured in weber (Wb), so that 1 Wb = 10^8 magnetic lines. Magnetic density is flux per unit area (Wb/m^2). The inactivation of microorganisms requires magnetic flux densities of 5–50 T (tesla); 1 T = 1 Wb/m^2 (Barbosa-Cánovas *et al.*, 1998).

The influence of static and/or oscillating magnetic fields on living organisms became evident in the early twentieth century with the observation of protoplasmic streaming in cells, which was accelerated or retarded according to the direction of applied magnetic fields. Homogeneous magnetic fields do not have any effect on the morphology, growth or reproduction of microorganisms; but heterogeneous ones do, being able to translocate free radicals such as OH• or O• and produce metabolic interruptions. A characteristic example is the inhibition of budding in yeast cells (Barbosa-Cánovas *et al.*, 1998). Biological membranes exhibit strong orientation in a magnetic field because of their intrinsic anisotropic structure. Orientation of cell membranes parallel or perpendicular to the applied magnetic field depends on the overall anisotropy of the biomolecules, mainly proteins associated with the membrane. Resonating peptide bonds possess diamagnetic anisotropy and therefore tend to orientate parallel to an external magnetic field, as does the plane of carbon–carbon double bonds. In addition, cell division rates can be altered by changes in ion flux across the plasma membrane following the application of a magnetic field (Maret and Dransfield, 1985).

Several experiments have shown that a strong SMF or a moderate OMF has the potential to inactivate both vegetative and spore forms. A single pulse with a flux density of between 5 T and 50 T and a frequency of 5–500 kHz reduces the number of microorganisms by at least 2 log cycles. The technology of inactivating microorganisms by placing foodstuffs in magnetic fields has significant potential use, such as stopping a fermentation process at the right time, or the improvement of quality and shelf-life extension of pasteurized foods (Barbosa-Cánovas *et al.*, 1998). Microorganisms inactivated by magnetic fields have included *S. thermophilus* in milk, *Saccharomyces* spp. in yogurt and orange juice, and bacterial spores in dough (Hofmann, 1985).

Generally speaking, preservation of foods with magnetic fields involves sealing the food in a plastic bag and subjecting it to 1–100 pulses in an OMF with a frequency of between 5 kHz and 500 kHz, at a temperature of 0–50 °C for a total exposure time of between 25 μs and 10 ms. Frequencies higher than 500 kHz are less effective, and also tend to increase the temperature significantly. Exposure time results from the product of the number of pulses and the duration of each pulse. The duration of each pulse includes 10 oscillations, after which the substantially decayed magnetic field has a negligible effect. No special preparation of food is required before treatment by OMF, although metal packages cannot be used. These treatments are carried out at atmospheric pressure with no detectable changes in quality (Barbosa-Cánovas *et al.*, 1998).

The main technological advantages associated with magnetic fields are minimal thermal denaturation of nutritional and organoleptic properties, reduced energy requirements for adequate processing, and the potential for treatment of foods inside a flexible film package (which preclude post-process contamination). Yet there are some important issues regarding this technique that should be addressed before it becomes commercially feasible, including a thorough understanding of the mechanism of microbial inactivation, the precise correlation between OMF operational parameters and microbial inactivation, and the long-term health effects on magnetic-field machine operators (Barbosa-Cánovas *et al.*, 1998).

9.6 Ultraviolet radiation

UV is a form of non-ionizing radiation having a wavelength of between 200 and 400 nm. It is usually divided into long-wave UV (UVA), with a 320–400 nm wavelength range; medium-wave UV (UVB), with a 280–320 nm range; and short-wave UV (UVC), with a 200–280 nm range. Low-pressure mercury gas (< 10 Torr) able to emit UV at 254 nm is commonly used for UV radiation, although there are also medium-pressure UV lamps (*ca.* 1000 Torr) that emit radiation between 185 and 1367 nm (Bintsis *et al.*, 2000).

UVC is lethal to most microorganisms, including bacteria, viruses, protozoa, mycelial fungi, yeasts and algae. Lethality is mostly related to the alteration of microbial DNA by dimer formation. The main types of photoproduct in UV-irradiated DNA are pyrimidine dimers, pyrimidine adducts and DNA-protein cross-links.

Once the DNA has been damaged, microorganisms can no longer reproduce. Temperature has little (if any) influence on the microbicidal action of UV radiation, but moisture exerts a strong effect. When bacteria are suspended in air, an increase in relative humidity results in a greatly reduced death rate, especially beyond 50 % RH. Similarly, bacteria in a liquid medium are more resistant (Bintsis *et al.*, 2000).

9.6.1 Disinfection of surfaces

Packaging materials can be sterilized by arranging appropriate UVC lamps over conveyors. It is key to the success of this process to have clean materials, as any dirt will absorb the radiation and protect any microorganisms present. As an example, during the manufacture of aseptically filled UHT dairy products, UVC sterilization has been applied to cartons and caps of high-density polyethylene bottles (Kuse, 1982; Nicholas, 1995).

UVC can also be employed to treat the actual surface of foods. Thus it has been used to control *B. stearothermophilus* in thin layers of sugar, or *Pseudomonas* spp. on meat surfaces. It is important to mention that sometimes UVC treatment leads to the development of off-flavors in meat and milk, due to the absorption of ozone and nitrogen oxides as well as from direct photochemical effects on the lipid fraction. Filtering UVC by covering the product with a thin layer of inert gas before irradiation can reduce this undesirable effect. Other applications include the inactivation of mold spores from the surface of baked goods (Sharma, 1999), reduction of total aerobic and mold numbers from the surface of eggs (Kuo *et al.*, 1997), and the extension of the shelf-life of fresh fish by reduction of initial contamination (Huang and Toledo, 1982).

Wright *et al.* (1999) considered using UVC for reducing *E. coli* O157:H7 in unpasteurized cider. Cider containing a mixture of acid-resistant strains of the pathogen (6.3 log CFU/ml) was treated using a thin-film UVC disinfection unit with output at 254 nm. Dosages ranged from 9402 to 61 005 μW.s/cm^2. Treatment significantly reduced *E. coli* O157:H7, with a mean reduction of 3.81 log CFU/ml. This reduction was affected by the level of background flora in cider, but ultimately the technique proved effective for reducing and possibly eliminating the pathogen.

9.6.2 Disinfection of air

UVC lamps are typically used in hospitals to create a curtain or barrier through which air must pass before reaching patients sensitive to infection (Bintsis *et al.*, 2000). For the handling of sensitive foodstuffs, a system combining laminar flow of air to remove particles of size > 0.1 μm and UVC radiation to kill any remaining live microorganisms has been suggested as a feasible means to supply clean sterilized air, particularly in food cold-storage areas (Decupper, 1992; Shah *et al.*, 1994). The microbiological quality of mechanically peeled fruits and vegetables is improved when UV-treated air is blown through the peeling unit, countercurrent to the flow of product (Dornow, 1992).

9.6.3 Disinfection of liquids

UVC is one of the simplest and most environmentally friendly ways of destroying a wide range of microorganisms in water (Gray, 1994). Is has been used to disinfect sewage effluent, drinking water and water for swimming pools, and the combination of UVC and ozone has a very powerful oxidizing effect (WHO, 1994). In most cases UVC can disinfect without any significant change in color, flavor, odor or pH, so that both microbiological safety and appropriate organoleptic quality of water are ensured. The standard is a 99.999 % reduction of microorganisms with a treatment time of less than 1 minute (Urakami *et al.*, 1997). The situation is different in the food industry, where a simple reduction in the water supply microbial load may be sufficient. In this respect, the brewing industry has become a major UVC user (Egberts, 1990). The treatment of opaque liquids is, however, a problem, considering the poor penetration of UVC. Yet there have been some indications that the US Food and Drug Administration may allow pathogen elimination from fruit juices using UVC, as long as the flow of the juice is turbulent rather than laminar and the temperature is continuously kept below 5 °C (Bintsis *et al.*, 2000).

The increased use of UVC radiation as a drinking-water treatment has instigated studies of the injury repair potential of microorganisms following treatment. Zimmer and Slawson (2002) challenged the repair potential of an optimally grown non-pathogenic laboratory strain of *E. coli* after UVC radiation from low- and medium-pressure lamps. Following irradiation, samples were incubated at 37 °C under photo-reactivating light or in the dark. Samples were analyzed for up to 4 hours following incubation, using a standard plate count. These researchers found that *E. coli* was capable of undergoing photo-repair following exposure to the low-pressure UVC source, but no repair was detected after exposure to the medium-pressure UVC source. Eventually, minimal injury repair was observed in the latter case at very low doses of 3 mJ/cm^2. This study clearly indicated differences in repair potential according to the UVC source used.

UVA is far less effective as a biocidal agent than UVC. For example, the incident energy required to bring about 50 % reduction in microbial viability is 5 J/m^2 using UVA, whereas UVC can achieve the same result with only 10^{-5} J/m^2 (Bintsis *et al.*, 2000). The mode of action of UVA on microbial cells is significantly different from that of UVC (Moss and Smith, 1981).

9.7 Light pulses

Pulsed light is a method of food preservation that involves the use of intense, short-duration pulses of broad-spectrum light, ranging from ultraviolet to near infrared. The material to be sterilized is exposed to at least one pulse of light having an energy density in the range of 0.01–50 J/cm^2 at the surface, using a wavelength distribution such that at least 70 % of the electromagnetic energy is distributed in the 170–2600 nm range.The process inactivates a wide range of microorganisms, including bacterial and fungal spores. Filtering of the spectrum eliminates wavelengths that may adversely affect food flavor and quality. Duration of the pulses used ranges from 1 μs to 0.1 s, at a rate of 1–20 flashes per second. For most applications, a few flashes applied in a fraction of a second provide high levels of microbial inactivation. Thus, the process is very fast and suitable for high throughput (Barbosa-Cánovas *et al.*, 1998).

This technology is mainly applicable in sterilizing or reducing the microbial population on the surfaces of packaging materials, packaging and processing equipment, foods and medical devices, as well as many other surfaces. As light pulses penetrate many transparent packaging materials, wrapped items can also be treated. Applications within the food industry include aseptic processing, liquid foods, solid foods (e.g. fish and meat products) and baked goods. With regard to aseptic processing, packaging materials are traditionally sterilized with hydrogen peroxide, but residues of this chemical may be undesirable. Light pulses have the potential to reduce or even eliminate chemical disinfectants and preservatives (Barbosa-Cánovas *et al.*, 1998).

On a smooth, non-porous surface, light pulses can lower the vegetative and spore populations of microorganisms by about 9 and 7 log cycles, respectively. On porous and complex surfaces such as meat, approximately 1–3 log cycle reductions are obtained (Barbosa-Cánovas *et al.*, 1998). In the case of meat products, thin slices will allow light penetration through the food material. Based on the same principle, white bread slices treated through the packaging material maintained a fresh appearance for more than 15 days, whereas untreated slices became moldy. Prepared and processed meat products, such as sausages and ground meat patties, can be treated to increase their shelf-life under refrigeration without the necessity for freezing. Similarly, vegetables such as tomatoes and potatoes, fruits such as apples and bananas, and prepared food products such as pastas and rice entrees can be treated to increase shelf-life, with minimal changes in nutritional quality (Rice, 1994).

A variety of microorganisms, including *E. coli*, *S. aureus*, *B. subtilis* and *S. cerevisiae*, have been inactivated by using between 1 and 35 pulses of light with intensities ranging from 1 to 12 J/cm^2. Greater inactivation can be obtained when full-spectrum light rather than glass-filtered light spectra are used. Thus, it appears that the UV component of light is essential to inactivate microorganisms using light pulses (Dunn *et al.*, 1991). The antimicrobial effect is primarily related to light absorption by highly conjugated carbon-to-carbon double-bond systems in proteins and nucleic acids, and evidently a similar mechanism to that of conventional continuous UV sources is involved. However, it has also been suggested that pulsed light causes an instantaneous lethal heating of the cell, leading to lysing or rupture of the cell wall (Dunn *et al.*, 1995; McDonald *et al.*, 2000; Wekhof, 2000).

A comparison of the disinfection rates due to pulsed light with those under conventional UV exposure suggests that doses for sterilization by the former are at least one order of magnitude lower than those of the latter. *B. subtilis*, for example, was sterilized (99.999 % destruction) by about 42 600 μW.s/cm^2 of UV, but the same level of inactivation required a dose of only 4500 μW.s/cm^2 under pulsed light. Clearly, pulsed light results in an apparent synergy of the pulsed energy quanta as compared to the relatively continuous stream of lower density conventional UV quanta (Wekhof, 1991; Dunn, 2000).

Pulsed light can be used to enhance product shelf-life and safety. Reductions of 2 log cycles were achieved with respect to *Salmonella* on chicken wings, *Listeria* on hot dogs, and both pathogens on primal and retail beef cuts (Dunn *et al.*, 1995). Curds of cottage cheese were inoculated with *Pseudomonas* and subjected to light with an energy density of 16 J/cm^2 and pulse duration of 0.5 ms. After only two flashes, viability was reduced by 1.5 log cycles. The temperature at the surface of the curd closest to the light source increased by 5 °C. Sensory evaluation using a trained panel showed no effect on cheese taste as a result of the light treatment (Dunn *et al.*, 1991). Pulsed light was also very effective in eliminating microbial contamination (mainly *S.* Enteritidis) from the surface of shell eggs. As much as an 8 log-cycle reduction was obtained. In fact, the inactivation effect was not limited to the surface but also extended to a certain degree into the egg shell pores (Dunn, 1996).

In vegetables and fruits, pulsed light was able to inactivate polyphenol oxidase, the enzyme that causes browning (Dunn *et al.*, 1991). Spores of *Bacillus cereus* and *Aspergillus niger* were inactivated by pulsed light from the surface of different packaging materials, whereas higher levels of inactivation of bacterial spores were achieved in water. In general, mold spores are more resistant to pulsed light than are bacterial spores (Dunn *et al.*, 1991; Dunn, 1996). DNA damage, such as the formation of single strand breaks and pyrimidine dimers, was induced in yeast cells by Takeshita *et al.* (2003) after irradiation by pulsed light. The effect was essentially the same as that observed with continuous UV light.

The generation of the pulsed light requires a considerable amount of energy. Thus, the power consumption of a typical pulsed light system is about 1000 W while similar results can be achieved with a conventional UV system drawing only 10 W of total power. Applications are consequently limited to situations where the benefits of achieving rapid sterilization outweigh the cost of pulse generation (Dunn *et al.*, 1997).

9.8 Ultrasound

Ultrasound refers to the result of an event generating pressure waves with a frequency of 20 kHz or more. In general, ultrasound equipment uses frequencies ranging from 20 kHz to 10 MHz. High-power ultrasound at lower frequencies (20–100 kHz) has the ability to create cavitation. This feature has value in inactivating microorganisms (Butz and Tauscher, 2002).

Investigation of ultrasound as a potential microbial inactivation method began in the 1960s (Earnshaw *et al.*, 1995). The mechanism of microbial killing is mainly thinning of cell membranes, localized heating, and production of free radicals (Fellows, 2000). During the sonication process, longitudinal waves are generated when a sonic wave meets a liquid medium, thereby creating regions of alternating compression and expansion (Sala *et al.*, 1995). These regions of pressure change cause cavitation, and gas bubbles are formed in the medium. These bubbles have a larger surface area during the expansion cycle, which increases the diffusion of gas, and the bubbles expand. When the ultrasonic energy is no longer sufficient to retain the vapor phase in the bubble, rapid condensation occurs. The condensed molecules collide violently, creating shock waves that form localized areas of very high temperature and pressure (up to 5500 °C and 50 000 kPa). The pressure changes yield the main bactericidal effect in ultrasound treatments (Piyasena *et al.*, 2003). The cavitation threshold of a medium (i.e. the minimum oscillation of pressure that is required to produce cavitation) depends on the dissolved gas, hydrostatic pressure, specific heat of the liquid, the nature of gas in the bubble, and the tensile strength of the liquid. Another essential variable is temperature, which is inversely proportional to the cavitation threshold. In addition, we now know that the ultrasonic frequency must be below 2.5 MHz, as cavitation will not occur above this value (Rahman, 1999).

Ultrasound is much more effective when combined with pressure (manosonication), heat (thermosonication), or both (manothermosonication). The enhanced mechanical disruption of cells is the explanation for enhanced killing (Sala *et al.*, 1995). It has also been suggested that microbial inactivation by ultrasound is more effective when combined with other decontamination techniques such as heating, chlorination or extreme pH (McClements, 1995).

Pagán *et al.* (1999) studied the application of ultrasound for the inactivation of *L. monocytogenes*. Ultrasonic treatment alone (20 kHz and 117 μm) at ambient temperature was not very effective, with a *D*-value of 4.3 minutes. However, the combination with pressures of 200 and 400 kPa lowered the *D*-values to 1.5 and 1.0 minutes, respectively. An amplitude increase of 100 μm decreased the resistance of the pathogen to manosonication by a factor of 6. Temperatures up to 50 °C did not have any significant effect on inactivation; but once they exceeded this threshold, an enhanced effect was noted. These authors also found that growth temperature affected heat resistance of *L. monocytogenes*. Cultures grown at 37 °C were found to be twice as heat-resistant as those grown at 4 °C; however, the cell growth temperature did not change the effect of manosonication itself. In addition, lower pH values resulted in greater inactivation rates, whereas greater sucrose concentrations increased *D*-values.

Salmonella Enteritidis, *S.* Typhimurium and *S.* Senftenberg were investigated by Manas *et al.* (2000) for their resistance to heat treatment, manosonication and manothermosonication in liquid whole eggs and citrate phosphate buffer solution. With manosonication (117 μm, 200 kPa, 40 °C), *S.* Enteritidis, *S.* Typhimurium and *S.* Senftenberg had *D*-values of 0.76, 0.84 and 1.4 minutes in whole egg, and 0.73, 0.78 and 0.84 minutes in citrate phosphate buffer, respectively. In comparison, *D*-values at 60 °C were 0.068, 0.12 and 1.0 minutes for the buffer, and 0.12, 0.20 and 5.5 miniutes for whole egg, respectively. A linear increase in ultrasonic wave amplitude resulted in an exponential increase in the inactivation rate of the manosonic treatment. When manothermosonication (117 μm, 200 kPa, 60 °C) was applied an additive effect resulted, with a reduction of 3 log cycles for the most resilient of the pathogens, *S.* Senftenberg.

The effects of ultrasound on *E. coli* in an aqueous medium, using a frequency of 24 kHz with varying intensities, were examined by Scherba *et al.* (1991). A significant reduction of the bacterial population was achieved, which increased with treatment time; however, intensity did not affect the killing rate. Utsonomiya and Kosaka (1979) noted that the initial temperature, medium, and pH influenced the survival of *E. coli* treated at 700 kHz. When the pathogen was suspended in saline at 32 °C, survival rates of 0.83 % and 0.2 % were obtained after 10 and 30 minutes of treatment, respectively; however, at an initial temperature of 17 °C, these values increased to 37.9 % and 8.1 %. No inactivation occurred in milk. When 10 % orange juice was added to milk, only 0.3 % survival was found at pH 2.6, whereas the survival was 100 % at pH 5.6.

The use of ultrasound to inactivate *E. coli* in biofilms could be beneficial to the food and water-bottling industries. For example, Johnson *et al.* (1998) reported that the combination of 70 kHz with gentamicin sulfate, an antibiotic, reduced *E. coli* numbers in a biofilm by up to 97 % in 2 hours. The main reason for the enhanced killing was increased diffusion of the antibiotic through the cell membrane, as the lipopolysaccharide layer of the outer cell membrane was believed to be destabilized by ultrasound. However, there are obvious concerns about utilizing a combination of ultrasound with an antibiotic in food processing. Rather, these types of combinations should be more suitable for the removal of biofilms from medical devices.

Raso *et al.* (1998a) carried out comparisons between manosonication and manothermosonication for the inactivation of spores of *B. subtilis*, and found that heat treatment provided by the latter made the inactivation process more effective.

Ultrasound, whether used alone, with thermal challenge or in combination with pressure treatment, could be an important element in food-processing technology because it is more effective and energy-efficient compared to conventional heat treatments.

9.9 High pressure

The technology of high-pressure processing (HPP) has been known for more than a century, but relatively recent scientific and technical progress has led to a renaissance regarding its utilization. A number of pressure-treated food products are already present in the Japanese, French, Spanish and American markets. HPP is a flexible technique that can subject liquid and solid foods, with or without packaging, to pressures between 100 MPa and 800 MPa, at temperatures that range from below 0 °C to beyond 100 °C, with exposure times that can range from a few seconds to more than 20 minutes. Food treated in this way keeps its original freshness, color, flavor and taste, as HPP acts instantaneously and uniformly throughout a mass of food, independently of size, shape and composition (Butz and Tauscher, 2002). Because pressure is uniform throughout the food, preservation is uniform, with no particle escaping treatment. In addition, and unlike thermal treatment, HPP is not time/mass dependent (Barbosa-Cánovas *et al.*, 1998).

Compression can increase the temperature of foods approximately by 3 °C per 100 MPa, and may also shift the pH of the food. Nevertheless, pressure pasteurization is also feasible at room temperature. Water activity and pH are critical process factors in the inactivation of microbes by HPP. Besides destruction of microorganisms, there are further influences of pressure on food materials, namely protein modification or denaturation, enzyme activation or inactivation, changes in enzyme-substrate interactions, and changes in the properties of polymerized carbohydrates and fats (Butz and Tauscher, 2002).

There are three main ways of generating high pressures (Deplace and Mertens, 1992):

- Direct compression, where the medium is pressurized with the small-diameter end of a piston and the large-diameter end is driven by a low-pressure pump. This method allows very fast compression, but the limitations of the high-pressure dynamic seal between the piston and the vessel internal surface restricts its use to small-diameter laboratory or pilot plant systems.
- Indirect compression, where a high-pressure intensifier pumps a pressure medium from a reservoir into a closed high-pressure vessel until the desired pressure is reached. This is the most widespread industrial procedure.

- Pressure-medium heating, where high pressure is generated by expansion of the pressure medium through high temperature. Thus, heating of the pressure medium is used when high pressure is applied in combination with heat treatment of food; it requires accurate temperature control within the entire internal volume of the pressure vessel.

Microbial inactivation is probably due to a number of factors, including protein conformation changes and membrane perturbation, leading to cell leakage. Moderately high pressures decrease the rate of growth and reproduction, whereas high and very high pressures cause microbial inactivation. The threshold pressures for retardation of reproduction and/or inactivation are dependent on the microorganism and species. In general, Gram-positive organisms are more resistant to pressure than Gram-negative ones (Barbosa-Cánovas *et al.*, 1998; Farkas, 2001). Those biochemical reactions in which reactants undergo either a decrease or increase in free volume are the ones most affected by HPP, because pressure causes either a decrease in the available molecular space or an increase in chain interactions. On the other hand, reactions involving formation of hydrogen bonds are favored by high pressure because bonding results in a decrease in volume. As a result of HPP, proteins are denatured (disruption of hydrophobic and ion-pair bonds), whereas nucleic acids are baroresistant. However, DNA transcription and replication are disrupted by high pressure due to the involvement of enzymes which can be inactivated. Different enzymes have considerable differences in barosensitivity. Membrane phospholipids suffer conformational changes, which disrupts membrane permeability (Hedén, 1964; Hoover *et al.*, 1989; Farr, 1990; Knorr, 1993).

The mode of action of pressure on bacterial spores is still not fully understood. Bacterial spores are killed directly by pressures higher than 1000 MPa, and they are sensitive to pressures of between 50 and 300 MPa. It is generally agreed that at such pressures spores germinate, followed by death of the germinated spore. However, it is not known whether pressure induces spore activation similar to reversible heat activation or triggers germination irreversibly. Ultimately, temperature and pressure ranges for germination and inactivation depend on the spore species (Smelt, 1998).

The pH of a food material plays a very important role in determining the extent to which HPP affects microorganisms. Yeasts and molds are quite resistant to low pH, and a pH < 4.0 does little to sensitize these organisms to pressure. By comparison, vegetative bacteria are quite sensitive to a combination of pressure and low pH. Low a_w protects cells against pressure, but microorganisms injured by pressure are generally more sensitive to low a_w. The net effect of a_w is not always easy to predict (Smelt, 1998; Tewari *et al.*, 1999). Microorganisms are particularly sensitive to nisin during or after pressure treatment. Masschalk *et al.* (2001) have proposed a mechanism of pressure-promoted uptake of nisin to explain the sensitization.

HPP can be combined with other processing techniques to enhance microbial inactivation. Raso *et al.* (1998b) studied the inactivation of *Y. enterocolitica* by combining ultrasonication, pressure and heat. The lethal effect of ultrasonication (20 kHz, 150 μm) increased with rising pressure until maximum inactivation occurred at an optimum pressure of 400 kPa. Pagán *et al.* (1998) studied the possibility of germinating *Bacillus* spores using HPP, then inactivating the germinated cells with a PEF treatment.

They found that germination of more than 5 log cycles of spores was initiated by pressurization, and that while the germinated cells did become sensitive to a subsequent heat treatment, they were not sensitized to PEF application below 40 °C.

Shigehisa *et al.* (1991) reported complete destruction of *S.* Typhimurium in beef at 300 MPa after 10 minutes at 25 °C. Carlez *et al.* (1994) reported that HPP caused reduction of vegetative mesophilic and psychrotrophic contaminants, and destruction of coliforms and *S. aureus*. Murano *et al.* (1999) considered combining HPP and temperature to inactivate *L. monocytogenes* in pork patties. Samples were inoculated with 10^9 CFU/ml of the pathogen, vacuum-packed, and treated with a number of combinations of high pressure and temperature. Lowest *D*-values, ranging from 0.37 to 0.63 minutes, depending on the *Listeria* strain, were obtained using 414 MPa for 6 minutes at 50 °C.

Erkmen and Karatas (1997) studied the effect of HPP on *S. aureus* in milk at pressures in the range of 50–350 MPa for up to 12 minutes at a constant temperature (20 ± 2 °C). No survival was found at 350 MPa for 6 minutes and 300 MPa for 8 minutes. Ponce *et al.* (1998) investigated the inactivation of *L. innocua* inoculated in liquid whole egg using HPP by subjecting the food to different combinations of pressure, temperature and time. Reductions of greater than 5 log cycles were obtained at 2 °C for 15 minutes using 450 MPa. Reduction values were greater than at room temperature – an effect that was explained as being due to the greater susceptibility of some proteins to denaturation at low temperatures. The effect of HPP on the survival of a pressure-resistant strain of *E. coli* O157:H7 in orange juice was investigated by Linton *et al.* (1999) over the pH range 3.4–5.0. The juice was inoculated with 10^8 CFU/ml of the pathogen and subjected to pressure treatments of 400, 500 and 550 MPa at 20 °C and 30 °C. A pressure treatment of 550 MPa for 5 minutes at 20 °C produced a 6 log cycle inactivation at pH 3.4, 3.6, 3.9 and 4.5, but not at pH 5.0. Combining pressure with mild heat (30 °C) did result in a 6 log cycle reduction at pH 5.0. Thus, it was concluded that microbiological safety of orange juice is achievable through appropriate combinations of HPP, temperature and time.

HPP is a very promising non-thermal food preservation method, not only for its inherent ability to cause proper microbial inactivation, but also due to the fact that the rheological and functional properties of foods remain unaltered (Barbosa-Cánovas *et al.*, 1998). Although much research has been done, there is still a great deal to be discovered regarding critical limits of the process and the extent to which this might ensure appropriate treatment of food materials. HPP commercialization will depend on its economic viability (Tewari *et al.*, 1999).

9.10 Food irradiation

Food irradiation is a non-thermal technology that involves exposing prepackaged or bulk foodstuffs to gamma rays, X-rays or electrons. The most common method is gamma radiation from a radioisotope source, typically cobalt 60. If electrons or X-rays are utilized, they are generated by an electron accelerator. The cost of gamma radiation is competitive with that of other methods of food preservation (Barbosa-Cánovas *et al.*, 1998).

The degree of chemical and physical change produced when food is exposed to high energy radiation is determined by the energy absorbed. In irradiation processing, it is described as the absorbed dose, measured in units of kilogray (kGy), where 1 gray (Gy) has an energy absorption equivalent of 1 J/kg. The following terms are used to describe the application of radiation in foods (Barbosa-Cánovas *et al.*, 1998):

- *Radicidation* is the application of ionizing radiation sufficient to reduce the number of specified viable non-spore-forming pathogenic bacteria to such a level that none is detectable in the treated food when examined by recognized bacteriological testing methods: it is a treatment with relatively low doses (0.1–8 kGy).
- *Radurization* is the application of ionizing radiation sufficient to cause a substantial reduction in the numbers of specific viable spoilage microorganisms: it involves doses of about 0.4–10 kGy to improve the shelf-life of a product.
- *Radappertization* is the application of ionizing radiation sufficient to reduce the number and/or activity of viable microorganisms (with the exception of viruses) to such a level that very few, if any, are detectable by recognized bacteriological or mycological testing methods applied to the treated food. No spoilage or toxicity of microbial origin must be detectable no matter how long or under what conditions the food is stored after treatment with doses of about 10–50 kGy to bring about virtually complete sterilization.

In 1983, on the basis of international agreement, the Joint Food and Agriculture Organization/World Health Organization Codex Alimentarius Commission accepted food irradiation as a safe and effective technology for the treatment of food and adopted a Codex General Standard for Irradiated Foods with an associated Code of Practice. Although the use of irradiation continues to grow worldwide, negative reactions in various countries have restricted its expansion. Introduction of commercial applications is rather slow because many governments require extensive data to support the wholesomeness of irradiated food, which leads to lengthy regulatory and approval-granting processes (Barbosa-Cánovas *et al.*, 1998). It is essential to emphasize that food irradiation is neither a miracle food preservation method nor a sinister technology. It has advantages and limitations like any other food preservation method (Castleman, 1993).

There are three main advantages associated with the use of irradiation in foods (Barbosa-Cánovas *et al.*, 1998): the ability to replace chemical treatments that are increasingly coming under suspicion or are even being banned; the capacity to slow the rate of food deterioration, thus extending shelf-life; and flexibility in application to packages or bulk units, in the frozen state or at room temperature.

Ionizing radiation is lethal for bacteria. The critical target for inactivation is the DNA, resulting in the loss of ability to reproduce, although alteration of membrane properties has also been indicated as a significant mechanism of inactivation. The proportion of a bacterial population that survives a given dose of irradiation depends on the intrinsic sensitivity of the microorganism, the stage of its growth cycle, the amount of irradiation damage inflicted, and the microorganism's potential for repair. Irradiation sensitivity differs with species and among strains, although the range of resistance among strains of a single species is usually small enough to be considered negligible (Ingram and Roberts, 1980; Moseley, 1990).

Bacterial spores are more resistant than their corresponding vegetative cells by a factor of 5–15 (Moseley, 1989). In general, the irradiation resistance of molds is equivalent to that of vegetative bacteria. Yeasts are more resistant than molds, and as resistant as bacterial spores. Viruses are even more irradiation-resistant than bacteria, so that irradiation treatments that destroy bacteria will not reliably inactivate viruses. Although there is no definitive explanation for microbial resistance to ionizing radiation, it is evident that the mechanism involves enzymatic activity able to repair radiation damage of nucleic acids (Ingram and Roberts, 1980; Jay, 1996).

Gram-negative bacteria involved in the spoilage of refrigerated fresh meats are more sensitive to irradiation than are lactic acid bacteria. For this reason, the use of low doses of irradiation that would inactivate those spoilage organisms, but not pediococci and/or lactobacilli, have potential application in fermented sausage production (Monk *et al.*, 1995). Of the Gram-negative pathogens, *Salmonella* is considered the most resistant. Thus, irradiation processes designed to eliminate this pathogen will also eliminate *Escherichia*, *Yersinia*, *Aeromonas* and *Campylobacter* species (Radomyski *et al.*, 1994). Irradiation is also very effective for eliminating *S. aureus* from meat products (Monk *et al.*, 1995).

Irradiation inactivation of *L. monocytogenes* is possible, even at relatively high concentrations (10^5 CFU/g), without inducing noticeable modifications to the taste, odor or textural properties (Ennahar *et al.*, 1994). As in heat processing, the 12D reduction concept is applied to the process of radappertization designed to kill *C. botulinum* spores (Monk *et al.*, 1995).

Aziz and Moussa (2002) studied the effect of gamma irradiation on the production of mycotoxins in fruits. Irradiation at doses between 1.5 and 3.5 kGy significantly decreased the total fungal counts compared to un-irradiated controls. After 28 days of storage at refrigeration temperatures, the un-irradiated fruits contained high concentrations of mycotoxins as compared with samples irradiated with 3.5 kGy. Mycotoxin production decreased with increasing irradiation dose, and mycotoxins were not detected after doses $\geq$ 5.0 kGy.

Radiation treatments at doses of 2–7 kGy can effectively eliminate potentially pathogenic non-spore-forming bacteria, including *Salmonella* spp., *E. coli* O157:H7, *S. aureus*, *Campylobacter* spp. and *L. monocytogenes*, without affecting sensory, nutritional and textural qualities. Candidates for radiation decontamination include poultry and red meats, egg products, fish and seafoods, and fresh fruits and vegetables. A unique feature is that irradiation can be applied to food in the frozen state. With today's demand for high-quality, convenient foods, irradiation in combination with other processes holds promise for enhancing the safety of many minimally processed foods. Radiation decontamination of dry ingredients, herbs and enzyme preparations with doses of 3–10 kGy has proven an alternative to fumigation with microbicidal gases. Radiation treatments at doses of 0.15–0.7 kGy appear to be feasible to control foodborne parasites (Farkas, 1998).

To promote worldwide introduction of food irradiation, it is necessary to develop national and international legislation and regulatory procedures to enhance confidence among trading nations that foods irradiated in one country and exported are irradiated under acceptable standards of wholesomeness, irradiation dose and hygienic practice (Barbosa-Cánovas *et al.*, 1998). Identifying irradiated food as such (i.e. properly informing consumers) is essential to reassure the public that consumer rights are protected. This is not trivial, because the effects produced by irradiation are often small and may be similar to changes produced by other means of food preservation (Brynjolfsson, 1989; Johnston and Stevenson, 1990). With regard to labeling, some countries require irradiated foods to be labeled with the characteristic green radura symbol and appropriate descriptive words; other nations demand just the radura symbol and no descriptive words, while others do not require any special identification. A universal labeling system is therefore imperative (Morehouse, 2002).

10 Plant hygiene and contamination during food processing

10.1 Introduction

Bacteria may be transferred to food by the production environment and personnel, either directly or by cross-contamination through surfaces, equipment, utensils and/or hands that have not been properly cleaned or disinfected. Cleaning and disinfection are two separate but closely related concepts. Cleaning is removing dirt and a portion of the microorganisms present, whereas disinfection is treating the surfaces in such a way that the remaining microorganisms are killed or reduced to an acceptable level. Cleaning comes always first, otherwise the subsequent disinfection will be less effective. Zoning – i.e dividing the production area into dry and wet and/or high-, medium- and low-care areas – is also useful in preventing product contamination. Nowadays zoning has evolved into a complex set of measures including equipment layout and design, air filtration, personnel hygiene, routes for personnel movement, and appropriate cleaning and disinfection procedures. Considering the cost involved, it is evident that zoning is only useful if applied logically (i.e. embedded in Good Manufacturing Practices, GMP).

Adoption and use of GMP and control through HACCP, coupled with equipment that is easier to keep clean as well as air-conditioned processing environments, have changed the profile of problematic organisms faced by the industry, but have not eliminated them. For example, outbreaks of foodborne illness caused by *Staphylococcus aureus* are significantly lower than 30 years ago; however, problems caused by psychrotrophic organisms are likely to increase as we continue to demand longer refrigerated shelf-life from perishable products. It is of interest to consider the fact that a subset of psychrotrophs that are alkali-tolerant could gain a selective advantage in food-processing environments where only alkaline cleaners and sanitizers are used. These psychrotrophic alkalitrophs include the pathogens *Listeria monocytogenes*, *Yersinia enterocolitica* and the potential pathogen *Aeromonas hydrophila*. Although not psychrotrophs, *Campylobacter* spp. are alkalitrophic and may be afforded the same selective advantage. Following rigorous cleaning, routine sequential exposure of food-contact surfaces to both alkali- and acid-based cleaners/sanitizers is imperative to control these organisms.

Introduction of potential pathogens via raw materials can lead to the establishment of foci or niches where these organisms may persist in the processing plant. Examples include hollow rollers on conveyors, cracked tubular equipment port rods, small spaces or gaps between close-fitting metal–metal or metal–plastic parts, valves and switches, or even saturated insulation. Failure to find and remove organisms from these niches may mean that these 'house flora' are periodically shed and organisms find their way into processed product. This type of event can lead to periodic foodborne illness outbreaks on an irregular basis over weeks or months that may involve large or small clusters of people scattered over the region served by the food plant. This pattern is typical of the type of problem caused by *L. monocytogenes* when it has established itself in a niche on equipment surfaces. Twelve such instances were reported in a variety of food plants (meat, dairy, fish) located in five European countries and the US between 1975 and 2000 (Tompkin, 2000).

Colonization of food-processing facilities and equipment by undesirable organisms that can withstand challenges generated by food processing is important. Listeriosis in Switzerland was traced to consumption of smear-ripened soft cheese. The same organism had survived on wooden shelves in 12 cheese-ripening cellars from which cheese was supplied. It was evident that the organism is able to adapt to different plant environments, survive for years, and contaminate products (Davidson and Harrison, 2002).

In a recent study, Autio *et al.* (2002) examined restriction endonuclease patterns or pulsotypes of *L. monocytogenes,* isolated from a variety of foods produced in 41 different plants in 10 different European countries, using pulsed field gel electrophoresis. Some of the pulsotypes were repeatedly recovered from the same product made by the same producer, which suggested persistence of the strain in the processing plant. In contrast, other pulsotypes were repeatedly found in products from different producers, suggesting that persistent 'house strains' may not always be producer-specific but have a wide geographic distribution. It has become generally accepted over the last 25 years that the preferential mode of existence for bacteria is not as free-living planktonic (free-floating) cells but as complex cellular communities at environmental interfaces. Such communities are referred to as biofilms (Characklis and Marshall, 1990).

10.2 Biofilms

10.2.1 Definition and implications for the food industry

Formally, biofilms are defined as 'collections of microorganisms and their associated extracellular products at an interface and generally attached to a biological or non-biological substratum' (Palmer and White, 1997). There are two main concerns regarding biofilms in the food industry: the presence of biofilms interfering with food-processing operations, and the potential for biofilms to serve as a reservoir for contamination of food with organisms of spoilage or safety concerns (Gill, 1998).

On most of the occasions where biofilms are a nuisance, the terms 'biofouling' and 'microbial-influenced corrosion' are applied.

10.2.1.1 Biofouling

Biofouling refers to the undesirable formation of a layer of living microorganisms and their decomposition products as deposits on the surfaces in contact with liquid media. Biofilm growth in pipes can reduce fluid flow rate and carrying capacity, which ultimately results in clogging. Biofilm formation on heat exchange surfaces can rapidly reduce heat transfer rates. The efficiency of ultrafiltration and reverse osmosis membranes may also be reduced by biofilm growth resulting in pore clogging (Kumar and Anand, 1998).

10.2.1.2 Microbial-influenced corrosion

Microbial-influenced corrosion is a process whereby the corrosion of metal surfaces is increased by the presence of biofilms. Corrosion results from a potential difference being created between sites on the surface that are covered by biofilm and those exposed to the surrounding liquid environment. This phenomenon can also result in the production of acid or of sulfides by sulfate-reducing bacteria (Little *et al.*, 1990).

10.2.1.3 Biofilm accumulation

Common places for biofilm accumulation are floors, wastewater pipes, vents in pipes, rubber and teflon seals, conveyor belts, stainless steel surfaces, etc. (Blackman and Franck, 1996). Herald and Zottola (1988) observed that *L. monocytogenes* was able to attach to stainless steel through attachment fibrils. The pathogen also attached to glass, polypropylene and rubber (Mafu *et al.*, 1990), and produced a sanitizer-resistant biofilm on glass, stainless steel and rubber surfaces (Ronner and Wong, 1993). Numbers of bacteria recovered from these surfaces were high, and dependent on the length of exposure time. It was also found that hydrophobic and electrostatic interactions were responsible for the attachment of *L. monocytogenes* to these surfaces (Mafu *et al.*, 1991). With regard to food surfaces, studies have shown the attachment of different microorganisms to poultry (Lilliard, 1988) and beef (Butler *et al.*, 1979) surfaces. These organisms, mainly coliforms, have not only been associated with slaughtering processes but are also responsible for cross-contamination of carcasses (Anand *et al.*, 1989).

10.2.2 Biofilm development

Biofilm formation and growth are complex, dynamic processes in which a number of steps can be identified; these are described below.

10.2.2.1 Conditioning of the surface

The formation of a biofilm occurs on virtually any submerged surface in any environment in which the bacteria are present. In food-processing environments, bacteria with organic and inorganic molecules (like proteins and minerals from milk and meat) get adsorbed to the surface, forming a conditioning film. These organic and inorganic substances, together with the microorganisms, are transported to the surface by diffusion or, eventually, by a turbulent flow of the liquid. The conditioning film leads to a higher concentration of nutrients compared to the fluid phase. Nutrient transfer rates are higher in a biofilm than in the aqueous phase.

Conditioning also alters the physicochemical properties of the surface, including surface free energy, hydrophobicity and electrostatic charges (Dickson and Koohmaraie, 1989; Jeong and Frank, 1994; Hood and Zottola, 1997).

Nevertheless, there appears to be no evidence that microorganisms always attach to a conditioned surface. In fact, the microtopography of the food-contact surface is equally important to favor bacterial retention, particularly if the surface consists of deep channels and crevices to trap bacteria (Kumar and Anand, 1998). It is also established that adsorption of certain proteins to surfaces plays an important role in the microbial adhesion *per se*. For example, albumin has been found to be inhibitory for the adhesion of *L. monocytogenes* to silica surfaces (Al-Makhlafi *et al.*, 1995). Milk proteins such as casein and β-lactoglobulin are also able to inhibit the attachment of *L. monocytogenes* and *S.* Typhimurium (Helke *et al.*, 1993). On the other hand, whey proteins seem to favor the attachment of milk-associated microorganisms to stainless steel, rubber and glass surfaces (Speers and Gilmour, 1985).

10.2.2.2 Adhesion of cells

The attachment of microorganisms to the conditioned surface may be active or passive, depending on bacterial motility or transportation of the planktonic cells by gravity, diffusion or fluid dynamic forces from the surrounding fluid phase. Adhesion is also dependent on nutrient availability in the surrounding medium, and the growth stage of the bacterial cells themselves (Kumar and Anand, 1998).

Two stages can be identified in this process: a reversible adhesion followed by an irreversible one. Initial weak, long-range interactions developed between bacterial cells and substratum are called reversible adhesion, and involve van der Waals forces, electrostatic forces and hydrophobic interactions. During this stage, bacteria still show Brownian motion and can easily be removed by fluid shear forces (e.g. merely by rinsing). In irreversible adhesion, short-range forces are involved: dipole–dipole interactions, hydrogen, ionic and covalent bonds, and hydrophobic interactions. The contact between bacteria and surface takes place mainly through bacterial appendages such as flagella, fimbriae, pili and fibrils, and removal of cells requires much stronger forces such as scrubbing or scraping (Marshall *et al.*, 1971; Jones and Isaacson, 1983; Hancock, 1991; Kumar and Anand, 1998). Spores exhibit a greater rate of adhesion than vegetative cells due to their relatively high hydrophobicity and hair-like structures on the spore surface. After surface adhesion has taken place, spores may germinate and the resulting vegetative cells multiply and produce exopolysaccharides (EPS) (Rönner *et al.*, 1990; Husmark and Rönner, 1992).

The temperature and pH of the contact surface have an influence on the degree of adhesion of microorganisms. *Yersinia enterocolitica* adhered better to stainless steel surfaces at 21 °C than at 35 °C or 10 °C (Herald and Zottola, 1988), and *Pseudomonas fragi* showed maximum adhesion to stainless steel surfaces at the pH range 7–8, optimal for its cell metabolism (Stanley, 1983).

10.2.2.3 Formation of microcolonies

The irreversibly attached bacterial cells proliferate by using the nutrients present in the conditioning film and the surrounding fluid environment. This leads to the formation of microcolonies that enlarge and coalesce to form a layer of cells covering the surface. The attached cells produce additional polymers (exopolysaccharides, EPS), which help in the anchorage of the cells to the surface and in stabilizing the colony from environmental fluctuations (Characklis and Marshall, 1990).

10.2.2.4 Formation of biofilm

Biofilm formation is a direct consequence of the continuous attachment of bacterial cells to the substratum and subsequent growth, along with associated EPS production. Biofilm composition is heterogeneous due to colonization by different microorganisms with different nutritional requirements. Further increase in the size of a biofilm takes place by the deposition or attachment of other organic and inorganic solutes and particulate matter from the surrounding liquid phase (Melo *et al.*, 1992; Costerton *et al.*, 1994).

10.2.2.5 Detachment and dispersal of biofilm

As the biofilm ages, the attached bacteria, in order to survive and colonize new niches, must be able to detach and disperse from the biofilm. The daughter cells can become individually detached or be sloughed off. Sloughing is a discrete process whereby periodic detachment of relatively large particles of biomass occurs. This can be due to various factors such as fluid dynamics, shear effects of the bulk fluid, presence of certain chemicals in the fluid environment, or altered surface properties of the bacteria or substratum. The released bacteria may be then transported to new locations and restart the biofilm process (Rittman, 1989; Applegate and Bryers, 1991; Marshall, 1992).

10.2.3 Biofilm structure and properties

The present model of biofilm structure is that of 'matrix-enclosed microcolonies interspersed with less dense regions that include highly permeable water channels'. These water channels penetrate throughout the biofilm and provide direct access to oxygen and nutrients from the bulk fluid, as well as allowing the removal of metabolic waste (Costerton *et al.*, 1994).

Studies found that the density of the bottom layers was 5–10 times that of the surface. This variation in density was paralleled by a decrease in porosity (84–93 % at the top vs 58–67 % at the base) and mean pore size (1.7–2.7 μm at the top layers vs 0.3–0.4 μm at the bottom layers). Living cells constituted a much greater proportion of the total biomass in the top layers (91 %) than in lower layers (31–39 %). These results were obtained from a mixed biofilm community under aerobic conditions provided with a non-restrictive nutrient source. Results of experiments conducted under more restrictive conditions suggest that more complex structures can be expected in response to restrictive growth conditions (Wolfaardt *et al.*, 1994; Zhang and Bishop, 1994).

It has been recognized for some time that microbial communities are capable of unique metabolic activity that isolated cells cannot conduct (Gill, 1998). Examples of this phenomenon can be seen in the metabolism of xenobiotic compounds (Lappin *et al.*, 1985) and the degradation of straw cellulose (Kudo *et al.*, 1987), but from a more general perspective, in the very synthesis of EPS and their biofilm functionality. EPS are a diverse group of polysaccharides that are secreted by many bacteria into the surrounding environment. Many organisms will convert a very significant proportion of available carbon into EPS, with rates of conversion of 50 % or higher (Sutherland, 1983). EPS are of great importance in the formation and structure of biofilms, composing a high proportion of the total biomass and serving a fundamental role in substratum attachment and in constituting a matrix for further cell attachment (Gill, 1998).

EPS production is stimulated by a variety of environmental conditions, including specific ion concentrations, aeration and low nutrient availability. Terms such as glycocalyx, slime, capsule and sheath have been used to refer to the EPS associated with biofilms. EPS production not only plays a role in initial adhesion and anchorage, but also in protecting bacteria from dehydration, as EPS can retain water several times its own mass and only slowly get desiccated. In addition, EPS help in the process of trapping and retaining nutrients, and protect cells against antimicrobial agents (Characklis and Cooksey, 1983; Roberson and Firestone, 1992; Ophir and Gutnick, 1994; Rinker and Kelly, 1996).

Biofilms are complex communities of organisms in which a variety of interspecies interactions may take place. It has been suggested that these interactions could influence the final development of the mature biofilm by inhibiting the attachments of some species while recruiting others (James *et al.*, 1995). Biofilms formed by mixed species are often thicker and more stable than monospecies biofilms. The biofilm formed between *L. monocytogenes* and *P. fragi* (the latter as the primary colonizing organism) was much more extendsive than the respective individual biofilms (Sasahara and Zottola, 1993).

It is well established that bacterial biofilms exhibit greater resistance to antimicrobial agents than the individual cells in suspension. It is difficult to establish that any single mechanism causes the resistance; rather, the combined mechanisms create resistant populations (Mustafa and Liewen, 1989; Frank and Koffi, 1990; Krysinsky *et al.*, 1992). Biofilm formation protects the innermost cells. Antimicrobial resistance is linked to the three-dimensional structure of the film, because the resistance is lost as the structure is disrupted (Hoyle *et al.*, 1992; Boyd and Chakrabarty, 1995).

Antimicrobial agents are generally far more effective against actively growing cells, which means that good disinfectants for planktonic cells are not necessarily suitable for biofilm cells. Typically, cells within the film receive less oxygen and fewer nutrients than those at the biofilm surface. In this condition, microorganisms may present altered growth and physiological states, resulting in increased resistance to antimicrobial agents (Gilbert *et al.*, 1990; Holah *et al.*, 1990; McFeters *et al.*, 1995). In mixed biofilms competition for nutrients results in nutrient deficiency, which also plays a major role in increased resistance of biofilms to disinfectants (Berg *et al.*, 1982; Jones and Pickup, 1989). Some studies with foodborne bacteria have indicated that resistance is more substantial in older biofilms (more than 24 hours) than in younger ones (Anwar *et al.*, 1990; Frank and Koffi, 1990; Wirtanen and Mattila-Sandholm, 1992).

Bacterial biofilms can develop increased resistance towards different antibiotics by production of antibiotic-degrading enzymes such as β-lactamases. In biofilms, many similar hydrolytic enzymes are produced and they become trapped and concentrated within the biofilm matrix, consequently exhibiting enhanced protective properties (Nickel *et al.*, 1985; Widmer *et al.*, 1990; Anwar *et al.*, 1992; Vergeres and Blaser, 1992).

10.2.4 Control and removal of biofilms

An effective cleaning and sanitation program will prevent biofilm formation, but it is a difficult task, and a thorough cleaning procedure should be developed and included in the food operation from the very beginning (Kumar and Anand, 1998). Food-processing equipment design is essential to achieve better cleanability of the food-contact surfaces once bacterial adhesion has occurred. With new surfaces, there is not much difference in cleanability among glass, nylon, polyvinyl or stainless steel, but the latter exhibits better hygienic properties with time by resisting damage caused by the cleaning process itself. It is important to realize that the application of sanitizers can and does cause surface corrosion (Dunsmore *et al.*, 1981; LeClercq-Perlat and Lalande, 1994). Proper choice of equipment, material and accessories, and correct construction, process layout and process automation are essential (Mattila-Sandholm and Wirtanen, 1992).

10.2.4.1 Chemical methods

It is asserted that before application of a disinfectant, it is essential to eliminate as much 'soil' and as many microorganisms as possible. Indeed, microorganisms become far more sensitive to disinfectants once they have been detached from the surface to which they were adhering. The mechanical or chemical breakage of the EPS matrix is therefore vitally important (Kumar and Anand, 1998).

Chelators such as EDTA have proven quite effective in destabilizing the outer membranes of bacterial cells by binding calcium and magnesium ions (Camper *et al.*, 1985). Some detergents are bactericidal, and some disinfectants may even depolymerize EPS, thus enabling the detachment of biofilms from surfaces. Examples are peracetic acid, chlorine, iodine, and hydrogen peroxide (Kumar and Anand, 1998). For example, the detergent monolaurin (50 μg/ml) combined with heat treatment at 65 °C for 5 minutes completely destroyed the biofilm formed by *L. monocytogenes*. A synergistic interaction between monolaurin and organic acids like acetic acid also caused a pronounced reduction of this pathogen (Oh and Marshall, 1995, 1996). Cetylpyridinium chloride (CPC) was reported to be valuable in the poultry-processing industry to reduce attachment of *Salmonella* on poultry skin (Breen *et al.*, 1997).

The impregnation of materials with biocides has been shown to play a major role in resisting bacterial colonization for as long as the antimicrobial agent is released from the surface. For example, antifoulant paints containing silver have been effective in controlling mixed biofilms in which *Legionella pneumophila* is present (Rogers *et al.*, 1995). Food-packaging materials containing antimicrobial compounds have gained practical importance in recent years for the control of spoilage and pathogenic microorganisms on food surfaces. These compounds are able to migrate to the food surface and eliminate microbial contamination.

10.2.4.2 Physical methods

Control of biofilms includes physical methods such as super-high magnetic fields, ultrasound treatments, high-pulse electrical fields (on their own and in combination with organic acids), and low-energy electrical fields (on their own and as enhancers of biocides) (Kumar and Anand, 1998). The biocidal effect of these procedures has been attributed to iontophoresis – that is, the generation of ions from chlorine-containing components such as NaCl, $CaCl_2$ and NH_4Cl, which are able to kill both Gram-positive and Gram-negative cells (Davis *et al.*, 1994). Another possible bioelectrical effect is that the electric current applied drives the charged molecules and antibiotics into the cells through the biofilm matrix, thus increasing mass transfer rates (Rajnicek *et al.*, 1994).

10.2.4.3 Biological methods

Newer strategies for the control of biofilms include the adsorption of bioactive compounds like bacteriocins onto food-contact surfaces for the inhibition or adhesion of bacteria. Bacteriocins are proteinaceous bactericidal agents. Nisin, a well-known and frequently applied antimicrobial peptide, has proven to be an effective inhibitor of many food pathogens and spoilage bacteria, including spore-formers (Kumar and Anand, 1998). There are reports showing that food-contact surfaces where nisin was adsorbed had lowered incidence of surface contamination by *L. monocytogenes* (Bower *et al.*, 1995). Similarly, the application of lactic acid cultures and their cell-free extracts has also been reported to selectively inhibit different spoilage and pathogenic microflora on the surface of poultry (Anand *et al.*, 1995).

Enzymes are very effective for degrading EPS and thus removing biofilms. The effectiveness of specific enzymes varies according to the type of microflora making up the biofilm (Kumar, 1997). Endoglycosidases able to degrade polymers, such as glycoproteins, were developed and used effectively in buffer and detergent solution for removing *S. aureus* and *E. coli* from glass and cloth surfaces (Lad, 1992). The most effective way to deal with biofilms is cleaning and sanitation to prevent the formation of conditioning films and to remove attached cells before colony growth can occur (Gill, 1998). Potthof *et al.* (1997) described the development of an in-place cleaning system for the dairy industry using a combination of enzymes and surfactants.

10.2.5 Beneficial aspects of biofilms

In many natural environments, maintenance of water quality is brought about by the microbial metabolism in biofilms. Bacteria present in these biofilms biodegrade many of the toxic compounds and minimize the buildup of pollutants, thus acting as pollutant moderators (Fuchs *et al.*, 1996). These systems that use mixed microbial consortia have found application in fluidized beds and trickling filters for sewage and wastewater management, as well as in water purification plants and waste gas treatment. The organic nutrient-trapping capability of biofilms helps in reducing the organic content of wastewaters (Kanekar and Sarniak, 1995; Pedersen *et al.*, 1997; Raunkjaer *et al.*, 1997). Biofilms have also received considerable attention from the viewpoint of bioremediation of industrial effluents (Nigam *et al.*, 1996) and in the nitrification process for treatment of high-strength, nitrogen-fertilizer wastewater (Beg *et al.*, 1995; Cecen and Orak, 1996).

It is evident that biofilms represent a natural form of cell immobilization. Immobilized microorganisms have been successfully employed in bioreactors to improve the productivity and stability of fermentation processes, including acetic acid and ethanol manufacture (Macaskie *et al.*, 1995; Pakula and Freeman, 1996). Even the lactic acid bacteria and bifidobacteria that colonize the gastrointestinal tract, protecting it from pathogenic bacteria, are assumed to exert their action through formation of biofilms (Vanbelle *et al.*, 1989).

11 Predictive microbiology

11.1 Introduction

In traditional practice, foods were preserved for storage by reduction of the water content, addition of solutes and/or reduction of the pH sufficient to ensure the bacteriological stability of the product at all temperatures. The spreading availability of refrigerated storage led to a reduced usage of preservatives, and this has been reinforced by a growing distrust of preservatives among consumers. Consumer demands dictate not only the reduced use of preservatives, but also the replacement of frozen by chilled products and the minimization of any thermal processing (Leistner, 1992).

There is now realization that several preservative actions, none of which alone is sufficient to stabilize a product microbiologically, may in combination be adequately inhibitory because of synergistic effects (Grant and Patterson, 1995). However, alterations in an established formula and/or process for a preserved food, which reduce the microbiological stability of the product, have inevitably raised questions as to the possible growth of pathogenic bacteria in the product (Russell and Gould, 1991). The usual approach to establishing the microbiological safety of minimally preserved and processed foods is challenge testing, in which the food is inoculated with pathogenic bacteria that might contaminate the commercial product, followed by monitoring of the behavior of the inoculated organisms in product stored under abusive as well as recommended conditions (Notermans and in't Veld, 1994).

Modeling the behavior of microorganisms began in about 1920, with the development of methods for calculating thermal death times. This revolutionized the canning industry. A resurgence in predictive modeling began in the 1980s, driven by a proliferation of refrigerated and limited-shelf-life foods, the development of hurdle technology, and the advent of personal computers. Modeling techniques have become standard tools for designing experimentation and describing results (Whiting and Buchanan, 2002).

11.2 Modeling

Modeling in food microbiology assumes that the growth or inactivation of microorganisms in a food or model system is predictable within the limits of normal biological variability, and that it can be described by mathematical equations. It commonly assumes that the measured behavior of an appropriate model system predicts the microorganism's behavior in foods with corresponding levels of environmental factors. Models should be validated with a number of selected tests (Whiting and Buchanan, 2002).

Modeling in food microbiology has taken an empirical approach. Empirical though these models may be, they are based on established linear and non-linear regression techniques. Growth data and/or model parameters are fitted to equations by using interactive least-squares computer algorithms. Assumptions regarding randomness, normal distributions, interpolations with tested ranges rather than extrapolations outside the ranges, parsimony and stochastic specifications must be made in microbial modeling, as they are for any other statistical application of regression (Whiting and Buchanan, 2002).

The responses of bacteria to environmental conditions of particular interest (with respect to the behavior of pathogens in perishable foods) concerns, for growth-permitting conditions, the time for resolution of the lag phase and the rate of increase of the numbers of viable cells (Zwietering *et al.*, 1990). For conditions that are non-permissive of growth, or are intended to be rapidly lethal, the behavior of interest is usually the rate of decrease of the number of viable cells (Cerf *et al.*, 1996). For some organisms, the duration of the lag phase and rate of increase may, for considerations of product safety, properly refer to the concentration of a toxin rather than to cell numbers, as there may be no simple relationship between the two under all circumstances (Ikawa and Genigeorgis, 1987).

Data on which to base lag duration and growth rate models have generally been collected by monitoring the growth of bacteria during their batch cultivation in liquid media. A common procedure has been to grow the test organism to the stationary phase in a non-defined medium such as tryptone soy broth or brain heart infusion. The stationary phase culture is suitably diluted with fresh growth medium, and the diluted culture is used to inoculate flasks containing the growth medium modified as required by the addition of acidulants, humectants and/or other preservatives at appropriate concentrations. Each culture is then held at one of several selected temperatures under aerobic or anaerobic conditions, or under an atmosphere containing a specified fraction of CO_2 (McClure *et al.*, 1994). When modeling of the combined effects of multiple environmental variables is intended, fractional or complete factorial designs may be employed in deciding the compositions, as well as different incubation temperatures and atmospheric conditions for the individual media, to define the behavior of the microorganism within the range chosen for each environmental variable (Whiting, 1997). Growth has usually been monitored by spread or pour plating, spiral plating, or measurement of the turbidity of each culture. However, other procedures, such as determination of the time required for the culture to attain a specified turbidity or conductivity, have been employed. Data obtained by different procedures can give somewhat different estimates of lag times or growth rates for the same culture (Dalgaard *et al.*, 1994).

It has been the established practice to determine the growth rate by fitting a line to a plot of the log transformed data against time, with the assumption that the growth is exponential. The lag time is then usually taken as being the time (t) at which the backward-extrapolated growth curve is at the log data value for the culture at $t = 0$. However, alternative procedures for estimating the lag time from a log plot have also been employed (Buchanan and Cygnarowicz, 1990). It has become common practice to use equations, such as the Gompertz equation, to fit sigmoidal curves to log-transformed growth data. The lag time is therefore taken as the time at which the extrapolated tangent is the log data value for the culture at t = 0, or at some other point otherwise estimable from curve parameters. (Gibson *et al.*, 1988).

Models can be conceived as having three levels (Whiting and Buchanan, 1997):

- The primary level, which describes changes in microbial numbers with time. A typical example is a growth model that estimates the change in log CFU/ml with time. Another example is the model that describes the decreasing counts with time during thermal processing. Microbial numbers may be estimated by counting, turbidity, conductance, or biochemical assays. Formation of a microbial toxin or other metabolic product with time constitutes another type of primary-level model.
- The secondary level, which describes how the parameters of the primary model change with changes in environmental or other factors. These equations may be based or Arrhenius or square-root relationships, particularly if temperature is the primary factor of concern, as is often the case when specific groups of foods are being modeled. When more factors are included in the model (e.g., a_w, pH, concentrations of antimicrobials), polynomial regression equations are needed, although they tend to be more difficult to interpret.
- The tertiary level, which makes predictions or estimations. Environmental values of interest are entered into the secondary-level model to obtain specific parameter values for the primary model, which is then solved for increasing periods of time to obtain the growth or inactivation curve expected from that specific combination of environmental values. Tertiary systems are essentially represented by applications software, varying in complexity from an equation on a spreadsheet to expert systems or risk-assessment simulations.

Because of their empirical nature, models predict bacterial behavior only uncertainly when the conditions are marginal for growth. This is particularly unfortunate for product formulation purposes, as here the requirement is often that a product be formulated to assure that there will be no growth of organisms, with adverse effects on human health, that may contaminate the product. To deal with such requirements, it has been proposed that probabilistic models, rather than kinetic ones, are needed. A probabilistic model would give, for specific sets of conditions, the probability of growth occurring, but not the rate of growth. Such models should allow narrow definition of the boundaries of a range of safe formulation/storage conditions, and so permit confident minimizing of preservatives and preserving treatments (Ratkowsky and Ross, 1995). However, published models of this type seem confined to descriptions of growth and toxin production by *Clostridium botulinum* (Roberts and Jarvis, 1983; Lund *et al.*, 1990).

The predictive power of any model will always be constrained by the complexities of food–microbe interactions. Often, models for an organism are based on the growth of only one strain, or at most a few strains, in a homogeneous broth. Various strains of the same organism may, however, behave very differently (for example, by exhibiting different rates of growth under identical conditions of cultivation) (Barbosa *et al.*, 1994). In addition, many food systems do not present a homogeneous environment. For example, on the larger scale, the outside of a product such as a hamburger patty will, if it is not vacuum-packed, present an aerobic environment for bacterial growth, while the environment within is anaerobic (Gill and Jones, 1994). On the smaller scale, and particularly if the patty is formed of coarsely ground meat of high fat content, contact between particles of fat will provide regions of high pH within an environment where the bulk pH is close to pH 5.5, which is normal for post-rigor muscle (Grau and Vanderlinde, 1993). Similarly, with any emulsified product, the environment for bacterial growth at an oil/water interface is likely to differ considerably from that of the bulk medium, not only with respect to pH but also with respect to the availability of nutrients or even oxygen (Robins *et al.*, 1994).

Even if the food does present an essentially homogeneous environment, it may contain micronutrients that are not present in the general purpose broths commonly used for generating data for models (Gill *et al.*, 1997). Then, some organisms may grow more rapidly and/or grow at lower temperatures on the food than in the broth. Also, the effects of factors such as pH and a_w may differ between a food and a broth because the chemical nature of acidulants and humectants as well as the purely physical factors can affect the growth of bacteria (Adams *et al.*, 1991). An additional difficulty arises in modeling the duration of the lag phase, in that the lag duration may depend upon the conditions that induced the lag. Despite that, most models of the lag phase have been based on the cultivation of organisms in which growth has ceased apparently as a result of carbon limitation. It is then not surprising that lag phase models generally appear rather unreliable for predicting the initiation of bacterial growth in foods (McKellar *et al.*, 1997).

Verification of model reliability has usually taken the form of comparing literature data for the growth of an organism in foods with the growth predicted by the model. A major difficulty with such an exercise is that parameters, such as pH and a_w, have to be assumed for a food (Sutherland and Bayliss, 1994). Because a substantial quantity of the data available from food often does not fit well with model predictions, it has been frequently suggested that a model be regarded as satisfactory provided that it underestimates observed growth relatively rarely, although it may often grossly overestimate growth. Overestimation of growth is then referred to as a 'safe failure' of the model (Jones *et al.*, 1994).

Bibliography

Aalto, T. R., M. C. Firman and N. E. Rigler (1953). *p*-Hydroxybenzoic acid esters as preservatives. I. Uses, antibacterial and antifungal studies, properties and determination. *J. Am. Pharm. Assoc.* **42**, 449–458.

Adams, M. R. (2001). Why fermented foods can be safe. In: M. R. Adams and M. J. R. Nout (eds), *Fermentation and Food Safety*, pp. 39–52. Aspen Publishers, Inc., Gaithersburg, MD.

Adams, M. R, and M. O. Moss (1995). *Food Microbiology*. The Royal Society of Chemistry, Cambridge.

Adams, M. R, and M. O. Moss (2000). *Food Microbiology*, 2nd edn. The Royal Society of Chemistry, Cambridge.

Adams, M. R., C. L. Little and M. C. Easter (1991). Modeling the effect of pH, acidulant acid and temperature on the growth rate of *Yersinia enterocolitica. J. Appl. Bacteriol.* **71**, 65–71.

Ahokas, J., H. El-Nezami, P. Kankaanpää *et al.* (1998). A pilot clinical study examining the ability of a mixture of *Lactobacillus* and *Propionibacterium* to remove aflatoxin from the gastrointestinal tract of healthy Egyptian volunteers. *Revue Méd. Vét.* **149**, 568–572.

Al-Makhlafi, H., A. Nasir, J. McGuire and M.A. Daeschel (1995). Adhesion of *Listeria monocytogenes* to silica surfaces after sequential and competitive adsorption of bovine serum albumin and β-lactoglobulin. *Appl. Environ. Microbiol.* **61**, 2013–2015.

Anand, S. K ., C. M. Mahapatra, N. K. Pandey and S. S. Verma (1989). Microbial changes on chicken carcasses during processing. *Indian J. Poultry Sci.* **24**, 203–209.

Anand, S. K., N. K. Pandey, S. S. Verma and R. Gopal (1995). Influence of some lactic cultures on microbial proliferation and refrigerated shelf stability of dressed chicken. *Indian J. Poultry Sci.* **30**, 126–133.

Andres, C. (1982). Mold/yeast inhibitor gains FDA approval for cheese. *Food Proc.* **43**, 83.

Anonymous (1976). *Plant Sanitation for the Meat Packaging Industry*. Office of Continuing Education, University of Guelph and Meat Packers Council of Canada.

Anonymous (1996). *Sanitizers for Meat Plants*. Diversey Corp., Wyandotte, MI.

Anonymous (1997). Guide to sanitizers. *Prepared Foods* **81**, 13–34.

Anwar, H., M. Dasgupta and J. W. Costerton (1991). Testing the susceptibility of bacteria in biofilms to antibacterial agents. *Antimicrob. Agents Chemother*. **34**, 2043–2046.

Anwar, H., J. L. Strap and J. W. Costerton (1992). Eradicating biofilm cells of *Staphylococcus aureus* with tobramycin and cephalexin. *Can. J. Microbiol.* **38**, 618–625.

Appendini, P. and J. H. Hotchkiss (2002). Review of antimicrobial food packaging. *Innov. Food Sci. Emerg. Technol.* **3**, 113–126.

Applegate, D. H. and J. D. Bryers (1991). Effects of carbon and oxygen limitation and calcium concentrations on biofilm recovery processes. *Biotechnol. Bioeng.* **37**, 17–25.

Arnold, R. R., M. Brewer and J. J. Gauthier (1980). Bactericidal activity of human lactoferrin: sensitivity of a variety of microorganisms. *Infect. Immun.* **28**, 983–899.

Aryanta, R. W., G. H. Fleet and K. A. Buckle (1991). The occurrence and the growth of microorganisms during the fermentation of fish sausage. *Intl J. Food Microbiol.* **13**, 143–156.

Auffray, Y., E. Lecesne, A. Hartky and P. Boulibonnes (1995). Basic features of the *Streptococcus thermophilus* heat shock response. *Curr. Microbiol.* **30**, 87–91.

Autio, T., J. Lunden, M. Fredriksson-Ahomaa *et al.* (2002). Similar *Listeria monocytogenes* pulsotypes detected in several foods originating from different sources. *Intl J. Food Microbiol.* **77**, 83–90.

Axelsson, L. (1998). Lactic acid bacteria: classification and physiology. In: S. Salminen and A. von Wright (eds), *Lactic Acid Bacteria: Microbiology and Functional Aspects*, pp. 1–72. Marcel Dekker, New York, NY.

Aytac, S. A. and Z. Y. Ozbas (1994). Survey on the growth and survival of *Yersinia enterocolitica* and *Aeromonas hydrophila* in yogurt. *Milchwissenschaft* **49**, 322–325.

Aziz, N. H., S. R. Mahrous and B. M. Youssef (2002). Effect of gamma-ray and microwave treatment on the shelf-life of beef products stored at 5 °C. *Food Control* **13**, 437–444.

Bacus, J. N (1986). Fermented meat and poultry products. In: A. M. Pearson and T. R. Dutson (eds), *Advances in Meat Research*, Vol. 2, *Meat and Poultry Microbiology*, pp. 123–164. AVI Publishing, Westport, CT.

Banks, J. G. and R. G. Board (1982). Sulfite-inhibition of *Enterobacteriaceae* including *Salmonella* in British fresh sausage and in culture systems. *J. Food Prot.* **45**, 1292–1297.

Banwart, G. J. (1989). *Basic Food Microbiology*. AVI-Van Nostrand Reinhold, New York, NY.

Baranowski, J. D., P. M. Davidson, C. W. Nagel and A. L. Branen (1980). Inhibition of *Saccharomyces cerevisiae* by naturally occurring hydroxycinnamates. *J. Food Sci.* **45**, 592–594.

Barbosa, W. B., L. Cabedo, H. J. Wederquist *et al.* (1994). Growth variation among species and strains of *Listeria* in culture broth. *J. Food Prot.* **57**, 765–775.

Barbosa-Cánovas, G. V., U. R Pothakamury, E. Palou and B. G. Swanson (1998). *Nonthermal Preservation of Foods*. Marcel Dekker, New York, NY.

Barbosa-Cánovas, G. V., M. M. Góngora-Nieto, U. R. Pothakamury and B. G. Swanson (1999). *Preservation of Foods with Pulsed Electric Fields*. Academic Press, San Diego, CA.

Batt, C., M. Solberg and M. Ceponis (1980. Inhibition of aflatoxin production by carrot root extract. *J. Food Sci.* **45**, 1210–1213.

Beg, S. A., M. M. Hassan and M. A. S. Chaudhry (1995). Multi substrate analysis of carbon oxidation and nitrification in an upflow packed-bed biofilm reactor. *J. Chem. Technol. Biotechnol.* **64**, 367–378.

Berg, J. D., A. Matin and P. V. Roberts (1982). Effect of the antecedent growth conditions on sensitivity of *Escherichia coli* to chlorine dioxide. *Appl. Environ. Microbiol.* **44**, 814–818.

Beuchat, L. R. (1994). Antimicrobial properties of spices and their essential oils. In: R. G. Board and V. Dillon (eds), *Natural Antimicrobial Systems in Food Preservation*, pp. 167–179. CABI Publishing, Wallingford.

Beuchat L. R and R. E. Brackett (1991). Inhibitory effects of raw carrots on *Listeria monocytogenes. Appl. Environ. Microbiol.* **56**, 1734–1742.

Beuchat, L. R. and D. A. Golden (1989). Antimicrobials occurring naturally in foods. *Food Technol.* **43**, 134–142.

Beumer, R. R. (2001). Microbiological hazards and their control: bacteria. In: M. R. Adams and M. J. R. Nout (eds), *Fermentation and Food Safety*, pp. 141–157. Aspen Publishers, Inc., Gaithersburg, MD.

Bintsis, T., E. Litopoulou-Tzanetaki and R. K. Robinson (2000). Existing and potential applications of ultraviolet light in the food industry – a critical review. *J. Sci. Food Agric.* **80**, 637–645.

Blackman, I. C. and J. F. Frank (1996). Growth of *Listeria monocytogenes* as a biofilm on various food-processing surfaces. *J. Food Prot.* **59**, 827–831.

Bodnaruk, P. W. and F. A. Draughon (1998). Effect of packaging atmosphere and pH on the virulence and growth of *Yersinia enterocolitica* on pork stored at 4 °C. *Food Microbiol.* **15**, 129–136.

Bonestroo, M. H., B. J. M. Kusters, J. C. de Wit and F. M. Rombouts (1993). The fate of spoilage and pathogenic bacteria in fermented sauce-based salads. *Food Microbiol.* **10**, 101–111.

Boutibonnes, P., V. Bisson, B. Thammavongs *et al.* (1995). Induction of thermotolerance by chemical agents in *Lactococcus lactis* subsp. *lactis* IL 1403. *Intl J. Food Microbiol.* **25**, 83–94.

Bower, C., J. McGuire and M. Daeschel (1995). Supression of *Listeria monocytogenes* colonization following adsorption of nisin onto silica surfaces. *Appl. Environ. Microbiol.* **61**, 992–997.

Boyd, A. and A. M. Chakrabarty (1995). *Pseudomonas aeruginosa* biofilms: role of the alginate exopolysaccharide. *J. Ind. Microbiol.* **15**, 162–168.

Breen, P. J., H. Salari and C. M. Compadre (1997). Elimination of *Salmonella* contamination from poultry tissues by cetylpyridinium chloride solutions. *J. Food Prot.* **60**, 1019–1021.

Brik, H. (1991). Natamycin. In: K. Flory (ed) *Analytical Profiles of Drug Substances.* Academic Press, New York.

Brock, T. D. (1967). Life at high temperatures. *Science* **158**, 1012–1019.

Brown, A. D. (1964). Aspects of bacterial response to the ionic environment. *Bacteriol. Rev.* **28**, 296–328.

Bruyneel, B., M. V. Woestyne and W. Verstraete (1991). Iron complexation as a tool to direct mixed clostridia-lactobacilli fermentations. *J. Appl. Bacteriol.* **69**, 80–85.

Brynjolfsson, A. (1989). Future radiation sources and identification of irradiated foods. *Food Technol.* **43(7)**, 84–89, 97.

Buchanan, R. L. and M. L. Cygnarowicz (1991). A mathematical approach toward defining and calculating the duration of the lag phase. *Food Microbiol.* **7**, 237–240.

Buchanan, R. L., D. G. Hoover and S. B. Jones (1983). Caffeine inhibition of aflatoxin production: mode of action. *Appl. Environ. Microbiol.* **46**, 1193–1200.

Bullerman, L. B., F. Y. Lieu and S. A. Seier (1977). Inhibition of growth and aflatoxin production by cinnamon and clove oils, cinnamic aldehyde and eugenol. *J. Food Sci.* **42**, 1107–1109.

Bunch, W. L., M. E. Matthews and E. H. Marth (1977). Fate of *Staphylococcus aureus* in beef soy loaves subjected to procedures used in hospital chill foodservice systems. *J. Food Sci.* **42**, 565–566.

Butler, J. L., J. C. Stewart, C. Vanderzant *et al.* (1979). Attachment of microorganisms to pork skin and surfaces of beef and lamb carcasses. *J. Food Prot.* **42**, 401–406.

Butz, P and B. Tauscher (2002). Emerging technologies: chemical aspects. *Food Res. Intl* **35**, 279–284.

Cai, P., M. A. Harrison, Y. W. Huang and J. L. Silva (1997). Toxin production by *Clostridium botulinum* type E in packaged channel catfish. *J. Food Prot.* **60**, 1358–1363.

Camper, A. K. and G. A. McFetters (1979). Chlorine injury and the enumeration of water-borne coliform bacteria. *Appl. Environ. Microbiol.* **37**, 633–639.

Camper, A. K., M. W. LeChevalier, S. C. Broadaway and G. A. McFeters (1985). Evaluation of procedures to desorb bacteria from granular activated carbon. *J. Microbiol. Methods* **3**, 187–198.

Carini, S., G. Mucchetti and E. Neviani (1985). Lysozyme: activity against clostridia and use in cheese products. *Microbiol. Alimen. Nutr.* **3**, 299–320.

Carlez, A., J. P. Rosec, N. Richard and J. C. Cheftel (1994). Bacterial growth during chilled storage of pressure treated mincemeat. *Lebensm. Wiss. Technol.* **27**, 48–54.

Carpenter, C. E., D. S. A. Reddy and D. P. Cornforth (1987). Inactivation of clostridial ferredoxin and pyruvate-ferredoxin oxidoreductase by sodium nitrate. *Appl. Environ. Microbiol.* **53**, 549–552.

Carter, M. J and M. R. Adams (2001). Microbiological hazards and their control: viruses. In: M. R. Adams and M. J. R. Nout (eds), *Fermentation and Food Safety*, pp. 159–173. Aspen Publishers, Inc., Gaithersburg, MD.

Castleman, M (1993). Radiant grub. *Sierra* **78**, 27–28.

Cecen, F. and E. Orak (1996). Nitrification of fertilizer waste water in a biofilm reactor. *J. Chem. Technol. Biotechnol.* **65**, 229–238.

Cerf, O. (1977). Tailing of survival curves of bacterial spores. *J. Appl. Bacteriol.* **42**, 1–19.

Cerf, O., K. R. Davey and A. K. Sadoudi (1996). Thermal inactivation of bacteria – a new predictive model for the combined effect of three environmental factors: temperature, pH and water activity. *Food Res. Intl.* **29**, 219–226.

CFR (2001). Code of Federal Regulations. National Archives and Records Administration. US Government Printing Office (available at http: //frwebgate.access.gpo.gov/cgi-bin/multidb.cgi).

Chang, J.-H., C.-C. Chou and C.-F. Li (2000). Growth and survival of *Escherichia coli* O157:H7 during the fermentation and storage of diluted cultured milk drink. *Food Microbiol.* **17**, 579–587.

Chang, R. K., J. O. Hearnsberger, A. P. Vickery *et al.* (1995). Sodium acetate and bifidobacteria increase shelf-life of refrigerated catfish fillets. *J. Food Sci.* **60**, 25–27.

Characklis, W. G. and K. E. Cooksey (1983). Biofilms and microbial fouling. *Adv. Appl. Microbiol.* **29**, 93–127.

Characklis, W. G. and K.C. Marshall (1991). Biofilms: a basis for an interdisciplinary approach. In: W.G. Characklis and K.C. Marshall (eds), *Biofilms*, pp. 3–16. John Wiley & Sons, Inc., New York, NY.

Chen, H. and D. G. Hoover (2003). Bacteriocins and their food applications. *Comp. Rev. Food Sci. Food Safety*, 82–100.

Chen, T. C., J. T. Culotta and W. S. Wang (1973). Effects of water and microwave energy precooking on microbiological quality of chicken parts. *J. Food Sci.* **38**, 155–157.

Chikindas, M. L. and T. J. Montville (2002). Perspectives for application of bacteriocins as food preservatives. In: V. K. Juneja and J. N. Sofos (eds), *Control of Foodborne Microorganisms*, pp. 303–321. Marcel Dekker, New York, NY.

Chipley, J. R. (1980). Effects of microwave irradiation on microorganisms. *Adv. Appl. Microbiol.* **26**, 129–145.

Chipley, J. R. (1993). Sodium benzoate and benzoic acid. In: P. M. Davidson and A. L. Branen (eds), *Antimicrobials in Foods*, 2nd edn, pp. 11–48. Marcel Dekker, New York, NY.

Chung, K. and C. Murdock (1991). Natural systems for preventing contamination and growth of microorganisms in foods. *Food Microbiol.* **10**, 361–365.

Clark, D. S and C. P. Lentz (1969). The effect of carbon dioxide on growth of slime producing bacteria on fresh beef. *Can. Inst. Food Sci. J.* **2**, 72–75.

Cleveland, J., T. J. Montville, I. F. Nes and M. L. Chikindas (2001). Bacteriocins: safe, natural antimicrobials for food preservation. *Intl J. Food Microbiol.* **71**, 1–20.

Code of Federal Regulations (1992). Title 21: Food and Drugs. Office of the Federal Register, National Archives and Records Administration, Washington, DC, USA.

Cole, M. B., K. W. Davies, G. Munro *et al.* (1993). A vitalistic model to describe the thermal inactivation of *Listeria monocytogenes. J. Ind. Microbiol.* **12**, 232–239.

Conley, A. J and J. J. Kabara (1973). Antimicrobial action of esters of polyhydric alcohols. *Antimicrob. Agents Chemother.* **4**, 501–506.

Cords, B. R. and G. R. Dychdala (1993). Sanitizers: halogens, surface-active agents and peroxides. In: P. M. Davidson and A. L. Branen (eds), *Antimicrobials in Foods*, 2nd edn, pp. 469–538. Marcel Dekker, New York, NY.

Costerton, J. W., Z. Lewandowski, D. de Beer *et al.* (1994). Biofilms, the customized microniche. *J. Bacteriol.* **176**, 2137–2142.

Cremer, M. L. and J. R. Chipley (1980a). Hospital ready-prepared type food service system: time and temperature conditions, sensory and microbiological quality of scrambled eggs. J. Food Sci. **45**, 1422–1424, 1429.

Cremer, J. L. and J. R. Chipley. (1980b). Time and temperature, microbiological and sensory assessment of roast beef in a hospital foodservice system. *J. Food Sci.* **45**, 1472–1477.

Cross, H. R., P. R. Durland and S. C. Seideman (1986). Sensory qualities of meat. In: P. J. Bechtel (ed.), *Muscle as Food*, pp. 279–320. Academic Press, Orlando, FL.

Curnutte, B. (1980). Principles of microwave radiation. *J. Food Prot.* **43**, 618–624.

Dahl, C. A., M. E. Matthews and E. H. Marth (1981). Cook–chill foodservice system with a microwave oven, injured aerobic bacteria during food product flow. *Eur. J. Appl. Microbiol. Biotechnol.* **11**, 125–130.

Dalgaard, P., T. Ross, L. Kamperman *et al.* (1994). Estimation of bacterial growth rates from turbimetric and viable count data. *Intl J. Food Microbiol.* **23**, 391–404.

Dallyn, H. (1994). Antimicrobial properties of vegetable and fish oils. In: V. M. Dillon and R. G. Board (eds), *Natural Antimicrobial Systems and Food Preservation*, pp. 205–221. CABI Publishing, Wallingford.

Daniels, J. A., R. Krishnamurth and S. S. H. Rizvi (1985). A review of effects of CO_2 on microbial growth and food quality. *J. Food Prot.* **48**, 532–537.

Dantzer, W. D., N. Hermawan and Q. H. Zhang (1999). Standardized procedure for inactivation of spores by pulsed electric fields (PEF). *IFT Annual Meeting Book of Abstracts, Session 83A-9*. Institute of Food Technologists, Chicago, IL.

Davidson, P. M (2001). Chemical preservatives and natural antimicrobial compounds. In: M. P. Doyle, L. R. Beuchat and T. J. Montville (eds), *Food Microbiology: Fundamentals and Frontiers*, 2nd edn, pp. 593–627. ASM Press, Washington, DC.

Davidson, P. M. and A. L. Branen (1981). Antimicrobial activity of non-halogenated phenolic compounds. *J. Food Prot.* **44**, 623–632.

Davidson, P. M. and C. H. Doan (1993). Natamycin. In: P. M. Davidson and A. L. Branen (eds), *Antimicrobials in Foods*, 2nd edn, pp. 395–408. Marcel Dekker, New York, NY.

Davidson, P. M. and M. A. Harrison (2002). Resistance and adaptation to food antimicrobials, sanitizers and other process controls. *Food Technol.* **56**, 69–78.

Davies, E. A., H. E. Bevis and J. Delves-Broughton (1997). The use of bacteriocin nisin as a preservative in ricotta-type cheeses to control the food-borne pathogen *Listeria monocytogenes. Lett. Appl. Microbiol.* **24**, 343–346.

Davis, B. D. (1991). Chemotherapy. In: B. D. Davis, R. Dulbecco, H. N. Eisen and H. S. Ginsberg (eds), *Microbiology*, 4th edn, pp. 201–228. J. B. Lippincott Co., Philadelphia, PA.

Davis, C. P., M. E. Shirtliff, N. M. Trieff *et al.* (1994). Quantification, qualification and microbial killing efficiencies of antimicrobial chlorine-based substances produced by iontophoresis. *Antimicrob. Agents Chemother.* **38**, 2768–2774.

Deans, S. G. and G. Richie (1987). Antibacterial properties of plant essential oils. *Intl J. Food Microbiol.* **5**, 165–180.

Decupper, J. (1992). Equipment for cold storage chambers for foods. French Patent Application FR2666742A1.

De Groene, E. M., A. Jahn, G. J. Horbach and J. Fink-Gemmels (1996). Mutagenicity and genotoxicity of the mycotoxin ochratoxin A. *Environ. Toxicol. Pharmacol.* **1**, 21–26.

Delaquis, P. J. and G. Mazza (1995). Antimicrobial properties of isothiocyanates in food preservation. *Food Technol.* **49**, 73–78.

Delves-Broughton, J., P. Blackburn, R. J. Evans and J. Hugenholtz (1996). Applications of the bacteriocin nisin. *Antonie van Leeuwenhoek* **69**, 193–202.

Deplace, G. and B. Mertens (1992). The commercial application of high pressure technology in the food-processing industry. In: R. Hayashi, K. Heremans and P. Masson (eds), *High Pressure and Biotechnology*, p. 469. Colloque INSERM John Libbey Eurotext, Ltd, Newtownabbey.

Dickens, J. A. and A. D. Whittemore (1994). The effect of acetic acid and air injection on appearance, moisture pick-up, microbiological quality and *Salmonella* incidence in processed poultry. *Poultry Sci.* **73**, 582–586.

Dickson, J. S. and M. Koohmaraie (1989). Cell surface charge characteristics and their relationship to bacterial attachment to meat surfaces. *Appl. Environ. Microbiol.* **55**, 832–836.

Dictionary.com (2003). Lexico Publishing Group, LLC.

Dillon, V. M. and P. E. Cook (1994). Biocontrol of undesirable microorganisms in food. In: V. M. Dillon and R. G. Board (eds), Natural Antimicrobial Systems and Food Preservation, pp. 255–296. CABI Publishing, Wallingford.

Dineen, S. S., K. Takeuchi, J. E. Soudah and K. J. Boor (1998). Persistence of *Escherichia coli* O157:H7 in dairy fermentation systems. *J. Food Prot.* **61**, 1602–1608.

Dixon, N. M. and D. B. Kell (1989). The inhibition by CO_2 of the growth and metabolism of microorganisms. *J. Appl. Bacteriol.* **67**, 109–136.

Domínguez, C., I. Gómez and J. Zumalacarregui (2001). Prevalence and contamination levels of *Listeria monocytogenes* in smoked fish and pâté sold in Spain. *J. Food Prot.* **64**, 2075–2077.

Domínguez, L., J. F. F. Garayazabal, E. R. Ferri *et al.* (1987). Viability of *Listeria monocytogenes* in milk treated with hydrogen peroxide. *J. Food Prot.* **50**, 636–639.

Doores, S. (1993). Organic acids. In: P. M. Davidson and A. L. Branen (eds), *Antimicrobials in Foods*, 2nd edn, pp. 95–136. Marcel Dekker, New York, NY.

Dornow, K. D. (1992). Process for mechanical peeling of fruits and vegetables in peeling machines. German Federal Republic Patent DE4037026C1.

Drosinos, E. H., C. Tassou, K. Kakiomenou and G.-J. E. Nychas (2000). Microbiological, physico-chemical and organoleptic attributes of a country tomato salad and fate of *Salmonella enteritidis* during storage under aerobic or modified atmosphere packaging conditions at 4 °C and 10 °C. *Food Control* **11**, 131–135.

Duncan, C. L. and E. M. Foster (1968). Effect of sodium nitrite, sodium chloride and sodium nitrate on germination and outgrowth of anaerobic spores. *Appl. Microbiol.* **16**, 406–411.

Dunn, J., T. Ott and W. Clark (1995). Pulsed light treatment of food and packaging. *Food Technol.* **49**, 95–98.

Dunn, J. E. (1996). Pulsed light and pulsed electric field for foods and eggs. *Poultry Sci.* **75**, 1133–1136.

Dunn, J. E. (2000). Unpublished data. Personal communication with W. J. Kowalski.

Dunn, J. E. and J. S. Pearlman (1987). Methods and apparatus for extending the shelf life of fluid food products. US Patent 4 695 472.

Dunn, J. E., D. Burguess and F. Leo (1997). Investigation of pulsed light for terminal sterilization of WFI filled/blow/fill/polyethylene seal containers. *J. Pharm. Sci. Technol.* **51**, 1110–1115.

Dunn, J. E., R. W. Clark, J. F. Asmus *et al.* (1991). Methods and apparatus for preservation of foodstuffs. US Patent 5 034 235.

Dunsmore, D. G., A. Twomey, W. G. Whittlestone and H. W. Morgan (1981). Design and performance of systems for cleaning product-contact surfaces of food equipment: a review. *J. Food Prot.* **44**, 220–240.

Earnshaw, R. G., J. Appleyard and R. M. Hurst (1995). Understanding physical inactivation processes: combined preservation opportunities using heat, ultrasound and pressure. *Intl J. Food Microbiol.* **28**, 197–219.

Egberts, G. (1991). UV sterilization of water in the brewery and beverage industries. *Brauerei Forum* **5**, 85–87.

Eklund, T. (1988). Inhibition of microbial growth at different pH levels by benzoic and propionic acids and esters of *p*-hydroxy-benzoic acids. *Intl J. Food Microbiol.* **2**, 159–165.

Ekstrand, B. (1994). Lactoperoxidase and lactoferrin. In: V. M. Dillon and R. G. Board (eds), *Natural Antimicrobial Systems and Food Preservation*, pp. 15–63. CABI Publishing, Wallingford.

Ellinger, R. H. (1972). Phosphates in food processing. In: T. E. Furia (ed.), *Handbook of Food Additives*, 2nd edn, pp. 617–807. CRC Press, Inc., Cleveland, OH.

Ellison, R. T. III and T. J. Giehl (1991). Killing of Gram-negative bacteria by lactoferrin and lysozyme. *J. Clin. Invest.* **88**, 1080–1194.

El-Nezami, H., P. Kankaanpää, S. Salminen and J. Ahokas (1998). Ability of dairy strains of lactic acid bacteria to bind a common food carcinogen, aflatoxin B_1. *Food Chem. Toxicol.* **36**, 321–326.

El-Ziney, M. G. and J. M. Debevere (1998). The effect of reuterin on *Listeria monocytogenes* and *Escherichia coli* O157:H7 in milk and cottage cheese. *J. Food Prot.* **61**, 1275–1280.

Encinas, J. P., C. J. González, M. L. García-López and A. Otero (1999). Numbers and species of motile aeromonads during the manufacture of naturally contaminated Spanish fermented sausages (longaniza and chorizo). *J. Food Prot.* **62**, 1045–1049.

Ennahar, S., F. Kunts, A. Strasser *et al.* (1994). Elimination of *Listeria monocytogenes* in soft and red smear cheeses by irradiation with low energy electrons. *Intl J. Food Sci. Technol.* **29**, 395–403.

Ennahar, S., T. Sashihara, K. Sonomoto and A. Ishizaki. T (2000). Class IIa bacteriocins: biosynthesis, structure and activity. *FEMS Microbiol. Rev.* **24**, 85–106.

Erkmen, O. and S. Karatas (1997). Effects of high hydrostatic pressure on *Staphylococcus aureus* in milk. *J. Food Eng.* **33**, 737–743.

Evrendilek, G. A., Q. H. Zhang and E. R. Richter (1999). Inactivation of *Escherichia coli* O157:H7 and *Escherichia coli* 8739 in apple juice by pulsed electric fields. *J. Food Prot.* **62**, 793–796.

Faith, N. G., N. S. Le Coutour, M. Bonnet Alvarenga *et al.* (1998). Viability of *Escherichia coli* O157:H7 in ground and formed beef jerky prepared at levels of 5 and 20 % fat and dried at 52, 57, 63, or 68 °C in a home style dehydrator. *Intl J. Food Microbiol.* **41**, 213–221.

Farber, J. M. (1991). Microbiological aspects of modified-atmosphere packaging technology – a review. *J. Food Prot.* **54**, 58–70.

Farber, J. M, E. Daley, R. Holley and R. Usborge (1993). Survival of *Listeria monocytogenes* during the production of uncooked German, American and Italian style fermented sausages. *Food Microbial.* **10**, 123-132.

Farkas, J. (1998). Irradiation as a method for decontaminating food, a review. *Intl J. Food Microbiol.* **44**, 189–204.

Farkas, J. (2001). Physical methods of food preservation. In: M. P. Doyle, L. R. Beuchat and T. J. Montville (eds), *Food Microbiology: Fundamentals and Frontiers*, 2nd edn, pp. 567–591. ASM Press, Washington, DC.

Farr, D. (1991). High pressure technology in the food industry. *Trends Food Sci. Technol.* **1**, 14–16.

Fellows, P. (2000). *Food Processing Technology: Principles and Practice*, 2nd edn. CRC Press, New York, NY.

Fennema, O. R., W. D. Powrie and E. H. Marth (1973). *Low-temperature Preservation of Foods and Living Matter*. Marcel Dekker, New York, NY.

Fleming, H. P., W. M. Walter Jr and J. L. Etchells (1973). Antimicrobial properties of oleuropein and products of its hydrolysis from green olives. *Appl. Microbiol.* **26**, 777–782.

Foegeding, P. M. and F. F. Busta (1991). Chemical food preservatives. In: S. S. Block (ed.), *Disinfection, Sterilization and Preservation*, 4th edn, pp. 802–832. Lea and Febiger, Philadelphia, PA.

Francis, G. A. and D. O'Beirne (1998). Effects of storage atmosphere on *Listeria monocytogenes* and competing microflora using a surface model system. *Intl J. Food Sci. Technol.* **33**, 465–476.

Francis, G. A. and D. O'Beirne (2001). Effects of vegetable type, package atmosphere and storage temperature on growth and survival of *Escherichia coli* O157:H7 and *Listeria monocytogenes. J. Ind. Microbiol. Biotechnol.* **27**, 111–116.

Francis, G. A., C. Thomas and D. O'Beirne (1999). The microbiological safety of minimally processed vegetables [Review article]. *Intl J. Food Sci. Technol.* **34**, 1–22.

Franco-Abuín, C. M., J. Rozas-Barrero, M. A. Romero-Rodríguez *et al.* (1997). Effect of modified atmosphere packaging on the growth and survival of *Listeria* in raw minced beef. *Food Sci. Technol. Intl* **3**, 285–290.

Frank, H. K. (1992). *Dictionary of Food Microbiology*. Technomic Publishing Co., Basel.

Frank, J. F. and R. A. N. Chmielewski (1997). Effectiveness of sanitation with quaternary ammonium compounds vs. chlorine on stainless steel and other food preparation surfaces. *J. Food Prot.* **60**, 43–47.

Frank, J. F. and R. A. Koffi (1991). Surface adherent growth of *Listeria monocytogenes* is associated with increased resistance to surfactant sanitizers and heat. *J. Food Prot.* **53**, 550–554.

Freese, E., C. W. Sheu and E. Galliers (1973). Function of lipophilic acids as antimicrobial food additives. *Nature* **241**, 321–327.

Fuchs, S., T. Haritopoulou and M. Wilhelmi (1996). Biofilms in freshwater ecosystems and their use as a pollutant monitor. *Water Sci. Technol.* **37**, 137–140.

Furr, J. R. and A. D. Russell (1972). Some factors influencing the activity of esters of p-hydroxybenzoic acid against *Serratia marcescens. Microbios* **5**, 189–195.

Galindo-Cuspinera, V., D. C. Westhoff and S. A. Rankin (2003). Antimicrobial properties of commercial annatto extracts against selected pathogenic, lactic acid and spoilage microorganisms. *J. Food Prot.* **66**, 1074–1078.

Galuska, P. J., R. W. Kolarik and P. C. Vasavada (1988). Inactivation of *Listeria monocytogenes* by microwave treatment. *J. An. Sci.* **67** (Suppl. 1), 139.

Genigeorgis, C. and H. Riemann (1979). Food processing and hygiene. In: H. Riemann and F. L. Bryan (eds), *Food-borne Infections and Intoxications*, 2nd edn, pp. 613–713. Academic Press, New York, NY.

Georgala, D. L. and A. Hurst (1963). The survival of food poisoning bacteria in frozen foods. *J. Appl. Bacteriol.* **26**, 346–358.

Gibson, A. M. and T. A. Roberts (1986a). The effect of pH, water activity, sodium nitrite and storage temperature on the growth of enteropathogenic *Escherichia coli* and salmonellae in laboratory medium. *Intl J. Food Microbiol.* **3**, 183–194.

Gibson, A. M. and T. A. Roberts (1986b). The effect of pH, sodium chloride, sodium nitrite and storage temperature on the growth of *Clostridium perfringens* and faecal streptococci in laboratory medium. *Intl J. Food Microbiol.* **3**, 195–210.

Gibson, A. M., N. Bratchell and T. A. Roberts (1988). Predicting microbial growth: growth responses of salmonellae in a laboratory medium as affected by pH, sodium chloride and storage temperature. *Intl J. Food Microbiol.* **6**, 155–178.

Gilbert, P., P. J. Collier and M. R. W. Brown (1991). Influence of growth rate on susceptibility to antimicrobial agents: biofilms, cell cycle, dormancy and stringent response. *Antimicrob. Agents Chemother.* **34**, 1865–1886.

Gill, A. O. (1998). Biofilms: towards a greater understanding of microbial life [Review paper for 'Topics in Food Science']. Department of Food Science, University of Manitoba, Winnipeg, MB, Canada.

Gill, C. O. (1995). MAP and CAP of fresh, red meats, poultry and offals. In: J. M. Farber and K. L. Dodds (eds), *Principles of Modified-atmosphere and Sous Vide Product Packaging*, pp. 105–136.. Technomic Publishing, Lancaster, PA.

Gill, C. O. and T. Jones (1994). The display of retail packs of ground beef after their storage in master packages under various atmospheres. *Meat Sci.* **37**, 281–295.

Gill, C. O., G. G. Greer and B. D. Dilts (1997). The aerobic growth of *Aeromonas hydrophila* and *Listeria monocytogenes* in broths and on pork. *Intl J. Food Microbiol.* **35**, 67–74.

Glass, K. A., J. M. Loeffelholz, J. P. Ford and M. P. Doyle (1992). Fate of *Escherichia coli* O157:H7 as affected by pH or sodium chloride and in fermented, dry sausage. *Appl. Environ. Microbiol.* **58**, 2513–2516.

González-Fandos, E., C. Olarte, M. Giménez *et al.* (2001). Behaviour of *Listeria monocytogenes* in packaged fresh mushrooms (*Agaricus bisporus*). *J. Appl. Microbiol.* **91**, 795–805.

González-Fandos, M. E., M. Sierra, M. L. García-López *et al.* (1999). The influence of manufacturing and drying conditions on the survival and toxinogenesis of *Staphylococcus aureus* in two Spanish dry sausages (chorizo and salchichón). *Meat Sci.* **52**, 411–419.

Goodnough, M. C. and E. A. Johnson (1991). Control of *Salmonella enteritidis* infections in poultry by polymyxin B and trimethoprim. *Appl. Environ. Microbiol.* **57**, 785–788.

Gould, G. W (1989). Heat induced injury and inactivation. In: G. W. Gould (ed.), *Mechanisms of Action of Food Preservation Procedures*, pp. 11–42. Elsevier Applied Science, London.

Gould, G. W. (2000). The use of other chemical preservatives: sulfite and nitrite. In B. M. Lund, T. C. Baird-Parker and G. W. Gould (eds), *The Microbiological Safety and Quality of Food*, pp. 200–213. Aspen Publishers, Inc., Gaithersburg, MD.

Gould, G. W. and N. J. Russell (1991). Sulphite. In: N. J. Russell and G. W. Gould (eds), *Food Preservation*, pp. 72–88. Blackie & Son, Ltd, Glasgow.

Gounot, A. M. and N. J. Russell (1999). Physiology of cold-adapted microorganisms. In: R. Margesin and F. Schinner (eds), *Cold-adapted Organisms.*

Ecology, Physiology, Enzymology and Molecular Biology, pp. 33–55. Springer Verlag, Berlin.

Gourama, H. and L. B. Bullerman (1988). Effects of potassium sorbate and natamycin on growth and penicillic acid production by *Aspergillus ochraceus. J. Food Prot.* **51**, 139–145.

Graham, A. F., D. R. Mason, F. J. Maxwell and M. W. Peck (1997). Effect of pH and NaCl on growth from spores of non-proteolytic *Clostridium botulinum* at chill temperatures. *Lett. Appl. Microbiol.* **24**, 95–100.

Grahl, T. and H. Maerkl (1996). Killing of microorganisms by pulsed electric fields. *Appl. Microbiol. Biotechnol.* **45**, 148–157.

Grant, I. R. and M. F. Patterson (1995). The potential use of additional hurdles to increase the microbiological safety of MAP and *sous vide* products. In: J. M. Farber and K. L. Dodds (eds), *Principles of Modified-atmosphere and Sous Vide Product Packaging*, pp. 263–286. Technomic Publishing, Lancaster, PA.

Grau, F. H. and P. B. Vanderlinde (1993). Aerobic growth of *Listeria monocytogenes* on beef lean and fatty tissue: equations describing the effects of temperature and pH. *J. Food Prot.* **56**, 96–101.

Graumann, P. and M. A. Marahiel (1996). Some like it cold: response of microorganisms to cold shock. *Arch. Microbiol.* **166**, 293–300.

Graumann, P. and M. A. Marahiel (1997). Effects of heterologous expression of CspB, the major cold shock protein of *Bacillus subtilis*, on protein synthesis in *E. coli. Mol. Gen. Genet.* **253**, 745–752.

Gray, N. F. (1994). *Drinking Water Quality – Problems and Solutions*. Wiley, Chichester.

Grijspeerdt, K. and L. Herman (2003). Inactivation of *Salmonella* Enteritidis during boiling of eggs. *Intl J. Food Microbiol.* **82**, 13–24.

Guan, T. Y. and R. A. Holley (2003). *Hog Manure Management, The Environment and Human Health*. Kluwer Academic/Plenum Publishers, New York, NY.

Guder, A., I. Wiedemann and H.-G. Sahl (2000). Post-translationally modified bacteriocins – the lantibiotics. *Biopolymers* **55**, 62–73.

Gundavarapu, S., Y. C., Hung, R. E. Brackett and P. Mallikarjunan (1995). Evaluation of microbiological safety of shrimp cooked in a microwave oven. *J. Food. Prot.* **58**, 742–747.

Han, Y. W. (1975). Death rates of bacterial spores: nonlinear survivor curves. *Can. J. Microbiol.* **21**, 1464–1467.

Hancock, I. C. (1991). Microbial cell surface architecture. In: N. Mozes, P. S. Handley, H. J. Busscher and P. G. Rouxhet (eds), *Microbial Cell Surface Analysis*, pp. 23–59. VCH Publishers, Weinheom.

Harmayani, E., J. N. Sofos and G. R. Schmidt (1991). Effect of sodium lactate, calcium lactate and sodium alginate on bacterial growth and aminopeptidase activity. *J. Food Safety* **11**, 269–284.

Harwig, J., P. M. Scott, B. P. C. Kennedy and Y. K. Chen (1973). Disappearance of patulin from apple juice fermented by *Saccharomyces* sp. *J. Inst. Technol. Aliment.* **7**, 295–312.

Hauschild, A. H. W. (1989). *Clostridium botulinum*. In: M. P. Doyle (ed.), *Foodborne Bacterial Pathogens*, pp. 112–189. Marcel Dekker, New York, NY.

Hébraud, M. and P. Potier (2000). Cold acclimation and cold shock in psychrotrophic bacteria. In: M. Inouye and K. Yamanaka (eds), *Cold Shock Response and Adaptation*, pp. 41–60. Horizon Press, Wymondham.

Heddleson, R. A. and S. Doores (1994). Factors affecting microwave heating of foods and microwave induced destruction of foodborne pathogens – a review. *J. Food Prot.* **57**, 1025–1037.

Heddleson, R. A., S. Doores, R. C. Anatheswaran *et al.* (1991). Survival of *Salmonella* species heated by microwave energy in a liquid menstruum containing food components. *J. Food Prot.* **54**, 637–642.

Heddleson, R. A., S. Doores, R. C. Anatheswaran and G. D. Kuhn (1994). Viability loss of *Salmonella* species, *Staphylococcus aureus* and *Listeria monocytogenes* heated by microwave energy in food systems of various complexity. *J. Food Prot.* **57**, 1025–1037.

Hedén, C. G. (1964). Effects of high hydrostatic pressure on microbial systems. *Bacteriol. Rev.* **28**, 14–29.

Helke, D. M., E. B. Somers and A. C. L. Wong (1993). Attachment of *Listeria monocytogenes* and *Salmonella typhimurium* to stainless steel and Buna-N in the presence of milk and milk components. *J. Food Prot.* **56**, 479–484.

Herald, P. J. and P. M. Davidson (1983). Antibacterial activity of selected hydroxycinnamic acids. *J. Food Sci.* **48**, 1378–1379.

Herald, P. J. and E. A. Zottola (1988). Attachment of *Listeria monocytogenes* to stainless steel surfaces at various temperatures and pH values. *J. Food Prot.* **53**, 1549–1552, 1562.

Hightower, L. E. (1991). Heat shock, stress proteins, chaperones and proteotoxicity. *Cell* **66**, 191–197.

Hoch, J. A. and T. J. Silhavy (1995). *Two-component Signal Transduction*. American Society for Microbiology, Washington, DC.

Hodgins, A. M., G. S. Mittal and M. W. Griffiths (2002). Pasteurization of fresh orange juice using low-energy pulsed electrical field. *J. Food Sci.* **67**, 2294–2299.

Hoek, K. S., J. M. Milne, P. A. Grieve *et al.* (1997). Antibacterial activity in bovine lactoferrin-derived peptides. *Antimicrob. Agents Chemother.* **41**, 54–59.

Hofmann, G. A. (1985). Deactivation of microorganisms by an oscillating magnetic field. US Patent 4 524 079.

Holah, J. T., C. Higgs, S. Robinson *et al.* (1991). A conductance-based surface disinfection test for food hygiene. *Lett. Appl. Microbiol.* **11**, 225–259.

Holley, R. A., G. Blank, A. Hydamaka and S. Cenkowski (2003). Unpublished data. Department of Food Science, University of Manitoba, Winnipeg.

Holley, R. A., M. D. Peirson, J. Lam and K. B. Tan (2004). Microbial profiles of commercial, vacuum packaged, fresh pork of normal or short storage life. *Intl J. Food Microbiol.* **97**, 53–62.

Holzapfel, W. H., R. Geisen and U. Schillinger (1995). Biological preservation of foods with reference to protective cultures, bacteriocins and food-grade enzymes. *Intl J. Food Microbiol.* **24**, 343–362.

Hood, S. K. and E. A. Zottola (1997). Growth media and surface conditioning influence the adherence of *Pseudomonas fragi*, *Salmonella typhimurium* and *Listeria monocytogenes* cells to stainless steel. *J. Food Prot.* **60**, 1034–1037.

Hoover, D. G., C. Metrick, A. M. Papineau *et al.* (1989). Biological effects of high hydrostatic pressure on food microorganisms. *Food Technol.* **43**, 99–107.

Houtsma, P. C., A. Heuvelink, J. Dufrenne and S. Notermans (1994). Effect of sodium lactate on toxin production, spore germination and heat resistance of proteolytic *Clostridium botulinum* strains. *J. Food Prot.* **57**, 327–330.

Hoyle, B. D., J. Alcantara and J. W. Costerton (1992). *Pseudomonas aeruginosa* biofilms as a diffusion barrier to piperacillin. *Antimicrob. Agents Chemother.* **36**, 2054–2056.

Huang, Y. W and E. Toledo (1982). Effect of high and low intensity UV irradiation on surface microbiological counts and storage-life of fish. *J. Food Sci.* **47**, 1667–1669, 1731.

Hughey, V. L. and E. A. Johnson (1989). Antibacterial activity of lysozyme against bacteria involved in food spoilage and foodborne disease. *Appl. Environ. Microbiol.* **53**, 2165–2170.

Hülsheger, H., J. Potel and E. G. Niemann (1981). Killing of bacteria with electric pulses of high field strength. *Radiat. Environ. Biophys.* **20**, 53–65.

Hülsheger, H., J. Potel and E. G. Niemann (1983). Electric field effects on bacteria and yeast cells. *Radiat. Environ. Biophys.* **22**, 149–162.

Husmark, U. and U. Rönner (1992). The influence of hydrophobic, electrostatic and morphologic properties on the adhesion of *Bacillus* spores. *Biofouling* **5**, 335–344.

Hyytia, E., S. Hielm. M. Mokkila *et al.* (1999). Predicted and observed growth and toxigenesis by *Clostridium botulinum* type E in vacuum-packaged fishery product challenge tests. *Intl J. Food Microbiol.* **47**, 161–169.

Ikawa, J. Y. and C. Genigeorgis (1987). Probability of growth and toxin production by nonproteolytic *Clostridium botulinum* in rockfish fillets stored under modified atmospheres. *Intl J. Food Microbiol.* **4**, 167–181.

Ingham, S. C. and H. E. Uljas (1998). Prior storage conditions influence the destruction of *Escherichia coli* O157:H7 during heating of apple cider and juice. *J. Food Prot.* **61**, 390–394.

Ingram, M. (1951). The effect of cold on microorganisms in relation to food. *Proc. Soc. Appl. Bacteriol.* **14**, 243.

Ingram, M. and T. A. Roberts (1980). Ionizing radiation. In: ICMSF (ed.), *Microbial Ecology of Foods*, Vol. 1, *Factors Affecting Life and Death of Organisms*, pp. 49–69. Academic Press, New York, NY.

Inhot, A. M., A. M. Roering, R. K. Wierzba *et al.* (1998). Behavior of *Salmonella typhimurium* DT104 during the manufacture and storage of pepperoni. *Intl J. Food Microbiol.* **40**, 117–121.

International Commission on Microbiological Specifications for Foods (1980) Temperature. In: ICMSF (ed.), *Microbial Ecology of Foods*, Vol. 1, *Factors Affecting Life and Death of Organisms*, pp. 1–37. Academic Press, New York, NY.

Jackson, L. S., S. K. Katta, D. D. Fingerhut *et al.* (1997). Effects of baking and frying on the fumonisin B_1 content of corn-based foods. *J. Agric. Food Chem.* **45**, 4800–4805.

James, G. A., L. Beaudette and J. W. Costerton (1995). Interspecies bacterial interactions in biofilms. *J. Ind. Microbiol.* **15**, 257–262.

Jay, J. M (1995). Food with low numbers of microorganisms may not be the safest foods. *Dairy Food Environ. Sanit.* **15**, 674–677.

Jay, J. M (1996). *Modern Food Microbiology*, 5th edn, New York, Chapman & Hall.

Jelen, P. (1982). Experience with direct and indirect UHT processing of milk – a Canadian viewpoint. *J. Food Prot.* **45**, 386–389.

Jen, C. M. C. and L. A. Shelef (1986). Factors affecting sensitivity of *Staphylococcus aureus* 196E to polyphosphate. *Appl. Environ. Microbiol.* **52**, 842–846.

Jeong, D. K and J. F. Frank (1994). Growth of *Listeria monocytogenes* at 21 °C in biofilms with microorganisms isolated from meat and dairy environments. *Lebensm. Wiss. Technol.* **27**, 415–424.

Jeyamkondan, S., D. S. Jayas and R. A. Holley (2000). Review of centralized packaging systems for distribution of retail-ready meat. *J. Food Prot.* **63**, 796–804.

Joerger, R. D., D. G. Hoover, S. F. Barefoot *et al.* (2000). Bacteriocins. In: L. Lederberg (ed.), *Encyclopedia of Microbiology*, Vol. 1, 2nd edn, pp. 383–397. Academic Press, San Diego, CA.

Johnson, E. A. (1994). Egg white lysozyme as a preservative for use in foods. In: J. S. Sim and S. Nakai (eds), Egg *Uses and Processing Technologies. New Developments*, pp. 177–191. CABI Publishing, Wallingford.

Johnson, E. A., J. H. Nelson and M. Johnson (1991). Microbiological safety of cheese made from heat-treated milk, part II: microbiology. *J. Food Prot.* **53**, 519–540.

Johnson, L. L., R. V. Peterson and W. G. Pitt (1998). Treatment of bacterial biofilms on polymeric biomaterials using antibiotics and ultrasound. *J. Biomat. Sci. Polymer Edn*, **9**, 1177–1185.

Johnston, D. E. and M. H. Stevenson (1991). *Food Irradiation and the Chemist*. The Royal Society of Chemistry, Cambridge.

Jones, G. W. and R. E. Isaacson (1983). Proteinaceous bacterial adhesions and their receptors. *CRC Crit. Rev. Microbiol.* **10**, 229–260.

Jones, G. W. and R. W. Pickup (1989). The effect of organic carbon supply in water on the antibiotic resistance of bacteria. *Aqua* **33**, 131–135.

Jones, J. E., S. J. Walker, J. P. Sutherland *et al.* (1994). Mathematical modeling of the growth, survival and death of *Yersinia enterocolitica*. *Intl J. Food Microbiol.* **23**, 433–447.

Judis, J. (1963). Studies on the mechanism of action of phenolic disinfectants. II. Patterns of release of radioactivity from *Escherichia coli* labeled by growth on various compounds. *J. Pharm. Sci.* **52**, 261–264.

Kabara, J. J. (1978). Fatty acids and derivatives as antimicrobial agents. A review. In: J. J. Kabara (ed.), *The Pharmacological Effect of Lipids*. I, pp. 1–14. American Oil Chemists Society, Champaign, IL.

Kabara, J. J (1993). Medium chain fatty acids and esters. In: P. M. Davidson and A. L. Branen (eds), *Antimicrobials in Foods*, 2nd edn, pp. 307–342. Marcel Dekker, New York, NY.

Kabara, J. J., R. Vrable and M. L. Ken-Jie (1977). Antimicrobial lipids: natural and synthetic fatty acids and monoglycerides. *Lipids* **9**, 753–759.

Kanatt, S. R., S. P. Chawla, R. Chander and D. R. Bongirwar (2002). Shelf-stable and safe intermediate-moisture meat products using hurdle technology. *J. Food Prot.* **65**, 1628–1631.

Kanekar, P. and S. Sarnaik (1995). Microbial process for the treatment of phenol bearing dye-industry effluent in a fixed film bioreactor. *J. Environ. Sci. Health* **A30**, 1817–1826.

Kato, N. and I. Shibashaki (1975). Enhancing effect of fatty acids and their esters on the thermal destruction of *E. coli* and *Pseudomonas aeruginosa. J. Ferment. Technol.* **53**, 802–807.

Kelch, F. and X. Bühlmann. (1958). Effect of commercial phosphates on the growth of microorganisms. *Fleischwirtschaft* **10**, 325–328.

Khalil, H. and R. Villota (1989). The effect of microwave sublethal heating on the ribonucleic acids of *Staphylococcus aureus. J. Food Prot.* **52**, 544–548.

Kim, H. J., Y. M. Choi, A. P. P. Yang *et al.* (1996). Microbiological and chemical investigation of ohmic heating of particulate foods using a 5 kW ohmic system. *J. Food Process. Preserv.* **20**, 41–58.

King, A. D., J. D. Ponting, D. W. Sanshuck *et al.* (1981). Factors affecting death of yeast by sulfur dioxide. *J. Food Prot.* **44**, 92–97.

Knabel, S. J., H. W. Walker and P. A. Hartman (1991). Inhibition of *Aspergillus flavus* and selected Gram-positive bacteria by chelation of essential metal cations by polyphosphates. *J. Food Prot.* **54**, 360–365.

Knorr, D. (1993). Effects of high hydrostatic pressure on food safety and quality. *Food Technol.* **47**, 156–161.

Kolsarici, N. and K. Candogan (1995). The effects of potassium sorbate and lactic acid on the shelf-life of vacuum-packed chicken meats. *Poultry Sci.* **74**, 1884–1893.

Koutchma, T. and H. S. Ramaswamy (2000). Combined effects of microwave heating and hydrogen peroxide on the destruction of *Escherichia coli. Lebensm. Wiss. Technol.* **33**, 30–36.

Krogh, P. (1987). Ochratoxins in food. In: P. Krogh (ed.), *Mycotoxins in Food*, pp. 97–121. Academic Press, London.

Krysinski, E. P., L. J. Brown and T. J. Marsichello (1992). Effect of cleaners and sanitizers on *Listeria monocytogenes* attached to product contact surfaces. *J. Food Prot.* **55**, 246–251.

Kudo, H., K. J. Cheng and J. W. Costerton (1987). Interactions between *Treponema bryantii* and cellulotypic bacteria in the *in vitro* degradation of straw cellulose. *Can. J. Microbiol.* **33**, 244–248.

Kuiper-Goodman, T. and P. M. Scott (1989). Risk assessment of the mycotoxin ochratoxin A. *Biomed. Environ. Sci.* **2**, 179–248.

Kulikoosky, A., H. S. Pankratz and H. L. Sandoff (1975). Ultrastructural and chemical changes in spores of *Bacillus cereus* after action of disinfectants. *J. Appl. Bacteriol.* **38**, 39–45.

Kumar, C. G. (1997). Studies on microbial alkaline proteases for use in dairy detergents. PhD Thesis, National Dairy Research Institute (Deemed University), Karnal, India.

Kumar, C. G. and S. K. Anand (1998). Significance of microbial biofilms in food industry; a review. *Intl J. Food Microbiol.* **42**, 9–27.

Kunene, N. F., J. W. Hastings and A. von Holy (1999). Bacterial populations associated with a sorghum-based fermented weaning cereal. *Intl J. Food Microbiol.* **49**, 75–83.

Kuo, F. L., J. B. Carey and S. C. Ricke (1997). UV irradiation of shell eggs: effect on populations or aerobes, moulds and inoculated *Salmonella typhimurium*. *J. Food Prot*. **60**, 639–643.

Kuse, D. (1982). UV-C sterilization of packaging materials in the dairy industry. *D. Milchwirtschaft* **33**, 1134–1137.

Labbe, R. (1989). *Clostridium perfringens*. In: M. P. Doyle (ed.), *Foodborne Bacterial Pathogens*, pp. 192–234. Marcel Dekker, New York, NY.

Lad, P. G. (1992). Endoglycosidases: new enzymes for cleaning. *Abstract from the 83rd AOCS Annual Meeting and Exposition, 10–14 May 1992, Toronto, Canada*.

Lappin, H. M., M. P. Greaves and J. H. Slater (1985). Degradation of the herbicide mecocrop [2-(2-methyl-4-chlorophenoxy) propionic acid] by a synergistic microbial community. *Appl. Environ. Microbiol*. **49**, 429–433.

LeChevallier, M. W., C. D. Cawthon and R. G. Lee (1988). Factors promoting survival of bacteria in chlorinated water supplies. *Appl. Environ. Microbiol*. **54**, 649–654.

LeClercq-Perlat, M. N. and M. Lalande (1994). Cleanability in relation to surface chemical composition and surface finishing of some materials commonly used in food industries. *J. Food Eng*. **23**, 501–517.

Lee, A., S. C. Smith and P. J. Coloe (1998). Survival and growth of *Campylobacter jejuni* after artificial inoculation onto chicken skin as a function of temperature and packaging conditions. *J. Food Prot*. **61**, 1609–1614.

Lee, R. M., P. A. Hartman, D. G. Olson and F. D. Williams (1994). Bacteriocidal and bacteriolytic effects of selected food-grade phosphates, using *Staphylococcus aureus* as a model system. *J. Food Prot*. **57**, 276–283.

Leistner, L. (1978). Hurdle effect and energy saving. In: W. K. Downey (ed.), *Food Quality and Nutrition*, pp. 553–557. Applied Science Publishers, London.

Leistner, L. (1992). Food preservation by combined methods. *Food Res. Intl* **25**, 151–158.

Leistner, L. (1994). Further developments in the utilization of hurdle technology for food preservation. *J. Food Eng*. **22**, 421–432.

Leistner, L. (1995a). Emerging concepts for food safety. In: 41st International Congress of Meat Science and Technology, 20–25 August, San Antonio, Texas, pp. 321–322.

Leistner, L. (1995b). Principles and applications of hurdle technology. In: G. W. Gould (ed.), *New Methods for Food Preservation*, pp. 1–21. Blackie Academic and Professional, London.

Leistner, L. (2000). Basic aspects of food preservation by hurdle technology [Review] *Intl J. Food Microbiol*. **55**, 181–186.

Leistner, L. and G. W. Gould (2002). *Hurdle Technologies: Combination Treatments for Food Stability, Safety and Quality*. Kluwer Academic/Plenum Publishing, New York, NY.

Leive, L. E. and B. D. Davis (1980). Cell envelope: spores. In: B. D. Davis, R. Dulbecco, H. N. Eisen and H. S. Ginsberg (eds), *Microbiology*, 3rd edn, pp. 71–110. Harper and Row, Philadelphia, PA.

Lewis, M. (1993). UHT processing: safety and quality aspects. *Food Technol. Eur.* **1993**, 47–51.

Leyer, G. J. and E. A. Johnson (1992). Acid adaptation promotes survival of *Salmonella* spp. in cheese. *Appl. Environ. Microbiol.* **58**, 2075–2080.

Liang, Z., G. S. Mittal and M. W. Griffiths (2002). Inactivation of *Salmonella* Typhimurium in orange juice containing antimicrobial agents by pulsed electric field. *J. Food Prot.* **65**, 1081–1087.

Liewen, M. B. and E. H. Marth (1985). Growth and inhibition of microorganisms in the presence of sorbic acid: a review. *J. Food Prot.* **48**, 364–375.

Lillard, H. S. (1988). Effect of surfactant on changes in ionic strength of attachment of *Salmonella typhimurium* to poultry skin and muscle. *J. Food Sci.* **53**, 727–730.

Lindsay, R. E., W. A. Krissinger and B. F. Fields (1986). Microwave vs conventional cooking of chicken: relationship of internal temperature to surface contamination by *Salmonella typhimurium. J. Am. Dietet. Assoc.* **86**, 373–374.

Lindström, M., M. Mokkila, E. Skyttä *et al.* (2001). Inhibition of growth of non-proteolytic *Clostridium botulinum* Type B in *sous vide* cooked meat products is achieved by using thermal processing but not nisin. *J. Food Prot.* **64**, 838–844.

Linton, M., J. M. J. McClements and M. F. Patterson (1999). Inactivation of *Escherichia coli* O157:H7 in orange juice using a combination of high pressure and mild heat. *J. Food Prot.* **62**, 277–279.

Little, B. J., P. A. Wagner, W. G. Characklis and W. E. Lee (1991). Microbial corrosion. In: W. G. Characklis and K. C. Marshall (eds), *Biofilms*, pp. 635–670. John Wiley & Sons, New York, NY.

Lund, B. M., A. F. Graham, S. M. George and D. Brown (1991). The combined effects of incubation temperature, pH and sorbic acid on the probability of growth of non-proteolytes type B *Clostridium botulinum. J. Appl. Bacteriol.* **69**, 481–492.

Lyver, A., J. P. Smith, I. Tarte *et al.* (1998). Challenge studies with *Listeria monocytogenes* in a value-added seafood product stored under modified atmospheres. *Food Microbiol.* **15**, 379–389.

Maas, M. R., K. A. Glass and M. P. Doyle (1989). Sodium lactate delays toxin production by *Clostridium botulinum* in cook-in-bag turkey products. *Appl. Environ. Microbiol.* **52**, 2226–2229.

Mabrouk, S. S and N. M. A. El-Shayeb (1981). The effects of garlic on mycelial growth and aflatoxin formation by *Aspergillus flavus. Chem. Mikrobiol. Technol. Lebensm.* **7**, 37–41.

Macaskie, L. E., R. M. Empson, F. Lin and M. R. Tollet (1995). Enzymatically-mediated uranium accumulation and uranium recovery using a *Citrobacter* sp. immobilized as a biofilm within a plug-flow reactor. *J. Chem. Technol. Biotechnol.* **63**, 1–16.

Mackey, B. M and C. M. Derrick (1986). Elevation of heat resistance of *Salmonella typhimurium* by sublethal heat shock. *J. Appl. Bacteriol.* **61**, 389–394.

Macura, D., A. M. McCannel and M. Z. C. Li (2001). Survival of *Clostridium botulinum* in modified atmosphere packaged fresh whole North American ginseng roots. *Food Research Intl* **34**, 123–125.

Mafu, A. A., D. Roy, J. Goulet and P. Hagny (1991). Attachment of *Listeria monocytogenes* to stainless steel, glass, polypropylene and rubber surfaces after short contact times. *J. Food Prot*. **53**, 742–746.

Mafu, A. A., D. Roy, J. Goulet and L. Savoie (1991). Characterization of physicochemical forces involved in adhesion of *Listeria monocytogenes* to surfaces. *Appl. Environ. Microbiol*. **57**, 1969–1973.

Maier, S. K., S. Scherer and M. J. Loessner (1999). Long-chain polyphosphate causes cell lysis and inhibits *Bacillus cereus* septum formation, which is dependent on divalent cations. *Appl. Environ. Microbiol*. **65**, 3942–3949.

Manas, P., R. Pagan, K. Raso, F. J. Sale and S. Condon. (2000). Inactivation of *Salmonella* Enteritidis, *Salmonella* Typhimearium and *Salmonella* Senftenberg by ultrasonic waves under pressure. *J. Food Protect*. **63**, 451–456.

Maret, G. and K. Dransfield (1985). Biomolecules and polymers in high steady magnetic fields. In: F. Herlach (ed.), *Topics in Applied Physics. Strong and Ultrastrong Magnetic Fields and their Applications*, pp. 143–204. Springer-Verlag, Berlin.

Marin, S., M. E. Guynot, P. Neira *et al*. (2002). Risk assessment of the use of sub-optimal levels of weak-acid preservatives in the control of mould growth on bakery products. Intl *J. Food Microbiol*. **79**, 203–211.

Marr, A. G and J. L. Ingraham (1962). Effect of temperature on the composition of fatty acids in *Escherichia coli. J. Bacteriol*. **84**, 1260–1267.

Marriott, N. G. (1999). *Principles of Food Sanitation*, 4th edn, Aspen Publishers, Inc., Gaithersburg, MD.

Marshall, D. L. and L. B. Bullerman (1986). Antimicrobial activity of fatty acid ester emulsifiers. *J. Food Sci*. **51**, 468–470.

Marshall, K. C. (1992). Biofilms: an overview of bacterial adhesion, activity and control at surfaces. *Am. Soc. Microbiol. News* **58**, 202–207.

Marshall, K. C., R. Stout and R. Mitchell (1971). Mechanisms of the initial events in the sorption of marine bacteria to surfaces. *J. Gen. Microbiol*. **68**, 337–348.

Marshall, V. M. E., W. M. Cole and A. J. Bramley (1986). Influence of the lactoperoxidase system on susceptibility on the udder to *Streptococcus uberis* infection. *J. Dairy Res*. **53**, 507–514.

Marwan, A. G. and C. W. Nagel (1986). Microbial inhibitors in cranberries. *J. Food Sci*. **51**, 1009–1013.

Masschalk, B., R. Van Houdt and C. W. Michiels (2001). High pressure increases bactericidal activity and spectrum of lactoferrin, lactoferricin and nisin. *Intl J. Food Microbiol*. **64**, 325–332.

Masson, P. L., J. F. Heremans and E. Schonne (1969). Lactoferrin: an iron-binding protein in neutrophilic leukocytes. *J. Exp. Medicine*, **130**, 643–658.

Mattila-Sandholm, T. and G. Wirtanen (1992). Biofilm formation in the food industry: a review. *Food Rev. Intl* **8**, 573–603.

McCarthy, S. A. (1996). Effect of sanitizers on *Listeria monocytogenes* attached to latex gloves. *J. Food Safety* **16**, 231–237.

McClements, D. J. (1995). Advances in the application of ultrasound in food analysis and processing. *Trends Food Sci. Technol*. **6**, 293–299.

McClure, P. J., C. de W. Blackburn, M. B. Cole *et al.* (1994). Modeling growth, survival and death of microorganisms in foods: the UK Food Micromodel approach. *Intl J. Food Microbiol.* **23**, 265–275.

McDonald, K. F., R. D. Curry, T. E. Clevenger *et al.* (2000). A comparison of pulsed and continuous ultraviolet light sources for the decontamination of surfaces. *IEEE Trans. Plasma Sci.* **28**, 1581–1587.

McFeters, G. A., F. P. Yu, B. H. Pyle and P. S. Stewart (1995). Physiological methods to study biofilm disinfection. *J. Ind. Microbiol.* **15**, 333–338.

McKellar, R. C., G. Butler and K. Stanich (1997). Modeling the influence of temperature on the recovery of *Listeria monocytogenes* from heat injury. *Food Microbiol.* **14**, 617–625.

McMeekin, T. A., K. Presser, A. D. Ratkowsky *et al.* (2000). Quantifying the hurdle concept by modeling the bacterial growth/no growth interface. *Intl J. Food Microbiol.* **55**, 93–8.

McMindes, M. K. and A. J. Siedler (1988). Nitrite mode of action: inhibition of yeast pyruvate decarboxylase (E.C.4.1.1.1) and clostridial pyruvate, oxidoreductase (E.C.1.2.7.1) by nitric oxide. *J. Food Sci.* **53**, 917–919.

Meinhold, N. M. (1991). Chlorine dioxide foam effective against *Listeria* – less contact time needed. *Food Proc.* **52**, 86–90.

Melo, L. F., T. R. Bott, M. Fletcher and B. Capdeville (1992). *Biofilms: Science and Technology*. NATO ASI Series E., Kluwer Academic Press, Dordrecht.

Meng, J. and C. A. Genigeorgis (1994). Delaying toxigenesis of *Clostridium botulinum* by sodium lactate in 'sous-vide' products. *Lett. Appl. Microbiol* **19**, 20–23.

Miller, R. K. and G. R. Acuff (1994). Sodium lactate affects pathogens in cooked beef. *J. Food Sci.* **59**, 15–19.

Mizobuchi, S. and Y. Sato (1985). Antifungal activities of hop bitter resins and related compounds. *Agric. Biol. Chem.* **49**, 339–403.

Monk, J. D., L. R. Beuchat and M. P. Doyle (1995). Irradiation inactivation of foodborne microorganisms. *J. Food Prot.* **58**, 197–208.

Morehouse, K. M. (2002). Food irradiation – US regulatory considerations. *Rad. Phys. Chem.* **63**, 281–284.

Morioka, Y., H. Nohara, M. Araki *et al.* (1996). Studies on the fermentation of soft salami sausage by starter culture. *Animal Sci. Technol.* **67**, 204–210.

Morris, E. O (1962). Effect of environment on micro-organisms. In: J. Hawthorn and J. M. Leitch (eds), *Recent Advances in Food Science*, Vol. 1, pp. 24–36. Butterworths, London.

Morris, S. L., R. C. Walsh and J. N. Hansen (1984). Identification and characterization of some bacterial membrane sulfhydryl groups which are targets of bacteriostatic and antibiotic action. *J. Biol. Chem.* **201**, 581–584.

Moseley, B. E. B. (1989). Ionizing radiation: action and repair. In: G. W. Gould (ed.), *Mechanisms of Action of Food Preservation Procedures*, pp. 43–70. Elsevier Applied Science, New York, NY.

Moseley, B. E. B. (1991). Radiation, micro-organisms and radiation resistance. In: D. E. Johnston and M. H. Stevenson (eds), *Food Irradiation and the Chemist*, pp. 97–108. The Royal Society of Chemistry, Cambridge.

Moss, M. O. (2001). Chemical hazards and their control: mycotoxins. In: M. R. Adams and M. J. R. Nout (eds), *Fermentation and Food Safety*, pp. 101–118. Aspen Publishers, Inc., Gaithersburg, MD.

Moss, S. H. and K. C. Smith (1981). Membrane damage can be a significant factor in the inactivation of *Escherichia coli* by near-ultraviolet radiation. *Photochem. Photobiol.* **33**, 203–210.

Mossel, D. A. A. and C. B. Struijk (1991). Public health implications of refrigerated pasteurized ('sous vide') foods. *Intl J. Food Microbiol.* **13**, 187–206.

Motarjemi, Y., A. Asante, M. R. Adams and M. J. R. Nout (2001). Practical applications: prospects and pitfalls. In: M. R. Adams and M. J. R. Nout (eds), *Fermentation and Food Safety*, pp. 253–273. Aspen Publishers, Inc., Gaithersburg, MD.

Murano, E. A., P. S. Murano, R. E. Brennan *et al.* (1999). Application of high hydrostatic pressure to eliminate *Listeria monocytogenes* from fresh pork sausage. *J. Food Prot.* **62**, 480–483.

Mustapha, A. and M. B. Liewen (1989). Destruction of *Listeria monocytogenes* by sodium hypochlorite and quaternary ammonium sanitizers. *J. Food Prot.* **52**, 306–311.

Muthukumarasamy, P., J. H. Han and R. A. Holley (2003). Bactericidal effects of *Lactobacillus reuteri* and allyl isothiocyanate on *Escherichia coli* O157:H7 in refrigerated ground beef. *J. Food Prot.* **66**, 2038–2044.

Nakamura, S., A. Kato and K. Kobayashi (1991). New antimicrobial characteristics of lysozyme–dextran conjugate. *J. Agric. Food Chem.* **39**, 647–650.

Nassar, A., M. A. Ismail and A.-E. Abdel-Hakeem (1995). The role of benzoic acid as a fungal decontaminant of beef carcasses. *Fleischwirtschaft* **75**, 1160–1162.

Nes, I. F. and H. Holo (2000). Class II antimicrobial peptides from lactic acid bacteria. *Biopolymers* **55**, 50–61.

Nes, I. F., D. B. Diep, L. S. Havarstein *et al.* (1996). Biosynthesis of bacteriocins in lactic acid bacteria. *Antonie van Leeuwenhoek* **70**, 113–128.

Newton, K. G. and C. O. Gill (1978). The development of the anaerobic spoilage flora of meat stored at chill temperatures. *J. Appl. Bacteriol.* **44**, 91–95.

Nicholas, R. (1995). Aseptic filling of UHT dairy products in HPDE bottles. *Food Technol. Eur.* **2**, 52–58.

Nickel, J. C., J. B. Wright, J. Ruseska *et al.* (1985). Antibiotic resistance of *Pseudomonas aeruginosa* colonizing a urinary catheter. *Eur. J. Clin. Microbiol.* **4**, 213–218.

Nieman, C. (1954). Influence of trace amounts of fatty acids on the growth of microorganisms. *Bacteriol. Rev.* **18**, 147–163.

Nigam, P., G. McMullan, I. M. Banat and R. Marchant (1996). Decolourisation of effluent from the textile industry by a microbial consortium. *Biotechnol. Lett.* **18**, 117–120.

Nissen, H. and A. Holck (1998). Survival of *Escherichia coli* O157:H7, *Listeria monocytogenes* and *Salmonella kentucky* in Norwegian, fermented dry sausage. *Food Microbiol.* **15**, 273–279.

Northolt, M. D. (1983). Pathogenic micro-organisms in fermented dairy products. *Neth. Milk Dairy J.* **37**, 247–248.

Notermans, S. and P. in't Veld (1994). Microbiological challenge testing for ensuring safety of food products. *Intl J. Food Microbiol.* **24**, 33–39.

Notermans, S., J. Dufrenne and B. M. Lund (1991). Botulism risk of refrigerated processed foods of extended durability. *J. Food Prot.* **53**, 1020–1024.

Nout, M. J. R. (2001). Fermented foods and their production. In: M. R. Adams and M. J. R. Nout (eds), *Fermentation and Food Safety*, pp. 1–19. Aspen Publishers, Inc., Gaithersburg, MD.

Nout, M. J. R., F. M. Rombouts and A. Havelaar (1989). Effect of accelerated natural lactic fermentation of infant food ingredients on some pathogenic micro-organisms. *Intl J. Food Microbiol.* **8**, 351–361.

Nychas, G. J. E. and J. S. Arkoudelos (1991). Staphylococci: their role in fermented sausages. *J. Appl. Bacteriol.* **69**, 167S–188S.

Nykänen, A., K. Weckman and A. Lapveteläinen (2000). Synergistic inhibition of *Listeria monocytogenes* on cold-smoked rainbow trout by nisin and sodium lactate. *Intl J. Food Microbiol.* **61**, 63–72.

Oh, D. H. and D. L. Marshall (1995). Destruction of *Listeria monocytogenes* biofilms on stainless steel using monolaurin and heat. *J. Food Prot.* **58**, 251–255.

Oh, D. H. and D. L. Marshall (1996). Monolaurin and acetic acid inactivation of *Listeria monocytogenes* attached to stainless steel. *J. Food Prot.* **59**, 249–252.

Ollinge-Snyder, P., F. El-Gazzar, M. E. Matthews *et al.* (1995). Thermal destruction of *Listeria monocytogenes* in ground pork prepared with and without soy hulls. *J. Food Prot.* **58**, 573–576.

Ophir, T. and D. L. Gutnick (1994). A role of exopolysaccharide in the protection of microorganisms from desiccation. *Appl. Environ. Microbiol.* **60**, 740–745.

Ostergaard, A., P. K. B. Embarak, C. Wedell-Neergaard *et al.* (1998). Characterization of anti-listerial lactic acid bacteria isolated from Thai fermented fish products. *Food Microbiol.* **15**, 223–233.

Ough, C. S. (1993). Sulfur dioxide and sulfites. In: P. M. Davidson and A. L. Branen (eds), *Antimicrobials in Foods*, 2nd edn, pp. 343–368. Marcel Dekker, New York, NY.

Ozbas, Z. Y. and S. A. Aytac (1996). Behaviour of *Yersinia enterocolitica* and *Aeromonas hydrophila* in skim milk during fermentation by various lactobacilli. *Z. Lebensm. Forsch.* **202**, 324–328.

Pace, B. and L. L. Campbell (1967). Correlation of maximal growth temperature and ribosome heat stability. *Proc. Natl. Acad. Sci. USA* **57**, 1110–1116.

Pagán, R., S. Esplugas, M. M. Góngora-Nieto *et al.* (1998). Inactivation of *Bacillus subtilis* spores using high intensity pulsed electric fields in combination with other food conservation technologies. *Food Sci. Technol. Intl* **4**, 33–44.

Pagán, R., P. Manas, I. Alvarez and S. Condon (1999). Resistance of *Listeria monocytogenes* to ultrasonic waves under pressure at sublethal (manosonication) and lethal (manothermosonication) temperatures. *Food Microbiol.* **16**, 139–148.

Pakula, R. and A. Freeman (1996). A new continuous biofilm bioreactor for immobilized oil-degrading filamentous fungi. *Biotechnol. Bioeng.* **49**, 20–25.

Palmer, R. J. and D. C. White (1997). Developmental biology of biofilms: implications for treatment and control. *Trends Microbiol.* **5**, 435–440.

Paquet, A. and K. Rayman (1987). Some N-acyl-amino acid derivatives having antibotulinal properties. *Can. J. Microbiol.* **33**, 577–582.

Parsell, D. A. and R. T. Sauer (1989). Induction of heat shock-like response by unfolded protein in *Escherichia coli*: dependence on protein level not protein degradation. *Genes Dev.* **3**, 1226–1232.

Paster, N., B. J. Juven and H. Harshemesh (1988). Antimicrobial activity and inhibition of aflatoxin B_1 formation by olive plant tissue constituents. *J. Appl. Bacteriol.* **64**, 293–297.

Payne, K. D., P. M. Davidson, S. P. Oliver and G. L. Christian (1991). Influence of bovine lactoferrin on the growth of *Listeria monocytogenes. J. Food Prot.* **53**, 468–472.

Pearson, L. J and E. H. Marth (1991). Behavior of *Listeria monocytogenes* in the presence of methylxanthines – caffeine and theobromine. *J. Food Prot.* **53**, 47–50.

Pedersen, A. R., S. Moller, S. Molin and E. Arvin (1997). Activity of toluene-degrading *Pseudomonas putida* in the early growth phase of a biofilm for water gas treatment. *Biotechnol. Bioeng.* **54**, 131–141.

Pelroy, G. A., M. E. Peterson, P. J. Holland and M. W. Ecklund (1994a). Inhibition of *Listeria monocytogenes* in cold-process (smoked) salmon by sodium lactate. *J. Food Prot.* **57**, 108–113.

Pelroy, G. A., M. E. Peterson, R. Paranjpye *et al.* (1994b). Inhibition of *Listeria monocytogenes* in cold-process (smoked) salmon by sodium nitrite and packaging method. *J. Food Prot.* **57**, 114–119.

Perigo, J. A. and T. A. Roberts (1968). Inhibition of clostridia by nitrite. *J. Food Technol.* **3**, 91–94.

Perkin, A. G., F. L. Davies, P. Neaves *et al.* (1980). Determination of bacterial spore inactivation at high temperatures. In: G. W. Gould and J. E. L. Corry (eds), *Microbial Growth and Survival in Extreme Environments*, pp. 173–188. Academic Press, London.

Pitotti, A., R. Zironi, B. Adal and A. Amati (1991). Possible application of lysozyme in wine technology. *Med. Facul. Landbouwetenschap* **56**, 1697–1699.

Pittet, A. (1998). Natural occurrence of mycotoxins in foods and feeds – an updated review. *Revue Méd. Vét.* **149**, 479–492.

Piyasena, P., E. Mohareb and R. C. McKellar (2003). Inactivation of microbes using ultrasound: a review. *Intl J. Food Microbiol.* **87**, 207–216.

Ponce, E., R. Pla, M. M. Mor *et al.* (1998). Inactivation of *Listeria innocua* inoculated in liquid whole eggs by high hydrostatic pressure. *J. Food Prot.* **61**, 119–122.

Portocarrero, S. M., M. Newman and B. Mikel (2002). *Staphylococcus aureus* survival, staphylococcal enterotoxin production and shelf stability of country-cured hams manufactured under different processing procedures. *Meat Sci.* **62**, 267–273.

Post, F. J., W. S. Coblentz, T. W. Chou and D. K. Salunhke (1968). Influence of phosphate compounds on certain fungi and their preservative effect on fresh cherry fruit (*Prunus cerasus* L.). *Appl. Microbiol.* **11**, 430–435.

Pothakamury, U. R., A. Monsalvez-González, G. V. Barbosa-Cánovas and B. G. Swanson (1995). High voltage pulsed electric field inactivation of *Bacillus subtilis* and *Lactobacillus delbrueckii*. *Rev. Españ. Ciencia y Tecnol. Aliment*. **35**, 101–107.

Pothakamury, U. R., H. Vega, Q. Zhang *et al*. (1997). Effect of growth stage and temperature on inactivation of *E. coli* by pulsed electric fields. *J. Food Prot*. **59**, 1167–1171.

Potthof, A., W. Serve and P. Macharis (1997). The cleaning revolution. *Dairy Ind. Int*. **64**, 25, 27, 29.

Powers, J. J., D. Somaatmadja, D. E. Pratt and M. K. Hamdy (1960). Anthocyanins. II. Action of anthocyanin pigments and related compounds on the growth of certain microorganisms. *Food Technol*. **14**, 626–632.

Preita, J., M. A. Moreno, S. Díaz *et al*. (1994). Survey of patulin in apple juice and children's apple food by the diphasic dialysis membrane procedure. *J. Agric. Food Chem*. **42**, 1701–1703.

Qin, B., U. R. Pothakamury, G. V. Barbosa-Cánovas and B. G. Swanson (1995). Food pasteurization using high intensity pulsed electric fields. *Food Technol*. **49**, 55–60.

Qin, B., G. V. Barbosa-Cánovas, B. G. Swanson *et al*. (1998). Inactivating microorganisms using a pulsed electric field continuous treatment system. *IEEE Trans. Ind. Applic*. **34**, 43–50.

Qin, B., U. R. Pothakamury, G. V. Barbosa-Cánovas and B. G. Swanson (1996). Nonthermal pasteurization of liquid foods using high-intensity pulsed electric fields. *Crit. Rev. Food Sci. Nutr*. **36**, 603–627.

Radomyski, T., E. A. Murano, D. G. Olson and P. S. Murano (1994). Elimination of pathogens of significance in food by low-dose irradiation: a review. *J. Food Prot*. **57**, 73–86.

Rahman, M. S. (1999). Light and sound in food preservation. In: M. S. Rahman (ed.), *Handbook of Food Preservation*, pp. 673–686. Marcel Dekker, New York, NY.

Rajnicek, A. M., C. D. McCaig and N. A. R. Gow (1994). Electric field induced growth of *Enterobacter cloacae*, *Escherichia coli* and *Bacillus subtilis* cells: implications for mechanisms of galvanotropism and bacterial growth. *J. Bacteriol*. **176**, 702–713.

Raso, J., A. Palop, R. Pagán and S. Condon (1998a). Inactivation of *Bacillus subtilis* spores by combining ultrasonic waves under pressure and mild heat treatment. *J. Appl. Microbiol*. **85**, 849–854.

Raso, J., R. Pagán, S. Condon and F. J. Sala (1998b). Influence of temperature and pressure on the lethality of ultrasound. *Appl. Environ. Microbiol*. **64**, 465–471.

Raso, J., M. M. Góngora, M. L. Calderón *et al*. (1999). *Resistant Micro-organisms to High Intensity Pulsed Electric Field Pasteurization of Raw Skim Milk*. IFT Annual Meeting Technical Program. Institute of Food Technologists, Chicago.

Ratkowsky, D. A. and T. Ross (1995). Modeling the bacterial growth/no growth interface. *Lett. Appl. Microbiol*. **20**, 29–33.

Raunkjaer, K., P. H. Nielsen and T. H. Jacobsen (1997). Acetate removal in sewer biofilms under aerobic conditions. *Water Res*. **31**, 2727–2736.

Reis-Tassinari, A. and M. Landgraf (1997). Effect of microwave heating on survival of *Salmonella typhimurium* in artificially contaminated ready-to-eat foods. *J. Food Safety* **17**, 239–248.

Reiter, B. and B. G. Harnulv (1984). Lactoperoxidase antibacterial system: natural occurrence, biological functions and practical applications. *J. Food Prot.* **47**, 724–732.

Rheeder, J. P., W. F. O. Marasas, P. G. Thiel *et al.* (1992). *Fusarium moniliforme* and fumonisins in corn in relation to human esophageal cancer in Transkei. *Phytopathology* **82**, 353–357.

Rice, J. (1994). Sterilizing with light and electrical impulses. *Food Proc.* **July**, 66–70.

Rice, K. M. and M. D. Pierson (1982). Inhibition of *Salmonella* by sodium nitrite and potassium sorbate in frankfurters. *J. Food Sci.* **47**, 1615–1617.

Richards, M. and R. M. Macrae (1964). The significance of the use of hops in regard to the biological stability of beer. II. The development of resistance to hop resins by strains of lactobacilli. *J. Inst. Brew.* **70**, 484–488.

Riemann, H. (1963). Safe heat processing of canned, cured meats with regard to bacterial spores. *Food Technol.* **17**, 39–42, 45–56.

Riley, R. T. and W. P. Norred (1999). Mycotoxin prevention and decontamination - a case study on maize. *Food Nutr. Agric.* **23**, 25–32.

Rinker, K. D. and R. M. Kelly (1996). Growth physiology of the hyperthermophilic archaeon *Thermococcus litoralis*: development of a sulfur-free defined medium, characterization of an exopolysaccharide and evidence of biofilm formation. *Appl. Environ. Microbiol.* **62**, 4478–4485.

Rittman, B. E. (1989). Detachment from biofilms. In: W. G. Characklis and P. A. Wilderer (eds), *Structure and Function of Biofilms*, pp. 49–58. John Wiley & Sons, New York, NY.

Robach, M. C. (1980). Use of preservatives to control microorganisms in food. *Food Technol.* **34**, 81–84.

Roberson, E. B. and M. K. Firestone (1992). Relationship between desiccation and exopolysaccharide production in a soil *Pseudomonas* species. *Appl. Environ. Microbiol.* **58**, 1284–1291.

Roberts, T. A. and B. Jarvis (1983). Predictive modeling of food safety with particular reference to *Clostridium botulinum* in model cured meat systems. In: T. A. Roberts and F. A. Skinner (eds), *Food Microbiology, Advances and Prospects*, pp. 85–95. Academic Press, London.

Roberts, T. A., L. F. J. Woods, M. J. Payne and R. Cammack (1991). Nitrite. In: N. J. Russell and G. W. Gould (eds), *Food Preservatives*, pp. 89–111. Blackie & Son Ltd, Glasgow.

Roberts, T. A., A. C. Baird-Parker and R. B. Tompkin (eds) (1996). *Microorganisms in Foods, 5, Microbiological Specifications of Food Pathogens*. Blackie Academic and Professional, London.

Robins, M., T. Brocklehurst and P. Wilson (1994). Food structure and the growth of pathogenic bacteria. *Food Technol. Eur.*, **1994**, 31–36.

Robson, J. N. (1976). Some introductory thoughts on intermediate moisture foods. In: R. Davies, G. G. Birch and K. J. Parker (eds), *Intermediate Moisture Foods*, pp. 32–42. Applied Science Publishers, London.

Rogers, J., A. B. Dowsett and C. W. Keevil (1995). A paint incorporating silver to control mixed biofilms containing *Legionella pneumophila*. *J. Ind. Microbiol.* **15**, 377–382.

Rong-Yu, Z., Y.-W. Huang and L. R. Beuchat (1996). Quality changes during refrigerated storage of packaged shrimp and catfish fillets treated with sodium acetate, sodium lactate or propyl gallate. *J. Food Sci.* **61**, 241–244.

Ronner, A. B and A. C. L. Wong (1993). Biofilm development and sanitizer inactivation of *Listeria monocytogenes* and *Salmonella typhimurium* on stainless steel and Buna-N rubber. *J. Food Prot.* **56**, 750–758.

Rönner, U., U. Husmark and A. Henriksson (1991). Adhesion of *Bacillus* species in relation to hydrophobicity. *J. Appl. Bacteriol.* **69**, 550–556.

Rørvik, L. M. (2000). *Listeria monocytogenes* in the smoked salmon industry. *Intl J. Food Microbiol.* **62**, 183–190.

Rose, A. H. (1968). Physiology of microorganisms at low temperatures. *J. Appl. Bacteriol.* **31**, 1–11.

Rose, D. (1995). Advances and potential for aseptic processing. In: G. W. Gould (ed.), *New Methods of Food Preservation*, pp. 283–303. Blackie Academic & Professional, Glasgow.

Rosen, C.-G. (1972). Effects of microwaves on food and related materials. *Food Technol.* **26**, 36–40, 55.

Ross, A. I. V., M. W. Griffiths, G. S. Mittal and H. C. Deeth (2003). Combining nonthermal technologies to control foodborne microorganisms [Review]. *Intl J. Food Microbiol.* **89**, 125–138.

Rowan, N. J., S. J. MacGregor, J. G. Anderson *et al.* (2001). Inactivation of *Mycobacterium paratuberculosis* by pulsed electric fields. *Appl. Environ. Microbiol.* **67**, 2833–2836.

Rowe, J. J., J. M. Yabrough, J. B. Rake and R. G. Eagon (1979). Nitrite inhibition of aerobic bacteria. *Curr. Microbiol.* **2**, 51–58.

Russell, N. J. (1989). Functions of lipids: structural roles and membrane functions. In: C. Ratledge and S. C. Wilkinson (eds), *Microbial Lipids*, Vol. 2, pp. 279–365. Academic Press, London.

Russell, N. J. (2002). Bacterial membranes: the effects of chill storage and food processing. An overview. *Intl J. Food Microbiol.* **79**, 27–34.

Russell, N. J. and G. W. Gould (1991). Factors affecting growth and survival. In: N. J. Russell and G. W. Gould (eds), *Food Preservatives*, pp. 13–21. Blackie Academic and Professional, Glasgow.

Sala, F. J., J. Burgos, S. Condon *et al.* (1995). Effect of heat and ultrasound on microorganisms and enzymes. In: G. W. Gould (ed.), *New Methods of Food Preservation*, pp. 176–204. Blackie Academic & Professional, London.

Sasahara, K. and E. A. Zottola (1993). Biofilm formation by *Listeria monocytogenes* utilizes a primary colonizing microorganism in flowing systems. *J. Food Prot.* **56**, 1022–1028.

Sauer, C. J., J. Majkowski, S. Green and R. Eckel (1997). Foodborne illness outbreak associated with a semi-dry fermented sausage product. *J. Food Prot.* **60**, 1612–1617.

Scherba, G., R. M. Weigel and W. D. O'Brien (1991). Quantitative assessment of the germicidal efficacy of ultrasonic energy. *Appl. Environ. Microbiol.* **57**, 2079–2084.

Schlesinger, M., M. Ashburner and A. Tissieres (1982). *Heat Shock from Bacteria to Man*. Cold Spring Harbor Laboratory Press, Cold Spring Harbor, N.Y.

Schlievert, P. M., J. R. Deringer, M. H. Kim *et al.* (1992). Effect of glyceroyl monolaureate on bacterial growth and toxin production. *Antimicrob. Agents Chemother.* **36**, 626–631.

Schnepf, M. and W. E. Barbeau (1989). Survival of *Salmonella typhimurium* in roasting chickens cooked in a microwave, convention microwave and conventional electric oven. *J. Food Safety* **9**, 245–252.

Scott, D. (1988). Antimicrobial enzymes. *Food Biotechnol.* **2**, 119–132.

Scott, P. M. and G. A. Lawrence (1996). Determination of hydrolysed fumonisin B_1 in alkali processed corn foods. *Food Add. Contam.* **13**, 823–832.

Seiler, D. A. L and N. J. Russell (1991). Ethanol as food preservative. In: N. J. Russell and G. W. Gould (eds), *Food Preservatives*, pp. 153–171. Blackie & Son Ltd, London.

Shah, P. B., U. S. Shah and S. C. B. Siripurapu (1994). Ultraviolet irradiation and laminar air flow systems for clean air in dairy plants. *Indian Dairyman* **46**, 757–759.

Sharma, G. (1999). Ultraviolet light. In: R. K. Robinson, C. Batt and P. Patel (eds), *Encyclopedia of Food Microbiology*, 3rd edn, pp. 2208–2214. Academic Press, London.

Shay, B. J. and A. F. Egan (1987). The packaging of chilled red meats. *Food Technol. Aust.* **39**, 283–285.

Shelef, L. A. (1994). Antimicrobial effects of lactates: a review. *J. Food Prot.* **57**, 445–450.

Shelef, L. A. and J. A. Seiter (1993). Indirect antimicrobials. In: P. M. Davidson and A. L. Branen (eds), *Antimicrobials in Foods*, 2nd edn, pp. 539–570. Marcel Dekker, New York, NY.

Shelef, L. A. and Q. Yang (1991). Growth suppression of *Listeria monocytogenes* by lactates in broth, chicken and beef. *J. Food Prot.* **54**, 283–287.

Shibasaki, I. (1982). Food preservation with non-traditional antimicrobial agents. *J. Food Safety* **4**, 35–58.

Shigehisa, T., T. Ohmori, A. Saito *et al.* (1991). Effects of high hydrostatic pressure on characteristics of pork slurries and inactivation of microorganisms associated with meat and meat products. *Intl J. Food Microbiol.* **12**, 207–216.

Sholberg, P. L. and A. P. Gaunce (1995). Fumigation of fruit with acetic acid to prevent postharvest decay. *Hort. Sci.* **30**, 1271–1275.

Shtenberg, A. J. and A. D. Ignatev (1970). Toxicological evaluation of some combinations of food preservatives. *Food Cosmet. Toxicol.* **8**, 369–380.

Simpson, R. K., R. Whittington, R. G. Earnshaw and N. J. Russell (1999). Pulsed high electric field causes 'all or nothing' membrane damage in *Listeria monocytogenes* and *Salmonella typhimurium*, but membrane H^+-ATPase is not a primary target. *Intl J. Food Microbiol.* **48**, 1–10.

Singleton, V. L. and P. Esau (1969). *Phenolic Substances in Grapes and Wine and their Significance*. Academic Press, New York, NY.

Siragusa, G. R. and M. G. Johnson (1989). Inhibition of *Listeria monocytogenes* growth by the lactoperoxidase-thiocyanate-H_2O_2 antimicrobial system. *Appl. Environ. Microbiol.* **55**, 2802–2805.

Sloan, A. E., P. T. Waletzko and T. P Labuza (1976). Effect of order-of-mixing on a_w-lowering ability of food humectants. *J. Food Sci.* **41**, 536–540.

Smelt, J. P. P. M. (1998). Recent advances in the microbiology of high pressure processing. *Trends Food Sci. Technol.* **9**, 152–158.

Smith, K, G. S. Mittal and M. W. Griffiths (2002). Pasteurization of milk using pulsed electrical fields and antimicrobials. *J. Food Sci.* **67**, 2304–2308.

Sofos, J. N. (1984). Antimicrobial effects of sodium and other ions in food: a review. *J. Food Safety* **6**, 45–78.

Sofos, J. N. (1986). Use of phosphates in low sodium meat products. *Food Technol.* **40**, 52, 54, 58, 60, 62, 64, 66, 68–69.

Sofos, J. N. (1989). *Sorbate Food Preservatives*. CRC Press, Boca Raton, FL.

Sofos, J. N (1993). Current microbiological considerations in food preservation. *Intl J. Food Microbiol.* **19**, 87–108.

Sofos, J. N (1994). Antimicrobial agents. In: J. A. Maga and A. T. Tu (eds), *Handbook of Toxicology*, Vol. 1, *Food Additive Technology*, pp. 501–529. Marcel Dekker, New York, NY.

Sofos, J. N and F. F. Busta (1980). Alternatives to the use of nitrite as an antibotulinal agent. *Food Technol.* **34**, 244–251.

Sofos, J. N. and F. F. Busta (1992). Chemical food preservatives. In: A. D. Russell, W. B. Hugo and G. A. J. Ayliffe (eds), *Principles and Practice of Disinfection, Preservation and Sterilization*, 2nd edn, pp. 351–386. Blackwell Scientific Publications, London.

Sofos, J. N. and F. F. Busta (1993). Sorbic acid and sorbates. In: P. M. Davidson and A. L. Branen (ed.), *Antimicrobials in Foods*, pp. 49–94. Marcel Dekker, New York, NY.

Sofos, J. N., J. A. Maga and D. L. Boyle (1988). Effect of ether extracts from condensed wood smokes on the growth of *Aeromonas hydrophila* and *Staphylococcus aureus*. *J. Food Sci.* **53**, 1840–1843.

Sofos, J. N., L. R. Buechat, P. M. Davidson and E. A. Johnson (1998). *Naturally occurring Antimicrobials in Food*. Task Force Report No. 132. Council for Agricultural Science and Technology, Ames, IA.

Somaatmadja, D., J. J. Powers and M. K. Hamdy (1964). Anthocyanins. VI. Chelation studies on anthocyanins and other related compounds. *J. Food Sci.* **29**, 655–660.

Speers, J. G. S. and A. Gilmour (1985). The influence of milk and milk components on the attachment of bacteria to farm dairy equipment surfaces. *J. Appl. Bacteriol.* **59**, 325–332.

Stanley, P. M. (1983). Factors affecting the irreversible attachment of *Pseudomonas aeruginosa* to stainless steel. *Can. J. Microbiol.* **29**, 1493–1499.

Stephens, P. J., M. B. Cole and M. V. Jones (1994). Effect of heating rate on the thermal inactivation of *Listeria monocytogenes*. *J. Appl. Bacteriol.* **77**, 702–708.

Stumbo, C. R., K. S. Purohit, T. V. Ramakrishnan *et al.* (1983). *Handbook of Lethality Guides for Low-acid Canned Foods*, Vols 1 & 2. CRC Press, Boca Raton, FL.

Sutherland, I. W. (1983). Microbial exopolysaccharides – their role in microbial adhesion in aqueous systems. *CRC Crit. Rev. Microbiol.* **10**, 173–201.

Sutherland, J. P. and A. J. Bayliss (1994). Predictive modeling of growth of *Yersinia enterocolitica*: the effects of temperature, pH and sodium chloride. *Intl J. Food Microbiol.* **21**, 197–215.

Tagg, J. R., A. S. Dajani and L. W. Wannamaker (1976). Bacteriocins of Gram-positive bacteria. *Bacteriol. Rev.* **40**, 722–756.

Takeshita, K., J. Shibato, T. Sameshima *et al.* (2003). Damage of yeast cells induced by pulsed light irradiation. *Intl J. Food Microbiol.* **85**, 151–158.

Tang, Z. and S. Cenkowski (2000). Dehydration dynamics of potatoes in superheated steam and hot air. *Can. Agric. Eng.* **42**, 43–49.

Tang, Z. and S. Cenkowski (2001). Equilibrium moisture content of spent grains in superheated steam under atmospheric pressure. *Trans. ASAE* **44**, 1261–1264.

Tarr, H. L. A. (1941). The action of nitrites on bacteria. *J. Fish Res. Board Can.* **5**, 265–275.

Taylor, M. (2001). Microbiological hazards and their control: parasites. In: M. R. Adams and M. J. R. Nout (eds), *Fermentation and Food Safety*, pp. 175–217. Aspen Publishers, Inc., Gaithersburg, MD.

Terebiznik, M., R. Jagus, P. Cerutti *et al.* (2002). Inactivation of *Escherichia coli* by a combination of nisin, pulsed electric fields and water activity reduction by sodium chloride. *J. Food Prot.* **65**, 1253–1258.

Tewari, G., D. S. Jayas and R. A. Holley (1999). High pressure processing of foods: an overview. *Sci. Aliments* **19**, 619–661.

Tiina, M. and M. Sandholm (1989). Antibacterial effect of the glucose oxidase–glucose system on food poisoning organisms. *Intl J. Food Microbiol.* **8**, 165–174.

Tompkin, R. B. (1993). Nitrite. In: P. M. Davidson and A. L. Branen (eds), *Antimicrobials in Foods*, 2nd edn, pp. 191–262. Marcel Dekker, New York, NY.

Tompkin, R. B. (2000). Managing *Listeria monocytogenes* in the food processing environment. *Can. Meat. Sci. Assoc. News*, **December**, 4–8.

Tranter, H. S. (1994). Lysozyme, ovotransferrin and avidin. In: V. M. Dillon and R. G. Board (eds), *Natural Antimicrobial Systems and Food Preservation*, pp. 65–97. CABI Publishing, Wallingford.

Tsai, S. and C. Chou (1996). Injury, inhibition and inactivation of *Escherichia coli* O157:H7 by potassium sorbate and sodium nitrite as affected by pH and temperature. *J. Sci. Food Agric.* **71**, 10–12.

Tsigarida, E., P. Skandamis and G.-J. E. Nychas (2000). Behaviour of *Listeria monocytogenes* and autochthonous flora on meat stored under aerobic, vacuum and modified atmosphere packaging conditions with or without the presence of oregano essential oil at 5 °C. *J. Appl. Microbiol.* **89**, 901–909.

Tsuchido, T. and M. Takano (1988). Sensitization by heat treatment of *Escherichia coli* K-12 cells to hydrophobic antibacterial compounds. *Antimicrob. Agents Chemother.* **32**, 1680–1683.

Urakami, I., M. Yoshikawa, J. Udagawa and T. Sugahara (1997). Ultraviolet light disinfection of natural mineral water. *J. Antibact. Antifung. Agents Jpn.* **25**, 697–701.

Utsonomiya, Y. and Y. Kosaka (1979. Application of supersonic waves to foods. *J. Faculty Appl. Biol. Sci., Hiroshima Uni.* **18**, 225–231.

Uyttendale, M., I. Taverniers and J. Debevere (2001). Effect of stress induced by suboptimal growth factors on survival of *Escherichia coli* O157:H7. *Intl J. Food Microbiol.* **66**, 31–37.

Valenta, H. and M. Goll (1996). Determination of ochratoxin A in regional samples of cow's milk from Germany. *Food Add. Contam.* **13**, 669–676.

Vanbelle, M., E. Teller and M. Focant (1989). Probiotics in animal nutrition: a review. *Arch. Animal Nutr.* **7**, 543–567.

van Kraijj, C., W. M. de Vos, R. J. Siezen and O. P Kuipers (1999). Lantibiotics: biosynthesis, mode of action and applications. *Nat. Prod. Rep.* **16**, 575–587.

Vanos, V and O. Bindschedler (1985). The microbiology of instant coffee. *Food Microbiol.* **4**, 19–33.

Vanos, V., S. Hofstaetter and L. Cox (1987). The microbiology of instant tea. *Food Microbiol.* **4**, 187–197.

Veeramuthu, G. J., J. F. Price, C. E. Davis *et al.* (1998). Thermal inactivation of *Escherichia coli* O157:H7, *Salmonella senftenberg* and enzymes with potential as time-temperature indicators in ground turkey thigh meat. *J. Food Prot.* **61**, 171–175.

Vega, H., U. R. Pothakamury, F. J. Chang *et al.* (1996). High voltage pulsed electric fields, pH, ionic strength and the inactivation of *E. coli. Food Res. Intl* **29**, 117–121.

Vergeres, P. and L. Blaser (1992. Amikacin, ceftazidine and flucloxacillin against suspended and adherent *Pseudomonas aeruginosa* and *Staphylococcus epidermidis* in an *in vitro* model of infection. *J. Infect. Dis.* **165**, 281–289.

Wakabayashi, H., W. Bellamy, M. Takase and M. Tomita (1992). Inactivation of *Listeria monocytogenes* by lactoferricin, a potent antimicrobial peptide isolated from cow's milk. *J. Food Prot.* **55**, 238–240.

Walker, J. R. L. (1994). Antimicrobial compounds in food plants. In: V. M. Dillon and R. G. Board (eds), *Natural Antimicrobial Systems and Food Preservation*, pp. 181–204. CABI Publishing, Wallingford.

Wang, L.-L. and E. A. Johnson (1997). Control of *Listeria monocytogenes* by monoglycerides in foods. *J. Food Prot.* **60**, 131–138.

Wederquist, H. J., J. N. Sofos and G. R. Schmidt (1994). *Listeria monocytogenes* inhibition in refrigerated vacuum packaged turkey bologna by chemical additives. *J. Food Sci.* **59**, 498–500.

Wederquist, H. J., J. N. Sofos and G. R. Schmidt (1995). Culture media comparison for the enumeration of *Listeria monocytogenes* in refrigerated vacuum packed turkey bologna with chemical additives. *Lebensm. Wiss. Technol.* **28**, 455–461.

Weinberg, E. D. (1978). Iron infection. *Microbiol. Rev.* **42**, 45–46.

Wekhof, A. (1991). Treatment of contaminated water, air and soil with UV flashlamps. *Environ. Progr.* **10**, 241–247.

Wekhof, A. (2000). *Pharma+Food* (January issue). Hüttig Verlag, Heidelberg.

Whiting, R. C. (1997). Microbial data base building: what have we learned? *Food Technol.* **51**, 82–86.

Whiting, R. C. and R. L. Buchanan (1997). Predictive modeling. In: M. P. Doyle, L. R. Beuchat and T. J. Montville (eds), *Food Microbiology: Fundamentals and Frontiers*, pp. 728–739. ASM Press, Washington, DC.

Whiting, R. C. and R. L. Buchanan (2002). Predictive modeling and risk assessment. In: M. P. Doyle, L. R. Beuchat and T. J. Montville (eds), *Food Microbiology: Fundamentals and Frontiers*, 2nd edn, pp. 813–831. ASM Press, Washington, DC.

WHO (World Health Organization) (1994). Ultraviolet radiation. Environmental Health Criteria 160, World Health Organization, Geneva.

Widmer, A. F., R. Frei, Z. Rajacic and W. Zimmerli (1991). Correlation between *in vivo* and *in vitro* efficacy of antimicrobial agents against foreign body infections. *Antimicrob. Agents Chemother*. **35**, 741–746.

Wills, E. D. (1956). Enzyme inhibition by allicin, the active principle of garlic. *Biochem. J.* **63**, 514–520.

Wirtanen, G. and T. Mattila-Sandholm (1992). Effect of the growth phase of foodborne biofilms on their resistance to a chlorine sanitizer. Part II. *Lebensm. Wiss. Technol*. **25**, 50–54.

Wolfaardt, G. M., J. R. Lawrence, R. D. Robarts *et al.* (1994). Multicellular organization in a degradative biofilm community. *Appl. Environ. Microbiol*. **60**, 434–446.

Woods, L. F. J. and J. M. Wood (1982). The effect of nitrite inhibition on the metabolism of *Clostridium botulinum. J. Appl. Bacteriol*. **52**, 109–110.

Woods, L. F. J., J. M. Wood and P. A. Gibbs (1981). The involvement of nitric oxide in the inhibition of the phosphoroclastic system in *Clostridium sporogenes* by sodium nitrite. *J. Gen. Microbiol*. **125**, 399–406.

Wright, J. R., S. S. Sumner, C. R. Hackney *et al.* (1999). Efficacy of ultraviolet light for reducing *Escherichia coli* O157:H7 in unpasteurized apple cider. *J. Food Prot*. **63**, 563–567.

Wu, Y. C., B. Kimura and T. Fujii (1999). Fate of selected food-borne pathogens during the fermentation of squid shiokara. *J. Food Hyg. Soc. Jpn* **40**, 206–210.

Wyman, D. P. (1996). Understanding active chlorine chemistry. *Food Qual*. **2**, 77–82.

Yousef, A. E and E. H. Marth (1989). Stability and degradation of aflatoxin M_1. In: H. P. Van Egmond (ed.), *Mycotoxins in Dairy Products*, pp. 127–161. Elsevier Applied Science, London.

Yu, S. L. and C. C. Chou (1987). Survival of some food poisoning bacteria in pickled vegetables stored at different temperatures. *Food Sci. China* **14**, 296–305.

Zaika, L. L. and A. H. Kim (1993). Effect of sodium polyphosphates on growth of *Listeria monocytogenes. J. Food Prot*. **56**, 577–580.

Zajak, M., L. Bjorck and O. Claasson (1981). Antibacterial effect of the lactoperoxidase system against *Bacillus cereus* (from milk). *Milchwissenschaft* **38**, 417–418.

Zhang, T. C. and P. L. Bishop (1994). Density, porosity and pore structure of biofilms. *Water Res*. **28**, 2267–2277.

Zimmer, J. L. and R. M. Slawson (2002). Potential repair of *Escherichia coli* DNA following exposure to UV radiation from both medium- and low-pressure UV sources used in drinking water treatment. *Appl. Environ. Microbiol*. **68**, 3293–3299.

Zwietering, M. H., I. Jongenburger, F. M. Rombouts and K. van't Riet (1991). Modeling of the bacterial growth curve. *Appl. Environ. Microbiol*. **56**, 1875–1881.

Food safety

19

Ivone Delazari, Hans P. Riemann and Maha Hajmeer

1 Overview

It has become increasingly common to address food safety problems along a 'farm-to-fork' or 'stable-to-table' continuum. 'Harvest' of vegetables, milk and eggs occurs on the farm; whereas the analogous event for meat occurs at the abattoir. Seafoods may be harvested at sea or on farms. Although it is certainly true that some food-borne disease agents originate at the farm level, it also appears that most measures that guarantee consumer safety become available postharvest. Therefore, this chapter begins with a brief survey of available preharvest food safety measures but devotes most attention to postharvest interventions. Some redundancy of effort may result, but the emphasis on postharvest food safety is inspired by our perception that measures applied in this phase are likely to cost less and accomplish more.

2 Preharvest food safety

2.1 Introduction

Harvest means 'to gather a crop', and *preharvest* food safety thus relates to plants in the field; the corresponding term related to animals is 'animal production food safety'. For the sake of simplicity, the term *preharvest* will be applied in this chapter to animals as well as to plants.

Preharvest food safety applies to the growing, packaging and marketing of raw fruits and vegetables, and to animal production on farms where animals and animal

Foodborne Infections and Intoxications 3e
ISBN-10: 0-12-588365-X
ISBN-13: 978-012-588365-8

products such as eggs and milk are produced; it also applies to fishing and fish farming. In the food-processing industry preharvest food safety is used to secure acceptable raw material, but critical control points (CCPs) are more often associated with the processing than with raw material treatments. However, the food service industry and consumers receive large amounts of unprocessed foods; and while it cannot be expected that these groups will develop hazard analysis-critical control points (HACCP) plans, they should at least stay informed about what CCPs mean, since they will perform the terminal food preparation in the kitchen.

Safety measures for fruits and vegetables have received much attention recently because an increasing proportion of foodborne disease outbreaks has been associated with these products. This increase may be due to several factors, such as an increased consumption of raw vegetables, import of products that have been produced under suboptimal safety conditions, packaging methods that extend shelf-life but do not improve safety, application of contaminated manure and water, and marginal acidity of fruit juices.

The measures that can be taken are sometimes straightforward, such as applying a critical control point – e.g. pasteurization of juices. In most cases, solutions are less clear. Good agricultural practices (GAPs) are recommended to prevent contamination from various sources, but more specific and effective procedures still remain to be identified.

Efforts to keep food animals healthy are not new; farmers have always tried to keep their animals well because healthy animals are better producers and unhealthy animals may be condemned at inspection. The problem is that colonization of animals with a variety of human pathogens does not manifest itself as an animal health problem, and only laboratory tests can reveal the presence of zoonotic pathogens. The tactics available in animal production food safety are the same as those applied in animal disease control and eradication; and they include detection and slaughter, quarantine, biosecurity and farm hygiene. Vaccination plays a minor role, and mass treatment such as competitive exclusion of pathogens has found only limited commercial use.

Quality assurance plans have been developed for different animal production systems; they are based on GAPs that are dictated by common sense, but (like those for vegetable and fruit production) they fail to identify effective interventions. Sometimes there is no clear distinction made between quality and safety.

Some coordinated programs supported by public agencies have sufficient support and manpower to maintain the effective laboratory-based surveillance needed for control based on detection and slaughter. Collection of appropriate surveillance data in such systems also makes it possible to conduct epidemiological studies to identify risk factors associated with the presence of foodborne pathogens. This in turn may generate more effective intervention procedures.

In the case of seafood preharvest food safety, efforts are limited to avoiding harvest from unsafe waters. Fish farming faces similar food safety control issues to other animal production systems. It should be made clear that preharvest and animal production food safety efforts are not substitutes for postharvest food safety. Some pathogens, such as *C. perfringens* and *S. aureus*, are ubiquitous and can not be controlled by preharvest intervention.

2.2 Historical aspects

A few zoonotic foodborne diseases have historically been controlled by preharvest activities. Prime examples are brucellosis and tuberculosis, both of which cause animal production losses and have a narrow host range that makes them amenable to eradication. Still, the direct protection of human beings was accomplished by postharvest activities such as meat inspection and cooking of meat, and pasteurization of milk in the case of tuberculosis. Certain zoonotic parasite infections, such as trichinosis and taeniasis, have traditionally been controlled by combinations of preharvest and postharvest activities. These limited and highly focused activities have been successful, partially because of the limited host range of the pathogens, but it is clear that while preharvest activities help to reduce incidence/prevalence of infections in food animals, it is the postharvest activities that most directly protect the consumers and will do so irrespective of preharvest activities.

Hazards in food can be physical (e.g. metal, glass, bone), chemical (e.g. drugs and pesticide residues) and microbial (bacterial and non-bacterial), depending on the nature of the product. Microbial hazards, specifically bacterial, have been of a greater public health concern compared to the physical and chemical ones. Even under the microbial category the focus seems to be on bacterial pathogens, although problems have occurred (and still do) with non-bacterial agents such as viruses, parasites, protozoa, and yeasts and molds. The massive waterborne cryptosporidiosis outbreak of 1993 is one example of a non-bacterial problem (Corso *et al.* 2003). About 403 000 illnesses were reported and 100 AIDS patients died in that incident.

Most enteric bacteria are harmless, but some are pathogenic and can pose a public health threat. Enteric pathogens, such as *Salmonella* and *Campylobacter jejuni*, can affect both animals and humans. *Shigella* is a human-specific microorganism. Enterohemorrhagic *Escherichia coli* (e.g. *E. coli* O157:H7) will colonize the intestinal tract of cattle with no clinical signs of the disease, but it can be deadly to humans. Until *E. coli* O157:H7 was discovered, pathogenic *E. coli* was thought to be human-specific. Enteric pathogens have a broad range of hosts, and are difficult to eradicate. Preventive measures can be implemented preharvest, but total control is difficult in such an open environment. Additionally, a better understanding of the ecology of these organisms in the farm environment is needed. This understanding will facilitate control of the pathogens.

Enteric pathogens are mostly transmitted via the fecal–oral route. This can be through contaminated food or water. Secondary transmission, such as person-to-person contact, is also possible. Most human cases of enteritis are foodborne. Certain food animals serve as reservoirs of pathogens, such as *E. coli* O157:H7 in cattle and *Salmonella* in cattle, pigs and poultry, while *Campylobacter* is most often associated with poultry. Animals colonized with pathogens might be shedding the microorganisms for some time (possibly weeks to months); humans are shedders too. Young animals are more prone to infections, are more likely to be ill and tend to shed more pathogens, perhaps because a competing intestinal flora has not yet been fully established.

The intestine is a very complex ecosystem that includes host defense mechanisms and the normal intestinal microflora. Numerous microbial species are present, and many are difficult to culture. Knowledge of the microbial flora in the intestinal tract is limited because of the technical problems associated with its study. The study of pathogens in isolation, the autecology of pathogens, has limited value because it does not identify the nature of interactions with other species.

It should be pointed out again that the enteric pathogens causing foodborne infections in man are not natural inhabitants of the intestinal tract of food animals. The animal host may not exhibit any clinical disease, but will clear the infection in much the same way as humans do.

2.3 Contemporary problems

Many human pathogens are zoonotic and multihost. There is thus a tremendous opportunity for interspecies transmission among hosts, and in many cases the transmission is fecal-to-oral, often carried by food. It must be realized that many of these infections do not cause disease in their animal hosts; therefore they are difficult to detect and control. Some foodborne disease agents have a broad range of hosts and several reservoirs; some are ubiquitous and do not require a living host, such as *B. cereus, S. aureus*, and *C. botulinum*; others, such as *Shigella* and some viruses, are human-specific. The ubiquitous and human-specific agents are as likely to cause postharvest as preharvest contamination/infection, and cannot be controlled by preharvest food safety actions. They must be targeted by sanitation standard operating procedures (SSOPs) in food preparation establishments, including homes, or through the ultimate critical control point, which is adequate heating just before consumption.

2.3.1 Animals

Although many of the foodborne disease agents are zoonotic, they seldom cause disease in animals and there is therefore no incentive for control from an animal health point of view. Pets have similar prevalence of infection with *Salmonella* and other foodborne agents to food animals. Pets may be an important but largely unrecognized source of foodborne pathogens; by nature, food contamination from pets is postharvest rather than preharvest. Animal products are seldom consumed raw, and the ultimate critical control point (CCP) is generally heat treatment; the risk of subsequent contamination must be controlled by SSOP. Of special interest are zoonotic agents with a low infectious dose, such as *E. coli* O157:H7, because control by detection and slaughter is not effective.

2.3.2 Plants

Plants do not form a part of the zoonotic cycle, but may be contaminated with zoonotic agents from animals and humans. Plants are often consumed raw (fruits, fruit juices, lettuce, sprouts and other vegetables); in many cases, where heat treatment is not acceptable for quality reasons, the only limited steps that can be taken are at the preharvest stage. These steps consist of attempts to reduce contamination during growing, harvesting and packaging. However, the methods are not very effective because so little is known about the ecology of the agents involved.

2.4 Sources of infection/contamination

2.4.1 Animals

Control measures for zoonotic agents with a narrow host range, such as *Brucella* spp., *Mycobacterium bovis*, and *Taenia saginata* and *solium*, are well established and have been largely successful in industrialized countries. This, however, is not the case with multihost agents such as *Salmonella, E. coli* strains and others; these agents colonize many animal species, often without causing overt disease. The agents spread among animals, often through the fecal–oral route. Adult animals may be fairly resistant to infection, but this is not the case with young animals. Young chicks may become infected after exposure to 1–5 CFU of *Salmonella* (Milner and Schaeffer, 1952; Pivnick and Nurmi, 1982; Schleifer *et al.*, 1984), while chickens over 4 weeks of age were not infected by 10^2 CFU of *Salmonella* (Sadler *et al.*, 1969) and very few adult hens became infected after exposure to 10^4 CFU of *Salmonella* (Baskerville *et al.*, 1992; Gast and Beard, 1993). The normal intestinal flora of adults reduces the risk of infection with enteric pathogens (Draser and Hill, 1974); the mechanism of inhibition of pathogens is not known, but may be due to accumulation of fatty acids produced by the normal microflora or physical exclusion (Savage, 1983). The effect of adult flora was demonstrated by Nurmi and Rantala (1973), who showed that feeding adult intestinal flora to young chicks provided partial protection against *Salmonella*. This has been called 'competitive exclusion'. The spread of *Salmonella* is generally via the fecal–oral route, but hens may also become infected via aerosol. *Salmonella* is not a normal inhabitant of the intestinal tract, and chickens will generally stop shedding after a few weeks. Infection with *Campylobacter* shows a different pattern; the infection is rare in young chicks before 2–3 weeks of age, but after that the agent may spread rapidly to practically all birds, and they may remain infected for at least 3–4 weeks (National Advisory Committee, 1997). The spread of zoonotic agents among animals is often initiated by spread from adults to young; in birds this can occur through infected or surface-contaminated eggs, and the egg is a focus for preharvest intervention. Zoonotic agents like *Salmonella*, when present in sewage, may survive and have been shown to cause infection of domestic birds (Kinde *et al.*, 1996a, 1996b, 1997); the organisms readily multiply in highly diluted manure slurries, despite the presence of a dense population of indigenous microorganisms (Himathongkham *et al.*, 2000).

Survival of *Salmonella* and other enterics in the environment is reduced by lowering the water activity (Mallinson *et al.*, 1993; Himathongkham and Riemann, 1998), but at a water activity of 0.96–0.97 (relative humidity 96–97 %) they multiply when minute amounts of organic matter are present (Himathongkham *et al.*, 2000). Water left after cleaning a facility may support growth, and the effect of wet cleaning of poultry houses has been questioned (Gast and Beard, 1993; *Salmonella* Enteritidis Analysis Team, 1995). Wet cleaning of pig facilities was found to increase the risk of *Salmonella* infection in pigs (Stege *et al.*, 1997a). Increased ammonia concentration in the environment (e.g. in litter) reduces the survival of enteric pathogens (Fanelli *et al.*, 1970; Kumar *et al.*, 1971; Turnbull and Snoeyenbos, 1973; Opara *et al.*, 1992; Himathongkham and Riemann, 1998).

Interspecies transmission of zoonotic agents occurs frequently, and animals like rats and mice that co-inhabit animal facilities may be infected with *Salmonella* (Henzler and Opitz, 1992; Kinde *et al.*, 1996b). A number of risk factors for *Salmonella* or *E. coli* infections have been indicated based on epidemiological case-control studies; these include dry pig feed (versus wet, fermented feed) (Stege *et al.*, 1997a; Tielen *et al.*, 1997), large pig-herd size (Christensen and Carstensen, 1997), use of broad-spectrum antibiotics or transport of pigs between herds (Berends and Snijders, 1997), and feeding barley to feedlot cattle (Hancock *et al.*, 1997); none of these risk factors seems to have been confirmed by controlled intervention studies. Most of the indicated risk factors are of moderate magnitude, and elimination of any single one will not eliminate infection in animals.

Salmonella contamination of feed at the point of production has not been shown to be an important factor (Stege *et al.*, 1997b) but can be greatly reduced by heat treatment at controlled humidity levels (Himathongkham *et al.*, 1996); and the effect can be enhanced by adding propionic acid to the feed (Matlho *et al.*, 1997).

2.4.2 Plants

An increasing proportion of foodborne outbreaks is traced to contaminated fruits or vegetables (Tauxe, 1997); the pathogens involved include hepatitis A virus, *Salmonella, E. coli* O157:H7 and *Cyclospora*. The increase may be partially due to an increased consumption of raw vegetables and fruits. In the US, the *per capita* consumption of fruits and vegetables increased almost 50 % between 1976 and 1996 (Beuchat and Ryo, 1997), but it is not known how much is eaten in the raw state.

Fruits and vegetables in the field can be contaminated from a number of sources (Beuchat and Ryo, 1997), including feces from man and animals, contaminated 'green' or inadequately composted manure, contaminated irrigation water, contaminated water used for pesticide application, contaminated insects, and contaminated seeds. The relative importance of these sources is not known, but it would seem that conditions resulting in direct contact with feces or feces-contaminated water are among the most important. Little is known about the possibility of actual multiplication of foodborne pathogens under field conditions. Grains and other crops are also exposed to contamination with pathogens, but most of these crops undergo some processing before being eaten and therefore pose a minor risk.

Most vegetables have little or no killing effect on foodborne pathogens; organic acids in fruits have a detrimental effect on pathogens, but the acidity is not always sufficient, as evidenced by outbreaks of *E. coli* O157:H7 infections caused by unpasteurized apple juice (Tauxe, 1997).

There do not appear to be any surveillance data providing information about the prevalence of foodborne pathogens on raw fruits and vegetables, or about the relative importance of sources of these pathogens. The potential for contamination must be considered significant, and both pre- and postharvest safety actions are needed. Postharvest handling of fruits and vegetables would appear to lend itself to HACCP procedures, but, except for processing such as pasteurization of juice, CCPs for effective elimination of pathogens remain to be identified. Dipping vegetables (alfalfa sprouts) in various disinfectants may result in a 5–7 log cycle reduction of *Salmonella*, while dipping of fruit (cantaloupe) resulted in less than 1 log cycle reduction (Beuchat and Ryo, 1997). There is clearly a need for research and development in preharvest food safety.

Raw fruits and vegetables may in some cases permit multiplication of zoonotic agents, although little is known about the ability of pathogens to multiply in/on vegetables and fruits in the field; however, the initial contamination is directly or indirectly from hosts infected with such agents. The relative importance of sources of contamination, such as contaminated water, 'green' manure etc., is seldom known.

2.5 Food safety programs

2.5.1 Animals

For the purpose of eradication of foodborne disease agents from animal populations, the methods are the same as those developed over the years to eradicate livestock diseases. New methods have not replaced the old, but rather merely increased the number of combinations of methods available for control of infections in animals. Methods include the following:

- *Slaughter of infected and exposed animals.* This method has been successfully used in control of tuberculosis and brucellosis, and is also being used in control of, for example, *Salmonella* infections and paratuberculosis, especially in breeding flocks/herds. It can be an expensive measure to use on a mass scale.
- *Quarantine to prevent transport/import of infected animals.* Officially regulated quarantine has been used for many years to control a variety of livestock infections. Its main effect has probably been to discourage importation because it is expensive. Quarantine is not effective against diseases with long incubation periods and no reliable laboratory diagnostic procedures, such as bovine spongiform encephalopathy (BSE) and paratuberculosis. With the increase in free trade, quarantine is likely to be replaced with risk analysis and establishment of recognized disease-free zones. Quarantine is used to some extent by individual farmers to control *Salmonella* infections, for example in dairy herds, where replacements are kept in isolation until there is evidence that they are not infected; this must be based on laboratory testing.
- *Mass screening based on laboratory tests or direct tests on animals.* This tactic is used to assure freedom from defined infectious agents; it is based on laboratory testing or direct tests on animals, and played an indispensable role in the eradication of brucellosis and tuberculosis. A serological screening test is used in the Danish control program for *Salmonella* infection in pigs (Chapter 3). Mass screening is a measure that must be used in conjunction with other control methods.
- *Farm hygiene, including cleaning/disinfection after slaughter of infected/exposed animals.* This is a potentially important step in the control of enteric foodborne agents in farm animals. Farm hygiene is defined in very general terms; actions that are known or perceived to reduce the risk of presence of pathogens are recommended. These actions are seldom directed specifically against proven risk factors, because only few risk factors are known.

- *Vaccination*. Much effort has been spent to develop effective vaccines against foodborne disease agents that colonize farm animals. Some *Salmonella* vaccines have shown promise (Hassan and Cartis, 1990; Methner *et al.*, 1997), but none have so far been shown to prevent infection of herds/flocks. Experimental vaccination of calves against *Cryptosporidium parvum* has shown some beneficial effect, but failed to prevent infection (Harp and Goff, 1998).
- *Mass treatment*. There are no examples of mass treatments that will effectively cure animals of important foodborne disease agents.

Combinations of the methods mentioned above have been used to eradicate some foodborne zoonoses that have a fairly narrow reservoir and also cause animal losses. Brucellosis and tuberculosis in cattle have been eradicated in many countries by the slaughter of infected animals, quarantine, mass screening, farm hygiene and, to some extent, vaccination. More recently, attempts are being made by affected countries to eradicate BSE, which is suspected to be able to cause human disease (Chapter 11).

However, most foodborne zoonoses, such as *Salmonella, E. coli* and *Listeria*, have a broad host range and often do not cause overt disease in animals. These agents cannot be eradicated, but some countries manage to keep the infections low in selected livestock populations by applying the methods used in disease eradication (Halgaard *et al.*, 1997; Wierup and Woldtroll, 1988). It is not known whether these efforts are cost-effective. Nevertheless, with increased international food trade, the countries that have governmental programs in place may enjoy a competitive advantage.

The approaches to preharvest food safety in animal populations have mostly taken a different direction. Commodity groups and agencies have developed model plans to be implemented by individual farms; none of the plans aim at eradication of foodborne disease agents. The plans differ with respect to the requirements for documentation of food safety activities; some require monitoring and verification records, similar to HACCP plans, and auditing by personnel from public agencies, and sometimes limited laboratory testing is required. Few or none of these plans have been validated with respect to efficacy. Few if any would qualify as true HACCP plans, one difficulty being identification of CCPs where infection/contamination with foodborne pathogens can be eliminated.

The plans are generally called 'quality assurance plans' rather than food safety plans; this may cause confusion, since the plans have nothing to do with the real or perceived organoleptic quality of the final product. When no documentation is required, plans may best be termed 'Good Agricultural Practices' (GAPs).

In summary, infection of farm animals with zoonotic foodborne disease agents is common, multiple risk factors are involved, and no CCP that will eliminate infection/contamination has been identified. It appears that the best approach to control zoonotic foodborne pathogens at the farm level, short of eradication efforts, is a set of GAPs that take into account the risk factors for infection/contamination. There is a need for research to identify, measure and confirm, through controlled trials, the risk factors involved.

2.5.2 Plants

When dealing with plants, the primary preharvest food safety concern is with produce; such crops grow close to the ground, with increased risk of contamination, and are often eaten raw. Grains and other crops are also exposed to foodborne pathogens, but most of these crops undergo some processing before being eaten and therefore pose a minor risk. The first step in control of foodborne pathogens associated with produce is to develop and implement GAPs. This must be done before development of a HACCP system is attempted; without GAPs, there can be no HACCP. GAPs should include (but are not limited to) the following:

- *Toilet facilities*. Provision and use of toilet facilities and facilities for hand-washing or hand disinfection for field workers are critical. Testing workers for carriage of foodborne pathogens must be done at high frequency to be of any benefit.
- *Decontaminated manure*. Manure used for fertilizer should be decontaminated by composting, long-term (several weeks) stacking, or drying, with or without gassing with ammonia (Himathongkham and Riemann, 1998).
- *Clean water*. Clean, if necessary decontaminated, water should be used for irrigation and pesticide application.
- *Control of wildlife*. Wildlife droppings may contain foodborne pathogens (Beuchat and Ryo, 1997). This source of contamination is difficult to control. Elimination or reduction of wildlife populations would probably meet public resistance; fencing is effective only against some wildlife species and is expensive.
- *Insects*. Insects are not considered to be reservoirs for foodborne pathogens, but may serve as vectors; they probably represent a minor threat.
- *Seeds for sprouting*. In the production of salad sprouts, which is often done in hydroponics, reduction of pathogens by treatment of seeds is a possibility. Dipping in solutions of various disinfectants may result in a 5–7 log reduction of *Salmonella* (Beuchat and Ryo, 1997).

In summary, as for animal production food safety, there is a need for identification of risk factors and methods to control them. For the time being, plant food safety must be based on GAPs dictated by common sense.

Additional detailed information on GAPs can be found in section 5 of this chapter.

2.6 Examples of animal-production food safety programs

2.6.1 Salmonella control program in slaughter herds

The Danish control program for *Salmonella* in slaughter pig herds consists of serological testing and corrective actions in terms of management if the prevalence is above a certain level (Halgaard *et al.*, 1987; Chapter 3). Depopulation occurs only when *Salmonella* outbreaks in humans can be traced back to a farm. The monitoring of this state-industry cooperative program is based on ELISA testing for antibodies to *Salmonella* in meat juice. All Danish herds producing more than 100 slaughter pigs per year are tested at random, and about 800 000 tests are performed on 16 000 herds per year. The ELISA testing is highly automated, and the data are fed into a central computer system. Based on the test results, the herds are assigned to one of three levels: level I herds have no or few reactors, level II herds have a moderate number of reactors, and level III herds have a high number of reactors and are considered to be a risk. Level II and III herds are required to seek advice on how to reduce the

Salmonella sero-prevalence, and if no action is taken there is a penalty. Level III herds are slaughtered under tightened hygienic conditions. During the surveillance period, from 1995 to 1997, the proportion of herds at levels I, II and III were 93.7–95.7 %, 2.9– 4.4 % and 1.2–3.3 %, respectively.

2.6.2 Egg quality assurance plans

There are plans that are generally called quality assurance plans, although they have little to do with the organoleptic quality of products. It would seem that commodity groups hesitate to use the term 'safety'; however, the substitution of 'quality' for 'safety' may cause confusion.

The methods applied by the California Egg Quality Assurance Program are farm hygiene and exclusion of *Salmonella* by controlling the inputs (birds, feed) and limiting the rodent population. Environmental testing for *Salmonella*, once for each flock, has been included in the plan. A 1996 environmental survey indicated that approximately 98 % of the poultry houses were *S.* Enteritidis-negative (Riemann *et al.*, 1998); similar results were found in a 1999 survey (Castellan *et al.*, 2004). The Pennsylvania Egg Quality Assurance Plan differs from the California plan mainly by requiring more environmental testing.

The California and Pennsylvania plans deal with layers that do not begin egg production until 20–23 weeks of age. Lateral transmission is more important than vertical transmission. If, in spite of the control effort, salmonellae are introduced in numbers that cause infection, an epidemic may spread through the flock and result in shedding that will subside after a few weeks. There is presently no proven method to prevent such a spread.

Similar quality assurance plans have been developed for swine and beef production, and for other agricultural commodities. Documentation in terms of monitoring and verification reports is not always required, and such plans may best be called GAP plans.

3 Postharvest food safety

It has become common to divide food safety into preharvest and postharvest food safety. Such fragmentation is not really warranted, since both are parts of the same food safety system. However, since the technical procedures and the degree of perfection are quite different, we followed this convention and here address food safety measures that are applicable postharvest.

3.1 Introduction

Food safety in this context means application of methods to secure food that is safe to eat; it is a warning system that will alert those who are responsible to any potential problems. Traditionally, the safety of food, especially meat and poultry, was controlled by inspection of the final product. This limited detectable defects to those that can be seen, smelled or felt. As science and technology progressed, inspection was supplemented with chemical and microbiological tests of samples of food, and elaborate sampling schemes were developed. However, the inefficiency of sampling and testing made it impractical to rely on such control measures to assure a desired degree of food safety. Several factors contributed to the inefficiency. First, because of the necessarily limited number of samples that can be tested, attribute sampling plans could not guarantee a high level of microbiological safety of the food. Second, sampling plans assume a random distribution of microorganisms, but microbial dispersion may not be random in foods such as vegetables, meat, grains, etc.; the uncertainty about distribution patterns makes it difficult to obtain a representative sample. Third, sampling of end products for microbiological testing is expensive due to the destruction of large amounts of product. Finally, the delayed feedback from end-product testing makes it almost meaningless (Jarvis, 1989). Regulatory agencies still depend on sampling for making decisions, and advances in biotechnology that increase test specificity dramatically have contributed to the popularity of sampling and testing. However, the laws of probability that determine sampling outcomes have not changed.

Recently, regulatory agencies have adapted the Hazard Analysis and Critical Control Point system (HACCP) as a part of food inspection (seafood, fruit juice, meat and poultry in the US), the emphasis being on detecting safety hazards during the production or manufacturing process itself, rather than depending on inspection of the finished product. However, there is little documentation of how well HACCP functions in a regulatory setting. HACCP is not a stand-alone system. Other programs, such as Good Manufacturing Practices (GMP) and SSOP, are needed; in fact, there can be no HACCP without GMP.

HACCP shifts the focus of control to monitoring procedures, and is a preventive system for hazard control. In HACCP, testing is used only for verification purposes, and the need to test end products to assure the safety of individual lots is substantially reduced (Kvenberg and Schwalm, 2000). HACCP focuses on the essentials – the critical steps called CCPs – and it ignores non-essentials; these are covered by the HACCP prerequisites such as GMP and SSOP. HACCP is more cost-effective than product testing, and it has several other advantages – such as keeping the personnel working on the production line involved in food safety, and providing them with food safety education. HACCP was first implemented in the food-processing industry, where CCPs were fairly easy to identify. Application of HACCP was later mandated in meat and poultry slaughter plants, and in the seafood and fruit juice industries, and is also applied to some extent to the production of food crops and animals. These applications of HACCP are more questionable because effective interventions are difficult to achieve, and without effective intervention there cannot be CCPs or HACCP; GMP or GAP would apply better in such situations.

Food safety is the responsibility of the food production and processing industries, the distribution and retailing sectors, and the consumer. Regulatory agencies are also an important part of the food safety system but, because the creation of agencies and the way they perform often occurs piecemeal and is influenced by prevailing politics, their attention to some of the important components of the food system can be limited. The structure of traditional governmental agencies, divided as they are into separate departments controlled by an established hierarchy, may also represent an obstacle to a unifying concept that may be perceived as a threat to those occupying domains within the existing organizational structure.

Systems analysis has become an accepted approach to other human health problems (World Health Organization (WHO), 1976). 'Systems approach' is a generic term that covers the body of theory and practice, of which systems analysis forms one part. Basically, the systems approach is concerned with entities perceived as sets of interacting parts. A system is not only the sum of its parts; it also includes the interaction between the parts. It is important to know how interrelationships operate, how they are managed, and how information flows. Systems-analysis methods seek to define the relationships existing in a system and between it and other systems. It attempts to calculate the effects of altering the elements or the way in which they interact. System analysis permits the use of a common logic and vocabulary across organizational and disciplinary lines, and this is no small part of its usefulness. Systems analysis has not played an important role in food safety development. With the introduction of mandatory HACCP plans the industry's role in food safety has been formally recognized, but it is not clear how the industry (including transporters, retailers, public eating places, etc.) should interact with various governmental responsibilities toward the overall goal of food safety.

3.2 History and contemporary situation

The migration of rural people into the new environment of urban life resulted in a change in their basic needs and behavior. The concentration of both people and enterprises in cities created new challenges in terms of food supply and food quality (Aebi, 1981). Food had to be transported from production sites to cities, and for some products methods of preservation were needed. Also, food production and processing techniques evolved as eating habits underwent changes in many countries (Codex Alimentarius Commission, 1994). Nowadays, more people eat a larger number of meals in public places, and the time spent in the home kitchen has decreased as more and more people depended on pre-prepared foods requiring little or no preparation at home (Aebi, 1981).

The increase in demand for a variety of foods has the consequence that large populations become increasingly dependent on the commercial development and production of foods that are closer to their final stages of preparation. This has induced the food industry to increase the production of 'convenience' foods, which means ready-to-eat and/or minimally processed foods; and the industry is becoming more complex and more sophisticated. To meet the demand for food supplies that can reach geographical areas far from the place of production and processing, food formulations, transport and storage are continuously being modified. This is done to increase safety and shelf-life, improve functional quality, and reduce the time consumers spend on final preparation. As the complexities of food composition, processing and distribution have increased, the chances of mishaps (including foodborne diseases) has also increased, presenting new difficulties and challenges.

Providing food is an enormous and complex operation that includes food production in the field and at sea, industrial processing, storage, distribution, display at sales points, and preparation for consumption. Food production and processing for consumption, which were earlier controlled by the consumers themselves – notably by housewives – are now largely in the hands of the food industry, and those in homes and the food service industry who do the final preparation of meals may be less experienced than previously and less able to decide on safe handling and preparation methods.

The assurance of a safe and wholesome food supply requires the attention of many groups of people with different skills and interests, in both the public and the private sectors. Consumers expect the food they eat to be safe and suitable for consumption. Unsafe food can affect human health. Food spoilage is wasteful and costly, and adversely affects trade and consumer confidence (Codex Alimentarius Commission, 1994). Effective controls are therefore necessary; and a cooperative effort to ensure the safety of the food supply is clearly needed.

There are several essential partners who have to cooperate: producers, processors, distributors, vendors, consumers and health authorities. Each partner has a role to play and must also respect the efforts of others.

Health authorities should function as an impartial referee, enforcing the law regardless of the interest and intentions of each partner. Government authorities have three basic duties: to help protect consumers from illness or injury caused by food, to help maintain confidence in internationally marketed food by adoption of Codex Alimentarius' codes and guidelines (Codex Alimentarius Commission, 1994), and to provide education for food handlers and consumers so they recognize the critical roles they play in maintaining food safety. Governments also support food safety research in their own laboratories and through universities.

International organizations play an increasingly important role in food safety. These include the World Health Organization (WHO) and the Food and Agriculture Organization (FAO), which establish expert committees on food safety and sponsor Codex Alimentarius' involvement in setting food safety standards. The International Commission on Microbiological Specifications for Foods publishes informational material and makes technical recommendations on food safety.

Producers, sellers and food handlers must realize that consumers' health must be a top priority, and that strict food legislation and an efficient food control system are mandatory (Aebi, 1981). Food producers have the responsibility to maintain healthy flocks and herds, and to supply safe crops. Food sources should be managed according to GMP, sometimes called good production practices (GPP) or GAP, to avoid contamination reaching levels that would render end products potentially harmful to human health or unsuitable for human consumption (Codex Alimentarius Commission, 1994).

Food industry plant management has the responsibility to provide food of a high degree of safety. Also, the industry has to provide consumers with information that enables them to protect their food from contamination or spoilage. This is done through clear and detailed label instructions that tell the consumers how to store, handle and prepare the food correctly (Codex Alimentarius Commission, 1994). Labeling should include a complete listing of food constituents, including allergenic substances.

Companies that transport food have the duty to keep and maintain food environment conditions that are specified by the producers and processors, from the point of origin to the final destination. Additionally, it is imperative that the vehicles are used only for the transport of harmless products.

Food handlers, retailers and consumers have the duty to follow label instructions and keep food under the environmental conditions specified by producers and processors. Consumers and food handlers also have to apply good handling practices and hygienic procedures when preparing food for serving. Human contact with food is generally greater at the service and preparation levels than in food production and processing, and mandates good personal hygiene.

3.3 Postharvest food safety systems

3.3.1 HAZOP, HAZAN, and HACCP

The systematic process control represented by HACCP was developed by the food processing industry and later considered in other sectors of the food chain; it was based partially on the safety and quality systems already existing in other industries. Different types of safety plans have been developed. HAZOP stands for Hazard and Operability Studies, and was pioneered by the chemical industry as a method of identifying hazards and problems that could prevent efficient and safe operations (Kletz, 1992). Ideally, HAZOP is carried out during the design phase, before the construction of a plant, by a team where members can stimulate each other and build upon each others' ideas, to help ensure that a sufficient number of detailed points have been considered during design. The complexity of modern plants makes it difficult to see what can go wrong unless a HAZOP team goes through the design systematically. A HAZOP team will evaluate a large number of situations, but only very few of these undergo a full analysis; this may happen if the team suggests a very expensive solution to guard against an unlikely hazard. HAZOP soon spread to the oil industry, and later to food processing. In the food industry the emphasis has been on identifying ways in which contamination could occur, rather than on the operating and safety problems that are of concern to other industries.

HAZAN stands for Hazard Analysis; it is also called risk analysis, risk assessment, probabilistic risk assessment or quantitative risk assessment. Risk analysis is increasingly being applied in food safety (see Chapter 2). It is not itself a safety system, but is used in conjunction with various safety systems.

HACCP is a risk-management system that includes risk assessment. The term HACCP stands for Hazard Analysis-Critical Control Points; it was pioneered by the food processing industry and designed to avoid the occurrence of food safety problems. A hazard analysis serves as a basis for identifying critical control points (CCPs), which are operations that must be controlled to assure food safety. Critical limits (CLs) are established for each CCP; if the CLs are exceeded, defined corrective actions must be carried out. Monitoring documents certify that the processes are under control. Verification of the HACCP system is done by examination of monitoring documents, and may include laboratory testing. Validation is an evaluation of the soundness of the scientific basis of the HACCP system. A HACCP team creates the HACCP plan for each individual product line. The participation of the plant personnel in the creation of the HACCP plans also improves morale and cooperation, and serves as an educational tool.

3.3.2 The history of HACCP

At the end of the 1950s, the National Aeronautics and Space Administration (NASA) in the US, planning to send astronauts on space flights, presented the Pillsbury Company with the challenge of producing the first food to be eaten at zero gravity (Bauman, 1990). The scientists faced two problems. The first was the unknown behavior of food and food particles at zero gravity and their possible effect on the atmosphere, and the electronic and mechanical components of equipment. The group solved this problem by developing 'bite-sized food' with a coating that prevented shattering by holding particles firmly together. In this way, damage to the spacecraft environment, atmosphere and components was avoided (Baumann, 1990; Stevenson, 1995).

Food safety and the possibility of the astronauts becoming ill during their performance in space was the second problem, and the most difficult to solve (Bauman, 1990). The risk would increase, as NASA was planning for longer-lasting space missions. While food scientists were well acquainted with the pathogenic microorganisms associated with foods, the usual industry methods for food safety control were not developed enough to inspire total confidence in food supply for the space program (Vail, 1994). The challenge was to produce space foods completely free of pathogenic microorganisms and their toxins (Sperber, 1991; Stevenson, 1995). The existing quality control procedures and inspection techniques were deemed insufficient (Baumann, 1990), and it was decided to base food safety on preventive measures that would include control of the processing of food, as well as of raw materials, ingredients, and environment and employee involvement in processing. The 'Zero Defect Approach' used by NASA in the evaluation and testing of hardware for the Space Program was initially explored. The Zero Defect Approach was based on non-destructive tests applied to each component, and on having several back-up systems for each operation (Vail, 1994; Stevenson, 1995).

The 'Mode of Failure' concept was also considered. This concept had been developed by the US Army Natick Laboratories, and was based on the study and evaluation of each stage of a process – searching, questioning and determining, throughout the process, what might go wrong (a hazard), what to expect from possible failure and how it could happen (a hazard analysis), and, finally, where it would happen (CCPs). In that way, detailed knowledge about product and processes was acquired that would permit the prediction of what could potentially go wrong and how and where it might happen. Measures could be established by the identification of CCPs and evidence provided as to whether or not the process was under control; corrective measures could be developed in advance to re-establish normality in case of failure (Stevenson, 1990, 1995; Vail, 1994). The system, now called HACCP, presumes that product quality and safety can be designed and developed as an integral part of a process.

The HACCP concept was presented at the first conference for Food Protection in 1971, and it served as a basis for the Food and Drug Administration's (FDA) regulation for the canning of low-acid food. Additionally, the FDA applied the HACCP system to develop its investigation activities. HACCP soon attracted the attention of several large food production enterprises and other sectors of the food industry (Stevenson, 1990, 1995).

In the 1980s, official agencies in the US, debating microbial criteria for foods, concluded that the application of standards to end products had severe limitations as a means of preventing microbiological hazards (US National Advisory Committee on Microbiological Criteria for Foods, 1991). The agencies asked the National Academy of Sciences (NAS) to consider and develop principles and general concepts for microbiological criteria for foods in relation to public health (Adams, 1990; Stevenson, 1990). In 1985, an NAS subcommittee suggested the use of HACCP in official food protection programs and in food production (Adams, 1990; Stevenson, 1990). Based on this recommendation, the US National Advisory Committee on Microbiological Criteria for Foods (NACMCF) was created in 1987. This committee has been working on establishing the bases for the implementation of HACCP in food safety programs, including the standardization of the procedures to facilitate development and application of the system, whether used by official agencies or in the food industry (US National Advisory Committee on Microbiological Criteria for Foods and US Department of Agriculture, USDA, 1993).

It is commonly agreed today, among official agencies involved with food safety inspection and public health, that HACCP is the preferred tool for assuring food safety. There is also agreement that HACCP may improve the existing food control programs and the regulatory basis of food standards. In 1996, HACCP was mandated by USDA-FSIS in meat and poultry slaughter and processing facilities in an attempt to instigate more strict food safety measures. Later it was mandated by FDA as a control system for seafood and, recently, for juices. Currently, there is an interest in using HACCP to control the safety of live animal production as well as plant production (i.e. preharvest food safety).

The World Health Organization considers HACCP to be an efficient system for better control in food production. It advises that the HACCP concept should be incorporated into international and national food legislation as a means of improving the efficacy of food inspection (WHO, 1993). Also, the Codex Alimentarius Commission (CAC) has adopted the HACCP system as essential for ensuring food safety (Codex Alimentarius Commission, 1994).

The concept of 'failure analysis' is occasionally brought up as something beyond HACCP. It is clear that this concept is an integral part of HACCP, and when failures have occurred, it has sometimes been because not all the links in the food chain were covered by HACCP. This indicates the need for a systems approach to cover the total food safety system. For example, an outbreak of salmonellosis with ice cream as the vehicle occurred because a tank truck that had previously been used to transport raw liquid eggs had not been adequately cleaned and disinfected (Hennesy *et al.*, 1996). This can be classified as a systems failure. Systems failures have not been addressed well enough in food safety. Only large, integrated firms have the capability to ensure that HACCP operates throughout the chain through which their products pass.

Care should be exercised in the use of HACCP as a regulatory tool. The effectiveness of HACCP depends on the commitment of management to fully support the system, and the willingness and enthusiasm of plant personnel in planning and operating HACCP. If these voluntary aspects are lost because of an imposed 'command and control' situation, the system is doomed to failure. It should also be recognized that the effectiveness of CCPs differs widely among HACCP plans; for example, the present seafood HACCP is far less efficient than HACCP applied in the food canning industry.

3.3.3 HACCP principles

The intent of the HACCP system is to prevent potential hazards that could affect food safety. It provides a more specific approach to the control of hazards than that provided by traditional inspection and quality control procedures (WHO, 1980). The purpose of HACCP is to assure food safety through the development, implementation and effective management of a hazard control program (Vail, 1994). It is based on seven functions also known as the seven principles of the HACCP system.

Principle 1: Conduct a hazard analysis

The aim of a hazard analysis is to identify hazards and associated risks, and develop appropriate preventive measures. Hazards and associated risks must be assessed at field/sea level, at processing, storage, distribution, retail sale, and preparation for eating. This provides a broad understanding and clear comprehension of a food system and its components, and is the basis for evaluation of risks from the contaminants throughout the entire food chain.

- *The hazard concept*. A hazard is a biological, chemical or physical agent in, or condition of, food with the potential to cause an adverse health effect on the human being (Codex Alimentarius Commission, 1997). A hazard must be of such a nature that its prevention, elimination or reduction is crucial for food safety. Hazards that are not significant are controlled by hygienic procedures or by GMPs. Hazards vary from one establishment to another, even when the same kind of food is produced, because of differences in the sources of raw materials and ingredients, formulation, processing equipment and methods, storage conditions, and levels of technology that can affect hazards. For these reasons, a hazard analysis must be carried out in each plant on all existing products and on any new product that a processor intends to manufacture. Hazard assessment is based on a combination of experience, epidemiological data and information in the technical literature.
- *Classification of hazards*. Hazards can be classified into three broad groups: chemical, physical and microbiological. The potential risk of each hazard is assessed based on its likelihood of occurrence and its severity.

Foods are made up of chemicals, and many chemicals can be toxic depending upon dosage levels (Rhodehamel, 1992). Chemical hazards may cause foodborne diseases, but the frequency is small compared to those caused by microbiological hazards. As such microbiological hazards are often a priority in food safety decisions, although chemical hazards remain significant safety issues for food processors. Chemical hazards consist of the presence of a chemical agent in foods that can cause harm to the consumer health. This includes naturally occurring substances, such as allergens and toxins, and contamination of food with chemical substances, their residues or their degradation products (e.g. preservatives, flavorings, pesticides, antibiotics, sanitizers, paints and solvents) at levels that can harm the consumer. This relates to the production, manufacturing, transportation, storage, display and usage of food products. Many chemicals are routinely used in food production and processing, and they do not cause any harm if used properly; however, some can cause severe illness when improperly used, and the hazard analysis must consider the possible use of allowed chemicals in an unsafe manner. Chemicals may be present inadvertently in animal and plant foods. Their possible presence in food creates a need for the establishment of tolerance levels in order to protect public health (Rhodehamel, 1992; Katsuyama, 1995a). There are three main sources of chemical contaminants in food plants: chemicals entering the plant with the raw materials or ingredients, chemicals used in the plant to support manufacturing processes, and chemicals intended for cleaning and sanitation purposes. Potential sources of chemical hazards must be addressed and controlled within the framework of a food company's HACCP. Chemical hazards may be classified as shown in Table 19.1.

Physical hazards are physical agents that are present in a food and can harm the consumer. Examples include fragments of stones, insect fragments, wood splinters, pieces of glass, hard fragments of plastics, carton boards, metal and others. Physical hazards characteristically affect only one or a few persons, since they are generally present in only a few packages. Physical hazards reflect the level of control in food-processing operations. A plant that has a large number of recorded occurrences of foreign material in its products normally would need to improve its GMPs and provide training for the personnel to develop a better sense of responsibility. Differentiation should be made between foreign bodies capable of physically injuring a consumer and those that present only an esthetic problem. In general, physical hazards do not cause a disease but an injury – such as a broken tooth or a cut in the

Table 19.1 Chemical Hazards Classification (World Health Organization, 1998. Guidance of regulatory assessment of HACCP. Report of a joint FAO/WHO Consultation on the Role of Governmental Agencies in Assessing HACCP. Geneva 2–6 June. WHO Food Safety Unit.)

I Hazards associated with poor practices in primary production
  1. In vegetable foods
     Pesticide residues, their metabolites, heavy metals
     Mycotoxins, especially in grains
  2. In foods of animal origin
     Pesticide residues, their metabolites, heavy metals, and veterinary drugs
II. Hazards associated with poor practices in food processing
  1. By intentional but incorrect addition: colorants and additives (nitrite, sulfite, etc)
  2. By unintentional addition: detergents, disinfectants, and pesticides
III. Hazards naturally associated with certain foods
  1. Toxic, naturally occurring substances
     In seafood: ciguatera, tetrodotoxin, and saxitoxins
     In plant foods: cyanogenic plants, myristicin in nuts, and solanine in potatoes
     In grains: mycotoxins
  2. Allergenic, naturally occurring substances
     Foods of animal origin: cow.s milk and dairy products, eggs, mollusks and crustaceans
     Plant foods: soybeans, wheat, nuts and peanuts
IV. Hazards associated with environmental pollution
  Heavy metals and chemical residues, including pesticide residues and dioxins.
V. Associated with sabotage or terrorism
  Unpredictable; the dimension of hazards may be enormous and the hazards may be of great severity

Table 19.2 Physical hazards classification

I. Associated with agricultural crops and storage
   Insect fragments in grains and their products[a]

II. Associated with incorrect/poor practices
   1. In agricultural foods and incoming material/produce: feathers, stones, rodent hairs, etc.
   2. In the production of foods of animal origin: hypodermic needles used in veterinary treatment, etc.
   3. In food processing: sharp metal fragments, glass, etc.

III. Associated with poor maintenance
   1. Poor maintenance of buildings and facilities: sharp fragments of tiles, fragments of metals, etc.
   2. Poor maintenance of equipment: sharp fragments of metal, acrylic, glass from thermometers, etc.

IV. Associated with poor hygienic practices
   1. Housekeeping: insects and their visible fragments, rodent droppings, etc.
   2. Personal: hairs, metal clips, metal pin, writing pen caps, etc.

[a] Tolerance level should be applied.

mouth. Some claims have been made that physical hazards cause a psychological impact that disturbs the relationship between individuals and their food. Such claims can almost never be substantiated (Katsuyama, 1995b). By definition, HACCP deals with food safety. Therefore, only physical contaminants that can cause injuries should be addressed. Table 19.2 shows one way to classify physical hazards.

Extraneous materials associated with crops are inevitable, and tolerance levels should be set. GMPs should be introduced to avoid improper practices during the production and harvesting of plant or animal foods. Poor maintenance may result in the introduction of foreign materials into foods from plant buildings and equipment. Poor hygienic practices, whether in housekeeping or among the employees, can introduce physical hazards into food that suggest the possibility for other kinds of hazards of public health concern and indicate the need for intervention by official authorities. Flies are a real menace to the maintenance of sanitary conditions, and their presence indicates the existence in the plant environment of putrefying and decaying matter in which they breed. The presence of rodent droppings is clear evidence of infestation with rodents (Eisenberg, 1981). Poor employee habits and behavior can also introduce foreign bodies into the food. In some operations, food may be handled by people who are ignorant of the basic principles of personal hygiene (Eisenberg, 1981). In this case there is a need for training employees to improve their lifestyle. Some foreign materials can contaminate food products unpredictably as a result of sabotage or terrorism; the personnel management must be alert to this problem (Katsuyama, 1995b).

Microbiological hazards are contamination of foods, at unacceptable levels, with microbial agents of foodborne diseases such as viruses, bacteria, fungi or parasites (Council for Agricultural Science and Technology, 1994). Microbiological hazards may be classified as shown in Table 19.3.

Table 19.3 Microbiological hazards classification

I. Microorganisms, invasive or not, associated with foodborne infections
   1. Bacteria: *Campylobacter jejuni*, *Listeria monocytogenes*, *Salmonella*, *Shigella* and others
   2. Viruses: Hepatites A virus, noroviruses, etc.
   3. Parasites: *Cryptosporiudium parvum*, *Giardia lamblia*, *Toxoplasma gondii*, etc

II. Microorganisms associated with foodborne toxicoinfections
   Bacteria: *Clostridium botulinum* (infant botulism), *C. perfringens*, toxin-producing *E. coli*, *Vibrio cholerae*, etc

III. Microorganisms associated with food intoxications
   Bacteria: *Bacillus cereus* (emetic type), *C. botulinum*, and *Staphylococcus aureus*

From Council for Agricultural Science and Technology, 1994.

Although the HACCP system was developed to identify all potential hazards that compromise food safety, regardless of their nature, microbiological hazards present the greatest menace to consumers and should therefore receive priority in any HACCP plan (Smith *et al.*, 1990; WHO, 1993). The nature of biological hazards is discussed throughout this book.

Microbiological hazards are the most important for several reasons. They are capable of causing foodborne disease outbreaks that affect large numbers of people. Microorganisms are frequently, if not always, present as food contaminants, and many are able to multiply in foods. Some food-processing methods, such as meat and poultry slaughter processing, cannot completely eliminate pathogenic microorganisms.

- *Hazard analysis*. Hazard analysis requires technical expertise to identify all hazards correctly, evaluating their severity and predicting the associated risks. Incorrect predictions will not provide the necessary confidence and may increase the costs of production or indeed human health risks (WHO, 1993). The main purpose of the hazard analysis is to identify the actual and potential hazards associated with food ingredients, processes, transportation, storage, display conditions and final usage. Hazard analysis includes investigation of the effects of a variety of factors on food safety. It is based on the information from the technical literature, epidemiological surveys and product history, including claims reports, and it should evaluate the seriousness of contamination and the associated risks. It is necessary to perform a hazard study revision when changes are made in the process, whether in terms of raw material supply, formulation, processing, packaging, distribution or final use. Hazard analysis is best conducted through the use of structured hazard analysis procedures, in order to have all potential hazards effectively evaluated (Notermans *et al.*, 1994). Hazard analysis must consider situations that may be beyond the food processor's immediate control, such as food distribution, retailing, and consumer handling. Information from these sources can help the food processor to design food products with a higher degree of safety, or to improve safety instructions on the label for retailer and consumer use. During the hazard analysis, safety concerns should be differentiated from quality concerns.

Hazard identification is the first step in hazard analysis. Only hazards that could be present in the food under study should be addressed. Hazards can be identified by reviewing publications on foodborne disease outbreaks and epidemiological surveys, conducting a comparison with similar processes and products, and identifying vehicles of foodborne pathogens and factors that contributed to the occurrence of outbreaks and sporadic cases. Surveying the internal data of the company can also help to identify hazards. Consumer claims are an excellent source of information; records of consumer complaints indicate the weak points related to the food and food processing, and are important tools to identify potential hazards, since they reflect the performance of food product at the point of consumption. If the target for HACCP is a new food product, past records for similar foods can serve as guidelines. Another internal source of information is laboratory test data. The most useful data are those related to shelf-life studies at the recommended temperature of display and at abuse temperatures. Also, the routine laboratory analysis data for the food should be used; they may directly point to the agents of concern. The description of the food product is a valuable document in helping to identify hazards. The description should include information on ingredients, raw materials, composition, display specifications and labeling, as shown in Figure 19.1. The intended use must be based on normal consumer practices and the types of populations for which the food is intended. The use of a hazard-decision tree (Notermans *et al.*, 1994), as shown in Figure 19.1, to identify hazards is advisable; however there are steps/questions in the decision tree that do not apply in some situations. Experience should prevail in such situations.

Estimation of the consequences of hazards should be carried out after the hazards have been identified. This is a fairly complicated procedure, and is based on several considerations. Factors that have an impact on hazards, including intrinsic and extrinsic factors affecting the persistence of the hazard, should be reviewed, as well as the parameters allowing microbial growth and toxin formation in foods (WHO, 1993). Raw materials and ingredients should also be evaluated for potential contamination, and the ability to support or inhibit microbial growth at each step should be studied throughout the food chain. Consumer habits and the expected preparation (cooking, roasting, defrosting, reconstitution) of the food product should be evaluated (Committee on Communicable Diseases Affecting Man, 1991). The probable identity of the final consumer should also be considered. The projected consumers can be the general population or a particular segment of the population (such as children or elderly persons), or the food can be intended for institutions, restaurants, hospitals, nurseries, etc. An important factor to be considered is consumer susceptibility to infection. Children, elderly persons, pregnant women, hospital patients and immunocompromised persons (organ transplants, cancer therapy, AIDS) are the most susceptible. It is therefore important to prioritize hazard analysis of food intended for these categories of people. It should be borne in mind that the at-risk population is increasing (Table 19.4).

Only under specific circumstances, such as in hospital food deliveries, is it possible to predict who the final consumer will be. Supermarkets provide food for all segments of the population – sick people in institutions or homes, healthy people, young children and old people. The food industry must therefore assume that both healthy and vulnerable individuals will buy its products, unless the products are clearly forbidden, by labeling, for susceptible persons.

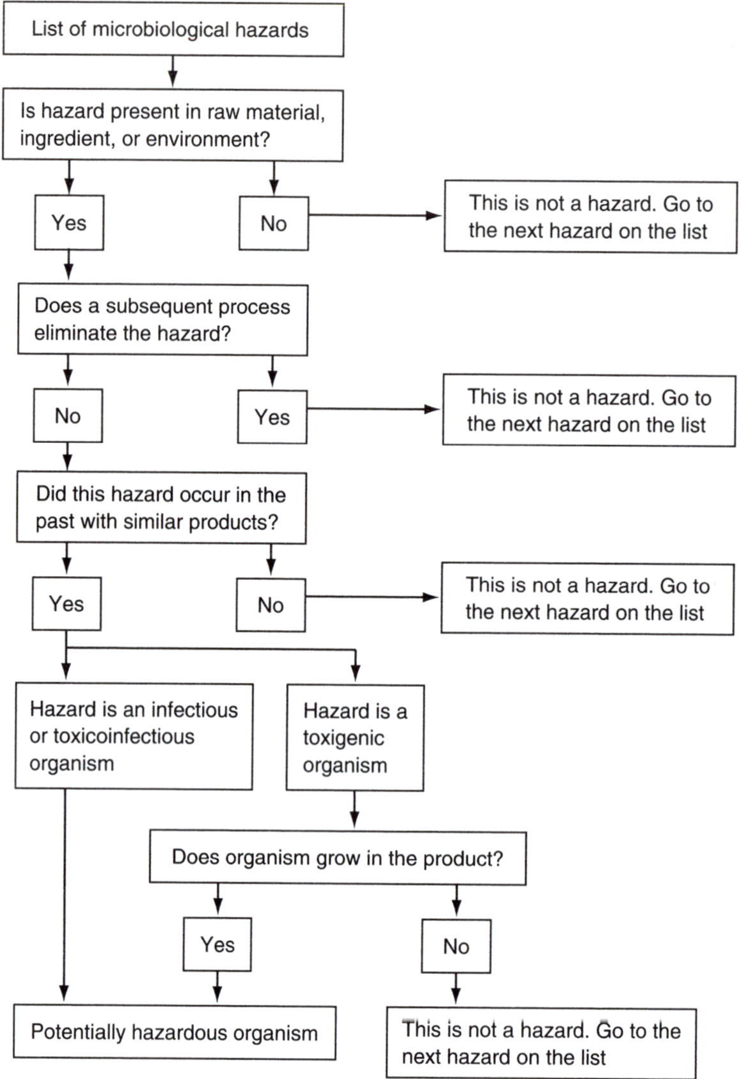

Figure 19.1 Hazard decision tree.

The severity of a hazard can be understood as the degree of consequences that can result from it. After a hazard has been identified, it can be assigned a severity value. Obviously, an event that may present a threat to life is much more serious than an event that causes only a moderately severe disease, which again is more important than an event that has only a minor public health effect (WHO, 1980; International Commission on Microbiological Specifications for Foods, 1988). Hazard severity also depends on the frequency of occurrence, duration of the disease, potential spread, possibility of asymptomatic carriers and probability of sequelae. Based on these considerations, the severity of a microbiological hazard can be classified as shown by the examples in Table 19.5.

Table 19.4 Population sensitive to foodborne disease in the US

Population category	Individuals	Year
Pregnant women	5 657 900	1989
Neonates	4 002 000	1989
Elderly persons (over 65 years of age)	29 400 000	1989
Residents in homes or related care facilities	1 553 000	1986
Cancer patients (non-hospitalized)	2 411 000	1986
Organ-transplant patients	110 270	1981–1989
AIDS patients	135 000	1993

US Department of Commerce, 1991; US Department of Health and Human Services, 1993. In Council for Agricultural Science and Technology, 1994.

Table 19.5 Severity of microbiological hazards

Hazards	Examples of microorganisms involved
Severe, direct	This type of hazard is characterized by the potential to cause life-threatening diseases. Examples include *Clostridium botulinum*, *C. perfringens* type C, *Listeria monocytogenes*, *Shigella dysenteriae*, shigatoxin-producing *Escherichia coli*, *Vibrio vulnificus*.
Moderate, with potential for extensive spread	Moderate hazards – usually do not cause serious disease in normal human adults. Examples include: *Salmonella typhimurium* and other *Salmonella* spp., *Shigella* spp, *Vibrio parahaemolyticus*, enteropathogenic *E. coli*, *Campylobacter jejuni*, *Cryptosporidium parvum*, etc.
Moderate, with limited spread	Moderate hazards, limited spread – those contaminations represented by toxigenic microorganisms that produce food intoxication but are not extensively disseminated or transmitted from person to person or person to food.
Low indirect	Microganisms included in this classification are those not usually related to foodborne disease and therefore do not cause human health concern. They are indicator organisms that reveal faults in food production and handling conditions, and include non-pathogenic *E. coli*, *Enterobacter aerogenes*, etc.

Based on International Commission on Microbiological Specifications for Foods, 1974.

Severe, direct hazards are characterized by their potential for causing life-threatening diseases (International Commission on Microbiological Specifications for Foods, 1974; Scott and Moberg, 1995). Moderate hazards 'with potential for extensive spread' means the potential for transmission through the environment and/or person-to-person transmission. Moderate hazards 'with limited spread' are represented by toxigenic microorganisms that produce food intoxication but are not extensively disseminated or transmitted from person-to-person (International Commission on Microbiological Specifications for Foods, 1974). Low indirect hazards are represented by microorganisms not usually related to foodborne disease and therefore of little public health concern. Broadly speaking, they are indicator organisms that reveal faults in food processing conditions. High numbers of coliform bacteria and enterococci may indicate poor hygiene and/or fecal contamination (International Commission on Microbiological Specifications for Foods, 1974); however, they are more often related to the environment than to direct fecal contamination of food.

Hazards associated with operations should be listed for each step of the food chain, from the primary production to the final use of the food, along with risk evaluation and estimates of the levels of hazard in terms of severity. Operations in food production, processing and preparation present hazards associated with ingredients, operations, storage or manipulation (Harrigan and Park, 1991). For this reason, relevant information on handling – related to industrial processing, retailing, and consumer use and each step in the food chain – must be analyzed to determine whether contaminants can reach the product (Committee on Communicable Diseases Affecting Man, 1991). To be included in the list, a hazard must be of such a nature that its prevention, elimination or reduction to an acceptable level is essential to food safety. Low-risk hazards should not be considered in a HACCP plan, and must be treated by regular GMP procedures. In order to identify microbial hazards, it is necessary to know how processing and handling conditions will affect microbial activities, as well as the degree of likelihood that a microorganism will be present in the ingredients and environment. The potential for abuse – for example, exposure to elevated temperatures or lengthy storage time during distribution or by the consumer – should also be considered. The effect of abuses is reflected in microbial growth during distribution or consumer handling. Precautions against effects of abuse include (but are not limited to) properties such as pH values, water activity, presence of inhibitors, and process efficacy (WHO, 1993). The product 'mode of failure' through transport, distribution and use by the consumer should be determined, and appropriate precautions taken to prevent failures. Hazard analysis starts with the development of a flow diagram of the steps in processing, and should ideally include steps before and after the processing (Figure 19.2). The purpose of the flow diagram is to provide a simple and clear description of the phases involved in the processing. The flow diagram should cover aspects of incoming raw material and ingredients, operations and processing conditions, including additives, processing time, temperature and other factors affecting hazards. Personnel in the processing area should check the flow diagram in order to assure that it is correct. This is also the best way to assure that the process is in agreement with GMP; if deviations from GMP occur, they have to be corrected before implementing HACCP.

Each ingredient and raw material must be evaluated with respect to its potential as a source of contamination or its ability to inhibit microbial growth (Committee on Communicable Diseases Affecting Man, 1991). Microorganisms and other hazardous substances may contaminate raw materials and cause them to deteriorate in the field, in storage or during transport to the food-processing plant. Such contaminations are common; it should be assumed that they are always present, and food-handling systems must be based upon that assumption (Brickey, 1981). Other hazardous substances may also contaminate raw material and, in order to keep contaminants below levels that would render end products potentially harmful, the safety of ingredients used should be verified by consulting the product formulation or by observing product preparation (Committee on Communicable Diseases Affecting Man, 1991). Foods in their natural state may harbor contamination from various sources. Raw beef, pork and poultry meats are frequently contaminated with *Salmonella*, *Campylobacter*, *Clostridium perfringens*, *Yersinia enterocolitica*, *Escherichia coli*, *Listeria monocytogenes* and

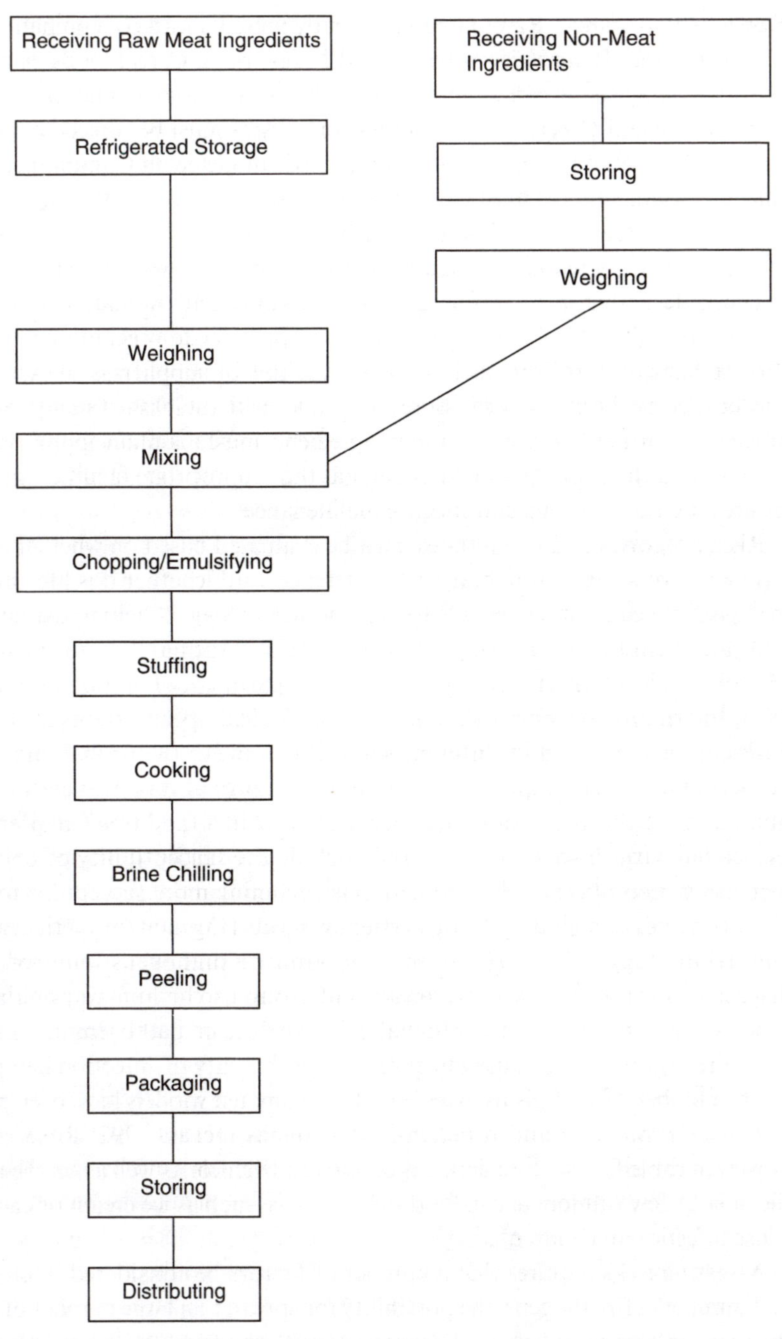

Figure 19.2 Flow diagram of hotdog-manufacturing process.

Staphylococcus aureus. Raw seafood frequently presents with contamination by non-O1 *Vibrio cholerae*, *V. parahaeomolyticus* and *V. vulnificus*. Cereal foods, particularly rice and spices, are usually contaminated by *Bacillus cereus*, and vegetables frequently carry *C. botulinum* and *C. perfringens*. All these organisms must be considered, in relation to the risks presented by the food under study (Committee on Communicable Diseases Affecting Man, 1991). It must be ensured that raw material and other ingredients are not produced in areas harboring excessive harmful substances, and do not come from sources with the potential for unacceptable contamination with fecal material. It should be checked whether there are any specific points at which raw materials and ingredients may be at high risk of contamination, and appropriate precautions must be taken (Codex Alimentarius Commission, 1994). Auditing of suppliers is an excellent aid, and may be used by the industry to assure compliance with established standards. Employees of suppliers of raw material and other ingredients must maintain appropriate personnel hygiene, and the supplier should document that appropriate facilities are available to ensure necessary cleaning and effective maintenance.

Risk categories and magnitudes must be evaluated based on whether a hazard will have a low or a high probability of occurrence, and whether it is life-threatening or may possibly cause many people to become sick. This will help to define the hazards and rate them based on risks and severity (US National Advisory Committee on Microbiological Criteria for Foods and US Department of Agriculture, 1993; WHO, 1993; International Commission on Microbiological Specifications for Foods, 1997). Risk can be expressed in different ways; it can mean the probability of a hazard occurring in a certain quantity of food, or the probability of a certain number of human infectious (or toxic) doses being present in a food (see Chapter 2). Factors associated with the magnitude of risk include the susceptibility of consumers. An increasing percentage of the population is becoming more susceptible to pathogenic microorganisms, including those carried by foods (Council for Agricultural Science and Technology, 1994). The dose of foodborne pathogens required to produce disease is not precisely known for any subgroup of the human population. Some experts reject the notion of a minimal infective dose of pathogens and maintain that even one organism can cause infection, the probability of infection being dependent on the number of organisms ingested. Dose–response models have been proposed for *Salmonella*, *Shigella* and other microorganisms (Haas, 1992; Rose *et al.*, 1996). However, the infective dose depends on the food vehicle as well as on the resistance of the host. A few salmonellae in food rich in lipids, such as ice cream or chocolate, may cause infection in children.

Assessing risks requires that a number of factors be considered, such as the types and numbers of pathogens, the possibility for spread to a large number of people, and intrinsic as well as extrinsic characteristics of the food. Risk assessment must be objective. The assessment must take into account the hazard as well as the degree of public health consequence of its occurrence. It is necessary to have a good comprehension and understanding of how certain hazards happen, and from where they come. Risk assessment involves the documentation and analysis of scientific evidence in order to evaluate risk and identify factors that influence the risks; this information is needed by risk managers (Sterrenberg and Notermans, 2000). A moderate hazard that carries a high risk may be more important than a more severe hazard that has little public health significance (International Commission on Microbiological Specifications for Foods, 1988). Application of predictive microbiology is playing an increasing role in quantitative risk assessment. Examples include the quantitative risk assessment model for *S.* Enteritidis in pasteurized liquid eggs (Whiting and Buchanan, 1997), the *Salmonella* risk assessment modeling program for poultry (Oscar, 1998), and the ground-beef hamburger study by Cassin *et al.* (1998). Risk assessment is discussed in Chapter 2.

Preventive measures are factors or actions of physical, chemical or biological nature to control identified hazards. More than one hazard may be controlled by one specific preventive measure, and more than one control measure may be required to control a specific hazard (Codex Alimentarius Commission, 1997). Preventive physical measures may be a time–temperature requirement, a moisture level or a water activity value. A preventive chemical measure may be titratable acidity, residual free chlorine level, salt concentration, or levels of nitrite and other chemical additives. The use of starter microorganisms added to food for fermentation purposes may represent a biological preventive measure. Each preventive measure has an associated critical limit, which defines the margin or safety limit for each CCP (US National Advisory Committee on Microbiological Criteria for Foods, 1992; WHO, 1993). By hazard analysis and identification of preventive measures, three goals are achieved: first, hazards and preventive measures are identified; second, this knowledge may be used to correct or improve processes or product performance related to food safety; and third, identification of CCPs is facilitated (WHO, 1993). It should be emphasized that the safety of a food product depends on the control of hazards by the use of appropriate measures applied on an operating processing line. Thus, these measures should be validated to guarantee their efficiency in yielding a safe food. Validation of the control measures will be discussed below.

Principle 2: Identification of critical control points

A CCP is a point, step or procedure at which control can be applied and a food safety hazard can be prevented, eliminated or reduced to acceptable levels. It is important to identify potential CCP(s) in food preparation. CCPs may be cooking, chilling, sanitation procedures, product formulation control (pH, salt and water activity), and employee and environmental hygiene, including prevention of cross-contamination. Different facilities preparing the same food may differ in the risk of hazards, depending on the operation. Information obtained during hazard analysis, together with that from the process flow diagram, helps to identify steps that are critical. Several stages may exist in the food-processing system where biological, physical or chemical hazards can be controlled. However, there will be only few situations, operations, locales or points where the loss of control could result in a potentially unsafe food. CCPs are those few, crucially important, points. All identified hazards must be controlled at specific points in the food system involving the entire food chain (Katsuyama and Stevenson, 1995).

- *Determination of a CCP* in the HACCP plan may be facilitated by the application of a decision tree, which outlines a logic-reasoning approach (WHO, 1993; Codex Alimentarius Commission, 1997). The CCP decision tree is based on a sequence of questions to be answered in an orderly fashion for each of the possible points to be controlled, as shown in Figure 19.3. All hazards and associated risks should be considered, and additional questions asked – Do the food-processing operations result in a possible exposure to contamination? Is there a possible failure to inactivate contaminants? Is there an opportunity for microbial development? Is the food going to be prepared in large batches and kept, before processing, for a long period under conditions favorable for microbial growth? (Committee on Communicable Diseases Affecting Man, 1991; WHO, 1993).

It is important to have detailed information about the food-processing operation. This can be obtained by interviewing people involved in processing and by observing operations during processing, starting with ingredient sources and following through shipping, transportation, storage, retail and consumption of the food product (Committee on Communicable Diseases Affecting Man, 1991). A specific point can

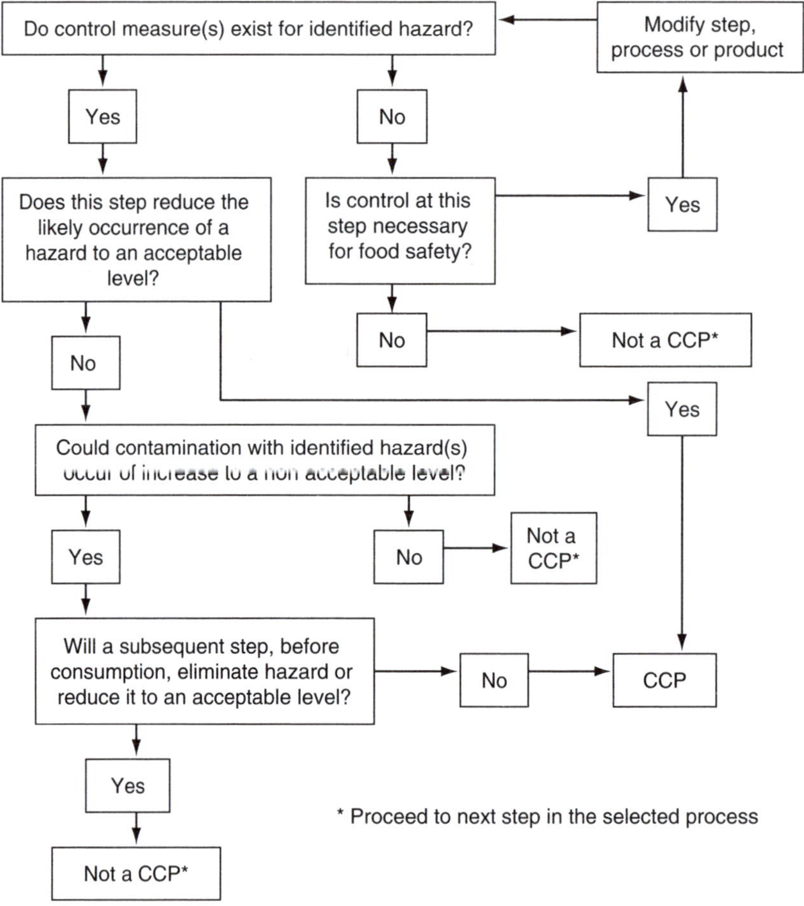

Figure 19.3 CCP decision tree.

only be considered a CCP if there is a significant risk for occurrence of a disease and/or failure of product integrity as a result of lack of control in the operation. In comparison, control points (CPs) are those points where loss of control does not result in an unsafe product. The CPs are addressed in the GMPs. A CCP does not have to be established for each step. If a hazard cannot be effectively controlled at one specific CCP, control should be achieved by a subsequent procedure. If not, a change must be made in the operations, or another supplier of raw material chosen. Sometimes, the control at one or two CCPs will eliminate all of the most important hazards. This is the case with heat processing or irradiation of hermetically packaged food. In other situations, a combination of control measures can be used at successive CCPs – for example, cooking, refrigeration and re-heating. In other situations, like raw beef and poultry meat processing without irradiation treatment, the *Salmonella* hazard may be reduced but cannot be eliminated. CCPs, such as appropriate levels of chlorine or other disinfectant levels in washing procedures and preventing microbial growth by appropriate storage temperature, will at best reduce contamination on the carcasses. Good manufacturing practices play an important role in the operations of a food-processing plant. Examples are the cleaning and disinfection of work surfaces for processing meat, hygienic practices of the workers, and keeping the environment cool and dry. These actions are applied to CPs and are defined in the GMPs or SSOPs. HACCP can be compared to critical surgery, which cannot be performed successfully unless the environment in the surgical room is clean and disinfected (SSOPs), the instruments are sterilized and appropriate for the specific surgery (GMPs), and the personnel are qualified for the event.

- *Criteria in selecting a CCP* must consider whether processing will result in an increase, decrease or persistence of a microbial hazard. The same rationale must be applied for chemical and physical hazards. The importance and complexity of the operations to which the food product is submitted before and during industrial processing, and the potential for abuses during transport and storage, should also be taken into account. All steps in a process that are critical to the safety of food should be identified. When this has been done, control procedures should be implemented at these steps and monitoring procedures developed and applied with appropriate frequency to ensure continuing effectiveness. Facilities, equipment and operations should be evaluated for contamination, and provision made for adequate cleaning, disinfection and maintenance (Codex Alimentarius Commission, 1994). The potential for abuses by the consumer through mishandling of food must be evaluated in order to define procedures for additional food protection measures or to develop specific labels instructing food users.

Quality-control programs are sometimes called HACCP, HACCP-like or HACCP-based. It must be emphasized that HACCP is a food *safety* system. Also, for each CCP there is a corresponding procedure to control an identified hazard. If a quality control system has no defined and effective intervention procedures, it has no CCP and does not qualify as HACCP. Operations in which a hazard is reduced, such as cleaning, sanitizing and hygienic food handling, hardly qualify as CCPs. This includes washing and steam pasteurization or steam vacuum of carcasses combined with hot-water wash or chemical intervention, which may cause a limited reduction of surface contamination (Delazari *et al.*, 1998; Castillo *et al.*, 1999). These operations are better covered in prerequisites (GMPs, SSOPs). It should be clear that, in a processing line, many steps work in reducing risks by reducing the hazard levels, but only a few of these steps will truly be CCPs. The HACCP team should have flexibility in choosing CCPs but must be prepared to defend its decisions.

Principle 3: Establishment of critical limits for preventive measures at each CCP

Critical limits (CLs) are the parameter levels within which the operation must flow in order to assure conditions of normality related to food safety. In other words, they are the boundaries for safety for each CCP, and may be limits with respect to temperature, time, meat patty thickness, water activity, pH, available chlorine, etc. Critical limits may be derived from regulatory standards or guidelines, experiments, and expert opinion. Also, they may be based on professional experience with the food system or on the technical literature (US National Advisory Committee on Microbiological Criteria for Foods, 1992), on pathogen challenge tests and on shelf-life tests. Use of pathogen modeling and predictive microbiology in both defining CCPs and setting critical limits has become increasingly common (McMeekin *et al.*, 1993; Elliott, 1996; Walls and Scott, 1997; Whiting and Buchanan, 1997). Critical limits may refer to pH value, titratable acidity, water activity, time and temperature in heat processing and storage, residual available chlorine, level of nitrite added, etc. (US National Advisory Committee on Microbiological Criteria for Foods, 1991; Weddig, 1999). Only people who know what levels of preventive measures will effectively prevent the specified hazards can set CLs. The CLs must be met for each preventive measure associated with CCPs. In some cases, more than one CL can be applied at a particular CCP (Codex Alimentarius Commission, 1997). Variability is inherent in food processing; therefore, specification of CLs should include realistic tolerances. The tolerances should be determined from the routine measurement data (Committee on Communicable Diseases Affecting Man, 1991). An establishment may choose to set operating limits that are stricter than the defined CLs; this will reduce the number of violations of CLs (Weddig, 1999).

Principle 4: Establishment of procedures to monitor CCP

Once the control measures and criteria have been established, adequate procedures for monitoring each CCP must be defined. Monitoring is a planned frequency of observations or measurements to determine whether or not a CCP is under control, and also produces records for future use in the verification process. The HACCP monitoring system is thus documented evidence of process conditions, by programmed tests, measurements and observations at each CCP. Most of the monitoring procedures for the CCPs need to be performed quickly, 'on line', to permit corrective actions without delays. There is no time to perform laboratory tests; microbiological methods in particular are not suitable as monitoring tools, except under special circumstances, because of time limitations (US National Advisory Committee on Microbiological Criteria for Foods, 1992). The monitoring procedures must be able to detect loss of control at the CCP (Codex Alimentarius Commission, 1997).

Monitoring makes it possible to perform a trace-back through the whole process. If monitoring shows a tendency toward loss of control (i.e. if deviations from CLs are frequent or the preventive measure is not properly controlling the CCP), a corrective action (CA) must be taken immediately to bring the process back under control. A monitoring document must show when a deviation from CLs occurred and what CA measure was performed. The monitoring procedures must be very precise, consistent and effective because of the potential consequences of a process that is not properly controlled. Monitoring can be performed on a continuous or discontinuous basis. On a continuous basis, use of automated equipment and sensors is recommended whenever possible. Examples of continuous monitoring are the recording of autoclave time–temperature in heat processing of foods, temperature in a storage room, free chlorine in the water entering poultry chiller tanks, and pH measurements during food fermentation. Continuous monitoring is not always possible. In that case, monitoring must be performed on a discontinuous basis with well-defined intervals between tests or measurements (US National Advisory Committee on Microbiological Criteria for Foods, 1992).

- *Methods of monitoring* are based on the type of hazard and the available tools and instruments. Monitoring can be done by observation, use of sensory evaluation of food characteristics, use of physical and chemical attributes, and sometimes by microbial analysis (Committee on Communicable Diseases Affecting Man, 1991).

  Monitoring by observation is frequently used in HACCP. Examples are supplier auditing, approved supplier checklist, or observation of standardized operations.

  Chemical tests must be rapid enough to permit sufficient time for correction of deviations as they occur on line. Common chemical measurements are salt or sugar concentration, titratable acidity, and added sodium nitrite in pasteurized canned meat, etc. (Committee on Communicable Diseases Affecting Man, 1991).

  Physical measurements should be rapid, and preferably done on a continuous basis. Important measurements are time and temperature, water activity, pH and moisture level (Committee on Communicable Diseases Affecting Man, 1991).

  Microbiological measurements are not practical for monitoring, because most require several hours or even several days. They could be useful for ingredients or raw materials if there is enough room and time to hold a lot until the microbiological results are available, or it is possible to have the results available in advance. It is preferable to request microbiological test results from the supplier.

  It must be stressed that monitoring should be done at CCPs and is related to the process but not the food, because HACCP is a process-control system and not a product-control procedure.
- *Visiting supplier installations or auditing supplier* production data is a way to monitor raw material. If the supplier employs HACCP, verification documents are useful to minimize laboratory analysis on receiving goods. Supplier guarantee letters are another way to assure raw material acceptability. Guarantee letters should document raw material origin and sources, and the supplier laboratory analytical report. Depending upon the type of food and process, analysis of incoming raw

material and ingredients may be unnecessary. An example is a case where the food has a pH level that will kill or prevent growth of pathogens, such as pickled products. Another example is food with a water activity low enough to prevent microbial growth, such as some cakes and candies (US National Advisory Committee on Microbiological Criteria for Foods, 1992).

- *Monitoring of cleaning and disinfection* can be organized in different ways. Normally, cleaning and disinfection are operations falling under GMPs or SSOPs; as such, they are prerequisites and may not be included in HACCP. In either case, monitoring of cleaning operations should be used to determine whether the operations are performed correctly, even when it is part of the SSOPs data sheet. It is important to monitor the concentration of sanitizer and pH, and time of exposure of equipment to the disinfecting solution. Surfaces and utensils that have been cleaned can be evaluated for cleanliness by quick methods such as measurement of residual levels of adenosine triphosphate.
- *Responsibility for monitoring* is very important, and must be delegated carefully. A reasonable choice would be employees such as supervisors, line workers, maintenance and mechanical engineering personnel (at the time of checking equipment for start up), or employees responsible for cleaning and disinfection. Individuals selected for monitoring a procedure should be trained in the techniques used for monitoring (US National Advisory Committee on Microbiological Criteria for Foods, 1992). They should completely understand the monitoring purposes and importance, have easy access to monitoring instruments, and be consistent and exact in their records. Deviations detected by monitoring should be immediately corrected, assuring a rapid return to normal conditions. The monitoring responsibility should be delegated to employees in such a way that all work shifts are covered and supplied with instruments for measuring and with tools for corrections when they are needed. Failure in this regard frequently causes a breakdown in HACCP. Each person responsible for monitoring a CCP should be listed, and must sign each of his or her monitoring sheets.

Principle 5: Establishment of corrective actions

The HACCP system was developed to identify potential hazards and to establish strategies to prevent their occurrence. However, ideal circumstances do not always prevail, and process deviation will occur. When a deviation exceeding an established CL occurs, a corrective action (CA) plan should follow. First, the non-conforming product must be disposed of in an appropriate way. Second, the cause of the non-conformity must be corrected immediately, to ensure that the CCP is back under control. Records should be kept of all CAs taken (Codex Alimentarius Commission, 1997).

Because of differences in CCPs for different foods and processes, and because of the diversity of possible deviations, specific action plans should be developed for each CCP. Only individuals with complete knowledge of the process, the product and the HACCP plan should be in charge of development of CAs (US National Advisory Committee on Microbiological Criteria for Foods, 1992). The rapid response to the detection of a CCP temporarily out of control is an attribute of HACCP. If a deviation is not corrected in time, the establishment should put a hold on the process until necessary analysis and CA become effective. Actions will vary according to the process, and may include reheating or other types of reprocessing, decreasing water activity, adjusting levels of additives, label alteration, and lot rejection or lot destruction.

Recall is the last CA the industry can count on. A successful recall requires the collaboration of a number of different entities, including food growers, manufacturers, retail outlets, and state and federal agencies (Wong *et al.*, 2000). It is advised that enterprises have recall teams and recall plans in place to rapidly stop consumption of a suspected food product. Recall should start as soon as evidence requires it. Different actions should be applied, depending upon recall category. Recalls may be classified as shown in Table 19.6.

Table 19.6 Guidelines for a product withdrawal

	Recall type I	Recall type II	Recall type III
Problem	Situation where there is a probability that the use of or exposure to the product will cause serious, adverse health consequences or death	Situation where there is a probability that use can cause a temporary or medically reversible effect; likelihood of severe consequences is remote	Situation in which the use or exposure to a violative product, is not likely to cause consequences to health
Example	Botulism, high-lethality poison	*Salmonella, Staphylococcus aureus*; low-lethality poisons; glass contamination	Adulteration with harmless/inert substances; spoilage with non-pathogenic organisms; less weight than declared, inferior quality
Range	Consumers, retail, warehousing and transit	Retail, warehouse and transit	Transport and direct deliveries
Communication	Urgent and immediate warning through news media in all distribution areas where the product was delivered	Warning through news media for groups at high risk	Usually no public communication
Effectiveness of recall	Objective is 100% return; auditing at retail points	As complete as possible; auditing in the largest area possible	Warehouse supervision

From Varnan and Evans, 1991.

Principle 6: Establishment of verification procedures

Verification is designed to ensure that the HACCP plan is being implemented properly, and may include verification of prerequisites of CCPs (Stevenson and Gombas, 1999). Verification of prerequisites is done primarily by auditing the GMP and SSOP procedures. Verification can be done by the quality assurance personnel involved in the HACCP team. Verification may consist, depending on the purpose, of a simple check of the monitoring data of CCP performance or a detailed analysis of the HACCP plan. Monitoring records and documents are checked in a routine verification, and laboratory testing may be part of the verification process. Also, the checking and testing for calibration and accuracy of the measurement instruments used for monitoring can be done during verification. Verification is a crucial procedure in HACCP. It includes activities such as inspections, and may use microbiological and chemical tests to confirm the effectiveness of control measures, the conditions of the product as found in the marketplace, and reviews of customer complaints (International Life Sciences Institute, 1997).

One of the first steps in a verification procedure is to check that at each CCP hazards are eliminated or prevented; this will generally require laboratory support. An example is shown in Table 19.7; this may also serve as a validation that a sequence of steps, each of which contributes to a limited decrease of pathogens and reduces hazards. Once CCPs have been verified, application of routine testing for pathogens in the end product become superfluous (Swanson and Anderson, 2000).

Periodic audits are very useful to demonstrate to industry management boards and others that the processes are under control, the HACCP plan and procedures are being followed, established standards are being met, resources are being used efficiently, and enterprise objectives are being met. Auditing a food process uses the same tools as verification and evaluates the total HACCP plan and GMP procedures, including documentation for each step of the process and the critical points. It starts with suppliers' accreditation and certification, as well as documentation of raw materials and other supplies. Auditors normally request laboratory analysis of the food item under audit as evidence that the system is generating safe food. Periodic auditing is required in addition to the more frequent verification.

Table 19.7 Verification process.[a] Microbiological testing using indicator organism. Process prouction of chicken carcasses.

Escherichia coli (CFU/g)[b]	Process stage	Product temperature (°C)[c]	Time (minutes)[d]
	Reception of birds		
	Hanging	42	2-3
	Bleeding	42	2-3
1.0×10^3	Scalding	16	2 3
3.0×10^4	Plucking (picking)	43	3-3.5
1.0×10^3	Evisceration	42	3-3.5
3.2×10^2	Washing	40	1-1.5
2.4×10^2	Cooling	20	15
1.8×10^2	Cutting and packaging	20	14-20
6.8×10^1	Freezing	−20	
2.3×10^1	Storage	−20	

[a] Similar procedure is done for validation of preventive (control) measures
[b] Colony Forming Units/gram
[c] Temperature measured at the end of the operation
[d] The indicated times correspond to the range from the start to the end of the operation.

Principle 7: Establishment of record-keeping procedures

The HACCP plan must be recorded in detail. Data sheets must be developed based on specific recording needs. Each step of the implementation must be recorded, and documents must be available for verification and for evaluation by government inspection services.

The approved HACCP plan and associated records should be kept on file at the establishment or at a location with easy access when needed. Monitoring records for CCPs include a list of all CCPs, with names of persons in charge of the monitoring. Without monitoring records, there is no evidence that the criteria for CCPs are being met. Records of time and temperature during storage, transport and distribution should be kept in order to document that no product was distributed after its shelf-life had expired; these documents make it possible to check, during verification, that the total process was effectively under control. If the occurrence of deviations and/or stops in processing is frequent, something is seriously wrong with the process or the equipment. The situation should be investigated to improve control. Records are also needed to document approval of processes, and approval of changes in formulation and processes. Records are important sources for internal use, and for official agencies conducting audits or verification actions. They are also important in the case of foodborne diseases alleged to have been transmitted by the food and in cases of litigation.

Records of training of personnel in HACCP techniques are best kept by the use of certificates proving that the employees attended specific courses on the subject, according to their duties in the HACCP plan. Training and retraining must be an ongoing program in the company. The purpose of retraining is not to repeat the initial training; it is to add to previously acquired knowledge, or can be an evaluation of how well the employees have learned. Education and training make personnel understand and realize how their work relates to the overall operations in the plant. The personnel must understand that the way they perform their task will affect the overall process. The instructor can be the manager or the supervisor. What is important in the training is that the instructor transfers the knowledge in a simple way, not using scientific jargon. For that reason, outside experts might be used to train higher-level staff, and the staff then transmit the knowledge to the plant personnel.

Validation

Validation is an important part of HACCP. When all the seven principles have been implemented, validation will provide information as to whether the system effectively leads to a safe product. There are two types of validation; one of the control measures, and the other of the HACCP system.

- *Validation of control measures*. This is the process of ensuring that a defined set of control measures is capable of achieving appropriate control over a specific hazard(s) in a food product for a specific CCP (Codex Alimentarius Commission, 2001). It is typically conducted before the initiation of a new food safety system, to assure that the system is capable of achieving the desired food safety outcome. During validation of preventive actions, measurements can be done on the food product to confirm both preventive measures and critical limits. The HACCP team personnel can perform validation of preventive measures.
- *Validation of the HACCP system*. Initial HACCP system validation includes a review of the scientific documentation used in developing the HACCP plan, and a general review of the scientific literature. It will check whether CCPs are valid,

and whether control measures and CLs are appropriate (Kvenberg and Schwalm, 2000). HACCP system validation generally covers the whole HACCP implementation process. The validation should assure that all necessary resources were applied for the HACCP team formation and training, all hazards were properly identified, all CCPs were correctly selected, and effective control measures were chosen based on solid technical and scientific evidence. The previous history of the food, compared to the situation after HACCP was introduced, should present a difference in safety aspects. Once the HACCP system has been validated, the plant is certified to operate under HACCP. Validation should not be done by a member of the team involved in carrying out the HACCP plan, but by an independent expert(s). Revalidation is needed when deviations from CLs occur frequently, or when there are significant changes in the product, process or packaging (Kvenberg and Schwalm, 2000).

3.3.4 Nature of HACCP

- *The HACCP team* has the responsibility for development of the HACCP plan. The plan is a graphic representation of the process as a whole. In order to provide a total picture of the process, it is necessary to incorporate in the HACCP plan all operations including those controlled by GMPs, as shown in Table 19.8. The HACCP team should include only people belonging to the plant, and should be multidisciplinary. HACCP cannot be a 'command and control' procedure, and it is a mistake and counterproductive to have CCPs dictated by outside 'experts'. The people in each plant must define the CCPs after appropriate training, as they are the ones that have intimate knowledge of the processes. They have to develop their own HACCP plan, and by doing this they become part of the plan and will often find improved ways to perform operations during the development of HACCP. There is a tendency to develop the so-called 'generic' HACCP plans. Generic plans sometimes contain very useful background information (US National Advisory Committee on Microbial Criteria on Foods and US Department of Agriculture, 1993), but they should not be a substitute for the plan(s) developed by the employees of a plant. HACCP plans must be developed by the people who are going to implement them, and the role of third parties should be to provide background information, 'train the trainers' and perform validation studies.
- *Management's* commitment determines the success of the HACCP system. The commitment is reflected in allocation of the needed resources, funds and personnel. The manager has to be well informed about the HACCP concepts and the benefits of HACCP to food safety and quality assurance, and in terms of workers' motivation and involvement. The company's posture in quality matters is a factor to be considered in general policies, and must be published throughout the company.
- *Responsibilities*. The responsibility for carrying out the HACCP plan must be delegated to a person who will be specially trained to perform that activity. The professional most appropriate for that is the process manager. In that way, it is assured that the top level of management has a real interest in the plan.

Table 19.8 HACCP plan[a] for frozen chicken (illustration purposes only)

Stage of process	Potential hazards	Stage classification	Control measures	Critical limits	Corrective measures	Monitoring & frequency	Verification
Catching birds	Spread of *Salmonella*, *C. jejuni*, *C. coli*	CP	1. Number of birds per cage 2. Fasting	1. ≤ 6 birds 2. ≥ 6 h before slaughter	1. Remove or add birds 2. Delay loading	1. Observe every cage 2. Check every record on fasting	1. Check operator log 2. Check operator log weekly
Incoming birds		CP	1. Temperature 2. Waiting time	1. ≤ 16°C 2. ≤2 h	1. Turn on water spray 2. Put on line immediately	1. Automatic record 2. Record arrival time for each lot	1. Check calibration 2. Check operator log
Bleeding		CP	1. Cutting blade 2. Clean blade 3. Bleeding time	1.Carotids and jugular cut through 2. Continuous washing 3. ≥ 30 s	1. Sharpen blade 2. Changing used blade 3. Correct line speed	1. Check blade sharpness twice a shift 2. Check need to change blade 3. Record speed	1. Check number of used blades/day
Scalding		CP	1. Water temperature 2. Water flow 3. Exposure time	1. ≥ 56°C 2. ≥ 1 l/bird 3. ≥ 3 min	1. Increase temperature 2. Increase water flow 3. Correct line speed	1. Continuous recording 2. Continuous recording 3. Check speed twice shift	1. Check calibration 2. Check operator log 3. Check operator log
Defeathering		CP	1. Free chlorine 2. Water flow	1. ≥ 10 ppm 2. ≥ 1 l/bird	1. Increase chlorine level 2. Increase water flow	1. Continuous record 2. Check every 2 hr	
Evisceration	Spread of *Salmonella*, *C. jejuni*, *C. coli*	CCP	1. Sanitize instruments 2. Intestine intact 3. Washing 4. Water flow	1. 83°C, 5 min 2. 100% free from visible intestinal contamination 3. 19 ppm free chlorine 4. ≥ 1 l/bird	1. Increase temperature 2. Remove contaminated carcass from the line[b] 3. Increase chlorine level 4. Increase flow	1. Check temperature 2. Observe operation 3. Continuous record 4. Continous record	1. Check calibration 2. Check operator log 3. Check operator log 4. Check operator log

Table 19.8 HACCP plan[a] for frozen chicken (illustration purposes only)—*continued*

Stage of process	Potential hazards	Stage classification	Control measures	Critical limits	Corrective measures	Monitoring & frequency	Verification
Final washing	Possible reduction of microbial hazards	CP	1. Water temperature 2. Water flow 3. Added chlorine	1. ≤ 16°C 2. ≥ 3 l/bird 3. ≥ 10 ppm	1. Add ice to the tank 2. Increase water flow 3. Correct concentration	1. Automatic record 2. Automatic measure 3. Automatic record	1. Check operator log 2. Check flow weekly 3. Check CL once a day
Chilling	Possible reduction of microbial hazards	CP	1. Water temperature 2. Water flow 3. Added chlorine	1. ≤ 2°C 2. ≥ 3 l/bird 3. 15 ppm	1. Add ice to the tank 2. Increase water flow 3. Increase concentration	1. Automatic record 2. Automatic measure 3. Automatic record	1. Check operator log 2. Check flow weekly 3. Check CL once a day
Packaging	Possible reduction of microbial hazards	CP	1. Room temperature 2. Relative humidity	1. ≤ 12°C 2. ≤ 65%	1. Decrease temperature 2. Increase air circulation, exhaust	1. Automatic record 2. Automatic record	
Freezing	Possible reduction of microbial hazards	CP	Tunnel temperature	−40°C	Adjust temperature	Automatic record	Check calibration
Storage	Little expected effect on hazards	CP	Temperature	−28°C	Adjust temperature	Automatic record	Check calibration
Transport	Little expected effect on hazards	CP	Truck temperature	−28°C	Turn on cooling air system	Automatic record	Check calibration

[a] Including control points on GMP

[b] Special washing of contaminated carcasses with hyperchlorinated water, assuring removal and decontamination before returning carcasses to the line.

The quality assurance manager is normally charged with advising and with verification and validation procedures, if not directly involved in the development of the HACCP plan. It should be emphasized, though, that a clear distinction between HACCP and quality assurance should be made. HACCP deals with food safety and not quality.

- *Training*. There is a need for uniform training to develop an appreciation for and understanding of HACCP and its implementation. This need exists in industry as well as in regulatory agencies. There are three specific objectives of HACCP training: it should provide a common understanding of the practical implications of HACCP in food safety; impart the practical skills and knowledge necessary for HACCP application; and provide the stimulus for further development and harmonization of HACCP (Mayes, 1994).

  After training, the person in charge of the HACCP plan must assemble a HACCP team. This team should be multidisciplinary and, whenever possible, include a microbiologist, a food engineer, a food chemist, a production supervisor, a sanitarian, a cleaning supervisor, a mechanical/electrical engineer, and the person responsible for process control.

  It is important to make sure that all members have the qualifications to understand fully the training they will receive. It is advisable that the team has access to external advice from an expert, especially a food microbiologist or other authority on food safety. Team members must be able to recognize hazards, to define levels of severity and associated risks, to recommend control criteria and CA, and to design procedures to monitor and verify the efficacy of the HACCP plan.

  The HACCP team has responsibility for the development of each step of the HACCP plan. The tasks should be uniformly distributed among team members, and a document should specify who is responsible for what in the HACCP plan by listing the name of each followed by a description of his or her specific duty.
- *Updating HACCP*. HACCP is an open-ended project and should be continuously improved; for this reason, steps must be taken to promote and maintain interest. It is advisable to create a library on HACCP. This is one of the measures that will keep the HACCP team united around the subject. Making each member the sponsor of one subject in the library promotes the desire to look for literature reviews of the subject. Also, monthly meetings should be scheduled for discussion of HACCP problems seen in the plant during the month. This does not eliminate the need for other meetings when required. The HACCP team has to have enough freedom to invite scientists from universities, food research institutes or governmental agencies to give presentations or lead discussions on various subjects of interest. The HACCP team has to have its own program in continuing education, and a budget for attending meetings related to the discipline. A programmed schedule for HACCP-team technical meetings, where each team member is encouraged to present a paper of his or her own and to publish it in the company's internal magazine(s) or in scientific journals on quality control, is another way to keep the HACCP team actively involved in the subject. Members are normally distinguished as the company sentinels for food safety.

3.3.5 HACCP and SSOPS

The cleaning and sanitation of food plants is part of the GMPs and is generally done according to SSOPs, often by an outside contractor. SSOPs are seldom included in the HACCP plan, although it may make sense to do that, especially in the case of slaughter and other plants that handle raw products but do not have a single CCP that is effectively eliminates hazards. Agencies differ with respect to what they want to include into HACCP. The FDA seafood HACCP differs from the FSIS meat and poultry HACCP (Bernard, 1997); it might be better to leave such decisions to the industry.

3.3.6 HACCP and ISO9000

HACCP is a systematic approach for assuring the production and processing of safe foods. The focus in the HACCP program is on food safety, not quality. The ISO9000 series system (ISO9000) of the International Organization for Standardization is not directed towards safety; it is set to provide common standards of quality during the production or manufacturing of products in order to assure that two or more trading partners (nationally or internationally) agree on the quality of the product.

Both ISO and HACCP are preventive. The ISO system provides confidence to the customer that the company has been working under a quality system that can and will result in a satisfactory product or service. The HACCP system provides the assurance to the management, to official agencies and to the consumer that the CCPs are under control and that the plan is adequate to control hazards. Definition of responsibilities and authority within the quality system is required by ISO. CLs for the preventive measures at specific CCPs have to be met in HACCP; standards are to be met in ISO.

The International Organization for Standardization, composed of private and governmental bodies, established the ISO9000 series in 1987. These are consensus standards and, by their nature, contain quality principles widely accepted in authoritative texts and by expert groups. They provide a yardstick for suppliers of goods and services to measure their capability to supply quality products. Conformance may be certified by the supplier or by a third party registrar. Governments and private organizations in over 50 countries have now adopted the ISO standards. There are five standards in the ISO9000 series. They include ISO9000, which contains general guidance for the selection and use of the remaining four. The ISO9001 is the most comprehensive, and includes design development and servicing capabilities in addition to the elements in ISO9002. The ISO9002 includes the basic elements of quality assurance, such as management responsibility, control system review, training, etc., along with the quality control elements contained in ISO9003. The ISO9003 relates to finished product quality control, and includes detection and control of defective product during final inspection and testing. The ISO9004 contains guidance for procedures to develop quality management systems, and addresses, among other things, product liability and safety issues. The ISO9000 is a quality management system the objectives of which are to prevent and detect non-conforming products during the stages of production and delivery to the customer. It is a group of international standards directed at and oriented towards the quality management of an organization. The ISO standards are generic, and assess the uniformity and repeatability of actions and procedures by providing the overall organizational discipline that is necessary for an effective quality management system.

It involves the firm as a whole; all departments and sectors must be consistent in the standardized way they perform their functions. The HACCP system involves only the sections and departments that directly or indirectly affect food safety. Industry has generally focused on ISO standards to provide a framework for the establishment and auditing of quality management systems, whereas regulatory authorities have focused on HACCP systems as tools to improve the safety and wholesomeness of food (Hathaway, 1995).

The ISO has as a focus the total system, and requires that the company has a quality system that meets the goals of the ISO standards. It aims to enforce the principle that procedures are always performed in the same manner, through documentation of 'the way to do' and 'how to do'. Auditing is used to prove conformation. Deviations from normality have to be corrected.

To apply ISO9000 in a food plant without HACCP is difficult. It is easier to implement ISO9000 in a food plant that is already using HACCP, since HACCP becomes part of the ISO system. Inspection and testing should be applied to the operations in the production line, not based only on product testing.

3.3.7 Benefits of HACCP

The high degree of interest in food safety has forced governments and food industries to search for means to achieve safety of the food supply. Most of the regulatory agencies have changed their strategy to achieve food safety, emphasizing the prevention of health hazards. In parallel, the food industry has adopted proactive preventive measures in the management of food, resulting in the adoption of HACCP, which has become an internationally recognized approach for prevention of faults in food safety. Implementing HACCP in the food industry encompasses and results in a series of benefits. It serves to develop a team spirit in multidisciplinary teams; helps to integrate the different specialists and their know-how, as engineers, chemists, veterinarians, sanitarians and others share their knowledge; and promotes a global vision of the food processes. It has been observed that the degree of communication among personnel increases when HACCP is implemented in a company.

In an HACCP-protected food industry, new projects, no matter whether related to food safety or not, tend to be developed by multidisciplinary teams that function in a similar way to HACCP teams. Also, a series of questions develop during a 'brainstorming' session on the project. This may permit the discovery in advance of possible faults, thus avoiding unpleasant surprises and unwanted problems at the end.

HACCP makes it possible to be in concordance with international market requirements. Since 1980, the WHO and the Codex Alimentarius, the reference agency for food safety of the World Trade Organization, have been advising governments and food industries to adopt HACCP (WHO, 1980; Tauxe, 1997). Based on these recommendations, governments are intensifying their actions on food inspection and food control.

HACCP improves confidence in food processing. Food plant personnel are able to visualize the process through a sequence of essential data that confirm the final product safety. It also assists in the search for improvement of operations. It creates motivation and inspires a search for technical updating through discussions in study groups.

HACCP provides for better use of laboratory time that would otherwise be used for final product testing. Laboratories can be better used in the control of the food plant environment, the evaluation of cleaning operations, the evaluation of personnel hygiene, and other relevant factors affecting food production. Finally, adoption of HACCP leads to a client and consumer confidence conquest. Years ago food safety was not so much on the mind of people as it is today, when it has become a subject of priority for consumers, producers, distributors and governmental agencies on the five continents (Brundtland, 2001). This has been intensified by the prospects of possible acts of terrorism; HACCP is an important tool for prevention of such acts.

An opinion poll carried out by Environics International in 10 countries (Australia, Brazil, Canada, China, England, Germany, India, Japan, Mexico and the US) requested 10 000 citizens to indicate one of the five following groups that they considered to be most responsible for food safety: government, producers, consumers, farmers and retailers. Of those interviewed, 36 % answered that producers were responsible for food safety, 30 % the government, 15 % the consumers, 6 % the farmers, 5 % the retailers, and 6 % all/combinations of these (Miller, 2001). HACCP can, to varying degrees, be applied by all groups, and is the most important management tool in food safety.

3.3.8. HACCP glossary

Acceptable level Refers to the presence of a hazard that does not pose the likelihood of causing an unacceptable health risk.

Auditing Systematic and independent examination/evaluation to determine whether quality activities and results comply with planned arrangements and whether these arrangements are implemented effectively and are suitable to achieve objectives (ISO8402, 1995).

CCP decision tree A sequence of questions to be answered orderly, to determine whether or not a control point is a critical one (CCP) (US National Advisory Committee on Microbiological Criteria for Foods, 1992).

Continuous monitoring Continuous recording of data (US National Advisory Committee on Microbiological Criteria for Foods and US Department of Agriculture, 1993).

Control The management of conditions to maintain conformity with established criteria (US National Advisory Committee on Microbiological Criteria for Foods and US Department of Agriculture, 1993).

Control point (CP) Any point, step or procedure in a specific food system where loss of control does not result in an unacceptable health risk to the consumer (US National Advisory Committee on Microbiological Criteria for Foods, 1991).

Corrective measures Actions or procedures to correct a situation temporarily out of control; must be applied whenever a deviation occurs (US National Advisory Committee on Microbiological Criteria for Foods, 1992).

Criteria Pre-established requirements that serve as support for a decision or judgment (US National Advisory Committee on Microbiological Criteria for Foods, 1992).

Critical control point (CCP) A locale, practice, procedure or process at which a control measure must be applied relating to one or more factors, to eliminate, prevent or reduce a hazard to an acceptable level (US National Advisory Committee on Microbiological Criteria for Foods, 1992).

Critical defects Deviations occurring at CCPs that can result in a hazard and consequently affect consumer health (US National Advisory Committee on Microbiological Criteria for Foods and US Department of Agriculture, 1993).

Critical limits (CLs) Criteria to be met for each preventive measure associated with a CCP (US National Advisory Committee on Microbiological Criteria for Foods, 1992).

Deviation Failure to reach a pre-established critical limit or a failure to stay within critical limits.

Food chain The sequence of operations dealing with food, starting from the field or sea and continuing through industrial processing, storage, transport, distribution, retailing, until the final preparation for consumption.

Food contamination The presence of objectionable levels of organisms, chemicals, foreign bodies, taints, or unwanted, diseased and decomposed material (Codex Alimentarius Commission, 1994).

Food safety Food that does not cause illness or injury to consumers (Codex Alimentarius Commission, 1994).

Food suitability Non-spoiled food and its fitness for normal human consumption – not associated with public health.

Good manufacturing practices (GMPs) Principal factors involved in achieving food hygiene. Covers every aspect of food production, employee training, plant design, equipment specification and cleaning, quality assurance evaluation, and distribution of food products (Gould, 1994). Includes hygienic practices, cleaning and disinfection program. Describes ways to avoid contamination in food manufacturing environment and is the basis on which HACCP is established.

Hazard analysis A process evaluation to determine where the contamination of a food product could reach a level that would be unacceptable, and to determine the potential for the persistence, increasing, or development of contamination (US National Advisory Committee on Microbiological Criteria for Foods, 1992).

Hazard analysis-critical control points decision tree Graphical presentation of a sequence of decisions to determine whether or not a CP is a CCP. The decision tree is nothing more than a sequence of questions to be answered orderly.

Hazard analysis-critical control points (HACCP) plan A document describing the activities developed in accordance with the principles of HACCP to ensure control of hazards that are significant for food safety in the product under consideration and its intended use (WHO, 1998).

Hazard analysis-critical control points system Result of the implementation of the HACCP (US National Advisory Committee on Microbiological Criteria for Foods, 1992).

Hazard analysis-critical control points team Multidisciplinary professional group, in charge of the development of the HACCP plan (US National Advisory Committee on Microbiological Criteria for Foods, 1992).

Hazard A hazard is a biological, chemical, or physical agent in, or conditions of, food with the potential to cause an adverse health effect on the human being (Codex Alimentarius Commission, 1997).

Monitoring Planned sequence of observations or measurements to evaluate whether or not a specific CCP is effectively under control and to produce exact recorded data for posterior use in the HACCP verification process (US National Advisory Committee on Microbiological Criteria for Foods, 1992). Monitoring is checking that the procedures at the CCPs are properly carried out (WHO, 1980), and is a routine task, performed continuously or several times during a day's operation.

Preventive measures Actions of physical, chemical or microbiological nature, used to control an identified hazard (US National Advisory Committee on Microbiological Criteria for Foods, 1992).

Quality assurance Strategic management function that establishes policies concerning quality, adopts programs to meet the established goals, and provides confidence that these measures are being effectively applied (Stauffer, 1994)

Quality control A tactical function to perform those programs identified by Quality Assurance as necessary to accomplish quality goals (Stauffer, 1994).

Risks Estimates of the probability for the occurrence of a specific hazard (US National Advisory Committee on Microbiological Criteria for Foods, 1992).

Risk assessment Evaluation of food hazards in terms of their degree of seriousness, including the potential for sequelae related to vulnerable populations, factors involved in dose response, and intrinsic and extrinsic circumstances affecting hazards. The costs related to specific hazards must be considered.

Risk categories The ranking of risks based on the food hazard characteristics.

Sensitive ingredients Ingredients recognized as being associated with a hazard, serious enough to be of concern (US National Advisory Committee on Microbiological Criteria for Foods, 1992).

Severity The seriousness, the dimension and magnitude of a hazard, is based on the degree of consequences in the case of its occurrence (US National Advisory Committee on Microbiological Criteria for Foods, 1992).

Significant risk A moderate probability of causing a health risk (US National Advisory Committee on Microbiological Criteria for Foods, 1991).

Validation Evaluation of the scientific and technical information/background to be determined if the implemented HACCP plan, will effectively control hazards (US National Advisory Committee on Microbiological Criteria for Foods, 1998).

Verification The adaptation of methods, procedures, tests and other evaluations, in addition to monitoring, to determine conformity with the HACCP plan. This is primarily an industry responsibility; however some verification activities can also be undertaken during regulatory assessments (WHO, 1998).

4 Comments and conclusions

All of the stages in the food chain are associated with food safety. In food production, the responsibility lies with the farmer who is in charge of using approved chemicals, biological substances and drugs, and of controlling animal diseases to avoid adverse effects on food quality.

In the food-processing industry, the responsibility lies with the management, which must assure the use of approved technologies and procedures to obtain a safe food product.

In the storage facilities, the responsibility for keeping the food at proper temperatures and controlling other environmental conditions lies with the managers who have the authority over the process. In the transport stage, the responsibility also lies with the managers, who should ensure adequate training of the drivers to keep the temperature under control and to assure food safety during transport. This includes cleaning and disinfection of vehicles between hauls. At the sale and retail level, the management is responsible for keeping food under the conditions requested in the instructions on its label.

In the food service industry, the duty lies with the managers to provide training for the personnel in personal hygiene, food hygiene and the safe handling of food. Finally, at home the person preparing the food has the responsibility for following the label instructions, and for hygienic and the safe handling of food in the kitchen.

Government regulations alone cannot assure food safety. Food production, processing and service industries and consumers are very important participants.

Bibliography

Adams, C. E. (1990). Use of HACCP in meat and poultry inspection. *Food Technol.* **44**, 169–170.

Aebi, H. E. (1981). Food acceptance and nutrition: experiences, intentions and responsibilities. In: J. Solms and R. L. Hall (eds), *Criteria of Food Acceptance. How Man Chooses What He Eats*, p. 461. Forster Verlag, A.G. Forster Publishing, Zurich.

Baskerville, A., T. J. Humphrey, R. B. Fitzgeorge *et al.* (1992). Airborne infection of laying hens with *Salmonella enteritidis* phage type 4. *Vet. Rec.* **130**, 395–398.

Bauman, H. E. (1990). HACCP: concept, development and application. *Food Technol.* **44**, 156–158.

Berends, B. R. and J. M. A. Snijders (1997). Risk factors and control measures during slaughter and processing. In: S. Bech-Nielsen and J. P. Nielsen (eds), *Proceedings of the 2nd International Symposium on Epidemiology and Control of* Salmonella *in Pork*, pp. 36–44. Federation of Danish Pig Producers and Slaughterhouses, Copenhagen.

Bernard, D. (1997). HACCP. National Food Processors Information Letter No 3387, 28 February.

Beuchat, L. R. and J,-H. Ryo (1997). Produce handling and processing practices. *Emerg. Infect. Dis.* **3**, 459–465.

Brickey, P. M. (1981). Food sanitation and the FD&C Act. In: J. R. Gorham (ed.), *Principles of Food Analysis for Filth, Decomposition and Foreign Matter*, 2nd edn, pp. 7–9. US Food and Drug Administration Technical Bulletin No 1, US Food and Drug Administration, Washington, DC.

Brundtland, G. H. (2001). *Food Safety – A Worldwide Challenge*. Office of the Director General, World Health Organization, Uppsala, 14 March.

Cassin, M. H., A. M. Lammerding, E. C. Todd *et al.* (1998). Quantitative risk assessment for *Escherichia coli* O157:H7 in ground beef hamburgers. *Intl J. Food Microbiol.* **5**, 21–44.

Castellan, D. M., H. Kinde, P. H. Kass *et al.* (2004). Descriptive study of California layer premises and analysis of risk factors for *Salmonella enterica* serotype enteritidis as characterized by manure drag swabs. *Avian Dis.* **48**, 550–561.

Castillo, A., L. M. Lucia, K. J. Goodson *et al.* (1999). Decontamination of beef carcass surface tissue by steam vacuum alone and combined hot water and lactic acid sprays. *J. Food Prot.* **62**, 146–151.

Christensen, J. and B. Carstensen (1997). The effect of herd size on the sero-prevalence *Salmonella enterica* in Danish swine herds. In: S. Bech-Nielsen and J. P. Nielsen (eds), *Proceedings of the 2nd International Symposium on Epidemiology and Control of* Salmonella *in Pork*, pp. 192–195. Federation of Danish Pig Producers and Slaughterhouses, Copenhagen.

Codex Alimentarius Commission (1994). Food and Agriculture Organization of the United Nations, World Health Organization. Joint FAO/WHO Food Standards Programme. Codex Committee on Food Hygiene. Considerations of the Draft Revised International Code of Practice – General Principles of Food Hygiene, 22nd Session, Washington DC, USA, 17–21 October.

Codex Alimentarius Commission (1997). Hazard Analysis and Critical Control Points (HACCP) system and guidelines for its application. Revision 1. Annex to Codex Alimentarius Commission, Recommended International Code of Practice, p. 34 – General Principles of Food Hygiene, Revision 3.

Codex Alimentarius Commission (2001). Discussion Paper on proposed Draft Guidelines for the Validation of Food Hygiene Control Measures. Joint FAO/WHO Food Standards Programme. Codex Committee on Food Hygiene, 34th Session, Bangkok, Thailand, 8–13 October.

Committee on Communicable Diseases Affecting Man (1991). *Procedures to Implement the Hazard Analysis Critical Control Point System.* International Association of Milk, Food and Environmental Sanitarians, Inc., Ames, IA.

Corso, D. S., M. H. Kramer, K. A. Blair *et al.* (2003). Cost of illness in the 1993 waterborne *Cryptosporidium* outbreak, Milwaukee, Wisconsin. *Emerg. Infect. Dis.* **4**, 426–431.

Council for Agricultural Science and Technology (1994). *Foodborne Pathogens: Risks and Consequences.* Task Force Report No. 122. CAST, Ames, IA.

Delazari I., S. T. Iaria, H. Riemann *et al.* (1998). Decontaminating beef for *Escherichia coli* O157:H7. *J. Food Prot.* **61**, 547–550.

Draser, B. S. and M. J. Hill (1974). *Human Intestinal Flora.* Academic Press, San Diego, CA.

Eisenberg, W. V. (1981). Sources of food contaminants. In: J. R. Gorham (ed.), *Principles of Food Analysis for Filth, Decomposition and Foreign Matter*, pp. 11–25. FDA Technical Bulletin No. 1. Food and Drug Administration, Washington, DC.

Elliott, P. H. (1996). Predictive microbiology and HACCP. *J. Food Prot.* (Suppl.), 48–53.

Fanelli, M. J., W. W. Sadler and J. R. Brownell (1970). Preliminary studies of persistence of salmonellae in poultry litter. *Avian Dis.* **14**, 131–141.

Gast, R. K. and C. W. Beard (1993). Research to understand and control *Salmonella enteritidis* in chickens and eggs. *Poultry Sci.* **72**, 1157–1163.

Gould, W. A. (1994). *Current Manufacturing Practices: Food Plant Sanitation*, pp. 1–5. CTT Publications Inc., Baltimore, MD.

Haas, C. N. (1992). Estimation of risks due to low doses of microorganisms: a comparison of alternative methodologies. *Am. J. Epidemiol.* **118**, 573–582.

Halgaard, C., A. C. Nielsen and J. C. Ajufo (1997). The Danish control program for *Salmonella* in slaughter pigs. In: S. Bech-Nielsen and J. P. Nielsen (eds), *Proceedings of the 2nd International Symposium on Epidemiology and Control of* Salmonella *in Pork*, pp. 260–263. Federation of Danish Pig Producers and Slaughterhouses, Copenhagen.

Hancock, D. D., D. H. Rice, L. A. Thomas *et al.* (1997). Epidemiology of *E. coli* O157:H7 in feedlot cattle. *J. Food Prot.* **60**, 462–465.

Harp, J. A. and T. P. Goff (1998). Strategies for the control of *Cryptosporidium parvum* infection in calves. *J. Dairy Sci.* **81**, 289–294.

Harrigan, W. F. and R. W. A. Park (1991). *Making Safe Food, A Management Guide for Microbiological Quality*. Academic Press, New York, NY.

Hassan, J. O. and R. Curtis (1990). Control of colonization by virulent *S. typhimurium* by oral immunization with avirulent delta cya delta crp *S. typhimurium. Res. Microbiol.* **141**, 839–850.

Hathaway, S. (1995). Harmonization of international requirements under HACCP based food control systems. *Food Control* **6**, 267–276.

Hennesy, T. W., C. W. Hedberg, L. Slutsker, and K. E. White (1996). A national outbreak of *Salmonella enteritidis* infections from ice cream. *N. Engl. J. Med.* **334**, 1281–1286.

Henzler, D. J. and H. M. Opitz (1992). The role of mice in the epidemiology of *Salmonella enteritidis* on chicken layer farms. *Avian Dis.* **36**, 625–631.

Himathongkham, S. and H. Riemann (1998). Destruction of *S. typhimurium*, *E. coli* O157:H7 and *Listeria monocytogenes* in chicken manure by drying and gassing with ammonia. *FEMS Microbiol. Letts* **171**, 179–182.

Himathongkham, S., M. das Gracas Pereira and H. Riemann (1996). Heat destruction of *Salmonella* in poultry feed: effect of time, temperature and moisture. *Avian Dis.* **40**, 72–77.

Himathongkham, S., S. Dragoi, S. Nuanualsuwan *et al.* (2000). Growth and survival of *Salmonella typhimurium, E. coli* O157:H7 and *Listeria monocytogenes* in the poultry farm environment (unpublished data).

International Commission on Microbiological Specifications for Foods (1974). *Microorganisms in Foods. 2. Sampling for Microbiological Analysis: Principles and Specific Applications*. University of Toronto Press, Toronto.

International Commission on Microbiological Specifications for Foods (1988). *Microorganisms in Foods. 4. Application of Hazard Analysis Critical Control Points (HACCP) System to Ensure Microbiological Safety and Quality.* Blackwell Scientific Publications, London.

International Life Sciences Institute (1997). *A Simple Guide to Understanding and Applying the Hazard Analysis Critical Control Point Concept*, 2nd edn. ILSI Europe Council Monographs, Brussels.

International Organization for Standardization (1986). ISO 8402: Quality Vocabulary. ISO, Geneva.**127

Jarvis, B. (1989). *Statistical Aspects of the Microbiological Analysis of Foods. Progress in Industrial Microbiology*. Elsevier Science Publishers, New York, NY.

Katsuyama, A. M. (1995a). Chemical hazards and controls. In: K. E. Stevenson and D. T. Bernard (eds), *HACCP, Establishing Hazard Analysis Critical Control Points, A Workshop Manual*, 2nd edn, pp. 5.1–5.2. National Food Processors Association, The Food Processors Institute, Washington, DC.

Katsuyama, A. M. (1995b). Physical hazards and controls. In: K. E. Stevenson and D. T. Bernard (eds), *HACCP, Establishing Hazard Analysis Critical Control Points, A Workshop Manual*, 2nd edn, pp. 6.1–6.6. National Food Processors Association, The Food Processors Institute, Washington, DC.

Katsuyama, A. M. and D. E. Stevenson (1995). Hazard analysis and identification of Critical Control Points. In: K. E. Stevenson and D. T. Bernard (eds), *HACCP, Establishing Hazard Analysis Critical Control Points, A Workshop Manual*, 2nd edn, pp. 8.1–8.14. National Food Processors Association, The Food Processors Institute, Washington, DC.

Kinde, H., D. H. Read, A. Ardans *et al.* (1996a). Sewage effluent likely source of *Salmonella enteritidis* phage type 4 infection in a commercial chicken layer flock in southern California. *Avian Dis*. **40**, 672–676.

Kinde, H., D. H. Read, R. P. Chin *et al.* (1996b). *Salmonella enteritidis* phage type 4 infection in a commercial layer flock in southern California; bacteriological and epidemiological findings. *Avian Dis*. **40**, 665–671.

Kinde, H., M. Adelson, A. Ardans *et al.* (1997). Prevalence of *Salmonella* in municipal sewage treatment plant effluents in Southern California. *Avian Dis*. **41**, 392–398.

Kletz, T. (1992). *HAZOP and HAZAN, Identifying and Assessing Process Industry Hazards*, Institute of Chemical Engineers, London.

Kumar, M. C., M. D. York, J. R. McDowell and B. S. Pomeroy (1971). Dynamics of *Salmonella* infection in fryer roaster turkeys. *Avian Dis*. **15**, 221–232.

Kvenberg, J. F. and D. J. Schwalm (2000). Use of microbial data for hazard analysis critical control point validation – Food and Drug Administration perspective. *J. Food Prot*. **63**, 810–814.

Mallinson, E. T., L. E. Carr, S. W. Joseph and L. E. Stewart (1993). Newer *Salmonella* control and identification techniques for farms and food plants. In: *Proceedings of the 97th Annual Meeting of the US Animal Health Association, Richmond, Virginia*, pp. 474–476.

Matlho, G., S. Himathongkham, H. Riemann and P. Kass (1997). Destruction of *Salmonella enteritidis* in poultry feed by combination of heat and propionic acid. *Avian Dis*. **41**, 58–61.

Mayes, T. 1994. HACCP training. *Food Control* **5**, 121–132.

McMeekin, T. A., J. N. Olley, T. Ross and D. A. Ratkowsky (1993). *Predictive Microbiology: Theory and Application*. John Wiley & Sons, New York, NY.

Methner, U., P. A. Barrow, G. Martin and H. Meyer (1997). Comparative study of the protective effect against *Salmonella* colonization in newly hatched chickens using live, attenuated *Salmonella* vaccine strains, wild-type *Salmonella* strains or a competitive exclusion product. *Intl J. Food Microbiol.* **35**, 223–230.

Miller, D. (2001). Global Consumer Trends. CIES Food Business Forum. International Food Safety Conference. Geneva, Switzerland, 19–20 September.

Milner, K. C. and M. F. Shaffer (1952). Bacteriologic study of experimental *Salmonella* infection in chicks. *J. Infect. Dis.* **90**, 81–96.

National Advisory Committee on Microbiological Criteria for Foods (1997). Generic HACCP application in broiler slaughter and processing. *J. Food Prot.* **60**, 579–604.

Notermans, S., M. H. Zwietering and G. C. Mead (1994). The HACCP concept: identification of potentially hazardous micro-organisms. *Food Microbiol.* **11**, 203.

Nurmi, E. and M. Rentala (1973). New aspects of *Salmonella* infection in broiler production. *Nature* **241**, 210–211.

Opara, O. O., L. E. Carr, E. Russek-Cohen *et al.* (1992). Correlation of water activity and other environmental conditions with repeated detection of *Salmonella* contamination on poultry farms. *Avian Dis.* **36**, 664–671.

Oscar, T. P. (1998). Salmonella Risk *Assessment Modeling Program for Poultry (SCRAMPP)*. USDA, ARS, Beltsville, MD.

Pivnick, H. and E. Nurmi (1982). The Nurmi concept and its role in control of *Salmonella* in poultry. In: R. Davis (ed.), *Developments in Food Microbiology*, p. 41. Applied Science Publishers, Barking.

Rhodehamel, E. J. (1992). Overview of biological, chemical and physical hazards. In: M. D. Pierson and D. A. Corlett (eds), *HACCP Principles and Applications*. Chapman & Hall, New York, NY.

Riemann, H., S. Himathongkham, D. Willoughby *et al.* (1998). A survey for *Salmonella* by dragswabbing manure piles in California egg ranches. *Avian Dis.* **42**, 67–71.

Rose, J. B., C. N. Haas and C. P. Gerba (1996). Linking microbial criteria for foods with quantitative risk assessment. In: J. J. Sheridan, R. L. Buchanan and T. J. Montville (eds), *HACCP: An Integrated Approach to Assuring the Microbial Safety of Meat and Poultry*. Food & Nutrition Press, Westport, CT.

Sadler, W. W., J. R. Brownell and M. J. Fanelli (1969). Influence of age and inoculum level on shed pattern of *Salmonella typhimurium* in chickens. *Avian Dis.* **13**, 793–803.

Salmonella enteritidis Analysis Team (1995). *Salmonella enteritidis* Pilot Project, Progress Report. US Department of Agriculture, Washington DC.

Savage, D. C. (1983). Association of indigenous microorganisms with gastrointestinal surfaces. In: D. J. Hentges (ed.), *Human Intestinal Flora in Health and Disease*, pp. 55–74. Academic Press, San Diego, CA.

Schleifer, J. H., B. J. Juven, C. W. Beard and N. A. Cox (1984). The susceptibility of chicks to *Salmonella Montevideo* in artificially contaminated poultry feed. *Avian Dis.* **28**, 497–503.

Scott, V. N. and L. Moberg (1995). Establishing Hazard Analysis Critical Control Points systems. In: K. E. Stevenson and D. T. Bernard (eds), *HACCP, Establishing Hazard Analysis Critical Control Points, A Workshop Manual*, 2nd edn, pp. 4.14–4.25. National Food Processors Association, The Food Processors Institute, Washington, DC.

Smith, J. P., C. Toupin, B. Gagnon *et al.* (1990). A Hazard Analysis Critical Control Point approach (HACCP) to ensure microbiological safety of sous vide processed meat products. *Food Microbiol.* **7**, 177–198.

Sperber, W. H. (1991). The modern HACCP system. *Food Technol.* **45**, 116, 118, 120.

Stauffer, J. E. (1994). Quality *Assurance of Food: Ingredients, Processing and Distribution*. Food & Nutrition Press Inc., Westport, CT.

Stege, H., J. Christensen, D. L. Baggesen *et al.* (1997a). Subclinical *Salmonella* infection in Danish finishing pig herds; risk factors. In: S. Bech-Nielsen and J. P. Nielsen (eds), *Proceedings of the 2nd International Symposium on Epidemiology and Control of* Salmonella *in Pork*, pp. 148–152. Federation of Danish Pig Producers and Slaughterhouses, Copenhagen.

Stege, H., J. Christensen, D. L. Baggesen and J. P. Nielsen (1997b). Subclinical infection in Danish pig herds: The effect of *Salmonella* contaminated feed. In: S. Bech-Nielsen and J. P. Nielsen (eds), *Proceedings of the 2nd International Symposium on Epidemiology and Control of* Salmonella *in Pork*, pp. 81–84. Federation of Danish Pig Producers and Slaughterhouses, Copenhagen.

Sterrenberg, P. and S. Notermans (2000). Risk analysis: theory and reality. *Proceedings of the 46th International Congress of Meat Science and Technology*, Buenos Aires, Argentina, 27 August–1 September, Vol. 2, 6.II–L3, pp. 690–697.

Stevenson, K. E. (1990). Implementing HACCP in the food industry. *Food Technol.* **44**, 179–180.

Stevenson, K. E. (1995). Introduction to Hazard Analysis Critical Control Points systems. In: K. E. Stevenson and D. T. Bernard (eds), *HACCP, Establishing Hazard Analysis Critical Control Points, A Workshop Manual*, 2nd edn, pp. 1.1–1.5. National Food Processors Association, The Food Processors Institute, Washington, DC.

Stevenson, K. E. and D. E.Gombas (1999). Verification procedures. In: K. E. Stevenson and D. T. Bernard (eds), *HACCP: A Systematic Approach to Food Safety. A Comprehensive Manual for Developing and Implementing a Hazard Analysis and Critical Control Points Plan*, pp. 99–104. National Food Processors Association. The Food Processors Institute. Washington, D.C.

Swanson, K. M. J. and J. E. Anderson (2000). Industry perspectives on the use of microbial data for hazard analysis critical control point validation. *J. Food Prot.* **63**, 815–818.

Tauxe, R. V. (1997). Emerging foodborne diseases; evolving public health challenge. *Emerg. Infect. Dis.* **3**, 425–434.

Tielen, M. J. M., F. W. van Shie, P. J. van der Wolf *et al.* (1997). Risk factors and control measures for subclinical *Salmonella* infections in pig herds. In: S. Bech-Nielsen and J. P. Nielsen (eds), *Proceedings of the 2nd International Symposium on Epidemiology and Control of* Salmonella *in Pork*, pp. 32–40. Federation of Danish Pig Producers and Slaughterhouses, Copenhagen.

Turnbull, P. C. and G.H. Snoeyenbos (1973). The role of ammonia, water activity and pH on the salmonellacidal effect of long-used poultry litter. *Avian Dis.* **17**, 72–86.

US Department of Commerce (1994). Statistical abstracts of the United States National Databook. Bureau of the Census. In: *Foodborne Pathogens: Risk and Consequences*. Council for Agricultural Science and Technology. Task Force Report No. 122, Ames, IA.

US Department of Health and Human Services, Centers for Disease Control and Prevention. 1994. HIV/AIDS Surveillance Report No. 5. In: *Foodborne Pathogens: Risk and Consequences*. Council for Agricultural Science and Technology. Task Force Report No. 122. Ames, IA.

US National Advisory Committee on Microbial Criteria for Foods (1992). Hazard Analysis Critical Control Points system. *Intl J. Food Microbiol.* **16**, 1–23.

US National Advisory Committee on Microbial Criteria for Foods and US Department of Agriculture (1993). Generic HACCP for raw beef. *Food Microbiol.* **10**, 726–775.

US National Advisory Committee on Microbiological Criteria for Foods (1991). Recommendations of the US National Advisory Committee on Microbiological Criteria for Foods: I. HACCP principles, II. Meat and poultry, III. Seafood. Guide to principles of Hazard Analysis Critical Control Points system. *Food Control* **2**, 202–211.

Vail, R. (1994). Fundamentals of HACCP. *Cereal Foods World* **39**, 393–395.

Walls, I. and V. N. Scott (1997). Use of predictive microbiology in microbial food safety risk assessment. *Intl J. Food Microbiol.* **36**, 97–102.

Weddig, L. M. (1999). Critical limits. InL K. E. Stevenson and D. T. Bernard (eds), *HACCP, A Systematic Approach to Food Safety. A Comprehensive Manual for Developing and Implementing a Hazard Analysis Critical Control Points System*, pp. 85–88. National Food Processors Association, The Food Processors Institute, Washington, DC.

Whiting, R. C. and R. L. Buchanan (1997). Development of a quantitative risk assessment model for *Salmonella enteritidis* in pasteurized liquid eggs. *Intl J. Food Microbiol.* **36**, 111–125.

Wierup, M. and M. Wold-Troell (1988). Epidemiological evaluation of the salmonella-controlling effect of a nation wide use of competitive exclusion culture in poultry. *Poultry Sci.* **67**, 1026–1033.

Wong, S., D. Street, S. I. Delgado and K. C. Klentz (2000). Recalls of foods and cosmetics due to microbial contamination reported to the US Food and Drug Administration. *J. Food Prot.* **63**, 1113–1116.

WHO (World Health Organization) (1976). *Application of Systems Analysis to Health Management*. Report of a WHO Expert Committee. Technical Series 596. WHO, Geneva.

WHO (World Health Organization) (1980). *Report of the WHO/ICMSF Meeting on Hazard Analysis Critical Control Point System in Food Hygiene*. Geneva, 9–12 June.

WHO (World Health Organization) (1993). *Training Considerations for the Application of the Hazard Analysis Critical Control Point System to Food Processing and Manufacturing*. Division of Food and Nutrition, Food Safety Unit, WHO, Geneva.

Recommended websites for food safety information

Centers for Disease Control and Prevention, at http://www.cdc.gov
Food and Drug Administration (FDA) Center for Food Safety and Applied Nutrition, at http://vm.cfsan.fda.gov/list.html
US Department of Agriculture (USDA) Food Safety and Inspection Service, at http://www.usda.gov/agency/fsis/homepage.htm
USDA/FDA Foodborne Illness Education Information Center, at http://www.nal. usda.gov/foodborne/index.html
US Environmental Protection Agency (EPA), at http://www.epa.gov

Index

Food Science and Technology

International Series

Maynard A. Amerine, Rose Marie Pangborn, and Edward B. Roessler, *Principles of Sensory Evaluation of Food*. 1965.

Martin Glicksman, *Gum Technology in the Food Industry*. 1970.

Maynard A. Joslyn, *Methods in Food Analysis*, second edition. 1970.

C. R. Stumbo, *Thermobacteriology in Food Processing*, second edition. 1973.

Aaron M. Altschul (ed.), *New Protein Foods:* Volume 1, *Technology, Part A*—1974. Volume 2, *Technology, Part B*—1976. Volume 3, *Animal Protein Supplies, Part A*—1978. Volume 4, *Animal Protein Supplies, Part B*—1981. Volume 5, *Seed Storage Proteins*—1985.

S. A. Goldblith, L. Rey, and W. W. Rothmayr, *Freeze Drying and Advanced Food Technology*. 1975.

R. B. Duckworth (ed.), *Water Relations of Food*. 1975.

John A. Troller and J. H. B. Christian, *Water Activity and Food*. 1978.

A. E. Bender, *Food Processing and Nutrition*. 1978.

D. R. Osborne and P. Voogt, *The Analysis of Nutrients in in Foods*. 1978.

Marcel Loncin and R. L. Merson, *Food Engineering: Principles and Selected Applications*. 1979.

J. G. Vaughan (ed.), *Food Microscopy*. 1979.

J. R. A. Pollock (ed.), *Brewing Science*, Volume 1—1979. Volume 2—1980. Volume 3—1987.

J. Christopher Bauernfeind (ed.), *Carotenoids as Colorants and Vitamin A Precursors: Technological and Nutritional Applications*. 1981.

Pericles Markakis (ed.), *Anthocyanins as Food Colors*. 1982.

George F. Stewart and Maynard A. Amerine (eds.), *Introduction to Food Science and Technology*, second edition. 1982.

Malcolm C. Bourne, *Food Texture and Viscosity: Concept and Measurement*. 1982.

Hector A. Iglesias and Jorge Chirife, *Handbook of Food Isotherms: Water Sorption Parameters for Food and Food Components*. 1982.

Colin Dennis (ed.), *Post-Harvest Pathology of Fruits and Vegetables*. 1983.

P. J. Barnes (ed.), *Lipids in Cereal Technology*. 1983.

David Pimentel and Carl W. Hall (eds.), *Food and Energy Resources*. 1984.

Joe M. Regenstein and Carrie E. Regenstein, *Food Protein Chemistry: An Introduction for Food Scientists*. 1984.

Maximo C. Gacula, Jr., and Jagbir Singh, *Statistical Methods in Food and Consumer Research*. 1984.

Fergus M. Clydesdale and Kathryn L. Wiemer (eds.), *Iron Fortification of Foods*. 1985.

Robert V. Decareau, *Microwaves in the Food Processing Industry*. 1985.

S. M. Herschdoerfer (ed.), *Quality Control in the Food Industry*, second edition. Volume 1—1985. Volume 2—1985. Volume 3—1986. Volume 4—1987.

F. E. Cunningham and N. A. Cox (eds.), *Microbiology of Poultry Meat Products*. 1987.

Walter M. Urbain, *Food Irradiation*. 1986.

Peter J. Bechtel, *Muscle as Food*. 1986.

H. W.-S. Chan, *Autoxidation of Unsaturated Lipids*. 1986.

Chester O. McCorkle, Jr., *Economics of Food Processing in the United States*. 1987.

Jethro Japtiani, Harvey T. Chan, Jr., and William S. Sakai, *Tropical Fruit Processing*. 1987.

J. Solms, D. A. Booth, R. M. Dangborn, and O. Raunhardt, *Food Acceptance and Nutrition*. 1987.

R. Macrae, *HPLC in Food Analysis*, second edition. 1988.

A. M. Pearson and R. B. Young, *Muscle and Meat Biochemistry*. 1989.

Marjorie P. Penfield and Ada Marie Campbell, *Experimental Food Science*, third edition. 1990.

Leroy C. Blankenship, *Colonization Control of Human Bacterial Enteropathogens in Poultry*. 1991.

Yeshajahu Pomeranz, *Functional Properties of Food Components*, second edition. 1991.

Reginald H. Walter, The Chemistry and Technology of Pectin. 1991.

Herbert Stone and Joel L. Sidel, *Sensory Evaluation Practices*, second edition. 1993.

Robert L. Shewfelt and Stanley E. Prussia, *Postharvest Handling: A Systems Approach*. 1993.

R. Paul Singh and Dennis R. Heldman, *Introduction to Food Engineering*, second edition. 1993.

Tilak Nagodawithana and Gerald Reed, *Enzymes in Food Processing*, third edition. 1993.

Dallas G. Hoover and Larry R. Steenson, *Bacteriocins*. 1993.

Takayaki Shibamoto and Leonard Bjeldanes, *Introduction to Food Toxicology*. 1993.

John A. Troller, *Sanitation in Food Processing*, second edition. 1993.

Ronald S. Jackson, *Wine Science: Principles and Applications*. 1994.

Harold D. Hafs and Robert G. Zimbelman, *Low-fat Meats*. 1994.

Lance G. Phillips, Dana M. Whitehead, and John Kinsella, *Structure-Function Properties of Food Proteins*. 1994.

Robert G. Jensen, *Handbook of Milk Composition*. 1995.

Yrjö H. Roos, *Phase Transitions in Foods*. 1995.

Reginald H. Walter, *Polysaccharide Dispersions*. 1997.

Gustavo V. Barbosa-Cánovas, M. Marcela Góngora-Nieto, Usha R. Pothakamury, and Barry G. Swanson, *Preservation of Foods with Pulsed Electric Fields*. 1999.

Ronald S. Jackson, *Wine Science: Principles, Practice, Perception*, second edition. 2000.

R. Paul Singh and Dennis R. Heldman, *Introduction to Food Engineering*, third edition. 2001.

Ronald S. Jackson, *Wine Tasting: A Professional Handbook*. 2002.

Malcolm C. Bourne, *Food Texture and Viscosity: Concept and Measurement*, second edition. 2002.

Benjamin Caballero and Barry M. Popkin (eds), *The Nutrition Transition: Diet and Disease in the Developing World*. 2002.
Dean O. Cliver and Hans P. Riemann (eds), *Foodborne Diseases*, second edition. 2002.
Martin Kohlmeier, *Nutrient Metabolism*. 2003.
Herbert Stone and Joel L. Sidel, *Sensory Evaluation Practices*, third edition. 2004.
Jung H. Han, *Innovations in Food Packaging*. 2005.
Da-Wen Sun, *Emerging Technologies for Food Processing*. 2005.
Hans Riemann and Dean Cliver (eds), *Foodborne Infections and Intoxications*, third edition. 2006.